15th International Conference on the Strength of Materials 2009

(ICSMA 15)

Journal of Physics: Conference Series Volume 240

Dresden, Germany
16-21 August 2009

Editors:

Werner Skrotzki
Carl-Georg Oertel
Horst Biermann
Martin Heilmaier

ISBN: 978-1-61738-806-4
ISSN: 1742-6588

Printed from e-media with permission by:

Curran Associates, Inc.
57 Morehouse Lane
Red Hook, NY 12571

Some format issues inherent in the e-media version may also appear in this print version.

Printed by Curran Associates, Inc. (2010)

For permission requests, please contact the Institute of Physics
at the address below.

Institute of Physics
Dirac House, Temple Back
Bristol BS1 6BE UK

Phone: 44 1 17 929 7481
Fax: 44 1 17 920 0979

techtracking@iop.org

Additional copies of this publication are available from:

Curran Associates, Inc.
57 Morehouse Lane
Red Hook, NY 12571 USA
Phone: 845-758-0400
Fax: 845-758-2634
Email: curran@proceedings.com
Web: www.proceedings.com

TABLE OF CONTENTS

PLENARY

ATOMISTIC AND MICROSTRUCTURAL ASPECTS OF PLASTIC DEFORMATION

ATOMISTIC AND MICROSTRUCTURAL ASPECTS OF FRACTURE

ATOMISTIC AND MICROSTRUCTURAL ASPECTS OF FRACTURE

HIGH TEMPERATURE DEFORMATION AND CREEP

MECHANICAL PROPERTIES RELATED TO PHASE TRANSFORMATIONS

LARGE AND SEVERE PLASTIC DEFORMATION

NANO- AND MICROSCALE PHENOMENA IN PLASTICITY AND FRACTURE

STRENGTH ISSUES IN BIOLOGICAL SYSTEMS AND BIOMATERIALS

MECHANICAL BEHAVIOUR OF GLASSES AND NON-CRYSTALLINE SOLIDS

MULTISCALE MODELLING AND EXPERIMENTAL VALIDATION

INSIGHT THROUGH NEW EXPERIMENTAL METHODS

OTHER NEW DEVELOPMENTS RELATED TO THE FIELD

The 15th International Conference on the Strength of Materials (ICSMA 15) took place in Dresden, Germany, August 16–21, 2009. It belongs to the triennial series of ICSMA meetings with a long tradition, starting in 1967 - Tokyo, 1970 - Asilomar, 1973 - Cambridge, 1976 - Nancy, 1979 - Aachen, 1982 - Melbourne, 1985 - Montreal, 1988 - Tampere, 1991 - Haifa, 1994 - Sendai, 1997 - Prague, 2000 - Asilomar, 2003 - Budapest, 2006 - Xian. ICSMA 15 was hosted by the Dresden University of Technology, Institute of Structural Physics.

Following the tradition of this conference series, it was the main focus of ICSMA 15 to promote and strengthen the fundamental understanding of the basic processes that govern the strength of materials. Nonetheless, it was the aim to forge links between basic research on model materials and applied research on engineering materials of technical importance. Thus, ICSMA 15 provided a forum for the presentation and discussion of research on the mechanical properties of all materials which are of interest to materials scientists and engineers from many different areas. The topics covered by ICSMA 15 were:

1.Atomistic and microstructural aspects of plastic deformation
2.Atomistic and microstructural aspects of fracture
3.Adhesion and interfacial strength
4.Cyclic deformation and fatigue
5.High temperature deformation and creep
6.Mechanical properties related to phase transformations
7.Large and severe plastic deformation
8.Nano- and microscale phenomena in plasticity and fracture
9.Strength issues in biological systems and biomaterials
10.Mechanical behaviour of glasses and non-crystalline solids
11.Multiscale modelling and experimental validation
12.Insight through new experimental methods
13.Other new developments related to the field

While there was large interest in the new topics 7 and 8, contributions to topic 9 were much less than expected.

ICSMA 15 attracted 352 scientists from 30 countries with one fourth of the participants being students. This is a very good ratio showing that we could attract the young generation. There have been 272 oral and 135 poster presentations. It is our pleasure to thank the members of the International ICSMA Committee for their valuable help, especially for proposing and choosing the 18 plenary speakers. 187 papers were submitted for publication in the proceedings, 167 were accepted after reviewing. We would like to express our thanks to all referees for their efficient and prompt efforts. We acknowledge particularly support from the German Research Society (DFG), the Saxon Ministry for Science and Art and the City of Dresden. We are also grateful for industrial support from PLANSEE Metall GmbH, Goodfellow GmbH, MTS Systems GMB, Nagoya University and IOP Publishing. Finally we thank all members of the Local Organizing

Committee, Intercom Dresden and Conwerk / Laboratory Ten for the excellent organization of ICSMA 15 and the very pleasant collaboration.

During the conference the International ICSMA Committee decided to convene the next conference in Bangalore, India, in 2012. We wish the organizers of ICSMA 16 great success and look forward to meeting you in Bangalore.

Werner Skrotzki (Technische Universität Dresden)
Carl-Georg Oertel (Technische Universität Dresden)
Horst Biermann (Bergakademie Technische Universität Freiberg)
Martin Heilmaier (Technische Universität Darmstadt)
Guest Editors
Dresden, June 23, 2009
(* Corresponding author; e-mail address: werner.skrotzki@physik.tu-dresden.de)

All papers published in this volume of *Journal of Physics: Conference Series* have been peer reviewed through processes administered by the proceedings Editors. Reviews were conducted by expert referees to the professional and scientific standards expected of a proceedings journal published by IOP Publishing.

Low and high temperature deformation mechanisms in TiAl alloys

A. Couret [1,2]

[1] CNRS ; CEMES (Centre d'Elaboration de Matériaux et d'Etudes Structurales) ; BP 94347, 29 rue J. Marvig, F-31055 Toulouse, France
[2] Université de Toulouse ; UPS ; F-31055 Toulouse, France

couret@cemes.fr

Abstract. The deformation mechanisms involving ordinary dislocations are studied at various temperatures, in single-phased and lamellar TiAl alloys. *In situ* straining and heating experiments are performed in the transmission electron microscope (TEM) as well as *post-mortem* studies of deformed samples. It is shown that the dislocations can move by glide or climb depending on the temperature range and that some dynamic strain aging occurs at intermediates temperatures. In lamellar alloys, the crossing of interfaces is found to play a determining role.

1. Introduction

During the last two decades, gamma titanium aluminides (γ-TiAl) have attracted considerable interest among industrial and research institutions for structural applications at high temperatures. This is due to a unique combination of properties, i.e. reduced weight, high temperature strength and good oxidation resistance. According to the phase diagram, γ-TiAl alloys can solidify in a single phase state only when Al content exceeds the composition of the peritectic point corresponding to the reaction: L + α → γ. On the other hand, two phased alloys can be achieved with compositions in the (α + γ) two phase region. In particular, slow cooling of α grains provides lamellar grains, which are formed by an irregular piled-up of γ ($L1_0$) and α_2 ($D0_{19}$) lamellae. Depending on the respective proportion of γ and lamellar grains, the microstructures were classified in four groups: near-γ, duplex, near lamellar and fully lamellar [1]. The plasticity of these γ TiAl alloys is mainly localised in the γ phase. Near-γ microstructures and fully lamellar ones which have generally large grains suffer of a poor ductility at room temperature in such way that the best compromise between ductility, strength and fracture toughness is achieved for duplex and refined lamellar alloys. In this context, to understand the mechanical properties of these various γ-TiAl alloys, it appears necessary to study the elementary deformation mechanisms activated in the γ phase and the deformation spreading in the lamellar microstructure.

The γ phase of TiAl alloys has the $L1_0$ tetragonal crystal structure which is a derivative of the fcc structure (Fig. 1).The c/a ratio is about 1.02. As per the <uvw] mixed brack notation, all permutations amongst u and v are allowed while the third index is fixed. Due to the tetragonality of the structure, the <110] and <101] directions are not equivalent to each other. Two types of perfect dislocations are usually considered in literature: a/2<110] ordinary dislocations similar to the dislocation of the fcc

structure and a<101] superlattice dislocations similar to the dislocations of the $L1_2$ ordered structure. A few studies have been performed in single crystals on these perfect dislocations [2-5]. They have shown that the corresponding resolved shear stresses are strongly dependent on the aluminium content and exhibit a so-called yield stress anomaly. Twinning due to the glide of a/6<112] Shockley dislocations in (111) plane is also observed [6]. Note that a/6<211] Shockley dislocations are not operative because their glide would produce a complex stacking fault. Except in single crystals, a<101] superlattice dislocations are scarcely observed in comparison with a/2<110] ordinary dislocations and twinning is the second operative mode.
This paper is devoted to the study of the deformation mechanisms involving ordinary dislocations at different temperatures. It will also be studied how they are moving in the lamellar microstructure. *In situ* and *post-mortem* TEM experiments were performed on $Ti_{48}Al_{48}Cr_2Nb_2$ and $Ti_{50.8}Al_{47}Cr_1Si_{0.2}$ alloys.

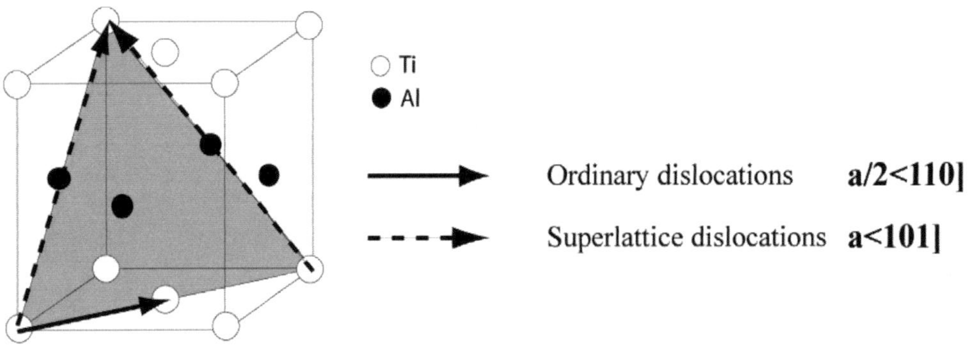

Figure 1. Atomic arrangement and Burgers vector of perfect dislocations in the $L1_0$ structure

2. Glide at low temperature

Fig. 2 shows ordinary dislocations moving during an *in situ* experiment performed at room temperature. These dislocations are elongated along their screw direction and they are anchored at many pinning points. The stereographic analysis of the slip traces left by the dislocations has shown that they are moving in their {111} glide plane. Such elongated and anchored configurations have been observed in deformed samples by *post mortem* TEM studies [7,8]. The dislocations glide by jumps between these elongated positions. Video recordings indicate that the jump times are always smaller than a fiftieth of second and that the duration of the locking varies between 0.02s and 1s.

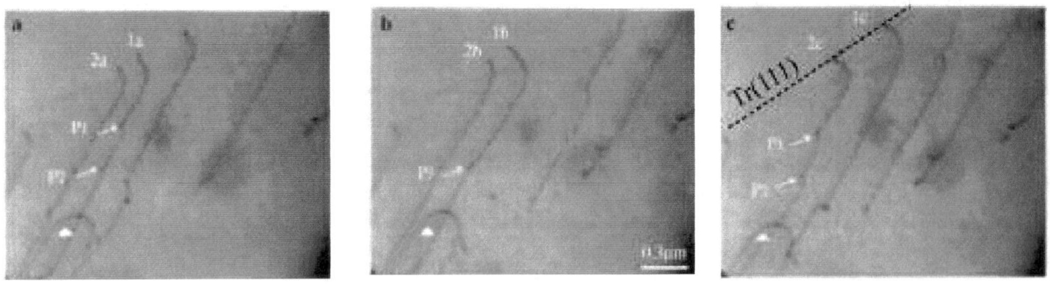

Figure 2. Two ordinary dislocations 1 and 2 gliding in (111) planes. P1, P2 and P3 are pinning points. ($Ti_{48}Al_{48}Cr_2Nb_2$ alloy).

As illustrated in Fig. 3, Shockley dislocations involved in twinning are also anchored at small pinning points. In various TiAl alloys, the separating distance between pinning points anchoring moving ordinary and twinning dislocations has been measured to be 0.15 μm. This demonstrates that the pinning is not due to cross-slip, as proposed in Refs. [7,8] since the Shockley dislocations are not able to cross slip because their the Burgers vector belong to only one {111} plane. Moreover, Fig. 2 shows that successive dislocations 1 and 2 can anchor at the same location (points P1 and P2). From these observations, it is concluded that these pinning points are due to fixed extrinsic obstacles that are randomly distributed in the crystal. This explanation is consistent with the observation of anchoring points on superlattice dislocations [9] and with the measurement of a decrease of their density during annealing treatments [5]. On the basis of atom probe analysis of TiAl alloys [10], the pinning points have been proposed to have a chemical origin, as a local enrichment in titanium, a segregation of interstitials oxygen atoms and/or the ordering of a small region [11].

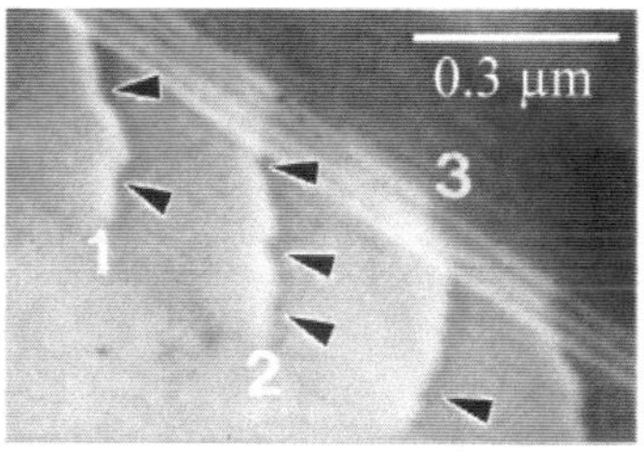

Figure 3. Three Shockley (1, 2 and 3) dislocations gliding on adjacent twinning planes. Triangles indicate pinning points. ($Ti_{48}Al_{48}Cr_2Nb_2$ alloy)

Fig. 4 displays three successive positions of an ordinary dislocation. In view (a), it is anchored at point (P). Between views (a) and (b), the screw segments (S) jumps. Between views (b) and (c), the macrokink at the pinning point glides quickly, forming a long screw segment which jumps and locks in a straight configuration. This rectilinear shape of the screw dislocations indicates the existence of a strong frictional force acting on the screw segments whereas no evidence of the same kind has been obtained for the non-screw segments. This frictional force has been attributed to the core spreading of screw segments in consistency with atomistic calculations [12,13].

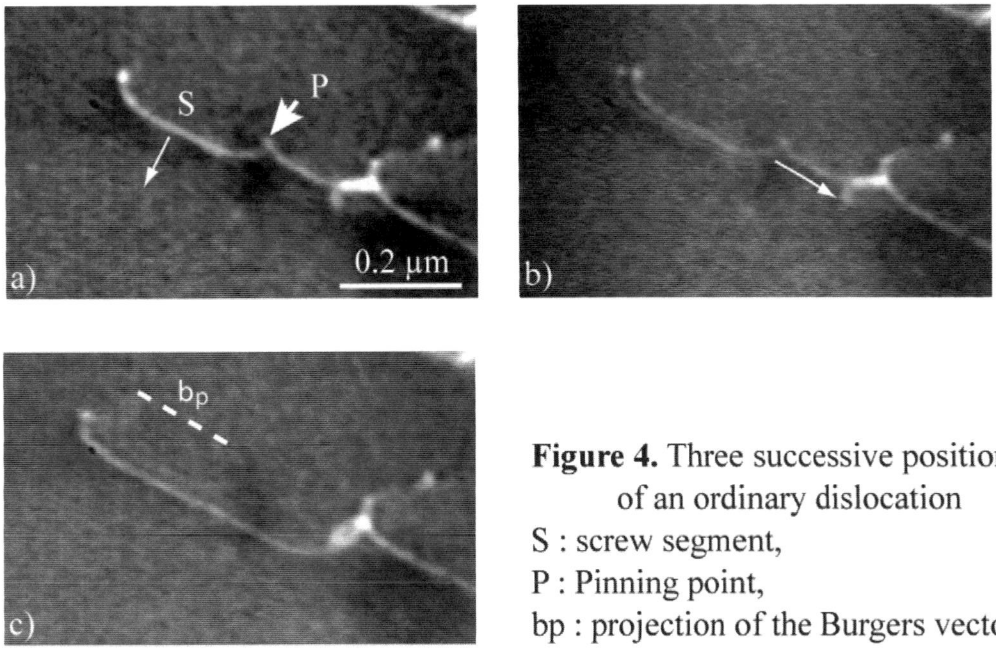

Figure 4. Three successive positions of an ordinary dislocation
S : screw segment,
P : Pinning point,
bp : projection of the Burgers vector.

3. Glide at intermediate temperatures (400°C-600°C)

At intermediate temperatures (T ≥ 400°C), glide ordinary dislocations are still elongated along their screw orientation and anchored at many pinning points (Fig. 5a). However, they are highly concentrated in bands. During *in situ* experiments (Fig. 5), movements are faster and more heterogeneous than at room temperature. When the strain is applied, the deformation occurs suddenly by bursts of dislocations, namely tens or hundreds of dislocations move instantaneously and quickly, and stop (Fig. 5b). Then, these dislocations will stay immobile and will not move again. Further strain application will generate a new burst.

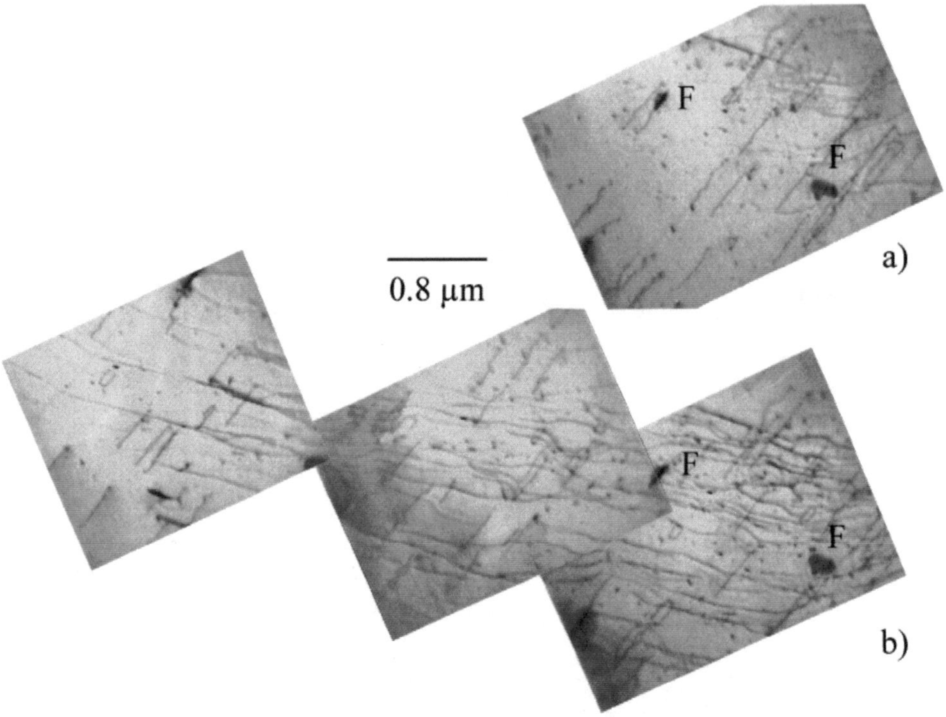

Figure 5. *In situ* observation of the formation of a band of ordinary dislocations at 400°C. F are fixed points in the crystal. ($Ti_{48}Al_{48}Cr_2Nb_2$ alloy)

4. Mixed climb at high temperature

At high temperatures, ordinary dislocations look very different to those described above. As shown on Fig. 6, these dislocations exhibit no preferential direction and are not anchored at pinning points. The planes of some dislocation loops were determined by tilt experiments associated with a stereographic analysis. Fig. 7 shows one example of such determination. The dislocation is tilted around the dashed line joining the two dislocation extremities, which is thus the trace of the intersection between the loop plane and the surface. From the variation of the apparent width of the loop, its plane is determined with an experimental uncertainty of ± 5° [14]. In more than fifty percents of cases, this plane deviates significantly from the (111) glide planes. However, this deviation was never found to be greater than 20°. This indicates that the movement occurs by a mixed climb mechanism, which involves climb and glide components.

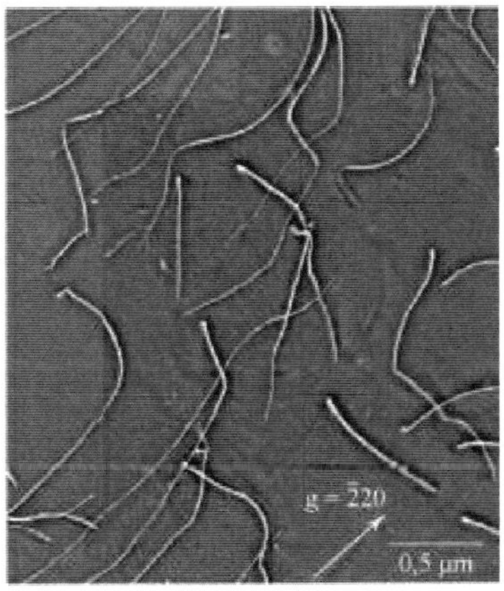

Figure 6. Ordinary dislocations in a sample crept at 750°C under 150 MPa. ($Ti_{48}Al_{48}Cr_2Nb_2$ alloy).

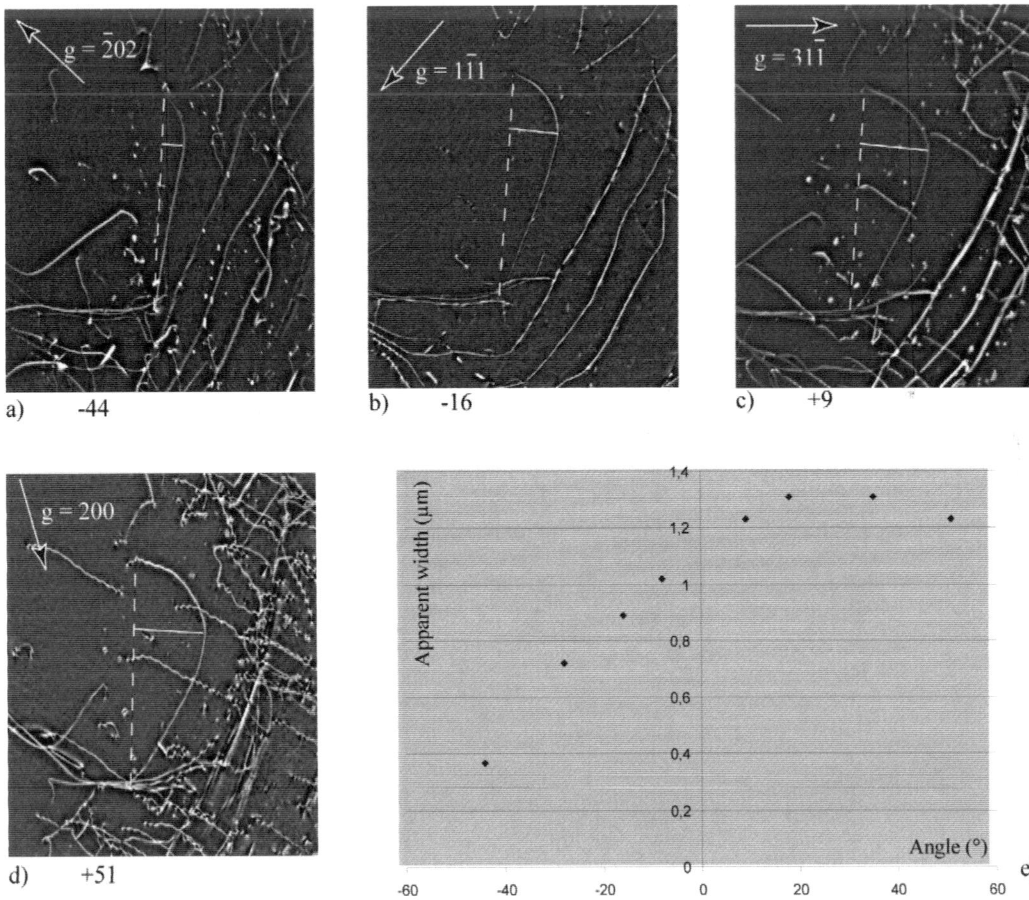

Figure 7. Tilts experiment performed on an ordinary dislocations, in a thin foil cut in a sample crept at 750°C under 150 MPa. ($Ti_{48}Al_{48}Cr_2Nb_2$ alloy).

In situ heating experiments were performed at 750°C on a thin foil cut from a sample crept (2% of strain) at the same temperature (Fig. 8). During this sequence, rectilinear non-screw segments are observed to move viscously, at a very low velocity as indicated by the time information. It has been evidenced that this movement in thin foil occurs in pure climb plane [14].

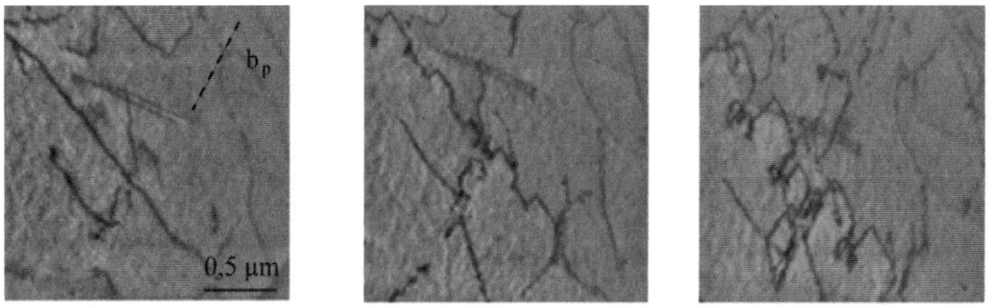

a) 0s b) 2mn 11s c) 5mn 59s

Figure 8. Heating *in situ* experiments performed on a thin foil cut from a sample crept (2% of strain, 750°C and 150 MPa). ($Ti_{48}Al_{48}Cr_2Nb_2$ alloy).

5. Ordinary dislocations in the lamellar microstructure

In consistency with the formation mechanism of γ lamellae during cooling [15-17], the interface planes of the lamellar microstructure are $\{111\}_\gamma$ and $(0001)_{\alpha 2}$ planes. Due to the tetragonality of the structure, the $L1_0$ lamellae can take six orientation variants corresponding to six possible orientations. Between these lamellae, three interface relationships can be found: ordered domain, twin and pseudo-twin [16-17]. Statistical analyses of the lamellar microstructure have shown a predominance of twin related interfaces and that the width of α_2 lamellae is smaller than that of γ lamellae [18].

Fig. 9 shows lamellar zones deformed at 600°C [19]. In both cases, the majority of the lamellae were deformed by ordinary dislocations, the Burgers vector of which is either parallel (view a), or inclined (view b) to the interface plane. These dislocations are elongated along their screw orientation and anchored as described previously. Some loops seem to have been emitted at the interfaces. Few lamellae (view b) are deformed by a combination of twins and ordinary dislocations.

Black arrows in view b (Fig. 9) indicate that ordinary dislocations are localized near areas where twins have impacted the γ/α_2 interface. This suggests that the activation of ordinary dislocations was influenced by the accumulation of stress due to the pile-up of twinning dislocations at interfaces. It has been shown elsewhere that the propagation of deformation across α_2 lamellae occurs by an elastically-mediated transfer under the effect of these dislocation piled-up [20].

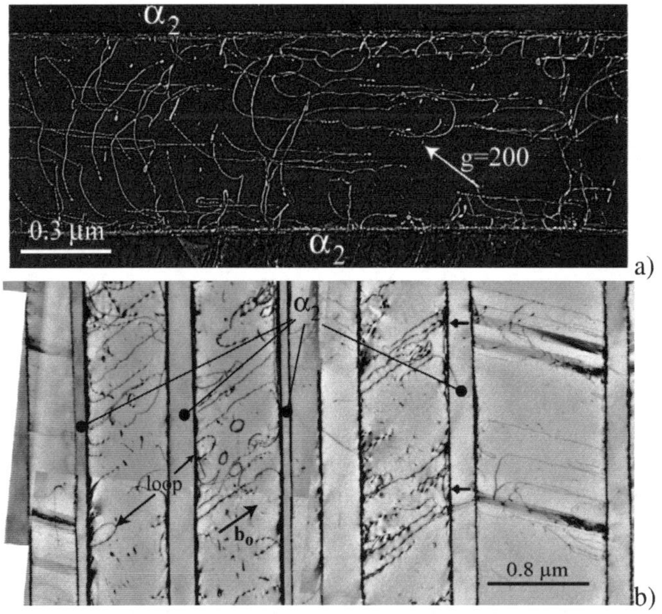

Figure 9. Deformation microstructures in two lamellar zones of a $Ti_{50.8}Al_{47}Cr_1Si_{0.2}$ alloy deformed at 600°C

Fig. 10 shows an example of the transmission of the deformation at interfaces at room temperature. It can be seen that pile-ups of dislocations are formed at the interfaces and that the transmission occurs. Several situations differing by the incident system, the interface relationship and the orientation of the stress were analysed in details [21,22]. In general, part of the incident shearing is transferred under the effect of the internal stress due to dislocation pile-up. The crossing is easier at twin interfaces and when the incident system is formed by a twin. It is promoted by favourable geometric conditions based on orientations of the Burgers vector, of the dislocation lines and of the glide planes.

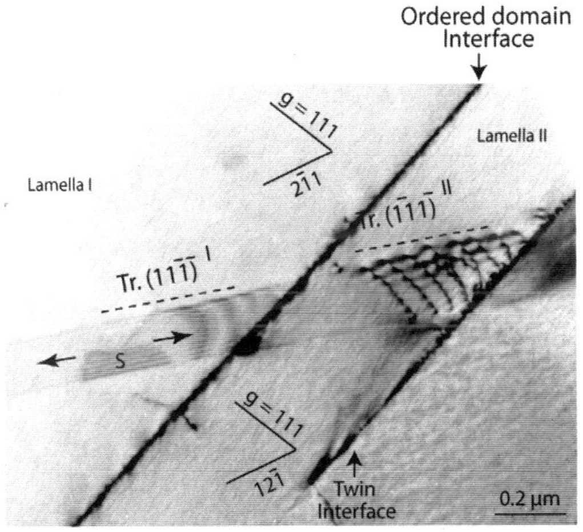

Figure 10. Transmission of the deformation at interfaces in a lamellar alloy. ($Ti_{48}Al_{48}Cr_2Nb_2$ alloy)

6. Discussion

At low temperature, the movement of ordinary dislocations is impeded by the extrinsic pinning points and the frictional force due to the core spreading. In such a case, the pinning points will not control the dislocation movement as long as their separating distance d_s is larger than a critical value d_c given by [23]:

$$d_c = \frac{2T}{\tau b}\cos(\frac{\theta_c}{2}),$$

where T is the line tension, θ_c the critical angle between the two anchored segments when the obstacle breaks, τ the resolved shear stress and b the Burgers vector (b = 0.28 nm). In fact, the dislocations never stop at pinning points as long as there is at least one rectilinear segment able to move. The sequence of events of Fig. 4 illustrates this point since it shows that the pinning point (P) overcoming results of the reaching of the critical configuration by the moving screw segment (S). For the case of ordinary dislocations of TiAl, it has been found that d_s is twice d_c [24]. As a conclusion, the ordinary dislocation glide is controlled by the frictional force, in consistency with the experimental observations showing rectilinear screw segments free of pinning points (see for instance Fig. 4).

The fast and catastrophic behavior observed at intermediate temperatures is correlated to plastic instabilities measured at the same temperature during tensile tests at constant strain rate [5]. It was observed in several intermetallics. It has been attributed to dynamic strain ageing [9,25] and correlated to yield stress anomalies. For ordinary dislocations, the coupling of an easy cross-slip and of a strong frictional force leads to easy multiplication processes by open loops, as identified for instance in titanium [26]. During the deformation at imposed constant strain rate, this intensive dislocation multiplication is at the origin of the deformation by bursts since it induces a decrease of the dislocation velocity leading to a decrease of the stress and eventually to a strong locking of the dislocations due to the frictional force.

At high temperature, during creep experiments (750°C) these ordinary dislocations are generally moving by a mixed climb mechanism, for which the moving plane is close but distinct from the glide plane [14]. The corresponding elementary process is the nucleation and lateral propagation of a jog pair, in consistency with the rectilinear shape of non screw segments. On the one hand, this movement allows the dislocations to escape from the frictional force which is operative only when they are lying along their screw direction in (111) planes. On the other hand, in terms of dislocation displacement, it is more efficient than pure climb for one diffusion event, in consistency with activation parameters measurements [14]. As a matter of fact, the moving distance for one elementary jump is proportional the jog height which is the distance between adjacent (111) planes in the inclined moving plane and this distance is higher for mixed plane due to the inclination of the moving plane.

7. Concluding remark

In the present paper, it has been shown that the frictional force acting on ordinary dislocations of TiAl plays a determining role in the entire temperature range. It controls the movement at low temperature, produces the strong locking of the dislocations which induce instabilities at intermediate temperature and is at the origin of the activation of mixed climb. In lamellar areas, it is also operative but the controlling process is the crossing of the interface.

References

[1] Kim YW and Dimiduk DM 1991 *JOM* **43** 40
[2] Kawabata T, Kanai T and Izumi O 1985 *Acta Metall. Mater.* **38** 1381
[3] Inui H, Matsumuro M, Wu DH and Yamaguchi M 1997 *Phil. Mag.* **A 75** 395
[4] Nakano T, Hagihara K, Seno T, Sumida N, Yamamoto M and Umakoshi Y 1998 *Phil. Mag. Lett.* **78** 385
[5] Grégori F, 1999 *Ph D Thesis* Université de Paris VI
[6] Farenc S, Coujou A and Couret A 1993 *Phil. Mag.* **A 67** 127
[7] Viguier B, Hemker KJ, Bonneville J, Louchet F and Martin J L 1995 *Phil. Mag.* **A 71** 1295
[8] Sriram S, Dimiduk D M, Hazzledine P and Vasudevan V K 1997 *Phil. Mag.* **A 76** 965
[9] Haussler D, Bartsch M, Aindow, Jones I P and Messerschmidt U 1999 *Phil. Mag.* **A 79** 1045
[10] Menand A, Huguet A and Nérac-Partaix A 1996 *Acta. Metall. Mater.* **44** 4729
[11] Zghal S, Menand A and Couret A. 1998 *Acta Metall. Mater.* **46** 5899
[12] Mahapatra R, Girshick, Pope D and Vitek V 1995 *Scripta Metall. Mater.* **33** 1921
[13] Simmons J P., Rao S I and Dimiduk D M. 1997 *Phil. Mag.* **A 75** 1299
[14] Malaplate J, Caillard D and Couret A 2004 *Phil. Mag.* **A 84** 3671
[15] Denquin A and Naka S 1996 *Acta Metall. Mater.* **44** 343
[16] Inui H, Oh M H, Nakamura A and Yamaguchi M 1992 *Phil. Mag.* **A 66** 539
[17] Zghal S, Naka S and Couret A 1997 *Acta Metall. Mater.* **45** 3005
[18] Zghal S, Thomas M, Naka, S, Finel A and Couret A 2005 *Acta Mater.* **53** 2653
[19] Singh JB, Molénat G, Sundararaman M, Banerjee S, Saada G, Veyssière P and Couret A 2006. *Phil. Mag.* **A 86** 2429
[20] Singh JB, Molénat G, Sundararaman M, Banerjee S, Saada G, Veyssière P and Couret A 2006. *Phil. Mag. Lett.* **86** 47
[21] Zghal S, Coujou A and Couret A 2001. *Phil. Mag.* **A 81** 345
[22] Zghal S and Couret A 2001. *Phil. Mag.* **A 81** 365
[23] Caillard D and Couret A 2002 *Mater. Sci. Eng.* **A322** 108
[24] Couret A 1999 *Phil. Mag.* **A 79** 1977
[25] Molénat G, Couret A and Caillard. 1997 *Mater. Sci. Eng.* **234** 66
[26] Farenc S, Caillard D and Couret A 1993 *Acta Metall. Mater.* **41** 2701

Stress-induced phase transformations studied by in-situ transmission electron microscopy

ML Nó[1] , A Ibarra[1], D Caillard[2] and J San Juan[3]

[1] Dpto de Física Aplicada II, Facultad de Ciencia y Tecnología, UPV/EHU. Apdo 644, 48080 Bilbao, Spain

[2] CEMES/CNRS, BP 4347, 29 rue J. Marvig, 31055 Toulouse Cedex, France

[3] Dpto de Física de la Materia Condensada, Facultad de Ciencia y Tecnología, UPV/EHU. Apdo 644, 48080 Bilbao, Spain

maria.no@ehu.es

Abstract. In this work, we carry out a detailed study, by in-situ Transmission Electron Microscopy (in-situ TEM), focused on two single-crystals of Cu-Al-Ni shape memory alloys with different transformation temperatures. The first single crystal is in beta phase at Room Temperature (RT) and has been cycled under stress, by super-elastic effect, inside the TEM. Two different mechanisms for the nucleation of β'_3 and γ'_3 martensite phases were observed: a) Martensite can nucleate on dislocations during super-elastic tests and when withdrawing the stress, the reverse transformation takes place by the disappearance of the martensite plate on the dislocation. b) During mechanical cycling martensite plates nucleate in other plates. The second single crystal is in martensite phase at RT, and when the stress is applied different mechanisms are observed: a) Reorientation and interface motion of the plates under the external applied stress, b) nucleation of mobile dislocations inside the martensite. A quantitative analysis of the experimental results, having into account the images and the diffraction patterns, has been realized and different mechanisms have been proposed to explain the experimental results.

1. Introduction

It is well known that TEM is a very powerful technique to study the microstructure of materials because it can give information on the real and reciprocal space from the same sample region. In particular in-situ TEM under cooling and heating is commonly used to study phase transformations in small volumes. With a similar idea, the study of structural phase transformations able of being induced under stress by loading and unloading cycles inside the microscope could be very useful. In particular the in-situ TEM study of the super-elastic and pseudo-elastic effects exhibited by Shape Memory Alloys (SMA) must give information about the sites for martensite nucleation (volume, dislocations etc.), the kind of martensites, the plates favored by the stress, the interface planes, austenite-martensite and martensite-martensite, as well as the microstructure evolution taking place during cycling.

Up today very few studies, concerning the super-elastic and pseudo-elastic effects of SMA [1], have been realized by in-situ TEM [2-7]. However, in-situ TEM experiments could give a unique information at nano-scale, which at present is becoming of paramount importance to understand and master the super-elastic and pseudo-elastic behavior observed in micro and nano devices, where even size effects have been recently observed [8, 9]. So, the objective of this work is to obtain microscopic information at nano-scale about the mechanisms taking place during super-elastic and pseudo-elastic effects on Cu-Al-Ni SMA, by using different in-situ TEM experiments.

2. Experimental

Two single crystals with compositions Cu-27.96%Al-3.62%Ni(%at) (sample I) and Cu-27.14%Al-4.54%Ni(%at) (sample II) were grown by the Stephanov method. From these single crystals, tensile samples 5mm x 1mm in cross-section and 25 mm in gauge length between the two griping heads, were spark-machined with $[001]_{L21}$ and $(0\text{-}10)_{L21}$ as tensile axis and plane of the sample respectively (Samples IA and IIA). Other samples 50mmx5mmx1mm with the same orientation than the tensile samples were also cut (samples IB and IIB) for TEM observations. All the samples were annealed at 1173K for 1800s in argon atmosphere and quenched in iced water in order to prevent precipitation and to obtain the β_3 metastable $L2_1$ (Fm-3m, a=5.82nm) ordered phase [10]. Furthermore samples I and II were aged during 4h and 24h respectively at 453K to increase the transformation temperatures [11]. Moreover the surfaces were mechanically polished to remove the damaged surface and the oxide layer produced during thermal treatments.

Thin foil samples for TEM were obtained from samples IB and IIB by mechanical polishing, until a thickness between 100μm and 200μm and punching in 3mm diameter discs. Then, they were electro-polished in a Struers Tenupol-5, using a mixture of 33% NHO_3 in methanol at about 258K. The TEM analysis of annealed samples was performed in a Philips CM200 with Gatan double-tilt and double-tilt heating holders.

The transformation temperatures were measured at 5% and 95% of the transformed volume fraction, calculated from the integral of the normalized entropy change associated to the martensitic transformation [12], measured by Differential Scanning Calorimetry (DSC) at a temperature rate of 10K/min in a Perkin Elmer DSC7. The transformation temperatures β_3 <-> γ'_3 remain below RT for sample I, (M_s = 233K, M_f = 231K, A_s = 271K, A_f = 276K), while for sample II the transformation temperatures β_3 <-> β'_3+ γ'_3 are above RT (M_s = 329K, M_f = 317K, A_s = 325K, A_f = 363K). Consequently super-elastic and pseudo-elastic effect tests can be carried out at room temperature for samples I and II respectively.

The super-elastic and pseudo-elastic effects have been studied, on samples IA and IIA respectively, at RT with a 4467 Universal Instron test machine at a strain rate of $6.67\text{x}10^{-4}\text{s}^{-1}$, and using an extensometer to measure the strain.

Rectangular 3 x 1mm^2 samples with the same orientation as IA tensile samples (tensile axis $[001]_{L21}$, foil normal $[0\text{-}10]_{L21}$) were also cut by spark cutting from sample IB to perform the super-elastic cycling inside the TEM. In addition, samples with the same shape and orientation were cut from sample IA after 54 previous cycles in the testing machine in order to continue the in situ super-elastic cycling at RT in the microscope.

Finally samples with the same rectangular shape were cut from sample IIB in order to observe pseudo-elastic behavior when the stress is applied in-situ at the TEM.

All the in situ TEM samples were mechanically polished to 40μm thickness and final electro-polished to electron transparency. Samples IA and IB were mounted in a single tilt Gatan tensile holder with $[001]_{L21}$ in the tensile direction (in coincidence with the goniometer tilt axis) and $[0\text{-}10]_{L21}$ perpendicular to the foil. The holder was introduced in a JEOL JEM2010 TEM and a digital video recorder was used to register the in-situ observations. The observed areas have been chosen having into account the simulation realized by Coujou et al. [13] in order to have the tensile stress in the same direction that the external applied stress $\sigma=[001]_{L21}$.

3. Super-elastic effect: Results and discussion

A preliminary study of the microstructure before thermo-mechanical tests was carried out on sample I. Antiphase boundaries (APB), corresponding to B2 and $L2_1$ order, have been observed by TEM, as for the quenched samples [14]. The microstructure of dislocations (Figure1a) is constituted by: a) prismatic loops in the foil plane $(0\text{-}10)_{L21}$ with **b** = (1/2) $[0\text{-}10]_{L21}$, b) straight edge dislocations with **b**= (1/2) $[100]_{L21}$, **b**= (1/2) $[001]_{L21}$ and vector line $[001]_{L21}$ and $[100]_{L21}$ respectively, coming from prismatic loops on planes $(100)_{L21}$ and $(001)_{L21}$, c) mixed dislocations with vector line $\langle 111\rangle_{L21}$ and **b** = (1/4) $\langle 111\rangle_{L21}$, d) other mixed dislocations. A detailed analysis of this microstructure has been carried out by LACBED [15]. Figure 1b shows the stress-strain curves for sample IA, after 34 and 54 cycles for a maximum strain about 5.5%. It can be observed that there is no remaining deformation after the completely repetitive super-elastic cycles. However when the stress is applied on sample IIA in the $\beta'_3 + \gamma'_3$ martensitic phase the pseudo-elastic behaviour takes place by the interfaces motion during loading-unloading and a 7.5% remaining deformation is observed, which will be recovered when the sample is heated at a temperature higher than A_f.

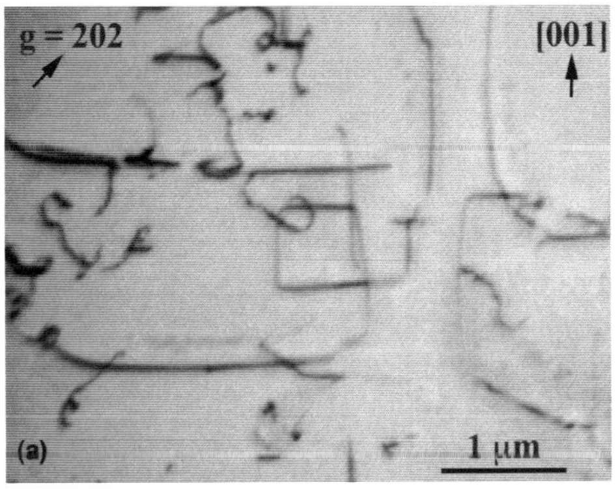

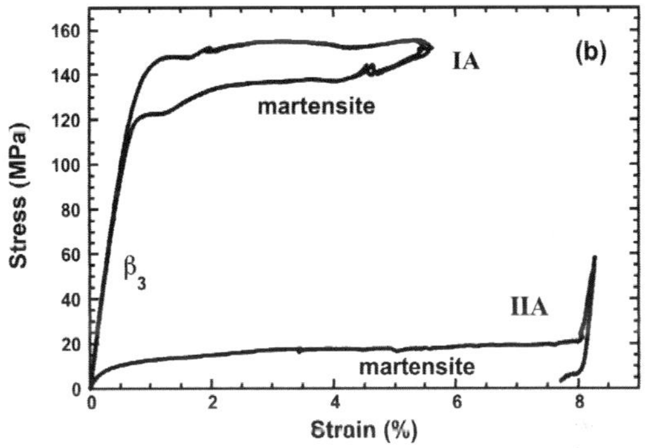

Figure 1a. Microstructure of sample IB just before applying the stress. TEM 200KV Bright field. Figure 1b. Curves IA: Super-elastic effect of sample IA. Cycles 34 and 54 are represented. Curve IIA: Pseudo-elastic effect for sample IIA in martensite phase.

3.1. Martensite nucleation on dislocations

When the stress is applied on sample IB it can be observed, during the loading cycles 3 to 13, the nucleation and growing of different martensite plates on dislocations. These events correspond to the direct transformation β_3->martensite. When the stress is withdrawn, the reverse transformation martensite->β_3 takes place and the martensite plate disappears on the dislocation. Figure 2 shows the last 34 seconds of the reverse transformation martensite-β_3 during unloading of the eighth cycle σ-ε. As shown in Figure 2e, the martensite plate disappears just in the dislocation.

The line vector $\mathbf{u}=[1\text{-}1\text{-}2]_{L21}$ of the dislocation was determined by trace analysis method for different tilts, but the Burgers vector cannot be univocally determined by the extinction method due to the high anisotropy of the Cu-Al-Ni alloys (A≈13 [16]). It cannot either be determined by weak beam or large convergent beam techniques with the holder and microscopes employed. The extinction method gives three possible Burgers vectors: $\mathbf{b}=(1/4)[1\text{-}1\text{-}1]_{L21}$, $\mathbf{b}=(1/4)[1\text{-}11]_{L21}$ and $\mathbf{b}=(1/2)[001]_{L21}$ but in any case the dislocation glide plane is $(\text{-}1\text{-}10)_{L21}$.

Figure 3 shows, under a different diffraction condition, the nucleation and growth of the martensite plate on the same dislocation, during the loading of the tenth cycle σ-ε. The Select Area Diffraction Pattern (SADP) realized on the plate at the end of the tenth cycle (Figure 3e), shows that it is γ'_3 martensite (Pmmn, a=0.53424nm, b=0.42244nm, c=0.43896 [17, 5]). Its basal plane $(020)_{\gamma'3}$ comes from the $(202)_{L21}$ plane of the β_3 phase. The shear direction $[001]_{\gamma'3}$, near the $[\text{-}101]_{L21}$, and the normal plane takes about 45° with the stress direction $[001]_{L21}$. If the equivalence with dislocations is made, the Schmidt factor is near 0.5 and this plate is preferentially favored by the stress.

The previous results show that γ'_3 martensite can be nucleated on mixed dislocations, in a similar way than β'_3 martensite, which was shown to be able of nucleating in dislocations [5]. Along σ-ε cycling, the dislocation is not modified during martensite nucleation, what means that the dislocation is compatible with both phases β_3 and γ'_3. Among the three possible Burgers vectors for the dislocation, $\mathbf{b}=(1/4)[1\text{-}1\text{-}1]_{L21}$ is the only one belonging to the basal plane of the γ'_3 martensite and we can assume that this is the Burgers vector of the dislocation. Consequently the dislocation has almost a screw character, which will be promoted by the applied stress in the $[001]_{L21}$ direction. In this situation the atomic shearing along $[10\text{-}1]_{L21}$ introduced by the dislocation is 0.133a√2 (Figure 4a and 4b) when the dissociation of the $<111>_{L21}$ screw dislocations in B2 ordered alloys is considered [18,19]. Figure 4c shows γ'_3 martensite with their basal plane $(020)_{\gamma'3}$ parallel to the $(202)_{L21}$ (Figure 4a) as was experimentally observed. In this situation the shearing produced between the β_3 and γ'_3 structures is 0.137a√2. Consequently, in the core of dislocation pre-exists the atomic shearing that γ'_3 martensite requires to be nucleated, and when the applied stress has a high Schmidt factor, for a specific shear direction and basal plane, the martensite plate is nucleated in the dislocation core.

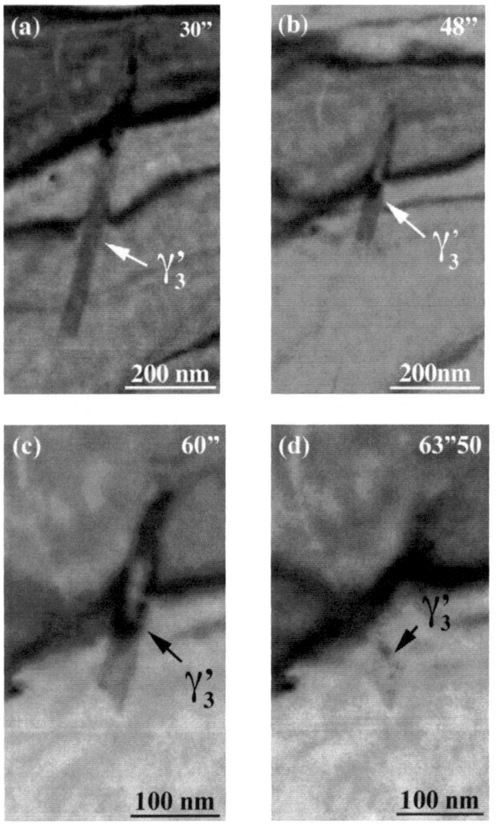

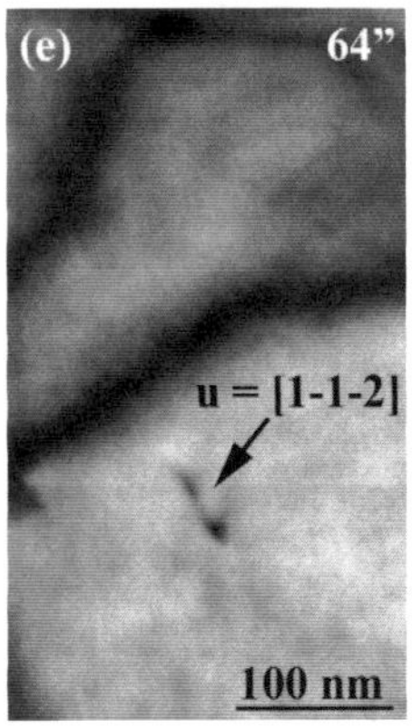

Figure 2. Eighth mechanical cycle on the sample, when the external stress is withdrawn.
a-d) γ'_3-β_3 reverse transformation. The size of the martensitic plate decreases versus time (indicated in pictures.
e) The martensite plate disappears on dislocation **u**.
TEM 200KV Bright field. **B**≈$[0\text{-}10]_{L21}$

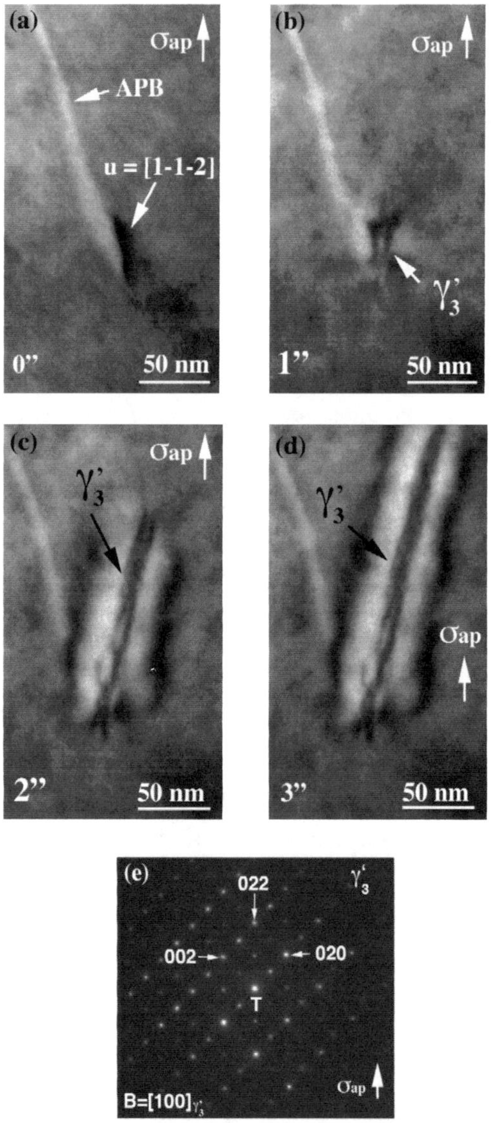

Figure 3. Tenth mechanical cycle on the sample. The external stress is applied.
a-d) β_3-γ'_3 direct transformation. Nucleation and growth of the plate versus time, which is indicated on the pictures. **B**≈$[1\text{-}20]_{L21}$
e) SADP of the martensite plate. **B**≈$[100]_{\gamma'3}$
TEM 200KV Bright field.

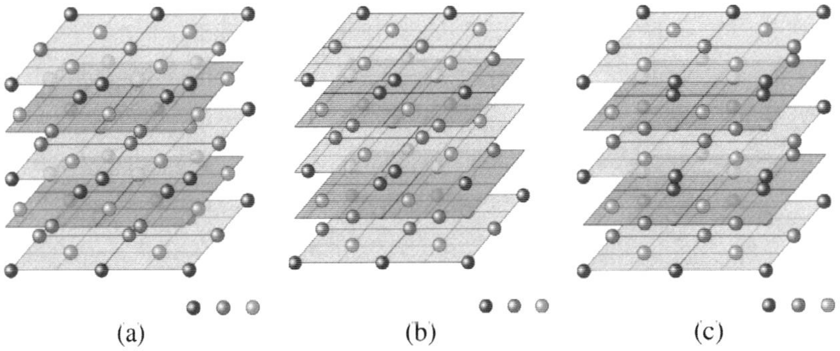

Figure 4. a) β_3 ($L2_1$ ordered structure. The (101) planes have been represented.
b) Shearing produced by $(1/4)[1\text{-}1\text{-}1]_{L21}$ screw dislocation on the (101) plane.
c) γ'_3 martensite. The basal planes have been represented.

3.2. Martensite nucleation on other martensite

Figure 5a shows a martensite plate (A1) nucleated when the stress is applied on sample IA, after the 54 previous σ-ε cycles from Figure 1b. The SADPs in Figures 5b and 5c were taken on martensite plate A1 and β_3 phase respectively. The martensite plate A1 can be indexed as β'_3 (C2/m, a=1.38017nm, b=0.52856nm, c=0.43987, β=113.60° [4]) and it can be observed that the basal plane $(600)_{\beta'3}$ comes from the $(202)_{L21}$ plane. Considering the nomenclature used by Saburi and Wayman [20] this plate will be named as variant 4 ($\beta'_{3\text{-}4}$). The streaks along $[600]^*_{L21}$ direction in the reciprocal space are perpendicular to the defects observed on the basal planes in the Bright field (BF) image (Figure 5a). Having into account the twinning planes for β'_3 martensite and the SADP of Figure 5b, these streaks can only be interpreted as stacking faults in the basal plane. The transformation β_3-> $\beta'_{3\text{-}4}$ continues growing, under the stress effect, along the $[701]_{L21}$ direction, and other plate (A'1) is nucleated in the interface between the β_3 phase and the $\beta'_{3\text{-}4}$ martensite plate A1 (Figure 5a and 5d). The SADP corresponding to the A'1 plate (Figure 5e) show that this plate is also β'_3 martensite with stacking faults in its basal plane $(600)_{\beta'3}$ coming from the $(-202)_{L21}$ plane. It corresponds to variant 3 from Sabury and Wayman nomenclature [20] and will be named as $\beta'_{3\text{-}3}$. The analysis of the SADP taken on the interface $\beta'_{3\text{-}4}$- $\beta'_{3\text{-}3}$ [21] shows that the interface plane is the twinning plane $(201)_{\beta'3\text{-}4}||(\text{-}20\text{-}1)_{\beta'3\text{-}3}$. In Figure 5f the stereographic projection for β_3 phase, $\beta'_{3\text{-}3}$ and $\beta'_{3\text{-}4}$ martensites, as well as the applied stress direction, have been represented. For $\beta'_{3\text{-}4}$ martensite it can be observed an angle near to 45° between the shear direction $[00\text{-}1]_{\beta'3\text{-}4}$ and the stress direction $[001]_{L21}$. Also an angle of about 45° exists between the normal plane $(600)_{\beta'3\text{-}4}$ and the stress direction. Equivalent results for $\beta'_{3\text{-}4}$ martensite are observed in Figure 5f. Consequently we can assume that the nucleation of martensite is favored by the stress, similarly to the mobility of dislocations, when the shear direction is considered as the Burgers vector and the basal plane as the gliding plane of dislocations [21]. Directions $[7\ 0\ 1]_{L21}$ and $[70\text{-}2]_{L21}$ correspond to the intersection between the $\beta'_{3\text{-}4}$ and $\beta'_{3\text{-}3}$ martensite habit plane and the $\approx(0\text{-}10)_{L21}$ sample plane respectively, showing that both martensites take part of the self-accommodating groups I or II from Saburi et al. [22]. Variants 5 and 6 of the same self-accommodating groups are not favored by the stress and they are not observed. The explanation is given by the fact that the normal to the basal planes and their shear directions are at 90° with respect the applied stress direction and consequently these basal planes are not favored by the stress. Finally when unloading the sample, the reverse transformation from $\beta'_{3\text{-}4}$ and $\beta'_{3\text{-}3}$ martensites to β_3 phase was experimentally observed, in agreement with the macroscopic super-elastic tests (Figure 1b).

When a stress is applied on sample IB, new B martensite plates were nucleated and propagated on the sample. They are β'_3 martensite plates and parallels between them [4]. When deformation increases new plates are nucleated at the interface β'_3- β_3, and these new martensite plates can be γ'_3 type [21]. Figure 6a shows a martensite plate A nucleated on one of the two B plates, growing until it reaches the other plate B, and the interfaces A-B are indicated by arrows. The SADPs in Figure 6b and 6c correspond to A plate (β'$_3$ martensite) and B plates (γ'$_3$ martensite) respectively. A detailed analysis of the interface plane has been realized in [21] for an equivalent interface β'$_3$- γ'$_3$.

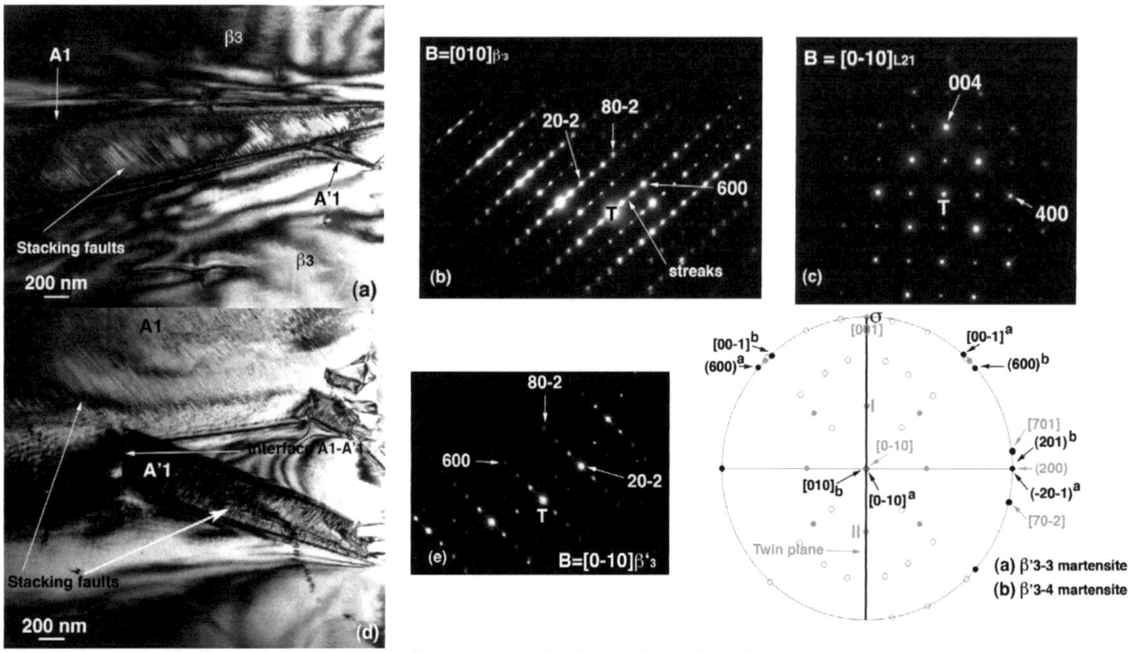

Figure 5. TEM 200KV. Martensite variants A1 and A'1 nucleated under the stress σ. a) BF B≈[0-10]$_{L21}$. b) A1 martensite plate SADP. c) β$_3$ phase SADP. d) BF B≈[0-10]$_{L21}$. e) A'1 martensite plate SADP. f) Stereographic projection: Black-a β'$_{3\text{-}3}$ martensite, black-b β'$_{3\text{-}4}$ martensite, grey β$_3$ phase.

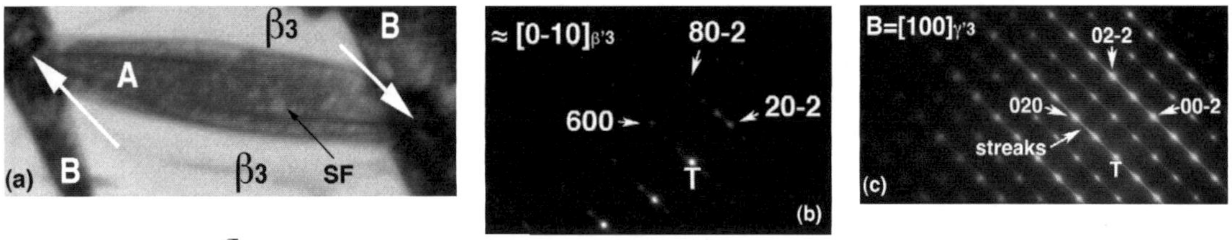

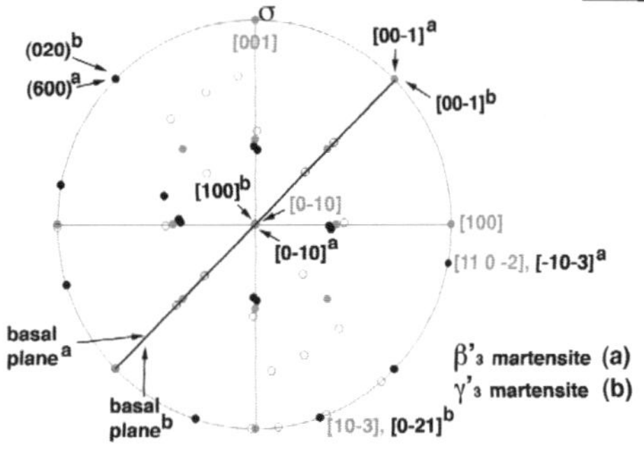

Figure 6. TEM 200KV. Martensite variants A and B nucleated under the stress σ. a) BF B≈[0-10]$_{L21}$. b) A martensite plate SADP. c) B martensite plate SADP. f) Stereographic projection: Black-a β'$_3$ martensite, black-b γ'$_3$ martensite, grey β$_3$ phase.

The stereographic projection for both martensites and the β$_3$ phase is represented in Figure 6d. The interface plane A-B is the basal plane $(600)_{\beta'3}||(020)_{\gamma'3}$. Both plates are favored by the stress because the shear direction $[00\text{-}1]_{\beta'3}||[00\text{-}1]_{\gamma'3}$ for both plates takes 45° with the stress direction [001]$_{L21}$ and an angle of 45° exist between the normal planes $(600)_{\beta'3}||(020)_{\gamma'3}$ and the stress direction. Directions [11 0-2]$_{L21}$ and [1 0-3]$_{L21}$ correspond to the intersection between the β'$_3$ and γ'$_3$ martensite habit planes and the $\approx(0\text{-}10)_{L21}$ β_3 plane respectively, showing that both martensites belong to different self-accommodating groups having into account the Saburi et al. [22] nomenclature. A and B plates belong to groups I (or II) and V (or VI) respectively. Once again, when the sample is unloaded the reverse transformation $\beta'_3 + \gamma'_3$ to β_3 takes place.

We can summarize that for sample I and $[001]_{L21}$ stress direction, two β'_3 plates of martensite with a twin relationship can be nucleated. Moreover β'_3 and γ'_3 plates can also be nucleated when the interface plane is the basal plane. In both cases the shear direction and the normal basal planes must be favored by the stress ($\approx 45^{o}$ between shear direction and normal basal planes with the stress direction).

4. Pseudo-elastic effect: Results and discussion

Sample II is in martensite phase at RT and so different mechanisms are observed when the stress is applied. Figure 7 shows the nucleation of dislocations on the surroundings of an interface between two martensite plates, then they move through a γ'_3 plate. This kind of phenomena, although it has been scarcely observed, indicates that γ'_3 martensite can be plastically deformed by creation and motion of dislocations when their Schmidt factor is high enough. Due to the ability of the γ'_3 martensite for easily twinning it can be suggested that the dislocation mechanisms must be produced when twinning is not favored by the stress.

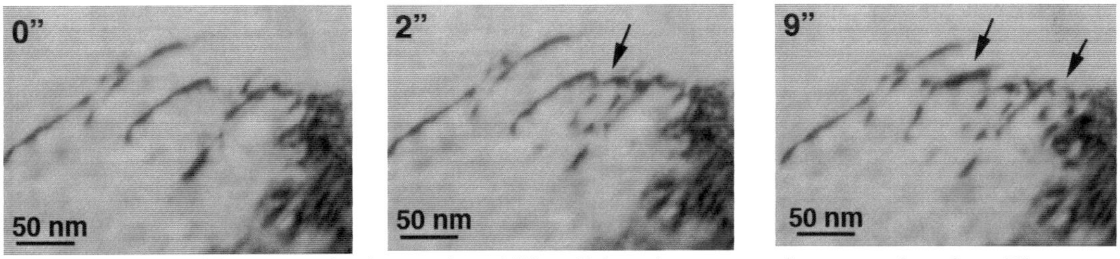

Figure 7. TEM 200KV. Nucleation and mobility dislocations on a γ'_3 martensite plate. The arrows show the positions where a dislocation is moving.

On other hand, during loading it can also be observed how the different γ'_3 and β'_3 martensite plates are reoriented by the effect of the stress. In Figure 8 it is show how β'_3 martensite plate is the more favored by the stress and then the interface between both plates moves by steps due to the presence of the stacking faults in the basal planes of both martensites. The interface position and shape change versus time and can be observed by comparing the figures 8a, 8b and 8c.

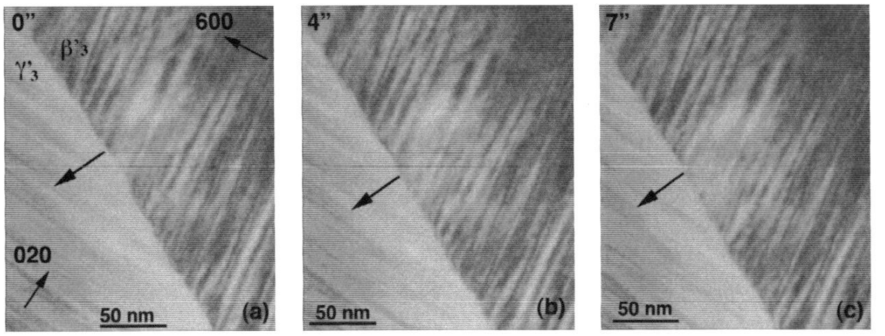

Figure 8. TEM 200KV γ'_3-β'_3 interface. The arrow shows the mean motion direction of the interface versus time. The interface is not flat and moves by steps.

5. Conclusions

The experimental results discussed on the previous sections show us that in-situ TEM is a good choice to analyze the microscopic mechanisms taking place during super-elastic and pseudo-elastic effects.

In particular during super-elastic effect on Cu-Al-Ni SMA, it has been observed that both γ'_3 and β'_3 martensite plates can be nucleated on dislocations as well as in the interfaces between other plates and the beta phase. In all cases the nucleation of plates must be preferentially favored by the stress and for that, the shear direction and the normal to the basal plane must be at about 45° from the stress direction in order to have a high Schmidt factor. In these conditions, the core of dislocations near to the screw character and with the adequate orientation versus the stress can be favorable nucleation points for both kind of martensite variants γ'_3 and β'_3. When σ is applied in the $[001]_{L21}$ direction only two kinds of interfaces are favored by the stress: a) The twin plane $(201)_{\beta'3\text{-}4} \,||\, (\text{-}20\text{-}1)_{\beta'3\text{-}3}$ for variants 3 and 4 from the self-accommodating groups I or II, b) The basal plane $(600)_{\beta'3} \,||\, (020)_{\gamma'3}$ as interface plane, but in this last case β'_3 and γ'_3 martensites are variants from different self-accommodating groups, I (or II) for β'_3 and V (or VI) for γ'_3.
During pseudo-elastic effect the plastic deformation of the γ'_3 phase has been observed by nucleation and mobility of dislocations inside the martensite plates. In addition we have also observed the stepped mobility of the γ'_3-β'_3 interfaces due to the presence of numerous stacking faults in both structures.

Acknowledgments
This work was supported by the MAT2008-02759 and the Consolider-Ingenio2010 CSD2009-00013 projects from the Spanish Ministry of Science and Innovation, the IT-434-07 Consolidated Group and project ACTIMAT-08 from ETORTEK program of the Basque Government.

References

[1] Shape Memory Materials, ed. K. Otsuka and C.M. Wayman. *Cambridge University Press* 1998
[2] Delaey L, Guenin G, Murakami Y, Gobin PF, Van Tendeloo G and Van Landuit J 1985 *Phys. Stat. Solidi* A **89** 457
[3] Kröger A, Wernhardt R, Somsen Ch, Eggeler G and Wieck A 2006 *Mater. Sci. Eng. A* **438-440** 513
[4] Ibarra A, San Juan J, Bocanegra EH, Caillard D and Nó ML 2006 *Mater. Sci. Eng. A* **438-440** 787
[5] Ibarra A, Caillard D, San Juan J and Nó ML 2007 *Appl. Phys. Lett.* **90** 101907-1
[6] Tirry W and Schrivers D 2008 *Mater. Sci. Eng. A* **481-482** 420
[7] Kröger A, Dziaszyk S, Frenzel J, Somsen Ch, Dlouhy A and Eggeler G 2008 *Mater. Sci. Eng. A* **481-482** 452
[8] San Juan J, Nó ML and Schuh CA 2008 *Adv. Mater.* **20** 272
[9] San Juan J, Nó ML and Schuh CA 2009 *Nature Nanotechnol.* **4** 415
[10] Pérez-Landazábal JI, Recarte V, Pérez-Sáez RB, Nó ML, Campo J and San Juan J 2002 *Appl. Phys. Lett.* **8** 1794
[11] Recarte V, Pérez-Sáez RB, Nó ML, and San Juan J 1999 *J. Mater. Res.* **14** 2806
[12] Ortín J, Mañosa LL, Friend CM, Planes A and Yoshikawa M 1992 *Phil. Mag. A* **65** 461
[13] Coujou A, Lours Ph, Roy NA, Caillard D and Clement N 1990 *Acta Metall. Mater.* **38** 825
[14] Morniroli JP, Nó ML, Rodriguez PP, San Juan J, Jezierska E, Michel N, Poulat S and Priester L 2003 *Ultramicroscopy* **98** 9
[15] Rodriguez PP, San Juan J, Morniroli JP and Nó ML 2002 *Proc. Int. Conf. on Electron Microscopy (Durban)* vol 1 (ISBN 0-620-29294-6) p 647
[16] Suezawa M and Sumino K 1976 *Scripta Metall.* **10** 789
[17] Ye J, Tokonami M and Otsuka K 1990 *Met. Trans. A* **21** 2669
[18] Vailhe C and Farkas D 1997 *Acta Mater.* **45** 4463
[19] Vailhe C and Farkas D 1997 *J. Mater. Res.* **12** 2559
[20] Saburi T and Wayman CM 1979 *Acta Mater.* **27** 979
[21] Nó ML, Ibarra A, Caillard D and San Juan 2010, *to be published*
[22] Saburi T, Wayman CM, Takata K and Nenno S 1980 *Acta Mater.* **28** 15

Severe plastic deformation and phase transformations

X. Sauvage, A. Chbihi, X. Quelennec

University of Rouen, Groupe de Physique des Matériaux, CNRS (UMR 6634), Avenue de l'Université - BP 12, 76801 Saint-Etienne du Rouvray, France

xavier.sauvage@univ-rouen.fr

Abstract. Large densities of crystalline defects are created in severely deformed metallic alloys, pushing systems far from the thermodynamic equilibrium. Thus, strain induced phase transformations may occur and affect the mechanical properties like in the well known white etching layer on the surface of rail tracks. Here we briefly report about different phenomena reported in the literature such as solid state amorphization, super saturated solid solutions, precipitate dissolution and disordering. Phase separation and precipitation in nanostructured metallic alloys by severe plastic deformation prior to aging are also discussed.

1. Introduction

Deformation mechanisms in metallic alloys have been widely investigated and are nowadays generally well understood. On the basis of this knowledge, material scientists can design microstructures for an optimal mechanical behavior under service condition. The understanding of the evolution of microstructures subjected to plastic deformation is also extremely important, especially for alloys used in metal forming processes. High levels of plastic deformation at low homologous temperature (T<1/3 Tm) are usually considered to be negative because of the resulting poor ductility, internal stresses and potential damages. There are however some situations where severe plastic deformation (SPD) is unavoidable, like on the surface of rail tracks. A so-called white etching layer (WEL) is commonly observed on the surface which illustrates the detrimental effects of SPD [1]. The WEL is a severely deformed layer resulting from the rolling friction of train wheels under high contact pressure. It is extremely hard but brittle and susceptible to crack formation. The mechanism for the formation of this phase from the pearlitic parent matrix has long puzzled metallurgists as it exhibits many features of quenched martensite but does not involve prior austenitisation. It has been shown that it is due to the transformation of the original pearlite into a nanocrystalline structure where the cementite phase is completely decomposed [2, 3]. It is interesting to note that similar steels are used to make tire cords by drawing. However, in this later case the SPD is used to process wires with a yield stress up to 3GPa or more [4, 5, 6]. These unique mechanical properties result from nanostructuring: during drawing cementite lamellae are aligned along the wire axis and the interlamellar spacing is typically reduced in a range of 20 to 50 nm for a total true strain of about 3.5. However, for higher strains, one has to deal also with strain induced phase transformations like on rail surfaces, and higher yield stress might be obtained only at the expense of ductility. These two examples show some of the possibilities offered by SPD, including nanostructuring and obtaining new structures with unique properties, in addition there is the related scientific issues: what are the exact nanostructuring mechanisms and how original microstructures are transformed? This later point includes of course SPD-induced phase transformations; but, also phase transformation following SPD processing. These two cases are briefly reviewed in the present paper.

2. SPD processes

More than 50 years ago, the relationship between the grain size and the yield stress of materials was established by Hall and Petch [7, 8]. The smaller the grain size, the higher the yield stress, thus nanostructured alloys look to be a promising way to achieve record mechanical strength. Although the exact physical mechanisms of plastic deformation in nanoscaled materials are still under debate [9, 10], the processing of bulk nanostructured materials (BNM) for applications as structural materials is a field involving numerous research teams worldwide. There are however two major issues that are not yet solved: the lack of ductility that is usually observed in such materials [11, 12] and the development of an industrial process that allows the fabrication of large quantities. Various strategies have been proposed like elecrodeposition, compaction of ball milled powders or severe plastic deformation (SPD) processes. This later approach looks very promising especially because materials free of porosity are obtained. A large number of techniques have been designed, most of them being modification of existing metal forming processes: Accumulative Roll Bonding (ARB) [13], Continuous Cyclic Bending (CCB) [14] and Constrained Groove Pressing (CGP) [15] were designed from the original rolling process, Accumulated Cold Drawing (ACD) [16, 17] was design from the original drawing process, Equal Channel Angular Pressing (ECAP) [18, 19], Cyclic Extrusion Compression (CEP) [20] and Twist Extrusion (TE) [21] were designed from extrusion. Only the High Pressure Torsion (HPT) [22, 23] that was originally designed for geological applications, is not inspired by an industrial process. New techniques are often proposed [24, 25], but all of them rely on the idea that a high hydrostatic pressure is necessary to avoid cracking, for a review see [26, 27]. It is also noted that metallic alloys could also be processed to a large level of plastic deformation at low homologous temperature ($T<1/3T_m$) with conventional processes like rolling or drawing, as the heavily drawn pearlitic steel wires mentioned in the introduction.

Commercially pure alloys processed by SPD exhibit a grain size that is typically in a range of 100 to 500 nm [26]. However, some steels [28, 29], intermetallic alloys [30, 31] or composite materials may exhibit a grain size as small as 10 or 20 nm [32, 33]. Nanostructuring mechanisms and the specific characteristics of the new grain boundaries resulting from SPD have been discussed by many authors [26, 27, 34-36]. As mentioned in the introduction, multi-phase nanostructural alloys may be of great benefit for the nanostructuring process: this is the case for drawn pearlitic steels, or more generally for metal matrix filamentary composites processed by drawing [37-39]. With a similar approach, it is also well known that one can easily prepare nanoscaled, multilayered structures using ARB with dissimilar metals [40-42]. On the other hand, in Mg or Al based alloys, several authors have shown that second phase particles promote the grain size reduction and also the thermal stability of the alloys [43-47].

3. Phase transformations induced by SPD

During SPD, large densities of defects like dislocations or boundaries are created; pushing systems far away from the thermodynamic equilibrium state, and these features may affect their stability [48]. The White Etching Layer at the surface of rail tracks mentioned in the introduction is a typical example: together with the strong grain size reduction, original carbides are dissolved [2, 3]. In this case, the frictional heating may promote this transformation, but it is interesting to note that the decomposition of similar carbides (Fe_3C) has been widely reported in similar steels processed by HPT [28, 29, 49], ball milling [50], and in heavily drawn pearlitic steels 30 years ago [51]. First revealed by Mössbauer spectroscopy, it was confirmed later by Atom Probe Tomography (APT) analyses (Fig. 1) [52-56].

The exact underlying mechanisms are still under debate, especially because the nanostructures and the carbon distribution are not yet fully understood. Some authors contend that the carbide dissolution leads to carbon super-saturated solutions in the ferrite with a crystallographic structure similar to that of quenched martensite [55, 57]. Others believe that carbon atoms are more likely distributed around dislocations [29, 58, 59]. In fact, both the driving force and the kinetics of the carbide decomposition are still controversial. Some authors point out the possible effect of moving dislocations that cross cementite lamellae. They may trap carbon atoms of the lamellae and then redistribute them inside the ferrite phase [51, 59]. Others argue that the driving force for the decomposition could be the dramatic increase of the Fe3C/α-Fe interfacial energy resulting from the lamellae thinning. This feature may lead to a strong increase of the carbon solubility in the ferrite through the well known Gibbs-Thomson effect [6, 54]. It is interesting to note that the dissolution of second phase particles has also been reported in FeNi based alloys processed by HPT [61, 62] and in an Al-Cu alloy processed by ECAP [63]. In these two cases, dislocations are though to play a major role by shearing precipitate and dragging solute atoms.

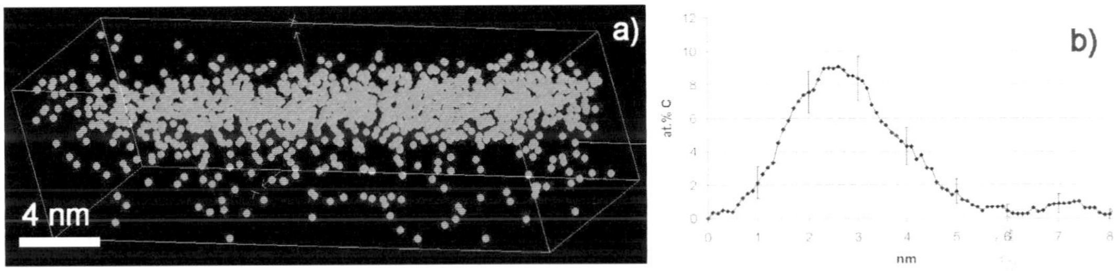

Fig. 1: Cold drawn pearlitic steel (Fe-3.6at.%C, true strain 3.5) (a) 3D reconstructed volume of a small volume analyzed by APT. Only carbon atoms are plotted to exhibit a carbon rich lamella. (b) Composition profile computed across the lamella showing the decomposition of Fe_3C (from [60])

Like irradiation, large strains by HPT may also destroy long range order in some intermetallics (TiAl [30], Ni_3Al [64], Cu_3Au [65]). This disordering appears progressively with increasing strain and is usually attributed to the huge density of crystallographic defects created during the plastic deformation. It is interesting to note that in the NiTi intermetallic the crystalline structure is mostly amorphized after SPD by rolling [66], HPT [67, 68] or drawing [69]. This phenomenon may be attributed to some shear instabilities that would induce the sudden collapse of the crystalline lattice [66, 70, 71]. Considering the large compressive hydrostatic stress involved in the above mentioned SPD processes, such a phase transformation where the final product exhibits a lower density is surprising. This paradox might be explained however by an asymmetric decrease of the repulsive and of the attractive part of the bonding energy of atoms [70]. Anyway, strain induced amorphization is extremely interesting because annealing at low temperature (typically about 300°C), gives rise to the nucleation of a huge density of crystallites and hence for a reasonable time a fully crystalline structure with a mean grain size smaller than 50 nm [68]. Strain induced solid state amorphization has also been observed in some so-called "immiscible" systems like CuNb [72] or CuZr [73]. In this later case the phase transformation is a first order transition. The amorphous phase nucleates and grows in a strain induced super saturated solid solution previously formed. Thus the driving force is usually attributed to the higher enthalpy of mixing comparing to the heat of formation of the amorphous phase [71]. It is however important to note that all strain induced super saturated solid solutions do not transform into an amorphous phase [73-75]. The physical mechanisms of the formation of such non equilibrium solid solutions has been widely investigated for ball milled powder (for a review, see [76]). Several mechanisms have been proposed, most of them relying on thermodynamic arguments: during milling, nanoscaled domains are formed leading to an interfacial energy sufficiently high to overcome the positive heat of mixing [77, 78]. However, such an approach does not provide any atomistic mechanism to account for the kinetics of the reaction. Eckert and co-authors argued that pipe diffusion along dislocations may play a significant role [79]. The role of dislocations is also underlying the so-called "kinetic roughening" model of Bellon and Averback [80]. Atoms would be shifted across interfaces by the shear of atomic glide planes. The final state would be determined by the balance between these forced jumps and decomposition due to thermal diffusion. Recent APT measurements on the Cu-Ag system seem to validate this approach [81]. There are also some models incorporating both kinetic and thermodynamic effects where the ballistic events are introduced as excess entropy [48, 82] but microstructural features are not taken into account. With a pure kinetic approach, other authors proposed a theory based on enhanced mobility resulting from deformation induced point defects [83]. Some model experiments were performed on the Cu-Fe system processed by HPT with deformation parameters fully under control (strain rate, total accumulated strain, temperature, no contamination). As previously reported [84, 85], it was confirmed that HPT may lead to the formation of super saturated solid solutions like ball milling (Fig. 2) and concentration gradients revealed by APT seem to confirm that strain induced point defects may play a significant role in the formation of super saturated solid solution [86].

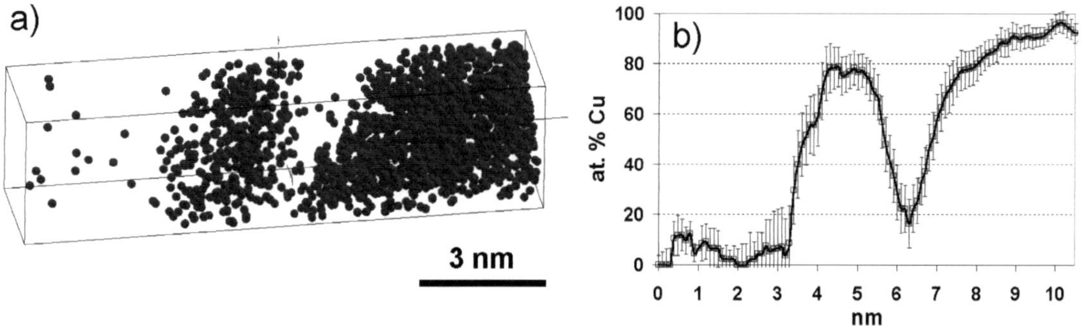

Fig. 2: Cu-Fe composite processed by HPT (a) 3D reconstructed volume of a small volume analyzed by APT, only Cu atoms are displayed to show a nanoscaled region. (b) Copper concentration profile across Cu/Fe interfaces showing the strain induced intermixing and resulting ssss (from [86])

4. Phase transformations in SPD materials

In all previous cases, materials were processed in the phase separated state, but another approach could be to process by SPD before phase separation in order to combine both strengthening by grain size reduction effect and the traditional precipitate hardening. Moreover, as proposed by some authors, a fine distribution of nanoscaled precipitates may act as dislocation sources and enhance significantly the ductility of BNM [87, 88]. Several authors tried to apply this approach for some aluminium alloys [89-91], but controlling the precipitation kinetics in SPD metals is a critical issue because there might be competition between recovery, recrystallisation and grain growth, while heterogeneous precipitation along dislocations and/or grain boundaries is very likely to occur [92]. A fastest precipitation kinetic is also usually reported in aluminium alloys and all classical metastable phases no longer appear [93-96]. In a model FeAuPd alloy, it was also observed that the decomposition process is dramatically affected by the nanoscaled structure resulting from HPT processing [97]. In the coarse grain homogenized alloy, the aging treatment leads to the nucleation and growth of the bcc α-Fe and fcc AuPd phases by a discontinuous precipitation process resulting in a typical lamellar structure. In the nanoscaled structure (grain size in a range of 50 to 100nm) aging leads to the decomposition of the solid solution into an equi axed microstructure with a very limited grain growth (Fig. 3). The decomposition process is thought to be very different because of the fast grain boundary diffusion that significantly affects the redistribution of elements necessary to grow the two phases. It is also interesting to note that the resulting two phase nanostructure exhibits a much higher hardness than the coarse grained alloy.

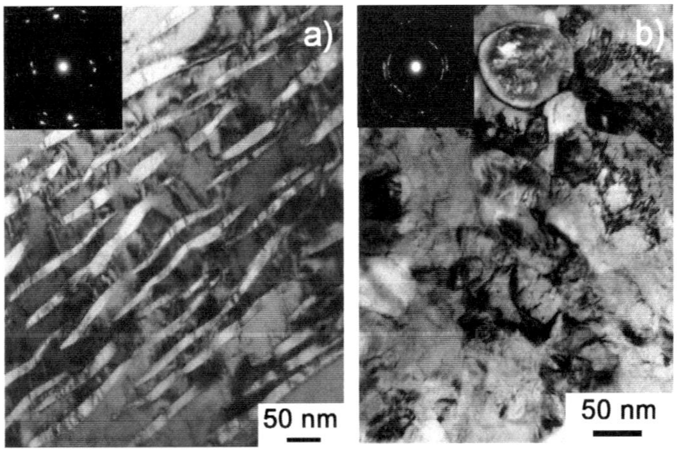

Fig. 3: TEM bright field images and corresponding SAED patterns of a Fe50Au25Pd25 alloy aged at 450°C (a) typical lamellar structure in the non-deformed state, (b) nanoscaled equi-axed structure in the alloy processed by HPT prior to aging. (from [97])

5. Conclusions

Severe plastic deformation strongly affects the original microstructures of metallic alloys, leading in some cases to nanoscaled structures. This approach has been used for several decades for nanostructuring cold drawn pearlitic steels that exhibit remarkable mechanical strength. More recently, various SPD techniques were designed to strengthen metals through grain size reduction. As discussed in the present paper, it is however important to note that the huge amount of crystalline defects created during the plastic strain could lead to some phase transformations that may affect the mechanical properties. For example, for large drawing strains, Fe_3C lamellae are dissolved in pearlitic steels, thus the original α-Fe/Fe_3C interfaces that are supposed to play a critical role in the strengthening mechanism are modified. Indeed, these interfaces cannot be sheared by dislocations and the latter are confined in thin α-Fe channels. However, surprisingly, even for large drawing strain where the proportion of dissolved carbides is large, the yield strength does not decrease and only the ductility seems to be affected. The underlying physical mechanisms are still unknown.

To obtain multiphase nanostructures, the other approach is to process by SPD homogenized metallic alloys where a super saturated solid solution is obtained by quenching. Once the nanostructure is obtained, phase separation is achieved by an aging treatment. This way, if heterogeneous precipitation can be avoided and if recrystallisation mechanisms are not too fast, an homogeneous distribution of precipitates may nucleate within the nanoscaled grains leading to higher strength and ductility. Even if phase transformations in systems containing a large amount of defects proceed differently when compared to non-deformed alloys, this route seems promising to design nanostructured alloys with enhanced properties. Indeed, second phase particles may enhanced the thermal stability through grain boundary pinning and as observed in drawn pearlitic steels, creating a large density of heterophase boundaries may significantly enhanced the yield stress while keeping a reasonable ductility.

References

[1] S.B. Newcomb, W.M. Stobbs, *Mater. Sci. Eng.* **66** (1984) 195.
[2] H.W. Zhang, S. Ohsaki, S. Mitao, M. Ohnuma, K. Honoa, *Mater. Sci. Eng.* **A 421** (2006) 191.
[3] W. Lojkowski, M. Djahanbakhsh, G. Burkle, S. Gierlotka, W. Zielinski, H.J. Fecht, *Mater. Sci. Eng.* **A 303** (2001) 197.
[4] J.D. Embury, R.M. Fisher, *Acta Metall.* **14** (1966) 147.
[5] G. Langford, *Met. Trans.* **1** (1970) 465.

[6] J. Languillaume, G. Kapelski, B. Baudelet, *Acta Mater.* **45** (1997) 1201.
[7] E.O. Hall, *Proc. Phys. Soc.* **B 64** (1951) 747.
[8] N.J. Petch, *J. Iron Steel Instrum.* **173** (1953) 25.
[9] Y.T. Zhu, X. Liao, *Nature Mater.* **3** (2004) 351.
[10] H. Van Swygenhoven, P.M. Derlet, A.G. Frøseth, *Acta Mater.* **54** (2006) 1975.
[11] E. Ma, *Nature Mater.* **2** (2003) 7.
[12] R. Valiev, *Nature Mater.* **3** (2004) 511.
[13] Y. Saito, H. Utsunomiya, N. Tsuji, T. Saikai, *Acta Mater.* **47** (1999) 579.
[14] Y. Takayama, M. Yamaguchi, T. Tozawa, H. Kato, H. Watanabe, T. Izawa, Proc. of the 4th Int. Conf. on Recrystallization and Related Phenomena, *The Japan Inst. of Metals, Sendai, Japan* 1999, 321.
[15] D.H. Shin, J.-J. Park, Y-S. Kim, K.-T. Park, *Mater. Sci. Eng.* **A 328** (2002) 98.
[16] Levi FP, *J. Appl. Phys.* **31** (1960) 1469.
.[17] L. Thilly, F. Lecouturier, G. Coffe, S. Askenazy, *IEEE Trans. Appl Supercond.* **12** (2002) 1181.
[18] V.M. Segal, V.I. Reznikov, A.E. Drobyshevkij, V.I. Kopylov, *Metally* **1** (1981) 115.
[19] Z. Valiev, A.V. Korznikov, R.R. Mulyukov, *Mater. Sci. Eng.* **A 168** (1993) 141.
[20] J. Richert, M. Richert, *Aluminium* **62** (1986) 604.
[21] D. Orlov, Y. Beygelzimer, S. Synkov, V. Varyukhin, Z. Horita, *Mater. Trans.* **49** (2008) 2.
[22] P.W. Bridgman, *Phys. Rev.* **48** (1935) 825.
[23] R.Z. Valiev, N.A. Krasilnikov, N.K. Tsenev, *Mater. Sci. Eng.* **A 137** (1991) 35.
[24] L.S. Tòth, M. Arzaghi, J.J. Fundenberger, B. Beausir, O. Bouaziz and R. Arruffat-Massion, *Scripta Mater.* **60** (2009) 175.
[25] Y. Harai, Y. Ito, Z. Horita, *Scripta Mater.* **58** (2008) 469.
[26] R.Z. Valiev, , R.K. Islamgaliev, I.V. Alexandrov, *Progress Mater. Sci.* **45** (2000) 103.
[27] R.Z. Valiev, T.G. Langdon, *Progress Mater. Sci.* **51** (2006) 881.
[28] Y. Ivanisenko, I. MacLaren, X. Sauvage, R.Z. Valiev, H.-J. Fecht, *Acta Mater.* **54** (2006) 1659.
[29] X. Sauvage, Y. Ivanisenko, *J. Mater. Sci.* **42** (2007) 1615.
[30] A.V. Korznikov, O. Dimitrov, G.F. Korznikova, J.P. Dallas, A. Quivy, R.Z. Valiev, A. Mukherjee, *Nanostruct. Mater.* **11** (1999) 17.
[31] A.V. Korznikov, O. Dimitrov, G.F. Korznikova, J.P. Dallas, S.R. Idrisova, R.Z. Valiev, F. Faudot, *Acta Mater.* **47** (1999) 3301.
[32] X. Sauvage, P. Jessner, F. Vurpillot, R. Pippan, *Scripta Mater.* **58** (2008) 1125.
[33] I. Sabirov, R. Pippan, *Scripta Mater.* **52** (2005) 1293.
[34] J.A. Wert, X. Huang, G. Winther, W. Pantleon, H.F. Poulsen, *Materials Today* **10** (2007) 24.
[35] R. Pippan, F. Wetscher, M. Hafok, A. Vorhauer, I. Sabirov, *Adv. Eng. Mat.* **8** (2006) 1046.
[36] H.W. Zhang, N. Hansen, *J. Mater. Sci.* **42** (2007) 1682.
[37] V. Vidal, L. Thilly, F. Lecouturier, P.-O. Renault, *Acta Mater.* **54** (2006) 1063.
[38] E. Snoeck, F. Lecouturier, L. Thilly, M.J. Casanove, H. Rakoto, G. Coffe, S. Askénazy, J.P. Peyrade, C. Roucau, V. Pantsyrny, A. Shikov, A. Nikulin, *Scripta Mater.* **38** (1998) 1643.
[39] A.M. Russell, L.S. Chumbley, Y. Tian, *Adv. Eng. Mat.* **2** (2000) 11.
[40] L. Battezzati, C. Antonione, F. Fracchia, *Intermetallics* **6** (1995) 67.
[41] G. Wilde, H. Sieber, J.H. Perepezko, *Scripta Mater.* **40** (1999) 779.
[42] X. Sauvage, D.G. Dinda, G. Wilde, *Scripta Mater.* **56** (2007) 181.
[43] K. Matsubara, Y. Miyahara, Z. Horita, T.G. Langdon, *Acta Mater.* **51** (2003) 3073.
[44] H. Hasegawa, S. Komura, A. Utsunomiya, Z. Horita, M. Furukawa, M. Nemoto, T.G. Langdon, *Mater. Sci. Eng.* **A 265** (1999) 188.
[45] H. Akamastu, T. Fujinami, Z. Horita, T.G. Langdon, *Scripta Mater.* **44** (2001) 759.
[46] S. Lee, A. Utsunomiya, H. Akamatsu, K. Neishi, M. Furukawa, Z. Horita, T.G. Langdon, *Acta Mater.* **50** (2002) 553.
[47] M. Furukawa, Z. Horita, M. Nemoto, T.G. Langdon, *Mater. Sci. Eng.* **A 324** (2002) 82.
[48] P. Pochet, E. Tominez, L. Chaffron and G. Martin, *Phys. Rev.* **B 52** (1995) 4006.

[49] Y. Ivanisenko, W. Lojkowski, R.Z. Valiev, H.J. Fecht, *Acta Mater.* **51** (2003) 5555.
[50] S. Ohsaki, K. Hono, H. Hidaka, S. Takaki, *Scripta Mater.* **52** (2005) 271.
[51] V.N. Gridnev, V.G. Gavrilyuk, *Phys. Met.* **4** (1982) 531.
[52] F. Danoix, D. Julien, X. Sauvage, J. Copreaux, *Mater. Sci. Eng.* **A 250** (1998) 8.
[53] M.H. Hong, W.T. Reynolds, T. Tarui, K. Hono, *Met. Trans.* **30A** (1999) 717.
[54] X. Sauvage, J. Copreaux, F. Danoix, D. Blavette, *Phil. Mag.* **A 80** (2000) 781.
[55] K. Hono, M. Ohnuma, M. Murayama, S. Nishida, A. Yoshie, T. Takahashi, *Scripta Mater.* **44** (2001) 977.
[56] X. Sauvage, N. Guelton, D. Blavette, *Scripta Mater.* **46** (2002) 459.
[57] A. Taniyama, T. Takayama, M. Arai, T. Hamada, *Scripta Mater.* **51** (2004) 53.
[58] V.G. Gavriljuk, *Mater. Sci. Eng.* **A 345** (2003) 81.
[59] X. Sauvage, X. Quelennec, J.J. Malandain, P. Pareige, *Scripta Mater.* **54** (2006) 1099.
[60] X. Sauvage, W. Lefebvre, C. Genevois, S. Ohsaki, K. Hono, *Scripta Mater.* **60** (2009) 1056.
[61] V.V. Sagaradze, S.V. Morozov, V.A. Shabashov, L.N. Romashev, R.I. Kuznetsov, *Phys. Met. Metall.* **66** (1988) 111.
[62] V.V. Sagaradze, V.A. Shabashov, *Nanostruct. Mater.* **9** (1997) 681.
[63] M. Muruyama, K. Hono, Z. Horita , *Mater. Trans. – JIM* **40** (1999) 938.
[64] A.V. Korznikov, G. Tram, O. Dimitrov, G.F. Korznikova, S.R. Idrisova, Z. Pakiela, *Acta Mater.* **49** (2001) 663.
[65] C. Rentenberger, H.P. Karnthaler, *Acta Mater.* **56** (2008) 2526.
[66] J. Koike, D.M. Parkin, M. Nastasi, *Phil. Mag. Lett.* **62** (1990) 252.
[67] A.V. Sergueeva, C. Song, R.Z. Valiev, A.K. Mukherjee, *Mater. Sci. Eng.* **A 339** (2003) 159.
[68] S.D. Prokoshkin, I.Y. Khmelevskaya, S.B. Dobatkin, I.B. Trubitsyna, E.V. Tatyanin, V.V. Stolyarov, E.A. Prokofiev, *Acta Mater.* **53** (2005)2703.
[69] K. Tsuchiya, Y. Hada, T. Koyano, K. Nakajima, M. Ohnuma, T. Koike, Y. Todaka and M. Umemoto, *Scripta Mater.* **60** (2009) 749.
[70] M. Tang, S. Yip, *Phys. Rev. Lett.* **75** (1995) 2738.
[71] E. Ma, *Scripta Mater.* **49** (2003) 941.
[72] X. Sauvage, L. Renaud, B. Deconihout, D. Blavette, D. H. Ping and K. Hono, *Acta. Mater.* **49** (2001) 389.
[73] S. Ohsaki, S. Kato, N. Tsuji, T. Ohkubo, K. Hono, *Acta Mater.* **49** (2001) 389.
[74] X. Sauvage, R. Pippan, *Mater. Sci. Eng.* **A 410–411** (2005) 345.
[75] X. Sauvage, C. Genevois, G. Da Costa, V. Pantsyrny, *Scripta Mater.* **61** (2009) 660.
[76] C. Suryanarayama, *Progress Mater. Sci.* **46** (2001) 1.
[77] A. Yavari, P. Desré, T. Benameur, *Phys. Rev. Lett.* **68** (1992) 2235.
[78] C. Gente, M. Oehring and R. Bormann, *Phys. Rev.* **B 48** (1993) 13244.
[79] J. Eckert, J. Holzer, C. Krill and W. Johnson, *J. Appl. Phys.* **73** (1993) 2794.
[80] P. Bellon and R. Averback, *Phys. Rev. Lett.* **74** (1995) 1819.
[81] F. Wu, D. Isheim, P. Bellon and D. Seidman, *Acta Mater.* **54** (2006) 2605.
[82] E. Ma, M. Atzmon, Mat. *Chem. Phys.* **39** (1995) 249.
[83] B. Khina, I. Soplan and G. Lovshenko, *J. Mater. Sci.* **39** (2004) 5135.
[84] V.A. Teplov, V.P. Pilugin, V.S. Gaviko, E.G. Chernyshov, *Phil. Mag.* **B 68** (1993) 877.
[85] V.V. Stolyarov, R. Lapovok, I.G. Brodova, P.F. Thomson, *Mater. Sci. Eng.* **A 357** (2003) 159.
[86] X. Sauvage, F. Wetscher, P. Pareige, *Acta Mater.* **53** (2005) 2127.
[87] S. Cheng, Y.H. Zhao, Y.T. Zhu, E. Ma *Acta Mater.* **55** (2007) 5822.
[88] E. Ma, *JOM* **58** (2006) 49.
[89] W.J. Kim, J.K. Kim, T.J. Park, S.I. Hong, D.I. Kim, Y.S. Kim and J.D. Lee, *Met. Mater. Trans.* **A 33** (2002) 3155.
[90] J.K. Kim, H.K. Kim, J.W. Park and W.J. Kim, *Scripta Mater.* **53** (2005) 1207.
[91] Y.H. Zhao, X.Z. Liao, Z. Jin, R.Z. Valiev, Y.T. Zhu, *Acta Mater.* **52** (2004) 4589.
[92] A. Cerri, P. Leo, *Mater. Sci. Eng.* **A 410-411** (2005) 226.

[93] G. Sha, Y.B. Wang, X.Z. Liao, Z.C. Duan, S.P. Ringer, T.G. Langdon, *Acta Mater.* **57** (2009) 3123.
[94] J. Cizek, I. Prochazka, B. Smola, I. Stulikova, R. Kuzel, Z. Matej, V. Cherkaska, R.K. Islamgaliev, O. Kulyasova, *Mater. Sci. Forum* **503-504** (2006) 149.
[95] B.B. Straumal, B. Baretzky, A.A. Mazilkin, F. Philipp, O.A. Kogtenkova, M.N. Volkov, R.Z. Valiev, *Acta Mater.* **52** (2004) 4469.
[96] Z. Horita, K. Ohashi, T. Fujita, K. Kaneka, T.G. Langdon, *Adv. Mater.* **17** (2005) 1599.
[97] X. Sauvage, A. Chbihi, D. Gunderov, E.V. Belozerov, A.G. Popov, *J. Mater. Sci.* **43** (2008) 7293.

Journal of Physics: Conference Series **240** (2010) 012004 doi:10.1088/1742-6596/240/1/012004

Microstructural and mechanical challenges in biomedical NiTi

Martin Franz-Xaver Wagner
Institute for Materials, Ruhr-University Bochum,
Universitaetsstr. 150, D-44801 Bochum, Germany

martin.wagner@rub.de

Abstract. The mechanical behaviour of NiTi shape memory alloys superficially resembles that of certain biomaterials, such as bones or tissues: By virtue of a reversible martensitic phase transformation, NiTi alloys can recover relatively large strains; uniaxial stress-strain curves exhibit constant stress-plateaus (at several hundreds of MPa, depending on alloy composition and testing temperature) associated with the phase transition. These novel functional properties, in combination with high mechanical strength in ultra-fine grained NiTi and good biocompatibility, are utilized in various implants and medical devices. Yet – and quite similar to hierarchically structured biomaterials – the deformation behaviour of NiTi is intricately linked to distinct deformation processes on several length scales, and there remain significant gaps in our understanding of the microstructure-property relations. In the present paper, recent experimental and theoretical results from first-principles calculations, micromechanical modelling and nanoindentation are discussed with a focus on the role of inelastic deformation processes, twin boundaries and the interaction of plastic deformation and stress-induced phase transformations. These novel findings challenge our understanding of the fundamental mechanical properties of NiTi. They highlight the importance of inelastic deformation mechanisms for the overall mechanical properties and strength of NiTi.

1. Introduction

NiTi shape memory alloys are the most important shape memory alloys today, mostly because their superior mechanical properties have led to the development and successful market launch of products and devices in the field of biomedical applications [1-4]. In particular implants such as stents are in direct contact with biological materials (tissues, bones), and therefore biocompatibility is an important topic in NiTi research in the chemistry and medical communities. It has been noted before [4] that shape memory alloys phenomenologically share some characteristic features in terms of mechanical behaviour, in particular non-linear behaviour associated with extended constant stress-plateaus in uniaxial tensile testing, thermo-mechanical hysteresis and the ability to recover substantial amounts of strain (in pseudoelastic NiTi, of the order of 10%). One may argue that there is also a microstructural analogy that extends beyond a merely superficial resemblance in mechanical behaviour: Biomaterials are characterized by highly hierarchical microstructures, where different elements and substructures on various length scales fulfil several functions [5-7]. At least to some extent, the stress-induced martensitic transformation (which is fully reversible during unloading and thus gives rise to the shape memory effect) results in the formation of similarly complex substructures. Martensite nucleating in the austenitic matrix forms twins in order to minimize strain energy related to the accompanying shape change (transformation strain), and the resulting martensitic microstructures consist of increasingly finer martensite plates and needles which are often internally twinned [1]. In ultrafine grained NiTi, twin variants with widths as small as a few nanometers are observed [8,9]. Research on the mechanical properties of pseudoelastic NiTi shape memory alloys is challenging because the deformation behaviour is not only directly affected by the stress-induced phase transformation. Additional deformation mechanisms encompass several length scales and also need to be considered in detail to obtain a better understanding of the pseudoelastic deformation of NiTi.

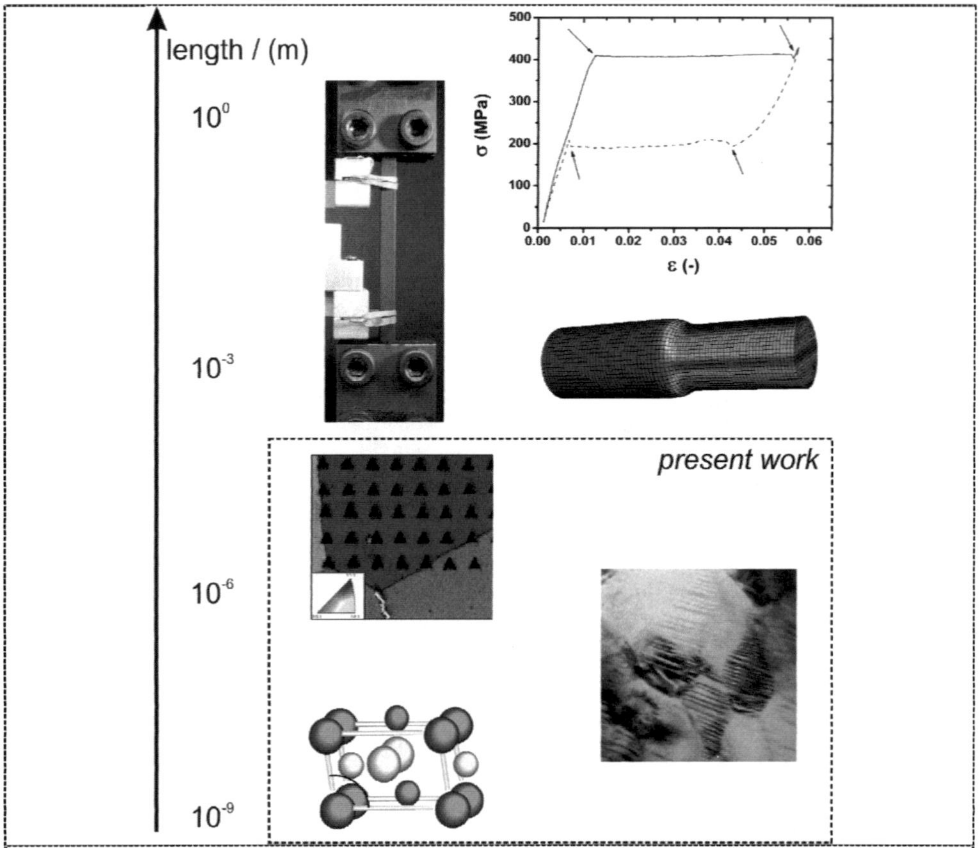

Figure 1. Schematic representation of some of the relevant length scales associated with various deformation mechanisms in pseudoelastic NiTi. This paper is focused on transformation-induced, elastic and plastic deformation at the atomistic and microstructural length scale.

Figure 1 provides a schematic overview of those length scales that are typically important for the mechanical behaviour of materials. The macroscopic mechanical properties of NiTi (pseudoelastic stress-strain curve on the upper right; the beginning and end of the forward and reverse phase transformations is highlighted by arrows) are typically investigated by conventional tensile testing. However, localized deformation (formation and propagation of Lüders-like transformation bands) occurs in thin tensile NiTi specimens, and the distinct stress- and strain-states, as well as the distribution of volume fractions of the different phases, in the transition regions between fully martensitic and fully austenitic regions play a key role in stabilizing this mode of deformation [10-14]. Direct experimental investigations of micro-scale mechanical properties and transformation processes can for instance be performed by nanoindentation, micro-pillar testing and in-situ transmission electron microscopy (TEM) studies [15-18]. On the atomistic length scale, the fundamental thermo-elastic properties of the crystal lattices are directly related to the electronic structure and lattice dynamics of the different phases [19-21].

Despite a vast body of literature on NiTi shape memory alloys, the aforementioned aspects of their mechanical behaviour are not yet understood completely. They are subject of intensive ongoing research activities. In this paper, several results from recent theoretical and experimental studies on different length scales will be briefly reviewed and discussed. Special emphasis is placed in (1) the elastic anisotropy of B19' martensite, (2) the contribution of inelastic (but reversible) deformation mechanisms to the macroscopic mechanical behaviour, (3) elastic compatibility stresses at martensitic twin boudaries, and (4) the interaction of stress-induced phase transformation and conventional plastic deformation during nanoindentation of NiTi.

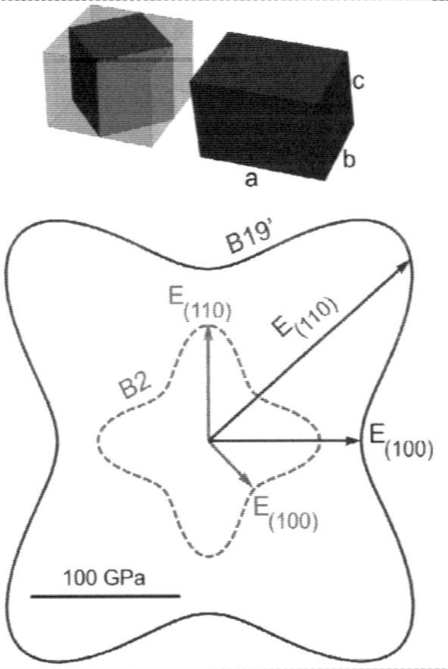

Figure 2. Direction-dependent moduli for B2 austenite and B19' martensite for uniaxial deformation along directions lying in a shared (001)-plane (see representation of unit cells and orientation relationship in the upper half of the figure). A complete, three-dimensional representation of the direction-dependent modulus of B19' is given in [22].

2. Elastic anisotropy in twinned NiTi martensites

Following a (rather idealized) bottom-up approach, one would like to completely capture the mechanical properties of an individual martensite variant, then study the mechanics of twinned microstructures, the properties of austenite/martensite interfaces, the interaction of distinct deformation processes with microstructural elements like grain boundaries and precipitates, and finally derive an accurate model for the resulting macroscopic mechanical behaviour, properly taking localized deformation of macro-scale specimens into account (see also figure 1). One most simple and first step is to study the elastic properties of the individual phases in NiTi. The anisotropic elastic behaviour of a crystal is fully determined by Hooke's law

$$\sigma_i = c_{ij} \epsilon_j \tag{1}$$

where the elastic constants c_{ij} relate stresses σ and strains ε. Resonant ultrasound spectroscopy (RUS) has been used to experimentally determine the elastic constants of B2 NiTi austenite (cubic structure: 3 independent c_{ij}), [23], and also for single variant specimens of 2H martensite (orthorhombic symmetry: 9 independent c_{ij}) in a CuAlNi shape memory alloy [24]. Unfortunately, this experimental approach is not feasible for B19' NiTi martensite because of the very fine twins that are present even after the application of high stresses (which generally favour the formation of fewer martensite variants and which have been used in [24] to obtain single variant RUS specimens of CuAlNi). The full set of 13 independent elastic constants of monoclinic B19' NiTi has, however, been recently estimated by a theoretical approach [22]. Ab-initio quantum-mechanical calculations allow the application of arbitrary strain states to a crystal lattice, and from the changes of free energy (and from the Hellmann-Feynman forces), the corresponding stress states and all elastic constants can be determined. The results are strictly valid only at zero K (i.e., finite temperature effects are neglected), and the high computational effort typically restricts such calculations to the simulation of perfect crystal lattices (i.e., important microstructural features like dislocations, phase or grain boundaries, cannot be easily incorporated in these calculations). Moreover, when studying the elastic properties of the different phases of NiTi, additional difficulties arise (further details, both technical and on the interpretation of the results, are presented in [22,25]): As B2 austenite is only stabilized by entropical contributions at higher temperatures, the corresponding crystal structure is mechanically unstable at zero K, and while they qualitatively agree with those measured by RUS at higher temperatures, the elastic constants of B2 NiTi determined from ab initio calculations reflect this instability. Furthermore, in the calculations special shear stresses are needed to stabilize B19' martensite which in their absence reverts into an orthorhombic B33 structure that has not been observed in experiment [21]. But despite these issues, the ab initio simulations were found to be useful in determining the elastic constants of B19' NiTi, and several interesting conclusions can be drawn.

The ab initio calculations indicate that the NiTi martensites (B19' as well as the theoretical B33 structure) are elastically anisotropic. To compare the anisotropic elastic properties of austenite and martensite, it is instructive to study the elastic modulus that would be measured if a single crystal specimen were subjected to uniaxial loading in a certain crystallographic direction. Figure 2 represents the direction-dependent elastic moduli for B2 austenite (determined from experimental elastic constants after [23]) and B19' martensite for the case where B2 and B19' share one (001)-plane and the direction of loading is contained in that plane (see also the schematic representation of the cubic and monoclinic unit cells in the upper half of figure 2; the monoclinic unit cell of B19' is uniquely characterized by the lattice constants/ monoclinic angle a = 0.4108 nm, b = 0.4646 nm, c = 2.898 nm, β = 97.8°, [26]). The distance of each curve from the origin is proportional to elastic modulus in that crystallographic direction. Figure 2 clearly demonstrates that B19' martensite is elastically stiffer than austenite. In the case of purely elastic deformation of twinned B19' martensite, Voigt-Reuss-Hill averaging predicts a macroscopic Young's modulus of ~120 GPa [22], which is considerably larger than that of austenite (~60-80 GPa). This finding apparently contradicts most experimental data, where the macroscopic Young's modulus of martensite is usually reported in the range of ~20-50 GPa (i.e., smaller than the theoretical value by a factor of three, and, more importantly, much smaller than that of austenite). Yet, the 'true' elastic modulus determined from ab initio calculations is in good agreement with recent neutron diffraction studies [27]; a high modulus confirms some long-held predictions [28] that additional deformation mechanisms play an important role in softening the macroscopic stress-strain response of martensitic NiTi. In fact, when the elastic constants of B19' are used as input for an advanced micro-mechanical model, an analysis of the evolution of the volume fractions of different martensite variants clearly shows that *detwinning* (growth of one favoured variant that consumes a less favoured twin variant), *reorientation* (growth of entirely new variants from an already twinned martensitic structure), or *additional transformation* of residual austenite can occur during uni- and biaxial loading [29]. In particular at higher stresses, these inelastic (but reversible) deformation mechanisms contribute by more than 50% to macroscopic strains. Therefore, despite the high elastic stiffness introduced in the micromechanical model, the macroscopic stress-strain behaviour is in excellent agreement with experimental data [30].

Further mechanical analysis of B19' martensite requires consideration of the special mechanical properties of twinned structures. While such analyses have been performed to a great level of detail and in excellent agreement with experimental observations before [31], the effect of the elastic anisotropy of B19' on the stress-states at twin boundaries has not been studied before simply because the elastic constants were unknown. In a first (and strongly simplified) attempt to study B19' twin boundaries, consider a stack of thin martensite twins as shown schematically in figure 3. When such a twinned structure is subjected to external loading, individual variants will deform differently because of elastic anisotropy and because of their different orientation (shown schematically by the monoclinic unit cells in figure 3) with respect to the external loads. As the thin twin variants are coherently joined at the twin boundaries, additional stresses (so-called compatibility stresses) arise at each twin boundary. A simple analytical model of these purely elastic compatibility stresses in (infinitely large) bicrystals has been presented by Gemperlova et al [32]. The special symmetry of twin boundaries can be used to further simplify their expressions for the compatibility stresses at the interface. The most important twins in coarse grained NiTi, <110> type II twins, are characterized by an irrational habit plane $K_I = (-1,1,0.72053)$ and by a shearing direction $\eta_I = [1,1,0]$, [33]. When the matrix of elastic constants is rotated so that η_I points in the 1-direction indicated in figure 3, and K_I is parallel to the 2-direction, the compatibility stresses σ^C in the first variant (e.g., the blue variant in figure 3) according to the model of Gemperlova et al. (and using Voigt's notation, [34]) for an arbitrary applied load case σ^A are given by

$$\sigma_1^C = -0.156 \cdot \sigma_5^A - 0.1804 \cdot \sigma_6^A \quad (2)$$

$$\sigma_3^C = -0.679 \cdot \sigma_5^A - 1.187 \cdot \sigma_6^A \quad (3)$$

$$\sigma_5^C = -0.054 \cdot \sigma_1^A - 0.304 \cdot \sigma_2^A + 0.297 \cdot \sigma_3^A - 0.209 \cdot \sigma_4^A \quad (4)$$

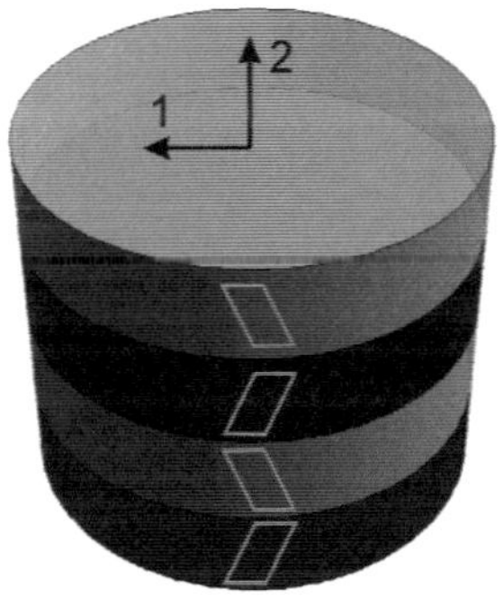

Figure 3. Compatibility stresses arise at twin boundaries between individual variants (blue or green; different crystal orientations are indicated by the orange unit cells) because of the anisotropic deformation of individual variants. In the simple scenario shown schematically here, stresses and elastic constants are defined in the Cartesian coordinate system with the 2-direction perpendicular to thee twin boundaries.

The stresses in the second variant (green in figure 3) have the same magnitude, but an opposite sign. In polycrystalline NiTi, the interaction with an adjacent austenitic or partly martensitic matrix does obviously impose more complicated boundary conditions than those considered in the simple toy scenario depicted in figure 3. Moreover, the approach of Gemperlova et al. is strictly only valid at the twin boundary and in infinite bodies. While further work is required to develop these considerations into a more realistic model of twin boundary compatibility stress in martensitic NiTi, it is noteworthy that equations (2)-(4) demonstrate that the elastic compatibility stresses in B19' type II twins are of a similar order of magnitude as the applied stresses. Because this scenario is purely elastic, the compatibility stresses are superimposed to the applied stresses. The resulting stress state may therefore well affect dislocation nucleation at twin boundaries or provide an additional driving force for reorientation or detwinning.

3. Nanoindentation and plastic deformation

Given the many different deformation mechanisms that can be activated in NiTi as well as the excellent shape recovery of pseudoelastic alloys, it is perhaps not too surprising that plastic deformation by conventional dislocation slip has not been studied in detail by the shape memory community. Instead, the focus of research has been mainly placed on crystallographic and thermodynamic aspects of the reversible phase transformation. However, dislocations do indeed play an important role in determining the mechanical properties of NiTi in several respects: During cyclic mechanical or thermal loading, the gradual accumulation of dislocations and stabilized martensite results in functional fatigue (changing phase transformation temperatures, decreasing plateau stresses). This degradation of functional properties limits the service life of NiTi applications and is not fully understood today [35,36]. Moreover, nanoindentation experiments indicate that, while pseudoelastic recovery during unloading can clearly be observed, strain gradients related with this micro-scale testing method also result in a more pronounced plastic deformation below the indenter tip. During nanoindentation, the applied force and the indentation depth are continuously recorded. As a measure of pseudoelastic recovery, one can define the ratio of residual indentation depth and maximum indentation depth at maximum load (remnant depth ratio: ρ). ρ-values near zero indicate perfect shape recovery during unloading (i.e., no residual indent remains after indentation), and large ρ-values are related to plastic deformation, [37].

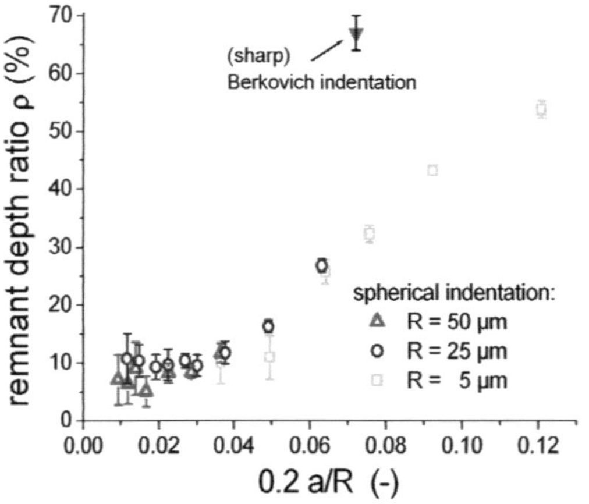

Figure 4. Remnant depth ratio during nanoindentation of pseudoelastic NiTi as a function of representative strain. Spherical indentation data [37] exhibit increasing ρ-values with increasing strain. Some residual deformation is observed even at small strains.

In figure 4, typical remnant depth data from experiments with different spherical indenter tips and maximum loads are summarized as a function of representative strain below the indenter tip. The representative strain, 0.2 a/R, where a is the contact radius and R is the radius of the indenter tip, provides a simple yet effective estimate of the complex strain state below the indenter. It increases with increasing indentation depth in the case of spherical indentation. Indentation with sharp (Berkovich) tips is associated with a constant representative strain of ~7.2% for all indentation depths/maximum loads because of the self-similar shape of the pyramidal indenter tip. Figure 4 demonstrates that the large representative strain associated with sharp indentation cannot be provided by the stress-induced phase transformation alone. Instead, dislocation-mediated plasticity must occur, and this is reflected by large ρ-values of ~65 % which are consistently observed for a wide range of indentation depths. An increased dislocation density has been reported below a Berkovich indent in a pseudoelastic NiTiFe alloy [38].

Spherical indentation allows to systematically vary the ratio between pseudoelastic and plastic deformation, figure 4. An increasing dislocation density below spherical indents with increasing indentation depth has also been documented [39]. But, most interestingly, even under optimum conditions (large indenter radii, small indentation depths and therefore small residual strains), some plastic deformation is evident from the corresponding ρ-values (which are typically ~5-10 %, figure 4). It is well known that nanoindentation is associated with large strain gradients [40]. Obviously, these strain gradients can only be accommodated by the formation of (geometrically necessary) dislocations even when the material can in principle accommodate large strains by a stress-induced phase transformation. The nanoindentation results presented here show that there is an intimate interaction between plastic deformation and transformation when pseudoelastic NiTi is deformed, in particular when strain gradients are present. As a closing thought, it is noted that future research on shape memory alloys may therefore allow extending conventional strain gradient plasticity models to more general classes of materials.

4. Summary and conclusions

The mechanical behaviour of NiTi shape memory alloys is affected by processes on various length scales. The present paper summarizes some aspects of the deformation mechanisms on the nano- and micro-scales. Recent ab initio simulations have demonstrated that B19' NiTi martensite is elastically anisotropic, and that – in contrast to macroscopic observations – the 'true' elastic modulus of B19' does exceed that of B2 austenite. While theoretically stiffer than austenite, martensitic NiTi is characterized by a considerably lower modulus of elasticity during macroscopic testing because additional deformation mechanisms occur when twinned NiTi martensite is subjected to external loading. The most important inelastic (but reversible) deformation processes include additional transformation, detwinning, or reorientation of twinned martensite. The effect of these deformation mechanisms has been demonstrated and analyzed by using the ab initio-derived elastic constants of B19' NiTi in a micromechanical model. These simulations of uni- and biaxial load cases indicate that inelastic deformation processes considerably contribute to the macroscopic stress-strain behaviour in polycrystalline NiTi. Elastic anisotropy moreover results in compatibility stresses at twin boundaries of B19' martensite. These stresses may well affect the twin boundary movement and/or dislocation nucleation at stresses beyond the pseudoelastic plateau. Finally, dislocation-mediated plastic deformation does not only affect the cyclic behaviour of NiTi – it becomes evident (and, in some cases, dominant) during nanoindentation of NiTi. While indentation with spherical indenters (where representative strains depend on the indentation depth and on the tip radius) allows studying the pseudoelastic behaviour of NiTi in small testing volumes, plastic deformation cannot be prevented completely because the strain gradients inherent in nanoindentation testing can only be accommodated by dislocation slip. Sharp (Berkovich) indenters are associated with a characteristic strain that requires considerable plastic deformation irrespective of indentation depth. One of the key challenges for future micromechanical/ microstructural studies lies in a detailed analysis of the complex interactions between dislocations (in both austenite and martensite) and the stress-induced, twinned martensite. Nanoindentation, in combination with in-situ or post-mortem microstructural analysis techniques, may represent a pathway for future systematic studies that address this need.

Acknowledgments
Challenging discussions with, and generous support of, many colleagues at the Ruhr-University in Bochum and at the Ohio State University, Columbus, USA, are gratefully acknowledged, as is financial support by the German Research Foundation (Deutsche Forschungsgemeinschaft, DFG) in the framework of the interdisciplinary research center SFB459 ("Shape Memory Technology") and through the Emmy Noether program, as well as by the Alexander von Humboldt Foundation and by the North-Rhine Westfalian Academy of Science and Arts.

References

[1] Otsuka K and Wayman C M 1999 *Shape Memory Materials* (Cambridge, UK: Cambridge University Press)
[2] Duerig T, Pelton A and Stoeckel D 1999 *Mater. Sci Eng.* A **273-275** 149
[3] Pelton A R, Russell S M and DiCello J 2003 *JOM* **55** 33
[4] Morgan N B and Friend C M 2001 *J. Phys. IV* **C11** 325
[5] Fratzl P and Weinkamer R 2007 *Prog. Mater. Sci.* **52** 1263
[6] Fratzl P 2008 *Nature Mater.* 2008 **7** 610
[7] Koester K J, Ager III J W and Ritchie R O 2008 *Nature Mater.* 2008 **7** 672
[8] Waitz T and Karnthaler H P 2004 *Acta Mater.* **52** 5461
[9] Waitz T, Pranger W, Antretter T, Fischer F D and Karnthaler H P 2008 *Mater. Sci. Eng.* A **481-482** 479
[10] Shaw J A and Kyriakides S 1995 *J. Mech. Phys. Solids* **43** 1243
[11] Favier D, Louche H, Schlosser P, Orgeas L, Vacher P and Debove L 2007 *Acta Mater.* **55** 5310
[12] Li Z Q and Sun Q P 2002 *J. Plasticity* **18** 1481

[13] Olbricht J, Schaefer A, Wagner M F X and Eggeler G 2007 *Proc. Shape Memory and Superelastic Technologies*, Tsukuba, Japan 47
[14] Schaefer A and Wagner M F X: Strain mapping at propagating interfaces in pseudoelastic NiTi. *Proc. of European Symposium on Martensitic Transformations 2009, Prague, Czech Republic, submitted.*
[15] Pfetzing J, Schaefer A, Somsen C and Wagner M F X 2009 *Int. J. Mater. Res.* **100** 936
[16] Frick C P, Orso S and Arzt E 2007 *Acta Mater.* **55** 3845
[17] Norfleet D M, Sarosi P M, Manchiraju S, Wagner M F X, Uchic M D, Anderson P M and Mills M J 2009 *Acta Mater.* **57** 3549
[18] Olbricht J, Wagner M F X, Condo A, Dlouhy A, Grossmann C, Kroeger A, Somsen C and Eggeler G 2008 *Int. J. Mater. Res.* **99** 1150
[19] Parlinski K and Parlinska-Wojtan M 2002 *Phys.Rev.*B **66** 064307
[20] Huang X, Bungaro C, Godlevsky V and Rabe K M 2001 *Phys.Rev.*B **65** 014108
[21] Huang X, Ackland G J and Rabe K M 2003 *Nature Mat.* **2** 307
[22] Wagner M F X and Windl W 2008 *Acta Mater.* **56** 6232
[23] Ren X, Miura N, Zhang J, Otsuka K, Tanaka K, Koiwa M, Suzuki T and Chumlyakov Y I 2001 *Mater. Sci. Eng.* A **312** 196
[24] Sedlak P, Seiner H, Landa M, Novak V, Sittner P and Manosa L 2005 *Acta Mater.* **53** 3643
[25] Wagner M F X and Windl W 2009 *Scripta Mater.* **60** 207
[26] Kudoh Y, Tokonami M, Miyazaki S and Otsuka K 1985 *Acta Mater.* **33** 2049
[27] Rajagopalan S, Little A L, Bourke M A M and Vaidyanathan R 2005 *Appl. Phys. Lett.* **86** 081901
[28] Liu Y and Xiang H 1998 *J. Alloys Compd.* **270** 154
[29] Heinen R, Hackl K, Windl W and Wagner M F X 2009 *Acta Mater.* **57** 3856
[30] Helm D and Haupt P 2003 *Int. J. Solids Struct.* **40** 827
[31] Waitz T, Antretter T, Fischer F D, Simha N K and Karnthaler H P 2007 *J. Mech. Phys. Solids* **55** 419
[32] Gemperlova J, Paidar V and Kroupa F 1989 *Czech. J. Phys.* **39** 427
[33] Otsuka K and Ren X 2005 *Prog. Mater. Sci.* **50** 511
[34] Nye J F 1985 *Physical Properties of Crystals* (Oxforf, UK: Oxford University Press)
[35] Eggeler G, Hornbogen E, Yawny A, Heckmann A and Wagner M 2004 *Mater. Sci. Eng* A **378** 24
[36] Wagner M F X, Nayan N and Ramamurty U 2008 *J. Phys.* D **41** 185408
[37] Pfetzing J, Wagner M F X, Frenzel J, Somsen C and Eggeler G: Thermomechanical constraints on pseudoelasticity during nanoindentation of binary and ternary NiTi(Fe) alloys. *Proc. of International Conferenc eon Martensitic Transformations 2008, Santa Fe, NM, in print.*
[38] Zheng H, Rao J, Pfetzing J, Frenzel J, Somsen C and Eggeler G 2008 *Scripta Mater.* **58** 743
[39] Pfetzing J, Wagner M F X, Simon T, Schaefer A, Somsen C and Eggeler G: TEM investigation of the microstructural evolution during nanoindentation of NiTi. *Proc. of European Symposium on Martensitic Transformations 2009, Prague, Czech Republic, submitted.*
[40] Nix W D and Gao H 1998 *J. Mech. Phys. Solids* **46** 411

Deformation mechanisms in ARB processed aluminium alloy AA6016 at low temperatures

L Hollang, J Scharnweber, K Reuther and W Skrotzki

Institut für Strukturphysik, Technische Universität Dresden , D-01062 Dresden, Germany

werner.skrotzki@physik.tu-dresden.de

Abstract. The deformation behaviour of the aluminium alloy AA6016 at low temperatures was investigated in the initial state and after 4 and 8 cycles of accumulative roll bonding (ARB). Tensile tests at 25 K, 77 K, 180 K and 296 K were performed at constant strain rate 10^{-4} s^{-1}. Stress relaxation experiments performed during the tensile tests were used to determine the experimental rate sensitivity λ as a function of the flow stress σ. In all cases $\lambda(\sigma)$ is found to be linear revealing that the Cottrell-Stokes law holds. The effect of grain size on σ can be adequately described through an additive athermal stress contribution σ_d, which is the higher the higher the degree of pre-deformation is. Moreover, the temperature dependence of the strain rate sensitivity $m(T)$ indicates that the rate controlling mechanism in the initial state is local single slip. The ARB states deviate from the single slip behaviour already at 25 K. The reason probably is the occurrence of additional thermally activated slip processes in the ARB states.

1. Introduction

In the last decade severe plastic deformation by accumulative roll bonding (ARB), first introduced by Saito et al. [1], turned out to be a very promising technique regarding "large scale production" of ultrafine grained (UFG) metals and alloys with grain sizes ranging from 1 µm to about 100 nm [2 - 4]. Compared to their conventional large grained counterparts, UFG materials show a markedly enhanced strength (e.g. [5]) with only a slightly decreasing and in some aluminium alloys even increasing ductility [6 - 8].

The influence of ARB processing on the tensile behaviour at ambient temperatures has been described for various UFG materials, e.g. for the technically relevant age-hardening aluminium alloy AA6016 [9]. Investigations at low temperatures are less frequent. Principally, investigations at low temperatures permit to study quantitatively the influence of ARB processing on the temperature dependent transition from pure dislocation slip controlled plasticity to diffusion controlled plasticity in aluminium AA6016.

This study aims at a qualitative investigation of the relevant deformation mechanisms in AA6016 at and below room temperature. For this purpose samples taken from three ARB sheets with different degree of deformation were subjected to tensile tests with stress relaxation experiments at temperatures between 25 K and 296 K. An application of thermal activation theory (e.g. [10]) then permits to distinguish between pure athermal ("Hall-Petch" like) and thermal ("nano" deformation mechanisms) effects of ARB processing on the flow stress [11].

2. Experimental

In this investigation samples of the initial state and after four and eight ARB cycles were tested. Prior to ARB the plates with a thickness of 1 mm were solution heat treated in a furnace for 1h at 520°C and subsequently quenched in water. The ARB-process with a thickness reduction of 50% per rolling pass was applied to aluminium plates with a width of 100 mm and a length of approximately 300 mm using a four high rolling mill (Carl Wezel, Mühlacker). Prior to rolling the plates were pre-warmed at 230°C for four minutes. After each cycle the surfaces of the air-cooled bonded plates were wire brushed to remove the oxide layer. The ARB process was done up to a maximum of eight cycles, which corresponds to a von Mises strain of 6.4 and leads to a plate with $2^8 = 256$ bonded layers each having a thickness of about 3.9 µm. The conditions of the ARB processing are described in more detail in [12]. The chemical composition of the aluminium alloy AA6016 is listed in Table 1.

Table 1. Chemical composition of AA6016 (mean values)

	Si	Cu	Fe	Mn	Mg	Cr	Zn	Ti	other	Al
wt.%	1.0-1.5	0.52	0.5	0.2	0.25-0.6	0.1	0.2	0.15	0.15	balance

Figure 1 shows the microstructures of the sheets investigated, determined by electron back scatter diffraction (EBSD) measurements in a Zeiss Ultra 55 scanning electron microscope using 40 nm step size. The average grain size determined by the line intersection method is about 16 µm in the initial state. After four and eight ARB cycles the grain size is d_{RD} = (660 ± 170) nm and (380 ± 40 nm) nm along the rolling direction (RD) and d_{ND} = (400 ± 70) nm and (180 ± 40) nm along the normal direction (ND), respectively. In all cases grain boundaries with misorientation angles ≥ 3° were taken into account.

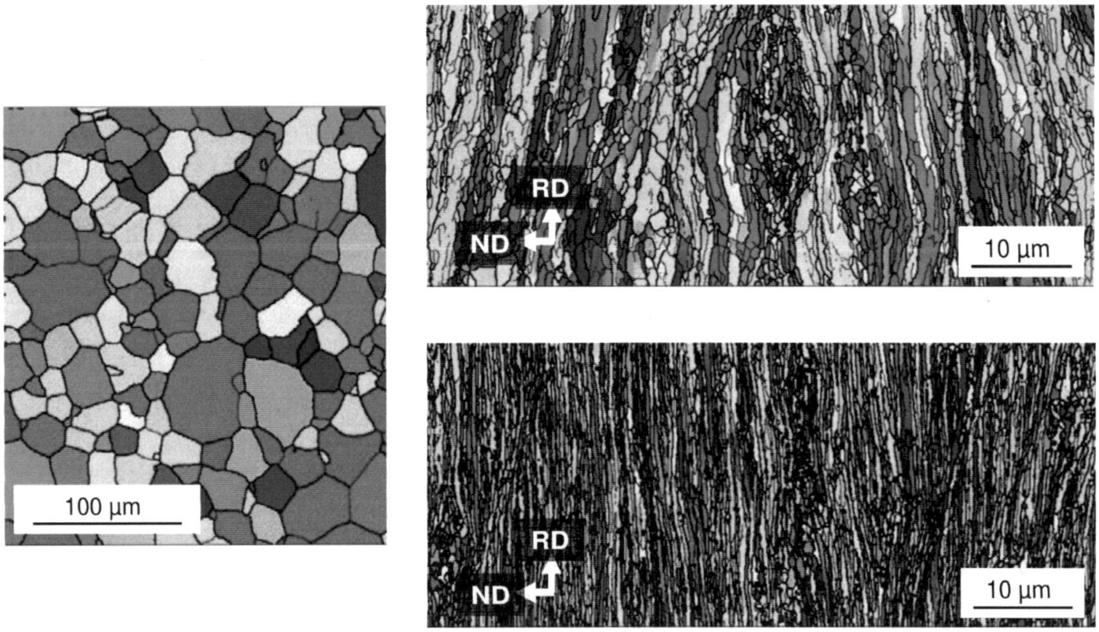

Figure 1. Aluminium alloy AA6016 in the initial state (left) and after 4 (upper right) and 8 (lower right) ARB cycles. Rolling direction (RD) and normal direction (ND) are indicated.

Journal of Physics: Conference Series **240** (2010) 012148 doi:10.1088/1742-6596/240/1/012148

From the ARB sheets flat specimens with a gauge length of about 15 mm and a cross section of 3.5×1 mm^2 were cut by spark erosion. The tensile direction was parallel to RD. The deformation experiments were carried out in an open continuous flow cryostat (He coolant) permitting to keep the temperature constant within less than 0.05 K [13]. The cryostat was mounted in an Instron 4502 load frame. The experiments were performed at constant crosshead displacement rate providing an initial strain rate of 9×10^{-4} s^{-1}. The tensile tests were performed at 25K, 77K, 180K and 296K until necking or fracture occurred. To calculate the flow stress σ and total strain ε the diminution of the cross section of the specimens was taken into account.

The thermal activation analysis is based on stress relaxation experiments performed during the tensile tests. Details about the data acquisition and evaluation as well as the underlying theory are described elsewhere [11].

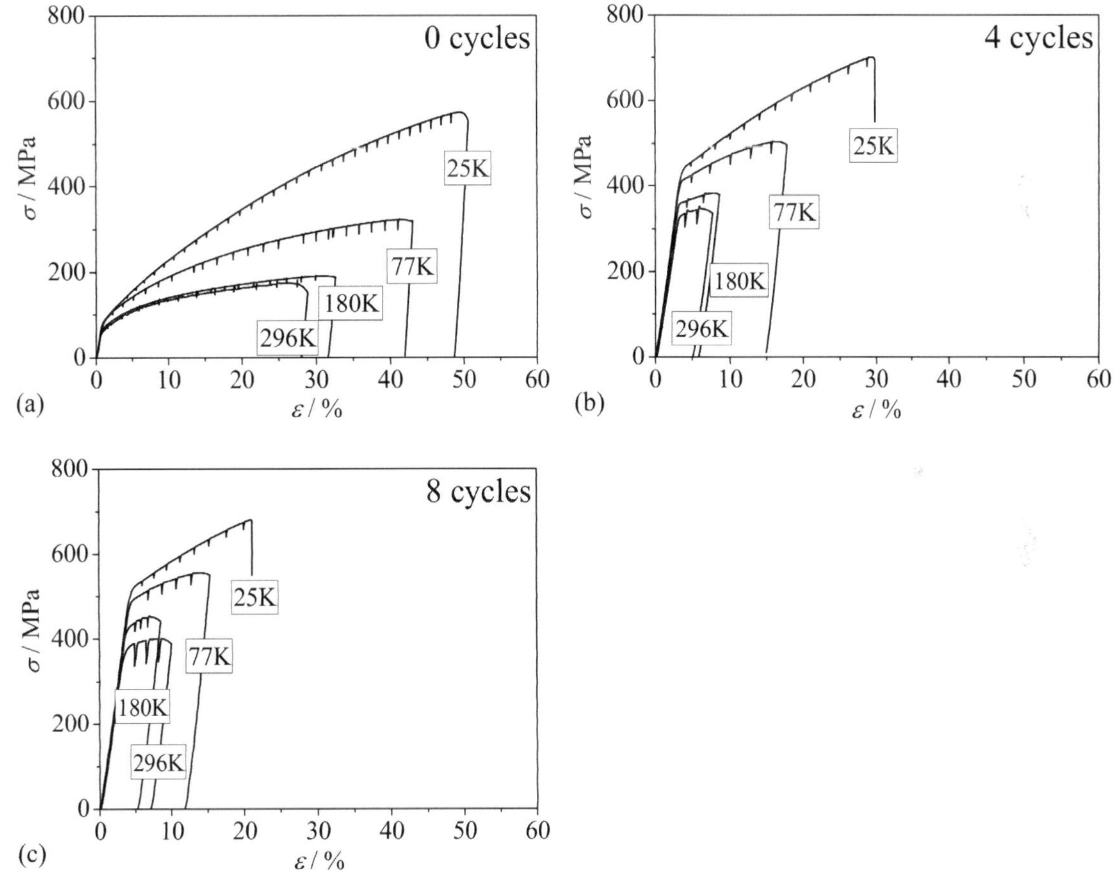

Figure 2. Stress–strain curves for samples deformed at different temperatures: (a) initial state, (b) four and (c) eight ARB cycles.

3. Results and Discussion

The stress–strain curves are shown in Figure 2. The stress drops are the result of the stress relaxation experiments. First of all, the curves show pronounced temperature dependence. Generally, it holds that for a given material the flow stress and the ductility are the higher the lower the deformation temperature is. On the other hand, at constant temperature the flow stress increases and the ductility decreases with increasing number of ARB cycles. All deformation experiments were terminated if pronounced necking took place. However, a special behaviour is observed at 25K. Here the specimens of the ARB processed material fractured suddenly after the onset of necking. Some other parameters obtained from the true stress–true strain curves are discussed in detail in [14].

Figures 3a-c show the corresponding hardening coefficient $\theta = (\partial\sigma/\partial\varepsilon)_T$ as function of flow stress. In all cases the θ values were determined by linear extrapolation of the stress–strain curves in front of a stress relaxation experiment. In the initial state θ almost linearly decreases with increasing stress, at least at high stresses. This clear indicates so-called stage III strain hardening of face-centred cubic polycrystals [15]. The upper limit $\theta = \sigma$ of the Considère criterion [16] for plastic instability valid for rate-insensitive materials is illustrated by the dashed line in figures 3a-d. Actually, in the initial state necking starts at all temperatures at this point.

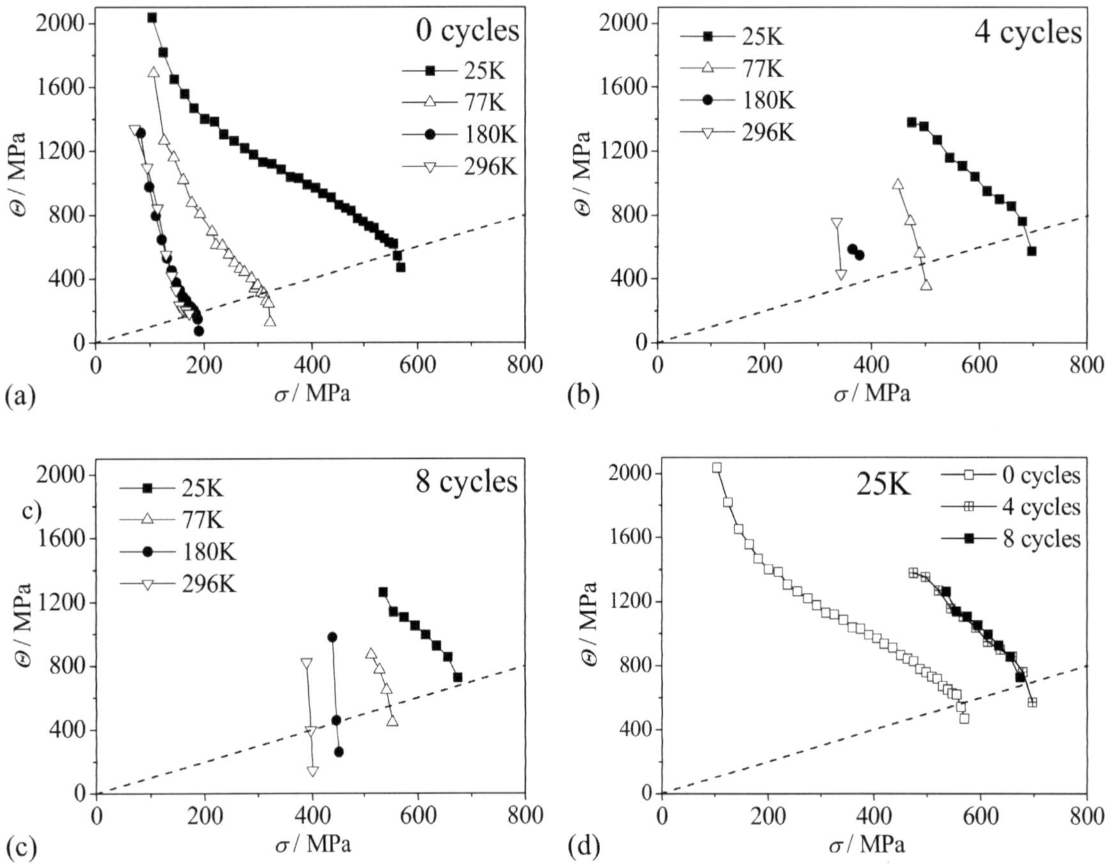

Figure 3. Hardening coefficient θ vs. flow stress σ at different temperatures: (a) initial state, (b) four and (c) eight ARB cycles, and (d) $\theta(\sigma)$ at 25 K in different ARB states. The dashed line marks $\theta = \sigma$.

The ARB samples deformed at 77 K and 25 K behave similar if compared with the initial state. Obviously, ARB processing of aluminium AA6016 at elevated temperatures (about 200°C) is not sufficient to exhaust the strain hardening capacity of the material at much lower temperatures. Figure 3d for all three ARB states shows the relationship $\theta(\sigma)$ at 25 K, that is at maximum ductility. It is interesting to note, that the $\theta(\sigma)$ curves after four and eight ARB are shifted to much higher stress levels, where they coincide extremely well. This indicates that most of the athermal Hall-Petch like hardening occurs during the first four ARB cycles, where most of the obstacles for dislocation motion were initially introduced (new grain boundaries, bonding layers, agglomerates of vacancies, etc.). Naturally, Taylor hardening by dislocation storage and grain fragmentation plays an important role at any stage of the ARB process. However, the balance between "pure" Taylor hardening and dynamic recovery processes may be the reason for the "moderate" microstructural refinement between four and eight ARB cycles where, probably, all structure lengths were further downsized simultaneously until saturation is reached. It seems that saturation will be reached in aluminium AA6016 not far beyond eight ARB cycles.

At 296 K and 180 K the ARB samples show a rather limited ductility, but the region of stable homogenous deformation seems to be somewhat extended beyond $\theta = \sigma$. This can be attributed to an increased strain rate sensitivity of the ARB states, which delays the onset of necking. In this case the Hart criterion [17] for plastic instability $\theta \leq \sigma(1 - m)$ is valid. m is the strain rate sensitivity of the material and at 296 K is about 0.01 in the initial state, but 0.4 after eight ARB cycles (cf. figure 5).

Figures 4a-c show the experimental rate sensitivity λ as function of flow stress σ of all materials investigated. There is a strictly linear relationship in all cases where strain hardening is sufficiently pronounced. In the ARB states this is the case only at 77 K and 25 K; at higher temperatures there is a lack of strain hardening. The linear relationship clearly indicates so-called Cottrell-Stokes behaviour, i.e. the increase of stress with strain during the experiment is totally caused by dislocation multiplication, movement, and storage.

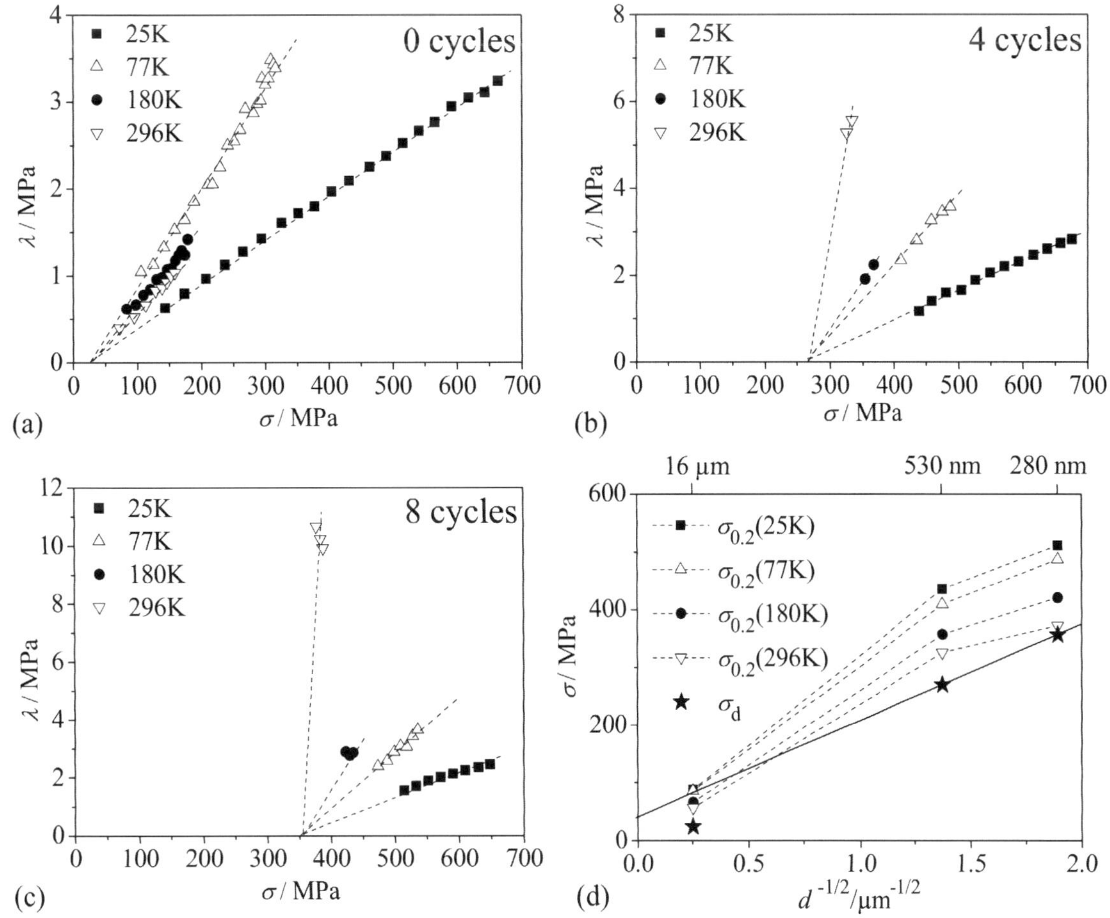

Figure 4. Experimental rate sensitivity λ vs. σ for samples deformed at different temperatures: (a) initial state, (b) four and (c) eight ARB cycles. (d) shows the offset stresses (here called σ_d, see text) together with the flow stress after 0.2% strain at different temperatures as a function of the inverse square root of the arithmetic average of grain size $d = (d_{RD} + d_{ND})/2$.

In all three materials the straight lines intersect more or less at one point at the stress axis. The resulting offset stress is 24 MPa in the initial state and 270 MPa and even 360 MPa after four and eight ARB cycles, respectively. Obviously, the offset stress is athermal by nature and adds to the Cottrell-Stokes stress caused by dislocation multiplication, movement, and storage mentioned above. The sum of both is the total flow stress σ measured in the tensile tests [11].

In the initial state the small athermal offset stress of 24 MPa is probably due to slight precipitation hardening [18, 19]. During the first four ARB cycles the athermal stress component increases rapidly and reaches already 270 MPa. This strong increase of the athermal offset stress during ARB can be attributed mainly to the influence of grain size d on the flow stress σ and is, therefore, called σ_d in the following.

Even between four and eight cycles, microstructural refinement still proceeds, but the increase of σ_d is not as pronounced as in the first four cycles (from 270 MPa up to 360 MPa only). The impression of a moderate strengthening during higher ARB cycle numbers is confirmed by observations of the process of grain refinement performed by EBSD studies [9]. It is remarkable, that in aluminium AA6016 after eight ARB cycles more than 50% of the flow stress measured in tensile tests at 25 K is due to σ_d. For comparison, in pure ECAP nickel the contribution of σ_d to σ at 25 K is only 15 % [11]. In this respect ARB processing seems to be very efficient for strengthening of aluminium alloys.

It cannot be excluded, however, that the ARB processing also affects the volume fraction, size, and distribution of precipitates and, thus, the athermal stress component. The incorporation of oxides from the (former) surface layers in the material may also be an important athermal strengthening mechanism in ARB aluminium alloys like AA6016 [20]. In this sense σ_d is an upper limit for the athermal influence of grain size in ARB processed AA6016.

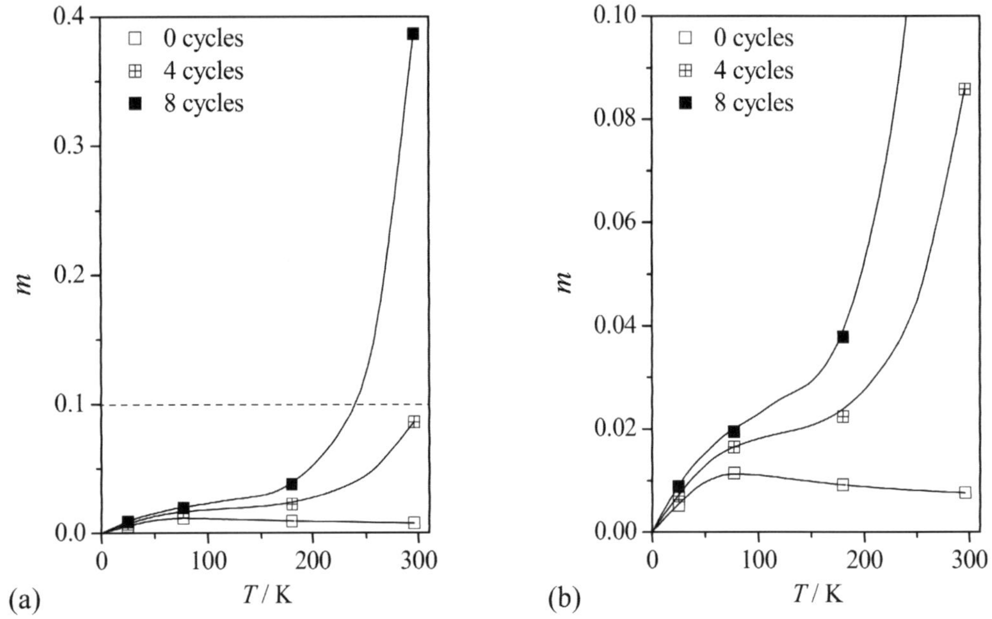

Figure 5. (a) Strain rate sensitivity m after different numbers of ARB cycles as a function of deformation temperature T. The right figure (b) is a blow-up of the lower part of (a).

Figure 4d shows σ_d (the offset stress in the initial state is included) together with the flow stress after 0.2% strain at different temperatures as a function of the inverse square root of the arithmetic average of grain size $d = (d_{RD} + d_{ND})/2$. The full line illustrates the expected Hall-Petch like behaviour of the ARB states after four and eight cycles assuming the same amount of oxide and precipitation hardening in both states. Probably, the initial state deviates to lower stresses (24 MPa) since here the precipitation hardening is not fully developed if compared with the ARB states. It is very likely that the too small offset stress in the initial state is the result of under-aging due to the missing pre-warming treatment (230° for five minutes prior to ARB, see e.g. [20]) and, of course, due to the lack of ARB induced oxide incorporation and precipitation hardening.

As can be seen from figure 4d the flow stress determined at 0.2% strain is only a poor estimate for the Hall-Petch stress since it always contains a thermal stress component that is the higher the lower the deformation temperature is. Thus, σ_d is a lower bound for the pure athermal grain size induced stress and, probably, is a better quantity for the "true" Hall-Petch stress. The slope of the full straight line in figure 4d is $k_{HP} = 0.17$ MPa m$^{1/2}$.

Figure 5 shows the temperature dependence of the strain rate sensitivity m, that is, of the slope of the straight lines in figure 4. In the initial state the usual behaviour expected from forest cutting theory is observed (e.g. [11, 21, 22]). m first strongly increases at low temperatures, passes a maximum at intermediate temperatures and then decreases towards high temperatures. This behaviour is typical for dislocation dominated plasticity [22] and may be attributed to local single slip [11, 23].

The ARB states deviate towards much higher strain rate sensitivities already at 25 K. As for pure ECAP nickel [11] it holds that the deviations are the more pronounced the higher the degree of pre-deformation is. The deviations indicate the appearance of additional, unusual deformation mechanisms in ARB aluminium AA6016 already at 25 K where, probably, diffusion plays no role in the deformation process. Therefore, we suppose that the thermally activated interaction of dislocations with grain boundaries is responsible for the increase of m at low temperatures. However, at 296 K m is up to forty times (after eight cycles) higher if compared with the initial state. There is no doubt that this is a hint for the dominance of diffusion controlled (recovery) processes taking place at room temperature [24, 25]. The behaviour of m above room temperature is discussed by other authors, e.g. in Ref. [26] for UFG aluminium of commercial purity and in Ref. [27] for the UFG aluminium alloy AA6061, both processed by equal channel angular pressing.

4. Conclusions

i) ARB processing of aluminium AA6016 is a very efficient technique to produce a microstructure of sub-micron scale exhibiting unique unusual deformation behaviour.

ii) After ARB processing at about 200°C the aluminium AA6016 is heavily pre-deformed and, therefore, shows a limited ductility at room temperature. Towards lower temperatures the ductility strongly increases due to an increase of strain hardening capacity. At constant temperature the remaining strain hardening capacity decreases with increasing number of ARB cycles.

(iii) In AA6016 most of the Hall-Petch like hardening during ARB processing occurs within the first four ARB cycles. At higher cycle numbers, the ongoing microstructural refinement by Taylor hardening competes with dynamic recovery processes.

(iv) The microstructurally saturated state in aluminium AA6016, where strain hardening and dynamic recovery balance each other, seems to be not far beyond eight ARB cycles.

(v) The athermal stress component σ_d obtained from the $\lambda(\sigma)$ relationship represents the "true" temperature independent Hall-Petch stress.

(vi) ARB processing strongly increases the strain rate sensitivity of aluminium AA6016 at room temperature and moderately at very low temperatures. This clearly indicates the appearance of unconventional deformation processes in ARB AA6016 at all temperatures. However, whereas at low temperatures probably dislocation–grain boundary interaction only slightly affects the plastic behaviour of ARB processed AA6016, diffusion controlled deformation mechanisms probably govern the plasticity of the material at room temperature.

Acknowledgments
The ARB aluminium alloy was kindly provided by Dr. H.-W. Höppel, University Erlangen-Nürnberg.

Journal of Physics: Conference Series **240** (2010) 012148 doi:10.1088/1742-6596/240/1/012148

References

[1] Saito Y, Utsunomiya H, Tsuji N and Sakai T 1999 *Acta Mater.* **47** 579.
[2] Valiev RZ, Korznikov AV and Mulyukov RR 1993 *Mater. Sci. Eng.* A **168** 141.
[3] Tsuji N, Saito Y, Utsunomiya H and Tanigawa S 1999 *Scripta Mater.* **40** 795.
[4] Valiev RZ, Islamgaliev RK and Alexandrov IV 2000 *Prog. Mater. Sci.* **45** 103.
[5] Valiev RZ, Estrin Y, Horita Z, Langdon TG and Zehetbauer M 2006 *JOM* **58** 33.
[6] Kim W, Kang SB, Tsuji N and Minamino Y 2005 *Acta Mater.* **53** 737.
[7] Höppel HW, May J and Göken M 2004 *Adv. Eng. Mat.* **6** 781.
[8] Tsuji N, Shiotsuki K and Saito Y 1999 *Mater. Trans.* **40** 765.
[9] Scharnweber J, Skrotzki W, Oertel C-G, Brokmeier HG, Höppel HW and Topic I 2009 *Adv. Eng. Mat.*, accepted.
[10] Kocks UF, Argon AS and Ashby MF1975 *Prog. Mater. Sci.***19** 1.
[11] Hollang L, Hieckmann E, Brunner D, Holste C and Skrotzki W 2006 *Mater. Sci. Eng.* A **424** 138.
[12] Topic I, Höppel HW, Staud D, Merklein M, Geiger M and Göken M 2008 *Adv. Eng. Mat.* **10** 1101.
[13] Brunner D and Diehl J 1992 *Z. Metallkd.* **83** 828.
[14] Scharnweber J, Hollang L, Reuther K, Dey SR, Oertel CG and Skrotzki W 2010 *J. Phys. Conf. Series,* this issue.
[15] Kocks UF and Mecking H 2003 *Prog. Mater. Sci.* **48** 171.
[16] Considère AG 1885 *Annales des Ponts et Chausses (Ser. 6)* **9** 574
[17] Hart EW 1967 *Acta Metall.* **15** 351.
[18] Andersen SJ, Zandbergen HW, Jansen J, Traeholt C, Tundal U and Reiso O 1998 *Acta Mater.* **46** 3283.
[19] Dutta I, Allen SM 1991 *J. Mat.Sci. Lett.* **10** 323.
[20] Barlow CY, Nielsen P, Hansen N 2004 *Acta Mater.* **52** 3967.
[21] Gibbs GB 1969 *Mater. Sci. Eng.* **4** 313.
[22] Wielke B 1977 *Z. Metallkd.* **68** 199.
[23] Essmann U, Rapp M, Wilkens M 1968 *Acta Metall.* **16** 1275.
[24] Blum W, Pöllmann T, Hofmann U, Weidinger P 1996 *phys. stat..sol. (a)* **157** 329.
[25] Les P, Zehetbauer M, Stüwe HP 1996 *phys. stat. sol.(a)* **157** 265.
[26] May J, Höppel HW, Göken M 2005 *Scripta Mater.* **53** 189.
[27] Priftaj AV, Böhner A, May J, Höppel HW, Göken M 2008 *Mater. Sci. Forum* **584-586** 741.

On the development of a pseudo micro-truss intermetallic microstructure in a high pressure die cast AZ91 alloy

A.V. Nagasekhar[1], **C.H. Caceres**[1], **C. Kong**[2]

[1] ARC Centre of Excellence for Design in Light Metals
Materials Engineering, The University of Queensland, Brisbane, QLD4072, Australia
[2] Electron Microscope Unit, UNSW Analytical Centre
University of New South Wales, Sydney, NSW2052, Australia

a.nagasekhar@uq.edu.au

Abstract. The three dimensional features of the intermetallic microstructure that develop across the thickness of a 1 mm thick casting has been studied using SEM and dual beam FIB. The intermetallics form a closely interconnected spatial network which resembles a scaffold or micro-truss structure near the casting surface. The degree of interconnection decreases, and the overall scale of the microstructure is coarser at the core of the casting. The contribution of this pseudo micro-truss structure to the overall strength of the casting is discussed.

1. Introduction

The yield strength of high pressure die cast (hpdc) AZ91 alloy increases with a decrease in section thickness [1-6]. The average grain size of the cast material decreases with decreasing section thickness, and the increased strength of the thinner sections is normally ascribed to Hall-Petch strengthening effects [1, 7, 8-11]. However, the decrease in casting thickness may affect the other contributions to the strength, particularly dispersion strengthening [5, 9], so the underlying reasons for the increased strength of the thinner sections are not completely clear. It has been suggested [5] that the closely interconnected structure of the intergranular intermetallics formed during the solidification behaves similarly to a ceramic cellular solid [5], akin to a pseudo micro-truss structure. Highly segmented ceramic structures exhibit increased strength due to a Weibull modulus effect [10], and it seems likely that this additional strengthening effect adds to the thickness effects on the strength [5].
In order to assess this possibility, it is necessary to determine the extent of the interconnection in the intermetallic structure as well as its relative scale, i.e., the size of the intermetallic cells. With these goals in mind, 2D and 3D microstructural studies of the intermetallic structure have been carried out at two locations of the cross section of a cast to shape tensile specimen of hpdc AZ91 alloy.

2. Experimental

A tensile specimen of AZ91 (1 mm thick) alloy from earlier studies [4, 5] was used for this study. Several sections were obtained from the gauge length and metallographically polished by standard methods, down to 0.5 μm colloidal silica. One of the polished sections was etched with glycol. Figure 1 shows the sites of interest in the cross-section. These sites were selected as they represent the regions of minimum (core) and maximum (corner) solidification rates. Earlier work indicates that the core and

corner regions exhibit the lowest and highest hardness values on the cross section [4, 5] as well as the coarsest and finest scale of Mg_{17} Al_{12} intermetallics [11], respectively.
The 2D analysis was carried out on a JEOL 6460 LA Scanning Electron Microscope, on the etched cross-section (plane x-z in Fig 1), whereas the serial sectioning for the 3D analysis was performed on a FEI xT Nova NanoLab 200 DualBeam FIB system, on the un-etched cross-section. The procedure involves digging a small square trench, and subsequently removing slices of given thickness from one of the trench's walls. See Fig. 1 for details on the geometry of the sectioning. Further details on the sectioning and imaging procedure have been published elsewhere [12]. Specific details to the sequential sectioning are given in Table I. The Amira® 3D reconstruction software [13] was used to create 3D images from the sequence of secondary electron images.

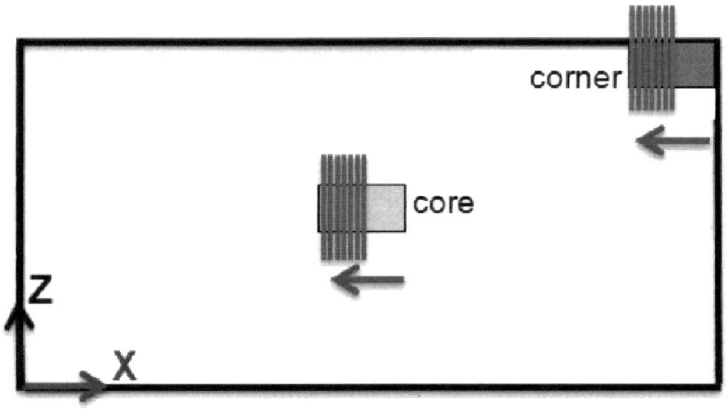

Figure 1. Identification of the sites for the microstructural study. The overall dimensions of the cross section were 10 mm by 1 mm. The sequential sectioning for the FIB study was carried out on y-z sections as indicated by the parallel lines and arrows. The x, z directions match those of Figures 2 and 3.

Table I. Details of the FIB sequential sectioning for the 3D images.

Location	Resolution	Slice thickness (μm)	HFW (μm)
Core	1024x884	0.18	36.6
Corner	1024x884	0.2	51.2

3. Results
SEM secondary electron images (x-z plane in Fig 1) from the core and corner regions are shown in Fig. 2. The bright regions are the $Mg_{17}Al_{12}$ intermetallic. The scale of the intermetallic appears relatively coarse at the core, Fig. 2a, whereas it is very fine at the corner, Fig. 2b. Some amount of interconnection is apparent in both microstructures, but especially at the corner.
Figure 3 shows the reconstructed 3D images of the intermetallic. Each reconstructed image in Fig. 3 has been created from a series of 100 SEM micrographs. Comparison with sectioning perpendicular to those depicted in Fig 2 shows that the staggered appearance is an artefact of the sequential slicing. The solidification direction lies along the z-direction at the core, and away from the x-z vertex at the corner. It is seen that the intermetallics exhibit a high degree of interconnection at the corner, indeed resembling a micro-truss structure. On the contrary, at the core the degree of interconnection is significantly lower and the scale of the microstructure is much larger, as anticipated from Fig. 2. The

latter is to be expected since the core contains many externally solidified large grains which keep apart the intermetallic-rich regions that form as the small grains solidify inside the die cavity.

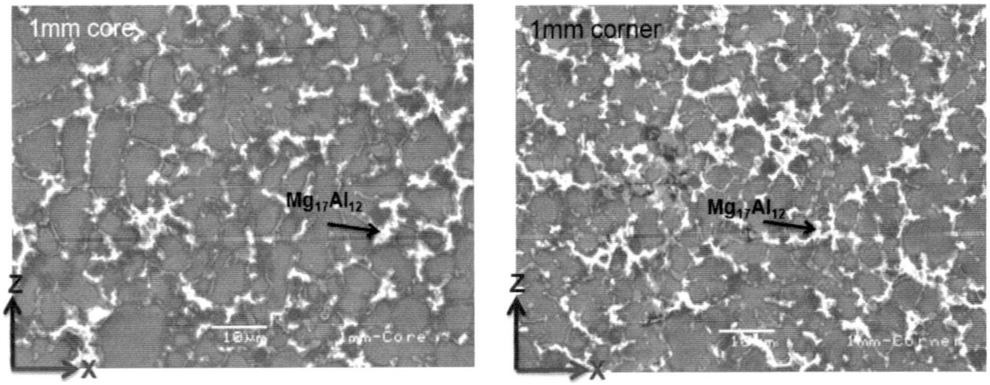

Figure 2. SEM secondary electron images showing the distribution of intermetallic at the core (a) and corner (b), respectively, of the specimen studied.

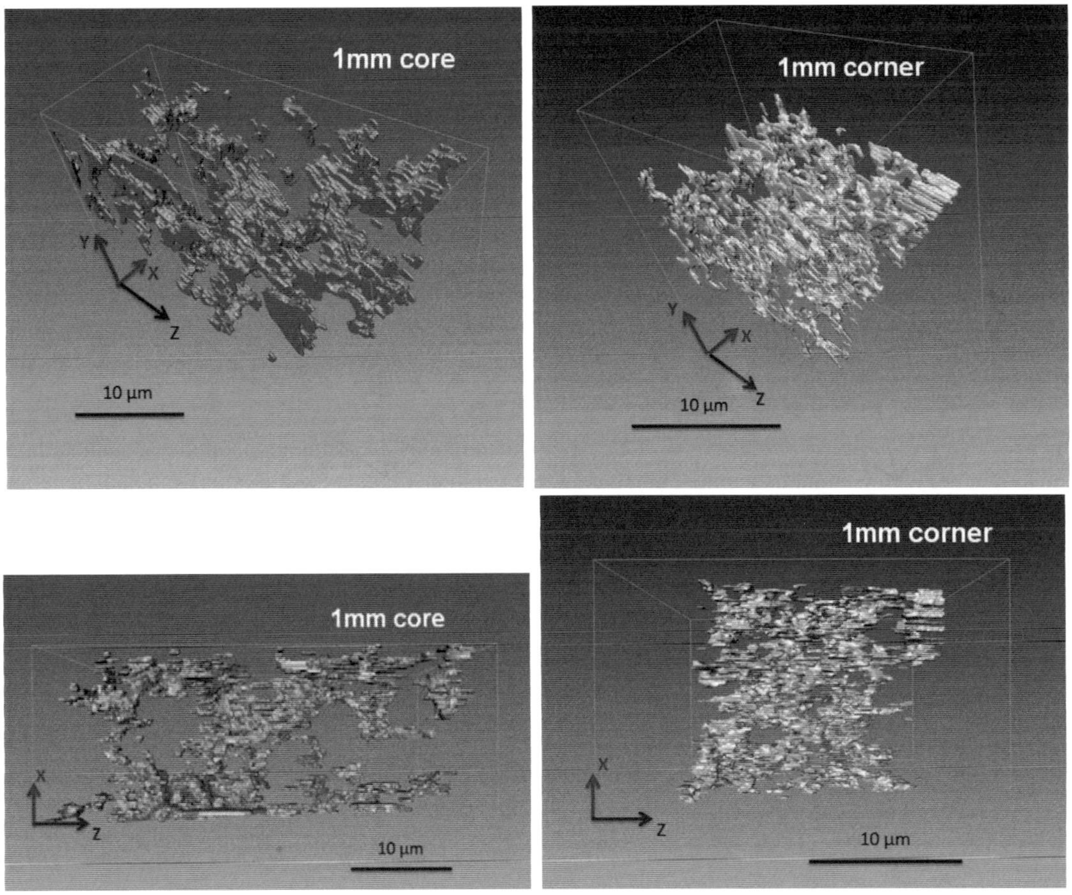

Figure 3. Different views of the intermetallic structure at the core and corner regions of the specimen of Fig. 2.

4. Discussion

Detailed quantitative analyses by Cáceres *et al.* [4, 5, 14] have shown that the strength of the as-cast structure of hpdc alloy AZ91 stems from three dominant mechanisms, namely: Hall-Petch hardening, solid solution hardening and dispersion hardening. The small grain size provides the largest strengthening, followed by solid solution hardening. The volume fraction of intermetallic is small (less that 10%) and dispersion hardening results in a minor reinforcement when the reinforcing particles are isolated and enable plastic deformation to bypass them after a small plastic strain. When a significant amount of interconnection is present, as in the present case, the intermetallic frame can be expected to behave like a rigid scaffold: plastic deformation can only occur by cracking the whole intermetallic structure. Extensive interconnection should thus result in a more rigid and harder material than when the intermetallic particles are isolated from each other, and it seems safe to ascribe the increased strength of thinner sections to the pseudo micro-truss structural reinforcement which in thin castings covers a larger share of the cross section [5]. It is noted that similar reinforcing effects stemming from the interconnected intermetallic network have recently been suggested with regards to the creep resistance of a thixomolded MRI 230D alloy [15].

5. Conclusions

The intermetallic microstructure at the core and corner regions of the cross section of hpdc AZ91 alloy specimens, of thickness 1 mm, has been characterised in 2D and 3D by using SEM and Dual Beam FIB. 2D images show that the scale of the intermetallics is relatively coarse at the core but appears very fine at the corner. 3D reconstructed images show that the intermetallic structure is finer and more profusely interconnected at the corner regions, resembling a micro-truss structure. The difference is scale and degree of interconnection is largely due to the presence of large dendritic grains at the core region. The highly interconnected intermetallic microstructure is likely to account for a measurable fraction of the alloys' strength, especially in thinner sections in which it is more prevalent.

Acknowledgements

We acknowledge the AMMRF for partial financial assistance.

References

[1] Sequeira W P, G L Dunlop and M T Murray 1997 *3rd International Magnesium Conference* ed G W Lorimer pp 63-73 London (The Institute of Metals, London)
[2] Bowles A L, J R Griffiths and C J Davidson 2001 *Magnesium Technology 2001* Ed: J Hryn pp 161-168 New Orleans (The Minerals, Metals and Materials Society (TMS), Warrendale, PA)
[3] Gokhale A M and G R Patel 2001 *Magnesium Technology 2001* Ed: J Hryn pp 195-199 New Orleans (The Minerals, Metals and Materials Society (TMS) Warrendale, PA)
[4] Cáceres C H, J R Griffiths, A R Pakdel and C J Davidson 2005 Mater. Sci. Eng. **A 402** 258-268
[5] Cáceres C H, W J Poole, A L Bowles and C J Davidson 2005 Mater. Sci. Eng. **A 402** 269-277
[6] Wei Y-h, L-f Hou, L-j Yang, B-s Xu, M Kozuka and H Ichinose 2009 J. Mater. Processing Tech. **209** 3278-3284
[7] Aghion E, N Moscovitch and A Arnon 2007 Mater. Sci. Eng. **A 447** 341-346
[8] Sakkinen D J 1994 *SAE Technical Paper 940779* pp 558-569 (Warrendale, PA: SAE)
[9] Aune T, H Westengen and T Ruden 1993 *SAE Technical Paper 930418* pp 51-57 (Warrendale, PA: SAE)
[10] Ashby M F and Y J M Brechet 2003 Acta Mater. **51** 5801-5821
[11] Nagasekhar A V, C H Caceres and M A Easton 2009 *Light Metals Technology 2009* Ed: M S Dargusch and S M Keay pp 479-482 Gold Coast, Australia (Trans Tech Publications)
[12] Nagasekhar A V, C H Caceres and C Kong 2009 *Magnesium Technology 2009* Ed: *E A Nyberg*, et al. pp 101-104 San Francisco (The Minerals Metals and Materials Soceity (TMS) Warrendale, PA)
[13] *Amira 3.1.* http://www.amiravis.com/
[14] Cáceres C H, J R Griffiths, C J Davidson and C L Newton 2002 Mater. Sci. Eng. **A 325** 344-355
[15] Amberger D, P Eisenlohr, M Göken 2009, ICSMA-15, *J. Phys.: Conf. Series,* this issue.

Journal of Physics: Conference Series **240** (2010) 012006 doi:10.1088/1742-6596/240/1/012006

Deformation band formation, characteristics, history

H J McQueen

Mechanical and Industrial Engineering
Concordia University, Montreal H3G 1M8, Canada

Abstract

Deformation bands (DMB) are crystal slabs that develop slip on systems different from neighbours; they rotate in different directions developing intervening transition boundaries (TB) that rise rapidly in misorientation. In polycrystals, grains divide into DMB, slipping on 2 or 3 systems to provide the 5 required components with minimum energy, as confirmed through OIM boundary misorientations and band rotation poles. TEM exposed the subgrains but seldom TB, except in single crystals as layers of cells. As the TB extend and align into layer bands, the microtextures are similar in cold and hot working. In hot working, TB are quite narrow and able to migrate; at large strains, lengthening and rotating TB (like grain boundaries) are responsible for rapid accumulation of high angle facets of many subgrains. Nevertheless sub-boundaries persist defining steady-state cellular dimensions and flow stress.

1. Introduction

When considering the plastic deformation of single-phase crystalline metals, there are three categories of constraints although the third one is effective only at high temperatures ($T>0.5T_M$ melting K). The dislocations move in glide, cross-slip or climb, according to the lattice slip systems to develop a substructure (cell walls (CW) below $0.5T_M$ and subgrain boundaries (SGB) above that) that is fairly uniform throughout the grain but with increased density near grain boundaries (GB) [1,2]. If there are no cracks, the deformation must conform to plastic stability requiring 5 independent components of the strain tensor. In polycrystals, a grain can lower its strain energy by dividing into volumetric regions called deformation bands (DMB) with intervening transition boundaries (TB). As the DMB slip on separate sets of only 2 slip systems, they rotate differently (net slip vector into elongation axis); the dislocations from the 4 systems interact in the TB that rises more rapidly in misorientation ψ than walls within a DMB. At high T, where boundaries are mobile to varying degrees, GB, TB and SGB migrate and rearrange as required to minimize surface energy [1,2].

The presentation of cold working proceeds from single crystals to polycrystals as observed by x-rays (XRD) optical and transmission electron microscopy (TEM) as understanding grew of strain hardening [3,4]. Examination of TB during annealing has thrown much light on static recovery (SRV) and nucleation of recrystallization (SRX). The role of DMB in texture formation from x-rays was confirmed through micro textures from TEM selected area diffraction (SAD) and by electron back scattering SEM-EBS (diffraction D or imaging I). Finally, the characteristics of TB in hot working are explained with the previous techniques aided by orientation imaging microscopy (OIM).

Dislocation mechanisms in single- and poly-crystals

On initially polished surfaces with varied orientations in single slip crystals, the slip bands at strains $\varepsilon \approx 0.05$ developed kinks transverse to the bands (XRD asterism), arising from bending constraints or internal perturbations (particles) [3-7]. As ε rose to 0.1-2, the kinks broadened into DMB, exhibiting secondary slip bands; this could also be related to rotational overshoot in orientation for slip on a system. In crystals oriented for poly-slip, bands start almost immediately developing large rotations; in channel compression forming textural regions expected [89]. In rolled single crystals of Fe Si and of Al alloys, the DMB were clearly defined with alternate rotations by etching [10]. In TEM of single crystals, the DMB were easily located and confirmed by SAD an alternating rotation direction [10], as could be predicted by slip patterns [10,11]. The TB, often referred to as transition bands, generally consisted of layers of very fine cells (sharper and smaller ~0.05 μm than the cell ~0.5 μm), with walls of ~5° with total ψ rising towards 35° as more layers formed with rising ε [10].

In polycrystals, the grains broke up into DMB from very low strains and rotated through large angles that were characteristic of the initial orientation of the grain relative to the imposed stress axes (consistent with single crystal rotations [10]) [3]. In Al-0.5Cu and Al-0.5Cu-1Si, compression to 0.69 [6,7] produced neat alternating DMB in the single-phase alloy at 20°C [5] but at -196°C, the DMB were finer and more irregular as in Cu-Si particle alloy at 20°C [68B]. None of the TEM micrographs illustrated TB but the substructure in the DMB was described as low energy dislocation structures [6,7], thus evidence of the first constraint acting alone. The TB are clearly observed in orientation imaging microscopy (OIM) and are permanent even on reversing the strain [12-14].

For cold worked polycrystals deformed by rolling through strains of 1 to 4, TEM does not definitely pinpoint TB in the small field of view. Cell structure (~0.5 μm) undergoes modifications due to micro bands cutting through it, forming blocks (~5 μm) with denser walls (BW). Some of these, also called geometrically necessary boundaries (GNB), rise fairly rapidly in misorientation, if neighboring blocks slip on systems different, whereas the cell walls (called incidental dislocation boundaries, IDB) rise slowly [15-17]. The cells maintain an aspect ratio of about 1.2 indicating that CW (IDB) rearrange, thus being transitory (non-permanent) [1,2]. Because of misalignment of slip systems in neighboring grains, dislocations form at GB smaller cellular structures, with denser walls than in grain centers. Additional dislocations are also induced by stress concentrations from the neighboring grain. While ψ may reach high values, the pattern of distribution at GB is quite different from the layers of fine cells in TB. The TB (and some BW) attain permanency and lengthen like GB, as grains deform and rotate into the elongation axis, separating bands a few cells thick with the same orientation; each is one component of the deformation texture. Micro-texture measurements showed that such regions with narrow strong textures created the X-ray texture designated as a spread of an ideal orientation averaged across the specimen [24].. From low ε and high ε layer bands; DMB are initially 20-100 μm wide, depending on grain size and impurity effects [1,18,19].

Annealing after cold work

In annealing after cold straining up to about 0.5, nucleation is observed in optical microsopy to occur primarily along GB. As ε rises above that, nucleation occurs more frequently along internal interfaces that are likely TB[17,19]. In single crystals with only TB nucleation, TEM clarified that dislocations from cell interiors and some weaker CW move to some surrounding walls that rise in ψ and appear sharper [10,11]. This process, called coalescence by Hsun Hu, continues with a cell enlarging and increasing in ψ, until the energy differential induces rapid migration [10]. Variations on the mechanism in proximity to GB include bulging of the original GB related to differences in intersecting walls [19]. In studies of SRX texture formation, nuclei originate in the layer bands, depending on their cell size and perfection and on boundary structure that is related to their micro-texture. The variations in rate of SRX of a deformation micro-texture band and not its initial volume fraction, controls its strength in the SRX texture [17,19].

Hot working change in constraints

Above $0.5T_M$, the mobilityof the boundaries SGB, TB and GB become significant principally at high strains [1,2,20]. From the start in hot working at constant T and strain rate $\dot{\varepsilon}$, dislocation behavior is changed through dynamic recovery (DRV) that, above a strain ε_S, results in steady state deformation with constancy of flow stress σ_S and equiaxed subgrain size d_S (= 1.27 W_S wall spacing), as well as of dislocation spacing in SGB, S_W (= b_V/ψ, Burgers vector) and in subgrains $\rho_i^{-0.5}$ (ρ_i Frank network density) [20,21]. All substructure parameters are dependent on each other and on σ_S/G (G, shear modulus). The subgrains were observed initially by XRD, etching in ferritic alloys and by polarized optical microscopy (POM) in Al. Through POM, DMB were reported in hot worked Ni, uranium and Al [22]; however, the contrast alternation in a few Al grains slightly above ε_S arises from rows of subgrains that reorganize at higher ε, where DMB are much wider in both POM and OIM [12-14].

Through TEM (imaging and SAD), the characteristics of SGB (d_S, W_S, S_W, ψ, $\rho_i^{-0.5}$) were confirmed but TB as layers of fine cells were not seen; not more than one boundary of high ψ >15° was noticeable in each view field, due to SAD or to foil tilting for ε < 3 [1,12-14,20,21]. In hot rolling to ε = 2.3 (90% reduction in one pass (20 s^{-1})) of Al, Cu, Ni and austenitic stainless steels (γ-Fe), elongated grains exhibited deformation textures similar to cold working of Cu (change from brass type at low T in γ-Fe) [23-25]. In Cu and 7030 brass, there were internal grain markings, possibly TB that gave rise to nucleation along with GB upon slower cooling [23].

Single crystals deformed hot did develop high-ψ TB that did not lead to nucleation in Al on continued straining or holding [26], although such TB did give rise to dynamic recrystallization (DRX) in Cu or Ni at similar T/T_M [27]. In hot torsion of polycrystals of Cu, Ni and γ-Fe (C, HSLA and stainless), nucleation of discontinuous dDRX took place at relatively low critical ε_C (1 - 0.2, as T rose, $\dot{\varepsilon}$ fell). The new grains reached final size as a continuous necklace along GB and then in repeated necklaces until initial grains consumed [28,29]. Compared to single crystals, dDRX occurs before TD's reach a suitable structure for nucleation. In steady state after work softening, nucleation appears to be random at GB of grains produced by dDRX, maintaining the texture unchanged [25].

Under high strains, the third constraint alters the microstructures compared to cold working. The GB and the TB become serrated as they migrate down SGB in order to balance the surface tensions (minimize surface energy) [1,2,12-14,18,20-23,28,29]. However, the SGB constantly rearrange to maintain subgrains equiaxed. There is continual slow rearrangement of the GB and TB in the course of which there is a net migration into grains or DMB that have smaller subgrains, i.e., higher Taylor factor [1,2,25]. This causes a change in torsion texture and a gradual softening across $\Delta\varepsilon \approx 10$ [18]. The micro textures have been measured to ε = 60 by scanning TEM Kikuchi patterns and agrees with the macro texture from XRD and is quite different from the dDRX texture in Cu and Ni [18,25]. At strains where layer banding was attained, the pattern of ψ values indicated that between each original GB, there were several TB, the number rising with strain [18].

Since the serrations have a wavelength and amplitude near $d_S/2$, those on neighboring grains start to meet as the average grain thickness approaches $2d_S$ [1,2,17,18,20,21,30]. This shortens the grains affected and thickens the neighboring ones; this is accelerated at the thin ends of pancaked grains [2,20,21]. In result, the layer bands always remain thicker than d_S and the microstructure in POM or in OIM takes on a spheroidal appearance with size near d_S [21,36], with σ_S being maintained; both being similar to those of specimens of 20 X larger initial grain size [18]. A good name for this is grain-defining gDRV [2], although grain shortening with an internal substructure was called geometric *gDRX (* no nucleation or growth) [18].

Since through constant rearrangement of the transitory SGB maintaining the subgrains constant in size and equiaxed despite rising strain, more subgrains must come into contact with the permanent GB and TB that are lengthening as the grains pancake [12,20,25,30] (like marbles filling a balloon that is being flattened), a count of all subgrain facets show that the fraction of boundaries >15° rises as the fraction of SGB decline [1,12-14]. TEM shows that ψ or S_W of SGB remain constant to ε = 16 [31] and that at ε = 60, there are many such SGB. OIM statistics for ε = 0.2 to 6 exhibit such a change while fraction with ψ=10 - 15 remains unchanged [12,14]. Moreover, when the ε is reversed, 0.2 to 0.5, then to 0, the fractions transfer from high to low with d_S and σ_S having constant magnitude [13]. This contradicts the theory of continuous cDRX that the SGB continuously increase in ψ to account for the rising fraction with ψ>15 [2,12,14,20].

Conclusion

In 20°C straining, the plastic equilibrium constraint gives rise to multi-layered transition boundaries between deformation bands quite different from walls due to dislocation behavior alone. Such TB have much greater density of dislocations in multi layers, rising rapidly in misorientations and on annealing, provide nucleation sites. The deformation bands rotate to give rise to the various components of the deformation texture. When above 0.5 T_M, DRV alters dislocation behavior, deformation bands still create a similar texture. The transition boundaries are much narrower (like GB), but still have high misorientation and permanence and are easily defined by OIM. Balancing surface tensions causes serrations of the non-transitory boundaries and a spheroidal microstructure no smaller than the steady state subgrain size best named grain defining dynamic recovery (gDRV).

References

[1] H.J. McQueen and W. Blum, *Al Alloys Physical and Mechanical Properties*, (ICAA6), T. Sato, ed., Japan Inst. Metals (1998), pp. 99-112.
[2] H.J. McQueen, E. Evangelista, M. Cabbibo and G. Avramovic-Cingara, *Can. Metal Quart.*, 47 (2008), 71-82.
[3] C.S. Barrett and L.H. Levenson, *Trans. AIME*, 137 (1990), 112-127.
[4] B. Jaoul, *Étude de la Plasticité et Application aux Métaux, Dunod*, Paris, 1965.
[5] R.E. Smallman and C.S. Lee, *Mater. Sci. Eng.* A184 (1994), 97.
[6] S.S. Kulkarni, E.A. Starke Jr. and D. Kuhlman Wilsdorf, *Acta Mater.*, 46 (1998), 5283-5301.
[7] D. Kuhlman Wilsdorf, *Acta Mater.*, 47 (1999), 1697-1712.
[8] G.Y.Chin, *Inhomogeneity Plastic Deformation*, ASM, Metals Park, OH (1973), pp. 83-112.
[9] M.C. Theyssier, B. Chenal, J.H. Driver, N. Hansen, *Phys. Stat. Sol.*, (a) 149 (1995), 367-378.
[10] H. Hu, *Recovery and Recrystallization of Metals*, Gordon and Breach, NY (1963), pp.311-364.
[11] A. Berger, P.J. Wilbrandt, P. Haasen, *Acta Metall.*, 31 (1983), 1433-1443.
[12] G. Avramovic-Cingara, H.J. McQueen and D.D. Perovic, *Light Metals/Metaux Legers 2004*, D. Gallienne, R. Ghomaschi eds., Met. Soc., CIM, Montreal (2004), pp. 141-152.
[13] G. Avramovic-Cingara, H.J. McQueen, *Mater. Res. Forum*, 519-523 (2006), pp. 1659-1664.
[14] G. Avramovic-Cingara and H.J. McQueen, Aerospace Materials Manufacturing, Emerging Techniques, *Met. Soc.*, CIM Montreal, M. Jahazi, et al., eds. (2006), pp. 173-186.
[15] D.A. Hughes, Q. Liu, D.C. Chrzan and N. Hansen, *Acta Mater.*, 45 (1997), 105-112.
[16] D.A. Hughes and A. Godfrey, *Hot Deformation of Al Alloys II*, T.R. Bieler, et al., eds. TMS, AIME, Warrendale, PA. (1998), pp. 23-36.
[17] R.D. Doherty, D.A. Hughes, F.J. Humphreys, J.J. Jonas, D. Juul-Jansen, M.E. Kassner, W.E. King, T.R. McNelley, H.J. McQueen and A.D Rollett, *Mater. Sci. Eng.*, 238 (1998), 219-274.
[18] J.K. Solberg, H.J. McQueen, N. Ryum and E. Nes, *Phil. Mag.*, 60 (1989), 447-471.
[19] F.J. Humphreys, M. Hatherly, *Recrystallization, Related Phenomenon*, Pergamon, Oxford 1995.
[20] H.J. McQueen, W. Blum, *Mater. Sci. Eng.*, A290 (2000), 95-107.
[21] W. Blum, Q. Zhu, R. Merkel and H.J. McQueen, *Z. Metallkde*, 87 (1996), 14-23.
[22] F.N. Rhines, *Inhomogeneity of Plastic Deformation*, ASM Metals Park, OH. (1973), pp. 251-284.
[23] H.J. McQueen, *Trans. Japan Inst. Met.*, 9 Suppl. (1968), 170-77.
[24] H.J. McQueen and H. Mecking, *Z. Metallkde*, 1987, 78, 387-395.
[25] H.J. McQueen, *Proc. ICOTOM 12*, J.A. Szpunar,ed., NRC Res. Pub.,Ottawa (1999), pp.836-841.
[26] M.E. Kassner, *Metal. Trans.*, 20A (1989), 2182-2185.
[27] P. Karduck, G. Gottstein and H. Mecking, *Acta Metal.*, 31 (1983), 1525-1536.
[28] H.J. McQueen, E. Evangelista and N.D. Ryan, *Recrystallization ('90) in Metals and Materials*, T. Chandra, ed., TMS-AIME, Warrendale, PA. (1990), pp. 89-100.
[29] C.M. Sellars, *Phil. Trans. Roy. Soc.*, A288 (1978), 147-158.
[30] T. Petterson and E. Nes, *Met. Trans.* 34A (2003), 2727-2736; 2737-2744.
[31] M.E. Kassner and M.E. McMahon, *Met. Trans.*, 18A (1987), 835-846.

Theoretical analysis of dislocation splittings in $MoSi_2$

V. Paidar[1], M. Čák[2], M. Šob[3,4], and V. Vitek[2]

[1]Institute of Physics, Academy of Sciences of the Czech Republic, Na Slovance 2, CZ-182 21 Praha 8, Czech Republic
[2]Department of Materials Science and Engineering, University of Pennsylvania, 3231 Walnut Street, Philadelphia, PA 19104, U.S.A.
[3]Department of Chemistry, Faculty of Science, Masaryk University, Kotlářská 2, CZ-611 37 Brno, Czech Republic
[4]Institute of Physics of Materials, Academy of Sciences of the Czech Republic, Žižkova 22, CZ-616 62 Brno, Czech Republic

paidar@fzu.cz

Abstract. Due to a lower symmetry of the tetragonal $C11_b$ structure when compared with the cubic BCC lattice, the 1/2<331] dislocation cores are split asymmetrically contrary to the 1/2<111> BCC dislocations. This has essential impact on their behaviour and, consequently, on mechanical properties. Various types of dislocation dissociations are analyzed in the frame of anisotropic elasticity with the help of the data from *ab initio* calculations of γ-surfaces for generalized stacking faults.

1. Introduction

The intermetallic compound $MoSi_2$ crystallizes in the body-centred-tetragonal $C11_b$ structure. The $C11_b$ lattice of this binary system is similar to the BCC lattice but the dimension c in the tetragonal direction is not $3a$ but is close to $a\sqrt{6}$ [1]. The bonds between all nearest Mo and Si atoms have the same length for the ideal c/a ratio equal to $\sqrt{6}$ and internal structure parameter $\Delta = 0$ [1]. Various slip systems can be activated in $MoSi_2$ for different sample orientations and temperatures [2-10] but for the compressive deformation along the tetragonal axis <001], the plastic deformation is exclusively controlled by the mobility of the 1/2<331] dislocations on the {013) planes. These dislocations are important for ductility of molybdenum disilicide in general and this is why they deserve a special attention. The 1/2<331] dislocations are in fact inherited from the 1/2<111> dislocations of the BCC crystals. Orientation dependence of the critical resolved shear stress clearly indicates that the dislocation properties of 1/2<331] dislocations are governed by their cores.

The first step in the dislocation core structure examination is determination of the existence of stacking faults and related calculations of the energies of generalized stacking faults. These planar defects are decisive for the dislocation core structure and their properties can be assessed using a suitable model of interatomic forces. Due to a rather complex interatomic bonding in silicides, a quantum-mechanical density functional based *ab initio* method was employed in the calculations of γ-surfaces, i.e. the energies of generalized stacking faults as functions of displacement vectors [11] that answers the above question.

Here we report the first results of our theoretical study. Our analysis of possible dislocation splittings utilizes anisotropic linear elasticity and is based on the *ab initio* calculations of the γ-surface.

2. Dissociations of the 1/2<331] screw dislocations

Due to a long 1/2<331] Burgers vector, the dislocation energy is high, and therefore, the activity of such dislocations with this Burgers vector is rather unexpected. Nevertheless, the most important factor is dislocation mobility that can be enhanced by appropriate dislocation dissociation.

Let us start with the planar dislocation splittings on the (013) plane that is related to the (011) BCC plane with large atomic density. There are three local minima on the γ-surface corresponding to three different non-crystallographic stacking faults (SF) on the (013) plane [11]. The vectors connecting these minima with the deepest minimum corresponding to the ideal crystal (IC) are possible Burgers vectors of partial dislocations. Three dislocation reactions with the Burgers vectors of partials that deviate least from the $[3\,\bar{3}\,1]$ direction are

$$\mathbf{b} = \mathbf{b1} \text{ –SF2– } \mathbf{b2}, \tag{1}$$
$$\mathbf{b} = \mathbf{b3} \text{ –SF1– } \mathbf{b4} \text{ –SF3– } \mathbf{b5}, \tag{2}$$
$$\mathbf{b} = \mathbf{b3} \text{ –SF1– } \mathbf{b6} \text{ –SF2– } \mathbf{b7} \text{ –SF3– } \mathbf{b5}. \tag{3}$$

The Burgers vectors **b1** to **b7** are defined in tables 1-3 and marked in figure 1 on the contour plot of the (013) γ-surface.

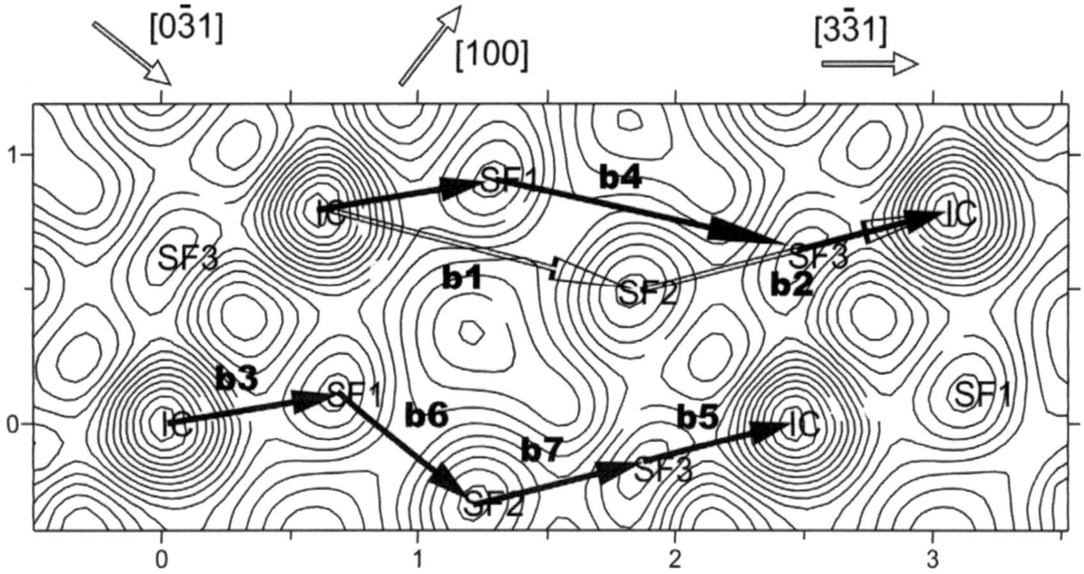

Figure 1. Contour plot of the (013) γ-surface and dissociations of the 1/2 $[3\,\bar{3}\,1]$ dislocation on the (013) plane. Three types of dissociations are depicted: Twofold **b1+b2**, threefold **b3+b4+b5** and fourfold **b3+b6+b7+b5** dissociations. Minima of the γ-surface corresponding to the non-perturbed ideal crystal are marked as IC and the positions of three stacking faults as SF1, SF2 and SF3.

Table 1. Twofold planar dissociation of the 1/2 $[3\,\bar{3}\,1]$ dislocation on (013).

Burgers vector	**b1**	**b2**
γ-surface minima	IC-SF2	SF2-IC
angle to $[3\,\bar{3}\,1]$	14.5°	14.5°

Table 2. Threefold planar dissociation of the 1/2 $[3\bar{3}1]$ dislocation on (013).

Burgers vector	**b3**	**b4**	**b5**
γ-surface minima	IC-SF1	SF1-SF3	SF3-IC
angle to $[3\bar{3}1]$	9.3°	12.6°	14.5°

Table 3. Fourfold planar dissociation of the 1/2 $[3\bar{3}1]$ dislocation on (013).

Burgers vector	**b3**	**b6**	**b7**	**b5**
γ-surface minima	IC-SF1	SF1-SF2	SF2-SF3	SF3-IC
angle to $[3\bar{3}1]$	9.3°	37.8°	14.5°	14.5°

The dissociation widths can be determined from the equations for the balance between the surface tensions of stacking faults and the dislocation interaction forces. In the framework of anisotropic elasticity, these forces depend on the orientation of dislocation line, the Burgers vectors and the elastic constants [12]. For the twofold dissociation, the width $r1$ of stacking fault ribbon is given by

$$\gamma 2 = F12/r1, \tag{4}$$

where $F12$ is the force parameter between the partials and $\gamma 2$ is the stacking fault energy.
For the threefold dissociation, two widths $r2$ and $r3$ are given by

$$\gamma 1 = F34/r2 + F35/(r2+r3), \tag{5}$$
$$\gamma 3 = F54/r3 + F53/(r2+r3), \tag{6}$$
$$F43/r2 + \gamma 3 = F45/r3 + \gamma 1. \tag{7}$$

And finally, for the fourfold dissociation, three widths $r4$, $r5$ and $r6$ are given by

$$\gamma 1 = F36/r4 + F37/(r4+r5) + F35/(r4+r5+r6), \tag{8}$$
$$\gamma 3 = F57/r6 + F56/(r5+r6) + F53/(r4+r5+r6), \tag{9}$$
$$F63/r4 + \gamma 2 = F67/r5 + F65/(r5+r6) + \gamma 1, \tag{10}$$
$$F75/r6 + \gamma 2 = F76/r5 + F73/(r4+r5) + \gamma 3. \tag{11}$$

The single-crystal elastic constants of tetragonal $MoSi_2$ corresponding to the temperature of 0 K were taken from [13] (c11=410 GPa, c33=514 GPa, c12=115 GPa, c13=87.5 GPa, c44=207 GPa and c66=200 GPa) and the c/a ratio of 2.447 (a = 0.3206 nm, c = 0.7846 nm) was used in the calculations of the force parameters Fik for dislocation interaction forces Fik/r where ik denotes different pairs of partial dislocations. The energies of three stacking faults on the (013) plane determined by *ab initio* calculations are $\gamma 1$ = 1.22 J/m^2, $\gamma 2$ = 1.06 J/m^2, and $\gamma 3$ = 1.12 J/m^2 [11].

The separation $r1$ of the **b1** and **b2** partials that divide the total Burgers vector into two parts was found to be about 12a. The threefold dissociation **b3**+**b4**+**b5** is slightly wider, both separations $r2$ and $r3$ are about 7a. Finally, the widths $r4$, $r5$ and $r6$ of the stacking fault ribbons for the fourfold dissociation **b3**+**b6**+**b7**+**b5** are 4a, 7a and 5a, respectively. The partial dislocation separations are expressed in units of the lattice constant a (0.3206 nm) that determines the $C11_b$ lattice dimensions. Dislocation dissociation causes an energy decrease that grows with the number of partial dislocations. For the twofold dissociation the energy gain is about 6.3×10^{-9} Jm^{-1}, for the threefold dissociation it is 6.8×10^{-9} Jm^{-1} and for the fourfold dissociation 8.3×10^{-9} Jm^{-1}.

The direction of [3 $\bar{3}$ 1] is parallel to the intersection of three planes with large atomic densities, namely (013), ($\bar{1}$ 03) and (110). Those three planes are analogous to the {110} planes in the BCC lattice into which the most common BCC Burgers vector 1/2 <111> can spread. The calculations of the γ-surface on the (110) plane found that the stacking fault energy on this plane is considerably smaller, $\gamma 0 = 0.357$ J/m^2, than on (013) [11]. The displacement vector is either 1/4 [3 $\bar{3}$ 1] or 1/2 [1 $\bar{1}$ 1] leading in both cases to the same crystallographic stacking fault on (110). Nevertheless, the Burgers vector 1/2 [1 $\bar{1}$ 1] is less likely to participate in the dissociation of 1/2 [3 $\bar{3}$ 1] since it is inclined by 30° to the [3 $\bar{3}$ 1] direction, and the corresponding dislocation energy is thus much higher than for the dissociations with 1/4[3 $\bar{3}$ 1].

Assuming the dissociation **b**/2+**b**/2 with the low stacking fault energy *γ0* we get the splitting width of about 39*a*. However, such a dissociation is not possible on the (013) plane as there is no local minimum for the displacement 1/4 [3 $\bar{3}$ 1]. Similarly as in the case of 1/2<111> screw dislocation in a BCC structure, the screw dislocation 1/2 [3 $\bar{3}$ 1] may dissociate in a non-planar manner on the three planes (013), ($\bar{1}$ 03) and (110) with the Burgers vectors **b3** and **b3*** on {013) and ($\bar{1}$ 03), respectively, together with **b**/2 on (110). A small Burgers vector **bc3** (12.8% of **b**/2) will remain in the dislocation centre at the intersection of three planes. An alternative non-planar dissociation is into **b5** and **b5*** on {013) and ($\bar{1}$ 03), respectively, together with **b**/2 on (110). The central dislocation **bc5** is even slightly smaller (11.5% of **b**/2) in this case. The Burgers vectors **b3*** and **b5*** lying on ($\bar{1}$ 03) are analogous to the vectors **b3** and **b5** lying on (013); they have the same lengths but different orientations.

Due to a low value of the energy *γ0* the non-planar threefold dissociations are much wider than the planar ones. Neglecting the small central dislocations the separation of the partial **b**/2 from the central dislocation is about 80*a*, while the widths of *γ1* or *γ3* on the {013) planes are about 2*a* in both cases. When the interaction forces with the central dislocation are taken into account the separation of **b**/2 slightly increases to 82*a* for the **b3**+**b3***+**b**/2 reaction while it remains nearly the same for the **b5**+**b5***+**b**/2 reaction. These dislocation splittings practically look as dissociations into two partials with the Burgers vectors **b**/2 and with the core of the partial at the intersection as spread into the two planes of the {013) type. The energy gain is larger for the **b3** and **b3*** partials (13.0 10^{-9} Jm^{-1}) than for the **b5** and **b5*** partials (10.8×10^{-9} Jm^{-1}). Nevertheless, these energetical gains due to splitting on (110) are larger than those for any planar dissociation (1), (2) or (3).

3. Dissociations of the 1/2<131] screw dislocations

There are other Burgers vectors of the dislocations on the (013) plane that are shorter than 1/2<331]. We will not discuss here the dislocations <100] and <110] lying on the (001) basal plane as those dislocations are not active for loading along the hard <001] tetragonal direction. Similarly, the dislocations <001] and 1/2<1 $\bar{1}$ 1] on the (110) plane will not be considered for the same reason. Notice that the dislocation <001] have a comparable length with 1/2<331]. Nevertheless, there are also dislocations **bi** = 1/2<131] inclined to the direction of <331] on the {013) plane that are shorter than **b** = 1/2<331].

Two dislocation reactions with the Burgers vectors close to the [1 $\bar{3}$ 1] direction can be envisaged:

bi = **b1** –SF2– **b8**, (12)

bi = **b3** –SF1– **b6** –SF2– **b8**. (13)

These Burgers vectors are defined in tables 4-5 and in figure 2. The dissociation widths can be again determined from the balance between the surface tensions of stacking faults and the dislocation interaction forces. For the twofold dissociation of 1/2 [1 $\bar{3}$ 1], the width *r7* is given by

$$\gamma 2 = F18/r7. \qquad (14)$$

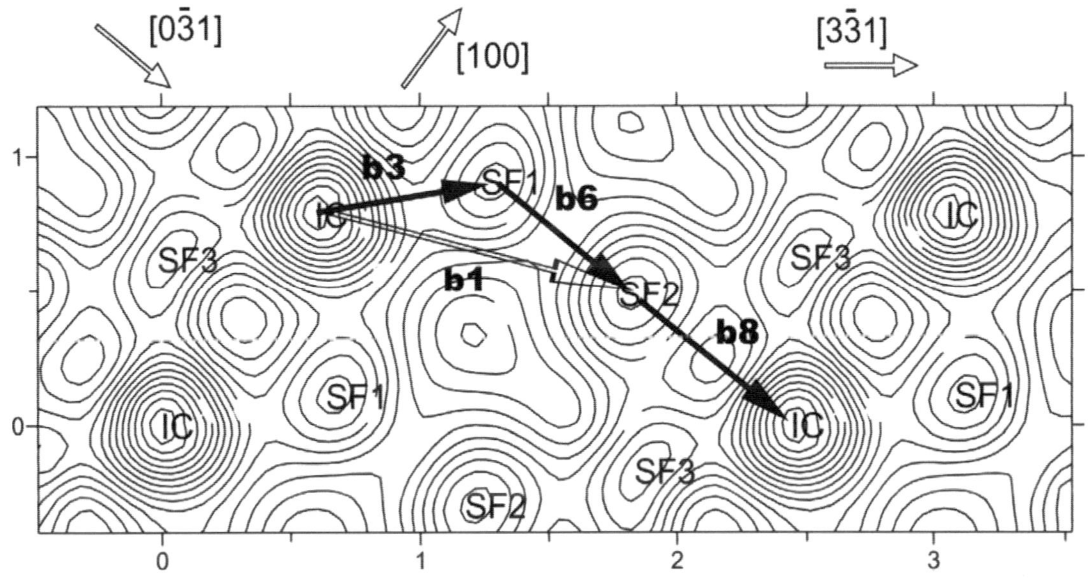

Figure 2. Dissociations of the 1/2 $[1\bar{3}1]$ dislocation on the (013) plane. Two types of dissociations are depicted: Twofold **b1+b8** and threefold **b3+b6+b8** dissociations.

Table 4. Twofold planar dissociation of the 1/2 $[1\bar{3}1]$ dislocation on (013).

Burgers vector	**b1**	**b8**
γ-surface minima	IC-SF2	SF2-IC
angle to $[1\bar{3}1]$	8.8°	14.5°

Table 5. Threefold planar dissociation of the 1/2 $[1\bar{3}1]$ dislocation on (013).

Burgers vector	**b3**	**b6**	**b8**
γ-surface minima	IC-SF1	SF1-SF2	SF2-IC
angle to $[1\bar{3}1]$	32.6°	14.5°	14.5°

Finally, for the threefold dissociation of 1/2 $[1\bar{3}1]$, two widths *r8* and *r9* are given by

$$\gamma 1 = F36/r8 + F38/(r8+r9), \quad (15)$$
$$\gamma 2 = F86/r9 + F83/(r8+r9), \quad (16)$$
$$F63/r8 + \gamma 2 = F68/r9 + \gamma 1. \quad (17)$$

The separation *r7* of the **b1** and **b8** partials that divide the Burgers vector **bi** into two parts was found to be about 8*a*. The threefold dissociation **b3+b6+b8** is again slightly wider, the separations *r8* and *r9* are about 3*a* and 7*a*, respectively. The energetical gains for these planar dissociations are 3.5 and 4.0×10^{-9} Jm^{-1}, respectively, what is much smaller than for the non-planar dissociations.

4. Discussion

The main difference between the two dislocations **b**=1/2<331] and **bi**=1/2<131] is that the screw dislocation **bi** inclined by 23.3° to the $[3\bar{3}1]$ direction cannot spread into the (110) plane and cannot thus profit from the energy decrease due to the low energy stacking fault on (110).

The lengths of all the Burgers vectors **b3, b6, b7, b5, b8** are comparable, the important difference is the deviation of **b6** from the $[3\,\bar{3}\,1]$ direction and of **b3** from the $[1\,\bar{3}\,1]$ direction. Due to this deviation the respective dislocation interaction forces are weaker and the splitting **b3**+**b6**+**b8** is asymmetric contrary to the splitting **b3**+**b4**+**b5** that is symmetric (cf. tables 5 and 2, the dissociation widths being 3.1*a* and 7.3*a* vs. 7.2*a* and 7.1*a*). Assuming the same energies of the stacking faults, $\gamma 1 = \gamma 2 = 1.14$ J/m^2, the separations *r8* and *r9* are 3.5*a* and 6.6*a*, respectively, while for the calculated energies they are 3.1*a* and 7.3*a*, respectively. It shows that there is certain effect of the stacking fault energy difference but since this difference is relatively small (15%), the deviation of the **b3** Burgers vector from the direction of the screw dislocation (32.6°) has a more pronounced effect on the asymmetrical character of dissociation.

The reported widths of dislocation splitting are accessible by modern electron microscopes and so it is a challenge for experimentalists to test our theoretical predictions. For this purpose the weak-beam technique is considered to be very promising [14].

Due to the asymmetrical splittings a strong anisotropic behaviour of the 1/2<331] dislocations can be expected. There are two non-planar dissociations, **b3+b3*+b**/2 and **b5+b5*+b**/2 that have similar nature and more detailed investigation would be necessary to describe their behaviour. Since the dislocation mobility is of primary importance, the effect of applied stresses has to be considered.

5. Conclusions

In addition to numerous planar dislocation splittings on the {013) planes, the non-planar core structures were reported that can play an important role in plastic deformation of $C11_b$ disilicides, in particular they may cause strong orientation dependence of mechanical properties of single crystals. This dislocation behaviour is much more complex than that of BCC 1/2<111> dislocations.

Acknowledgments

This research was supported by the Grant Agency of the Academy of Sciences of the Czech Republic, Project No. IAA100100920 (VP and MŠ), by the Ministry of Education of the Czech Republic, Project Nos. OC09014 (VP), MSM0021622410 and OC09011 (MŠ), and by the Academy of Sciences of the Czech Republic, Project Nos. AV0Z10100520 (VP) and AV0Z20410507 (MŠ). Partial support was provided by the Department of Energy, BES Grant No. DE-PG02-98ER45702 (MČ and VV).

References

[1] Villars P 1997 *Pearson's Handbook, Desk Edition, Crystallographic Data for Intermetallic Phases* (ASM International)
[2] Maloy S A, Heuert A H, Lewandowski J J and Mitchell T E 1992 *Acta Metall. Mater.* **40** 3159
[3] Maloy S A, Mitchell T E and Heuert A H 1995 *Acta Metall. Mater.* **43** 657
[4] Ito K, Inui H, Shirai Y and Yamaguchi M 1995 *Phil. Mag. A* **72** 1075
[5] Ito K, Moriwaki M, Nakamoto T, Inui H and Yamaguchi M 1997 *Mater. Sci. Eng. A* **233** 33
[6] Ito K, Yano T, Nakamoto T, Moriwaki M, Inui H and Yamaguchi M 1997 *Prog. Mater. Sci.* **42** 193
[7] Guder S, Bartsch M, Yamaguchi M and Messerschmidt U 1999 *Mater. Sci. Eng. A* **261** 139
[8] Inui H, Nakamoto T, Ishikawa K and Yamaguchi M 1999 *Mater. Sci. Eng. A* **261** 131
[9] Inui H, Ishikawa K and Yamaguchi M 2000 *Intermetallics* **8** 1131
[10] Dietzsch C, Bartsch M and Messerschmidt U 2005 *Phys. Stat. Sol. a* **202** 2249
[11] Čák M, Šob M, Paidar V and Vitek V 2009 in *Advanced Intermetallic-Based Alloys for Extreme Environment and Energy Applications* (Mater. Res. Soc. Proc. Vol. 1128) paper 1128-U07-10
[12] Hirth J P and Lothe J 1968 *Theory of Dislocations* (New York: McGraw-Hill)
[13] Tanaka K, Onome H, Inui H, Yamaguchi M and Koiwa M 1997 *Mater. Sci. Eng. A* **239** 188
[14] Williams DB and Carter CB 2009 *Transmission Electron Microscopy* (Springer)

On the dissociation of dislocations in semiconductor crystals

Guy Vanderschaeve[1, 2] and Daniel Caillard[1]

[1]CEMES/CNRS BP 94347, 31055 Toulouse Cedex, France
[2]Université de Toulouse, INSA, 31077 Toulouse Cedex, France

caillard@cemes.fr

Abstract. An analysis of moving dissociated dislocations subjected to both a quasi-static viscous friction force and a periodic lattice resistance is performed: Metastable configurations are expected, the range of the dissociation distances being asymmetrically distributed with respect to the equilibrium dissociation distance. The data of earlier investigations on the influence of high external stress on the dissociation of dislocations in silicon and germanium, as determined by transmission electron microscopy, are re-examined. It is shown that in most cases the range of observed dissociation widths is consistent with the predictions of the model

1. Introduction

From a theoretical analysis of the energy of dislocations that are not freely mobile in their dissociation plane, it was recently shown that metastable dissociations are expected [1]. The mean value of the measured partial separation may be higher than the equilibrium distance, as a consequence, the surface energy of the fault connecting the two partials may be underestimated.

In covalent materials such as semiconductors, dislocations are moving through a periodic Peierls-type potential. At low temperature dislocations in semiconductors align themselves along the low energy <110> Peierls valleys. Then perfect dislocations have either 60° or screw character and partial dislocations have either 90° or 30° characters, each with a specific atomic core structure.

In earlier studies [2, 3] the influence of high external stress on the dissociation of dislocations in Si and Ge was investigated by considering that, in addition to the applied forces, moving dislocations are subjected to a quasi-viscous friction force (proportional to their velocity). Then, measuring the dissociation widths under stress can give useful information on the mobilities of dislocations in these materials: it was concluded that partial dislocations have different mobilities, depending not only of their character (90° or 30°), but also on their position (leading or trailing) with respect to the dislocation motion. If the former point is readily explained by considering that 90° and 30° partials have different atomic core structure, the latter result is rather unexpected, because the core structure of a given partial does not depend on its position with respect to the stacking fault ribbon.

In this paper, we investigate the influence of the periodic Peierls-type lattice friction force. It is shown that metastable dissociation configurations are possible, the range of which is asymmetrically distributed with respect to the equilibrium dynamic dissociation distance. In most cases the range of observed dissociation widths is consistent with the predictions of the model. A main conclusion is that the mobility of a partial dislocation (which is defined as the ratio velocity/mean friction force) depends only of its character (30° or 90°) and not of its position (leading or trailing) with respect to dislocation motion.

2. Dynamic dissociation of dislocations

In a strained crystal each partial dislocation P_l and P_t (l and t means leading and trailing, respectively) experiences four forces (per unit length): (i) the attractive force exerted by the stacking fault ribbon (equal to the stacking fault energy γ), (ii) the repulsive elastic interaction kb^2/d, with b the Burgers vector of the perfect dislocation, d the partial separation and k the interaction constant which depends on the elastic constants of the crystal and on the dislocation character, (iii) the glide component F_i (i = l or t) of the applied force, which causes the dislocation movement and (iv) a friction force R_i which opposes to the dislocation movement. According to Wessel and Alexander [2] R_i is proportional to the mean partial velocity v_i, which defines the mobility M_i of the partial ($v_i = M_iR_i$). The dynamical dissociation width d_e is obtained by considering that in the stationary regime of velocity, the net force acting of each partial is 0. According to the present authors [4, 5] these results can be analysed by introducing an "effective" stacking fault energy Γ, so that the elastic energy E (per unit length) of a dislocation dissociated in two partials separated by a distance d writes as:

$$E = \Gamma d - kb^2 \ln d + (\text{self energy terms}) \tag{1}$$

$$\Gamma = \gamma + \frac{\tau b}{2}\left(f - \frac{1-\mu}{1+\mu} \right) \tag{2}$$

with τ the stress component acting in the glide plane on the total dislocation, $f = (F_t - F_l)/\tau b$ a geometrical factor given by the orientation of the stress axis and $\mu = M_t/M_l$ the mobility ratio. Then:

$$d_e = kb^2/\Gamma \tag{3}$$

to be compared to the equilibrium dissociation width in the absence of applied forces d_0:

$$d_0 = kb^2/\gamma \tag{4}$$

Then, according to [2, 3] the equilibrium dissociation width in stressed semiconductor crystals is determined not only by the orientation of the stress axis, but also by the ratio μ of the mobilities of partial dislocations.

The previous analysis does not take into account the periodic character of the Peierls potentiel. Let us assume that each partial experiences, in addition to the previous forces, a periodic lattice friction which derives from a Peierls type potential. The corresponding energy can be simply described by a cosine function:

$$-A_i \cos(2\pi x_i/h) \tag{5}$$

where x_i are the coordinates of the partials P_i in the slip plane (h being the distance between the Peierls valleys) and A_i is temperature dependent (A_i is the depth of the Peierls potential at 0K, and A_i decreases as T increases). That means that the force which opposes to the dislocation movement is the sum of two terms: a mean friction force (R_i) and a periodically varying force.

The total energy $E(x_l, x_t)$ of the dissociated dislocation is then (omitting the partials self energy terms which do not depend on the partial separation):

$$E = \Gamma(x_l - x_t) - kb^2 \ln(x_l - x_t) - \Sigma_i A_i \cos(2\pi x_i/h) \tag{6}$$

Minimising the energy function with respect to x_t and x_l, and taking account of the balance of forces in the stationary state of steady motion, yields to the dissociation width $d = (x_l - x_t)$.

When $A_i = 0$ (i.e. the Peierls friction is negligible), the partial separation is $d_e = kb^2/\Gamma$ (cf Eq. (3); however when $A_i \neq 0$ there are several solutions and metastable dissociation configurations (each with a different energy E_m) are possible. It turns out that E_m is minimum for $d = d_e$; i.e. the most probable dissociation distance (for which the total energy is minimum) is the one given by equation (3).

The range of metastable dissociation widths is controlled by the weakest Peierls valley[1] so that, labelling A_m which of the two quantities A_i is the lowest one, one should have:

$$-\frac{2\pi A_m}{h} \leq \frac{kb^2}{d} - \Gamma \leq \frac{2\pi A_m}{h} \tag{7}$$

[1] This point was dismissed in previous papers by the present authors [4, 5], where Equation (7) was wrongly written as: $-\frac{\pi(A_l + A_t)}{h} \leq \frac{kb^2}{d} - \Gamma \leq \frac{\pi(A_l + A_t)}{h}$

It is noteworthy that the range of possible dissociation widths is asymmetrically distributed with respect to d_e: it is shorter for $d < d_e$ than for $d > d_e$. Therefore the mean value of the measured partial separations may be higher than the equilibrium distance. Inded, from equation (7):

$$d_{min} = d_e \frac{1}{1+\dfrac{2\pi A_m}{\Gamma h}} \leq d \leq d_{max} = d_e \frac{1}{1-\dfrac{2\pi A_m}{\Gamma h}} \tag{8}$$

Note that we have:

$$\frac{2\pi A_m}{\Gamma h} = \frac{d_{max} - d_e}{d_{max}} = \frac{d_e - d_{min}}{d_{min}} = \frac{d_{max} - d_{min}}{d_{max} + d_{min}} \tag{9}$$

Equation (9) makes it possible to check the consistency of experimental data with the model: d_{min} and d_{max} are the limits of the metastable dissociation range (assuming that the whole dissociation range is detected experimentally); d_e is identified as the most frequently measured dissociation distance d_F. Note that, although the range of metastable dissociation distances d is asymmetrically distributed with respect to d_e, their inverse $1/d$ is symmetrically distributed with respect to $1/d_e$.

3. Are the experimental data consistent with the model?

The dislocation microstructure resulting from a two-step deformation process, the second step being a low temperature-high stress deformation regime, was investigated in [2, 3]:

(a): Si, T = 420 °C; τ = 256 MPa;
(b) Si P-doped Si (labelled Si(P)), T = 360 °C; τ = 296 MPa ;
(c) Ge, T = 220 °C ; τ = 294 MPa ;
(d) Ge, T = 220 °C ; τ = 343 MPa

The histograms of experimental data present a rather wide scatter: Qualitatively consistent with the model, the asymmetry predicted by the previous analysis is noticed in more than half of the studied configurations (see Table 1): the range of observed dissociation widths (d) is larger for $d > d_F$ than for $d < d_F$, where d_F is the most frequent dissociation distance. Consequently, the mean dissociation distance $\langle d \rangle$ is higher than d_F. It should be pointed out that for each configuration, the number of dissociation distance measurements is relatively low (10-20), which could be insufficient to get a correct statistics. (For comparison, in the study on dissociation of superdislocations in the intermetallic alloy Ti_3Al, about 100 measurements were performed on a given configuration [6]).

More quantitative agreement could be obtained by checking if the experimental results are consistent with Equation (9). According to the predictions of the model, d_F is put equal to d_e. Moreover we assume that the measured minimum and maximum dissociation distances, d^e_{min} and d^e_{max}, respectively, are those corresponding to the limits of the dissociation range.

Table 1 shows the corresponding ratio for 60° dislocations: When the asymmetry with respect to d_F is observed, the equalities (9) are reasonably well satisfied. (It should be keep in mind that the error bar on the dissociation distances measured by the TEM weak beam technique is at least of 0.5 nm, which leads to rather high relative error bars on d^e_{min} ($d^e_{min} \approx$ 1-1.5 nm); of course the incidence of this experimental uncertainty on d_F and d^e_{max} is less important). Similar agreement is noticed for screw dislocations.

So, most of the experimental data on dislocation dissociation in low-temperature highly stressed Si and Ge crystals are consistent with the occurrence of metastable configurations originating from the influence of the periodic Peierls potential. As a consequence, to determine the mobility ratio μ using equations (2, 3), one should use the most frequent dissociation width d_F and not the mean experimental dissociation width $\langle d \rangle$. This will be done in the next section.

4. The mobility of partial dislocations

Let us first consider the data on screw dislocations (dissociated into two identical 30° partials). It turns out that they can be interpreted in a consistent way without the assumption of different mobilities for leading and trailing 30° partials. Indeed putting $M_l = M_t$ (i.e. μ = 1) in Eq. (2), leads to $d_F = d_e$, within the experimental errors [4, 5].

Table1. 60° dislocations: Comparison between the experimental data and the prediction of the model

crystal (dislocation)	asymmetry/d_F	$\frac{d_F - d^e_{min}}{d^e_{min}}$	$\frac{d^e_{max} - d_F}{d^e_{max}}$	$\frac{d^e_{max} - d^e_{min}}{d^e_{max} + d^e_{min}}$
Si[(a)] (30°/90°)	yes	(0 - 0.80)	(0 - 0.57)	(0.13 - 0.37)
Si[(a)] (90°/30°)	no	/	/	/
Si(P)[(b)] (30°/90°)	no	/	/	/
Si(P)[(b)] (90°/30°)	no	/	/	/
Ge[(c)] (30°/90°)	yes	(0 - 0.50)	(0.40 - 0.67)	(0.25 - 0.50)
Ge[(c)] (90°/30°)	yes	(0.33 - 1.5)	(0.44 - 0.60)	(0.50 - 0.67)
Ge[(d)] (30°/90°)	yes	(0 - 1)	(0.33 - 0.57)	(0.33 - 0.56)
Ge[(d)] (90°/30°)	yes	(0.25 -1)	(0.50 - 0.62)	(0.50- 0.63)

[(a) (b) (c) (d)] : see text.

For 60° dislocations, let us assume again that the mobility of a given partial depends only of its character (30° or 90°) and not of its leading/trailing position. That means that the ratio (Equation (9) which depend only of A_m, should be the same for both kinds of 60° dislocations, (30°/90°) and (90°/30°), respectively. This is exactly what is observed in Ge (Table 1) for both sets of strained samples. This is a strong indication that the partial mobility depends only of its character and not on its leading/trailing position. (The comparison could not be made for Si since the data for (90°/30°) dislocations are not consistent with the model).

The mobility ratio μ can be estimated either by comparing d_0 (the dissociation distance in the absence of applied forces) and d_F (the most frequent dissociation distance in strained crystals). Unfortunately, stacking fault energies in elemental semiconductors are rather scattered amongst the available reports [7].This, combined with the relative error bar on dissociation widths measurements leads to a rather rough estimate. One finds:

- Si: $3 \leq M_{90}/M_{30} \leq 6$
- Ge: $1.5 \leq M_{90}/M_{30} \leq 5$

5. Conclusion

Taking into account the periodic nature of the lattice friction in covalent materials leads to a possible range of metastable dissociated dislocation configurations. In the frame of this model, we have re-examined the literature data on the dissociation of dislocations in highly stressed elemental semiconductors. The main conclusion is that it is not necessary to postulate that the mobility of a Shockley partial depends on its leading/trailing position, to account for the experimental results.

References

[1] Paidar V and Caillard D 1994 *Phil. Mag. Lett.* **69** 305
[2] Wessel K and Alexander H 1977 *Phil. Mag.* **35** 1523
[3] Gottschalk H 1979 *J. Physique* **40** C6-127
[4] Vanderschaeve G and Caillard D 2005 *Phys. Stat. Sol. (a)* **202** 939
[5] Vanderschaeve G and Caillard D 2007 *Mater. Sci. Eng.* **462** 418
[6] Legros M and Caillard D 2001 *J. Microscopy* **203** 90
[7] George A and Rabier J 1987 *Rev. Phys. Appl.* **22** 941

Journal of Physics: Conference Series **240** (2010) 012009 doi:10.1088/1742-6596/240/1/012009

The intermittency of plasticity in an Al3%Mg alloy

Y Bougherira[1], D Entemeyer[1], C Fressengeas[1], N P Kobelev[2], T A Lebedkina[2] and M A Lebyodkin[1]

[1] Laboratoire de Physique et Mécanique des Matériaux, Université Paul Verlaine-Metz, CNRS, Ile du Saulcy, 57045 Metz, France
[2] Institute of Solid State Physics, Russian Ac. Sci., 142432 Chernogolovka, Russia

Youcef.bougherira@univ-metz.fr

Abstract: Statistics of acoustic emission accompanying plastic deformation and of stress serrations caused by the Portevin-Le Chatelier effect are studied during tension of an Al3%Mg alloy at room temperature. Power-law distributions of acoustic emission reflecting self-organization of dislocations and intermittency of plastic flow are found, irrespective of the strain rate, both before and after the critical strain for the onset of the serrated flow. In contrast, several regimes including both power-law and peaked distributions are observed at the macroscopic scale of stress serrations, depending on the applied strain rate.

1. Introduction

Recent studies of jerky flow in dilute alloys, also referred to as the Portevin–Le Chatelier (PLC) effect, proved that it is often characterized by scale invariance through power law distributions of jerks reflecting self-organization of the dislocation ensembles (see review [1] and references therein). Statistical and dynamical analyses of serrated stress-time series revealed such complex nonlinear phenomena as self-organized criticality (SOC) [2] and chaotic dynamics [3]. Independently, power-law statistics were found for bursts of acoustic emission (AE) and local strain rate recorded during deformation of pure crystalline solids [4, 5], which bears evidence to an intermittent, avalanche-like character of the macroscopically homogeneous plastic "flow", albeit at sizes of the local strain-rate jumps much smaller than during jerky flow. These observations suggest that self-organization phenomena are of a general nature in dislocation ensembles, and may become apparent at various plastic event scales [6].

Due to huge collective effects resulting in the macroscopic stress serrations, the PLC effect presents a distinct advantage for the investigation of self-organization of dislocation ensembles. Indeed, simultaneous application of the traditional stress-time recording techniques and of more sensitive methods, such as AE or local extensometry, allows for studying deformation processes in a large range of event scales. Moreover, the existence of a critical strain for the occurrence of jerky flow offers an opportunity to study collective effects both during the seemingly smooth flow below the critical strain and the unstable regime beyond it. Hence a comparative study of intermittent plasticity in dilute alloys and in pure solids can be documented.

To our knowledge, the statistical study of the PLC effect over such diverse event scales has not been performed so far. In the present work we report on the statistical analysis of AE accompanying plastic deformation of an AlMg alloy – a classical material exhibiting the PLC effect – and compare the AE statistics with that of stress serrations.

2. Experimental technique

Tensile specimens with a gauge part 25 x 6.8 x 2.5 mm^3 in size were cut from a polycrystalline cold-rolled sheet of Al3%Mg alloy, annealed at 400°C during 2 h, and quenched into water. Uniaxial tensile tests were conducted at room temperature with constant crosshead velocity corresponding to the initial imposed strain rate $\dot{\varepsilon}_a$ from 2×10^{-5} s^{-1} to 6×10^{-3} s^{-1}. The data acquisition rate was chosen in a range from 2 Hz to 500 Hz, depending on the strain rate.

AE was recorded during tensile tests by a computer-controlled LOCAN 320 system. The piezoelectric transducer was attached to the surface just above the deforming part of the sample with the aid of grease. The total gain and the threshold voltage for the identification of the starting point of an AE event were respectively 80 dB and 27db.

3. Macroscopic behavior

Figure 1 shows examples of stress-strain curves observed at three different strain rates. Together with local extensometry measurements presented elsewhere, these data show that the studied alloy exhibits the well-known types of the PLC behavior. Type *A* regime is observed at high strain rates. It is characterized by irregular stress fluctuations associated with deformation bands propagating quasi-continuously along the tensile axis. At lower strain rates, these irregular deformation curves are replaced by more regular oscillations of type *B* (see also Fig. 3) which result from static deformation bands sequentially nucleated ahead of each other. Such correlated nucleation is conventionally referred to as "hopping propagation". Further decrease in strain rate leads to sharp type *C* serrations, which are usually attributed to randomly nucleated bands, although the analysis of the serrations bears evidence to some correlation [7]. It can also be seen from Fig. 1 that the macroscopic stress jumps occur after some critical strain ε_{cr} exhibiting strain rate dependence.

The statistical behavior of the stress serrations in the investigated alloy is also consistent with literature data [1]. Scale-free power-law distributions are found for the amplitude of type *A* serrations, in agreement with the SOC hypothesis for type *A* behavior. In the case of type *B* curves, the histograms are characterized by large peaks with irregular shape reflecting the occurrence of characteristic scales. Two distinct scales of stress drops, leading to essentially bimodal distributions (cf. [8]), are distinguished for type *C*: the sequence of deep drops is accompanied by serrations with essentially lower amplitudes, usually preceding them. The latter are generally considered as stochastic fluctuations and disregarded in the theories of the PLC effect

An unexpected feature is observed for these small-scale drops. By calculating the distributions separately for the two kinds of events (Fig. 2), we found power-law statistics at small scale, which proves the nonrandom nature of the small serrations. The power-law exponent α varies roughly between 1 and 1.5, similar to α-values determined for type *A* serrations. It should be noted that small drops with the same amplitude also occur before the critical strain for type *C* instability, indicating that plasticity is strongly heterogeneous at this scale all over the deformation curve. This conclusion is also confirmed by the observation of stress undulations before the critical strain for type *B* instability.

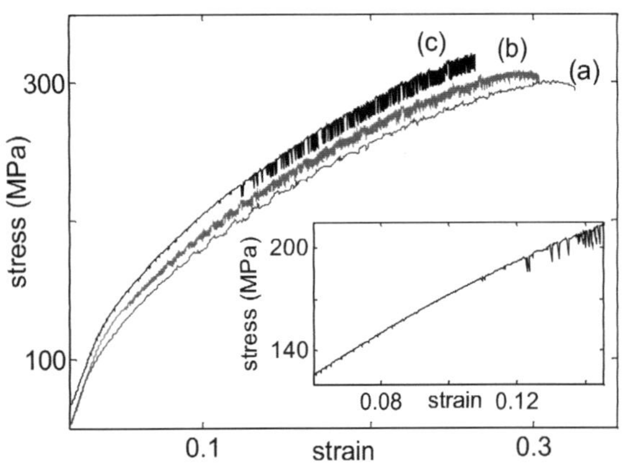

Figure 1. True stress-true strain curves recorded at room temperature and various strain rates: (a) 6×10^{-3} s^{-1}, type *A* behavior; (b) 2×10^{-4} s^{-1}, type *B*; (c) 2×10^{-5} s^{-1}, type *C*.

Inset: Portion of a stress-strain curve at 2×10^{-5} s^{-1}, which displays small-amplitude stress drops before the onset of the macroscopic instability.

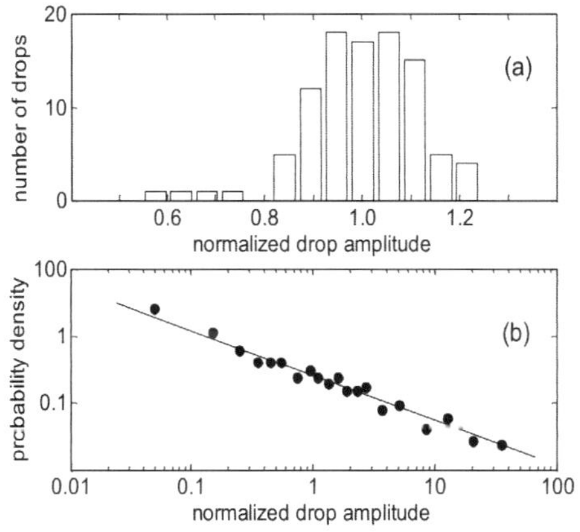

Figure 2. Distribution of the amplitudes of (a) large stress drops and (b) low-amplitude serrations for $\dot{\varepsilon}_a = 2\times10^{-5}\ s^{-1}$.

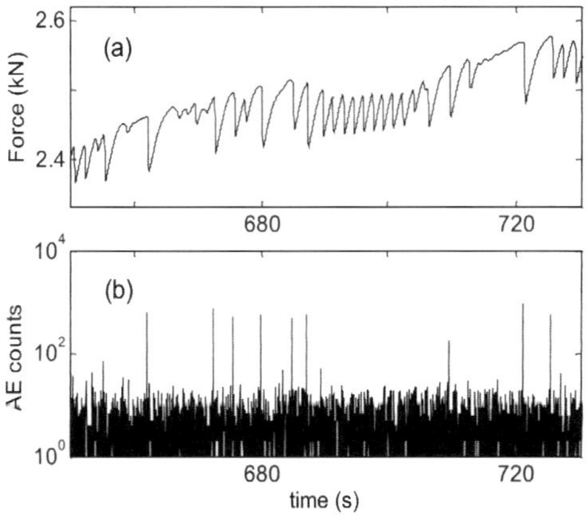

Figure 3. Synchronous records of (a) the deformation curve and (b) AE counts; $\dot{\varepsilon}_a = 2\times10^{-4}\ s^{-1}$.

4. Acoustic emission

Figure 3 represents synchronous records of a type *B* deformation curve and the concomitant AE count. Such behavior is typical of the PLC effect and suggests distinguishing "continuous" vs. burst-like AE components [9]. The continuous AE also consists of discrete bursts but owes its name to its random-like appearance, so that it is customarily attributed to uncorrelated motion of dislocation pile-ups. This kind of AE is observed from the beginning of plastic deformation. Bursts in the AE count rate and event duration are noticed for some type *B* stress drops only (Fig. 3). They accompany almost every deep stress drop of type *C*, but are hardly discernible during type *A* serrations. Note that the amplitude of the AE events practically does not display any increase at the instant of the stress drops. Thus, the bursts in AE count rate and duration are likely due to overlapping of numerous AE events.

As justified in Ref. [10], a physically based measure of the plastic activity in the sample is provided by the AE energy which can be evaluated as $E \sim A^2$, where A is the amplitude of the acoustic events. Statistical analysis of this quantity reveals unexpected behavior. Namely, the calculated distributions are found to obey power laws in all deformation conditions. This result refutes the hypothesis of randomness and seems to be a generic property of the intermittency of plastic flow. The power-law exponent α_{AE} depends on strain rate and generally changes in the course of deformation. However, as shown in Fig. 4, it is possible to determine time intervals during which it remains statistically invariant within some least square error. In Fig. 4, such intervals are designated by rectangles, the height of which displays the respective error.

During unstable flow, α_{AE} usually takes on rather high values, ranging roughly from 2 to more than 3. Close values of α_{AE} are observed for various specimens in type *A* regime (Fig. 4a), whereas it noticeably changes with deformation, and from sample to sample, at lower strain rates, particularly, in type *B* regime (Figs. 4b). This observation agrees with conclusions on the variable sensitivity of the dislocation dynamics to the microstructure in various deformation conditions, drawn in Ref. [11] from the analysis of stress serrations. It is worthy of notice that α_{AE} typically takes on lower values below $\varepsilon_{cr.}$ In particular, it appears to be less than 2 for type *A* instability, thus approaching the value of 1.5-1.6 reported for smoothly deforming pure materials [4, 5, 10].

5. Conclusions

The statistical analysis of both stress serrations and AE signals during deformation of a dynamically strain ageing alloy reveals a complex multiscale behavior of dislocation ensembles. The following major conclusions can be made.

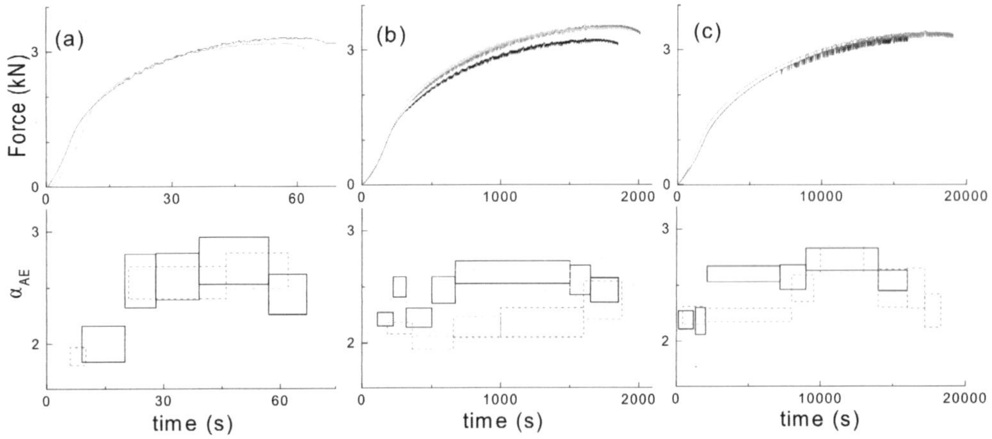

Figure 4. Force vs. time curves and variation of the power-law index α_{AE} for the AE energy distribution. (a) $\dot{\varepsilon}_a = 6\times10^{-3}$ s^{-1}; (b) $\dot{\varepsilon}_a = 2\times10^{-4}$ s^{-1}; (c) $\dot{\varepsilon}_a = 2\times10^{-5}$ s^{-1}.

(1) A persistent scale invariance feature, as reflected by power laws, is found for the statistical distributions of AE energy. The power-law statistics is also observed for low-amplitude stress serrations, corresponding to an intermediate scale between AE and macroscopic stress drops. These observations bear evidence to an inherently intermittent nature of plastic flow associated with an avalanche-like dislocation dynamics. At the same time, the power law is not universal in the present experiments, in contrast to a situation often met in pure crystals [5, 6], but its slope depends on the deformation conditions. Understanding these variations requires further investigation of the role of the dislocation aging, grain boundaries, as well as of the effect of the finite time resolution leading to overlapping of the AE events.

(2) The power-law statistics of the AE confirms the hypothesis of SOC-like behavior in the conditions of type *A* instability, which was previously based on the observation of the power-law statistics of stress serrations [1], whereas it differs from the peaked distributions characterizing type *B* and type *C* stress drops. Thus, distinct statistics coexist in this case: whereas the power-law distributions and scale-free behavior uncovered by AE monitoring correspond to the motion of dislocation groups, the peaked histograms reveal the existence of characteristic scales at the macroscopic scale of the deformation curve. The latter behavior may occur through synchronization of dislocation groups, a process reminiscent of the collective synchronization of coupled oscillators [12].

References

[1] Kubin L P, Fressengeas C, Ananthakrishna G, 2002 *Dislocations in solids*, vol **11**, Nabarro F.R.N, Duesbery M S, Editors 101
[2] Bak P, Tang C, Wiesenfeld K 1988 *Phys. Rev.* A **38** 364
[3] Abarbanel H D I, Brown R, Sidorowich J J and Tsimring L Sh 1993 *Rev. Mod. Phys.* **65** 1331
[4] Miguel M C, Vespignani A, Zapperi S, Weiss J and Grasso J R 2001 *Nature*, **410** 667
[5] Weiss J, Richeton T, Louchet F, Chmelík F, Dobron P, Entemeyer D, Lebyodkin M, Lebedkina T , Fressengeas C and McDonald R J 2007 *Phys.Rev.* B **76** 224110
[6] Zaiser M 2006 *Advances in Physics* **55**, 185
[7] Lebyodkin M A and Lebedkina T A 2008 *Phys. Rev.* E **77** 26111
[8] Lebedkina T A and Lebyodkin M A 2008 *Acta Mater.* **56** 5567
[9] Chmelík F, Ziegenbein A, Neuhäuser H and Lukáč P 2002 *Mater. Sci. Eng.* A **324** 200
[10] Weiss J, Grasso J R, Miguel M C, Vespignani A and Zapperi S 2001 *Mater. Sci. Eng.* A **309** 360
[11] Bharathi M S, Lebyodkin M A, Ananthakrishna G, Fressengeas C, Kubin L P 2002 *Acta Mater.* **50** 2813
[12] Strogatz S H 2000 *Physica* D **143,** 1

Development of a method to determine Burgers vectors from atomistic data

Jun Hua, Alexander Hartmaier

Micromechanical Modelling of Macroscopic Material Behaviour, Interdisciplinary Centre for Advanced Materials Simulation (ICAMS), Ruhr-University Bochum, 44801 Bochum, Germany

Alexander.Hartmaier@rub.de

Abstract. Large-scale molecular dynamics simulations have been widely used to investigate the mechanical behaviour of materials. But complex datasets, involving the positions of millions of atoms, generated during the simulations make quantitative data analysis quite a challenge. This paper presents a novel method to determine not only dislocations in the crystal, but also to quantify their Burgers vectors. This is achieved by combining geometrical methods to determine the atoms lying in the dislocations cores, like for example the common neighbour analysis or the bond angle analysis, with the slip vector analysis. The first methods are used to filter out the atoms lying in undisturbed regions of the crystal; the latter method yields the relative slip of the remaining atoms and thus indicates the Burgers vector of those atoms lying in the dislocation cores. The validity of the method is demonstrated here on a single edge dislocation in a relatively small sample. Furthermore a way will be sketched how this analysis can be used to determine densities of statistically stored and geometrically necessary dislocations, respectively. Hence, this method can be expected to provide valuable input for strain gradient plasticity models.

1. Introduction

Atomic scale computer simulation such as Molecular Dynamics (MD) is a powerful tool for revealing the processes underlying the plastic response of a crystal and for providing valuable insights into its material behavior. The important goals of atomistic studies of confined geometries, like nanoindentation, are to identify the atomistic mechanisms in the early states of plastic deformation under the indenter, to characterize the defect structures and even to directly compare force-indentation behaviour with experiments.

The analysis of atomic bonds is widely used in large-scale molecular dynamics simulations to characterize the state of a local crystal structure. Most local structure analysis methods such as the centrosymmetry parameter (CSP) [1] and the common neighbor analysis (CNA) [2] use such abstractions to characterize the local state of solid materials. Ackland and Jones [3] take advantage of the bond angular distribution (BAD) function to distinguish the crystal types of body centered cubic (bcc), face centered cubic (fcc) and hexagonal close packed (hcp). This BAD analysis is faster and gives fewer false positive results (atom lying in a defect is identified as bulk atom) and negative results (bulk atom identified as defect atom) than previous methods. The slip vector analysis (SVA), developed by Zimmerman et al [4], compares the current atomic configuration to a reference state and thus quantifies the local atomic slip in the ensemble. The slip vector is a measure of local plastic deformation and is related to the Burgers vectors of a lattice dislocation that caused this plastic slip. This method has been shown to be capable of detecting different contributions to plastic slip from atomistic simulations [5].

In our approach, we combine SVA and BAD to yield the magnitudes and directions of plastic slip of atoms situated in dislocation cores. The objective of our research work is to provide an atomic level insight into plastic deformation mechanisms during nanoindentation and utilize all the information of the atomic data to make the calculation of signed dislocation densities and dislocation density tensors possible.

2. Burgers Vector Analysis

The slip vector developed by Zimmerman [4] is defined as

$$\mathbf{s}^{(k)} = -\frac{1}{n_s}\sum_{k\neq l}^{n}\left(\mathbf{x}^{(kl)} - \mathbf{X}^{(kl)}\right) \tag{1}$$

where atom l is a nearest neighbor of atom k, n_s is the number of the slipped neighbors, $\mathbf{x}^{(kl)}$ and $\mathbf{X}^{(kl)}$ are the position vector difference between atom k and l in the current and reference configurations, respectively. The reference configuration denotes the arrangement of atomic positions at the beginning of the deformation step.

The slip vector is a measure of the local plastic deformation since it denotes the average displacement of an atom in its current position compared to the initial position and relative to its nearest neighbors. However, the slip vectors of atoms lying on opposite sides of the slip plane point in opposite directions. Hence, we have to develop a way to uniquely define the direction of the slip vector, before we can associate it with plastic slip in the crystal. To accomplish this, we identify the atoms situated in dislocation cores with the help of the bond angle analysis method [3] and only analyze the slip vector within the dislocation core. Furthermore, by identifying those atoms with a higher coordination number than bulk atoms, we can identify the atoms in the core of an edge dislocation that lie on the side of the slip plane were it intersects the inserted half-plane. In fcc crystals, we typically find atoms with coordination numbers of 13 on this side of the slip plane. On the opposite side of the slip plane, in contrast, we find atoms with a smaller coordination number than bulk atoms. By using this property of edge dislocations that allows us to define an "upper" and "lower" side of the slip plane, we are able to define a unique direction of the plastic slip that is characterized by the slip vector of atoms in the dislocation core.

To demonstrate the method, a copper crystal (97,296 atoms) containing a single edge dislocation as shown in figure 1 is studied here. The size of the crystal prior to deformation is 40x6x17 unit cells which correspond to a volume of 202.8x53.0x106.4 Å^3. For simplicity, and since we only want to demonstrate the analysis method, we use rather simple geometry, where the top surface is a crystallographic (111) plane with free boundary conditions, the bottom surface is fixed, and periodic boundary conditions are applied to the sides of the simulation box. The simulations using an embedded atom method (EAM) potential for Cu developed by Mishin et al. [6] are conducted in the NVE (constant number of atoms, constant volume, constant energy) ensemble. The MD simulations were performed using a version of the IMD parallel molecular dynamics code [7] in which we implemented the BAD and SVA methods. For visualization, the atomistic configuration viewer Atomeye [8] and the Tecplot software were used.

Before mechanically loading the simulation box, the dislocation has been properly relaxed, until all atoms found the energetic minimum position, which corresponds to a simulation a T=0K. As it is well-known, a perfect edge dislocation in the fcc structure is dissociated into a leading and a trailing Shockley partial gliding on the (111) slip plane and linked by a stacking fault. Here, the dissociation reaction for copper is given by:

$$\frac{a}{2}[\bar{1}\,10]=\frac{a}{6}[\bar{1}\,2\,\bar{1}]+\frac{a}{6}[\bar{2}\,11] \tag{2}$$

The lattice constant of copper is a=3.615 Å. The Burgers vector magnitude of copper is b_{partial} = 1.48 Å for a {111}<112> partial dislocation and b_0 = 2.56 Å for a perfect {111}<110> dislocation.

Finally the relaxed ensemble is loaded with a shear stress of 300 MPa in a dynamic simulation with a time step is of 0.001 ps for 90 iteration steps. During this simulation, the edge dislocation glides on the (111) plane under the action of the applied shear stress. Figure 1(a) shows the atomistic configuration of the lower half of the simulation box. In figure 1(b) the result of the slip vector ananlysis for the atoms lying below and above the slip plane are shown, while in figure 1(c) only the slip vector of the atoms lying in the dislocation core is represented, by combining the BAD and the SVA method. Furthermore, in this figure the sign of the slip vectors lying above the slip plane have been flipped, such that all vectors point in the same direction. The vectors shown in figure 1(c) thus represent the Burgers vector of the dislocation on the atomic scale.

In figure 1(b), there are four groups of atoms represented by the slip vectors: (i) atoms lying in the wake of the dislocation with a full Burgers vector as slip vector (dark blue), (ii) atoms in the stacking fault region with a partial Burgers vector as slip vector (light blue), (iii) atoms in the leading partial dislocation core, and (iv) atoms in the trailing partial dislocation core (green, red, and orange). Since in each group the slip vector is almost the same, here we only choose one atom of a group to analyze. A closer analysis shows that the atoms in the stacking fault region have slip vectors of (-1.19, 0.61, 0.58) in the crystallographic orientations of $[\bar{1}\,10]$, $[\bar{1}\,\bar{1}\,2]$ and $[1\,1\,1]$. The magnitude of this slip vector is 1.46 Å which is very close to the magnitude of the partial Burgers vector b_{partial}. The direction of this slip vector is close to (-1.0, 0.5, 0.5) which is actually the leading Shockley partials of $[\bar{2}\,11]$. So in fact, this part can be identified as Burgers vector of the leading Shockley partial dislocation. The slip vectors of atoms in the dislocation core are not in such good agreement with the expected full Burgers vectors. This is seen by the analysis of an atom in the trailing partial core, which has the slip vector components (-1.41, 1.19, 0.32). The direction of this slip vector is reasonably close to the (-1.0, 1.0, 0) vector, i.e., the complete Burgers vector. The magnitude of this slip vector is 1.87 Å and thus smaller than b_0, which shows the lattice slip is still incomplete in the dislocation core.

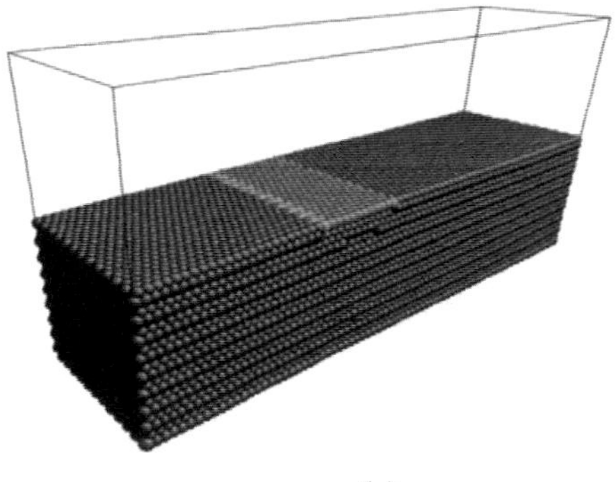

(a)

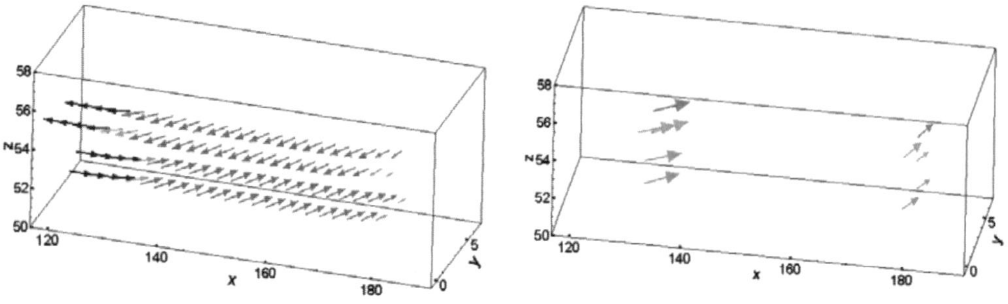

(b) (c)

Figure 1. The atomistic configuration of an edge dislocation and the corresponding Burgers vector analysis. In figure 1(a) a cut through the atomistic configuration is shown, where the atoms are color coded according to the BAD analysis (dark red: surface, dark blue: fcc, light blue: stacking fault (hcp), green: 12 neighbors, neither fcc, nor hcp, orange: 11 neighbors); Figure 1(b) shows the results of the slip vector analysis for the atoms lying in the slip plane of the dislocation; in figure 1(c) only the slip vectors of the atoms lying in the dislocation core are shown, and the sign of the slip vectors of the atoms on top of the slip plane have been flipped. All slip vectors are normalized with the norm of the Burgers vector b_0.

3. Summary and Outlook

Through the above analysis, we have successfully identified Burgers vectors purely from atomistic data of a sample containing an edge dislocation in fcc copper. This is accomplished by combining the slip vector analysis and the bond angle distribution method. The method has already been tested for more complex situation, as for example the dislocation structure underneath a nanoindenter in fcc copper and also in bcc tungsten. These results will be published in forthcoming work. This method, hence, opens the way to calculate signed dislocation densities and dislocation density tensors directly from atomistic data. The knowledge of the dislocation density tensor, in turn, allows plastic strains and work hardening to be calculated for each slip system independently. Thus, the condensed information gained on the atomic scale can be incorporated into continuum mechanics methods through Orowan law or Taylor work hardening rules in a multiscale modeling approach.

References

[1] Kelchner C L, Plimpton S J and Hamilton J C 1998 *Phys. Rev. B* **58**, 11085
[2] Honeycutt J D and Andersen H C 1987 *J. Phys. Chem.* **91** 4950
[3] Ackland G J and Jones A P 2006 *Phys. Rev. B* **73** 054104
[4] Zimmerman J A, Kelchner C L, Klein P A, Hamilton J C and Foiles S M 2001 *Phys. Rev. Lett.* **87** 165507
[5] Vo N Q, Averback R S, Bellon P, Odunuga S and Caro A 2008 *Phys. Rev. B* **77** 134108
[6] Mishin Y, Mehl M J, Papaconstantopoulos D A, Voter A F and Kress J D 2001 *Phys. Rev. B* **63**, 224106
[7] http://www.itap.physik.uni-stuttgart.de/~imd/userguide/imd.html, University of Stuttgart, Institute of Theoretical and Applied Physics (ITAP).
[8] Li J 2003 *Modelling Simul. Mater. Sci. Eng.* **11**, 173

TEM in situ straining experiments in Fe at low temperature

D. Caillard

CEMES-CNRS, 29 rue J. Marvig, BP4347, 31055 Toulouse cedex, France

caillard@cemes.fr

Abstract. In situ TEM straining experiments were carried out in pure Fe, to investigate the origin of the discontinuity observed at 250K in the temperature variation of the deformation activation parameters. The results show that the motion of screw dislocations is steady at 300K, in agreement with a kink-pair mechanism, but jerky at 110K. This change has been attributed to a transition from a kink-pair mechanism to a locking-unlocking mechanism, similar to that observed previously in Ti.

1. Introduction

The low-temperature mechanical properties of body centred cubic (BCC) metals exhibit all the characteristic features of the kink-pair mechanism, controlling the motion of screw dislocations with extended cores (see reviews by Kubin [1], Christian [2], Suzuki et al [3], Duesberry [4], Taylor [5]). In particular, there is a rapid decrease of the flow stress at increasing temperature, and activation volumes are fairly small.

In iron single crystals, the investigations of Quenel et al [6], Kuramoto et al [7], and Brunner and Diehl [8], show that the stress versus temperature curve exhibits a discontinuity at around 250K, to which corresponds a peak in the variation of the activation volume. These discontinuities indicate a change of controlling mechanism at 250K. Such a behavior is not specific to pure iron, as it is also observed in Nb and Nb alloys, Mo, as well as in hexagonal close packed metals deformed in prism slip (see a review in [9]).

In BCC metals, discontinuities have been ascribed to several possible effects: i) the occurrence of anomalous slip (in Nb and Mo, not in Fe where anomalous slip has never been reported), ii) a change in the kink-height, provided the Peierls potential has a "camel hump" shape (Guyot and Dorn [10], Takeuchi [11], Aono et al [12], Koizumi et al, [13]), iii) a transition between the low-stress and high-stress regimes of the kink-pair mechanism (Statham et al [14], Kubin and Jouffrey [15], Brunner et al [16], Groger and Vitek [17]), iv) a change of the kink-pair plane from $\{110\}$ to $\{112\}$ (Brunner and Diehl [18, 19]). These interpretations are however not fully convincing, because they have never been verified experimentally, and because (i) is restricted to Nb and Mo, (iv) cannot be transposed to hexagonal close-packed metals, the camel-hump potential assumed in (ii) is not reproduced by recent ab initio calculations [20], and the sole modification of kink-pair mechanism, proposed in (ii)-(iv), seems incompatible with the huge variation of pre-exponential term.

The aim of the present study is thus to perform in situ TEM experiments in pure Fe at different temperatures, in order to detect potential changes in the dynamical behaviour of dislocations that could be related to the discontinuity at 250K.

2. Experimental procedure

A large-grain polycrystal of pure iron, containing 2.5 to 4 ppmwt (ppm in weight) of C, N, O, and Si (15 ppmwt as a whole), and less than 2 ppmwt of other elements, has been provided by J. Le Coze, Ecole des Mines of St. Etienne, France. Microsamples were prepared as described in [21], and strained in a JEOL 2010HC transmission electron microscope. Two Gatan straining holders were used, one working at room temperature, and the other (lent by R. Schaeublin, PSI, Switzerland) between 110K and room temperature. The microsamples were deformed by series of constant strain-rate deformations followed by relaxations, at an average strain-rate of the order of 10^{-6} to 10^{-5} s^{-1}. The videos were recorded by a Megaview III camera, at a frame rate of 25 images per second, and analysed frame by frame.

3. Results

Fig. 1 shows a dislocation source anchored at one extremity, in S, in a microsample strained at room temperature. The screw segment labelled 1 moves to the top-left between t = 0 and t = 3.60s (note the anchoring in P). When both opposite screw segments 1 and 2 are sufficiently widely separated, the curved edge portion moves rapidly to the top, emerges at the foil surface, and leaves the slip trace tr (t = 3.60s to t = 3.64s). The velocity of this motion is too high to be measured with the time-resolution of the camera (0.04s between two successive frames). Dislocation 1 moves on to the top-left, whereas dislocation 2 moves in the opposite direction (t = 3.80s to t = 6.68s). Then, the second curved edge portion escapes rapidly at the opposite free surface, and the rotation resumes.

Several tens of successive rotations of the dislocation source have been observed at 300K, with the same characteristics, summarized as follows:

i) the velocity of screw dislocations is considerably lower than that of edge and other non-screw segments, which results in a large density of moving straight screw dislocations.

ii) slip traces are generally wavy, and do not correspond to any low-index slip plane. In some cases, however, cross-slip is less abundant, and slip occurs preferentially in {110} planes.

iii) screws move steadily, at a velocity which increases with their length.

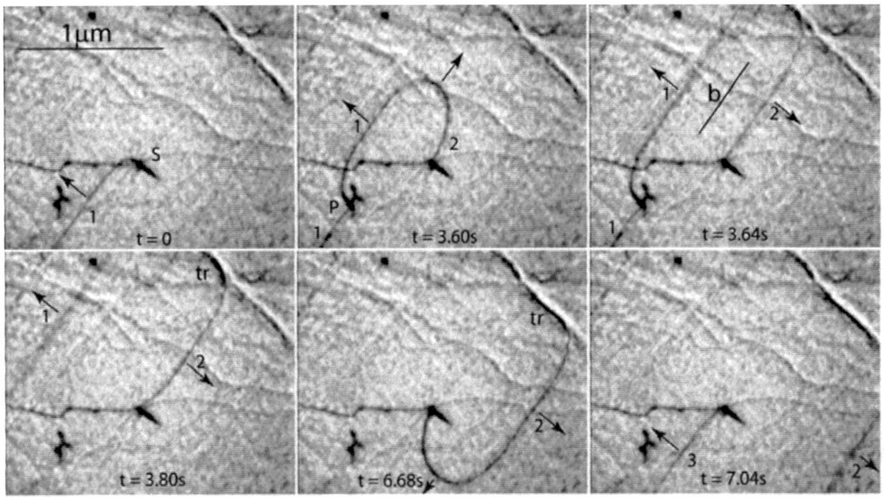

Fig. 1: Dislocation source in pure Fe, at 300K. The anchoring point is S, the slip trace is tr, and the Burgers vector direction is b.

At low temperature, screw dislocations are much straighter than at room temperature, and the edge parts are more highly curved. Their kinetics is also very different from the above one. Fig. 2 shows a group of dislocations under stress at 110K. One dislocation (noted d) moves very quickly over ~ 20nm between the two first frames of the sequence. Then, nothing occurs during almost one second, till the same dislocation jumps again. Other dislocations have the same jerky motion in the following of the sequence (not shown).

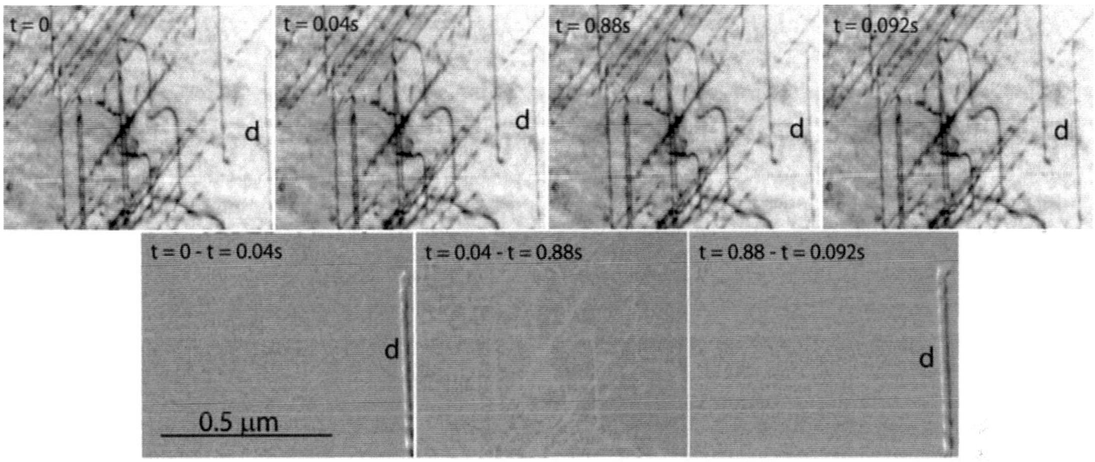

Fig. 2: Dislocation motion in pure Fe at 110K. The subtractions of the different successive images underline the jerky motion of dislocation d.

4. Discussion

Except for the activation of anomalous slip, which has not been observed in Fe, the explanations proposed for the discontinuity in the mechanical properties at 250K are all based on a change in the properties of the kink-pair mechanism. This mechanism is characterized by jumps of screw dislocations between adjacent Peierls valleys, which results in the steady motion of rectilinear dislocations at the scale of in situ TEM observations.

The observations at 300K are fully consistent with the kink-pair mechanism. The jerky motion observed at 110K is, however, difficult to describe in the same way. Indeed, a simple evaluation shows that it cannot be due to any variation of the local stress, e.g. due to elastic interactions with neighbouring dislocations. It is not either due to interactions with extrinsic obstacles like impurity atoms, because the jump length decreases with decreasing temperature. Under such conditions, the change of kinetics occurring between 110K and 300K reveals a change of controlling mechanism, which probably corresponds to the discontinuities discussed in the introduction.

A similar change in the kinetics of screw dislocations occurs in titanium deformed in prismatic slip around 300K, in correlation with a similar peak in the activation area versus stress or temperature. It has been ascribed to a change in the mechanism of dislocation motion, from the kink-pair one above 300K, to the locking-unlocking one below 300K [22, 23]. The locking-unlocking mechanism is a variant of the kink-pair mechanism, where dislocations extracted from Peierls valleys can glide freely over a large number of inter-atomic distances, before being locked again. Since the experimentally measured free-glide distance decreases with increasing temperature, there is a transition from the locking-unlocking mechanism to the kink-pair one, when this distance reaches the kink-pair height. As shown in [9], this transition naturally induces a peak in the stress or temperature dependence of the activation area, and a discontinuity in the stress versus temperature curve.

In Fe, we propose that the discontinuities measured at 250K have the same origin as in Ti, i.e. they are related to a transition between the locking-unlocking mechanism, below 250K, and the kink-pair one above.

5. Conclusions

In situ straining experiments in pure Fe have revealed different kinetics of screw dislocations at 300K and 110K:

- At 300K, screw dislocations move steadily, at a velocity which depends on their length, as expected from the kink-pair mechanism
- At 110K, screw dislocations have a jerky motion, with jumps longer than several tens of nanometers, separated by locked positions that last up to several seconds. This jerky motion is inconsistent with the kink-pair mechanism.

This change of behaviour has been compared to that studied previously in titanium, and ascribed to a transition from a kink-pair mechanism to a locking-unlocking one. It is likely to be responsible for the discontinuities in the activation parameters of the deformation, observed at 250K.

Acknowledgments: The author warmly thanks R. Schaeublin (PSI Willigen) for the loan of the low-temperature straining holder, Jean Le Coze (Ecole des Mines St Etienne) for the elaboration of the iron ingot, and prof. Jacques Friedel (Orsay University) for his very stimulating comments. This work has been supported by the European Fusion Development Agreement.

References

[1] Kubin L 1982 *Reviews on the Deformation Behaviour of Metals* vol. IV n°3, ed P Feltham (Tel Aviv: Scientific Pub. Div. Freund Pub. HouseLtd) p 181.
[2] Christian JW 1983 *Met. Trans.* **14A** 1237.
[3] Suzuki T, Takeuchi S, and Yoshinaga H 1985 *Dislocation Dynamics and Plasticity* vol. 12, (Berlin: Springer Series in Materials Science, Springer-Verlag).
[4] Duesberry MS 1989 *Dislocations in Solids* vol. 8, ed. FRN Nabarro (Amsterdam: North Holland) p 67.
[5] Taylor G 1992 *Prog. Mater. Sci.* **36** 29.
[6] Quenel DJ, Sato A, and Meshii M 1975 *Mater. Sci. Eng.* **18** 199.
[7] Kuramoto E, Aono Y, and Kitajima K 1979 *Scripta Metall.* **13** 1039.
[8] Brunner D, and Diehl J 1997 *Phys. Stat. Sol. (a)* **160** 355.
[9] Caillard D, and Martin JL 2003 *Thermally Activated Mechanisms in Crystal Plasticity* Pergamon Materials Series, vol. 3, (Amsterdam: Pergamon).
[10] Guyot P, and Dorn JE 1967 *Can. J. Phys.* **45** 983.
[11] Takeuchi S 1979 *Phil. Mag. A* **39** 661.
[12] Aono Y, Kitajima K, and Kuramoto E 1981 *Scripta Metall.* **15** 275.
[13] Koizumi H, Kirschner HOK, and Suzuki T 1993 *Acta Metall. Mater.* **41** 3483.
[14] Statham CD, Koss DA, and Christian JW 1972 *Phil. Mag.* **26** 1089.
[15] Kubin LP, and Jouffrey B 1973 *Phil. Mag.* **27** 1369.
[16] Brunner D, Diehl J, and Seeger A 1984 *Structure and Properties of Crystal Defects*, ed. V Paidar and L Lejcek, (Amsterdam: Elsevier Publ. Co) p. 175.
[17] Gröger R, and Vitek V 2008 *Acta Mater.* **56** 5426.
[18] Brunner D, and Diehl J 1991 *Phys. Stat. Sol. (a)* **124** 155 and 455.
[19] Brunner D, and Diehl J 1991 *Phys. Stat. Sol. (a)* **125** 203.
[20] Ventelon L, and Willaime F 2006 *J. Computer-Aided Materials Design* **94** 85.
[21] Pettinari F, Couret A, Caillard D, Molénat G, Clément N, and Coujou A 2001 *J. Microscopy* **203** 47.
[22] Farenc S, Caillard D, and Couret A 1993 *Acta Metall. Mater.* **41** 2701.
[23] Farenc S, Caillard D, and Couret A 1995 *Acta Metell. Mater.* **43** 3669.

Dynamic strain aging in materials with a high Peierls relief: competition of diffusion and dragging impurities

B V Petukhov
Shubnikov Institute of Crystallography, Russian Academy of Sciences, Leninskii pr. 59, Moscow, 119333 Russia

petukhov@ns.crys.ras.ru

Abstract. A model of the dynamic interaction of dislocations with the impurity subsystem of crystals that have high Peierls barriers has been developed. It is justified that the impurity kinetics during atmosphere formation includes two stages. The initial stage leads to a negative strain rate and positive temperature dependence of the yield stress of the material in some range.

1. Introduction

The effect of impurities on the dislocation dynamics significantly modifies the mechanical properties of crystalline materials. When the dislocation and impurity mobilities are comparable, plastic flow instability often arises (the Portevin–Le Chatelier effect [1]). The dynamic aging of dislocations is also considered as a possible reason for the temperature and strain rate anomalies of the yield stress of materials ([2,3], etc.).

To date, the dynamic interaction of impurities with dislocations has been well studied for fcc crystals (see for reviews [4,5]) and much less for bcc metals, semiconductors, ceramics, and other materials in which dislocation motion is controlled by the Peierls barriers. The key problem of the theory of dynamic strain aging of materials is the variation of the impurity content in dislocation lines. The consideration of this problem in literature is too simplified; in many cases, the kinetic law is just postulated as a simple extrapolation between the initial diffusion power-law Cottrell-Bilby kinetics of impurity segregation in a dislocation line and the final approximation to equilibrium [6]. However, this approach may underestimate the impurity content at dislocations, because it does not take into account the impurity dragging. To understand this phenomenon better, it would be useful not to postulate but to derive a kinetic law from the underlying physical mechanisms. This is the purpose of the present study.

1. Model of short-range dislocation-impurities interaction

At different velocities of dislocations, different spatial scales of their interaction with impurities manifest themselves. We are interested in the case of sufficiently fast motion, when distances of the order of few lattice parameters around a dislocation are significant. The dislocation motion on the atomic scale (between valleys of the periodic crystal lattice) occurs, as is known, via thermally activated jumps [1,2]. A sharp change of the dislocation position in the slip plane leads to redistribution of the impurity atmosphere in each step, the redistribution kinetics depends on the structure of the potential profile for impurity migration in the vicinity of dislocation core. The simplest one-dimensional shape of such relief is schematically shown in figure 1.

Far from a dislocation, the migration relief is periodic and the time τ_0 of impurity displacement by the lattice parameter is unperturbed. In the vicinity of dislocation core, the profile is distorted by the interaction in such a way that there is a deep minimum corresponding to the most energetically favorable impurity position with respect to the dislocation, which is indicated by the letter *d* in figure 1. The attraction to this position reduces barrier *E* for impurity transition into it and decreases the time of the corresponding thermally activated jump $\tau = \tau_* \exp(E/kT)$ in comparison with the time of reverse transition, $\tau_1 = \tau_{*1} \exp(E_1/kT)$, or the time of migration far from the core, $\tau_0 = \tau_{*0} \exp(E_0/kT)$, thus, $\tau << \tau_0, \tau_1$. The meaning of the activation energies E_1, and $E = E_1 - \Delta E$ is explained in figure 1.

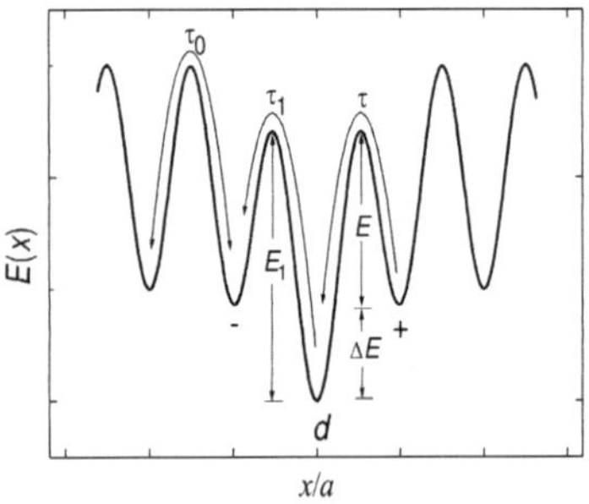

Figure 1. Schematic diagram of the migration relief of impurities near a dislocation core. The states directly in the core (*d*) and the neighboring states in the slip planes (+) and (–) are selected, with indication of the impurity transition times between these states and the crystal bulk

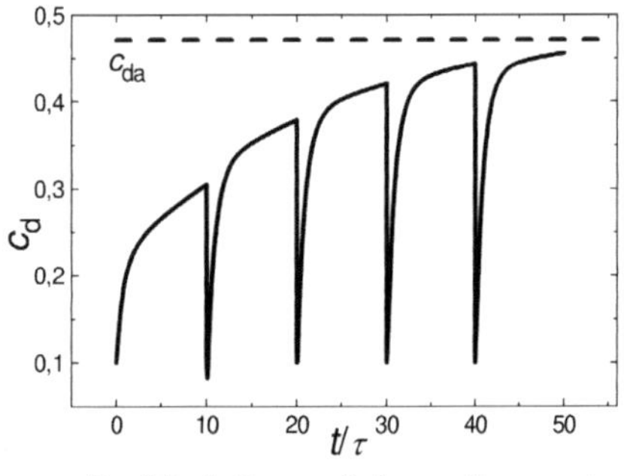

Figure 2. Variation of impurity content at an initially fresh dislocation in course of sequential jumps. The dashed line marks the steady amplitude value of impurity content established at $t \to \infty$ ($c_0 = 0.1$, $t_a = 10\tau$).

We focus on calculating the occupancy of the state *d*, which is a trap for impurities, the other states in the simplest approximation will be considered as weekly perturbed and interpreted as some reservoir with a constant concentration c_0. An exception is the (+) and (–) states (adjacent to *d* in the slip plane) because, due to the dislocation motion with possible entrainment of impurities, the impurity concentration is significantly redistributed between these states.

Let us analyze the evolution of the impurity concentration in the dislocation core c_d. We will also introduce the designations c_+ and c_- for the impurity concentrations in the (+) and (–) states. We can write the equations for impurity redistribution between these states and the crystal bulk during the waiting time t_a for jump to the next valley, when the dislocation is at rest in one of the potential valleys:

$$\frac{dc_d}{dt} = (c_+ + c_-)(1 - c_d)/\tau - c_d(2 - c_+ - c_-)/\tau_1, \tag{1}$$

$$\frac{dc_\pm}{dt} = - c_\pm(1 - c_d)/\tau + c_d(1 - c_\pm)/\tau_1 - c_\pm/\tau_0 + c_0(1 - c_\pm)/\tau_0. \tag{2}$$

The first terms on the right-hand sides of these equations describe the flux of impurities from the adjacent (+) and (–) states to the core (*d*) per unit time, the second terms describe the reverse flux. The third and fourth terms on the right-hand side of (2) describe the exchange by impurities between the (+) and (–) states and the crystal bulk. Equations (1) and (2) take into account that transitions may occur only to unoccupied states: for example, the transition frequency to the state *d* is proportional to the probability $1 - c_d$ that this site is free, etc.

Due to the large difference in the times of impurity transitions to the dislocation core (τ) and the neighboring lattice sites (τ_1 and τ_0), the evolution of the impurity content occurs in two stages with significantly different durations. During a time of about τ, only the fastest transition of impurities from the nearest states (±) to the core (*d*) occurs, without the participation of more distant impurities. And only then, for a time of about τ_0, exchange with impurities from the reservoir occurs and the equilibrium concentration is established. The presence of the pronounced initial stage, describing the behavior of far from equilibrium impurity content, results in a number of peculiar effects, which are considered below.

3. Entrainment of impurities at steady dislocation motion

On the microscopic level, the dislocation motion over the Peierls barriers occurs via successive thermally activated jumps by a lattice period a with jump waiting times t_a, thus, the average dislocation velocity is $V_d = a/t_a$. The impurities in the dislocation core behave differently, depending on the migration barrier height. At a sufficiently high binding energy ΔE, as opposed to the barrier height E, impurities are predominantly entrained by the dislocation to the next potential valley. At the reverse relation ($\Delta E < E$), impurities are predominantly dropped away the dislocation and remain in the initial valley. This condition gives an approximate quantitative criterion for distinguishing by convention high- and low-mobile impurities. A complete analysis of model predictions is given in [7]. Let us shortly describe the behavior of low-mobile impurities ($\Delta E < E$).

Using equations (1), (2), one can easily calculate kinetics of dynamic aging of a dislocation, as is shown in figure 2. Accumulation of impurities due to the dragging is clearly seen, which illustrates the necessity to modify the Louat equation [6] for the considered case. The variation of amplitude value c_{da} of the impurity content attained in a step versus the duration of the waiting time of a jump t_a, or, equivalently, versus $V_d=a/t_a$, is illustrated in figure 3. This dependence has a simple qualitative explanation. Low-mobile impurities at a short transition time t_a behave as static ones, retaining the unperturbed content c_0. With an increase in the transition time, the amplitude content c_{da} of low-mobile impurities in the dislocation core monotonically increases to the equilibrium value

$$c_{deq}=\frac{c_0\exp(\Delta E/kT)}{1+c_0\exp(\Delta E/kT)}. \qquad (3)$$

It follows from equations (1), (2) that initial deviation of c_{da} off the bulk value c_0 at the onset of dynamic aging is described by expression

$$c_{da}\approx c_0\exp(t_a/\tau). \qquad (4)$$

Clearly, this exponential increase with t_a is much faster than the determined by the diffusion power-like increase. As $t_a=a/V_d$, (4) predicts strong negative rate dependence of the impurity contribution to the yield stress. As usual, the strain rate anomaly is accompanied by the temperature anomaly. Accounting for the Arrhenius temperature dependence of τ in (4), the total dependence of the impurity content in the core on T may be represented as in figure 4: first c_{da} increases with the increasing temperature due to the increase in the impurity mobility, furthermore, it decreases with a decrease in the equilibrium content, when the temperature increases.

For a simplified qualitative description of the temperature peak $c_{da}(T)$, one can use some interpolation formula. For example, for c_0<<1 a satisfactory approximation is given by the formula

$$c_{da}(T)\approx\frac{c_0}{c_0+\exp(-\Delta E/kT)+\exp(-t_a/\tau)}. \qquad (5)$$

This dependence is shown in figure 4 by a dashed line c_{int} to compare it with the result of the numerical solution of equations (1) and (2) and analytical approximations describing the low- and high-temperature branches of the peak.

4. Conclusion

An analytical theory of the dislocation dynamic aging in materials with the Peierls mechanism of the plastic deformation is developed. The theory takes into account the impurity dragging effect, which dominates over the bulk diffusion of impurities to dislocation core, thus predicting the enrichment of the impurity content at dislocations. The theory elucidates the origin of anomalies of the plastic deformation, such as negative strain rate and positive temperature dependences of the yield stress.

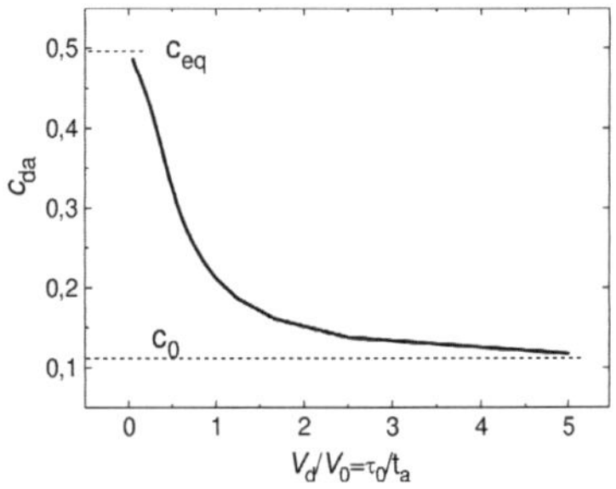

Figure 3. Dependence of impurity content in dislocation core on dislocation velocity V_d (or on waiting time $t_a=a/V_d$) at the parameters $c_0 = 0.1$ and $\tau_1 = \tau_0 = 10\tau$.

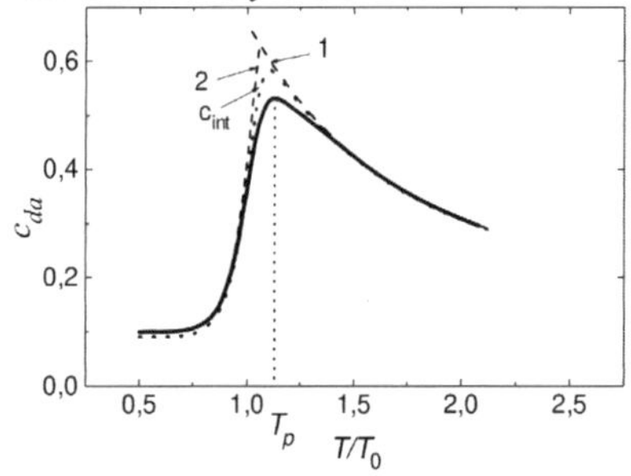

Figure 4. Temperature dependence of the impurity content c_{da} in the dislocation core for the parameters $c_0 = 0.1$, $T_0 = 0.1E/k$, $E_1 = 1.3E$, and $t_a = 4.4\times10^4\tau_*$. The dashed lines show: *1* the equilibrium dependence (3), and *2* the dependence characteristic of the first stage (4). The dotted line is the interpolation dependence c_{int} (5).

The predicted increase in the content of low-mobile impurities as the temperature increases is responsible for various specific features in the temperature dependence of the yield stress $\sigma(T)$ in impure materials: from inflections in $\sigma(T)$ to the anomalous increase in $\sigma(T)$ in a certain temperature range. Both these features were repeatedly observed in experiments with different "Peierls" materials: in bcc metals and (in the case of prismatic slip) in hcp metals Ti, Zr, and Be and intermetallic compounds (see review in [2]). One may suppose that a sufficiently large increase of the impurity content in dislocation cores with increase of temperature (predicted by (4)) can completely block the dislocation motion and cause an interesting phenomenon known as the "inverse brittle–ductile transition," at which a material loses its plasticity not upon cooling (as usual) but, vice versa, upon heating, as was observed in GaAs [8], or SrTiO3 [9].

References

[1] Hirth J P and Lothe J 1980 *Theory of Dislocations* (McGraw-Hill, New York)
[2] Caillard D and Martin J L 2003 *Thermally Activated Mechanisms in Crystal Plastisity* (Pergamon, Amsterdam)
[3] Messerschmidt U, Guder S, Haüssler D, and Bartsch M 2001 Mater. Res. Soc. Proc. **652** Y11.6.1
[4] Ananthakrishna G 2007 Phys. Rep. **440** 113
[5] Estrin Y and Kubin L P 1995 *Continuum Models for Materials with Microstructure*, ed H. B. Mühlhaus (Wiley, Chichester) p 395.
[6] Louat N 1981 Scr. Metall. **15** 1167
[7] Petukhov B V 2009 Cryst. Reports **54** 82
[8] Suzuki T, Tokuoka T, Yonenaga I, and Kirchner H O K 2000 Scripta Mater. **43** 645
[9] Gumbsch P, Taeri-Baghbadrani S, Brunner D, Sigle W, and Rühle M 2001 Phys. Rev. Lett. **87** 085505

Journal of Physics: Conference Series **240** (2010) 012013 doi:10.1088/1742-6596/240/1/012013

Al-matrix composite materials reinforced by Al-Cu-Fe particles

J Bonneville, G Laplanche, A Joulain, V Gauthier-Brunet and S Dubois

Université de Poitiers, PHYMAT, CNRS-UMR 6630, SP2MI, F-86962 Chasseneuil

joel.bonneville@univ-poitiers.fr

Abstract. Al-matrix material composites were produced using hot isostatic pressing technique, starting with pure Al and icosahedral (i) Al-Cu-Fe powders. Depending on the processing temperature, the final reinforcement particles are either still of the initial i-phase or transformed into the tetragonal ω-$Al0_{0.70}Cu_{0.20}Fe_{0.10}$ crystalline phase. Compression tests performed in the temperature range 293K – 823K on the two types of composite, i.e. Al/i and Al/ω, indicate that the flow stress of both composites is strongly temperature dependent and exhibit distinct regimes with increasing temperature. Differences exist between the two composites, in particular in yield stress values. In the low temperature regime ($T \leq 570K$), the yield stress of the Al/ω composite is nearly 75% higher than that of the Al/i composite, while for $T > 570K$ both composites exhibit similar yield stress values. The results are interpreted in terms of load transfer contribution between the matrix and the reinforcement particles and elementary dislocation mechanisms in the Al matrix.

1. Introduction

Icosahedral quasicrystalline (i-QC) materials exhibit very attractive mechanical properties such as high hardness, high elastic modulus and elevated yield stress. However, they all suffer from a serious drawback for industrial applications due to their extreme brittleness up to high temperatures. A possibility to take advantage of their high mechanical properties is to use them as reinforcement particles in composite materials.

The mechanical properties of Al-based composite materials reinforced by i-QC particles are little known, even unclear. Tsaï *et al.* [1] were the first to produce Al-metal matrix composites (MMCs) reinforced by Al-Cu-Fe i-QC particles. Using a hot isostatic pressing (HIP) technique, they highlighted that for a processing temperature of 673 K the i-QC phase of the reinforcement particles was preserved, while at 873 K a phase transformation of the i-QC reinforcement particles occurred. In the latter case, the i-QC phase was transformed into the tetragonal crystalline ω-phase with the $Al_{0.70}Cu_{0.20}Fe_{0.10}$ composition. Tsaï et al [1] also investigated the macro-hardness of both MMCs, i.e. Al/i and Al/ω MMCs, using Vickers indentations. They found for the two MMCs that the hardness H_v increases with increasing volume fraction, V_f, of the reinforcement particle. In addition, they obtained that for V_f ranging between 5% and 15% Al/i and Al/ω MMCs exhibit similar hardness, while for V_f = 20% and 25% H_v is higher for the Al/i MMC than for the Al/ω MMC. The latter result was later confirmed by Kaloshkin *et al.* [2], but surprisingly these authors also found that the flow stress measured by compression tests was higher for the Al/ω MMC than for the Al/i MMC.

The aim of the present study is therefore to examine in detail the consequence of the phase transformation of the reinforcement particles on the mechanical properties of Al/Al-Cu-Fe MMCs. For this, Al/Al-Cu-Fe MMCs with reinforcement particles of the i-QC phase and of the ω-phase were produced using a hot isostatic pressing technique with an initial mixture of pure Al and i-QC Al-Cu-Fe powders. Crystallographic and chemical composition investigations were performed on the as-synthesised MMCs to identify the resulting phases. Compression tests were performed on the two types of MMC at constant strain-rate in the temperature range 293K-823K. The mechanical properties of both MMCs are reported and the results are interpreted in terms of possible strengthening sources.

2. Experimental procedures

Experimental details concerning i-QC powder processing and Al/Al-Cu-Fe MMC synthesis have already been reported elsewhere [3, 4]. They are here briefly summarised. A HIP method is used for producing both MMCs, which consists in hot pressing a green compact composed of a mixture of pure Al and i-QC Al-Cu-Fe powders. For the two composites, the initial volume fraction of the i-QC Al-Cu-Fe powder was adjusted in order to obtain a final volume fraction of 40% of Al-Cu-Fe reinforcement particles [3]. The main difference concerns the sintering temperatures, which is set to 673K for preserving the i-QC phase of the reinforcement particles and to 823K for producing the i-QC to ω-phase transformation. Pure Al samples were also produced by the same HIP process at 823K for a direct estimate of the reinforcement particle strengthening efficiency.

The crystallographic structure of the MMCs was determined by XRD experiments using a Bruker D501 diffractometer. Chemical composition was investigated by Energy Dispersive X-Ray Spectroscopy (EDXS) in a JEOL 5600LV scanning electron microscope. Deformation tests were performed in compression at a nominal strain rate of $1.4 \times 10^{-4}\ s^{-1}$ over the temperature range 293K-823K.

3. Experimental Results

3.1. Crystallographic features

Figure 1 shows two XRD profiles obtained on the two Al/Al-Cu-Fe MMCs after the HIP process at T = 673 K (fig. 1a) and T = 823 K (fig. 1b). All peaks are indexed in fig. 1a according to the Al face-centred cubic (fcc) structure and the i-QC Al-Cu-Fe structure and in fig. 1b according to the Al-fcc structure and the ω-Al-Cu-Fe structure. No other peak corresponding to an eventual third phase is detected in each profile that is, at the level of XRD detection, each MMC can be considered as two-phase materials. EDXS analyses confirm XRD patterns of figure 1a and 1b. The chemical composition of the reinforcement particles are $Al_{0.649}Cu_{0.226}Fe_{0.125}$ for profile 1a and $Al_{0.7}Cu_{0.2}Fe_{0.1}$ for profile 1b, which correspond to the i- and ω-phase respectively. It must be noticed that a third composition close

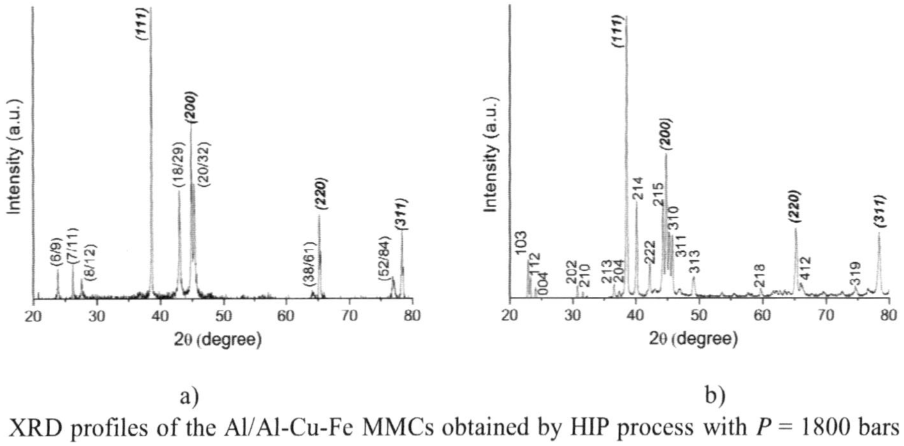

a) b)

Fig. 1 XRD profiles of the Al/Al-Cu-Fe MMCs obtained by HIP process with P = 1800 bars and at: (a) T = 673 K (the (N,M) notation is used for indexing the i-QC phase) and (b) T = 823 K.

to $Al_{0.732}Cu_{0.058}Fe_{0.210}$ is detected by EDXS in some places inside the reinforcement particles of both MMCs. The latter phase is not detected by XRD due to its low volume fraction, but certainly corresponds to the λ-$Al_{13}Fe_4$ phase already reported in [5]. A low Cu content is also detected by EDXS at some places in the Al matrix of the two MMCs.

3.2. Mechanical properties

Figure 2 shows typical stress-strain curves of the two MMCs and pure Al at room temperature (RT). The shape of the stress strain curves is similar over all the investigated temperature range. After yielding, plastic deformation consists in a short transient with strong hardening that is followed by a plastic stage of almost steady state conditions. The stress drops on the Al and Al/i curves are due to load relaxation experiments. The conventional $\sigma_{0.2\%}$ proof stress, defined at 0.2% plastic strain, extracted from these curves is reported in fig. 3 as a function of temperature. A clear strengthening is

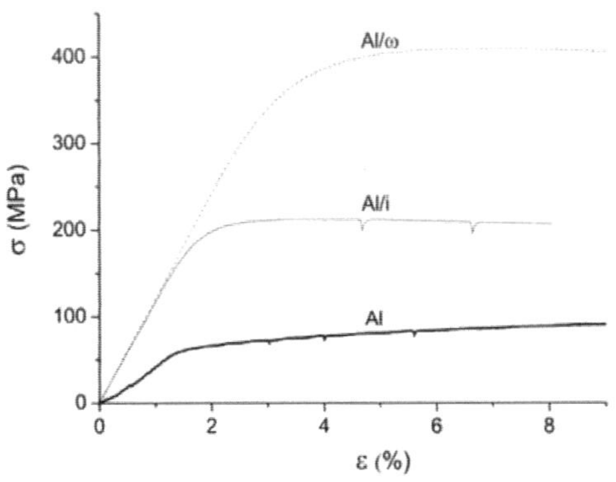

Fig. 2 Stress-strain curves at RT of pure Al, Al/i and Al/ω materials

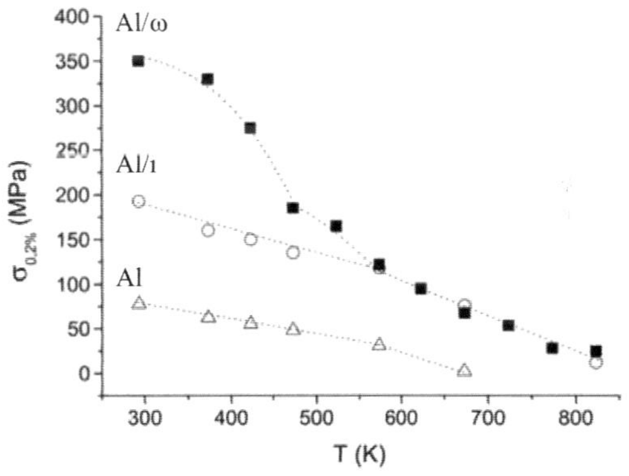

Fig. 3. $\sigma_{0.2\%}$temperature dependence of pure Al, Al/i and Al/ω materials.

observed for the two MMCs as compared with pure Al. The temperature dependence of $\sigma_{0.2\%}$ allows us to distinguish at least two temperature regimes. In the low temperature regime, below 550K, the Al/ω MMC exhibits higher $\sigma_{0.2\%}$ values, which however decrease more rapidly with increasing temperature than those of the Al/i MMC. Above 550K, $\sigma_{0.2\%}$ behaves similarly for the two MMCs. It is remarkable that a break in the $\sigma_{0.2\%}$ temperature dependence is observed at approximately 550 K for pure Al as well.

4. Discussion

Using a HIP process we produced two Al-MMCs reinforced by Al-Cu-Fe particles. Depending on the synthesis temperature, the reinforcement particles are either of the initial i-QC phase or of the ω-tetragonal phase in agreement with Tsai et al. [1]. This is also coherent with the high temperature (T>773K) phase diagram of Al-Cu-Fe [6, 7], which indicates that the i-QC phase does not coexist with the Al phase. In principle, the phase transformation occurring at T = 823 K from the i-QC to the ω-phase only requires Al diffusion from the Al particles toward the i-QC particles and is accompanied by a volume expansion of the initial i-QC reinforcement particles. The phase transformation occurs with a migration of the Al/Al-Cu-Fe interfaces. Little Cu diffusion from the i-QC particles into the Al matrix is also revealed for the two MMCs, which may lead to matrix solid solution hardening.

The $\sigma_{0.2\%}$ stresses are larger for the two MMCs than for pure Al and exhibit with temperature for all materials two distinct temperature regimes. Several contributions can be responsible for the large flow stress strengthening of the two MMCs as compared to pure Al [8]. In particular, load transfer between matrix and reinforcement particles is considered as the major cause of strengthening for this type of

MMCs [9]. Load transfer is also very dependent from the type of interface between the matrix and the reinforcement particles. At low temperature, the interfaces are either of Al-fcc/i-QC type or Al-fcc/ω-tetragonal type depending on the considered MMC. It seems reasonable to assume that interfaces between quasiperiodic and periodic structures have different properties than interfaces between two periodic structures, which may partially explain the low temperature difference in $\sigma_{0.2\%}$ between the two MMCs. The stiff decrease of $\sigma_{0.2\%}$ with temperature for the Al/ω MMC can be hardly attributed to load transfer only, because it would require a drastic change in the interface nature, which is not expected at low temperature. Other contributions, such as geometrically necessary dislocations, grain sizes and thermal expansion coefficients, are not sufficient for explaining the recorded difference in $\sigma_{0.2\%}$ between Al/i-QC and Al/ω MMCs [10]. Transmission electron microscopy observations, not presented here, clearly demonstrate that plasticity takes place in the soft Al matrix [3, 11]. They also highlight that dislocation obstacles are of different nature in the two composites. Only pinning points are observed on dislocations in the Al/i-QC composite while small ω-particles impede dislocation movement in the Al/ω composite, with sizes ranging between 100 nm and 300 nm. This may account for the large stress increase of Al/ω MMC in comparison to Al/i-QC MMC [11]. At high temperature, similar recovery processes take place in the both matrices. Then, the difference between MMCs and pure Al mainly results from load transfer. Al/i-QC interfaces evolve with increasing temperature to Al/ω interfaces in agreement with the similar measured $\sigma_{0.2\%}$ values for both composites.

5. Conclusion

Two Al-MMCs were synthesized using a HIP technique. Depending on processing temperature, the initial i-QC phase of the reinforcement particles is preserved (T = 673K) or transformed into the ω tetragonal phase (T = 823K). A strong strength improvement is recorded for the two MMCs in comparison with pure Al. In addition at low temperature, $\sigma_{0.2\%}$ is larger for Al/ω MMC than for Al/i-QC MMC, which may be partly ascribed to better load transfer from the matrix to the reinforcement particles, resulting from better interfaces in the Al/ω MMC. The temperature dependence of $\sigma_{0.2\%}$ necessitates however to take into account dislocation mechanisms into the Al matrix. At high temperature, load transfer seems to be the major strengthening contribution for the two MMCS.

Acknowledgments

Région 'Poitou-Charentes' is acknowledged for financial support through a PhD fellowship.

References

[1] A. P. Tsaï, K. Aoki, I. Akihisa, T. Masumoto : J. Mater. Res. **8**, 5-7 (1993).
[2] S. D. Khaloshkin, V. V. Tcherdyntsev, A. I. Laptev, A. A. Stepashkin, E. A. Afonina, A. L. Pomadchik, V. I. Bugakov: J. Mater. Sci. **39,** 5399-5402 (2004).
[3] T. El Kabir, A. Joulain, V. Gauthier, S. Dubois, J. Bonneville: J. Mater. Res. **23**, 904 (2008).
[4] G. Laplanche, A. Joulain, J. Bonneville, R. Schaller, T. El Kabir: submitted to J. Alloys Comp. (2009).
[5] E. Giacometti, J. Fikar, N. Baluc, J. Bonneville, Phil. Mag. Lett. **82**, 183 (2002).
[6] F. Faudot, A. Quivy, Y. Calvayrac, D. Gratias, H. Havemel, Mater. Sci. Eng. A **133**, 383 (1991).
[7] D. Gratias, Y. Calvayrac, J. Devaud-Rzepski, F. Faudot, H. M., A. Quivy, J. Non-Cryst. Solids **153-154**, 482 (1993).
[8] T. W. Clyne and P. J. Withers: An Introduction to Metal Matrix Composites, Cambridge University Press (1993).
[9] F. Tang, I. E. Anderson, T. Gnaüpel-Herold, H. Prask, Mater. Sci. Eng. A **383**, 362 (2004).
[10] T. El Kabir, PhD dissertation, University of Poitiers (2007).
[11] G. Laplanche, A. Joulain, J. Bonneville, V. Gauthier-Brunet, S. Dubois, T. El Kabir, accepted for publication in J. Mater. Res. (2009).

Universal Scaling of the Temperature Dependence of the Strength of Crystals Governed by the Peierls Mechanism

S Takeuchi[1] and T Suzuki[2]
[1]Science University of Tokyo, Shinjuku-ku, Tokyo 162-8601, Japan
[2]Professor Emeritus, Institute of Industrial Science, The University of Tokyo, Meguro-ku, Tokyo, 153-8505, Japan

E-mail: takeuchi@rs.noda.tus.ac.jp

Abstract. The temperature dependences of the critical resolved shear stress (CRSS) governed by the Peierls mechanism in pure NaCl type crystals, those in pure bcc transition metals, those by dissociated dislocations in covalent crystals of the diamond and the zinc blende structures and those by perfect dislocations at low temperatures in zinc blende crystals have been demonstrated to be roughly scalable with respect to the non-dimensional normalization of the CRSS by the shear modulus G and the temperature by Gb^3/k_B, where b is the strength of the Burgers vector and k_B the Boltzmann constant. Furthermore, CRSS vs. T relations have been shown to be scaled universally by normalizing respectively the CRSS by the estimated Peierls stress τ_P and the temperature by the kink-pair energy parameter of $(\tau_P/G)^{1/2}(bd)^{3/2}G/k_B$, where d is the period of the Peierls potential.

1. Introduction

It is considered that for a group of crystals with the same crystal structure and with a similar bonding nature, the deformation behavior is analogous to each other. From the point of predicting the strength of crystals, various attempts have been made to scale the strength by materials parameters. Ashby and Brown have succeeded in scaling the temperature dependence of the Vickers hardness values of single crystals of different groups with a considerable accuracy [1]. They have also shown that the high temperature creep rate for particular classes of solids can be represented by common formulae [2].

In this paper, we try to scale the critical resolved shear stress (CRSS) of crystals, the strength of which is governed dominantly by the Peierls mechanism. It has been shown experimentally that the estimated Peierls stresses of pure crystals with the same crystal structure possess a similar value when normalized by the shear modulus [3-5], and that the Peierls stresses of different crystal structures obey quite roughly the classical theoretical relation obtained by Peierls and Nabarro [6,7]. It has been established that the low temperature yield strengths of pure bcc metals and pure NaCl type ionic crystals and those at high temperatures of tetrahedrally coordinated crystals are governed dominantly by the Peierls mechanism. For the tetrahedrally coordinated cubic crystals, it has been shown that the activation enthalpy of dislocation motion can be scaled by $K_T b^3$, where K_T is the line tension factor and b is the strength of the Burgers vector. [8]. For the same crystals, a homologous relation has been shown to hold for the CRSS vs. the temperature relation by Kirchner and Suzuki [9] and for the hardness vs. the temperature relation by Yonenaga and Suzuki [10]. Recently, we have shown that the temperature dependences of the CRSS of <110>{110} slip in NaCl type ionic crystals are distinctly divided into a high Peierls stress group and a low Peierls stress group, whereas those of <110>{001}slip in NaCl type crystals can be scaled almost universally [11].

In the present paper, we try to scale all the CRSS data in a wide temperature range of crystals governed by the Peierls mechanism in a unified manner, based on a single dislocation velocity equation deduced from the dislocation theory in the framework of the line tension approximation.

2. Kink pair formation enthalpy

For the smooth kink case where the Peierls potential of the second kind (Peierls potential for the kink motion) is much lower than that of the first kind, the dislocation velocity is determined by the rate of the kink pair formation along straight dislocation segments lying in Peierls potential valley. For covalent, tetrahedrally coordinated crystals, the Peierls potential of the second kind is considered to be appreciable, but it has been shown experimentally that even in these covalent crystals, the velocity of the dislocation is also governed solely by the rate of the kink pair formation on dislocation segments lying in Peierls valley [8]. Thus, well above the athermal stress level where the backward kink-pair formation can be neglected, the dislocation velocity governed by the Peierls mechanism can generally be written by an Arrhenius type equation with the activation enthalpy of the kink-pair formation. Since the pre-exponential factor of the Arrhenius equation is much less sensitive to stress compared with the exponential factor, the temperature dependence of the CRSS is determined essentially by the condition

$$\frac{\Delta H(\tau)}{k_B T} = m, \tag{1}$$

with a m value in the range of 25±10 depending on the type of crystal and the slip system. Here, $\Delta H(\tau)$ is the activation enthalpy and $k_B T$ has the usual meaning. The functional form of $\Delta H(\tau)$ depends on the shape of the Peierls potential; $\Delta H(\tau=0)$ is equal to the kink-pair formation energy E_{kp} or twice the kink energy, and $\Delta H(\tau=\tau_P)$ is zero, where τ_P is the Peierls stress. Correspondingly, the temperature dependence of the CRSS, τ_c, should satisfy $\tau_c=\tau_P$ at $T=0$ and $\tau_c=0$ above a critical temperature T_0 given by

$$T_0 = E_{kp}/(mk_B). \tag{2}$$

In the framework of the line tension approximation of the dislocation, kink pair formation enthalpy has been formulated by Celli et al. [13] and Dorn and Rajnak [14]. The kink pair formation enthalpy corresponding to the saddle point configuration of a dislocation overcoming the Peierls potential depends on the shape of the Peierls potential [14]. The kink pair formation energy has been shown to be expressed as [12]

$$E_{kp} = B\zeta^{1/2}(db)^{3/2}(\tau_P K_T)^{1/2}, \tag{3}$$

where B is a numerical constant which depends on the shape of the Peirels potential, ζ is the pre-factor of the dislocation self-energy expression given by $(1/4\pi)\ln(R/r_0)$, d the period of the Peierls potential and K_T the line tension factor. The value of $\ln(R/r_0)$ is usually taken as 10 and $\zeta \approx 0.8$. For the elastically isotropic case, K_T is about $2G$ and $G/2$ for screw and edge dislocations, respectively. Koizumi et al. have calculated the numerical constant B for a variety of Peierls potential shapes, from the sinusoidal shape through a flat-top shape to a camel-hump shape, and showed that B is rather insensitive to the shape of the potential and is in the range of 1.0 ± 0.1 [15]. Thus, writing $db=k_1 b^2$, $K_T=k_2 G$ and $\tau_P=\alpha G$, the kink pair energy E_{kp} is generally written as

$$E_{kp} = 0.9 k_1^{3/2} k_2^{1/2} \alpha^{1/2} G b^3 \equiv k\alpha^{1/2} G b^3, \tag{4}$$

where k_1 depends on the slip system and the dislocation character, and k_2 also depends on the dislocation character. From equations (2) and (4) one obtains

$$T_0 = \frac{k\sqrt{\alpha}}{m}\left(\frac{Gb^3}{k_B}\right) \quad \text{or} \quad \alpha = \left(\frac{mk_B T_0}{kGb^3}\right)^2. \tag{5}$$

3. Universal scaling of the CRSS vs. *T* relation

3.1. Non-dimensional scaling of the CRSS vs. T Relation

The CRSS vs. *T* relations which we deal with in this paper are those for {110}<110> and {001}<110> slips in NaCl type ionic crystals, <111> slip in bcc transition metals and {111}<110> slip in zinc blende crystals. In this subsection, we show how CRSS vs. *T* relations can be scaled with respect to the non-dimensional normalizations of CRSS by *G* and *T* by Gb^3/k_B.

We have reported recently the results for NaCl type ionic crystals [11]. There exist two groups of crystals for {110}<110> slip, i.e., the high Peierls stress group with τ_P/G=0.0013±0.0002 and the low Peirels stress group with τ_P/G=0.0004±0.0001. Meanwhile, the normalized CRSS vs. the normalized temperature plots for {001}<110> slip fall in a single zone though considerably wide.

It has already been shown that for each of bcc metals the temperature dependence of the CRSS can well be interpreted by the theory of the Peierls mechanism of the smooth kink regime [16]. When we plot the normalized stress vs. the normalized temperature relations for seven pure bcc transition metals (for the data source, see [12]), the data points fall in a single zone, though the data are considerably scattered, as shown in figure 1(a).

In recent years, the CRSSs of {111}<110> slip in III-V compounds of the zinc blende crystals have been obtained for a wide temperature range in compression experiments under a hydrostatic confining pressure [17-19]. In figure 1(b), we show the normalized CRSS vs. the normalized temperature relations for four zinc blende crystals together with the data zone of the ambient pressure experiments for eleven zinc blende crystals, Si, Ge and diamond, compiled by Kirchner and Suzuki [9]. It is found that all the data fall in a common zone, but the zone consists of two parts, the high temperature, low stress zone with a steep temperature dependence and the low temperature, high stress zone with a less steep temperature dependence. The low stress part has long been established to be due to glide of dissociated dislocations, while the high stress part is most probably due to glide of non-dissociated, perfect dislocations [17].

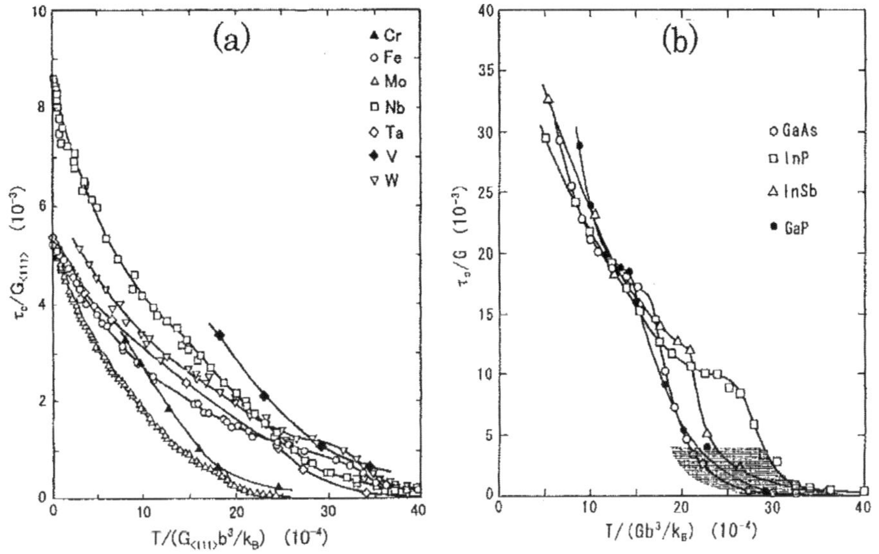

Figure 1. Normalized CRSS vs. normalized temperature relations for seven bcc transition metals (a), and those for III-V zinc blende crystals (b). The shaded zone in (b) indicates the data point zone for eleven zinc blende crystals, Si, Ge and diamond compiled by Kirchner and Suzuki [9].

Table 1. Character of dislocation governing the deformation, k_1, k_2 and k values (see text), m value in equation (2), average $\alpha(\equiv\tau_P/G)$ value and average value of $T_0/(Gb^3/k_B)$ for five types of slip systems.

Slip system	Disl. Type	k_1	k_2	k	m [ref]	α_{av}	$[T_0/(Gb^3/k_B)]_{av}$
NaCl{110}<110>: High τ_P	Edge	1	2^{-1}	0.64	30 [20]	0.0013(exp)	0.00077(calc)
NaCl{110}<001>: Low τ_P	Screw	$2^{-1/2}$	2	0.76	30 [20]	0.0004(exp)	0.00051(calc)
NaCl{001}<110>	Screw	1	2	1.27	20 [21]	0.0075(calc)	0.0055(exp)
bcc transition metal	Screw	$2^{3/2}/3$	2	1.17	25 [12]	0.006(exp)	0.0036(calc)
Zinc blende: Dissociated	30°-partial	2^{-1}	13/8	0.41	35 [22]	0.089(calc)	0.0035(exp)
Zinc blende: Perfect	Screw	1	2	1.27	35 [22]	0.045(exp)	0.0077(calc)

Thus, for five types of slip, (a) {110}<110> slip in NaCl type ionic crystals, (b) {001}<110> slip in NaCl ionic crystals, (c) <111> slip in bcc transition metals, (d) {111}<110> slip by dissociated dislocations in zinc blende crystals and (e) {111}<110> slip by perfect dislocations in zinc blend and diamond structure crystals, the CRSS vs. T relations can be scaled approximately with respect to the non-dimensional normalization of CRSS by G and the temperature by Gb^3/k_B, though there exist two groups of crystals for (a).

3.2. *Average values of $\alpha(\equiv\tau_P/G)$ and $T_0/Gb^3/k_B)$ for five types of slip*

If we assume that the line tension approximation holds for the five types of slips treated in the previous subsection, equation (4) should be satisfied for every case. k_1 value is determined from the character of the dislocation controlling the deformation rate. Throughout the present analysis, we assume elastic isotropy to determine k_2 value. After determining k ($\equiv 0.9k_1^{3/2}k_2^{1/2}$) in equation (4) for each slip, we determine the average α value from the experimental average value of $[T_0/(Gb^3/k_B)]_{av}$ and the experimental m value, or conversely the average $T_0/(Gb^3/k_B)$ value from the experimental average value of α and m value from equation (5). In table 1, we list the dislocation type, k_1, k_2, k, m values with their references in parentheses, experimental (exp) or calculated (calc) values of α_{av} and experimental or calculated values of $[T_0/(Gb^3/k_B)]_{av}$.

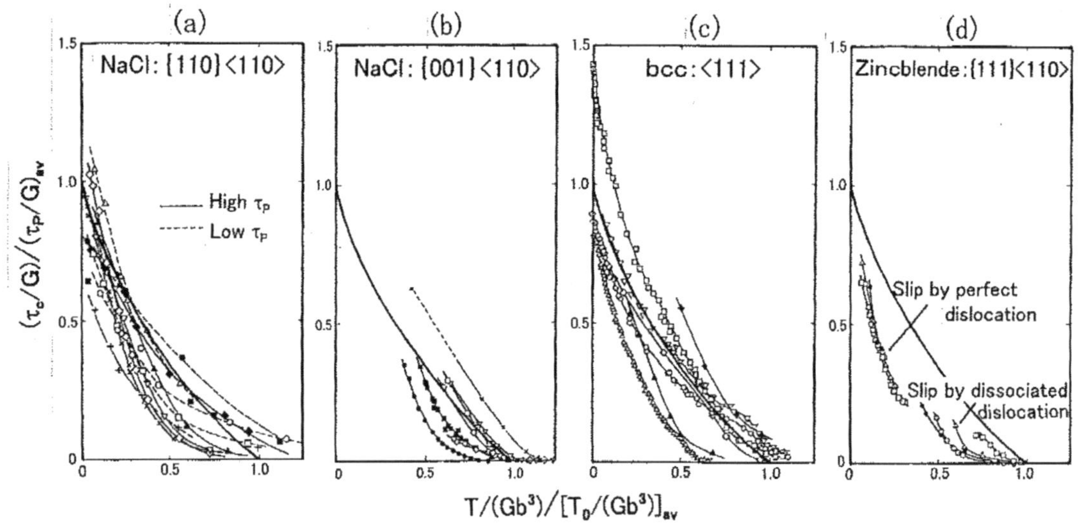

Figure 2. Universal scaling plots of CRSS vs. T relations for five types of slip: (a) {110}<110> slip in NaCl type crystals, (b) {001}<110> slip in NaCl crystals, (c) <111> slip in bcc transition metals, (d) {111}<110> slip by dissociated dislocations and that by perfect dislocations in zinc blende crystals.

3.3. Universal scaling plot

Using the average values of α and $T_0/(Gb^3/k_B)$ for each slip, we show in figure 2(a) to (d) the universal scaling plot for five types of slip. The thick solid line in each figure in figure 2 indicates the theoretical relation for the sinusoidal Peierls potential.

4. Discussion

Even a group of crystals having a common structure and a similar bonding nature, each crystal has its own characteristics different from the others. For example, ionic radius ratios are widely different among NaCl type ionic crystals, and ionicities are at variance among tetrahedrally coordinated crystals [23]. Nevertheless, we have shown that the five types of slip governed by the Peierls mechanism can be scaled by materials parameters within ±40% accuracy for each group, except {110}<110> slip in NaCl type crystals. This fact means that the Peierls stress of a crystal is essentially determined by the crystals structure. The validity of the scaling relation encourages us to predict the strength of crystals governed by the Peierls mechanism with a considerable accuracy. We have interpreted that the existence of two groups in {110}<110> slip in NaCl type crystals is related to the special situation in that the dislocation core of this slip system is in a critical state for the dissociation, as discussed in a previous paper [11].

The results of the universal scaling plots in figure 2 for the five types of slip system are generally satisfactory. This fact indicates that the line tension model of the kink pair formation energy is generally applicable for a wide range of the Peierls stress. It should be noted that in the present universal scaling process, we need to know experimentally the character of the dislocation controlling the deformation and the m value (equation (1)), but no adjustable parameter is needed.

In the present scaling, we have assumed elastic isotropy for the estimation of K_T value in equation (3). However, due to elastic anisotropy, the actual K_T values for bcc metals for example differ by almost a factor of two in many cases [12]. For a more refined analysis, we should obviously use exact K_T value calculated from elastic constants of each crystal. Actually, if we use the exact K_T values for bcc metals, it has been shown that τ_c/τ_P vs. $k_BT/(db)^{3/2}(\tau_PK_T)^{1/2}$ plots fall in a quite narrow zone [16].

The results for zinc blende crystals in figure 2(d), the experimental data both at temperature and low temperature regions are rather far below the theoretical curve for the sinusoidal Peierls potential. Edagawa et al. [19] have made a computer simulation of a perfect screw dislocation motion in a model zinc blende crystal overcoming the Peierls potential and showed that in a high stress range the dislocation motion is controlled by overcoming of a camel-hump Peierls potential and hence the CRSS vs. T relation is expected to be far below that expected from a sinusoidal Peierls potential, in agreement with figure 2(d). In the low stress range, a partial dislocation of the two partials having the lower mobility controls the dislocation mobility. In III-V compounds, 30°-β-partial dislocation is known to have the lowest mobility (see, [8]) and controls the deformation rate. If the other partial has a much lower mobility, then the effective stress acting on the lower mobility dislocation should be much higher than the applied stress; if the stress to move the other partial is negligibly small, the effective stress acting on the 30°-β-parital is to be double of the applied stress. The much lower level below the thick solid curve of the data for the dissociated dislocation glide must be due to an effective stress larger than the applied stress due to the interaction between partials.

5. Conclusion

1) The CRSS vs. temperature relations for the same class of crystals governed by the Peierls mechanism are roughly scalable (within ±40%) with respect to materials parameter such as shear modulus and the strength of the Burgers vector.
2) CRSS vs. temperature relation of any crystal governed by the Peierls mechanism can be scaled universally by normalizing the CRSS by the estimated τ_P and the temperature by the observed critical temperature T_0 which is related to τ_P/G by $T_0=[k(\tau_P/G)^{1/2}](Gb^3/k_B)$, where constants k and m

can be experimentally determined.

3) Although CRSS data are not available at low temperatures for crystals with high Peierls stress, the validity of the universal scaling relation makes us possible to predict τ_P value with some accuracy based on the T_0 value obtained from high temperature CRSS data.

References

[1] Ashby M F and Brown A M 1981 *2nd Risø Conference on Materials Science*, ed N Hanson, A Horsell, T Leffers and H Lilholt p 1
[2] Brown A M and Ashby M F 1980 *Scripta Metall.* **14** 1297
[3] Takeuchi S and Suzuki T 1988 *Strength of Metals and Alloys*, ed P O Kettunen, T K Lepisto, M E Lehtonen (Oxford: Pergamon Press) pp 553-554
[4] Suzuki T and Takeuchi S 1988 *Rev. Phys. Appl.* **23** 685
[5] Wang J N, 1996 *Mater. Sci. Eng.* A **206** 259
[6] Peierls R E 1940 *Proc Phys. Soc.* **52** 23
[7] Nabarro F R N 1947 *Proc. Phys. Soc.* **59** 256
[8] Maeda K and Takeuchi S 1988 *Dislocations in Solids* ed F R N Nabarro and M S Duesbery (Amsterdam: Elsevier) vol 10 pp 444-504
[9] Kirchner H O K and Suzuki T 1998 *Acta Mater.* **46** 305
[10] I. Yonenaga I and T. Suzuki T 2002 *Philos. Mag. Lett.* **82** 535-
[11] Takeuchi S, Koizumi H and Suzuki T 2009 *Mater. Sci. Eng.* A **521-522** 90
[12] Suzuki T and Takeuchi S 2001 *Crystal Lattice Defects and Dislocation Dynamics* ed R A Vardanian (New York: Nova Science Publishers) pp 1-70
[13] Celli V, Kabler M, Ninomiya T and Thomson R 1963 *Phys. Rev.* **131** 58
[14] Dorn J E and Rajnak S 1964 *Trans. Met. Soc. AIME* **230** 1052
[15] Koizumi H, Kirchner H O K and Suzuki T 1993 *Acta metall. Mater.* **41** 3483
[16] Suzuki T, Kamimura Y and Kirchner H O K 1999 *Philos. Mag.* A **79** 1629
[17] Suzuki T, Yasutomi T, Tokuoka T and Yonenaga I 1999 *Phys. Status Solidi* (a) **171** 47.
[18] Suzuki T, Yasutomi T, Tokuoka T and Yonenaga I 1999 *Philos. Mag.* A **79** 2637
[19] Edagawa K, Koizumi H, Kamimura Y and Suzuki T 2000 *Philos. Mag.* A **80** 2591
[20] Suzuki T, Koizumi H 1985 *Dislocations in Solids* ed H Suzuki, T Ninomiya, K Sumino and S Takeuchi (Tokyo: University of Tokyo Press) pp 159-164
[21] Suzuki T, Strotzki W and Haasen P 1981 *Phys. Status Solidi* (b) **103** 763
[22] Boivin P, Rabier J and Garem H 1990 *Philos. Mag.* A **61** 619
[23] Phillips J C 1973 *Bonds and Bands in Semiconductors* (New York: Academic Press)
[24] Suzuki T, Kamimura Y and Kirchner H O K 1999 *Philos. Mag.* A **79** 1629

Journal of Physics: Conference Series **240** (2010) 012015 doi:10.1088/1742-6596/240/1/012015

Analysis of deformation behaviors of ultrafine grained Cu-30%Zn with bimodal grain-size distribution

Takatoshi Morimitsu[1], Naoki Takata[2], Daisuke Terada[3] and Nobuhiro Tsuji[3]

[1]Department of Adaptive Machine Systems, Graduate School of Engineering, Osaka University, 2-1 Yamadaoka, Suita, 565-0871, Japan
[2]Department of Metallurgy and Ceramics Science, Graduate School of Science and Engineering, Tokyo Institute of Technology, 2-12-1-S8-8, Ookayama, Meguro-ku, Tokyo, 152-8552, Japan
[3]Department of Materials Science and Engineering, Graduate School of Engineering, Kyoto University, Yoshida Honmachi, Sakyo-ku, Kyoto, 606-8501, Japan

takatoshi.MORIMITSU@ams.eng.osaka-u.ac.jp

Abstract Ultrafine grained (UFG) metals have remarkably high strength in comparison with that of coarse grained metals. However, the UFG metals often exhibit low tensile ductility. It has been reported that the UFG metals having bimodal grain-size distributions consisting of coarse grains and fine grains perform large elongation keeping high strength. However, the exact deformation mechanism of the bimodal UFG metals has not yet been clarified. Deformation behaviors of a bimodal UFG metal were studied in the present paper. A sheet of Cu-30%Zn was highly deformed by accumulative roll-bonding (ARB) process and then annealed. The annealed specimen showed a fully-recrystallized UFG structure with a bimodal grain size distribution. To clarify deformation behaviors of the bimodal metal, an identical region was observed by the electron backscattered diffraction (EBSD) method at various tensile strains. The grain average misorientation (GAM) increased with increasing the tensile strain. The increase in GAM of the coarse grains was larger than that of the fine grains, suggesting the difference in local deformation and strain-hardening behaviors between different grain sizes. The enhanced uniform elongation of the bimodal UFG specimen was considered to be associated with the difference in strain-hardening between the coarse grains and fine grains.

1. Introduction

Ultrafine grained (UFG) metals with grain sizes smaller than 1μm have remarkably high strength in comparison with that of coarse-grained pure and even alloyed metals. However, the fine grained materials often exhibit low tensile ductility compared with the coarse-grained counterparts. On the other hand, it has been reported that the UFG materials having bimodal grain-size distributions consisting of coarse and fine grains perform large elongation keeping high strength [1, 2]. It has been explained that the fine grains maintain high strength of the material while the coarse grains produce large elongation. However, the exact deformation mechanism of the bimodal UFG materials has not yet been clarified. Thus, we investigated the contribution of the different grain sizes on tensile deformation behaviors, using a fully-recrystallized bimodal UFG material in this study.

2. Experimental Procedures

Chemical composition of a Cu-30%Zn used in the present experiments is shown in Table 1. The Cu-30%Zn sheet with a size of 200 × 40 × 1 mm^3 was highly deformed by the accumulative roll-bonding (ARB) process [3] with lubrication at room temperature up to 6 cycles, corresponding to an equivalent strain of 4.8, for making an UFG structure. After the ARB process, the Cu-30%Zn sheet was annealed at 250 °C for 5.4 ks to fabricate a fully recrystallized bimodal UFG structure.

An identical region of the tensile specimen was observed by the EBSD method in order to investigate deformation behaviors of the bimodal metal. The tensile test was stopped at 3.4%, 15% and 24.2% elongation, and the identical region was analyzed by EBSD. The grain average misorientation (GAM) [4, 5] is calculated in the following process. First, the misorientation between neighboring pairs of each measured point within the grain is calculated. Second, these misorientations are averaged in each grain surrounded by high-angle grain boundaries. The increase in GAM is considered to correspond to the development of the dislocation substructures within the grains. Consequently, the GAM is expected to be a parameter that indicates the amount of local strain.

Table 1. Chemical composition of the Cu-30%Zn (JIS2600P) studied (wt%).

	Cu	Pb	Fe	Zn
JIS2600P	69.26	0.0017	0.0073	Bal.

3. Results and Discussion

3.1. Microstructure and Mechanical properties

Figure 1 shows a grain boundary map of the Cu-30%Zn specimen ARB processed by 6 cycles and subsequently annealed at 250 °C for 5.4 ks. The specimen shows a fully recrystallized UFG structure composed of submicrometer UFGs and relatively coarse grains with several μm grain sizes. The mean grain sizes (in area) of the fine grains and coarse grains were 0.25 μm^2 (circle approximated diameter of 0.56 μm) and 10.7 μm^2 (circle approximated diameter of 3.7 μm), respectively. This fully-recrystallized bimodal structure is significantly different from the bimodal structure in the Cu specimen reported in [1, 2] that was composed of as-SPD (severe plastic deformation) processed UFG matrix and partially recrystallized coarse grains.

Figure 2 shows engineering stress-strain curves of the starting material, as-ARB processed specimen, and the specimen annealed at 250 °C for 5.4 ks after the ARB process (bimodal specimen). The tensile strength and the uniform elongation of the starting sample were 369 MPa and 53.4%, respectively. The tensile strength of the as-ARB specimen increased to 819 MPa, while the uniform elongation greatly decreased to 2.2%. The bimodal specimen showed the tensile strength of 516 MPa and uniform elongation of 24.4%. It can be concluded that the bimodal UFG structure performs a large tensile elongation keeping high strength.

3.2. Observations and Analysis of an identical region by EBSD

Figure 3 shows the IPF maps of the identical region in the tensile specimen obtained from the EBSD measurement for the Cu-30%Zn having the bimodal UFG structure. In the 16% tensile-tested specimen (figure 3(b)), the grains are elongated to the tensile direction (RD). The gradation of the color in the grains in figure 3(b) indicates that misorientation has appeared within individual grains. The color gradation in the coarse grains is more significant than that in the fine grains, which suggests that the coarse grains have larger misorientation than the fine grains.

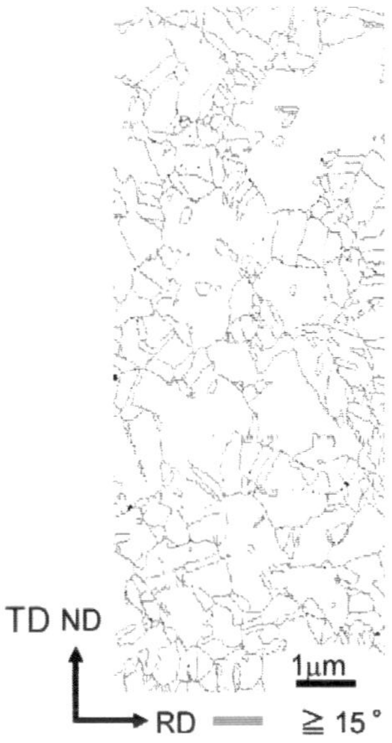

Figure 1. High angle grain boundary map obtained from the EBSD measurement of the Cu-30%Zn specimen ARB processed by 6 cycles and subsequently annealed at 250 °C for 5.4 ks.

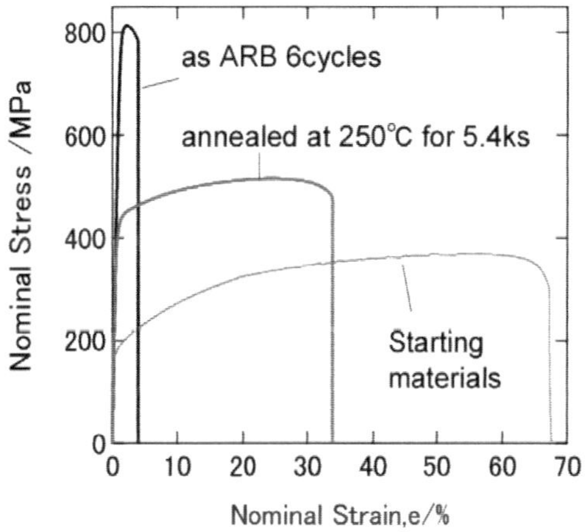

Figure 2. Nominal stress-strain curves of the starting material (Cu-30%Zn alloy), the as-ARB processed specimen, and the bimodal UFG specimen (ARB processed and then annealed at 250 °C for 5.4 ks).

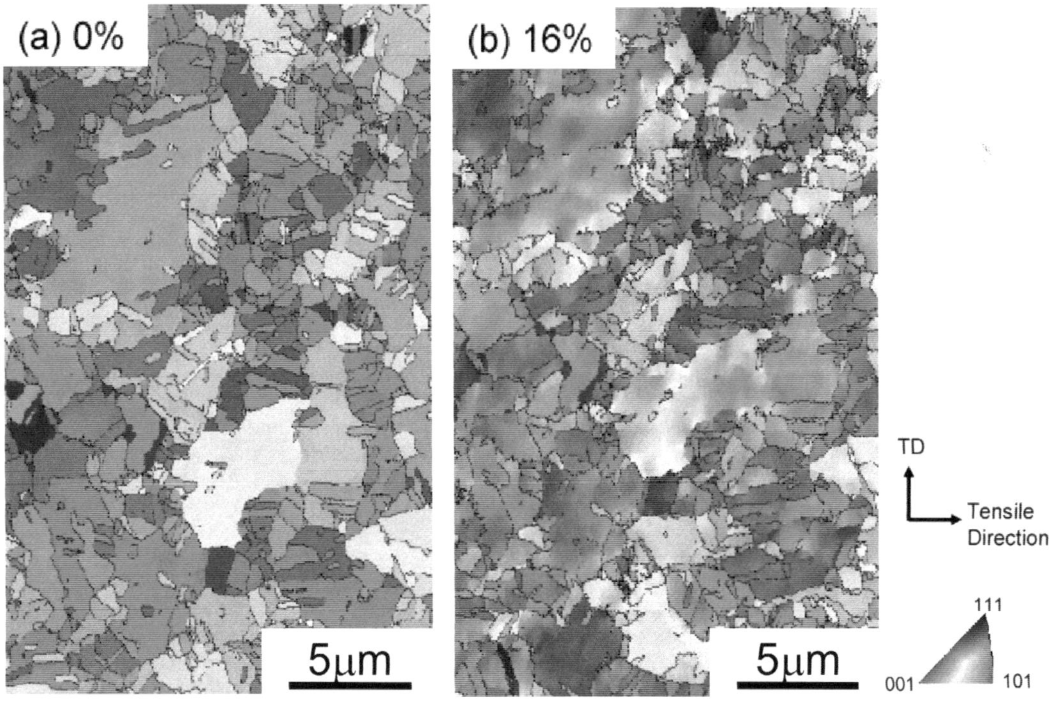

Figure 3. Inverse pole figure maps obtained from the EBSD measurement of the identical area in the bimodal UFG specimen (a) before tensile test and (b) after 16.0% tensile elongation.

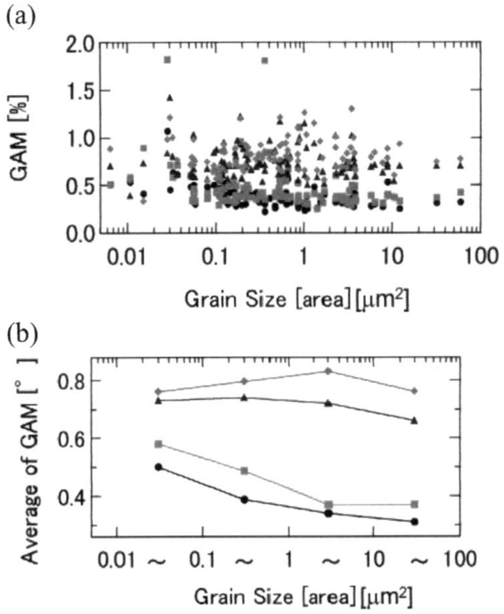

● Before tensile ■ 3.4% ◆ 24.2% ▲ 16.0%

Figure 4. (a) GAM of the grains within the observed area shown in figure 3 as a function of the grain size (in area). The GAM values for the bimodal specimen before the tensile test, tensile-deformed to 3.4%, 24.2% and 16.0% are indicated in different marks with different colors, as is shown above. (b) Average GAM as a function of grain size (in area). GAM values were averaged within the ranges; 0.01 ~ 0.1 μm^2, 0.1 ~ 1 μm^2, 1 ~ 10 μm^2 and 10 ~ 100 μm^2.

Figure 4 (a) shows GAM values of the grains in the identical observed area shown in figure 3 as a function of the grain size (in area). The GAM increased with increasing the tensile strain for all grain sizes. In figure 4 (b), the GAM values were averaged within four grain size ranges; 0.01~0.1 μm^2, 0.1~1 μm^2, 1~10 μm^2 and 10~100 μm^2. The increase in the average GAM of the coarse grains (10~100 μm^2 range) is 0.45°, which is obviously larger than that of the fine grains in the 0.01~0.1 μm^2 range (0.26°). The larger increase in GAM is considered to correspond to the more significant development of the dislocation substructures within the grains. The result thus suggests that the coarse grains are more strain-hardened than the fine grains in the present bimodal UFG structure. The enhanced uniform elongation of the bimodal UFG specimen might be associated with the difference in strain-hardening between the coarse grains and fine grains.

4. Summary

A fully recrystallized bimodal UFG structure was obtained in a Cu-30%Zn alloy through ARB and subsequent annealing processes. The EBSD measurement of an identical region in the tensile specimen indicated that the development of misorientation within the coarse grains is larger than that within the fine grains during tensile deformation. This suggests the difference in strain-hardening between the coarse grains and fine grains, which are considered to be attributed to the large elongation keeping high strength in the bimodal UFG specimen.

References

[1] Wang Y, Chen M, Zhou F and Ma E, 2002, *Nature* 419, 912.
[2] Wang Y M and Ma E, 2004, *Acta Mater.* 52, 1699-1709.
[3] Saito Y, Utsunomiya H, Tsuji N and Sakai T, 1999, *Acta Mater.* 47, 579.
[4] Wright S I, 1999, *Proc.12th Int. Conf. on Textures of Materials* 1, 104-109.
[5] Schwartz A J, Kumar M and Adams B L, 2000, Springer, *Electron Backscatter Diffraction in Materials Science*, 247-264.

Microstructure engineering from metallic powder blends for enhanced mechanical properties

P Langlois[1], N Fagnon[1] and G Dirras[2]

[1] LIMHP, Université Paris 13, CNRS;
Institut Galilée, 99 av. J.-B. Clément, 93430 Villetaneuse, France

[2] LPMTM, Université Paris 13, CNRS;
Institut Galilée, 99 av. J.-B. Clément, 93430 Villetaneuse, France

dirras@univ-paris13.fr

Abstract. The present work focuses on the transformation of high-purity Ni powder blends of controlled volume fractions (40 and 60 %) of nanometre-sized (100 nm) and micrometre-sized (544 nm) particles into bulk samples as part of a strategy for producing ultrafine-grained materials usefully exhibiting both strength and ductility. The process involved cold isostatic pressing at 1.5 GPa and sintering. The resulting bulk samples had relative densities near 95 %, were texture-free, and exhibited two different grain size distributions with an average value of 600 ± 30 nm. The mechanical properties were investigated by compression and microhardness tests, both at room temperature, and compared to the behaviour of a sample processed from micrometre-sized powder only. Samples prepared from the blends exhibited high yield stresses of 440 and 550 MPa after compression, and they did sustain work hardening. Tests conducted before and after compression up to 50 % deformation showed the same relative amount of hardness increase around 20 %, which was three times lower than that of the monolithic sample for which a decrease of the average grain size close to 26 % was measured.

1. Introduction

Powder-metallurgy-based methods are versatile routes for producing in-demand microstructures of various types ; centimetre-sized and fully dense, ultrafine-grained as well as nanostructured materials can indeed be processed in respect to the bottom-up approach through techniques such as hot isostatic pressing (HIP) [1] or differential hydrostatic extrusion combined to cold isostatic pressing (CIP) and sintering [2]. Those methods are therefore of prime interest for addressing the dual problem of improving both the strength and the ductility of these materials. Different strategies have yet been proposed to offset the low tensile ductility accompanying their high flow stress [3] ; one of them deals with the incorporation of coarser grains, which are prone to store dislocations, in a fine-grained matrix. The bimodal grain structure can be achieved by heterogeneous grain growth in severely deformed materials ; that way is already investigated for Ni [4]. Another way to produce such structures, which has been scarcely studied yet [5], relates to the sintering of blends of nanocrystalline and microcrystalline powders ; the present work enters that scope and focuses on the transformation of high-purity Ni powder blends of controlled volume fractions of nanometre-sized and micrometre-sized particles into bulk samples.

2. Experimental procedure

The experimental procedure has consisted in previously identifying the characteristics of each subsequent-blend component under the whole range of CIP and in processing the samples through the mixture of two starting powders, their compaction by CIP, their sintering, and their characterization along with the one of a reference sample submitted to the common process ; characterizations have consisted in testing the sintered samples under compression and in proceeding, both before and after compression, with the last two items of the following detailed account of the experimental procedure :

- The starting nickel powders were commercial high-purity powders : the micrometre-sized powder was an Alfa Aera spherical Ni powder APS 3-7 μm of 99.9 % purity and the nanometre-sized powder was an Argonide Corporation spherical Ni powder with an average grain size of 100 nm and a 99.9 % purity, synthesized by electro-explosion of wires. The latter powder had previously been internally characterized in a separate earlier work, including the determination of its chemical composition by inductively coupled plasma spectroscopy [6].
- Micrometre-sized and nanometre-sized powders have been mixed in a three-dimensional shaker mixer (Turbula® type T2F) for 6 h at room temperature (RT) with two different controlled volume fractions (40 and 60 %).
- CIP has been performed on laboratory equipment specifically designed and built (multi-ring HP cylinder reinforced by shrinkage) for containing fluid pressures up to 2 GPa at RT.
- Sintering has reproducibly been performed under 500 °C during 1 h.
- Compression tests have been carried out on cylindrical specimens having a diameter of 5 mm and a length of 6.5 mm with an Instron universal testing machine (model 1195) at a strain rate of ~ 10^{-4} s^{-1}. Deformation has been exerted up to 50 % for all the tested specimens.
- Vickers hardness measurements have been performed with a Struers microhardness tester (Duramin 20 model).
- The as-processed microstructures have been investigated by electron backscattering diffraction (EBSD) conducted on a conventional LEO S360 scanning electron microscope. Post-data treatment was carried out by using the OIM version 4 software from TexSEM Laboratories.

3. Results and discussion

CIP constitutes a consolidation step during which most of the densification shall take place prior to any further ultimate-densification processing route such as HIP, and it therefore requires special attention. Its results are illustrated for each of the subsequent-blend components by figure 1 according to the Shapiro-Kolthoff's plot primarily adopted for powder die compaction [7]. This plot is consistent with the integration of the differential equation that Heckel developed by assuming similarity to a first-order chemical reaction with porosity and pressure substituting for concentration and time respectively [8] :

$$\frac{dD}{dp} = \mathrm{K}(1 - D) \tag{1}$$

where D is the relative green density, p the applied pressure and K a constant. Equation (1) stands for the underlying assumption that the rate of change in density with respect to pressure is directly proportional to the remaining porosity. By integration, it gives

$$\ln\frac{1}{1-D} = \mathrm{K}p + \mathrm{A} \tag{2}$$

where A is a constant. It is agreed that such a simple model as equation (2) can for the best fit the data in either the initial or the final stage of the densification process ; Heckel claimed on his own that constant K should relate to plastic deformation for ductile materials whereas constant A would represent the close packing achieved at low pressures from a loose array as a result of rearrangement before appreciable amounts of inter-particle bonding occur. Although continuum and mechanistic modelling approaches have introduced many more equations that fit compaction data better under particular circumstances [9], the Shapiro-Kolthoff's plot remains phenomenologically convenient for fitting at once the consecutive CIP stages of consolidation through rearrangement and deformation that actually give rise to two distinct linear-relationship domains as yet observed for die-compacted Fe powders for instance [10]. Figure 1 also emphasizes the two distinct "compactibilities" of the starting powders depending on their average grain size ; analyses had shown that the micrometre-sized powder corresponds to spherical clusters of particles whose actual average size is 544 nm. It is worth noting that, despite a lower "compactibility", the nanometre-sized powder sample CIPed at 2 GPa attained a relative green density of 92 %.

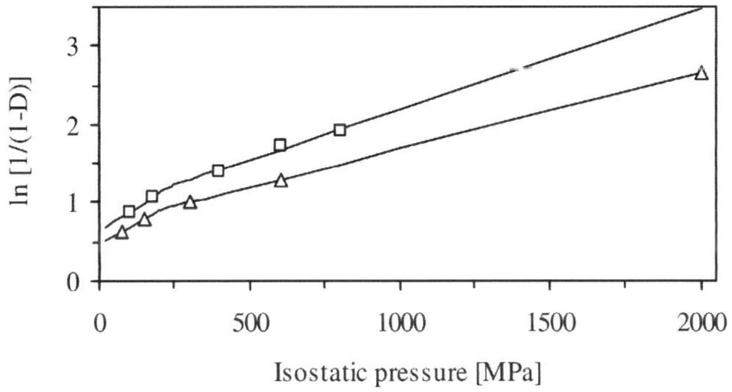

Figure 1. Shapiro-Kolthoff's plot of relative green density D versus isostatic pressure applied for compacting micrometre-sized (squares) and nanometre-sized (triangles) Ni powders.

In order to limit the number of parameters used for this experiment, CIP has been performed at 1.5 GPa only for processing all the samples along with sintering. Table 1 summarizes the main characteristics of the resulting bulk samples that were texture-free as confirmed by the random EBSD-analysed grain orientations and structurally homogeneous as confirmed by the quite uniform hardness distributions, and that exhibited different grain size distributions as a consequence of the dislocations induced during CIP in the second linear-relationship domain and of their arrangement into large-angle boundaries during sintering. In regard to residual porosity, deformation was appropriately limited to compressive testing ; the resulting main characteristics are summarized in table 2.

Table 1. Characteristics of the samples as sintered prior to compression.

Composition ([a])	Relative density (%)	Vickers hardness	Grain size[b] (nm)
0:100	94.62	125 ± 4	848
40:60	93.10	186 ± 8	623
60:40	94.12	204 ± 19	579

[a] Initial nanometre-sized to micrometre-sized powder volume percentage ratio.
[b] Average value based on the EBSD analysis of the grain size distribution.

Table 2. Characteristics of the samples as deformed after compression.

Composition	Yield stress (MPa)	Vickers hardness	Grain size (nm)
0:100	250	212 ± 10	629
40:60	440	225 ± 9	589
60:40	550	240 ± 6	568

The mechanical properties are in direct line with the seminal Hall-Petch relationship which states that the flow stress, which can be related to hardness under certain circumstances, increases when the grain size of metallic polycrystals decreases [11]. The microstructure-dependent yield stress itself increases significantly and is accompanied by an impressive deformation whereas a reduced strain hardening contributes, for the samples prepared from the blends, to a strength level finally similar to the one of the reference sample. Figure 2 emphasizes the respective evolutions of hardness and grain size in regard to initial blend composition. Analyses conducted before and after compression show relative amounts of hardness increase of 21.6 and 17.7 % for the samples prepared from the blends but no significant grain size reduction, compared to a 69.6 % increase and a 25.8 % reduction for the monolithic sample respectively, which also denote an enhanced structural stability.

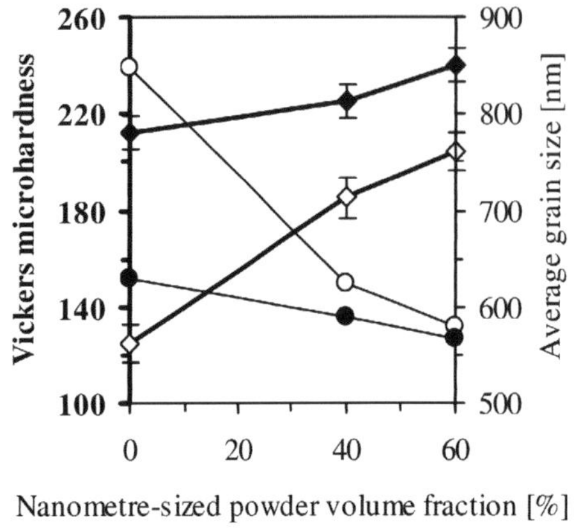

Figure 2. Influence of initial Ni blend composition on hardness (diamonds) and grain size (circles) of sintered samples before (open symbols) and after (filled symbols) compression.

4. Concluding remarks

Samples prepared from the blends exhibited high yield stresses of 440 and 550 MPa after compression, depending on the grain size distribution, and did sustain work hardening. Analyses conducted before and after compression also denoted through hardness and grain size an enhanced structural stability. Yet, ductility should be evaluated through tensile testing that requires full densification. The optimization of the process is therefore underway, either by varying the presently fixed parameters or by combining HIP or differential hydrostatic extrusion to CIP, and remains aimed at getting a better trade-off between strength and ductility.

Acknowledgments

The authors are grateful for the support provided by French ANR research funding agency through Blanc 2009 programme to the MIMIC project.

References

[1] Billard S, Fondère J-P, Bacroix B and Dirras G F 2006 *Acta Mater.* **54** 411-21
[2] Langlois C, Hÿtch M J, Langlois P, Lartigue-Korinek S and Champion Y 2005 *Metall. Mater. Trans. A* **36** 3451-60
[3] Wang Y M and Ma E 2004 *Acta Mater.* **52** 1699-709
Fedorov A A, Gutkin M Y and Ovid'ko I A 2002 *Scripta Mater.* **47** 51-5
[4] Zhao Y H, Topping T, Bingert J F, Thornton J J, Dangelewicz A M, Li Y, Liu W, Zhu Y T, Zhou Y Z and Lavernia E L 2008 *Adv. Mater.* **20** 3028-33
[5] Wang Y, Chen M, Zhou F and Ma E 2002 *Nature* **419** 912-5
Witkin D, Lee Z, Rodriguez R, Nutt S and Lavernia E 2003 *Scripta Mater.* **49** 297-302
[6] Gubicza J, Bui H Q, Fellah F and Dirras G F 2009 *J. Mater. Res.* **24** 217-26
[7] Shapiro I and Kolthoff M 1947 *J. Phys. Colloid Chem.* **51** 483-93
[8] Heckel R W 1961 *Trans. Metall. Soc. AIME* **221** 671-5
Heckel R W 1961 *Trans. Metall. Soc. AIME* **221** 1001-8
[9] Piehler H R and DeLo D P 1997 *Prog. Mater. Sci.* **42** 263-76
Panelli R and Ambrozio Filho F 2001 *Powder Technol.* **114** 255-61
[10] Dominguez O, Phillippot M and Bigot J 1995 *Scripta Metall. Mater. Sci.* **42** 263-76
[11] Hall E O 1951 *Proc. Phys. Soc. London* **B-64** 747-53
Petch N J 1953 *J. Iron Steel Inst.* **174** 25-8

Journal of Physics: Conference Series **240** (2010) 012017 doi:10.1088/1742-6596/240/1/012017

Investigation of the role of grain boundary on the mechanical properties of metals

Nousha Kheradmand, Afrooz Barnoush and Horst Vehoff
Saarland University, P. O. Box 151150, D-66041 Saarbrücken
n.kheradmand@matsci.uni-sb.de

Abstract. Compression testing of micropillars was used to investigate the gain boundary effect on the strength of metals which is especially interesting in ultra fine grained and nanocrystalline metals. Single and bicrystal micropillars of different sizes and crystallographic orientations were fabricated using a focused ion beam system and the compression test was performed with a nanoindenter. A reduction of the pillar size as well as the introduction of a grain boundary results in an increase in the yield strength. The results show that the size and the orientation of different adjoining crystals in bicrystalline pillars have an obvious effect on dislocation nucleation and multiplication.

1. Introduction

The mechanical properties investigation of metals by microindentation shows that the hardness increases by decreasing the grain sizes. However, the trend of this change for grain sizes between 20 nm and 1 μm is altering [1]. This clearly reveals the local interaction between dislocations and grain boundaries in small grain materials. In this range strain rate sensitivity is also observed [2]. Yang et al [3] used nanoindentation to examine the role of dislocations-grain boundary interaction in this grain size range. They indented within only one grain in order to investigate the interaction between grain boundary and dislocations. Their results provided the interaction of dislocations with all grain boundaries surrounding the indented grain. In this paper we introduce a novel method based on micropillar compression testing [4], to investigate the interaction between dislocations and a specifically chosen single grain boundary.

2. Experiments

Micropillars were cut in a coarse grained pure nickel (99.99%) sample which was prepared according to the procedure described in [5]. The Orientation Imaging Map (OIM) shown in figure 1a used to choose the arbitrary orientations, in which single crystalline (figure 1c) and bicrystalline (figure 1e) pillars were cut using a Strata™ dual beam focused ion beam (FIB) system manufactured by FEI. These pillars, with different diameters, were compressed in a Hysitron® TriboIndenter™ nanoindentation system equipped with a diamond flat punch tip. The compression tests were performed in two ways: load control and displacement control, where a constant loading rate or strain rate was applied, respectively.

3. Results and discussion

3.1. Single crystalline micropillars (SCM)

The stress-strain curves of $\approx$ [153] oriented micropillars with diameters ≈ 7 and ≈ 3 μm are shown in figures 2a and 2b. The curves show following differences: 1) the larger pillar has a continuously elastic to plastic deformation. The smaller pillar, however, shows an elastic deformation followed by several strain bursts (pop-in) at constant loads. 2) The larger sample has a lower strength in comparison to the smaller micropillar which is in good agreement with previous observations [4, 6, 7].

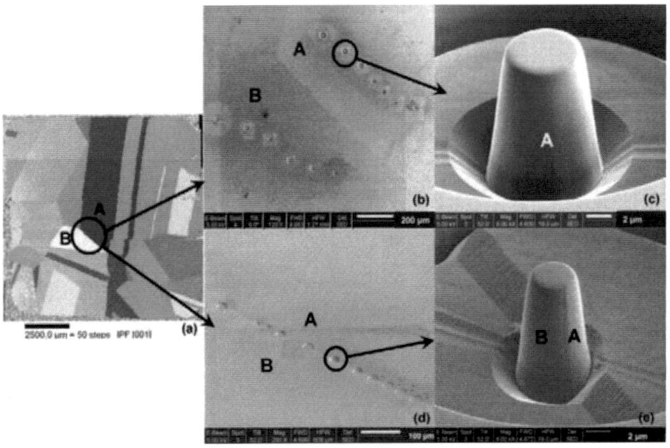

Figure 1: a) OIM of the coarse grained Ni sample used for producing micropillars. The orientations of grains are marked on the inverse pole figure. b,d) SEM images of grains (A and B) and the grain boundary, where the micro SC and BC pillars are produced. c,e) SEM images of SC and BC pillars. The SC Pillar is cut inside one grain (e.g. grain A) and BC pillar on the grain boundary between grains A and B.

A new approach in this work was the stepwise compression test. This means that the stress-strain curves are not resulted from one single compression test but they are sum of consecutively performed compression tests on a pillar with increasing maximum load in the load function. Between each step the sample was observed in SEM, as shown in figure 2b, which provided more detailed information from intermediate steps of pillar deformation. Additionally by means of stepwise loading it was possible to observe the plastic deformation initiation in micropillars.

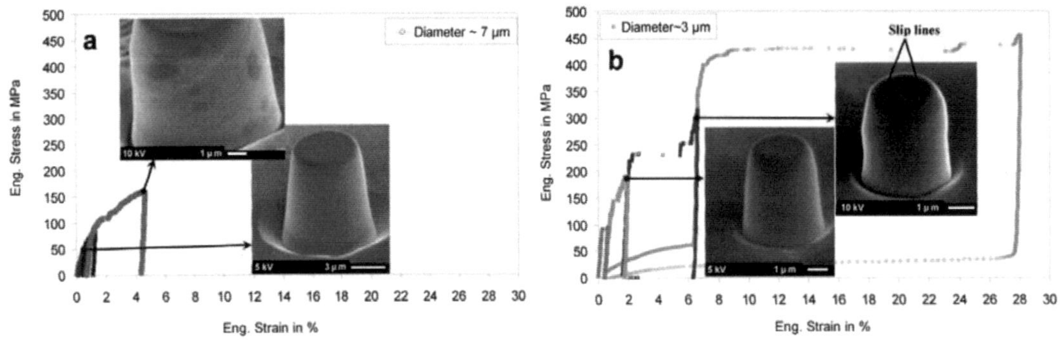

Figure 2: The stress-strain curves of SCMs of different sizes a)≈ 7 μm and b)≈ 3 μm with the SEM micrograph taken in intermediate steps of compression test.

3.2. Birystalline micropillars (BCM)

As mentioned the aim of this work was the investigation of the grain boundary effect on the plastic deformation of fine crystalline metals. Therefore the compression tests were performed on BCM's with different diameters in the range of 1 to 7 μm. The stress-strain curves for two different BCM's with ≈ 1 and ≈ 3 μm diameters are shown in figure 3. The observations show, as in the case of the SCM, by decreasing the size of the BCM the strength increases.

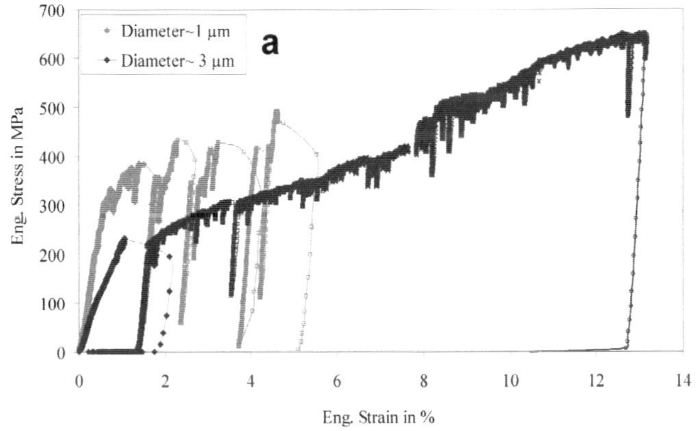

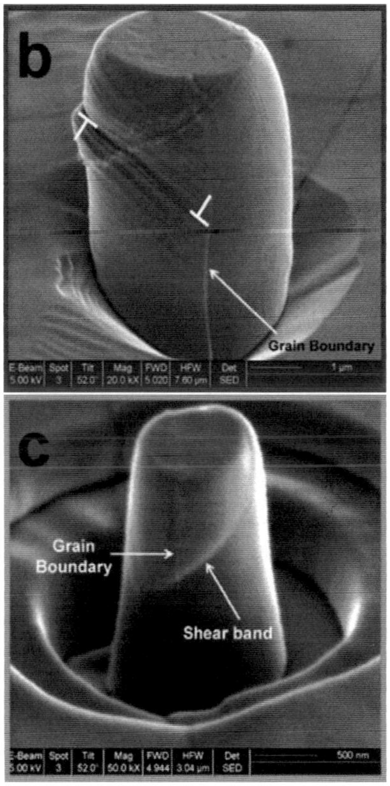

Figure 3: a) Stress-strain curves of BCM's with different sizes. b) Post-compression SEM image of a ≈ 3 μm BCM, a dislocation loop is shown schematically. One side of the loop reaches the free surface, which results in formation of steps. The other side of the loop reaches the grain boundary, which results in the pile-up of dislocations and bowing out of the grain boundary. c) Post-compression SEM image of a ≈ 1 μm BCM. The pillar deforms as a SCM and shear bands cross the grain boundary.

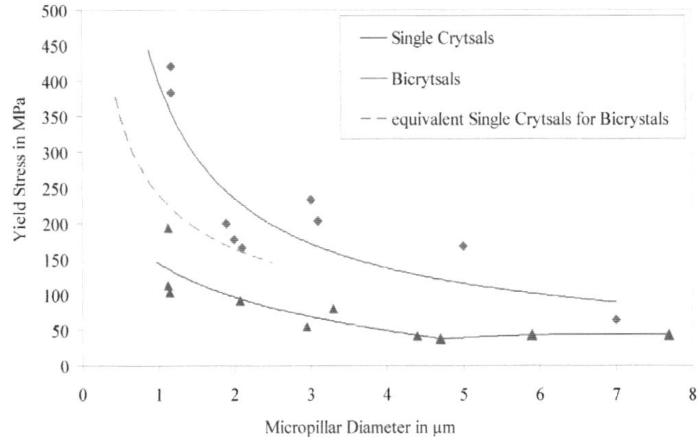

Figure 4: Yield stress vs. micropillar diameter of SCMs and BCMs. The curves show two parameters affecting the yield stress: 1) the pillar size and 2) the grain boundary in BCMs.

It is worth mentioning that the compression tests presented in figure 3 were performed under displacement control at a constant strain rate. Therefore, instead of pop-ins, as in the load control compression tests (Figure 2), here load drops are observed. Figure 4 summarizes the results of the compression test of all BCM's and SCM's examined in this study. It shows that there are two important parameters affecting the yield stress of pillars: 1) the pillar diameter and 2) the grain boundary. An explanation to the first effect can be found in [8]. A decrease of the pillar's diameter leads to a lower dislocation density due to the loss of dislocations through the free surface image stresses. The grain boundary effect is more complicated compared to the size effect. First of all the BCM's can be considered as two single crystals adjoining each other. In the first order approximation we can consider the pillars elastic isotropic and write the relation 1;

$$\epsilon_{BCM} = \epsilon_1 = \epsilon_2 \Rightarrow \sigma_{BCM} = \sigma_1 = \sigma_2 \tag{1}$$

where are the strains and are the stresses in BCM, grain 1 and grain 2, respectively. It means that by ignoring the effect of the grain boundary a BCM can be equivalated to two SCM with the following relation

$$d_{BCM} = d_{SCM}\sqrt{2} \tag{2}$$

where and are the diameters of BCM and equivalent SCM's, respectively. Since the Schmid factors of two grains are different, the one with a higher Schmid factor will tend to yield earlier. Therefore, the measured yield stress can be associated with the grain with the higher Schmid factor i.e. equivalent SCM containing this gain. Therefore, we can expect a shift to the left for BCM's yield stress curve in figure 4 by considering the BCM's as two single crystals adjoining each other. This is shown in figure 4 by the dash line "equivalent SCM for BCM". In spite of this shift, the curve dose not overlap with the curve belonging to the single crystals. It shows the strengthening effect of grain boundary for small sizes. The other interesting observation during the compression tests on BCM's was the difference in their plastic deformation behaviors as their sizes changed. As shown in post-compression SEM image of $\approx 3\ \mu$m BCM, the grain boundary was deformed under the loading. One can imagine that, the plastic deformation initiates in the grain with the higher Schmid factor (left side grain of BCM in figure 3b by the activation of a Frank-Read source. The activated dislocation loop starts to grow by further increasing of the load. One side of the dislocation loop reaches the free surface and results in the formation of steps, while the other side of the loop reaches the grain boundary and results in the pile-up of dislocations behind the grain boundary. Thus by increasing the load the pile up force on the grain boundary increases and results in the bowing of the grain boundary as shown in figure 3b. In the $\approx 1\ \mu$m BCM shown in figure 3c, however, the grain boundary seems to be absent and the whole pillar deforms like a single crystalline pillar and shear bands crosses the micropillar. This effect is still not fully understood and further experiments are required to clarify this size effect in small BCM's.

References

[1] Yang B and Vehoff H 2005 *Mater. Sci. Eng., A* **400-401** 467–470
[2] Vehoff H, Lemaire D, Schler K, Waschkies and Yang B 2007 *Int. J. Mat. Res.* **98** 259
[3] Vehoff H, Yang B, Barnoush A, Natter H and Hempelmann R 2008 *Z. Phys. Chem.* **222** 499–525
[4] Uchic M D and Dimiduk D M 2005 *Mater. Sci. Eng. A* **400-401** 268–278
[5] Barnoush A and Vehoff H 2008 *Corros. Sci.* **50** 259–267
[6] Maaß R, Van Petegem S, Van Swygenhoven H, Derlet P, Volkert C and Grolimund D 2007 *Phys. Rev. Lett.* **99** 145505
[7] Greer J R, Oliver W C and Nix W D 2005 *Acta Mater.* **53** 1821–1830
[8] Volkert C A and Lilleodden E T 2006 *Phil. Mag.* **86** 5567–5579

Strain localization in <111> single crystals of Hadfield steel under compressive load

E G Astafurova, G G Zakharova, E V Melnikov

Institute of Strength Physics and Materials Science, Siberian Branch of Russian Academy of Sciences, 634021, Akademichesky pr. 2/4, Tomsk, Russia

astafe@ispms.tsc.ru

Abstract. A study of strain localization under compression of <111> Hadfield steel single crystals at room temperature was done by light and transmission electron microscopy. At $\varepsilon<1\%$, macro shear bands (MSB) form that have non-crystallographic and complex non-linear habit planes and are the results of the interaction of dislocation slip on conjugate slip planes. Mechanical twinning was experimentally found inside the MSB. After the stage of MSBs formation, deformation develops with high strain hardening coefficient and corresponds to interaction of slip and twinning inside as well as outside the MSBs.

1. Introduction

Hadfield manganese steel is well known for its anomalous high work hardening, ductility, and strength [1–3]. It has austenitic structure (FCC) and low stacking fault energy (γ=23mJ/m^2) for a nominal composition of Mn 12-14 wt.% and C 1.0-1.3 wt.%. This combination makes steel successful for industrial applications such as impact hummers, railroad frogs, etc. The polycrystalline Hadfield steel is well-studied, but scientists have not got a single meaning on its hardening mechanism.

In the recent works, Hadfield steel single crystals were studied well under tension, and unusual strain hardening of this steel was attributed to pronounced twinning and strain ageing phenomenon [4,5]. The present paper examines the deformation macro- and microstructures of the <111>-oriented single crystals of Hadfield steel under compressive load to provide further development of our knowledge about their deformation mechanisms under different strain paths.

2. Methods

The material used in this study was the austenitic Hadfield steel with a composition of 13%Mn, 1.0%C in weight and Fe balance. The single crystals were grown by Bridgman technique in a He atmosphere. They were then homogenized for 24 h at 1373K. Electro-discharge machining was utilized to cut regular parallelepiped specimens with nominal dimensions of 4×4×7 mm. All specimens were mechanically ground, chemically etched to remove damaged surface layer. They then were solution-treated and water-quenched from 1373K after 1h. Before testing all samples were electrolytically polished. The compression test was performed with electromechanical Instron 3369 test machine at room temperature and a strain rate of 1.2×10^{-3} sec^{-1}. Experiments were repeated on five companion specimens to check repeatability. Olympus GX-71 microscope was utilized for optical analysis. For TEM investigations, the foils were prepared by mechanical grinding and jet electropolishing. Microstructure was examined in Philips CM 30 and Philips SEM 515 electron microscopes.

3. Results and Discussion

The microstructure and texture evolution of Hadfield steel under tension and compression at room temperature are determined mainly by mechanical twinning intensity and by a number of operating twin and slip systems [1-5]. The initial deformation mechanism of Hadfield steel single crystals of all orientations is slip, but it changes fast to twinning as the Hadfield steel possesses low stacking fault energy and high level of plastic flow stresses due to high carbon content [4,5]. Plastic strain corresponding to slip-to-twin transition depends on crystal axis orientation, and, in case of compressive load, it has the smallest value (ε<0,5%) in the single crystals oriented close to <001>-direction. Under compression of <111>-oriented crystals, slip can be accompanied by twinning but, nevertheless, it has to be the basic deformation mechanism.

The experimental true stress-strain curve of the <111>-oriented single crystal as presented in Figure 1 clearly indicates both a non-monotonic increase in stress during straining and non-homogeneous deformation of a specimen. The first part of the stress-strain curve, up to the strain of 15%, represents the work-hardening stage with serrated flow, stress drops and low strain hardening coefficient of $\theta=d\sigma/d\varepsilon$=800MPa. Stress drops are associated with the development of macroscopically visible shear bands (Fig. 1 a,b). After moderate strain, i.e. at the end of macro shear bands (MSB) formation, these bands occupy practically the whole volume of the crystal (Fig. 1 c). The second part of stress-strain curve is smooth, it has the high strain hardening coefficient of 1800MPa and is associated with deformation both inside and outside of MSBs.

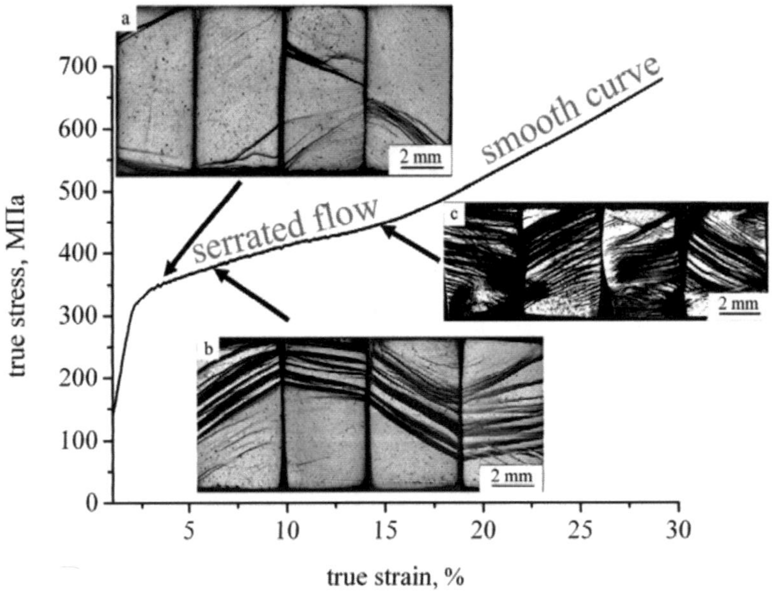

Figure 1. The stress-strain curves and the optical images of successive stages of the MSB formation on the sample scale. (a-c) crystal compressed to plastic strains of (a) 1,8% (b) 3,4% (c) 11,6%

An appropriate model for MSB boundary nucleation under multiple slip due to the formation of Lomer-Cottrell locks is proposed in previous papers [6,7]. On the basis of this model and our own experimental data, we assume the following stages of the <111> single crystals deformation. First, very low strains are controlled by multiple slip, as the <111> orientation is highly symmetrical one and the slip has homogeneous character (without pile-ups) [5]. The interaction of dislocations gliding on intersecting planes produces continuous non-crystallographic surfaces which consist of Lomer-Cottrell locks. These surfaces form habit planes for subsequent MSBs and do not coincide with {111}-type planes that are slip planes of fcc materials (Fig. 2a).

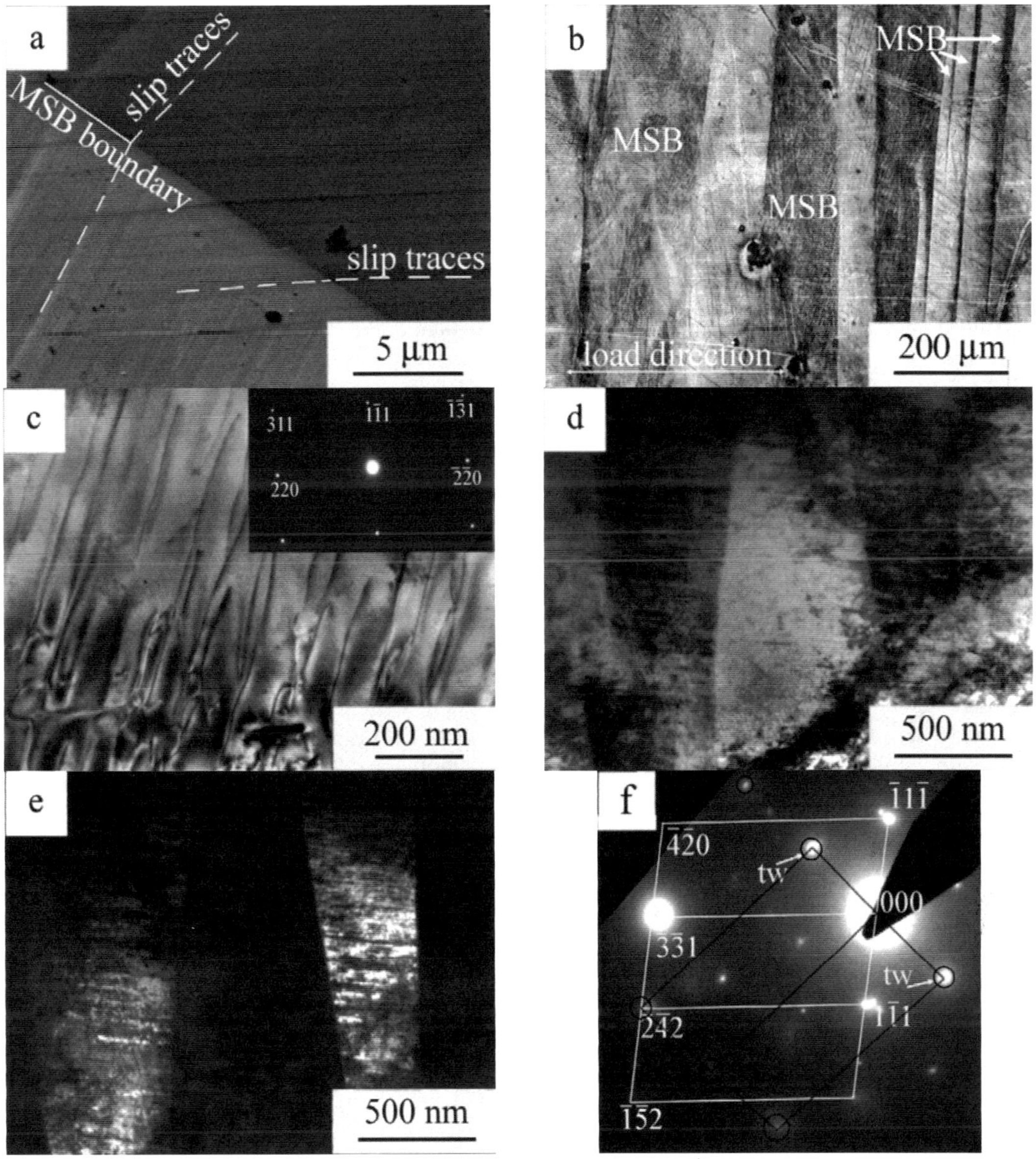

Figure 2. The micro- and macro structures of the MSB in compressed <111> single crystal: (a) SEM micrograph of slip lines on the MSB boundary, ε=11,6%; (b) optical image of singe crystal face side after repolishing and etching, ε=11,6%; (c,d) TEM bright field images showing dislocation structure and micro twins inside of the MSB, foils were cut in parallel with the MSB macrohabit plane, ε=5%; (e,f) TEM dark field image obtained in twin reflex and SADP to (d,e);

The macroscopically observed broad bands of 20-300μm (Figs. 1a,b; 2b) do occur in all specimens after the plastic strain of ε>0,5% and coincide with periodic stress relaxations. The decrease in stress (stress drop) is determined by softening mechanism inside the band due to lattice reorientation. The intensive slip inside MSBs causes crystal lattice rotation up to 10° compared to the matrix region as revealed by EBSD analysis. Both slip (Fig. 2 c) and twinning were found (Fig. 2 d-f) inside the MSB. It means that the lattice reorientation makes twinning possible inside the band, and following increase in stress after each drop is related to the interaction of slip and twinning inside the band. So, the inner structure of the band is getting stronger than that of surrounding material regions, and the next MSB forms. Although the curve is not smoothed, the amplitude of jerky flow is only a small fraction of the stress level. The stress oscillations cease at strain of 15%, and further deformation occurs in the whole specimen. This process can be easily observed both by optical microscopy and SEM.

No strain localization into MSBs has been found under compression of the Hadfield steel single crystals oriented along other crystallographic orientations (for instance, <001>, <123>, <011>) as twinning or single shear deformation starts from the very beginning of plastic flow. Moreover, change in type of dislocation structure from homogeneous dislocation distribution to planar dislocation configuration suppresses the MSB formation as it is observed during compression of the <111>-oriented specimens of Hadfield steel alloyed with aluminum. This also confirms the hypothesis that, for well-developed MSBs which form during compression in Hadfield steel single crystals, the multiple homogeneous slip is necessary.

4. Conclusions

In the present paper, <111> single crystals of conventional Hadfield steel (Fe-13Mn-1.0C, wt.%) were studied in order to shed light on the mechanisms of MSB formation under compressive load. The results obtained may be summarized as follows:

1. Shear banding is closely related to the structural anisotropy of pre-existing deformation mechanism. It is preceded by the formation of persistent non-crystallographic (shear habit plane is not a {111} plane) boundaries due to gliding dislocation interactions (Lomer-Cottrell locks). Formation of shear bands is strongly assisted by multiple slip, whereas, twinning, single slip or microlocalization (pile-ups of dislocations) will suppress MSB formation.
2. Rotation-induced mechanical instability within MSB leads to local reorientation of crystal lattice inside of the band and favors mechanical twinning in it.

Acknowledgments

The authors wish to thank Prof. Yu.I. Chumlyakov for his contributing discussions. This research was supported by the Grant of RF President (MK-1436.2008.8) and RFBR (project No.07-08-00064).

References

[1] Adler, P.H., Olson, G.B., Owen, W.S., 1986. *Met. Trans. A* **17A**: 1725-37.
[2] Raghavan K.S., Sastri A.S., Marcinkowski M.J., 1969. *Trans. Met. Soc. AIME* **245**: 1569-75.
[3] Dastur Y.N., Leslie W.C., 1981. *Met. Trans. A,* **12A**: 749-59.
[4] Karaman I., Sehitoglu H., Beaudoin A.J. et al., 2000. *Acta Mater.* **48**: 2031-47.
[5] Astafurova E.G., Kireeva I.V., Chumlyakov Yu.I., et al., 2007. *Int. J. Mat. Res.* **98**(2):144-49.
[6] Chang Y.W., Asaro R.J., 1981. *Acta. Metall.* **29**: 241-57.
[7] Dao M., Asaro R.J., 1994. *Scripta Mater.* **30**: 791-96.

Study of thermally activated dislocation motion in AJ51 and AE42 magnesium alloys

P Lukáč and Z Trojanová

Department of Physics of Materials, Faculty Mathematics and Physics, Charles University Prague, Ke Karlovu 5, CZ-121 16 Praha 2, Czech Republic

lukac@met.mff.cuni.cz

Abstract. Stress relaxation tests along stress-strain curves of two magnesium alloys AJ51 and AE42 are used for determining the activation volume at different test temperatures. The applied stress has been separated into its effective stress and internal stress components. The variations of the activation volumes with the effective stress are examined.

1. Introduction

Improvement of mechanical properties of Mg-Al based alloys has attracted many studies. Third alloying elements are used to improve the properties of the alloys. Among the alloying elements, rare earth elements, Ca, and Sr are promising additions. They also improve high temperature properties of the alloys. However, only a few studies have been reported where dislocation mechanisms during deformation of Mg alloys were clarified.

It is widely accepted [1] that the resolved shear stress necessary for dislocation motion in the slip plane can be divided into two components. In polycrystals, the resolved shear stress is related to the applied stress, σ. The same is valid for the components. Then:

$$\sigma = \sigma_i + \sigma^*, \tag{1}$$

where σ_i is the (internal) athermal contribution to the stress, resulting from long-range internal stresses impeding the plastic flow. The effective shear stress σ^* acts on dislocations during their thermally activated motion when they overcome short range obstacles. The plastic strain rate $\dot{\varepsilon}$ for a single thermally activated process can be expressed as:

$$\dot{\varepsilon} = \dot{\varepsilon}_0 \exp\left[-\Delta G\left(\sigma^*\right)/kT\right] . \tag{2}$$

Here $\dot{\varepsilon}_0$ is a pre-exponential factor, T is the absolute temperature and k is the Boltzmann constant. $\Delta G(\sigma^*)$ is the change in the Gibbs free enthalpy depending on the effective stress $\sigma^* = \sigma - \sigma_i$ and the simple form is

$$\Delta G(\sigma^*) = \Delta G_0 - V\sigma^* = \Delta G_0 - V(\sigma - \sigma_i). \tag{3}$$

Here ΔG_0 is the Gibbs free enthalpy necessary for overcoming a short-range obstacle without the applied effective stress and V = bdL is the activation volume where d is the obstacle width and L is the mean length of dislocation segments between obstacles. It should be mentioned that L may depend on the stress acting on dislocation segments. The stress relaxation technique has been demonstrated to be a useful method for estimating the activation volume. In a stress relaxation (SR) test, the specimen is deformed to a certain stress σ_0 and then the testing machine is stopped and the stress is allowed to relax. The stress decreases with time *t*. The specimen can be again reloaded to a higher stress (load) and the SR test may be repeated. The stress decrease with time during the SR can be described by the well known Feltham equation [3]:

$$\Delta\sigma(t) = \sigma(0) - \sigma(t) = \alpha \ln(\beta t + 1)\,, \tag{4}$$

where $\sigma(0) \equiv \sigma_0$ is the stress at the beginning of the SR at time $t = 0$, β is a constant and

$$\alpha = \frac{kT}{V}\,. \tag{5}$$

The aim of this work is to use the SR tests during plastic deformation of AJ51 and AE42 magnesium alloys at different temperatures in order to obtain the values of the activation volume and to identify thermally activated process(-es) responsible for the plastic deformation.

2. Experimental procedure

Magnesium alloys AJ51 (Mg-5Al-0.6Sr) and AE42 (Mg-4Al-2Nd) were used. Samples for tensile tests of a cylindrical form with a diameter of 5 mm and a gauge length of 25 mm were deformed in an INSTRON 5882 machine at a constant cross head speed giving an initial strain rate of $6.7\text{x}10^{-5}\text{s}^{-1}$ over a wide temperature range of 23 to 300 °C. Duration of the SR was 300 s. Components of the applied stress (σ_i, σ^*) were estimated using Li's method [4, 5]. The SR curves were fitted to the power law function in the form: $\sigma - \sigma_i = [a(m-1)]^{\frac{1}{1-m}}(t+t_0)^{\frac{1}{1-m}}$, where a, t_0 and m are fitting parameters. This relation was derived based on dislocation dynamics (a power low between dislocation velocity and stress) assuming both the internal stress and the density of mobile dislocations are constant during the SR test.

3. Results and discussion

A part of the AE42 true stress-true strain curve at 25 °C (points indicate the stresses at which the SR tests were performed) is shown in figure 1. The internal stress and effective stress are shown as functions of strain. It is obvious that the internal stress σ_i forms the substantial contribution to the applied stress σ_{ap}. Analogous dependences obtained at elevated temperatures show that the internal stress decreases with increasing temperature. Similar results were also obtained in the case of some Mg-Al-Ca and Mg-Al-Sr alloys [6, 7].

The apparent activation volumes (in b^3) estimated for AJ51 samples deformed in tension and compression at various temperatures are shown in figure 2 as a function of the effective stress σ^*. From figure 2 it can be seen that all values of V/b^3 lie at one line, "master curve", done by the equation:

$$V = kT\frac{\partial \ln \dot{\varepsilon}/\dot{\varepsilon}_0}{\partial \sigma^*} = \frac{\Delta G_0 pq}{\sigma_0^*}\left[1-\left(\frac{\sigma^*}{\sigma_0^*}\right)^p\right]^{q-1}\left(\frac{\sigma^*}{\sigma_0^*}\right)^{p-1}, \tag{6}$$

where p and q are phenomenological parameters reflecting the shape of a resistance profile and σ_0^* is the effective stress at 0 K. The fit of the experimental values of equation (6) gives for the activation enthalpy $\Delta G_0 = 0.95 \pm 0.05$ eV for the AJ51 alloy. It should be mentioned that in the case of other Mg-Al-Sr alloys, all values of the apparent activation volume lie on the master curve [6, 7].

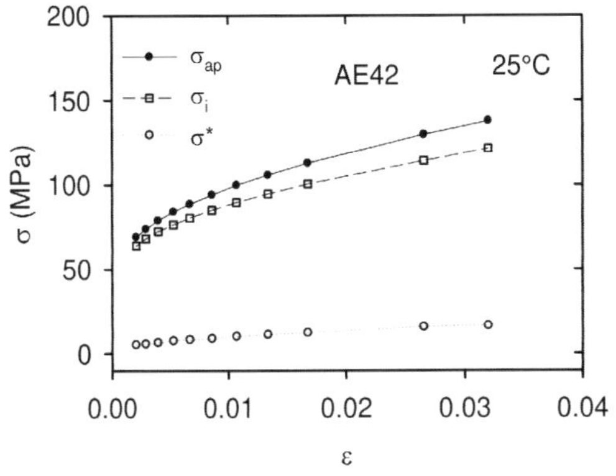

Figure 1. A part of the true stress true strain curve. Points at the curve indicate the points at which the SR tests were performed.

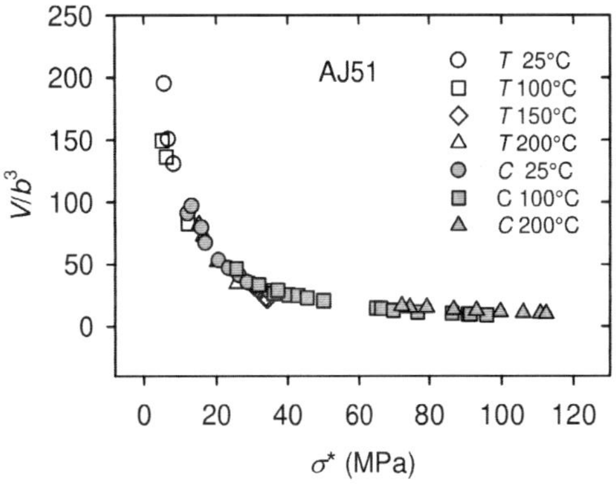

Figure 2. Plot of the apparent activation volume against σ^* estimated for four deformation temperatures in tension (*T*) and compression (*C*) for the AJ51 alloy [7].

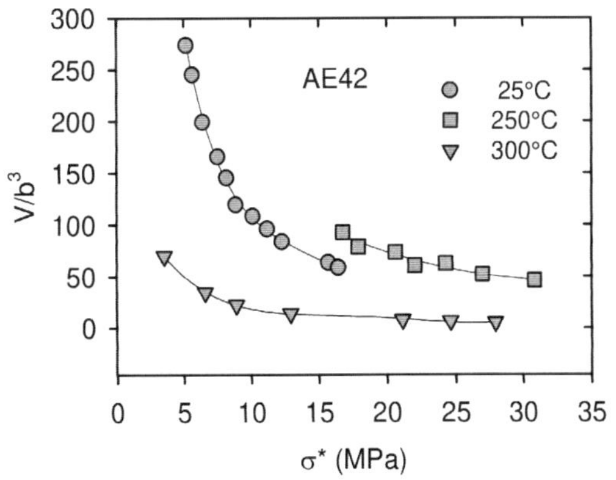

Figure 3. The apparent activation volume depending on the effective stress, estimated for three temperatures.

The apparent activation volumes for AE42 estimated at three temperatures are plotted against the thermal stress in figure 3. It can be seen that three different curves were obtained in this case. Equation (6) is not fulfilled in the case of the AE42 alloy. This is probably due to strain ageing at intermediate temperatures. Strain ageing phenomena (serrated yielding, post-relaxation effect,) were also observed in some Mg alloys containing rare earth [8]. The values of an increase in the flow stress $\Delta\sigma$ after the SR test observed at different strains are plotted against temperature in figure 4. It can be seen that the maximum of the ageing effect is at a temperature of 150 °C. $\Delta\sigma$ is zero at temperatures lower than 50 °C and higher than 250 °C. In an alloy, the flow stress may be considered as a sum of two contributions:

$$\sigma = \sigma_f + \sigma_d \tag{10}$$

with σ_f relating to a friction imposed by the solutes-dislocation interaction and σ_d relating to the dislocation-dislocation interaction [9, 10]. Hong [11, 12] have shown that irrespective of the mechanisms of dynamic strain ageing, the frictional stress σ_f could be approximately described by the following equation:

$$\sigma_f = \alpha_2 Gb\delta c \exp\left(-\frac{(T-T_0)^2}{B} \right), \tag{11}$$

where α_2 is a constant, δ is the atomic size misfit parameter, G is the shear modulus, c is the solute concentration and B is the width of the statistical normal distribution about the temperature T_0 where the maximum of the solute–dislocation interaction force occurs. This temperature was found, in the case of AE42 alloy, to be in the vicinity of 150 °C. At this temperature the interaction of mobile solute atoms with moving dislocations may reach its maximum value. The isolated solute atoms are not significant obstacles for the thermally activated dislocation motion at higher temperatures. They may diffuse to the staking fault and influence double cross slip from basal to non-basal planes.

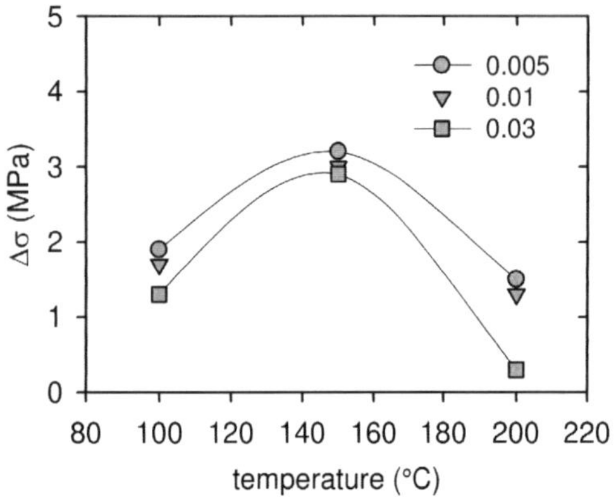

Figure 4. Temperature dependence of the flow stress increase after SR for various strains.

The values of the activation volume and the activation enthalpy may help to identify thermally activated processes considering some of the common short–range barriers to dislocation motion. The values and the observed stress dependence of the activation volume indicate that the double cross slip (and formation of jogs) may be the rate-controlling mechanism.

4. Conclusions

Stress relaxation tests are used to estimate the activation volume and to determine the internal stress and effective stress. In the case of the AJ51 alloy, the activation volume is a function of the effective stress independent of temperature because the microstructure is constant in the temperature range studied. Dynamic strain ageing influenced the deformation behaviour of AE42 magnesium alloy; the variation of the activation volume with the effective stress is different for different temperatures.

Acknowledgements.

The authors are grateful for support offered by the Grant Agency of the Academy of Sciences of the Czech Republic under grant IAA201120902. A part of this research was also supported by the Ministry of Education, Youth and Sports of the Czech Republic (Research Project 1M2560471601 "Eco-centre for Applied Research of Non-ferrous Metals").

References

[1] Caillard D and Martin J L 2003 *Thermally Activated Mechanisms in Crystal Plasticity* (Amsterdam: Pergamon Press)
[2] Dotsenko V I 1979 *Phys. Stat. Sol. (b)* **93** 11
[3] Feltham P 1963 *Phys. Stat. Sol.* **3** 1340
[4] Li J C M 1967 *Canad. J. Phys.* **45** 493
[5] De Batist R and Callens A 1974 *Phys. Stat. Sol. (a)* **21** 591
[6] Trojanová Z and Lukáč P 2009 *Inter. J. Mat. Res.* **100** 270
[7] Trojanová Z, Lukáč P and Kainer K U 2007 *Adv. Eng. Mater.* **9** 370
[8] Trojanová Z, Lukáč P, Kainer K U and Gärtnerová V 2005 *Adv. Eng. Mater.* **7** 1027
[9] Balík J and Lukáč P 1989 *Czech J. Phys.* B **39** 447
[10] Balík J and Lukáč P 1989 *Czech J. Phys.* B **39** 1138
[11] Hong S I 1989 *Mater. Sci. Eng.* A **110** 125
[12] Hong S I 1987 *Mater. Sci. Eng.* **91** 137

Kinematics of deformation bands in an austenitic FeMnC TWIP steel

J P Chateau[1], T A Lebedkina[2], M A Lebyodkin[3] , A Jacques[1] and S Allain[4]

[1] Institut Jean Lamour, Ecole des Mines, Parc de Saurupt, CS14234, 54042 Nancy Cedex, France
[2] Institute of Solid State Physics, Russian Ac. Sci, 142432 Chernogolovka, Russia
[3] LPMM, Université de Metz / CNRS, Ile du Saulcy, 57045 Metz Cedex 01, France
[4] Arcelormittal Maizières Research SA, Voie Romaine, BP 30320, 57283 Maizières-lés-Metz, France

Jean-Philippe.Chateau@mines.inpl-nancy.fr

Abstract. Tensile tests on a Fe22Mn0.6C steel at room temperature and different strain rates show serrations on the curves similar to Portevin-Le Chatelier (PLC) serrations of type *A*, associated with negative strain rate sensitivity. Propagation of deformation bands have been observed by high-rate extensometry over more than two orders of magnitude of the applied strain rate. This constitutes a remarkable difference with the PLC effect which shows a transition to static bands (type *B* or *C*) when the applied strain rate decreases. In this steel, bands moving as slow as a few tenth of mm/s are observed instead of static bands, which is two orders of magnitude lower than what is reported for type *A* PLC bands. This emphasises a strong correlation between plastic events, also confirmed by multifractal analysis of the tensile curves. Twinning which is responsible of the high strain hardening rate of this steel at room temperature is discussed as one of mechanisms of correlation between instabilities.

1. Introduction

The plastic deformation mechanisms of the Fe22Mn0.6C (wt%) steel depend on temperature and corresponding variation of the stacking fault energy [1,2]. Twinning occurs at room temperature and progressively disappears when the temperature increases. The tensile curves exhibit fluctuations characteristic of plastic instabilities and which morphology depends on temperature. The mechanisms of these instabilities are not known yet, with an exception above 423K/473K where dynamical strain aging of dislocations by carbon atoms may lead to the Portevin-Le Chatelier effect [3,4]. At lower temperatures, the bulk diffusion of carbon is too slow to account for this mechanism [5]. At room temperature, this steel displays negative strain rate sensitivity (SRS) at strain rates below 10-1 s-1. While this property is generally ascribed to the PLC effect, it is not a peculiar feature of this mechanism. Furthermore, at this temperature, twinning enters into competition with dislocation glide and promotes a high strain hardening rate, effect known as twinning-induced plasticity (TWIP), which leads to simultaneously high mechanical resistance and deformability [6,7]. Twins can also be at the origin of pronounced stress fluctuations [8]. To characterise these instabilities, we study the local strain distribution during tensile deformation using high-rate extensometer technique [9] and examine the correlation between instabilities by a multifractal analysis of the tensile curves [10].

2. Local strain measurement

Flat tensile specimens of a 3 μm grain size Fe22Mn0.6C steel with a gauge section of 60mm× 12.6mm×1.25mm were tested in tension at room temperature with a constant crosshead speed. The nominal value of the imposed strain-rate $\dot{\varepsilon}_a$ varied from 2.08×10^{-5} s^{-1} up to 2.5×10^{-2} s^{-1}. One side of the specimen is painted black and ten 1 mm wide white marks are painted normal to the longitudinal axis, distant by 1mm [11]. The sequence of pairs of black and white stripes forms a set of ten 2 mm wide extensometers in the centre of the gauge length. The positions of the black/white transitions are followed during the tensile test using a high resolution 1D CCD camera with a recording frequency of 10^3 Hz and a pixel size of 1.3μm and fixed relative to the fixed crosshead.

The local strain is determined from the displacements measured by the CCD camera:

$$\varepsilon_i(t) = \ln \frac{x_{i+2}(t) - x_i(t)}{x_{i+2}(0) - x_i(0)}$$

where x_i is the coordinate of the ith transition and t is the time with reference taken at 0.

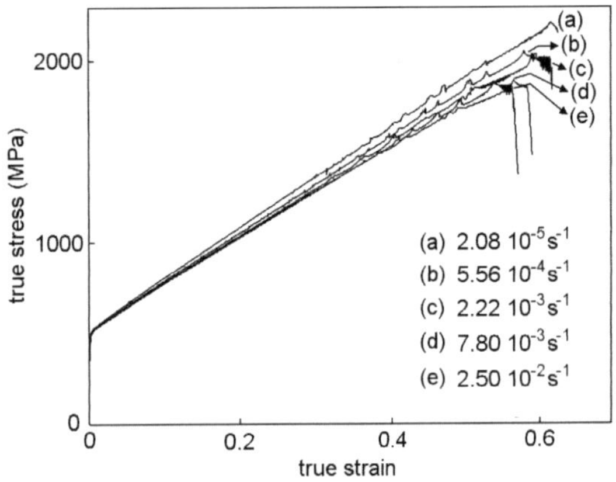

Figure 1. True stress-strain curves for various $\dot{\varepsilon}_a$ values: high strain hardening rate (TWIP effect), presence of serrations and negative SRS.

The macroscopic true stress-strain curves in Figure 1 show the typical behaviour of this steel: high fracture stress and strain, a high strain hardening rate due to TWIP effect, stress fluctuations and a negative SRS. The latter varies from a few MPa at the beginning of deformation to 30MPa at $\varepsilon = 0.5$, which corresponds to what was observed in C-Mn steels under PLC conditions [3,4].

Figure 2 illustrates the measurement of synchronised local displacements and strains with stress fluctuations: a stress drop (Figure 2.a) corresponds to an instantaneous acceleration of the displacements of the entire set of extensometers (Figure 2.b), as marked by the dotted lines. This shows that a strain localisation appears between the fixed crosshead and the set of extensometers. Then the displacements slow down one extensometer after another (inclined dotted line in Figure 2.b), which results in local strain jumps, marked by dashed lines in Figure 2.c. The occurrence of strain steps followed by intervals of slower deformation, plus the delay between the steps and the initial acceleration suggest the formation of a band moving through the set of extensometers. This is confirmed by the magnification of the local strain curve illustrated on Figure 3, which shows that the deformations of adjacent extensometers are delayed in time.

In the case of the PLC effect, moving bands (type *A*) are only observed at the highest applied strain rates, typically above 10^{-3} s^{-1}. When the strain rate is decreased, a transition to hopping of correlated static bands is observed (type *B*), associated with oscillations on the stress-strain curves. At the lowest strain rates, typically 10^{-5} s^{-1}, the correlations between the static bands disappear (type *C*). The main difference in the studied steel is the persistence of the propagation mode over the entire applied strain rate range, with an exception for $\dot{\varepsilon}_a = 5.56\times10^{-4}$ s^{-1} where intermittent static bands superimposed to propagating bands were also observed and for $\dot{\varepsilon}_a = 2.08\times10^{-5}$ s^{-1} where the recording time was too short to access a full propagation sequence. The band propagation velocity V_b can be determined as it is presented in Figure 3 where one can see a linear dependence with $\dot{\varepsilon}_a$ at a given applied strain, even at the lowest strain rates. This is consistent with the power law observed in the case of type *A* PLC bands, but with a higher exponent (1 instead of 0.8) [12,13]. However, bands moving as slow as a few tenths of mm/s are observed, what is far less than it is reported in the case of type *A* PLC bands (several mm/s).

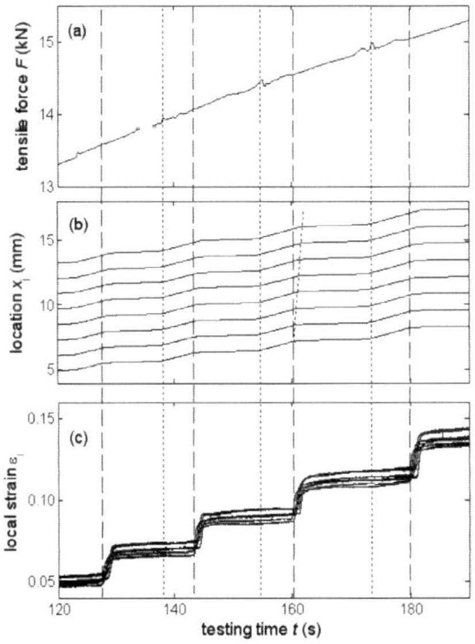

Figure 2. Typical experimental record for $\dot{\varepsilon}_a = 2.22\times10^{-3}$ s^{-1}. (a) Tensile force. (b) Positions of the black/white transitions. (c) Local strain.

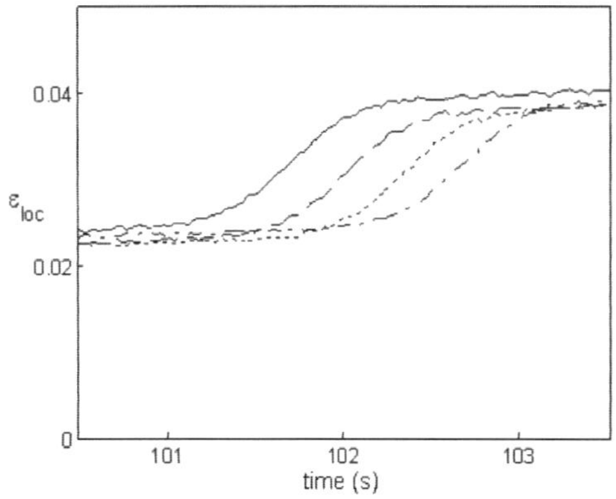

Figure 3. Magnification of Figure 2.c for four juxtaposed extensometers.

3. Multifractal analysis

The transitions from type *A* to type *C* when $\dot{\varepsilon}_a$ decreases is linked to correlations between the plastic deformation events, which generally lead to self-similar or fractal properties of deformation curves [13]. In order to test the self-similarity of the stress fluctuations, the tensile curve of a specimen tested at room temperature and at $\dot{\varepsilon}_a = 7\times10^{-4}$ s^{-1} has been time differentiated numerically. The signal is then composed of successive peak values ψ_k. By using a time scale of N intervals δt, a local measure $\mu_i(\delta t)$ of the signal in the ith interval can be defined as the sum of all the ψ_k values in this interval, normalised by the sum of all the ψ_k values in the N intervals. Introducing the functions Z_q:

$$Z_q = \sum_{i=1}^{N} \mu_i^{\,q}$$

where q is a real number, then [10]:

- in the case of a purely stochastic signal, Z_q is proportional to δt^{q-1},
- if the signal has a self-similar character, Z_q is proportional to $\delta t^{(q-1)D_q}$, where the D_q depend on q and are called generalised fractal dimensions.

Figure 5.a shows that the signal has a strong self-similar character over 2 orders of magnitude of δt, despite the small number of recorded points (<3000). At room temperature, the stacking fault energy is equal to 20 mJ/m^2 and many microtwins are formed during plastic deformation [2]. As they are similar to dislocation pile-ups, they induce high stress concentrations at interfaces which are able to trigger further plastic events in their vicinity. This can be at the origin of strong correlations. In comparison, the same analysis on a sample deformed at 673K, where no twinning occurs, shows a random signal of fluctuations due to noise (Figure 5.b). Tests performed at 423K show stress fluctuations on part of the tensile curves similar to type *B* PLC. Further multifractal analyses are needed to compare the results with previous studies in the case of the PLC effect [13].

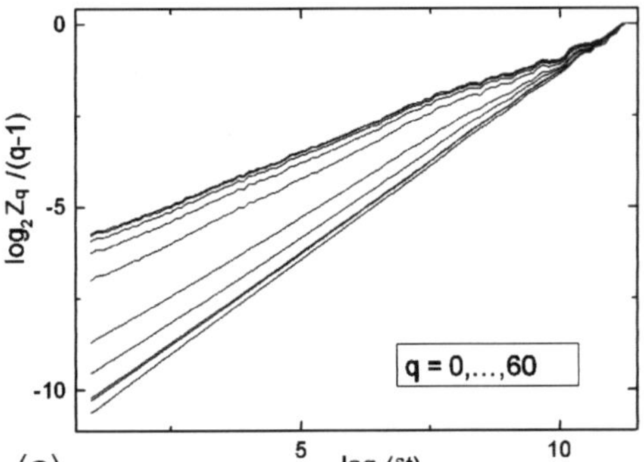

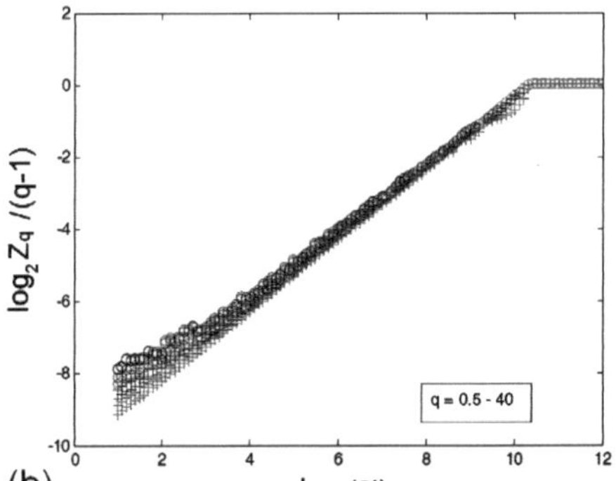

Figure 5. Result of the multifractal analysis. Evolution of functions Z_q with δt, (a) with TWIP effect at room temperature, showing straight lines with different slopes D_q and (b) with no TWIP effect at 673 K, showing only one straight line with a slope equal to 1.

4. Conclusion

The propagation of localisation bands has been observed by high rate extensometry at room temperature in the Fe22Mn0.6C steel over more than two orders of magnitude of the applied strain rate. This is relevant with type *A* bands observed at high strain rate in the case of the PLC effect. Contrary to this effect, no transition to static bands occurs when the applied strain rate decreases: the band velocity simply decreases linearly down to very low values. The correlations between plastic events seem to be sufficient to extend the propagation mode to the lowest strain rates. Twinning induces a high strain hardening rate in this steel at room temperature. It may also play an important role in the complex mechanism producing the plastic instabilities and different from a classical dynamic strain aging.

References

[1] Schumann V H 1972 *Neue Hütte* **17** 605-609
[2] Allain S, Chateau J P, Dahmoun D and Bouaziz O 2004 *Mater. Sci. Eng.* A **387-389** 272-276
[3] Belotteau J, Berdin C, Forest S, Parrot A and Prioul C 2006 Local Approach to Fracture (EMMC9: Ecole des Mines de Paris) 187-192
[4] Wagner D, Roubier N and Prioul C 2006 *Mater. Sci. Technol.* **22** (3) 301-307
[5] Allain S, Cugy P, Scott C, Chateau J P ,Rusinek A and Deschamps A 2008 *Int. J. Mater. Res.* **99** (7) 734-738
[6] Grässel O, Krüger L, Frommeyer G and Meyer L W 2000 *Int. J. Plast.* **16** 1391-1409
[7] Allain S, Chateau J P and Bouaziz O 2002 *Steel Res.* **73** 299–302
[8] Klassen-Nekludova M V 1964 *Mechanical Twinning of Crystals* (Consultants Bureau: NY)
[9] Weiss J, Richeton T, Louchet F, Chmelik F, Dobron P, Entemeyer D, Lebyodkin M, Lebedkina T, Fressengeas C and McDonald R J 2007 *Phys. Rev.* B **76** 224110 (8).
[10] Halsey T C, Jensen M H, Kadanoff L P, Procaccia I and Shraiman B I 1986 *Phys. Rev.* A **33** 1141-1151
[11] Lebedkina T A, Lebyodkin M A, Chateau J P, Jacques A and Allain S 2009 *Mater. Sci. Eng.* A **519** 147–154
[12] Ziegenbein A, Hähner P and Neuhäuser H 2000 *Comput. Mater. Sci.* **19** 27-34
[13] Lebyodkin M A and Lebedkina T A 2006 *Phys. Rev.* E **73** 036114

Deformation behavior of pure titanium at a wide range of strain rates

Y Tanaka, M Kondo, N Miyazaki, R Ueji

Kagawa University Faculty of Engineering, 2217-20 Hayashicho, Takamatsu, Kagawa 7610396, Japan

tanaka@eng.kagawa-u.ac.jp

Abstract. Plastic deformation behavior of pure titanium has been investigated using a tensile test at a various range of strain rates. Flow stress curves reveal that the work hardening behavior depends on the strain rates. The higher strain rates, the more work hardening. Dislocation cell structure has been developed after the deformation at a low strain rate. At a high strain rate, in addition to that, microstructure of twinning is well evolved. Much occurrence of twinning at the high strain rate leads the pronounced work hardening.

1. Introduction

Outstanding properties of titanium such as excellent strength-to-weight ratio, corrosion resistance, prominent biocompatibility are giving itself wide spread applications. The deformation behavior of titanium with an hcp structure can be more complex than that of common cubic metals. Slips with *a*-type Burgers vectors are the most favorable system. Despite the requirement of five independent slip system for polycrystalline to adapt an arbitrary plastic strain, only four slip systems exist. The occurrence of deformation twin plays an important role.

Regarding work hardening behavior, titanium displays substantial rate sensitivity [1] and exhibits several different stages, as reported from compression material tests [2-5]. Formation of deformation twin increase the strain hardening, and a number of attempts have been paid to elucidate the structure of the deformation twin [6-8]. Almost all of the previous investigations are discussed based on compression tests. The purpose of this study is conducting tensile tests at a various range of strain rates and clarifying the tensile plastic deformation behavior regarding to the strain rate sensitivity.

2. Experiments

In this investigation, Grade 1 quality CP-Ti ingots were used. The ingots were cold-rolled from 9 to 1 mm in thickness. Tensile test pieces were cut out from the cold-rolled sheet using an electric discharge machine in a way to coincide the tensile direction with the rolling direction. The test pieces were subjected to heat treatments in vacuum furnaces at a temperature of 923 K for 10 min, 1 h and 10 h. The test pieces were recrystallized by annealing for 1 h and 10 h, and yielding equiaxed grain sizes were 14.3 μm and 25.8 μm, respectively. The test pieces annealed for 10 min were partially recrystallized and we could not evaluate the grain size of them. The tensile test with various strain rates at room temperature were performed using a dynamometer sensing block type high deformation speed material testing machine (HMH206, Saginomiya Seisakusho Inc., Jpn). The rate-dependent mechanical behavior of the CP-Ti was studied over a strain rate range of 10^{-3} to 10^{3} /s. TEM samples were sectioned from the position near a grip end in order to avoid sectioning samples from the necking region. TEM thin foils were prepared by electro-polishing to perforation. These foils were examined using a JEOL JEM-3010 transmission electron microscope operating at 300 kV.

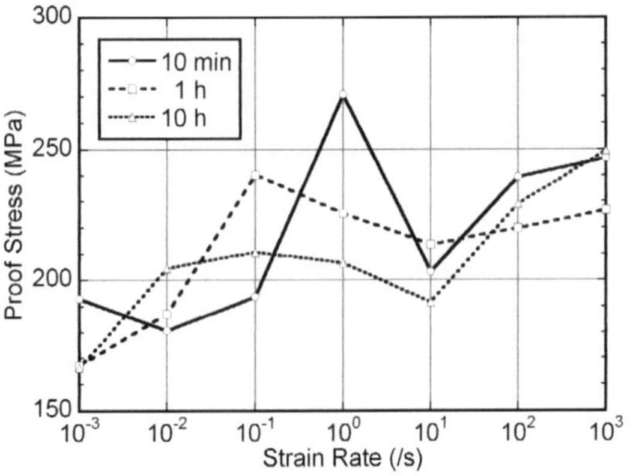

Figure 1. Relation between the proof stress and the strain rate obtained by the tensile tests.

3. Experimental results

3.1. Mechanical behavior

Figure 1 reveals the dependence of the proof stress on the strain rate. Even though the existence of wide dispersion, it seems to be a slight correlation between the strain rate and the proof stress for all of the heat treatment periods. The annealing period gives no effect on the change of the proof stress, or the wide range of dispersion of the proof stress may mask the grain size effect.

The flow stress curves are shown in figure 2. Wavy curves for a 10^3 /s strain rate have arisen from stress vibration which occurs when an elastic wave caused by an impact from a hummer in the test machine is propagated through the test sample and detected by the load cell. The wavy shape of the curves is due to the inherent problem of the test machine and is not essential for the mechanical behavior. The true stresses at some constant strain increase with the strain rate for all of the heat treatment periods. Significant difference between the heat treatment periods cannot be recognized. Only in the case of a low strain rate, the sample annealed for 10 min, partially recrystallized, obtain higher tensile strength and elongation (figure 2(a)). For the medium strain rate (10^0 /s) the elongation becomes lower than those for the lower strain rates. For further high strain rates, however, the elongation increases with the strain rates. This phenomena have been found not only hexagonal crystal but also cubic iron (both bcc and fcc structures) [9], and the reason of the reduction in elongation for the medium strain rate is still vague for all kind of crystal structure.

3.2. Microstructure

As anticipated from the flow stress curves, dimple structure has been seen on the fractured surfaces after the tensile test for all of the specimens by SEM observation. No significant difference in the dimple structure was detected regarding the effects of the strain rates and the annealing periods.

Using TEM observations, we have found the difference in the microstructures between low and high strain rate deformations. Figure 3(a) shows the microstructure after the tensile test with the low strain rate of 10^{-3} /s. A large number of dislocations can be seen, and the parallel band that is marked

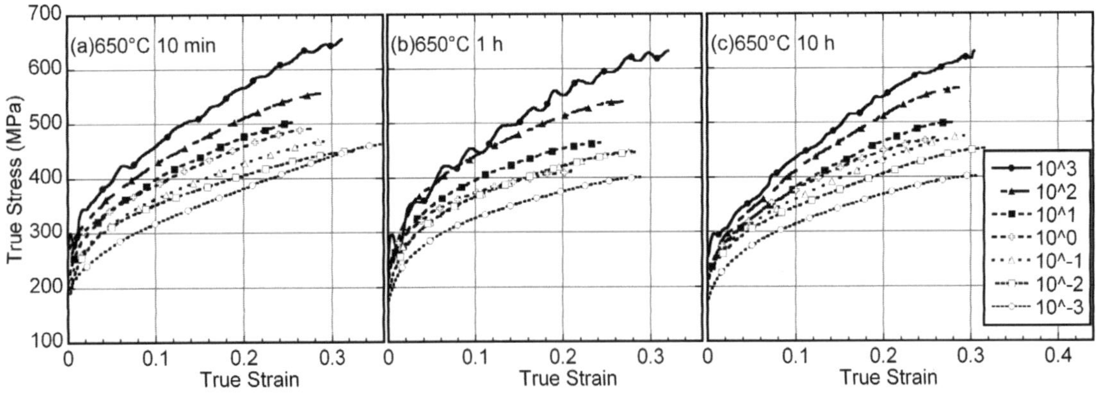

Figure 2. Flow stress curves measured in various strain rates indicated in an inset.

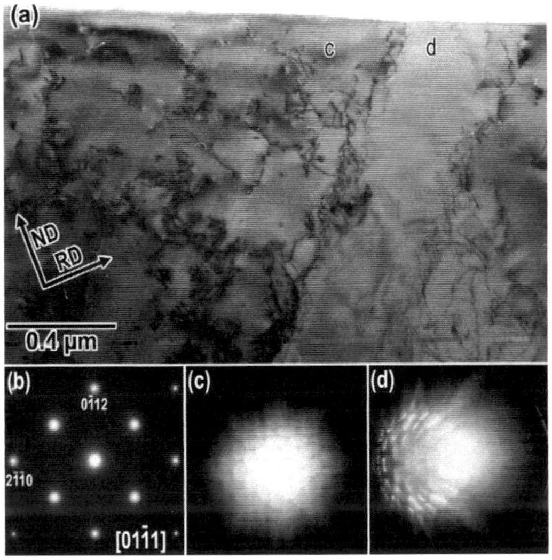

Figure 3. (a) TEM micrograph of the deformation microstructure after the tensile test of 1 h annealed titanium with the strain rate of 10^{-3} /s, (b) $[01\bar{1}1]$ zone axis SAED pattern, (c, d) CBED patterns taken from the positions "c" and "d" as indicated in figure 3(a).

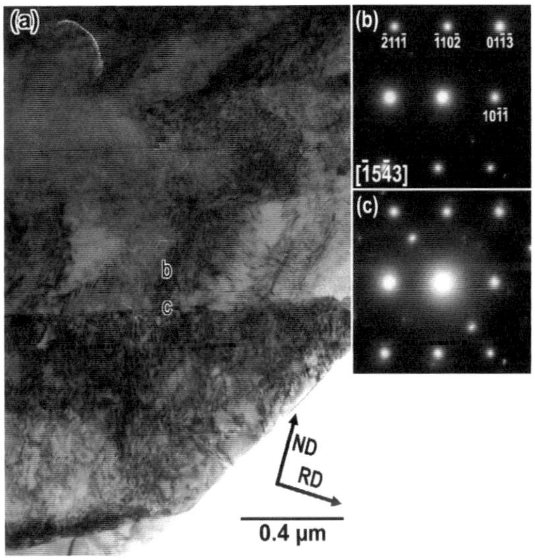

Figure 4. (a) TEM micrograph of the deformation microstructure after the tensile test of 1 h annealed titanium with the strain rate of 10^{3} /s, (b) $[\bar{1}5\bar{4}3]$ zone axis SAED pattern taken from the positions "b" in figure 4(a), (c) SAED pattern taken from the twin boundary.

"d" in the right side of the TEM image seems to be a twin plate at first glance. The crystal orientation relation between "c" and "d" is not twin but almost the same orientation. Kikuchi lines in figure 3(d) are moved from the position in figure 3(c) to the $[2\bar{1}\bar{1}0]$ direction. The movement of the Kikuchi pattern equals to $2\boldsymbol{G}_{2\bar{1}\bar{1}0}$ that means the tilt angle between them is 0.8°. The boundary dividing the crystal orientation is a small angle tilt boundary composed of tangled dislocations, and the microstructure after the low strain rate deformation is mainly dislocation cell structure. Deformations twins occur with a low frequency. A remarkable point on the diffraction patterns is that the movement of Kikuchi pattern is just lateral $[2\bar{1}\bar{1}0]$ direction which contains only a-axes direction in the hexagonal crystal. The tangled dislocations making the small angle boundary have proven to be with a-type Burgers vectors only.

For the deformation at a high strain rate, many dislocations and small angle boundaries have been seen as well as the case of the deformation at low strain rate. In addition, twin boundaries have been observed. Figure 4(a) shows the microstructure after the tensile deformation at the strain rate of 10^{3} /s. Dense dislocations are formed in the twin plate and horizontal twin plates are seen as indicated "c" in figure 4(a). This twin plane is almost parallel with the direction of a $10\bar{1}\bar{1}$ diffraction spot. Using the relation between Miller-Bravais index and 4-dimensional index [10], we can calculate the index of the twin plane. The twin is formed on a $(1\bar{1}02)$ plane. It is well known that the $(1\bar{1}02)$ twin compensate the tensile stress along the c-axis. Those twins will obstruct the motion of dislocations. Therefore, much work hardening behavior has been measured at the high strain rate tensile test.

4. Discussion

Many papers pointed out that the behavior of work hardening in compression test shows a distinctive 3-stage pattern in the change of work hardening rate [2-5]. Many difficulties attended our attempt to differentiate the flow stress curves due to the wavy fluctuation for high strain rates. On the other hand, we can analyze the stress flow curves simply based on a Hollomon's equation. n-values (work hardening exponent) obtained from the Hollomon's equation have been consistent with uniform

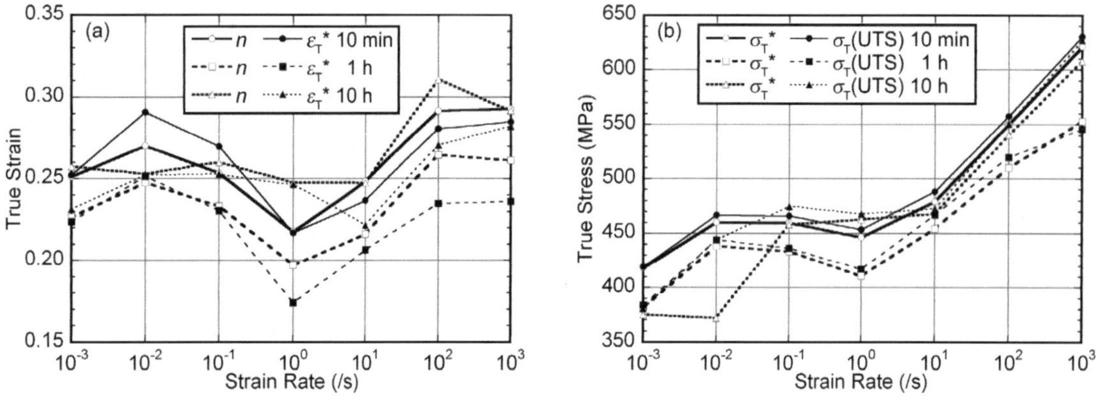

Figure 5. (a) Effect of the tensile strain rate on the work hardening exponent, *n*, and the true strain, ε_T^* satisfying the necking condition, (b) effect of the strain rate on the true stress, σ_T^* and the ultimate tensile true stress.

elongations derived from the criterion of plastic instability. The exponent, *n* and ε_T^*; the true strain when starting plastic instability, are shown in figure 5(a). As mentioned before, elongations decrease for the medium strain rate. When the plastic instability is held, the work hardening rate coincide with the true stress, σ_T^*, which should correspond to the ultimate tensile strength (UTS). Relation between the tensile strengths and the strain rate is shown in figure 5(b). At a high strain rate, the tensile strength significantly increases, since the deformation twins have a role of the obstacles against the dislocation motion.

5. Conclusion

Findings from the tensile tests at a various range of strain rates and the following TEM observations after deformation are summarized as follows.

1. The higher tensile strain rates, the more work hardening.
2. The microstructure developed after a low-rate deformation mainly exhibits the dislocation cell structure.
3. The microstructure developed after a high-rate deformation revels the deformation twinning in addition to the dislocation cell structure.
4. Analyses of the flow stress curves are successfully performed based on the Hollomon's equation.
5. The formation of twinning at a high-rate deformation seems to enhance the work hardening.

Acknowledgments

This research is partially supported by the funds for Kagawa University Encouragement 2007 and Young Scientists 2009.

References

[1] Chichili D R, Ramesh K T and Hemker 1998 *Acta mater.* **46** 1025.
[2] Nemat-Nasser S, Guo W G and Cheng J Y 1999 *Acta mater.* **47** 3705.
[3] Cheng J and Nemat-Nasser S 2000 *Acta mater.* **48** 3131.
[4] Salem A A, Kalidindi S R and Doherty R D 2003 *Acta mater.* **51** 4225.
[5] Salem A A, Kalidindi S R, Doherty R D and Semiatin S L 2006 *Metall. Mater. Trans. A* **37A** 259.
[6] Song S G and Gray III G T 1995 *Acta metall. mater.* **43** 2325, 2339.
[7] Serra A, Bacon D J and Pond R C 1999 *Acta mater.* **47** 1425.
[8] Barnett M R 2007 *Mater. Sci. Eng. A* **464** 1.
[9] Ueji R private communications.
[10] Frank F C 1965 *Acta Cryst.* **18** 862.

Effect of microstructural morphology on the mechanical properties of titanium alloys

A. Dehghan-Manshadi[1], M.H. Reid and R.J. Dippenaar
Faculty of Engineering, University of Wollongong, Wollongong 2500, NSW Australia

alidm@uow.edu.au

Abstract. Different morphologies of α+β microstructures were obtained in a commercial Ti-6Al-4V alloy by cooling at different rates from the single β-phase region into the two phase region. The effect of such morphologies on mechanical properties was studied using hot compression tests in a Gleeble thermomechanical simulator. A variety of complex morphologies could be obtained since the cooling rate has a significant influence on the β to α phase transformation and the resulting morphological development. While most of the β phase transformed to colonies of α at high cooling rates, it was possible to obtain a complex mixture of α colonies, grain boundary α and lamellar structure by decreasing the cooling rate. These complex morphologies each exhibited distinctive mechanical properties and characteristic dynamic phase transformation behaviour during deformation as a function of strain rate.

1. Introduction

Titanium alloys exhibit mechanical and physical properties which fit aerospace applications in which light weight, good wear and corrosion resistance are required and provide significant economic benefit in these applications. However, these unique properties are strongly affected by the chemical composition, microstructure, deformation and heat treatment history. Ti-6Al-4V (Ti64) as the most popular α+β titanium alloy used in the aerospace industry, has attracted significant research interest and an enormous amount of work has been done regarding the improvement of its mechanical properties through heat treatment, hot deformation or a combination of these processes. Much of the research centred on improving the microstructure and attaining appropriate mechanical properties. Earlier work has shown that the initial microstructure such as the fraction of α and β phases, the morphology and thickness of α-lathes as well as the size of α colonies (i.e. the geometrical arrangement of α and β-phases) have significant effects on the mechanical properties of the Ti64 alloy. For example, Lütjering [1] has shown that among the all microstructural characteristics, the α colony size has the most significant influence on the mechanical properties. It has furthermore been shown that parameters such as the cooling rate from the β- phase field, the initial β grain size and the presence of interstitial impurities (oxygen and carbon), can affect the geometrical arrangement within the microstructure of the Ti64 alloy. The cooling rate seems to be one of the most important parameters affecting microstructural development. Whereas slow and intermediate cooling rates lead to nucleation and growth of α-lamellae into the initial β grains through a diffusion controlled process, higher cooling rates render a martensitic transformation [1, 2].

[1] To whom any correspondence should be addressed.

The influence of different initial microstructures on the mechanical properties of a variety of α+β titanium alloys containing a single morphology, has been extensively studied [3, 4]. However, the case is more complicated when a mixture of different morphologies with different responses to hot deformation exists in the initial structure. The present project was therefore designed to study the effect of cooling rate, from the single β-phase into the two-phase α+β region, on the geometrical arrangement of the constituent phases and in addition, to determine the effect of the resulting arrangements on the mechanical properties of a commercial Ti64 alloy.

2. Experimental procedure

12.7 mm diameter rods of commercially available Ti-6Al-4V alloy (0.06% C, 0.2% Fe, 0.15% O, 5.8% Al, 4.1% V and balance of Ti) were used in this study. Samples for the compression tests, 12.7mm diameter and 19 mm in length were cut from the as-received rods. Hot compression tests were carried out in a Gleeble 3500 thermomechanical simulator. Samples were heated to 1323 K at a rate of 5 Ks^{-1} and held for 180 s, aiming to achieve a single phase β microstructure. To achieve different morphologies of α+β structures from this initial β phase, samples were cooled to 1073 K at different cooling rates (10, 5, 2, 1 or 0.5 Ks^{-1}). Following cooling to 1073 K, some samples were directly quenched in water to study the microstructural development as a function of cooling rate. While other samples were deformed at a strain rate of 1 s^{-1} or 0.1 s^{-1} followed by water quenching to assess the influence of cooling rate on the resulting mechanical properties. Microstructural development was studied by optical and scanning electron microscopy following standard metallographical practice.

3. Results and Discussion

Figure 1 shows the microstructure of Ti64 samples after soaking at 1323 K and cooling to 1073 K at different rates. The microstructure developed by cooling at a rate of 5 ks^{-1} consists of colonies of α within the prior coarse β-grains of about 350 μm (Figure 1a). On some selected β grain boundaries α-plates nucleated and grow towards the β-grain interior (acicular α). This acicular structure of α was presumably formed during cooling from the β transition temperature to 1073 K. Also, a thin layer (<1 μm) of grain boundary α (GB-α) was observed in some regions of prior β-grain boundaries. By decreasing the cooling rate to 2 Ks^{-1}, the microstructure changed to a mixture of α-colonies, α-lamellas with an average size of 40 μm and α-plates (i.e. acicular-α) of 1 μm thick as well as some GB-α (Figure 1b). Slower cooling caused the formation of even more complex structures consisting of large lamellar-α, acicular-α, GB-α as well as some large equiaxed or elongated α-grains (Figures 1c and d). The thickness of both the acicular and GB-α phases is increased by decreasing the cooling rate. Figure 2 shows the influence of cooling rate on the thickness of α-phase as measured from SEM images. It is clear that a decrease in the cooling rate from 10 to 0.5 ks^{-1} leads to an increase in the thickness of GB-α from 1 μm to more than 10 μm. Figure 1 shows that the volume fractions of both morphologies of α are also increased by decreasing the cooling rate. It stands to reason that during slow cooling, more time is available for the diffusional transformation of β to α leading to an increase in the size and volume fraction of the different α- morphologies.

These experiments clearly demonstrated that a wide range of microstructures with different morphologies and sizes of α-plates can be brought about by changing the cooling rate and it is expected that these different morphologies can not only influence the mechanical properties of the as-quenched specimens but perhaps more importantly, have a determining effect on the mechanical properties following hot deformation. To further probe this premise, hot deformation tests were carried out at different strain rates and examples of typical shapes of flow curves are shown in Figure 3. The flow curves show a rapid increase to the peak followed by a continuous flow softening beyond the peak. The peak stress as a function of cooling rate (i.e. different microstructures) is superimposed in Figure 2. At both strain rates, an increase in the cooling rate leads to an increase in peak stress up to a cooling rate of 2 Ks^{-1} and then remains constant. Lütjering [1] has shown that for a fully lamellar structure, the flow stress remain constant at cooling rates between 1 $Ks^{-1}$and 100 Ks^{-1}, and only at cooling rates higher than 100 Ks^{-1} a rapid increase in flow stress is observed due to a decrease in the α-colony size. The present work has shown that at cooling rates between 2 Ks^{-1} and 5 Ks^{-1}, where most of the microstructure is consists of α-colonies (Figure 1a and b), the flow stress is essentially constant (Figure 2) in agreement with Lütjering. However, a gradual decrease in cooling rate below 2 Ks^{-1} leads to a rapid decreased in peak stress. Figures 1c and d, confirm that the microstructure following slow cooling comprise a complex mixture of different α-phase morphologies. It is well known that these different morphologies have different strengths and deform by different mechanisms. Colonies of α are expected to have the highest strength due to their unique orientation. By contrast, β-layers are very soft at high temperature while grain boundary-α should be softer than α-colonies, but will be harder than the equiaxed α-phase. Therefore, mixtures of these phases in the initial microstructures result in complex mechanical behaviour upon further deformation at high temperature. Moreover, dissimilarity between the restoration mechanisms of the different α-morphologies during hot deformation will lead to even more complex mechanical response. While a kinking process, which has less effect on stress, is proposed for the restoration process of α-colonies [5], it is expected that lamellar and GB-α will be restored by dynamic recovery and recrystallization. These proposed phenomena were studied in more detail by analyzing the as-deformed microstructures.

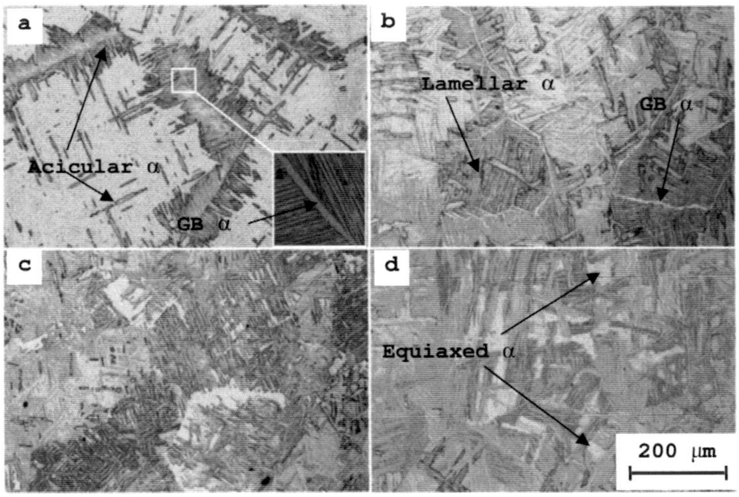

Figure 1: Microstructure of Ti64 alloy after re-heating to 1323 K and cooling to 1073 K at different cooling rates **a)** 5 Ks^{-1}, **b)** 2 Ks^{-1}, **c)** 1 Ks^{-1} and **d)** 0.5 Ks^{-1}

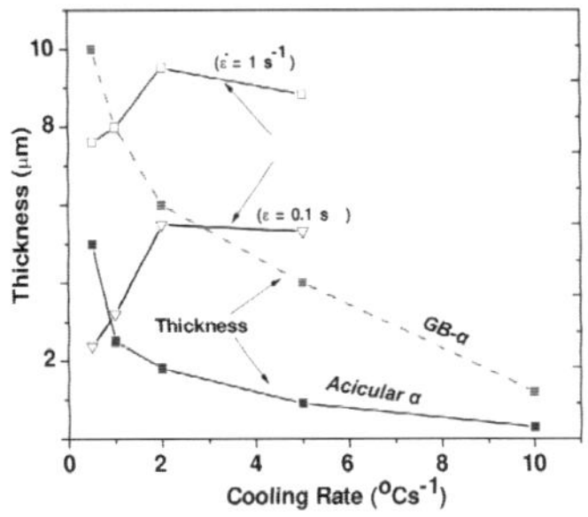

Figure 2. The thickness of α-phases and peak stress as functions of cooling rate

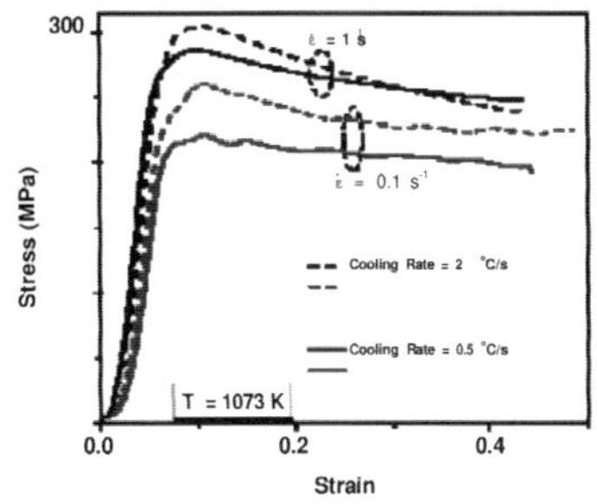

Figure 3. The flow curves of samples deformed at 1073 K at different strain rates

Figure 4 shows the optical microstructure of samples with different initial microstructures resulting from different cooling rates to 1073 K and then isothermally deformed at a true strain of 0.5. While the lamellar α-structure globulized, the grain boundary and equiaxed-α phases were plastically deformed and elongated (pancake structure) along the direction of deformation. The result of globularization is a massive decrease in the average grain size of the α+β structure and the effect of this globularized structure on mechanical properties as shown in Figure 3 is a decrease in the flow stress (a continuous softening in flow stress). The influence of globularization of the lamellar structure on decreasing the flow stress is clear for samples cooled at low rates. As these samples have more and larger α-lamellar phase, the globularization of this phase caused a significant decrease in the flow stress (Figure 2). The GB-α and equiaxed grains of α showed a different restoration mechanism compared to lamellar-α. A trace of dynamic recrystallization was frequently observed in these α morphologies by the formation of small grains in the vicinity of the initial grain boundaries (Figure 4c) and it is reasonable to assume that the occurrence of such a dynamic recrystallization is the main reason for decrease in peak stress in slow cooled samples since they contain a larger fraction of GB and equiaxed α.

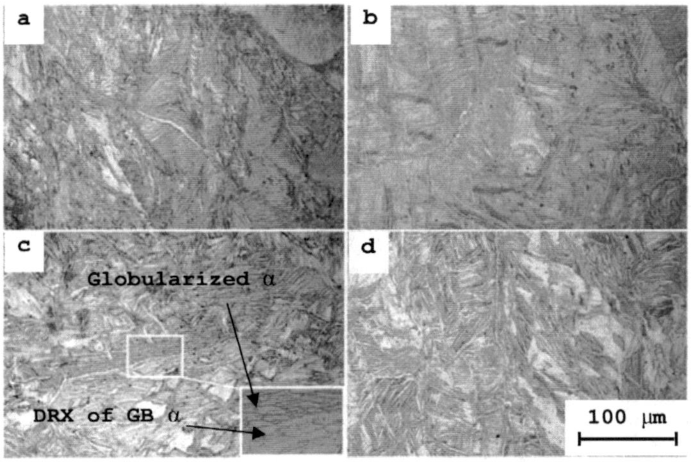

Figure 4: The microstructure of Ti64 alloy deformed at 1073 K after cooling at different rates of **a)** 5 ks^{-1}, **b)** 2 ks^{-1}, **c)** 1 ks^{-1} and **d)** 0.5 ks^{-1} $(\dot{\varepsilon} = 1s^{-1})$

An interesting observation is the presence of multiple peaks in the flow curves following deformation at low strain rates (Figure 3). This phenomenon has been observed during the hot deformation of steel under very low strain rates and/or very high temperatures and has been attributed to the occurrence of discontinuous dynamic recrystallization [6]. However, more accurate analyses of deformed microstructures are necessary to verify that discontinuous dynamic recrystallization is responsible for the appearance of multiple peaks in the present case.

4. Conclusions

A wide variety of simple and complex microstructures were obtained following cooling of a commercial Ti-6Al-4V alloy from the single β-phase region to the α+β region after different cooling rates. Different initial microstructures responded differently to hot deformation. While, lamellar structures globularized, grain boundary-α and equiaxed α-grains deformed plastically. Also, traces of dynamic recrystallization were observed following hot deformation in microstructures containing grain boundary α-phase.

5. References

[1] Lutjering G 1998 *Mater. Sci. Eng.* A **243** 32.
[2] Filip R, Kubiak K, Ziaja W and Sieniawski J 2003 *J. Mater. Proc. Tech.* **133** 84.
[3] Seshacharyulu T, Medeiros S C and Frazier W G 2000 *Mater. Sci. Eng.* A **284** 184.
[4] Venkatesh B D, Chen D L and Bhole S D 2008 *Mater. Sci. Eng.* A **506** 117.
[5] Seshacharyulu T, Medeiros S C, Frazier W G and Prasad Y V R K 2002 *Mater. Sci. Eng.* A **325** 112.
[6] Dehghan-Manshadi A and Hodgson P D 2007 ISIJ Int. **47** 1799.

Journal of Physics: Conference Series **240** (2010) 012023 doi:10.1088/1742-6596/240/1/012023

Precipitation hardening of a FeMnC TWIP steel by vanadium carbides

J P Chateau[1], A Dumay[1], S Allain[2] and A Jacques[1]

[1] Jean Lamour Institute, Ecole des Mines, Parc de Saurupt, CS14234, 54042 Nancy cedex, France
[2] Arcelormittal Maizières Research SA, Voie Romaine, BP 30320, 57283 Maizières-lés-Metz, France

Jean-Philippe.Chateau@mines.inpl-nancy.fr

Abstract. A fine precipitation of spherical vanadium carbides is obtained in a Fe22Mn0.6C base steel during the final recrystallisation heat treatment. Precipitates formed in recrystallised grains have a cube-cube orientation relation with the matrix, confirmed by Moiré patterns observed in TEM. The theoretical size for loss of coherency is below the nm, much lower than the precipitates' size. Deformation contrasts were observed around the precipitates and their residual coherency was measured. It was shown to decrease when the carbides' size increases, to vanish above 30 nm. The net increase of the yield stress was estimated to be 140 MPa. Precipitation hardening by vanadium carbides do not alter the strain hardening rate by TWIP effect, as they do not seem to act as obstacles for the propagation of microtwins.

1. Introduction

The Fe22Mn0.6C TWIP steel has a low stacking fault energy and deforms by twinning at room temperature in addition to dislocation glide. Microtwins of secant variants form in each grain and define subgrain cells, which size decreases along with strain [1]. The microtwins stored in stacks [2] act as strong obstacles for dislocations and progressively decrease their mean free path. The resulting dynamical Hall&Petch effect leads to high strain hardening rate and an outstanding compromise between elongation (> 50%) and mechanical resistance (> 1 GPa). On the other hand, the yield stress of this steel is rather low, below 400 MPa depending on the grain size, but can be increased by precipitation hardening. If the hardening effect of intragranular precipitates by blocking dislocations is well known, their interaction with mechanical twins has not been systematically studied. It is important that the presence of fine precipitates do not also block the expansion of the microtwins, in order to keep the high hardening rate. We characterize the precipitation of vanadium carbides by TEM, their interaction with twins and their resulting effect on the mechanical properties.

2. Studied material

The studied material is an austenitic Fe22Mn0.6C (wt%) base grade with 0.21 at% of vanadium furnished by ArcelorMittal, hot rolled at a temperature higher than 900°C and air quenched at 20°C/s. At this stage the steel is fully recrystallised and only 50 ppm of vanadium has precipitated, as predicted by models [3]. Sheets are then cold rolled down to a thickness of 1.2 mm and undergo a final recrystallisation and vanadium carbide precipitation optimized heat treatment [4]. Recrystallisation is

Journal of Physics: Conference Series **240** (2010) 012023 doi:10.1088/1742-6596/240/1/012023

complete and the final average grain size is 1.7 μm. TEM samples were prepared by mechanical (tripod) and ion (PIPS) polishing. The presence of fine intragranular spherical precipitates is confirmed by TEM observations. Their diameter varies from 2 to 18 nm with an average value $\langle d \rangle = 6.8$ nm and about 3 % of larger precipitates, up to 100 nm in size, are observed.

3. Lattice relation between austenite and vanadium carbides

3.1. Theoretical approach

The vanadium carbide phase has a stoichiometric VC composition and a B2 structure (NaCl type): f.c.c. vanadium structure with carbon atoms in each octahedral site. The lattice parameter is $a_{VC} = 0.4165$ nm. A cube-cube orientation relation in γ-austenite has been reported in the literature [5]: $(001)_\gamma // (001)_{VC}$ - $[100]_\gamma // [100]_{VC}$. The lattice parameter of Fe22Mn0.6C is $a_\gamma = 0.3605$ nm and the lattice mismatch with VC is high: $\delta = \Delta a / a_\gamma = 15.4$ %.

If we consider a semi-coherent precipitate with a lattice mismatch $\delta^* < \delta$ relaxed by interface dislocations, its total energy is:

$$E_p = 4\pi R^2 e_s + E_{el}^{VC}(\delta^*) + E_{el}^{\gamma}(\delta^*)$$

where $e_s = e_0 + e_{dis}$ is the energy per unit surface of the interface and E_{el}^{VC} and E_{el}^{γ} are the elastic energies stored respectively in the precipitate and the infinite matrix. e_0 is the specific surface energy of a coherent interface and e_{dis} is the energy of the interface dislocations, given by [6]:

$$e_{dis} = \frac{\mu_\gamma b}{2\pi^2}\left\{1 + \beta - (1+\beta^2)^{1/2} - \beta \ln\left[2\beta(1+\beta^2)^{1/2} - 2\beta^2\right]\right\} \quad with \quad \beta = \frac{\pi(\delta - \delta^*)}{1 - \nu_\gamma}$$

$\mu_\gamma = 60$ GPa and $\nu_\gamma = 0.3$ are respectively the shear modulus and Poisson ratio of the matrix, assumed equal to the ones of the precipitate in this expression and b = 0.255 nm is the Burgers vector. In the case of a coherent precipitate, $\delta^* = \delta$ and $e_{dis} = 0$. The elastic energy can be treated as in the case of an Eshelby spherical inclusion, with a radial displacement field and an eigenstrain applied to the precipitate equal to $-\delta^*$. The general solution for the displacement field in spherical coordinates is:

$$u_r = \mathrm{A}r + \mathrm{B}r^{-2}$$

where A and B are constants specific to the precipitate and the matrix. After elastic relaxation of the eigenstrain, the strain induced by the precipitate is $\delta^{*\prime}$. By solving the problem with the boundary conditions: $u_r(0) = 0$ and $u_r(R) = \delta^{*\prime} - \delta^*$ for the precipitate and $u_r(R) = \delta^{*\prime}$ and $u_r(\infty) = 0$ for the matrix, the continuity of the radial stress at the interface leads to:

$$\delta^{*\prime} = \delta^*\left[1 + 4\mu_\gamma(3\lambda_{VC} + 2\mu_{VC})^{-1}\right]^{-1} = 0.75 \cdot \delta^*$$

where $3\lambda_{VC} + 2\mu_{VC} = 750$ GPa, calculated from [6]. The elastic energies are thus:

$$E_{VC} = 2\pi(3\lambda_{VC} + 2\mu_{VC})(\delta^* - \delta^{*\prime})^2 R^3 \quad \text{and} \quad E_\gamma = 8\pi\mu_\gamma \delta^{*\prime 2} R^3$$

Coherency is energetically favorable for small radii and semi-coherency for large radii. By an iterative calculation on δ^* and R, the critical radius for the loss of perfect coherency is $R_{crit} = 0.275$ nm, which is very small due to the high initial lattice mismatch, and $\delta^* = 2.9$ %. No perfectly coherent precipitate should be observed and semi-coherent precipitates should have a small residual coherency.

3.2. TEM observations

Spherical precipitates are observed, as shown in Figure 1. In two wave conditions and certain values of the Bragg vector ***s***, Moiré patterns perpendicular to the diffraction vectors ***g*** are observed in small precipitates (Figure 1.a) with a wave length of 1.6 nm for ***g*** = <111>. Interferences between beams diffracted by parallel planes with different intereticular distances d_{VC} and d_γ leads to fringes distant by $\lambda_{Moiré} = d_{VC} d_\gamma (d_{VC} - d_\gamma)^{-1} = 1.56$ nm in the case of unstrained {111} planes in the two phases. This is consistent with the cube-cube orientation relation and small deformations due to residual coherency.

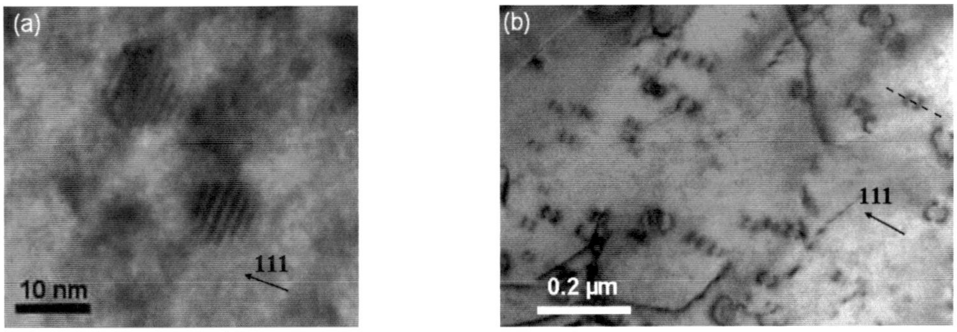

Figure 1. Two wave conditions TEM bright field with different Bragg vectors. a) Moiré patterns in the precipitates. b) Deformation contrasts with a coffee bean shape around precipitates.

In other *s* conditions, deformation contrasts are observed around the precipitates (Figure 1.b), which confirms their semi-coherent character. The coffee bean shape of these contrasts is characteristic of a radial displacement field ***u*** around the precipitates: in the plane normal to ***g***, ***g***.***u*** = 0 and the contrast vanishes. The displacement scales like r^{-2} and the width of the contrast increases with the deformation due to residual coherency. By measuring the width at which the intensity around the precipitate is 80% of the continuous intensity of the micrograph (see Figure 2.a) and using the method and tables proposed by [8], we measured the residual coherency of several precipitates. As shown in Figure 2.b, the residual coherency of small precipitates decreases when their size increases. A linear regression predicts that precipitates lose their semi-coherent character above 30 nm in diameter.

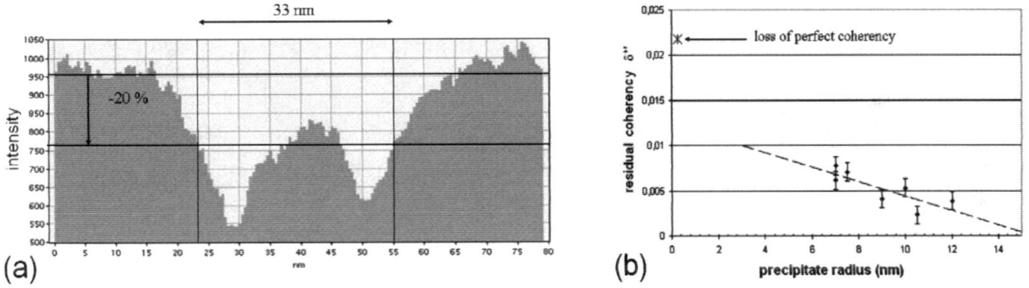

Figure 2. a) Intensity profile through a precipitate, along a line parallel to the diffraction vector (e.g. dashed line on Figure 1.b). The diameter of the precipitate is 14 nm. b) Measured deformation δ^*' due to residual coherency as a function of the size of the precipitates.

Larger precipitates show neither Moiré pattern nor deformation contrast. As precipitation and the final recrystallisation occur during the same heat treatment, these large precipitates formed in not recrystalised grains and are randomly oriented with the matrix and incoherent after the recrystilsation is complete. Small precipitates nucleated in recrystallised grains have the cube-cube orientation and a semi-coherent character below 30 nm in size.

4. Mechanical properties

Figure 3.a shows the true stress-strain curves of the present steel and of the reference Fe22Mn0.6C carbide free steel with a similar grain size (2.5 µm). Knowing that the contribution of the grain size *D* to the yield stress is $0.4\mu_\gamma(b/D)^{1/2}$ in FeMnC steels, the estimated effect of VC is an increase of the yield stress by 140 MPa. The theoretical hardening by incoherent precipitates is given by [9]:

$$\Delta\sigma_{\text{th}} = M \cdot 0{,}81\left(\frac{6f_{\text{V}}}{\pi}\right)^{1/2}\frac{\mu_\gamma b}{<d>}$$

where M ~ 3 is the Taylor factor, and f_v is the volumetric fraction of precipitates. If all the vanadium had precipitated, f_v = 0.32% and the resulting hardening would be $\Delta\sigma_{th}$ = 430 MPa. This shows that an important quantity of vanadium is still in solid solution and/or that the hardening efficiency of the large precipitates is poor, while they consummate a high amount of vanadium.

Figure 3.a also shows that the strain hardening rate is not affected by the presence of carbides: the entire curve is simply shifted to higher stresses by the amount of the initial hardening. The presence of precipitates does not alter the TWIP effect. Carbides do not seem to block microtwins, as shown on Figure 3: a microtwin continues on both sides of a precipitate. Although unknown, the crossing mechanism is likely to be of Orowan type because the friction stress of carbides is very high and their residual coherency is low. But, as the microtwin is 13 dense plane thick, it leaves 13 Shockley loops at the interface, which induces high stresses and should promote the emission of secondary dislocations.

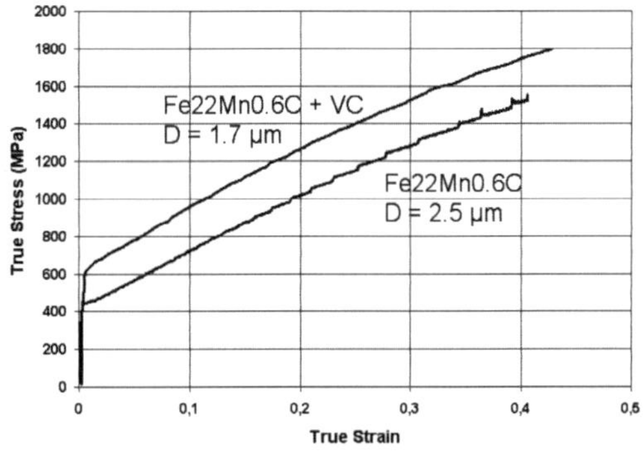

Figure 3. Compared true stress-strain curves of the precipitation hardened steel and the reference steel.

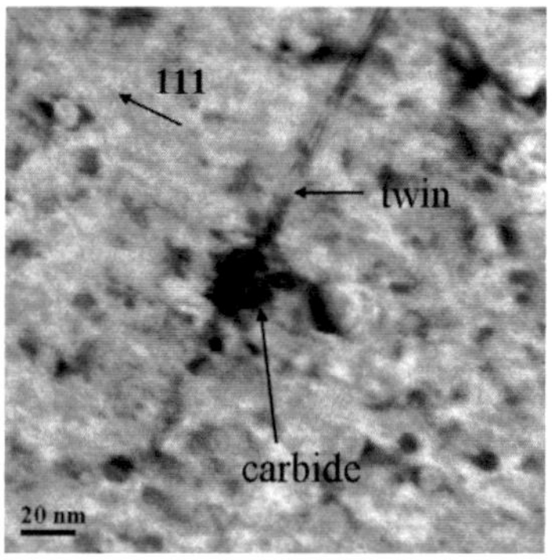

Figure 5. TEM micrograph of a carbide on the path of a microtwin seen edge-on.

5. Conclusion

A fine precipitation of vanadium carbides increases the yield stress of a Fe22Mn0.6C by 140 MPa, without decreasing the strain hardening rate. The hardening is performed by small precipitates formed in recrystallised grains. They are semi-coherent below 30 nm in size and have a cube-cube orientation relation with the matrix. Larger precipitates formed in non recrystallised grains do not participate much to hardening. The presence of carbides does not seem to block the propagation of microtwins and do not alter the hardening efficiency by TWIP effect.

Acknowledgements

The authors would like to thank Colin Scott for providing the samples with the optimised precipitation state and the stimulating discussions on this topic.

References

[1] Allain S, Chateau J P and Bouaziz O 2004 *Mater. Sci. Eng.* A **387-389** 143-147
[2] Allain S, Chateau J P, Dahmoun D and Bouaziz O 2004 *Mater. Sci. Eng.* A **387-389** 272-276
[3] Maugis P and Gouné M 2005 *Acta Mater.* **53** 3359-3367
[4] Scott C, Cugy P and Allely C 2008 Patent #WO2008007192 (A2)
[5] Yazawa Y, Furuhara T and Maki T 2004 *Acta Mater.* **52** 3727-3736
Silcock J M 1996 *Acta Mater.* **14** 687
[6] Jesser W A 1968 *Phil. Mag.* **19** 993
[7] Zhang X, Comins J D, Every A G and Stoddart P R 1998 *Int. J. Refract. Metal. & Hard Mat.* **16** 303-308
[8] Ashby M F and Brown L M 1963 *Phil. Mag.* **8** 1649
Edington J W 1975 *Practical Electron Microscopy in Material Science* (Philips Technical Library) **T3** 81-84
[9] Ardell A J 1985 *Met. Trans.* A **16A** 2131-2165

Deformation twinning in polycrystalline NiMnGa alloys

R Chulist[1*], M Pötschke[2], T Lippmann[3], C-G Oertel[1] and W Skrotzki[1]

[1] Institut für Strukturphysik, Technische Universität Dresden, D-01062 Dresden, Germany

[2] Institut für Metallische Werkstoffe, Leibniz-Institut für Festkörper- und Werkstoffforschung, D-01069 Dresden, Germany

[3] Institut für Werkstoffforschung, GKSS Forschungszentrum Geesthacht, D-21502 Geesthacht, Germany

robert.chulist@physik.tu-dresden.de

Abstract. The texture and local orientation after compression of a NiMnGa polycrystal and a bicrystal with 5M modulated structure were measured with synchrotron radiation and electron backscatter diffraction (EBSD), respectively. Compression of both samples leads to motion of those twin boundaries changing the volume fraction of particular martensitic variants in such a way that the shortest axis (*c*-axis) becomes preferentially aligned parallel to the compression axis. EBSD directly confirms the motion of twin boundaries within individual grains.

1. Introduction

Magnetically actuated shape memory alloys such as NiMnGa have recently attracted special attention due to large strains achieved by the reorientation of martensitic twin variants in which twin boundaries are highly mobile. The rearrangement of the martensitic variants under a magnetic field resulting from the magnetocrystalline anisotropy leads to a magnetic field induced strain (MFIS) [1 - 3]. Until now, MFIS has been mainly reported for NiMnGa single crystals [4 - 6]. However, since for broad technical applications growth of single crystals is quite time consuming and cost intensive, it is necessary to investigate polycrystalline samples on their suitability for MFIS.

The present work focuses on the rearrangement of twin variants during compression of a polycrystalline sample. Additionally, to confirm the motion of twin boundaries a bicrystal was investigated. To examine the deformation-induced orientation changes in NiMnGa alloys diffraction of synchrotron radiation and electron backscatter diffraction (EBSD) in the scanning electron microscope (SEM) has been applied. The texture changes observed are discussed with respect to movement of twin boundaries.

2. Experimental

The polycrystalline sample studied was fabricated by directional solidification: ($Ni_{50}Mn_{29}Ga_{21}$, 5M). Directional solidification was carried out under Ar atmosphere. A prealloy was prepared by induction melting of pure elements (Ni 99.98, Mn 99.8, Ga 99.999) [7]. The casting process was performed in an Indrudet S investment casting machine followed by furnace-cooling to room temperature. The cast sample had a diameter and length of 10 mm and 50 mm, respectively. Coarse columnar grain growth was achieved by using a hot mould with a cooled copper plate at the bottom. The mould was heated in a separate furnace to a temperature close to melting. The rectangular samples cut from the cast ingot about 1 cm from the bottom had a width and height of 4 mm and 15 mm, respectively. The bicrystal ($Ni_{50}Mn_{29}Ga_{21}$, 5M) was grown by the Bridgman method. The bicrystal with dimensions of (5×5×3) mm^3 was cut by spark erosion from an as-grown ingot and mechanically and electropolished for EBSD analysis. It is characterized by a general high angle grain boundary and different martensitic variants within the two grains (Fig. 1a). The investigated region was chosen under the aspect that the grain boundary has been acted as barrier for twin boundary motion.

The polycrystalline sample was compressed at room temperature up to a permanent strain of 3% without control of strain rate. Before deformation the longest lattice parameter of the 5M structure was preferentially aligned parallel to the compression axis. The texture was measured by diffraction of high-energy synchrotron radiation (100 keV) using beam line HARWI-II at DESY in Hamburg, Germany. Due to the high penetration depth of high-energy X-rays, texture measurements in transmission allow for collecting orientation data from a large sample volume, i.e. in a coarse-grained sample with a good grain statistics. Processing of the texture data was done with the Labotex software. The bicrystal was deformed in an Instron 8562 testing maschine at room temperature under a compressive stress of 30 MPa at a strain rate of 10^{-3} s^{-1} up to a permanent strain of 4.5%. The local orientation measurements within individual grains were done in the SEM (Zeiss Ultra 55) by EBSD (HKL company). Before deformation the bicrystal was trained. The training route consisted in successively compressing the sample along all three axes in order to achieve maximum strain and to reduce the twinning stress and work-hardening rate [8].

All planes and directions mentioned in this paper are given in the cubic coordinate system which is related to the cubic axes of the parent $L2_1$ phase.

3. Results and discussion

X-ray diffraction shows that the as-cast polycrystal simultaneously contains two martensitic structures, 5M and NM. Both phases have a strong <100> fibre texture along the growth direction (GD) [9]. It is known that only the 5M and 7M martensites in NiMnGa have a uniaxial magnetic anisotropy and therefore show MFIS. As shown by Gaitzsch et al. [10], the type of structure is not only determined by the composition but also by thermal treatment. This has been confirmed on a directionally solidified $Ni_{50}Mn_{29}Ga_{21}$ alloy, where the portion of 5M martensite increased by annealing at 1000°C for 48 h (Ar/H_2 atmosphere). After annealing the texture is still a fibre texture with <100> parallel to GD. However, in contrast to the as-cast state the variant selection is such that [001] parallel to GD is less preferred. The reason for this may be different internal stresses along the radial direction, i.e. tension or compression, respectively.

The axial compression test of the annealed polycrystal was carried out in the sample direction with longest lattice parameter of the 5M structure (two tetragonal *a*-axes) preferentially aligned parallel to the compression axis. Due to the preferred motion of those twin boundaries which produce most compressive strain the texture changes in such a way that the shortest axis, i.e. the *c*-axis, becomes preferentially aligned parallel to the compression axis. This can be clearly seen in inverse pole figures by the change of <100> fibre intensities. Fig. 1 shows inverse pole figures of GD taken of the undeformed and compressed sample, respectively. The <100] directions represent the variant '*a*' with the longer lattice parameter of the tetragonal 5M structure and [001], variant '*c*', with the shorter lattice parameter along GD. The numbers given are the percentages of the calculated volume fractions of the variants mentioned above. Due to compression the volume fraction of the variant with the shorter lattice parameter parallel to the compression axis increases at cost of the other variant. This clearly confirms the motion of twin boundaries in polycrystalline NiMnGa alloys.

The bicrystal was compressed at ambient temperature up to 30 MPa. The stress-strain curve shows a plateau region up to 4% at about 2 MPa and after that a strong increase. One grain (Fig. 2) with the <100> axes nearly parallel to the external edges of the sample was chosen for investigation The orientation of two twin-related martensitic variants measured with EBSD are presented in Figs. 2b and c. The {100) pole figures represent variant '*a*' and (001), the shortest lattice parameter, variant '*c*'.

After compression along the Y direction the light grey area spreads out to the position marked with black arrows (Fig. 3b). This directly confirms the motion of twin boundaries within individual grains in a NiMnGa bicrystal. Most interestingly, the motion takes place close to a grain boundary. However, in spite of a high compressive stress, this area is not completely detwinned (Fig. 3b). There are dark grey regions left with the *c*-axis perpendicular to the compression axis. This indicates that the grain boundary indeed impedes twin boundary motion preventing a complete reorientation. In other words, it acts as barrier for twin boundary motion.

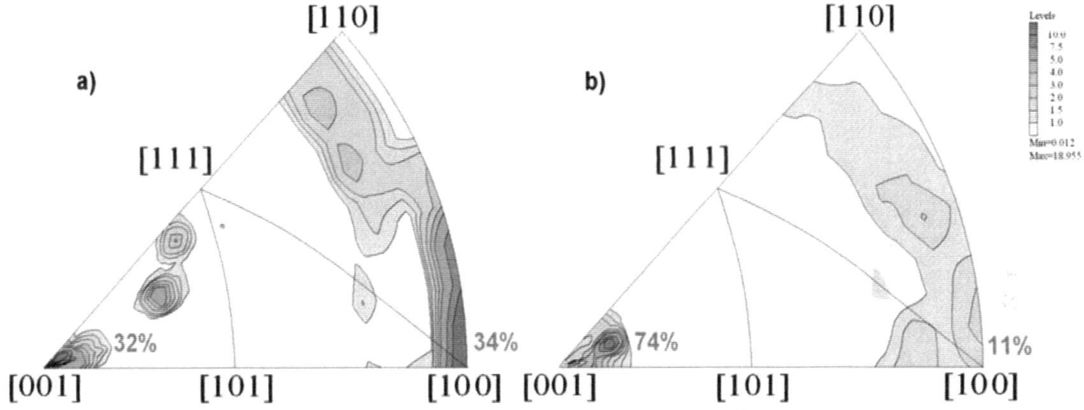

Figure 1. Inverse pole figures of the GD of the tetragonal 5M phase in sample I taken of the undeformed (a), and 3% along GD plastically compressed sample (b). The volume fraction of different orientations along GD with a spread of 20° is given.

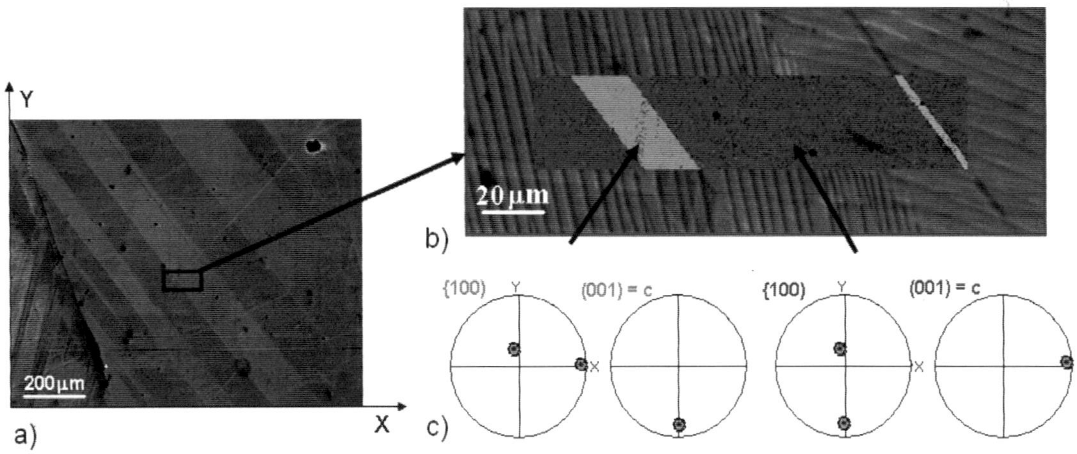

Figure 2. BSE image of an undeformed NiMnGa bicrystal (a) and EBSD mapping taken from the marked area (b). The dark and light grey coloured areas represent orientations with the *c*-axis nearly parallel to X and Y, respectively (c). The black spots are not indexed.

4. Conclusion

Compression of an annealed polycrystalline 5M NiMnGa sample confirms the motion of those twin boundaries which produce most compressive strain. Due to this preferred motion the texture changes in such a way that the shortest axis (*c*-axis) becomes preferentially aligned parallel to the compression axis. The change of the local orientation of a trained 5M bicrystal after compression studied with EBSD in the SEM also directly confirms twin boundary motion close to a grain boundary.

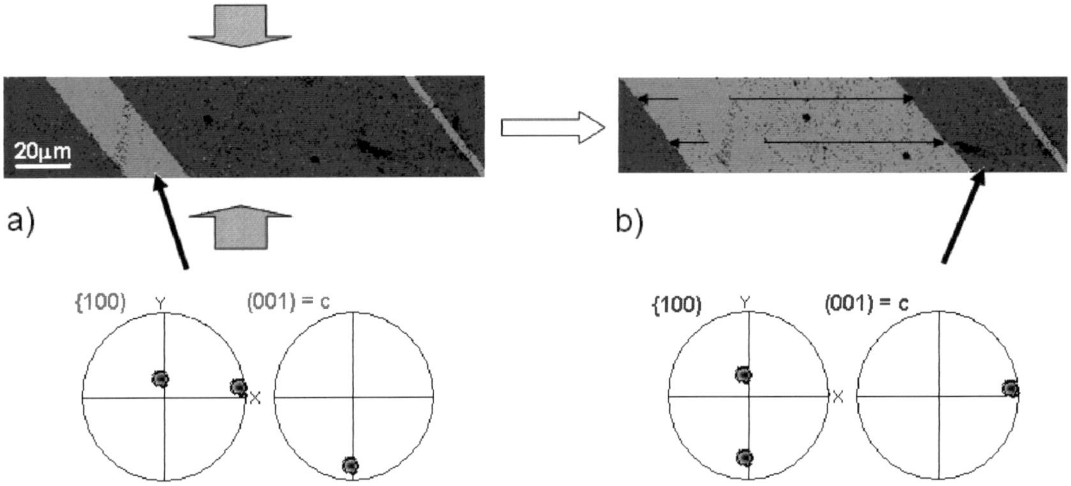

Figure 3. EBSD mapping of an undeformed NiMnGa bicrystal area (a) and after compression (b). The dark and light grey coloured areas represent orientations with the *c*-axis nearly parallel to X and Y, respectively. The black arrows show the direction of twin boundary motion. The black spots are not indexed.

Acknowledgements

The work was carried out within the priority program SPP 1239 "Magnetic shape Memory: Change of microstructure and shape of solid materials by external magnetic fields" of the German Research Society (DFG) under contract Sk 21/22-1, 2.

References

[1] Ullakko K, Huang JK, Kantner C, O`Handley RC and Kokorin VV 1996 *Appl. Phys. Lett.* **69** 1966
[2] Murray SJ, Marioni M, Allen SM, O`Handley RC and Lograsso TA 2000 *Appl. Phys. Lett.* **77** 886
[3] Müllner P, Chernenko VA, Wollgarten M and Kostorz G 2002 *Appl. Phys. Lett.* **92** 6708
[4] Müllner P, Chernenko VA and Kostorz G 2003 *J. Magn. Magn. Mater.* **267** 325
[5] Sozinov A, Likhachew AA, Lanska N and Ullakko K 2002 *Appl. Phys. Lett.* **80** 1746
[6] Heczko O, Sozinov A and Ullakko K 2000 *IEEE Trans. Magn.* **36** 3266
[7] Pötschke M, Gaitzsch U, Roth S, Rellinghaus B and Schultz L 2007 *J. Magn. Magn. Mater.* **316** 383
[8] Gaitzsch U, Pötschke M, Roth S, Rellinghaus B and Schultz L 2007 *Scripta Mater.* **57** 493
[9] Chulist R, Pötschke M, Böhm A, Brokmeier H-G, Garbe U, Lippmann T, Oertel C-G, and Skrotzki W 2008 *MRS Proc.* **1050E**, Warrendale, PA, BB09-03
[10] Gaitzsch U, Roth S, Rellinghaus B and Schultz L 2007 *J. Magn. Magn. Mater.* **305** 275

The Taylor relation in compression deformed Ge single crystals

K Nyilas[1], C Dupas[2], T Kruml[3], T Ungar[1] and JL Martin[2]

[1]Department of Materials Physics, Eötvös University Budapest, H-1518 POB 32, Budapest, Hungary
[2]IPMC, FSB, Ecole Polytechnique Fédérale de Lausanne, CH 1015 Lausanne, Switzerland
[3]Institute of Physics of Materials, Academy of Sciences, CZ 61662 Brno, Czech republic

kruml@ipm.cz

Abstract. Ge single crystals are deformed in compression at 850K and the same strain rate to various extents of strains. In each sample, the internal stress is measured through stress reduction tests and the dislocation densities by X-ray measurements. Data about these two parameters follow fairly well the Taylor-Saada relation, provided a correction term is added. It probably corresponds to dislocations which are seen by X-rays, though they do not contribute to crystal hardening.

1. Introduction

As a dislocation glides through a Germanium crystal, it has to overcome two types of “obstacles" : localized obstacles consist of the lattice resistance to glide (Peierls Nabarro force) while the others are due to long distance elastic interactions with other dislocations. Only the former ones are overcome with the help of thermal activation. They necessitate a shear stress τ^* called the thermal stress. The second ones necessitate a stress τ_i, called the internal stress. It has been proposed [1] that the applied shear stress τ can be decomposed according to:

$$\tau = \tau^* + \tau_i \tag{1}$$

The point of this study is to relate values of τ_i measured through stress reduction experiments in compression deformed Germanium single crystals at one temperature T and (shear) strain rate $\dot{\gamma}$. Then check the Taylor-Saada relation that claims that τ_i is proportional to the square root of the dislocation density ρ.

$$\tau_i = \alpha G b \sqrt{\rho} \tag{2}$$

There is a large amount of investigations about the plasticity of covalent crystals over the past 40 years (see reviews by e.g. [2, 3]). To the best of our knowledge, preliminary work about the determination of τ^* and τ_i and the origin of the internal stresses in Ge single crystals can be found in [4-6].

Similarly, the parameter ρ which is usually measured through the observation of etch pits with an optical microscope at specimen surfaces, is estimated here by appropriate X-ray techniques. High resolution double crystal diffractometer dedicated for line profile analysis is used to determine active slip systems and dislocation densities along with the fluctuation of dislocation densities as a function of deformation in Ge single crystals.

2. Experimental procedures

2.1. Material

Intrinsic Ge single crystals were cut in a parallelepipedic shape by a diamond saw in the [−123] single slip orientation with (1-11) and (54-1) lateral faces. The primary slip system is [−101] (111). Such an orientation ensures a much larger Schmid factor on the primary system as compared to the secondary one [011] (-1-11) (0.47 and 0.38 respectively). Care was taken to remove the dislocation rich surface layers by mechanical and chemical polishing.

2.2. Mechanical testing

Stresses τ, strains γ and strain rates $\dot{\gamma}$ are resolved with respect to the primary slip system.

The compression specimens were deformed in a computer controlled Schenck RMC 100 set up, equipped with a furnace, under a flux of He. The testing temperature was 850K at shear strain-rate $\dot{\gamma}$ of $1.6 \times 10^{-4}\ s^{-1}$. At the end of the test, the Ge samples were cooled under load so as to freeze the stressed dislocation structures. In an attempt to estimate the internal stresses along the monotonic curves, dip tests were performed using a procedure first proposed by [7]. At given stress along the stress strain curve, fast unloading is applied followed by a short creep experiment (15 seconds). The amount of stress decrease that corresponds to a zero creep rate equals the effective stress under these conditions [5,7]. For this purpose, computer programs have been designed for machine control and data acquisition. Details about the procedure can be found in [5].

2.3. X-ray investigations

The investigated face of the specimens is (54-1). The diffraction profiles were measured by a special double crystal diffractometer with negligible instrumental broadening. A fine focus rotating copper anode, Nonius FR 591, was operated as a line focus at 40 kV and 70 mA. The symmetrical 220 reflection of a Ge monochromator was used in order to have wavelength compensation in the lower angle region. The profiles were registered either by a linear position sensitive gas flow detector, OED 50 Braun, Munich, or by two dimensional imaging plate detectors of 100 mm height and 200 mm length with 50 μm pixel size.

3. Results

A typical stress strain curve is presented in Fig. 1. It exhibits the well known multiplication yield point (UYP) at the beginning, while normal hardening sets in after the lower yield point (LYP). Given the extremely low dislocation density in covalent crystals, intense multiplication phenomena operate at the onset of deformation. An evaluation of multiplication laws can be found in [8]. An example of the stress decomposition into effective and internal parts is illustrated in Fig. 1, showing that under the present conditions, τ^* ($\cong$ 6.1 MPa) is constant along the curve, within the uncertainty on the measurement. This is expected for the Peierls stress, under such conditions. τ_i is an increasing function of strain, due to dislocation hardening, above the lower yield point, $3.5\text{MPa} < \tau_i < 8\text{MPa}$ in the case of Fig. 1.

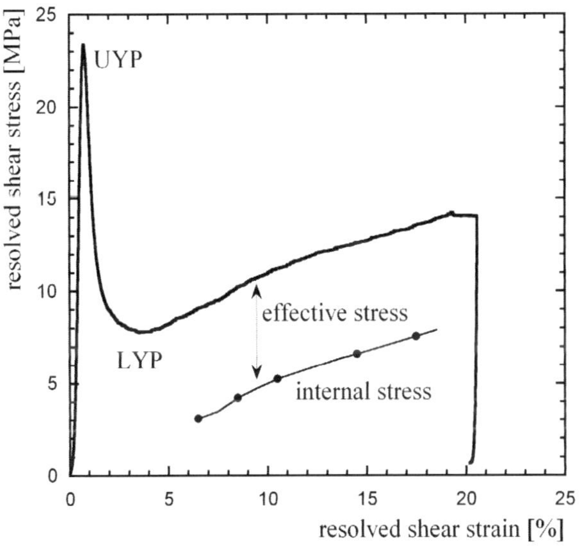

Figure 1. Stress-strain curve, effective and internal stress along the curve. T = 850K, $\dot{\gamma}$ = 1.6 x 10^{-4} s^{-1}.

Two other specimens were deformed at the same temperature and strain rate up to different levels of total strain (see Table 1). The above X-ray experiments were used to measure the dislocation densities in the three specimens after deformation. Table 1 summarizes the results.

Table 1. Results of mechanical tests and dislocation density measurements at the end of deformation.

γ_p (%)	τ (MPa)	τ* (MPa)	τ_i (MPa)	ρ (m^{-2})
7.0	10.4	6.0	4.4	2.6×10^{13}
20.5	14.0	6.0	8	2.0×10^{13}
30.8	36.6	6.0	30.6	6.5×10^{13}

4. Discussion

In an attempt to check whether the data about τ_i and ρ of Table 1 fit the Taylor-Saada relation, they have been plotted in Fig. 2 which shows τ_i as a function $\sqrt{\rho}$. Given the uncertainty on the measurement of the internal stress and especially the dislocation densities, a linear relation between τ_i and $\sqrt{\rho}$ is acceptable in agreement with the Taylor formula. Inspection of Fig.2 yields the following remarks:

i) The slope of the line on Fig. 2 equals αGb according to equation (2). With G = 60 GPa and b = 4.0×10^{-10} m, the slope of 7.09 Pa m measured on Fig. 2 yields a value of α = 0.295. This can be compared to previous attempts at estimating the coefficient α. Lavrentev [9] lists measured values of α in various situations. A value between 0.2 and 0.33 is reported for Cu single crystals at 300K deformed in stages I to III. Berner and Alexander [10] propose a theoretical value of $\alpha = \frac{1}{2\pi(1-\nu)} =$ 0.23 for Ge single crystals and a measured one of 0.32, using the etch pit technique for density measurements. The present data are in fair agreement with these estimations, given the uncertainty of measurements.

ii) Extrapolating the straight line of Fig. 2 yields a zero value for τ_i that corresponds to a certain dislocation density ρ_0 so that relation (2) should be written:

$$\tau_i = \alpha G b(\sqrt{\rho} - \sqrt{\rho}_0) \quad (3)$$

The density ρ_0 corresponds to dislocations which are seen by X-rays but do not contribute to hardening the crystal. Dislocation dipoles could be a possible type of such dislocations since they distort locally the lattice but do not create long distance stress-fields.

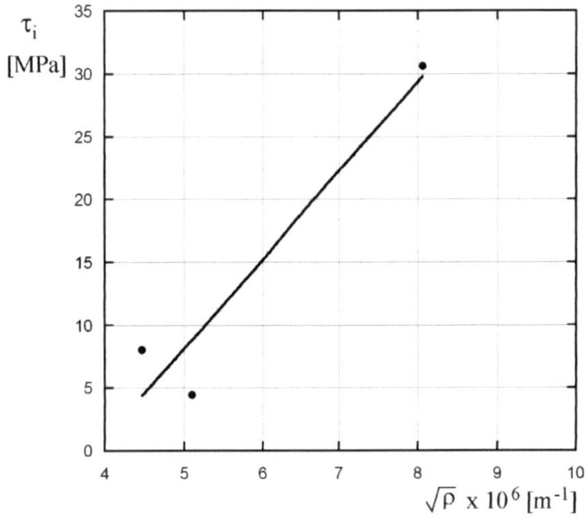

Figure 2. Internal stress τ_i as a function of the square root of the dislocation density ρ for the three samples of Table 1.

5. Conclusions

The two independent experimental techniques which have been used here to measure respectively the dislocation density and the internal stress in samples deformed to various extents at the same temperature and strain rate yield a relation between these parameters which strongly supports the Taylor relation, as old as 1934. The coefficient α is in fair agreement with previous theoretical estimations or measurements through other means. A correction term has to be added to the formula. It corresponds very likely to dislocations that are seen by X-rays but do not contribute to crystal hardening.

Acknowledgements
The financial support was provided by Fond National Suisse and Academy of Sciences of the Czech republic (AV0Z20410507).

References

[1] Seeger A, Diehl J, Mader S and Rebstock H 1957 *Phil. Mag.* **2** 323
[2] Alexander H and Haasen P 1968 *Solid State Physics* **22** 28
[3] George A and Rabier J 1987 *Rev. Phys. Appl.* **22** 941
[4] Dupas C, Zuodar N, Coddet O, Kruml T and Martin JL 2002 *J. Phys.: Condens. Matter* **14** 12989
[5] Kruml T, Coddet O, Saada G and Martin JL, 2003 *Phil. Mag. Lett.* **83** 651
[6] Nyilas K, Dupas C, Kruml T, Zsoldos L, Ungar T and Martin JL 2004, *Mater. Sci. Eng.* **A387-389** 25
[7] Gibbs GB 1966 *Phil. Mag.* **13** 317
[8] Kruml T, Dupas C and Martin JL 2006 *Acta Mater.* **54** 4721
[9] Lavrentev FF 1980 *Mater. Sci. Eng.* **46** 191
[10] Berner K and Alexander H 1967 *Acta Metall.* **15** 933

Journal of Physics: Conference Series **240** (2010) 012026 doi:10.1088/1742-6596/240/1/012026

Deformation mechanisms of hard oriented NiAl single crystals

M Scharnweber, C-G Oertel and W Skrotzki

Institut für Strukturphysik, Technische Universität Dresden, 01062 Dresden, Germany

werner.skrotzki@physik.tu-dresden.de

Abstract. Hard oriented NiAl single crystals (<100> deformation axis) have been compressed at temperatures between 296K and 963K. At about 600K the yield stress changes from a plateau to a steep fall. Slip line investigations show that this coincides with a transition from {112}<111> to {110}<110> slip. At low temperatures slip of <111> dislocations is determined by the Peierls mechanism while in the plateau region the deformation mechanism is still not understood. For the deformation by {110}<110> slip the activation enthalpy measured supports the deformation mechanism proposed by Mills et al. [1] based on the diffusion-assisted motion of macro-kinks in <110> edge dislocations which are found to be decomposed into <100> edge dislocations. The slip transition temperature is discussed with regard to the brittle-to-ductile transition temperature of polycrystalline NiAl.

1. Introduction

NiAl is an intermetallic compound with a number of favourable properties that make it a promising candidate for high temperature, e.g. turbine engine applications. NiAl has a brittle-to-ductile transition temperature (BDTT) of about 600K for polycrystals and hard oriented single crystals (<100> deformation axis) [1, 2]. This corresponds to the temperature where besides primary slip on {110}<100> secondary slip changes from {112}<111> to {110}<110> [2 - 4]. Based on TEM investigations revealing a complicated core structure of dislocations with <110> Burgers vector, Mills et al. [1] proposed a model for dislocation motion on {110}<110> slip systems. It is the aim of the present work to analyze the strain rate and temperature dependence of secondary slip in NiAl above the BDTT and check whether the model by Mills et al. is appropriate in terms of activation parameters to explain the fall of yield stress of hard oriented single crystals at medium temperatures.

2. Experimental

The single crystal used in this work (16mm in diameter and 30mm long) exhibited a stoichiometric composition with a maximum deviation of ±1at.% and a <100> growth direction which was verified by Laue back reflection technique. Compression samples of $2.1\times2.1mm^2$ in cross section and 4mm in length were cut by spark-erosion and subsequently annealed at 1273K for 60h under vacuum (heating/cooling rate about $1K\cdot min^{-1}$). The {100} oriented side faces were ground using SiC paper of different grit size to remove the surface layer affected by the erosion process leading to the final sample size of $1.8\times1.8\times4mm^3$. The base and top surface were only slightly ground with fine SiC paper to remove the roughness caused by the erosion process, thus minimizing the change in orientation and planarity. Additionally, for slip line investigations two side faces per sample were electropolished three times for 10s with cooling in between using a solution of 10% perchloric acid and 90% ethanol at a temperature of 263K and a voltage of 20V.

All compression tests were carried out in an Instron testing system under vacuum with the deformation temperature being obtained by heating with a rate of 12K·min^{-1}. Subsequently, a temperature dependent holding time of at least 1h was inserted to reach thermal equilibrium. The compression tests were performed with an initial strain rate of 3.6×10^{-5} s^{-1}. To study the effect of strain rate, after reaching a plastic strain of approx. 1% the strain rate was doubled and then reset to the original value at a plastic strain of approx. 2% with the test being stopped at a plastic strain of approx. 3%. The accuracy of stress and strain is about 3%. The absolute accuracy of the temperature measurement is 10K.

To analyze the mechanical data it is assumed that above the BDTT the material obeys a power law constitutive relationship

$$\dot{\varepsilon} = A\sigma^n \exp\left(-\frac{\Delta H}{kT}\right) \quad (1)$$

between strain rate $\dot{\varepsilon}$ and stress σ [5] (k = Boltzmann constant), with the stress exponent n given by

$$n = \frac{d\ln\sigma}{d\ln\dot{\varepsilon}} \quad (2)$$

and the apparent activation enthalpy given by

$$\Delta H = nk\frac{d\ln\dot{\varepsilon}}{d\ln(1/T)}. \quad (3)$$

In order to investigate the slip lines and determine the activated slip systems two neighbouring surfaces of the deformed samples were inspected by optical microscopy.

3. Results and discussion

Figure 1 shows an optical micrograph of the surface of a sample deformed at room temperature (RT). The slip lines under an angle of 63.5° to the compression axis are caused by {112}<111> slip. At low temperatures and high strain rates slip of <111> dislocations has been observed on {110}, too [6]. Figure 2 shows a sample deformed at 723K with horizontal slip lines which can be related to {110}<110> slip as there are no slip lines under an angle of 45° on the neighbouring sample faces ruling out the possibility of {110}<111> slip. The slip steps seen are produced by <110> screw dislocations. The Burgers vectors of <110> edge dislocations on {110} slip planes are parallel to the {100} side faces, thus they do not produce slip steps. In summary, the slip line investigations in the temperature range investigated show quite a sharp slip transition at about 600K.

The temperature dependence of the 0.2% yield stress is shown in figure 3. Similar curves have already been obtained by other researchers. The curves show a steep drop from a plateau at about 600K coinciding with the slip transition. Strain rate change tests show that there is practically no strain rate sensitivity in the plateau region, while at higher temperatures the mean stress exponents (average over a certain temperature interval) decrease from 4.3 to 3.7 (figure 4). Simultaneously, the apparent activation enthalpies increase from 1.5eV to 2.9eV.

It was the aim of the present study to analyze secondary slip in NiAl. Below about 600K the active secondary slip system is {112}<111>. Deformation via {112}<111> slip is temperature and strain rate dependent below room temperature. The activation volume at RT is about $20\,b^3_{<111>}$ (about $3\,b^3_{<111>}$ at RT [7] and below [8]) which together with the very long Burgers vector (0.5nm) of the moving dislocations indicate that the most likely deformation mechanism is the Peierls mechanism.

Between RT and about 600K the deformation via {112}<111> slip is neither temperature nor strain rate dependent. Frequently kinking takes place. According to Takeuchi [9] the yield stress plateau may be explained by the Peierls mechanism of narrowly dissociated <111> screw dislocations ($1.6\,b^3_{<111>}$). However, this contradicts the strain rate insensitivity observed here as well as the compact core structure of <111> screws predicted by molecular static calculations using an embedded atom potential ($\leq 1.2\,b^3_{<111>}$ [10]).

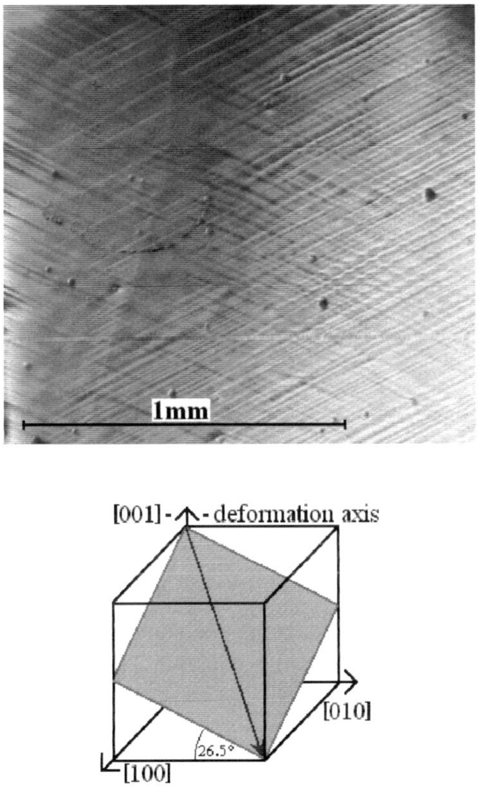

Figure 1. Slip lines caused by {112}<111> slip at 296K after compression up to a true strain of 3.1% and corresponding schematic representation of one slip system (plane plus Burgers vector (arrow)).

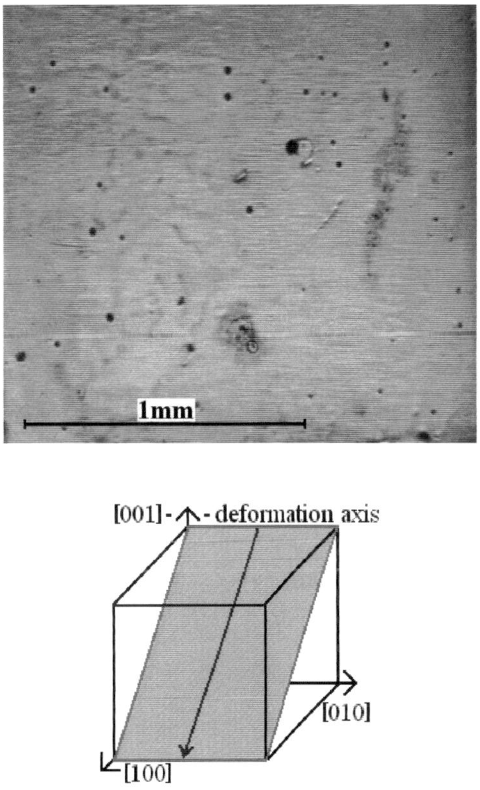

Figure 2. Slip lines caused by {110}<110> slip at 723K after compression up to a true strain of 2.7% and corresponding schematic representation of one slip system (plane plus Burgers vector (arrow)).

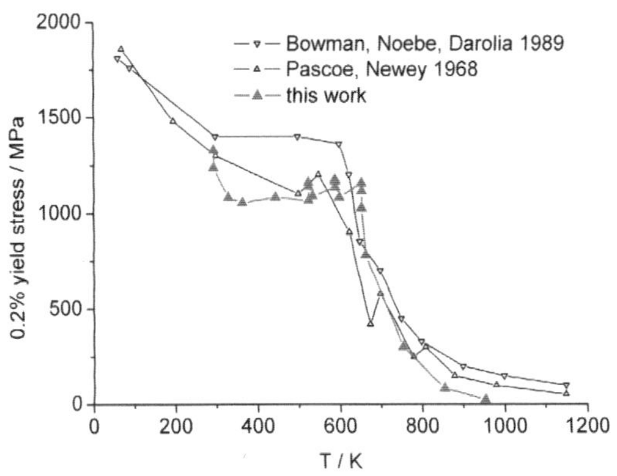

Figure 3. 0.2% yield stress vs. temperature (literature data are taken from [2]).

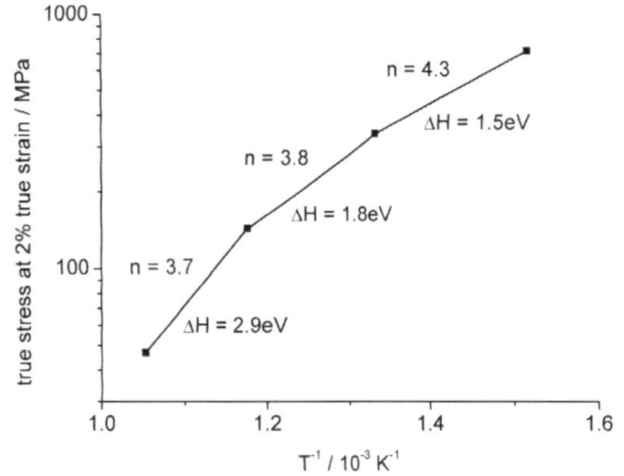

Figure 4. Apparent activation enthalpies from a plot of logarithmic true stress vs. reciprocal temperature using the mean values given for the stress exponent *n*.

Above about 600K the active secondary slip system is of the {110}<110> type. Mills et al. [1] proposed an interesting model for the movement of <110> dislocations in NiAl. According to TEM studies as well as atomistic calculations <110> edge dislocations tend to decompose into separate <100> edge dislocations with orthogonal Burgers vectors. This decomposition effectively locks the edge portions of <110> dislocations. Based on experimental evidence, it is suggested that motion of these dislocations is instead controlled by lateral movement of non-edge segments (macro-kinks). The model proposed requires the constriction of the decomposed edge segments involving the transfer of vacancies from one <100> dislocation to the other via pipe diffusion. The constitutive equation developed is in accordance with equation (1), with $n = 3$ and $\Delta H = 1.5$eV being the activation enthalpy for pipe diffusion (half of that for Ni lattice diffusion, 3.1eV [2]). These values compare quite well with those estimated in this work at about 700K (figure 4). In addition an activation volume of $2\,b^3_{<100>}$ was obtained suggesting that the climb of <100> edge dislocations is the process limiting the deformation [11]. With increasing temperature the activation enthalpy observed approaches that of Ni lattice diffusion suggesting a change in diffusion mechanism. A quantitative comparison of stress or strain rate is difficult because the constant A depends on the height of the macro-kinks and the distance over which the <100> edge dislocations have decomposed, parameters which may be temperature and stress dependent.

It should be mentioned that another locking mechanism has been proposed for 35° <110> dislocations on {110} based on a TEM analysis of the dislocation core structure [12, 13].

Concerning the BDTT of polycrystalline NiAl [3] it is conspicuous that this temperature compares quite well with that of hard oriented NiAl single crystals [14], respectively the slip transition observed here. Moreover, the strain rate dependence of the BDTT of NiAl polycrystals yields an activation enthalpy of 1.3eV [15] indicating that the BDTT may also be determined by pipe diffusion mediated motion of macro kinks in decomposed <110> edge dislocations. As the ductility of a polycrystalline material requires the compatible deformation of its crystallites, secondary slip has to be activated in order to provide sufficient independent slip systems. Thus, below the BDTT the stress needed to fulfill this condition by activating {112}<111> slip systems is much higher than the fracture stress resulting in a brittle behaviour. Above the BDTT due to the onset of diffusional processes secondary slip changes to {110}<110> leading to a much lower stress needed for its activation. Hence, the requirements for strain compatibility in polycrystals that is ductility are fulfilled.

References

[1] Mills M J, Srinivasan R and Daw M S 1998 *Phil. Mag. A* **77** 801
[2] Noebe R D, Bowman R R and Nathal M V 1993 *Int. Mater. Rev.* **38** 193
[3] Noebe R D, Bowman R R and Nathal M V 1996 *Physical Metallurgy and Processing of Intermetallic Compounds* ed N S Stoloff and V K Sikka (New York: Chapman & Hall) p 212
[4] Miracle D B 1993 *Acta Metall. Mater.* **41** 649
[5] Ilschner B 1973 *Hochtemperaturplastizität* (Berlin: Springer)
[6] Maloy S A, Gray III G T and Darolia R 1995 *Mater. Sci. Eng. A* **192/193** 249
[7] Sun Y Q, Taylor G, Darolia R and Hazzledine P M 1995 *Mat. Res. Soc. Symp. Proc.* **364** 261
[8] Kitano K and Pollock T M 1993 *Structural Intermetallics* ed R Darolia and J J Lewandowski et al. (Warrendale; PA: The Minerals, Metals & Materials Society) p 591
[9] Takeuchi S 2004 *J. Alloys Comp.* **378** 61
[10] Schroll R, Vitek V and Gumbsch P 1998 *Acta Mater.* **46** 903
[11] Scharnweber M 2008 *Diploma thesis, TU Dresden*
[12] Yang N and Sun Y Q 2004 *Acta Mater.* **52** 5615
[13] Messerschmidt U, Häussler D, Bartsch M and Sauthoff G 1999 *Intermetallics* **7** 455
[14] Field R D, Lahrman D F and Darolia R 1991 *Acta Metall. Mater.* **39** 2951
[15] Noebe R D, Cullers C L and Bowman R R 1992 *J. Mater. Res.* **7** 605

Journal of Physics: Conference Series **240** (2010) 012027 doi:10.1088/1742-6596/240/1/012027

Tensile properties of UFG aluminium alloy AA6016 at low temperatures

J Scharnweber, L Hollang, K Reuther, SR Dey, C-G Oertel and W Skrotzki

Institut für Strukturphysik, Technische Universität Dresden , D-01062 Dresden, Germany

werner.skrotzki@physik.tu-dresden.de

Abstract. The ultrafine grained aluminium alloy AA6016 produced by accumulative roll bonding (ARB) up to eight cycles was deformed in tension at 25K, 77K, 180K and 296K at a constant strain rate. The stress-strain curves were analyzed with regard to the evolution of yield stress σ_p, ultimate stress σ_m and fracture strain ε_f as a function of temperature for zero, four and eight ARB cycles processed aluminium plates. For all materials σ_p, σ_m, σ_m-σ_p and ε_f decrease with increasing temperature. With increasing number of ARB cycles, σ_p and σ_m increase, while σ_m-σ_p and ε_f were found to decrease. Based on the experimental results the influence of temperature and number of ARB cycles is discussed.

1. Introduction

Since its introduction by Saito et al. [1] accumulative roll bonding (ARB) has become a successful severe plastic deformation (SPD) method to produce ultrafine grained (UFG) metals and alloys with grain sizes ranging from 1 μm to about 100 nm [2 - 4]. Compared to their conventional large grained counterparts, UFG materials show an enhanced strength [e.g. 5] with only a slightly decreasing and in some aluminium alloys even increasing ductility [6 - 8].

The influence of ARB processing on the tensile behaviour at ambient temperatures has been described for various UFG materials, e.g. for the technically relevant age-hardening aluminium alloy AA6016 [9]. Investigations at other temperatures are less frequent. Therefore, in this study AA6016 plates of different degree of ARB deformation were subjected to tensile tests below room temperature in order to investigate the influence of temperature on the strength and ductility.

2. Experimental

In this investigation samples of the initial state and after four and eight ARB cycles were tested. Prior to ARB the plates with a thickness of 1 mm were solution heat treated in a furnace for 1h at 520°C and subsequently quenched in water. The grain sizes of the completely recrystallized initial material in rolling and normal direction (RD and ND) are about d_{RD} = (18.2 ± 2.2) μm and d_{ND} = (16.2 ± 3.0) μm. The ARB-process with a thickness reduction of 50% per rolling pass was applied to aluminium plates with a width of 100 mm and a length of approximately 300 mm using a four high rolling mill (Carl Wezel, Mühlacker). Prior to rolling the plates were pre-warmed at 230°C for four minutes. After each cycle the surfaces of the air-cooled bonded plates were wire brushed to remove the oxide layer. The ARB process was done up to a maximum of eight cycles, which corresponds to a von Mises strain of 6.4 and leads to a plates with 2^8 = 256 bonded layers each having a thickness of 3.9 μm. The conditions of the ARB processing are described in more detail in [10]. The grain sizes after four and eight ARB cycles are d_{RD} = (0.7 ± 0.4) µm and d_{ND} = (0.6 ± 0.2) µm and d_{RD} = (0.5 ± 0.1) µm and d_{ND} = (0.2 ± 0.1) µm, respectively [9]. Only high angle grain boundarys with a misorientation angle higher than 15° were considered for the grain size determination. The chemical composition of the aluminium alloy AA6016 is listed in Table 1.

Table 1. Chemical composition of AA6016 (mean values).

wt.%	Si	Cu	Fe	Mn	Mg	Cr	Zn	Ti	other	Al
AA6016	1.0-1.5	0.52	0.5	0.2	0.25-0.6	0.1	0.2	0.15	0.15	balance

The tensile tests were performed at 25K, 77K, 180K and 296K using an Instron 4500 deformation machine at a constant strain rate of $9.2\times10^{-4}s^{-1}$ until necking occurred. The low temperature deformation is described in detail in [11, 12]. The tensile direction was parallel to the rolling direction. To calculate the true stress and true strain the diminution of the cross section of the specimen was taken into account.

3. Results and Discussion

Figure 1 reveals the strong influence of both deformation temperature and numbers of ARB cycles on the stress-strain curves. All curves show work hardening which increases with decreasing temperature. At 25K the specimens of the ARB processed material fractured directly after onset of necking.

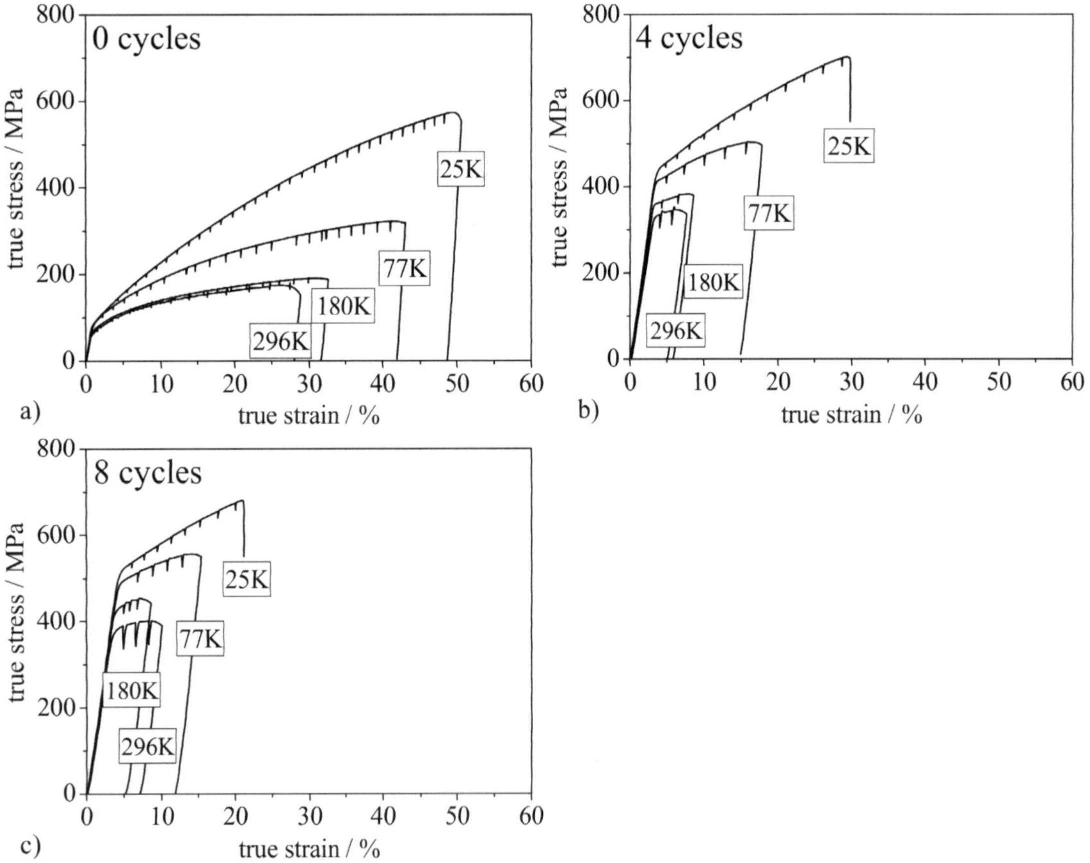

Figure 1. Stress-strain curves for samples deformed at different temperatures: (a) initial material, (b) four and (c) eight ARB cycles.

As can be seen in figure 2, yield stress σ_p and ultimate stress σ_m are similarly affected by the ARB process. For the initial as well as the ARB processed material there is an almost linear decrease of σ_p with temperature (figure 2a). A higher deformation degree shifts the yield stress to higher values and leads to a stronger temperature dependence. Compared to the initial state eight ARB cycles increase the yield stress by about 600% at all temperatures applied. It is widely accepted that this increase with the deformation degree is due to dislocation and grain boundary hardening. In [9] it was shown that the main grain size reduction of AA6016 has taken place until the fourth cycle. Accordingly, about 83% of σ_p is already reached after four cycles.

The decrease with temperature is stronger for σ_m than for σ_p (figure 2b). Compared to the initial material, eight ARB cycles increase σ_m at 25K only by 22%, while at 296K it is 130%.

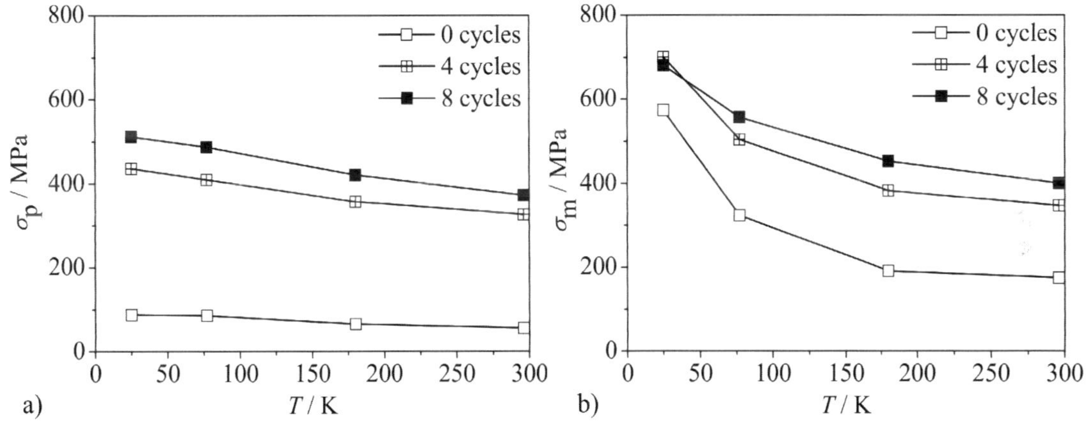

Figure 2. (a) σ_p and (b) σ_m for different numbers of ARB cycles versus deformation temperature.

The difference σ_m-σ_p shown in figure 3b characterizes the strain hardening capacity of the material. If during tensile deformation certain regions of the specimen deform stronger than others, more dislocations are generated there in order to achieve the higher deformation. The enhanced dislocation hardening in these regions stabilizes and hence homogenizes the deformation process of the specimen. Thus, the strain hardening capacity of the material governs its ductility (figure 3b). Accordingly, both σ_m-σ_p and the fracture strain ε_f are influenced similarly by deformation degree and deformation temperature: At lower temperatures both strain hardening capacity and ductility non-linearly decrease with temperature. The strain hardening capacity is limited by the saturation dislocation density ρ_s. If during tensile deformation ρ_s is reached, no more dislocations can be produced to stabilize the deformation of the specimen leading to its inhomogeneous deformation and hence to necking. Due to faster recovery caused by enhanced climb and cross slip of edge and screw dislocations, respectively, the saturation dislocation density and accordingly strain hardening capacity and ductility decrease with increasing deformation temperature. The reason that strain hardening is observed even at 296K is that ARB processing and tensile testing was done at elevated and ambient temperatures, respectively.

ARB processing strongly reduces the high strain hardening capacity and ductility of the annealed initial state. This reduction correlates with the degree of deformation: after eight cycles σ_m-σ_p at 25K and 296K is only 35% and 23% of that of the initial material, respectively. As mentioned above, the strain hardening capacity during tensile testing is related to the difference between the dislocation density before testing and at saturation. Accordingly, the highly deformed ARB specimens show lower strain hardening than the annealed initial material.

For the two highest temperatures, 180K and 296K, σ_m-σ_p is constant from the fourth to the eighth cycle. Assuming that at temperatures of 180K and higher, ρ_s is already reached during the first four ARB cycles, the specimens after four and eight cycles would have the same dislocation density before the tensile testing and hence the same strain hardening capacity.

The maximum fracture strain of 46% is reached in the initial material at 25K. ε_f reduces to 28% at 296K. After eight cycles ε_f at 25K and 296K is reduced by a factor of 3 and 4, respectively. Again, the main decrease has occurred during the first four ARB cycles. The slight increase of ε_f at 296K after four cycles may be related to the increase of the strain rate sensitivity [9].

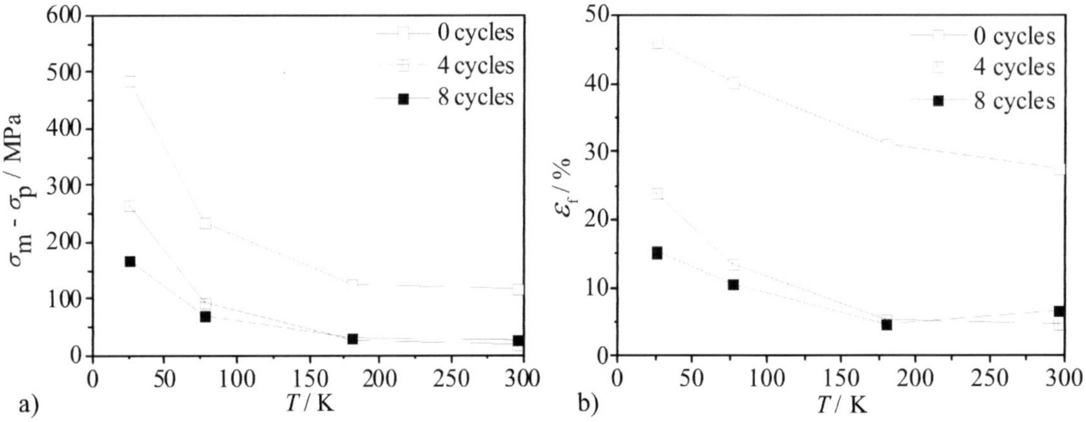

Figure 3. (a) σ_m-σ_p and (b) ε_f for different numbers of ARB cycles versus deformation temperature.

During the tensile tests additional relaxation experiments were performed (stress drops in figure 1). They yield information about the deformation mechanisms controlling the plastic behaviour at the particular deformation temperatures. They are analysed in detail in [13].

4. Conclusions

Between 25K and 296K the yield stress and ultimate stress decrease with increasing temperature, while there is an increase with the number of ARB cycles due to dislocation and grain boundary hardening. Moreover, the ductility decreases with increasing temperature and increasing number of ARB cycles (except at 296K) due to decreasing strain hardening capacity.

Acknowledgments

The ARB aluminium alloy was kindly provided by Dr. H.-W. Höppel, University Erlangen-Nürnberg.

References

[1] Saito Y, Utsunomiya H, Tsuji N and Sakai T 1999 *Acta Mater.* **47** 579
[2] Valiev RZ, Korznikov AV and Mulyukov RR 1993 *Mater. Sci. Eng.* A **168** 141
[3] Tsuji N, Saito Y, Utsunomiya H and Tanigawa S 1999 *Scripta Mater.* **40** 795
[4] Valiev RZ, Islamgaliev RK and Alexandrov IV 2000 *Prog. Mater. Sci.* **45** 103
[5] Valiev RZ, Estrin Y, Horita Z, Langdon TG and Zehetbauer M 2006 *JOM* **58** 33
[6] Kim W, Kang SB, Tsuji N and Minamino Y 2005 *Acta Mater.* **53** 737
[7] Höppel HW, May J and Göken M 2004 *Adv. Eng. Mat.* **6** 781
[8] Tsuji N, Shiotsuki K and Saito Y 1999 *Mater. Trans.* **40** 765
[9] Scharnweber J, Skrotzki W, Oertel C-G, Brokmeier HG, Höppel HW and Topic I 2009 *Adv. Eng. Mat.*, submitted
[10] Topic I, Höppel HW, Staud D, Merklein M, Geiger M and Göken M 2008 *Adv. Eng. Mat.* **10** 1101
[11] Hollang L, Hieckmann E, Brunner D, Holste C and Skrotzki W 2006 *Mater. Sci. Eng.* A **424** 138
[12] Reuther K 2008 Diploma thesis, Technische Universität Dresden
[13] Hollang L, Scharnweber J, Reuther K and Skrotzki W 2010, this issue

The hyperbolic sine relation and dip test data in constant strain rate compression experiments

T Kruml[1], O Coddet[2] and JL Martin[3]

[1]Institute of Physics of Materials, Academy of Sciences, CZ 61662 Brno, Czech republic
[2]Platit A.G., CH 2540 Grenchen, Switzerland
[3]IPMC, FSB, Ecole Polytechnique Fédérale de Lausanne, CH 1015 Lausanne, Switzerland

kruml@ipm.cz

Abstract. Single crystals of Ni_3Al, intrinsic Ge, Cu, Cu with 5at%Al and 7.5at%Al are submitted to constant strain rate compression tests and stress reduction experiments. Curves of the initial creep rate as a function of the amount of stress reduction are analysed in terms of a hyperbolic sine relation based on the thermal activation of the dislocation velocity. It is shown that this relation satisfactorily fits the curves of the two first crystals. However, this is not observed for the three last ones. Reasons for this behaviour are proposed as well as a method for the effective stress evaluation.

1. Introduction

Stress reduction experiments have been proposed [1,2] to determine the thermal component τ^* of the applied stress τ. τ^* is the stress necessary for a dislocation to overcome localized obstacles. In a simple description, τ and τ^* are related by [3]:

$$\tau = \tau^* + \tau_i \quad (1)$$

where the internal stress τ_i represents the athermal elastic interaction with other distant dislocations.

Stress reduction experiments are usually performed in creep, see e.g. [2,4,5], much more seldom under constant strain rate tests, see e.g. [1,6], where they are more complicated to perform. The procedure can be briefly described as follows: at strain γ on the stress-strain curve, the stress is suddenly reduced by an amount $\Delta\tau$, then the sample is allowed to creep under the stress $\tau-\Delta\tau$, for a short period of time. The initial creep rate $\dot{\gamma}_{pi}$ is recorded and observed to be positive, negative or zero depending on the magnitude of $\Delta\tau$. At zero creep rate $\Delta\tau = \tau^*$, for strain γ. The curve of $\dot{\gamma}_{pi}$ as a function of $\Delta\tau$ can be described by a hyperbolic sine function, as proposed by Milicka [5]:

$$\dot{\gamma}_{pi} = A \sinh [V(\tau^*-\Delta\tau)/kT] \quad (2)$$

In terms of the thermal activation theory, A is a coefficient that depends on temperature through an activation energy and V is the activation volume, A and V being constant during the stress dip. Relation (2) assumes i) a constant mobile dislocation density during the fast stress reduction and ii) a dislocation velocity that is thermally activated, i.e. depends on stress and temperature through the hyperbolic sine law [7]. This latter dependence is valid whatever the stress, even at very low values when dislocations exhibit forward and backward motions, provided these are likely to occur. Relation

(2) indicates that $\dot{\gamma}_{pi}$ is positive for $\Delta\tau$ values smaller than τ^*, is zero for $\Delta\tau = \tau^*$ and negative for $\Delta\tau < \tau^*$, as expected.

The point of this study is to compression deform various single crystals, record the $\dot{\gamma}_{pi}(\Delta\tau)$ curves and to assess the Milicka law (relation (2)). The crystals which have been selected are known to deform through different microscopic mechanisms. The corresponding data have already been published [8] but are reanalyzed here through different procedures.

More precisely, for a given material at given strain, a curve is plotted representing experimental data of $\dot{\gamma}_{pi}$ as a function of $\Delta\tau$. Then the following questions will be addressed:

i) How accurately does the hyperbolic sine relation fit this curve?

ii) How well can an extrapolation of this curve guide the experimentalist in the choice of $\Delta\tau$ values. Indeed, the extrapolation starts from the first low $\Delta\tau$ values, could indicate an approximate value of τ^*, and suggest a choice of $\Delta\tau$ values larger than τ^*, thus saving time. In the result section, dotted blue curves refer to a fit considering all the data points, red ones refer to a few first points at low $\Delta\tau$ values.

2. Experimental methods

2.1 Materials

The single crystals under investigation consist of Ni_3Al with 3at%Hf tested at 573K (0.35 of melting point temperature T_M), intrinsic Ge tested at 750K (0.62 T_M), Cu tested at 295K (0.22 T_M), Cu with 5at%Al and 7.5at%Al tested at 295K (0.22 T_M), referred to as Cu5Al and Cu7Al respectively. These crystals are known to deform according to different mechanisms. They were cut in the [−123] single slip orientation.

2.2. Mechanical testing

Stresses τ, strains γ and strain rates $\dot{\gamma}$ are resolved with respect to the primary slip system.

Details about compression tests, stress reduction experiments and computer programs designed for machine control and data acquisition can be found in [8]. Shear strain rates $\dot{\gamma}$ were in the range of 10^{-4} s^{-1}. Tests are performed at relatively low homologous temperatures, except for Ge.

3. Results

A typical curve $\dot{\gamma}_{pi}(\Delta\tau)$ is presented in Fig. 1 for Ni_3Al deformed at 573 K. The blue dotted curve is a fit of relation (2) with all the data. Given the experimental uncertainty, the fit is considered to be satisfactory. It yields a value $\tau^* = 75.7$ MPa. The red curve is a fit of relation (2) with the 5 first data points. It is very close to the blue curve yielding a very similar value of $\tau^* = 77.8$ MPa.

Fig. 2 illustrates a similar attempt for Ge at 750K. The red curve is a fit considering the 4 first data points. It yields $\tau^* = 12.6$ MPa. The blue curve is very close to the red one and yields $\tau^* = 12.8$ MPa.

In these two cases, τ^* could be obtained from about 5 dip tests (i.e. 5 specimens) only, at low $\Delta\tau$ values and extrapolating with relation (2).

Fig. 3 shows a very different behavior for Cu at 295 K. There is a wide range of $\Delta\tau$ values for which $\dot{\gamma}_{pi}$ is close to zero and never negative. Fitting relation (2) with all data points (dotted blue curve) is still acceptable, yielding a τ^* value of 5.4 MPa. However, when the 4 first data points are considered, a very different curve is obtained as well as a different value of $\tau^* = 3.0$ MPa.

Cu-5Al (Fig. 4) exhibits a behaviour similar to Cu. The fitting curves are superimposed at low $\Delta\tau$ values but differ for $\Delta\tau > 7$ MPa. They yield quite different values of τ^* (8.8 MPa for the whole set of data and 4.5 MPa for the 6 first points).
Cu-7Al (Fig. 5) exhibits a peculiar behaviour. Some negative values of $\dot{\gamma}_{pi}$ are observed. The two fitted curves are very similar up to $\Delta\tau$ = 11MPa. They yield τ^* = 6.8 MPa when the 7 first points are considered and 7.8 MPa otherwise.

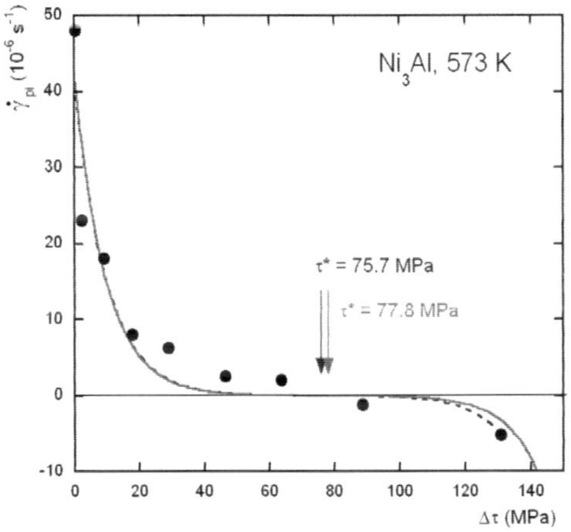

Figure 1. Reduced strain rate $\dot{\gamma}_{pi}$ as a function of the stress reduction $\Delta\tau$ for a Ni_3Al single crystal at 573 K and 620 MPa. Relation (2) fits all data points for the dotted blue curve, the 5 first ones for the red one. The corresponding values of τ^* are indicated.

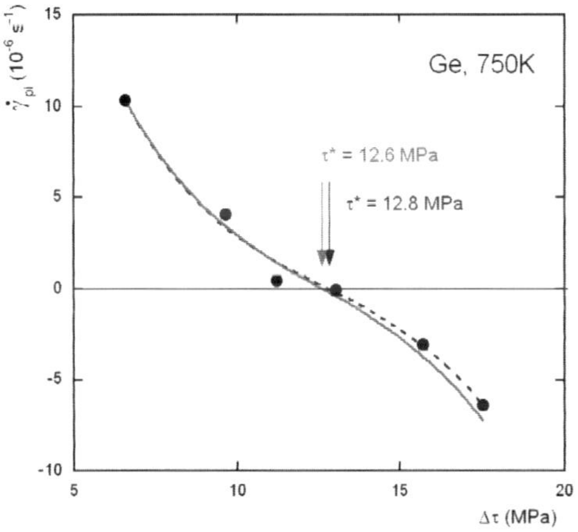

Figure 2. Same as Fig. 1 for a Ge single crystal at 750 K and 24 MPa. The red curve considers the 4 first data points.

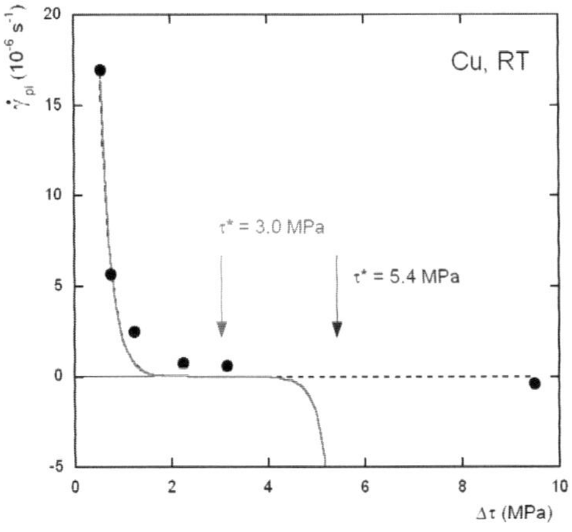

Figure 3. Same as Fig. 1 for a Cu single crystal at 295 K and 47 MPa. The red curve considers the 4 first data points.

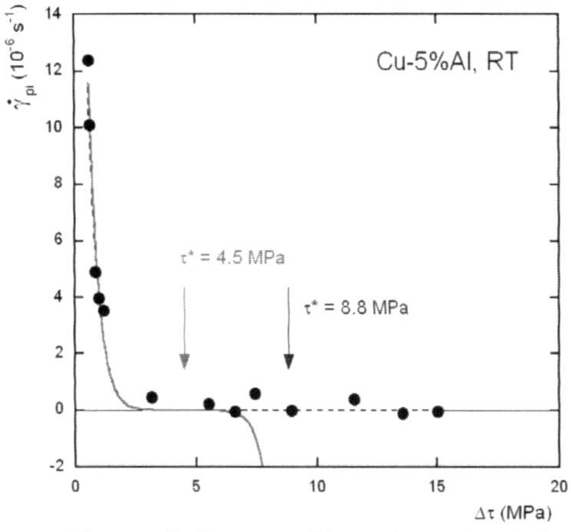

Figure 4. Same as Fig. 1 for a Cu-5Al single crystal at 295 K and 15 MPa. The red curve considers the 6 first data points.

4. Discussion

Table 1 summarizes the results. The estimated values of τ^* refer successively to the dotted blue and red curves of the figures. A comparison of the curves on the figures shows that Ni_3Al and Ge behave similarly according to relation (2). τ^* can be determined from the whole curve or the few first points. However, the slopes of the curves for $\dot{\gamma}_{pi} = 0$ are different. Taking the derivative of (2) for $\Delta\tau = \tau^*$ yields a slope of -AV/kT, which has different values for the two crystals under different conditions. Cu and Cu-5Al exhibit no clear evidence of negative $\dot{\gamma}_{pi}$ over a large range of $\Delta\tau$ values. The behaviour of Cu-7Al is rather similar. It is thought that the forest mechanism is operating in Cu and Cu-5Al so that the backward jumps of dislocations are unlikely thus excluding negative $\dot{\gamma}_{pi}$ values. In this case V depends on stress and therefore relation (2) is not justified.

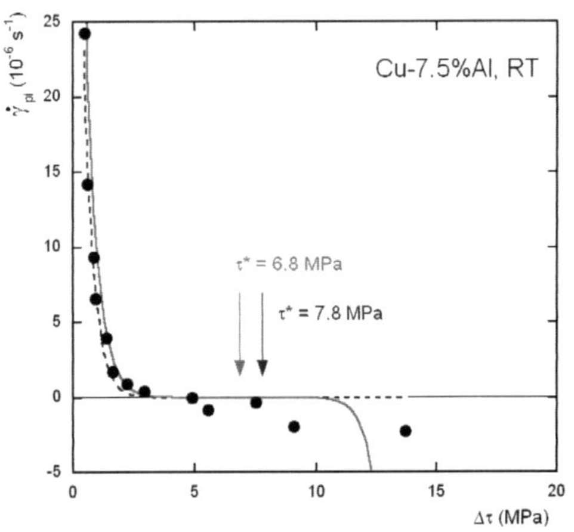

Figure 5. Same as Fig. 1 for a Cu-7Al single crystal at 295 K and 30 MPa. The red curve considers the 7 first data points.

Table 1. Comparison of mechanical test data for the different materials.

sample	T K	τ (MPa)	τ^* (MPa)	τ^* (MPa)
Ni_3Al	573	620	75.7	77.8
Ge	750	24	12.8	12.6
Cu	295	47	5.4	3.0
Cu-5Al	295	15	8.8	4.5
Cu-7Al	295	30	7.8	6.8

5. Conclusions

The hyperbolic sine relation accounts for the dip test data in Ni_3Al (573 K, 620 MPa) and Ge (750 K, 24 MPa). It can be used safely to determine τ^* and save samples and experimental time. However, for Cu, Cu-5Al the forest mechanism seems to be rate controlling, negative values of $\dot{\gamma}_{pi}$ predicted by Milicka's relation are not observed. Cu-7Al exhibits an intermediate behaviour.

Acknowledgements

The financial support was provided by Fonds National Suisse and Academy of Sciences of the Czech republic (AV0Z20410507). Discussions with J. Bonneville, D. Caillard and G. Saada are acknowledged.

References

[1] Gibbs GB 1966 *Phil. Mag.* **13** 317
[2] Solomon AA and Nix WD 1970 *Acta Metall.* **18** 863
[3] Seeger A, Diehl J, Mader S and Rebstock H 1957 *Phil. Mag.* **2** 323
[4] Blum W and Finkel A 1982 *Acta Metall.* **30** 1705
[5] Milicka K 1999 *Acta Mater.* **46** 1831
[6] Kruml T, Coddet O, Saada G and Martin JL 2003 *Phil. Mag. Lett.* **83** 651
[7] Caillard D and Martin JL 2003 *Thermally Activated Mechanisms in crystal Plasticity* (New York: Pergamon) p 6
[8] Kruml T, Coddet O and Martin JL 2005 *Z. Metallkde.* **96** 6

Mechanical properties of 15%Mn steel with fine lamellar structure consisting of ferrite and austenite phases

R Ueji[1], Y Okitsu[2], T Nakamura[1], Y Takagi[1], Y Tanaka[1]

[1] Faculty of Engineering, Kagawa University, 2217-20, Hayashi-cho, Takamatsu, Kagawa, 761-0396, Japan
[2] Honda R&D Co. Ltd., 4630, Shimotakanezawa, Haga-machi, Haga-gun, Tochigi, 321-3393, Japan

ueji@eng.kagawa-u.ac.jp

Abstract. New steel with fine lamellar structure consisting of austenite and ferrite was developed. 15mass%Mn-3%Al-3%Si steel sheet was used in this study. First of all, the effect of the cooling rate on the microstructure was examined. The cooling at the slower speed of 100 deg/hour created the dual phase structure consisting of both austenite and ferrite. The additional rolling developed the fine lamellar duplex structure. Improvement of both the tensile strength and elongation was achieved by rolling. The strength increases furthermore by the rolling up to larger reduction. The 90% rolled sheet shows high tensile strength around 1000MPa with large elongation (15%-20%). These results indicate that the multi-phased structure with controlled lamellar morphology is beneficial for the management of both high strength and large ductility.

1. Introduction

Austenitic steel is one of the advanced steels due to its extremely large ductility. For example, twining induced plasticity (TWIP) steel with a large amount of manganese in addition to some aluminium and silicon (e.g. 25-30mass%Mn-3%Al-3%Si) was studied for the automotive application [1,2]. Previous study [1] has shown that TWIP steel exhibits high tensile strength with adequate elongation even at high strain rates, so that TWIP steel is desired to be applied for automotive use. However, the TWIP steel need expensive costs due to large amount of Mn. When the Mn content is reduced, the austenite becomes unstable and other phases such as ferrite and ε-martensite can appear. These phases are regarded as harder phases than austenite. Consequently, the combination of austenite and other phases is useful for the commercial production and even for strengthening. However, few studies have been conducted to clarify the process how the duplex structure in these steels can evolve. Consequently, the aim of this study is to clarify the basic concept for the process for the medium Mn duplex structured steel with good mechanical properties.

2. Experimental procedures

The hot rolled steel sheet with chemical composition of 15%Mn-3%Al-3%Si-bal.Fe was used. The initial thickness is 12mm. Thermomechanical treatments examined in this study is illustrated in Fig.1. First of all, the effect of the cooling rate (a) was studied. The sample was reheated to 1200°C followed by cooling at various cooling media; water quench, air-cooling and furnace-cooling. Subsequently, the combination of the reheating and multi-pass rolling with various total reductions was also examined

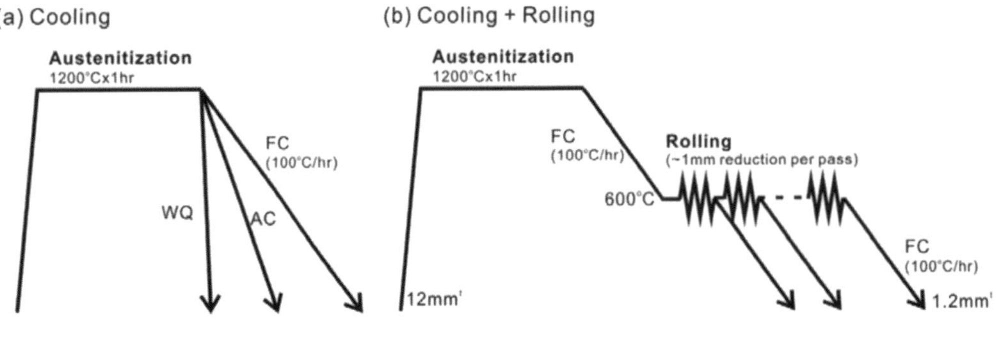

Fig. 1. Thermomechanical procedures examined in this study

(b). The rolling temperature was fixed at 600°C, at which the rolling can be conducted without any cracking/failure. The tensile test was conducted at a strain rate of $10^{-3}s^{-1}$ and the tensile axis was parallel to the rolling direction (RD). Microstructural observations were also conducted by transmission electron microscope (TEM) etc.

3. Results and discussion

3.1. Cooling-rate effects

Figure 2 shows the optical microstructures of the samples cooled from 1200°C by water-quench (a) or in the furnace (b). The cooling rate in the furnace (b) is 100 deg./hour. The water-quenched sample (a) shows the coarse grained ferrite structure (~ a few hundred μm) with several cracks introduced during the cooling. Similar ferrite structure can be found in the air-cooled sample. On the other hand, the furnace-cooled sample (b) shows duplex structure consisting of ferrite (white parts) and austenite (black parts). Both parts have elongated morphologies in several different directions. No crack exists in the furnace-cooled sample.

Nominal stress – nominal strain curves of the samples cooled at various cooling rates are shown in Fig.3. The water-quenched and the air-cooled samples fractured in brittle manner due to cracking during the cooling. When the cooling rate is slower, the elongation can be improved. The stress – strain curve of the furnace-cooled sample shows plastic deformation with work hardening, although fracture takes place immediately at the maximum of the nominal stress and no local elongation (necking) can be found. This feature suggests that the furnace-cooled sample still keeps brittle property.

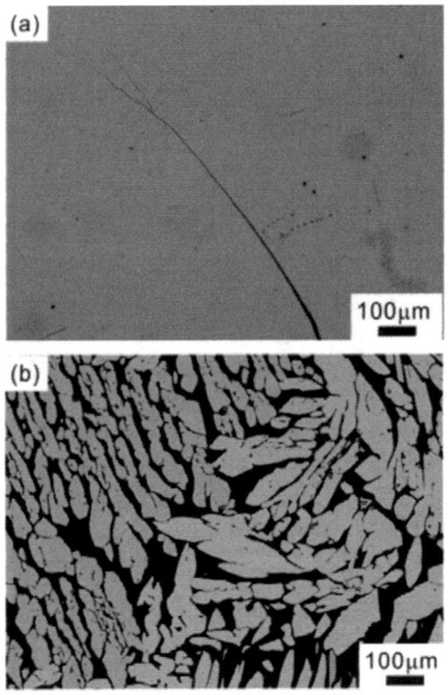

Fig. 2. Optical microstructures of the 15Mn-3Al-3Si steels reheated to 1200°C followed by water-quenching (a) or furnace-cooling (b).

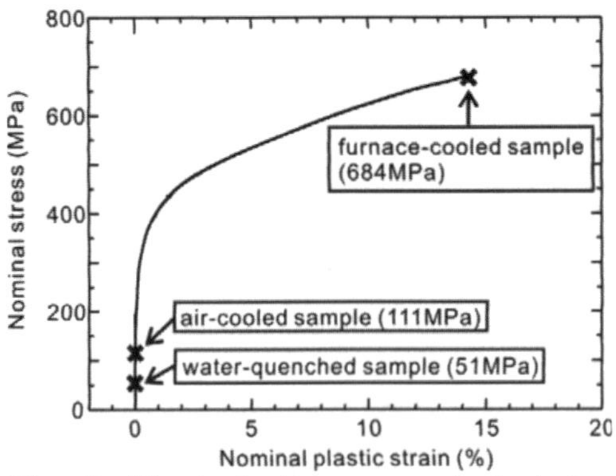

Fig. 3. Nominal stress - nominal strain curves of the 15Mn-3Al-3Si steels cooled at various rates from 1200°C.

3.2. Rolling effects

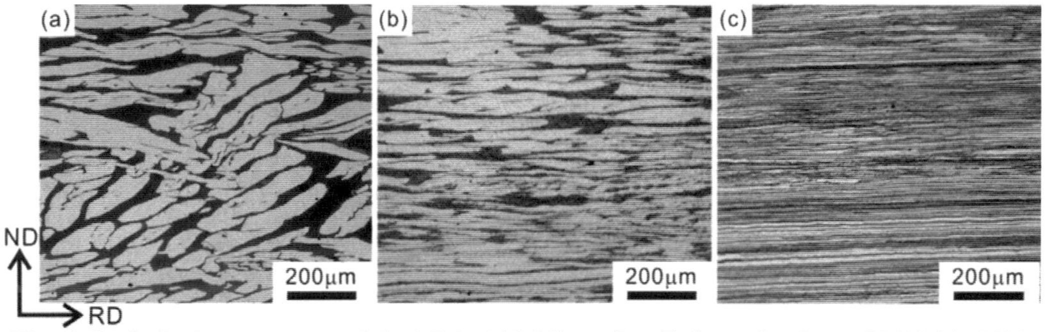

Fig. 4. Optical microstructures of the 15Mn-3Al-3Si steels rolled to reductions of 50% (a), 75% (b) or 90% (c) at 600°C during the furnace cooling from 1200°C.

Figure 4 shows the optical microstructure of the samples rolled at 600°C up to the various reduction in thickness during the furnace cooling. In this figure, the horizontal and vertical directions are parallel to the RD and the nominal direction (ND), respectively. The 50% rolled sample has the microstructure similar to that observed in the just furnace-cooled sample (Fig. 2(b)) although both ferrite and austenite parts are elongate along the RD. This morphology changes to a lamellar structure when the rolling reduction reaches up to 75% (Fig. 4(b)). The layers of the lamellar structure become thinner with increase of the rolling reduction. The 90% rolled sample exhibits the uniform and fine lamellar structure consists of the ferrite and the austenite layers of mean thickness of a few μm. According to the point-counting measurements of the consisting phases, the area fraction of ferrite slightly increases. The fractions of ferrite in the 0%, 25%, 50%, 75% and 90% rolled samples are 39%, 39%, 42%, 45% and 53%, respectively. These results indicate that the phase transformation from ferrite to austenite proceeds mainly before the rolling and one of the significant effect introduced by the rolling can be regarded as the alternation of the morphology.

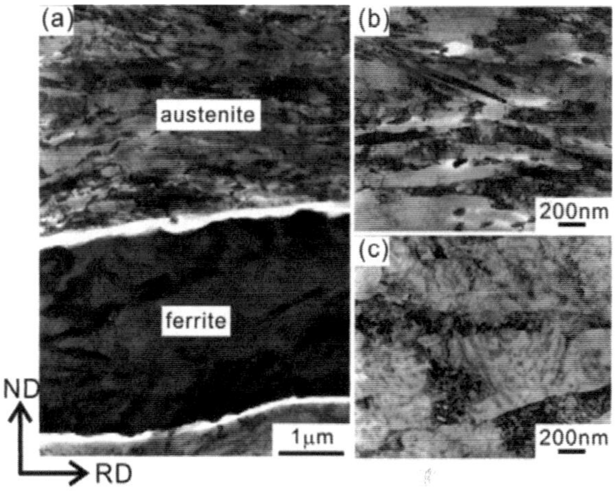

Fig. 5. TEM microstructure of the 15Mn-3Al-3Si steel rolled to 90% reduction at 600°C. The entire microstructure (a) consists of ferrite (b) and austenite (c) layers.

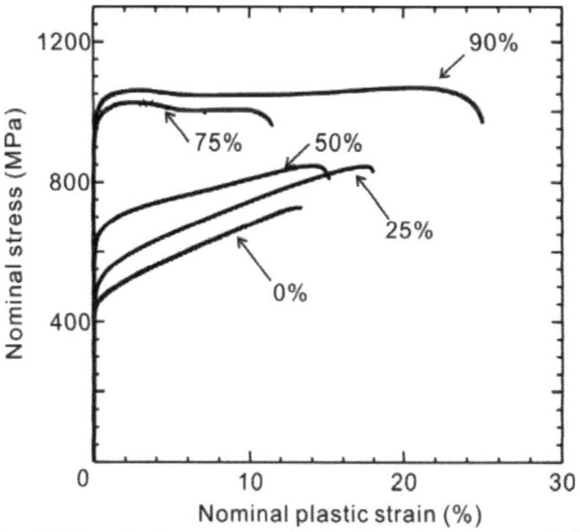

Fig. 6. Nominal stress - nominal strain curves of the 15Mn-3Al-3Si steels rolled to various reductions at 600°C during the furnace-cooling from 1200°C.

Figure 5 shows TEM microstructures of the 90% rolled sample. The analysis of selected area diffraction patterns clarified the existence of both austenite and ferrite layers as denoted in (a). The elongated dislocation structure (cell and/or subgrain) was also observed in both austenite (b) and ferrite (c) layers. Such dislocation microstructure cannot be found in the just furnace cooled samples, so that these evolved during 600°C rolling. Additionally, fine Fe_3Al (or Fe_3Si) particle can be found in ferrite layers (c).

Figure 6 shows the stress – strain curves of the samples rolled at 600°C to various reductions. Data of the furnace-cooled sample without rolling (0%) is also denoted. The yield stress becomes larger with larger reduction. Concerning of the elongation, although the simple tendency is difficult to be determined, it should be noted that rolling hardly brings worse effect to the elongation. It should be noted that some of the rolled sheets show large elongations as is in the 90% rolled sample. Additionally, all the rolled samples show gentle drops of the nominal stress corresponding to necking prior to the fracture. These features suggest that the rolling contributes both to the large strength and preferable ductility in the 15Mn-3Al-3Si steel.

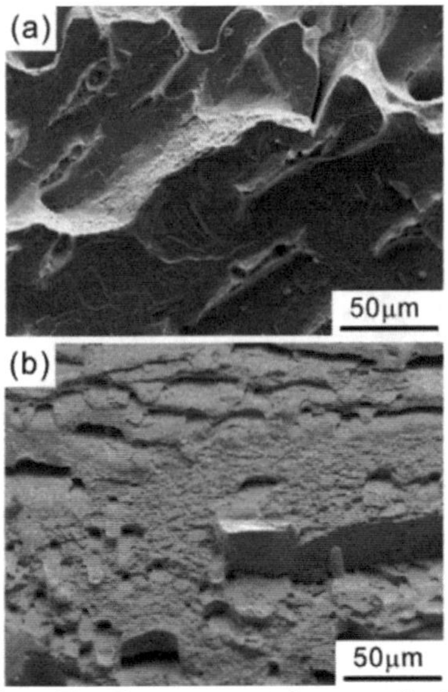

Fig. 7. Fracture surfaces of the 15Mn-3Al-3Si steels cooled without rolling (a) or 90% rolled during cooling (b).

In order to clarify the effect of the rolling, the fractography was conducted. The SEM images of fracture surfaces on the samples just furnace-cooled (a) or 90% rolled at 600°C (b) are shown in Fig.7. The furnace-cooled sample exhibits the mixture of the brittle river pattern and the dimple pattern in the ductile manner. On the other hand, the rolled surface exhibits the dimple pattern predominantly. This result indicates that the brittle fracture is prohibited by rolling. The reason of this inhibition can be considered as following: the origin of the brittle fracture of the furnace-cooled steel could be the ferrite with large amount of Al and Si since these elements can accelerate brittle fracture as it is well known in the Fe-Si ferrite alloy for a transformer in electronic power industry. This brittle fracture is partially due to mechanical twinning in ferrite [3]. Twinning has strong orientation dependence [2] and the microstructure (grain size, dislocation structure) also affect twinning. In case of the rolled 15Mn-3Al-3Si steel, the ferrite layers can contain strong <110>//RD texture. <110> is less preferable tensile orientation for twinning [3]. Additionally, the ferrite layer has many dislocation cells (or subgrains) which can also prohibit the twinning. One can finally conclude that the rolling can provide difficult structure for twinning.

4. Summary

This study demonstrated the efficiency of thermomechanical process on the microstructural control in 15Mn-3Al-3Si steel. The lamellar structure consisting of austenite and ferrite duplex phases can be evolved by the combination of slow-rate cooling and warm temperature rolling. This duplex steel has preferable mechanical properties.

Acknowledgments

Mr. Imanaga in Honda R&D Co. and Prof. Kunishige in Kagawa Univ. helped our experiments and gave us useful comments. RU acknowledges the financial supports by the Industrial Technology Research Grant Program from New Energy and Industrial Technology Development Organization (NEDO) of Japan (project ID 07A26002a) and the fund for Kagawa University Young Scientists 2009.

References

[1] Grässel O, Krüger L, Frommeyer G and Meyer L W 2000 *Int. J. Plast.* **16** 1391
[2] Ueji R, Tsuchida N, Terada D, Tsuji N, Tanaka Y, Takemura A, Kunishige K 2008 *Scripta Mater.* **59** 963
[3] Narita N and Takamura J 1992 *Dislocations in Solids* **9** 135

Fracture analysis in the modified split-cantilever beam using the classical theories of strength of materials

A Szekrenyes

Budapest University of Technology and Economics, Department of Applied Mechanics, Budapest, Muegyetem rkp. 5, Building MM, H-1111

szeki@mm.bme.hu

Abstract. The modified split-cantilever beam is a specimen type applied in fracture mechanics to measure the mode-III fracture properties of composites. In this paper an improved beam model is developed, which is based on the superposition of four different effects. The analytical model is validated by numerical calculations using the finite element software ANSYS and it is demonstrated that the agreement is very good between the analytical and the numerical models. Experimental measurements are also performed on unidirectional glass/polyester composite specimens, and a very good agreement is obtained between the results of the analytical model and the experiments. Apart from the excellent accuracy of the beam model it is shown that the modified split-cantilever beam has an important role in fracture mechanics, because it applies the same specimen geometry as the standard mode-I and mode-II tests and it is suitable to investigate the mode-III fracture properties in a quite extended crack length range.

1. Introduction

In the last decade more and more attention was focused on the investigation of the mode-III interlaminar fracture mechanisms of laminated composite materials [1,2]. This indicates that – apart from the mode-I and mode-II fractures - the mode-III is also important for the complete fracture characterization of the material. However, the mode-III fracture involves several difficulties, which do not take place under mode-I, mode-II and mixed-mode I/II tests. One of them is that – to the best of the author's knowledge – a pure mode-III fracture can not be produced. One of the mode-III specimen types is the modified split-cantilever beam (MSCB).

The aim of the present work is to improve the efficiency of the MSCB specimen [3,4], to provide an accurate closed-form solution for the compliance and the energy release rate and to demonstrate its applicability for the reduction of the experimental data. The MSCB specimen maintains the traditional beam-like geometry. Another reason for developing closed-form solution and performing experiments is that it is possible to combine it with mode-I double cantilever-beam (DCB) [5] and mode-II end-notched flexure test (ENF) [6,7]. Although these tests have several drawbacks, their simplicity is a great advantage.

2. Analysis – beam theories

The MSCB specimen is demonstrated in Fig. 1, where it is shown that the rigs subject the specimen by four grub screws inducing a scissor-like load of the sample [5,6]. The specimen is treated as a slender beam and the bending, shear, the so-called Siant-Venant effect and the free torsion of the specimen are equally accounted for [8]. The compliance (defined as $C=\delta/P$, δ is the displacement at roller C, P is the applied load at roller C in Fig. 1) of the MSCB specimen is [8]:

$$C=\frac{8a^3}{b^3hE_{11}}\left[f_{EB1}+f_{TIM1}+f_{FT1}+f_{S-V1}\right], \tag{1}$$

where:

$$f_{EB1}=1-3\left(\frac{s_1+s_2}{a}\right)+3\left(\frac{s_1+s_2}{a}\right)^2-\frac{s_1(s_1+s_2)(s_1+2s_2)}{a^3}, \tag{2}$$

$$f_{TIM1}=0.3\left(1-\frac{s_2^2-s_1^2}{as_1}\right)\left(\frac{b}{a}\right)^2\left(\frac{E_{11}}{G_{13}}\right),\quad f_{FT1}=0.19\frac{1}{\varsigma}\left(1-\frac{s_1}{a}\right)\left(\frac{b}{a}\right)^2\left(\frac{E_{11}}{G_{12}}\right), \tag{3}$$

$$f_{S-V1}=0.48\left(\frac{a-(s_1+s_2)}{a}\right)^2\left(\frac{b}{a}\right)\left(\frac{E_{11}}{G_{13}}\right)^{\frac{1}{2}},\quad \varsigma=1-0.63\mu\frac{h}{b},\quad \mu=\left(\frac{G_{13}}{G_{12}}\right)^{\frac{1}{2}}. \tag{4}$$

where a is the crack length, b is the specimen width, h is the half thickness, s_1 and s_2 are the distances between the grub screws (see Fig. 1), E_{11} is the flexural modulus of the specimen, G_{12} and G_{13} are the shear moduli of the specimen, furthermore, f_{W1} is from the Winkler-Pasternak foundation, f_T is from transverse shear and f_{S-V1} accounts for the Saint-Venant effect [8].

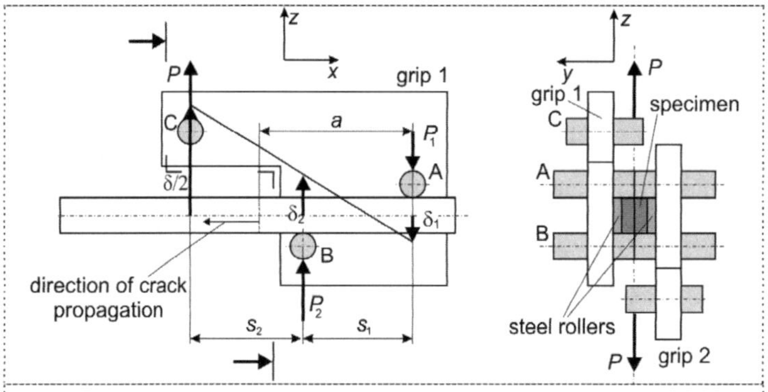

Figure 1. 2D illustration of the MSCB specimen.

The energy release rate of the MSCB specimen is [8]:

$$G_{MSCB}=\frac{12P^2a^2}{b^4hE_{11}}\left[f_{EB2}+f_{TIM2}+f_{FT2}+f_{S-V2}\right], \tag{5}$$

where:

$$f_{EB2}=1-2\left(\frac{s_1+s_2}{a}\right)+\left(\frac{s_1+s_2}{a}\right)^2,\quad f_{TIM2}=0.1\left(\frac{b}{a}\right)^2\left(\frac{E_{11}}{G_{13}}\right), \tag{6}$$

$$f_{FT2}=0.06\frac{1}{\varsigma}\left(\frac{b}{a}\right)^2\left(\frac{E_{11}}{G_{12}}\right),\quad f_{S-V2}=0.32\left(1-\frac{s_1+s_2}{a}\right)\left(\frac{b}{a}\right)\left(\frac{E_{11}}{G_{13}}\right)^{\frac{1}{2}}. \tag{7}$$

3. Comparison with numerical and experimental results

The analytical model (Eqs. (1)-(7)) is compared to finite element results to prove the accuracy of the model. The details of the FE models and the calculations are given in [8]. Fig. 2 shows the compliances at rollers A, B and C (Fig. 1) calculated from analysis and numerical models. The agreement is excellent and proves that the beam models captures very well the deformation of the specimen.

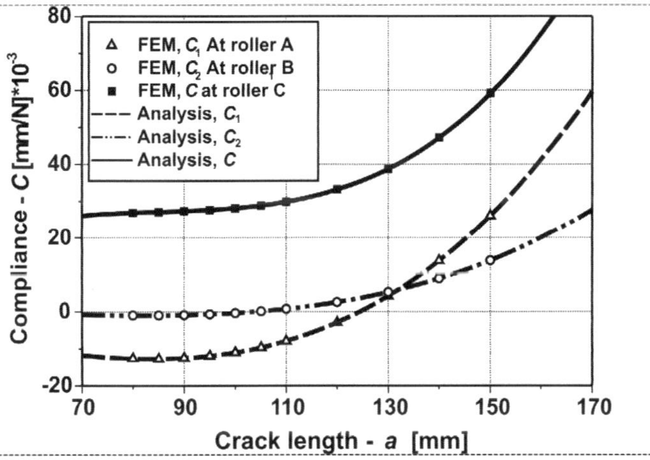

Figure 2. Comparison of the compliances by analysis and FE computations

The next step was to apply the model to data reduction in experimental tests. For this reason glass/polyester specimen were manufactured and tested using two rigs like the one, shown in Fig. 1. During testing the displacement of the specimen at roller C was measured by using a mechanical dial gauge, the load was recorded by using the scale of the testing machine. The measurements were performed by varying the crack length in the range of a=80 to 150 mm. The compliance versus crack length data was fitted by a third order polynomial and was plotted in Fig. 3 as well as the analytical model (Eq. (1)). The agreement is excellent, indicating the accuracy of the measurements and the analytical model. Table 1 shows the energy release rates calculated by two methods: the analytical model and the compliance calibration (CC) method [1, 2]. In this case the largest difference is 18.9 % between the results by beam model and CC, which can be explained by the sensitivity of the experimental compliance curve to the curve fit process. Overall the analytical model is assessed to be more accurate than the CC method.

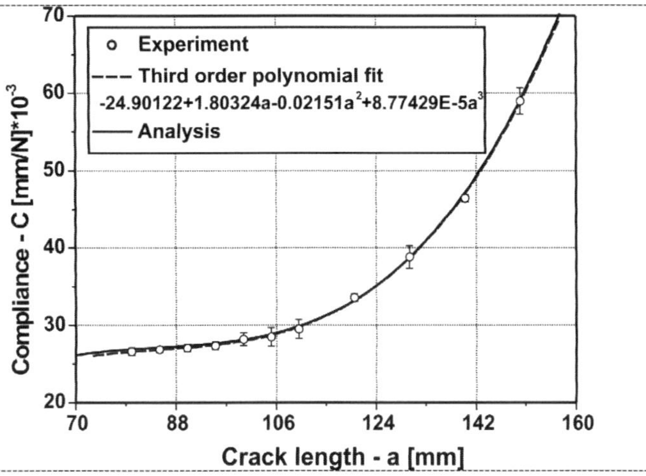

Figure 3. Experimentally measured and calculated compliances.

Table 1. Energy release rates calculated by the analytical model and the compliance calibration.

a [mm]	80	85	90	95	100	105	110	120	130	140	150
$G_C{}^{a}$	120.9	110.5	130.2	156.6	199.7	244.3	269.1	381.1	488.0	576.2	659.0
Scatter	±8.31	±2.7	±9.2	±3.01	±3.51	±4.91	±3.04	±6.86	±2.56	±8.44	±0.34
$G_{Beam}{}^{b}$	138.8	101.3	105.6	127.5	171.0	220.4	253.7	382.6	509.9	618.7	721.5
scatter	±6.91	±7.6	±8.2	±4.8	±5.01	±8.61	±8.83	±5.68	±2.86	±6.33	±8.34
diff^{ab}[%]	-14.8	8.4	18.9	18.6	14.4	9.8	5.7	-0.4	-4.5	-7.4	-9.5

$G_C{}^{a}$ – ERR by compliance calibration method [J/m^2]
$G_{Beam}{}^{b}$ – ERR by beam theory (Eq. (5)) [J/m^2]

4. Conclusions

In this paper the modified-split cantilever beam was revisited and an accurate analytical solution was developed. The compliance and the energy release rate of the specimen were calculated and were compared to results by numerical analysis and experiments. In both cases the agreement was excellent with the analytical model confirming the applicability and accuracy of the analytical model. The possibility to develop closed-form solution for fracture mechanical specimens is important, since this can make the data evaluation very fast. On the other hand the modified-split cantilever beam specimen is the only mode-III configuration, which is suitable to investigate the effect of crack length on the critical energy release rate at crack initiation in an extended crack length range.

5. Acknowledgements

This paper was supported by the János Bolyai Research Scholarship of the Hungarian Academy of Sciences and the National Science and Research Fund (OTKA) under Grant No. T34040.

References

[1] Brunner A J and Flüeler P 2005 Prospects in fracture mechanics of ''engineering'' laminates *Eng. Fract. Mech.* **72** 899–908
[2] Brunner A J, Blackman B R K and Davies P 2008 A status report on delamination resistance testing of polymer–matrix composites *Eng. Fract. Mech.* **75** 2779–2794
[3] Sharif F, Kortschot M T and Martin R H 1995 Mode III Delamination Using a Split Cantilever Beam *Composite Materials: Fatigue and Fracture – Fifth Volume* ASTM STP 1230, Ed.: Martin R H, ASTM, Philadelphia 85-99
[4] Trakas K and Kortschot M T 1997 The relationship between critical strain energy release rate and fracture mode in multidirectional carbon-fiber/epoxy laminates In: *Composite Materials: Fatigue and Fracture* – Sixth Volume ASTM STP 1285, 1997, Ed.: E. A. Armanios, ASTM, 283-304
[5] Szekrényes A 2009 Interlaminar fracture analysis in the G_I-G_{III} plane using prestressed transparent composite beams. *Composites Part A – Applied Science and Manufacturing* (accepted for publication)
[6] Szekrényes A 2007 Delamination fracture analysis in the G_{II}-G_{III} plane using prestressed transparent composite beams. *Int. J. Solids Struct.* **44** 3359-3378
[7] Szekrényes A 2007 Application of prestressed composite beams in fracture mechanics, *Per Polyt. Mech. Eng.* **51/2** 89-97
[8] Szekrényes A 2009 Improved analysis of the modified split-cantilever beam for mode-III fracture. *Int. J. Mech. Sci.* (accepted for publication)

Detection of nano-particles by dynamic dislocation-defect analysis

Shig Saimoto

Mechanical and Materials Engineering, Queen's University, Kingston, ON, K7L 3N6, Canada

saimoto@me.queensu.ca

Abstract. Dynamic dislocation-defect analysis is derived from precise determination of the activation volume υ and distance *d*. Upon precipitation of nano-particles of Al_6Fe in nominally pure aluminum, the Haasen plots can reveal their existence and a new activation distance plot can determine their magnitudes for specific obstacles from which particle size can be assessed.

1. Introduction

Nano-sized precipitates can be produced in nominally pure aluminum (Al) due to the low solubility of iron (Fe) but a special processing is required to overcome the low Fe diffusivity. This methodology first observed in electrical grade Al has been pursued and culminated in a quantitative kinetics of grain growth in very dilute Al-Fe alloys [1]. To achieve Fe solute levels below 1 ppm atomic, long-term anneals at 230 °C segregated Fe solutes to dislocations followed by Al_6Fe precipitation [2]. Subsequent anneals at 275 and 300 °C showed that three different levels of purity were attained [1] according to the phase diagram determined by precision strain rate sensitivity (PSRS) [3]. Furthermore the ultra-fine precipitates act as Zener drag on grain boundaries to limit grain growth resulting in continuous recrystallization (CReX) as described elsewhere [4]. Following this microstructural evolution, the thermodynamic response during tensile testing at low temperature was investigated. The Haasen plot determined by PSRS results in precise determination of the activation volume υ such that it was shown that the up strain-rate change is sensitive to the strong obstacles, primarily the forest dislocations, whereas the down-rate change incorporates the weaker defects termed deformation debris [5]. On the other hand, the less thermally activatable obstacles such as fine particles give rise to a steeper slope in the Haasen plot during the initial (yield) or micro-strain region and its analysis has been described [3]. Most recently it was shown that a plot of the inverse work-hardening coefficient θ versus the ratio of mean slip distance λ to the activation volume results in a slope which depicts the ratio of the activation distance *d* to the Burgers vector b [6]. The latter plot designated the activation distance diagram (ADD) together with the Haasen one and the log λ versus log flow stress τ plot comprise the primary relations for dynamic dislocation analysis (DDDA) as shown below:

Mean slip distance $$\lambda = \phi \frac{b}{2}\frac{\mu^2}{\tau}\frac{\partial \gamma}{\partial \tau}\Big|_{T,\dot{\gamma}} = \phi \frac{b}{2}\frac{\mu}{\tau}\frac{\mu}{\theta_{II}} \qquad \tau = \frac{\alpha \mu b}{\ell} \tag{1}$$

Activation volume $$v = kT\left(\frac{\partial \ln \dot{\gamma}}{\partial \tau}\right)\Big|_{T,\Sigma} = bd\ell \qquad S = m/T = \frac{1}{T}\frac{\partial \ln \tau}{\partial \ln \dot{\gamma}}\Big|_{T,\Sigma} = \frac{k}{\tau v} \tag{2}$$

Activation distance ratio $$d/b = (k/S)(\alpha b^3 \mu)^{-1} \cong \frac{\theta_{II}/\theta}{b^2 \lambda / \nu} \quad (3)$$

where μ is the shear modulus at test temperature T; θ_{II}, the Stage II work-hardening slope for [001] crystal; φ = f(α), a calibrating factor [6]; k, the Boltzmann constant and others are defined in text or retain their usual meaning. The ADD of equation (3) is unique in that the inter-obstacle length ℓ which inversely correlates to τ intrinsically cancels since υ is defined as $db\ell$ and derived from the stress drop. Hence the only modeling parameter which is encompassed in this evaluation is α, the strength of obstacle parameter, which ranges from 0 to 1.0. For the dislocation-forest (D-F) interaction this value for Al is taken to be 0.4 [6]. Thus the ADD plot gives rise to the thermodynamic signature of the obstacles which are statistically being intersected by the mobile dislocations and elucidates the interpretation of the Haasen plot; whereas λ is a linear function of α and is dependent on the choice of σ_0 used to define the flow stress τ from $(\sigma - \sigma_0)/M$ where σ_0 is the proof stress at 0.02 % strain and M, the Taylor factor of 3.

In this report, the purest Al matrix achieved by annealing at 230 °C for 100 h (RC1 in Ref. [1]) were tensile tested at 78 K. To validate that fine precipitates exist, the DDDA of specimens which were annealed for CReX at 255 °C for 1, 3 and 12 h was compared to that of super-pure Al of large (300 μm) grain size. The results clearly show that particles in the nm range are present.

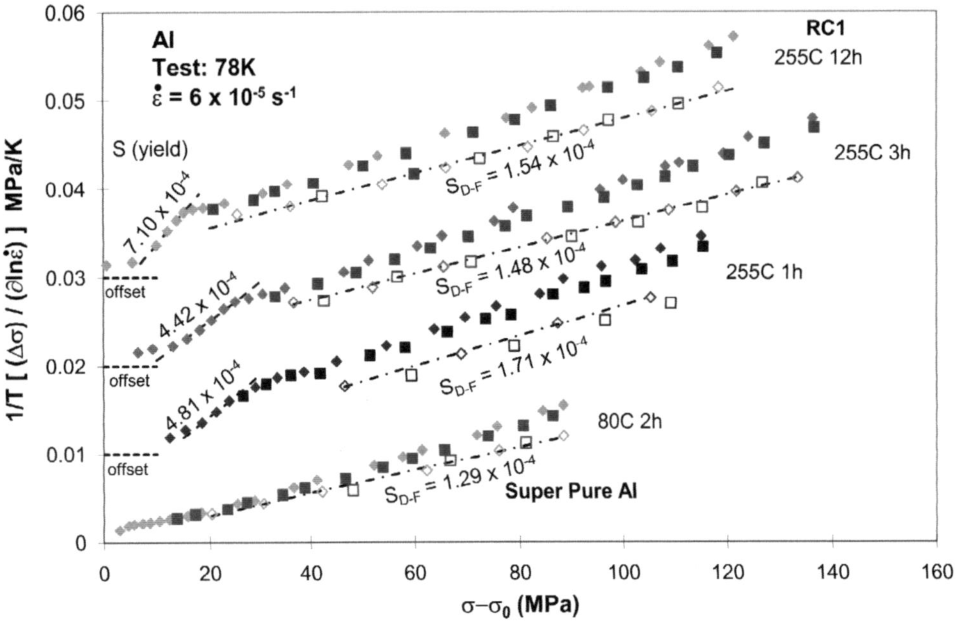

Figure 1. Haasen plot with the strain rate sensitivity slopes as referred to in the text.

2. Experimental approach.

The 27 mm as-cast slab of chemical composition (Al-8Cu-43Fe-55Mn-5Ni-67Si-112Ti in ppm atomic) was homogenized for 48 h at 600 °C; processing details are found elsewhere [1]. Tensile specimens with gauge section of about 40 x 2.4 x 0.9 mm^3 with an etched-reduced section were prepared prior to annealing at 255 °C in a salt bath. The PSRS tests [3] were performed with strain rate changes of ¼ and 1/10 down from 6 x 10^{-5} s^{-1} with sustained strain intervals of about 0.5% prior to up-change determinations. The microstructure was examined by electron channeling contrast in a scanning electron microscope after electro-polishing coupons after various heat treatments. In this examination, the one-μm sized CReX grains could be clearly recorded but no particles could be detected by this means. As reported [1], grain growth in the prepared sheets underwent a two stage process whereby the initial one was very slow. Thus even after 12 h at 255 °C, the grain size was about 2-3 μm. Super-pure Al was rolled after pre-cooling at 78 K but grain size smaller than 300 μm could not be attained by annealing at 80 °C for 2 h.

3. Results and Discussion

Figure 1 shows the Haasen plots derived from tensile testing of RC1 specimens aged at 255 °C for times t of 1, 3 and 12 h compared to that of super-pure Al which was recrystallized at 80 °C for 2 h. This figure shows that compared to the super-pure Al, the RC1 series manifests an initial steeper slope in the yielding process, the magnitude of which seems to increase with ageing time. The σ_0 for RC-1 series was larger compared to the super pure Al due to the fine grain size and the expected presence of nano-particles. Previous analysis [3] had shown that the strain rate sensitivity for less thermal particles S_p can be calculated from $(S_{D\text{-}F})^2/S$ and is tabulated in table 1. The particle size is inversely related to S_p and the plot in figure 2 reveals a coarsening kinetics of t^{κ} where theoretically κ is 0.33 for three dimensional ageing, 0.25 for interfacial and 0.20 for linear reactions. Although the derived value is 0.243 suggesting ageing within the grain boundaries, the particles which affect σ_0 are located within the grains. For particle size calculation, the parameter α is involved as indicated in table 1 and it is expected to be larger than that for dislocations of the assumed 0.4 and its maximum is 1.0. Thus the change in particle size could affect the magnitude of α. For this reason log S_p versus log t was plotted, in case this effect of α become obvious. The slopes corresponding to 0.33 and 0.20 is also shown to depict the possible range. The supposition is that Fe solute segregate to dislocations via Cottrell-Bilby [7] mechanism and precipitate by means of pipe diffusion during the preliminary anneal at 230 °C. During subsequent CReX the grain boundaries are stabilized by the pre-set precipitates which would coarsen by grain boundary diffusion but the intra-granular particles, if connected by the residual dislocation network after recrystallization, will coarsen by pipe diffusion. The current data is insufficient to resolve this subtle difference.

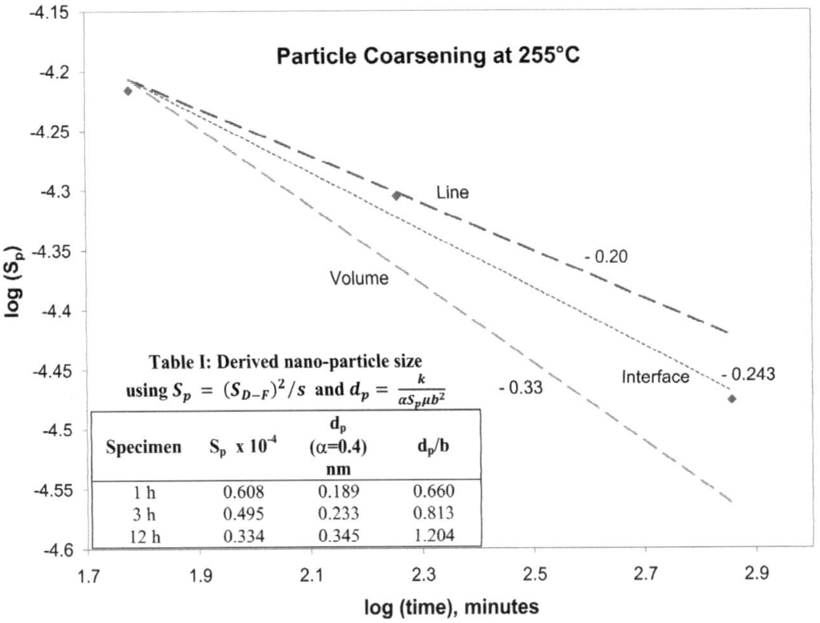

Table I: Derived nano-particle size using $S_p = (S_{D-F})^2/s$ and $d_p = \frac{k}{\alpha S_p \mu b^2}$

Specimen	S_p x 10^{-4}	d_p (α=0.4) nm	d_p/b
1 h	0.608	0.189	0.660
3 h	0.495	0.233	0.813
12 h	0.334	0.345	1.204

Figure 2. Log plot of coarsening kinetics with inset of Table 1 depicting derived values.

The role of fine particles should also manifest itself in the log λ versus log τ plot as shown in figure 3. For the super-pure case, yielding as expected takes place with the grown-in dislocations widely spaced in the 10 μm range and decreases as the dislocation density increases with strain. However for the RC1 cases, the opposite effect is observed in the micro-strain region whereby the mean slip distance increases from a minimum λ_m. For dislocation motion through randomly dispersed obstacles, the mean slip distance increases till slip becomes homogeneous [8]. This process of percolation [9] means that λ_m is smaller than the mean inter-obstacle spacing. Examination of this model indicates that a typical value for ℓ is about 2 λ_m and if $\alpha = 0.4$ is used with $\mu = 36{,}522$ MPa, the resulting τ is 13.2 MPa or a σ_0 of 39.6 MPa compared to the measured one of 65.2 MPa, ignoring grain-size effects. Since the expected α is larger than 0.4, the enforced agreement will change the operative α to become 0.66; which is not unreasonable. However for the 12 h-age case with $\lambda_m = 0.25$ μm and hence ℓ of 0.5 μm, results in calculated σ_0 of 40.8 MPa compared to the measured 70.12 MPa. The increase in α also imply an increase in the λ_m/ℓ ratio. Taking the case of very strong obstacles with $\alpha = 0.81$ [8] and $\lambda_m = \ell$, the calculated σ_0 of 101.6 MPa is found. Although both α and λ_m/ℓ are both directly related to the strength of obstacle defined by the internal angle between the bowing dislocation components around the obstacle, the models [8,10] has various geometrical terms which makes the predictions ambiguous.

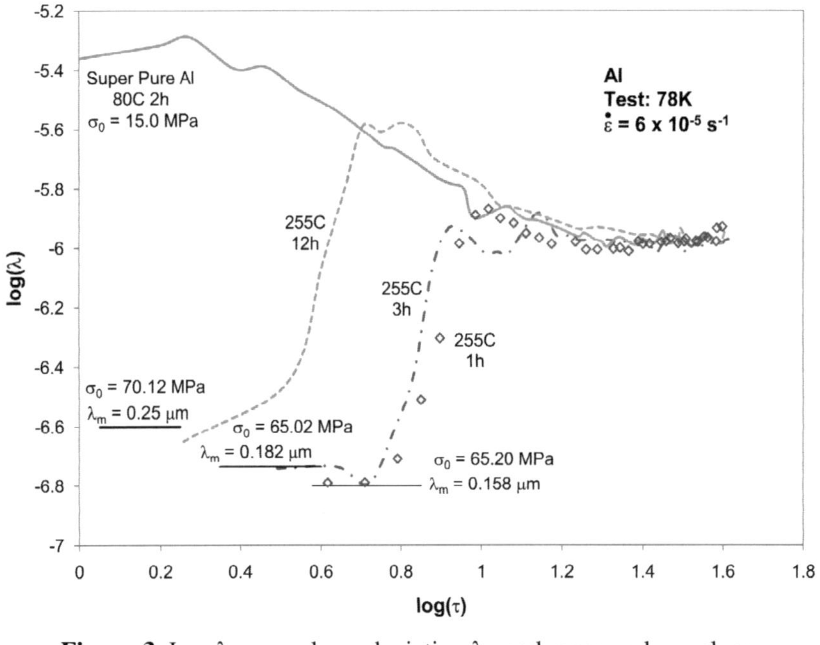

Figure 3. Log λ versus log τ depicting λ_m and measured σ_0 values.

The third aspect of the DDDA is the ADD plot shown in figure 4. The initial slope in units of *d*/b is that from the micro-strain region and the second designated slope is the least square fit from the 1/10 up-change data-points, the response of which is attributed to dislocation-forest interaction [5]. The D-F *d*/b values for all the specimens are near the 0.25 expected for the Peierls dislocation. Previously the value in Stage II for [001] Al crystal tested at 78 K was 0.35 which decreased to 0.22 during Stage III [6]. Hence the observed lower slope for the down-change data-points indicates the incorporation of debris as for Stage III. Thus the determinations in the higher strain regions are fairly self-consistent. In the micro-strain region for RC1, an effect is observed for the 12 h-age case wherein a zig-zag pattern is evident. This type of pattern was observed in age-hardening alloys [11] and may be due to punching out of loops around particles which are too strong to be cut through. Thus the passing dislocations intersect these very small weak loops with lower value of α [8] than the 0.4 used for data analysis. On the other hand the smaller zig-zag effect for the super-pure case must be due to the debris-loops left behind from the low temperature recrystallization temperature and hence manifest slopes similar to RC1 aged for 1 and 3 h. The lower α values for loops [8] means that a correction to the initial slope is required, which increases the *d*/b determined by using $\alpha = 0.4$ since λ is linearly related to α.

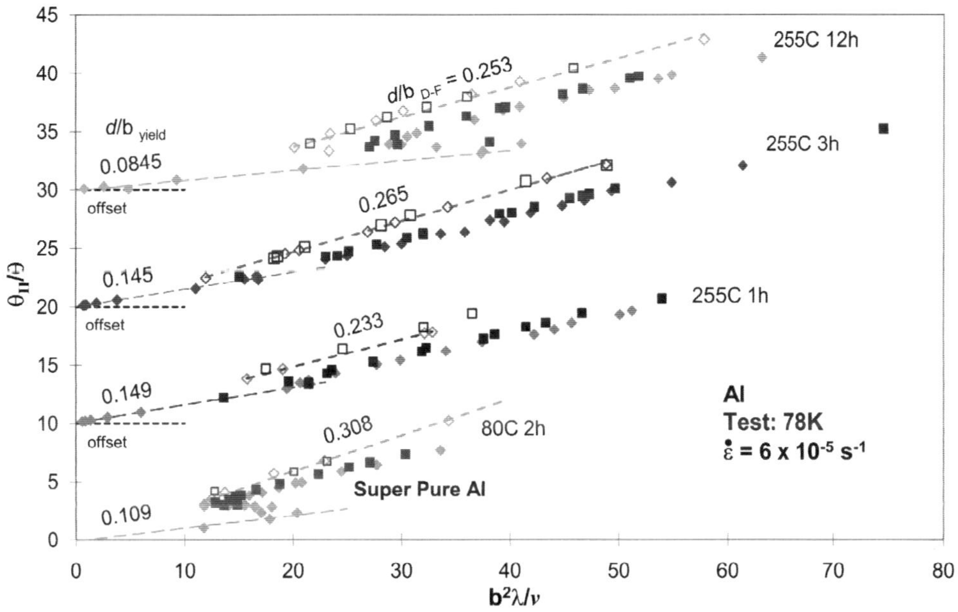

Figure 4. θ_{II}/θ versus $b^2\lambda/v$ ADD plot derived from equation (3) depicting the *d*/b values for the yield region and the dislocation-forest (D-F) interactions. The unfilled symbols are for up-change and filled ones for down-change. The square symbols indicate ¼ change whereas the diamond the 1/10 ones; the latter are used for all least square fits. The RC1 plots are offset for clarity.

4. Conclusion

The dynamic dislocation-defect analysis examines the interrelation of τ with θ, α, λ, υ and *d*. The activation distance can only be determined dynamically as illustrated to characterize the dislocation-defect intersection with thermal assistance. Using the dislocation-forest as a calibration factor, the obstacle species can be compared and their thermodynamic response categorized. In this manner, the formation of nano-sized Al_6Fe precipitates were confirmed upon suitable heat treatment to getter the Fe solutes by segregation to dislocations.

Acknowledgements

The author thanks the Natural Science and Engineering Research Council of Canada for continuous support and Ms. J. I. Cooley for computational assistance.

References

[1] Saimoto S and Jin H 2007 *Mater. Sci. Forum* **550** 339
[2] Yamamoto A *et al.* 1998 *Al Alloys:Their Physical and Mechanical Properties,* vol 2, ed T Sato *et al.* (Tokyo:Japan Inst. Light Metals) p 639
[3] Diak B J, Upadhyaya K R and Saimoto S 1998 *Prog. Mater. Sci.* **43** 223-363
[4] Jin H and Saimoto S 2003 *Mater. Sci. Technol.* **19** 1197
[5] Saimoto S, Cooley J, Larsen H and Scholler C 2009 *Phil. Mag.* **89** 853
[6] Saimoto S 2006 *Phil. Mag.* **86** 4213
[7] Cottrell A J and Bilby B A 1949 *Proc. R. Soc.* A **62** 49
[8] Foreman A J E and Makin M J 1967 *Canadian J. Phys.* **45** 511
[9] Kocks U F and Mecking H 2003 *Prog. Mater. Sci.* **48** 171-273
[10] Saimoto S, Saka H, Imura T, 1977 *Scripta Metall.* **11** 903
[11] Saimoto S, Gabryel C, Cooley J and Mishra R K 2008 *Al alloys*, vol 1, ed J Hirsch, B Skrotzki and G Gottstein (Weinheim:Wiley-VCH) p 815

High-temperature plastic deformation of fine-grained Y-doped $BaCeO_3$ polycrystals

C Vaquero-Aguilar[1], C Real[2] and M Jiménez-Melendo[1]

[1] Departamento de Física de la Materia Condensada, Universidad de Sevilla. Aptdo. 1065. 41080 Sevilla, Spain
[2] Instituto de Ciencia de Materiales de Sevilla, Centro Mixto CSIC-US, Av. Américo Vespucio 49. 41092 Sevilla, Spain

melendo@us.es

Abstract. The high-temperature plastic deformation of $BaCe_{0.95}Y_{0.05}O_{3-\delta}$ polycrystals with average grain size of 0.50 μm has been studied in compression between 1000 and 1250°C in air at different initial strain rates. The stress-strain curves display yield drop at strains close to 5%, followed by steady state or strain-softening stages. Large ductilities were achieved at the higher temperatures, without appreciable changes in grain shape and size. Mechanical data and microstructural observations are consistent with a flow mechanism by grain boundary sliding.

1. Introduction

Perovskite-structured oxides of formula ABO_3 (where A and B denote site cations) have received increasing attention in the last years because of their potential as membranes or electrolytes in solid oxide fuel cells (SOFCs) [1]. In particular, doped perovskites as $BaCeO_3$, $KTaO_3$ and $SrZrO_3$ exhibit high oxygen-ion and electronic conductivity, as well as excellent chemical stability over a large range of temperatures and oxygen partial pressures. These oxides have been also found to be proton conducting at high temperatures when doped with trivalent ions occupying the B sites in the structure [2-4].

Many studies have addressed the processing, structure, defect chemistry and conductivity of doped perovskite compounds [2-4]. Nevertheless, comparatively few studies have focused on the high-temperature mechanical properties of these protonic conducting materials [5], despite the utility of such investigations for ultimate component performance (deformation, mechanical degradation and failure). Additionally, they can yield valuable information about the processes involved in processing (sintering and grain growth) of perovskite ceramics.

In this paper, we present results of studies on the high-temperature mechanical behavior of polycrystalline $BaCe_{0.95}Y_{0.05}O_{3-d}$ fabricated by solid state reaction. The creep curves obtained at constant cross-head speeds in air at elevated temperatures and microstructural information are presented.

2. Experimental methods

2.1. Sample preparation

$BaCe_{0.95}Y_{0.05}O_{3-\delta}$ (BC5Y) powders were prepared by solid-state reaction of $BaCO_3$, CeO_2 and Y_2O_3 powders (Aldrich, purity > 99.0%). The required stoichiometric ratio was ball-milled with 20 mm-diameter agate balls for 1 h, calcined for 10 h at 1200°C in air and ball-milled again for 1 h. X-ray diffraction analysis confirmed that the resulting powder was pure orthorhombic perovskite phase. The powder was uniaxially pressed at 150 MPa into 13 mm-diameter pellets and then isostatically cold pressed at 210 MPa. The pellets were sintered in air in two steps: at 600°C for 2 h to eliminate water and hydroxides formed during intermediate stages, followed by sintering at 1550°C for 10 h. The final density of the samples, determined using Archimedes' method, was about 95% of the theoretical value (6310 kg/m^3).

2.2. Mechanical tests

Specimens of 5 x 3 x 3 mm in size were cut from the sintered pellets with a low-speed diamond saw and used for mechanical testing. Compression tests were carried out using a screw-driven universal testing machine in the temperature range between 1000 and 1250°C in air at constant cross-head speeds v_t between 1 and 500 μm/min, corresponding to initial strain rates $\dot{\varepsilon}_0$ (= v_t/l_o, with l_o the initial specimen length) between 4x10^{-6} and 2.0x10^{-3} s^{-1}. The specimens were sandwiched between CSi pads in order to reduce the friction with the alumina rods of the deformation machine; no appreciable reaction occurred between CSi pads and BC5Y specimens. The recorded data, load vs time, were analyzed in $\sigma - \varepsilon$ curves, where σ is the true stress and ε the true strain (= $\ln(l_o/l)$, with l the instantaneous specimen length). Mechanical tests were discontinued after reaching a total true strain of 50%.

2.3. Microstructural studies

The microstructure of as-sintered and strained BC5Y specimens was characterized by optical and scanning electron microscopy (SEM). To reveal grain boundaries, cross sections of the samples (parallel to the compressive stress axis in deformed specimens) were conventionally polished and then thermally etched at 1200°C for 2 h in air.

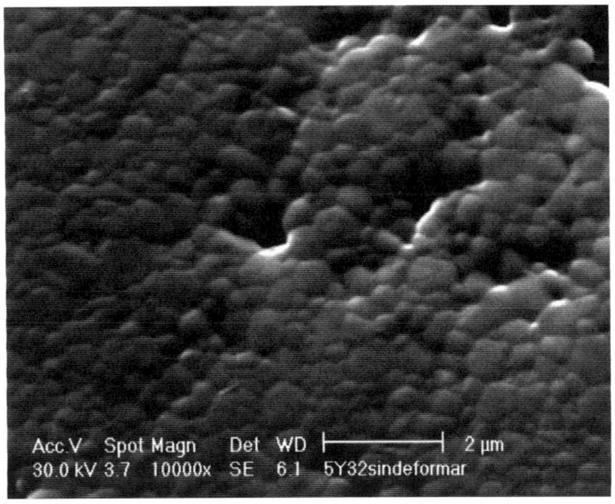

Figure 1. SEM micrograph of as-sintered BC5Y polycrystal

3. Results and discussion

The microstructure of as-sintered specimens is shown in figure 1. Grains are equiaxed, with an average grain size (taken as the planar equivalent diameter) of 0.50 μm. The porosity is intergranular, with pore sizes less than 1 μm.

The samples exhibit a remarkable ductility at temperatures above 1000°C. Below this temperature, the specimens fractured catastrophically without any plastic deformation. Figure 2(a) shows the $\sigma - \varepsilon$ curves obtained at 1250°C for various initial strain rates. Several features should be pointed out: (i) All the curves exhibit yield strength, followed by a near steady-state flow behavior (except for the highest strain rate imposed, where the onset of degradation can be clearly seen); (ii) The magnitude of the yield drop (the difference between upper and lower yield stresses) increases when increasing the strain rate; and (iii) The peak stress shifts towards lower strains as the strain rate decreases. Figure 2(b) displays the $\sigma - \varepsilon$ curves at two different temperatures, 1150 and 1250°C, for selected initial strain rates. The magnitude of the yield drop and the position of the peak stress increase when decreasing the temperature. Also, the compound failed at correspondingly lower strain rates. In this regard, a decrease in temperature is equivalent to an increase in strain rate.

Mechanical data were analyzed with the high-temperature steady-state equation [6]:

$$\dot{\varepsilon} = A\,\sigma^{n} \exp(\;Q / RT)$$

where A is a constant, n is the stress exponent, R is the universal gas constant, T is the absolute temperature and Q is the activation energy for flow. An apparent stress exponent of 1.7 was deduced from the fit of the plot of log $\dot{\varepsilon}$ vs log σ (measured in the steady state regime at $\varepsilon = 0.2$) at a fixed temperature. Similarly, an apparent activation energy of 450 kJ/mol was determined from a plot of log σ vs 1/T at given initial strain rates.

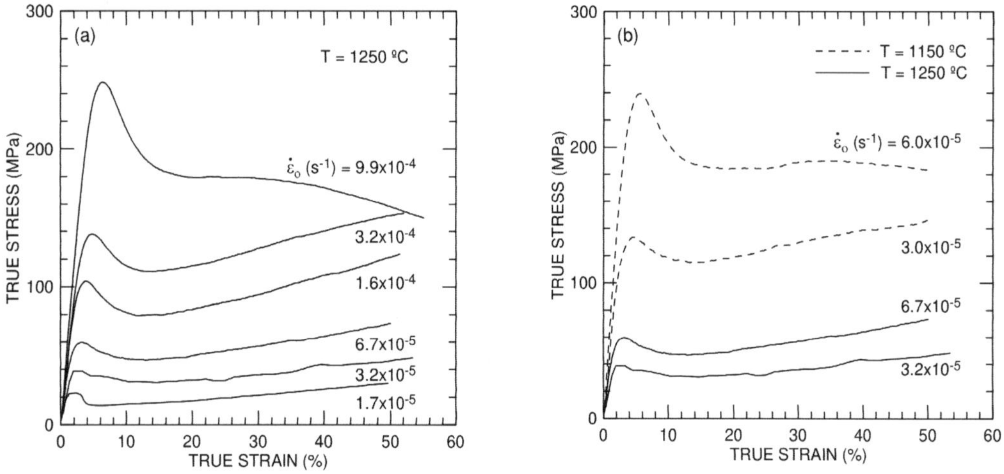

Figure 2. True stress vs true strain curves for BC5Y: (a) at 1250°C with different initial strain rates; (b) at 1150 and 1250°C for selected initial strain rates

Microstructural observations of strained samples revealed that no appreciable changes in grain size and shape occurred during deformation, despite the large strains attained. The density of the samples also remained practically unaltered after testing.

The lack of differential microstructural features in the deformed samples with respect to the as-sintered ones clearly indicates that grain boundary sliding is the main mechanism of deformation. The measured stress exponent, close to 2, is also consistent with this mechanism [6]. This behavior has been widely reported in other fine-grained superplastic ceramics, particularly yttria-stabilized tetragonal zirconia (YTZP) and dual phase Al_2O_3/ZrO_2 composites [6,7]. In these superplastic materials, however, no yield strength was found. Dokko and Pask [8] reported the presence of yield drop followed by steady state or strain-hardening stages in MgO polycrystals with grain sizes ranging from 10 to 30 μm. Such a behavior was explained in terms of dislocation multiplication necessary for initially promoting grain boundary sliding.

Regarding the present study, there are two works devoted to the high-temperature mechanical characterization of yttrium-doped barium cerate $BaCe_{1-x}Y_xO_{3-\delta}$, with x varying from 0.05 to 0.20 [9,10]. In these works, the grain sizes ranged from 1.8 μm for x = 0.05 to 4.5 μm for x = 0.20, with densities < 90% of theoretical values. Samples were typically deformed to total strains < 10% at constant cross-head velocity primarily in high-purity Ar at temperatures between 1200 and 1400°C. A stress exponent close to 1 and an activation energy for flow of 350 kJ/mol were reported, which were temptatively ascribed to grain boundary sliding accommodated by cation lattice diffusion. The lack of cation lattice and grain-boundary diffusion data, however, precluded the exact determination of the strain-rate controlling species.

The transition in stress exponent from n = 1 towards n = 2 as the grain size decreases has been reported in other ceramic oxides [6]; the critical grain size seems to be about 1 μm, in agreement with the present results. A detailed understating of the exact mechanism of superplastic flow has not been thoroughly developed. Grain boundary sliding must be accompanied by an accommodation process of the stresses generated during the sliding. This process is, in ultimate instance, the rate-controlling mechanism; otherwise, extensive cavitation would cause premature degradation and failure of the material. Although many different theoretical models have been developed to explain the superplastic behavior of fine-grained materials, based on diffusion and/or dislocation slip [11], none of them is able to explain successfully the body of experimental data. This is presently a matter of intense debate.

4. Conclusions

$BaCe_{0.95}Y_{0.05}O_{3-\delta}$ polycrystals were prepared by solid-state reaction. After sintering, the samples exhibit a very fine-grained microstructure of equiaxed grains with average size of 0.5 μm, and densities of about 95% of the theoretical value. The compound shows a high ductility above 1000°C, without noticeable changes in grain shape and size even at strains of 50%. The stress-strain curves obtained at constant cross-head speeds display systematically a yield drop followed by an extended steady state stage. The apparent stress exponent is close to 2. Microstructural observations and mechanical data are consistent with a superplastic behavior due to grain boundary sliding.

Acknowledgments

This work was supported by the Spanish Ministry of Science and Education, Project No. MAT2006-13005-C03-03.

References

[1] Norby T 2001 *Nature* **410** 877
[2] Iwahara H, Esaka T, Uchida H and Maeda N 1981 *Solid State Ionics* **3-4** 359
[3] Nowick A S and Du Y 1995 *Solid State Ionics* **77** 137
[4] Ma G, Matsumoto H and Iwahara H 1999 *Solid State Ionics* **122** 237
[5] Salama K, Majkic G and Balachandran U 2005 *Defect and Diffusion Forum* **242-244** 43
[6] Jiménez-Melendo M, Domínguez-Rodríguez A and Bravo-León A 1998 *J. Am. Ceram. Soc.* **81** 2761
[7] Jiménez-Melendo M, Clauss C, Domínguez-Rodríguez A, De Portu G, Roncari E and Pinasco P 1998 *Acta Mater.* **46** 3995
[8] Dokko P C and Pask J A 1979 *J. Am. Ceram. Soc.* **62** 433
[9] Goretta K C, Park E T, Guan J, Balachandran U, Dorris S E and Routbort J L 1998 *Solid State Ionics* **111** 295
[10] Park E T, Goretta K C, de Arellano-López A R, Guan J, Balachandran U, Dorris S E and Routbort J L 1999 *Solid State Ionics* **117** 323
[11] Nieh T G, Wadsworth J and Sherby O D 1997 *Superplasticity in Metals and Ceramics* (Cambridge: Cambridge University Press)

Brittle-to-ductile transition in polycrystalline aluminum containing gallium in the grain boundaries

A Itoh, J Izumi, K Ina and H Koizumi

Deaprtment of Physics, Meiji University, Higashi-mita, Tama-ku, Kawasaki, 214-8571 Japan

koizumi@isc.meiji.ac.jp

Abstract. It is well known that aluminum/gallium couple causes liquid metal embrittlement. Gallium atoms penetrate the grain boundaries of polycrystalline aluminum and degrade it. Polycrystalline aluminum specimens were contacted with a small droplet of gallium for 24 h. After gallium was removed from the surface of the specimens, tensile tests were performed between 77 K and 313 K. The specimens are ductile below 230 K and brittle above 303 K, the melting temperature of gallium. Between 280 K and 300 K, the maximum stress is larger in the specimens heated from 77 K than in those cooled from 313 K. This thermal history dependence of the maximum stress is considered to be attributed to the solidification of supercooled gallium in the grain boundaries.

1. Introduction

Polycrystalline aluminum becomes brittle if it is stressed in liquid gallium. This is a typical example of liquid metal embrittlement, which refers to the loss of ductility of normally ductile metals when stressed whilst in contact with a liquid metal. In the case of aluminum and gallium, it is also known that fracture stress and fracture strain are reduced above the melting point of gallium (29.8°C), even if it is stressed after gallium is removed from the surface of aluminum specimens. The fracture stress reduces with an increase in the contact time. This phenomenon is attributed to the penetration of gallium into the polycrystalline aluminum, which reduces the strength of grain boundaries [1]. The rapid penetration of gallium atoms has been confirmed by radiotracer [2], transmission electron microscope [3–5], and X ray techniques [6, 7].

Polycrystalline aluminum containing gallium in the grain boundaries shows a large ultrasonic attenuation above the melting temperature of gallium, while the attenuation is nearly the same as in polycrystalline aluminum without gallium at low temperatures [8]. A change in ultrasonic attenuation occurs over a small temperature range, and the temperature depends on the thermal history. The change in the ultrasonic attenuation may be accompanied by changes in some mechanical properties of the same specimen. In this paper, we describe the fracture properties of polycrystalline aluminum containing gallium between 77 K and 313 K.

2. Experimental

Specimens were spark cut from a 3 mm thick polycrystalline aluminum plate to the dimensions shown in figure 1. The purity of the specimens was 99.999% and the average grain size was 0.1 mm. After the specimens were chemically polished and annealed at 700 K, a small drop of liquid gallium saturated with aluminum was placed on the surface of the specimens at 313 K. To ensure a contact between aluminum and liquid gallium, an aluminum surface covered with the liquid gallium was scratched with a needle. After keeping the specimens at 308 K for 24 h, they were immersed in liquid nitrogen and then gallium was removed from the surface with emery paper. Since the penetration velocity of gallium atoms into the grain boundaries of polycrysltalline aluminum is several micrometers per second [1, 2, 4], 24 h is long enough for the grain boundaries to be invaded by gallium over the entire thickness of the specimen.

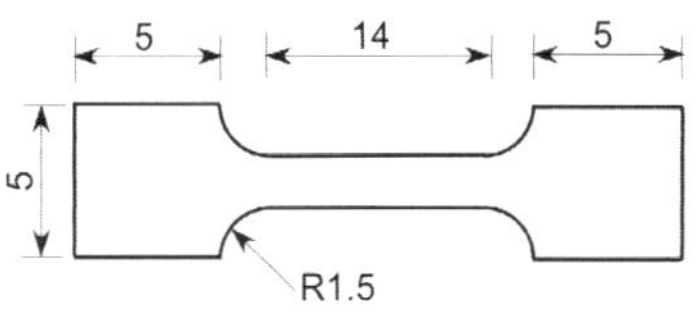

Figure 1. Tensile specimen with thickness of 3 mm.

Tensile tests were performed at a strain rate $\dot{\epsilon}$ of 1.2×10^{-2}/s or 1.2×10^{-3}/s. Two types of thermal histories were examined. Some of the specimens were immersed in ethanol at 313 K and then cooled to the test temperature, and the others were immersed in liquid nitrogen and then heated to the test temperature. The fracture surface was examined using a scanning electron microscope.

3. Results

The typical stress strain curves obtained at $\dot{\epsilon} = 1.2\times10^{-2}$/s are shown in figure 2. The nominal stress and nominal strain are shown in this and the following figures. At 313 K, both the maximum stress σ_{max} and the fracture strain ϵ_f are considerably reduced by gallium, while at 77 K, σ_{max} is not affected, but ϵ_f is somewhat reduced. At 281 K, σ_{max} is larger in the specimen heated from 77 K than that cooled from 313 K. At this temperature, both σ_{max} and ϵ_f depend on the thermal history. At 242 K, σ_{max} is almost the same for the heated and cooled specimens; however, ϵ_f seems to depend on the thermal history.

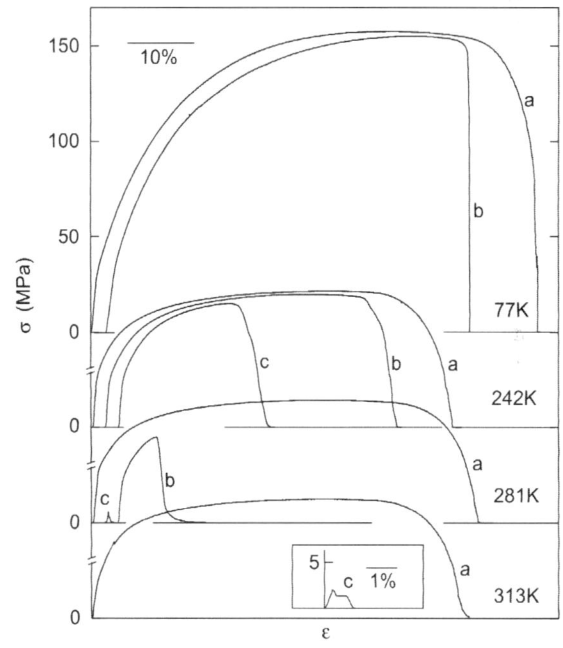

Figure 2. Stress strain curves of polycrystalline aluminum without gallium (a) and with gallium in the grain boundaries stressed after heating from 77 K (b) and cooling from 313 K (c). Temperatures indicated in the figure are the test temperatures.

The temperature dependence of σ_{max} and ϵ_f measured at $\dot{\epsilon} = 1.2\times10^{-2}$/s are shown in figure 3. The broken curves represent σ_{max} and ϵ_f of polycrystalline aluminum without gallium. Both σ_{max} and ϵ_f decrease with an increase in temperature. The effect of gallium is noticeable in σ_{max} above 230 K. The maximum stress σ_{max} of the specimens heated from 77 K decreases gradually, and above 300 K, it is about the same as that measured in a polycrystalline aluminum specimen stressed in liquid gallium. On the other hand, σ_{max} of specimens cooled from 313 K shows an abrupt increase at 280 K. Between 300 K and 280 K, σ_{max} depends on the thermal history. The fracture strain ϵ_f also decreases with temperature; however, the hysteresis is less clear because of the scattering of data.

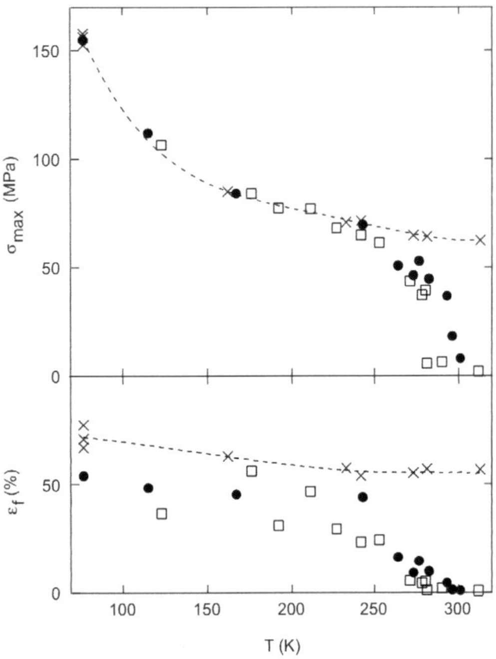

Figure 3. Maximum stress σ_{max} (upper) and fracture strain ϵ_f (lower) as functions of measuring temperature. $\dot{\epsilon} = 1.2 \times 10^{-2}$/s.
$\times$ polycrystalline aluminum without gallium.
● polycrystalline aluminum with gallium in the grain boundaries heated from 77 K.
□ polycrystalline aluminum with gallium in the grain boundaries cooled from 313 K.

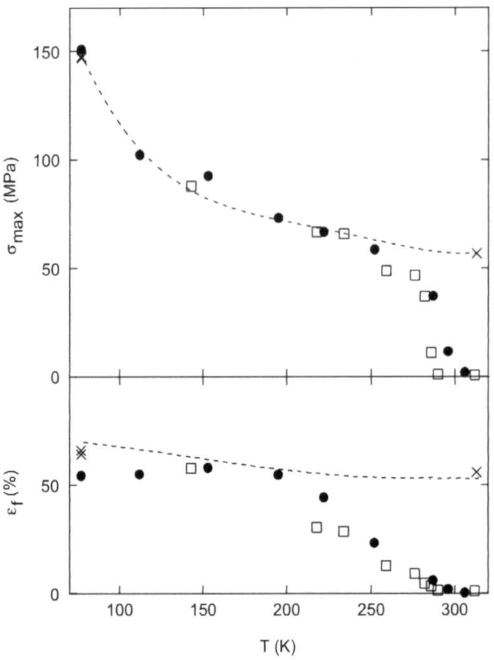

Figure 4. Maximum stress σ_{max} (upper) and fracture strain ϵ_f (lower) as functions of measuring temperature. $\dot{\epsilon} = 1.2 \times 10^{-3}$/s. Symbols are the same as those shown in figure 3.

Figure 4 shows σ_{max} and ϵ_f obtained at $\dot{\epsilon} = 1.2 \times 10^{-3}$/s. Similar tendencies as shown in figure 3 are observed for both σ_{max} and ϵ_f. The maximum stress σ_{max} of the specimens cooled from 313 K changes abruptly around 285 K, which is higher by 5° than that shown in figure 3.

Figure 5 shows the SEM of fracture surfaces. When the maximum stress is low, most of the fracture surface consists of smooth areas, suggesting intergranular fracture. When the maximum stress is large, the area of transgranular fracture increases. Secondary cracks are often observed perpendicular to the fracture surface.

4. Discussion

Since aluminum specimens without gallium are ductile in all the temperature range tested, transition from ductile fracture at low temperatures to brittle fracture at high temperatures observed in the specimens with gallium is attributed to gallium in the grain boundaries of polycrystalline aluminum.

The degradation consists of two components: a gradual decrease in σ_{max} above 230 K, and very small σ_{max} above 280 K or 285 K, which is only observed in the specimens cooled from above the melting temperature of gallium. The second component is, therefore, considered to be attributed to supercooled Ga in the grain boundaries, and an abrupt change in σ_{max} is caused by the solidification of supercooled gallium. Although at atmospheric pressure, stable α-gallium melts at 303 K, gallium can be easily supercooled, especially in small particles [9].

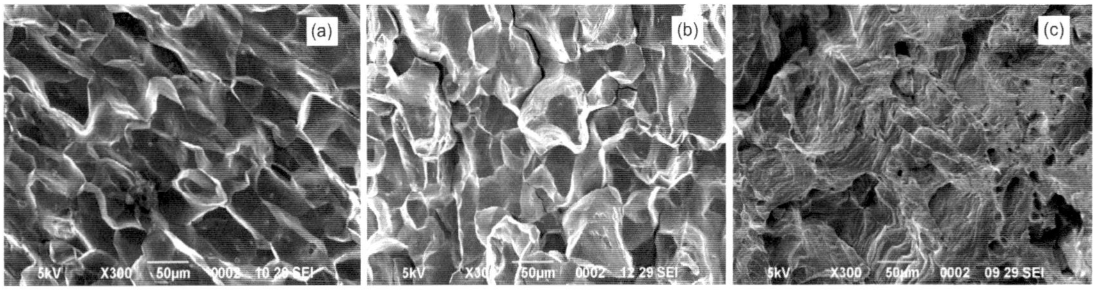

Figure 5. SEM of fracture surfaces of (a) a polycrystalline aluminum specimen cooled from 313 K and tensile tested at 281 K, (b) a specimen heated from 77 K and tested at 281 K, (c) a specimen heated from 77 K and tested at 242 K. The strain rate was 1.2×10^{-2}/s.

The thickness of gallium layers in some grain boundaries is measured to be of the order of 1 μm [6], and gallium in these grain boundaries may have similar properties as small particles have. Ultrasonic attenuation measured in polycrystalline aluminum containing gallium shows an abrupt decrease in attenuation around this temperature when the specimen is cooled from 308 K, while the attenuation is small up to 300 K when the specimen is heated from 200 K [8]. This temperature dependence of the ultrasonic attenuation also supports the solidification of supercooled gallium. The transition temperature of the ultrasonic attenuation and the abrupt change in $\sigma_{\max}$ cannot be compared directly, because both temperatures depend on various factors such as content of gallium, stress and changing rate of temperature.

Between 230 K and 280 K, a gradually decreasing component of $\sigma_{\max}$ is independent of the thermal history. SEM micrographs show that most of the fracture surface consists of smooth grain surfaces when the maximum stress is small, while the fracture surface is mixture of intergranular and transgranular fracture surfaces when the maximum stress in large. This suggests that in this temperature range the number of weakened grain boundaries increases with temperature, even when gallium in the grain boundaries may not be in the liquid state. The penetration of gallium depends on the nature of grain boundaries of aluminum [5]; therefore, it can be expected that each grain boundary becomes weak at a different temperature.

Penetration of liquid metal atoms into the grain boundaries of a solid metal has been confirmed in several solid metal and liquid metal systems which cause liquid metal embrittlement, such as copper/bismuth [10] and silver/mercury [1] systems. In these systems the similar brittle-to-ductile transition may be observed near the melting point of the liquid metal.

References

[1] Ina K and Koizumi H 2004 *Mater. Sci. Eng.* A **387-389** 390
[2] Roques-Carmes C, Aucouturier, M. and Lacombe P 1973 *Metal Sci. J.* **7** 128
[3] Hugo R C and Hoagland R G 1998 *Scripta Mater.* **38** 523
[4] Hugo R C and Hoagland R G 1999 *Scripta Mater.* **41** 1341
[5] Hugo R C and Hoagland R G 2000 *Acta Mater.* **48** 1949
[6] Ludwig W and Bellet D 2000 *Mater. Sci. Eng.* A **281** 198
[7] Pereiro-López E, Ludwig W, Bellet D, Cloetens P and Lemaignan C 2005 *Phys. Rev. Lett.* **95** 215501
[8] Miyagi T and Koizumi H 2006 *Mater. Sci. Eng.* A **442** 184
[9] Bosio L, Defrain A and Epelboin I 1966 *J. Physique* **27** 61
[10] Joseph B, Barbier F and Aucouturier M 1999 *Scripta Mater.* **40** 893

Journal of Physics: Conference Series **240** (2010) 012034 doi:10.1088/1742-6596/240/1/012034

Evaluation of quasi-static fracture characteristics considering surface conditions of silicon nitride for a space component

Y Kakimoto[1,2], Y Motoyashiki[2], K Kitazono[1], E Sato[2]

[1] Tokyo Metropolitan University Department of Aerospace Engineering, Hino, Tokyo 191-0065, Japan
[2] Institute of Space and Astronautical Science, Japan Aerospace Exploration Agency (ISAS/JAXA) ,Sagamihara, Kanagawa 229-8510, Japan

kakimoto.yuuki@ac.jaxa.jp

Abstract. ISAS/JAXA is now planning to adopt a thruster made of monolithic silicon nitride (SN282 manufactured by Kyocera Co.) onto a Venus exploration probe, PLANET-C, in replacement of conventional niobium heat-resistant alloy. Silicon nitride is still brittle and requires precise analysis on multiaxial thermal stresses induced during firing, though it has high toughness among other structural ceramics. This study evaluated quasi-static fracture characteristics of SN282 considering the surface conditions through compression-torsion biaxial fracture tests as well as the conventional four-point-bending tests. The samples were applied to the mechanical tests either as-ground or after annealing at 1300℃ in air for 1 h, which formed an oxidation layer of more than 250nm on the specimen surface. Symmetry four-point-bending tests showed that annealing improves flexure strength and reduce the difference caused by grinding directions. Biaxial stress fracture tests showed the high compressive stress makes the influence of facial crack insensitive.

1. Introduction

A thruster is a small rocket engine to control attitude or orbit of a spacecraft such as a planetary exploration probe or a satellite. Thruster materials are required oxidation resistance and low density. In addition particularly, propulsion power of a thruster is improved by using higher heat resistance material. Although a conventional thruster has been made of niobium heat-resistant alloy, this alloy must have oxidation resistant coating. Therefore, now ISAS/JAXA is planning to adopt a thruster made of monolithic silicon nitride (SN282 manufactured by Kyocera Co.) onto a Venus exploration probe, PLANET-C, in replacement of niobium heat resistant alloy [1].

SN282 has been developed for a ceramics turbine, exhibits good resistance to oxidation and has larger heat-resistance than niobium heat-resistant alloy. However, it is still brittle and requires precise analysis on multiaxial thermal stresses induced during firing, though it has high toughness among other structural ceramics. It is well known that surface condition strongly effects on mechanical quasi-static strength of ceramics. In addition, for brittle materials, fracture analysis is determined by combination of fracture statistics theory [2] and mixed mode crack propagation criteria [3] [4].

This study evaluated quasi-static fracture characteristics of SN282 silicon nitride considering the surface conditions and multiaxial stress condition through compression-torsion biaxial fracture tests as well as the conventional four point bending tests.

2. Experimental Procedure

2.1 Test machine and specimens

The experiments were conducted using a compression-torsion combined test machine, which is based on a tension/compression tester (AG1 manufactured by Shimadzu) and a retrofitted torque cell with a torque motor. Materials were commercial gas-pressure sintered silicon nitride, SN282. This material is toughened with microstructures tailored to achieve elongated grain structure.

2.2 Symmetric four-point-bending test

Symmetric four-point-bending tests were performed by using a four-point flexure fixture with 10 mm inner and 30 mm outer spans in the ambient temperature and atmosphere condition. The specimens were machined into flexure test specimens of 3 mm × 4 mm × 40 mm bars. Two kinds of finishing were applied: ground parallel or perpendicular to the longitudinal direction of the specimen with #400 grind stones. Half of them were annealed at 1300℃ in air for 1 h prior to testing. Flexure strength was determined as follows:

$$\sigma = \frac{3P(L-l)}{2\omega t^2} \tag{1}$$

where P is the load, L and l are the outer and inner spans, respectively, and ω and t are the specimen width and depth, respectively.

Brittle materials have a large variety of flexure strength between each specimen. Therefore, flexure strength was determined using a total 30 test specimens through evaluating the Weibull plot.

2.3 Biaxial stress fracture test

Biaxial stress fracture tests were performed in compression-and-torsion mode in ambient temperature and atmosphere condition. This specimen and loading mode are shown in Figs. 1 (a) and (b), respectively. These samples were machined and finished perpendicular to the specimen axes. At first, the compression force, σ, was applied. Next, torsion, τ, was introduced on the specimen with the compression force being kept.

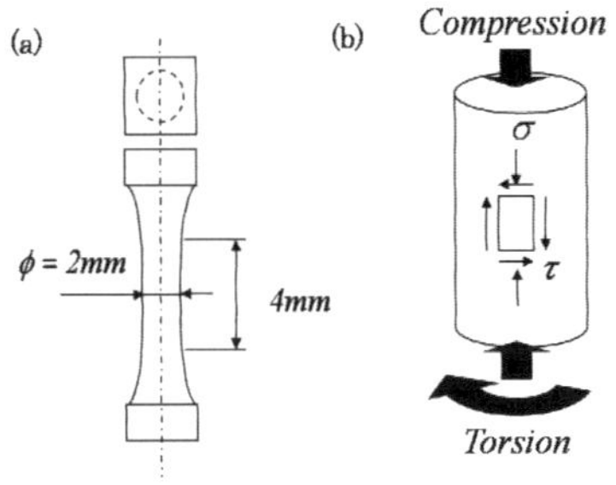

Fig.1. (a)Biaxial fracture test specimen geometry, (b)stress condition under the biaxial fracture test.

The maximum, intermediate and minimum principal stresses σ_1, σ_2 and σ_3 at the moment of fracture were conducted by the following equations:

$$\sigma_1 = \frac{\sigma}{2} + \sqrt{\frac{\sigma^2}{4} + \tau^2}\,, \tag{2}$$

$$\sigma_2 = 0\,, \tag{3}$$

$$\sigma_3 = \frac{\sigma}{2} - \left(\sqrt{\frac{\sigma^2}{4} + \tau^2}\right). \tag{4}$$

Since torsion causes non-uniform stress distribution, the fracture stresses are corrected considering effective volume ,V_e, by a following formula (5) (6) (7)

$$\sigma_1 = \left(\frac{\sigma}{2} + \sqrt{\frac{\sigma^2}{4} + \tau^2}\right)\left(\frac{V_e}{V}\right)^{\frac{1}{m}}, \tag{5}$$

$$\sigma_3 = \left(\frac{\sigma}{2} - \sqrt{\frac{\sigma^2}{4} + \tau^2}\right)\left(\frac{V_e}{V}\right)^{\frac{1}{m}}, \tag{6}$$

$$V_e = 2V\int_0^{r_0} \left(\frac{\dfrac{\sigma}{2} + \sqrt{\dfrac{\sigma^2}{4} + \left(\tau\dfrac{r}{r_0}\right)}}{\dfrac{\sigma}{2} + \sqrt{\dfrac{\sigma^2}{4} + \tau^2}}\right) r\,dr\,. \tag{7}$$

3. Experimental results and discussion

The two-parameter Weibull plots of flexure strengths of SN282 are shown in Fig. 2, where the variables F and σ are failure probability and flexure strength, respectively.

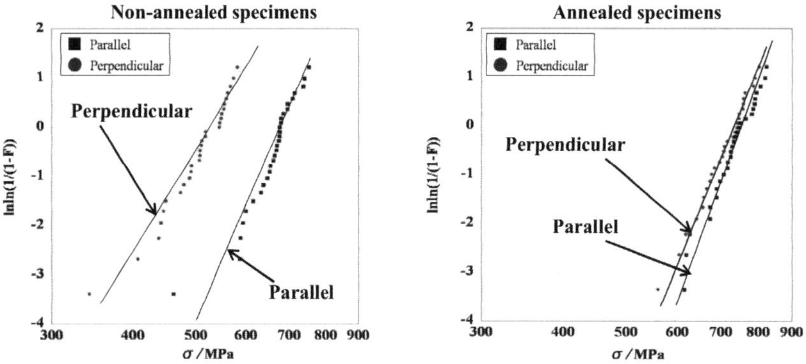

Fig.2. Two parameter Weibull plots of flexure strength of SN282 test specimens with two kinds of finishing obtained by symmetric four-point-bending test.

For non-annealed specimens, Weibull modulus, m, and characteristic strength, σ_A, were m = 11 and σ_A = 630 MPa for the specimen ground parallel to the longitudinal direction, and m = 9 and σ_A = 508 MPa for the specimen ground perpendicular to the longitudinal direction. On the other side, for the annealed specimens, m and σ_A were $m = 14$ and $\sigma_A = 731$ MPa for ground parallel, and $m = 13$ and $\sigma_A = 708$ MPa for ground perpendicular. These results show that annealing improves flexure strength of SN282 and reduces the difference of finishing directions.

Auger Electron Spectroscopy (AES) was executed to measure the surface conditions of the annealed and non-annealed specimens. For the non-annealed specimens, oxygen remains 8 at.% to 50 nm in depth from the specimen surface. In contrast, for the annealed specimens oxygen remains 34 at.% to 250 nm in depth. Therefore, an oxidation layer was formed on the surface of the annealed specimens and the oxidation layer functions to infill surface micro-flaws.

For biaxial stress fracture test, as shown Fig. 3, a compression force, σ, is applied on the specimen, with varying σ_3. Next, a torsion, τ, is introduced on the specimen, with varying both σ_1 and σ_3. For a low compressive stress field, annealing has a large effect on the fracture criteria. On the other side, for a high compressive stress field, annealing has little effect on the fracture criteria. AES results showed that annealing generated an oxidation layer on the specimen surface which infiltrates facial micro-flaws of the specimen. Therefore, it is considered that the high compressive stress makes the influence of facial micro-flaws insensitive.

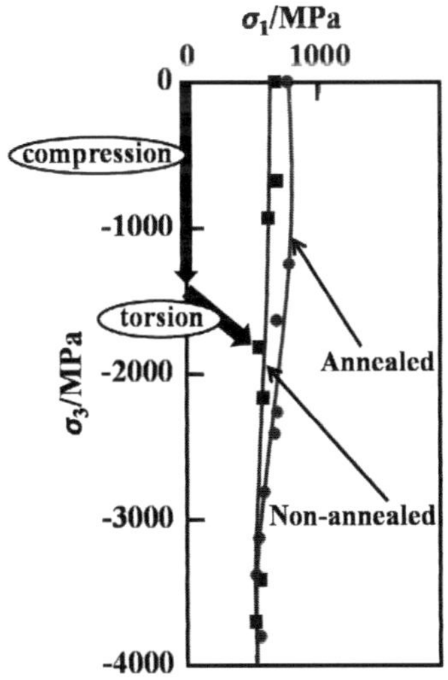

Fig.3. Biaxial stress fracture test result for annealed and non-annealed specimens.

4. Conclusions

Quasi-static fracture characteristics of SN282 silicon nitride considering the surface conditions and maltiaxial stress condition were evaluated, through compression-torsion biaxial fracture tests as well as the conventional four point bending tests. Four-point-bending tests showed that annealing improves flexure strength and reduce the difference of grinding directions. Biaxial stress fracture tests showed that the high compressive stress makes the influence of facial micro-flaws insensitive.

References

[1] K. Uesugi, E. Sato, S.Sawai, K. Furuhara, H. Mishima, Y. Nonaka and M.Kondo, *Proc 6th IAA Int. Conf. on Low-Cost Planetary Missions*, (2005) 177.
[2] W. Weibull : *Ingenioersvetenskapsaked.*, Handl., (1939), 151.
[3] S. B. Batford and H. L. Heinisch : *J. Amer. Cer. Soc.*, **61**, (1978), 355.
[4] A. G. Evans and R. L. Jones : *J. Amer. Cer. Soc.*, **61**, (1978), 156.

Journal of Physics: Conference Series **240** (2010) 012035 doi:10.1088/1742-6596/240/1/012035

Fractal analysis of the multiple fracture patterns in the low-carbon steel

M. Tyutin, L. Botvina

Baikov Institute of Metallurgy and Materials Science, Russian Academy of Sciences, Laboratory of structural steels and alloys, Leninski prospect, 49, 119991 Moscow, Russian Federation

marattut@mail.ru

Abstract. We studied the process of damages accumulation at tension of low carbon steel specimens with lateral notch. The fractal dimension of microcrack patterns in a plastic zone and angular parameter of microcracks distribution have been estimated. It is established that these parameters change in opposite phase. The analogy to the similar dependences estimated according to seismic activity is drawn.

1. Introduction

The authors of [1-3] were admittedly the first to establish self-similarity of multiple fracture patterns for a wide range of materials (metals, polymers, and rocks) under conditions of tension, creep, fatigue, and superplasticity. The self-similarity was confirmed when a unique curve for the distribution of defects was plotted based on the normalization, which corresponds to the relation

$$\frac{N}{N_{\max}} \sim \left(\frac{l}{l_{\max}}\right)^{\alpha} \tag{1}$$

The power law describing the distribution of faults in the Earth's crust found by many researchers (in particular, by the authors of [4]) confirms self-similarity of the microcrack pattern and Eq. (1) at the global level of development of the multiple fracture process. The tangent of the inclination angle of these cumulative distributions was accepted in [5] as the fractal dimension (D) of the microcrack patterns observed in the Earth's crust. It was assumed that fractal dimension can be estimated from relation

$$D = \frac{3b}{c}, \tag{2}$$

where b is the exponent in Guttenberg-Richter power equation ($N \sim E^{-b}$) for the relation between the number of seismic events (N) and their energy (E) (or magnitude $M = \log E$) and c is a constant equal to ~3/2. This means that the fractal dimension of the fracture pattern is proportional to the seismic parameter b (b value) estimated from the tangent of the inclination angle of the distribution of accumulated number of seismic events by magnitude.

The objective of this study included estimation of fractal dimension and parameter b of the real pattern of fractures of metal specimens and establishment of the correlation between them.

Let us introduce the following notations for easier description of the results: b_C, b_{AE} and b_S are exponents in the power laws describing cumulative distributions of microcracks (or faults), acoustic signals, and seismic events, respectively; D_C, D_{AE}, and D_S are fractal dimensions estimated from the data of the analysis of real damage, acoustic, or seismic responses to this damage, respectively.

Relation (1) was confirmed in [7, 8], where acoustic parameter b_{AE} was estimated on the rock specimens by the method of the acoustic emission. However, in the first case [7], fractal dimension was accepted as the angular coefficient of cumulative curves of microcracks distribution by length b_C; in the second case [8], as the fractal (correlation) dimension of the epicenters of acoustic events D_{AE}. In both cases, positive correlation of the estimated parameters was found (although the authors of [8] observed it only at the initial stage of the deformation of samples) corresponded in the accepted notations to relations:

$$b_C = 3\frac{b_{AE}}{c} \text{ and } D_{AE} = 3\frac{b_{AE}}{c}.$$

2. Results and discussion

In this work, we estimated b_C, the parameter of real microcracks pattern based on the method of replicas obtained at the polished surface of samples at different stages of loading. We also estimated fractal dimension D_C (corresponding to these patterns) measured by the least squares using relation [9]: $N = \lim_{R\to 0} BR^{-D}$, where N is the number of squares with side R containing black pixels (microcracks were colored black) and B is a constant. Variation in the area of the covered square is no less than two orders of magnitude. The cumulative curve of the distribution of microcrack length (L) used for determining parameter b_C corresponded to the power relation ($N = BL^{-bc}$) over the interval of variation in the microcracks length no less than by 1.5 orders of magnitude.

The investigations were carried out with flat notched specimens of model material (low-carbon steel) under conditions of static and cyclic loading. The thickness of the specimens ranged from 4 to 16 mm. The replicas allowed us to study the zone of the localization of fracture and the microcracks patterns in the zone at each stage of the process. After this, the replicas were analyzed using an optical microscope equipped with digital video camera, and the obtained patterns of multiple fractures were processed by the image analysis program. The maximal number (1500 - 10000) of microcracks measured on each sample depended on the degree of damage of the material at the each stage of loading. The results of measurements were used for plotting cumulative curves of the distribution of microcracks by length in the following coordinates: total number of microcracks (N) with the length greater and equal to the current length versus current length of microcracks (L).

Figure 1 shows a typical deformation curve of samples with highlighted points, corresponding to them zones of fracture localization, and patterns of microcracks in zones. The curves of variations in parameters b_C and D_C with deformation are also shown. It is seen that the zone of localization of fractures is already formed at the linear region of the deformation curve. At this initial stage of loading numerous slip bands that are developed both at the boundaries of the structural elements and within them were found. Microcracks are absent. Increase of strain leads to increase of the density of slip bands and appearance of defect of the next rank - pores formed in the slip bands or places of stress concentrators (scratches at the polished surface of the sample). When the yield stress is exceeded and the related size of the localization zone increases, the defect of the third rank appears as microcracks formed by the merging of pores along the slip bands.

Figure 1 also suggests a decrease in the b_C / D_C ratio with the increase in strain accompanied by the variation in parameters b_C and D_C in the opposite phase, unlike in-phase correlation of these parameters corresponding to Eq. (2). Such variation in parameters b_S and D_S was observed in [11, 12]. In [8], such variation in b_{AE} and D_{AE} was observed at the final stage of the fracture of specimens.

In this case, decrease in parameter b_C is caused by merging of microcracks, while increase in the fractal dimension D_C is caused by increase in the total area of fractures related to increase in the length and opening of microcracks. Investigation of fractures of specimens with different thicknesses confirmed the noted variation of these parameters with deformation and demonstrated that the dependence of the fractal dimension on parameter b_C corresponds to a linear dependence $D_C = 2.86 - 0.67b_C$ (correlation coefficient 0.95) for all tested samples. A similar linear dependence was also found in the analysis of interrelation between the correlation fractal dimension D_S and parameter b_S estimated from the data of seismic activity: $D_S = 2.3 - 0.73\ b_S$ (correlation coefficient 0.76) [11] and $D_S = 2.7 - 1.39\ b_S$ (correlation coefficient 0.85) [12].

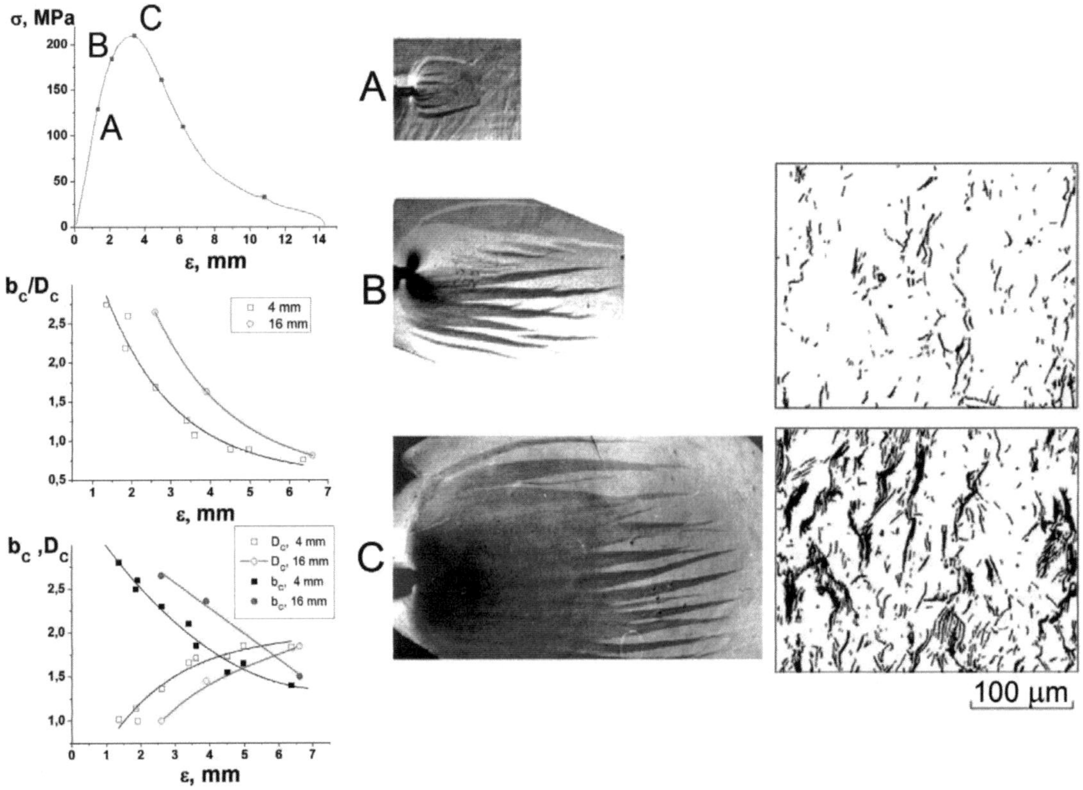

Fig. 1. Deformation dependences of stress in net-section of samples with notch, parameters b_C and D_C and their ratio $\frac{b_C}{D_C}$, zones of localization of fractures in the notch tip, and patterns of microcracks in zones.

Thus, it was established that:

- The kinetics of multiple fractures observed in steel specimens is similar to one on a global level of fracture process development. It confirms the ideas of self-similarity of fracture process of solids.
- For the multiple fracture patterns the dependence of the fractal dimension on b parameter corresponds to the reverse linear relation D = 2.86 - 0.67b.

Acknowledgments
The work was supported by the Russian Foundation for Basic Research (projects No. 09-05-01166 and 09-01-09279-моб_з).

References

[1] Botvina L R and Barenblatt G I 1985 *Problemy prochnosti (in Russian)* **12** 17-24
[2] Barenblatt G I and Botvina L R 1983 *Izv. AN USSR (in Russian)MSS* **2** 88-92
[3] Botvina L R 1989 Kinetika razrusheniya konstrukcionnyh materialov (Moscow: Nauka)
[4] Gudmunson A 1987 *Tectonophysics* **139** 295-308
[5] Aki K 1981 *Earthquaqe Prediction: An International Review, Maurice Ewing Ser.* **4** 566-74
[6] Botvina L R 2004 *Int. J. Fracture* **128** 133-7
[7] Hatton C G, Main I G and Meredith P G 1993 *J. Struct. Geol.* **15** 1485-95
[8] Ponomarev A V, Zavyalov A D, Smirnov V B and Lockner D A 1997 *Tectonophysics* **277** 57-81
[9] Feder J 1988 *Fractals* (New York: Plenum Press)
[10] Botvina L R, Jarkova N A, Tyutin M R, Petersen T B and Budueva V G 2005 *Deformation and Fracture of Materials (in Russian)* **3** 2-8
[11] Hirata T 1989 *J. Geophys. Res.* **94** 7507-14
[12] Őnsel A O, Main I, Alptekin Ő and Cowie P 1996 *Tectonophysics* **257** 189-202

Characterisation of transgranular crack propagation by EBSD in a soft steel

A. Boumaiza[1,2], N. Rouag[1], T. Baudin[3], Z. Larouk[1], R. Penelle[3]

[1] Laboratoire Microstructures et Défauts, Université Mentouri Constantine, Algeria
[2] Département de Physique, Université Farhat Abbas Jijel, Algeria
[3] Laboratoire de Physico-Chimie de l'Etat Solide UMR CNRS 8648, Paris-Sud Orsay, France

nadrou@wissal.dz (N. Rouag)

Abstract

This study is concerned with crack propagation in a soft steel sheet during drawing. The drawability is considered in relation with the structural anisotropy. Most existing studies on crack propagation are based on the global mechanical properties. However, microstructural inhomogeneity can lead to micro-crack formation. Micro-texture can affect crack propagation and stops in soft steel during drawing. The EBSD technique is used to show that the adjustment of the grain orientation from the initial recrystallization component {111}<112> towards the deformation orientation {111}<110 > incites a trans-granular crack inside a {111}<112> grain in a globally ductile material.

1. Introduction

Industrial metals always undergo various external stresses during fabrication and under service conditions. When they possess a heterogeneous microstructure, the usual mechanical properties do not allow accurately determining the risks of cracking that often occurs throughout manufacturing and service. In the plastic deformation studies, the microstructure is a basic factor of predicting the mechanical material behaviour. The presence of microstructural inhomogeneities will certainly affect the agreement between predictions based on homogeneous deformation and the experimental observations [1-3].Crack propagation and stops is an important topic in the formability field [4-6]. It is often required in safety analysis to consider that a crack might be initiated and lead to its propagation. It is important to know whether or not the material will be able to stop the crack before it goes through the entire structure, during its response to the external stresses. To answer this question, it is essential to consider the combination of "micro and macro" effects [7-9]. Undeniably, the phenomenon of localization is a result of both the external solicitation (loading, piece geometry) and the microstructure properties. It is this duality that makes the success of the micro-macro approaches which are based on changes of dimension scales. Therefore, such approaches allow a more comprehensive study, closer to the material behaviour during use.

The global behaviour of industrial metals is essentially controlled by the local micro-stress concentrations. Strain localization that appears may locally lead to a strong decrease of the ductility and therefore to a material rupture. The EBSD technique (Electron Back Scattering Diffraction) allows the correlation between the microstructure characteristics and the distribution of plastic deformation and stresses [10,12].

EBSD is presently used predominantly in nearly all metallurgy research approaches and has become a common technique used in the characterization of polycrystalline materials [13-16]. Since its advent, the method has become increasingly useful in the analysis of grain size and distribution, grain boundary disorientations and texture analysis of materials. The EBSD technique makes it possible to

Journal of Physics: Conference Series **240** (2010) 012036 doi:10.1088/1742-6596/240/1/012036

extract both local and global information about the material structure; it allows a complete microstructure characterization from local measurements.

The initial texture of the material is an essential parameter related to its mechanical behavior. In the metallurgical considerations of formability, the texture analysis links the crystallographic characteristics of the material to stress distribution [17-19]. The strong drawing deformations are accompanied by a noticeable modification of the initial texture. The low carbon steels generally possess a γ fiber texture {111}<uvw>, favourable to deep drawability [20-22]. In the present study, the investigated sheet possesses such fiber with a main component {111}<112>. After drawing, this fibre is kept with a change of the main component to bcc orientation deformation {111}<110>. The industrial problem is the initiation and propagation of cracks during forming. The average mechanical parameters of the sheet (Re=230 $N.mm^{-2}$, Rm=311 $N.mm^{-2}$, Λ%=36, HV= 140 $Kg.mm^{-2}$) are correct and do not allow an appreciation of the risk of cracking.

2. Experimental procedure

The material used in the present study is a primary recrystallized steel sheet of 1.5mm thickness, containing 0.07% C, 0.03% Si, 0.38% Mn, 0.025% P, 0.012% S, 0.02% Al, 0.007% N (wt%).

The samples ($1cm^2$) examined by EBSD are polished on standard emery papers up to grade 1200 and with the diamond paste up to 0.5μm. They are subsequently electropolished in a solution of: 25vol. perchloric acid (d=1.67), 235 vol. acetic acid (d=1.05) and 250vol. monobutyl ethylene glycol ether (d = 0.9), at (-5 ≈ -2) °C, using an applied voltage of 25V with 0,5A. The characterization of the global matrix texture is performed by X-Rays analysis and locally completed by the EBSD technique near the crack. Mapping was also performed in the area around the crack. However, it was difficult to obtain good patterns and thus to determine the crystallographic grain orientation close to the crack. Therefore, the misorientation profile across the crack is used in the present study in addition to mapping.

3. Observations

The preliminary micro-graphical analysis showed that the central crack progress with a zigzag by a globally ductile rupture of the material with some secondary fissures around the main line (figure 1).

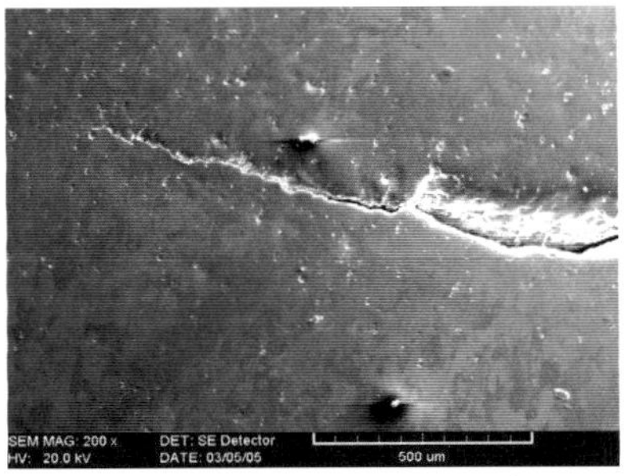

Figure 1
SEM micrograph of the analysed crack

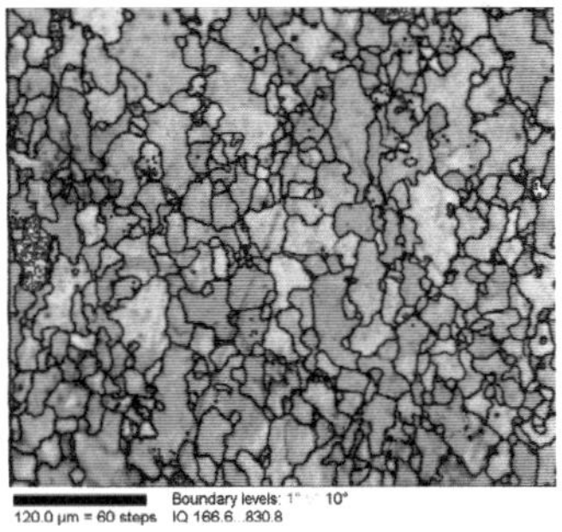

Figure 2
Initial inhomogeneous microstructure

Figure 2 shows the initial microstructure. The average grain size is about 150 μm^2, but the noticeable inhomogeneity does not make it a reliable parameter. The grain size distribution is very heterogeneous; there is a gathering together of small and large grains shared in clusters. These two grain populations behave differently during forming. This risk is not taken into account when only the average size is considered, while the grain distribution is heterogeneous. In that clustered matrix, the distribution of stresses will be also heterogeneous [1-3]. The existence of these different populations can lead to local thinning or micro-cracking. The reduction in grain size is accompanied by a strengthening the plastic stress flow. Therefore, more global plastic deformation is increased, more important deformation is localised in the small grain clusters, where the risk of crack appearing is high. The existence of a heterogeneous distribution of grain sizes involves the existence of important stresses locally, which can incite local thinning or micro-cracks initiation.

Figure 3 shows the local orientation characterization around the crack. We can note the existence of a fiber texture {111}<uvw >, with a variation from <110> to <112> for <uvw > orientation. This local characterization gives a similar component texture to the one obtained by global analysis.

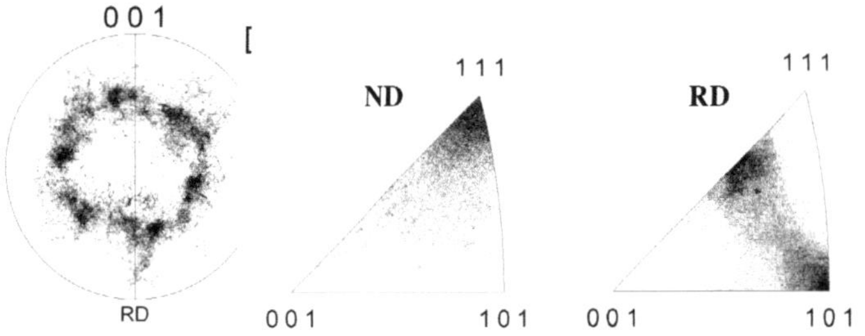

Figure 3
{111}<uvw> orientations of grains surrounded the crack

Figure 4 shows a misorientation profile across the crack. Except for the pic that corresponds to the crack position, the misorientation profile shows a perfect continuity of the observed evolution. This continuity of the observed evolution clearly indicates that it is the same grain along the considered profile, on the two sides of the crack. The right side corresponds to orientation {111} <112> whereas the left side includes orientations from {111} <112> towards {111} <110>.

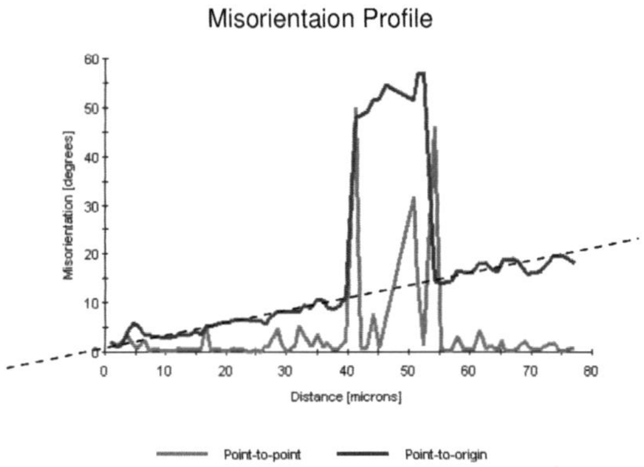

Figure 4
Misorientation profile across the crack showing a perfect continuity of the observed evolution.

4. Discussion

There is a continuous variation of site orientations from {111}<112> towards {111}<110 >. This observation confirms that the observed crack is trans-granular and not inter-granular, as will let it expected by the globally ductile character of the considered material. It could be suggested that the apparently local brittle-mode of fracture is actually the result of grain reorientation. In deformation mechanism by slip, each grain undergoes a lattice rotation, it is certain that the grain reorientation from the initial orientation {111} <~112 > towards the bcc deformation orientation {111} <110 > implies an important micro-constraint state, which is able to initiate a trans-granular evolution of the crack, if the slip planes present some sufficiently weak angles with the propagation direction. Figure 5 shows the global evolution of texture main component from initial orientation {111} <112> to {111}<110> for deformed samples by tensile tests with different angles α of tensile axis rapport the rolling direction for all tensile axis (α=0°,45°,90°). An analysis of different tensile cracking samples is performed in order to confirm the possibility of fragile micro-zone existence in the considered soft steel. Figure 6 shows the existence of a cleavage micro-zone, merged in a global ductile neighbourhood generally observed on most the analyzed samples.

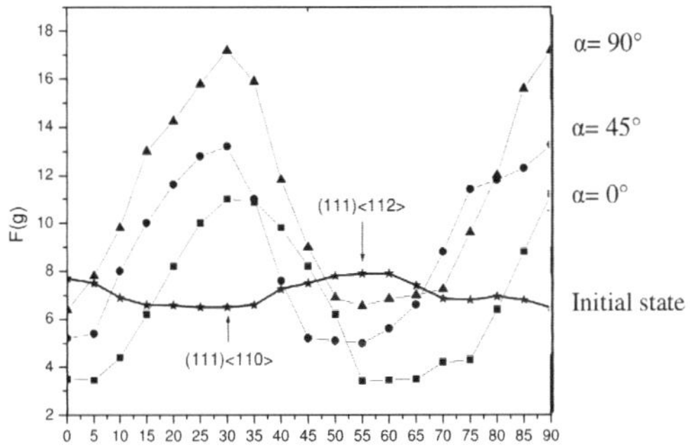

Figure 5
Evolution of texture main component from {111}<112> for the initial state towards {111}<110 > for all tensile axis (α=0°,45°,90°).

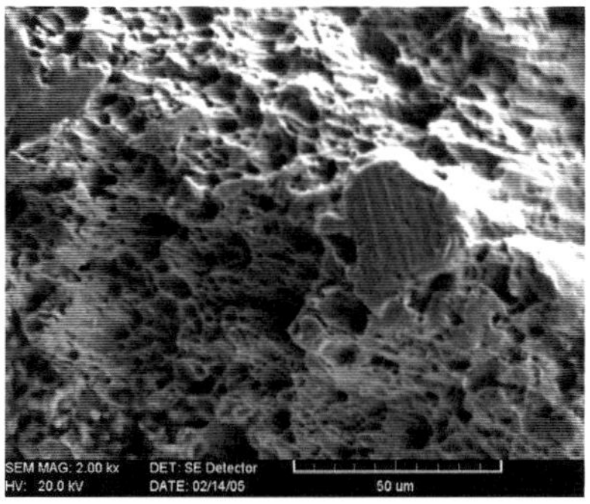

Figure 8
SEM micrograph showing the presence of cleavage region merged in a global ductile neighbourhood.

Conclusion

This study considers the texture effect on the propagation of cracks in soft steel during forming. The misorientation profile across the crack is used in EBSD in addition to mapping of the neighbourhood of the crack. The misorientation profile permits to clearly show adjustment of the grain orientation from {111}<112 > initial orientation towards the bcc deformation orientation {111}<110 >. This process can explain the possibility of transgranular crack propagation in a globally ductile material. In a heterogeneous structure the risk of cracking seems to increase in the presence of small grains {111}<112 > clusters.

References

1 Jago RA, Hansen N, Acta Metall. 1986; 34:1711-1720.
2 Ritchie RO, Thompson AW, Metall. Trans. A 1985; 16:233-248.
3 Thompson AW, Acta Metall. 1977; 25:83-86.
4 Bouyne E, Flower HM, Lindley TC, Pineau A, Scripta Mater. 1998; 39:295–300.
5 Pardoen T, Delannay F, Eng. Fract. Mech. 2000;65:455-466.
6 Kamp N, Parry MR, Singh KD, Sinclair I, Acta Mater. 2004; 52:343–353
7 Kuhlmann-Wilsdrof D, Acta Mater. 1999; 47:1697-1712.
8 Raabe D, Zhao Z, Park SJ, Roters F, Acta Mater. 2002;50:421–440.
9 Raabe D, Sachtleber M, Zhao Z, Rotters F, Zaefferer S, Acta Mater. 2001; 49: 3433–3441.
10 Dingley DJ, Randle V, J. Mater. Sci. 1992;27:p4545-4566.
11 Gourgues AF, Mater.Sci. Technol. 2002; 18,2:119-133.
12 Venegas V, Caleyo F, Gonzalez JL, Baudin T, Hallen JM, Penelle R, Scripta Mater. 2005;52:147-152.
13 Pouchou JL, L'analyse EBSD, Département des Matériaux Métalliques et Procédés, Editions du groupe GN ; 2004.
14 Michael JR, Microsc. Microanal. 2002;8(Suppl. 2): 724-725.
15 Randle V, Acta Mater. 2004; 52:4067–4081.
16 Klimek L, Pietrzyk B, J. Alloys Comp. 2004; 383:17-23.
17 Hu YM, Floer W, Krupp U, Christ HJ, Mater. Sci. Eng. A 2000; 278:170-180.
18 Stanford N, Dunne D, Ferry M, Mater. Sci. Eng. A 2003; 348:154-162
19 (a) Hwang KC, Lee S, Lee HC, Mater. Sci. Eng. A 1998 ; 254: 282-304
(b) Hwang KC, Lee S, Lee HC, Mater. Sci. Eng. A 1998 ; 296-304.
20 Barnett MR, Jonas JJ, ISIJ Int. 1999;39:856–73.
21 Tomitz A, Kaspar R, ISIJ Int. 2000;40: 927–31.
22 Asensio J, Romano G, Martinez VJ, Verdeja JI, Pero-Sanz JA, Mater. Charact. 2001;47:119-127.

Effect of residual stress and surface roughness on the fatigue behaviour of aluminium matrix composites

M. Smaga and D. Eifler
Institute of Materials Science and Engineering,
University of Kaiserslautern, D-67653 Kaiserslautern, Germany

smaga@mv.uni-kl.de

Abstract. In this investigation the fatigue properties of specimens manufactured with different turning parameters were investigated in stress-controlled constant amplitude tests at ambient temperature. The change of feed rate and depth of cut lead to a change in the near surface microstructure. Hence the fatigue properties were influenced significantly due to different surface roughness and surface residual stress resulting from the unequal turning processes. The cyclic deformation behaviour of AMC225xe is characterised by pronounced initial cyclic hardening. Continuous load increase tests allow a reliable estimation of the endurance limit of AMC225xe with one single specimen on the basis of cyclic deformation, temperature and electrical resistance data.

1. Introduction

The reinforcement of aluminium alloys with ceramic particles provides a new family of composite materials during the last years, which is applied in structural components subjected to high cyclic loads [1]. The machining process of metal matrix composites is more complex due to the inherent differences in metallurgical characteristics of MMCs compared to the machining of monolithic metallic materials. In the literature we can find systematic experimental investigations of the monotonic and cyclic deformation behaviour [2-4] as well as information about machining process [5-7] of MMCs. However, the available literature on machining of metal matrix composites shows limited effort towards a simultaneous study of the effects of turning parameters on surface characteristics and fatigue properties, in which ultrafine SiC reinforcement particles have been embedded in an Al-Cu-Mg matrix alloy.

2. Material and experimental setup

The investigated aluminium matrix composite AMC225xe exists of the high quality aerospace aluminium alloy AA 2124 reinforced with 25 vol.-% fine 2-3 μm silicon carbide particles SiC_p (Fig. 1). The AMC225xe was produced by a powder metallurgical processing route, followed by open die forging and T6 heat treatment. Uniaxial stress-controlled fatigue tests were performed at ambient temperature with a load ratio of R=-1 using a frequency of 5 Hz. Stress-strain-hysteresis loops, the change in temperature ΔT and the change in electrical resistance ΔR were measured during the fatigue tests [8]. The residual stresses were measured using the uniaxial sine square technique and the diffraction of copper $K\alpha$ radiation from the (311) planes of the aluminum matrix of the MMC.

Fig. 1: Microstructure of AMC225xe (a) homogeneous and (b) inhomogeneous distribution of SiC_p

3. Results

The fatigue tests were performed on specimens prepared by turning using a high performance polycrystalline diamond based cutting tool with constant cutting speed of 100 m/min, a feed rate (v) in the range from 0.02 to 0.08 mm/rev as well as the depth of cut (a_p) in the range from 0.4 to 1.2 mm. Specimens with carefully polished surface were used as a reference. To estimate the endurance limit of AMC225xe for specimens with polished surface with one single specimen, continuous load increase test (LIT) was performed (Fig. 2a). At the beginning of the test, the plastic strain amplitude $\varepsilon_{a,p}$, the change in temperature ΔT and the change in electrical resistance ΔR are nearly zero. Significantly increasing $\varepsilon_{a,p}$, ΔT and ΔR values are detected for $\sigma_a \geq 300$ MPa. This stress amplitude $\sigma_{D\ (LIT)}$ is in a very good accordance with the endurance limit determined in conventional constant amplitude tests until $N=4{\cdot}10^6$ cycles (Fig. 2b). The S-N curve for specimens with polished surface was described according to Basquin by $\sigma_a=664{\cdot}N_f^{-0.045}$. As can be seen in Fig. 2b, for the stress amplitude $\sigma_{D\ (LIT)}=300$ MPa the extrapolated Basquin equation yields a fatigue life of about $N_f \sim 10^8$ cycles. The test with $\sigma_a=300$ MPa and f=5 Hz was interrupted at $4{\cdot}10^6$ cycles for experimental reasons.

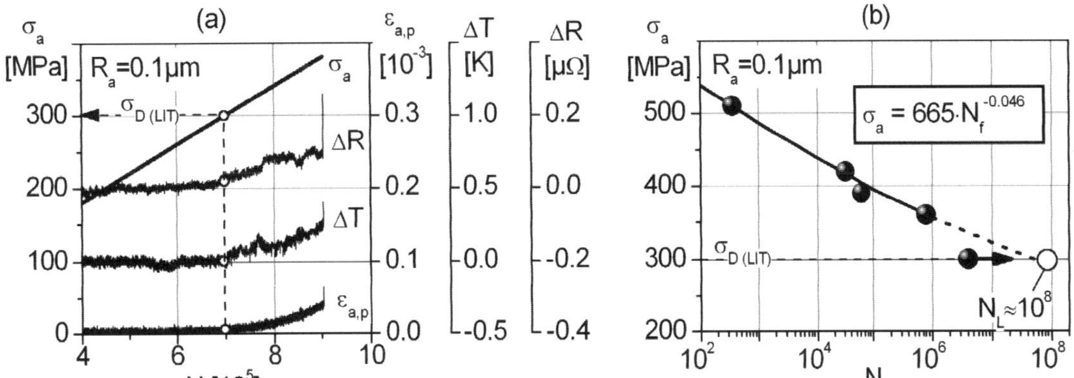

Fig. 2: Development of plastic strain amplitude $\varepsilon_{a,p}$, change in temperature ΔT and change in electrical resistance ΔR in a continuous load increase test and S-N-curve for polished surface (R_a=0.1µm)

The cyclic deformation behaviour of AMC225xe is characterised by pronounced initial cyclic hardening, exemplarily shown for specimens with turning parameter v=0.05 mm/rev and a_p=0.4 mm in Fig. 3a. With increasing stress amplitude the initial plastic strain amplitudes $\varepsilon_{a,p}$ increase and consequently the lifetimes decrease, although cyclic hardening is more pronounced. For a detailed characterisation of the cyclic deformation behaviour and to show the difference between tension and compression stress-strain response under cyclic loading and exemplarily the development of a stress-strain hysterese at the first cycle is plotted in Fig. 3b for σ_a=390 MPa. In the tension half cycle the first cyclic plastic deformation occurs at $\sigma\approx$300 MPa. In the compression half cycle plastic deformation already occurs at the significantly lower stress of $\sigma\approx$200 MPa. This documents that there exists a huge difference in fatigue strength between the tensile and the compressive part of the loading cycles. The difference in plastic strain response between tensile and compressive loading can be caused by several factors, e.g. thermal residual stress distribution in the composite or closing of pores. Figure 3c shows the change of residual stresses in the specimen surface during fatigue tests with a stress amplitude σ_a=390 MPa (0.05 mm/rev and 0.4 mm). The first tension half cycle leads to a higher reduction of the compressive stresses from -80 MPa to -44 MPa in comparison to the tension half cycle at 90% N_f ($N=1\cdot10^4$), which reduces the residual stresses only about 4 MPa from -42 MPa to -38 MPa. The cyclic loading leads to a continuous reduction of the residual stresses from -80 MPa at N=0 to -46 MPa at 10% N_f. These residual stresses were measured after the compression half cycle at the end of the cycle. The residual stresses, which were developed during the turning processes, were reduced at about 50 % due to initial plastic deformation up to 10 % life time. After this phase they are stable up to specimen's failure and diminish to a negligible value.

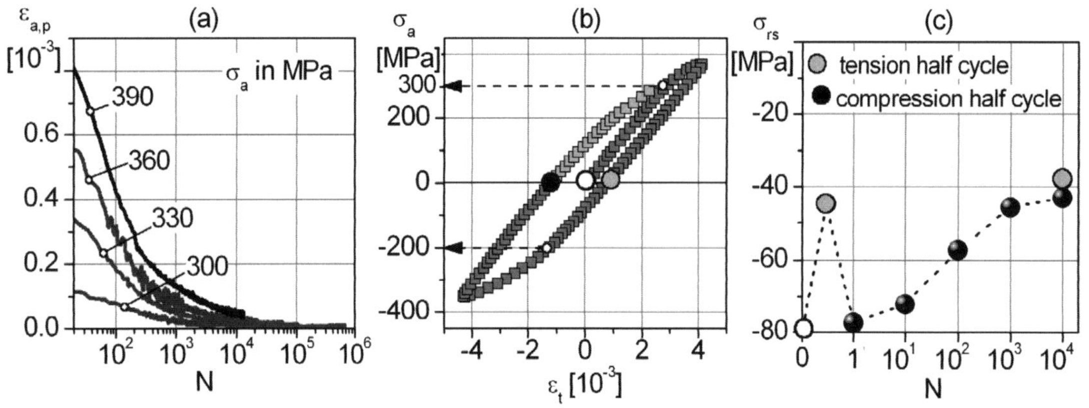

Fig. 3: Cyclic deformation curves (a), stress-strain-hystreses (b) and change of surface residual stresses during a fatigue test with σ_a=390 MPa (c) for specimens with R_a=0.4 µm

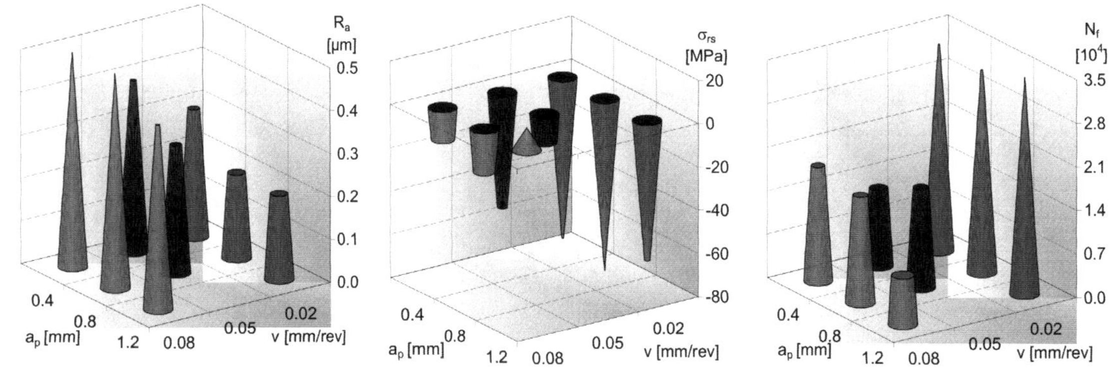

Fig. 4: Effect of the change of feed rate and depth of cut on surface roughness (a), residual stress (b) and fatigue life (c)

Figure 4 shows the effect of turning parameters on surface roughness (a), residual stresses (b) and fatigue behaviour (c) of aluminium matrix composites. The feed rate is the most important parameter, which significantly influences surface roughness as well as residual stress and consequently the fatigue life. With a specific combination of feed rate and depth of cut the same surface roughness and different residual stresses, e.g. v=0.02 / a_p=0.4 and 0.05 / 0.8 can be achieved. For the same surface roughness R_a=0.3 μm the change of residual stresses from -13 MPa to -74 MPa leads to an increase of fatigue life by about a factor of two. Obviously the best fatigue properties are obtained with specimens with small surface roughness and higher compressive residual stresses. The increase of feed rate from 0.02 to 0.08 mm/rev and the simultaneous increase of cutting depth from 0.4 to 1.2 mm lead to a decrease of the fatigue life by a factor of four. Figure 5 shows the correlation between the fatigue life of specimens loaded with σ_a=390 MPa and the surface residual stresses developed during turning.

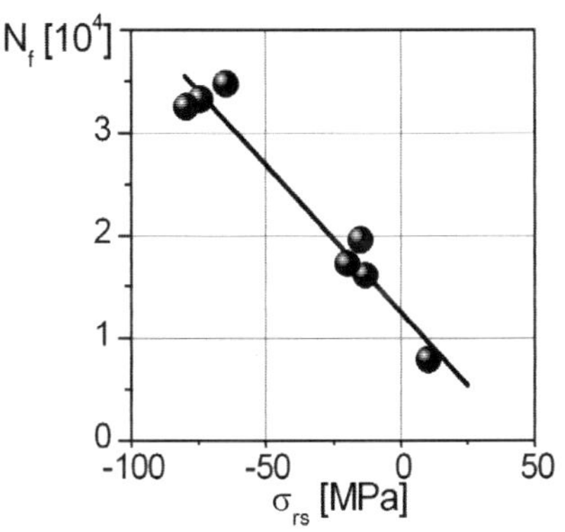

Fig. 5: Correlation between fatigue life and surface residual stresses

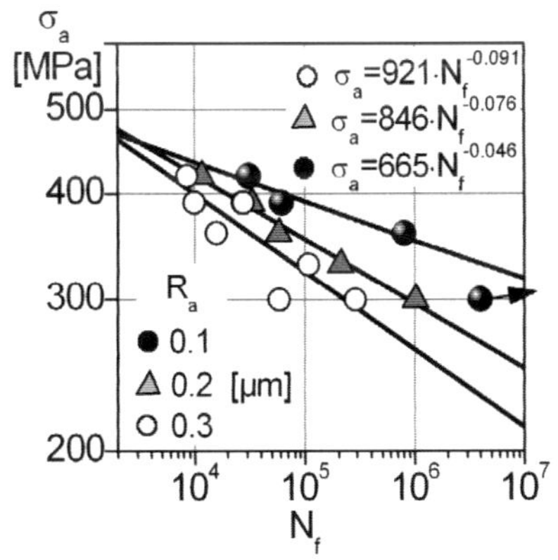

Fig. 6: S-N-curves

The fatigue life for specimens turned with two different feed rates v=0.02 mm/rev (R_a=0.2 μm) and 0.8 mm/rev (R_a=0.3μm) at constant depth of cut a_p=0.2 mm, compared to polished reference specimens (R_a=0.1 μm) is shown in Fig. 6 in form of S-N-curves. All S-N-curves converge in the LCF-range and diverge in to the HCF-range, where the surface roughness and residual stresses play a very important role in terms of fatigue life and lead to the huge difference in fatigue life according to the Basquin equation. Additionally the specimens turned with v=0.08 mm/rev show higher scattering in the number of cycles to failure due to higher scattering in the surface roughness.

4. Conclusions

The cyclic deformation behaviour of AMC225xe is characterised by pronounced initial cyclic hardening. The surface residual stresses are reduced by 50 % due to initial plastic deformation up to 10 % fatigue life and are nearly constant after this phase up to specimen's failure. The feed rate is the most important machining parameter, which significantly influences surface roughness as well as residual stresses and consequently the fatigue life. The extrapolated Basquin equation yields a fatigue life of polished specimens in range of $N_f \sim 10^8$ cycles for the stress amplitude $\sigma_{D\,(LIT)}$=300 MPa estimated in LIT. The fatigue behaviour of specimens with different surface properties is similarly in the LCF-range and different in the HCF-range, where the surface roughness and residual stresses play a very important role.

5. References

[1] Llorca J, 2002 *Prog. Mater. Sci.* **47** 283
[2] Koh SK, Oh SJ, Li C, Ellyin F, 1999 *Int. J. Fatigue* **21** 1019
[3] Hall JN, 1994 *Mater. Sci. Eng. A* **183** 69
[4] Srivatsan TS, Jr DL, Soni K.K, 1993 *Int. J. Fatigue* **15** 231
[6] Li XP, Seah WKH, 2001 *Wear* **247** 161
[7] Palanikumar K, Karthikeyan R, 2006 *Mach. Sci. Technol.* **10** 417
[8] Starke P, Walther F, Eifler D, 2007 *Fatigue Fract. Eng. Mater.* **30** 1044

Journal of Physics: Conference Series **240** (2010) 012038 doi:10.1088/1742-6596/240/1/012038

Cyclic stress-strain behaviour under thermomechanical fatigue conditions – Modeling by means of an enhanced multi-component model

H J Christ[1] and V Bauer[2]

[1] Institut für Werkstofftechnik, Universität Siegen, D-57068 Siegen, Germany
[2] Wieland Werke AG, Graf-Arco Str. 36, D-89072 Ulm, Germany

hans-juergen.christ@uni-siegen.de

Abstract. The cyclic stress-strain behaviour of metals and alloys in cyclic saturation can reasonably be described by means of simple multi-component models, such as the model based on a parallel arrangement of elastic-perfectly plastic elements, which was originally proposed by Masing already in 1923. This model concept was applied to thermomechanical fatigue loading of two metallic engineering materials which were found to be rather oppositional with respect to cyclic plastic deformation. One material is an austenitic stainless steel of type AISI304L which shows dynamic strain aging (DSA) and serves as an example for a rather ductile alloy. A dislocation arrangement was found after TMF testing deviating characteristically from the corresponding isothermal microstructures. The second material is a third-generation near-gamma TiAl alloy which is characterized by a very pronounced ductile-to-brittle transition (DBT) within the temperature range of TMF cycling. Isothermal fatigue testing at temperatures below the DBT temperature leads to cyclic hardening, while cyclic softening was found to occur above DBT. The combined effect under TMF leads to a continuously developing mean stress. The experimental observations regarding isothermal and non-isothermal stress-strain behaviour and the correlation to the underlying microstructural processes was used to further develop the TMF multi-composite model in order to accurately predict the TMF stress-strain response by taking the alloy-specific features into account.

1. Introduction

Engineering components and structures that operate at elevated temperatures, e.g. in electric power generation equipment, gas turbines or chemical reactors, are often subjected not only to isothermal cyclic or static loading but also to thermal transients during start-up and shut-down stages resulting in plastic deformations in the highly stressed regions. In many high-temperature applications this combination of cyclic thermal loading and cyclic mechanical loading leads to thermomechanical fatigue (TMF) and may even be considered to be the primary life-limiting damage contribution. In laboratory testing, most commonly TMF tests are carried out (in addition to isothermal LCF tests) applying the two extreme TMF modes of in-phase (IP) and out-of-phase (OP) temperature-strain phasing.

The present study deals with the cyclic stress-strain response under TMF loading conditions and the description of this behaviour by means of material-specific expansions of a simple multi-component model. Two extremely different metallic materials, an austenitic stainless steel of type AISI304L and a third-generation near-γ TiAl alloy (in duplex condition), were selected, in order to illustrate the broad variety of TMF-induced changes as compared to isothermal LCF. A reliable simulation of the non-isothermal stress-strain path is a valuable contribution to many TMF life prediction methods. More details on this aspect and information on the materials studied and the experimental techniques applied can be found in [1-4].

2. Application of Masing's model to thermomechanical fatigue

It has been shown in an earlier work [5] that the so-called pseudo-isothermal thermomechanical fatigue behaviour (PITMF behaviour) can be calculated from the hysteresis loops of isothermal fatigue tests performed at different temperatures by applying a multi-component model originally proposed by Masing [6]. The basic assumption of this model is that a material consists of a parallel arrangement of elements which deform in a linear elastic-perfectly plastic manner. The distribution of the element-specific yield stresses σ_{if} of the elementary volumes is expressed by means of a yield stress distribution density function $f_p(\sigma_{if})$ which can be determined from a single branch of a hysteresis loop.

The stress-strain path can easily be calculated on the basis of the second integral function $G_p(\sigma_{if})$ of $f_p(\sigma_{if})$ or by using the model in a discretised manner (e.g. on the basis of 30 elementary volumes). In the case of anisothermal conditions, the effect of temperature on the yield stress distribution density function must be included and the calculation is carried out in small strain and temperature increments. PITMF behaviour means that the material is assumed to behave under varying temperature at each temperature as it does in the corresponding isothermal test. Hence, it defines a reference behaviour which helps to identify TMF-specific deformation features on the basis of a comparison of the simulation result and the experimental observation.

3. Stress-strain behaviour of the austenitic stainless steel

The isothermal LCF behaviour of AISI304L is strongly affected by the occurrence of dynamic strain aging (DSA) in an intermediate temperature range between 250°C and 500°C. DSA manifests itself in a change of the saturation dislocation arrangement, as clearly seen in figure 1. Both, at low and high temperatures a cell or subgrain dislocation arrangement forms, whereas a disorderly planar-type arrangement was observed in an intermediate temperature range. The more planar dislocation slip of DSA increases the saturation stress amplitude at constant plastic strain amplitude (figure 2).

Figure 1. Dislocation arrangement of isothermal fatigue tests at different temperatures and a plastic strain amplitude of 0.5%.

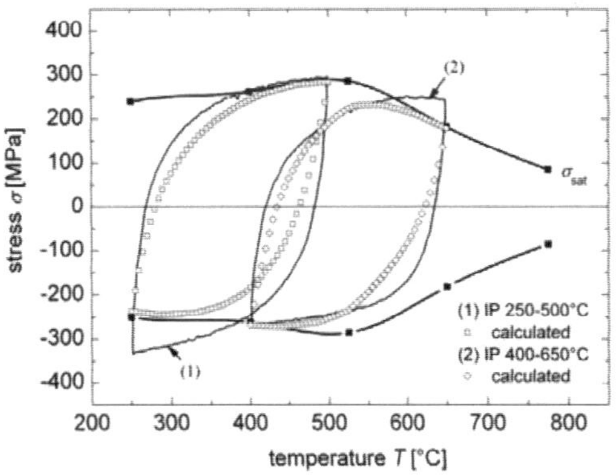

Figure 2. Calculated and measured hysteresis loops of two IP TMF tests compared to the isothermal saturation stress amplitude.

Figure 2 shows the measured and the calculated (PITMF) hysteresis loops for two IP tests together with a representation of the saturation stress amplitudes of isothermal tests *vs* temperature. Peak and valley stresses of the calculated loops correspond very nicely to the saturation stress amplitude values σ_{sat} of the respective isothermal test. However, outside the temperature range of DSA, the load reversal points of the experimentally observed loops deviate significantly from the σ_{sat} course. The impeded dislocation mobility in the DSA region strongly affects the adjusting TMF dislocation arrangement and leads to higher strength both at higher and lower temperature.

Additional tests were carried out with temperature changes from the saturation state within the DSA regime to a temperature above or below the DSA temperature range. Figure 3 shows the cyclic deformation curve of a test, during which the temperature was changed from 500°C (inside the DSA regime) to 250°C (above DSA), and compares the stress amplitude with the values from isothermal testing at 250°C. The strengthening effect of predeformation in the DSA temperature range is clearly visible. This effect can be incorporated into the stress-strain response modeling by using the yield stress distribution density function from hysteresis loops obtained in respective temperature change tests instead of those from loops of isothermal tests (without pre-cycling in the DSA regime). Figure 4 compares the hysteresis loop of cyclic saturation of an IP test with the predicted hysteresis loop documenting a reasonable agreement. For higher maximum temperatures, low strain rates and superimposed dwell times creep and stress relaxation has to be taken additionally into account in the model.

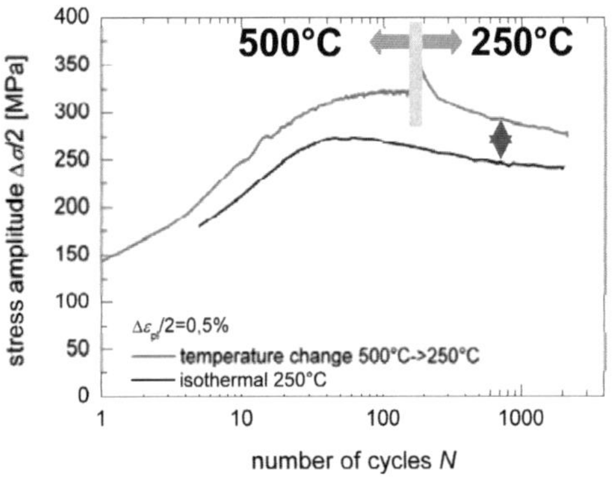

Figure 3. Comparison of the stress response during a temperature change test (500°C→ 250°C) with the isothermal response at 250°C.

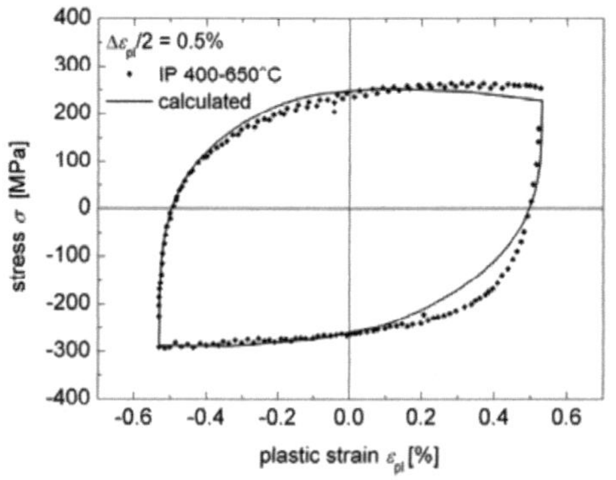

Figure 4. Comparison of a measured hysteresis loop of a IP test with a calculated loop considering the DSA prehistory.

4. Stress-strain behaviour of the third-generation near-γ TiAl alloy

Isothermal fatigue tests performed on the near-γ TiAl alloy studied showed a rather pronounced cyclic saturation independent of testing temperature. However, a more detailed analysis of the course of the plastic strain amplitude in the total-strain-controlled tests revealed a slight cyclic hardening for T < 550°C and cyclic softening at T > 650°C. As a consequence of this behaviour a continuously developing mean stress σ_m results under TMF conditions, while the stress amplitude $\Delta\sigma/2$ remains fairly constant (see figure 5 and 6).

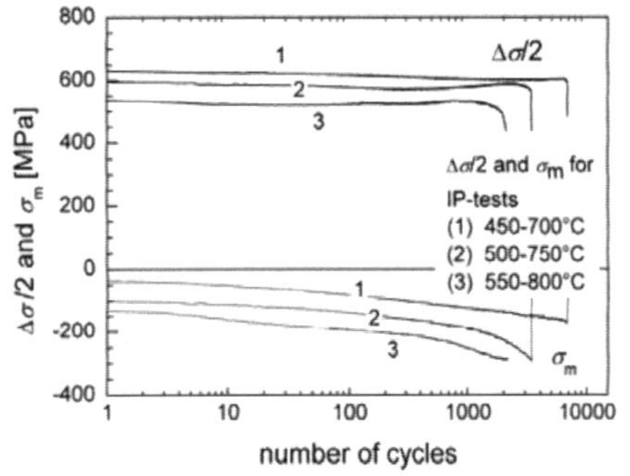

Figure 5. Development of stress amplitude and mean stress with number of cycles for three in-phase TMF tests at a mechanical strain amplitude of 0.6%.

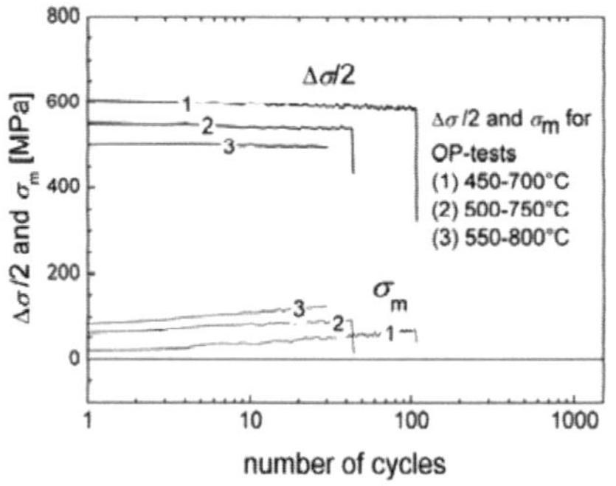

Figure 6. Development of stress amplitude and mean stress with number of cycles for three out-of-phase TMF tests at a mechanical strain amplitude of 0.6%.

In order to model this TMF mean stress development, simple power-law equations were deduced from the isothermal cyclic-deformation curves. In the calculation of the cyclic stress-strain response of TMF, cyclic softening and hardening was iteratively considered cycle by cycle at the maximum and minimum temperature, respectively. For this purpose the isothermal softening/hardening law was applied to the corresponding peak and valley stresses calculated by means of the multi-component model in the first instance. This procedure is schematically shown in figure 7, which also compares the predicted and the experimentally observed mean stress values. In figure 8 the hysteresis loops of IP TMF loading with a temperature range of 500-750°C are shown for the second cycle and the cycle at half life $N_f/2$ and compared with the corresponding results from the simulation. As a consequence of the similarity in the deformation processes during isothermal and thermomechanical cyclic deformation the predicted results for the first few cycles coincide well with the measured loops. With advancing number of cycles the negative mean stress during IP testing decreases and the hysteresis loop shifts downwards. This shift is reasonably predicted on the basis of the isothermal softening and hardening behaviour.

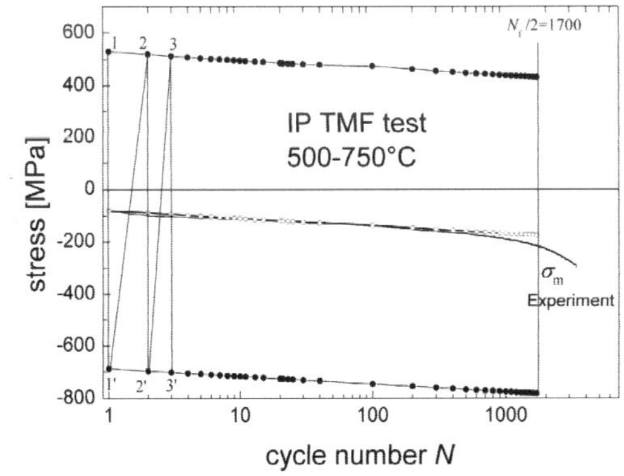

Figure 7. Development of maximum, minimum and mean stress with cycle number; comparison of the theoretical approach with the experimental result.

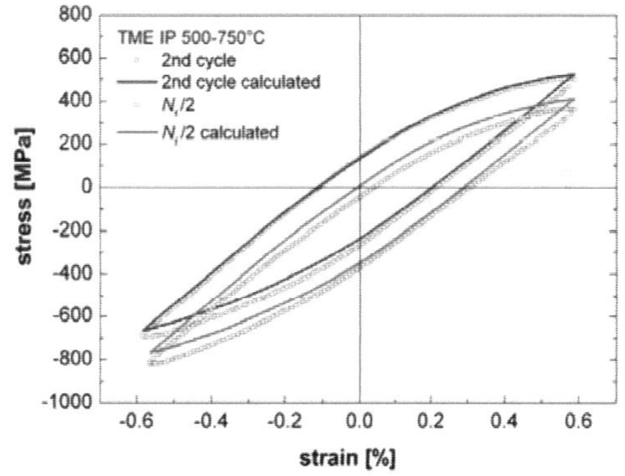

Figure 8. Comparison of the measured hysteresis loops of an IP test for the second cycle and at half life with the calculated loops considering mean stress development.

5. Conclusions

Both materials studied show a cyclic stress-strain behaviour under TMF conditions that deviates characteristically from that observed under isothermal conditions. This phenomenon can be attributed to the occurrence of dynamic strain aging in the case of AISI304L, while cyclic hardening and softening of the TiAl alloy at low and high temperatures, respectively, lead to a continuously developing mean stress at varying temperature. The multi-component model (i) helps to identify the TMF-specific behaviour and (ii) can easily be expanded in such a way that these effects are considered.

References

[1] Bauer V 2007 *Verhalten metallischer Konstruktionswerkstoffe unter thermomechanischer Belastung – Experimentelle Charakterisierung und modellmäßige Beschreibung* (Doctorate Thesis, Universität Siegen, Aachen: Shaker Verlag GmbH)

[2] Christ H J and Bauer V 2008 *Proc. of the Hael Mughrabi Honorary Symposium: Plasticity, Failure and Fatigue in Structural Materials – from Macro to Nano (TMS 2008 Annual Meeting&Exhibition, March 9-13, 2008, New Orleans, U.S.A)* ed K J Hsia, M Göken, T Pollock, P D Portella, N R Moody (TMS, Warrendale) pp 49-54

[3] Bauer V and Christ H J 2009 *Intermetallics* **17** 370-377

[4] Christ H J 2007 *Mater. Sci. Eng. A* **468-470** 98-108

[5] Maier HJ and Christ H J 1996 *Scripta Mater.* **34** 609-615

[6] Masing G 1923 *Wissenschaftliche Veröffentlichungen aus dem Siemens-Konzern* **3** 231-239

Mechanical properties of B-modified Ti-6Al-4V alloys

Indrani Sen
Department of Materials Engineering, Indian Institute of Science
Bangalore 560 012, India

indrani@platinum.materials.iisc.ernet.in

Abstract. Minor addition of B to the Ti-6Al-4V alloy reduces the prior β grain size by more than an order of magnitude. TiB formed in-situ in the process has been noted to decorate the grain boundaries. This microstructural modification influences the mechanical behavior of the Ti-6Al-4V alloy significantly. In this paper, an overview of our current research on tensile properties, fracture toughness as well as notched and un-notched fatigue properties of Ti-6Al-4V-xB with x varying between 0.0 to 0.55 wt.% is presented. A quantitative relationship between the microstructural length scales and the various mechanical properties have been developed. Moreover, the effect of the presence of hard and brittle TiB has also been studied.

1. Introduction

Ti and its alloys are widely used in aerospace industries owing to their unique combination of low density and excellent mechanical properties such as strength and toughness. Among the Ti alloys, Ti-6Al-4V (referred hereafter as Ti64) is the most popular. In the as-cast condition, this alloy has grain size of the order of a few mm. Hence, expensive thermo-mechanical processing is employed to refine the microstructure where necessary. Recently it has been found that minor amount of B additions in the hypoeutectic range, markedly reduces the as-cast prior β grain size by an order of a magnitude [1]. For instance addition of 0.1 wt.% B reduces the grain size from ~ 1.7 mm to 150 μm [2]. Apart from the microstructural refinement, B addition to Ti64 alloy has significance in terms of producing in-situ TiB needles of width ~ 2-5 μm and aspect ratio ~ 10. Interestingly, TiB posses the appropriate balance of thermo-mechanical stability and similar thermal expansion co-efficient [3].

Since Ti64 is an aerospace grade alloy; its fracture and fatigue properties are of paramount importance from a designer's point of view. The present study was initiated to examine these in detail. This includes characterization of the elastic and plastic deformation behavior of the alloy under uniaxial tension, plane strain fracture toughness, fatigue crack growth and the un-notched fatigue behavior in terms of cyclic stress-strain curves as well as high cycle fatigue endurance limit. All these properties are rationalized in terms of microstructural modification that occurs due to B-addition to the Ti64 and TiB formation.

2. Materials and Experiments

Five grades of Ti-6Al-4V-xB alloys with x = 0.0, 0.04, 0.09, 0.30 and 0.55 wt.% are examined in this study. These alloys have been induction skull melted to achieve homogeneity in composition throughout the cast structure, followed by hot isostatic pressing at 900°C and 100 MPa pressure for two hours to reduce the cast porosity. Microstructural characterization of the alloys was performed on specimens that were metallographically polished and etched with Kroll's reagent.

Room temperature quasi-static tensile tests were performed on cylindrical dog-bone shaped specimens with 5 mm gage diameter and 25 mm gage length, subjected to a nominal strain rate of 0.001/s. Four specimens were tested for each composition. An extensometer of gage length 25 mm was mounted on the specimen surface to monitor the strains. Yield strength (σ_Y) and ultimate tensile strength (σ_U) as well as elastic modulus (E) were thus measured from the tensile tests. Ductility as measured by the elongation to failure, ε_f has also been recorded.

To characterize the cyclic stress-strain response, incremental step test (IST) method was utilized on highly polished dog-bone shaped specimens of gauge diameter and length 4.5 and 14 mm respectively. Maximum strain amplitude used was ± 1.0 or ± 1.2%. It has been found that a strain limit of ± 1.0 gives better stabilization. Highly polished hour glass shaped specimens of gauge diameter 4 mm are used for rotating-bending high cycle fatigue tests performed at a frequency of 100 Hz. 30-40 specimens were tested for each composition to obtain the endurance limit (EL) on the basis of a lifecycle of 10^7.

Fatigue crack growth experiments were performed on Chevron notched full C[T] specimens of width (W) 50 mm and thickness (B) 25 mm. The samples were fatigue pre cracked at a at constant cyclic load amplitude ΔP of 20,000N with the load ratio, R (minimum to maximum load) as 0.1. A frequency of 10 Hz is applied throughout the test with a sinusoidal waveform. After the precracking, the crack is grown further with the standard load-shedding method to determine the threshold stress intensity factor range (ΔK_0). Experimentally ΔK_0 is considered as the value of ΔK for which crack growth per cycle i.e. da/dN becomes less than 10^{-9} m $cycle^{-1}$. After reaching the ΔK_0, the specimens are observed in optical microscope to track the crack growth behavior in the near-threshold regime.

Plane-strain fracture toughness tests were performed after the initial precracking according to the ASTM standard procedure E399-90. For this, a crack opening displacement gage with a 2 mm travel is mounted on the knife edges, fixed on the specimens to measure the crack extension. Valid K_{IC} could not be achieved for Ti64-0.0B as well as Ti64-0.04B. However, for the rest of the B-modified alloys, K_{IC} is achieved. Fractography of the fatigue crack growth and fracture toughness specimens are done by scanning electron microscope to identify the fracture modes for the different alloys.

3. Results and Discussions

3.1.1. Microstructures

Representative microstructure of the B-modified alloys examined in this study is shown in Figure 1. It shows Widmanstätten structure with alternate lamellas of (hcp) α and (bcc) β oriented along particular directions within individual colonies. Such colonies are again distributed within the prior β grains. The TiB phase forms needle shaped particles of aspect ratio ~10 at the grain boundaries forming a necklace like arrangement. Prior β grain size (*D*) has been found to reduce by 96% with 0.55 wt.% B addition (Figure 2). Moreover, with B addition to Ti64, reductions in the aspect ratio of the laths as well as randomization of the colonies were also seen.

3.2. Tensile properties

Yield strength, σ_y, ultimate tensile strength, σ_u and tensile ductility, ε_f are determined from the quasistatic tensile stress-strain plots. Both σ_y and σ_u has been found to increase by 12-16% respectively with B addition (Figure 3). Interestingly this is found to be correlated to the inverse square root of grain size according to the Hall-Petch relation (Figure 4). However, E has been found to show a drastic enhancement with 0.04 wt.% B to Ti64, followed by a reduction. The overall E of the system has been found to increase linearly with B addition, except for 0.04 wt.% B addition [4]. On the other hand ε_f has been found to increase initially with up-to 0.09 wt.% B addition primarily due to drastic grain refinement (Figure 3). Further B additions, however produces higher volume fraction of hard TiB needles in the matrix and in turn reduces the global ductility of the system. Thus with 0.55

wt.% B addition, ε_f becomes almost similar to that of B-free alloy. In this sense, it seems practical to consider addition of ~0.1 wt.% B as the optimum amount of B that will give the highest benefits in terms of tensile properties.

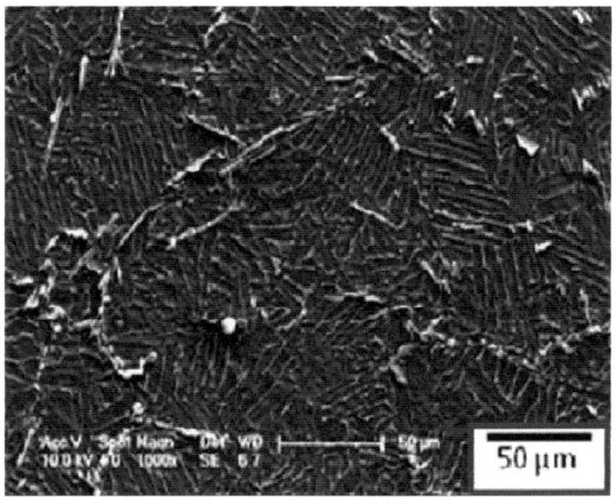

Figure 1. Representative scanning electron microstructure of Ti64-0.30B

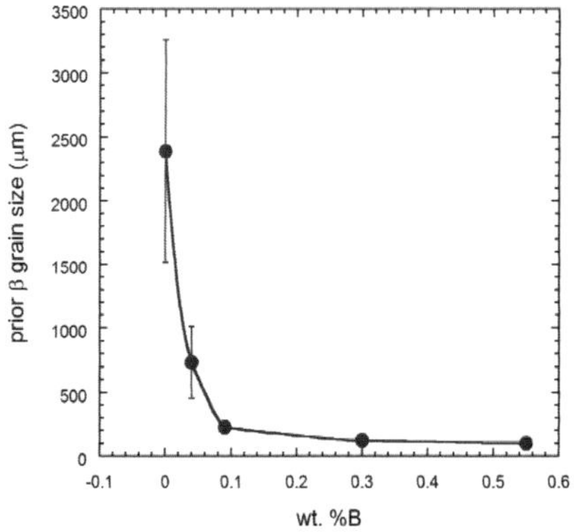

Figure 2. Variation of prior β grain size with B addition to Ti64.

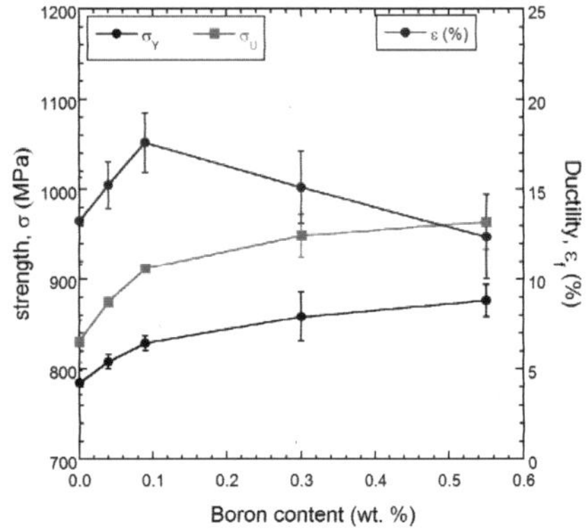

Figure 3. Variation of of σ_Y (MPa), σ_U (MPa) and ε_f (%) with B addition to Ti64.

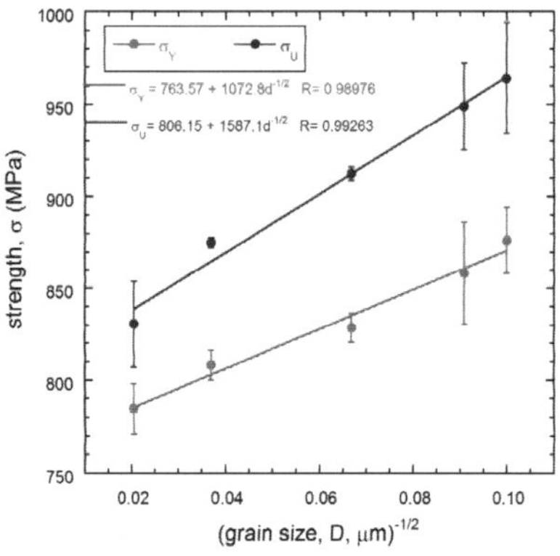

Figure 4. Variation of of σ_Y (MPa) and σ_U (MPa) with the inverse square root of grain size, *D*

3.3. Cyclic stress-strain

Cyclic softening is observed in all the alloys examined in this work. However, B modification has been found to increase the degree of softening. One possible reason could be the dislocation pile up at the α-β phase boundary, as well as dislocation-TiB particle entanglement. This activity in term will enhance the internal stress of the system. As a result, the applied stress necessary for further deformation will reduce and the material will soften cyclically. The more the amount of B is added, thicker will be the laths that can accommodate larger number of dislocation in a pile up and also larger will be the volume fraction of TiB and hence more will be the entanglement. Therefore cyclic softening has been found to increase with B addition to Ti64 [5].

3.4. High cycle fatigue

Endurance limit has been calculated from the high cycle fatigue experiments on the basis of limiting life of 10^7. It has been found that endurance limit increases with the increase in B addition to the system. This has been attributed to the grain size refinement with B addition. Since crack initiation is of prime importance for un-notched fatigue, it is the size of the grain that will have key role to play in controlling the endurance limit [5].

3.5. Fatigue crack growth

The values of ΔK_0 and the Paris exponent, *m*, for the alloys are calculated and ΔK_0 has been found to decrease with B addition up to 0.30B. However, a sudden increase is noted for the Ti64-0.55B alloy. Optical micrographs of fatigue crack near the threshold regime show that crack branching is a common feature for both the 0B and B-modified alloys. Deflection of the cracks by the lath boundaries as well as cutting the laths by the cracks to move forward on achieving higher stress concentration are other possible micro mechanisms, which enhance crack path tortuousity [2]. The crack path tortuousity was seen to decrease gradually with addition of B to Ti64. Since the crack path is trans-granular in nature, this can be related to the reduction of grain size. This is because of the fact that the growth of the crack gets arrested as the plastic zone size ahead of the crack tip reaches a critical microstructural size scale. A stress intensity factor range of the order of ΔK_0 is then needed to overcome the barrier and resume the crack growth. ΔK_0 is therefore directly related to the microstructural size of the alloy according to the relation ΔK_0 vs. $\sigma_Y{*}D^{0.5}$.

3.6. Fracture Toughness

Plane strain fracture toughness, K_{IC} (K_Q for 0.0B and 0.04B) has been found to decrease sharply by 66% with 0.55 wt.% B addition to Ti64. It is evident that with B addition, the resistance of Ti64 to fracture in the presence of a sharp crack and under severe tensile constraint decreases markedly. K_{IC} is thus believed to represent lower bound value of fracture toughness. According to RKR model [6], K_{IC} is directly proportional to σ_Y and a microscopic size parameter according to the following relation.

$$K_{IC} \propto \sqrt{D \times \sigma_y}$$

This relation shows a very good correspondence with a value of 0.99. Since with B addition, the increment in σ_y is only 12 % in comparison to the reduction in *D* by 96%, microstructural refinement is found to dominantly control the fracture toughness of the alloy. However, although B appears to be detrimental in terms of fracture resistance of Ti64, a value of 56 MPa $m^{1/2}$ is obtained for 0.09 wt.% B which is still in practical application limit and hence 0.1 wt.% B can be considered optimum to get maximum grain refinement, strength increment as well as ΔK_0 and K_{IC} still in applicable limit.

4. Conclusions

B addition to as cast-Ti64 causes significant grain refinement by an order of magnitude. This leads to enhancement in strength, ductility and elastic modulus as well as endurance limit. Degree of cyclic softening also increases with B addition. However, fracture toughness and fatigue threshold stress intensity factor range get reduced due to grain refinement. Considering all the above properties, it seems 0.1 wt.% B to Ti64 should be optimum in reducing the grain size to the maximum limit, enhancing strength and ductility while possessing the fracture and fatigue parameters within the practical application limits.

5. References

[1] Tamirisakandala S, Bhat R B, Tiley J.S and Miracle D B 2005 *Scripta Mater.* **53** 1421
[2] Sen I, Tamirisakandala S, Miracle D B and Ramamurty U 2007 *Acta Mater.* **55** 4983
[3] Sen I, Maheshwari L, Tamirisakandala S, Miracle D B and Ramamurty U 2009 *Mater. Sci. Eng. A* **518**; 162
[4] Sen I and Ramamurty U (2009) *submitted to Scripta Mater.*
[5] Sen I unpublished Ph.D thesis work
[6] Ritchie R O, Knott J F and Rice J R 1973 *J. Mech. Phys. Solids* **21** 39

Journal of Physics: Conference Series **240** (2010) 012040 doi:10.1088/1742-6596/240/1/012040

Prehistory effects on the VHCF behaviour of engineering metallic materials with different strengthening mechanisms

M Zimmermann , C Stöcker , C Müller-Bollenhagen , H-J Christ

Institut für Werkstofftechnik, Universität Siegen, D-57068 Siegen, Germany

martina.zimmermann@uni-siegen.de

Abstract. Engineering materials often undergo a plastic deformation during manufacturing, hence the effect of a predeformation on the subsequent fatigue behaviour has to be considered. The effect of a prestrain on the microstructure is strongly influenced by the strengthening mechanism. Different mechanisms are relevant in the materials applied in this study: a solid-solution hardened and a precipitation-hardened nickel-base alloy and a martensite-forming metastable austenitic steel. Prehistory effects become very important, when fatigue failure at very high number of cycles ($N > 10^7$) is considered, since damage mechanisms occur different to those observed in the range of conventional fatigue limit. With the global strain amplitude being well below the static elastic limit, only inhomogeneously distributed local plastic deformation takes place in the very high cycle fatigue (VHCF) region. The dislocation motion during cyclic loading thus depends on the effective flow stress, which is defined by the global cyclic stress-strain relation and the local stress distribution as a consequence of the interaction between dislocations and precipitates, grain boundaries, martensite phases and micro-notches. As a consequence, no significant prehistory effect was observed for the VHCF behaviour of the solid-solution hardening alloy, while the precipitation-hardening alloy shows a perceptible prehistory dependence. In the case of the austenitic steel, strain-hardening and the volume fraction of the deformation-induced martensite dominate the fatigue behaviour.

1. Introduction

Many modern applications, such as in transportation, require fatigue lives beyond the classical fatigue limit of 10^7 cycles. Recent studies have revealed that crack initiation in the VHCF regime is initiated by other damage mechanisms than can be observed for low and high cycle fatigue behaviour (e.g. [1]). Fatigue life is dominated by crack initiation rather than crack growth [2] with local microstructural heterogeneities acting as crack nucleation sites. These heterogeneities can be defects like pores and inclusions [3] as well as precipitates or a second phase [4].

As classical damage mechanisms do not apply to the specific fatigue behaviour in the VHCF regime, the question arises whether acknowledged strengthening mechanisms, such as strain hardening, precipitation-hardening or martensite transformation, actually contribute to a cyclic strength enhancement in this regime. In the case of the nickel-base superalloy Nimonic 80A, it was demonstrated that the overaged condition shows a slightly better fatigue behaviour than the peak-aged condition [5], thus acting in contrast to its behaviour in the low cycle and high cycle fatigue regime. Moreover, those strengthening mechanisms are rarely used singularly but appear as combined processes throughout a manufacturing process, such as the prestraining of a metastable austenitic steel. In this case the prestraining evokes a strength enhancement by means of a deformation-induced phase transformation from fcc austenite into bcc martensite [6] as well as strain hardening due to an increase of the dislocation density.

2. Material and experimental details

The material studied in the research work presented was chosen so as to typify different strengthening mechanisms: a solid-solution hardened nickel-base alloy Nimonic 75, a precipitation-hardened nickel-base alloy Nimonic 80A and a martensite-forming metastable austenitic steel AISI 304. Nimonic 80A was tested in the peak-aged and overaged condition (according to the heat treatments applied). Prestraining was carried out as monotonic tensile deformation by means of a servohydraulic testing machine at room temperature for the nickel-base alloys and at temperatures below room temperature for the austenitic steel.

The nickel-base alloys were prestrained up to a total strain of 8%. The martensite volume fractions (MVF) of 26% and 54% in AISI304 was gained by a plastic strain of 15% and deformation start temperatures of -20 and -70°C, thus adjusting the quasi-static mechanical properties. Prestrained specimens were electro-chemically polished after predeformation in the case of the austenitic steel and before prestraining for the nickel-base alloys. Some of the overaged specimens of Nimonic 80A were also electro-chemically polished after prestraining. Undefined micro-notches were introduced in some of the austenitic samples by soldering thermocouples to the specimens' surfaces.

For the fatigue experiments a resonance pulsating high-frequency test system (f ~ 100 Hz), an ultrasonic fatigue testing machine (f ~ 20 kHz) and a servohydraulic test system (f ~ 760 Hz) were used. Tests were conducted under symmetrical tension-compression loading (R = -1) in ambient air. More details on the materials and the experimental techniques applied can be found in [5, 6].

3. Results and discussion

The quasi static strength of the analysed material was quantified by tensile tests. Table 1 gives an overview of the results for the yield strength and the fracture strain of the nickel-base alloys. The prestraining leads to a pronounced increase of the yield strength for the precipitation hardened nickel-base alloy. The enhancement is even more pronounced for the austenitic steel (from 284 MPa to 652 MPa for 26% MVF and even 752 MPa for 54% MVF). As can be seen from table 1 and figure 1 this strength increase comes along with a pronounced decrease of the ductility for those materials. In contrast the predeformation effect on the yield strength of the solid-solution hardening nickel-base alloy Nimonic 75 is rather small and though the ductility decreases slightly it is still quite high compared to Nimonic 80A.

Table 1. Static mechanical properties before and after prestraining (pa = peak-aged, oa = overaged).

		Nimonic 80A pa	Nimonic 80A oa	Nimonic 75
prestrain 0%	Yield Strength [MPa]	976	537	302
	A [%]	26	34	48
prestrain 8%	Yield Strength [a] [MPa]	1201	811	436
	A [%]	18,5	26,3	40,2

[a] True yield strength.

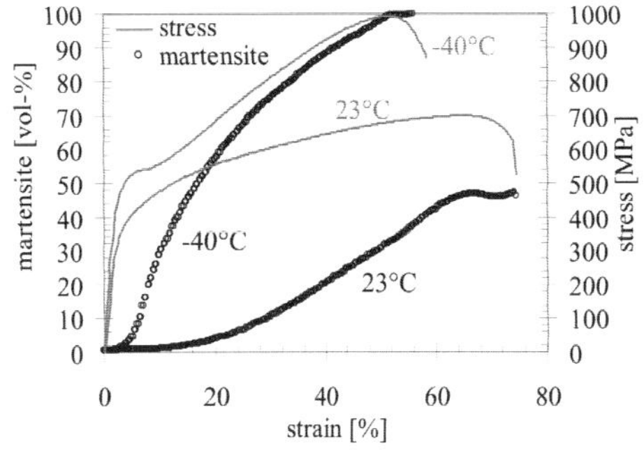

Figure 1. Stress strain curves and MVF for AISI304 at different testing temperatures.

Journal of Physics: Conference Series **240** (2010) 012040 doi:10.1088/1742-6596/240/1/012040

Figure 2a and 2b shows a comparison of the dislocation arrangement found in overaged Nimonic 80A in the annealed starting condition without plastic deformation and after a prestraining of 8%. The prestraining leads to a considerable increase of the dislocation density. A comparable effect of a high dislocation density due to prestraining can likewise be assumed for the other materials. In addition to strain hardening, the austenitic steel undergoes a phase transformation from austenite into ε-martensite and α'-martensite as can be seen in figure 2c. Therefore, the strengthening effect in AISI304 is a combination of strain hardening and phase transformation.

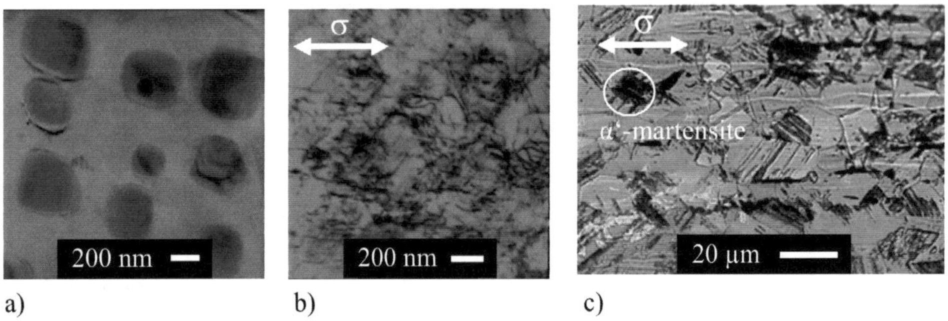

Figure 2. Microstructure before a) and after b) prestraining with a significant increase of dislocation density in overaged Nimonic 80A and c) martensite transformation for AISI304 after prestraining.

The prestraining not only leads to changes in the quasi-static mechanical properties but also induces a surface roughening in the tensile test specimens. For the nickel-base alloys this resulted in a change of surface roughness from Rz = 1,4 μm to Rz = 6,6 μm. As this surface condition is representative for the major application examples, where no further surface treatment ensues the original forming process, it was maintained for the majority of fatigue specimens except for the austenitic steel and some overaged specimens of Nimonic 80A. This allows a distinction between the effect of strengthening mechanism alone and superimposed micro-notch effects on the VHCF behaviour.

Figure 3 depicts the fatigue testing results for the analysed materials without and with predeformation. The most intense influence of a prestrain can be found for the austenitic steel (Fig. 3a). The combination of strain hardening and martensite formation induces a fatigue strength enhancement of more than 200 MPa with an astonishingly small discrepancy between 26% and 54% martensite volume fraction. Hence, the major strengthening seems to be caused by strain hardening rather than phase transformation (as the change in MVF is mainly generated by different deformation starting temperatures and 15% plastic strain). A comparison of the results for the austenitic steel reveals that although the undeformed material has a much lower fatigue strength, it possesses a true fatigue limit which so far could not be proven for the specimens with 54% MVF. Moreover with an increasing amount of MVF the scatter in the fatigue life results seems to increase tremendously from 26% to 54% MVF.

The nickel-base alloys are lesser affected by a prestraining prior to fatigue testing (Fig. 3b). Nonetheless for the precipitation-hardening Nimonic 80A the fatigue strength decreases by 30-40 MPa compared to its original strength in the VHCF range for the peak-aged (pa) and overaged condition (oa). On the contrary no significant influence of prestraining can be found for the solid-solution hardening alloy Nimonic 75 as well as for those specimens of overaged Nimonic 80A which had undergone a second electro-chemical polishing.

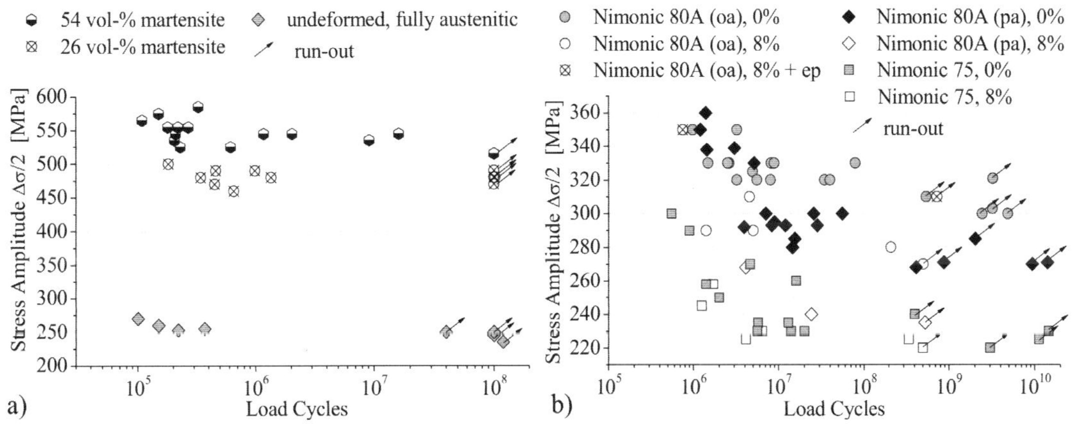

Figure 3. Fatigue results of a) AISI304 and of b) Nimonic 75 and 80A without and with prestraining.

A comparison of the different fatigue results presented in the given study shows that the strengthening mechanism alone does not define the fatigue behaviour of the materials in the VHCF range. An increase in strength does not necessarily lead to a higher fatigue limit in the VHCF regime and whether an increased dislocation density enhances the fatigue limit due to an alteration of the dislocation structure to a more stable state is also not predictable on the basis of the quasi static behaviour. What seems to underlie all of the fatigue results is a change in ductility after prestraining. Keeping in mind that the global strain amplitude during VHCF testing ranges far below the elastic limit, local stress raisers like micro-notches gain importance. This evokes the question whether not the strengthening mechanism itself than rather its effect on the notch sensitivity dominates the fatigue behaviour in the VHCF range.

The given assumption is underlined by the fact that surface polishing of the Nimonic 80A specimens after predeformation erases the prestraining effect on the fatigue behaviour. This would also explain why Nimonic 75, with its rather high ductility even after prestraining, does not show any prehistory dependence of its VHCF behaviour. Even in the case of the austenitic steel the missing true fatigue limit can be explained with an increased notch sensitivity, as some specimens with 54% martensite volume fraction failed due to crack initiation around a "stress raising" inclusion starting from the interior of the specimen, where the brittle martensite phase prevails. Moreover, tests with soldered thermocouples confirmed the high notch sensitivity of the prestrained austenitic steel and its influence on the VHCF behaviour.

4. Conclusions

Cyclic strength enhancement by means of acknowlegded strengthening mechanisms does not necessarily lead to a beneficial VHCF behaviour. With the applied global strain well below the elastic limit and local plastic strain only inhomogeneously distributed the influence of micro-notches acting as local stress raisers gain importance. Hence, prehistory effects depend above all on subsequent changes in the ductility as well as the introduction of surface roughening due to forming processes.

References

[1] Stanzl-Tschegg S E and Schönbauer B 2008 Proc. 17th ECF (Brno) p890
[2] Mughrabi H 2006 *Int. J. Fatigue* **28** 1501-1508
[3] Murakami Y, Nomoto T, Ueda T 1999 *Fatigue Fract. Eng. Mater. Struct.* **22** 581-590
[4] Chai G 2006 *Int. J. Fatigue* **28** (11) 1533–1539
[5] Zimmermann M, Stöcker C, Christ H-J 2008 *TMS Annual Meeting* (New Orleans) ed Hsia K J, Göken M, Pollock T, Portella P D, Moody N R (TMS Publications, Pensylvania) p63
[6] Müller-Bollenhagen C, Zimmermann M, Christ H-J 2009 *Int. J. Fatigue* doi:10.1016/j.ijfatigue.2009.05.07

Journal of Physics: Conference Series **240** (2010) 012041 doi:10.1088/1742-6596/240/1/012041

Evolution of dislocation structure and fatigue crack behavior in Fe-Si alloys during cyclic bending test

Kohsaku Ushioda[1], Shoji Goto[2], Yoshinari Komatsu[3], Akinori Hoshino[4] and Shigeto Takebayashi[1]

[1] Technical Development Bureau, Nippon Steel Corp, 20-1 Shintomi Futtsu 293-8511
[2] Graduate School of Engineering and Resource Science, Akita University, now Professor Emeritus
[3] Graduate School of Engineering and Resource Science, Akita University
[4] Graduate School of Engineering and Resource Science, Akita University, now Furuya Metal Co., Ltd.

ushioda.kohsaku@nsc.co.jp

Abstract. The evolution of dislocation structures was investigated by means of TEM in Fe-Si alloys with 0, 0.5 and 1.0 mass% Si during a cyclic bending test in conjunction with fatigue crack behavior. The addition of Si increased the fatigue strength. In steel without Si the cell structure develops, whereas in steel with 1%Si the vein structure evolves, which is considered to lead to the increased fatigue strength. The cell structure in 0%Si steel is postulated to be caused by the easy cross slip of dislocations, whereas the vein structure in the steels with Si is inferred to be caused by the difficulty in cross slip presumably due to the decrease in stacking fault energy. Furthermore, the steel containing Si shows a dislocation free zone (DFZ) along grain boundaries. A transgranular fracture takes place in 0%Si steel, while in 1%Si steel many intergranular cracks were observed just beneath the top surface, which was thought to be caused by the fact that a) strains are dispersed within grains owing to the vein structure and b) micro cracks are initiated and propagated along a DFZ.

1. Introduction

Efficiency and safety are required for automobiles, ships and so forth, while still achieving weight reduction. Increasing the strength of steels is a very effective means of achieving these requirements. The relationship between the strengthening mechanism and fatigue behavior of steels is extremely important. However in the case of practical steels, the strengthening mechanism is not completely understood as more than one mechanism is simultaneously involved. This study focused on Si in Fe, which is typically used as a solid solution hardening element. Previous studies [1- 4] have reported that Si in Fe presumably lowers the stacking fault energy to suppress cross slip which was considered to result in planar dislocation structures under fatigue testing [1]. However no study has associated the evolution of dislocation structures with fatigue crack behavior, which is the focus of the present study.

2. Experimental procedure

The specimens used were Fe-Si alloys having the chemical compositions and average grain size shown in Table 1. Ti-bearing ultra-low carbon steels (Ti-IF) with respectively added Si of 0, 0.5 and 1.0% were processed through hot rolling, cold rolling and annealing [4].

Table 1 Chemical compositions (mass%) and grain size (μm) of steels used for fatigue testing

Steels	C	Si	Mn	P	S	Al	Ti	N	Grain Size
Ti-IF1	0.0025	0.0021	<0.001	0.0013	0.0005	0.0460	0.0360	0.0007	24
Ti-IF2	0.0023	0.4800	<0.001	0.0008	0.0005	0.0460	0.0360	0.0006	20
Ti-IF3	0.0032	0.9600	<0.001	0.0009	0.0005	0.0440	0.0360	0.0008	21

The fatigue test was conducted on 1mm thick recrystallized specimens. The test machine was a planar cyclic bending type [4]. The frequency was 7 Hz. The stress ratio, R, was equal to −1 and the applied stress amplitude was within a range of 160MPa to 380MPa. The point at which the bending moment lowered by 50% from the saturation point was defined as a fracture. The shapes and dimensions of the specimens for the tensile test and fatigue test were the same as those described in the literature [4]. The evolution of the dislocation structure was investigated with TEM. The evolution and propagation behaviors of fatigue cracks were evaluated by measuring the length of cracks per area as a function of the depth from the top surface.

3. Experimental results

3.1. Tensile and fatigue test

The typical stress and strain curves obtained indicate that the Si addition increased the work hardening rate [4]. Figure 1 shows the results of S-N curves. It is clear that the fatigue strength increased with an increase in the added Si amount.

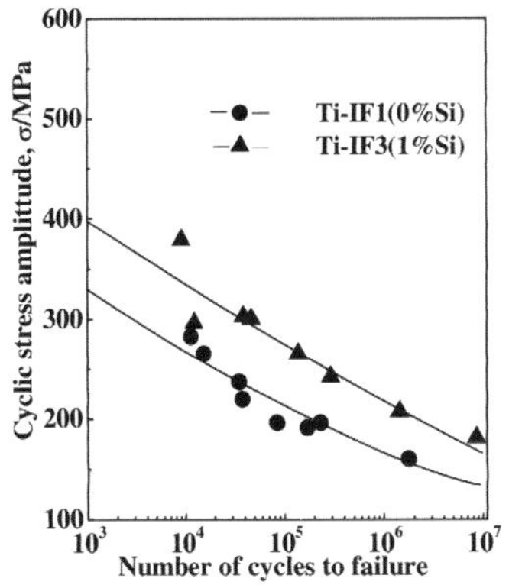

Fig.1 Influence of Si addition in Ti-IF steels on S-N curves.

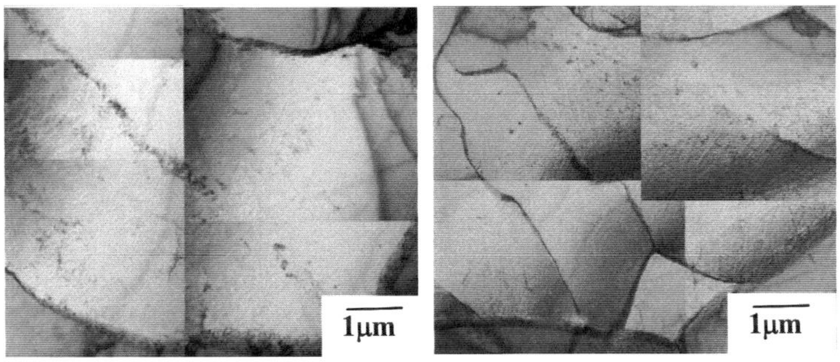

Fig.2 TEM micrographs of Ti-IF1 (0%Si) tested under a) a low (σ=162.8MPa, N=1,673,800) and b) a high (σ=267.8MPa, N=15,100) stress amplitude.

3.2. Evolution of dislocation structure

The dislocation structures fatigue-tested at the low-stress amplitude in steel containing 0%Si were carefully observed with TEM. Figure 2 a) shows dislocation cell structure with a relatively large size of 5 to 10μm. The center of the photo shows the initiation of the cell formation process. In another area, many comparatively refined cells with a heterogeneous structure can be observed in some positions. When the stress amplitude rose, the cell structures became clear and the cell size was reduced to 4 to 5μm. As the stress amplitude was further increased, the cell size became smaller and was reduced to 3 to 4μm (Fig.2 b). The cell structures became discontinuous in the initial grain boundaries. It is inferred that the cells were formed independently for respective grains and cells were blocked by the grain boundaries.

When 1%Si was added, the dislocation structure was changed remarkably. Figure 3 a) shows dislocation structures with low stress amplitude and that the planar dislocations were evenly scattered and tangled with each other in some locations. As the stress amplitude increased, the dislocations aggregated. The structure consisted of a mixture of bundled dislocation regions and dislocation-free regions (channels with a width of 0.3-1μm), which had an interval of 1μm. No dislocation cell structure was observed. In addition, particular structures were observed in which no dislocation was present along the initial grain boundaries within a distance of 0.5μm (Fig.4). It must be noted that this type of structure, namely a dislocation free zone (DFZ), was observed near all of the adjacent grain boundaries in the grain. Figure 3 b) shows the dislocation structures after applying further increased stress. A typical vein structure or a labyrinth-like structure was observed. Again, in this case a DFZ was confirmed.

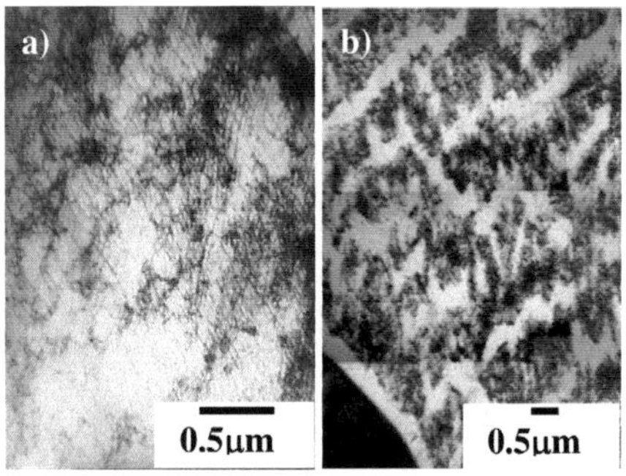

Fig.3 TEM micrographs of Ti-IF3 (1%Si) tested under a) a low (σ=210.0MPa, N=1,353,200) and b) a high (σ=305.3MPa, N=35,600) stress amplitude.

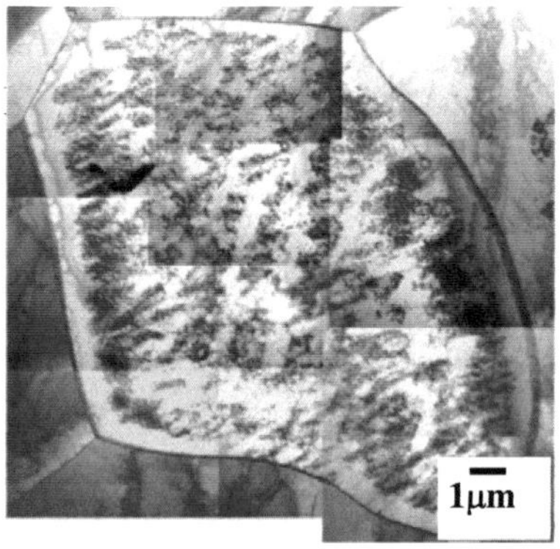

Fig.4 TEM micrograph of Ti-IF3 (1%Si) tested under a middle stress amplitude showing a DFZ along grain boundaries. (σ=267.8MPa, N=129,600) [4]

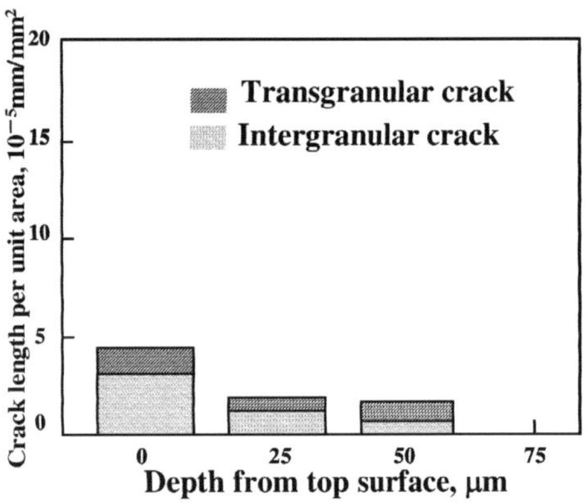

Fig.5 Crack length per unit area as a function of depth from the top surface of the specimen tested under a high stress amplitude of Ti-IF3 (1%Si) (σ=364.2MPa, N=19,200) [4]

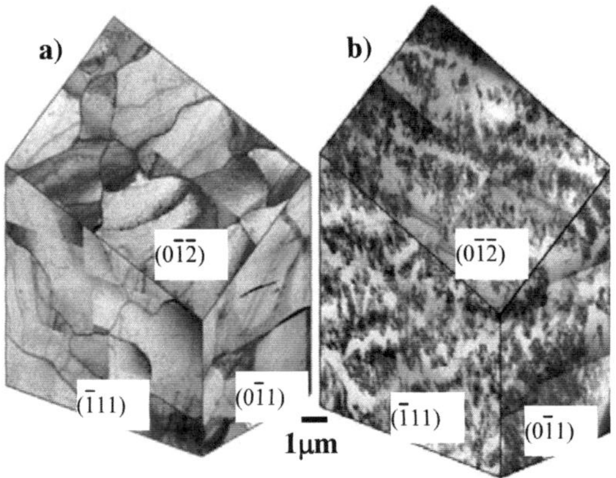

Fig.6 Three dimensional representation of dislocation structures of specimens tested under a high stress amplitude of a) Ti-IF1(0%Si) (σ=267.8MPa, N=15,100) and b) Ti-IF3 (1%Si) (σ=305.3MPa, N=35,600) [4]

3.3. Observation of fatigue-fractured surfaces

A fatigue-fractured surface in the top surface region was observed with SEM. Many striations were observed in 0%Si steels, while grain boundary fractures were observed in 1%Si steels [4]. Figure 5 shows the length of fatigue cracking per unit area of the 1.0%Si steels. The length of fatigue cracking is the longest at the top surface and such cracks are mainly located along the grain boundaries.

4. Discussion

The 3-dimensional description of the dislocation structure in Fig. 6 reveals that dislocation cell structures were observed in the 0%Si steel and vein structures with bundle-like dislocations were observed in the1%Si steel. The dislocation on the {011} slip plane expanded most conspicuously.

The mechanism can be attributed to the assumption [1, 2] that the Si-free steels, having comparatively high stack fault energy, resulted in stable dislocation cell structures owning to the cross slip of dislocations. On the other hand, in the Si-added steels, presumably having lower stacking fault energy, it was more difficult for dislocations to cross slip leading to vein structures. It is worthwhile mentioning that the periodically arranged dislocation walls observed in the vein structures are similar to the ladder structure found in Cu [6], which is formed by the activation of only a primary slip system, but at the same time they may be referred to as a labyrinth structure, which is formed by the activation of more than two slip systems as reported in Fe-3%Si [5] alloy and Cu [6]. The present study has not clarified the crystallographic relationship between a slip system and a dislocation wall. Therefore, further research is required.

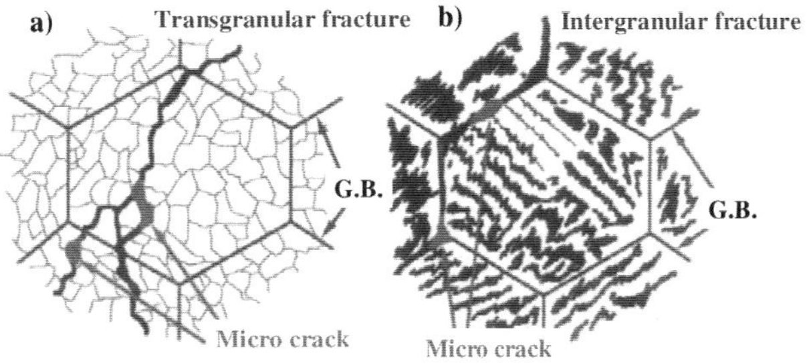

Fig.7 Schematic representations of the fatigue crack behavior in the specimens of a) 0%Si and b) 1%Si [4]

Concerning the interaction between dislocations and grain boundaries, Luoh et al. [7] carried out a detailed study on Cu polycrystal [7] taking into account the grain boundary characteristics. They reported [7] that a DFZ formed along grain boundaries when the vein structures developed, whereas a DFZ did not form in the cell structures, which is very similar to the results of the present study.

For the consideration of the relationship between the dislocation structures and fatigue crack behavior in terms of dislocation arrangement, a model was represented schematically in Fig. 7. It is thus inferred that in 0%Si steels an obvious cell structure was related to the fractured surface with a striation. In the case of 1.0%Si steels, the strain within grains is dispersed uniformly due to a vein structure, while in the grain boundaries the strain can be concentrated due to a DFZ, leading to the formation of micro cracks and their propagation along a DFZ.

5. Conclusions

The results obtained are as follows:

(1) Si addition increases fatigue strength.

(2) The dislocation cell structures are evolved in 0%Si steels, whereas in 1.0%Si steels vein structures are formed. It is also inferred that in Si-free steels cross slip was easy to perform, while in Si-containing steels cross slip became difficult.

(3) In addition, DFZs along initial grain boundaries were observed in Si-containing steels.

(4) Transgranular fractures occurred in 0%Si steels, while in 1.0%Si steels intergranular cracks were observed. Dislocation structure was considered closely related to fracture behavior.

References

[1] M. Klesnil and P. Lukas 1980 Fatigue of Metallic Materials, Elsevier Scientific Publishing Company, 30

[2] J. L. Walter and E. F. Koch 1965 Trans. Metall. Soc. AIME **233** 1209

[3] B. Scstak, V. Novak and S. Libovicky 1988 Phil. Mag. A **57** 353

[4] K. Ushioda, S. Goto, Y. Komatsu, A. Hoshino and S. Tekebayashi 2009 ISIJ Int. **49** 312

[5] H. Mori, M. Tokuwame and T. Miyazaki 1979 Phil. Mag. A **40** 409

[6] F. Ackermann, L. P. Kubin, J. Lepinnoux and H. Mughrabi 1984 Acta. Metall. **32** 715

[7] T. Luoh and C.P. Chang 1996 Acta. Mater. **44** 2683

Low-cycle fatigue behaviour and microstructure of copper and alpha-brass under biaxial load paths

S. Henkel[1], J. Fischer[1], L. Balogh[2], T. Ungar[2] and H. Biermann[1]
[1]TU Bergakademie Freiberg, Institute for Materials Engineering, Gustav-Zeuner-Strasse 5, D-09599 Freiberg, Germany
[2]Eötvös University, Department of General Physics, P.O. Box 32, 1518 Budapest, Hungary

henkel@ww.tu-freiberg.de

Abstract. The low-cycle fatigue behaviour of copper and α-brass CuZn30 was investigated in uniaxial and biaxial tests. Planar biaxial fatigue tests were carried out using cruciform samples with proportional stain paths with and without phase shift between the two axes. Microcharacterisation was performed by electron microscopy as well as by high-resolution X-ray line profile analysis. The biaxial cyclic stress-strain curves show good agreement with the uniaxial ones using the von Mises equivalent strain hypothesis. The dislocation densities and microhardness values of the biaxial case, however, show significantly lower values compared to the uniaxial case at equivalent von Mises stresses.

1. Introduction

Life prediction for highly and complex loaded engineering components like aircraft components is always a question of assignability. Often, it is impossible to infer the component behaviour from uniaxial tests. Therefore, starting from uniaxial tests a chain of coupon tests, component relating and full scale tests is necessary. The testing of biaxial planar specimens is one chain link. It gives the possibility to study the damage mechanism of low cycle fatigue behaviour of metals in planar stress state on a clearly arranged sample with fixed main stress axes.

2. Experimental details

The investigated materials - commercially pure (99.9964%) copper and α-Brass CuZn30 (69.2% copper and 30.7% zinc)- are well known model materials for wavy and planar dislocation slip. Both materials were available as hot-rolled plates with a thickness of 15mm. The properties from the static tensile test, Young's modulus and grain size are given in table 1.

Table 1. Properties of the investigated materials.

Material	R_m [MPa]	$R_{p0,2}$ [MPa]	A_5 [%]	Young's modulus [GPa]	Grain size [μm]
Copper	217	78	48	126	90
CuZn30	339	113	65	110	33

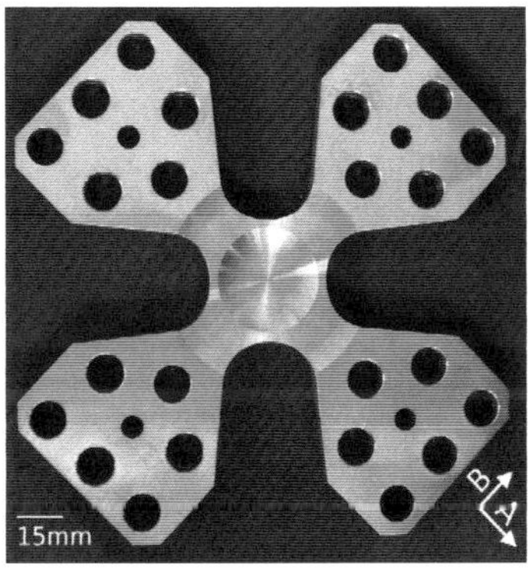

Figure 1. Cruciform sample with load axes directions A and B.

The tests were performed on a biaxial 250 kN servohydraulic test rig (Instron) using cruciform samples with a thinned central area of diameter 15 mm and 1.6 mm thickness (figure 1). The samples were milled and polished mechanically. Total strain controlled load was applied with a frequency of 0.3Hz using a biaxial orthogonal extensometer. The amplitude in both loading axes was equal while the phase between the loading axes was varied as 0°, 22.5°, 45° and 90°. Cruciform samples give no direct access to stresses, therefore finite element analyses was widely used to identify them [3]. In this work the starting of the elastic unloading after changing the ramp direction was used to calculated stresses from local strains by Hookes law. Figure 2 shows the drive signal without phase shift. For cases with phase shift an additional unloading was applied to the drive signal to get elastic unloading conditions in both axes (figure 3). For comparision uniaxial tests were carried out on a 100 kN servohydraulic testing system using conventional LCF samples at equal von Mises strain amplitudes. Dislocation densities were measured by a high-resolution double-crystal diffractometer [4] on samples deformed to half-life.

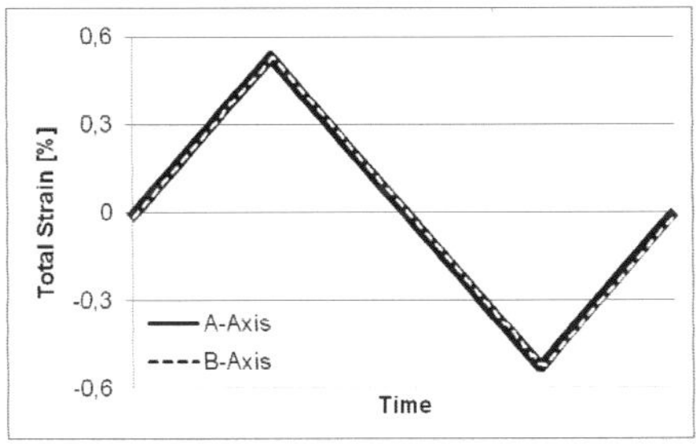

Figure 2. Synchronous strain

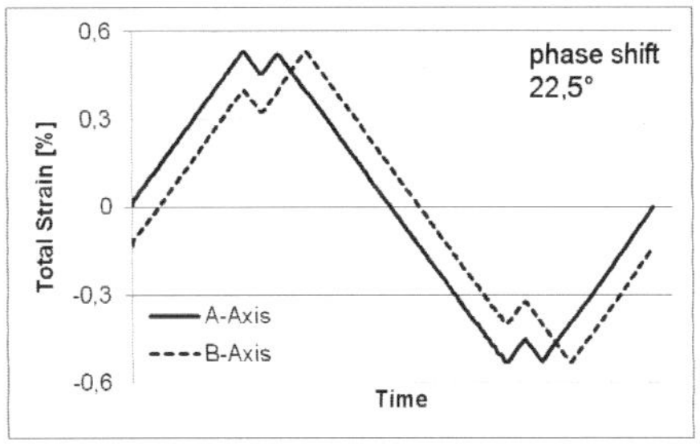

Figure 3. Asynchronous strain

3. Results

The fatigue lives are quite similar in uniaxial and biaxial synchronous cases for the same von Mises strain amplitudes, cf. table 2. The lower lifetime with increasing phase shift is supposed to result from notch effects of the sample geometry. In these cases the cracks started always from outside the central measurement area. This could be caused by rotating the main stress axes in these areas under phase shift loads. Figures 4 and 5 show the cyclic deformation curves of uniaxial and biaxial synchronous fatigue tests. In both materials initial cyclic hardening is observed followed by cyclic softening or cyclic saturation and/or secondary hardening, in accordance with literature [2], [1]. Interestingly, the biaxial tests show good agreement with the uniaxial tests. The small differences may be due to uncertainties in the stress determination procedure.

Table 2. Lifetime [cycles] until technical crack for α-brass and copper

material	strain amplitude in the axes	synchronous equivalent strain	uniaxial	biaxial phase shifts			
				0°	22.5°	45°	90°
α-brass	0.28%	0.5%	18 000	13 200	13 000	8 000	10 000
	0.53%	1.0%	4 000	3 500	2 000	2 000	1 500
copper	0.09%	0.15%		50 000	67 500		
	0.16%	0.3%	13 000	15 000	10 500		

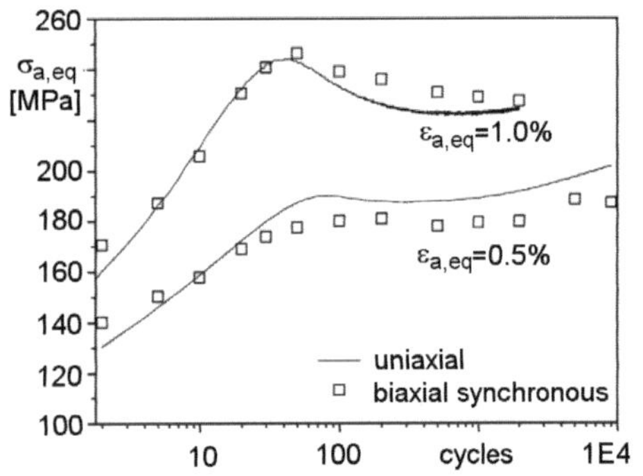

Figure 4. Cyclic hardening, α-brass

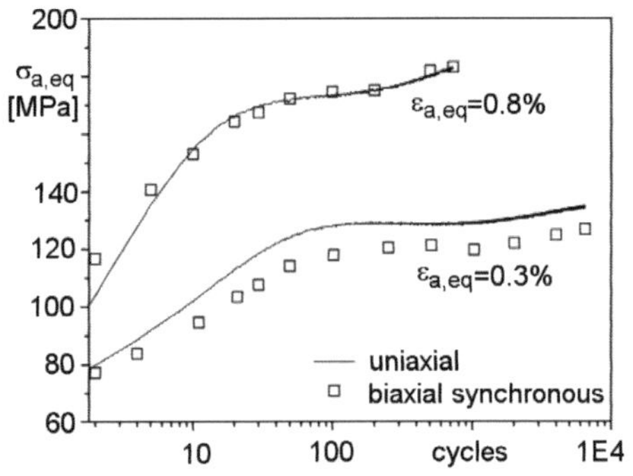

Figure 5. Cyclic hardening, copper

The dislocation densities are in the biaxial cases significantly lower than in the uniaxial cases (figures 6 and 7). In all cases of copper samples a dislocation cell stucture was observed. α-brass shows planar dislocation structures, however, in the asynchronous case they are more irregular, as shown in the TEM images given in figure 8 for the uniaxial, biaxial synchronous and biaxial asynchronous (22.5°) cases, respectively, at a von Mises equivalent strain amplitude of 1%.

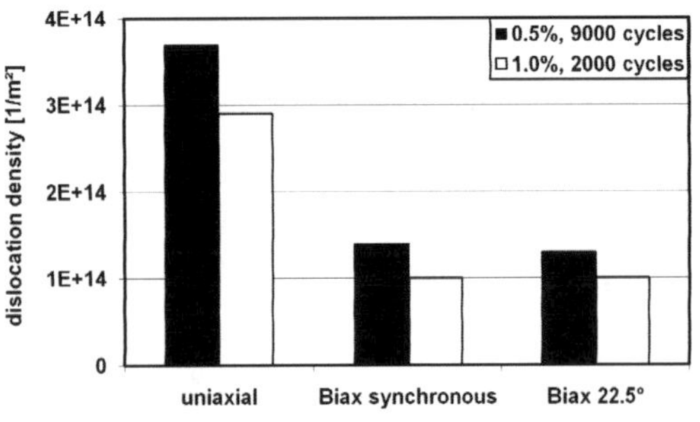

Figure 6. Dislocation densities, α-brass

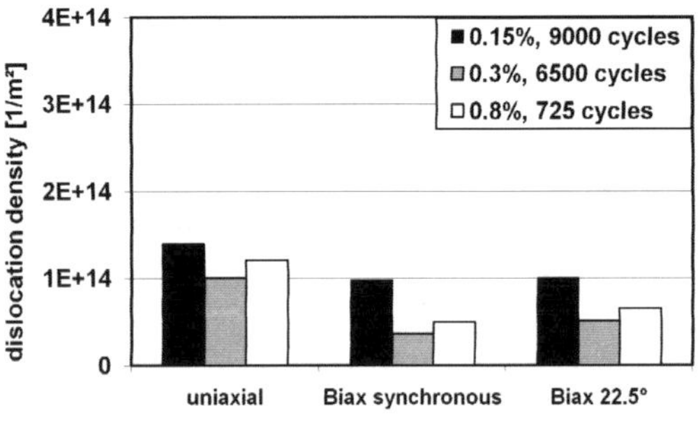

Figure 7. Dislocation densities, copper

Figure 8. TEM investigations of α-brass, $\Delta\epsilon_{a,eq} = 1\%, N = 2000$ cycles, a) uniaxial , b) biaxial synchronous, c) biaxial asynchronous fatigue

4. Summary and conclusions

The strain and stress amplitudes in uniaxial and synchronous biaxial cases using the von Mises equivalent strain hypothesis are in good agreement. Also the fatigue lives are in agreement because life time is mainly controlled by the grains with slip systems of high Schmid factor (high shear stresses). However, dislocation densities are smaller in the biaxial cases. This may result from the lower number of slip systems with high Schmid factor than in the uniaxial case due to the reduced symmetry of the cruciform sample. Thus, at the same von Mises stress, the mean value of the resolved shear stresses of the active glide systems is lower, giving lower dislocation densities under biaxial loading conditions. Asynchronous and synchronous tests give comparable values of dislocation densities. In addition, lifetimes in asynchronous load cases are affected by sample geometry because of rotating main stress axes in the notched areas of the samples. In these cases the starting cracks initiate always from outside the measurement area.

Acknowledgments

Many thanks go to Dr. H.-A. Kuhn, *Wieland-Werke AG* for providing the test material. Special thank goes to János L. Lábár from Eötvös University Budapest for the TEM studies.

References

[1] Bayerlein M 1991 *Wechselverformungsverhalten und Ermüdungsrissbildung von vielkristallinem Kupfer*. Ph.D. thesis Uni. Erlangen-Nürnberg
[2] Carstensen J 1998 *Structural Evolution and Mechanisms of Fatigue in Polycrystalline Brass* Ph.D. thesis Risø-R-1005(EN), Risø National Laboratory, Denmark
[3] Lecompte D, Smits A, Sol H, Vantomme J and Van Hemelrijck D 2007 *Int. J. Solids Struct.* **44** 1643–56
[4] Ungar T 2004 *Scripta Mater.* **51** 777–781

Journal of Physics: Conference Series **240** (2010) 012043 doi:10.1088/1742-6596/240/1/012043

Simulation of fatigue crack growth under large scale yielding conditions

Christoph Schweizer, Thomas Seifert and Hermann Riedel

Fraunhofer-Institute for Mechanics of Materials IWM, Woehlerstrasse 11, D-79108 Freiburg, Germany

christoph.schweizer@iwm.fraunhofer.de

Abstract. A simple mechanism based model for fatigue crack growth assumes a linear correlation between the cyclic crack-tip opening displacement ($\Delta CTOD$) and the crack growth increment (da/dN). The objective of this work is to compare analytical estimates of $\Delta CTOD$ with results of numerical calculations under large scale yielding conditions and to verify the physical basis of the model by comparing the predicted and the measured evolution of the crack length in a 10%-chromium-steel. The material is described by a rate independent cyclic plasticity model with power-law hardening and Masing behavior. During the tension-going part of the cycle, nodes at the crack-tip are released such that the crack growth increment corresponds approximately to the crack-tip opening. The finite element analysis performed in ABAQUS is continued for so many cycles until a stabilized value of $\Delta CTOD$ is reached. The analytical model contains an interpolation formula for the J-integral, which is generalized to account for cyclic loading and crack closure. Both simulated and estimated $\Delta CTOD$ are reasonably consistent. The predicted crack length evolution is found to be in good agreement with the behavior of microcracks observed in a 10%-chromium steel.

1. Introduction

Fatigue life of metallic materials is typically controlled by the growth of microcracks when plastic strain amplitudes become large (large scale yielding conditions). Experiments by Neumann [1] suggested a correlation between da/dN and $\Delta CTOD$ for cracks under mode 1 loading. In this paper an analytical estimate for $\Delta CTOD$ is presented and compared to results of finite element calculations. Further, the evolution of microcracks in a 10%-chromium turbine steel is investigated using the replica technique. The predictions of the model are found to be consistent with the observed crack growth behavior.

2. Analytical model

The J-integral and $CTOD$ are related by [2]

$$CTOD = d_N \frac{J}{\sigma_0}, \tag{1}$$

where σ_0 is the yield stress. The constant d_N depends on the hardening behavior of the material. For power-law hardening the crack-tip fields are HRR-type fields [3, 4], and d_N is a function of the Ramberg-Osgood hardening exponent N and was tabulated by Shih [5]. For cyclic loading, all parameters in eq. (1) are replaced by their cyclic counterparts. The resulting cyclic J-integral (ΔJ) has a solid mathematical foundation, if the material exhibits Masing behavior [6]. Following Kumar *et al.* [7] and Heitmann *et al.* [8], ΔJ can be approximated by the sum of an elastic (or small scale yielding) and a plastic contribution. Assuming a small penny-shaped crack under plane strain and mode 1 loading conditions, one obtains

$$\Delta CTOD = \frac{d_N}{\sigma_{cy}} \left(\frac{4}{\pi} \frac{\Delta\sigma_{eff}^2 (1-\nu^2)}{E} a + \frac{6}{\pi} \frac{\Delta\sigma \, \Delta\epsilon_{pl}}{\sqrt{1+3N}} a \right), \tag{2}$$

where σ_{cy} denotes the cyclic yield stress. The second term in brackets was derived by He and Hutchinson [9] in the fully plastic limit for a power-law hardening material. Crack closure is taken into account by the use of an effective stress range $\Delta\sigma_{eff} = \sigma_{max} - \sigma_{op}$ in the elastic part of ΔJ, as was proposed by Heitmann *et al.* [8]. σ_{max} and σ_{op} are the maximum and crack opening stress, respectively. All quantities in eq. (2) can be determined from stress-strain hysteresis loops except Poisson's ratio ν and σ_{op}. The latter could either be estimated with empirical formulas given in [8, 10, 11] or taken from finite element simulations.

3. Fatigue crack growth simulation

For the simulation of fatigue crack growth, an axisymmetric model with an internal crack and an initial crack length to width ratio of $a/W = 1/35$ is used. This corresponds to a small penny-shaped crack in a large, practically infinite body. Two different finite element meshes are used: the first with an element to crack length ratio in front of the crack-tip of 1/2000, the second with a ratio of 1/200. The material is described by a rate-independent plasticity model with nonlinear kinematic hardening, which is available in ABAQUS version 6.8. Two sets of material parameters are employed and $\nu = 0.3$ is used in both sets. Both give an almost perfect fit to a hysteresis loop, which was measured at room temperature for a 10%-chromium steel (see fig. 1). In order to investigate the effect of hardening behavior on $\Delta CTOD$, one set of material parameters is fitted such as to describe power-law hardening and Masing behavior with a hardening exponent of $N = 0.16$ up to twenty percent plastic strain range (red symbols in fig. 2). The second parameter set leads to completely saturated stresses above one percent strain range and will be refered to as the low-hardening model (blue symbols). Large deformation theory is used for all simulations. The crack is propagated by releasing several nodes during

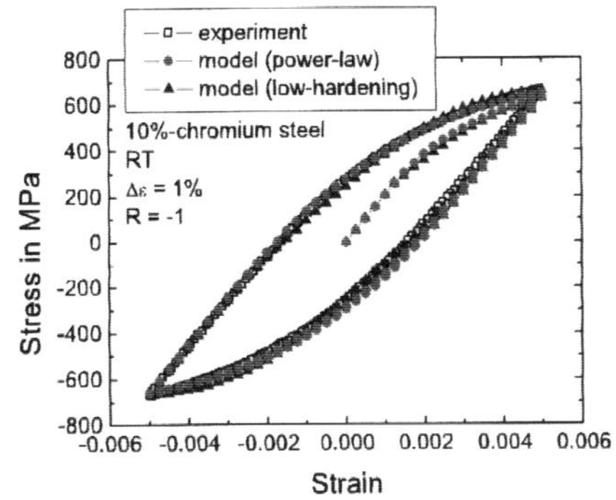

Figure 1. Material model for both parameter sets in comparison to a hysteresis loop.

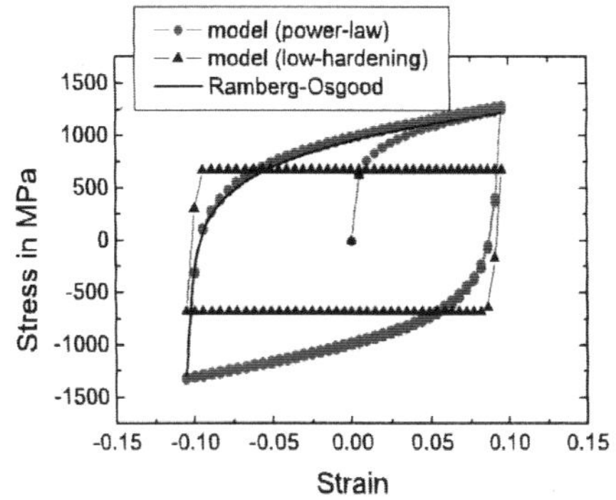

Figure 2. Material model for both parameter sets at large strain amplitudes.

the tensile part of the cycle using the ABAQUS *DEBOND option. The node release scheme is somewhat arbitrary, but by trial and error the crack growth increment per cycle can be chosen to be in the same order as $\Delta CTOD$. The proposed method allows to study the effect of mesh size without changing the crack growth rate. All simulations are continued until the normalized crack-tip opening displacement ($\Delta CTOD/a$) reaches a constant plateau. $\Delta CTOD$ is evaluated at the distance behind the crack tip where the crack opening angle is 45°. The results for both meshes are shown in fig. 3. Smoother curves are obtained with the finer mesh. The low-hardening model leads to significantly larger crack-tip openings than the power-law. In fig. 4 the effect of varying the crack growth rate by a factor of two is shown for the power-law model. Apparently, the result is not sensitive to the node release scheme. Further, fig. 3 and fig. 4 show analytical estimates for $\Delta CTOD/a$ based on eq. (2) with σ_{op} and σ_{max} taken from the finite element simulation. σ_{op} is here defined as the macroscopic stress at which the stresses in front of the crack-tip become tensile for the first time. The analytical model is capable of capturing the effect of hardening as fig. 3 shows. Since the power-law model seems to be more reasonable for a high-strength steel, the stabilized value $(\Delta CTOD/a)_{pow} = 0.005$ from the power-law in fig. 3 is used to model fatigue crack growth in the following section.

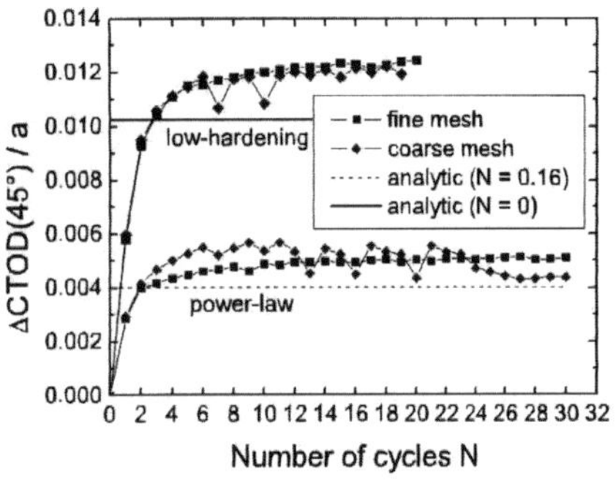

Figure 3. Effect of mesh size and hardening behavior on $\Delta CTOD/a$ for both parameter sets.

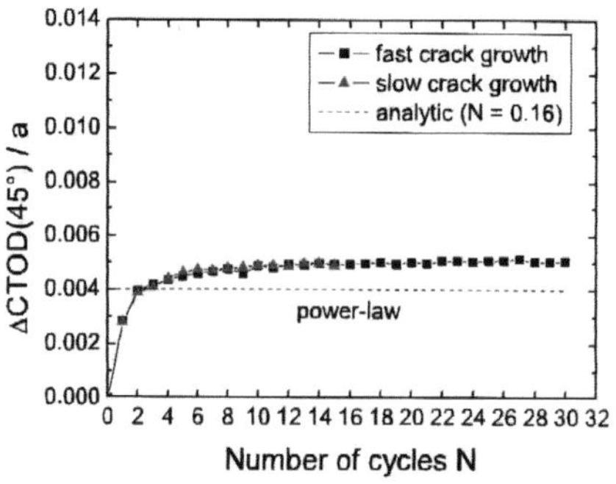

Figure 4. Effect of the node release scheme on $\Delta CTOD/a$ for the power-law hardening material.

4. Prediction of microcrack growth

The replica technique was used to follow the crack length evolution of two cracks in strain controlled fatigue tests on cylindrical specimens of the 10%-chromium steel at room temperature. Both cracks were approximately semicircular and nucleated after three and nine percent of the total lifetime. The simple relationship

$$da/dN = \beta \Delta CTOD \tag{3}$$

serves as a mechanism-based model for fatigue crack growth. $\Delta CTOD$ in eq. (3) is now replaced by $1.25(\Delta CTOD/a)_{pow}a$, where the factor 1.25 accounts for the fact that a semicircular surface crack rather than an embedded penny-shaped crack is considered [12]. Integration from an initial crack length a_0 to an arbitrary crack length a yields the exponential crack growth law:

$$a = a_0 e^{1.25\beta(\Delta CTOD/a)_{pow}(N-N_0)}. \tag{4}$$

a_0 is set to the first crack length observed after N_0 cycles for each crack. Hence the only unknown parameter in eq. (4) remains β, which is treated as a fitting parameter. The results are presented in fig. 5 and fig. 6. The model prediction is better for the crack in fig. 6, which grew perpendicularly to the applied load throughout the whole lifetime. The crack in fig. 5 was noticably affected by microstructure and showed very pronounced stage 1 crack growth until 50% of the lifetime. For both cracks β is close to 0.3, a value which was found in experimental and numerical studies on fatigue crack growth [13, 14].

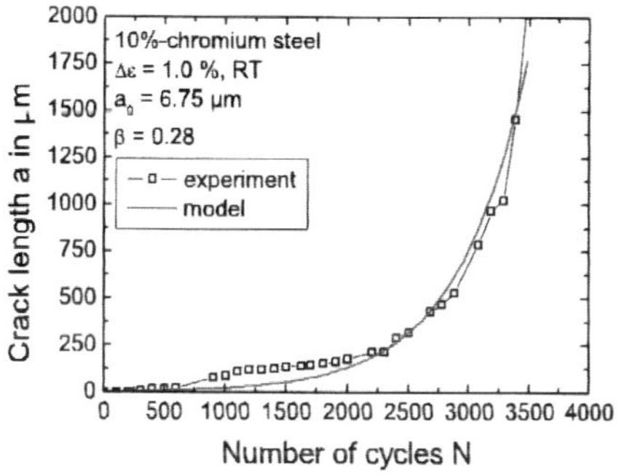

Figure 5. Predicted versus measured crack length evolution for the main crack.

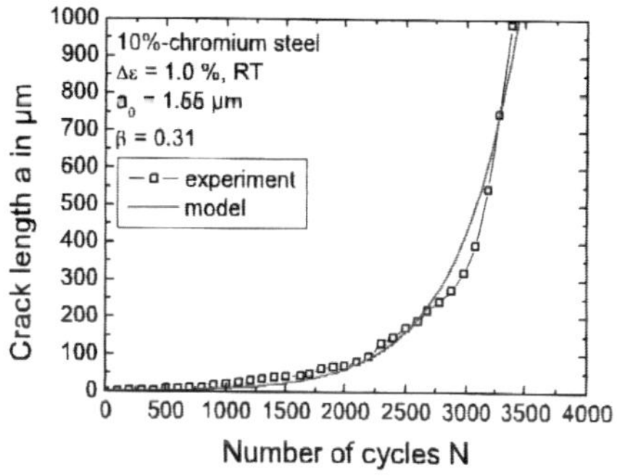

Figure 6. Predicted versus measured crack length evolution for crack 1.

5. Summary and conclusions

In the present work an analytical estimate for the cyclic crack-tip opening displacement is presented and fatigue crack growth simulations under large scale yielding conditions are conducted. Numerically calculated and estimated $\Delta CTOD$ are reasonably consistent for both material models considered, if opening stresses are taken from the simulation. The results from the fatigue crack growth simulations are used to model the evolution of small cracks in a 10%-chromium steel. A simple model, which is based on a linear correlation between $\Delta CTOD$ and da/dN is capable of predicting the crack length evolution observed. The model needs a single fitting parameter, β, which is found to have a reasonable value in comparison with physical models of crack growth by crack-tip blunting.

References

[1] Neumann P 1973 *Acta Metall.* **22** 1155–65
[2] Shih C F 1981 *J. Mech. Phys. Solids* **29** 305–26
[3] Hutchinson J W 1968 *J. Mech. Phys. Solids* **16** 13–31
[4] Rice J R and Rosengren G F 1968 *J. Mech. Phys. Solids* **16** 1–12
[5] Shih C F 1983 *Brown University Report MRL E-147*
[6] Wuethrich C 1982 *Int. J. Fracture* **20** 35–7
[7] Kumar V, German M D and Shih C F 1981 *Report NP-1931 on Project 1237-1 for Electric Power Research Institute, Palo Alto, California*
[8] Heitmann H H, Vehoff H and Neumann P 1984 *Advances in Fracture Research 84 - Proc. of ICF6* vol 5 (ed S R Valluri et al, Oxford and New York: Pergamon Press Ltd.) pp 3599–606
[9] He M Y and Hutchinson J W 1981 *J. Appl. Mech.* **48** 830–40
[10] Newman J C 1984 *Int. J. Fract.* **24** 131–35
[11] Schijve J 1981 *Eng. Fract. Mech.* **14** 461–5
[12] Riedel H 1987 *Fracture at High Temperatures* 1st ed (Berlin, Heidelberg, New York: Springer-Verlag)
[13] Bichler C and Pippan R 2007 *Eng. Fract. Mech.* **74** 1344–59
[14] Tvergaard V 2004 *J. Mech. Phys. Solids* **52** 2149–66

Ultrasonic fatigue of a high strength steel

M Koster, G Wagner, D Eifler

Institute of Materials Science and Engineering, University of Kaiserslautern, Germany

koster@mv.uni-kl.de

Abstract At the Institute of Materials Science and Engineering at the University of Kaiserslautern an ultrasonic testing system for the fatigue assessment of metallic materials in the very high cycle fatigue (VHCF) regime was developed. The ultrasonic testing system allows to control the test and to measure detailed fatigue data. The achieved results can be used to describe the cyclic deformation behaviour of wheel steels at ultrasonic frequencies. In load increase tests (LIT), the critical stress amplitude can be determined, which leads to a defined change of process parameters like generator power, dissipated energy and specimen temperature. With SEM investigations it was proved that the change of the process parameters correlates with irreversible changes in the microstructure. It can be shown that the stress amplitude, leading to first irreversible changes in the microstructure, strongly depends on the depth position within the original wheel rim. New and basic results on the fatigue mechanisms of high strength steels in the VHCF-regime can be achieved.

1. Introduction

During the 19th century, the classical description of metal fatigue was developed. In general, the main aspects of the lifetime-oriented fatigue behaviour of metals are described in Woehler curves which are generally used in the most industrial fields for the design of reliable constructions under cyclic load. But during the last decade, new research works indicate that the common comprehension of the Woehler curves has to be revised in the VHCF range [1, 2].

The assumption that the fatigue limit of bcc-metals is reached mostly after $2 \cdot 10^6$ cycles was disproved. In the literature a decrease of the maximum tolerable stress amplitude after 10^8 cycles is described for different metals. A possibility to achieve these results in an acceptable time range yields the application of ultrasonic oscillation systems with a frequency of 20 kHz. By using an ultrasonic oscillation system, the test time for $N = 10^9$ cycles can be theoretically reduced to less than one day, in the experimental realisation it requires about two weeks, since the loading sequence is divided into ultrasonic stimulation and sufficient long interruptions to ensure that the temperature increase of the specimen remains below 10 °C in the crack-free state [2, 3].

The following results, achieved by using non-contact and high-resolution measuring techniques, give a first insight to the very complex characterization of the fatigue assessment of a railway wheel steel in the VHCF-regime by using piezoelectric testing facilities.

2. Experimental setup

To extend the knowledge of the cyclic deformation behaviour in the VHCF range a commercial ultrasonic oscillation system (UOS) was developed at the Institute of Materials Science and Engineering to meet the special requirements of VHCF investigations (Figure 1 a):

2.1. The VHCF-testing facility

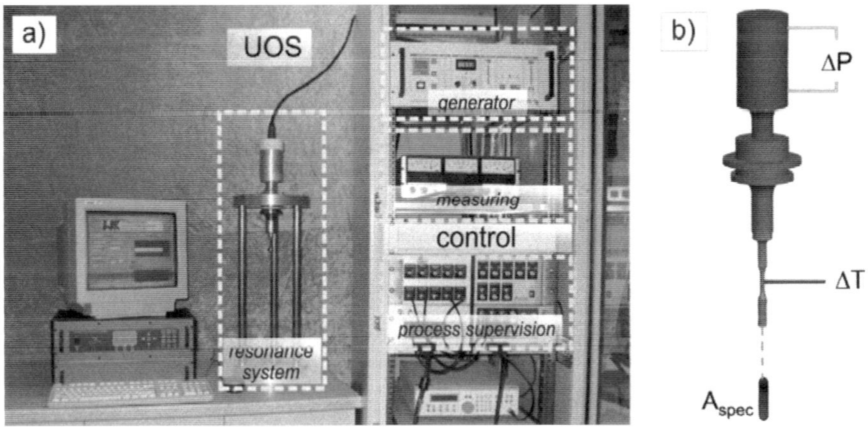

Figure 1: VHCF testing facility (a) and applied measuring techniques (b)

The VHCF testing facility is divided in two functional groups: The ultrasonic oscillation system (UOS) and the control unit.

The UOS is mounted in a customized frame that guarantees by its open design the access to the necessary measurement components. The process control is realized by a specially developed control unit. In order to guarantee a high process stability with a sufficiently high resolution of the ultrasonic signal, the control unit is subdivided into a process supervision device and a measuring device. The process supervision defines the loading sequences and records relevant process data. The measuring device records relevant process data with a sufficiently high recording speed, allowing a detailed analysis of the ultrasonic signal.

2.2. Measuring techniques

The measurement of important process parameters like generator power, specimen's oscillation amplitude and temperature with a very high resolution allows a detailed evaluation of the cyclic deformation behaviour on the basis of deformation-induced changes in the microstructure and an improved understanding of the fatigue mechanisms in the VHCF regime.

2.3. Test material

The fatigue behaviour of railway wheel steels has been investigated in the Very High Cycle Fatigue regime until 10^9 cycles with a frequency of 20 kHz at room temperature. The specimens are machined from original wheels of the unalloyed medium carbon steel SAE 1050, with a carbon content of 0.5 wt-%, widely used for high-speed passenger traffic [4]. The industrial heat treatment consists of austenising, spraying a cooling liquid on the rim, the so-called 'rim chilling', and finally annealing, leading to a typical distribution of the hardness in the wheel (Figure 2). As a result of this process the hardness decreases with increasing distance from the rim surface and the ferrite fraction and cementite lamellae spacing increase.

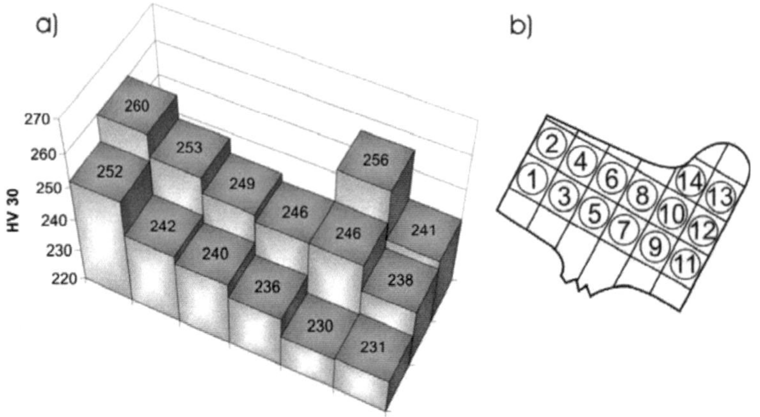

Figure 2: Distribution of the Vickers hardness (a) and specimen positions (b)

For a detailed microstructure-oriented evaluation of the cyclic deformation behaviour, specimens were machined from defined cross-section areas of the rim in rolling direction.

3. Results of load increase tests in the VHCF-regime

Load increase tests allow to estimate the endurance limit with only a few specimens. LITs are commonly used for fatigue investigations in the high-cycle-fatigue range (HCF) with servo-hydraulic and resonance vibration testing systems. For the realization of LITs with piezoelectric testing systems, the VHCF-testing facility was improved in order to apply defined varying loads.

The load increase tests start at a damage-free stress level of 230 MPa with a stepwise increase of the stress amplitude of 5 MPa every 10^6 cycles. In order to describe the fatigue assessment the maximum generator power and the dissipated energy during every ultrasonic pulse were determined (Figure 3).

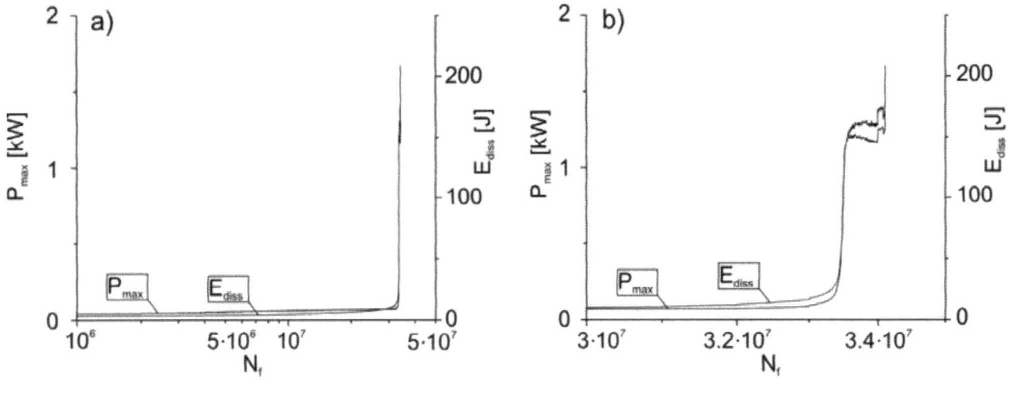

Figure 3: Maximum generator power and energy dissipation during a LIT a) overview b) detail

Over 95% of the test time, the curves for the maximum generator power and energy dissipation don't show any significant changes (Figure 3 a). The first increase of the parameters occurs after $3.2{\cdot}10^7$ cycles, indicating first irreversible changes in the microstructure (Figure 3 b). After a distinct increase of the investigated parameters after $3.3{\cdot}10^7$ cycles, there is a range, where the course of the parameters is nearly constant. Specimen failure occurs in the following loading step, caused by fatigue cracks that start preferentially in the softer phase, ferrite and grow along the phase boundaries between ferrite and pearlite (Figure 4).

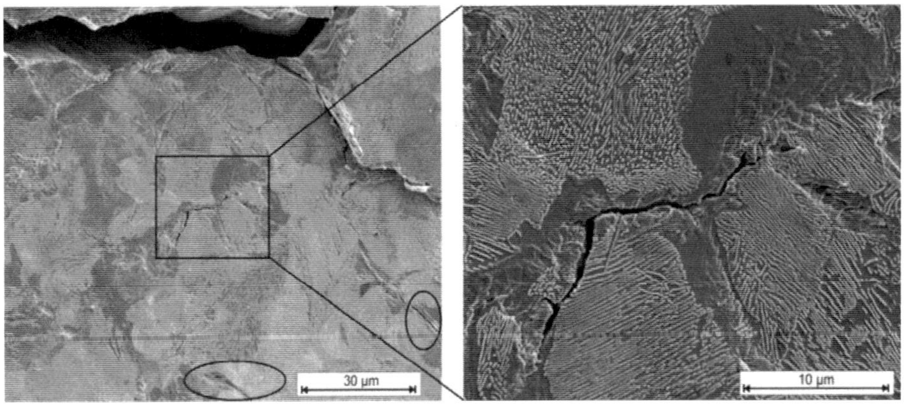

Figure 4: SEM investigations of a crack resulting from a LIT

A series of LITs shows the sensitivity of the applied measuring techniques (Figure 5). The stress amplitude leading to changes in the microstructure and finally to the failure of the specimens correlates closely with the Vickers hardness, depending on the specimen position in the wheel cross section.

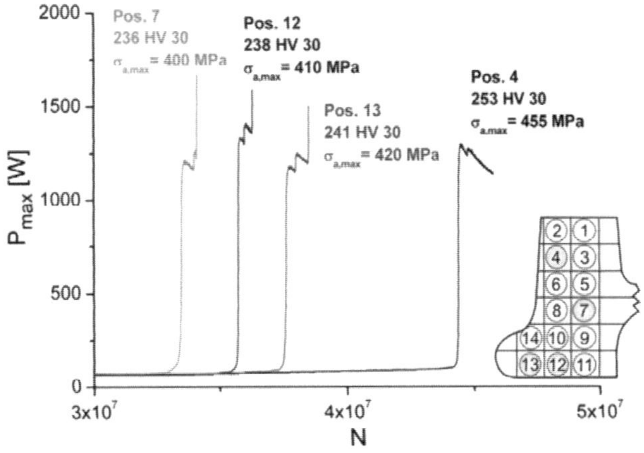

Figure 5: LITs: maximum load amplitude for different specimen as a function of the Vickers hardness

4. Conclusion

Load increase tests have been successfully performed with the railway wheel steel SAE 1050 by using UOS. The applied measuring techniques, especially the maximum generator power and the dissipated energy during an ultrasonic pulse are best suited to indicate changes in the microstructure up to $3{\cdot}10^6$ cycles before failure. Based on the results of the LITs, constant amplitude tests are performed for an appropriate description of the cyclic deformation behaviour and damage mechanisms in the VHCF-regime.

References

[1] Mughrabi H 2006 *Int. J. Fatigue* **28** 1501-8
[2] Bathias C 1999 *Fatigue Fract. Eng. Mater. Struct.* **22** 559-65
[3] Zettl B, Mayer H, Ede C and Stanzl-Tschegg S 2006 Int. J. Fatigue **28** 1583-9
[4] Wagner V, Ebel-Wolf B, Walther F and Eifler D 2007 *VHCF-4 Proceedings* (Warrendale: TMS) 137-42

Cyclic plasticity and internal dislocation structure in two-phase alloy

J Polák, M Petrenec, T Kruml and J Tobiáš

Institute of Physics of Materials, Academy of Sciences of the Czech Republic
Žižkova 22, 616 62 Brno, Czech Republic

polak@ipm.cz

Abstract. Austenitic-ferritic stainless steel has been cycled in symmetrical push-pull regime with constant strain rate and different strain amplitudes. Hysteresis loops were recorded and analysed. The plot of the second derivative of the half-loop vs. relative strain shows one peak for low amplitude straining corresponding to the cyclic deformation of softer phase, austenite. With increasing strain amplitude second peak appears and corresponds to the cyclic deformation of the harder phase, ferrite. Transmission electron microscopy study revealed parallel bands in austenite and initial virgin dislocation structure in ferrite at low amplitude while cell structure in austenite and secondary wall structure in ferrite at high amplitude.

1. Introduction

Cyclic straining results in the modification of the internal structure of the material and the changes of its cyclic stress-strain response follow. The relation of the cyclic stress-strain response and internal dislocation structure has been studied extensively in single-phase materials [1] where the dislocation structure determines the internal stress amplitude at particular strain amplitude. In two-phase alloys in which one or both phases are deformed plastically the dislocation structure starts to develop always in the softer phase [2] and with increasing strain amplitude in both phases. Generally, the parallel arrangement of phases and also of individual microvolumes can be supposed and therefore the statistical theory [1] can be applied for the description of the hysteresis loop.

In this contribution the hysteresis loop and the internal dislocation structure in two-phase austenitic-ferritic duplex steel have been studied in cycling with different strain amplitudes at room temperature and the relation of probability density function of the internal critical stresses to dislocation arrangement in both phases is demonstrated.

2. Experimental

Austenitic-ferritic SAF 2507 type duplex stainless steel was supplied by Sandvik, Sweden as rods of 30 mm in diameter. The chemical composition (in wt. %) was: 0.02 C, 23.0 Cr, 7.0 Ni, 3.8 Mo and 0.27 N, the rest Fe. The structure of the steel was formed by the islands of austenite elongated in the rolling direction embedded in a ferritic matrix. The volume fraction of austenite was 53%, the volume fraction of ferrite was 47%. Cylindrical specimens had the gauge length 12 mm and the diameter 8 mm. They were cycled in a computer controlled electrohydraulic testing machine using longitudinal extensometer having the gage length of 8 mm with constant strain rate $2.5\text{x}10^{-3}$ s^{-1}. Hysteresis loops with high number of data points were continually recorded and stored for further analysis. The second derivatives of the tensile or compressive half-loops were obtained using smoothing and numerical derivative procedures.

Thin foils for the transmission electron microscopy were prepared using standard procedure from sections taken parallel to the specimen axis. They were studied in a transmission electron microscope STEM Philips CM-12 operating at 120 kV using a double tilt holder and a MegaView II digital camera.

3. Results and discussion

Cyclic straining of austenitic-ferritic duplex steel with constant strain amplitudes results in a short initial hardening followed by long term softening with a tendency to saturation [3]. Figure 1 shows saturated tensile hysteresis half-loops in relative coordinates. In this representation the details of the shape of the loop are not evident and the shape of the half-loops is approximately determined by the shape of the half-loop with the highest strain amplitude. The shape of the loop and its evolution in cyclic straining can be found by plotting the second derivative of the half-loop (divided by the half of the square of the effective elastic modulus E_{eff}) vs. the relative strain or vs. fictive stress (relative strain multiplied by the half of the elastic modulus). E_{eff} is the effective elastic modulus evaluated for the tensile hysteresis half-loop from the slope of the half-loop where the second derivative reaches the minimum. This plot can be used for the determination of the effective stress and for the probability density distribution of the internal critical stresses [1].

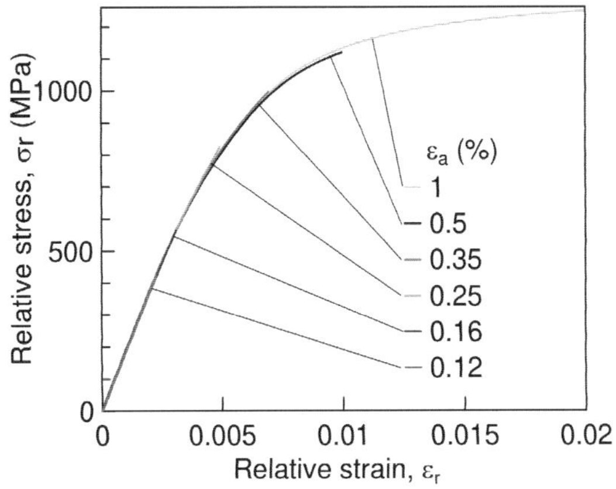

Figure 1. Tensile half-loops in relative coordinates.

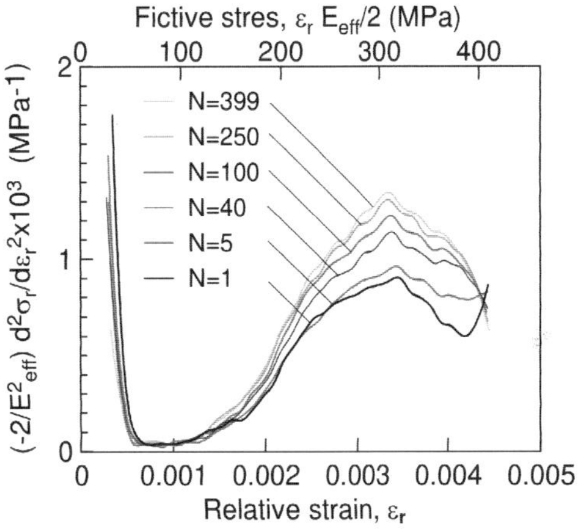

Figure 2a. Evolution of the second derivative of the tensile half-loop with the number of cycles (a) $\varepsilon_a = 2.5\text{x}10^{-3}$, ($\varepsilon_{ap} = 2\text{x}10^{-4}$).

Figure 2 shows this plot for three strain amplitudes and different cycle numbers. In cycling with the lowest strain amplitude (figure 2a) sudden initial drop corresponds to the relaxation of the compressive effective stress to zero. Later the effective stress is built up in the tensile direction and single peak corresponds to the cyclic straining of the austenitic phase. It represents the probability density function of the internal critical stresses in the deformed austenitic phase. The height of the peak increases which corresponds to the localization of the cyclic strain in the slip bands with increasing dislocation density (see later). Cycling with medium strain amplitude (figure 2b) results in the appearance of a second not fully developed peak. The evolution of the first peak with the number of cycles is similar to that in cycling with the lowest strain amplitude. High amplitude cycling (figure 2c) produces two distinct peaks, the first corresponding to cyclic straining of austenite and the second to cyclic straining of ferrite. The small peak in between both peaks was found to be connected with the subtle variation

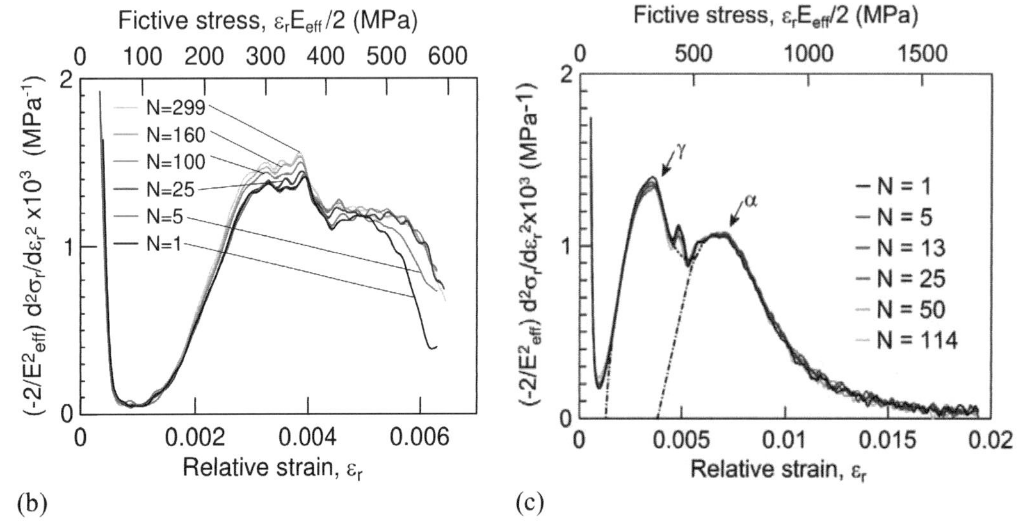

Figure 2. Evolution of the second derivative of the tensile half-loop with the number of cycles (b) $\varepsilon_a = 3.5\text{x}10^{-3}$, ($\varepsilon_{ap} = 7.2\text{x}10^{-4}$), (c) $\varepsilon_a = 1\text{x}10^{-2}$, ($\varepsilon_{ap} = 6.4\text{x}10^{-3}$).

of the strain rate due to imperfection in the cylinder and appears at relative strain corresponding to zero stress in all plots. No appreciable changes of both peaks during cyclic straining appeared.

Dislocation structure in a specimen cycled with low strain amplitude shows figure 3. The typical dislocation structure corresponding to cyclic loading is apparent only in austenitic grains. Dislocation structure in ferritic grains corresponds to the non-deformed state. It is typical for all grains observed. Dislocations in the austenite lie in parallel bands, forming pile-ups. This is a typical dislocation arrangement provided one slip system is activated and cross-slip is difficult. Low stacking fault energy is a reason for the splitting of the perfect dislocation into the two Shockley partials in the primary slip plane. In the ferritic grains, low density of randomly oriented dislocations is found (see figure 3). The

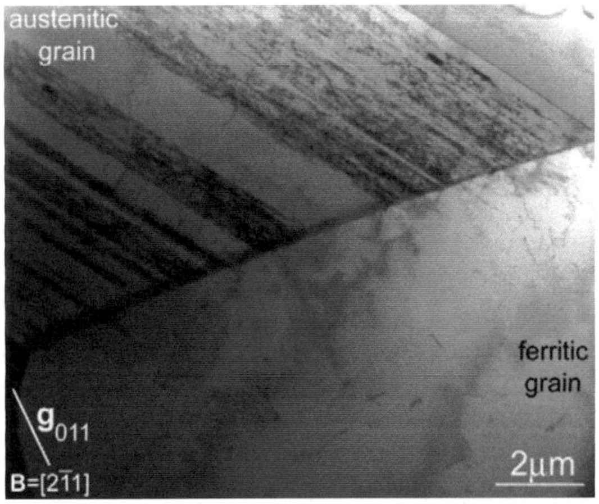

Figure 3. Dislocation structure in specimen cycled with constant plastic strain amplitude $1\text{x}10^{-4}$ for $5\text{x}10^{6}$ cycles.

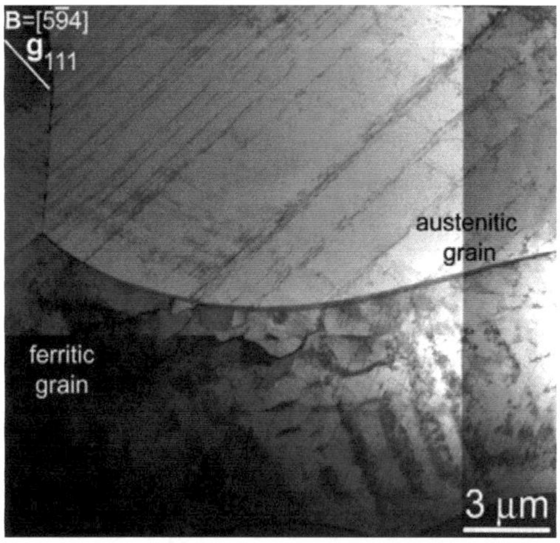

Figure 4. Dislocation structure in specimen cycled with constant plastic strain amplitude $2\text{x}10^{-3}$ to fracture ($N_f = 10140$ cycles).

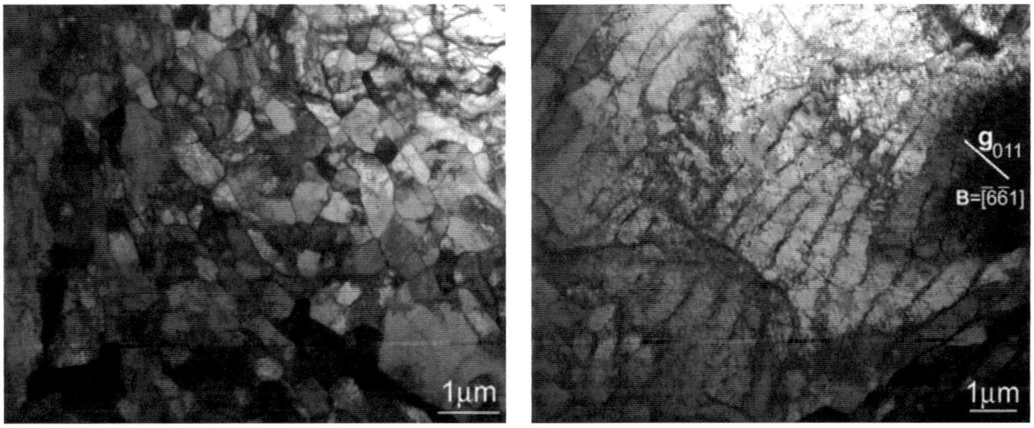

(a) (b)

Figure 5. Dislocation structure in specimen cycled with constant plastic strain amplitude 1×10^{-2} to fracture (N_f = 925 cycles), (a) austenitic grain, (b) ferritic grain.

dislocation density and arrangement is the same as in the virgin material. The applied stress is thus not high enough to induce a notable dislocation activity in majority of ferritic grains. This is in agreement with the analysis of the loop shape in low amplitude loading.

With increasing strain amplitude characteristic dislocation structure is formed both in austenitic and in ferritic grains. In austenitic grains parallel bands become thicker and their density increases until secondary slip is activated. In narrow twins ladder-like structure can be formed (see e.g. [4]). The dislocation structure in the ferritic grains is composed of vein-like matrix and well defined walls. Two or more slip systems are activated; the walls are produced by the interaction of the dislocations from two of most active slip systems. Screw dislocation segments are present in the channels. Characteristic dislocation structure in specimen cycled with the high strain amplitude shows figure 5. Characteristic cell dislocation structure in austenite was produced by multiple slip (figure 5a). Dislocations are concentrated to the cell walls and individual cells are disoriented by a small angle. In the ferritic grain (figure 5b) cells are formed in the lower left corner while secondary walls dominate the majority of the grain.

Due to much higher effective stress in the ferrite the second peak of the second derivative is shifted to higher fictive stress and both peaks of the probability density function of the internal critical stresses are well separated. The effective stresses 120 MPa and 340 MPa were evaluated for austenite and ferrite respectively. The maxima of the critical internal stresses of elementary volumes in the probability density plot with the highest occurrence correspond to 180 MPa in austenite and 290 MPa in ferrite. The major part of the cyclic stress in duplex steel for all plastic strain amplitudes is thus carried by ferritic phase. The dislocation structures of both phases vary appreciably with applied plastic strain amplitudes in agreement with the changes of the probability density function of the internal critical stresses.

Acknowledgements
The authors acknowledge the support by the research project No. AV0Z 20410507 and grant No. 1AA100480704 of the GA of the Academy of Sciences and grant No. 106/09/1954 of the GACR.

References
[1] Polák J 2003 Cyclic Deformation, Crack Initiation, and Low Cycle Fatigue *Comprehensive Structural Integrity* vol 4, ed I Milne, R O Ritchie and B Karihaloo (Amsterdam: Elsevier) pp 1-39
[2] Mateo A, Llanes L, Iturgoyen L and Anglada M 1996 *Acta Mater.* **44** 1143-53
[3] Polák J 2005 *Kovove Mater.* **43** 280-89
[4] Marinelli M C, Degallaix S and Alvarez-Armas I 2006 *Mater. Sci. Eng.* A **435-436** 305-8

Thermo-mechanical fatigue behavior of the intermetallic gamma-TiAl alloy TNB-V5 with different microstructures

M. Roth[1] and H. Biermann

TU Bergakademie Freiberg, Institute for Materials Engineering, Gustav-Zeuner-Strasse 5, D-09599 Freiberg, Germany
[1] now at IAV GmbH, Kauffahrtei 25, D-09120 Chemnitz, Germany

marcel.roth@iav.de

Abstract. The cyclic deformation and fatigue behavior of the γ-TiAl alloy TNB-V5 is studied under thermo-mechanical load for the three technically important microstructures Fully-Lamellar (FL), Near-Gamma (NG) and Duplex (DP), respectively. Thus, thermo-mechanical fatigue (TMF) tests were carried out with different temperature-strain cycles, different temperature ranges from 400°C to 800°C and with two different strain ranges. Cyclic deformation curves, stress-strain hysteresis loops and fatigue lives are presented. The type of microstructure shows a surprisingly small influence on the cyclic deformation and fatigue behavior under TMF conditions. For a general life prediction the damage parameter of Smith, Watson and Topper P_{SWT} is well suitable, if the testing and the application temperature ranges, respectively, include temperatures above the ductile-brittle transition temperature (approx. 750°C). If the maximum temperature is below that temperature, the brittle materials′ behavior yields a high scatter of fatigue lives and a low slope of the fatigue life curve and therefore the damage parameter P_{SWT} cannot be applied for the live prediction.

1. Introduction

Alloys on the basis of the intermetallic γ-TiAl phase are very attractive for modern lightweight high-temperature components, because they combine low density and excellent high-temperature strength [1-6]. Especially 3rd generation γ-titanium aluminides with 5 - 10 at% Nb (TNB-alloys) show a high potential for applications. Because of the low ductility at temperatures below the ductile-brittle transition (approx. 750°C) [1] and the resulting poor damage tolerance a good knowledge of the cyclic deformation and fatigue behavior is required. Moreover, during start-up and shut-down operations structural components in automotive and aerospace engines can be exposed to cyclic thermal and superimposed cyclic mechanical loadings. This thermo-mechanical fatigue (TMF) load can be decisive for the fatigue life of a component. Up to now only a few studies on the TMF behavior of γ-TiAl are published. Christ et al. studied the influence of temperature, environment and temperature-strain cycle on the TMF behavior of different γ-TiAl alloys [1-5]. Brookes et al. studied the TMF behavior of TNB-V5 under multiaxial axial-torsional fatigue [6]. First results of the authors` group on the TMF behavior of TNB-V5 with Near-Gamma microstructure were recently published elsewhere [7-9].

In the present paper the influence of temperature-strain cycle and temperature range on the cyclic deformation and fatigue behavior of the γ-TiAl alloy TNB-V5 under thermo-mechanical load is evaluated on three different microstructures, i.e. Near-Gamma (NG), Duplex (DP) and Fully-Lamellar

(FL). TMF tests were carried out with the temperature-strain cycles in-phase (IP), out-of-phase (OP), clockwise diamond (CD) and counter-clockwise diamond (CCD), temperature ranges between 400°C and 800°C and with two different strain ranges.

2. Experimental Details

The investigated alloy TNB-V5 with the nominal composition of Ti-45Al-5Nb-0.2C-0.2B (at-%) was available in three microstructures (Figure 1) designated as a) Near-Gamma b) Duplex and c) Fully-Lamellar, respectively, according to [10].

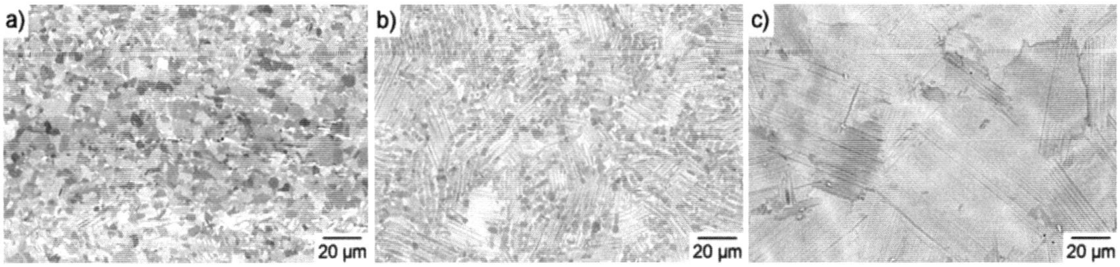

Figure 1: Near-Gamma (a), Duplex (b) and Fully-Lamellar (c) microstructure of TNB-V5, extrusion direction horizontal, scanning electron microscopy, back-scattered electron contrast.

For TMF tests electrolytically polished, compact cylindrical specimens with a gauge length of 19 mm and a gauge diameter of 6.5 - 7.0 mm were used. The TMF tests were carried out on a servohydraulic testing system under total-strain control mode at constant mechanical strain rates of $5.75 \times 10^{-5}\ s^{-1}$ to $1.02 \times 10^{-4}\ s^{-1}$ and at a strain ratio $R_\varepsilon = -1$. Two different mechanical strain amplitudes, $\Delta\varepsilon_{mech}/2 = 5.75 \times 10^{-3}$ and $\Delta\varepsilon_{mech}/2 = 6.50 \times 10^{-3}$, were applied. Three different temperature ranges were studied, i.e. 400°C - 650°C, 550°C - 800°C and 400°C - 800°C. All tests were performed in laboratory air and match the recommendations of the "Code-of-Practice for Strain-Controlled Thermo-Mechanical Fatigue Testing" [11]. Four different TMF temperature-strain cycles (in-phase (IP), out-of-phase (OP), clockwise diamond (CD) and counter-clockwise diamond (CCD)) were applied.

3. Results and Discussion

Figure 2a shows the stress-strain hysteresis loops for the different microstructures (NG, DP and FL) for IP loading as an example within the temperature range 400°C – 800°C with $\Delta\varepsilon_{mech}/2 = 5.75 \times 10^{-3}$. The Fully-Lamellar microstructure yields slightly higher stresses than the other microstructures. Near-Gamma and Duplex show comparable behaviour with almost identical plastic strain amplitudes which are larger than that of Fully-Lamellar microstructure. In Figure 2b the corresponding cyclic deformation curves are presented. All microstructures show similar characteristics of mean stresses which are compressive in case of IP loading. The microstructures Fully-Lamellar and Duplex offer comparable lifetimes, the microstructure Near-Gamma has the shortest fatigue time. However, it has to be pointed out that there is a surprisingly small influence of the type of microstructure on the cyclic deformation and fatigue behavior under thermo-mechanical loading conditions.

All three microstructures (NG, DP, FL) show a comparable influence of the temperature-strain cycle (IP, OP, CD, CCD) on the cyclic deformation and fatigue behavior. OP loading yields high positive and IP loading high negative mean stresses, whereas in CD and CCD tests significantly smaller positive and negative, respectively, or nearly no mean stresses are observed. The shortest fatigue lives are always found in OP tests, the longest in IP tests. CD and CCD testing yield similar fatigue lives intermediate between those of OP and IP tests, as shown earlier for the Near-Gamma microstructure [7-9].

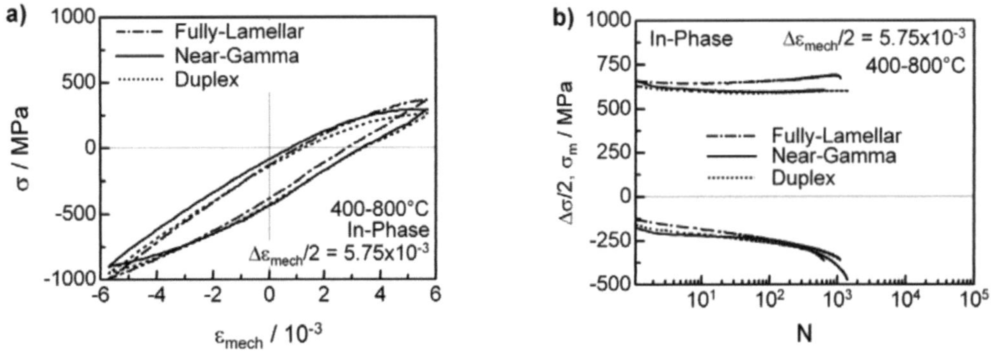

Figure 2: Stress-strain hysteresis loops at $N_F/2$ (a) and corresponding cyclic deformation curves (b) for the different microstructures.

Comparing the temperature ranges shows that the stress amplitudes and therefore the maximum tensile stresses at 400°C – 650°C are much higher than those at 400°C – 800°C and at 550°C – 800°C. The tests within the temperature range 400°C – 650°C show the smallest numbers of cycles to failure and the tests within the temperature ranges 550°C – 800°C and 400°C – 800°C show significantly higher fatigue lives (not shown in the present paper).

A suitable lifetime prediction for the γ-TiAl alloy TNB-V5 under thermo-mechanical loading can be achieved for the cycles including a maximum temperature of 800°C by using the damage parameter P_{SWT} (equation 1) suggested by Smith, Watson and Topper [12]

$$P_{SWT} = \sqrt{\sigma_{max}\varepsilon_a E} \tag{1}$$

Here, σ_{max} is the maximum tensile stress at half number of cycles to failure $N_F/2$, ε_a the mechanical strain amplitude ($\Delta\varepsilon_{mech}/2$) and E the Young´s modulus at the temperature of the maximum stress.

In Figure 3a the damage parameter P_{SWT} is plotted in a double-logarithmic diagram vs. the fatigue lives for all tests in the temperature ranges 550°C – 800°C and 400°C – 800°C and in Figure 3b the predicted fatigue lives are compared with the experimentally obtained ones for the temperature ranges including the maximum temperature of 800°C. If the testing and the application temperature ranges, respectively, include temperatures above the ductile-brittle transition temperature (the DBTT is approx. 750°C) the predicted fatigue lives deviate from the experimental fatigue lives in maximum by a factor of 3. If the maximum temperature is below the DBTT, the brittle materials´ behavior yields a high scatter of fatigue lives and a low slope of the P_{SWT} vs. N_F curve, cf. [7-9]. Therefore, this damage parameter cannot be applied for the live prediction.

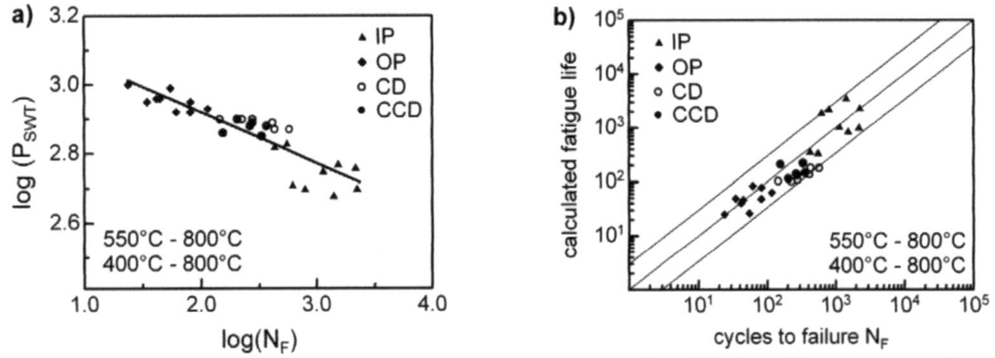

Figure 3: Damage parameter P_{SWT} vs. fatigue lives (a) and predicted fatigue lives vs. experimentally obtained fatigue lives (b) for all tests in the temperature ranges 550°C – 800°C and 400°C – 800°C.

4. Conclusions

The influence of microstructure, temperature-strain cycle and temperature range on the cyclic deformation and fatigue behavior of the γ-TiAl alloy TNB-V5 under thermo-mechanical load is evaluated. The most important results are:

The microstructures Fully-Lamellar and Duplex offer comparable lifetimes. The microstructure Near-Gamma has a slightly shorter lifetime. However, there is a surprisingly small influence of the type of microstructure on the cyclic deformation and fatigue behavior under TMF loading conditions.

All three microstructures (NG, DP, FL) show a comparable influence of the temperature-strain cycle (IP, OP, CD, CCD) on the cyclic deformation and fatigue behavior. The shortest fatigue lives are always observed in OP tests, the longest in IP tests. The fatigue lives correlate with the amount of maximum tensile stress and thus with the mean stresses.

In accordance to the assumption that the fatigue life is strongly influenced by maximum tensile stress, the tests within the temperature range 400°C – 650°C show the smallest numbers of cycles to failure and the tests within the temperature ranges 550°C – 800°C and 400°C – 800°C show significantly higher fatigue lives.

For a general life prediction the damage parameter of Smith, Watson and Topper P_{SWT} is well suitable, if the testing and the application temperature ranges, respectively, include temperatures above the DBTT. If the maximum temperature is below, the brittle materials′ behavior yields a high scatter of fatigue lives and therefore this damage parameter cannot be applied for the live prediction.

Acknowledgments
This work was supported by Deutsche Forschungsgemeinschaft (Bi 418/10). Further thanks are expressed to Dr. H. Kestler, Plansee AG, Reutte, Austria, for providing the material.

References

[1] Schallow P: Ph.D. Thesis. Univ. Siegen, VDI Verlag, Düsseldorf, 2005
[2] Fischer F O R: Ph.D. Thesis, Univ. Siegen, 2000
[3] Christ H-J, Fischer F O R and Maier H J 2001 *Mater. Sci. Eng. A* **319-321** 625
[4] Christ H-J 2007 *Maert. Sci. Eng. A* **468-470** 98
[5] Bauer V and Christ H-J 2009 *Intermetallics* **17** 370
[6] Brookes S P, Kühn H-J, Skrotzki B: Proc. 11th World Conference on Titanium, 2007.
[7] Roth M and Biermann H 2006 *Scripta Mater.* **54** 137
[8] Roth M and Biermann H 2007 *Mater. Sci. Forum* **539-543** 1559
[9] Roth M and Biermann H 2008 *Int. J. Fatigue* **30** 352
[10] Kim Y-W and Dimiduk D M 1991 *JOM* **43** 40
[11] Hähner P et al.: Validated Code-of-Practice for Strain-Controlled Thermo-Mechanical Fatigue Testing. EC – EUR 22281 EN – Directorate General, Joint Research Centre – Institute for Energy, 2006, Petten, NL.
[12] Smith K N, Watson P and Topper T H 1970 *J. Mater.* **5** 767

Scanning electron microscopy and computer modelling of microstructural changes in the vicinity of propagating short fatigue cracks in austenitic stainless steels

U Krupp[1], I Roth[2], HJ Christ[2], M Kübbeler[3], CP Fritzen[3], W Skrotzki[4]

[1] University of Applied Sciences Osnabrück, Fakultät Ingenieurwissenschaften und Informatik, D-49009 Osnabrück, Germany
[2] University of Siegen, Institut für Werkstofftechnik, D-57068 Siegen, Germany
[3] University of Siegen, Institut für Mechanik und Regelungstechnik, D-57068 Siegen, Germany
[4] Technische Universität Dresden, Institut für Strukturphysik, D-01069 Dresden, Germany

u.krupp@fh-osnabrueck.de

Abstract. The present paper deals with the in-situ observation of local fatigue damage processes by using a conventional servohydraulic testing machine and a new developed piezo-driven miniature testing system in the scanning electron microscope in combination with electron backscatter diffraction to correlate crack initiation and propagation with local microstructural features. In the case of metastable austenitic 304L steel it is shown and supported by finite element calculations that fatigue-induced strain localization causes the formation of α′ martensite. Martensite nuclei lying parallel to the slip planes in the austenite grains and twin boundaries can be considered as crack-initiation sites. Fatigue crack propagation causes an increasing martensite transformation rate ahead of the crack tip leading to the occurrence of crack closure effects. As shown in an earlier study, crack closure in the case of short cracks follows a transient regime, i.e. immediately after crack initiation the crack stays open almost during the complete fatigue cycles. The development of a short crack model based on the boundary element method aims to the prediction of crack initiation and short fatigue crack propagation rates.

1. Introduction

More than 90% of the fatigue life of structural stainless steel components loaded by high numbers of cycles is determined by crack initiation and short crack growth. During this phase of the fatigue damage process the material can not be considered as a homogeneous structure. The variation in grain geometry and crystallographic orientation causes an inhomogeneous distribution of mechanical stresses due to elastic anisotropy (cf. Fig. 1 and [1, 2]). As a consequence of ongoing cyclic loading, stress/strain localization leads to plastic deformation and in the case of metastable austenitic steels to a transformation from γ austenite to ε and α′ martensite. Eventually, accumulation of plastic strain and transformation-induced martensite results in fatigue crack initiation and crack propagation which during the early stage of the fatigue damage process is strongly influenced by the microstructure, like grain size and orientation distribution (cf. [2]). Due to a 2.57% volume increase and the higher strength of the martensite laths [3] there is an intrinsic crack tip shielding effect, i.e., the transformed plastic zone ahead of the crack tip may retard the crack propagation rate by (i) transformation-induced crack closure and (ii) by a higher yield strength (cf. [4, 5]).

Since the effect of martensitic transformation has a significant impact on the strength properties not only of metastable austenitic steels but also of TRIP steels (transformation-induced plasticity) being frequently used in the automotive industry, there is a strong need to implement it into modeling concepts relating the microstructure with the applied fatigue loading conditions.

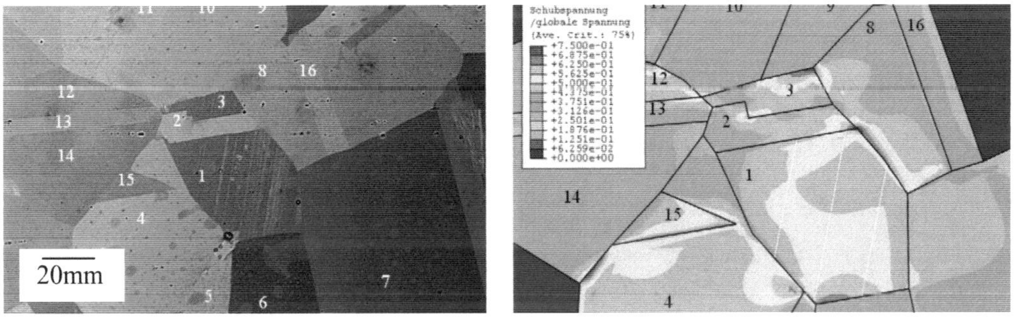

Fig. 1: Plastic strain localization (grains 1 and 3) in AISI 304L steel: (a) SEM micrograph (after fatigue loading) and corresponding inhomogeneous stress distribution, with stress maxima at the grain boundaries of grains 1 and 3 (FE calculation based on an elastic anisotropic material model and EBSD data)

2. Experimental

Fully reversed (R = -1) fatigue tests were applied to AISI304L metastable austenitic steel (0.03C, 18.1Cr, 8.8Ni, 1.9Mn, in wt%) under stress control and a stress amplitude slightly above the 10^7 cycle fatigue limit of $\Delta\sigma/2$ = 230MPa. Experiments were carried out (i) in-situ in a scanning electron micrsoscope (SEM) using a self-designed piezo-driven miniature testing system shown in Fig. 2, and (ii) in a MTS servohydraulic testing machine. To correlate fatigue damage and martensite transformation phenomena with the microstructure, fatigue experiments were accompanied by automated electron backscatter diffraction (EBSD) measurements of crystallographic orientations.

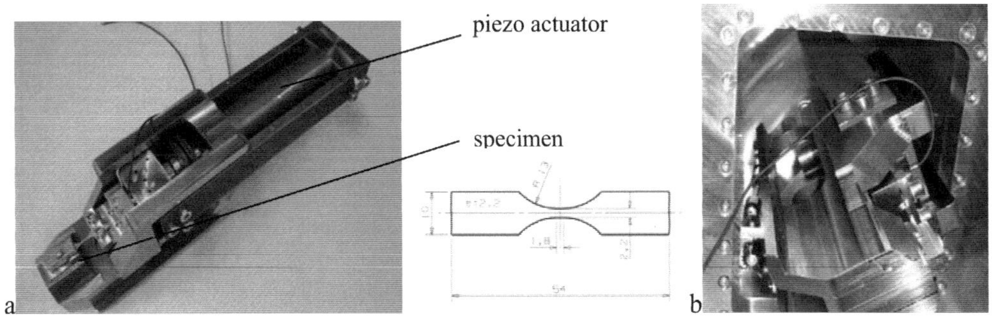

Fig. 2: In-situ fatigue testing system for integration into a SEM: (a) piezo actuator and specimen stage, (b) 70° tilt stage for EBSD measurements during fatigue testing

3. Influence of Martensite Formation during Early Fatigue Crack Propagation

Even small-strain fatigue loading is sufficient to create slip markings in many grains. This is due to the low friction stress in face-centred cubic stainless steels. The origin of the slip markings are grains of high elastic anisotropy stresses (cf. Fig. 1) and high Schmid factors (close to 0.5). The slip markings act as nuclei for the austenite martensite formation as is clearly revealed by EBSD measurements. Here, only α' martensite having the Kurdjumov-Sachs orientation relationship with the parent austenite phase, i.e. $(\bar{1}1\bar{1})_{\gamma} \parallel (101)_{\alpha'}$ and $[\bar{1}\bar{1}0]_{\gamma} \parallel [\bar{1}\bar{1}1]_{\alpha'}$ is detected. The metastable hexagonal ε martensite is assumed to vanish after a short transition period.

The initial crack density at small-strain fatigue is very small; interestingly, crack initiation is found to occur not only at the martensite lamellae but also at twin boundaries (cf. [1, 6]). Immediately after initiation, microcracks grow crystallographically along slip planes (stage 1), but after short time they change to normal stress controlled stage 1b growth (cf. [2, 7]), operated by two alternately active {111} slip planes. This stage of the crack propagation process strongly involves martensite formation as is shown in Fig. 3. The crack path appears in the EBSD in form of martensite that was formed in the plastic zone ahead of the crack tip. Extension of the plastic zone and hence crack propagation is blocked by the grain boundaries as indicated by the crack propagation rates that show an oscillating behavior. Since cracks are observed to stay open at zero load (Fig. 3b) in combination with martensite formation at the crack tip, transformation-induced crack closure is assumed to be active in such a way, that the crack tip is not shielded but remains open even during a certain part of the compressive half cycle.

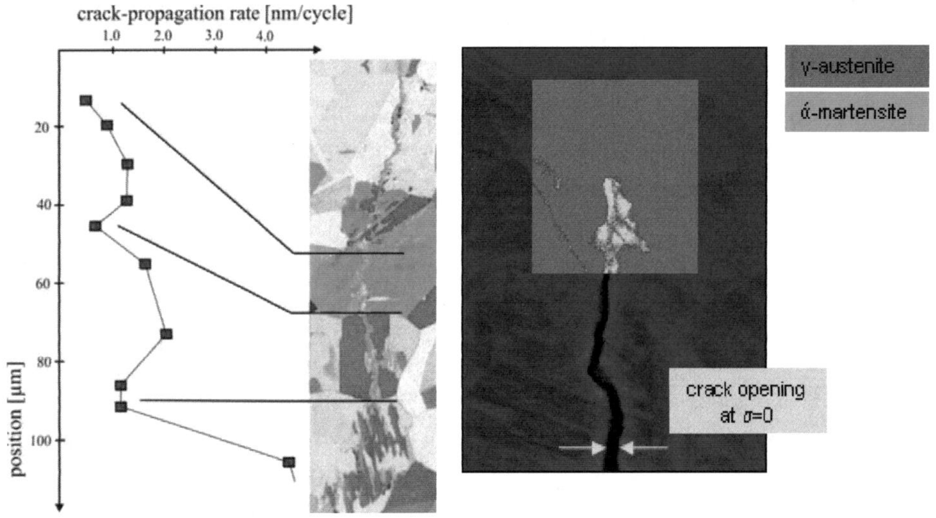

Fig. 3: Crack propagation in AISI304L metastable austenitic steel: (a) barrier effect of the grain boundaries (the lines correlate the crack propagation rate data with the EBSD grain orientation map) and (b) crack opening at σ = 0 due to martensite formation at the crack tip

4. Modeling Concept

Microstructural control of short fatigue cracks has been modeled by means of a boundary element (BE) approach (cf. [2, 8]), where slip band cracks and the adjacent plastic zones are meshed by displacement discontinuity elements. This approach allows for calculating the cyclic crack tip slide displacement (ΔCTSD) being the direct driving force for crack propagation. During crack advance the distance between the crack tip and the next grain boundary and consequently, ΔCTSD decreases. This decrease causes the crack propagation rate da/dN to slow down. At the same time the stress acting on slip systems of the neighboring grains increases. After reaching a critical stress, the plastic zone extends to the respective grains and ΔCTSD and da/dN increase. Hence, the BE concept is capable to reflect the oscillating propagation behavior of short cracks as a function of local microstructural features.

To implement both the effect of an inhomogeneous stress distribution, as is shown in Fig. 1, and the phenomenon of local plastic strain-induced martensite formation, a second set of BE was introduced to mesh the grain boundaries [9]. Both sets are superimposed as two sub-problems: crack in an infinite plate and crack-free grains.

Fig. 4 shows the application of the modified BE approach to a simple microstructure consisting of three grains of two different Young's moduli E_0 and E_1. In the case of a homogeneous plate ($E_1 = E_0$), ΔCTSD starts to decrease when approaching the grain boundary. At a normalized crack length of about 0.9, ΔCTSD exhibits a substantial jump since the critical stress for activation of the slip system in the neighboring grain is exceeded. The strong effect of an inhomogeneous distribution of the Young's modulus (elastic anisotropy) becomes evident for situations, where the crack propagates in an elastically softer ($E_1 = E_0/1.5$) or an elastically harder material ($E_1 = 1.5\ E_0$). As compared to the homogeneous materials, there are strong variations in ΔCTSD and the crack propagation rate, respectively.

In future work, the superposition concept will be applied to microstructures that are generated with experimental EBSD data and where effect of the austenite to martensite formation is included.

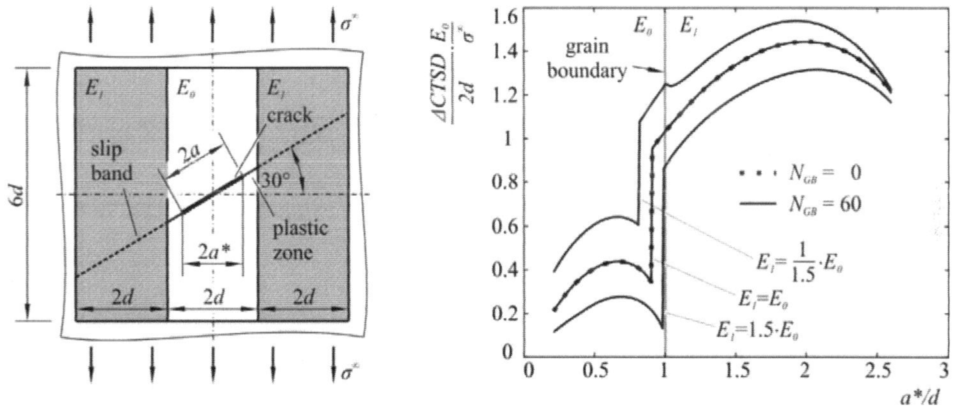

Fig. 4: Crack in an elastically inhomogeneous microstructure: (a) geometrical arrangement and (b) crack tip slide displacement vs. normalized crack length for various arrangements of elastically soft and hard grains (dashed line: BE generation only for the crack).

5. Conclusions

Fatigue damage of AISI304L metastable austenitic steel involves the formation of plastic strain-induced martensite, even at macroscopically very low strain amplitudes. Elastic anisotropy causes strain localization and fatigue crack initiation. As soon as a short crack starts to grow the plastic zone around its tip exhibits martensitic transformation. It is shown qualitatively that this transformation has a strong impact on the current crack propagation rate. This can be attributed to both the development of transformation-induced crack closure and a change in the crack growth mechanism itself, i.e. the crack propagates into high-strength martensite instead of soft austenite. A new superposition boundary element approach based on an existing short crack model accounts for crack propagation within an elastically and plastically inhomogeneous microstructure.

References

[1] Heinz A and Neumann P 1990 *Acta Metall. Mater.* **38** 1933

[2] Krupp U 2007 *Fatigue Crack Propagation in Metals and Alloys* (Wiley, Weinheim)

[3] Padilha AF and Rios PR 2002 ISIJ Int. **42** 325

[4] Mayer HR, Stanzl-Tschegg S, Sawaki Y, Huhner M and Hornbogen E 1995, *Fatigue Fract. Eng. Mater. Struct.* **18** 935

[5] Mei Z and Morris Jr. JW 1991 Eng. Fract. Mech. **39** 569

[6] Blochwitz C, Jacob S and Tirschler W 2008 *Mater Sci. Eng.* **A496** 59

[7] Düber O, Künkler B, Krupp U, Christ HJ and Fritzen CP 2006 *Inl. J. Fatigue* **28** 983

[8] Künkler B, Düber O, Köster P, Krupp U, Fritzen CP and Christ HJ 2008 *Eng. Fract. Mech.* **75** 715

[9] Kübbeler M, Roth I, Krupp U, Fritzen CP and Christ HJ 2009 *Eng. Fract. Mech.* in press

Cyclic deformation of extruded AM30 magnesium alloy in the transverse direction

D L Chen[1], A R Emami[1] and A A Luo[2]

[1]Department of Mechanical and Industrial Engineering, Ryerson University, 350 Victoria Street, Toronto, Ontario M5B 2K3, Canada

[2]General Motors Research and Development Center, Warren, MI 48090, USA

dchen@ryerson.ca

Abstract. Cyclic deformation characteristics of a recently developed AM30 Mg extrusion alloy in the transverse direction were evaluated under strain-controlled tests at different strain amplitudes. The alloy exhibited strong cyclic hardening especially at higher strain amplitudes. While the initial tensile Young's modulus was essentially the same in both transverse and longitudinal directions, the hysteresis loops were asymmetric in the longitudinal direction, but nearly symmetric in the transverse direction. This tension-compression asymmetry was associated with the presence of strong texture in the extruded Mg alloy. With increasing strain amplitude the mid-life hysteresis loops showed a clockwise rotation which was related to nonlinear or pseudoelastic deformation behavior. Fatigue crack initiation occurred at the specimen surface, and multiple initiation sites were observed at higher strain amplitudes. Crack propagation was basically characterized by the formation of characteristic fatigue striations.

1. Introduction

Concerns about greenhouse gas emissions and resulting global warming have spurred the development and application of lightweight magnesium (Mg) alloys in the automotive and aerospace industries. Currently the majority of Mg alloys used in automotive applications is cast alloys because of their high productivity [1-3]. To expedite the load-bearing structural applications, it is necessary to develop wrought Mg alloys with improved formability and higher strength than cast Mg alloys. For this purpose a new wrought Mg alloy, AM30, has been developed by General Motors Research and Development Center [4]. The design and life estimation of automotive components require low cycle fatigue (LCF) behavior of the materials selected. The cyclic deformation of this alloy in the longitudinal/extrusion direction has been recently reported [5,6], where a strong asymmetric tension-compression characteristic was observed due to twinning in the compressive phase and detwinning in the tensile phase. It is unclear if this alloy in the transverse direction would also exhibit a strong asymmetric characteristic. The present study was aimed to identify cyclic deformation behavior of AM30 extruded alloy in the transverse direction, in comparison with that in the extrusion direction.

2. Material and experimental procedure

The as-received AM30 Mg alloy, with a composition (in wt%) of 3.4 Al, 0.33 Mn, 0.16 Zn, 0.0026 Fe, 0.0006 Ni, 0.0008 Cu and the balance Mg, was extruded at an extrusion exit speed of 76.2 mm/s with an extrusion ratio of about 6. The extrusion temperature range was about 360~382°C and the alloy was air quenched after extrusion. The microstructure of the alloy after polishing and etching with Acetic-Picral solution (10 ml acetic acid, 4.2 g picric acid, 10 ml distilled water, and 70 ml ethanol (95%)) was observed using an optical microscope equipped with an image analysis system. Sub-sized LCF specimens oriented in the transverse direction with a gauge length of 25 mm and a width of 6 mm were machined and then ground along the loading direction with emery papers up to 4/0 to remove the machining marks and to achieve a smooth surface. Strain-controlled, push-pull type fatigue tests were conducted using a computerized Instron fatigue testing system, at a strain ratio of -1, a strain rate of 1×10^{-2} s^{-1} and room temperature of 25°C. Triangular loading waveform was applied during fatigue tests. The strain-controlled tests at lower strain amplitude levels were continued up to 10,000 cycles, then the tests were changed to load control at a frequency of 50 Hz. After the LCF tests, the fracture surfaces of the fatigued samples were examined using scanning electron microscope (SEM) to examine fatigue crack initiation site and propagation mechanisms.

3. Results and discussion

Figure 1 shows a typical microstructure of the AM30 alloy, consisting of basically recrystallized grains. However, the grain size was heterogeneous exhibiting fairly typical bimodal feature, since small and large grains were co-existent. Further large grains were observed at both top and bottom surfaces of the extruded plate [5].

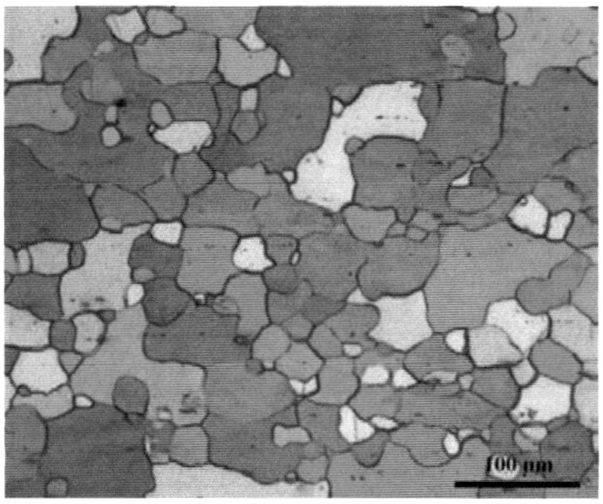

Figure 1. Microstructure of AM30 extruded Mg alloy.

The change of cyclic stress amplitude and plastic strain amplitude of AM30 alloy in the transverse direction with the number of cycles at different total strain amplitudes is shown in Figure 2. It is seen that the increase in the stress amplitude corresponded well to the decrease in the plastic strain amplitude, indicating the occurrence of cyclic hardening. At the low strain amplitude of 0.1% only very small cyclic hardening occurred. As the applied strain amplitude increased cyclic hardening became much more pronounced, as shown by the increasing slope in Figure 2(a). The cyclic hardening could also be seen in Figure 3(a) where the mid-life stress amplitude was much higher than that of the initial cycles. It is seen from Figure 3(b) that with increasing strain amplitude the mid-life hysteresis loops became larger and exhibited a clockwise rotation with increasing nonlinearity and smaller elastic modulus or pseudoelasticity. Pseudoelasticity might arise from several origins such as reversible movement of dislocations, twinning, and stress induced phase transformations [5]. Twinning pseudoelasticity was due to the reversible movement of twin boundaries, since the position of the twins in the deformed state was unstable, and a driving force could cause them to return back (i.e., detwinning occurred) upon unloading [5,7]. Caceres *et al.* [7] examined the change of elastic modulus in a cast AZ91 Mg alloy using tensile loading-unloading procedures, and observed that with increasing plastic strain up to 1-2% the elastic modulus decreased considerably, i.e., the pseudoelastic strain could cause a decrease of up to 70% in the elastic modulus of the alloy. This was attributed to the partial reversal of $\{10\bar{1}2\}$ twins upon unloading [5-7].

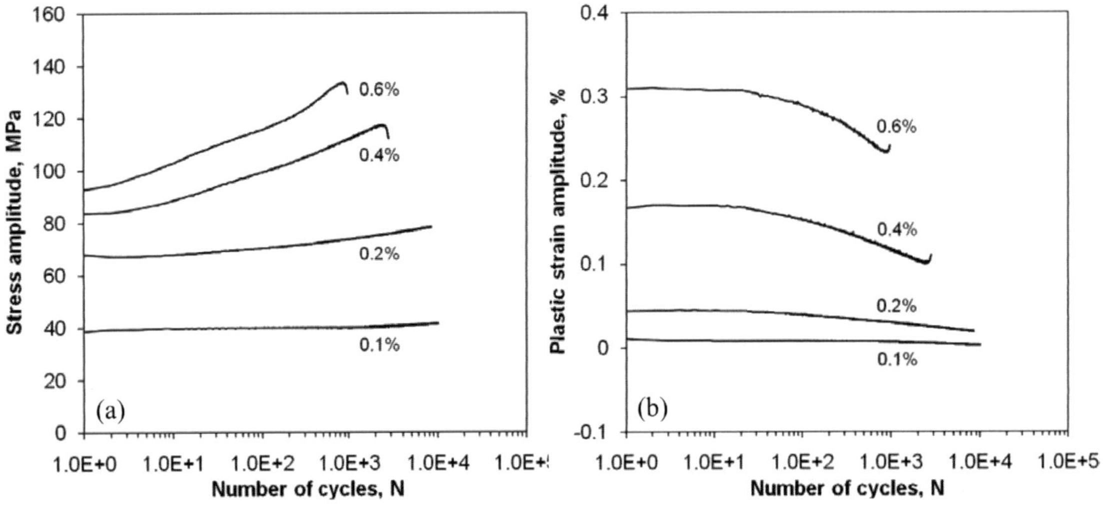

Figure 2. (a) Stress amplitude and (b) plastic strain amplitude vs. the number of cycles.

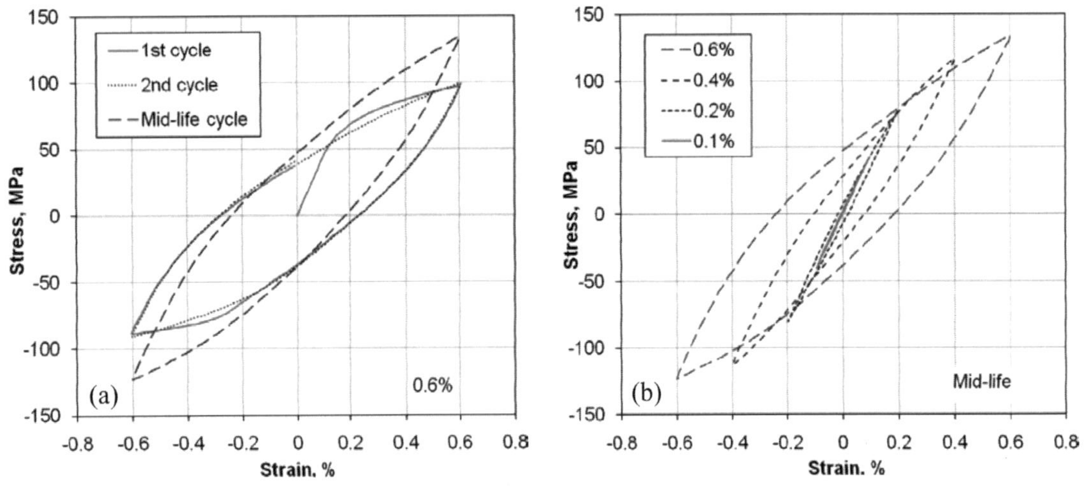

Figure 3. Hysteresis loops of AM30 Mg alloy fatigued in the transverse direction (a) at a given strain amplitude of 0.6%, and (b) at the mid life for different strain amplitudes.

It is of interest to see in Figure 4(a) that the hysteresis loop of the first cycle was very asymmetric in the longitudinal direction, but relatively symmetric in the transverse direction, even though the initial tensile Young's modulus (or slope) up to about 55 MPa was basically the same in both orientations. At the mid life the loops became narrower and taller in both orientations due to the occurrence of cyclic hardening but the asymmetry in the longitudinal direction still remained, as shown in Figure 4(b). This tension-compression asymmetry was primarily attributed to the presence of texture in the extruded Mg alloy. The main texture of the extruded AM30 plate has been determined to be $(11\bar{2}4)$ $\langle 22\bar{4}1\rangle$. The shape of hysteresis loops would depend on the relative ease of basal slip and twinning-detwinning together with the texture. In the longitudinal direction, twinning in compression and detwinning in tension occurred [5,6,8,9], while in the transverse direction basal slip and twinning would be expected to be basically equivalent in tension and compression in view of the nearly symmetric hysteresis loops. Fatigue crack initiated from the specimen surface, and multiple initiation sites at the specimen surface were observed at higher strain amplitudes. Crack propagation was mainly characterized by fatigue striation-like features.

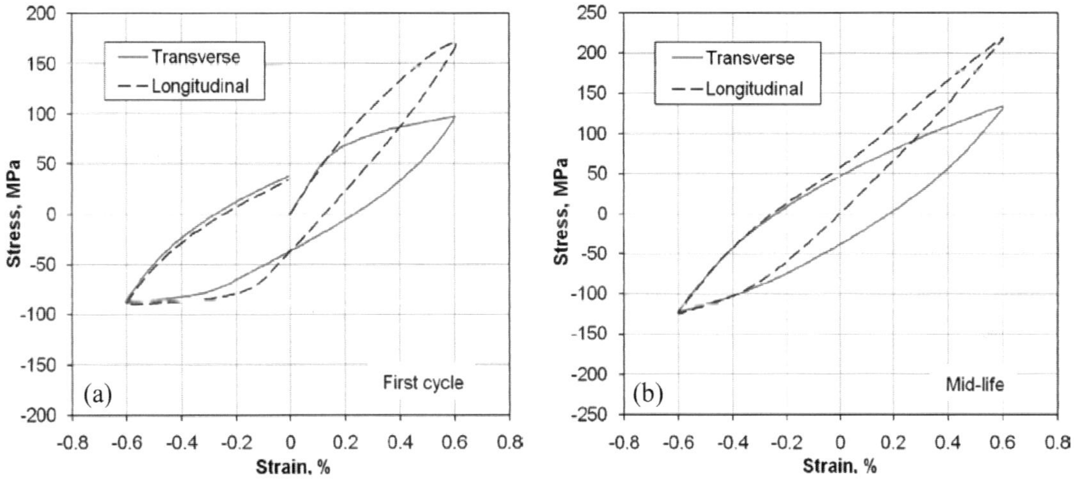

Figure 4. Comparison of hysteresis loops obtained at a strain amplitude of 0.6% between the transverse and longitudinal directions for (a) the first cycle and (b) the mid-life cycle.

4. Conclusions

The grain size of AM30 extruded alloy showed a bimodal microstructural feature. As cyclic deformation progressed the stress amplitude increased and plastic strain amplitude decreased. Cyclic hardening became increasingly stronger with increasing total strain amplitude. While the initial tensile Young's modulus up to about 55 MPa was essentially the same in both transverse and longitudinal directions, the hysteresis loop of the first cycle was very asymmetric in the longitudinal direction, but nearly symmetric in the transverse direction. The mid-life hysteresis loops exhibited a clockwise rotation with increasing strain amplitude, which was related to nonlinear or pseudoelastic behavior. The loops at the mid life in both orientations became narrower and taller due to the occurrence of cyclic hardening but asymmetry in the longitudinal direction still remained. This tension-compression asymmetry was attributed to the presence of strong texture in the extruded Mg alloy. Fatigue crack initiation occurred at the specimen surface, and multiple initiation sites were observed at higher strain amplitudes. Crack propagation was mainly characterized by the formation of fatigue striations.

References

[1] Benedyk J C 2005 *Light Metal Age* **63** 36-38
[2] Dieringa H and Kainer K U 2007 *Materialwissenschaft und Werkstofftechnik* **38** 91-96
[3] Nyberg E A, Luo A A, Sadayappan K and Shi W F 2008 *Adv. Mater. Processes* **166** 35-37
[4] Luo A A and Sachdev A K 2007 *Metall. Mater. Trans. A* **38** 1184-92
[5] Begum S, Chen D L, Xu S and Luo A A 2008 *Metall. Mater. Trans. A* **39** 3014-26
[6] Fan C L, Chen D L and Luo A A 2009 *Mater. Sci. Eng. A* **519** 38-45
[7] Caceres C H, Sumitomo T and Veidt M 2003 *Acta Mater.* **51** 6211-18
[8] Wu L, Agnew S R, Brown D W, Stoica G M, Clausen B, Jain A, Fielden D E and Liaw P K 2008 *Acta Mater.* **56** 3699-707
[9] Noster U and Scholtes B 2003 *Z. Metallkd.* **94** 559-63

Fatigue behavior and microstructure of an Al-Mg-Sc alloy at an elevated temperature

C Watanabe and R Monzen

Division of Innovative Technology and Science, Kanazawa University, Kakuma-machi, Kanazawa 920-1192, Ishikawa, Japan

chihiro@t.kanazawa-u.ac.jp

Abstract. Al-Mg-Sc alloy polycrystals bearing Al_3Sc particles with different sizes, i.e. 4, 6 and 11 nm in diameter, have been cyclically deformed at 423 K under constant plastic-strain amplitudes, and the microstructural evolution has been investigated in relation to the stress-strain response. Cyclic softening after initial hardening is found in specimens with small particles of 4 and 6 nm, but no cyclic softening takes place in specimens with larger particles of 11 nm. These features of cyclic deformation behavior are similar to the results previously obtained at room temperature. Transmission electron microscopy observations reveal that dislocations are uniformly distributed under all applied strain amplitudes in the specimens containing large particles of 11 nm, whereas slip bands are formed in the cyclically softened specimens bearing smaller particles. The cyclic softening is explained by a loss of particle strength through particle shearing within strongly strained slip bands. The 6 and 11 nm Al_3Sc particles have a stronger retardation effect on the formation of fatigue-induced stable dislocation structure than 4 nm particles at 423 K.

1. Introduction

The addition of a small amount of Scandium (Sc) to Al-Mg alloys causes a significant increase in the strength of the alloys, due to the existence of coherent, finely dispersed $L1_2$ Al_3Sc precipitate particles [1]. Also the Al_3Sc particles have modest coarsening rates at elevated temperatures, leading to the effective suppression of recrystallization and the stabilization of microstructures at high temperatures. Thus Sc-containing Al-Mg alloys are expected to be used in higher-temperature application compared to conventional structural Al alloys. Although some experiments have been performed on the mechanical properties of Al-Mg-Sc ternary alloys at elevated temperatures [1], there are no investigations of the high-temperature fatigue properties of these alloys, to the best of my knowledge.

In previous studies [2, 3], we examined the cyclic deformation behavior and dislocation microstructure under plastic-strain-controlled and stress-amplitude-controlled conditions at room temperature (RT), using an aged Al-Mg-Sc alloy with Al_3Sc particles of different diameters, i.e. 4, 6 and 11 nm, which correspond to under-age, peak-age and over-age conditions, respectively. The over-aged alloy showed cyclic hardening to saturation despite the fact that the Al_3Sc particles were coherent. On the other hand, cyclic softening occurred in the under-aged and peak-aged alloys.

In this work, a polycrystalline Al-1wt%Mg-0.27wt%Sc alloy bearing Al_3Sc particles with different sizes has been cyclically deformed at 423 K under various constant plastic-strain amplitudes, and the microstructural evolution has been investigated in relation to the stress-strain response. The results have also been compared with previous results obtained at room temperature. As in the previous studies [2, 3], we have selected three particle diameters, i.e. 4, 6 and 11nm.

2. Experimental Procedure

Specimens for fatigue tests were cut from hot-rolled polycrystalline Al-1wt%Mg-0.27wt%Sc alloy plates. These specimens had the stress axis parallel to the rolling direction. The specimens were solutionized at 905 K for 7.2×10^3 s and water quenched. Transmission electron microscopy (TEM) observations revealed that no precipitates existed in the solution treated specimens. To obtain spherical and coherent Al_3Sc particles, one set of solutionized specimens was aged at 573 K for 9.0×10^2 s, second set was aged at 573 K for 7.2×10^3, and third set was aged at 623 K for 6.48×10^4 s, corresponding to under-ageing, peak-ageing and over-ageing conditions, respectively. Hereafter, these under-aged, peak-aged and over-aged specimens will be referred to as specimens UA, PA and OA, respectively. The average diameter of Al_3Sc particles were 4 nm for specimen UA, 6nm for specimen PA and 11 nm for specimen OA. The average size was determined from over 200 particles for each taken by TEM. Judging from the Al-Sc equilibrium phase diagram [4], the volume fractions of the Al_3Sc particles in specimens UA, PA and OA are nearly identical, 0.007.

All mechanical tests were carried out at 423 K in air. Tensile properties were determined with an initial strain rate of 3.0×10^{-3} s^{-1}. Fully-reversed tension-compression fatigue tests were performed under plastic-strain-amplitude control using a triangular command signal with a constant strain rate of 10^{-3} s^{-1}.

The fatigued specimens were sliced into 3mm disks parallel to the stress axis and were mechanically ground down to 0.2 mm. Thin foils for TEM observations were prepared by electro-polishing. Microscopy was performed with JOEL-2000EX and JOEL-2010FEF microscopes operating at 200 kV.

3. Results & Discussion

Table 1 summarizes the tensile properties of specimens UA, PA and OA tested at 423 K. For comparison, the results obtained at RT are also indicated. At 423 K, the values of 0.2% proof stress of the specimens UA and OA are almost the same and smaller than that of the specimen PA. The values of 0.2% proof stress and tensile stress of each specimen are smaller than those obtained at RT, while the values of fracture strain of each specimen show a reverse tendency.

Figure 1 shows cyclic-hardening curves of specimens containing Al_3Sc particles of 4, 6 and 11nm, tested under a plastic-strain amplitude of 1.0×10^{-3} at 423 K. The data obtained at RT are also indicated. The stress amplitude σ is plotted against the cumulative plastic-strain $4N\varepsilon_{pl}$, where N is the number of fatigue cycles and ε_{pl} is the applied plastic-strain amplitude. The specimen OA exhibited a cyclic hardening to saturation under all applied plastic strain amplitudes, while in the specimens UA and PA a cyclic softening was observed after initial hardening. This result is similar to that obtained at RT [2], but the level of stress amplitudes at 423 K is lower than that at RT.

Table 1. Tensile properties of specimens UA, PA and OA obtained at 423 K and room temperature (RT).

Specimen	0.2% proof stress (MPa)		Tensile stress (MPa)		Fracture elongation (%)	
	423 K	RT	423 K	RT	423 K	RT
UA	127	130	150	170	24	21
PA	150	168	167	213	30	24
OA	121	125	147	174	27	21

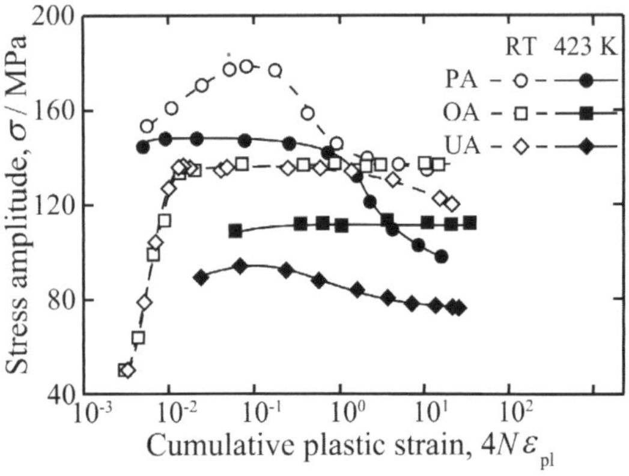

Figure 1. Cyclic-hardening curves of specimens UA, PA and OA, fatigued under a plastic-strain amplitude of 1.0×10^{-3} at 423K. Also shown are the data obtained at RT [2].

Figures 2(a), (b) and (c) depict the typical fatigue dislocation microstructures formed at 423K under $\varepsilon_{pl} = 1.0 \times 10^{-3}$ for specimens OA, UA and PA. In the specimen OA, dislocations were uniformly distributed under all applied plastic strain amplitudes, whereas both the specimens UA and PA, in which cyclic softening occurred, exhibited clearly developed slip bands along the trace of primary slip plane. These slip bands indicate the occurrence of very inhomogeneous deformation. In other words, strong strain localization took place within the slip bands. Such strong strain localization usually causes the destruction and re-dissolution of particles [3]. To check whether shearing of Al_3Sc particles occurred during fatigue tests at 423 K, the average size of the Al_3Sc particles within slip bands or in the matrix was measured by TEM using a specimen PA re-aged at 623K for 6.48×10^4 s following a fatigue test to failure. The resulting average diameters were about 9 nm within the slip bands and 12 nm in the matrix. This discrepancy in the particle size implies that the small Al_3Sc particles of 6 nm were cut by moving dislocation within the strongly strained slip bands. The cyclic softening in specimens UA and PA in figure 1 can then be explained by a decrease in particle strengthening through particle shearing or re-dissolution within the slip bands.

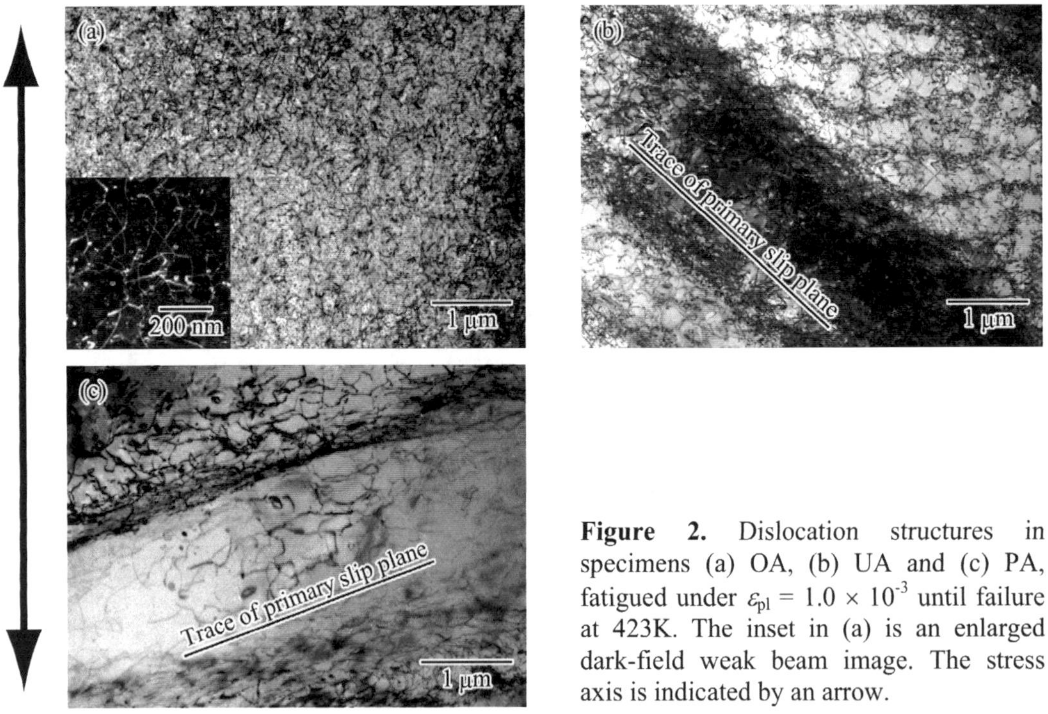

Figure 2. Dislocation structures in specimens (a) OA, (b) UA and (c) PA, fatigued under $\varepsilon_{pl} = 1.0 \times 10^{-3}$ until failure at 423K. The inset in (a) is an enlarged dark-field weak beam image. The stress axis is indicated by an arrow.

It is known that second-phase particles act as barriers against the formation and development of fatigue-induced stable dislocation structure by preventing the dislocation motion and re-arrangement during cyclic deformation [5]. Since the inter-particle spacing decreases with decreasing particles size, smaller particles have more significant retardation effect on the development of the dislocation structure if volume fraction of the particles is fixed [5]. All the specimens fatigued at RT showed no geometrically regular dislocation arrangements [2], and the relatively uniform dislocation distribution were observed except the slip bands. This result shows the effectiveness of Al_3Sc particles in preventing formation of the fatigue dislocation structures at RT. However, as shown in figure 2(b), not only the slip band but also the labyrinth structure was observed in the specimens UA fatigued at 423 K. Moreover, so-called cell structure was formed with the slip bands in a specimen UA fatigued under a high plastic-strain amplitude of 1.0×10^{-3}. On the other hands, the specimens PA and OA fatigued under all applied plastic-strain amplitudes exhibit no typical fatigue dislocation structures such as labyrinth or cell structure formed in the UA specimen. This indicates that dislocation re-arrangement is more difficult in the specimen PA and OA at 423 K despite the fact that the inter-particle spacings are larger than that in the specimen UA. In the specimen UA, moving dislocations may get over relatively easily small Al_3Sc particles by climb motion and cross slip at the high temperature of 423 K, and then re-arrange themselves to form stable structures. On the other hand, larger Al_3Sc particle in the specimens PA and OA act as effective obstacles to moving dislocations even at 423 K, and thus a relatively uniform dislocation distribution is formed.

It should be noted in figure 1 that, at 423 K, the stress level of the specimen UA is smaller than that of the specimen OA, although those of both the specimens are almost the same at RT. Since recovery seems to easily take place in the specimen UA fatigued at 423 K as mentioned above, the dislocation density in the specimen UA may be lower than that in the specimen OA. This may result in the lower stress level of the specimen UA.

4. Summary

Plastic-strain controlled low-cycle fatigue tests of Al-Mg-Sc alloy polycrystals with Al_3Sc particles of 4, 6 and 11 nm in diameter were performed at 423 K. The results and conclusions are summarized as follows.

(1) Specimens UA and PA bearing 4 and 6 nm Al_3Sc particles show cyclic softening, while specimen OA with 11 nm Al_3Sc particles show cyclic hardening to saturation.
(2) In the specimens UA and PA, slip band dislocation structures are observed, and, in the specimen OA, dislocations are uniformly distributed.
(3) Fatigue behaviors and dislocation structures are similar to those obtained at room temperature [3]. However, the stress levels tend to decrease at 423 K.
(4) The cyclic softening of specimens UA and PA are caused by shearing of small Al_3Sc particles within the slip bands.
(5) The Al_3Sc particles larger than 6 nm act as strong obstacles against the dislocation motion and re-arrangement, leading to the suppression of fatigue-induced dislocation structures at 423 K.

Acknowledgements

This work was supported in part by a Grant-in-Aid for Young Scientist (B) from Japan Society for Promotion of Science under Grant No. 19760063. The authors wish to acknowledge Furukawa-Sky Aluminum Corp. for supplying the alloy.

References

[1] Marquis EA, Seidman DN and Dunand DC 2003 *Acta Mater.* **51** 4751
[2] Watanabe C, Jin CY, Monzen R and Kitagawa K 2004 *Mater. Sci. Eng.* A **387-389** 552
[3] Watanabe C, Monzen R and Tazaki K 2008 *Int. J. Fatigue* 30 635
[4] Murray JL J. *Phase Equilib.* 1998 **19** 380
[5] Watanabe C, Fujii T, Onaka S and Kato M 2001 *Mater. Sci. Eng.* A **319-321** 506

Effect of hydrogen cathodic charging on fatigue fracture of type 310S stainless steel

A Kishi, N Takano

Advanced Materials Processing Research Laboratory,
Kanazawa Institute of Technology,
3-1 Yatsukaho, Hakusan, Ishikawa 924-0838, JAPAN

ntakano@neptune.kanazawa-it.ac.jp

Abstract. It is known that hydrogen embrittlement occurs in austenitic stainless steels like type 304 because of the strain induced martensitic transformation. Rotary bending fatigue tests of type 310S stainless steel with hydrogen cathodic charging were carried out to investigate the effect of hydrogen charging on the austenitic phase. High dissolved hydrogen decreases the fatigue strength. It also enhances the plasticity and induces ε_H martensitic transformation. Hydrogen may promote crack initiation in slip bands.

1. Introduction

Austenitic stainless steels have been utilized as materials for equipments related to hydrogen energy lately, because they have high tolerance for hydrogen embrittlement. It is, however, known that hydrogen embrittlement occurs in type 304 because of stress induced martensitic transformation [1]. It was reported that hydrogen embrittlement does not occur in type 310S, which is most stable austenitic steel [1]. It is difficult to measure the effect of hydrogen on tensile tests of the bulk, because hydrogen diffusivity in 310S is quite small and the affected region is limited to the surface layer. Inoue et al. [2] reported that the elongation of a thin sheet decreases under high hydrogen fugacity and the crack propagation is influenced by hydrogen induced martensitic transformation.

It is considered that the effect of hydrogen on fatigue tests is remarkable compared with tensile test, because the interaction between hydrogen and slip bands plays an important role in hydrogen embrittlement. It is reported [3] that hydrogen induces micro cracks along slip bands and enhances the fatigue crack growth in 310S. However, the effect of hydrogen on the fatigue strength has not been clarified. Rotary bending tests were carried out up to high cycle region and X-ray diffraction was used to investigate the effect of hydrogen on the fatigue strength of 310S austenitic stainless steel in the present paper.

2. Experimental procedures

Specimens were machined from φ8mm drawing rods of type 310S stainless steel, whose chemical composition is shown in Table 1. Dimensions of the fatigue specimen are shown in figure 1. Its surface was polished chemically by the aqueous

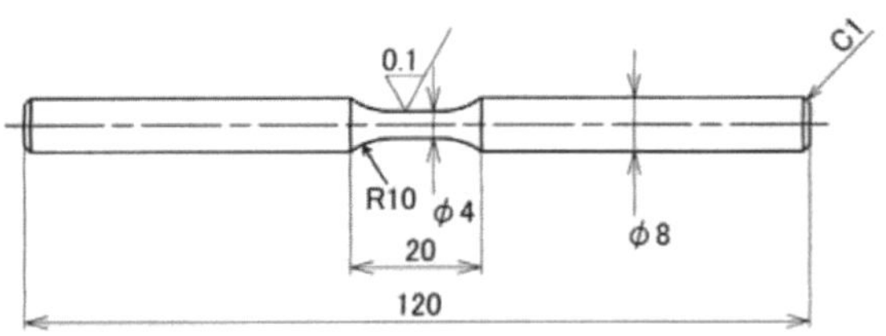

Figure 1. Dimensions of the fatigue specimen.

solution that contains 2.9mass% of hydrofluoric acid and 29mass% of hydrogen peroxide after polishing mechanically by alumina powder (0.1 μm).

Table 1. Chemical composition (mass%).

C	Si	Mn	P	Ni	Cr	S
0.02	0.3	1.07	0.028	19.26	24.1	0.001

Rotary bending fatigue tests were carried out using a water tank, as shown in figure 2, to charge hydrogen into a specimens. The tank was filled with an electrolytic solution that contained 0.5 kmol/m^3 sulfuric acid and 1.4 kg/ m^3 thiourea. Cathode current was passed through the specimen during a fatigue test. The temperature of the solution was held at 300 K. A frequency of cyclic stress was 60 Hz.

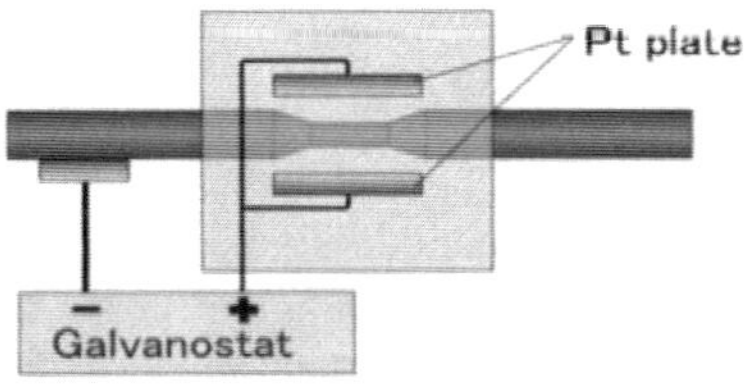

Figure 2. Schematic illustration of the apparatus of fatigue test with hydrogen charging.

For comparison, fatigue tests were also carried out in glycerin, which was used to keep a temperature 300 K under the hydrogen free condition. Moreover, to free the surface reaction from the electrolytic solution, fatigue tests of hydrogen pre-charged specimen were carried out in glycerin. The hydrogen pre-charge was run for 48 hours in the same electrolytic solution, in which the current density was 2590 A/m^2.

X-ray diffraction was employed with Cr K_α and Co K_α radiation to prove the existence of martensite phase and hydrides. In the measurement, an incident angle was fixed at 5 degree.

3. Results and discussion

3.1. Fatigue tests

S-N curves are shown in figure 3. The fatigue strength σ_a with hydrogen charging up to a current density 490A/m^2 was almost the same as that in glycerin without hydrogen. In the condition whose current density is 2590A/m^2, the fatigue strength decreased slightly. The decrease in strength was also obtained on the fatigue tests of hydrogen pre-charged specimen in glycerin. It takes almost 48hours for a fatigue test up to 10^7 cycles. It is considered that an amount of absorbed hydrogen is almost the same in the test with hydrogen charging and in glycerin after hydrogen pre-charged at high cycle if the hydrogen transport by dislocation is disregarded. This result implies that the fatigue strength is affected not by the surface reaction but by the dissolved hydrogen. Moreover, we confirmed that the fatigue strength deceased similarly, even if a specimen was polished by alumina powder after hydrogen pre-charging.

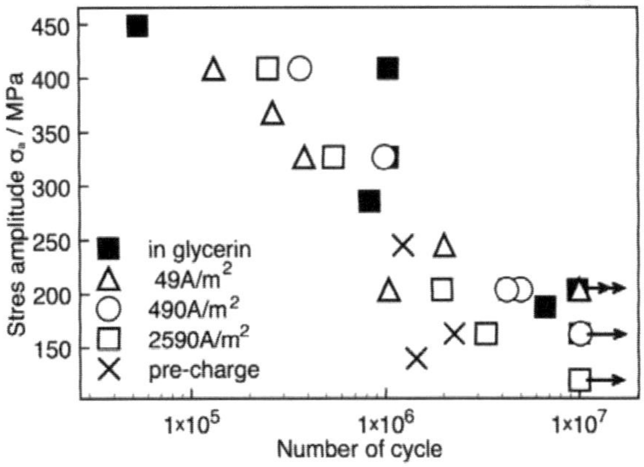

Figure 3. S-N curves. Values in the legend are cathode current density in charging with hydrogen. Arrows indicate unbroken specimens.

3.2. Fracture surfaces

A lot of traces of the slip bands were observed on the fracture surfaces, as shown in figure 4. The trace is almost straight on the fracture surface in glycerin (figure 4 (c)), while it is wavy on the fracture surfaces of specimens with hydrogen (figure 4 (a) (b)). This indicates that hydrogen enhances plastic deformation. There is no secondary crack on the fracture surfaces in glycerin, but there are many cracks with hydrogen. Such cracks were observed near the side surface, and hardly observed in the centre of specimens (figure 5). Moreover, cracks along slip bands were observed on the side surfaces of the specimens charged with hydrogen (figure 6). This corresponds to the observation of cracks by Mine et al. [3] The fracture surfaces of specimens with hydrogen charging and those tested in glycerin after hydrogen pre-charged are similar. This is in agreement with S-N curves.

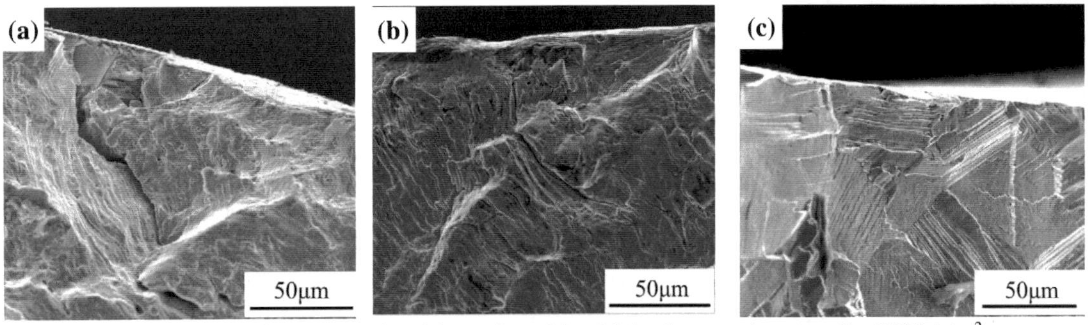

Figure 4. Fracture surfaces near the side surface (a) with hydrogen charging (i_c=2590A/m^2, σ_a=204MPa, N_f=1.9×10^6), (b) in glycerin after hydrogen pre-charged (σ_a=163MPa, N_f=2.2×10^6) and (c) in glycerin without hydrogen (σ_a=187MPa, N_f=6×10^6).

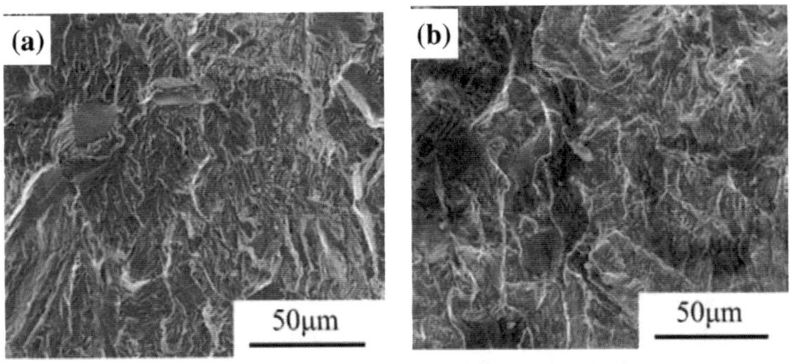

Figure 5. Fracture surfaces in the centre of the specimen (a) in glycerin without hydrogen (σ_a=187MPa, N_f=6×10^6) and (b) with hydrogen charging (i_c=2590A/m^2, σ_a=204MPa, N_f=1.9×10^6).

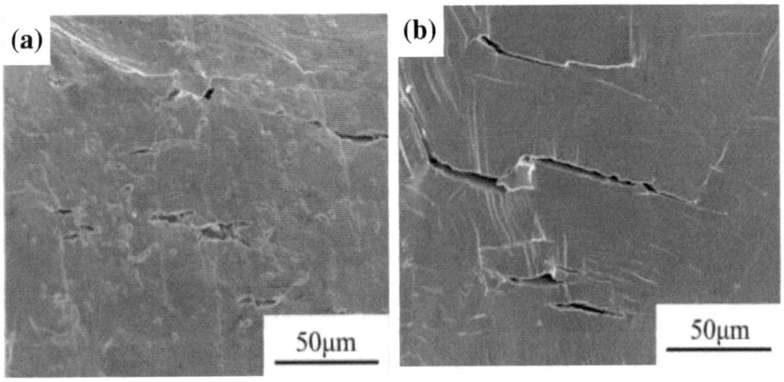

Figure 6. Side surfaces (a)with hydrogen charging (i_c=2590A/m^2, σ_a=204MPa, N_f=1.9×10^6) and (b) in glycerin after hydrogen pre-charged (σ_a=163MPa, N=2.2×10^6).

3.2 X-ray diffraction

No other phase except γ phase was detected on all fracture surfaces as shown in figure 7. That is, neither stress induced martensitic transformation nor hydride was detected. On the other hand, the martensite phase was detected on the surface of a specimen into which hydrogen was charged without stress (figure 8). It is not stress induced matentie ε but hydrogen induced ε_H. Kamachi [4] has presented that ε_H phase is transformed from γ phase when hydrogen is charged for long time. Any hydride was not detected on the surface as well as on the fracture surface. Then, it is considered that absorbed hydrogen did not induce hydrides but ε_H martensite phase and this phase disappeared during fracture process.

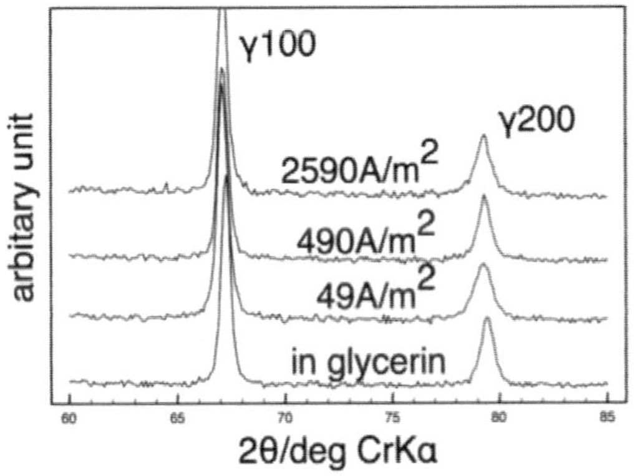

Figure 7. X-ray diffraction patterns on the fracture surface.

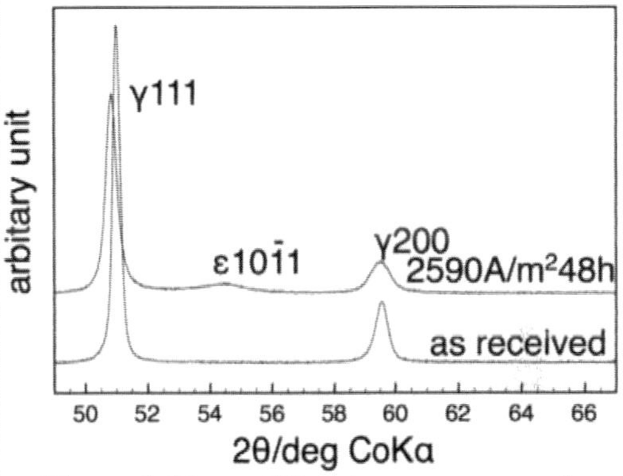

Figure 8. X-ray diffraction patterns on the surface of hydrogen pre-charged unloaded specimen and as received.

4. Discussion

High dissolved hydrogen may promote crack initiation in slip bands and decrease the fatigue strength. The effect of hydrogen on the crack initiation in slip bands is considered as follows. In the first, it is due to hydrogen-enhanced localised plasticity, which was observed on the fracture surface. In the second, it is brittle fracture of ε_H martensite induced by hydrogen, which was detected on the surface of hydrogen charged specimen without stress. In the third, it is hydrogen segregation in slip bands, which enhances the internal pressure. Polák et al. [5] proposed that the intrusions take an important role in fatigue crack initiation of stainless steel. The influence of hydrogen on fatigue fracture may be related to the mechanism. A further research is necessary which mechanism governs the fatigue behaviour of 310S in the hydrogen environment.

5. Conclusion

It is concluded that high dissolved hydrogen decreases the fatigue strength in type 310S stainless steel. Hydrogen may promote crack initiation in the slip bands under the cyclic stress condition. But, the mechanism of the crack initiation is under discussion.

References

[1] Benson R B Jr, Dann R K and Roberts Jr 1968 *Trans. Metall. Soc. AIME* **242** 2199
[2] Inoue A, Hosoya Y and Masumoto T 1979 *Trans. ISIJ* **19** 170
[3] Mine Y, Narazaki C, Matsuoka S and Murakami Y 2007 *Tetsu-to-Hagane* **93** 247
[4] Kamachi K 1978 *Trans. ISIJ* **18** 485
[5] Polák J, Man J, Vystavěl T and Petrenec M 2009 *Mater. Sci. Eng.* A **517** 204

Influence of grain orientation on evolution of surface features in fatigued polycrystalline copper: a comparison of thermal and uniaxial mechanical fatigue results

Markus Aicheler

EN-MME-MM Metallurgy and Metrology, CERN European Organization for Nuclear Research, CH-1211 Genve 23

markus.aicheler@cern.ch

Abstract. Surface state plays a major role in the crack nucleation process of pure metals in the High-Cycle-Fatigue (HCF) as well as in the Ultra-High-Cycle-Fatigue (UHCF) regime. Therefore, in studies dealing with HCF or UHCF, special attention is paid to the evolution of surface degradation during fatigue life. The accelerating structures of the future Compact Linear Collider (CLIC) under study at CERN will be submitted to a high number of thermal-mechanical fatigue cycles, arising from Radio Frequency (RF) induced eddy currents, causing local superficial cyclic heating. The number of cycles during the foreseen lifetime of CLIC reaches 2×10^{11}. Fatigue may limit the lifetime of CLIC structures. In order to assess the effects of superficial fatigue, specific tests are defined and performed on polycrystalline Oxygen Free Electronic (OFE) grade Copper, a candidate material for the structures.

Surface degradation depends on the orientation of near-surface grains. Copper samples thermally fatigued in two different fatigue experiments, pulsed laser and pulsed RF-heating, underwent postmortem Electron Backscattered Diffraction measurements. Samples fatigued by pulsed laser show the same trend in the orientation-fatigue damage behavior as samples fatigued by pulsed RF-heating. It is clearly observed that surface grains, oriented [1 1 1] with respect to the surface, show significantly more damage than surface grains oriented [1 0 0].

Results arising from a third fatigue experiment, the ultrasound (US) swinger, are compared to the results of the mentioned experiments. The US swinger is an uniaxial mechanical fatigue test enabling to apply within several days a total number of cycles representative of the life of the CLIC structures, thanks to a high repetition rate of 24 kHz. For comparison, laser fatigue experiments have much lower repetition rates. The dependence of surface degradation on grain orientation of samples tested by the US swinger was monitored during the fatigue life. Results are presented and compared to the ones arising from the two other test methods.

1. Introduction

The need for high accelerating gradients imposes considerable constraints on the materials of the accelerating structures of the future compact linear collider (CLIC) for high energy physics under study at CERN. The design accelerating fields (100 MV/m) will be the highest ever achieved in accelerating devices [1]. The surfaces exposed to high pulsed Radio Frequency (RF) currents will be subjected to cyclic thermal stresses, arising from cyclic Joule heating. This may well cause surface roughening and degradation by fatigue after the 10^{11} cycles foreseen after 20 years of operation [1]. The negative influence of increased surface roughness on RF-Performance or electrical breakdown resistance must be considered as a limiting factor. As OFE copper is traditionally the standard material for RF applications, it is considered in the present study as the basic structural material for the accelerating structure.

In both High–Cycle–Fatigue (HCF) and Ultra–High–Cycle–Fatigue (UHCF), the limiting factor for fatigue life is crack initiation. In contrast, during Low-Cycle-Fatigue (LCF), crack propagation plays the major role [2]. Therefore special attention is payed in HCF and in UHCF to the evolution of the surface during cycling when cycle numbers exceed 10^4 [3, 4].

This study focuses on surface degradation as a function of the orientation of surface grains. As shown in [5], copper exhibits highly anisotropic elastic properties. A significant dependence of surface degradation on local orientation can be observed. Two different fatigue experiments were presented in [5] and the results are discussed. Cycle numbers are significantly extended by introducing an additional fatigue technique, the ultra sound (US) swinger. Results will be discussed and compared to [5].

2. Material and experimental procedure

Material

All samples are made of high conductivity–oxygen free–copper min 99.99 % (Cu–OFE REF. UNS C10100 Grade 1) already used in [5]. The state without any heat treatment is "half-hard" (H02 according to ASTM B601). Samples in the "annealed" state were exposed to 1000 °C for 2 h in vacuum. For properties see [5].

Fatigue Devices

In the laser–fatigue device an Excimer–laser irradiates repetitively a disc like specimen under vacuum. The experimental setup is described in detail in [6, 5]. A repetition frequency of 200 Hz, a pulse length of 40 ns and an energy density 0.3 J/cm^2 (which induces a calculated temperature rise of $\Delta T = 280$ K [6]) are chosen in order to ensure that between each cycle the irradiated area can cool completely down to room temperature due to heat diffusion through the specimen into the specimen holder. Therefore there is no steady temperature rise in the specimen. The heating while irradiation induces thermal expansion which is suppressed by surrounding material in lateral direction as well as in depth. This cyclic biaxial stress results in fatigue damage. The estimated thermally induced cyclic compressive strain load is between $\epsilon_{th} = 0$ and 7 x 10^{-3}.

The RF-fatigue device consists of a mushroom cavity which produces well defined temporarily changing magnetic fields. A detailed description of the device can be found in [7, 5]. Pulsed RF magnetic fields (11,4 GHz called X–Band) induce eddy currents with a repetition rate of 60 Hz. During one pulse (1,5 μs) these eddy currents induce heating up to a calculated temperature change of $\Delta T_{max} = 110$ K for the current test setup, which corresponds to a maximal thermal strain of $\epsilon_{th} = 1.8$ x 10^{-3}. As for the laser device the cyclic heating produces cyclic compressive biaxial stress which results finally in fatigue damage. The magnetic field intensity and the cooling system are chosen in the way to ensure the cooling between pulses to be sufficient to reach an average steady state of minimum temperature of ≈ 60 °C.

The US swinger (USS) device cycles mechanically an hour glass shaped sample which is mounted on one end to a piezo–electric resonator. The experimental setup is described in detail in [8]. The sample undergoes cyclic uniaxial tension and compression loads with a load ratio $R = -1$. The calculated resulting stress for the present experiment is 60 MPa hence strain $\epsilon = 6$ x 10^{-4}. The sample dissipates heat during high frequency cycling. Therefore the free sample end is dipped in a water bath, which is constantly recirculated. On the resonator site the leverage is cooled through forced air convection, provided by a high power fan. The resulting steady state sample temperature is below 50 °C.

For details about the determination of grain orientations see [5]

3. Experimental Results

Laser-fatigued samples

An annealed sample has been submitted to laser fatigue. Fig. 1 shows the transition area between a [1 0 0] twin and a [1 1 1] grain after 5 x 10^5 cycles (= test end). A strong difference in roughness can be observed across the twin boundary. The bright contrast of the [1 1 1] grain within the irradiated area arises from roughening effects, since the surface roughness increased during the laser irradiation treatment for this orientation. The dark appearance of the twin [1 0 0] is due to its almost completely smooth surface apart from the adaptation area in the close vicinity of the twin boundary, illustrated with dashed lines. No surface features developed therefore during the thermal cycling for this orientation.

RF-fatigued samples

Fig. 2 shows the surface of the RF-fatigued sample in the area of highest cyclic temperature load after 1 x 10^7 cycles (= test end). RF tests were performed on half–hard samples, showing significantly smaller grains than in the annealed state for the laser discs. Only the grains with the main orientations [1 0 0] and [1 1 1] are indicated. The dark regions [1 0 0] correspond to smoother surfaces whereas bright regions correspond to clearly developed surface features.

USS-fatigued samples

Fig. 3 shows the surface of a sample submitted to the USS-fatigue device after 2 x 10^{10} cycles (fracture = test end). Dark spots are artifacts of electro polishing before cycling. One can see a central [1 0 0] oriented grain (G1) without surface degradation. It is surrounded by several [1 1 1] oriented grains which show different levels of degradation. There can be distinguished 3 groups of [1 1 1] degradation: grains G2 and G5 show severe surface roughening; grains G3 and G4 show moderate roughening and grain G6 shows no roughening. The in–plane orientation of the [1 1 1] grains differ significantly. For G2 and G5 the first Euler angle (which determines the in plane rotation) is 350°. For G3 and G4 the angle is 212° and 222° respectively, whereas for G6 the angle is 278°. All angles have a statistical accuracy of ± 4°.

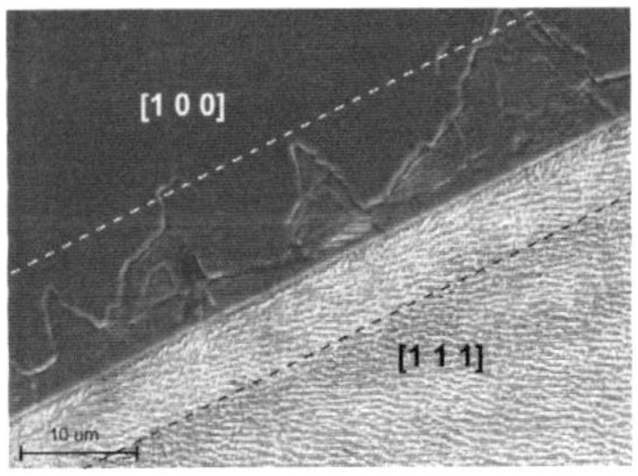

Figure 1. SEM picture of laser fatigued sample. No damage on [1 0 0] grain outside the transition region (dashed line) is observed. The [1 1 1] grain shows severe roughening.

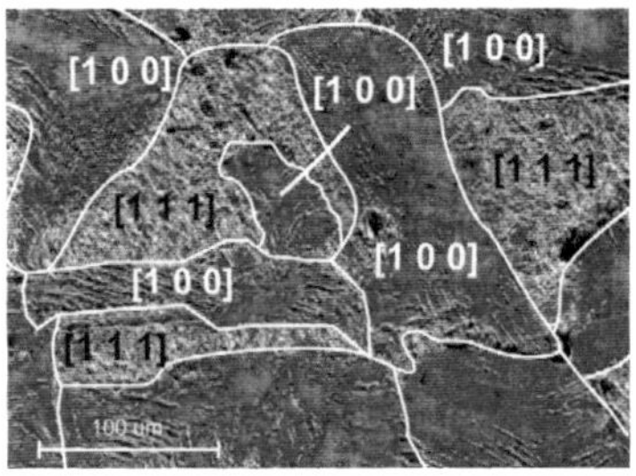

Figure 2. SEM picture of RF fatigued sample. Only grains with the main orientations are indicated. Bright and dark contrast correspond to rougher ([1 1 1]) and smoother ([1 0 0]) surface.

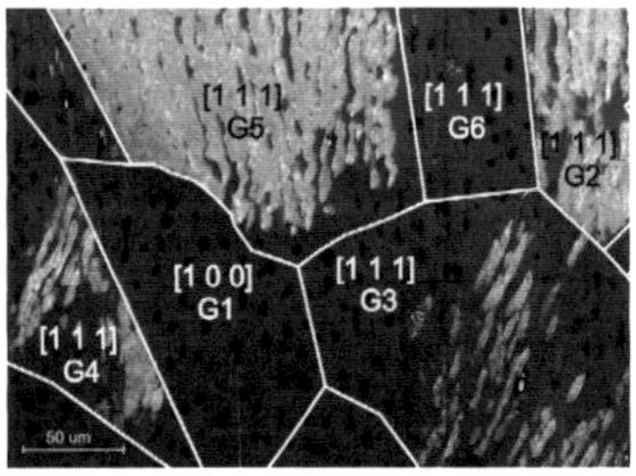

Figure 3. SEM picture of USS fatigued sample. Loading axis is horizontal. Dark spots are artifacts of electro polishing. Only grains with main orientations are indicated.

4. Discussion and Outlook

In [5] it is clearly shown that surface grains oriented in [1 0 0] submitted to thermal fatigue show much less surface degradation than grains in [1 1 1] (see also Fig. 1, 2). Although surface degradation features look quite similar (see Fig. 2, 3), in uniaxial mechanical loading this trend cannot be observed as clearly as in thermal fatigue. By calculating the Schmid factors (SF) for each of the 12 slip systems for different crystallographic orientations one can show that in biaxial case only out–of–plane orientations of surface grains change the SFs. In contrast, in uniaxial case the SF configuration is strongly depending on the in–plane orientations of surface grains. Therefore different surface degradation behaviour of grains differing only in the in–plane orientations can be expected.

While in thermal fatigue the degradation behaviour was clearly observed to be as reported in [5], in uniaxial testing also undamaged [1 1 1] grains were observed (Fig. 3). The group of grains G2 and G5 show severe surface damage whereas the group of grains G3 and G4 show moderate damage. The surface of grain G6 is completely smooth. Differences in the in–plane orientations are corrected, taking into account the three fold symmetry of a [1 1 1] direction. The corrected values show that only the grain G6 differs significantly (38°) whereas the the grains G2-G5 are oriented at 102° ± 10°.

The differences in extend of degradation of the [1 1 1] grains can therefore only be partly described by the in-plane orientation. A complete explanation of the dependency of degradation of the surface through thermal or mechanical cycling on grain orientation is still not established. To this purpose a more detailed Schmid factor analysis and active slip system orientation observation are foreseen.

5. Conclusion

The influence of the grain orientation on the fatigue behavior of surfaces is shown on the base of the main grain orientations [1 1 1] and [1 0 0]. In thermal fatigue both orientations showed very different developments during fatigue life. The [1 1 1] grains showed severe surface deformation whereas the [1 0 0] grains stayed almost unchanged smooth. Introducing a third fatigue device, the uniaxial USS-device, makes it more difficult to assign surface degradation directly to an out-of-plane orientation. The in-plane orientation of the grains becomes highly important because the Schmid factor configuration on the slip systems change significantly. In uniaxial loading it is shown that [1 1 1] oriented grains can show significantly lower surface degradation as compared to the biaxial loading arising from thermal fatigue experiments.

Acknowledgments

The author thanks Holger Neupert for the support at the laser experiment, the SLAC team for the results of the RF-experiment, Sergio Calatroni, Mauro Taborelli and Gonzalo Arnau Izquierdo for creative and fruitful discussions, Stefano Sgobba for the continous advice and the CLIC study in the frame of CERN for the financial support.

References

[1] Braun H 2008 Clic 2008 parameters CLIC Note 764 CERN
[2] Suresh S 1998 *Fatigue of Materials* 2nd ed (Cambridge, UK: Cambridge University Press)
[3] Mayer H 2006 *Int. J. Fatigue* **28** 1446–1455
[4] Stanzl-Tschegg S E, Mughrabi H and Schoenbauer B 2007 *Int. J. Fatigue* **29** 2050–2059
[5] Aicheler M, Sgobba S, Taborelli M, Calatroni S, Izquierdo G A and Neupert H 2010s *ubmitted to Int. J. Fatigue*
[6] Calatroni S, Neupert H and Taborelli M *Proceedings EPAC 2004, Lucerne, Switzerland*
[7] Nantista C, Tantawi S, Weisend J, Siemann R and Dolgashev V *Particle Accelerator Conference 2005, Knoxville, TN, USA*
[8] Heikkinen S 2008 *Study of High Power RF Induced Thermal Fatigue in the High Gradient Accelerating Structures* PhD dissertation Helsinki University

Influences of the manufacturing process chain design on the near surface condition and the resulting fatigue behaviour of quenched and tempered SAE 4140

M. Klein, D. Eifler
Institute of Materials Science and Engineering,
University of Kaiserslautern, D-67663 Kaiserslautern, Germany

klein@mv.uni-kl.de

Abstract. To analyse interactions between single steps of process chains, variations in material properties, especially the microstructure and the resulting mechanical properties, specimens with tension screw geometry were manufactured with five process chains. The different process chains as well as their parameters influence the near surface condition and consequently the fatigue behaviour in a characteristic manner. The cyclic deformation behaviour of these specimens can be benchmarked equivalently with conventional strain measurements as well as with high-precision temperature and electrical resistance measurements. The development of temperature-values provides substantial information on cyclic load dependent changes in the microstructure.

1. Introduction

The fatigue properties of metallic construction units are affected strongly by their near surface condition which is influenced by manufacturing processes and their individual parameters. Because of their complexity, the influence-cause-effect-relations of process chains are not understood satisfactorily until now [1-4].

The aim of the current investigations is the evaluation of the influence of the process chain design on the fatigue properties of metallic components and specimens. In particular, the microscopic change of the near surface area was regarded over the entire process chain from semi-finished parts to finished specimen.

2. Material and process chains

To estimate the influencing factors of process chain design and process parameters on the fatigue behaviour, specimens with tension screw geometry were manufactured (Fig.1a).

Five typical manufacturing process chains (PC), consisting of the processes heat treatment, turning and grinding, were applied. The studies are focused on the fatigue relevant “extension areas”. The investigated SAE 4140 was supplied in a quenched and tempered condition as round bars with a diameter of 25 mm. The microstructure is characterised by a tempered martensite microstructure with ferrite and fine dispersed Fe_3C carbides.

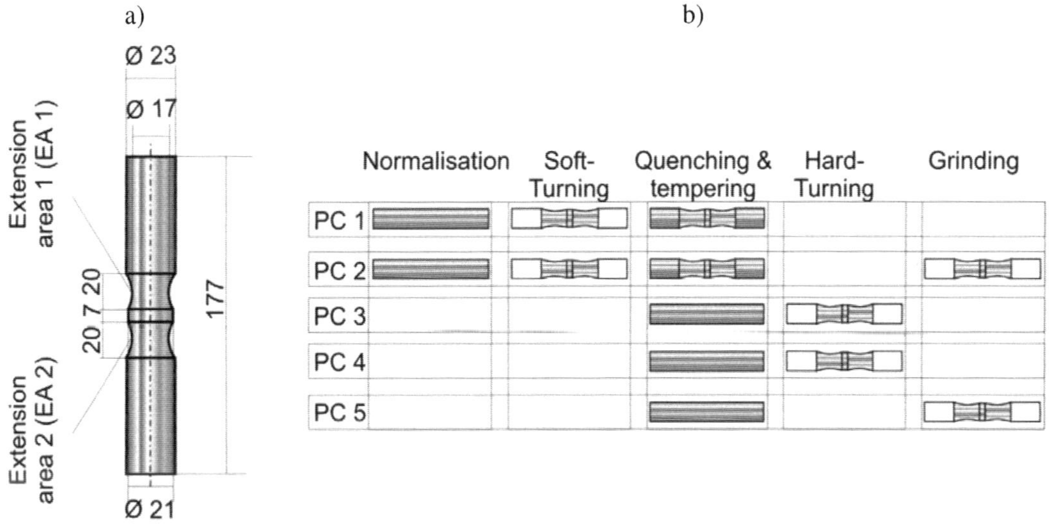

Figure 1. Specimen with tension screw geometry (a); Process chains PC 1-5 (b)

The overview matrix of the investigated process chains is given in Fig. 1b. Areas machined in the particular processes are marked in grey. After normalisation (T_a=860°C, t=120 min) the workpieces of PC 1 and 2 were soft-turned and in step three quenched and tempered (T_a=850°C, t=120 min; T=550°C, t=90 min), PC 1 ended with this heat-treatment-process. The workpieces of PC 2 were turned with a tolerance of 300 µm. PC 3-5 were started with quenching and tempering. The extension areas of PC 1, 3, 4 and 5 are machined with only one chipping process: soft turning / hard turning / grinding and consequently large chipping volume. In PC 2 the chipping volume was realised in two steps: soft turning and grinding.

3. Experimental setup

Light and scanning electron microscopy, surface roughness, micro hardness und residual stress measurements as well as fatigue tests were carried out. The experimental setup for the fatigue tests is shown detailed in [5]. In addition to stress-strain hysteresis measurements $\varepsilon_{a,p}$, deformation-induced changes in temperature ΔT [6-8] and changes in electrical resistance ΔR [9-12] were measured with high precision.

The mechanical stress-strain hysteresis measurements were carried out with an extensometer in the extension area two (EA 2). ΔT was detected with four thermocouples (T_1-T_4) fixed along the specimen. T_1 and T_2 were fixed in the middle of the extension areas EA 1 and 2 and T_3 and T_4 at the elastically loaded shafts. The local change in temperature for the extension areas 1 and 2 can be evaluated with the subtraction of the mean of T_3 and T_4 from T_1 and T_2 respectively. ΔR was measured integrally across both extension areas. The experiments yield cyclic deformation $\varepsilon_{a,p}$-N, cyclic temperature ΔT-N and cyclic electrical resistance ΔR-N curves representing the actual fatigue state.

The stress-controlled constant amplitude tests (CAT) were performed at the load ratio R=0 on a servohydraulic testing system at ambient temperature with a frequency of 5 Hz and triangular load-time functions.

4. Results

The surface roughness of PC 2-5 was determined in average of 4.5 µm. The values of PC 3-4 were slightly better than the values of PC 2 and PC 5. The as heat-treated surface of PC 1 with oxide layers was detrimental with respect to fatigue properties.

As a consequence of the grinding process, the near surface area of the specimens manufactured according to PC 2 is characterised by significant plastic deformations reaching to a depth of 3 µm.

Compressive residual stresses σ_{rs}=-330 MPa were measured (Fig. 3a). The grinding process in PC 5, inducing compressive residual stresses of -350 MPa, leads to changes in the microstructure to a depth of about 7 µm (Fig. 3b). The pronounced plastic deformations are caused by the larger chipping volume in PC 5. The microstructural changes due to the hard-turning process in PC 3 (Fig. 3c) can be divided into a primary (P) zone near the surface to a depth of 2 µm where the distance between the carbide lines is nearly zero and in the secondary (S) zone for further 3 µm where the carbide lines are strongly deformed. The feed rate (a_p) of 50 µm and the depth of cut (V) of 0.6 mm/rev. lead to a nearly residual stress free state. The double feed rate and half depth of cut applied in PC 4 result in the microstructure shown in Fig. 3d with tensile residual stresses of about +430 MPa. In comparison to PC 3, the primary zone is smaller and the secondary zone is significantly enlarged. The process type and the process parameters, respectively, are responsible for the different developments of the microstructural changes and the resulting amount and sign of the near surface residual stresses.

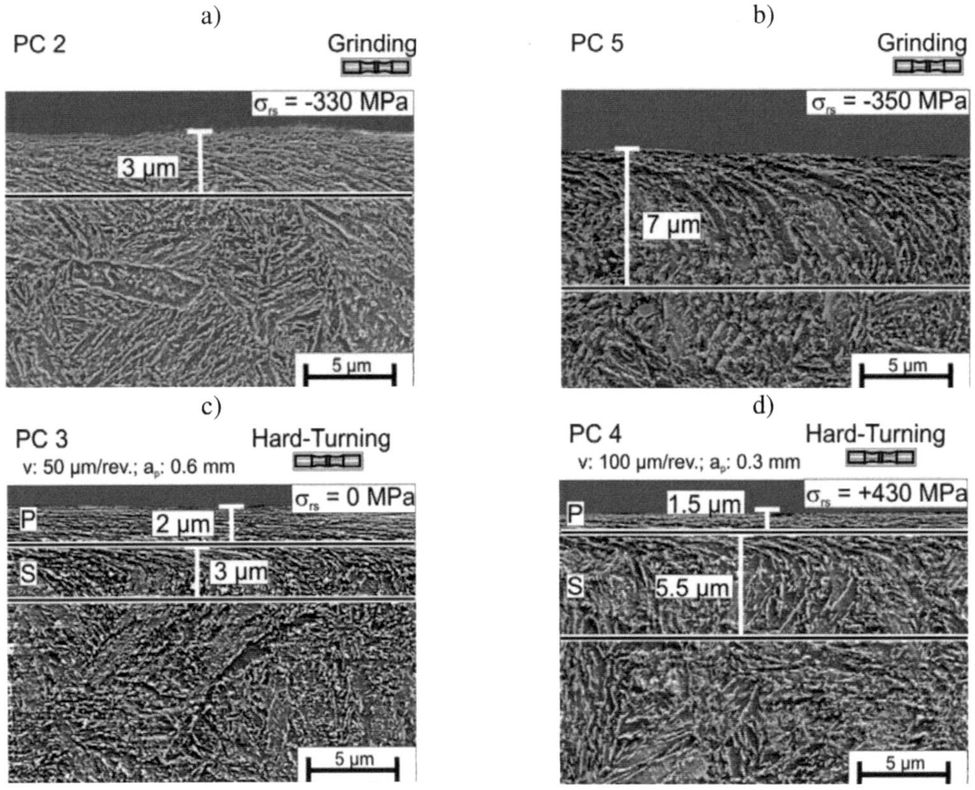

Figure 3. Changes of the near surface area in the extension areas due to the process chains PC 2 (a) and PC 5 (b) after grinding as well as PC 3 (c) and PC 4 (d) after hard-turning

Fig. 4a shows the development of $\varepsilon_{a,p}$ measured in the EA 2, ΔT measured individually in EA 1 and EA 2 and ΔR measured integrally across both extension areas of a specimen machined according to PC 4, for a CAT with σ_a=440 MPa. Fig. 4b gives an overview of ΔT-courses for specimens machined according to PC 3, PC 4 and PC 5 for σ_a=440 MPa measured in EA 2.

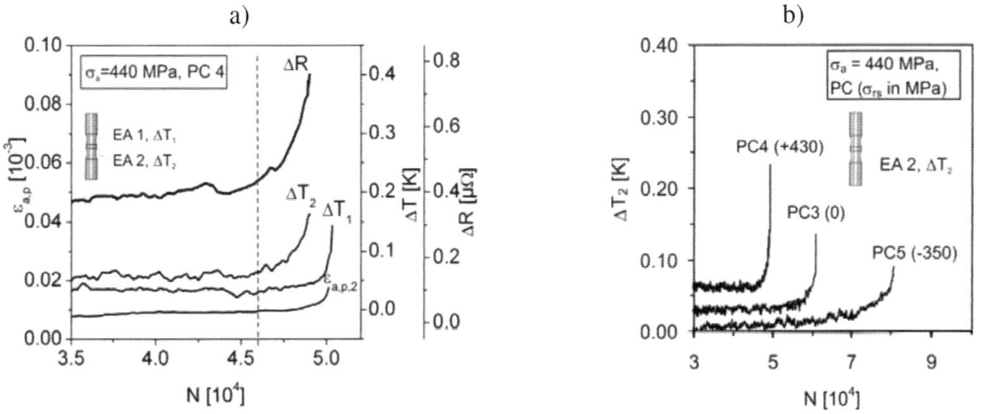

Figure 4. Development of plastic strain amplitude, change in temperature and change in electrical resistance in a constant amplitude test for a tension screw specimen of PC 4 (a); cyclic temperature curves and residual stresses for tension screw specimens of PC 3, PC 4 and PC 5 (b)

A significant increase of ΔT_2 and ΔR is measured at $4.6{\cdot}10^4$ cycles due to a proceeding fatigue damage in EA 2. The course of EA 1 (ΔT_1) increases at distinct higher number of cycles, right before failure because of thermal conduction. Obviously, ΔT- and ΔR-measurements indicate fatigue failure earlier than $\varepsilon_{a,p}$- measurements. This is caused by very small and localised plastic deformation volumes in high-strength steels leading to very small plastic strain amplitudes. The local application of ΔT and ΔR allows to identify precisely component areas which are critical for fatigue damage.

Specimens with higher (lower) ΔT-values reach lower (higher) number of cycles to failure. With increasing residual stresses from compression to tension, ΔT increases and the lifetime consequently decreases.

5. Conclusion

It was shown that machining processes and their parameters influence the near surface condition and consequently the fatigue behaviour in a characteristic manner. Significant changes in distance and arrangement of the carbide lines were observed as a result of turning as well as grinding processes. The process type and its parameters, respectively, lead to different developments of microstructure and the resulting amount and sign of residual stresses. The plastic strain amplitude, temperature and electrical resistance measurements can be equivalently used for fatigue assessment of specimens with tension screw geometry. The temperature and electrical resistance measurements seem to be more beneficial for the evaluation of fatigue life.

6. References

[1] Klein D, Thoben K D, Nowag L 2006 *Materialwissenschaft und Werkstofftechnik* **37** 19
[2] Tönshoff H K, Denkena B, Hessel D, Brandes A 2004 *CIRP J. Manuf. Syst.* 379
[3] Klocke F, Fallböhmer M, Kopner A, Trommer G 2000 *Rob. Comp.-Integr. Manuf.* 411
[4] Warnecke G, Eifler D, Aurich J C, Mauren F, Klein M 2005 *Res. and Develop. in Germany* 85
[5] Klein M, Walther F, Eifler D 2007 Proceedings Werkstoffprüfung 243
[6] Lukáš P, Klesnil M 1973 *Mater. Sci. Eng. A* 345
[7] Stärk K F 1982 *Z. Werkstofftech.* 333
[8] Harig H, Frank E 1981 Wire 227
[9] Polák J 1969 *Czech. J. Phys.* 315
[10] Charrier J R, Roux R 1991 *Nondestr. Test. Eval.* 113
[11] Piotrowski A, Eifler D 1995 *Materialwissenschaft und Werkstofftechnik* 121
[12] Starke P, Eifler D 2009 *Mat. Test.* ***51*** *261*

Surface crack nucleation and propagation in electrodeposited nanocrystalline Ni-P alloy during high cycle fatigue

Shigeaki Kobayashi[1], **Akiyuki Kamata**[1], **Tadao Watanabe**[2]

[1] Department of Mechanical Engineering, Faculty of Engineering, Ashikaga Institute of Technology, 268-1 Omae, Ashikaga, Tochigi 326-8558, Japan.
[2] Visiting Professor, Key Laboratory for Anisotropy and Texture of Materials, Northeastern University, Shenyang, 110004, China. Formerly, Tohoku University, Sendai, Japan.

skoba@ashitech.ac.jp

Abstract. The morphology of specimen surface after fatigue fracture was evaluated in connection with grain orientation distribution and grain boundary microstructure to reveal a mechanism of fatigue fracture in nanocrystalline materials. The electrodeposited and sharply {001} textured Ni -2.0 mass% P alloy with the average grain size of ca. 45 nm and high fractions of low-angle and Σ3 boundaries showed 2 times higher fatigue limit than electrodeposited microcrystalline Ni polycrystal. The surface features of fatigued specimen were classified into two different types of morphologies characterized as brittle fracture at the central area and as ductile fracture at the surrounding area.

1. Introduction

The fatigue properties of nanocrystalline materials have received a great interest of researchers [1-6], because fatigue property is particularly important for micro electro mechanical systems (MEMS) materials. However, unfortunately, there has been only a little experimental work on fatigue property and behavior of nanocrystalline materials up to now. The fatigue property and behavior of nanocrystalline materials has been discussed mostly by focusing only the effect of grain refinement. The study of the roles of other microstructural factors such as texture and grain boundary microstructure has been very scarce, although grain boundaries very likely play some key roles in fatigue crack nucleation and propagation depending on boundary character. Recently, we have found that the intergranular fatigue cracks nucleated predominantly at random boundaries in polycrystalline Al, while the fatigue cracks never occur at low-angle boundaries [7]. The low-Σ boundaries showed the higher resistance to fatigue crack nucleation than random boundaries [7]. Thus the effect of grain boundary microstructure on the fatigue fracture in the nanocrystalline materials may become more significant than that in microcrystalline materials, because the volume fraction of grain boundaries becomes remarkably high in nanocrystalline materials.

This work was performed to investigate the fatigue fracture process in electrodeposited nanocrystalline Ni-P alloy in connection with grain boundary microstructure. The formation of surface damage closely related to fatigue crack nucleation under cyclic stress condition were observed and discussed in comparison with the case of static tensile stress condition on the basis of the results of FE-SEM/EBSD analyses for grain boundary microstructure.

2. Experimental procedure

A nanocrystalline Ni -2.0 mass% P alloy was produced by electrodeposition using an electrolytic bath of 150 g/l nickel sulphate, 45 g/l nickel chloride, 80 g/l phosphoric acid and 0.5 g/l phosphorous acid with pH 1.5 at 338 K and a current density of 2.0 mA/mm^2 for 10.8 ks. Flat dog-bone shaped tensile and fatigue test specimens whose gage zone dimension was 5.0 mm long, 2.0 mm wide and 0.2 mm thick were cut out from the electrodeposited sheets mechanically stripped from the Ti substrate. The specimen showed a yield stress of 920 MPa, the tensile strength of 1550 MPa and the plastic elongation of 7.5 %.

X-ray diffractometer with monochromatic Cu-Kα radiation was utilized to analyze the initial microstructure of the specimen. The automated FE-SEM/EBSD/OIM was employed to quantitatively analyze of grain boundary microstructures on the surface of pre- and post-deformed specimens.

Fatigue tests were carried out by using a servo-hydraulic machine in air at room temperature. The sinusoidal load was applied at the stress amplitude between 290 MPa and 550 MPa at the stress ratio of 0.1 and the frequency of 10Hz.

3. Results and Discussion

3.1. Initial microstructure of electrodeposited nanocrystalline Ni-P alloy

Figures 1 (a)-(b) show an OIM micrograph with the inverse pole figure, the grain boundary characterization (GBC) micrograph with the average grain size and the grain boundary character distribution (GBCD) for the initial specimen, respectively. The OIM micrograph shows individual grain orientations in distinct colors corresponding to those indicated in the stereo-triangle of the inverse pole figure. In the GBC micrograph, low-angle (L, θ = 2°-15°), low-Σ and random (R) boundaries are shown by distinct colors shown in color bars. Although the Ni-P binary phase diagram [8] shows that the solubility limit of P into Ni matrix is 0.17 mass%, the secondary Ni_3P phase did not precipitate in the specimen. According to Boylan et al., the nanocrystalline Ni-P alloy is electrodeposited as a supersaturated solid solution [9].

From the OIM micrograph shown in figure 1 (a), it was found that the surface orientation of the initial specimen strongly localizes around {001}. The other grains tend to have surface orientations around {112}, {123}. The {001} grains had the grain size ranging from 10 nm to 180 nm and the average grain size of 56 nm. The grain size distribution of the {001} grains were more widely spreading than that of the grains with other surface orientations. It was found that the initial specimen had a considerably high fraction of low-angle and Σ3 boundaries, as shown in figure 1 (b). The fractions of low-angle and Σ3 boundaries were 30 % and 28 %, almost 13 and 15 times higher than that theoretically predicted for a random polycrystal shown in the brackets, respectively. Judging from OIM micrograph and GBC micrograph, Σ3 boundaries tend to lie between the grains with {001} surface and the grains with {112} or {123} surface. These Σ3 boundaries may be of nonequilibrium state, because these Σ3 boundaries showed sluggish and irregular shape.

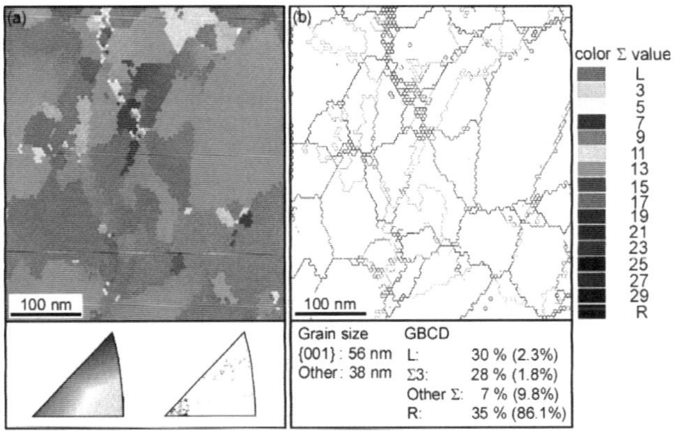

Figure 1. (a) OIM and (b) GBC micrographs of electrodeposited nanocrystalline Ni -2 mass% P alloy specimen.

3.2. High-cycle fatigue property of electrodeposited nanocrystalline Ni-P alloy

Figure 2 shows the relationship between the stress amplitudes and the logarithm of the number of cycles to fracture for the nanocrystalline Ni -2.0 mass% P alloy specimens. The data obtained from electrodeposited nanocrystalline pure Ni with an average grain size in the range 20-40 nm [2] and electrodeposited microcrystalline Ni with grain size around 1-7 μm [10] are shown together in this figure. The fatigue limit for the nanocrystalline Ni-P alloy in the present work was about 360 MPa. Accordingly, the fatigue limit is 2 times higher than that for the microcrystalline pure Ni, but slightly lower than that reported for the nanocrystalline pure Ni. Thus, it is evident that the fatigue limit of polycrystalline materials is improved by grain refinement to nanocrystalline level of grain size. However, the effect of nanocrystallization for improvement in the fatigue limit was lower than that for improvement in static tensile strength and hardness. Change in grain boundary microstructure and surface morphology under cyclic stress condition may affect the moderate fatigue limit of nanocrystalline Ni-P alloy.

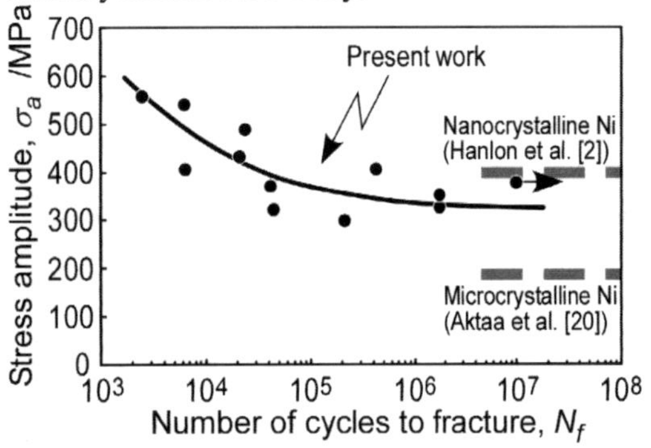

Figure 2. S-N curve of electrodeposited nanocrystalline Ni -2.0 mass% P alloy subjected to cyclic deformation of 290 MPa – 550 MPa in stress amplitudes at frequency of 10 Hz and at room temperature.

3.3. Surface morphology and microstructure after static tensile and dynamic fatigue tests

Figure 3 (a) is a SEM micrograph of the specimen surface after tensile fracture test. Figure 3 (b) and (c) are the higher magnification images of the areas (i) and (ii) in figure 3 (a), respectively. The stress direction is the horizontal in the micrographs. The specimen surface fractured by tensile stress was characterized as ductile fracture mode resulting in the formation of steps pattern which probably associated with formation of shear bands in the whole area of specimen. The spacing of neighboring shear steps was 1.0 μm – 5.0 μm. The surface cracks were nucleated and propagated along these shear steps. The fracture surface was also characterized as dimple pattern typical for ductile tensile fracture.

Figure 4 (a) is a SEM micrograph of the surface of fatigued specimen. Figure 4 (b) and (c) are the higher magnification images of areas (i) and (ii) in figure 4 (a), respectively. The stress direction is the horizontal in the micrographs. The morphology of fatigued surface can be classified into two distinct types. The area (i) around the crack source showed a brittle fracture mode. The specimen surface was rather flat and waving micro-cracks were observed in area (i). The nucleation of these micro-cracks was found to be the source of fatigue fracture. The fracture surface of the area (i) was planer and the fracture propagates toward the peripheral area of the specimen, and the adjacent area shows striation pattern of characteristic of fatigue fracture. On the other hand, the surrounding area (ii) showed a ductile tensile fracture mode. Steps which probably resulted from the formation of shear bands were observed. The spacing of neighboring shear steps was 1.0 μm – 5.0 μm. Several experimental studies of the formation of shear bands during cyclic deformation have been recently reported for nanocrystalline metals [11]. The surface cracks nucleated and propagated along these shear steps. Therefore, the formation of shear steps is very important in the fracture resulted from fatigue crack propagation in nanocrystalline Ni-P alloy. The OIM analysis of the surface of fatigued specimen suggested that the area among the shear steps was formed by group of grains with similar surface orientation and high fraction of low-angle boundaries. It is thought that these shear steps were formed by sliding at connected random boundaries *i.e.* cooperative grain boundary sliding by shear of groups of grains. The morphology of fracture surface of the area (ii) was characterized as dimple pattern typical for tensile ductile fracture. Moreover, the results of OIM analyses suggest that the grain growth which is associated with random boundaries, probably occurred up to the grain size about 300 nm during high cycle fatigue tests, though the grain boundaries surrounding fatigue deformed grains were not clear. Until recently, studies of the stress-induced grain growth under cyclic deformation in nanocrystalline materials is very scarce, although a moderate increase in grain size (the order of 30%) after cycle deformation has been reported for nanocrystalline copper [1]. The further studies are required for full understanding of mechanisms of room temperature grain growth under cyclic stress in nanocrystalline materials.

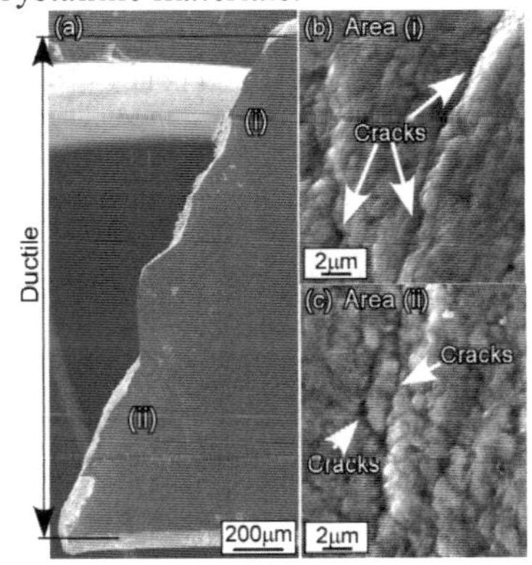

Figure 3. SEM micrographs of specimen surface after tensile test.

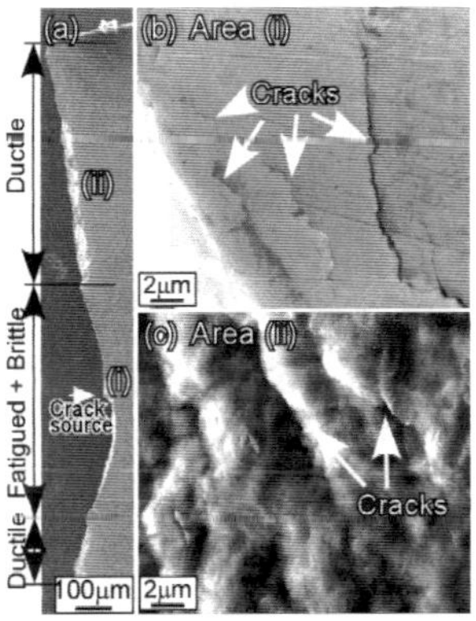

Figure 4. SEM micrographs of specimen surface after high-cycle fatigue test.

4. Conclusions

The electrodeposited and {001} textured nanocrystalline specimens with high fractions of low-angle (30 %) and Σ3 (twin) (28 %) boundaries showed the fatigue limit of 360 MPa. The surface features of fatigued specimen were classified into two different types of morphologies. The central area was characterized as brittle fracture and the adjacent area as ductile fracture mode resulting in the formation of steps pattern due to shear banding, On the other hand, the specimen surface after static tensile test was only characterized by ductile fracture. It was also found that stress-induced grain growth occurred. The fatigue fracture in nanocrystalline metals was found to be strongly affected by grain boundary microstructure, not simply by the grain size.

Acknowledgement

The authors would like to express their hearty thanks to Professor Tsurekawa of Kumamoto University for the provision of the EBSD/OIM analyses.

References

[1] Witney A B, Sanders P G, Weertman J R, Eastman J A 1995 *Scripta Metall. Mater.* **33** 2025
[2] Hanlon T, Kwon Y-N, Suresh S 2003 *Scripta Mater.* **49** 675
[3] Kumar K S, Swygenhoven H Van, Suresh S 2003 *Acta Mater.* **51** 5743
[4] Dao M, Lu L, Asaro R J, Hosson J T M De, Ma E 2007 *Acta Mater.* **55** 4041
[5] Yang Y, Imasogie B, Fan G J, Liaw P K, Soboyejo W O 2008 *Metall. Mater. Trans.* **39A** 1145
[6] Cheng S, Xie J, Stoica A D, Wang X-L, Horton J A, Brown D W, Choo H, Liaw P K 2009 *Acta Mater.* **57** 1272
[7] Kobayashi S, Inomata T, Kobayashi H, Tsurekawa S, Watanabe T 2008 *J. Mater. Sci.* **43** 3792
[8] Massalski T B, Murray J L 1986 *Binary Alloy Phase Diagrams*, Vol 2, ed L H Bannett amd H Baker (Materials Park, OH: ASM International) p 1739
[9] Boylan K, Ostrander D, Erb U, Palumbo G, Aust K T 1991 *Scripta Metall. Mater.* **25** 2711
[10] Aktaa J, Reszat J Th, Walter M, Bade K, Hemker K J 2005 *Scripta Mater.* **52** 1217
[11] Xie J, Wu X, Hong Y 2007 *Scripta Mater.* **57** 5

Extrusion and intrusion evolution in cyclically strained cast superalloy Inconel 738LC using confocal laser scanning microscope and AFM

K Obrtlík[1], M Juliš[2], J Man[1], T Podrábský[2] and J Polák[1]

[1] Institute of Physics of Materials, Academy of Sciences of the Czech Republic, Žižkova 22, 616 62 Brno, Czech Republic
[2] Institute of Material Science and Engineering, Faculty of Mechanical Engineering, BUT, Technická 2, 616 69 Brno, Czech Republic

obrtlik@ipm.cz

Abstract. Surface relief within persistent slip markings (PSMs) was studied using confocal scanning laser microscopy (CSLM) and atomic force microscopy (AFM) in cast nickel base superalloy Inconel 738LC cyclically strained in strain control at room temperature. Extrusion and intrusion topography and kinetics are documented. The dependence of extrusion height on the number of cycles is obtained. Two regimes of extrusion growth are identified. Average intrusion growth rate is assessed.

1. Introduction

The extrusion and intrusion formation within persistent slip markings (PSMs) of crystalline materials represents the first surface indication of fatigue damage in a material specimen or in a component [1]. Since fatigue cracks nucleate within the PSMs, experimental data on the topography of extrusions and intrusions and the kinetics of their evolution can reveal the details of fatigue damage mechanisms.

Surface relief within PSMs and its connection with crack initiation was studied extensively in a number of materials (see review [2]). Though fatigue cracks originate from intrusions, much more direct quantitative information has been obtained on extrusions. Quantitative data on intrusions have been acquired particularly using sectioning and plastic replica techniques. The intrusion growth rate is reported higher than that of extrusions [3].

Quantitative data on the surface relief within PSMs in nickel-base superalloys, which are designed for critical parts of gas turbines, are sporadic [4,5]. An extrusion height is reported [4,5] and limited data on extrusion growth rate can be inferred from [4]. Therefore, the present paper deals with the quantitative observation of surface relief evolution with focus to extrusions and intrusions kinetics in Inconel 738LC (In 738LC) cyclically strained under total strain control at room temperature using two advanced techniques, namely confocal scanning laser microscopy (CSLM) and atomic force microscopy (AFM).

2. Experimental

Inconel 738LC polycrystals were supplied by PSB Velká Bíteš, a.s. as conventionally cast rods in fully heat treated condition, i.e. after 1120 °C / 4 h / air cooling (AC) + 845 °C / 24 h / AC. Chemical composition and the details of microstructure are shown elsewhere [6]. The average grain size, found using the linear intercept method, was 3.6 mm.

Cylindrical button-end specimens had gauge length and diameter of 15 mm and 6 mm, respectively. They were fatigued in a symmetrical push-pull cycle in strain control at room temperature. The strain was measured and controlled using a sensitive extensometer with a 12 mm base. The strain rate of $2x10^{-3}$ s^{-1} and total strain amplitude of $\varepsilon_a = 0.55$ % were kept constant.

The surface relief was studied in a CSLM (Olympus LEXT OLS3000) and in an AFM (Acurex IIL, Topometrix). The maximum lateral resolution of the CSLM is 120 nm and the resolution in the direction of the beam is 5 nm (for the detailed description of the technique, see [6,7]). The CSLM software can provide the surface relief profiles. The AFM in contact imaging mode in the air was used to obtain constant-force topographic images. Further details has been reported elsewhere [3,6].

3. Results

To study the surface relief evolution the CSLM inspection of the specimen surface was performed at the following numbers of cycles: N = 30, 60, 100, 150, 200, 300. N = 300 corresponds approximately to 50 % of the fatigue life N_f. The AFM was used to observe both the specimen surface and the plastic replica at N = 300 cycles.

An example of AFM micrograph of the metallic surface is shown in figure 1. Three parallel PSMs consisting of extrusions and intrusions can be seen. The extrusion height varies considerably along a

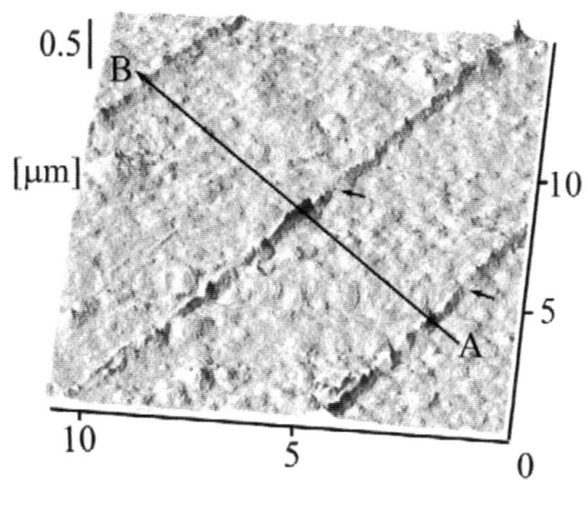

Figure 1. Three-dimensional AFM micrograph of surface relief. N = 300, $\varepsilon_a = 0.55$ %.

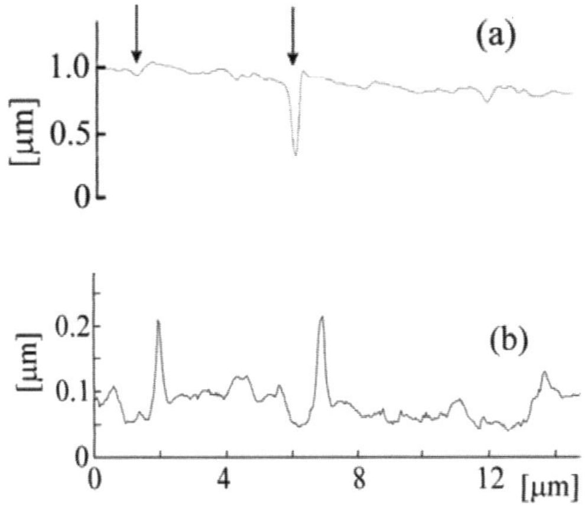

Figure 2. Surface relief profiles in the section perpendicular to the specimen surface (denoted AB in Figure 1). (a) CSLM, (b) AFM.

PSM. There are even PSM segments where extrusions were not distinguished. These locations are indicated in with small black arrows in figure 1. Surface profiles in the section perpendicular to the specimen surface and marked as AB in figure 1 are shown in figure 2. In the direction from the left to the right moderate and deep intrusions marked with the arrows and followed by extrusions are shown in figure 2a (CSLM technique). The typical surface relief obtained in the section AB with the AFM is shown in figure 2b. Shallow intrusions and clear extrusions can be seen in individual PSMs.

Figure 3 shows the growth kinetics of a hilly extrusion. The specimen was removed from the testing machine at preselected number of cycles and the extrusion height was measured using the CSLM. The extrusion height vs. number of cycles can be approximated with the linear dependence and the extrusion growth rate of 0.29 nm/cycle is obtained. The results presented in figure 3 confirm

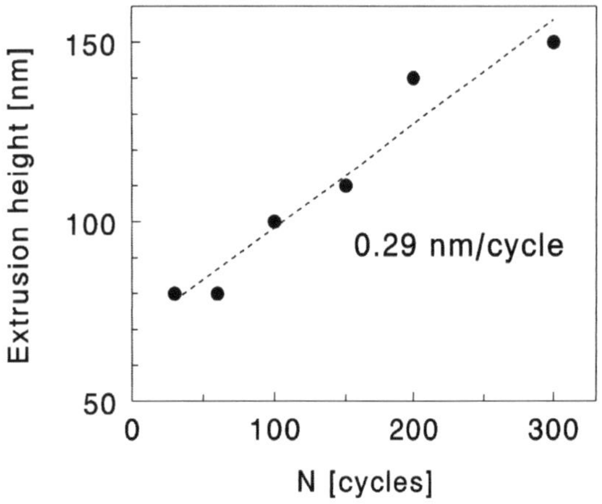

Figure 3. Extrusion height vs. number of cycles in cycling with constant total strain amplitude $\varepsilon_a = 0.55$ %.

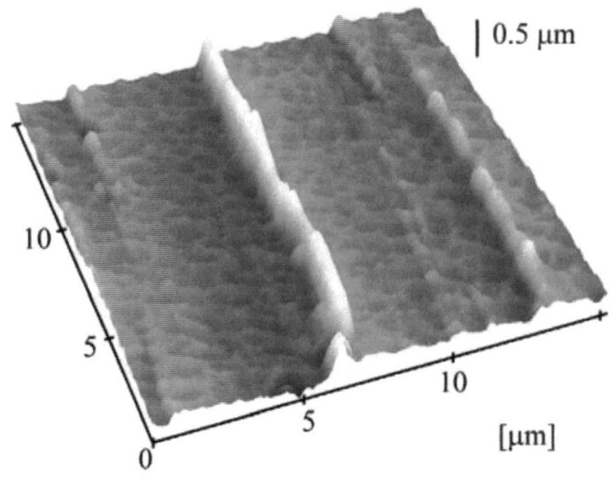

Figure 4. Surface relief obtained by AFM using plastic replica. N = 300 cycles (0.5 N_f).

quantitatively that slip bands in the In 738 LC superalloy are persistent.

Three-dimensional AFM micrograph of surface relief using plastic replica is shown in figure 4 in non-inverted format. Several PSMs with intrusions of various depths are apparent. The deepest intrusion can be found in the PSM located in the center of the image. Its maximum depth is 0.6 μm. The intrusion morphology and occurrence are in agreement with those found using the CSLM technique (compare figure 5).

Figure 5 shows inverted CSLM micrographs (i.e. intrusions appear as thin projections protruding from PSMs) in three-dimensional representation taken in the identical part of a metallic specimen at different numbers of cycles. At the number of cycles N = 23 % N_f, intrusions are not present continuously within the PSMs but there are locations where intrusions were not distinguished (see figure 5a). The intrusion morphology varies along a PSM. Thin almost ribbon-like or completely

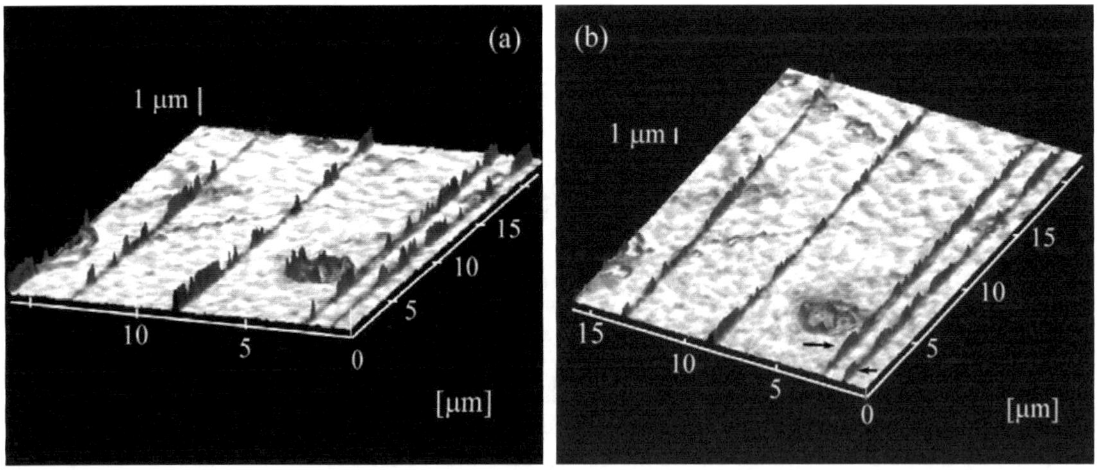

Figure 5. CSLM micrograph of surface relief in metallic specimen showing intrusion evolution (inverted image). (a) N = 140 cycles, (b) N = 300 cycles (0.5 N_f).

isolated peak-like intrusions can be observed. Figure 5 shows that the amount of intrusions increases with increasing number of cycles – see the location marked with black arrows in figure 5b. The average intrusion depth was measured in selected sections. It increases from 0.30 μm at N = 140 cycles to 0.61 μm at N = 300 cycles. These data allow assessing intrusion growth rate to 1.9 nm/cycle.

4. Discussion

Detailed quantitative study of the PSM topography and its evolution using the high resolution CSLM and AFM techniques revealed characteristic features of the surface relief of nickel-based In 738LC superalloy. The topography of PSMs comprises extrusions and intrusions analogous to those in simple FCC and BCC metals and alloys [1]. The dependence of extrusion height vs. number of cycles can be divided into two regimes. In the incipient regime up to about 5 % N_f the assessed extrusion growth rate is approximately one order of magnitude higher than the constant extrusion growth rate in the interval between 5 to 50 % N_f. The present data are in good agreement with the results obtained previously [3,8]. In the second regime, the extrusion growth rate was reported 0.2 to 0.4 nm/cycle for polycrystalline copper [8] and 0.35 nm/cycle for 316L steel [3]. Intrusion growth rate is reported 1.52 nm/cycle [3], which is higher than that of extrusions in agreement with the present results. However, the intrusion growth rate can be influenced by probable crack initiation at the tip of an intrusion.

Present results show that both the extrusion height and the intrusion depth vary significantly along PSMs after 140 or 300 cycles elapsed – see figures 1, 4 and 5. Therefore, local activity of a PSB leading to the surface relief formation within the corresponding PSM changes during cycling.

5. Conclusions

The detailed study of the evolution and topography of extrusions and intrusions within PSMs of cyclically strained Inconel 738LC using CSLM and AFM leads to the following conclusions:

(i) Slip markings consisting of extrusions and intrusions that appear early in the fatigue life are persistent.

(ii) Two regimes of extrusion growth are identified. Short initial regime is followed by constant growth rate period.

(iii) PSB activity leading to extrusion and intrusion formation varies locally during cycling.

(iv) CSLM proves to be a suitable technique for the direct observation of intrusion topography and crack initiation.

Acknowledgments

The financial support of the grants No. 106/07/1507 of the Grant Agency of the Czech Republic, Nos. 1QS200410502 and AV0Z 20410507 of the Academy of Sciences of the Czech Republic and No. 2E13800119 of the FME BUT is gratefully acknowledged.

References

[1] Polák J 2003 *Comprehensive Structural Integrity* vol 4, Ed I Milne and RO Ritchie et al, (Amsterdam: Elsevier) pp 1-39

[2] Man J, Obrtlík K and Polák J 2009 *Phil. Mag.* **89** 1295-1336

[3] Man J, Klapetek P, Man O, Weidner A, Obrtlík K and Polák J 2009 *Phil. Mag.* **89** 1337-1372

[4] Risbet M, Feaugas X, Guillemer-Neel C and Clavel M 2003 *Scripta Mater.* **49** 533-538

[5] Obrtlík K, Petrenec M, Man J, Polák J and Hrbáček K 2009 *J. Mater. Sci.* **44** 3305-3314

[6] Juliš M, Man J, Obrtlík K, Podrábský T and Polák J 2008 *Proc. 17th European Conf. on Fracture – ECF 17 (Brno, CZ, 2-5 Sept 2008)* ed J Pokluda, P Lukáš, P Šandera and I Dlouhý (Brno: ESIS Czech Chapter) CD pp 2314-2319

[7] *Olympus - CSLM LEXT OLS3000*. User`s manual on http://www.olympus.cz

[8] Mughrabi H, Bayerlein M and Wang R 1991 *Proc. 9th Int. Conf. on Strength of Metals and Alloys – ICSMA 9* ed D G Brandon, R Chaim and A. Rosen (London: Freund Publishing) pp 879-896

Fatigue properties of an 1421 aluminum alloy processed by ECAE

A Mogucheva, R Kaibyshev

Belgorod State University, Pobeda 85, Belgorod, 308015, Russia

mogucheva@bsu.edu.ru

Abstract. Fatigue properties and fatigue crack growth rate were examined in an Al-Mg-Li-Sc-Zr allow subjected to equal channel angular extrusion (ECAE) with rectangular shape of channels up to a total strain of ~4 at a temperature of 325°C followed by solution treatment with subsequent oil quenching with aging. After this processing the fraction recrystallized was ~80pct; the deformed microstructure remains essentially unchanged under solution treatment due to high density of Al_3Sc coherent dispersoids playing a role of effective pinning agents. It was shown that the fatigue limit of this material attained a value of ~185 MPa. Thermomechanical processing provided a decrease in fatigue crack propagation growth rate and an increase in the stress intensity factor, K_{1c}, in comparison with extruded bar. However, characteristics of crack propagation resistance did not attain values suitable for application of this alloy for critical aircraft components.

1. Introduction

Al-Li-Mg alloys are considered as advanced materials for airplanes due to excellent combination of low density and sufficient strength [1, 2]. However, crack propagation resistance of these alloys is not enough for commercial application for critical components of aircrafts. Nowadays, extensive work with Al-Li-Mg alloys is carried out to increase their values of the stress intensity factor, K_{1c}, and fatigue crack growth rate up to those obtained for the AA2524 aluminum alloy. Two approaches are used to improve crack propagation resistance of the Al-Li-Mg alloys. First approach is associated with optimization of chemical and phase composition of these alloys. Second path is in the development of thermomechanical processing (TMP) route which provides the manufacturing of semi-finished products with recrystallized structure. Authors [1] assumed that the formation of recrystallized grains can improve crack propagation resistance of Al-Li-Mg alloys in contrast to aluminum alloys belonging to the other systems. However, conventional TMP consisting of cold or warm rolling and final recrystallisation annealing could not be applied to the Al-Li-Mg alloys because of poor workability of these alloys at low temperatures that led to a premature fracture under working conditions. At the same time, recrystallization structure can be easily produced in Al-Li-Mg alloys by equal-channel angular extrusion (ECAE) at a temperature of ~300°C [3-5].The aim of the present study is to examine fatigue resistance of an Al-Li-Mg alloy with recrystallized structure. This alloy was subjected to standard heat treatment consisting of solution treatment followed by oil quenching and subsequent aging after ECAE.

2. Experimental

The experiments were conducted using the 1421 Al with a chemical composition of Al-5.1%Mg-2.1%Li-0.17%Sc-0.08%Zr (in weight pct). Details of manufacturing process and procedure of ECAE and techniques used for structural characterization as well were reported in previous works [3-5] in detail. The ECAE with rectangular shape of channels was carried out using route B_{cz} [5] at a temperature of 325°C up to a total strain of ~4. Thin foils were examined by using a Jeol-2010 TEM with a double-tilt stage and equipped with energy dispersive spectrum analysis (EDS) produced by Oxford Instruments, Ltd at an accelerating potential of 200 kV. The specimen surface was observed by scanning electron microscopy (SEM) Quanta 200 3D. Fatigue life characterization in terms of normal stress (S-N curve characterization) and measurement of fatigue crack growth rates were performed using a 100 kN servohydraulic Instron8801 testing machine under load control condition. Completely reversible push-pull (stress ratio, R=-1) sinusoidal load cycles at a frequency of 25 Hz were used for all the tests. Compact tension (CT) specimens 20 mm thick were cut in accordance with ASTM Standard E399-06. The specimens were loaded using a sinusoidal cycle with a frequency of 5 Hz and an asymmetry coefficient R = 0.1. The fatigue crack growth rate as a function of the stress-intensity factor range, ΔK, was obtained in accordance with ASTM standard E647-08.

3. Results and discussion

ECAE followed by the heat treatment results in the formation of recrystallized structure with volume fraction of recrystallized grains of ~80pct. (Fig. 1a); average size of recrystallized grains is ~2 μm, portion of high-angle boundaries (HAGB) is ~79pct., average misorientation is 30.7°. Average density of lattice dislocation within recrystalized grains is $\rho \sim 5\times10^{14}$ m^{-2}. Particles of S_1-phase with an average size of 0.4 μm are observed at grain boundaries and even within grain interiors (Fig.1b). Their volume fraction attends ~16 pct. There is a weak evidence for the formation of the δ' – phase dispersoids (Fig.1b); their volume fraction is negligible.

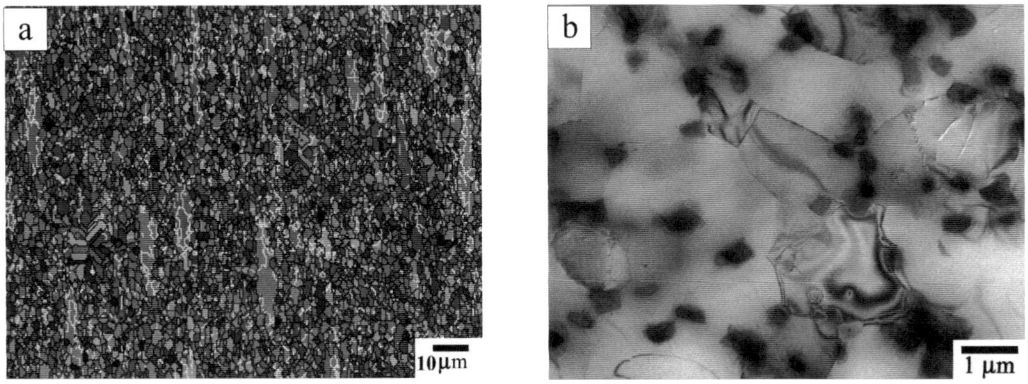

Figure 1. Typical microstructures of the 1421Al subjected to ECAE with $\varepsilon \sim 4$ followed by T6 heat treatment.

The S-N plot characterizing the fatigue life under testing with constant load amplitude is shown in figure 2. The S-N curve does not show any deflection points and apparent knee as well. The fatigue behaviors of the 1421 alloy in low cycle fatigue (LCF) and high cycle fatigue (HCF) range are almost the same. Linear dependence of S vs N remains virtually unchanged with transition from region of LCF to region of HCF. Therefore, the fatigue limit, σ_f, could not be defined. At 10^7 cycles, the fatigue strength was determined as 185 MPa. This value is relatively high for aluminum alloys.

In order to explain the usual fatigue behavior of the 1421 Al with recrystallized structure the fatigue crack growth rate (da/dN) on the stress intensity range, ΔK, was obtained (figure 3). Only two stages can be defined in figure 3. At stage I ($\Delta K \leq 6$ MPa×√m), any small increase in the ΔK value leads to dramatic increase in the fatigue crack growth rate. As a result, this rate attains a relatively high value of 0.1 μm/cycle even at ΔK ~6 MPa×√m. At higher values of stress intensity range, ΔK, a well-defined stage II is observed (figure 3). A fatigue growth rate of 0.65 μm/cycle is attained at ΔK ~14 MPa×√m. At ΔK ~ 15 MPa×m$^{1/2}$, fracture occurs rapidly; no transition to stage III was detected. The stress intensity factor of K_{1c} = 22.7 MPa×√m is evaluated for the present samples. It is worth noting that characteristics of crack propagation resistance of the present 1421 Al with recrystallized structure and those of the early version of this alloy (1420Al) with unrecrystallized structure are almost the same despite the fact that yield stress of the 1421 alloy was high than that of the 1420 alloy by a factor of 1.5 [1].

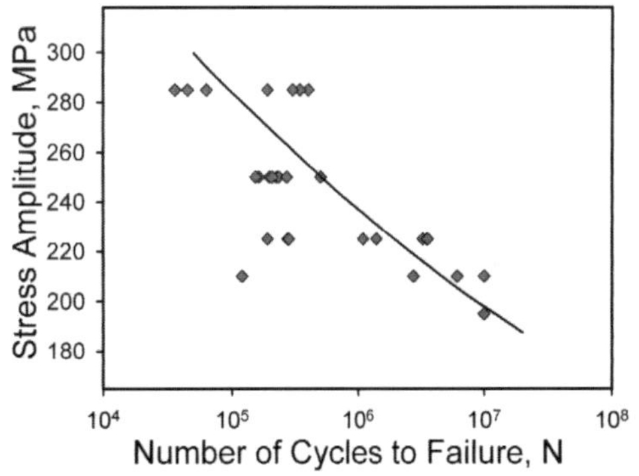

Figure 2. S-N curve for ECAE 1421 alloy.

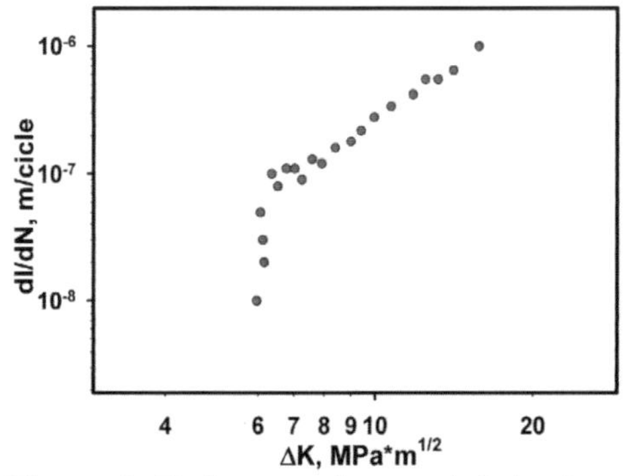

Figure 3. Fatigue crack growth behavior of 1421 alloy.

Figure 5. Optical micrograph (a); SEM fractographs (b, c, d).

Typical views of fracture surface of the flat rectangular sample with the edge crack are given in figure 4. Concentric lines, which are peculiar for fatigue failure can be observed on the sample fracture (figure 4a). The intergranular type of fatigue crack propagation with same separate sections of plastic fracture is seen in figure 4b. Crack nucleation appears also on the large particles of the primary phase $Al_3(Sc,Zr)$ on the samples after ECAE up to the strain of 4 (figure 4c). The fracture observations also suggest that the crack splitting frequently occurs in the material after ECAP that increases its service properties.

With low values of $\Delta K \sim 6$ MPa×m$^{1/2}$ the fracture morphology shows the intergranular character of material failure. That is indicated by the facet fracture with the facet size close to the grain size (figure 4a). Data in figure 4 clearly show that the fatigue crack gradually ceases to be sensitive to the grain boundaries. This is confirmed by the non structure-sensitive character of its spreading in relation to the grain size at the stage of linear growth (figure 3)

The results of the given research show that intergranular failure that is characterized by the low sinuosity of the fatigue crack of the samples occurs with low values of the stress intensity factor range (figure 4a). This causes lower resistance to the fatigue crack growth. With the high values of the stress intensity factor range the mixed type of failure occurs. The transition from intergranular to transgranular fatigue failure occurs at scarce non-recrystallized areas (figure 4d). Nevertheless, the intergranular failure is the dominant mechanism. Rough primary particles also cause brittle fracture (figure 4c).

4. Conclusions

The development of ultrafine grained microstructure by ECAE does not lead to significant improvement of the fatigue properties for the 1421 aluminum alloy as compared to relatively coarse grained microstructure evolved after conventional hot extrusion. The S-N curve does not show any deflection points and noticeable knees. Behavior of the 1421 alloy in LCF range and HCF range is almost the same. The stage III of unstable crack propagation can hardly be revealed on the relationship between the fatigue crack growth rate and the stress intensity range.

References

[1] Fridlyander I, Chuistova K, Berezina A, Kolobnev N 1992 *Aluminum - Lithium alloys. Structure and Properties* Naukova Dumka Kiev.

[2] Fridlyander I, Khokhlatova L, Kolobnev N, Rendix K, Tempus G 2002 *Metal. Sci. Heat Treat.* **1** 3.

3 Kaibyshev R, Shipilova K, Musin F, Motohashi Y 2005 *Mater. Sci. Eng.* **396** 341.

[4] Mogucheva A and Kaibyshev R 2008 *Phys. Met. Metall.* **106** 424.

[5] Kaibyshev R, Shipilova K, Musin F, Motohashi Y 2005 *Mater. Sci. Techno* . **21** 408.

Numerical prediction of the cyclic behaviour of metallic polycrystals and comparison with experimental data

M Sauzay[1], E Ferrié[1,2], A Steckmeyer[1]

1 CEA Saclay, SRMA, Bât. 455, 91191 Gif-sur-Yvette, France
2 now at SIMAP-GPM2, INP Grenoble, BP46, 38402 Saint-Martin d'Hères, France

maxime.sauzay@cea.fr

Abstract. Grain size seems to have only a minor influence on the cyclic strain strain curves (CSSCs) of metallic polycrystals of medium to high stacking fault energy (SFE). Many authors therefore tried to deduce the macroscopic CSSCs curves from the single crystals ones. Either crystals oriented for single slip or multiple slip were considered. In addition, a scale transition law should be used (from the grain scale to the macroscopic scale). The Sachs rule (homogeneous stress, single slip) or the Taylor one (homogeneous plastic strain, multiple slip) were usually used. But the predicted macroscopic CSSCs do not generally agree with the experimental data for metals and alloys, presenting various SFE values. In order to avoid the choice of a particular scale transition rule, many finite element (FE) computations are carried out using meshes of polycrystals including more than one hundred grains without texture. This allows the study of the influence of the crystalline constitutive laws on the macroscopic CSSCs. Activation of a secondary slip system in grains oriented for single slip is either allowed or hindered (slip planarity), which affects strongly the macroscopic CSSCs. The more planar the slip, the higher the predicted macroscopic stress amplitudes. If grains oriented for single slip obey slip planarity and two crystalline CSSCs are used (one for single slip grains and one for multiple slip grains), then the predicted macroscopic CSSCs agree well with experimental data provided the SFE is not too low (austenitic steel 316L, copper, nickel, aluminium).

1. Introduction

Following experimental investigations, the effect of grain size is weak in metallic materials such as nickel [1] and copper [2,3]. That is why it seems possible to deduce the macroscopic cyclic behaviour (polycrystal scale) from the single crystal one (grain scale). Several authors showed that the Sachs homogeneization model [4], which assumes that single slip occurs in each grain homogeneously, leads to reasonable predictions in the case of copper for plastic strain amplitudes lower than a few 10^{-4} [5,3]. This is in agreement with many observations showing that only one slip system is activated per grain provided the macroscopic plastic strain is low enough [2,3]. But, Gorlier showed that the Sachs model leads to large underestimation of the macroscopic stress amplitude whatever the applied strain for the more planar austenitic stainless steel 316L [6]. The Taylor model gives more accurate predictions for low strain even if its assumptions are not true (homogeneous plastic strain tensor which generally requires the activation of five slip systems per grain [7])! In fact, it was shown later that both Sachs and Taylor models lead to underestimations of the stress amplitude for aluminium and nickel [8].
In order to avoid the choice of particular homogeneization hypothesis, large scale polycrystalline finite element (FE) computations are carried out [9-11]. Each grain obeys particular crystalline elastic-plastic laws. The influence of two important slip mechanisms is studied for FCC metals at room temperature: single slip versus multiple slip cyclic hardening and slip planarity.

2. Crystalline constitutive laws and FE computations

Crystalline elasticity is used because of the cubic symmetry of the FCC structure. Twelve slip systems ({111}<110>) are considered in each grain. The macroscopic predictions are based on three kinds of crystalline cyclic hardening laws for better understanding which mechanisms are the most influent.

2.1. Crystalline plasticity laws adjusted on the CSSCs of single crystals oriented for single slip and allowing secondary slip

In copper [12] as in nickel [13-14], the CSSCs of crystals oriented for single slip are very close together because of their similar dislocation microstructures made of persistent slip bands. These orientations belong to a rather large domain of the standard crystallographic triangle (Fig. 1). During cycling at low plastic strain amplitudes, the corresponding crystals/grains are plastically activated because of their high Schmid factor values (close to 0.5, figure 1) whereas the ones oriented for multiple slip are still elastically deformed because of their low Schmid factor values. At low strain, the coefficients of the crystalline plastic hardening laws could therefore be adjusted using only the CSSCs of the well-oriented crystal (Schmid factor: 0.5). In addition, Laird and co-workers showed experimentally that the cyclic hardening of crystals is mainly due to cyclic increase of the kinematics stress (long range stress) whereas the isotropic stress (short range stress) is rather constant whatever the plastic strain amplitude [12,15]. A simple Armstrong-Fredericks type law is used (non-linear kinematics hardening leading to saturation). For each slip system, i=1,12, the kinematics resolved shear stress, x_i , is a non-linear function of the corresponding plastic slip, γ^p_i [11]. For each material, two parameters, C and α, should be adjusted as well as the initial critical shear stress, τ_0 [11]. The Activation of the ith slip system occurs if: $|\tau_i - x_i| = \tau_0$ with τ_i the corresponding resolved shear stress.

2.2. Additional effect of hardening laws based the CSSCs of crystals oriented for multiple slip

As it was shown a long time ago, crystals oriented for multiple slip, such as <100> or <111>, display stronger hardening than crystals oriented for single slip (plastic slip versus resolved shear stress curves). The case of nickel was extensively studied [13-14,16] as well as the one of copper. For these orientations, the dislocation microstructures are often made of labyrinths and cells/walls respectively, which differ from the persistent slip band microstructure usually observed for crystals oriented for single slip (figure 1). Therefore, taking into account the stronger cyclic hardening of the grains for which the loading direction is close to these crystallographic directions could lead to higher hardening at the polycrystal scale. Following the idea of Schwab and Holste [17], the grains are divided in several sets. Only two domains are used for the sake of simplicity. The standard crystallographic triangle is divided in two domains using a simple criterion. If the ratio between the secondary and primary resolved shear stresses, τ_2/τ_1, is lower than a critical value, r_{crit}, the grain is considered as being oriented for single slip (figure 2). If the ratio is higher, the grain is oriented for multiple slip. The critical ratio, r_{crit}, should be higher than 0.94 (value of this ratio for a well-oriented single crystal which is known to be oriented for single slip) and lower than 1 (sides of the standard crystallographic triangle) (figure 1). Finally, two different couples of non-linear kinematics hardening coefficients, (C_{single}, α_{single}) and ($C_{multiple}$, $\alpha_{multiple}$), should be adjusted using the single crystal CSS curves (figure 2). For the single slip oriented grains, the CSSC of the well-oriented crystal is used (Schmid factor: 0.5). For grains oriented for multiple slip, the <100> crystal curve is used because it has been published in the literature for all the metallic materials we study (aluminium, copper, nickel, 316L). The initial critical shear stress, τ_0, is considered to be the same for all grains, whatever their orientations.

2.3. Effect of slip planarity imposed to grains oriented for single slip

Following the observations reported in the literature, generally only one slip system is activated per grain, provided the macroscopic axial plastic strain amplitude is smaller than a few 10^{-3}. This is observed whatever the metallic material (Cu [5,3,18], Ni [16], 316L [19]). Let us first consider grains oriented for single slip. By choosing a very low hardening coefficient for the isotropic self-hardening, a negligible hardening of the isotropic primary resolved shear stress is predicted as observed by Laird and co-workers [12,15]. But a very high value of the latent hardening coefficient is chosen. Therefore, primary plastic slip induces a so high hardening on the secondary slip system that only single slip is predicted in both crystals and grains oriented for single slip. For grains oriented for multiple slip, weak self and latent isotropic hardening coefficients are used. The experimental single crystal CSS curves used for the adjustment of the crystalline plasticity laws have been published in the cited references (for example: Cu [3], Ni [13-14,16], 316L [6]).

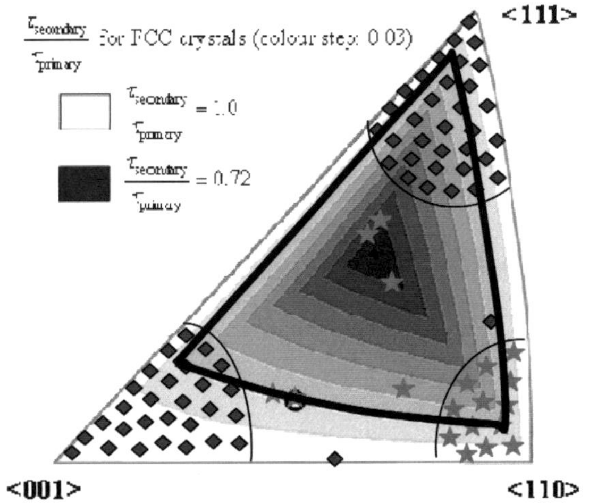

Figure 1. Isovalues of the ratio between secondary and primary shear stresses plotted in the standard triangle (courtesy from Th. Kruml). Bold line: ratio equal to 0.94, (value for well-oriented crystal, black circle). Orientations with PSBs (red stars) or labyrinths/cells (blue diamonds) (nickel [13-14,16]).

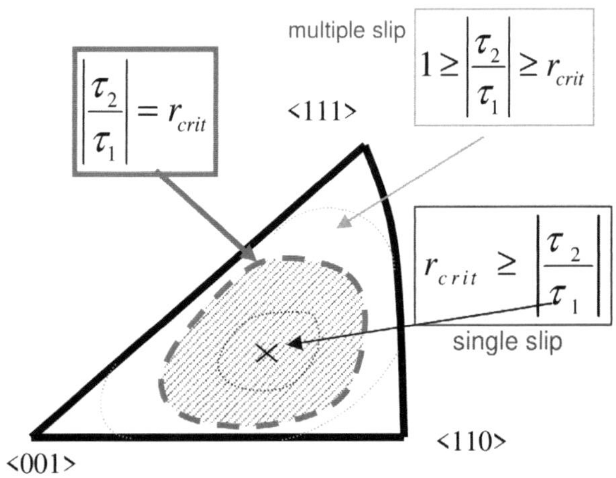

Figure 2. Two domains of orientations corresponding to either single slip or multiple slip. Standard crystallographic triangle. Two different couples of non-linear kinematics hardening coefficients, (C_{single}, α_{single}) and ($C_{multiple}$, $\alpha_{multiple}$), should be adjusted using the experimental single crystal CSS curves.

3. Comparison between predicted and experimental polycrystalline cyclic stress strain curves

Numerical homogeneization is applied using the FE method. Many details concerning the FE computations are given in [8,11] (FE code, boundary conditions, minimum number of grains and FEs, number of cycles required for getting cyclic saturation…). A typical mesh of 125 cubic grains is used.

3.1. Crystalline plasticity laws adjusted on the CSSCs of single crystals oriented for single slip and allowing secondary slip

The predicted CSSCs are generally close to the ones predicted by the Sachs model, whatever the considered metal or alloy [8,11]. This means that the macroscopic stress level is generally underestimated even if the plastic strain is in the 10^{-5} to 10^{-3} range (figures 3 and 4 for copper and 316L steel respectively, red curves with triangles). As no particular assumption is made concerning the homogeneization procedure, some physical hardening mechanisms occurring at the grain scale should have been neglected. As the hardening of grains oriented for multiple slip leads to a reduced increase of the predicted stress, the effect of slip planarity in grains oriented for single slip is now investigated.

3.2. Effect of slip planarity imposed to grains oriented for single slip

Hindering secondary slip in grains oriented for single slip leads to stronger macroscopic hardening whatever the applied macroscopic strain. The choice of the critical ratio value affects the predicted macroscopic stress only for axial plastic strain amplitude higher than about 10^{-3}. If the macroscopic plastic strain is lower, using the minimum (0.94) or maximum value (1) leads to close predicted macroscopic CSS curves because the grains belonging to the multiple slip family are not plastically deformed at low plastic strain. Following figures 3 and 4, imposing slip planarity by hindering secondary slip leads to reasonable predictions whatever the material (copper, 316L, aluminium) for plastic strain smaller than 10^{-3}. For higher amplitude, multiple slip in the corresponding grains should be taken into account in order to avoid too strong hardening due to slip planarity. Following figure 1, a value of the critical ratio, r_{crit}, close to 0.95 seems reasonable at least for nickel. If a higher value is used, many more grains/crystals displaying multiple slip dislocation microstructures such as labyrinths or walls would be in the single slip domain meaning that the partition is less accurate.

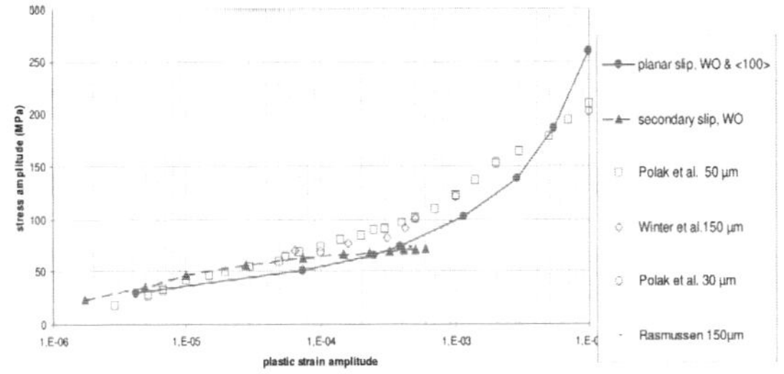

Figure 3. Predicted & experimental macroscopic CSS curves. Green symbols: experimental data. Red curve with triangles: basic model (well oriented grain (WO), secondary slip is allowed) (2.1 and 3.1). Red curve with circles: enhanced model (both single slip and multiple slip grains with r_{crit}=0.95, planar slip is imposed to well-oriented grains (2.2, 2.3 and 3.2). Copper.

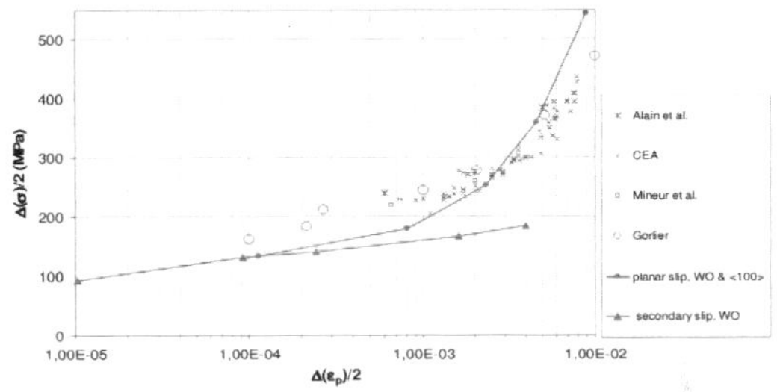

Figure 4. Predicted & experimental macroscopic CSS curves. Green symbols: experimental data. Red curve with triangles: basic model (well oriented grain (WO), secondary slip is allowed) (2.1 and 3.1). Red curve with circles: enhanced model (both single slip and multiple slip grains with r_{crit}=0.95, planar slip is imposed to well-oriented grains (2.2, 2.3 and 3.2). 316L

References

[1] Buque C, Bretschneider J, Schwab A and Holste C 2001 *Mater. Sci. Eng.* A **319-321** 631
[2] Pedersen OB, Rasmussen KV and Winter AT 1982 *Acta Metall.* **30** 57
[3] Mughrabi H and Wang R 1988. *Basic Mechanisms in Fatigue of Metals* ed P Lukas, J Polak (Elsevier) pp 1
[4] Sachs G 1928 *Z. Verein dt. Ing.* **72** 734
[5] Rasmussen KV and Pedersen O 1980 *Acta Metall.* **14** 1467
[6] Gorlier C 1984 (Thèse Ecole des Mines de Saint Etienne) France
[7] Taylor G 1938 *J. Inst. Metals* **62** 307
[8] Sauzay M 2009 Habilitation à Diriger des Recherches (Université Paris VI) France.
[9] Barbe F, Forest F and Cailletaud G 2001 *Int. J. Plast.* **17** 537
[10] Bennett V and McDowell D 2003 *Int. J. Fat.* **2** 27
[11] Ferrié E and Sauzay M 2008 *Mater. Sci. Forum* **567-568** 97
[12] Cheng AS and Laird C 1981 *Mater. Sci. Eng.* A **51** 111.
[13] Mecke K and Blochwitz C 1982 *Cryst. Res. Technol.* **17** 743
[14] Blochwitz C and Veit U 1982 *Cryst. Res. Technol.* **17** 529
[15] Laird C and Li Y 1994 *Mater. Sci. Eng.* A **186** 65
[16] Buque C, Bretschneider J, Schwab A and Holste C 2001 *Mater. Sci. Eng.* A **300** 254
[17] Schwab A and Hoste C 2002 *Acta Mater.* **50** 289
[18] Christ H-J, Mughrabi H and Wittig-Link C 1988. *Basic Mechanisms in Fatigue of Metals* ed P Lukas, J Polak (Elsevier) pp 83
[19] Obrtlik K, Kruml T and Polak J 1994 *Mater. Sci. Eng.* A **187** 1

Journal of Physics: Conference Series **240** (2010) 012057 doi:10.1088/1742-6596/240/1/012057

Fatigue properties of high Nb TiAl alloy

T Kruml, A Dlouhý, M Petrenec, K Obrtlík and J Polák

Institute of Physics of Materials, Academy of Sciences, CZ 61662 Brno, Czech Republic

kruml@ipm.cz

Abstract. Fatigue properties of the new generation of TiAl alloys with high Nb content are studied. Comparison with the previous alloy with 2at% Nb shows that the new alloy resists better to cyclic deformation at high temperatures. The microstructural observation proved that at 750 °C, the easiest deformation mode is the glide of ordinary dislocations followed by superdislocation glide and twinning. Nevertheless, all three modes seem to be active.

1. Introduction

While TiAl alloys are already used in some applications, their development continue since their potential of a material with low density and good high temperature properties (both strength and corrosion resistance) is limited by the brittleness and difficult machinability. The third generation of these alloys contains high Nb addition, typically 7-8 at. %.

In the most foreseen applications, the low-cycle and/or high cycle fatigue as well as the crack growth resistance of the material are important [1]. The data on low cycle fatigue properties of the high Nb content alloy are rather scarce. Christ et al. [2,3] observed stable cyclic behaviour, satisfactory strength and lifetime at high temperatures and environmental surface embrittlement which increases with temperature.

The first results measured in our laboratory have been reported recently [4]. The microstructure of the material and deformation conditions (symmetrical strain controlled tests) are similar to those used by Gloanec et al. [5] who studied previous generation of TiAl alloy with 2at%. Nb. Their results are used for comparing the behaviour of low and high Nb content alloys and are complemented by TEM observation of dislocation structure.

2. Experimental details

2.1. Material and specimens

The material was delivered in the form of a ingot of 70 mm in diameter and 1.2 m in length prepared by casting and subsequent hot isostatic pressing. Chemical composition of the material in at.% is: 44Al – 48Ti – 7.8Nb – 0.2Ni. The majority of the volume consists of fine lamellar microstructure of γ and α_2 phases. The grains of irregular but not elongated shape have the average size of about 1 mm. Some larger γ phase islands are present on the grain boundaries, the microstructure thus can be called as nearly lamellar (Fig. 1). The structure is not completely homogeneous across the diameter of the ingot; the fatigue specimens were prepared from the outer regions of the ingot with the axis parallel to the ingot axis. Cylindrical specimens with 6 mm in diameter were carefully mechanically and electrolytically polished.

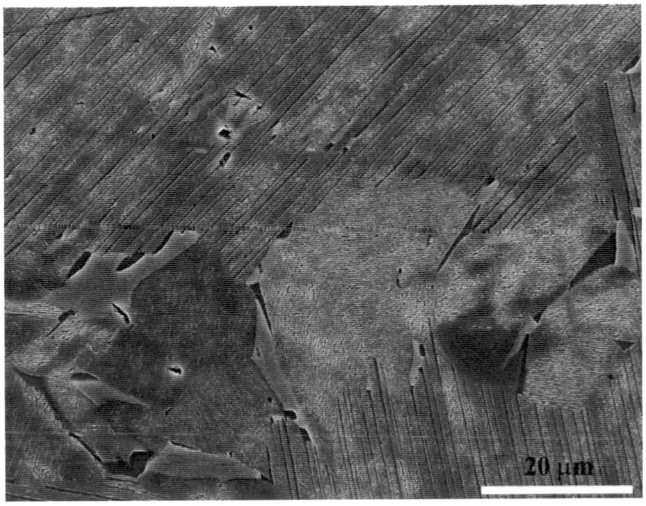

Figure 1. The microstructure of the experimental material.

2.2. Mechanical testing
Cyclic deformation tests were performed in a MTS 810 servohydraulic machine. Strain amplitude was kept constant as well as strain rate of $2x10^{-4}$ s^{-1}. The loading cycle was symmetrical (R_{ε} = -1).. Tests were performed at room temperature and at temperatures of 700 °C and 750 °C in air. Three thermocouples were used to monitor the homogeneity and stability of the temperature during testing.

2.3. Microscopy
Light microscopy and SEM was used for the inspection of the as-received microstructure and fatigue fracture surfaces. The thin foils for TEM were prepared from the slices cut perpendicularly to the specimen axis. The direction of the axis was conserved during the foil preparation, consisting from the mechanical polishing and final thinning in a double jet electropolishing apparatus. The foils were inserted in the Philips CM 12 TEM in the manner that the holder axis was parallel to the specimen axis. It was thus possible to relate crystallographic planes and directions of individual grains to the specimen geometry and to calculate Schmid factors for observed deformation modes.

3. Results and discussion

3.1. Cyclic response and fatigue life
In all three test temperatures, stable cyclic response of the material was observed. Almost the same behaviour was found by Christ and Bauer [3] who observed slight cyclic hardening for temperatures below 500 °C.. On the contrary, significant cyclic hardening was observed at RT in 2at.%Nb alloy [5].

At 750 °C, stable cyclic response is found in both types of alloy. Fatigue life curves are shown in Fig. 2. If total strain amplitude is considered, the fatigue life curves of both materials remarkably overlap (Fig. 2a). Nevertheless, the fatigue lifetime of the 8Nb alloy is one – two orders of magnitude longer if the saturated stress amplitudes are compared (see the derived Wöhler curve in Fig. 2b). The reason for this is the higher strength of the 8Nb alloy in comparison with the 2Nb alloy. The yield stress is higher and the cyclic deformation curve is shifted to higher stresses for the 8Nb alloy. It means that at the same stress amplitude, the plastic strain amplitude is lower for 8Nb and the fatigue life is longer. Both life curves at 750 °C change their slopes at about N_F ~ 2000. The two observed deformation regimes are characterized by different exponents of Manson-Coffin and Basquin laws [4]. The same sudden change in slope of the fatigue life curves was observed also at T = 700 °C. The lower slope of the Basquin curve at low stresses results in even more important improvement of fatigue life for low loading levels.

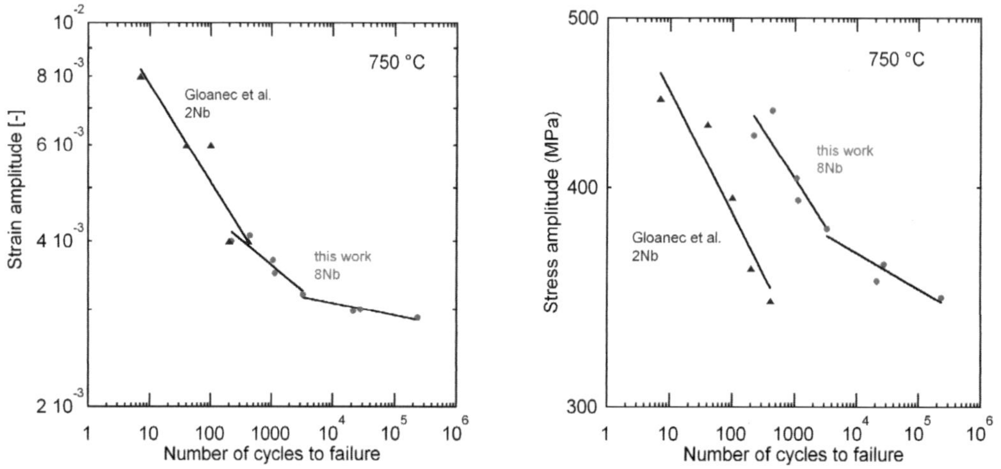

Figure 2. Fatigue life curves of the investigated material compared with previous generation of TiAl alloy [5].

3.2 Deformation mechanisms at 750 °C

Dislocation arrangement in γ phase containing also four twins has been analysed (Fig. 3a, $\vec{g} = 020$). The crystallographic orientation of the stress axis is close to [5 -10 3]. The investigated region was cut so that the foil plane is close to (655). The dislocation arrangement was observed using five different diffraction vectors which enabled finding Burgers vectors $\vec{b}$ of dislocations. In Fig. 3b ($\vec{g} = \bar{1}11$), the twins lying along $(\bar{1}11)$ plane are edge-on. Ordinary dislocations from the primary ordinary slip system $(1\,1\,1)$. $[1\,10]$ are present in high density. They have elongated screw segments. The Burgers vector of primary ordinary dislocation is assessed using invisibility diffraction conditions for $\vec{g} = \bar{1}\,\bar{1}\,1$ in Fig. 3c. If $\vec{g} = 002$ is applied, only superdislocations with non zero c component of $\vec{b}$ are visible. This condition allows separation of ordinary dislocations and superdislocations. It can be observed that superdislocations are present with a low density in the investigated volume.

It suggests that the main deformation mode is the slip of ordinary dislocation, in spite of the fact that the slip of superdislocations and twinning are significantly more favoured by the crystallographic orientation, as can be seen from the Schmid factors shown in Table 1. The two latter modes require higher flow stress but apparently it is possible to activate them. This is important for the appearance of the interlamellar fracture since according to the von Misses criterion, five independent deformation modes must be available to adapt deformations in between neighbouring grains or lamellas.

Table 1. Schmid factors of observed deformation modes.

deformation mode					
	plane	$(1\,11)$	$(1\,1\,1)$	$(1\,11)$	$(1\,11)$
	$\vec{b}$	$[0\bar{1}\,1]$	$[\bar{1}\,10]$	$[12\bar{1}]$	$[\bar{1}\,1\bar{2}]$
Schmid factor		0.475	0.366	0.37	0.44
comment		primary SD	primary OD	twinning	twinning

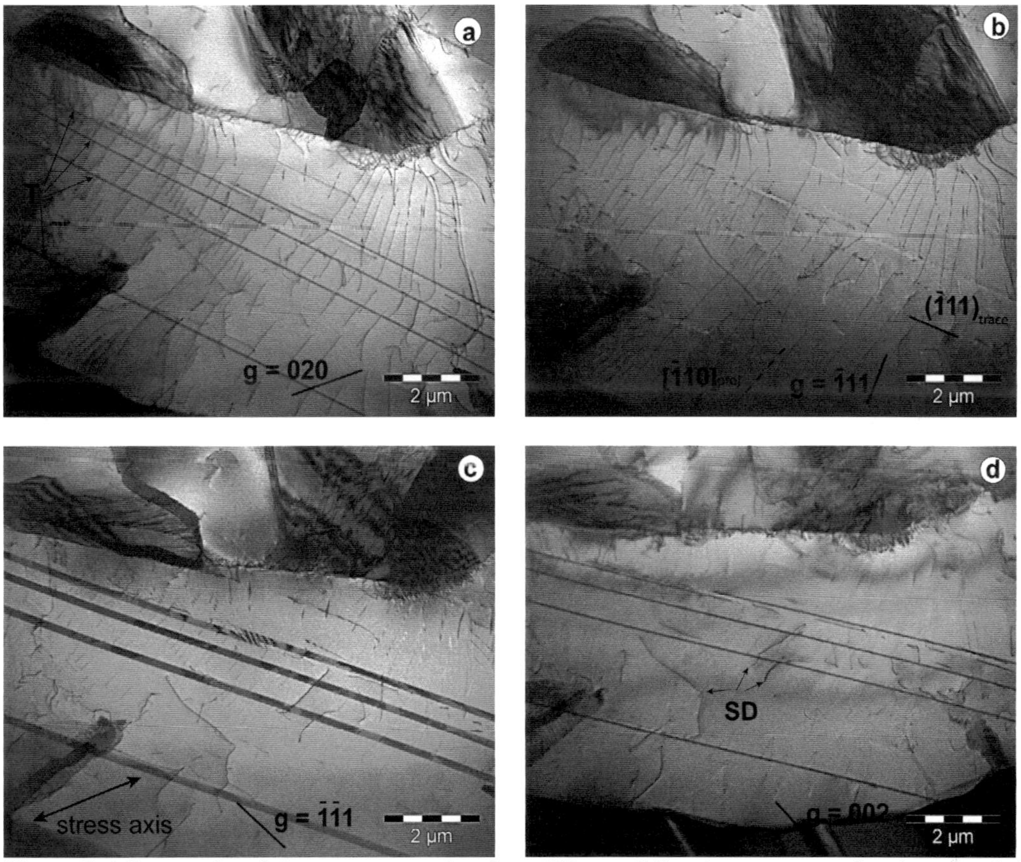

Figure 3. Micrographs of the same area using four different diffraction vectors g. Specimen cycled with $\varepsilon_a = 0.41\%$ at T = 750 °C up to the fracture ($N_F = 438$).

4. Conclusions

- The TiAl alloy with high Nb is cyclically stable in the investigated temperature interval (RT – 750 °C).
- The fatigue life curves of the low and high Nb alloys are similar in N_F vs. ε_a representation. In N_F vs. σ_a representation the fatigue life of the high Nb alloy is shifted more than one order of magnitude to longer fatigue life.
- Glide of ordinary dislocations seems to be the most easy deformation mode at 750 °C.

Acknowledgement. Financial support was provided by projects GACR 106/07/0762 and 106/08/1631.

References

[1] Henaff G and Gloanec AL 2005 *Intermetallics* **13** 543
[2] Heckel TK and Christ HJ 2008 *6th Int. Conf. on Low Cycle Fatigue*, eds. Portella PD et al., 579
[3] Christ HJ and Bauer V 2008 *6th Int. Conf. on Low Cycle Fatigue*, eds. Portella PD et al., 585
[4] Kruml T, Obrtlík K, Petrenec M and Polák J 2009 *Int. J. Mat. Res.* **100** 349
[5] Gloanec AL, Jouiad M, Bertheau D, Grange M and Henaff G 2007 *Intermetallics* **15** 520

Application of FIB technique to study of early fatigue damage in polycrystals

J Man[1], A Weidner[2], I Kuběna[1,3], T Vystavěl[4], W Skrotzki[5] and J Polák[1]
[1] Institute of Physics of Materials, AS CR, Žižkova 22, 616 62 Brno, Czech Republic
[2] Institute of Materials Engineering, TU Bergakademie Freiberg, Gustav-Zeuner-Str. 5 D-09596 Freiberg, Germany
[3] Institute of Materials Science and Engineering, Brno University of Technology, Technická 2, 616 69 Brno, Czech Republic
[4] FEI, Podnikatelská 6, 612 00 Brno, Czech Republic
[5] Institute of Structural Physics, TU Dresden, D-01062 Dresden, Germany

man@ipm.cz

Abstract. Focused ion beam (FIB) technique together with other advanced microscopic techniques were applied to study early microstructural changes leading to crack initiation in fatigued polycrystals. Dislocation structures of persistent slip bands (PSBs) and surrounding matrix were revealed in the bulk of surface grains by electron channelling contrast imaging (ECCI) technique on the FIB cross-sections. True shape of extrusions, intrusions and the path of initiated fatigue cracks were assessed in three dimensions by serial FIB cross-sectioning (FIB tomography). Advantageous potential of FIB technique and its other possible utilization in fatigue crack initiation studies in polycrystals are highlighted.

1. Introduction

Focused ion beam (FIB) technique represents a modern tool which has been used in various areas of materials research for studying or modifying material systems down to nanometre levels (for a review see e.g. [1]). In fatigue damage studies in polycrystals FIB has been used to study environment-assisted crack profiles in pipeline steel [2], the interaction of microcracks with selected interfaces in nickel-based superalloy [3] and in the study of short crack path in dual phase steel [4].

In comparison with the above papers the present work is directed to the study of early stages of fatigue damage, namely to the study of cyclic strain localization leading to fatigue crack initiation. The localization of the plastic strain in persistent slip band (PSB) is followed by the formation of persistent slip markings (PSMs) consisting of extrusions and intrusions. Numerous experimental studies proved the dominant role of PSMs in the process of fatigue crack initiation (for a review see e.g. [5,6]). Study of surface profiles of PSMs in individual cross-sections [7] and cumulative evolution of PSMs on two perpendicular surfaces [8] in 316L steel using FIB technique were performed recently. In this work the first results on dislocation structure of PSBs and three-dimensional (3D) form of PSMs and fatigue crack initiation obtained by FIB in conjunction with other microscopic techniques are reported.

2. Experimental

Flat specimens with a square cross-section of polycrystalline nickel and cylindrical specimens with a shallow notch of 316L austenitic stainless steel were cycled with controlled constant plastic strain amplitudes to various stages of fatigue life. Central parts of both specimen geometries were mechanically and electrochemically polished before fatigue tests. Further details concerning the materials, specimens and fatigue tests can be found elsewhere [7–9].

The studies with the FIB technique in polycrystalline nickel (specimen IV used in our recent study [9]) were performed on a Zeiss CrossBeam 1540 XB instrument. Dislocation structures were studied by electron channelling contrast imaging (ECCI) technique in high-resolution SEM–FEG (Zeiss Ultra 55 FESEM). For serial cross-sectioning of PSMs in 316L steel (FIB micro-tomography) a FEI Dual Beam Helios NanoLab was used. Thin protective platinum layer was deposited on a small area of interest across PSMs before ion milling. U-shaped trench with one high quality cross-sectional surface perpendicular to PSMs and to the specimen surface was produced by FIB. Subsequently series of surface relief profiles were obtained by periodical removing of material within U-shaped trench parallel with the above high quality cross-sectional surface by FIB using automated software FEI Slice and View. In total, 500 sections were milled (300 sections with 20 nm slice thickness followed by 200 sections with 12 nm milling step).

3. Results and discussion

3.1. Dislocation structure and the depth of PSBs

Recent study of the local cyclic plastic strain within PSBs in nickel polycrystals revealed the broad spectrum of strain amplitudes with the average value around 1.5% [9]. Among factors, which can influence the local plastic shear in PSBs, the depth of PSBs and their dislocation arrangement beneath the specimen surface are of a great importance.

Figure 1a shows a grain of polycrystalline nickel cycled with the plastic strain amplitude ε_{ap} = 5×10^{-4} for 30,000 cycles ($0.6N_f$). PSMs coming from two activated slip systems are apparent in this grain. The EBSD (electron backscattering diffraction) measurements [9] showed that this grain is demarcated by twin boundaries TB1, TB2 and TB3 and high angle grain boundary GB1, as indicated in figure 1b. The calculation of trace angles for the twin boundaries and PSMs both on the specimen surface and in the interior resulted in the construction of a tentative sub-surface profile of this individual grain and the possible PSB depth evaluation – see figure 1b.

To reveal the unknown trace of GB1 and the dislocation structure below the specimen surface combined FIB/ECCI technique was adopted. After finishing the half-cycle experiments [9] additional 2000 cycles with the same ε_{ap} were applied to specimen IV followed by a regular FIB cross-sectioning across GB1 (see figure 1b). Adoption of the BSE (back-scatter electron) contrast in SEM revealed the path of the grain boundaries GB1 and TB2 on the FIB cut below the specimen surface (figure 1b).

Dislocation arrangement both on the specimen surface and FIB-cut plane visualized by ECCI technique in SEM–FEG are documented in figure 2a. To improve the quality of ECCI micrographs the specimen was shortly electropolished which resulted in the slight rounding of FIB-cut edges and removing thin surface layer (see figure 2a). Detail of dislocation structure from the central area on the

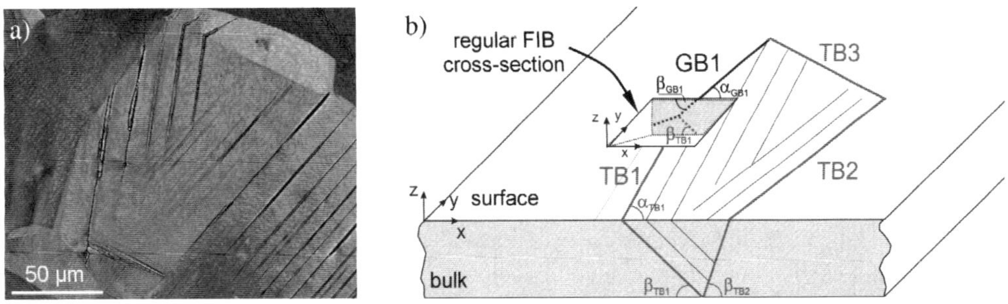

Figure 1. (a) Surface relief in the grain of polycrystalline nickel fatigued for 30 000 cycles with $\varepsilon_{ap} = 5\times10^{-4}$, ECCI. (b) Schema of a virtual cut through the grain shown in figure 1(a). Traces of grain boundaries found by EBSD and the position of FIB crater are indicated.

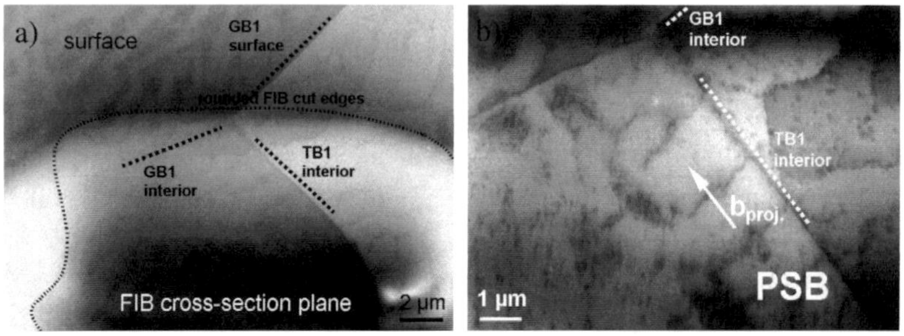

Figure 2. Dislocation structure revealed by ECCI in polycrystalline nickel fatigued for 32 000 cycles with ε_{ap} = 5×10^{-4}, SEM–FEG. (a) Overview of the dislocation arrangement both on the surface and FIB cross-section plane. (b) Detail of structure from FIB cross-section plane.

FIB cross-section plane shows figure 2b. A PSB lamella with ladder structure lying parallel to TB1 is clearly seen. This finding perfectly agrees with the surface relief observation of the PSM going along the intersection of TB1 with the specimen surface (cf. figures 1 and 2b).

3.2. The shape of PSMs and fatigue crack initiation – FIB tomography

After the documentation of the specimen surface 3D form of PSMs in individual grains of cyclically strained 316L steel were studied by FIB either in regular cross-sections or in U-shaped trench using automated milling procedure. Here we will present only the results obtained by the latter way.

Figure 3a shows parts of two surface grains of 316 steel fatigued with ε_{ap} = 1×10^{-3} for 3800 cycles (8.3%N_f). Well developed PSMs in the right grain are equidistantly displaced whereas less developed PSMs in the left grain are present only locally (see the upper left corner in figure 3a). An individual sharp PSM which developed parallel to a thin twin separating these grains (see below) was chosen for the subsequent 3D FIB micro-tomography study. U-shaped trench created for this purpose around the area of interest is shown in figure 3b.

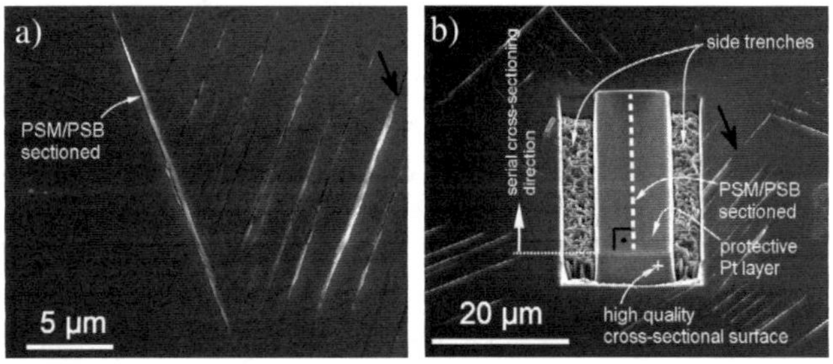

Figure 3. (a) SEM image of 316L steel fatigued for 3800 cycles with ε_{ap} = 1×10^{-3} before (a) and after the creation of U-shaped FIB trench (b). Both images were taken with the 52° specimen tilting. Small black arrow indicates the same position on both micrographs.

Figure 4 demonstrates an example of results obtained by FIB sectioning. The sectioning started from the lower end of PSM in direction to its centre. The profile of extrusion indicates that PSB is adjacent to a thin twin. The profile of the PSM changes. At the end of PSM only a thin extrusion was present. The extrusion widened and its height increased until in section 360 (figure 4b) the extrusion was well developed. The first indication of the starting crack was in section 390 (figure 4c). The growing crack follows twin boundary and later deviates from it (figure 4e). Figure 4e also reveals thin

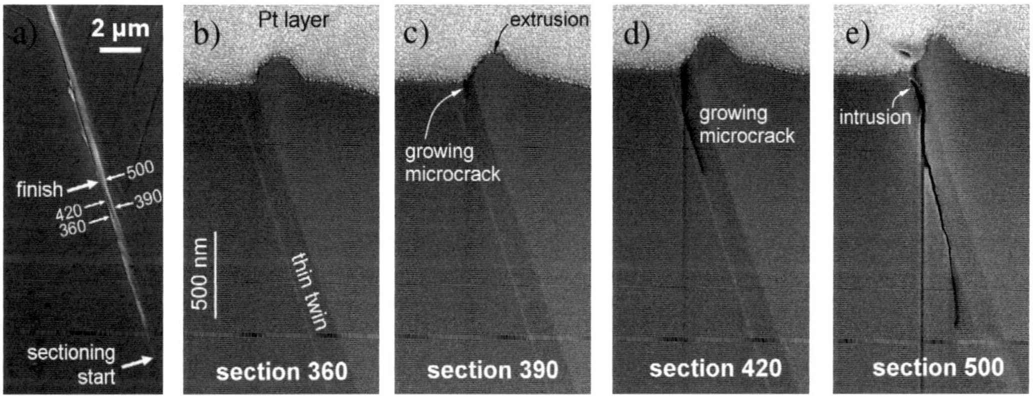

Figure 4. (a) Detail of PSM in 316L steel sectioned by FIB tomography (see figure 3a). (b–e) Profiles of PSM and initiated fatigue crack in sections indicated in figure 4a. All SEM–FEG micrographs were corrected for a 52° sample tilt.

intrusion accompanying the extrusion and running along twin boundary. This observation suggests that the fatigue crack started to initiate from an intrusion in the centre of the grain where the PSB depth is most pronounced. Further milling can prove or disprove this hypothesis.

4. Summary

The present work demonstrates that FIB, coupled with other advanced high-resolution techniques (SEM–FEG, ECCI, EBSD), represent an effective tool for systematic studies of the early fatigue damage and fatigue crack initiation in individual surface grains of cyclically strained polycrystals at submicroscopic level. Detailed data on the 3D form of PSMs and the initiation of fatigue cracks can be obtained. Dislocation structures corresponding to surface PSMs (with their possible 3D reconstruction) can be identified and the depth of PSBs (i.e. their length in the active slip direction) beneath the specimen surface can be evaluated. Another promising use of FIB technique in the early fatigue damage studies in polycrystals which has not been checked so far is the preparation of thin foils directly from specimen surface for transmission electron microscopy (TEM) analysis. Experimental data obtained by these modern high-resolution techniques will be invaluable for checking the validity and further advancement of theoretical models of surface relief formation leading to fatigue crack initiation.

Acknowledgments

Support of this work by the grants No. 106/06/1096 and No. 101/07/1500 from the Czech Science Foundation, by the research project No. AV0Z 20410507 and the grant No. 1QS200410502 from the AC SR and the grant Sk21/24-2 from the German Research Foundation (DFG) is acknowledged.

References

[1] Munroe P R 2009 *Mater. Charact.* **60** 2–13
[2] Wang Y-Z, Revie R W, Phaneuf M W and Li J 1999 *Fatigue Fract. Eng. Mater. Struct.* **22** 251–6
[3] Holzapfel C, Schäf W, Marx M, Vehoff H and Mücklich F 2007 *Scripta Mater.* **56** 697–700
[4] Motoyashiki Y, Brückner-Foit A and Sugeta A 2008 *Eng. Fract. Mech.* **75** 768–78
[5] Mughrabi H 2009 *Metall. Mater. Trans. A* **40** 1257–79
[6] Man J, Obrtlík K and Polák J 2009 *Phil. Mag.* **89** 1295–336
[7] Man J, Klapetek P, Man O, Weidner A, Obrtlík K and Polák J 2009 *Phil. Mag.* **89** 1337–72
[8] Polák J, Man J, Vystavěl T and Petrenec M 2009 *Mater. Sci. Eng. A* **517** 204–11
[9] Weidner A, Man J, Tirschler W, Klapetek P, Blochwitz C, Polák J and Skrotzki W 2008 *Mater. Sci. Eng. A* **492** 118–27

Increasing the fatigue limit of a high-strength bearing steel by a deep cryogenic treatment

E. Kerscher[1,2], K.-H. Lang[1]

[1]Institut für Werkstoffkunde I, Universität Karlsruhe (TH), D-76128 Karlsruhe, Germany
[2]now: Lehrstuhl für Werkstoffkunde, Technische Universität Kaiserslautern, D-67653 Kaiserslautern, Germany

kerscher@mv.uni-kl.de

Abstract. High-strength steels typically fail from inclusions. Therefore, to increase the fatigue limit of high-strength steels it is necessary to modify the inclusions and/or the surrounding matrix. The goal must be a higher threshold for crack initiation and/or crack propagation. One possibility to reach this goal seems to be a deep cryogenic treatment which is reported to completely transform the retained austenite as well as to facilitate the formation of fine carbides. Therefore, specimens were annealed before or after deep cryogenic treatment, which was carried out with different cooling and heating rates as well as different soaking times at -196° C. Hardness and retained austenite measurements and fatigue experiments were used to evaluate the different sequences of treatments mentioned above. The fatigue limit increases only after some of the sequences. The results show that the soaking times are not relevant for the fatigue limit but it is very important to temper the specimens before the deep cryogenic treatment. Also, repeated deep cryogenic treatments had a positive influence on the fatigue limit.

1. Cryogenic Cooling and Fatigue

High strength steels are often based on martensitic heat treatments so that it is possible that austenite remains besides the martensitic microstructure in the final material. The volume content and the stability of the retained austenite influence the fatigue limit of these steels significantly. Formerly, the cryogenic cooling was primarily used to reduce the content of retained austenite by transforming it into martensite by cooling the material below the secondary martensite start temperature. Unfortunately the higher martensite content often does not increase the fatigue limit due to microcracks, which were created by microstresses caused by the cooling below room temperature. [1, 2]

Recent reports show a significant influence of cryogenic treatments with controlled temperature rates and long soaking times on the wear resistance of tool steels [3]. Also a change in the carbides morphology by the cryogenic treatment is discussed [4, 5].

2. Experimental Setup

The heat treatment of all material states of the hypereutectic carbon steel (1.0 wt.-% C, 1.5 wt.-% Cr, German trade name 100Cr6) starts with austenitisation in a vacuum furnace at 840° C for 20 min followed by quenching to room temperature in oil. After this hardening different treatments of annealing and/or cooling are conducted. Figure 1 gives an overview of the different treatments. Not all of them were applied to every state: annealing is carried out at 180° C for 2 hours in some cases before and/or after deep cryogenic cooling. The deep cryogenic cooling is realized by cooling the specimens down to the temperature of liquid nitrogen (-196° C). The cooling down-process is either done by placing the specimens directly in liquid nitrogen or by a controlled process with a cooling rate of 1 K/min. Heating up is realized with the same rate in the second case. In the first case of fast cooling the specimens are taken out of the nitrogen and directly put on a table at room temperature. The nomenclature of the specimens indicates the elements given in Figure 1. For instance ASC100A is a state which is annealed, slowly cooled down to -196° C, soaked for 100 hours at -196° C, heated up to room temperature with 1 K/min and finally annealed, again. There was normally only one round of cooling and heating up, except in case of ASC65A, where the sequence was repeated 10 times with a soaking time of 65 hours in total. All heat treatments were carried out with flat fatigue specimens (for details of geometry see [6]).

The specimens were tested in deflection controlled bending fatigue tests. There were about 20 specimens for each state. The fatigue limit for an ultimate number of cycles of 10^7 was determined by using the staircase method [7]. The hardness was measured by Vickers indentation test. The retained austenite was calculated from X-ray-diffraction by the 6-line-method [8]. All fracture surfaces were observed in SEM and the non metallic inclusions were analyzed regarding their stress intensity factors [9].

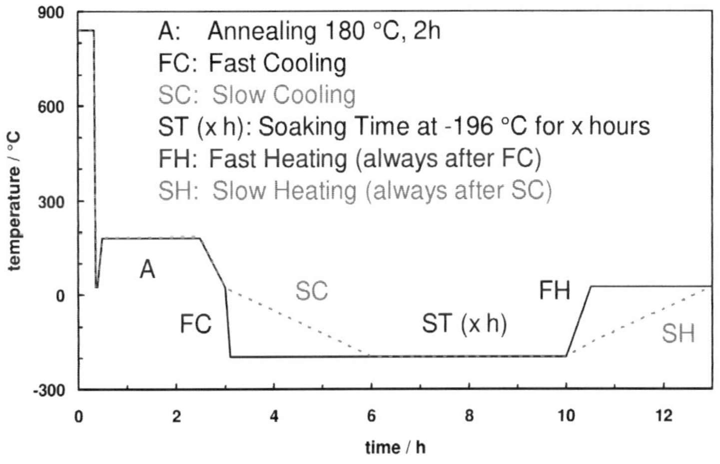

Figure 1. Schematic curve of the elements for the different paths at the heat treatments.

3. Results

3.1. Hardness and content of retained austenite after Cryogenic Treatment

The just hardened state had a hardness of 950 HV 10 and a content of 11.3 vol.-% of retained austenite. Deep cryogenic cooling of the hardened state did not change the hardness but reduced the retained austenite to 9.5 vol.-%. Final annealing reduced the hardness to 825 HV 10 and the content of retained austenite to 8.0 vol.-%.

Hardening and annealing led to 780 HV 10 and 8.5 vol.-% of retained austenite. A second annealing reduced the content of retained austenite significantly but changed the hardness just slightly to 770 HV 10. All variations of cooling between two annealing treatments did not change the resulting hardness values and the content of retained austenite.

3.2. Fatigue results

All states show a wide scatter in the number of cycles to failure (see Fig. 2) caused by the initiation from nonmetallic inclusions which had different morphologies and are located in different distances from the surface. The fatigue limit of the annealed and not cooled state is 940 MPa. A second annealing increases the fatigue limit slightly to 953 MPa. Slow and fast cooling directly after hardening and before annealing reduced the fatigue limit to about 890 MPa, which indicates that the annealing performed directly after hardening and before cryogenic cooling is evidently for higher fatigue limits, while the rate of cooling and heating does not seem to have any influence on the fatigue limit.

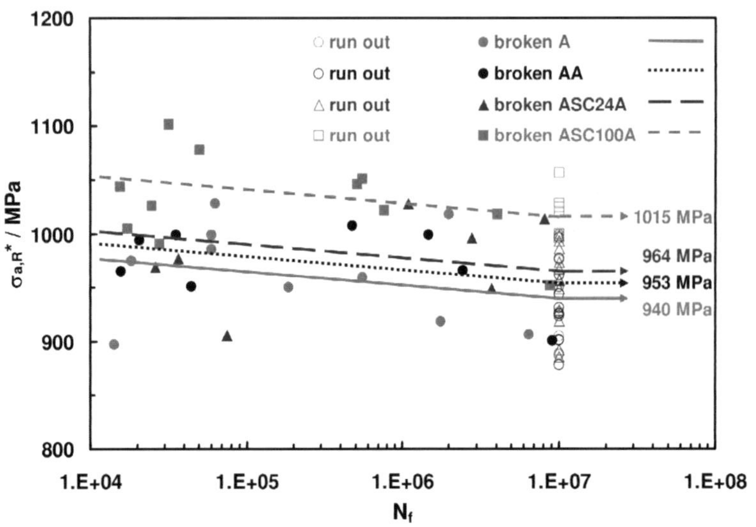

Figure 2. Woehler curve of the specimens without deep cryogenic treatment and with a slow cooling and heating rate and a soaking time of 24 and 100 hours.

Therefore, in two states the specimens were annealed, cooled fast to -196° C, soaked for 100 and 1000 hours and finally annealed after fast heating up to room temperature. Woehler curves show that a longer soaking time increases the fatigue limit by about 7 %. Yet, there is still a reduction of the fatigue limit by the fast cooling process, which can by seen by comparing the fatigue limit of state AA and AFC100A. The second one has still a lower fatigue limit than the just annealed state.

Evidently, the best fatigue limits are expected after annealing, slowly cooling, soaking a long time at -196° C, heating up slowly and annealing again. Starting with a soaking time of 24 hours the fatigue limit is increased linearly with the soaking time up to 100 hours. Unfortunately, a further increasing of the soaking time to 1000 hours did not further increase the fatigue limit.

Table 1. Results of different paths of the deep cryogenic treatments: given are the hardness, the retained austenite and the fatigue limits as well as the mean value of K_{max}/K_{mod}.

	Vickers hardness HV 10	*retained austenite vol.-%*	fatigue limit MPa	K_{max}/K_{mod}
only hardened	950	11.3	-	-
FC100	950	9.5	-	-
SC100	950	9.5	-	-
FC100A	825	8.0	888	0.95
SC100A	825	8.0	896	0.84
A	780	8.5	940	1.00
AA	770	4.0	953	1.06
AFC100A	770	4.0	923	1.09
AFC1000A	770	4.0	988	1.01
ASC24A	770	4.0	964	1.08
ASC65A (10 x SC)	770	4.0	990	1.15
ASC100A	770	4.0	1015	1.08
ASC1000A	770	4.0	988	1.10

3.3. Evaluation of the inclusions at the fracture surface

To exclude a statistical effect caused by different inclusion morphologies of the fractured specimens the stress intensity factors K_{max} at the critical site of the nonmetallic inclusions were calculated and compared to the expected value K_{mod} [9]. By this we get information about the load which is applied to the matrix directly at the inclusion by K_{max}. The higher the ratio K_{max}/K_{mod} the higher is the fatigue limit in the matrix directly surrounding the inclusion.

The value of 1 for the state A shows that the calculation works quite well for normally heat treated states. The values below 1 for FC100A and SC100A represent the reduced fatigue limit of these states. The second annealing without cooling increases the K_{max}/K_{mod} of about 6 %. Additional slow cooling increases K_{max}/K_{mod} slightly to 1.08. Surprisingly the state ASC65A shows an increase of 15 %. This is caused by the repeated cooling, soaking and heating up, which was done 10 times for this state.

4. Conclusion

The surroundings of inclusions and their modification are the key factor in determining the fatigue limit. We suppose that microcracks are initiated at these places by deep cryogenic treatments if the following conditions were not taken into account: the annealing directly after hardening is necessary to reduce the thermal eigenstresses around the inclusions, which result from the different thermal expansion coefficients of the inclusion and matrix, and the microstresses resulting from martensitic transformation. In consequence all states which are annealed before cooling exhibit $K_{max}/K_{mod} > 1$

The fast cooling may cause problems due to the thermal stresses up to 530 MPa tensile stresses which are induced between the interior (24°C) and surface (-196°C) of a specimen. They are high enough to induce microcracks.

However, increasing the fatigue limit of a high-strength bearing steel is possible by deep cryogenic cooling. Therefore, it is essential to anneal before and after deep cryogenic cooling and to cool down and heat up with slow cooling rates. Other treatments led to a decrease of the fatigue limit. The soaking time should have no influence since there is no classical diffusion of C in α-Fe possible.

The reason for the beneficial effect to the fatigue limit is yet unclear. Possibly, very fine carbides are formed, but investigations with light and electron microscope did not reveal any evidence for this theory.

References

[1] Erdmann-Jesnitzer F 1960 Beitrag zur Tieftemperaturbehandlung gehärteten Stahles *HTM* **15** 128-33

[2] Beumelburg W 1974 *Das Verhalten von einsatzgehärteten Proben mit verschiedenen Oberflächenzuständen und Randkohlenstoffgehalten im Umlaufbiege-, statischen Biege- und Schlagbiegeversuch* (Karlsruhe: Universität Karlsruhe)

[3] Mohan Lal D, Renganarayanan S and Kalanidhi A 2001 Cryogenic treatment to augment wear resistance of tool and die steels *Cryogenics* **41** 149-55

[4] Alexandru I and Bulancea V 2002 Effect of Cryogenic Cooling of Residual Stresses, Structure, and Substructure *Handbook of Residual Stress and Deformation of Steel* ed. G Totten, M Howes and T Inoue ASM International, 331-44

[5] Huang J Y, Zhu Y T, Liao X Z, Beyerlein I J, Bourke M A and Mitchell T E 2003 Microstructure of cryogenic treated M2 tool steel *Mater. Sci. Eng. A* **339** 241-44

[6] Kerscher E, Lang K-H, Vöhringer O and Löhe D 2008 Increasing the fatigue limit of a bearing steel by dynamic strain ageing *Int. J. Fatigue* **30** 1838-42

[7] Braam J J and van der Zwaag S 1998 A Statistical Evaluation of the Staircase and the ArcSin Methods for Determining the Fatigue Limit *J. Testing and Evaluation* **26** 125-31

[8] Macherauch E and Wohlfahrt H 1972 Prinzipien der quantitativen röntgenographischen Phasenanalyse (RPA) *HTM* **27** 230-32

[9] Kerscher E, Lang K-H and Löhe D 2008 Increasing the Fatigue Limit of a High-Strength Bearing Steel by Thermomechanical Treatment *Mater. Sci. Eng. A* **483-484C** 415-17

Fatigue crack growth in nodular cast iron - Influences of graphite spherical size and variable amplitude loading

T. Mottitschka[1], G. Pusch[1], H. Biermann[1], L. Zybell[2], M. Kuna[2]

[1]Institute for Materials Engineering, Gustav–Zeuner-Strasse 5, D-09599 Freiberg, Germany

[2]Institute of Mechanics and Fluid Dynamics, Lampadiusstrasse 4, D-09599 Freiberg, Germany

thomas.mottitschka@iwt.tu-freiberg.de

Abstract. Investigations of constant and variable amplitude loading, including a variety of overloads, were carried out experimentally as well as numerically in order to characterize the fatigue crack growth behaviour of nodular cast iron with a ferritic matrix. Under constant cyclic loading the crack growth rate, which depends on the graphite size can be described well by the NASGRO equation. The mean distance between the graphite particles and the shape factor also influence the fracture behaviour. The experimental investigations show that overloads yield to acceleration effects in fatigue crack growth. The calculation of the a-N curves based on different usual life prediction models yields non-conservative results. Therefore, a modified strip yield model is presented which allows the prediction of the crack growth acceleration after overloads in nodular cast iron.

1. Introduction

During use, a component is often subjected to variable amplitude loadings, which influence the lifetime. Therefore, fracture mechanics assessment of fatigue crack growth behaviour under variable amplitude plays an important role in many areas of applications of nodular cast iron especially in the case of components made for wind turbines. In order to quantify load history effects in ductile cast iron, fatigue crack growth experiments were carried out under constant and variable amplitude loading. The acceleration effects observed in fatigue crack growth of nodular cast iron significantly affect the accuracy of life predictions. Therefore, an extended strip yield model was developed to simulate the fatigue crack growth in nodular cast iron under variable amplitude loading.

2. Experimental setup

A ductile cast iron EN-GJS-400-18LT with a ferritic matrix and two specific adjusted graphite spherical sizes was studied. The cast test material variants G10 and G50 are produced from one slab by using different cooling rates to adjust specific graphite sizes d_G. Higher cooling rate leads to finer spherical graphite size (G10) using a chill-mould. Lower cooling rate leads to larger spherical graphite size (G50) by means of sand mould. The graphite morphology of both test materials was statistically evaluated by quantitative structural analysis. Stroppe et al. found that the mean distance λ between the graphite particles and the shape factor f influence the fracture behaviour of cast iron with a ferritic matrix [1]. The deviation from the spherical size is described by the shape factor (for an ideal sphere f=1). Table 1 gives the mechanical properties and the graphite morphology for both materials.

Table 1: Mechanical properties and graphite morphology of both tested materials [7]

Material	T [°C]	R_m [MPa]	$R_{p0,2}$ [MPa]	A_5 [%]	E [GPa]	d_G [µm]	λ [µm]	f [-]
G10	RT	388	247	23,5	176	13	21	0,75
G50	RT	374	242	22,9	179	41	58	0,53

The experiments were performed with SENB-specimens of dimensions 10 mm x 20 mm x 100 mm. The fatigue crack growth experiments under constant amplitude loading were carried out at constant stress ratios of R=0.1, 0.3 and 0.5 according to ASTM E 647-08 [2] using a resonance testing machine *RUMUL*. The compliance method was used to determine the crack length. The investigations of load history effects with variable amplitude loading were carried out using a MTS servo-hydraulic testing machine under load-controlled conditions, in which the crack length were measured by using crack gauges. Fatigue crack propagation under variable amplitude loading was realized by

- Constant cyclic load with various tensile overload ratios (R_{OL}=1.25…2.25)

  R-ratio during a single overload defined by $R_{ol} = \dfrac{P_{ol}}{P_{BL,\max}}$ (1)

  where P_{ol} is the maximum force during the overload and $P_{Bl,\max}$ is the maximum force of the baseline-level loading.
- Block loading (low-high-low sequences) with various block loading ratios (R_{block}=1.25…2.0)

  R-ratio during the block loading was defined by $R_{block} = \dfrac{P_{block}}{P_{BL,\max}}$ (2)

  where P_{block} is the maximum force during the block loading and $P_{BL,\max}$ is the maximum force of the baseline-level loading.

3. Experimental results and numerical simulation

3.1. Fatigue crack growth behaviour under constant and variable loading

The crack growth curves for both materials G10 and G50 are illustrated in Fig. 1 for all investigated R-ratios. In Fig. 1a the crack growth curves are only given for lower and middle crack growth rates, whereas Fig. 1b compares the complete crack growth curve of G10 and G50 for R=0.3. The fracture mechanical properties for both materials at various stress ratios are given in Table 2. Fatigue crack growth under constant amplitude loading can be described well by the NASGRO equation, as seen in Fig. 1b. This equation is an extended Paris equation, which describes all three ranges of the crack growth curve in dependency of the stress ratio. For further information it is referred to [3, 4]. The results in Table 2 and Fig. 1 show that fatigue crack growth depends on the graphite size. With increasing spherical graphite size, higher threshold values ΔK_{th} (see Table 2) and higher critical cyclic stress intensity factors ΔK_{fc} are obtained (see Fig. 1). Examples for the extensive analysis of the fatigue crack growth under variable amplitude loading to investigate load history effects on nodular cast iron are given in Fig. 2. For G10 the influence of a 1.5-fold single tensile overload is presented in Fig. 2a. No crack growth retardation occurs. The crack growth rate shortly increased up to about 1E-03 mm/cycle and then returned to the previous crack growth rate level. As a result of one tensile overload the crack required about 440 cycles to level off to its previous crack growth rate. Fig. 2b shows the influences of block loading (low-high-low sequence) with two different block loading ratios R_{block}=1.25 and 1.50 on G10, whereby higher block loading ratios lead to higher crack growth rates. The transition from a low to a high sequence shortly induces higher crack growth acceleration before the crack growth rates settle down. The transition from high to low does not yield retardation effects.

Table 2: Results of the fatigue crack growth behaviour under constant amplitude loading

Material	*R*-ratio	ΔK_{th} [$MPam^{1/2}$]	m	C
G10	0.1	10.1	6.0	5.1E-12
	0.3	8.1	5.0	1.5E-10
	0.5	5.1	5.0	2.8E-10
G50	0.1	10.2	6.1	3.3E-12
	0.3	8.5	5.0	2.0E-10
	0.5	5.5	4.6	3.1E-10

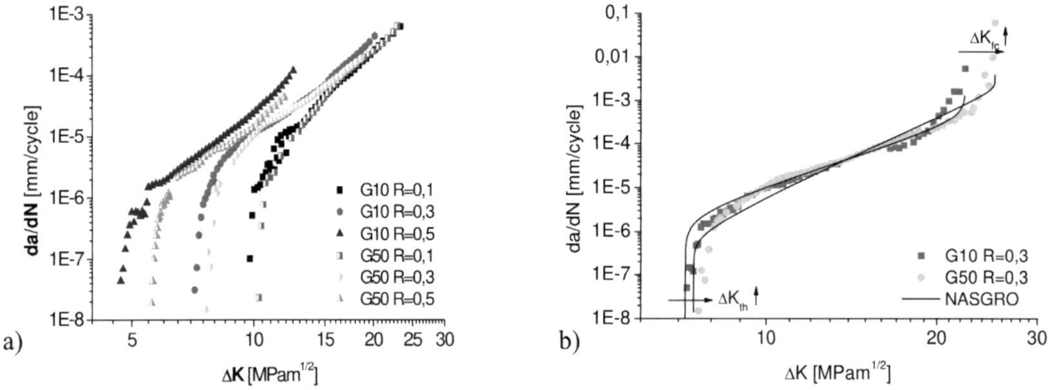

Fig. 1: Influence of stress ratio a) and graphite spherical size b) on fatigue crack growth

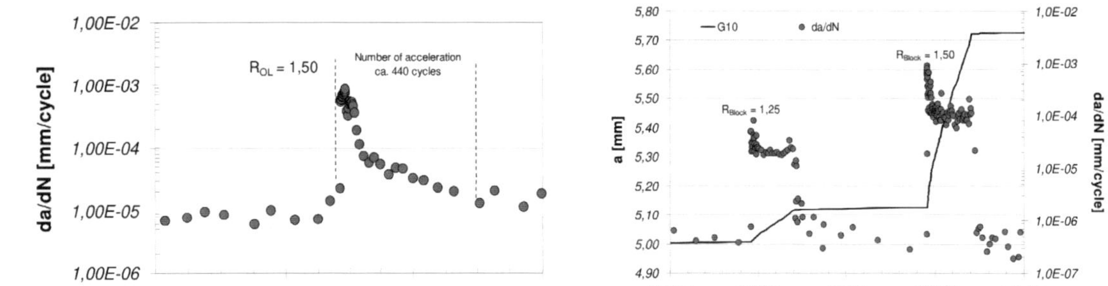

Fig. 2: Influence of a single tensile overload a) and block loading b) on fatigue crack propagation of G10

Scanning electron microscopy was applied to investigate the damage behaviour during crack propagation in nodular cast iron under variable amplitude loading. It could be shown that overloads lead to debonding the nodular graphite from the ferritic matrix and also micro crack initiation was observed on the specimen surfaces in a region forms ahead of the crack tip. This damage behaviour can lead to crack growth acceleration [4].

3.2. Numerical simulation of load history effects with an extended Strip Yield Model (SYM)

In order to be able to simulate the acceleration effects occurring during fatigue crack growth of nodular cast iron an extended strip yield model (SYM), based on the work of Newman [5] and Wang [6], was developed. The basic equations of the strip yield model arise from a modified Dugdale model, where the crack is fictitiously extended by the length of the plastic zone. Within the plastic zone there are elastic ideal-plastic bar elements of length L_i, which are separated by crack advance and which further carry the residual plastic deformation on the crack surfaces. The extension of the strip yield model regarding the acceleration effects is realised by an additional boundary condition for the plastic deformation L_i of the bar elements within the plastic zone. If a critical value L_i* is reached, the corresponding element is separated at once and the crack advances.

$$L_i \geq L_i^* : \qquad \sigma_i > 0 \;\Rightarrow\; \sigma_i = 0 , \quad \sigma_i < 0 \;\Rightarrow\; \sigma_i = 0 \tag{3}$$

The crack growth rate is calculated by the NASGRO equation, using parameters identified for EN-GJS-400-18LT by Hübner et al. [3]. Accelerated crack growth in nodular cast iron after single overloads can be described with the extended strip yield model, see Fig. 3. In order to indentify the microstructure dependent parameter L_i* for crack growth acceleration the evaluation of systematic experiments with single overloads is currently under progress.

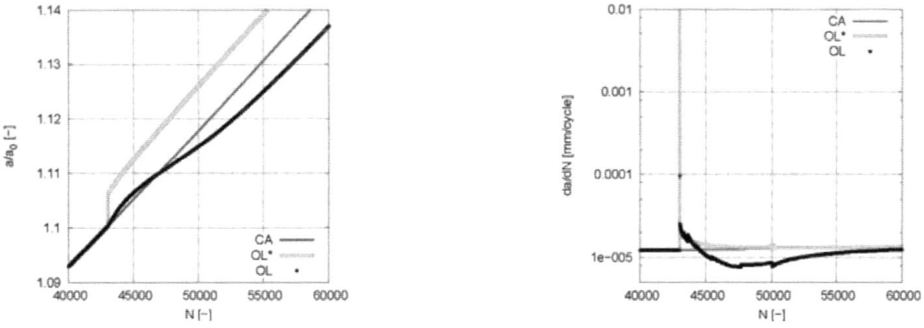

Fig. 3: Comparison of crack length a and crack growth rate da/dN versus cycle plot under constant amplitude loading (CA) and one integrated overload (OL-general SYM, OL*-extended SYM)

4. Conclusions

Within the scope of this work, systematic investigations of constant and variable amplitude loading were carried out by experiments as well as numerical studies in order to characterize the fatigue crack growth behaviour of nodular cast iron. The fatigue crack growth behaviour depends on the graphite size. Increasing graphite size leads to higher thresholds values ΔK_{th} and higher critical cyclic stress intensity factors ΔK_{fc}. The experimental investigations under variable amplitude loading always show acceleration effects in fatigue crack growth and no retardation effects have been obtained. The presented numerical simulations show that accelerated crack growth after a single tensile overload can be described with the extended strip yield model.

Acknowledgement

The authors are grateful to the German Research Foundation (DFG) for financial support.

References

[1] H. Stroppe, G. Pusch and A. Ludwig *Giessereiforschung* **54** (2002) 115-120
[2] ASTM E647-08
[3] P. Hübner and G. Pusch *Konstruieren und Giessen* **32** (2007) 34-37
[4] P. Hübner, H. Schlosser, G. Pusch, H. Biermann: *Int. J. Fatigue* **29** (2007) 1788-1796
[5] J.C. Newman Jr. *ASTM STP* **748** (1981) 53-84
[6] G.S. Wang, A.F. Blom *Eng. Fract. Mech.* **40** (1991) 507-533
[7] L. Zybell, M. Kuna, T. Mottitschka, M. Hoffmann, G. Pusch, H. Biermann *DVM-Bericht* **241** (2009) 73-88

Microstructural aspects of duplex steel during high cycle and very high cycle fatigue

H Knobbe[1], P Köster[1,2], U Krupp[3], H-J Christ[1], C-P Fritzen[2]

[1] Institut für Werkstofftechnik, Universität Siegen, Paul-Bonatz-Straße 9-11, D-57068 Siegen, Germany
[2] Institut für Mechanik und Regelungstechnik – Mechatronik, Universität Siegen, Paul-Bonatz-Straße 9-11, D-57068 Siegen, Germany
[3] Fakultät Ingenieurwissenschaften und Informatik, Fachhochschule Osnabrück, Albrechtstraße 30, D-49076 Osnabrück, Germany

helge.knobbe@uni-siegen.de

Abstract. Fatigue experiments were conducted up to $N=10^8$ cycles on a duplex stainless steel. The investigations revealed important information: (i) most of the slip markings on {111}-slip planes form during the early stage of fatigue (~10^4), although many grains stay active in developing more and/or longer slip bands (investigations with SEM/EBSD). It was found that the damage in these grains often has its origin at grain or phase boundaries, growing into the interior of the grains; (ii) in most of the recrystallisation twins slip markings are present in all parts of the grain leading to the assumption that just tilted grain boundaries are not very effective as microstructural barriers. In cases where the twin grain boundary is favorably oriented, it even can promote microcrack initiation; (iii) phase boundaries are most effective in restricting dislocation movement to the softer austenite, since no evidence of growing fatigue damage was found in ferritic grains. These findings lead to the conclusion that for the material studied an endurance limit exists despite occurring plastic deformation. This behaviour is a consequence of the barrier effect induced by phase boundaries against short crack propagation and holds true as long as no inclusions are present.

1. Introduction

During the last years much attention has been drawn to the investigation of metallic materials' fatigue behaviour in the very high cycle regime, implying numbers of cycles to fracture ranging from 10^6 up to 10^{10} or 10^{12} cycles [1,2]. In consequence of the first results obtained mostly on high strength steels and aluminium alloys the ongoing discussion whether a fatigue limit exists or not was raised. Several authors found that many different influencing factors such as microstructural condition, prehistory, notches and experimental setup may have a crucial effect on the fatigue life [3]. However, it was discovered that materials can fail below the conventional fatigue limit [4], but most of the work concentrates on single-phase materials [5]. Few papers are available covering the complex situation of two-phase materials such as two-phase titanium alloys and duplex steels [6-8]. The crack initiation sites can be shifted from the surface in case of HCF to subsurface for VHCF loads. The latter can occur at inclusions or non-defect sites. In this study, only inclusions were found to promote subsurface crack initiation in some rare cases.

2. Material and Experimental

The present material X2 CrNiMoN 22-5-3 consists of a strong phase (α ferrite) and a soft phase (γ austenite). The strength distribution strongly depends on the alloying condition, primarily the nitrogen content [9]. The chemical composition for the material studied is given in table 1.

Table 1. Chemical composition of duplex steel in wt. %.

Fe	Cr	Ni	Mo	Mn	Si	P	S	Cu	Ti	Co	N	C
bal.	21.91	5.58	3.08	1.83	0.53	0.023	0.002	0.11	0.03	0.07	0.19	0.02

The material was heat-treated in order to obtain a homogeneous microstructure. Heat treatment was done in laboratory air at 1250°C for 4h followed by controlled cooling to 1050°C in 3h leading to a phase fraction of 1:1. Afterwards the material was water-quenched to prevent the formation of brittle phases such as the σ phase. After heat treatment the surface oxide scale was mechanically removed.
Fatigue experiments were carried out at a stress ratio of R = -1 with a servohydraulic fatigue testing system (MTS) capable of frequencies up to 1000 Hz. The occurrence of excessive damping heat requires the use of compressed air cooling and restricts the testing frequency to 300 Hz. The cylindrical specimens had a diameter of 5mm with a gauge length of 10 mm in order to test a relatively large volume. All experiments were performed in stress control with a sinusoidal waveform and were stopped at $1x10^8$ cycles. The damage evolution was followed by interruption of some tests after reaching defined numbers of cycles. The specimens used for these experiments contained a shallow-notch in the middle of the gauge length in order to localize the damage. All specimens were electrolytically polished prior to testing to obtain a surface without residual stresses allowing to measure orientation data. The EBSD technique was used to determine the distribution of the austenite and ferrite phases and the crystallographic orientations.

3. Results

The S-N-data of the material is published in ref. [10]. In summary, three different situations were observed at stress levels in the order of the endurance limit at 10^8 cycles: (i) failure of the specimens in the classical high cycle fatigue regime (~10^6 cycles) with crack initiation at the surface, (ii) failure with crack initiation at non-metallic inclusions with fish eye formation (~10^7 cycles) and (iii) no failure after 10^8 cycles (run-out specimens). The existence of slip markings at run-out specimens together with the measurable temperature increase during the experiments lead to the assumption that the occurring plastic slip is present during the whole experiment but reversible to some extent. In order to gain more precise information on the evolution of fatigue damage with ongoing cyclic loading, different areas were selected on the shallow notch and analyzed by means of SEM after certain numbers of cycles. Figure 1 shows the evolution of slip traces in one austenite grain, marked by the orange circle. Two important observations can be made: (i) the cyclic deformation starts at or near the phase boundary in the upper part of the grain and grows into the interior of the grain (this may be explained by incompatibility stresses raised by the different strengths of both phases [11]); (ii) after reaching the phase boundary at the lower part of the grain more and more other parts of the grain are involved, so no saturation can be found up to $2x10^7$ cycles. This indicates a continuous plasticity during the whole experiment, which is in good agreement with the measured temperature increase. Orientation measurements by EBSD reveald that slip most frequently occurs on the {111} slip planes, but the phase boundaries obviously are strong enough to prevent the dislocations from crossing into the ferritic phase. Figure 2 shows another example of slip trace evolution for two interesting sites. The orange circle marks an austenite grain containing a twin boundary. This boundary is oriented in such a way that slip is induced on it. It is well known for pure austenitic steels that twin boundaries act as crack initiation sites [12]. However, another slip system is also activated. In this case the slip traces become more pronounced during the fatigue process, but they do not cross the twin boundary, because the plastic deformation on the boundary, together with the unfavorable orientation of the grain, prevents slip into the other part of the grain. The grain highlighted by the blue circle is also a twin but shows another situation. Here, slip markings have crossed the boundary very early in the fatigue process leading to the assumption that generally twin boundaries do not have a strong influence on cyclic slip. This is in good agreement with the idea of special (coincident site lattice, CSL) grain boundaries. In this case CSL-boundaries of type Σ3 are found, meaning that every 3rd atom contained in the boundary is shared by both sides of the twin, making it easy for the dislocations to move over the boundary. Furthermore this observation gives rise to the assumption that tilted boundaries are not very effective as barriers. In other words high twist angles are required in order to gain more efficiency against dislocation movement.

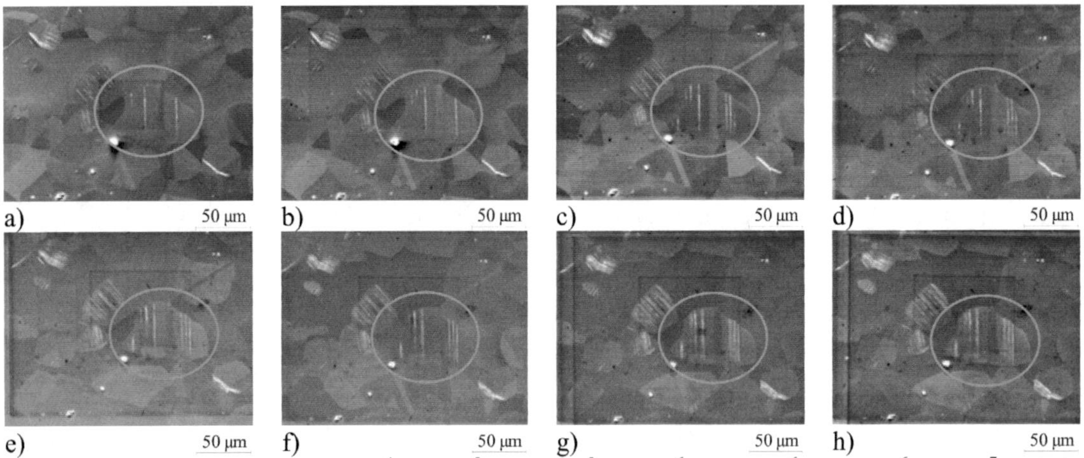

Figure 1. Slip traces after (a) $5x10^4$, (b) 10^5, (c) $5x10^5$, (d) 10^6, (e) $2x10^6$, (f) $5x10^6$, (g) 10^7 and (h) $2x10^7$ cycles showing that cyclic deformation starts at grain or phase boundaries.

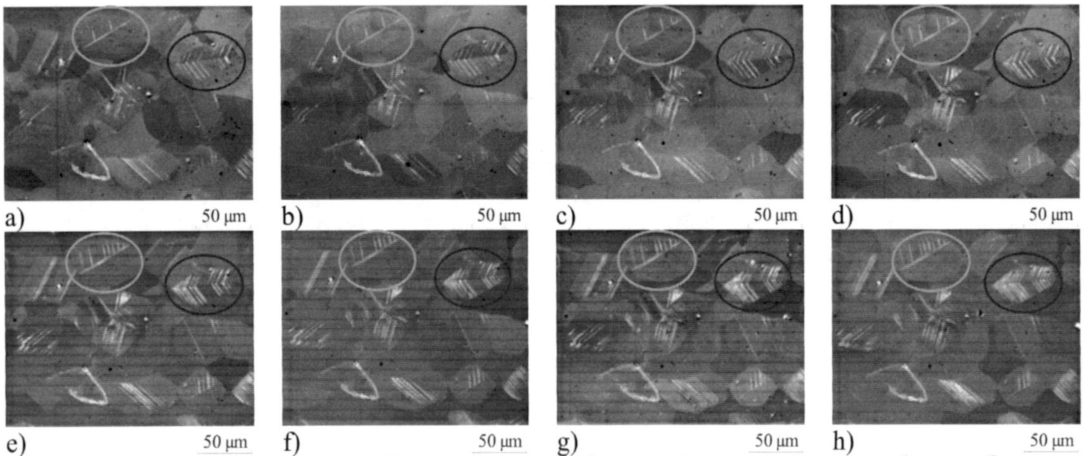

Figure 2. Slip traces after (a) $5x10^4$, (b) 10^5, (c) $5x10^5$, (d) 10^6, (e) $2x10^6$, (f) $5x10^6$, (g) 10^7 and (h) $2x10^7$ cycles showing that cyclic deformation occurs throughout the whole experiment.

Figure 3 indicates that areas near the phase boundaries do not only induce cyclic slip but can also promote microcrack initiation. These findings could be linked to incompatibility stresses together with elastic anisotropy. In this case a stage I crack has formed on a single slip system in the austenite, but obviously the crack is not able to grow (a,b). Figure 3 c) shows a detailed microgragh of the slip bands with considerably strong intrusions and extrusions. But still there are areas in between the slip traces that are not deformed up to this stage of fatigue. The observations above led to the assumption that these areas will be involved in the cyclic deformation with higher numbers of cycles. Finally, figure 3 d) gives an insight into the microcrack formation on the slip band. It is clearly visible that the crack started from an intrusion growing with a certain angle inside the material. This is typically observed for stage I microcracks [13]. The situation of non-growing cracks can be governed by two factors: (i) the phase boundary in the lower part (figure 3 a,b) simply arrests the crack. In other words the crack-driving force is not high enough to exceed the critical stress in the ferrite since this is the stronger phase; (ii) the incompatibility stresses which are superimposed to the external loading are released by the crack initiation. Once the crack has formed and the microstructural stresses are gone, the external stress is not high enough to cause further crack growth and thus an endurance limit can be defined, at least as long as no inclusions are present in the tested volume.

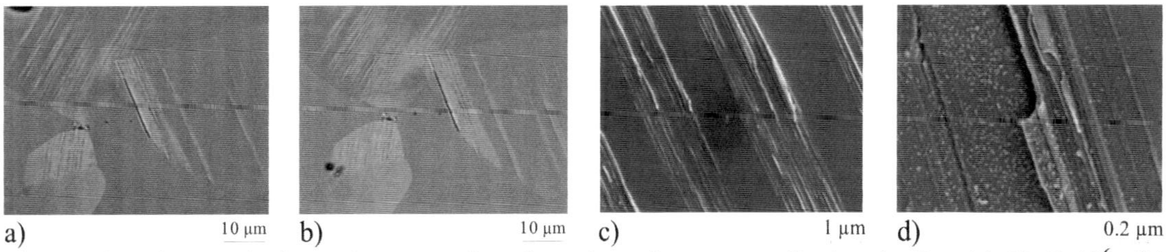

Figure 3. Microcrack formation near the phase boundary on a slip band after (a) N=5x10^6, (b) N=2x10^7, (c) N=2x10^7 (intrusions and extrusions in slip bands), (d) N=2x10^7 (microcrack initiation at intrusions).

4. Conclusions
The fatigue behaviour of duplex steel in the very high cycle fatigue regime was investigated with special emphasis on fatigue damage evolution. It was found that cyclic slip is present during the whole experiment in the softer austenite. Twin boundaries play an important role because they either act as damage initiation sites or promote slip across the boundary. However, microcracks predominantly form near phase boundaries due to incompatibility stresses. Such cracks do not grow further as soon as the microstructural stresses are reduced. This leads to the definition of an endurance limit despite ongoing plastic deformation if no inclusions are existent.

References
[1] Stanzl-Tschegg S and Mayer H 2001 *Proc. Int. Conf. on Fatigue in the Very High Cycle Regime* (Vienna: University of Agriculture Science Press)
[2] Sakai T and Ochi Y 2004 *Proc. Int. Conf. on Very High Cycle Fatigue* (Kusatsu: Ritsumeikan University)
[3] Allison J E, Jones J W, Larsen J M and Ritchie R O 2007 *Proc. Int. Conf. on Fatigue in the Very High Cycle Regime* (Ann Arbor: University of Michigen)
[4] Bathias C 1999 *Fatigue Fract. Eng. Mater. Struct.* **22** 559
[5] Lukáš P, Kunz L and Svoboda M 2006 *Proc. Int. Conf. on Fatigue* ed W S Johnson (Atlanta: Georiga Institute of Technology).
[6] Notkina E, Lütjering G and Gysler A 2001 *Proc. Int. Conf. on Fatigue in the Very High Cycle Regime* ed S Stanzl-Tschegg and H Mayer (Vienna: University of Agriculture Science Press).
[7] Chai G 2006 *Int. J. Fatigue* **28** 1611
[8] Chai G 2006 *Int. J. Fatigue* **28** 1533
[9] Johansson J and Odén M 2000 *Met. Mater. Trans. A* **31** 1557
[10] Knobbe H, Köster P, Krupp U, Christ H-J, Fritzen C-P, Cherif M A and Altenberger I 2007 *Proc. Int. Conf. on Fatigue in the Very High Cycle Regime* ed J E Allison, J W Jones, J M Larsen and R O Ritchie (Ann Arbor: University of Michigan).
[11] Baczmanski A, Braham C and Seiler W 2004 *Phys. Stat. Sol.* **201** 2886
[12] Blochwitz C and Tirschler W 2003 *Mater. Sci. Eng. A* **339** 318
[13] Forsyth P J E 1963 *Acta Metall.* **11** 70

Journal of Physics: Conference Series **240** (2010) 012062 doi:10.1088/1742-6596/240/1/012062

Analysis of creep behaviour of TiAl-8Ta intermetallic alloy

G Angella, M Maldini, V Lupinc

CNR – Istituto IENI – Milano – Via R. Cozzi 53, 20125 Milano, Italy

angella@ieni.cnr.it

Abstract. The creep behaviour of a high Ta containing γ-TiAl alloy was investigated in the temperature range of 700-850°C with applied stresses in order to have rupture times up to 3000 h. Decelerating primary and accelerating tertiary regimes dominated the experimental curves, whilst the secondary regime with constant minimum creep rates was absent. Reporting the experimental data in plot $\log\dot{\varepsilon}$ vs. ε, the tertiary data lie on lines with similar slopes at all temperature and load conditions, indicating that a damage mechanism, depending only on the accumulated creep strain causes the accelerating tertiary regime. Creep tests with step like changes in load and/or temperature changes were run and microstructure investigations were performed through X ray diffraction, scanning and transmission electron microscopy to have an insight on the nature of the damage mechanisms that control the accelerating tertiary regime.

Introduction

The dual phase $\alpha_2 + \gamma$ TiAl alloys have been widely investigated, since they are candidates to be used in aeronautic and land based gas turbine [1-3]. A route to produce the desired fine microstructure in TiAl alloys rich in heavy elements such as Nb and Ta is based on the massive transformation typical of such alloys [4-7]. The designed thermal and HIP treatments, following to the massive transformation, are imposed to obtain the final convoluted cross $\alpha_2 + \gamma$ lamellar microstructure that confer to the alloys an adequate room temperature ductility, the critical parameter for intermetallic alloy applications.

The physical and mechanical proprieties of a new Ti-46Al-8Ta (atom. %) have been studied in the integrated European project IMPRESS. In the present work the first creep characterisation results are reported with particular attention on the accelerating tertiary regime that dominates the creep curves of TiAl-8Ta, and on the damaging mechanisms that cause the accelerating regime through creep tests with changing conditions and microstructure investigations.

Material and experimental procedure

The alloy Ti-46Al-8Ta (atom. %) was provided by ACCESS (Aachen) in the form of cast and heat treated cylindrical bars with a diameter of 13 mm and length of 120 mm. Heat treatment consisted of hot isostatic pressing (HIP) at an applied pressure of 200 MPa, temperature of 1260 °C for 4 h, which was followed by solution annealing at 1360 °C for 1 h and air cooling. A final HIP ageing at an applied pressure of 150 MPa and temperature of 1260°C was imposed for 2 h, followed by cooling at a rate of 5 °C/min.

Constant load creep tests were performed on cylindrical samples with gauge length and diameter of 28.0 and 5.6 mm respectively. The gauge elongation was measured through capacitive transducers connected to extensometers clamped to the specimen gauge ledges. The specimens were heated through furnaces and the gauge temperature was controlled through three thermocouples in order to avoid temperature gradients along the specimen gauge.

The microstructure characterization was carried out through X ray diffraction (XRD), scanning electron microcopy (SEM) with energy dispersion spectroscopy (EDS) and transmission electron microscopy (TEM). The TiAl-8Ta specimens for SEM were polished and were not etched chemically, since the backscattered electron signal in SEM was adequate to collect quality micrographs.

Results and discussion

3.1. Creep test results

When the true creep rates $\dot{\varepsilon}$ are plotted against the true creep strains ε, the TiAl-8Ta curves do not exhibit a wide plateau corresponding to the conventional secondary regime, but decelerating primary and

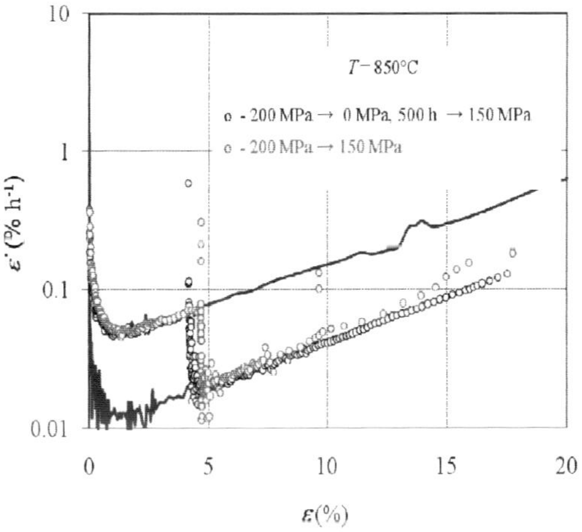

Figure 1. Creep data log $\dot{\varepsilon}$ vs. ε from load change tests at 850°C. Blue continuous lines are from constant load tests, reported for comparison.

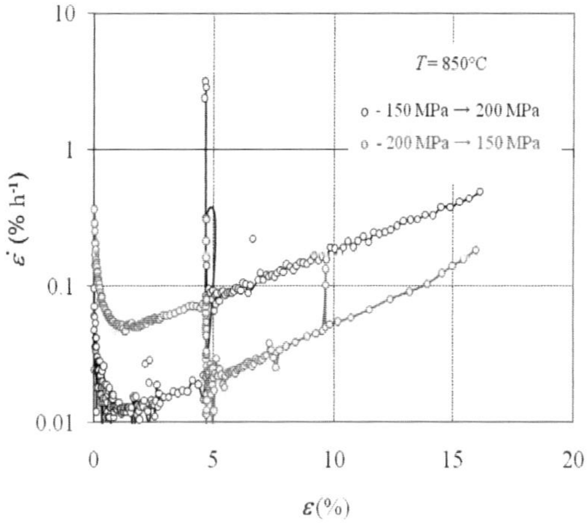

Figure 2. Creep data log $\dot{\varepsilon}$ vs. ε from load change tests at 850°C.

accelerating tertiary regimes dominate the experimental curves at all the present experimental conditions. Minima of the creep rates, i.e. $\dot{\varepsilon}_{min}$, are observable rather than steady state creep rate regimes. From the analysis of log $\dot{\varepsilon}_{min}$ vs. logσ plots, where σ is the nominal applied stress, different Norton exponents were obtained at different explored temperatures: 7.6 at 700°C, 6.6 at 750°C, 6.3 at 800°C and 4.9 from 850°C.

In log $\dot{\varepsilon}$ vs. ε plot, the experimental points representing the long accelerating creep, lie on straight lines with similar slopes at all tested temperatures and loads. Therefore, the tertiary creep can be described by the following relationship:

$$\dot{\varepsilon} = \dot{\varepsilon}_o \exp(m\varepsilon) \tag{1}$$

where $\dot{\varepsilon}_o$ is the back-extrapolated strain rate to $\varepsilon = 0$ and $m = 15.6 \pm 1.7$. In the log $\dot{\varepsilon}$ vs. ε plots, the experimental creep curves are congruent, i.e. it is possible to overlap the creep curves obtained at different stresses and temperatures by a translation along the $\dot{\varepsilon}$ axis. As a consequence, the shape of the creep curves is independent on the applied stresses and all the creep curves coincide if plotted as strain in function of the fraction life t/t_f where t_f is the time to fracture or, the time to reach a defined strain. The experimental results indicate that the damage producing the dominant accelerating stage depends on strain only and the creep strain *univocally* quantify both the damage and the life fraction of the material. Such a behaviour is generally not found in nickel base superalloys, where there is a systematic variation of the creep curve shape at different stresses [8].

In order to confirm that the damage producing the accelerating creep is independent on the applied stress and on the time spent at high temperature, creep tests with step like load changes were run at 850°C when the microstructure degradation was expected to be more significant. The alloy was first loaded with an initial stress of 200 MPa to about 5%, corresponding at about half of the creep life, after that the nominal stress was reduced to 150 MPa. A second test consisted of 200 MPa to about 5%, after that a period of 500 h in temperature with no applied load was imposed and finally a nominal load of 150 MPa was applied. The experimental results of the variable load creep tests between 200 and 150MPa are plotted as log $\dot{\varepsilon}$ vs. ε in figure 1 and confronted with the constant load creep curves: following the stress increment and after a small and short transient, the strain rates lie over the reference constant load creep curves at the final applied stress. A further test consisted of 150 MPa to about 5%, after that an increase of nominal load to 200 MPa was imposed. The data are reported in log $\dot{\varepsilon}$ vs. ε and compared to the 200 MPa + 150 MPa test in figure 2:

again, after reloading, no significant change of the strain rates and the tertiary creep slopes occurred. The experimental results confirm the damage is not depending on time and on applied stresses, but only on the accumulated creep strain. Moreover, the small and short transient following the reloading, after a long permanence at high temperature without load, show the static recovery of the dislocation microstructure is very low. The average of the activation volumes calculated through the above stress change creep tests resulted in about $6.9b^3$, where $b = 0.283$ nm with the Taylor factor assumed to be 3.06 [9].

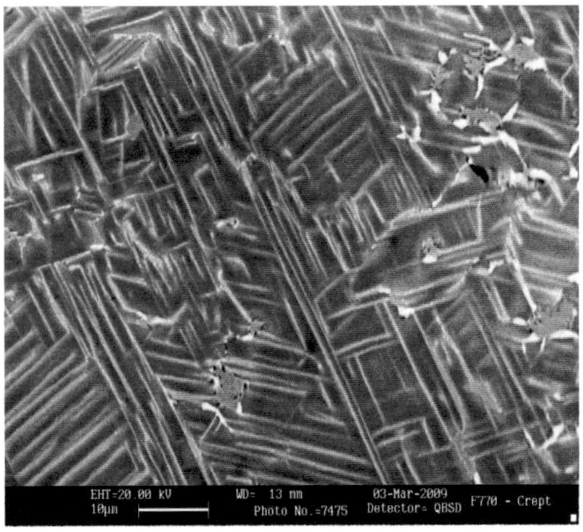

Figure 3. TiAl-8Ta microstructure after creep at 700°C: cross lamellar $\gamma + \alpha_2$ structure similar to the as received structure (α_2 bright). The grain boundaries present numerous Ta rich precipitates

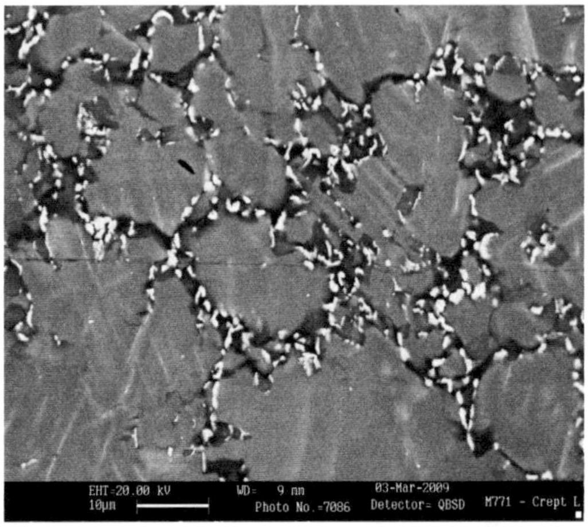

Figure 4. TiAl-8Ta microstructure after creep at 850°C: the α_2 lamellae are reduced in number and thickness; the grain boundaries present Ta rich precipitates and are depleted in Ta.

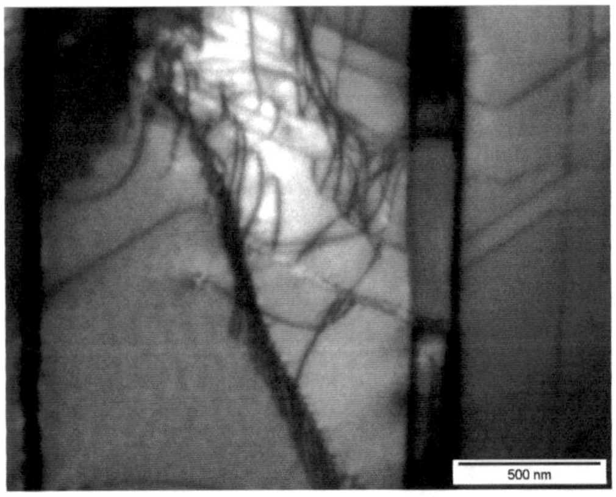

Figure 5. TEM micrograph of TiAl-8Ta after creep at 850°C. Dark thin lamellae are α_2 phase and the bright phase is γ: cut of α_2 phase is evident.

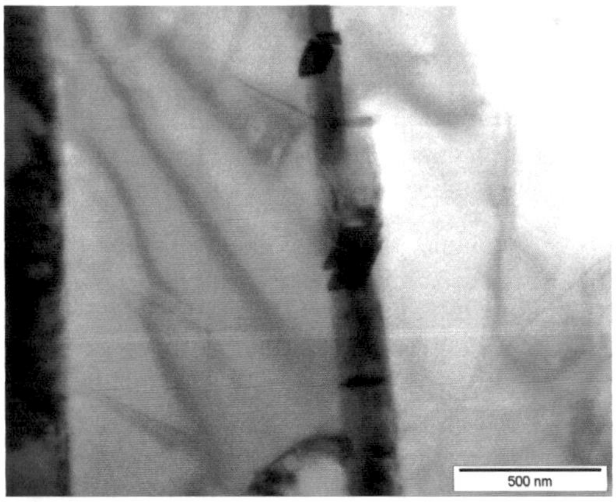

Figure 6. TEM micrograph of TiAl-8Ta after creep at 850°C. The planes of cut in α_2 phase are put in evidence.

Also creep tests with change of temperature from 850°C to 800°C and from 750°C to 700°C with constant load shown the same behaviour. The apparent activation energy of deformation resulted to be 356.5 kJ·mol^{-1} at 700°C and 365.4 kJ·mol^{-1} at 850°C, in both cases similar to the apparent energy for Al diffusion in Ti about 350 kJ·mol^{-1} in [8]. The found apparent energy of deformation and activation volume indicate that creep deformation in TiAl-8Ta in the present testing conditions is controlled by a dislocation climb [9].

3.2. Microstructure analysis

The microstructure of the alloy was examined before and after creep tests. Micrographs of the crept microstructure after about 1100 h at 700°C with nominal applied stress of 400 MPa and after about 500 h at 850°C/150 MPa are reported in figure 3 and 4 respectively. The microstructure of the crept alloy at 700°C is not significantly evolved with respect to the as received conditions, where cross lamellar $\gamma + \alpha_2$ structure is observable. Conversely after creep at 850°C the microstructure is considerably evolved with respect to the as-received material: the α_2 lamellae are reduced in number and thickness, a second phase rich in Ta precipitates profusely at the grain boundaries and the grain boundaries appear dark because of Ta depletion. Similar grain boundary degradation has been reported in TiAl alloys with significant addition of Nb [10]. The Ta rich phase was observed also in the as received alloy and at 700°C, however in smaller quantity. The XRD analysis confirmed the reduction of α_2 volume fraction at 850°C and indicated that the Ta rich phase was the ternary phase Ti_4TaAl_3, designated as τ with hexagonal crystal structure of the $B8_2$ [11]. At all the explored test conditions no sign of cavitations was detected.

Selected TEM micrographs of the TiAl-8Ta alloy crept at 850°C with an initial load of 100 MPa crept to rupture in about 3500 h are reported in figures 5 and 6. Several α_2 lamellae are still observable and they appear to be significantly cut, as shown in figure 6, where the planes of cut is put in evidence. The TiAl-8Ta crept at 700°C presented similar microstructure. The alloy did not present any sign of mechanical twinning that is very common in TiAl alloys [12-14].

The significant α_2 volume fraction reduction and grain boundary degradation during creep, particularly important in the tests performed at the highest temperatures, do not seem to be able to describe the experimental accelerating creep in all the temperature and stress conditions investigated. Furthermore, cavity formation at the grain boundaries, significant in other alloys for high temperature applications [8], has not been reported in the present alloy. A microstructure damage mechanism that is sensitive to strain only is not evident from microstructure results and further investigation is needed.

Conclusions

The experimental investigation on the creep behaviour of a high Ta containing γ-TiAl alloy has shown:

- the creep curves are dominated by the tertiary accelerating creep stage and no steady state was observed;
- the apparent activation energy of creep and activation volume indicate that creep deformation in TiAl-8Ta is controlled by a dislocation climb;
- the increment of the strain rate with the strain during the accelerating follows the same exponential strain softening law in all the explored temperature and stress conditions, in spite of widely differing rupture life;
- the accumulated creep strain is a good parameter to quantify the damage of the alloy;
- mechanical twinning has not been detected;
- the significant α_2 volume fraction reduction and grain boundary degradation during creep, specially in the tests performed at the highest temperatures, do not seem to be able to describe the experimental accelerating creep.

Acknowledgements

This work was financially supported by EU Integrated Project IMPRESS "Intermetallic Materials Processing in Relation to Earth and Space Solidification" under the contract No. NMP3-CT-2004-500635 and the alloy was provided by ACCESS (Aachen). Experimental assistance of E. Signorelli is gratefully acknowledged.

References

[1] Kim Y W 1994 *JOM-Journal of the Minerals, Metals & Materials Society* **46** 30
[2] Clemens H., Kestler H. 2000 *Adv. Eng. Mater.* **2** 551.
[3] Appel F, Wagner R 1998 *Mater. Sci. Eng.* A **22** 187
[4] Hu D, Wu X, Loretto M H 2005 *Intermetallics* **13** 914
[5] Clemens H, Bartels A, Bystrzanowski S, Chladil H, Leitner H, Dehm G, Gerlin R, Schimansky F P 2006 *Intermetallics* **14** 1380
[6] Hu D, Huang A J, Loretto M H 2007 *Intermetallics* **15** 327.
[7] Maldini M, Lupinc V 1997 *Mater. High Temp.* **14** 47
[8] Saage H, Huang A J, Hu D, Loretto M H, Wu X 2009 *Intermetallics* 17 32
[9] Appel F 2001 *Mater. Sci. Eng.* **A317** 115
[10] Bystrzanowski S, Bartles A, Clemens H, Gerling R, Schimansky F P, Dem G, Kestler H 2005 *Intermetallics* **13** 515
[11] IMPRESS WP5 annual report D-5-5 on "Microstructure Proprieties" 25th May 2009
[12] Appel F 2001 *Intermetallics* **9** 907
[13] Wang J G Hsiung L M Nieh T G 1998 *Scripta Mater.* **7** 957
[14] Hsiung L M, Nieh T G, Choi B W, Wadsworth J, *Mater. Sci. Eng.* A **329** 637

Synthesis and characterization of lamellar and fibre-reinforced NiAl-Mo and NiAl-Cr

T Haenschke[1], **A Gali** [2], **M Heilmaier** [3], **M Krüger** [4], **H Bei** [2], **E P George** [2,5]

[1] Department of Metallurgy and Materials, IRC, The University of Birmingham, Birmingham B15 2TT, UK

[2] Oak Ridge National Laboratory, Materials Science and Technology Devision, 1 Bethel Valley Road, Oak Ridge, TN 37831-6093, USA

[3] TU Darmstadt Materials Science, Petersenstraße 23, 64287 Darmstadt, Germany

[4] Otto-von-Guericke University Mechanical Engineering, Große Steinernetischstraße 6, 39104 Magdeburg, Germany

[5] The University of Tennessee, Materials Science and Engineering, Knoxville, TN 37996-2200, USA

m.heilmaier phm.tu-darmstadt.de

Abstract. Directionally solidified (DS) alloys of the eutectic systems NiAl-10Mo and NiAl-34Cr (at.%) are potential candidates for high-temperature structural applications. Here, these alloys were first arc-melted and drop-cast. Thereafter, they were directionally solidified (DS) at growth rates of 20 and 80 mm/h while rotating at a fixed rotation speed of 60 revolutions per minute. Specimens of the DS alloys were tested in three-point-bending and uniaxial compression to obtain mechanical properties, including the ductile to brittle transition temperature (DBTT). For the NiAl-Cr system DBTT was found to be around 300 °C. Microstructural observations revealed that in the section perpendicular to the growth direction a uniform distribution of fibres was observed. The expected decrease of the fibre diameter with increasing growth rate was not observed. Instead, the fibre diameter slightly increased with increasing crystal growth rates. First compression tests were performed to get insights into the creep behaviour of these fibre-reinforced microstructures.

1. Introduction

The intermetallic compound NiAl possesses attractive properties for high temperature structural applications, e.g. high melting point, high oxidation resistance, a comparatively low ductile to brittle transition temperature (DBTT) and good thermal and electrical conductivity. Nevertheless, NiAl suffers from poor ductility and fracture toughness at room temperature (RT) and low strength and creep resistance at elevated temperature. Hence, substantial efforts have been made to enhance the mechanical properties through grain refinement, micro- and macro-alloying and the incorporation of second phase particles [1, 2].

Another way to increase ambient and high temperature strength is to employ fibre reinforcement. To avoid interface and debonding problems, intermetallics that were in-situ reinforced with refractory metals were investigated earlier, especially, eutectics produced by directional solidification [2, 3].

2. Experimental procedures

Alloys having a nominal composition of NiAl-10Mo and NiAl-34Cr (at.%) were prepared. First the alloying elements were weighed, placed in a copper crucible, and arc-melted under argon atmosphere. The arc-melted buttons were flipped and remelted five times to ensure homogeneity, after which the alloys were drop cast in a cylindrical copper mould 10 mm in diameter. The total weight loss was less than 0.05%. The cast ingots were directionally solidified (DS) according to the procedure given in [4] using two different growth rates, 20 and 80 mm/h.

The DS rods were cut transverse to the growth direction using an electro-discharge machine or a circular diamond saw. The samples were embedded in epoxy resin or conductive bakelite and prepared by standard metallographic methods (grinding to grit size 2400, polishing to 1μm). The NiAl-Mo samples were etched with a solution of 80% hydrochloric and 20% nitric acids [4], the NiAl-Cr samples with a solution of 18% HCl, 8% H_2O_2 and balance H_2O [5].

The metallographically prepared samples were examined by optical and scanning electron microscopy in the secondary electron (SE) image mode. The fibre spacing, fibre diameter and fibre density were analyzed with ImageJ and MATLAB.

Bending and compression tests were carried out in a Zwick electromechanical testing device equipped with a Maytec vacuum furnace, the former using specimens with a rectangular cross section (length of 30 mm and a support length of 25 mm) and the latter using specimens with a rectangular cross section of about 2 mm x 3 mm and a height of about 4.5 mm. The strain rate was approximately 10^{-3} s^{-1}. Tests on the NiAl-Mo alloy were carried out in a protective atmosphere, whereas NiAl-Cr was tested in air, which demonstrated the promising oxidation resistance of the latter. The direction of compression corresponded to the growth direction of the crystals, whereas bending was carried out perpendicular to the fibre axis. Hence, the fibres were loaded in tension and compression on either side of the neutral axis.

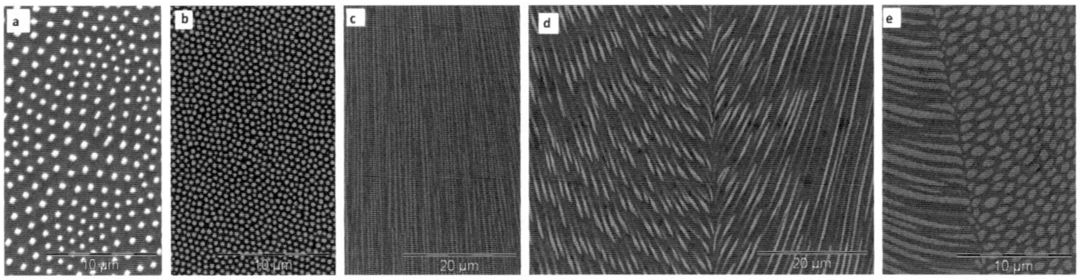

Figure 1. SEM images of sections perpendicular to the growth direction for NiAl-10Mo grown at 80 mm/h (a) and NiAl-34Cr 20 mm/h (b), cuts parallel to the growth direction showing well-aligned fibres in NiAl-10Mo 20 mm/h (c) and misalignments at colony boundaries in NiAl-10Mo 20 mm/h (d) and NiAl-34Cr 80 mm/h (e).

3. Results and discussion

The microstructure observed in the NiAl-X alloys perpendicular to the growth direction of the crystals shows the expected structure (figures 1 a and b), namely molybdenum and chromium fibres, respectively, embedded in a matrix containing mainly nickel and aluminum [4, 6, 7, 8]. The Mo fibres appear more uniform and are arranged in domains with a nearly hexagonal lattice, whereas the Cr-fibres seem to be less ordered and scatter more widely in diameter. In addition, the Cr-containing material exhibits a cellular solidification structure. The area fraction of the Cr-fibres is substantially higher than that of the Mo fibres, as expected from their compositions and the phase diagrams [2]. Table 1 summarizes characteristic properties of these structures. Some values are not consistent with the information available in the literature: for example, one expects an increasing fibre diameter with decreasing growth rate [8]; here the opposite was observed. Figures 1 c, d and e show the microstructure parallel to the growth direction. Figure 1 c illustrates clearly that fibres with an aspect ratio of at least 100 were achieved. However, grain boundaries are visible as well (figures 1 d and e). The presence of grains with different fibre orientations clearly demonstrates that a single crystalline NiAl-matrix was not achieved, which was also confirmed by EBSD-analysis. This might be a possible reason for the mismatch between expected and observed relationship between growth rate and fibre size/spacing in table 1.

Table 1. Characteristic values of the fibre structure of NiAl-X

Material [at.%]	Growth rate [mm/h]	Average fibre diameter [μm]	Area fraction [%]	Fibre density [μm^{-2}]	Average fibre spacing [μm]	Theoretical density [g/cm^3]
NiAl-9Mo	20	0.54	13	0.85	1.10	6.29
NiAl-9Mo	80	0.68	12	0.39	1.65	6.29
NiAl-34Cr	20	0.48	38	3.48	0.60	6.31
NiAl-34Cr	80	0.52	37	2.67	0.75	6.31

Figure 2(a) illustrates the temperature dependence of the strain at the outer tensile fibre in bending tests. At room temperature (RT) NiAl-Mo behaves in a completely brittle manner, whereas between 300 and 400°C it up to 5 times higher strain and 10 times higher ultimate stress values (Fig. 2(a)) were observed . NiAl-Cr grown at 20 mm/h more than doubles its strain value between 300 and 400°C, from 3.4 to 8.4%. Hence, the DBTT is around 300°C. All the other tested materials, excluding NiAl-Cr, 80 mm/h, show significant strains of more than 4% at 400°C. Therefore, the DBTT is expected to be around this temperature. Miracle [1] reported a DBTT of 200°C and 375°C for soft and hard oriented single NiAl crystals and between 325 and 625°C for polycrystalline NiAl. Hence, the DBTT of the present alloys is at the lower end of polycrystalline NiAl. It is also lower than the 675°C given by Bei and George [4] for NiAl-Mo tested in tension (as opposed to the bending tests performed here). Further, experiments have to be done to correlate the bending and tensile results, nevertheless, the values for DBTT obtained in this work are encouraging.

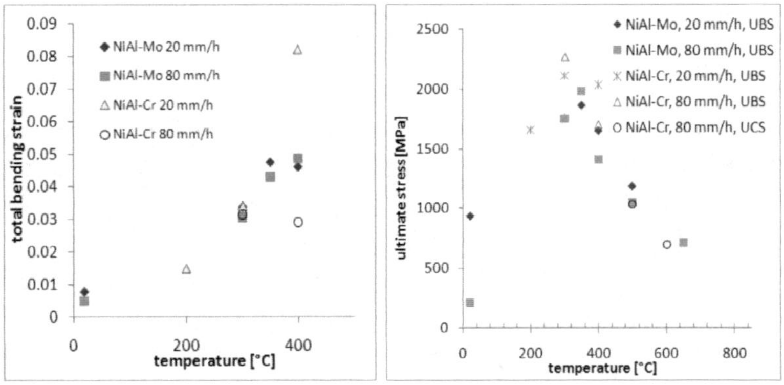

(a) total bending strain vs. temperature (b) ultimate stress vs. temperature

Figure 2. Results of the 3-point bending and compression tests

Figure 2(b) illustrates the temperature dependence of the achieved maximum stresses. In agreement with [1] the maximum stress peaks for all materials around the DBTT at approximately 2000 MPa and decreases towards both lower and higher temperatures to roughly

600 MPa around 600°C for NiAl-Mo grown at 80 mm/h and NiAl-Cr grown at the same speed, respectively. These strength values are consistent with [2, 8] and more than twice as high as that achieved by Zhong et al. [9] in tensile tests of ex-situ Al_2O_3 fibre-reinforced NiAl at 700°C, hence, confirming the notion that in-situ reinforcement is less prone to fibre delamination.

The creep performance achieved of the tested eutectic materials is shown in figure 3, which is a plot of creep rate vs. density-compensated creep stress. Stress exponents n between 7 and 10 were observed which is in good agreement with data reported by Miracle [1]. For comparison, the figure contains own measurements of compressive creep properties for a single crystalline Nickelbase superalloy CMSX4 in the [001] direction at 1050°C as well as values for polycrystalline NiAl at 750°C given in [10]. In comparison with polycrystalline NiAl, the creep performance of in-situ fibre reinforced NiAl is strongly enhanced, even though the former was tested at a much lower temperature. By contrast, the creep performances of our NiAl-X in-situ composites are already comparable with that of CMSX-4.

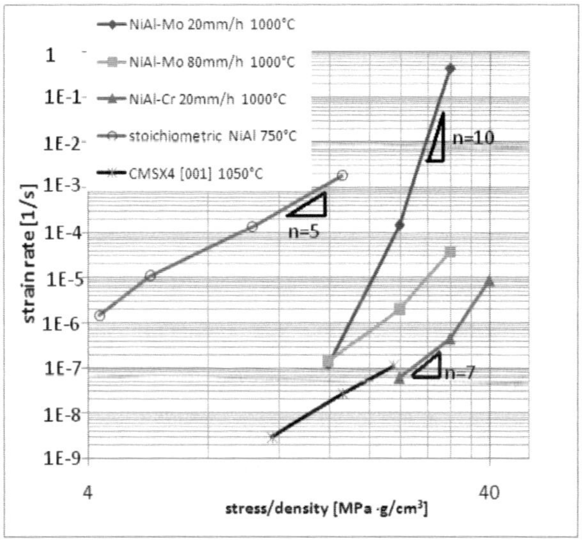

Figure 3. Strain rate vs. normalized stress for the compression creep tests carried out at 1000°C and for polycrystalline NiAl at 750°C [10] and single crystalline CMSX4 at 1050°C.

4. Conclusion

In summary, the tested materials possess comparatively low DBTTs of < 300 °C and it appears as if our data merge with the SX superalloy's creep response at application relevant low strain rates. These encouraging results have to be proven with further (long term) creep tests at even lower applied stresses, preferably using tensile specimens.

H Bei, A Gali and E P George were supported by the Division of Materials Sciences and Engineering, U.S. Department of Energy.

References

[1] Miracle, D. B. *Acta Metall. Mater.* **41**, 649.
[2] Noebe, R. D., Bowman, R. R., and Nathal, M. V. *Int. Mater. Rev.* **38**, 193.
[3] Bei, H., Pharr, M., and George, E. P. *J. Mater. Sci.* **39**, 3975.
[4] Bei, H. and George, E. P. *Acta Mater.* **53**, 69.
[5] Bei, H., George, E. P., Kenik, E. A., and Pharr, G. M. *Acta Mater.* **51**, 6241.
[6] Cline, H. E. and Walter, J. L. *Met. Trans.* **1**, 2907.
[7] Walter, J. L. and Cline, H. E. *Met. Trans.* **4**, 33.
[8] Walter, J. L. and Cline, H. E. *Met. Trans.* **1**, 1221.
[9] Zhong, Y., Chen, H., Hi, W., and Gottstein, G. *Mater. Sci. Eng.* **464**, 241.
[10] Wang, Y. L., Jones, I. P., and Smallman, R. E. *Intermetallics* **14**, 800.

Creep characterization of power-law materials through pseudo-steady indentation tests and numerical simulations

Hidenari Takagi [1], Ming Dao [2] and M. Fujiwara [1]

[1] Division of Applied Physics, College of Engineering, Nihon University, Fukushima 963-8642, Japan
[2] Department of Materials Science and Engineering, School of Engineering, Massachusetts Institute of Technology, Cambridge, MA 02139, USA

h_takagi@ge.ce.nihon-u.ac.jp

Abstract. A constant-indentation creep rate test (CICRT) has been carried out for an Al-Mg solid-solution alloy using a microindenter in the temperature range of 636-773 K. When a conical indenter is pressed into the specimen surface under a load condition of $F = F_0 \exp(\lambda t)$ (F the indentation load, F_0 the initial load, λ the loading rate parameter, t the loading time), the indentation pressure and indentation creep rate approach constant values of p_s and $\dot{\varepsilon}_{in(s)}$, respectively. The representative points of the deformation in the underlying material are defined on a contour line of the equivalent stress $\sigma_r = C_1 p_s$, where C_1 is the so-called constraint coefficient of $1/3$ reported by Tabor. The finite element simulation of a power-law material subjected to the CICRT shows that the relationship between the equivalent plastic strain rate $\dot{\varepsilon}_r$ at these points and $\dot{\varepsilon}_{in(s)}$ is $\dot{\varepsilon}_r = C_2 \dot{\varepsilon}_{in(s)}$ and that $C_2 \cong 1/3.6$ in the case of a creep stress exponent of 3.0. The constitutive equation of $\dot{\varepsilon}_r$ versus σ_r obtained from experimental data and the computed value of C_2 is in good agreement with that evaluated from conventional uniaxial creep tests.

1. Introduction

The instrumented indentation testing technique has been attracting attention as a method of determining mechanical properties in a selected region of a small specimen [1-4]. In this study, a constant-indentation creep rate test (CICRT) is carried out for an Al-Mg solid-solution alloy using a microindenter, and a finite element (FE) simulation of the CICRT is performed to examine the deformation behavior in the region beneath the indenter. We demonstrate here that the constitutive equation for the tensile creep of power-law materials can be robustly predicted by the CICRT. To this end, we discuss the following items:

- In the CICRT, the indentation creep rate and indentation pressure approach constant values after a transition period.
- Representative points in the deformed region that practically govern indenter velocity are defined, and the equivalent plastic strain rate at these points is obtained by FE simulation to clarify its relationship with the indentation creep rate.
- A constitutive equation for equivalent plastic strain rate versus equivalent stress at these points is derived and compared with that obtained from tensile creep.

2. Constitutive equation for pseudo-steady indentation creep

When a conical indenter is pressed into a specimen surface, it is known that the indentation pressure p and the flow stress of the specimen σ have the following relationship:

$$\sigma = C_1 p, \tag{1}$$

where C_1 is the so-called constraint coefficient of 1/3 [1, 5]. In this paper, the representative points of the deformation in the underlying material are defined on a contour line of the equivalent stress $C_1 p$. The equivalent stress at these points is referred to as the representative stress σ_r. Moreover, the equivalent plastic strain rate at these points is named the representative strain rate $\dot{\varepsilon}_r$. The indentation creep rate $\dot{\varepsilon}_{in}$ is given in terms of the indenter displacement u and the indenter velocity $\dot{u}$ as $\dot{\varepsilon}_{in} = \dot{u}/u$ [1]. For compatibility, a linear relationship must hold between $\dot{\varepsilon}_r$ and $\dot{\varepsilon}_{in}$;

$$\dot{\varepsilon}_r = C_2 \dot{\varepsilon}_{in}, \tag{2}$$

where C_2 is a constant that depends on the creep properties and the apex angle θ of the indenter. In the steady-state deformation of metallic materials at high temperatures, the equivalent plastic strain rate $\dot{\bar{\varepsilon}}$ and the equivalent stress $\bar{\sigma}$ obey the power-law creep; $\dot{\bar{\varepsilon}} = A_1 (\bar{\sigma}/E)^n$, where A_1 is a material constant that depends on test temperature T, E is Young's modulus, and n is the creep stress exponent. When the law is assumed to hold at the representative points, the constitutive equation for indentation creep is given by

$$\dot{\varepsilon}_{in} = A_2 \left(\frac{p}{E} \right)^n = A_3 \left(\frac{F}{Eu^2} \right)^n, \tag{3}$$

where F is the indentation load, and A_2 and A_3 are material constants that depend on T, n, and θ. Assuming that $F \propto u^2$ is applicable in the indentation creep test, F can be expressed by $F = F_0 \exp(\lambda t)$, where F_0 is the initial load, λ is the loading rate parameter, and t is the loading time. In this case, the time dependences of $\dot{\varepsilon}_{in}$ and p/E are given by [6]

$$\dot{\varepsilon}_{in} = \frac{\lambda}{2} \left(\frac{1}{1 - \exp(-\lambda n t)} \right), \tag{4}$$

$$\frac{p}{E} = \left[\frac{\lambda}{2A_2 (1 - \exp(-\lambda n t))} \right]^{1/n}. \tag{5}$$

Equations (4) and (5) indicate that $\dot{\varepsilon}_{in}$ and p/E approach constant values after a transition period. Specifically, $\dot{\varepsilon}_{in(s)} \cong \lambda/2$ and $p_s/E \cong (\lambda/(2A_2))^{1/n}$. When both parameters are kept constant, $\dot{\varepsilon}_{in(s)} = A_2 (p_s/E)^n$ holds. Since $\dot{\varepsilon}_r = C_2 \dot{\varepsilon}_{in(s)}$ and $\sigma_r = C_1 p_s$, the constitutive equation for creep at the representative points, i.e., that obtained from tensile creep tests is given as $\dot{\varepsilon}_r = A_1 (\sigma_r/E)^n$.

3. Experimental procedures and computational setup

Commercially available ingots of an Al-5.3 mol% Mg alloy were cut into parallelepipeds. They were carefully shaped so that the specimen surface (5×10 mm^2) becomes parallel to the corresponding bottom surface and then annealed in Ar gas at 773 K for 3.6 ks. Immediately before the tests, specimens were electropolished to remove the surface layer of up to about 40 μm in thickness. The CICRTs were carried out in Ar gas at temperatures ranging from 636 to 773 K using a microindenter. An indentation load F given by the function $F = F_0 \exp(\lambda t)$ was applied, where $F_0 = 0.29$N and $\lambda = 0.5\times10^{-3} - 8.0\times10^{-3}\,s^{-1}$. Details of the experimental procedures have been reported elsewhere [3, 4].

Computational simulations of CICRTs were performed using an FE package, ABAQUS Standard (SIMULIA Inc.), combined with our original subroutines. The conical indenter was modeled by a rigid body, and the cylindrical material was modeled using four-node bilinear axisymmetric quadrilateral elements. Each element was assumed to be subjected to elastic deformation ($E = 37.8$GPa [7]; Poisson's ratio, $\nu = 0.345$ [8]) and to obey power-law creep; $\dot{\bar{\varepsilon}} = A\bar{\sigma}^n$ ($A = 1\times10^{-6}\,MPa^{-3}s^{-1}$, $n = 3.0$)

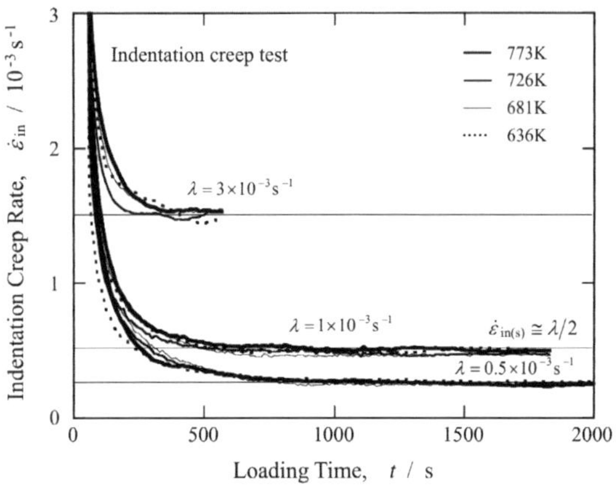

Figure 1 Time dependence of indentation creep rate.

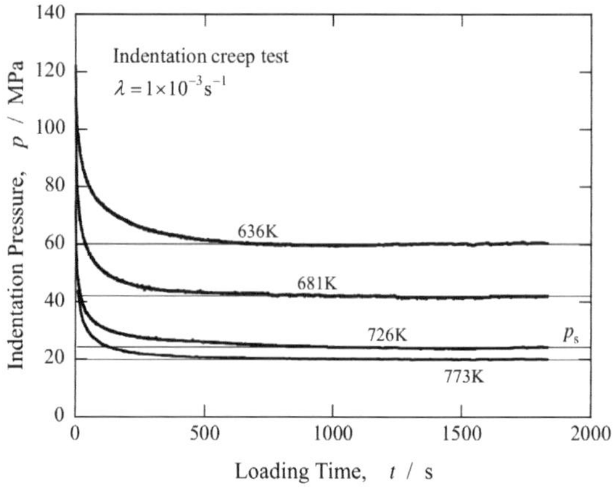

Figure 2 Time dependence of indentation pressure.

[4]. These material constants were those obtained for the Al-Mg solid-solution alloy when the creep rate was controlled by the viscous glide of dislocations. More details of the computational model setup can be found elsewhere [4, 6].

4. Results and discussion

4.1 CICRT using the microindenter

Figure 1 shows the time dependence of indentation creep rate $\dot{\varepsilon}_{in}$. $\dot{\varepsilon}_{in}$ sharply decreases immediately after loading then approaches a constant value $\dot{\varepsilon}_{in(s)} \cong \lambda/2$ after a transition period t_{trans}. Under the same value of λ, $\dot{\varepsilon}_{in}-t$ curves are similar regardless of test temperature T. When λ is smaller, it is found that t_{trans} increases. For example, $t_{trans} \cong 400$s for $\lambda = 3\times10^{-3}\,s^{-1}$, $t_{trans} \cong 700$s for $\lambda = 1\times10^{-3}\,s^{-1}$, and $t_{trans} \cong 1000$s for $\lambda = 0.5\times10^{-3}\,s^{-1}$. These characteristics agree with those of equation (4). Figure 2 shows the time dependence of indentation pressure p in the case of $\lambda = 1\times10^{-3}\,s^{-1}$. In addition to $\dot{\varepsilon}_{in}$, p also becomes a constant value p_s after t_{trans}. Moreover, it is found that p_s decreases with increasing T, and in this case t_{trans} is about 700s regardless of T. These characteristics are consistent with those described by equation (5).

4.2 Relationship between representative strain rate and indentation creep rate

The FE simulation revealed that $\dot{\varepsilon}_{in}$ and p approach constant values of $\dot{\varepsilon}_{in(s)}$ and p_s, respectively, after t_{trans}. The computational results are similar to the experimental results shown in Figures 1 and 2. When both parameters are kept constant, the indenter is surrounded by self-similar stress and strain fields as indentation creep proceeds [6]. This fact implies that the indenter tip can continuously detect the pseudo-steady indentation creep parameters at representative points, $\dot{\varepsilon}_r$ and σ_r.

Figure 3 shows the relationship of $\dot{\varepsilon}_r$ versus $\dot{\varepsilon}_{in(s)}$ obtained from the computational results. All data denoted by circles lie on a straight line that passes through the origin. The slope of this line corresponds to the proportionality constant C_2 of equation (2), $C_2 = 1/3.6$. Therefore, $\dot{\varepsilon}_r = \dot{\varepsilon}_{in(s)}/3.6$ holds for $n = 3.0$.

4.3 Predicting a constitutive equation for tensile creep

Circles in Figure 4 show the relationship between $\dot{\varepsilon}_{in(s)}$ and p_s/E obtained from the CICRTs at T=681 K for $\lambda = 0.5\times10^{-3} - 8\times10^{-3}\,s^{-1}$. All the circles lie on straight line A, and the constitutive equation for pseudo-steady-state indentation creep is expressed as $\dot{\varepsilon}_{in(s)} = 5.72\times10^{5}\left(p_s/E\right)^{3.0}$. Next, let us verify that the constitutive equation for tensile creep can be predicted through the CICRTs. First, when the values of p_s/E are multiplied by $C_1 = 1/3.0$, line A shifts to the left, becoming line B representing $\dot{\varepsilon}_{in(s)}$ versus σ_r/E. Second, when the values of $\dot{\varepsilon}_{in(s)}$ on line B are multiplied by $C_2 = 1/3.6$, line B shifts downward to become line C, expressed as $\dot{\varepsilon}_r = 4.18\times10^{6}\left(\sigma_r/E\right)^{3.0}$. This

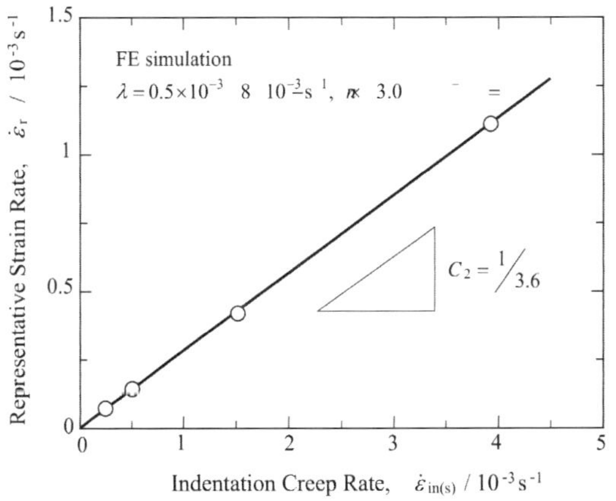

Figure 3 Representative strain rate versus pseudo-steady indentation creep rate.

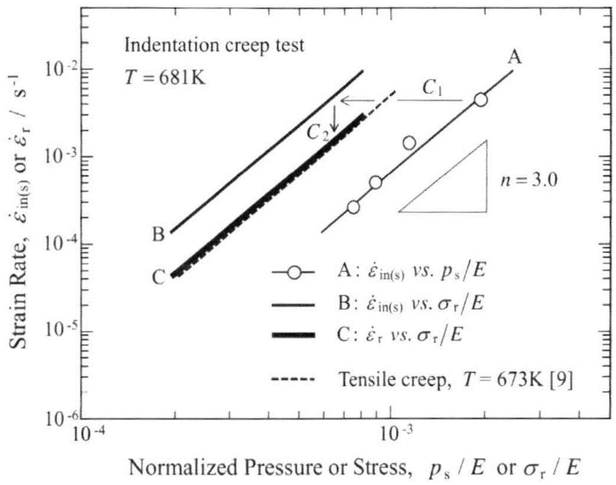

Figure 4 Logarithmic plots of strain rate versus normalized pressure or normalized stress.

line is in good agreement with the dotted line that represents the tensile creep test result for Al-Mg solid solution alloy reported in Reference [9]. This fact indicates that the constitutive equation for tensile creep can be robustly predicted by conducting CICRTs and using the known values of C_1 and C_2.

5. Conclusions

(1) The indentation creep rate and indentation pressure approach constant values after a transition period when indentation load given by the function $F = F_0 \exp(\lambda t)$ is applied.

(2) The FE simulation reveals that the representative strain rate $\dot{\varepsilon}_r$ is proportional to the indentation creep rate $\dot{\varepsilon}_{in(s)}$ as pseudo-steady indentation creep proceeds, that is, $\dot{\varepsilon}_r \cong \dot{\varepsilon}_{in(s)} / 3.6$ for the creep stress exponent of 3.0.

(3) The constitutive equation for creep predicted through CICRTs is in good agreement with that obtained from conventional uniaxial creep tests.

References

[1] Pollock H M, Maugis D and Barquins M *Microindentation Techniques in Materials Science and Engineering*, eds. P. J. Blau and B. R. Lawn (ASTM, Philadelphia, 1986) p. 47.
[2] Fischer-Cripps A C *Nanoindentation* (Springer-Verlag, New York, 2004).
[3] Fujiwara M 2002 *J. Jpn. Inst. Light Metals* **52**, 282.
[4] Takagi H, Dao M, Fujiwara M and Otsuka M 2003 *Phil. Mag.* **83**, 3959.
[5] Tabor D *The Hardness of Metals* (Oxford University Press, New York, 1951).
[6] Takagi H, Dao M and Fujiwara M 2008 *Acta Mech. Sol. Sinica* **21**, 283.
[7] Abe K, Tanji Y, Yoshinaga H and Morizumi S 1977 *J. Jpn. Inst. Light Metals* **27**, 279.
[8] Brandes E A and Brook G B *Smithells Metals Reference Book Butterworth-Heinemann* (Butterworth-Heinemann, Oxford, 1992).
[9] Horiuchi R and Otsuka M 1972 *Mater. Trans JIM* **13**, 284.

Effect of thermal and thermomechanical ageing on the microstructure and mechanical properties of a Ni-Cr alloy

S Chomette[1], J M Gentzbittel[1], J Calapez[1] and B Viguier[2]

[1] LITEN/DTH/LTH, CEA Grenoble, 17 rue des Martyrs, 38054 Grenoble cedex 9 France
[2] Institut Carnot CIRIMAT, ENSIACET-INPT, 4, allée Emile Monso - BP 74233 31432 Toulouse Cedex 4 France

jean-marie.gentzbittel@cea.fr

Abstract. Thermal ageing experiments at 850 and 950°C for 1000 h with and without an applied stress have been applied to INCONEL 617 alloy. The consequences of these treatments characterized in terms of microstructure evolution and the tensile behavior have been studied at room and elevated temperatures. It is shown that thermal ageing strongly reduces the ductility of the alloy at room temperature, whereas the saturation stress measured at elevated temperature is only slightly affected. Applying a small stress of 7MPa during thermal ageing has negligible effects on the tensile behavior.

1. Introduction

In Very High Temperature Reactors (VHTR), the Intermediate Heat eXchanger (IHX) imposes severe conditions (coolant temperature up to 1000°C) for long time (target: 20000h). Currently, the solid-solution nickel based alloy INCONEL 617 is a candidate material for the IHX and the hot ducts of the VHTR thanks to its oxidation and corrosion resistance and its high temperature mechanical properties. In the range of ageing temperature [700°C-1000°C], alloy 617 shows a peak of hardness within hundred first hours of ageing followed by hardness stabilization [1,2]. This peak has been related to the precipitation of intragranular Cr rich $M_{23}C_6$ carbides that appear at the beginning of thermal treatment, localized on dislocations in the slip lines [1,3,4]. For longer ageing time, these intragranular carbides are dissolved or coalesce. The consequences of such microstructural evolution on the mechanical properties of the INCONEL 617 have been widely studied in the literature for various types of mechanical testing: tensile testing [3,5-8], creep [1,4,9,10] and fatigue [11].

In the present paper, we report the results of tensile testing at room and elevated temperatures (850 and 950°C) and we compare the properties of annealed and heat treated alloy. Heat treatments were realised at 850 and 950°C and were complemented by thermomechanical treatment, consisting of ageing at the same temperatures under an applied stress of 7 MPa. This latter treatment is believed to be more representative of the life operation conditions of the alloy in the IHX. The effect of the imposed strain rate is also examined at high temperature for the three states of the alloy.

2. Experimental

A bar of INCONEL 617 with a diameter of 25.4 mm was purchased from Special Metals (chemical composition is given in Table 1). The material was hot forged and then solution annealed at 1175°C for 36 min followed by air cooling; this annealed state is further referred to "as received" (AR).

Table 1. Chemical composition of the Inconel 617 supplied by Special Metals (wt %)

Ni	**Cr**	**Co**	**Mo**	C	Ti	Al	Si	Mn	Fe	Cu
55,35	**21,49**	**12,00**	**9,23**	0,060	0,36	0,95	0,12	0,06	1,24	0,06

Parts of the initial bar were cut for thermal ageing (TA) treatments of 1000 h at 850°C or 950°C. Those TA sample were introduced in cold furnace working in air, the furnace was then warmed-up to ageing temperature, held for 1000 h and finally the samples were all furnace cooled. The samples for thermomechanical (TMA) treatments (part of the initial bar) were aged under a constant load at 850°C or 950°C for 1000 h corresponding to a stress of 7 MPa (normal service conditions in the IHX). The ageing duration of 1000 h has been chosen because all the major microstructural evolutions occur for shorter treatments. After thermal and thermomechanical ageings, part of the treated rods are mechanically polished and electrochemically etched for optical and scanning electron microscopy.

Cylindrical specimens with a gauge diameter of 4 mm and 20 mm gauge length were machined from the as received and aged alloy 617 rods. Tensile tests were performed in laboratory at room temperature and ageing temperature (850°C or 950°C). For high temperature testing, the sample was first heated and held for 1h at the testing temperature in order to obtain good thermal equilibrium of the whole load train. Tensile tests were conducted on screw driven electromechanical testing machines, using a displacement rate of the crosshead of 0.6 and 6 $mm.min^{-1}$ corresponding to an imposed strain rate of 5.10^{-3} and 5.10^{-4} s^{-1}, respectively. Deformation was measured directly on the gauge length using a MTS extensometer. The tensile curves are shown in engineering stress and strain coordinates.

3. Results and discussion

3.1. Microstructural evolution during thermal and thermomechanical treatments

The initial microstructure reveals γ austenitic grains (ASTM 2) containing homogeneously dispersed Ti(C,N) precipitates and alignment of Mo rich M_6C carbides parallel to the axis of the bar. Few intergranular Cr rich $M_{23}C_6$ secondary carbides are also present, principally at grain boundaries. After ageing at 850°C, numerous fine Cr rich carbides are observed on slip lines, while they have totally disappeared after 1000 h at 950°C. The decomposition of primary carbides (Ti(C,N) and M_6C) that transform to $M_{23}C_6$ appears in aged samples, more developed at 950°C [1]. $M_{23}C_6$ carbides are also observed at grain boundaries with a morphology which depends on the treatment temperature (dendritic and globular at 850°C, globular at 950°C [1,4]). At the scale of the optical and scanning electron microscope one cannot distinguish any difference of microstructure for the samples treated with and without stress. However, it is difficult to quantify and compare the amount of small $M_{23}C_6$ intragranular carbides in samples aged with and without stress at 850°C since their distribution depends strongly on the observed grain.

3.2. Tensile properties

3.2.1. Room temperature tensile tests. The curves of tensile tests at room temperature on the as received Inconel 617 and aged materials are presented in figure 1. Comparing the AR and TA material, the ageing treatment at 850°C slightly reduces the yield strength (YS) but it also increases the work hardening rate so that the ultimate tensile strength (UTS) of the aged sample is nearly the same as the AR one while the elongation to rupture and reduction in area have been strongly reduced. These differences can be related to the precipitation of the $M_{23}C_6$ carbides on slip lines and at grain boundaries. During ageing at 950°C, the intragranular carbides have dissolved, inducing a more important reduction of the YS and the UTS. According to figure 1, applying low stress during the ageing treatment at 850°C has little influence on the tensile response of the alloy. Only a slight reduction of tensile strength and ductility is observed. This phenomenon might be related to the coarsening of $M_{23}C_6$ carbides which could be enhanced by a promoted diffusion thanks to the stress during heat treatment.

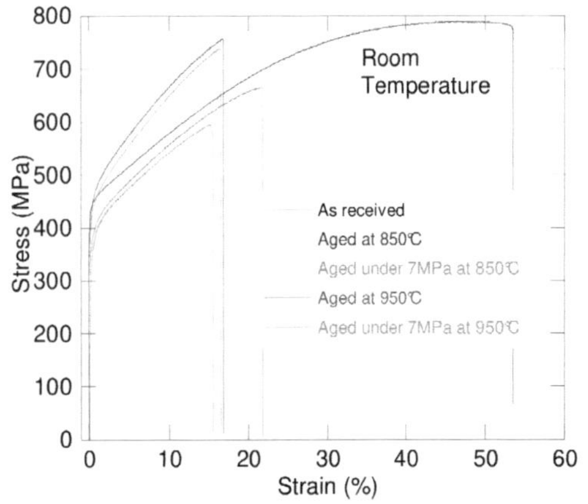

Figure 1. Room temperature engineering stress vs. strain curves of as received and treated Inconel 617 at a strain rate of 5.10^{-4} s^{-1}.

At 950°C the diffusion is more efficient. So, the reduction of the tensile response of the INCONEL 617 due to the stress during ageing treatment at 950°C is a bit more significant for YS and UTS values and further reduces the ductility, while no difference is observed on work hardening rate, as after treatments at 850°C.

3.2.2. High temperature tensile tests. Figure 2 shows the ensemble of results obtained by high temperature tensile testing of the as received and treated Inconel 617 at two strain rates: 5.10^{-3} s^{-1} and 5.10^{-4} s^{-1}. First of all, one notices that all the tensile curves obtained at high temperature are very similar for all the conditions that have been tested, and the material behaves at high temperature very differently from room temperature. Indeed, some strain hardening is only observed at 850°C for the higher imposed strain rate, 5.10^{-3} s^{-1}, at the beginning of the tensile curve. After the UTS, it has been verified that the engineering stress decrease does not correspond to the necking of the specimen but instead to a homogenous strain and a constant true stress, revealing a steady state deformation. For the lower strain rate at 850°C and for 950°C, this steady state deformation starts nearly at the plastic transition, sometimes associated to a more or yielding. Such behaviour has been reported in the literature [12]. It clearly appears in Figure 2 that this steady state stress level strongly depends on both temperature and strain rate. We have verified that these stress levels fit very well with the Norton plot obtained during constant load creep experiments performed at the same temperature [1], showing that similar mechanism operates over a wide plastic strain rate range (from 5.10^{-3} s^{-1} as shown in this paper down to 10^{-8} s^{-1} for creep tests [13]) .

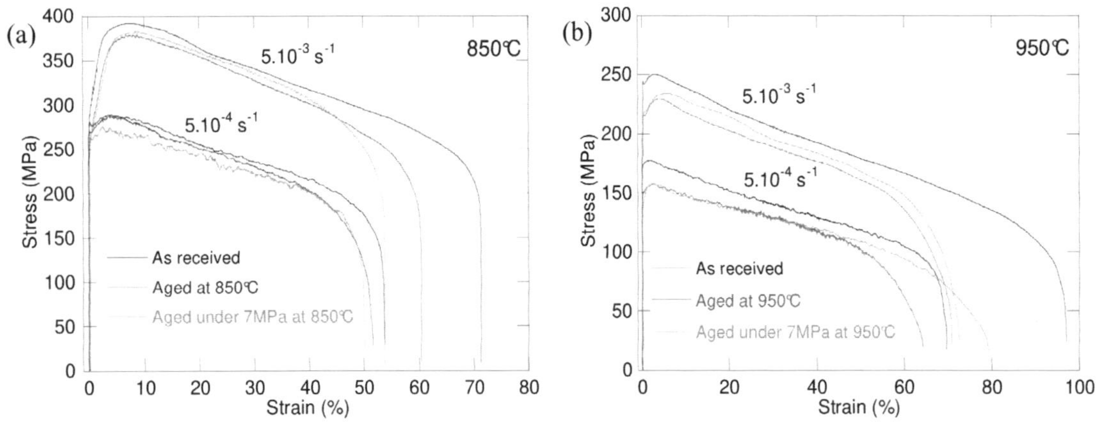

Figure 2. High temperature engineering stress vs. strain curves of as received and treated Inconel 617 at 5.10^{-3} s^{-1} and 5.10^{-4} s^{-1} imposed stain rate for two temperatures: (a) 850°C and (b) 950°C.

One can also notice the occurrence of plastic instabilities on tensile curves, especially for the lower imposed strain rates. The role of solutes atoms may be important in the high temperature regime.

For both 850 and 950°C testing temperature, the general effects of thermal ageing are similar: TA and TMA tend to slightly decrease the stress level, but in moderate way. A more marked effect is visible on the ductility of the alloy, especially for the higher strain rate (5.10^{-3} s^{-1}). Finally, comparing TA and TMA, it appears that applying a stress does not affect the tensile behavior as compared to samples that were aged by thermal treatment only.

4. Conclusions

This paper presents one of the first studies on the effect of stress ageing on tensile properties of the INCONEL 617. The low stress imposed during 1000 h heat treatments at 850°C and 950°C to simulate normal service conditions in the IHX does not induce major microstructural differences and tensile properties compared to alloy aged without stress under the same conditions. Ageing treatments are therefore sufficient to study the alloy for high temperature applications.

Precipitation of secondary $M_{23}C_6$ carbides occurred during ageing treatments in INCONEL 617; they promote damage evolution and reduce ductility at room temperature. The RT work hardening is increased after treatments at 850°C as compared to the as received and 950°C treated material due to different intragranular precipitation.

For high temperature tensile tests, deformation reaches a saturation stage which is only slightly affected by TA or TMA treatments.

References

[1] Chomette S, Gentzbittel J M and Viguier B 2009 *J. Nucl. Mater.* submitted JNM-D-09-00204.
[2] Wu W 2006 *PhD Thesis*, University of Cincinnati
[3] Kirchhofer H, Schubert F and Nickel H 1984 *Nucl. Tech.* **66** 139.
[4] Kihara S, Newkirk J B, Ohtomo A and Saiga Y 1980 *Met. Trans. A* **11** 1019.
[5] Susukida H, Tsuji I, Kawai H 1977 *Mitsubishi Tech. Bull.* **122** 1
[6] Donnelly M E 1984 *Advanced Gas Cooled Reactor Materials Program, General electric Company Report n°112*
[7] Burlet H, Gentzbittel J M, Lamagnere P, Cabet C, Blat M, Renaud D, Dubiez-Le Goff S, Pierron D 2007 *NEA NSC Workshop on Structural Materials for Innovative Nuclear Systems* (Karlsruhe)
[8] Breitling H, Dietz W and Penkalla H J 1988 *High Temperature Metallic Materials for Gas-Cooled Reactors* (Cracow, IAEA Specialists Meeting) p 96.
[9] Schneider K, Hartnagel W, Schepp P and Ilschner B 1984 *Nucl. Tech.* **66** 289.
[10] W. L. Mankins W L, J. C. Hosier J C and Bassford T H 1974 *Met. Trans.* **5** 2579
[11] Strizak J P, C. Brinkman C R and Rittenhouse P L 1981 *High Temperature Metallic Materials for Gas-Cooled Reactors* (Vienna/Austria, IAEA Specialists Meeting).
[12] Abd El-Azim M E, Mohamed K E and Hammad F H 1992 *Mech. Mater.* **14** 33
[13] Chomette S 2009 *PhD Thesis*, University of Toulouse

Co-Re-based alloys for high temperature applications: Design considerations and strengthening mechanisms

D. Mukherji, J. Rösler

Technische Universität Braunschweig, Institut für Werkstoffe, Langer Kamp 8, D-38106 Braunschweig, Germany

d.mukherji@tu-bs.de

Abstract. Cobalt-Rhenium base alloys are being developed for applications at temperatures beyond Ni-base superalloys. The high melting refractory element Re readily dissolves in Co and thereby changes the character of a Co-based alloy to a high melting point material. So far Co-Re system has not been investigated from the point of view of strengthening and oxidation and therefore different possibilities of strengthening mechanisms have been explored for our alloy development. Cr, a common element for oxidation resistance in many systems, is added along with Si in Co-Re alloys to improve oxidation behaviour at high temperatures.

1. Introduction

Ni-base superalloys are the dominant material class for gas turbine applications. However, the temperature capability of superalloys is limited by the possibility to raise further their melting point (which is around 1300° – 1400°C). There is an urgent need for the development of alloys for temperatures beyond the application range of Ni-superalloys. New Co-Re-based alloys are being developed at the Technische Universität of Braunschweig for applications aimed at +100°C metal temperature above Ni-base superalloys to meet this need. Addition of refractory metal Re, which has complete miscibility in the Co-Re system, considerably increases the melting temperature of Co-Re alloys. This opens a unique possibility, as it allows to steadily elevate the melting range in Co-alloys, changing the character of the material from that of a contemporary Co-based alloy to that of a high melting point material. This gives a good chance to tune the properties and find a suitable balance between the need for toughness and ductility on the one hand and strength at temperatures beyond the capability of Ni-based superalloys on the other hand.

Higher application temperatures however, also impose severe demand on the high temperature oxidation behaviour of alloys, which is generally met in Ni-superalloys. Oxidation resistance along with high temperature strength is therefore important in our alloy development consideration. The Co-Re system has been explored very little so far, in fact from strengthening and oxidation point of view not at all. Cr which promotes the formation of a protective Cr_2O_3 oxide layer in conventional iron, nickel and cobalt alloys is a logical choice as alloying addition to Co-Re alloys. It is however clear that chromium alone will not suffice for oxidation protection at very high temperatures. In this paper we present the alloy development strategy and the design considerations we adapted.

2. Alloy development strategy

Co-based alloys are presently used in gas turbines for static parts up to 1000°C. In contrast to the conventional cast Co-base superalloys, where the matrix phase has a face centred cubic (fcc) structure, Re addition stabilizes the hexagonal close packed (hcp) phase as the matrix in the new Co-Re alloys. This demands a new strategy in alloy development for the Co-Re alloys. Moreover, the Co-Re is a little studied alloy system and we therefore explored different strengthening possibilities for the alloy development. Primary aim was to explore different strengthening mechanisms in the Co-Re system exploiting the following possibilities:

- Precipitation hardening: Carbides precipitates ($M_{23}C_6$ and MC type)
- Solid-solution hardening: by Re atoms
- Second phase (composite) hardening: σ phase (Cr-Re rich TCP phase)
- Grain boundary strengthening: by B and Zr atoms addition

Thermodynamic calculations formed the basis for choosing the alloy compositions. Thermocalc [1] was used to determine the stability of phases in the multi-component Co-Re-base alloys. This information was used to fix the composition of the experimental alloys studied.

As mentioned, along with high temperature strength, oxidation resistance is an important parameter in the alloy development. We felt the necessity to study these two aspects of the alloy development separately, from a more fundamental point to understand the physical basics and mechanisms. The knowledge gained formed the basis for the development of technical alloys.

2.1. Microstructure, strength and ductility

The atom size of Re is ~10% larger than Co atom and it provides significant matrix strengthening (figure 1) by solid solutioning, in addition to increasing the alloys' melting temperature. Re in amounts of 17 to 30 at.% is added in the experimental alloys. Further, Cr and Ta are added in combination with C to derive precipitation strengthening through carbides. Microstructures of the CoRe1 (Co-17Re-23Cr-2.6C) and CoRe2 (Co-17Re-23Cr-1.2Ta-2.6C) alloys after solution heat treatment at 1450°C are shown in figure 2. In addition to the massive $Cr_{23}C_6$-type carbides (several micrometers in size) at the grain boundaries and within the grains reported in [2, 3], the matrix also contains lamellar $Cr_{23}C_6$-type carbides in the form of thin plates (~40 nm thick and inter-plate distance between 200 to 500 nm), oriented in parallel (figure 2 Left). In CoRe2 alloy, additionally, a very fine dispersion of TaC-type MC carbide forms (figure 2 Right). These carbide particles are extremely fine and are homogenously dispersed between the plate-like $M_{23}C_6$ type Cr-carbides. The microstructure of the heat treated Co-Re alloys show that the alloys are strengthened by $M_{23}C_6$ and MC carbides in different morphologies and on varying length scales.

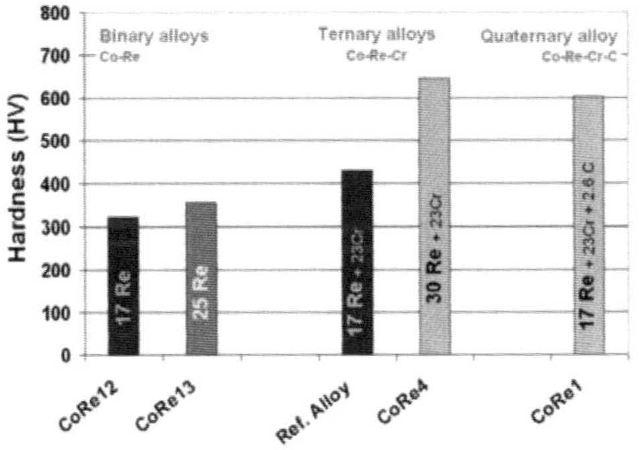

Figure 1. Hardness measurements in binary / multi-component Co-base alloys, with varying Re content.

In figure 1 it is also seen that the hardness increases much more in the ternary alloys containing Cr compared to the Co-Re binary alloys. This is because Cr also stabilizes the Cr-Re-rich σ phase which has very high hardness (microhardness > 800 HV). Further, when C is added to the Co-Re-Cr alloy (i.e. alloy CoRe1), lamellar carbides are present in the matrix in place of the σ phase, which also significantly increase the matrix hardness. This was clearly demonstrated by comparing microhardness values measured in differently heat treated CoRe1 alloy - when the lamellar carbides are present and absent in the matrix.

Generally, the topologically close packed σ phase is very hard and brittle. This is seen in figure 3 (Left), where the Co-Re alloy containing a significant volume fraction of the σ phase shows that cracks develop in the σ particles (bright phase) due to the hardness indentation. However, the surrounding matrix shows considerable plasticity (slip lines) and no cracks, thereby clearly indicating that in Co-Re alloys the presence of σ phase does not render the material inherently brittle. This gives us the possibility to consider a second phase (composite) hardening mechanism using the σ phase. In order to utilize this hardening concept, it is however, necessary to optimize the σ phase morphology and distribution.

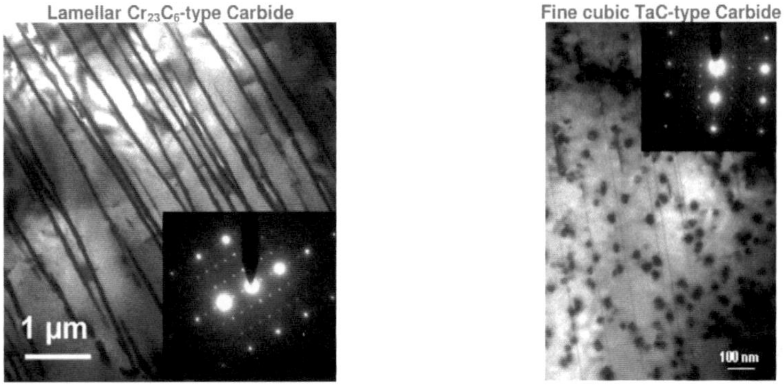

Figure 2. Different morphologies of $M_{23}C_6$ and MC type carbides in: CoRe1 (Left) and CoRe2 (Right).

An addition of 15 to 25 at.% Ni to Co-17Re-23Cr refines the σ particles (~ 500 nm), which are also sometimes present as thin lamellae (~ 200 nm thick). The σ phase is very homogenously dispersed. Furthermore, in this size scale the σ phase is no longer brittle (figure 3 Right). This is a potential microstructure for high temperature creep strengthening. The stability of σ phase was tested at high temperatures up to 1300°C.

Figure 3. (Left) Cracks in σ phase due to hardness indentation in Co-17Re-23Cr alloy. (Right) Finer σ phase in the Ni containing CoRe5 alloy (Co-17Re-23Cr-15Ni) show no cracks.

In any polycrystalline cast alloy, grain boundary cohesion is an important issue. In the case of Co-Re alloys, it was found that the grain boundaries are rather weak. Bend tests at room temperature showed e.g. that the Reference alloy (Co-17Re-23Cr) failed with brittle inter-granular fracture (figure 4 Left). To address this issue B and Zr was added in ppm levels up to 2000 ppm (by wt.) to the Reference alloy. The room temperature bend-test results in alloys containing B and Zr are shown in figure 4 (Right). The results show that B and Zr addition is effective in addressing the grain boundary cohesion issue. In fact, an addition of only 200 ppm B not only results in a significant increase in ductility, but the flow stress of the alloy is almost doubled (at RT). This trend is maintained also at high temperatures. Apparently, Zr also affects at high temperatures.

2.2. *Oxidation behaviour*

Oxidation mechanisms and scale formation is well understood in Co-Cr binary system [4] however, how Re addition affects the oxidation in Co-Cr alloys was not known and recently we investigated this [5, 6]. Generally, Re has a minor but negative effect and acts in two ways. Firstly, in combination with Cr, Re stabilizes the σ phase, which has an inferior oxidation behaviour compared to Co(Re) solid solution. Secondly, the element Re itself oxidizes readily forming volatile Re_2O_7, which is lost from the base metal. Knowing that chromium alone can not provide the needed oxidation protection, addition of Si (1-3 at.%) to the Co-17Re-Cr-base was investigated, where the Cr content was also varied between 23 and 30 at.%. Si addition showed promising results and matched the oxidation behaviour of the conventional Co-base alloy X40.

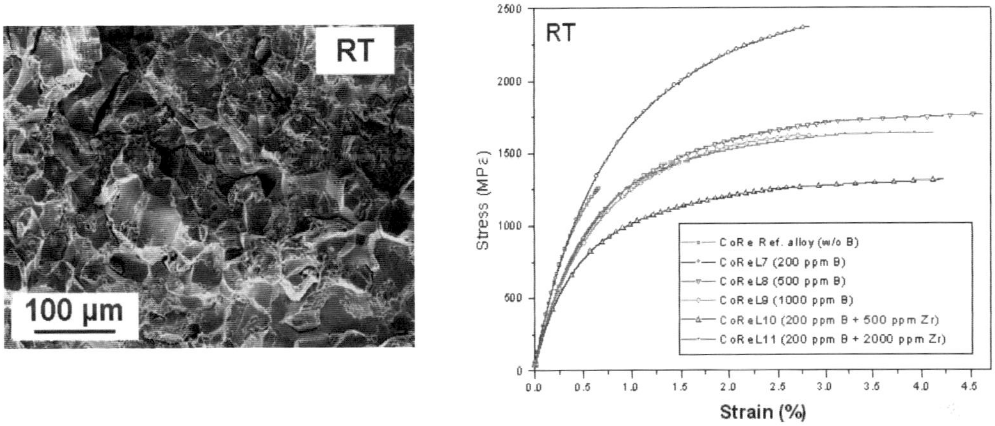

Figure 4. (Left) RT bend test results: Fracture surface in Co-17Re-23Cr alloy shows inter-granular fracture. (Right) Bend test results at RT.

Addition of Al in Co-Re-Cr-Si alloy can also be helpful, as at high temperatures the presence of Si and Al, can assist in the formation of a more protective SiO_2 and Al_2O_3-oxide scales. Additionally, combination of B with Si will also be explored, which is expected to form a glassy protective layer and thereby prevent oxidation.

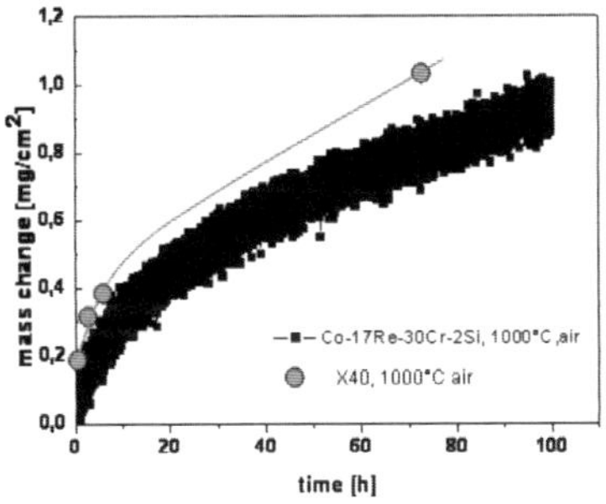

Fig. 5: Oxidation behaviour of Co-17Re-30Cr-2Si alloy at 1000°C compared to Co-base alloy X40.

Acknowledgements
We thank Ms Kruger and Prof. Heilmaier (TU Darmstadt) for the bend tests and Ms Gorr and Prof. Christ (Universität Siegen) for the oxidation tests. The financial support of Deutsche Forschungsgemeinschaft for the research group 727 ''Beyond Ni-base Superalloys'' is gratefully acknowledged.

References

[1] Saunders N 1996 *Superalloys 1996* eds. R.D. Kissinger *et al* (Warrendale: TMS)
[2] Rösler J, Mukherji D and Baranski T 2007 *Adv. Eng. Mater.* **9** 876.
[3] Mukherji D, Klauke M, Strunz P, Zizak I, Schumacher G, Wiedenmann A and Rösler J 2009 *Int. J. Mater. Res.* in print.
[4] Berthod P, Lemoine P and Ravaux J 2009 *J. Alloy. Comp.* **467** 227
[5] Klauke M, Mukherji D, Gorr B, Christ H–J and Rösler J 2009 *Int. J. Mater. Res.* **100** 104
[6] Gorr B, Trindade V, Burk S, Christ H-J, Klauke M, Mukherji D and Rösler J 2009 *Oxid. Metal* **71** 157

The influence of creep properties on crack propagation in thermal barrier coatings

Martin Bäker

Institut für Werkstoffe, Technische Universität Braunschweig, Langer Kamp 8, 38106 Braunschweig, Germany

martin.baekertu-bs.de

Abstract. Thermal barrier coatings are used to protect turbine blades from the high temperature of the process gas inside a turbine. They consist of a metallic bond coat and of a ceramic top coat with low thermal conductivity. During service, an additional oxide layer forms between bond coat and top coat that eventually causes failure. Finite element simulations show that the roughness of the interface between top and bond coat is crucial for determining the stress state. Lifetime models have been inferred that assume that cracks form in the peak positions at small oxide thickness and propagate when the oxide layer grows and the stress field shifts. A two-dimensional finite element model of crack propagation in the TBC layer is presented. Since the cracks propagate near a material interface and since plasticity may occur in the bond coat, standard tools of fracture mechanics for predicting the crack propagation direction are difficult to apply. This problem is circumvented in a very simple way by propagating short "test cracks" in different directions and optimising to find the crack direction with the maximum energy release rate. It is shown that the energy release rate and the crack propagation direction are sensitive to the details of the stress state and especially to the creep properties of the materials. Implications for failure models are discussed.

1. Introduction

Modern turbine blades are coated with thermal barrier coatings to partially protect the base material from the hot gas. The full potential of these coatings is difficult to exploit because they fail after some time by spallation for reasons that are not completely understood [1]. A thermal barrier coating system comprises a metallic bond coat that serves as corrosion protection, a ceramic top coat (the TBC in the narrower sense), and an oxygen layer that forms at the interface during service (TGO, thermally grown oxide). Usually, coatings spall upon cooling when this oxide layer reaches a critical thickness of the order of $10\,\mu$m [2, 3].

There are two causes of stress in the coating system: The difference of the coefficient of thermal expansion between the coating materials causes thermal stresses. Furthermore, the growing oxide layer has a larger volume than the bond coat it forms from, leading to growth stresses. It has been suggested [4] that cracks form initially at the peaks of an interface asperity because the stress is tensile here for a thin TGO. When the TGO grows, the stress state at the peak region becomes compressive, and the region of tensile stress shifts to the valley position. It has been assumed that cracks would follow the tensile region. However, this model takes not into account that a forming and growing crack would change the stress field. In this paper, a model of a thermal barrier coating system is presented that includes crack propagation. Creep

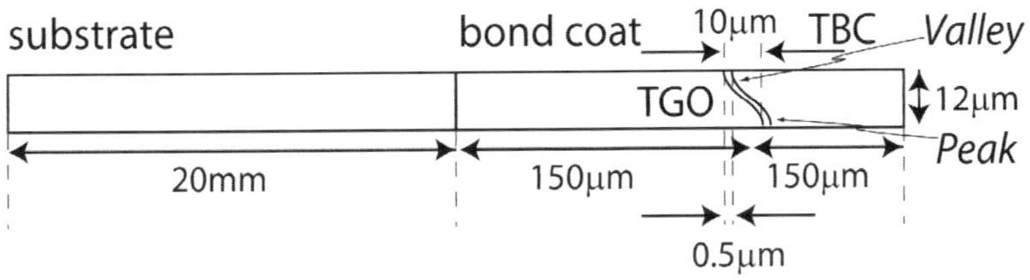

Figure 1. Model geometry. The model is cylindrically symmetric, with the axial direction being vertical. The "peak" and "valley" positions discussed in the text are marked.

Table 1. Material properties used in the model, taken from [5]. Norton creep with a creep law $\dot{\epsilon} = A\sigma^n$ was assumed for bond coat, TGO, and TBC.

	Substrate	Bond coat	Oxide	TBC
Young's Modulus (20 °C) [GPa]	184	200	400	48
Young's Modulus (1100 °C) [GPa]	145	110	325	22
Poisson number (20 °C)	0.3	0.3	0.23	0.1
Poisson number (1100 °C)	0.3	0.33	0.25	0.12
CTE (20 °C) [K^{-1}]	$12 \cdot 10^{-6}$	$13{,}6 \cdot 10^{-6}$	$8{,}0 \cdot 10^{-6}$	$9 \cdot 10^{-6}$
CTE (1000 °C) [K^{-1}]	$16 \cdot 10^{-6}$	$17{,}6 \cdot 10^{-6}$	$9{,}3 \cdot 10^{-6}$	$12{,}2 \cdot 10^{-6}$
Creep exponent n	no creep	3	1	1
Creep prefactor A (1000 °C) [$MPa^{-n}s^{-1}$]	no creep	$1{,}39 \cdot 10^{-7}$	varied	varied

has been shown previously to strongly influence the stress field [5]. Therefore, we also examine how crack propagation is influenced by the creep properties of the materials.

2. Finite element model

The model geometry (see Fig. 1) is that of a solid cylindrical disk which is thought to be infinitely extended in its axial direction, almost identical to the model described in [5]. The geometry was meshed using approximately 15000 four-node elements with selectively reduced integration and was implemented using the standard finite element program ABAQUS. Material data can be found in table 1. Growth of the TGO was implemented by using the swelling-option in ABAQUS. Isotropic swelling with a constant growth rate during each cycle was assumed.

To thermally load the system, the system temperature is isothermally raised from 20 °C to 1000 °C within 60 s. It is held at hot temperature for 120 s or 1 200 s, and cooled back in another 60 s to 20 °C.

After performing one thermal cycle, a crack of size 0.375 μm is initiated at the peak position of the TBC at the position of highest stress, directly above the TGO. This initial crack serves as starting point for the subsequent crack propagation calculation. Because the stress state is multi-axial and the crack is close to an interface, it is difficult to apply standard tools of fracture mechanics to predict the crack propagation direction. Therefore, we use a simple, but CPU-intensive, technique in which trial cracks are propagated in different directions, performing one finite element simulation per trial crack [7, 8]. The energy release rate during crack propagation is measured in each simulation by calculating the change in the energy stored in the system. It is assumed that the direction with the highest release rate is the direction in which the crack

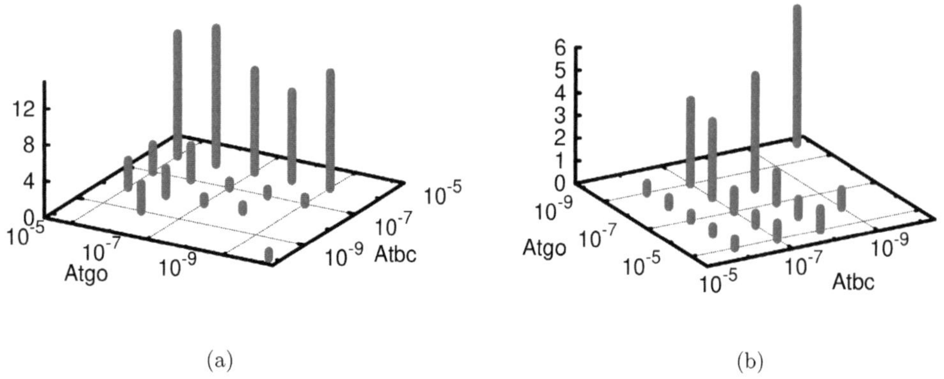

Figure 2. Crack propagation as a function of the creep strength of TGO and TBC (in MPa s^{-1}). Left: Total accumulated crack length (in μm) before the energy release rate becomes smaller the initial energy release rate. If the creep strength is low, cracks propagate throughout the simulation volume. Right: Initial value of the energy release rate in J/m^2. Note the change in perspective relative to Figure 2(a)

would propagate. If this is larger than the critical energy release rate G_c, the crack propagates in this direction and another crack increment is calculated. If it is smaller than G_c, the crack does not propagate and the state of the system is reset to the configuration before the crack increment. Throughout this paper, the trial crack length is fixed at 0.75 μm.

3. Results

To check the standard failure model described in the introduction, crack propagation simulations with a thin TGO layer were performed. After one thermal cycle with hold time 120 s, a crack was introduced at the peak position near the TGO/top coat interface. If the failure model is valid, a crack should be able to propagate and then stop. To study this, the simulation was run with a critical energy relase rate G_c of zero so that the crack would propagate unhindered until it has crossed the full simulation volume. The energy release rate was measured throughout to see whether the energy release rate becomes lower than the initial energy release rate (which is the condition for a stopping crack). The creep properties of top coat and TGO were varied.

Figure 2 shows the results results of the variation of the creep pre-factors. Figure 2(a) shows the crack length for which the energy release rate drops below its initial value. This is thus the minimum crack length before a crack might stop. In materials will low creep strength, the crack cannot stop because the energy release rate never becomes smaller than the initial value, whereas a high creep strength leads to a decrease of the energy relase rate with increasing crack length. Figure 2(b) shows the absolute value of the initial energy release rate. Although cracks in creep-soft materials would not stop once they have started, the absolute value of the energy release rate is rather small so that crack propagation would probably not begin.

These results show that the standard failure model can only be valid if the creep strength of the coating materials is sufficiently large. Furthermore, reducing the creep strength might be beneficial for the lifetime of the coatings because the energy release rate for crack propagation becomes smaller.

In the simulations shown above, the absolute values of the energy release rate after one thermal cycle shown in the previous section were small. In this section, the energy release rate

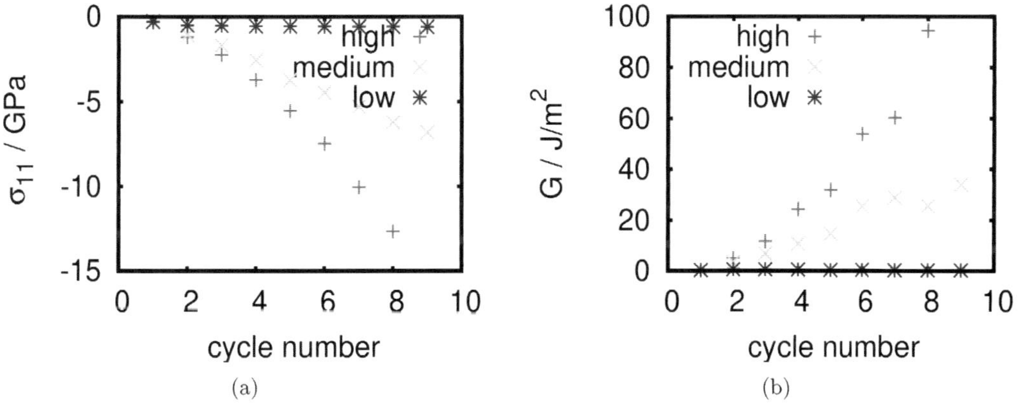

Figure 3. Left: Radial stress evolution at the peak position within the TBC for different values of the creep strength. Right: Energy release rate for cracks initiated in the peak position.
High strength $A_{\mathrm{TGO}} = A_{\mathrm{TBC}} = 10^{-10}\,\mathrm{MPa\,s^{-1}}$;
Medium strength $A_{\mathrm{TGO}} = A_{\mathrm{TBC}} = 10^{-9}\,\mathrm{MPa\,s^{-1}}$;
Low strength $A_{\mathrm{TGO}} = A_{\mathrm{TBC}} = 10^{-8}\,\mathrm{MPa\,s^{-1}}$.

of the first crack increment is measured after each of several thermal cycles (with hold time 1 200 s per cycle) to see whether it increases to more realistic values when the TGO grows. To do this, the critical energy release rate G_c was set to a very high value so that the crack could not propagate. Due to the TGO growth, the radial stresses at the peak position of the TBC become compressive with increasing number of cycles. This is shown in Figure 3(a). Nevertheless, as Figure 3(b) shows, the energy release rate increases despite the increase in compressive stresses. Thus, crack propagation occurs in mode II in this model. The results indicate the simple assumption of a failure model as sketched in the introduction may be wrong. However, crack propagation in mode II would require a larger energy release rate because friction between the crack surfaces would dissipate energy that is not available for creating new surface. In the future, the model will be improved by adding a friction coefficient on the crack surfaces.

References

[1] Stöver D, Funke D 1999 Directions of the development of thermal barrier coatings in energy applications *J. Mater. Proc. Technol.* **92** 195–202

[2] Singheiser L, Steinbrech R, Quadakkers W J and Herzog R 2001 Failure aspects of thermal barrier coatings *Materials at High Temperatures* **18** 249-259

[3] Rabiei A, Evans A 2000 Failure mechanisms associated with the thermally grown oxide in plasma-sprayed thermal barrier coatings, *Acta Mater.* **48** 3963–3976

[4] Freborg A M , Ferguson B L, Brindley W J and Petrus G J 1998 Modeling oxidation induced stresses in .thermal barrier coatings, *Mater. Sci. Eng. A* **245** 182–190

[5] Rösler J, Bäker M and Aufzug K 2004 A Parametric Study of the Stress State of Thermal Barrier Coatings – Part I: Creep Relaxation, *Acta Mater.* **52** 4809–4817

[6] Press W H, Flannery B P, Teukolsky S A, Vetterling W T1989 *Numerical Recipes* (Cambridge: Cambridge University Press)

[7] Bäker M 2008 Finite element crack propagation calculation using trial cracks, *Computational Materials Science* **43** 179–183

[8] Bäker M 2009 Simulation of crack propagation in mixed mode and at bimaterial interfaces using trial cracks, *Computational Materials Science* **45** 680–683

Journal of Physics: Conference Series **240** (2010) 012068 doi:10.1088/1742-6596/240/1/012068

Influence of microstructure on creep strength of MRI 230D Mg alloy

D Amberger[1], P Eisenlohr[2] and M Göken[1]

[1] Department of Materials Science and Engineering, Institute I: General Materials Properties WW I, University Erlangen-Nürnberg, Martensstraße 5, D-91058 Erlangen, Germany

[2] Max-Planck-Institut für Eisenforschung, Max-Planck-Straße 1, D-40237 Düsseldorf, Germany

dorothea.amberger@ww.uni-erlangen.de

Abstract. The low density of magnesium alloys makes them attractive for lightweight constructions. However, creep remains an important limitation of Mg alloys. To gain a more detailed understanding of the correlation between microstructure and creep properties in Mg alloys, creep tests have been performed on MRI 230D samples featuring various microstructures. For this purpose, the MRI 230D Mg alloy has been thixomolded into a plate with four steps of different height, which gives different microstructures in each step due to different cooling rates. With an increase in cooling rate (e.g., a decrease in step height) the interconnectivity of the eutectic phase increases at virtually constant volume fraction. The creep strength is found to decrease with decreasing interconnectivity of the eutectic phase. This implies that a eutectic phase morphology, which is highly interconnected, benefits the creep properties and should therefore be one goal in further developments for creep resistant Mg alloys.

1. Introduction

Magnesium, due to its low density, is attractive for lightweight constructions. However, next to other issues, the rather poor creep resistance of the most commonly used alloy AZ91 remains an important limitation to its widespread application. To overcome this issue, a series of creep resistant and cost competitive alloys has been developed. In many cases, Sr, Ca or rare earth elements (RE) have been added to improve the creep strength [1, 2, 3].

However, little is known about the correlation between microstructure and creep and on the mechanisms responsible for the improved creep behavior of those newly developed alloys. The difficulty in gaining an insight into these reasons lies in the fact that with an introduction of new alloying elements (for example Ca or Sr) several microstructural features change simultaneously and the effects of these changes on the mechanical properties are hard to separate. For Mg-Al-based alloys like AZ91 or AM60, with the introduction of, for example, Ca, the precipitate phase changes from $Mg_{17}Al_{12}$ to Al_2Ca, which is harder than $Mg_{17}Al_{12}$ (up to testing temperatures of about 200°C, [4]). Furthermore, not only the hardness and the chemistry of the second (intermetallic) phase changes, but also its morphology. The divorced eutectic, which is typical for AZ91 [5, 6] and shows rather isolated particles, changes into a network-like morphology of the eutectic phase [7, 8, 9]. Another effect is that the alteration of the elements and their concentrations in the α-Mg solid solution causes differences in solid solution strengthening.

In order to separate the above-mentioned influences on the creep strength we employ different cooling rates by means of casting stepped plates, which result in different morphology of the eutectic phase without significant changes in the chemistry.

2. Experimental

MRI 230D alloy was prepared by thixomolding at a melt temperature of 605°C (that means, in fully liquid state) by Neue Materialien Fürth GmbH (NMF, process details can be found, e.g., in [9, 10]). To produce different microstructures as a result of varying cooling rates, the alloy was thixomolded into stepped plates having a base area of 110 × 120 mm² and step heights of 2, 6, 10 and 14 mm (see also Figure 2). For the investigations, only the 6, 10 and 14 mm steps were used, due to geometrical restrictions of the sample size.

The concentration of alloying elements remains constant for all step heights (compare [11]), which is in accordance with the expectations as cooling rates in thixomolding are too high to allow for diffusion over macroscopic distances.

Microstructural investigation on polished and etched (0.5% HNO_3 in ethanol for 5 s) samples have been performed with a light microscope (LM, Leica DmRm 200) and a scanning electron microscope (SEM, Zeiss CrossBeam 1540 EsB) at 5 kV acceleration voltage. Precipitate volume fraction was measured as linear fraction from a set of parallel test lines superimposed onto the SEM images of the microstructure, i.e., by line intercept method.

Creep tests were performed at constant stress (= force per current cross section) and temperature in compression. Compressive stresses and strains conventionally have negative sign but are reported here in absolute values. Details of the creep test conduction can be found in [7, 11].

3. Results and Discussion

3.1. Microstructure

Light optical images (Fig. 1) of the investigated step heights reveal that all share a similar microstructure: a network of eutectic phase is enclosing the α-Mg solid solution.

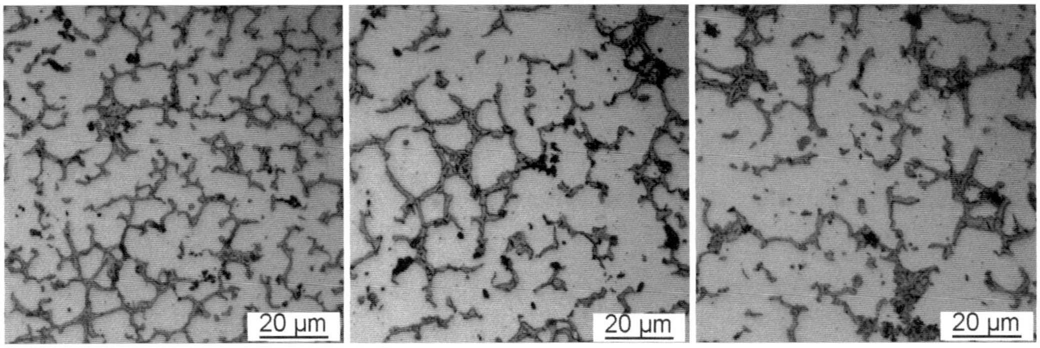

Figure 1: LM images of MRI 230D alloy at 6, 10, and 14 mm step (left to right)

However, a closer look shows that the network of the eutectic phase in the 6 mm step is finer and also seems to be more closed than the network of the other two steps. Between the 10 and the 14 mm steps a slight decrease in interconnectivity of the skeleton can be identified from the LM pictures (see also [11]). It can be concluded that the connectivity of the network of the eutectic phase increases with decreasing step height (increasing cooling rate), although no exact quantification of this connectivity is presently available.

The volume fraction of the eutectic phase, measured by the line intercept method and shown in Figure 2 remains constant within the margin of error among all step heights.

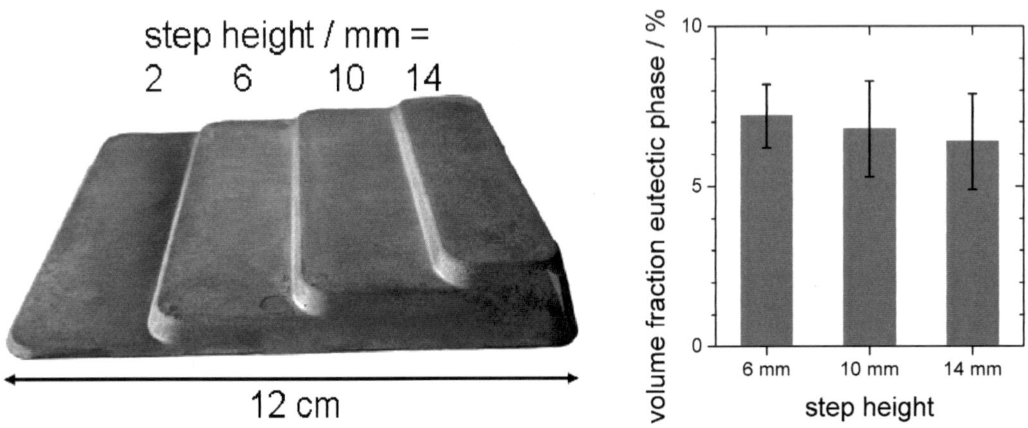

Figure 2: Stepped casting plate and volume fraction of the eutectic phase for the different step heights

Therefore differences in creep properties cannot be rationalized from different volume fractions of the eutectic phase. With regard to the fact that the chemical composition does not vary for the different steps [11] one can also conclude that the chemical composition of the eutectic phase as well as of the solid solution remains constant. Differences in solid solution hardening as well as hardness differences of the eutectic phase can thus be excluded – the differences in mechanical properties seem solely linked to the different morphology of the eutectic phase as seen in the LM images.

3.2. Creep properties

Figure 3 shows the creep curves (creep rate vs. creep strain) for all three steps tested at a temperature of 200°C and a compressive stress of 100 MPa.

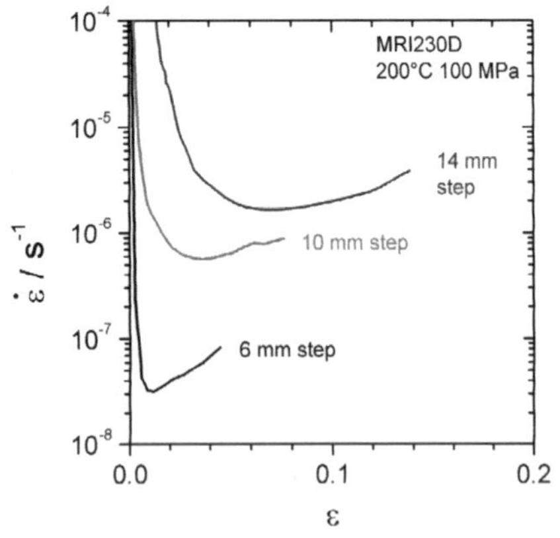

Figure 3: Compressive creep rate vs. creep strain for the different step heights of MRI 230D alloy measured at 200°C and 100 MPa

The 6 mm step exhibits by far the lowest creep rate of all tested step heights, its creep rate lying more than one order of magnitude lower than the creep rate of the 10 mm step and almost two orders of magnitude lower than the creep rate of the 14 mm step. The differences in creep rate between the 10 and the 14 mm step are far less pronounced, as the minimum creep rate of the 14 mm step is approximately a factor of 2 to 3 higher than the minimum creep rate of the 10 mm step.

Observing the chemical and microstructural results reported above, the lower creep strength is predominantly connected to the lower interconnectivity of the eutectic phase (resulting from lower cooling rate in thicker steps). Independently, a similar correlation has been reported [12] between the local room-temperature strength measured on cross-sections of AZ91 high-pressure die castings and the corresponding degree of eutectic phase interconnectivity observed in three-dimensional reconstructions of the microstructure. In line with both such results it was shown [7] that a destruction (by low-temperature prestraining) of the highly interconnected eutectic phase network leads to a massive deterioration of the creep strength.

4. Summary and Conclusion

The creep behavior of as-cast MRI 230D Mg alloy strongly depends on the structure of its eutectic phase network solidifying in the inter-dendritic channels. When this is highly interconnected a significantly higher creep strength results. The eutectic phase acts as a scaffold similar in structure to an embedded open-cell ceramic foam and thus provides strengthening in line with the expected behavior of such foams. Therefore, a conclusion with regard to further developments of creep-resistant Mg alloys would be to produce a highly interconnected network of second and harder phase within the soft α-Mg matrix.

5. Acknowledgement

Financial support from Deutsche Forschungsgemeinschaft (DFG) under contract number GO 741/9-3 is gratefully acknowledged. The authors would like to thank Neue Materialien Fürth GmbH, in particular Prof. Dr. R. F. Singer, Dr. A. Lohmüller and N. Kömpel, for supplying the thixomolded magnesium alloys and performing the GDOES measurements.

References

[1] M.O. Pekguleryuz, E. Baril 2001 *Mater. Trans.* **42** 1258-1267
[2] A.A. Luo 2004 *Int. Mater. Rev.* **49** 13-30
[3] M.O. Pekguleryuz, A.A. Kaya 2003 *Adv. Eng. Mater.* **5** 866-878
[4] L.L. Rokhlin 2007 personal communication
[5] S. Barbagallo, H.I. Laukli, O. Lohne, E. Cerri 2004 *J. Alloy Comp.* **378** 226-232
[6] S. Spigarelli, M. Cabibbo, E. Evangelista, M. Talianker, V. Ezersky 2000 *Mater. Sci. Eng.* **A 289** 172-181
[7] D. Amberger, P. Eisenlohr, M. Göken 2009 *Mater. Sci. Eng.* **A 510-511** 398-402
[8] R. Ninomiya, T. Ojiro, K. Kubota 1995 *Acta Metall. Mater.* **43** 669-674
[9] H. Eibisch, A. Lohmüller, N. Kömpel, R.F. Singer 2008 *Int. J. Mater. Res.* **99** 56-66
[10] A. Lohmüller, M. Scharrer, M. Hilbinger, R. Jenning, M. Hartmann, R.F. Singer 2005 *Giessereiforschung* **57** 2-9
[11] D. Amberger, P. Eisenlohr, M. Göken 2009 *Proc. 8th Int. Conf. Mg Alloys and their Applications (Weimar 2009)* ed K.U. Kainer (Weinheim: Wiley-VCH) in press
[12] A.V. Nagasekhar, C.H. Cáceres, C. Kong 2009 *J. Phys.: Conf. Ser.*, this issue

FEM simulation of TBC failure in a model system

P. Seiler[1], M. Bäker[1], T. Beck[2], M. Schweda[2] and J. Rösler[1]

[1] Institut für Werkstoffe (IfW), Technische Universität Braunschweig, Germany
[2] Institut für Energieforschung/ Werkstoffstruktur und -Eigenschaften (IEF 2), Forschungszentrum Jülich, Germany

p.seiler@tu-bs.de

Abstract. In order to study the behavior of the complex failure mechanisms in thermal barrier coatings on turbine blades, a simplified model system is used to reduce the number of system parameters. The artificial system consists of a bond-coat material (fast creeping Fecralloy or slow creeping MA956) as the substrate with a Y_2O_3 partially stabilized plasma sprayed zircon oxide TBC on top and a TGO between the two layers. A 2-dimensional FEM simulation was developed to calculate the growth stress inside the simplified coating system. The simulation permits the study of failure mechanisms by identifying compression and tension areas which are established by the growth of the oxide layer. This provides an insight into the possible crack paths in the coating and it allows to draw conclusions for optimizing real thermal barrier coating systems.

1. Introduction

Plasma sprayed thermal barrier coating systems are used on top of highly stressed components e.g. in gas turbines. They protect the underlying nickel-basis substrates from the high surrounding temperatures. A typical coating system consists of a nickel-basis substrate, the metallic bond-coat, and the thermal barrier coating (TBC). These coatings fail in service because of the thermally grown oxide (TGO) between the bond-coat and the TBC, and the complex load [1]. Furthermore, the interfacial roughness of the substrate and the creep properties play an important role in the failure mechanisms [2].

A model system was developed to reduces the influence parameters which leeds to the spallation of the coating system. The model system contains of the MCrAlY bulk material Fecralloy which is in this case used as the substrate of the coating system. The TBC is applied directly to this substrate. Therefore, all influences of a nickel-basis substrate can be neglected and a system with reduced coating layers can be analyzed in order to draw conclusions for optimizing real thermal barrier coating systems.

To study the creep influence of the bond-coat material itself a second substrate was used in the model system: the slow creeping oxide-dispersion-strengthened (ODS) material MA956.

The TGO as the third layer in the system grows between the substrate and the TBC by oxygen that diffuses through the TBC. This is caused by the high porosity of the TBC and the high ionic diffusivity of oxygen [3]. The oxidation is an inward growth of α-Al_2O_3 [1]. The growth of the TGO promotes the failure of the coating system. Therefore, the growth stresses are examined.

2. FEM model

The geometry is almost identical to the model described in [2] and [4]. Fig. 1 shows a sketch of the system. It was generated by a C++ preprocessor for ABAQUS [5] and consists of *CAX8RHT* elements. All layers in the model are free of stress at the beginning of the simulation. The initial TGO thickness was varied between 1 μm and 7 μm. The temperature of 1000°C was held for 2 h. The analysis of the stresses was performed at prominent positions at the interface (valley and peak positions in Fig. 1). The discussed stress values at this positions are the mean values of the radial stresses σ_{11} at the integration points.

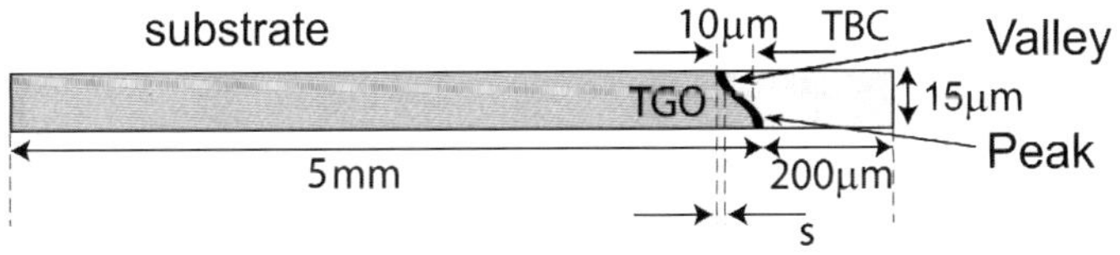

Figure 1. Sketch of the model system with the marked *valley* and *peak* positions.

The material parameters of the TGO and the TBC can be found in [2]. The measured parameters of Fecralloy and MA956 are shown in Tab. 1. The Poisson number ν and the thermal expansion coefficient α are the same as MCrAlY and can also be found in [2].

	Fecralloy	*MA956*
Young's modulus E, 20°C [GPa]	148.6	148.6
Young's modulus E, 1000°C [GPa]	81.7	81.7
density ρ [kg/m^3]	7.1×10^3	7.1×10^3
creep activation energy Q [kJ/mol]	486	452
creep exponent n [-]	5.29	41.0
creep prefactor A_0 [MPa^{-n}s^{-1}]	1.72×10^{-10}	3.46×10^{-124}
creep factor a [K^{-1}]	-0.012	0.10

Table 1. Material parameters of Fecralloy and MA956.

Creep was permitted in all layers (slow creep parameters for TGO and TBC from [2]). The creep rate in the TBC and in the TGO was calculated with a Norton creep law $\dot{\varepsilon} = A\sigma^n$. The creep rate of Fecralloy and MA956 is calculated by

$$\dot{\varepsilon} = \underbrace{A_0 \cdot \exp(aT) \cdot \exp\left(-\frac{Q}{RT}\right) \cdot}_{A} \sigma^n \tag{1}$$

with the activation energy Q, the creep exponent n and an additional factor a (Tab. 1). The creep parameters Q, A_0, n, and a are fitted values obtained from experimental creep tests. The resulting maximum growth stresses inside the TGO are in the range of 0.1 to 1 GPa and are of the same order of magnitude as shown in [6].

The TGO growth was simulated by simplifying the process as isotropic *swelling*. A Tammann law was used to describe the oxidation [7, p. 268]:

$$\dot{s} = \frac{1}{2} \cdot \frac{k'_p}{s^2} \tag{2}$$

with the current TGO thickness s, the parabolic oxidation constant k'_p and the growth rate $\dot{s}$. The parabolic oxidation constant of this simulation is $k'_p = 1.5 \times 10^{-17}$ m^2/s which describes the oxidation kinetics of α-Al_2O_3 at 1000°C [4]. The growth rate was calculated once at the beginning of the simulation and is constant for the further simulation.

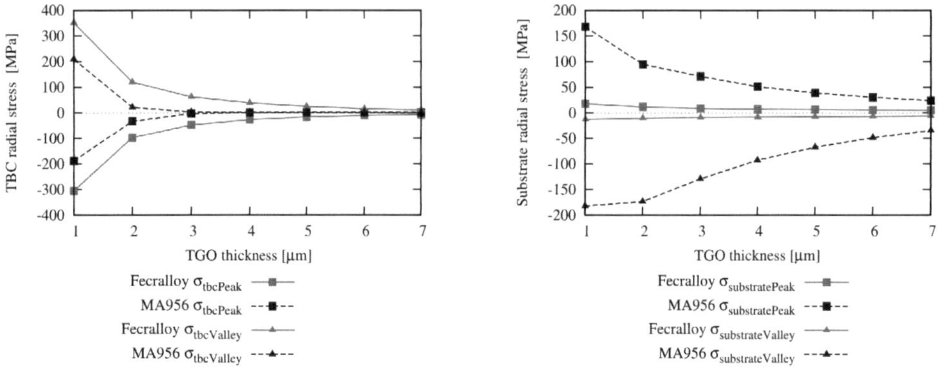

(a) Growth stresses inside the TBC. (b) Growth stresses inside the substrate.

Figure 2. Growth stresses of the TGO at 1000°C (hold time 2 h) in the Fecralloy and MA956 coating system against the initial TGO thickness.

3. Results

Fig. 2 shows the radial stress σ_{11} at the substrate/TGO and the TGO/TBC interfaces against the respective initial TGO thickness. The radial stresses decrease while increasing the initial TGO thickness because the growth rate of the TGO decreases while increasing the TGO thickness (Eq. 2). The TGO growth is governed by oxygen diffusion through the TGO [4]. The stresses in the MA956 substrate are greater than the stresses in the Fecralloy substrate for all TGO thicknesses. This relation changes in the TBC: The TBC stresses in the Fecralloy coating system are larger then the TBC stresses in the MA956 coating system.

Fig. 3 shows the distribution of radial stress after 2 h at 1000°C in Fecralloy and MA956. The reason for the stress maximum inside the MA956 substrate was discussed in [4]. The growth of the TGO in lateral direction results in compressive stresses in the peak and tensile stresses in the valley of the TBC. Furthermore, it results in a changed position of the TGO in which the valley of the TGO is shifted to the inside (in Fig. 3 to the left) and the peak to the outside (in Fig. 3 to the right). The comparatively fast creeping Fecralloy substrate can not resist the strain because of the resulting low stresses. The resistance of the MA956 substrate to the strain of the TGO is larger caused by the creep strength. This leads to higher magnitude of stresses inside the substrate and to lower stresses inside the TBC. Thus, it can be expected that the Fecralloy coating system fails most likely inside the TBC and MA956 coating systems inside the substrate or at the interface substrate/TGO.

The prediction of different failure mechanisms of the studied substrate materials was validated with experimental tests. Fig. 4(a) shows a micrograph of a Fecralloy sample with a cyclic grown TGO (no applied initial TGO). The crack path of the coating system is above the TGO inside the TBC layer. Fig. 4(b) shows the failure of the ODS strengthened MA956 sample, also with a cyclic grown TGO. The failure occurs at the TGO/substrate interface, which correlates with the distribution of stresses in Fig. 2. For this reason, the creep strength of the used substrate has a large influence on the position where the failure is induced and the resulting crack path, even if adequate MCrAlY materials are used as a bond-coat layer in technical thermal barrier coating systems.

The simulation of the growth stress shows that slow creeping substrates and bond-coats respectively reduce the lifetime caused by cracks under the TGO. On the other hand a more creep strengthened bond-coat reduces the stresses inside the TBC. MA956 samples with an artificial applied initial TGO by preoxidation or sputtering fail more likely under the TGO.

Journal of Physics: Conference Series **240** (2010) 012069 doi:10.1088/1742-6596/240/1/012069

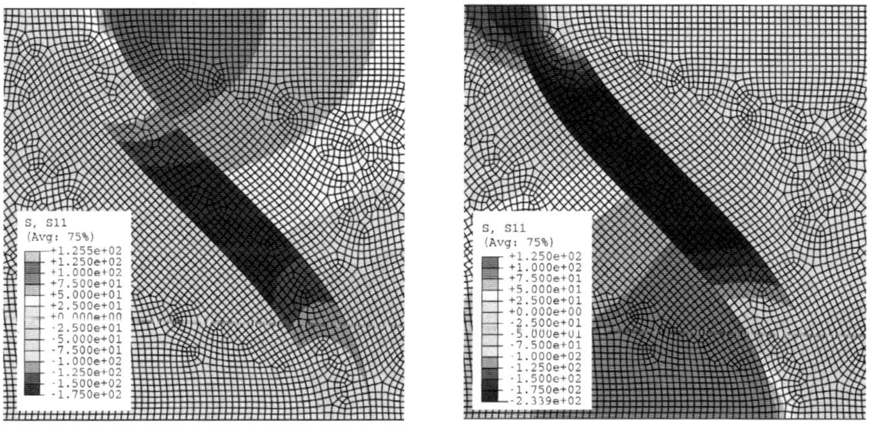

(a) σ_{11} stresses in the Fecralloy model. (b) σ_{11} stresses in the MA956 model.

Figure 3. Growth stresses of the TGO at 1000°C (hold time: 2 h, initial TGO thickness: 2 μm).

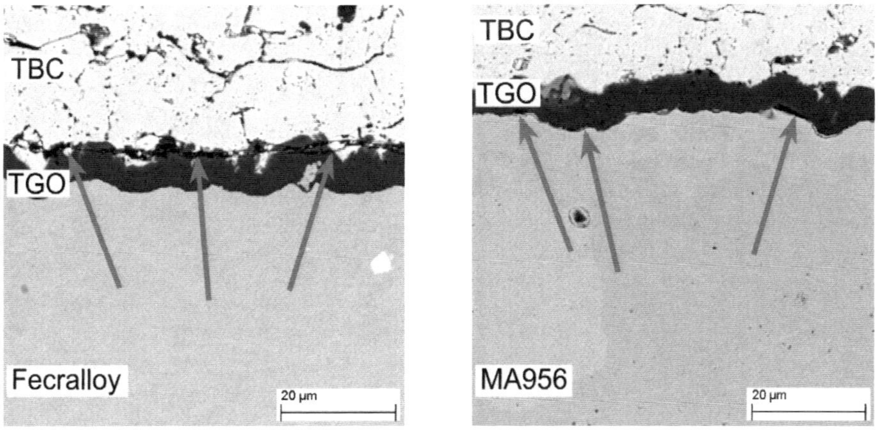

(a) Fecralloy with crack above the TGO. (b) MA956 with crack under the TGO.

Figure 4. Comparison of the failure in coating systems with Fecralloy and MA956 substrates (~300 cycles, 2 h at 1050°C each).

Fecralloy samples with an initial TGO fail only above the TGO. This suggests that there is an optimum of the bond-coat creep strength. These results contradict the considerations from Padture et al. [1] and Evans et al. [8]. They discuss that only slow creeping bond-coat materials benefit the lifetime of the coating systems.

References

[1] Padture N P, Gell M and Jordan E H 2002 *Science* **296** 280 – 284
[2] Bäker M, Rösler J and Heinze G 2005 *Acta Mater.* **53** 469–476 ISSN 1359-6454
[3] Fox A C and Clyne T W 2004 *Surf. Coat. Technol.* **184** 311–321 ISSN 0257-8972
[4] Rösler J, Bäker M and Aufzug K 2004 *Acta Mater.* **52** 4809–4817 ISSN 1359-6454
[5] Bäker M 2006 *Tagungsband ABAQUS Benutzerkonferenz*
[6] Schumann E, Sarioglu C, Blachere J R, Pettit F S and Meier G H 2000 *Oxid. Metals* **53**
[7] Bürgel R 2001 *Handbuch Hochtemperatur- Werkstofftechnik* 2nd ed (Braunschweig/ Wiesbaden: Vieweg)
[8] Evans A, Clarke D and Levi C 2008 *J. Eur. Ceram. Soc.* **28** 1405

Journal of Physics: Conference Series **240** (2010) 012070 doi:10.1088/1742-6596/240/1/012070

Dynamic polygonization in 9%Cr heat resistant steel

V Dudko[1], A Belyakov[1], V Skorobogatyh[2], I Schenkova[2], and R Kaibyshev[1]

[1] Belgorod State University, Pobeda 85, Belgorod 308015, Russia
[2] Central Research Institute for Machine-Building Technology, Sharikopodshipnikovskaya 4, Moscow 115088, Russia

dudko@bsu.edu.ru

Abstract. Structural changes in 9%Cr martensitic steel during creep were examined. The grip section of the crept specimen was characterised by a lath martensite structure, which hardly changed during the test. In contrast, quite different microstructure developed in the necking portion of the specimen. The structural changes were characterized by the evolution of relatively large equiaxed subgrains with remarkably lowered density of interior dislocations at places of initial martensite laths. The development of the well-defined subgrains in the necking portion was accompanied with a coarsening of second phase precipitations. The structural mechanism responsible for the microstructure evolution during creep is considered as a dynamic polygonization.

1. Introduction

Nowadays, tempered martensitic 9-11%Cr steels are considered as promising creep resistant structural materials for fossil power plants [1-4]. The creep properties of these materials at elevated temperatures depend on the stability of martensite laths [5, 6]. Structural changes leading to significant changes of original microstructures during hot deformation are commonly related to the development of dynamic recrystallization [7]. Two general mechanisms for dynamic recrystallization were proposed [8]. The discontinuous dynamic recrystallization was shown to occur in alloys with relatively low stacking fault energy, when the new grains nucleate and grow due to bulging and migration of original grain boundaries over long distances. Another mechanism of dynamic recrystallization takes place in materials with high ability to dynamic recovery and involves continuous development of new grains as a result of formation of strain-induced subboundary network and gradual increase in misorientations between subgrains. However, the mechanisms of microstructure evolution operating in martensitic steels during the creep at elevated temperatures have not been detailed. Efficiency of different second-phase particles for microstructure stabilization against dynamic evolution is still unclear. High dislocation densities in martensite laths may promote a local bulging of grain boundaries, leading to discontinuous recrystallization. On the other hand, the rearrangement of lath boundaries and interior dislocations may result in the development of new recrystallized grains that is similar to continuous recrystallization by subgrain coalescence. The aim of the present work is to study the mechanism of microstructure evolution in a 9%Cr steel during the creep at a temperature of 650°C, which is just above the service temperature of 620°C for this material.

2. Experimental

A P92-type steel, Fe-0.1C-0.17Si-0.54Mn-8.75Cr-0.21Ni-0.51Mo-1.60W-0.23V-0.07Nb (all in mass%), was fabricated by Chelyabinsk Metallurgical Plant, Russia. The steel was subjected to solution treatment at 1050°C followed by air cooling and then tempered at 730°C for 3 hours. The tensile specimens of ∅10 mm with the gauge length of 50 mm were crept up to rupture at 650°C under starting stresses of 118 MPa. The sample ruptured after 1271 hours was selected for detailed structural examination. Structural investigations were performed by using a Quanta 600 FEG scanning electron microscope equipped with an electron back scattering diffraction pattern (EBSP) analyser incorporating an orientation imaging microscopy (OIM) system, and a JEM-2100 transmission electron microscope. The subgrain sizes were measured on the TEM micrographs by the linear intercept method, including those at all the clear visible (sub)boundaries. The dislocation densities were evaluated by counting the individual dislocations in grain/subgrain interiors. Hardness measurements were carried out with the load of 5 N on the central section along the specimen axis.

3. Results and discussion

The starting martensite microstructure was obtained by air cooling of austenite with an average grain size of ~20 μm. A typical tempered fine structure is shown in figure 1a. The tempered martensite consists of lath blocks subdividing original austenite grains (figure 1). Parameters of microstructure is shown in table 1. The lath interiors are homogeneously filled with plate-shaped fine V(C,N) particles (figure 2a). The particles obey the Baker-Nutting type orientation relationship [9], *i.e.* $(100)_{V(C,N)} \parallel (100)_{\alpha}$, $[011]_{V(C,N)} \parallel [001]_{\alpha}$. The average longitudinal size and thickness of these fine carbonitrides are about 8 and 2 nm, respectively.

The hardness of initial tempered steel remarkably decreased after the creep test. However, the softening is not uniform along the crept specimen. The hardness gradually decreases from the portion close to the grip section to the specimen neck. The softening during creep correlates with the total creep strain.

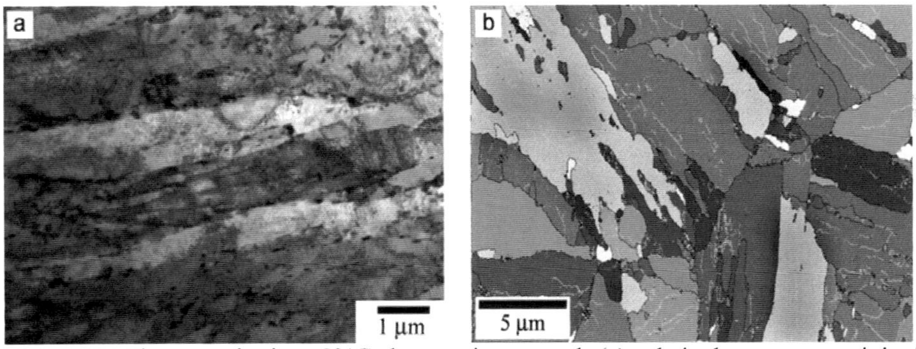

Figure 1. Tempered martensite in a 9%Cr heat resistant steel; (a) relatively coarse precipitations of carbides and Laves phases at boundaries of blocks and laths; (b) OIM map.

Table 1. Microstructure parameters of P92 steel.

	Transverse size of laths, nm	Densities of dislocation in the lath, m^{-2}	Average size of particles, nm	Portion of CSL boundaries, %	Hardness, MPa
Tempered microstructure	330	6.2×10^{14}	85	9	2680
Head of specimen	330	4.4×10^{14}	112	7	2340
Neck of specimen	740	10^{14}	211	3	1820

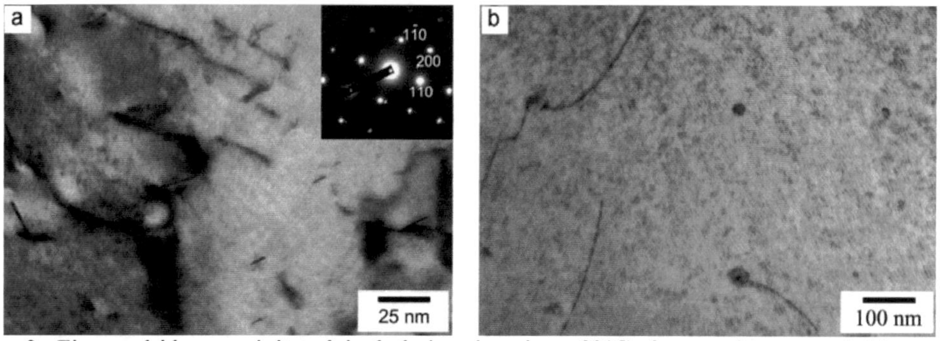

Figure 2. Fine carbides precipitated in lath interiors in a 9%Cr heat resistant steel (a) tempered martensite; (b) carbonitrides close to the neck portion of sample after the creep test.

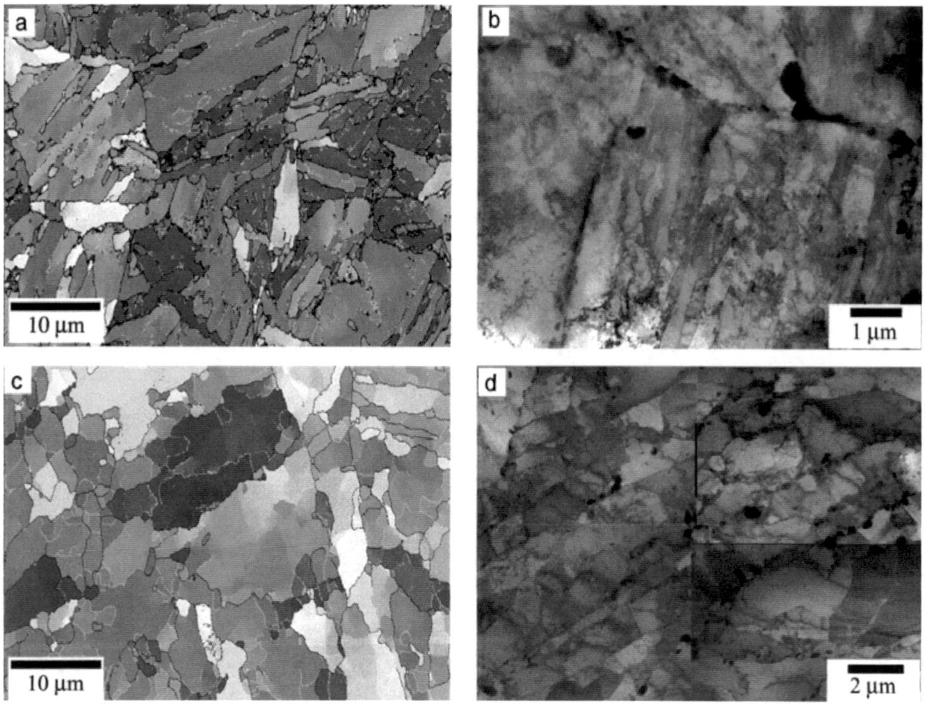

Figure 3. OIM maps and TEM photographs for (a, b) thread and (c, d) neck portions of a 9%Cr heat resistant steel specimen after the creep test.

The effect of hot plastic deformation on the structural changes in the tempered martensite of 9%Cr steel is clearly illustrated by OIM and TEM micrographs in figure 3. The grip section of the crept specimen did not experience a remarkable plastic deformation. Therefore, the structural changes in the grip section can be simply considered as a result of static recovery or recrystallization. The creep test did not change significantly the lath martensitic microstructure evolved in the grip section that looks like typical tempered martensitic microstructure (figures 3(a, b)). In contrast, the microstructure developed in the neck part of the crept specimen is quite different from the initial one. The microstructures in figure 3(c, d) look like typical dynamically recrystallized that is composed by almost equiaxed grains/subgrains with irregular boundaries. It should be noted that the fraction of special boundaries with high density of coincident sites decreased after the creep test (table 1). Coarsening of microstructure during the creep test is accompanied by a growth of fine V(C, N) precipitations. A few of round-shaped fine particles with an average size of about 30 nm can only be observed with in interiors of subgrain after creep test (figure 2b).

The development of relatively large subgrains in necking portion of crept specimen and the creep softening indicate that the plastic flow can significantly accelerate the coarsening of all second phase particles. The dislocation fluxes promote the diffusion processes leading to the particle coagulation during creep. The mobility of dislocation subboundaries increases due to decreasing the pinning forces. Also, the dislocation motion and interaction with the special boundaries result in the change of orientation relationship and, therefore, facilitate the boundary mobility. Then, the dislocation rearrangement within martensite substructure and the development of dynamic recovery result in the evolution of conventional hot worked substructure. The formation of new subgrains takes place continuously during the deformation due to rearrangements of interior dislocations; that is in situ without nucleation and growth of new recrystallizing grains. Any local bulging of high-angle grain boundaries scarcely accompanies the microstructure evolution. The fraction of high-angle grain boundaries does not rise and the (sub)boundary misorientation distributions do not significantly change during the process. Therefore, such evolutional process is similar to a continuous dynamic recrystallization, which frequently develops during hot working of metallic materials with high stacking fault energy [10]. In the present study the structural changes took place on the substructural level; namely, the martensitic laths gradually transformed to hot deformation subgrains. Thus, the considered mechanism of microstructure evolution should be discussed as a dynamic polygonization.

4. Conclusions

The structural changes depended significantly on the amount of plastic strain. The initial microstructure of tempered martensite with the transverse lath size of 330 nm was changed to relatively coarse substructure with the average subgrain size of about 740 nm in the neck portion of the crept specimen, where the experienced strain (reduction in area) was about 60%. In contrast, the martensitic microstructure in the grip section of the specimen scarcely changed during the test.

The subgrain coarsening in the necking part of the specimen during the creep test was accompanied by corresponding growth of second phase particles that originally precipitated at boundaries of blocks and laths. The average particles size increased from 85 to 211 nm during the test.

The mechanism of microstructure evolution in the neck portion of the crept specimen is considered as a dynamic poligonization.

Acknowledgements

The study was supported by Federal Agency for Science and Innovations under grant No. 02.523.12.3019. Authors are grateful to Mr. E. Kudryavtsev, Joint Research Center, Belgorod State University, for his assistance in specimen preparation and instrumental analysis.

References

[1] Ennis P J and Czyrska-Filemonowicz A 2003 *Sadhana* **28** 709
[2] Taneike M, Abe F and Sawada K 2007 *Nature* **424** 294
[3] Hald J 2008 *Int. J. Pressure Vessels and Piping* **85** 30
[4] Gupta G and Was G S 2008 *Metall. Mater. Trans. A* **39A** 150
[5] Sawada K, Takeda M, Maruyama K et al. 1999 *Mater. Sci. Eng. A* **A267** 19
[6] Abe F, Horiuchi T, Taneike M and Sawada K 2004 *Mater. Sci. Eng. A* **A378** 299.
[7] McQueen H J and Jonas J J 1975 *Treatise on Materials Science and Technology*, ed Arsenoult R J (Academic Press NY) p 393
[8] Sakai T and Jonas J J 2001 *Encyclopedia of Materials: Science and Technology* (Elsevier, Oxford) **7** 7079
[9] Wei F G, Hara T, and Tsuzaki 2004 *Phil. Mag.* **84** 1735
[10] Gourdet S and Montheillet F 2000 *Mater. Sci. Eng. A* **A283** 274

High temperature properties of an austenitic stainless steel

I Nikulin[1], R Kaibyshev[1], V Skorobogatykh[2]

[1] Belgorod State University, Pobeda 85, 308015 Belgorod, Russia

[2] Central Research Institute for Machine-Building Technology, Sharikopodshipnikovskaya 4, 115088 Moscow, Russia

nikulin_ilya@bsu.edu.ru

Abstract. Tensile properties of the 18Cr-9Ni-W-Nb-V-N austenitic stainless steel were studied at strain rates ranging from 6.7×10^{-6} to 1.3×10^{-2} s^{-1} in the temperature interval 20–740°C. It was found that this steel exhibits jerky flow at temperatures ranging from 530 to 680°C and an initial strain rate of 1.3×10^{-3} s^{-1}. This phenomenon was interpreted in terms of Portevin-Le Chatelier (PLC) effect occurring due to dynamic strain aging (DSA). PLC yields significant increase in high temperature strength of this steel due to extending of plateau on temperature dependence of yield strength (YS) and ultimate tensile strength (UTS) to higher temperatures. As a result, YS and UTS remain virtually unchanged with increasing temperature from 350 to 740°C. Role of additives of tungsten and vanadium in DSA and high temperatures strength of the austenitic stainless steel is discussed.

1. Introduction

Austenitic stainless steels, containing 18% of Cr and 8% of Ni, are widely used in high temperature applications. Additives of W, V, Nb and N provide an increment in their service temperature [1]. However, the origin of positive effect of these elements on high temperature strength is not yet clear. It has been reported that at high service temperatures the austenitic stainless steels are susceptible to dynamic strain aging (DSA) associated with interactions between solute atoms and mobile dislocations during plastic deformation. This phenomenon depends on temperature and strain rate. DSA may have a large impact on the mechanical properties such as strength, ductility, fatigue and creep life of high temperatures structural units [2-4]. The present study aimed to examination of DSA phenomenon in the 18Cr-9Ni-W-Nb-V-N steel and its effect on strength at high temperature. A role of alloying elements in high temperature strength of the 18Cr-9Ni-W-Nb-V-N austenitic steel is discussed.

2. Experimental procedures

The austenitic stainless steel with a chemical composition Fe–0.018%C–18.2%Cr–8.2%Ni–2.1%W–0.43%Nb–0.29%V–0.17%N–1.75%Mn–0.14%Si was manufactured by chill casting. The ingots were forged into rods at a temperature of ~1180°C, then solution treated at 1150°C for one hour and quenched into water. The flat tensile specimens having a gauge length of 25 mm and 7×3 mm^2 cross-section were cut from solution treated rods. These specimens were tensioned in the temperature interval of 20–740°C at strain rates ranging from 6.7×10^{-6} to 1.3×10^{-2} s^{-1}. An Instron universal testing machine (Model 5882) equipped with a three-zone split furnace was used. Temperature accuracy was within ±3°C. Each specimen was held at a testing temperature for about 20 minutes in order to reach a thermal equilibrium.

Specimens for metallographic examinations were etching in a solution of 30%H_2O+30% HCl+10%HNO_3. Metallographic analysis was carried out using Olympus GX70. For TEM examinations, samples were ground to about 0.1 mm. Discs with 3 mm diameter were cut and electropolished to perforation with a Tenupol-5 twinjet polishing unit using a 10%$HClO_4$ in CH_3COOH at 25 V. A Jeol JEM-2100 electron microscope with a double-tilt stage at an accelerating voltage of 200 kV was used for the thin foil examinations.

3. Results and discussion

Typical microstructure of solution treated steel consists of fully annealed austenitic grains containing annealed twins and small amount of fine second phase particles (figure 1). Dispersoids having average size of about 50 nm and located within austenitic grains was identified as a primary NbCrN. The average size of austenitic grains was about 16 μm. Energy-dispersive X-ray spectroscopy showed that alloying elements fully dissolves in matrix (figure 2).

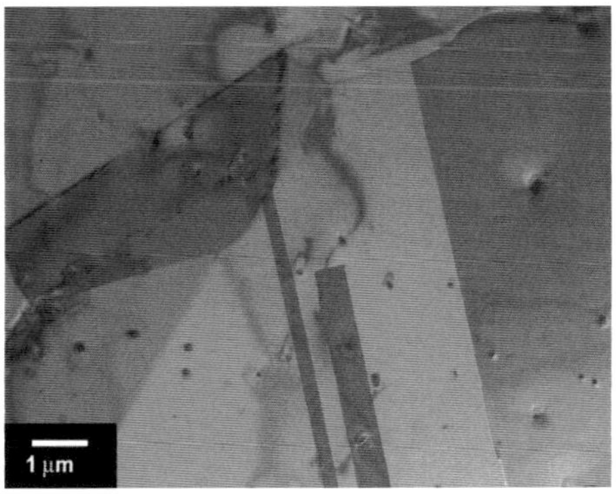

Figure 1. TEM micrograph showing typical microstructure of 18Cr-9Ni-W-Nb-V-N steel.

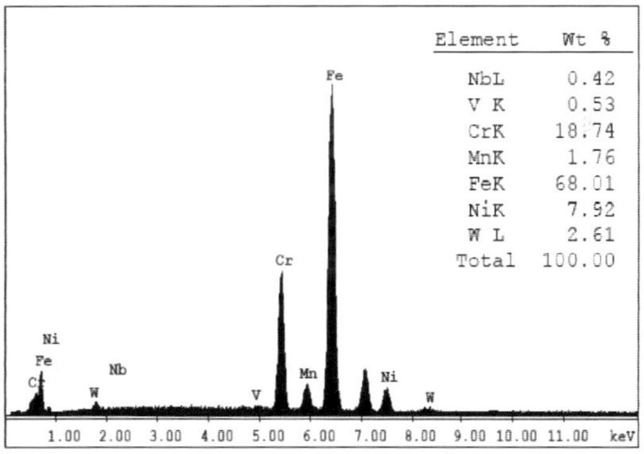

Figure 2. EDS spectra and quantification of elements taken from precipitations free matrix.

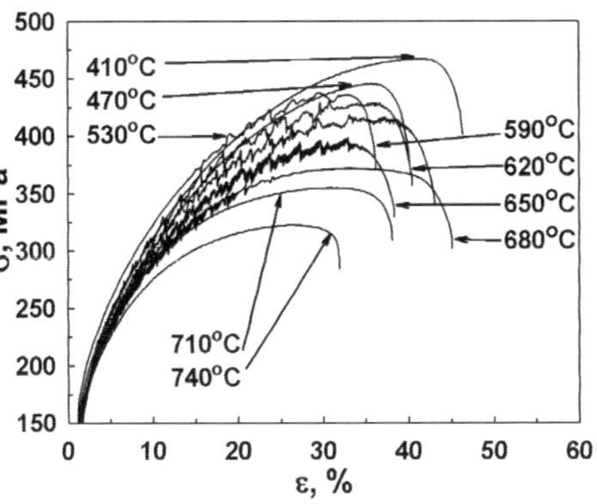

Figure 3. Typical stress–strain curves obtained at strain rate of 1.3×10^{-3} s^{-1}.

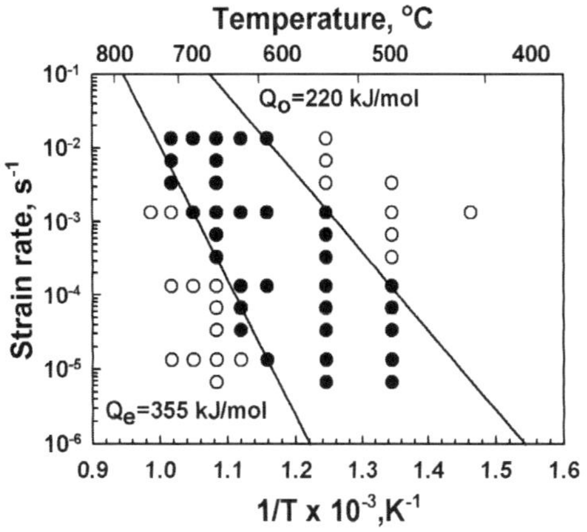

Figure 4. Test conditions where serrations occurred during tensile tests (open symbols: no serration, closed symbols: serration).

It was found that deformation behavior of the 18Cr-9Ni-W-Nb-V-N steel characterized by discontinuous plastic flow manifested as serrations on the stress–strain curves. The typical stress–strain curves obtained in the temperature interval from 410 to 740°C at a strain rate of ~ 1.3×10^{-3} s^{-1} are shown in figure 3. Different types of jerky flow observed on stress–strain curves in the temperature interval 530–680°C (figure 3) suggest the occurrence of various mechanisms of DSA. This assumption is supported by difference in activation energy for the onset and the end of jerky flow calculated as the slope of the boundaries delineating serrated flow regime in an $\ln\dot{\varepsilon}-1/T$ plot (figure 4). Values of activation energy were of 220 kJ/mol and 355 kJ/mol for the onset (Q_o) and the end (Q_e) of jerky flow, respectively. Several studies were dedicated to mechanism and activation energy for dynamic strain aging in austenitic stainless steel [3-6]. As a rule, diffusion of interstitial solutes and substitutional solutes to dislocations is considered to be the mechanisms responsible for jerky flow at low and high temperatures, respectively. However, the values of activation energy calculated for 18Cr-9Ni-W-Nb-V-N steel are considerably higher than that reported previously for the 18Cr-9Ni type austenitic steels at the same strain rate – temperature conditions. For example, in type 304 austenitic steels the activation energy for the onset and the end of jerky flow was found to be ~200 kJ/mol and ~290 kJ/mol, respectively [5]. The examined steel contains W, V, Nb and N in solid solution (figure 2). It is widely known [4] that nitrogen has a strong effect on the value of activation energy due to high interaction energy with chromium. However, Kim et el. [4] reported the value of activation energy for the onset and the end of serrations of 109 and 320 kJ/mol in 316 austenitic steel with nitrogen content of ~0.15%. Notably, that nitrogen content in the present steel is almost the same. It seems, that W, Nb and V dissolved in solid solution increase the values of activation energy due to strong interaction between mobile dislocations and solute atoms in the 18Cr-9Ni-W-Nb-V-N steel. This interaction impedes dislocation mobility affecting strength of this steel. However, additional examination of microstructure evolution during high temperature tensile test is necessary to define the actual role of these alloying elements. To confirm the influence of DSA on the strength the tensile properties of presented steel were examined at strain rate of 1.3×10^{-3} s^{-1} in the temperature interval 20–740°C.

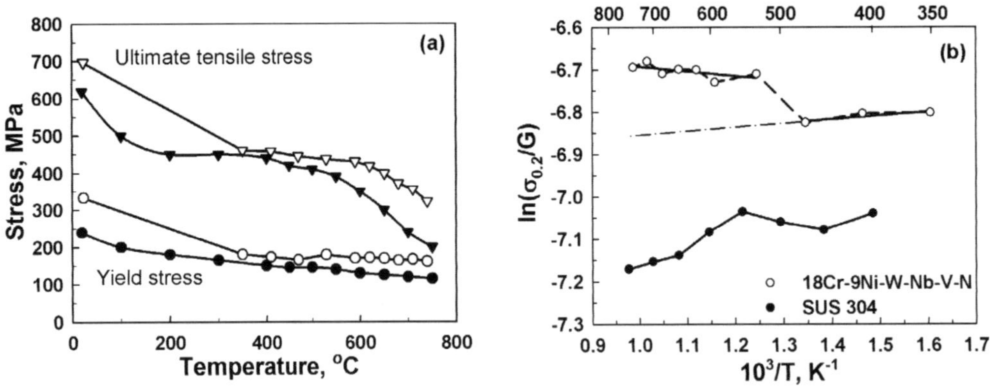

Figure 5. The variation of tensile properties with temperature at an initial strain rate 1.3×10^{-3} s^{-1} (a). Yield stresses normalized by the shear modulus *vs.* reciprocal temperature (b) (open symbols: 18Cr-9Ni-W-Nb-V-N steel, closed symbols: SUS304 steel [7]).

Yield stress (YS) and ultimate tensile stress (UTS) were plotted as a function of temperature in figure°5a. The temperature dependence of YS and UTS for SUS304 austenitic steel [7] is also presented for comparison. It is clearly seen that the 18Cr-9Ni-W-Nb-V-N steel demonstrates increased value of YS and UTS at all temperatures. The values of YS of SUS304 steel decrease with increasing temperature while these values for the 18Cr-9Ni-W-Nb-V-N steel remain virtually unchanged at temperatures > 350°C. The plateau of YS at the stress level of about 168 MPa was found (figure 5a).

In order to examine temperature dependence of yield stress for the 18Cr-9Ni-W-Nb-V-N steel in details the shear modulus compensated yield stress was plotted against the reciprocal absolute temperature in logarithmic scale (figure°5b). For the SUS304 the values of yield stress normalized on temperature dependence of shear modulus steel decrease with increasing temperature excepting short temperature interval where DSA effect was found. On the other hand, for the 18Cr-9Ni-W-Nb-V-N steel the normalized YS increases with increasing temperature from 530 to 740°C. Thus, positive temperature dependence of $\sigma_{0.2}/G$ takes place providing high strength of the steel examined in a wide temperature interval. It is obvious that this positive temperature dependence is attributed to the PLC effect.

In the temperature interval 350–680°C where DSA occurred, the values of UTS of the 18Cr-9Ni-W-Nb-V-N steel remain almost unchanged; plateau is observed at these temperatures (figure 5a). At high temperatures, UTS decreases extensively. It is seen in figure 5a, that at T > 530°C the present steel demonstrates values of UTS significantly higher than that for the SUS304 steel. It is known that the strong interaction between mobile dislocations and solute elements provides accumulation of lattice dislocations in DSA regime [8,9]. As a result the values of UTS increase at temperatures there DSA occurs. Therefore, it is possible to conclude that additional additives of W, Nb and V to 18%Cr-8%Ni steel lead to extension of the DSA regime to higher temperatures providing high value of UTS. As a result, the 18Cr-9Ni-W-Nb-V-N steel exhibits increased high temperature strength.

4. Conclusions

1. PLC effect was found in the 18Cr-9Ni-W-Nb-V-N steel with following typical attributes: jerky flow; the plateaus on temperature dependencies of YS and UTS; positive temperature dependence of $\sigma_{0.2}/G$.
2. The extending of DSA to higher temperature due to additions of W, Nb and V increases high temperature strength of the 18Cr-9Ni-W-Nb-V-N steel and provides high value of YS and UTS up to 740°C and 680°C, respectively.

Acknowledgements
This work was supported by Federal Agency for Science and Innovations under grant No. 02.523.12.3019. Authors acknowledge facilities and technical assistance from the Center of Common Facilities of Belgorod State University.

References

[1] Ishitsuka T, Mimura H 2002 *JSME* **45** 110
[2] Tjong S C, Zhu S M 1997 *Met. Mater. Trans.* **28A** 1347
[3] Hong S G, Lee S B 2005 *J. Nucl. Mater.* **340** 307
[4] Kim D W, Ryu W S, Hong J H, Choi S K. 1998 *J. Mater. Sci.* **33** 675
[5] Iino Y 1986 *Bulletin JSME*, **29** 355
[6] Jenkins C F, Smith G V 1969 *Trans. Metall. Soc. AIME* **245** 2149
[7] Abe F 2004 *Austenitic Stainless Steels: Creep and Rupture Data of Heat Resistant Steels, Creep Properties of Heat Resistant Steels and Superalloys* (Berlin: Springer)
[8] Hong S G, Lee S B 2004 *Int. J. Fatigue* **26** 899
[9] Ryu W S, Kim D W, Kim W G High temperature behaviors of type 316L(N) stainless steel *Transaction of the 15th International Conference on Structural Mechanics in Reactor Technology (Seoul, Korea, 15-20 August 1999)* pp 275-282

Journal of Physics: Conference Series **240** (2010) 012072 doi:10.1088/1742-6596/240/1/012072

New creep region and mechanism in hexagonal close-packed metals

T Matsunaga[1, 2], T Kameyama[2, 3], S Ueda[2, 4], E Sato[2]

[1] Department of Space and Astronautical Science, School of Physical Sciences, Graduate University for Advanced Studies, Sagamihara, Kanagawa 229-8510, Japan
[2] Institute of Space and Astronautical Science, Japan Aerospace Exploration Agency, Sagamihara, Kanagawa 229-8510, Japan
[3] Department of Materials Engineering, School of Engineering, University of Tokyo, Bunkyo, Tokyo 113-8656, Japan
[4] Department of Aerospace Engineering, Faculty of System Design, Tokyo Metropolitan University, Hino, Tokyo 191-0065, Japan

matsunaga.tetsuya@ac.jaxa.jp

Abstract. Only hexagonal close-packed (h.c.p.) materials show creep behaviour significantly at ambient temperature or less even below their 0.2% proof stresses with their stress exponents of 3.0 and their apparent activation energies of 20 kJ/mol. Transmission electron microscopy revealed dislocation arrays as a planar slip without any tangled dislocations inside each grain. Atomic force microscopy and electron backscatter diffraction pattern analyses brought about the occurrence of grain boundary sliding. The grain-size exponent was evaluated as 1.0, which means grain boundaries work as the barrier of the dislocation motion. Ambient-temperature creep of h.c.p. materials is schematically illustrated as that lattice dislocations move inside each grain without any obstacles and then pile up at grain boundaries. To continue the creep deformation, these dislocations are absorbed by grain boundaries to accommodate the internal stress and lead to grain boundary sliding.

1. Introduction

It has been revealed that several Ti alloys show significant creep behaviour at ambient temperature [1–6]. These works reported two types of deformation mechanisms: a planner slip in α-grains [1–3, 5]; time-dependent deformation twins [4]. Thompson et al. [1] evaluated apparent activation energy (Q) of 37 kJ/mol as dislocations overcame interstitial atoms. Neeraj et al. [5] and Hasija et al. [6] claimed that this creep was Andrade creep. However, the extra low Q value reported by ref. [1] contradicts that of Andrade creep which is 100 kJ/mol. Our previous works [7, 8] revealed that ambient-temperature creep appeared in not only Ti alloys but also hexagonal close-packed (h.c.p.) materials. Additionally, we introduced a new region, i.e. the ambient-temperature creep region [8], into the Ashby-type deformation mechanism map of annealed CP-Ti [9], where the Q value was 10 kJ/mol. According to these previous works [1–8], the deformation and the rate-controlling mechanism of ambient-temperature creep have not been understood clearly. In the present paper, transmission electron microscope (TEM) and atomic force microscope (AFM) observations were conducted to reveal the both mechanisms. Samples were typical pure h.c.p. metals, i.e. commercially pure titanium (CP-Ti), pure magnesium (Mg) and pure zinc (Zn).

2. Experimental Procedure

Samples were rolled sheets of CP-Ti (JIS grade 1) with grain size, *d*, of 75 μm, Mg (99.95%) with *d*=120 μm and Zn (99.995%) with *d*=100 μm. Another Zn sample with *d*=210 μm was produced by annealing at 483 K for 24 h.

Tensile tests were performed to evaluate 0.2% proof stresses ($\sigma_{0.2}$) at ambient temperature with the initial strain rate of 1×10^{-3} s^{-1}. Creep tests were conducted in tension at temperatures from 203 K to 873 K with loads of 0.5–0.9$\sigma_{0.2}$. The loading direction is parallel to the rolling direction in the both tests. Strains were measured using strain gauges and an optical displacement camera. Creep curves were fitted by the logarithmic creep equation [10] to evaluate steady state creep rates:

$$\varepsilon = \varepsilon_0 + \varepsilon_p \ln(1 + \beta_p t) + \dot{\varepsilon}_s t, \qquad (1)$$

where ε is the total true strain, ε_0 is the instantaneous strain, ε_p and β_p are parameters characterizing the primary creep region, $\dot{\varepsilon}_s$ is the extrapolated steady state creep rate and t is the elapsed time.

TEM observations were performed after creep tests lasting 4.3×10^5 s with 0.8$\sigma_{0.2}$ to clear a dislocation structure. AFM observations were executed before and after creep tests to observe grain boundary sliding (GBS). Specimens were polished mechanically using colloidal silica before the tests.

3. Results

The $\sigma_{0.2}$ at ambient temperature were 176, 61, 40 and 24 MPa for CP-Ti, Mg and Zn with *d*=100 and 210 μm, respectively. The right part of Fig. 1 is a log-log plot of modulus compensated applied stress, σ/E, and $\dot{\varepsilon}_s$ at 298 K: all specimens show the stress exponents, *n*, of about 3.0. It is noted that *E* values were obtained by tensile tests in previous reports, whereas they were the reference data [9, 11] in this study. A log-log plot of **b**/*d* vs. $\dot{\varepsilon}_s$ (the left part of Fig. 1) at σ/E=0.001, where **b** is the Burgers vector, shows that the *p* value is evaluated 1.0 for Zn.

The Arrhenius plots, Fig. 2, are shown for CP-Ti, Mg and Zn with *d*=100 μm at σ/E=0.004. These

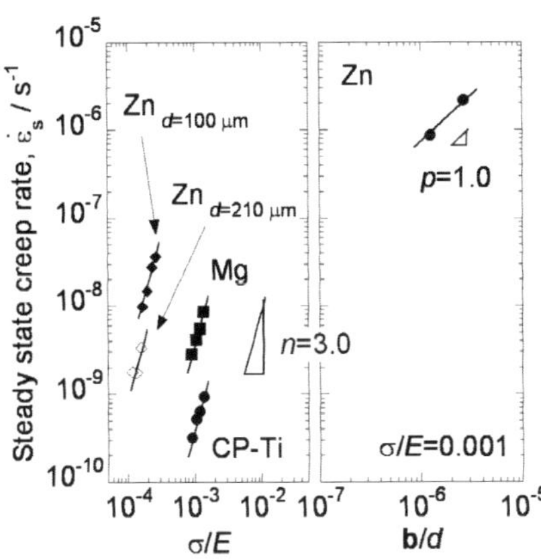

Figure 1. Right part is a double logarithmic plot of σ/E and $\dot{\varepsilon}_s$ at 298 K for all specimens; left part is that of **b**/*d* and $\dot{\varepsilon}_s$ for Zn.

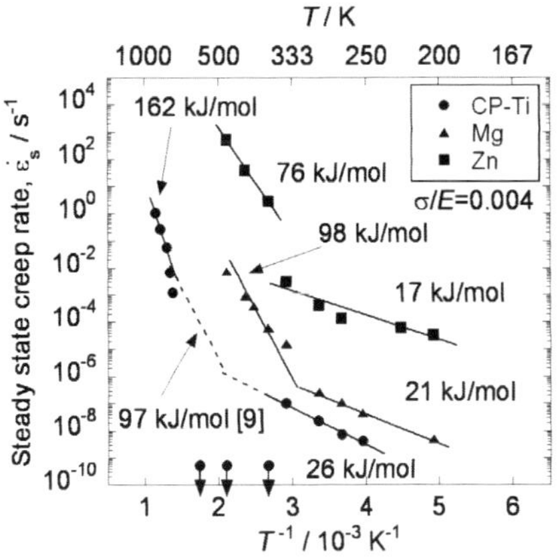

Figure 2. Arrhenius plots for CP-Ti, Mg and Zn with *d*=100 μm. The apparent activation energy of ambient-temperature creep is about 20 kJ/mol.

plots are divided into two regions. The higher-temperature regions, at T>330 K for Mg and T>353 K for Zn, have larger Q values of 76 and 98 kJ/mol. This region corresponds to the low-temperature dislocation creep region with dislocation-core diffusion. The Q value of CP-Ti, 166 kJ/mol, is higher than that of dislocation-core diffusion, 97 kJ/mol, reported previously [9] and might indicate lattice diffusion. Because it was influenced by strain aging [11], steady state creep rates at temperatures from 423 K to 573 K were not obtained. Another region with extremely low Q value of about 20 kJ/mol is the ambient-temperature creep region.

TEM images of CP-Ti, Mg and Zn with d=100 μm are shown in Fig. 3. Their total true strains were 0.003, 0.021 and 0.048 for CP-Ti, Mg and Zn, respectively. All images show same dislocation structure: dislocation arrays as a planar slip without any tangled dislocations inside a grain.

Figure 4 shows a creep curve and strain induced by GBS, ε_{GBS}, with $0.8\sigma_{0.2}$ for Zn with d=210 μm. ε_{GBS} was estimated by equation (2) [12];

$$\varepsilon_{GBS} = k\left(\frac{v}{d}\right) \tag{2}$$

where k is a geometrical averaging factor, v is the component of GBS perpendicular to the stress axis and to the specimen surface. In this study, the k value is 1.1 because the specimens were polished before the tests. The broken line in Fig. 4 is the fitting curve of $\dot{\varepsilon}_{GBS}$ using eq. (1), which revealed the extrapolated steady state $\dot{\varepsilon}_{GBS}$ as 2.1×10^{-10} s^{-1}; $\dot{\varepsilon}_{GBS}$ is only 6% of $\dot{\varepsilon}_s$.

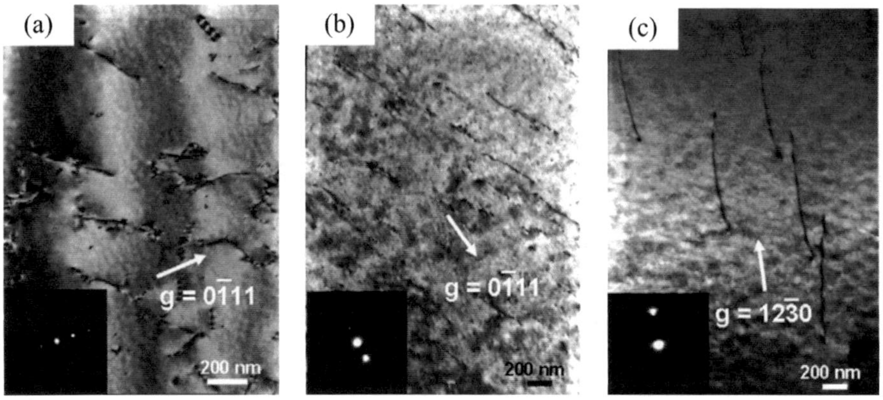

Figure 3. TEM images of (a) CP-Ti, (b) Mg and (c) Zn after creep test. The same dislocation structure, i.e. a planner slip without any tangled dislocations, was observed.

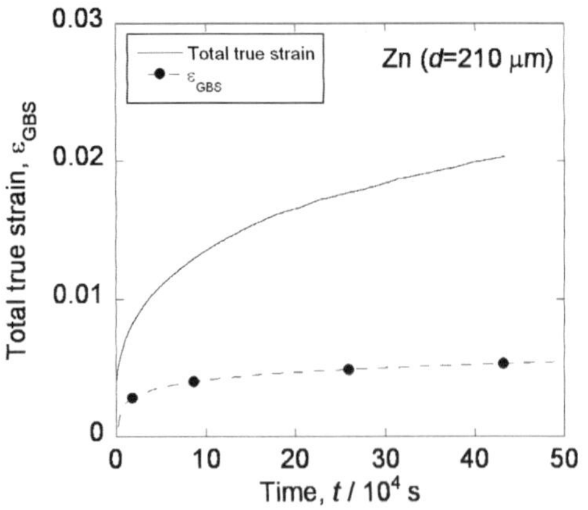

Figure 4. Creep curve and ε_{GBS} with $0.8\sigma_{0.2}$ of Zn with d=210 μm. The dotted line indicates the $\dot{\varepsilon}_{GBS}$ of 2.1×10^{-10} s^{-1}. This value is smaller than $\dot{\varepsilon}_s$ by one order.

4. Discussion

Although higher-temperature dislocation creeps do not show the p value, ambient-temperature creep show the grain-size dependency. In ambient-temperature creep region, lattice dislocations lying on a slip plane piled up at GBs easily and then these piled-up dislocations have to be accommodated to continue the creep. Slip-induced GBS was proposed as an accommodation process [13]. This process is as follows: (1) lattice dislocations pile up at GBs; (2) these dislocations resolve at GBs and are absorbed by them; (3) the dislocations slip on a boundary plane and produce GBS. In addition, the Q value of this mechanism is evaluated 15 kJ/mol for AZ31 [13]. Since the $\dot{\varepsilon}_{GBS}$ is smaller than $\dot{\varepsilon}_s$ by one order, main part of the strain is yielded by the dislocation motion inside grains. Summarizing the results, ambient-temperature creep is occurred by lattice dislocations as a planner slip without tangled dislocations inside each grain and slip-induced GBS to accommodate piled-up dislocations in series.

The constitutive relation for ambient-temperature creep is identified in the form of the modified Dorn equation including the grain-size dependency [14]:

$$\dot{\varepsilon}_s = AD_0 \exp(-Q/RT)\, \frac{Gb}{kT}\left(\frac{\sigma}{E}\right)^n \left(\frac{b}{d}\right)^p \tag{3}$$

where A is a constant, D_0 is a frequency factor, R is the gas constant and k is Boltzmann's constant.

5. Conclusions

To reveal the features of ambient-temperature creep, TEM, AFM and EBSD analyses were performed before and after creep tests. Typical h.c.p. metals show the distinguishing features differing from conventional dislocation creeps. The conclusions obtained from this work are shown bellow:

(1) The creep parameters were identified as $Q{\approx}20$ kJ/mol, n=3.0 and p=1.0.
(2) The dislocation structure, a planner slip without tangled dislocations, was observed.
(3) The mechanism of GBS is resulted from lattice-dislocation motion, i.e. slip-induced GBS.
(4) Constitutive relation was proposed as equation (3).

Acknowledgement

The authors appreciate funding and support from a Grant-in-Aid for Scientific Research and JSPS Fellows (00020141) and from the Light Metal Educational Foundation, Inc.

References

[1] Thompson A W and Odegard B C 1973 *Metall. Trans.* **A4** 899
[2] Odegard B C and Thompson A W 1974 *Metall. Trans.* **A5** 1207
[3] Imam M A and Gilmore C M 1979 *Metall. Trans.* **A10** 419
[4] Ankem S, Greene C A and Singh S 1994 *Scripta Metall.* **30** 803
[5] Neeraj T, Hou D H, Daehn G S and Mills M J 2000 *Acta Mater.* **48** 1225
[6] Hasija V, Ghosh S, Mills M J and Joseph D S 2003 *Acta Mater.* **51** 4533
[7] Yamada T, Kawabata K, Sato E, Kuribayashi K and Jimbo I 2004 *Mater. Sci. Eng. A* **387-389** 719
[8] Tanaka H, Yamada T, Sato E and Jimbo I 2006 *Scripta Mater.* **54** 121
[9] Frost H J, Ashby M F 1982 *Deformation Mechanism Maps* (Oxford: Pergamon Press) pp 43–52
[10] Garofalo F 1963 *Fundamentals of Creep and Creep-rupture in Metals* (New York, MacMillan) pp 18–19
[11] Boyer R, Weish G and Collings E W 1994 *Materials Properties Handbook: Titanium Alloys* (Ohio: ASM International) pp 233–236
[12] Bell R L and Langdon T G 1967 *J. Mater. Sci.* **2** 313
[13] Koike J, Ohyama R, Kobayashi T, Suzuki M and Maruyama K 2003 *Mater. Trans.* **44** 445
[14] Langdon T G and Mohamed F A 1978 *J. Mater. Sci.* **13** 1282

Re-examination of creep behaviour of high purity aluminium at low temperature

S Ueda[1 2], T Kameyama[2], T Matsunaga[2], K Kitazono[1], E Sato[2]

[1] Tokyo Metropolitan University, Department of Aerospace Engineering, Hino, Tokyo 191-0065, Japan

[2] Institute of Space and Astronautical Science, Japan Aerospace Exploration Agency, Sagamihara, Kanagawa 229-8510, Japan

ueda.shohji@ac.jaxa.jp

Abstract. The deformation behaviour of high-purity aluminium at low temperatures was investigated in order to re-examine Ashby-type deformation mechanism map. All specimens with different purities showed significant creep below room temperature. Under the same stress and temperature, the steady-state creep rate increased with increasing purity of the material. They showed stress exponents around 5.0 and apparent activation energies around 20 kJ/mol at temperatures below about 400 K, and 4.0 and 70-80 kJ/mol at temperatures above that temperature. The grain size had no effect in the low temperature region. From the microstructural observation, secondary slip system was observed. These features imply that pure aluminium deforms in the different mode from the ambient temperature creep of h.c.p. metals which has similar activation energy.

1. Introduction

Deformation mechanisms of metals are often plotted in Ashby-type deformation mechanism maps [1]. The map is composed by normalized stress, homologous temperature and strain rate, and the deformation mechanism is given as a function of these parameters. In the map of pure aluminium (Al) drawn by Frost and Ashby [1], creep deformation should proceed even below $0.4T_M$ (T_M : melting point). However, because of the lack of experimental data at low temperatures with low strain rates, low-temperature region has not been examined well. The map by Frost and Ashby [1] has been drawn based only on the experiment in the power-law breakdown region with $\dot{\varepsilon} > 10^{-7}$ s^{-1} [3, 4]. On the other hand, creep at ambient temperature was observed in hexagonal close-packed (h.c.p.) metals, and ambient-temperature creep was added in the deformation mechanism map of Ti [2]. In Al also, Ishikawa et al. [5] reported different creep behaviour from that by Luthy et al. [4] in the low temperature region. This study, therefore, aimed to re-examine the deformation behaviour in the low-temperature creep region of high-purity Al.

2. Experimental procedure

Creep tests were performed at temperatures from 253 K to 523 K on the samples of 99.999% (5N) Al, 99.99% (4N) Al and 99.5% (1050) Al. The average grain sizes were 140 μm for 5NAl, 100 μm for 4NAl and 50 μm for 1050Al. Creep strains were measured using strain gauges with resolution of 10^{-6} with the loading direction parallel to the rolling direction. The steady-state creep rate was estimated through fitting of the logarithmic creep equation [6]:

$$\varepsilon = \varepsilon_i + \varepsilon_p \ln(1 + \beta_p t) + \dot{\varepsilon}_s t \, ,$$

where ε is the total true strain, ε_i is the instantaneous strain, ε_p and β_p are parameters characterizing the primary creep region, $\dot{\varepsilon}_s$ is the extrapolated steady-state creep rate, and t is the elapsed time. The optical microscope observations were conducted on the samples before and after the creep test to figure out the microstructual development. The observed sample was polished mechanically by colloidal silica before the creep test.

3. Results

The results of the creep tests are shown in Fig.1 in a double logarithmic plot of strain rate and stress. All specimens showed significant creep above and below room temperature. 5NAl shows stress exponents of about 5.0 below 373 K ($0.4T_M$), and 4.0 above 473 K. 4NAl shows a similar behaviour, and 1050Al shows a slightly larger stress exponent than the materials with higher purity.

Figure 2 is the Arrhenius plot of the three materials at σ/E written in the figure. This plot shows two regions with different apparent activation energies, whose border is at about 400 K. All samples show apparent activation energies of 70-80 kJ/mol above this temperature, and 20-35 kJ/mol below this temperature.

Figure 3 shows a double logarithmic plot of creep rate vs. normalized stress for 5NAl with several grain sizes at 300 K; all data points locate on a single line. It is clearly concluded that the creep rate in the low-temperature region is independent of grain size. In addition, this creep phenomenon occurs above the 0.2% proof stress. This means that strain hardening took place before creep testing.

Figure 4 shows an optical micrograph of the deformed sample at room temperature to a strain of 1% at a rate of $10^{-9}s^{-1}$. Circles in Fig.4 show that secondary slip system acts. In addition, the other circles show that slip lines propagate into the adjacent grain.

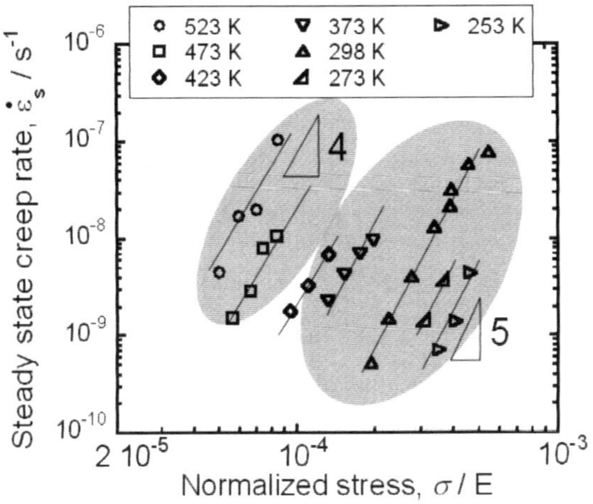

Figure 1. Steady state creep rate for various applied stress normalized by young's modulus of 5NAl.

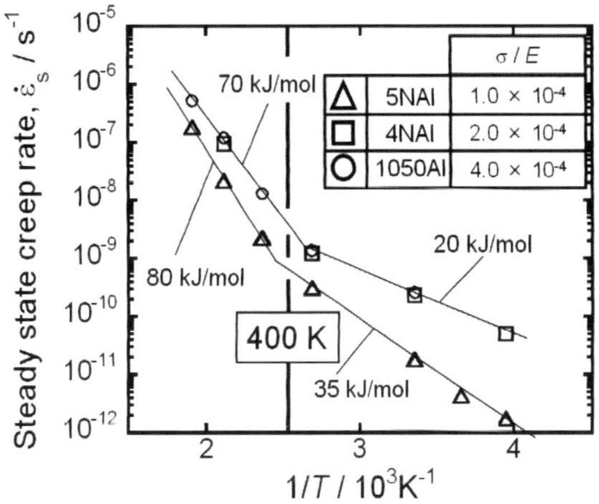

Figure 2. The Arrhenius plot for Aluminiums with different purity.

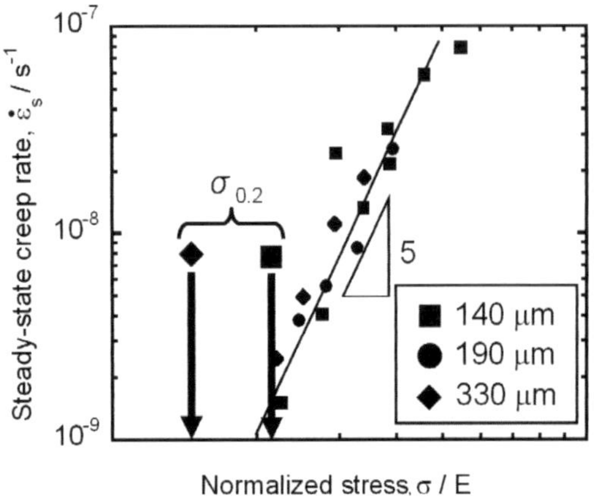

Figure 3. Grain size dependency of the creep behaviour for 5NAl sample crept at 300 K

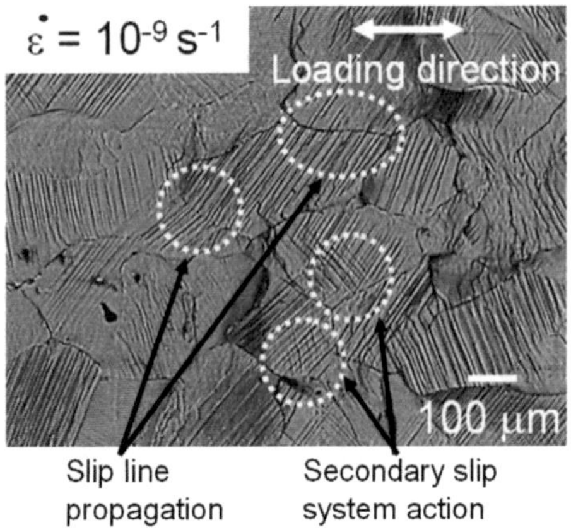

Figure 4. Optical micrograph of 5NAl sample after extension of 1% at 300 K under the strain rate of 10^{-9} s^{-1}.

4. Discussion

In Fig.5, the present data are compared to previous creep test data of high purity Al. Luthy et al. [4] and Ishikawa et al. [5] investigated the creep behaviour in the low temperature region. Servi and Grant [7] and Weertman [8] researched the creep behaviour in the high temperature region. As shown in Fig.4, above 523 K, the data of Weertman and Servi show an activation energy of about 140 kJ/mol which corresponds to the volume diffusion of Al. Present high-temperature data with 70-80 kJ/mol are connecting to their data smoothly. Below 400 K, the plot branches into two lines. One line is reported by Luthy with 80 kJ/mol, which is close to the pipe-diffusion activation energy of Al [9]. On the other hand, Ishikawa reported 20 kJ/mol below 400 K which agrees to the present study well.

Figure 6 plots some selected data of stress and strain rate by Luthy, Ishikawa and us. It is clear that Luthy estimated the activation energy by the data in the power-law breakdown region. Since the effect of temperature dependency of shear modulus is significant in the power-law breakdown region [1], Luthy might have estimated too high activation energies. On the other hand, Ishikawa and we evaluated the creep behaviour near and in the power law region, respectively. Therefore, the activation energy of low-temperature creep of pure Al is concluded to be 20 kJ/mol.

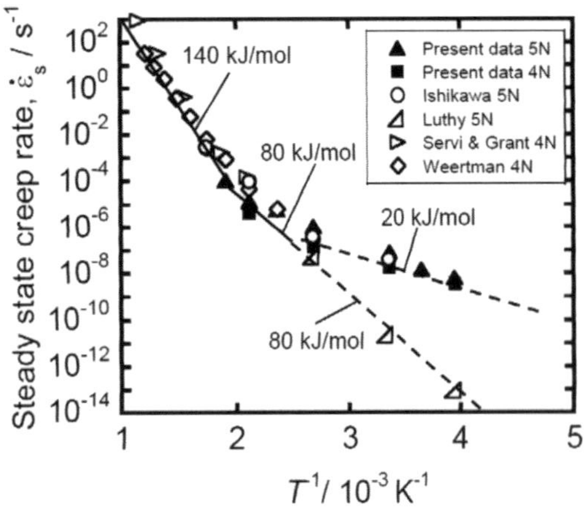

Figure 5. Arrhenius plot for the present data and creep data obtained from literature.

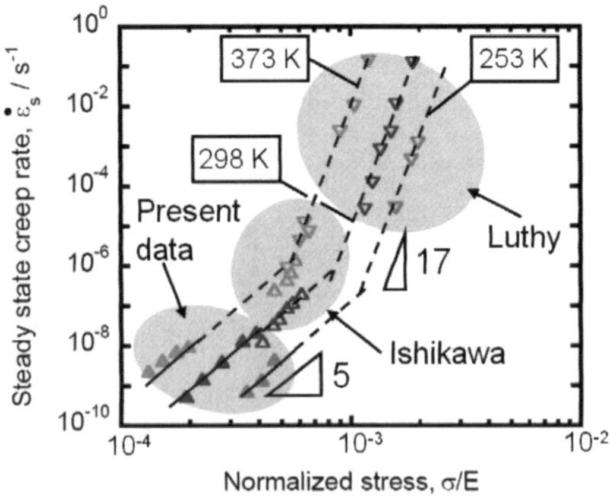

Figure 6. Double logarithmic plot of selected data from Luthy, Ishikawa, and us below 400 K.

The present study revealed that pure aluminium show creep with stress exponents of 4-5 and activation energies of 20-35 kJ/mol below 400 K. This implies that dislocations contribute to the creep deformation and diffusion hardly contributes. The ambient temperature creep of h.c.p. metals is one of such creep [10]; creep occurs at low temperatures with activation energy of about 20 kJ/mol. Therefore, now we discuss on the differences between the low temperature Al creep and the ambient temperature creep of h.c.p. metals.

First, considering intragranular deformation, h.c.p. metals show only the primary slip system activating and straightly aligned dislocation lines without any tangles. On the other hand, secondary slip line was observed in Al, and its creep proceeds under tangled dislocation structure. In Al, accommodations such as jog formation or cross slip are necessary inside of grains. Activation energies of cross slip and jog formation is about 100 kJ/mol [11] and 14 kJ/mol [12], respectively, and the former may acts as intragrain accommodation.

Next, considering grain boundary deformation, the von Mises law requires at least five slip systems to continue deformation for polycrystalline materials. Since h.c.p. metals have only two slip systems, accommodation has to occur at grain boundaries. However, since Al has 12 slip systems, strain is continuous at grain boundary and accommodation needs not.

5. Conclusion

The low-temperature creep behaviour of pure Al was investigated to re-examine the deformation mechanism map. The extremely low activation energy region was obtained, which has not been reported in the Ashby's map. The low temperature creep region has the features as follows;

(1) The stress exponent is about 4-5.
(2) The creep behaviour is independent of grain size.
(3) The creep occurs above the 0.2% proof stress.
(4) Secondary slip system acts.
(5) Slip lines propagate to the adjacent grain.

Acknowledgement
This research is supported by Light Metal Educational Foundation Inc. as Integrated Advanced Research.

References

[1] Frost H J, Ashby M F 1982 *Deformation Mechanism Maps* (Oxford Pergamon Press) pp. 1-29
[2] Tanaka H, Yamada T, Sato E, Jimbo I 2006 *Scripta Mater.* **54** 121
[3] Sherby O D, Trozera T A, Dorn J E 1956 *Proc. ASTM.* **56** 789
[4] Luthy H, Miller A K, Sherby O D 1980 *Acta. Metall.* **28** 169
[5] Ishikawa K, Maehara M, Kobayashi Y 2002 *Mater. Sci. Eng. A* **322** 153
[6] Garofalo F 1963 *Fundamentals of creep and creep-rupture in metals* (New York, MacMillan) pp18-19
[7] Servi I S, Grant N J 1951 *Trans. AIME* **191** 909
[8] Weertman J 1956 *J. Mech. Phys. Solids* **4** 230
[9] Volin T E, Balluffi R W 1971 *Acta. Metall.* **19** 263
[10] Matsunaga T, Takahashi K, Kameyama T, Sato E 2009 *Mater. Sci. Eng. A* **510-511** 356
[11] Schoeck G, Seeger A 1954 *Report of the Conference on Detects in Crystalline Solids* pp.340-346
[12] Harper J G, Shepard L A, Dorn J E 1958 *Acta. Metall.* **6** 509

Development of an experimental setup for testing the properties of γ / γ' superalloys

Siret Christophe[1], Viguier Bernard[2], Salabura Jean Claude[2], Andrieu Eric[2], Lesterlin Sandrine[3]

[1] TURBOMECA DT/MPE/MM, Avenue du 1er Mai, 40220 Tarnos, France
[2] Institut Carnot CIRIMAT, ENSIACET-INPT, 4, allée Emile Monso - BP 74233 31432 TOULOUSE Cedex 4 France
[3] TURBOMECA DT/MPE/MM, Avenue Joseph Szydlowski 64511 Bordes Cedex, France

Bernard.Viguier@ensiacet.fr

Abstract. Certification tests on turboshaft engines for helicopters can expose components as high pressure turbine blades to very high temperature during short time periods. To simulate these complex temperature and mechanical stress loadings and to study dimensional and microstructural stability under severe testing conditions, an experimental set-up has been recently developed. In this paper, we first present this new device and describe its performances. Then, the device is used to study the effect of heating procedure on creep results at 1200°C and rafting during primary creep on the single crystal nickel-based superalloy MC2.

1. Introduction

Most of the published studies dealing with single crystals made of nickel based superalloys have been performed with the aim of increasing thermal stability of the γ / γ' microstructure, improving mechanical properties under thermomechanical loadings and particularly the creep lifetime up to 1050°C [1-6]. In the case of helicopter twin engines turbines, one has to consider that in the event of one engine failure during take off or cruising power an emergency power requirement is needed to ensure the safety of the flight. During this short time, the blades can experience very high temperature. Even if the overheating duration is reduced, one has to take into account γ' dissolution leading to a decrease of the volume fraction on this strengthening phase [7-11]. The thermal shock related to this emergency situation is very rapid. The purpose of this study is at first to present an experimental set-up able to generate loadings procedures much closer to real loadings and secondly to present two examples of applications of interest regarding the helicopter turbine specific loadings.

2. Description of the experimental set-up

A conventional 20 kN electromechanical machine was used as the basis for the development of the experimental set-up described here after. A low thermal inertia loading frame was designed in order to minimize thermal gradient along the gage length of the sample under heating or cooling steps. A three zones radiation furnace has been developed allowing to impose maximum heating and cooling rates respectively equal to 3°C.s^{-1} and 10°C.s^{-1} in the range 1000 – 1200°C. The deformation of the sample was followed by using a laser extensometer. The accuracy of the strain measurement in the most severe testing conditions was assessed to be ± 2 µm. Finally, the global behavior of the experimental set-up is similar to a dilatometer in which mechanical stresses or strains can be superimposed.

3. Material of the study

The material chosen to test the experimental set-up was MC2 single crystal Ni based superalloy. The chemical composition of the alloy is given in table 1.

Table 1. Chemical composition of the MC2 (wt %)

Ni	Cr	Co	Mo	W	Al	Ti	Ta
Eq	8	5	2	8	5	1,5	6

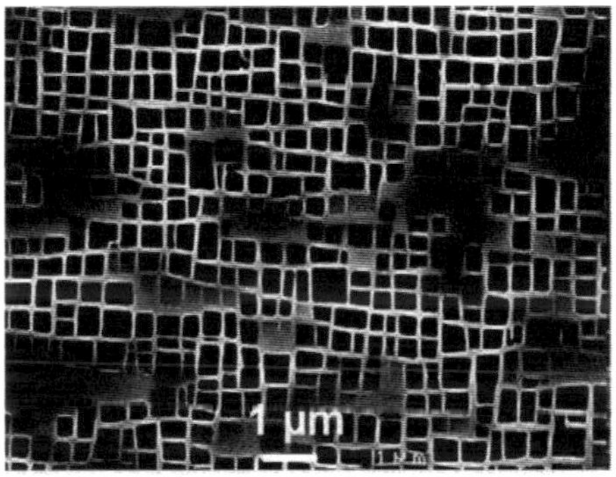

Figure 1 : MC2 optimized initial microstructure

It was given a classical two steps heat treatment (R1, R2) in order to optimize the microstructure. A representative γ - γ' microstructure of this alloy is presented in the figure 1. As mentioned in the introduction, the explored testing conditions have to trigger γ' phase dissolution. Thus, in figure 2, the evolution of the equilibrium γ' volume fraction for MC2 alloy up to the solvus of the γ' phase is presented [11]. In figure 3, as a matter of example, the γ' dissolution kinetics at 1200°C is also presented [12]. It is worth noting that, even at 1200°C, it takes about 60 min for the γ' volume fraction to reach the equilibrium value.

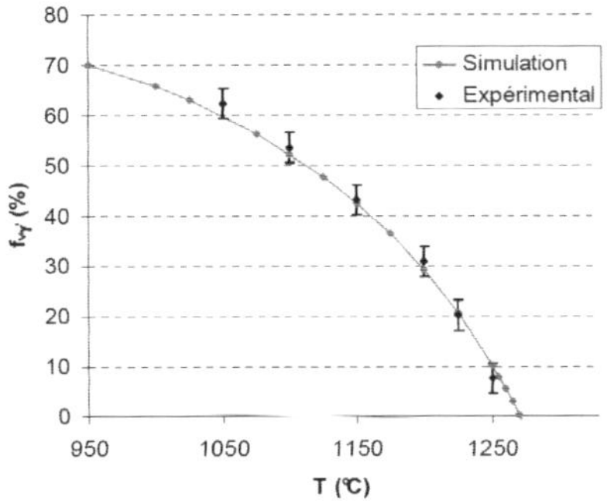

Figure 2 : Evolution of the equilibrium γ' volume fraction for MC2 alloy [11]

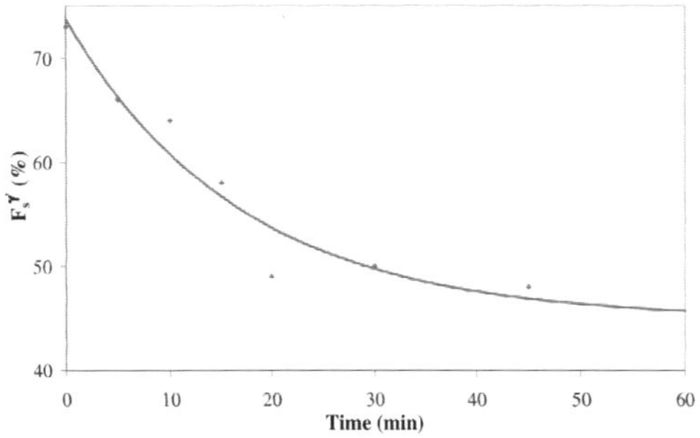

Figure 3 : γ' dissolution kinetics at 1200°C [12]

4. Examples of application

4.1. Effect of creep procedure

Two different testing procedures were carried on with the same final creep conditions (1200°C – 80MPa). The first one (Figure 4) simulates the conventional laboratory creep testing procedure in terms of heating rate (HR) and thermal stabilization (TS) whereas the second one better simulates the thermal loading which is expected when an emergency event is encountered (Figure 5).

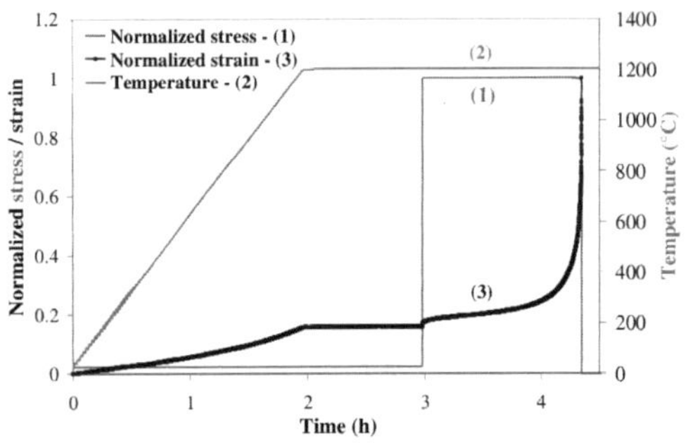

Figure 4 : Conventional laboratory creep testing procedure

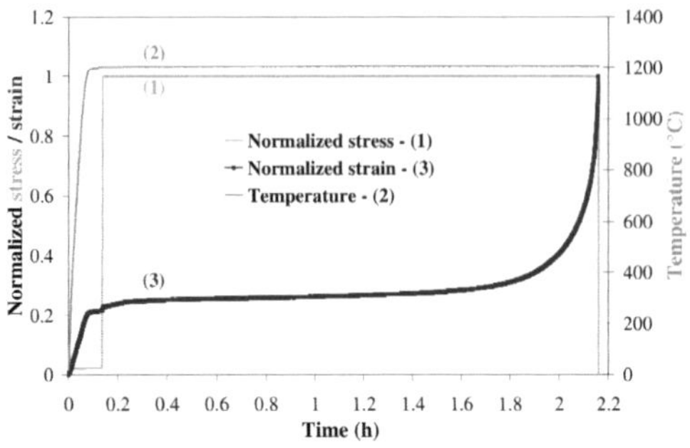

Figure 5 : Representative procedure of an emergency event

The complete thermomechanical history of both tests is presented in figures 4 and 5. It is clear that the life time corresponding to the second procedure is longer than the one corresponding to the first type. This trend is not really surprising, since during the slow heating and stabilization time used in procedure one, γ' precipitates have enough time to significantly dissolve. When dealing with creep rates, the Figure 6 illustrates clearly the effect of the initial γ' volume fraction on the primary and stationary creep stages.

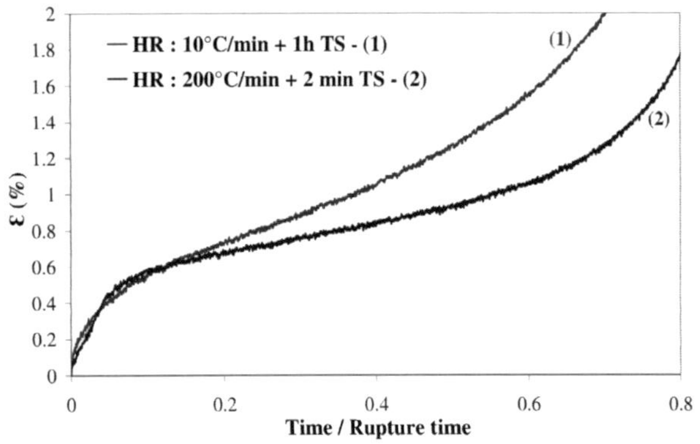

Figure 6 : Creep response as function of testing procedure

Here too, creep behavior related to the second procedure is logically better and secondary stage is evidenced whereas for the first loading procedure, secondary stage is more difficult to define. Fortunately, these results confirm that the creep resistance depends on the γ' volume fraction [13].

4.2. Rafting during primary creep

Due to the excellent accuracy of the strain measurement, it became possible to investigate the dimensional evolution of the sample during the rafting process [14-19]. The first testing condition which has been explored is 1050°C under 200 MPa. An inflexion of the creep curve is observed during the early stage of the creep deformation as presented in Figure 7. Interrupted creep tests on both sides of this inflexion point (30 min and 4 h) were carried out and the corresponding microstructures characterized. Micrographs included in Figure 7 clearly show that γ' rafting occurred during this period, as it has been observed on a different superalloy [3]. The question of the influence of an applied stress on this specific behavior raised. Then, other creep test conditions were explored, the results of which are presented on Figure 8. It appears that decreasing the applied stress seems to delay the occurrence of the inflexion. However this is certainly related to a correlated decrease of the strain rate, since the inflexion occurs for the same amount of creep strain (0.15 %) for the three tested creep conditions. This result highlights the role of plastic strain on rafting process and could be interpreted as proposed by other authors [3; 20-24] who identified the difference in terms of state of stress between perpendicular and parallel corridors as the driving force for rafting process and dislocation movement as the controlling kinetics parameters for coalescence.

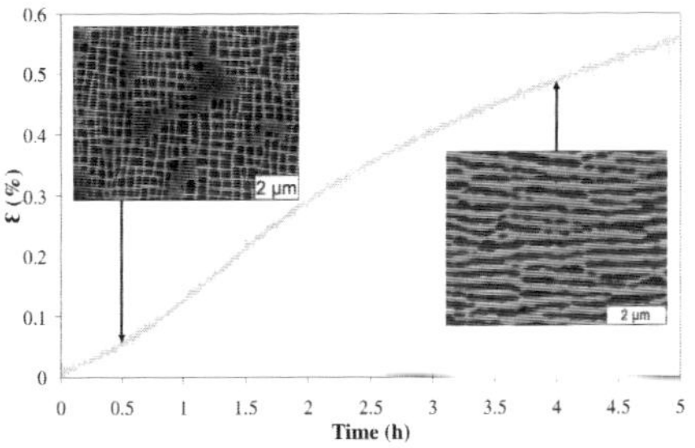

Figure 7 : Creep behavior during primary stage at 1050°C under 200MPa

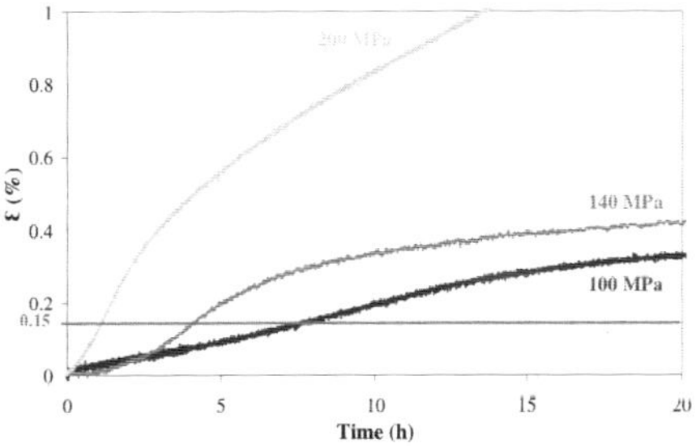

Figure 8 : Stresses effect on the shape of the creep response at 1050°C

5. Conclusion

The development of an experimental set-up allowing to test Ni based single crystal superalloy under testing conditions as closer as emergency events in helicopter engine allows the design of original experimental testing.

Creep properties in the temperature range corresponding to γ' dissolution or rafting as well as the study of dimensional evolutions during microstructural modifications are interesting topics both for industrial application and for academic questioning.

References

[1] A. Fredholm, and J.-L. Strudel, *Proceedings of the Petten International Conference, ed. J.B. Mariott, et al.* (London, U.K.: Elsevier Applied science, 1987), 9-18
[2] A. Royer, A. Jacques, P. Bastie, M. Veron, *Mater. Sci. Eng.* A **319-321** (2001) 800-804
[3] P. Caron, M. Benyoucef, A. Coujou, J. Crestou, N. Clément, *ISOMALM 2000, Chennai, India*, 2000, 148-156
[4] P. Caron, C. Ramusat, F. Diologent, in *Superalloys 2008, TMS, Seven Springs*, 2008, 159-167
[5] A. Epishin, T. Link, U. Brückner, and P. Portella, *Acta Mater.* **49** (2001), 4017-4023
[6] F. Diologent, P. Caron, *Mater. Sci. Eng.* A **385** (2004) 245-257
[7] M. Soucail, Y. Bienvenu, *Mater. Sci. Eng.* A **220** (1996) 215-222
[8] T. Grosdidier, A. Hazotte, A. Simon, *Scripta Metall. Mater.* **30** (1994) 1257-1262
[9] T. Grosdidier, A. Hazotte, A. Simon, *Mater. Sci. Eng.* A **256** (1998) 183-196
[10] A. Royer, P. Bastie, M. Veron, *Acta Mater.* **46** (1998) 5357-5368
[11] J. Cormier, PhD Thesis, University of Poitiers, ENSMA, France, 2006
[12] J. Cormier, X. Milhet, Jose Mendez, *J. Mater. Sci.* **42** (2007) 7780-7786
[13] T. Murakumo, T. Kobayashi, Y. Koizumi, H. Harada, *Acta Mater.* **52** (2004) 3737-3744
[14] J. K. Tien, S.M.Copley, *Met. Trans.* **2** (1971) 215-219
[15] A. Hazotte, J. Lacaze, *Scripta Metall.* **23** (1989) 1877-1882
[16] L. Muller, U. Glatzel, M. Feller-Kniepmeier, *Acta Metall. Mater.* **40** (1992) 1321-1327
[17] M. Veron, Y. Brechet, F. Louchet, *Acta Mater.* **44** (1996) 3633-3641
[18] M. Veron, P. Bastie, *Acta Mater.* **45** (1997) 3277-3282
[19] M. Veron, Y. Brechet, F. Louchet, *Scripta Mater.* **34** (1996) 1883-1886
[20] A. Racine, A. Hazotte, *J. Physique* **IV 3** (1993) 355-358
[21] T. M. Pollock, A. S. Argon, *Acta Metall. Mater.* **42** (1994) 1859-1874
[22] D. Blavette, L. Lettelier, A. Racine, A. Hazotte, *Microsc. Microanal. Microstruct.* **7** (1996) 185-193
[23] T. Ohashi, K. Hidaka, S. Imano, *Acta Mater.* **45** (1997) 1801-1810
[24] U. Hemmersmeier, M. Feller-Kniepmeier, *Mater. Sci. Eng.* A **248** (1998) 87-97

Hot working of high purity Fe-C alloys in the α-range

F Montheillet and J Le Coze

Ecole des Mines (SMS), CNRS UMR 5146,
158 cours Fauriel, 42023 Saint-Etienne Cedex 2, France

montheil@emse.fr

Abstract. The influence of carbon in solid solution on the stress-strain curves of α-iron was investigated using model alloys prepared from high purity iron. Uniaxial compression tests were carried out within the ferritic domain at temperatures between 700 and 880 °C. Oscillating stress-strain curves observed at high temperatures and low strain rates indicate that discontinuous dynamic recrystallization takes place. The macroscopic strain rate sensitivities m and apparent activation energies Q associated with the flow stress are not significantly modified by carbon additions. By contrast, the "mesoscopic" parameters h and r associated with strain hardening and dynamic recovery, respectively, are strongly dependent on the carbon content. Finally, an estimation of the grain boundary mobilities during dynamic recrystallization was carried out from the above rheological data.

1. Introduction

Commercial alloys, especially steels, usually contain a number of addition elements and impurities, which makes it difficult to analyze their respective effects on hot working properties. The aim of the present study was to investigate the influence of carbon in solid solution on the stress-strain curves of iron in the ferritic range.

Pure iron and a series of three Fe-C alloys containing 50, 100, and 200 wt ppm carbon were therefore prepared from electrolytic iron purified by induction melting under high purity hydrogen and argon, in a water cooled silver crucible [1, 2]. The ingots were then processed by hot forging and cold swaging, and cylindrical specimens were machined. Uniaxial compression tests were carried out within the range 700-880 °C at strain rates from 10^{-3} to 1 s^{-1}, up to strains of ~ 0.7. Stress-strain curves are first analyzed through the usual strain rate sensitivity m and apparent activation energy Q in Section 2. In a second step (Section 3), it is shown that parameters h and r derived from the curves using the Yoshie-Laasraoui-Jonas (YLJ) equation are much more sensitive to the carbon content. Finally, in Section 4, a model of discontinuous dynamic recrystallization (DDRX) recently published is used for estimating the associated grain boundary mobilities.

2. Analysis of the stress-strain curves

Fig. 1 shows four sets of flow curves for pure Fe, Fe-50 wt ppm C, Fe-100 wt ppm C, and Fe-200 wt ppm C deformed at 700 °C and various strain rates. A transition occurs from multiple peak behaviour at the lowest strain rate to single peak curves at larger strain rates, which indicates that discontinuous dynamic recrystallization takes place in these materials. This is in agreement with former observations by Glover and Sellars [3], but contrasts with the case of industrial ferritic alloys.

The yield stress σ_e increases with the amount of carbon, especially at 0.1 and 1 s^{-1}. In the same way, the peak stress σ_M, and the steady state stress (whenever it is reached) increase monotonically, although the peak strain ε_M seems to reach a maximum for 100 ppm C, which is likely to reflect lower DDRX kinetics.

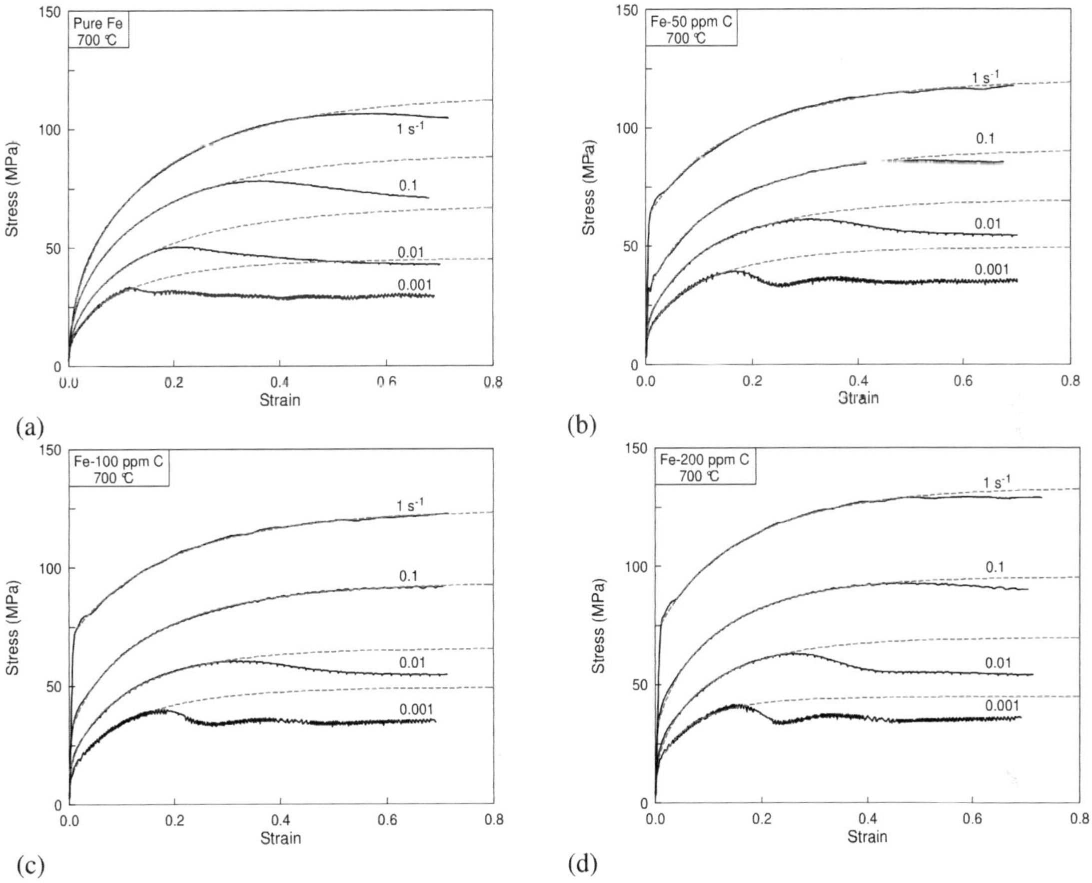

Figure 1. Uniaxial compression stress-strain curves for (a) pure iron, (b) Fe-50 ppm C, (c) Fe-100 ppm C, and (d) Fe-200 ppm C at 700 °C and various strain rates. Broken lines indicate the YLJ fits of the curves.

The classical macroscopic rheological parameters m (strain rate sensitivity of the flow stress) and Q (apparent activation energy) were determined from the above data. They both remain almost independent of the carbon content with average values m = 0.164 (700 °C) and Q = 324 kJ.mol^{-1} (700 °C and 0.01 s^{-1}). Note that this value is quite larger than Q = 280 kJ.mol^{-1} reported for self-diffusion in α-Fe.

3. Mesoscopic rheological parameters h and r

A more detailed description of the above stress-strain curves can be obtained when theoretical fits are applied. A first model was used for the combination of strain hardening and dynamic recovery, *viz.* the Yoshie-Laasraoui-Jonas (YLJ) equation:

$$\sigma_0 = \left[\sigma_\infty^2 - (\sigma_\infty^2 - \sigma_e^2)\exp(-r\varepsilon)\right]^{1/2} \tag{1}$$

where $\sigma_\infty = \alpha\mu b\sqrt{h/r}$ is a virtual steady state flow stress that would occur in the absence of recrystallization, and σ_e the yield stress (μ is the elastic shear modulus, b the Burgers vector, and $\alpha = 1$). The fitting procedure was extended over the strain range $[\varepsilon_e - \varepsilon_L]$, where ε_e is the elastic strain, and ε_L a given fraction of ε_M, that is expected to correspond with the onset of DDRX. Here a classical value $\varepsilon_L = (5/6)\varepsilon_M$ was chosen. The shapes of the curves depend basically on the two mesoscale parameters h and r, associated with strain hardening and dynamic recovery, respectively. The resulting fits are shown in Fig. 1 (broken lines). To a first approximation, for each of the investigated grades, the strain rate and temperature dependence of h and r can be written in the form:

$$h = h_0\left(\frac{\dot{\varepsilon}}{\dot{\varepsilon}_0}\right)^{m_h} \exp\left(\frac{m_h Q_h}{RT}\right) \tag{2}$$

and

$$r = r_0\left(\frac{\dot{\varepsilon}}{\dot{\varepsilon}_0}\right)^{-m_r} \exp\left(-\frac{m_r Q_r}{RT}\right) \tag{3}$$

(where $\dot{\varepsilon}_0 = 1\ s^{-1}$), which means that h becomes larger with increasing strain rate or decreasing temperature, while r varies in the opposite way.

As a general trend, the addition of carbon solutes in iron increases both h and r. The strain rate and temperature sensitivities of h are not significantly modified, whereas m_r grows and Q_r strongly drops with increasing carbon content, as shown in Figs 2 a and b. This indicates that dynamic recovery is strongly temperature dependent in pure iron, but much less when dislocation movements are impeded by the presence of solutes.

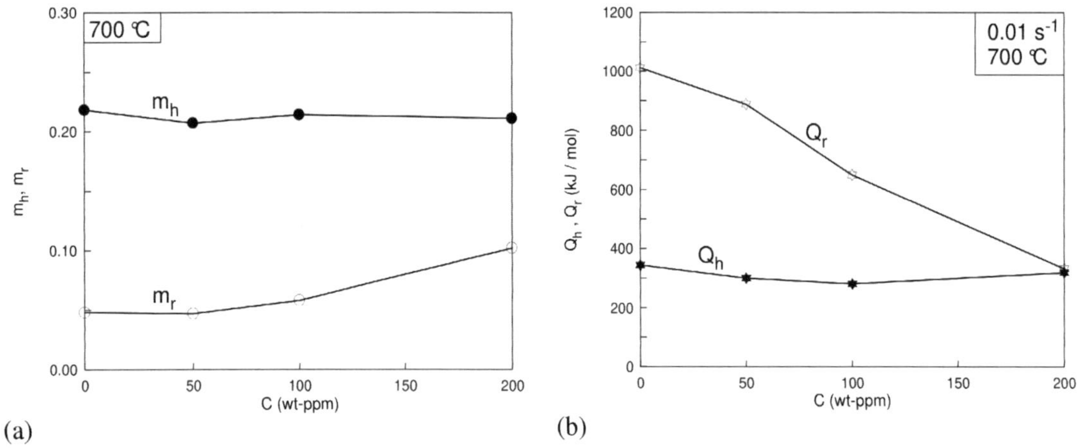

Figure 2. Influence of the carbon content on (a) the strain rate sensitivities and (b) the apparent activation energies associated with strain hardening (h) and dynamic recovery (r) of various Fe-C alloys.

4. Estimation of the grain boundary mobility Mτ

It has been shown in a recently published paper [4] that the grain boundary mobility pertaining to dynamic recrystallization can be estimated quite easily from experimental data. For that purpose, the (Swift type) power law (PW) equation was used (instead of YLJ) to fit the experimental curves:

$$\sigma_0 = K\left[\varepsilon + (\sigma_e/K)^{1/n}\right]^n \tag{4}$$

The latter is based on the mesoscopic differential equation:

$$\frac{d\rho}{d\varepsilon} = \frac{H^{\nu+1}}{\rho^{\nu}} \tag{5}$$

in which ρ is the dislocation density, and H and ν are two material parameters associated both with strain hardening and dynamic recovery. Furthermore, H and ν are simple functions of K and n.

The fitting procedure was conducted as above to determine these parameters from the experimental stress-strain curves. The grain boundary mobility is then given by:

$$M\tau = (2\nu + 3)\frac{(\nu+1)^{\nu+2}}{(\nu+2)^{\nu+1}} H^{\nu+1}\dot{\varepsilon}\frac{D_s}{(\sigma_s/\alpha\mu b)^{2(\nu+2)}} \tag{6}$$

where σ_s and D_s are the steady state flow stress and grain size, respectively, and τ is the line energy of dislocations. Since D_s was not available, it was estimated from σ_s using the well known Derby relationship $\sigma_s = k_D/D_s^{\,a}$. Values of Mτ predicted in this way are displayed in Fig. 3. They again suggest that DDRX kinetics reaches a minimum for 100 ppm C.

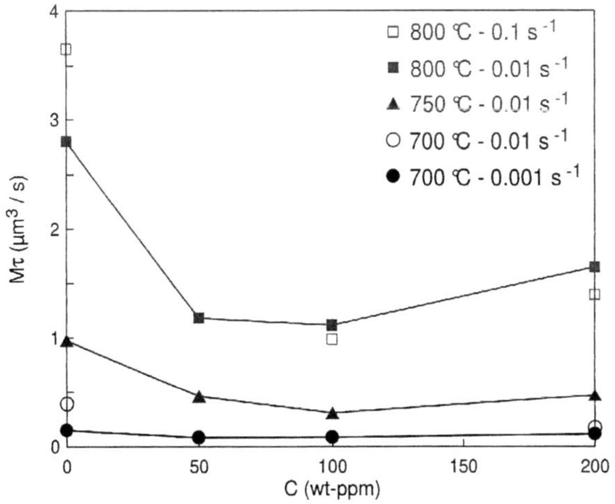

Figure 3. Dependence of grain boundary mobility during DDRX on the carbon content in pure iron.

5. Conclusions

- Specific effects of carbon solutes on the hot working behaviour of α-iron were brought into evidence from uniaxial compression stress-strain curves, using high purity base model materials.
- The dependence on carbon content of the main rheological parameters associated with (discontinuous) dynamic recrystallization, *viz.* h (strain hardening), r (dynamic recovery), and M (grain boundary mobility), was determined. The latter parameter strongly decreases up to ≈ 50 wt ppm C, and seems to exhibit a minimum for ≈ 100 wt ppm C.
- Microstructural investigations (*e.g.*, steady state grain size measurements) are currently being carried out to support the above observations by their physical counterparts.

References

[1] J. Le Coze, R. Tardy, A. Kobylanski and M. Biscondi. *Proc. UHPM-94*, Kitakyushu-City, Japan, p. 371 (1994).
[2] C. Desrayaud, S. Girard, J. Le Coze and F. Montheillet. *Mater. Trans.* **43**, 135 (2002).
[3] G. Glover and C.M. Sellars. *Met. Trans.* **4**, 765 (1973).
[4] F. Montheillet and J.J. Jonas. *ASM Handbook, vol. 22,* Modeling and Simulation : Processing of Metallic Materials, in press.

Morphology of the creep crack tip in P92 steel and its relation to microstructure

V A Yardley[1], T Matsuzaki[2], R Sugiura[2], A T Yokobori Jr.[2], S Tsurekawa[1], Y Hasegawa[3]

[1] Kumamoto University Graduate School of Engineering, Kurokami-machi 2-chome-39-1, Kumamoto 860-8555, Japan
[2] Tohoku University Graduate School of Engineering, Aramaki Aoba 6-6-01, Aoba-ku, Sendai 980-8579, Japan
[3] Steel Research Laboratories, Nippon Steel Corporation, Shintomi 20-1, Futtsu, Chiba 293-8511, Japan

yardley@kumamoto-u.ac.jp

Abstract. The region around the tip of a creep crack obtained using an interrupted test on a C(T) specimen of the 9 wt.% Cr power plant alloy P92 has been examined using electron backscatter diffraction to study the relationship between the crack morphology and the martensitic substructure. Nucleation of voids was found to occur primarily on prior austenite grain and packet boundaries. This agrees well with previous results in which minimum creep rate decreased with increasing prior austenite grain size for small prior austenite grains. Growth and coalescence of voids to form a continuous crack appears to occur by recovery, recrystallisation and eventual rupture of the bridging regions between the voids.

1. Introduction

Ferritic 9-12 wt. % Cr heat-resistant steels, such as the W-containing 9 wt.% Cr alloy P92 (NF616) [1,2] have a tempered lath martensitic microstructure. Martensite is formed by a diffusionless transformation from the parent austenite, and has a fixed crystallographic relationship with the austenite. However, because of crystal symmetry, a number of equivalent orientational variants satisfy this relationship and, while the prior austenite grain boundary network is conserved, each prior grain breaks up into substructures containing some or all of the allowed variants. Morphologically, the microstructure consists of highly dislocated laths, around 0.5 μm in width. Laths of the same, or closely related, orientational variant are grouped into elongated, straight-sided blocks, and blocks with parallel long axes occur together in regions called packets, within a prior austenite grain (PAG) [3-6].

This microstructure, together with the fine carbides formed after tempering, contributes to the long-term creep resistance of these steels under service conditions [7,8]. However, exposure to high temperatures and stresses can lead to the nucleation and growth over time of voids and cracks.

Safe plant operation thus requires accurate prediction of safe component lives under operating conditions and the choice of appropriate inspection intervals. The compact tension C(T) specimen with side groove is a standard geometry used in the determination of creep crack growth rates [9]. A C(T) specimen of P92 steel with fine PAG (≈30 μm diameter) has been examined using optical microscopy and SEM after testing to failure at 873K in air with a constant load of 311.8 MPa [10]. Creep curves measured during this test are shown in figure 1. The fracture surface showed a dimpled appearance characteristic of ductile fracture, and in a cross-section parallel to the tensile axis and direction of crack growth, the surface had a zig-zag or toothed shape, with the width of one 'tooth' approximately 500 μm. Lines of voids were observed running parallel to the local inclination of the crack surface and extending beyond the turning points of the zig-zags.

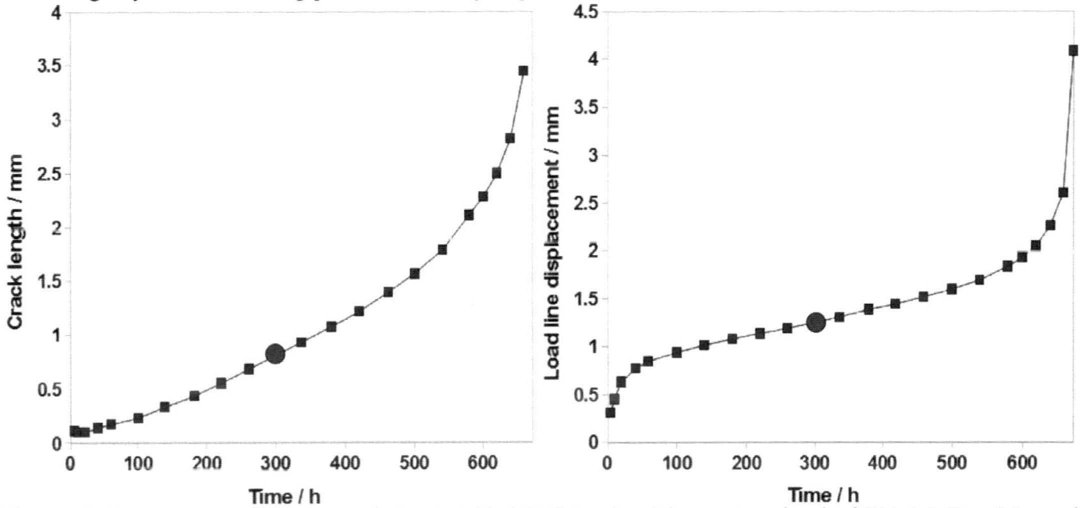

Figure 1. Creep curves for test carried out at T=873K in air with constant load of 311.8 MPa: (a) crack length; (b) load line displacement. Round marker indicates point at which test was interrupted.

In the present work, the creep test on fine-PAG P92 has been repeated but interrupted part-way through to obtain a metallographic specimen containing a growing crack. Using electron backscatter diffraction (EBSD), which enables mapping of the crystal orientation as a function of position [11], the relationship between the morphology of the crack and voids and the martensitic substructure, in terms of packets, blocks and prior austenite grains, has been investigated.

2. Experimental

The composition of the P92 steel is given in table 1. The material was normalised for 2 hours at 1343K and tempered for 2 hours at 1053K.

The creep tests were interrupted after 300 hours; it can be seen from figure 1(b) that at the point of interruption, marked with a round dot, the sample is in the steady-state creep regime. The area around the crack tip was sectioned in a plane parallel to the tensile axis and crack growth direction and prepared for EBSD observation using a final stage of polishing with 0.05 μm colloidal silica for around 2 hours [12]. A sample of the as-received, untested material was also observed using EBSD for comparison.

3. Results

3.1. As-received sample

As a result of the orientation relationship between martensite and its parent austenite, the martensitic variants also have fixed orientation relationships with one another, leading to a characteristic distribution of misorientation angles. The high-angle part of the misorientation angle distribution obtained using EBSD for the as-received sample is shown in

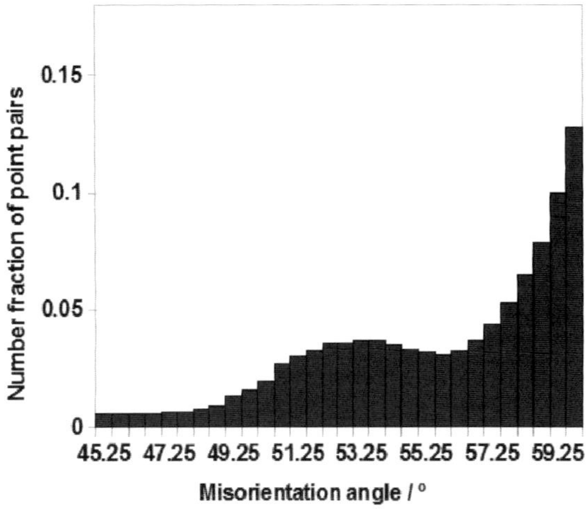

Figure 2. Distribution of misorientation angles θ in the interval $45^o < \theta < 60^o$ for the as-received sample

figure 2. A large peak can be observed at 60° and a smaller one at 54.5°, with a trough in between. The smaller peak is displaced by about 5° from the position it would take if the ideal Kurdjumow-Sachs orientation relationship [13] were exactly obeyed.

This misorientation angle distribution for the as-received specimen can be used for comparison to test whether recovery or recrystallisation have occurred in different regions of the interrupted test specimen.

Table 1. Composition of P92 sample, in mass %.

C	Si	Mn	P	S	Cr	Mo	W	Nb	V	N	B
0.09	0.16	0.47	0.01	0.001	8.72	0.45	1.87	0.06	0.21	0.05	0.002

3.2. Interrupted sample

The area around the crack tip in the interrupted test specimen can be divided into three regions according to microstructure and crystallographic characteristics.

Region I: Void nucleation region. Small voids of irregular size are found. Figure 3 is an EBSD image quality (IQ) map [14] of the void nucleation region, with the boundary of a prior austenite grain and the boundaries of the packets within it, as determined from crystallographic analysis, marked with white and black lines respectively. It can be seen that the majority of voids are located on the prior austenite grain boundary, with some others on packet boundaries and a few within packets. The misorientation profiles from both the bulk and the areas in the immediate vicinity of the voids are unchanged from that for the as-received sample. This suggests that the voids have formed without any gross microstructural recovery or plastic deformation.

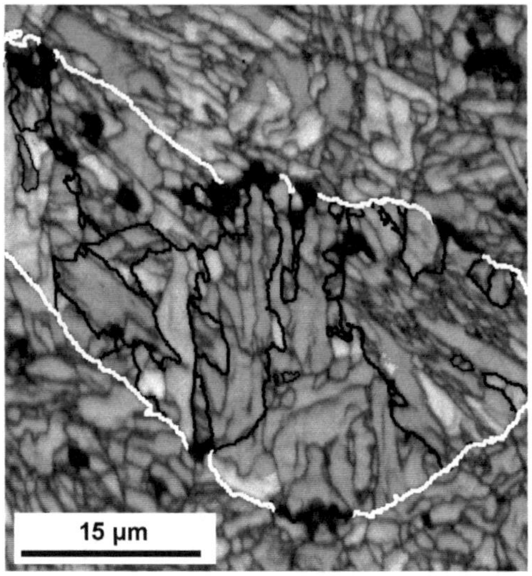

Figure 3: EBSD image quality map of the void nucleation region, with a prior austenite grain boundary marked in white and packet boundaries in black.

Region II: Void growth and bridging region. Large, circular voids are present. Some of these, from their shape, are clearly made up of two or more small voids which have coalesced. The voids are surrounded with layers of fine, equiaxed grains. Between the voids are bridging ligaments exhibiting grain rotation, as demonstrated by the crystal orientation map, and recovery and recrystallisation, as can be seen from the map of "image quality", characterising the clarity of the electron backscatter diffraction pattern. High-angle misorientation profiles from areas immediately around the voids show a loss of the characteristic two-peak martensitic profile.

Region III: Continuous crack region: Voids have coalesced to form a continuous crack extending from the edge of the sample. The surfaces of the crack are covered with a layer of recovered or recrystallised material, as demonstrated by the high-angle misorientation profile, but away from the immediate vicinity of the crack, the misorientation profile is similar to that of as-received material. Lines of 'failed' voids, which did not form a successful crack, are also present extending beyond the zig-zags of the continuous crack.

4. Discussion and conclusion

From the results presented above, a mechanism of fracture can be proposed. As a result of the specimen geometry, bands oriented at 45° to both the tensile axis and the overall crack growth direction experience the highest stress and have the greatest propensity for void formation. Voids nucleate mostly at prior austenite grain boundaries, with some at packet boundaries and within the packets. Nearer the crack tip, voids grow and coalesce by recovery, recrystallisation and eventual rupture of the bridging ligaments between them. This leads to the observed crack with a layer of recovered or recrystallised grains on its surface and the dimpled appearance of the fracture surface.

The zig-zag morphology is formed by nucleation and growth of lines of voids both above and below the plane of the crack. Out of the two lines of voids, the one in which bridging occurs first becomes part of the crack. The other remains as one of the observed 'failed' lines of voids. Zig-zags change direction after extending far enough away from the crack plane that they no longer experience sufficient stress for void coalescence.

It has been found experimentally in CrMoV martensitic steels that the minimum creep rate decreases with increasing PAG diameter for small PAG diameters [15]. The present results, showing void nucleation occurring on PAG boundaries, are consistent with this, since larger PAG give fewer possible void nucleation sites per unit volume. In addition, it has been found in plain-carbon steels that the packet size increases with increasing PAG size [16], so the number of available packet boundary sites for void nucleation will also decrease as PAG size increases. However, the relationship between PAG size and minimum creep rate breaks down for PAG diameters >50 μm [15], and increasing the PAG diameter beyond a critical value leads to the onset of brittle cracking along PAG boundaries [15,17]. Recent work suggests a similar dependence of fracture mode on PAG size in 12 wt.% Cr P122 steel [18]. Further EBSD observations of cracked 9-12 wt.% steel specimens with various thermal histories and thus different PAG sizes may help to test the existence of such a transition and, if it does exist, determine the critical PAG size corresponding to the transition and its dependence on such factors as composition and test temperature and stress.

Acknowledgements
The authors acknowledge the 129 committee of the Japan Society for the Promotion of Science (chairman: Prof. T. Yokobori) for support of this work and provision of samples.

References

[1] Nippon Steel Technical Report No. 50, 1991
[2] ASTM Standard A335M - 06 Standard Specification for Seamless Ferritic Alloy-Steel Pipe for High-Temperature Service, 2006
[3] Marder A R and Krauss G 1969 *Trans. ASM* **62** 957
[4] Maki T, Tsuzaki K and Tamura I 1980 *Trans. ISIJ* **20** 207-214
[5] Morito S, Tanaka H, Konishi R, Furuhara T and Maki T 2003 *Acta Mater.* **51** 1789-1799
[6] Kitahara H, Ueji R, Tsuji N and Minamino Y 2006 *Acta Mater.* **54** 1279-1288
[7] Maruyama K, Sawada K and Koike J 2001 *ISIJ Int.* **41(6)** 641-653
[8] Masuyama F 2001, *ISIJ Int.* **41(6)** 612-625
[9] ASTM E1457-07: Standard Test Method for Measurement of Creep Crack Growth Times in Metals, 2007
[10] Matsuzaki T, Sugiura R, Yokobori A T and Yardley V A 2008 EBSD study of effect of material microstructure in limiting creep crack growth in high-Cr heat-resistant steels *Japan Institute of Materials Autumn Meeting, Kumamoto, Japan*
[11] Adams B L, Wright S I and Kunze K 1993 *Metall. Trans. A* **24A** 819-831
[12] Nowell M W, Witt R A and True B W 2005 *Microscopy Today* **July** 44-48
[13] Kurdjumow G and Sachs G 1930 *Z. Physik* **64** 325-343
[14] Wright S I and Nowell M M 2006 *Microsc. Microanal.* **12** 72-84
[15] Kimura K, Ohi N, Shimazu K, Matsuo T, Tanaka R and Kikuchi M 1987 *Scripta Metall.* **21** 19-22
[16] Morito S, Yoshida H, Maki T and Huang X 2006 *Mater. Sci. Eng. A* **438-440** 237-240
[17] Hong S and Yu J 1989 *Scripta Metall.* **23** 1057-1062
[18] Sugiura R, Yokobori A T, Takamori S, Tabuchi M, Fuji A, Yoda M, Kobayashi K, and Yokobori T 2007 *Mater. Trans.* **48** 2928-2936

Influence of bondcoat creep and roughness on damage and lifetime of ZrO_2 TBCs for gas turbines under thermocyclic loads

M Schweda, T Beck, L Singheiser
Forschungszentrum Jülich GmbH, IEF-2, 52425 Jülich

t.beck@fz-juelich.de

Abstract. As a simplified model system for thermal barrier coatings (TBCs) in gas turbines, Y_2O_3 partially stabilized ZrO_2 TBCs were applied on FeCrAlY substrates. The creep strength of the substrate was varied by using an oxide-dispersion-strengthened (ODS) alloy and a non-ODS alloy with similar chemical composition. Defined interface profiles were produced before coating. Creep properties of the oxide layer between substrate and TBC were varied by either coating the test pieces with nanocrystalline PVD-alumina or with coarser grained naturally grown Al_2O_3 before plasma spraying the TBC. During thermal cycling (T_{max}=1050°C, T_{min}=60°C, dwell of 2 hours at T_{max}) periodic 2-d interfaces resulted in very low lifetime independent from substrate strength and interface oxide type. With stochastic 3-d interfaces lifetimes up to 900 cycles were reached, especially for the substrate with low creep strength combined with a coarse grained alumina interlayer.

1. Introduction

To protect blades and vanes in the first stages of gas turbines from overheating and hot corrosion, Y_2O_3 partially stabilized ZrO_2 (PYSZ) thermal barrier coatings (TBC) are used with NiCoCrAlY bondcoats (BC) forming a dense, thermally grown oxide (TGO) layer between TBC and BC. Thermal mismatch stresses between these components and TGO growth stresses promote cracks near the TBC/BC interface [1]. Creep of BC and TGO as well as the roughness of the BC/TBC-interface significantly affect the stress state in the TBC system and therefore determine its lifetime [2]. In the present study a simplified TBC system consisting of a plasma-sprayed PYSZ TBC applied on solid FeCrAlY substrates simulating the BC was investigated under thermocyclic loadings. The Ni base substrate was excluded to avoid additional thermal mismatch and interdiffusion. A conventional FeCrAlY substrate and an ODS alloy with similar chemical composition were used to investigate the influence of BC creep strength. The influence of interface roughness was analyzed by applying periodic 2-dimensional roughness profiles by high speed turning and stochastic 3-dimensional roughnesses by sand blasting. The influence of creep strength of the TGO layer between BC and TBC was investigated by applying either a nanocrystalline Al_2O_3 layer by sputtering or a naturally grown Al_2O_3 layer with grain sizes in the range of 1µm before plasma spraying of the TBC. All samples were subjected to thermocyclic fatigue with in-situ and ex-situ observation of delamination cracking.

2. Experimental Details

The FeCrAlY substrates were Fecralloy Eisen-Chrom™ and MA956™. The main alloying elements are 22% (20%) Cr, 4.8% (4.5%) Al, and 0.3% (0.5%) Y (weight.-%, numbers in brackets: MA956). The solid cylindrical specimens had a length of 25 mm, and a diameter of 8 mm. The peak-to-peak roughness of the interfaces was 10 µm, 15 µm and 20 µm with wavelengths between 40 and 45 µm. The fine grained Al_2O_3 layer applied by sputtering had an initial grain size of 100 up to 200 nm which increased during thermal cycling to ≈500 nm. The alumina interlayer applied by oxidation had initial grain sizes between 0.5 and 1.0 µm which did not change significantly during thermal cycling. The ZrO_2 / 7% Y_2O_3 TBC with 200µm thickness was applied at Universität Braunschweig, IfW, along 20 mm in the middle of the samples with smoothly decreasing thickness towards the ends of the test pieces in order to minimize edge effects. Thermal cycles were applied in air with T_{max}=1050°C, T_{min}=60°C, and a dwell of 2 hours at T_{max} using a resistance furnace with a pneumatic actuator periodically placing the sample inside and outside the furnace. The samples were examined by infrared impulse thermography to quantify delaminations under the TBC with a resolution of 0.2 mm. After failure by segmentation cracking of the TBC the samples were cross sectioned and examined by SEM.

3. Results and Discussion

1.1. Periodic 2-dimensional interface

Lifetimes were extremely short in the case of periodic 2-d interfaces. The samples with the highest peak-to-peak roughness, naturally grown Al_2O_3 interlayer and Fecralloy substrate (low creep strength) reached up to 40 cycles until macroscopic segmentation was observed. All other specimens failed at cycle numbers up to 10. In all cases delamination occurred mainly along the Al_2O_3/TBC interface with few cracks in the Al_2O_3 and at the Al_2O_3/substrate interface. Infrared thermography showed large longitudinal and tangential delaminations (Figure 1). These results allow the following conclusions:

- Tangential crack growth is very fast due to the fact that no roughness exists in this direction.
- The fast longitudinal crack growth indicates that the periodicity of the profile increases the crack growth rate compared to stochastic 3-dimensional profiles (see section 3.2).
- Increasing interface roughness rises lifetime because mechanical interlocking is essential for sufficient attachment of plasma sprayed TBCs to the substrate.

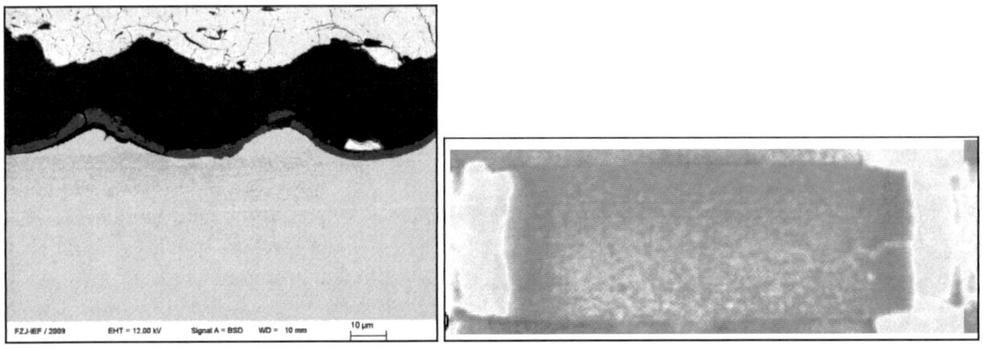

Figure 1. Delamination crack path (left) and thermography image (right, red areas indicate delamination) of a specimen with Fecralloy substrate, periodic 2-dimensional interface and naturally grown alumina

1.2. Stochastic, 3-dimensional interface

Figure 2 gives an overview of the lifetimes measured in the case of stochastic, 3-dimensional interfaces. The numbers inside the bars indicate the position of the delamination cracks determined by SEM at cross sections after failure. In addition to the samples with coarse grained or with nanocrystalline alumina layers applied before TBC application, in the intermediate roughness class two samples were tested without an initial Al_2O_3 layer. In this case, a naturally grown TGO with relatively high grain size (around 1 μm) is formed as in real-life TBC systems. The exemplarily chosen thermography image shows that crack propagation occurs mainly in longitudinal direction. It should be noted that a quantitative analysis of delamination crack growth using infrared thermography indicated an incubation phase with crack lengths below 0.2 mm followed by nearly linear or progressive crack growth as reported in [3]. From these results the following conclusions are drawn:

- Distinctly higher lifetimes are found in the case of Fecralloy-substrate (low creep strength) compared with MA956 (high creep strength).
- For high roughness, the fine grained Al_2O_3 interlayer leads to similar or higher lifetimes than the coarse grained Al_2O_3.
- The lifetime in the reference case with no initial alumina interlayer is nearly equal to the result for a coarse grained interlayer. This indicates that the attachment of the TBC is not adversely affected by preoxidation of the substrate before plasma spraying.
- In the case of smaller roughness, the fine grained Al_2O_3 results in distinctly lower life than the coarse grained variations (naturally grown and applied by pre-oxidation).
- Increasing roughness and increasing substrate creep strength tends to shift the crack to the Al_2O_3 and the Al_2O_3/substrate interface.

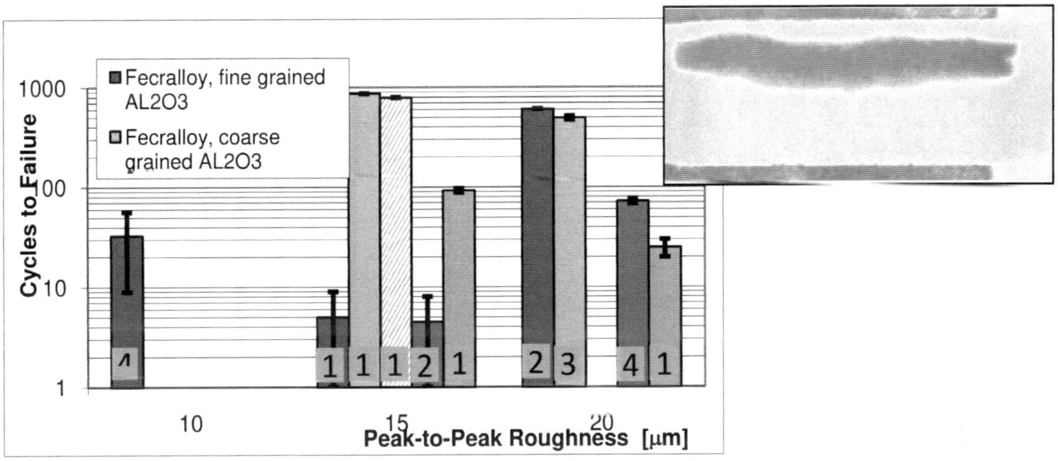

Figure 2. Lifetimes of the model TBC systems with stochastic 3-d interface. The numbers indicate the delamination crack path: (1) – TBC and TBC/Al_2O_3 interface; (2) – TBC/Al_2O_3 interface; (3) – TBC, TBC/Al_2O_3 interface and Al_2O_3; (4) – TBC/Al_2O_3 interface, Al_2O_3 and Al_2O_3/substrate interface. The inlay picture is a typical thermography image at end of life

Acknowledgement
The authors gratefully acknowledge funding of this work by the German Research Foundation (DFG)

References

[1] Evans, A.G., Clarke, D.R. and Levi, C.G. 2008, The Influence of Oxides on the Performance of Advanced Gas Turbines, *J. Europ. Ceram. Soc.* **28**, 1405-1419
[2] Busso, E.P., Quian, Z.Q., Taylor, M.P. and Evans, H.E. 2009, The Influence of Bondcoat and Topcoat Mechanical Properties on Stress Development in Thermal Barrier Coating Systems, *Acta Mater.* **57**, 2349-2361
[3] Trunova, O., Beck, T., Herzog, R., Steinbrech, R.W. and Singheiser, L. 2008, Damage Mechanisms and Lifetime Prediction of Thermal Barrier Coatings during Cyclic Oxidation – Part I: Experiments, *Surface Coatings Technol.* ***202***, 5027-5032

Journal of Physics: Conference Series **240** (2010) 012078 doi:10.1088/1742-6596/240/1/012078

Macroscopic localization of plastic deformation in single crystals Ni_3Ge with $L1_2$ superstructure

V A Starenchenko, S V Starenchenko[1], and Yu V Solov'eva

Tomsk State University of Architecture and Building, 2, Solyanaya sq., Tomsk, 634003, Russia

star@tsuab.ru

Abstract. The dislocation structure forming in the single crystal Ni_3Ge alloy during the deformation is studied in this paper. It is established that plastic deformation leads to the uniform strain stability loss in the Ni_3Ge alloy at increasing the temperature. The homogeneous dislocation structure loses its stability in local places, which are situated close to the zone of localization. The internal structure of the localization band of deformation consists of cleaned from dislocations fragments in size about 1μm that are disorientated from each other to 10-60°. Then the polycrystalline local band is formed, in which the plastic deformation reaches hundreds of percents.

1. Introduction

One of the main features of intermetallic compounds with the superstructure $L1_2$ is a rather homogeneous slip [1]. Homogeneous sliding is stable to the definite strain. Two channels of the loss stability of the uniform strain can be considered. The first is the catastrophic growth of cracks and the fracture of the material. The second one is connected with accumulating antiphase boundaries in slip bands and changing the type of deformation carriers from motion of superstructure dislocations to motion of single dislocations. It leads to appearance of rough slip traces [2]. It should be noted that the both channels are not mutually exclusive. Deformation of the intermetallic alloys, as a rule, is completed by brittle fracture but plasticity of alloys increases when the rough slip appears.

In this paper results of studying the macroscopic localization of plastic deformation and the uniform strain stability loss in the Ni_3Ge alloy are presented.

2. Experimental

A single crystal of the Ni_3Ge alloy was grown by the Bridgeman method in an argon atmosphere from Ni and Ge of 99.99% purity. Specimens shaped as parallelepipeds $3.0 \times 3.0 \times 6.0$ mm^3 in size were cut by the electric spark method from single crystal ingots. Samples were compressed on a test machine with a rate of $2\%min^{-1}$ at different temperatures (T=293, 523, 673, 873 K). Before tests specimens were homogenized for 40 h at T=1228 K. The directions of the compression axes of the investigated single crystals coincide with the [001] direction. The dislocation structure and the relief of the surface were studied by transmission electron

Journal of Physics: Conference Series **240** (2010) 012078 doi:10.1088/1742-6596/240/1/012078

microscopy and scanning one. The orientation of crystal faces was determined by the Laue and X-ray diffraction methods.

3. Results and Discussion

The alloy Ni_3Ge belongs to group of alloys with the $L1_2$ ordered structure and exhibits strongly expressed positive temperature dependence of mechanical characteristics [3, 4]. In certain temperature interval, increasing temperatures causes an increase of shear stresses, yield stress, and work-hardening rate.

The analysis of slip traces has shown, that sliding is homogeneous in orientation of the deformation axis [001] up to temperature T=873 K. The shift in lines of sliding reaches 50-100 Å. Lines of sliding 10000 - 15000 Å in length are distributed in regular intervals on a surface of a crystal. The distances between them are 300 – 400 Å. Roughening the slip, which is typical for severe deformation in alloys with the $L1_2$ superstructure, is not observed in the Ni_3Ge alloy.

Formed to the moment of fracture the dislocation structure is homogeneous as well. The dislocations are situated uniformly within the deformed crystal. They are combined mainly into dipole and multipole configurations. Measured characteristics of the dislocation structure and following statistical treatment have shown that the dislocation structure is described by the lognormal distribution.

At T=873 K the deformation is realized homogeneously as before by means of the fine slip up to $\varepsilon \approx 3 - 4$ %. The dislocation structure, which is formed at mentioned strains, is also homogeneous and described by the lognormal function. However the qualitative differences are observed in the homogeneous dislocation structure at different temperatures. It should be noticed that the dipole configurations and the rectilinear dislocations are absent at T=873 K.

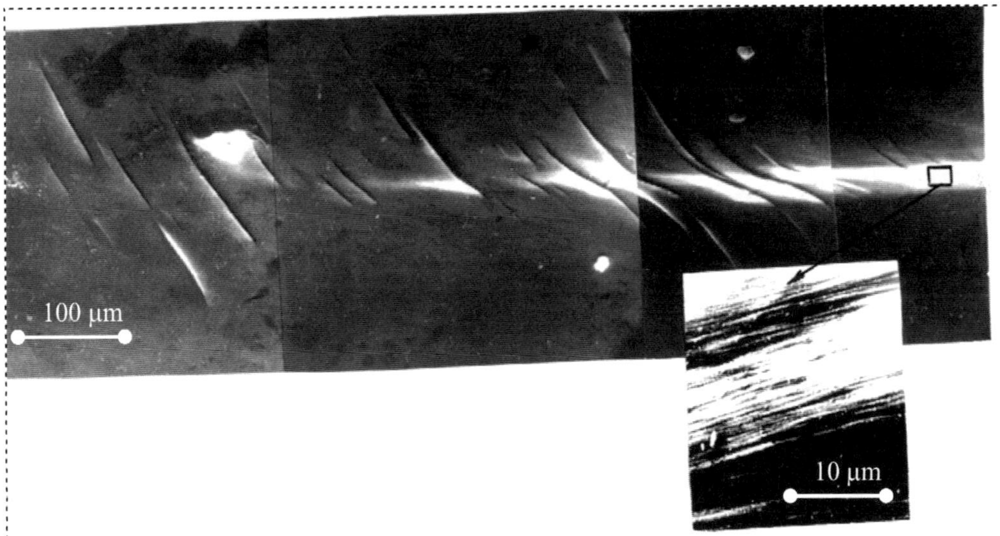

Figure 1. An initial stage of forming a band of superlocalization

The increase in temperature has displayed that plastic deformation is non-uniform beginning from $\varepsilon \approx 5$ % at the temperatures exceeding T=893 K. On the surface of the crystal there is a band passing through all crystal of width about 10^{-3} mm, in which deformation is localized. Originally, the band is formed from set of rough slip traces (figure 1).

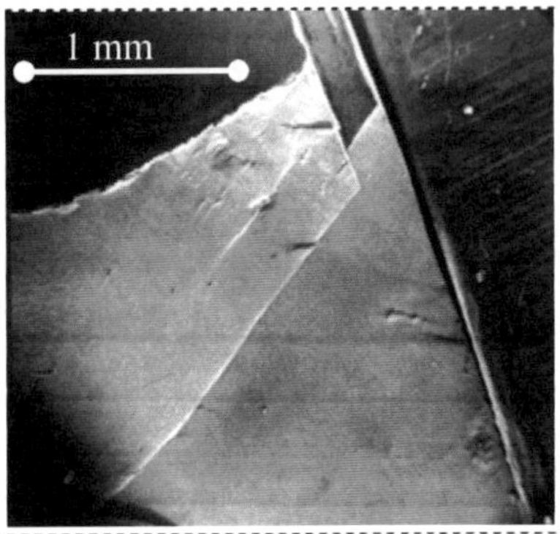

Figure 2. Displacement of parts of a crystal along the band of localization T=893 K

The evolution of the band caused by increase in the strain leads to even greater localization of deformation and displacement of parts of a crystal relative to each other on the localization deformation band. The single crystal is divided into two parts shifted along planes close to {111}. The further deformation develops along the division plane of parts of the crystal (figure 2).

Newly created surface has a complex rough structure, which consists of many fragments (figure 3*a*, *b*). Local deformation in the band, defined as $\Delta l/l \cdot 100$ %, where Δl – a value of displacement, l – a width of the band, is hundreds of percents.

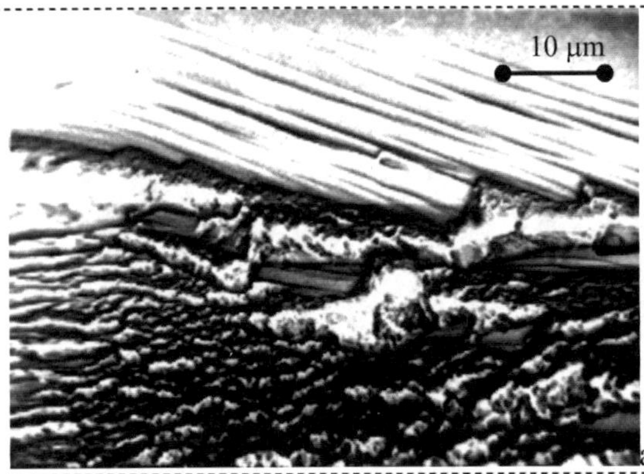

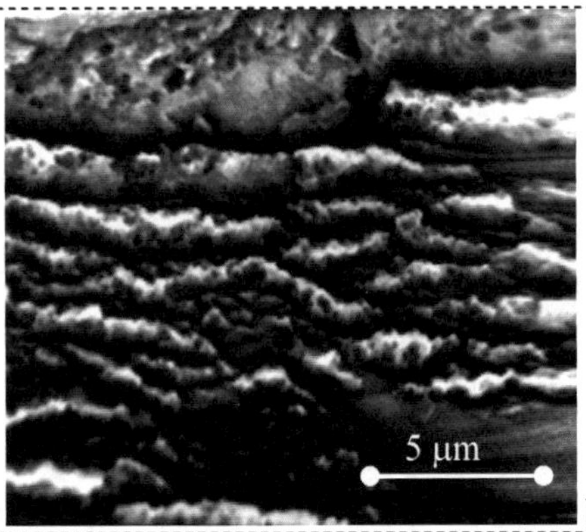

Figure 3. The structure of newly created surface at the superlocalization (the single crystal Ni_3Ge, T=893 K)

The drop in stresses in curve of strengthening corresponds to development of this process. To study the dislocation structure, foils have been prepared from different parts of the crystal: a) far from the band of localization, where deformation is homogeneous; b) near the zone of localization; c) the zone of localization. It turned out that the homogeneous dislocation structure with the high density of dislocations (2-3x10^{10} sm^{-2}) is formed in parts are placed far from the localization band. The homogeneous dislocation structure loses its stability close to the zone of localization. There appear areas of the disorientated band structures with 1-2° disorientations. Polycrystalline substructure, composed of the fragments with the low density of dislocations bounded by the high-angled boundaries, is detected directly in the band of localization. Disorientation between the fragments in this zone goes up to 10-60° (figure 4).

Sizes of observed fragments in the localization area fall into the interval 10^{-3} – 10^{-4} mm. The same sizes have roughness of the newly appearing surface that points to possible unity of their nature.

Raising the temperature leads to changing the shape of the localization band. The band of localization gets diffuse losing the sharp outline (figure 5).

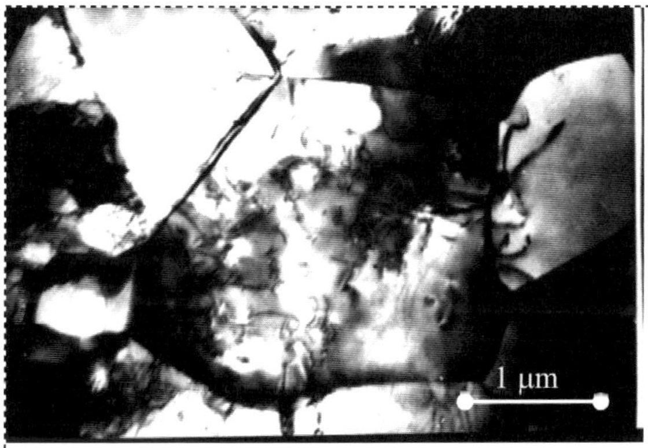

Figure 4. The dislocation structure in the superlocalization band (the single crystal Ni_3Ge, T=893 K)

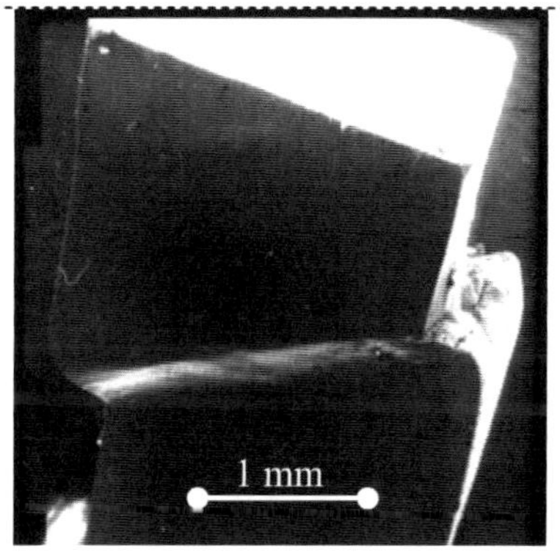

Figure 5. The superlocalization band of the deformation (the single crystal Ni_3Ge, T=973 K)

4. Conclusions

The investigation and the analysis of the deformation relief and dislocation structure for the single crystal Ni_3Ge, oriented along the [001] axes at the temperatures 893 - 973 K carried out in this work make it possible for the authors to offer the following stages of plastic deformation development. There are several stages of macroscopic localization process of the deformation.

1. The initial stage observed before deformations of 3-4 %, is the stage of the homogeneous slip. Dislocations are accumulated very fast up to density (2-3) $\times 10^{10}$ cm^{-2}.

2. The stage with zero coefficient of hardening is the stage of the homogeneous slip too; however the density of dislocations at this stage remains at constant level. The balance of processes of multiplication and annihilation of dislocations is observed at this stage.

3. The stage with negative coefficient of work-hardening is a stage of superlocalization of deformation. The density of point defects, dislocations, the level of the stress and the temperature are sufficient to form the localization band of deformation, which internal structure consists of fragments in size about 1μm that are disorientated from each other to 10-60 °. Internal areas of these fragments are cleaned from dislocations, and the polycrystalline local band is formed, in which the plastic deformation goes up to hundreds of percents.

References

[1] Taoka T, Sakata S 1990 *Acta Metall.* **5** 61.

[2] Popov L E, Koneva N A, Tereshko I V. 1979 *Deformatsionnoe uprochnenie uporyadochennykh splavov (Work-hardening the ordered alloys)* (Moscow: Metallurgia).

[3] Aoki A, Izumi O 1978 *J. Mater. Sci.* **13** 2313.

[4] Starenchenko V A, Colov'eva Yu V, Abzaev Yu A, Nikolaev V I, Shpeizman V V and Smirnov B I 1996 *Phys. Tvyordogo Tela* (*Phys. Solid State*) **38** 1668.

High temperature deformation behavior of the molybdenum alloy TZM

T Mrotzek[1], **U Martin**[2] **and A Hoffmann**[3]

[1] Plansee SE, 6600 Reutte, Austria
[2] Institute of Materials Science, Freiberg University of Mining and Technology, D-D-09599 Freiberg, Germany
[3] Plansee Metall GmbH, 6600 Reutte, Austria

martin@ww.tu-freiberg.de

Abstract. The molybdenum alloy TZM (Mo-0.5wt%Ti-0.08wt%Zr) is a commonly used constructional material for high-temperature applications. It is well known that molybdenum and its alloys develop a distinct subgrain structure and texture during hot deformation. These microstructural aspects have a significant effect on strength at elevated temperatures. It was observed that with proceeding primary recrystallization and therefore with disappearance of subgrains the yield strength drops almost to the level of pure molybdenum. The aim of the present work was to investigate and describe the strain hardening of hot deformed TZM on a microstructural basis. For this purpose sintered and prerolled TZM rods were recrystallized and each of them deformed to a specific degree of deformation afterwards.. Especially the evolution of disorientation distributions was analyzed by electron backscattering diffraction (EBSD) and used to describe the work hardening effect. The yield strength was determined by tensile tests between room temperature and 1473 K. By analyzing disorientation profiles the formation and evolution of geometrically necessary and incidental dislocation boundaries could be observed. A model developed by Pantleon was used to describe the work hardening of TZM.

1. Introduction

The molybdenum alloy TZM contains 0.5 wt.% Ti, 0.08 wt.% Zr and approximately 200 ppm C. It is commonly used as structural material in thermally loaded applications or as plate material in rotary X-ray tubes. Compared to pure molybdenum TZM exhibits a higher strength and creep resistance. Especially at elevated temperatures the primary strengthening mechanism was attributed to carbide particles. An earlier study [1] revealed that these carbides do not increase the strength by an Orowan-like dislocation pinning mechanism but are able to pin dislocation ensembles like cell walls or subgrain boundaries. Furthermore the particles help to increase the recrystallization temperature. Dislocation arrangements like cell walls or subgrain boundaries are able to act as glide barriers even at high temperatures and hence increase the yield strength of the material until recrystallization occurs.

A strong work hardening effect in TZM was reported earlier [2]. The tendency of pure molybdenum to form dislocation networks during hot deformation is well known. Results for Mo single and polycrystals were reported e.g. in [3]. The aim of this work is to provide a better microstructure-based understanding of the work hardening behaviour of TZM.

Work hardening is commonly expressed by plotting the work hardening rate versus the flow stress [4-6]. Normally a so called stage-III behaviour is observable. Depending on the strain path a stages IV and V may be present too [6]. For stage III there is the common understanding that hardening is caused by generation and storage of dislocations resulting in the formation of dislocation walls. Kocks developed a model for stage III which needs just one microstructural parameter, namely the overall dislocation density [6]. Unfortunately stage IV cannot be described by this model. Pantleon however picked up this approach incorporating the existence of excess dislocations [7]. With this modification the model is able to explain the formation and evolution of dislocation boundaries and describe work hardening up to stage IV. This approach was used to model the work hardening of TZM. For a detailed description of the model the interested reader is referred to [7,8,9].

The flow stress of molybdenum and its alloys below the transition temperature T_0 is governed by a high Peierls-Nabarro-potential [10,11]. This thermal part of flow stress has to be separated from the athermal contributions like work hardening. The description of the used procedure would exceed the scope of this paper and is given elsewhere [12].

2. Experimental

The investigated material was produced via a standard PM route. After sintering the alloying contents were 0.49 wt% Ti, 0.071 wt% Zr, 180 ppm carbon and 150 ppm oxygen. Afterwards, the sintered material was forged to a (logarithmic) strain of φ = 0.5 and completely recrystallized at 1900 K. This condition defines the non-deformed state and acts as a base line for the following experiments. It was named initial state (IS). Samples with increasing degree of deformation φ were produced from the IS material by hot rolling at 1470 K. Finally, samples for tensile tests and orientation mappings were machined from all states. All tensile tests were done at a constant strain rate of $2 \cdot 10^{-3}\ s^{-1}$ between room temperature and 1473 K.

In order to determine the microstructural parameters which characterize the hardening behavior the samples were investigated by EBSD. To achieve a sufficient success rate (fraction of indexed Kikuchi Maps) of more than 85% of the measured points, the specimens were electro polished. The measurements were done on a LEO Gemini 1530 SEM with an HKL Nordlys EBSD detector. Due to the small cross section of the acquisition area each sample was measured at three different points to ensure the representativeness of the measurement. The texture effect was determined by calculating the inverse pole figures and Taylor factors in tensile direction. The disorientations were calculated from the orientation mappings.

3. Results and Discussion

The samples were characterized by SEM/EDX to verify the precipitation of particles. Carbidic and oxidic particles were found. The carbide particles are mostly located at grain boundaries and appeared to be molybdenum carbides. Fine precipitated particles with potential to pin dislocations and cause a significant particle hardening effect were not found. Titanium is mostly soluted in the molybdenum matrix. Its hardening contribution is minor compared to the observed effects and will be neglected [13]. During the first stages of deformation the original grains were elongated and start to disorient. This may indicate the formation of dislocation boundaries (figure 1b). With increasing degree of deformation subgrains are formed within the original grains. According to literature [e.g. 14] the subgrain boundaries are of different type called incidental dislocation boundaries (IDB) and geometrically necessary boundaries (GNB) [14,15].

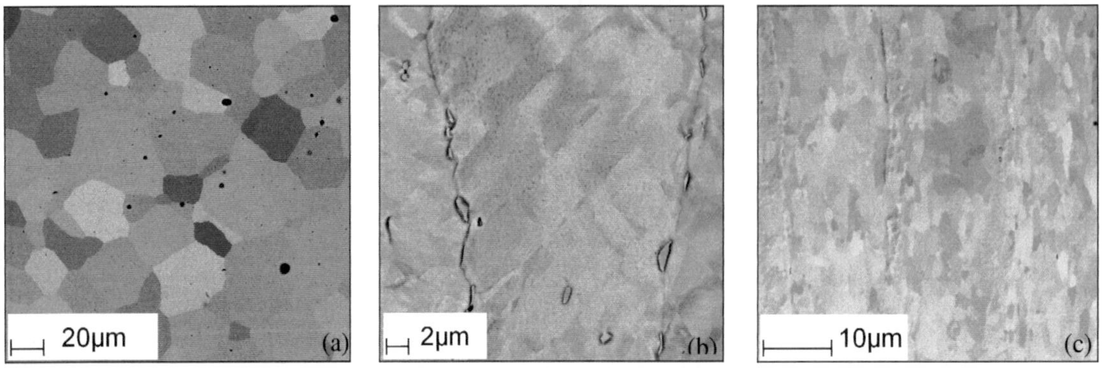

Figure 1. SEM micrographs of (a) IS and deformed states with (b) φ=0.5 and (c) φ=1.5, molybdenum carbides located at grain boundaries can be observed (b)

For a quantitative description of the substructure development the disorientation distributions of the different stages were calculated from the EBSD results. Even in the early stages of deformation the distributions reveal two maxima. A superposition of two lognormal distributions was fitted to the measured values. From the parameters of each function i the global maximum θ_i^{max} was calculated. The development of the values for θ_i^{max} is shown in figure 2. One maximum is in the range of two degrees, the other develops depending on the degree of deformation between 3° to 8° (figure 2). It is assumed that the first peak represents the IDBs whereas the broad second peak represents the GNBs.

For modeling the disorientation evolution Pantleon assumes that a fluctuation in statistical trapping of dislocations and an imbalance in the activation of slip systems on both sides of a boundary leads to a bias of dislocation fluxes. The imbalance of slip activation is Gaussian distributed with a standard deviation σ_{imb} [7,9]. The evolution of disorientations is described by:

$$\left\langle \theta^2 \right\rangle = \frac{Pb}{d^*}\gamma + \sigma_{imb}^2 \gamma^2 \quad (1)$$

For each stage of deformation the shear strain γ was calculated from φ using the related Taylor factor which was determined from EBSD orientation measurements [2]. The probability P of dislocation immobilization in a boundary is set equal to 1. Values for d^* and σ_{imb} were determined by fitting the right side of equation (1) to the square of θ_i^{max}.

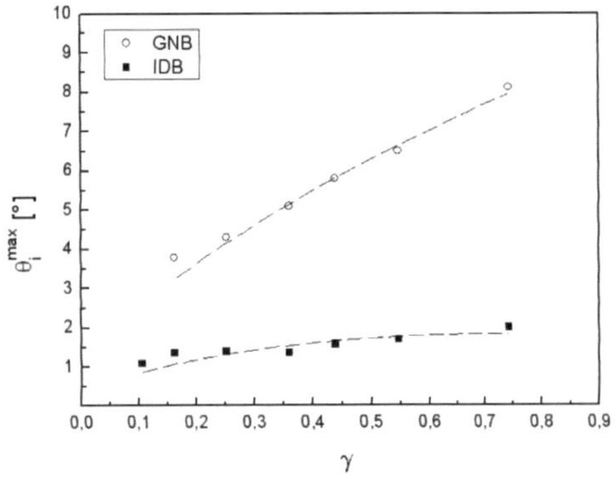

Figure 2. Development of θ_i^{max} with γ indicating the evolution of IDBs and GNBs during rolling at 1473 K

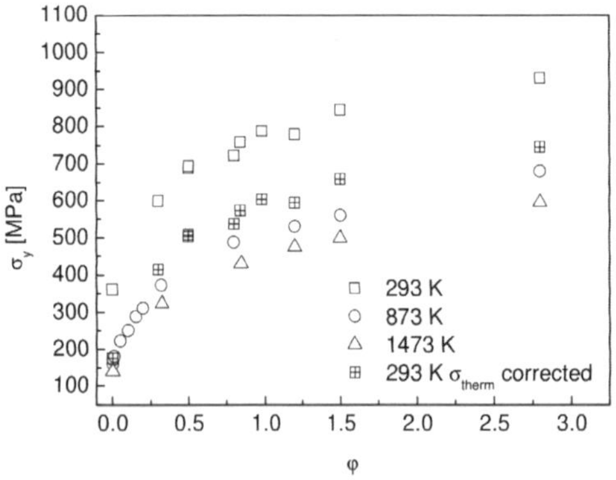

Figure 3. Yield strength of TZM at different temperatures and φ, indicated values for 293 K are reduced by the thermal stress component [12]

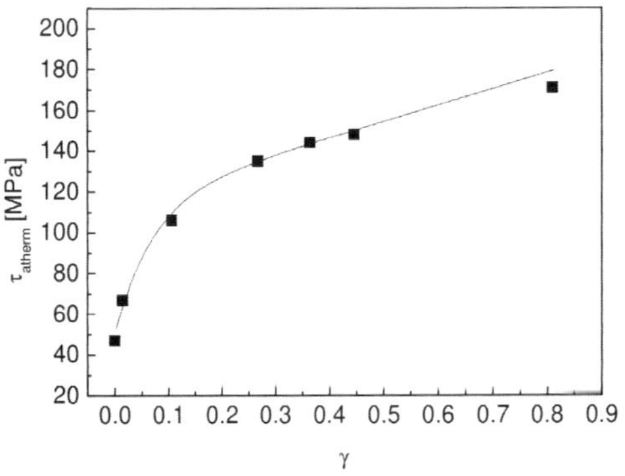

Figure 4. Athermal shear stress at yielding in tensile tests at 1473 K as function of hot rolling strain γ. solid line: modelled

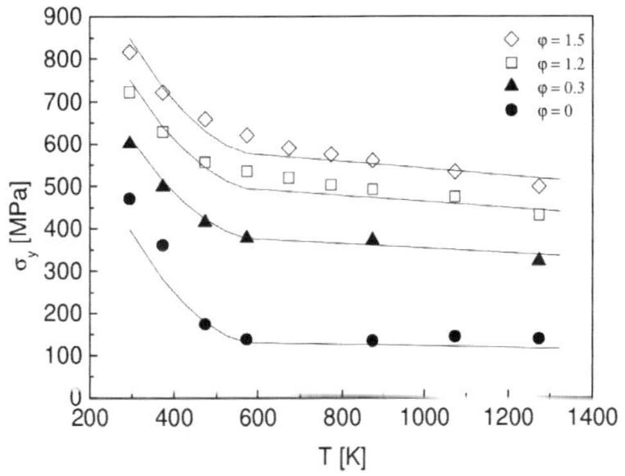

Figure 5. Tensile yield stress σ_y as function of temperature T for different rolling strains φ. solid lines: modelled

From the tensile tests the yield strengths of the hot worked TZM at different degrees of deformation φ and temperatures were obtained. The σ_y-φ plots reveal a strong work hardening effect (figure 3). Certainly the thermal stress contribution below a transition temperature T_0 must not be neglected. The transition temperature T_0 was determined to be 233°C (506K) at a strain rate of $2{\cdot}10^{-3}s^{-1}$ [12]. For this reason the stress values for $T<T_0$ were reduced by the thermal stress component (see figure 3) which was determined for this material as described in [12]. Using the model proposed by Pantleon for stages III and IV which is based on the development of GNBs and IDBs, the athermal shear stress τ_{atherm} at yielding as function of γ can be described in good agreement with the experimental values (figure 4). Based on these findings and incorporating texture effects, the temperature dependence of the shear modulus [13] as well as the thermal stress part, the development of the yield strength σ_y can be predicted for different temperatures and φ (figure 5). It can be concluded that even at temperatures up to 1300 K work hardening strongly enhances the strength of TZM. The observed strength increase is in the range of 500 to 600 MPa. The effect is mainly carried by geometrically neccessary and incidental dislocation boundaries and can be described by the model proposed by Pantleon in good agreement with experimental data.

References

[1] Mrotzek T, Hoffmann A and Martin U 2006 *Int. J. Refract. Met. H.* **24** 298
[2] Mrotzek T, Hoffmann A, Martin U and Oettel H 2007 *Mater. Sci. Forum* **539-543** 2725
[3] Luft A 1995 *Prog. Mater. Sci.* **35** 69
[4] Nabarro F R N, Basinski Z S and Holt D B 1964 Adv. *Physics* **13** 193
[5] Gil Sevillano J, van Houtte P and Aernoudt E 1981 *Prog. Mater. Sci.* **25** 69
[6] Kocks U F and Mecking H 2003 *Prog. Mater. Sci.* **48** 171
[7] Pantleon W 2004 *Mater. Sci. Eng. A* **387-389** 257
[8] Holt D L 1970 *J. Appl. Phys.* **41** 3197
[9] Pantleon W 2005 *Mater. Sci. Eng. A* **400-401** 118
[10] Peierls R 1940 *Proc. Phys. Soc.* **52** 34
[11] Nabarro F R N 1947 *Proc. Phys. Soc.* **59** 256
[12] Schimpf C, Mrotzek T and Martin U 2009 *Proc. 17th Int. Plansee Sem. (Reutte)* vol 2
[13] Wesemann I, Hoffmann A, Mrotzek T and Martin U 2009 *Proc.17th Int. Plansee Sem.(Reutte)*
[14] Hansen N 2005 *Adv. Eng. Mater.* **7** 815
[15] Kuhlmann-Wilsdorf D and Comins N R 1983 *Mater. Sci. Eng.* **60** 7

Strains and dislocation densities within a rafted superalloy: *in situ* investigation during mechanical testing at 1080°C

L Dirand[1], A Jacques[1], J P Chateau[1], T Schenk[1], O.Ferry[1] and P Bastie[2]

[1] Institut Jean Lamour (SI2M) (UMR CNRS-Nancy Université N° 7198) Ecole des Mines, parc de Saurupt 54042 Nancy cedex, France.
[2] Laboratoire de Spectrométrie Physique, (UMR CNRS-Université Joseph-Fourier N° 5588), 38402 Saint-Martin-d'Hères, France

laura.dirand@mines.inpl-nancy.fr

Abstract. During *in situ* investigation of the distribution of lattice parameters within a superalloy with a rafted microstructure by Three Crystal Diffractometry, both the γ and γ' peak positions and shapes are shown to change. While the peak positions can be used to measure the average stresses and strains within both phases, the changes in shape are related to these in the dislocation distribution within the material. It is shown that the density and order of the dislocation array at the γ/γ' interface and within the γ' phase give different contributions to the peak width, and that the widening of both peaks under high stresses is due to an increase of the dislocation density within the rafts.

1. Introduction

During the major part of their high temperature creep life (i.e. stage II of the creep curve), single crystal superalloys have a so-called rafted microstructure [1]: semi-coherent platelets, or rafts, of the $L1_2$ γ' phase perpendicular to the [001] tensile axis, are embedded in a disordered fcc γ matrix. Such a microstructure can be seen as the simplest two-phase material, i.e. a stack of alternating phases where dislocations move in both phases by glide or climb, and are trapped or transmitted at the γ/γ' interfaces. It might thus be possible to build physically-based constitutive laws.

In order to build such laws with some confidence, we need to know several parameters, such as the stresses within both phases, the strain rate of each phase, and if possible to have some insight about dislocation densities and their variations. Such data are however difficult to obtain from usual post mortem TEM investigations, as the volume fraction of each phase and the lattice parameters change, and the dislocation microstructure might relax during cooling,

It has been shown [2-5] that *in situ* diffraction methods using neutrons or high energy synchrotron radiation, especially Three Crystal Diffractometry (TCD) could give some insight on stresses and strain rates within the material. The basic result from TCD (analyzer scans) using a (200) reflection is the distribution of the lattice parameters within the material, along the [100] direction, i.e. perpendicular to the [001] tensile axis. In the case of a single crystal superalloy with a raft microstructure, it is a two-peak profile, corresponding to the γ' and γ phases. From the peak positions, the stresses and strains within both phases can be calculated [6].

Recent papers [7,8] suggested that the high temperature plastic strain of the γ' phase involved correlated climb of dislocations. The aim of the present paper is to investigate the effect of the experimental conditions on the shape of the TCD profiles, to link changes in these profiles and changes in the order of the interfacial dislocation array and in the dislocation densities within that phase, and the mechanical behavior of the rafts.

2. Experimental

A cylindrical AM1 single crystalline specimen ([001] tensile axis) with a 0.45 μm precipitate size was grown by SNECMA, and machined by ONERA. It was pre-strained in tension (1080°C, 120 MPa), until the onset of stage II of creep, in a high temperature tensile device adapted to in situ X-Ray tests, in order to obtain a raft microstructure, and stable lattice parameters. The specimen was loaded by successive steps up to 275 MPa, then unloaded to 120 MPa for 320 min, reloaded to 275, unloaded for 30 min, and reloaded for 15 min before cooling down under stress. The experiment lasted less than 24 hours, in order to limit the variations of raft and corridor thicknesses.

The distribution of the lattice parameters within the bulk of the specimens was measured *in situ* at ~ 300 s intervals by analyzing the diffraction profile of a (200) type reflection (i.e. perpendicular to the tensile axis), using scans of the analyzer crystal of the Triple Axis Diffractometer of the High Energy (BW5) beamline of Hasylab at 120 keV. The data were analyzed following the procedure described in [6], as the sum of three peaks: γ' peak (rafts), γ peak (corridors), and a low intensity γ peak (distorted zones at raft ends and dislocation cores). The main parameters of the fit are their positions, heights, and slopes (in logarithmic scale). The strain rate of the whole specimen was continuously measured during the test.

3. Results

3.1. Strain rate of the γ' phase

The Von Mises stresses and plastic strains in both phases were calculated using the procedure given in [6], taking into account the applied load and the elastic strain in the [001] direction, i.e. shifts of the TCD peaks position.

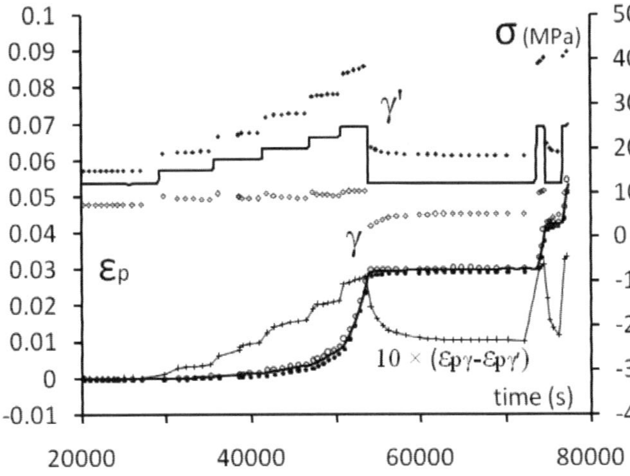

Figure 1. Von Mises stresses (top) and plastic strains (bottom) in the whole specimen (lines), in the γ ' (full symbols) and γ (empty symbols) phases. "Plus" symbols indicate the difference between the plastic strain of both phases, deduced from variations of $\delta_\perp$.

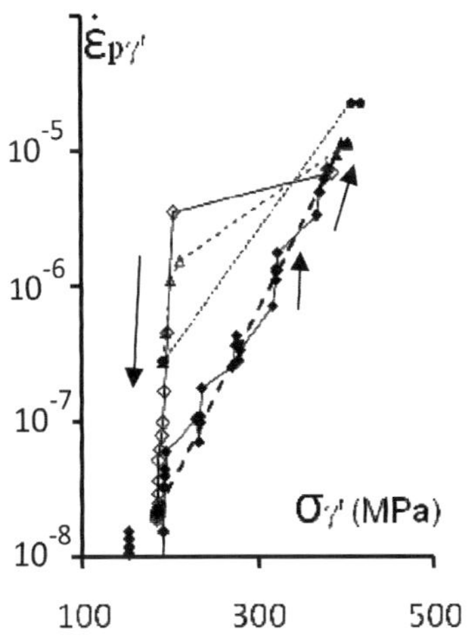

Figure 2. Stress dependence of $\dot{\varepsilon}_{p\gamma'}$. Diamonds, triangles, and dots (resp. full and empty) show the behavior during successive loadings (resp. unloadings) .

All internal stresses can be related to:

$$\delta_\perp = 2\left(a_{\gamma'100} - a_{\gamma 100}\right)/\left(a_{\gamma 100} + a_{\gamma' 100}\right) \tag{1}$$

where $a_{\gamma 100}$ and $a_{\gamma' 100}$ are the lattice parameters measured along [100]. The data are summarized in Figure 1 as a function of time, and the plastic strain rate of the γ' phase $\dot{\varepsilon}_{p\gamma'}$ is plotted vs. the Von Mises stress in Figure 2. Due to the high relative precision of the TCD measurements (a few 10^{-5}), the major sources of error are the specimen temperature and its elongation. While the uncertainty on the difference between the Von Mises stresses in the γ and γ' phases is ~ 10 MPa, the precision in their variations is within the MPa range. The same is true for the plastic strain: the precision in: $\varepsilon_{p\gamma}$ -$\varepsilon_{p\gamma'}$ is better than that in the plastic strain of the whole specimen. It should be noted (Fig. 2) that $\dot{\varepsilon}_{p\gamma'}$ increases under constant stress (arrows) and, after unloading, remains high before slowly returning to its initial value: $\dot{\varepsilon}_{p\gamma'}$ depends on the stress, but also on a parameter which varies on longer timescales.

3.2. Evolution of the TCD peak profiles

Figure 3 shows the definition of the width parameters w_γ and $w_{\gamma'}$ (inverse of the slope of the peaks on a Logarithmic scale) of the experimental peaks and Figure 4 their variation with time. Both parameters tend to increase with the applied stress, and decrease after unloading.

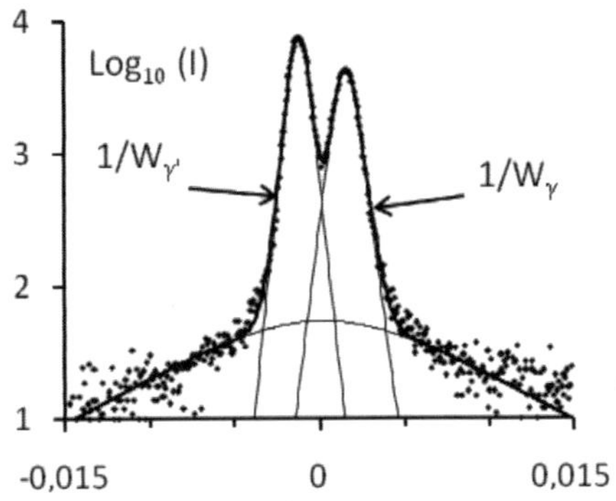

Figure 3. Definition of $w_{\gamma'}$ and w_γ.

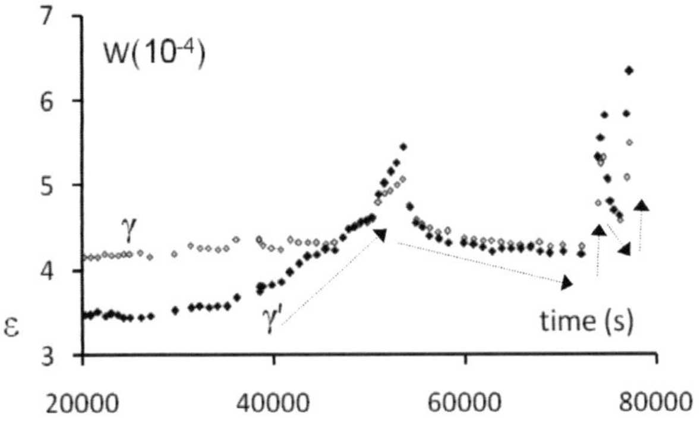

Figure 4. Evolution of $w_{\gamma'}$ and $w_{\gamma'}$ during the test.

4. Simulation and discussion

The short timescale of the variations of both the strain rate during unloading and the w parameters leaves changes in the dislocation microstructure (i.e. the dislocations trapped at the γ/γ' interface, and possibly dislocations moving within the rafts), as their only possible explanation.

A periodic square array of edge interface dislocations having all their Burgers vector $b_\perp$ within the interface plane, with a distance: $d = b_\perp / \delta_\perp$, and their additional lattice plane on the γ' side, will result in a two-peaks distribution of the lattice parameters within the material, as in Fig. 3, but with a very small width. However, some disorder (position fluctuations or presence of dislocations with a different Burgers vector) may result in long range fluctuations of the stress field, i.e. in wider peaks.

A first attempt to simulate the distribution of lattice parameters within the material was done, taking two arrays of dislocations with $[110]$ and $[1\bar{1}0]$ line directions for each interface, $b_\perp = 2^{-1/2}$.a, and a = 3.45 10^{-10} m, with a distance d chosen so that $\delta_\perp$ = 2.7 10^{-3}. Five interfaces perpendicular to the [001] axis were considered, at positions z = - 500 nm (+), z = - 200 nm (-), z = 0 nm (+), z = 300 nm (-), z = 500 nm (+), where + (-) indicates a γ (γ') to γ' (γ) transition. The volume fraction of γ' phase is 0.6. The strain due to both dislocation arrays was calculated at 200^3 points within a 2000^2 nm^2× 250 nm (from z = -100 nm to z = 150 nm). The resulting strain distribution is given by plot A (Fig. 5). Using the same fit procedure as for experimental peaks, width parameters w_γ = 1.58 10^{-4} and $w_{\gamma'}$ = 1.16 10^{-4} are found (point A in Fig.6). Due to the assumed periodicity of the γ/ γ' stacking, these values are lower than that found for an experimental profile recorded at the beginning of the experiment.

Part of the initial dislocations were then replaced by a pair of dislocations belonging to one of the eight slip systems with a non-zero Schmid factor, with random Burgers vectors orientation, except the edge component in the interface plane, and put at a random position within the same interface. These dislocations can be seen as dislocation segments left at an interface by a mobile dislocation gliding in the γ corridor. All the (w_γ, $w_{\gamma'}$) points (empty squares) are found on a straight line with a slope 0.68, with profile and point B standing for the replacement of 90% of the initial dislocations.

Adding dislocations with the same random Burgers vectors and line directions at random points within the γ' layers moves the position in the diagram (Fig. 6) along a line with a slope ≈ 1.67 (dots and triangles). Going from 0 to 4 10^{13} dislocations m^{-2} and changes profile B into C in Fig. 5, and point B to point C in Fig.6. This value is to be compared to the 1.61 slope measured for the experimental points (full diamonds). The changes in the peaks during the experiments might thus be linked to the introduction or annihilation of a few 10^{13} dislocations m^{-2} within the rafts.

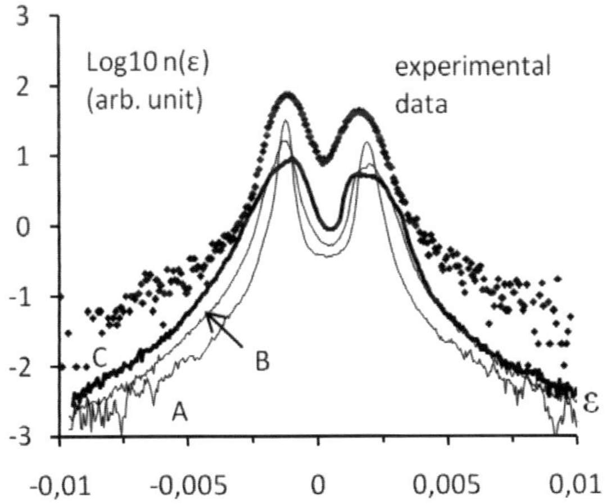

Figure 5. Distribution of the strain within the material, due to disorder at the interface (A to B) and dislocations within the rafts (B to C).

THE GRAPHIC ON THIS PAGE WAS NOT AVAILABLE ON THE E-MEDIA VERSION OF THIS PAPER.

Figure 6. Relative evolution of the w parameters with increasing interface disorder and/or increasing dislocation density within the rafts.

5. Conclusion

The strain rate of the γ' phase of a superalloy was measured during high temperature loading and unloading experiments. A correlation can be found between the long time response of the rafts and changes in the width parameters of the TCD profiles. These changes are themselves probably correlated with the introduction of a dislocation density of a few 10^{13} m^{-2} within the rafts.

Acknowledgements

The authors would like to thank M. Von Zimmermann (BW5 beamline) and the DESY staff for their help in the experiments, P. Caron (ONERA) and Y. Guedoux (Snecma) who provided the specimens.

References

[1] Veron M, Brechet Y and Louchet F 1996 *Acta Mater.* **44** 3633.
[2] Kuhn H A, Biermann H, Ungar T, Mughrabi H, 1991 *Acta Metall.* **39** 2783
[3] Royer A, Bastie P, Bellet D, Strudel J L, 1995 *Phil. Mag .A* **72** 669
[4] Biermann H, Strehler M, and Mughrabi H 1996 *Metall. Mater. Trans. A* **27**, 1003
[5] Jacques A and Bastie 2003 *Phil. Mag. A* **83**, 26, 3005
[6] Jacques A, Diologent F, Caron P and Bastie P 2008 *Mater. Sci. Eng. A* **483-484** 568
[7] Sarosi P M, Srinivasan, R, Eggeler G F, Nathal M V and Mills M J 2007 *Acta Mater.* **55** 2509
[8] Mompiou F, and Caillard D 2008 *Mater. Sci. Eng. A* **483-484** 143

Short-range ordering and mechanical properties of a Ni-20%Cr alloy

N R Dudova, R O Kaibyshev
Belgorod State University, Pobeda 85, Belgorod, 308015, Russia

dudova@bsu.edu.ru

Abstract. The mechanical behavior of a coarse-grained (100 μm) nickel-base alloy nichrome (Ni-20%Cr) was studied in compression at temperatures ranging from 150 to 1000°C. It was shown that in the temperature interval of 300-600°C this alloy demonstrates the following features of mechanical behavior: i) positive temperature dependence of yield stress; ii) jerky flow associated with the Portevin–Le Chatelier (PLC) effect; 3) very high value (115 MPa) of "threshold" stress at 650°C. These features of mechanical behavior can be related to short-range ordering (SRO). It was shown by differential scanning calorimetry that SRO takes place in this temperature range, causing PLC effect and positive temperature dependence of yield stress. In addition, SRO has persistency effect on yield stress and creep resistance.

1. Introduction

The short-range ordering (SRO) occurs in Ni-Cr-based alloys due to attractive forces operating between unlike atoms and results in abnormal temperature dependencies of physical and mechanical properties [1-3]. It is well known that SRO leads to a significant increase in electrical resistivity of Ni-Cr based alloys upon heating in the temperature range of 400-550°C, also known as the K state [3]. At the same time, poor attention was given to examination of effect of SRO on mechanical properties [1,2], in spite of its great practical importance for high temperature strength and creep resistance. The purpose of the present work is to report detailed information on deformation behaviour of dilute Ni-20%Cr alloy. Specific attention was paid to establish relationship between SRO and mechanical properties of the Ni-20%Cr alloy at high temperatures.

2. Experimental

The Ni-20%Cr alloy had the following chemical composition in weight pct.: Ni (base), 21%Cr, 1.1%Si, 0.6% Mn, 0.75% Fe, 0.31% Al, 0.08% Ti, 0.35% Cu, 0.05% C. A uniform structure with an average grain size of 100 μm was obtained by annealing of hot-rolled rod with 40 mm diameter for 2 h at a temperature of 1025°C with subsequent cooling in air. Mechanical compression tests were carried out using cylindrical samples with dimensions of $\varnothing 10\times 12$ mm^2 in the temperature range of 150-1000°C with initial strain rates ranging from 1.5×10^{-5} to 5×10^{-2} s^{-1}. Other details of mechanical and microstructural characterizations were described in previous work [4]. Values of "threshold" stresses were taken from previous study [5]. Differential scanning calorimetry was performed using an SDT Q600 (TA Instruments) device. Pure gold was used as inert reference material, the sample weight was less than 0.1 g and the experiments were carried out under an argon atmosphere.

3. Results and discussion

The calorimetric studies were undertaken in order to determine the temperature interval of SRO which depends on the exact chemical alloy composition [3]. Figure 1 shows differential scanning calorimetry (DSC) curves of the samples. Two exothermic peaks at about 200 and 550°C are distinguished in figure 1 (a) indicating the SRO in the quenched alloy. Four exothermic peaks are revealed in figure 1 (b) by the heating of the alloy subjected to severe plastic deformation. Two peaks are located at nearly the same temperatures as for quenched material. Moreover, at 370 and 625°C two additional exothermic peaks were found. Thus, DSC analysis shows that some SRO of the deformed alloy occurs at about 200 and 370°C and the more pronounced ordering transformation occurs in the temperature interval 450-610°C that complies with [3].

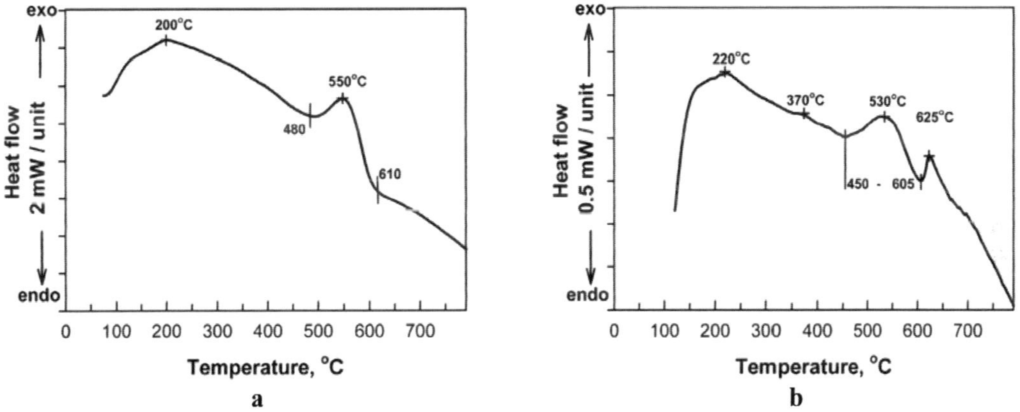

Figure 1. DSC thermogram of Ni-20%Cr alloy heated at 20°C/min (a) in the water quenched condition after solution treatment at 1100°C, (b) and after severe cold deformation by a Bridgman anvil to a strain of ~6.

Figure 2 shows true stress (σ) – true strain (ε) curves and values of the coefficient of strain rate sensitivity $d\log\sigma/d\log\dot{\varepsilon}$ in dependence of temperature. Three important features can be emphasized.

First, jerky plastic flow occurs at temperatures ranging from 300 to 600°C (figure 2 (a)). Such discontinuous flow is usually attributed to the Portevin–Le Chatelier (PLC) effect [6]. The PLC effect is known to be caused by the dynamic strain ageing, leading to periodic blocking and unblocking of dislocations by atmospheres of impurity atoms. The obtained negative strain-rate sensitivity of the flow stress (figure 2 (b)) is typical for the PLC effect [7]. It should be noted that the serrated σ-ε curves demonstrate extensive work hardening up to high strains while at T≥650°C the plastic flow attained the steady state.

In addition, non-typical behaviour for the PLC effect was revealed. The critical strain ε_c for the onset of jerky flow normally increases with increasing the strain rate and/or decreasing the temperature of deformation [7]. However, both the "normal" strain rate dependence (figure 3 (a)) and the "inverse" temperature dependence (figure 3 (b)) of critical strains were obtained for the present material. Such combination is a feature of materials strengthened by coherent precipitates [6,7]. Plastic flow becomes unstable in these alloys due to the softening induced by the shearing of the precipitates [6]. Surface observations of planar slip in the present alloy [4,5] support this conclusion. Furthermore, softening by planar slip was found [8] in SRO alloys as a result of cutting and disordering the SRO zones. In Ni-20 at.% Cr single crystal [9] SRO was reported to be correlated with planar slip which can initiate strain bursts in fatigue experiments. Thus, the SRO can result in jerky flow in the Ni-20%Cr alloy.

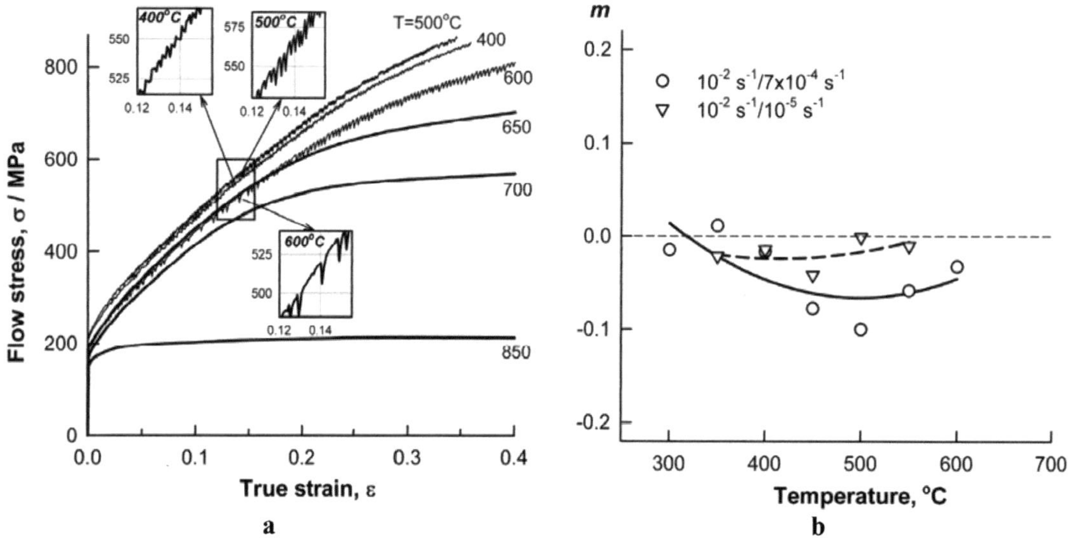

Figure 2. (a) True stress – true strain curves at temperatures of 400-850°C and strain rate of 7×10^{-4} s^{-1} and (b) temperature dependence of the coefficient, *m*, of strain-rate sensitivity of the flow stress.

The height of serrations and their extent in the σ-ε curves increase with increasing the deformation temperature. This is probably caused by an increase in the size of the areas in which SRO took place. Accordingly, the applied stress being necessary to shear these regions by mobile dislocations increases which leads to an increase in the amplitude of serrations from 10 MPa at 400°C to 25 MPa at 600°C. For destroying SRO areas with increased dimension it is required that a larger number of dislocations pass through them; this is reflected in an increase in strain between the stress peaks.

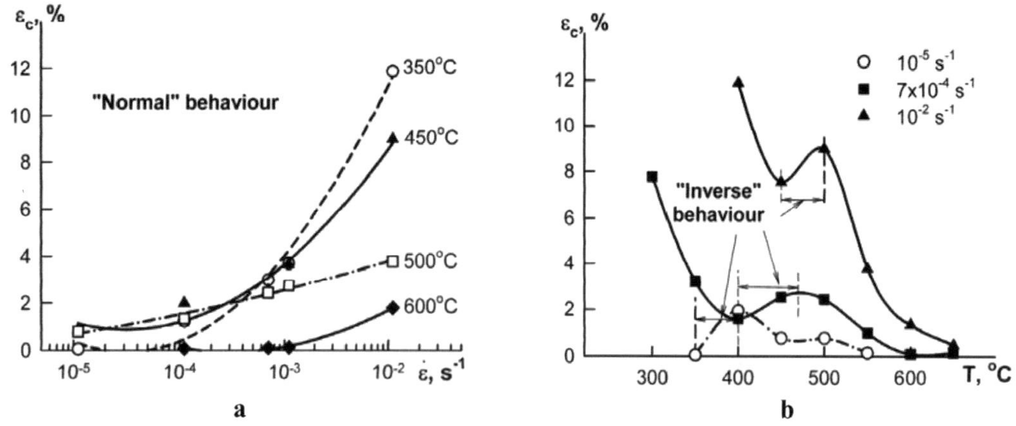

Figure 3. Variation of the critical strains with strain rate (a) and temperature (b) of deformation.

Second, the alloy exhibits a positive temperature dependence of the yield stress $\sigma_{0.2}$ in the temperature range **II** of 300-500°C (figure 4 (a)). Four regions can be distinguished. The temperature intervals **I** (150-300°C) and **III** (600-800°C) are characterized by a similar slope of the straight line (0.3 and 0.5, correspondingly) reflecting the decrease in stress with increasing temperature. In range **II** the temperature dependence is affected by large scatter of the experimental points. However, the slope of the average straight line takes on a negative value (about -0.15). It should be noted that such a (positive) temperature dependence of the flow stress is rarely observed in metals [10]. The temperature

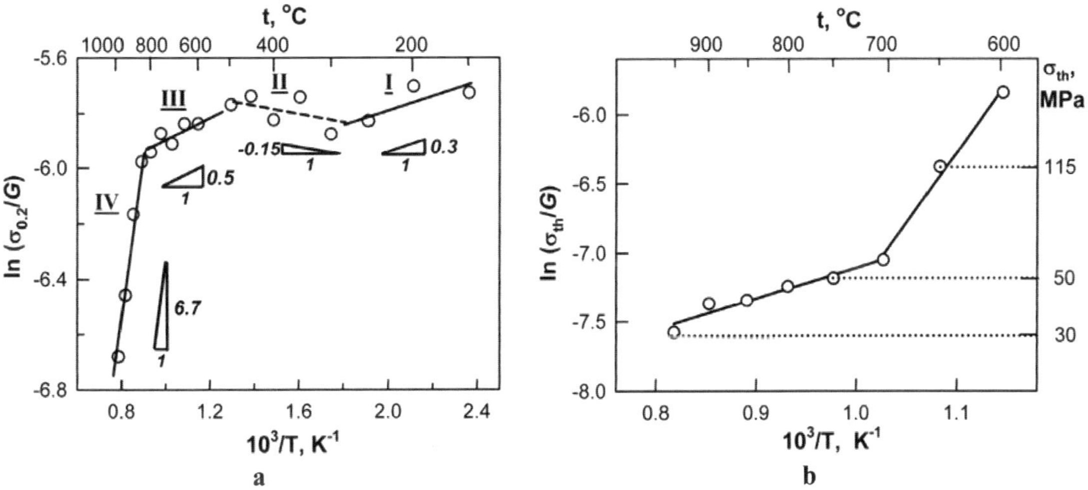

Figure 4. (a) Yield stress $\sigma_{0.2}$ and (b) "threshold" stress σ_{th} normalized by the shear modulus G(T) as functions of the reciprocal deformation temperature.

range **IV** (800-1000°C) is characterized by a significant decrease in the normalized yield stress. Therefore, there is evidence for the fact that SRO does not only increase the yield stress at T≤ 500°C but also results in high strength in the interval of 600-800°C where the alloy is in disordered state. Perhaps, this is caused by the stress necessary to drag solute atoms along with the dislocations. As it can be seen from figure 4 (a) the straight line **III** is located considerably above the straight line **I** extrapolated into the temperature range of 600-800°C.

Third, the SRO can result in high values of the "threshold" stresses σ_{th} below which the creep rate is extremely low [5]. The Ni-20%Cr alloy was shown to exhibit the "threshold" behaviour like a dispersion-strengthened material [5]. The absolute values of σ_{th} at T≤650°C higher than 115 MPa ensure excellent creep resistance of the Ni-20%Cr alloy.

4. Summary

Strengthening of the Ni-20%Cr alloy by short-range ordering is reflected in appearance of the Portevin–Le Chatelier effect in the temperature range of 300-600°C, a positive temperature dependence of the normalized yield stress $\sigma_{0.2}/G$ at 300-500°C and high values of "threshold" stress (>115 MPa) at T≤650°C.

Acknowledgments

The authors are grateful to Dr. I. Tarasova, Joint Research Center, Belgorod State University, for her assistance with thermal analysis.

References

[1] Marucco A 1995 *Mater. Sci .Eng. A.* **194** 225
[2] Metcalfe E, Nath B and Wickens A 1984 *Mater. Sci .Eng. A* **67** 157
[3] Lang E, Lupinc V and Marucco A 1989 *Mater. Sci .Eng. A* **114** 147
[4] Dudova N, Kaibyshev R and Valitov V 2008 *Phys. Met. Metallogr.* **105** 98
[5] Dudova N, Kaibyshev R and Valitov V 2009 *Phys. Met. Metallogr.* **107** 409
[6] Brechet Y and Estrin Y 1994 *Key Eng. Mater.* **97-98** 235
[7] Brechet Y and Estrin Y 1996 *Scripta Mater.* **35** 217
[8] Gerold V and Karnthaler H P 1989 *Acta Metall.* **37** 2177
[9] Wolf K, Gudladt H-J, Calderon H A and Kostorz G 1994 *Acta. Metall. Mater.* **42** 3759
[10] Couret A and Caillard D 1989 *Phil. Mag. A* **59** 801

Journal of Physics: Conference Series **240** (2010) 012082 doi:10.1088/1742-6596/240/1/012082

Work hardening and kinetics of dynamic recrystallization in hot deformed austenite

X Quelennec, E Martin, L Jiang, J J Jonas

Materials Engineering, McGill University, 3610 University Street, Montreal, Canada H3A 2B2.

john.jonas@mcgill.ca

Abstract. The work hardening behaviors of 4 steels deformed under dynamic recrystallization (DRX) conditions were analysed. The *h* and *r* parameters necessary to describe the work hardening or dynamic recovery (DRV) curves in the *absence* of DRX were determined from the experimental flow curves. The fractional softening X attributable to DRX, defined as the difference between the calculated DRV and experimental DRX curves, was used to derive the Avrami kinetics of DRX. Some conclusions are drawn about the effects of deformation temperature, strain rate and steel composition on the kinetics of DRX.

1. Introduction

The fractional softening X produced by dynamic recrystallization is of considerable importance for modeling the rolling load in steel mills. Most studies concerned with this topic have represented the DRX kinetics in terms of the recrystallized volume fraction [1] or the peak stress σ_p of the stress-strain curve [2]. In the present work, the net softening due to DRX is defined as the difference between the work hardening curve pertaining to the as-yet unrecrystallized grains and the experimental flow curve. A method for calculating the former is described and the differences are employed to derive the Avrami kinetics of DRX according to a method described earlier [3]. The DRX kinetics under industrial conditions can then be predicted and used to model the flow stress in rolling mills.

2. Experimental data

The stress-strain curves pertaining to 4 different steels deformed in compression under the conditions shown in Table 1 [4-6] were analyzed. The curves were sampled using a higher density of points before the peak stress. The 17 stress-strain curves are depicted in Figures 1 and 2. The yield stresses were defined in terms of a 2% offset and then the curves were fitted using the MATLAB™ software to an 8th order polynomial. The latter was required for the numerical calculations that followed.

Table 1. The compositions of the 4 steels and the associated deformation conditions of T and $\dot{\varepsilon}$.

Steel	Composition (wt%)	Ref.	$\dot{\varepsilon}$ (s^{-1})	T (°C)
A : SiMn low carbon steel	0.18C; 0.71Si; 1.11Mn; 0.018S; 0.007P	[4]	0.5	1050, 1000, 950, 900
			0.005	900, 850, 800
B : Nb low carbon steel	0.11C; 0.17Si; 1.23Mn; 0.038Nb; 0.006P; 0.005S; 0.004N	[5]	0.1	1050, 1000
C : Ti low carbon steel	0.12C; 0.18Si; 1.23Mn; 0.015Ti; 0.006P; 0.005S; 0.004N	[5]	0.1	1000, 950, 900
D : Mo-Cr high carbon steel	0.99C; 0.24Si; 0.31Mn; 1.44Cr; 0.12Cu; 0.02Mo; 0.05Ni; 0.01P; 0.003S	[6]	0.1	1150, 1100, 1050, 1000, 950

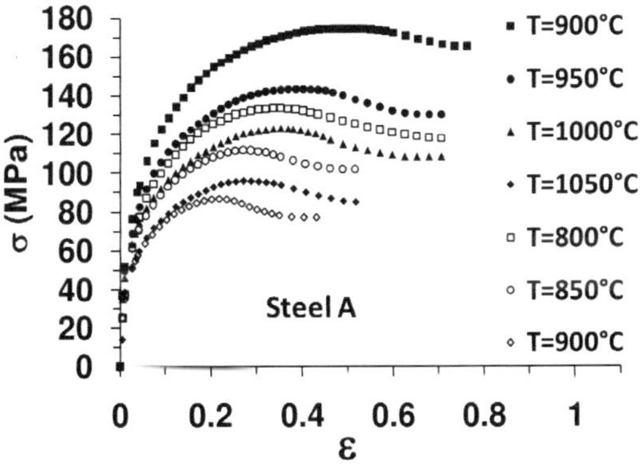

Figure 1. Seven flow curves determined on steel A. The closed and open symbols represent strain rates of 0.5 s^{-1} and 0.005 s^{-1}, respectively.

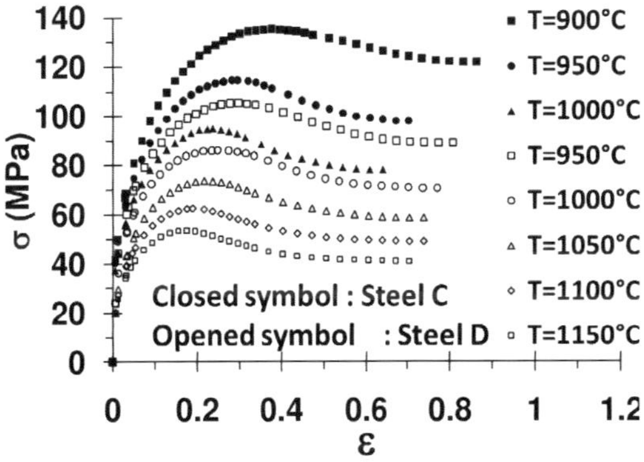

Figure 2. Eight flow curves determined on steels C (closed symbols) and D (open symbols). The steel B flow curves are not displayed here.

3. Analysis procedure

The work hardening behavior of the unrecrystallized regions was considered to be described by:

$$d\rho / d\varepsilon = h - r \times \rho \tag{1}$$

In this relationship [7], ρ is the current value of the dislocation density, ε the strain, h the athermal work hardening rate, and r the rate of dynamic recovery. Both are strain independent [3]. As shown earlier [3], the integration of equation (1) leads to the following description of the flow curve:

$$\sigma = \left(\sigma_{sat}^{\ 2} - \left(\sigma_{sat}^{\ 2} - \sigma_0^{\ 2}\right) \times \exp\left(-r \times \varepsilon\right)\right)^{\frac{1}{2}} \tag{2}$$

Here, σ_0 is the yield stress and σ_{sat} represents the saturation stress, which is attained at high strains. The latter depends on h and r according to $\sigma_{sat} = \mathrm{M} \times \alpha \times \mu \times b \times \sqrt{(h/r)}$, where M is the Taylor factor, α a material constant, μ the shear modulus and b the Burgers vector. On the basis that the flow stress σ is given by:

$$\sigma = M \times \alpha \times \mu \times b \times \sqrt{\rho} \tag{3}$$

Equation (1) can be expressed in terms of the flow stress, for which purpose we specify that:

$$d\rho / d\varepsilon = (M \times \alpha \times \mu \times b)^{-2} \times d\sigma^2 / d\varepsilon = 2 \times (M \times \alpha \times \mu \times b)^{-2} \times \sigma \times d\sigma / d\varepsilon \tag{4}$$

By employing equations (3) and (4), equation (1) can be rewritten as follows:

$$2 \times \sigma \times \theta = r \times \sigma_{sat}^2 - r \times \sigma^2 \tag{5}$$

where θ = dσ/dε is the current value of the work hardening rate. The coefficient r is then obtained from the slopes ($-r$) of the $2 \times \theta \times \sigma$ vs. σ^2 plots of the experimental data and h from the vertical intercepts ($r \times \sigma_{sat}^2 = (\mathrm{M} \times \alpha \times \mu \times b)^2 \times h$). This formalism indicates that the rate of storage energy (i.e. dρ/dε) depends on the current value of the stored energy ρ.

4. Work hardening parameters

Some selected $2 \times \theta \times \sigma$ vs. σ^2 plots are presented in Figure 3. The critical stresses σ_c for the initiation of DRX were determined by the double differentiation method [8]. It can be seen that the trend line for

Journal of Physics: Conference Series **240** (2010) 012082 doi:10.1088/1742-6596/240/1/012082

σ_0 passes through the origin, whereas those for σ_c and σ_p do not. This indicates that DRX cannot take place until a minimum driving force, that is, a minimum dislocation density, has been provided.

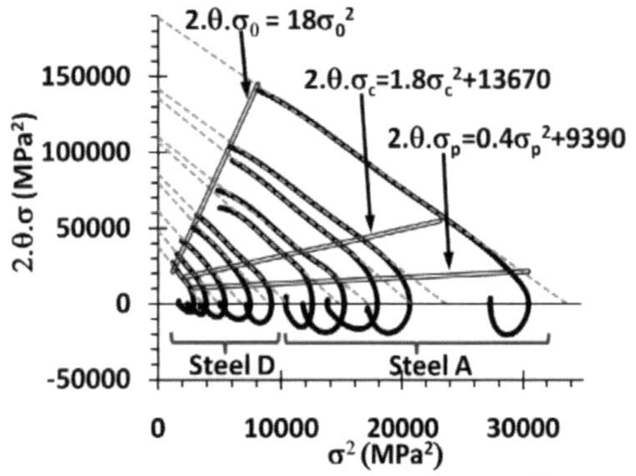

Figure 3. Some $2 \times \theta \times \sigma$ vs. σ^2 plots pertaining to steels A and D (black lines). The linear behaviors of the work hardening regime have been extrapolated to provide both horizontal and vertical intercepts.

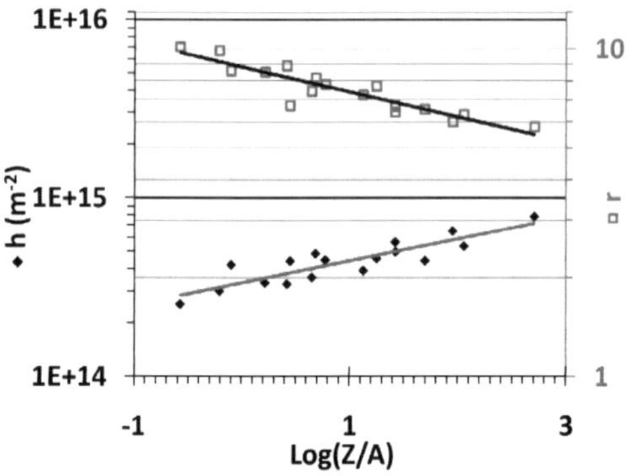

Figure 4. Dependences of r (open symbols) and h (closed symbols) on Z/A as calculated from the 17 experimental $2 \times \theta \times \sigma$ vs. σ^2 curves.

The dependences of r and h on Z are illustrated in Figure 4. Here, allowances have been made for the temperature dependences of b [9] and μ [10] and M and α have been set equal to 3 and 0.5, respectively. The other material constants employed are listed in Table 2. The latter were used to calculate the Z/A scale employed in the preparation of Figure 4.

Table 2. Values of α', n, A and Q as provided in Refs. 4-6 for the 4 steels. The starred values of n and α' were not specified and were estimated here.

Steel	n	α' (MPa^{-1})	A	Q ($kJ.mol^{-1}$)	Ref.
A	4.5*	0.012*	?	?	[4]
B	5.31	0.0097	8.8×10^{14}	423.4	[5]
C	5.31	0.012	3.1×10^{14}	361.1	[5]
D	4.73	0.013*	?	351.0	[6]

The r and h trend line equations are given by $r = 8.225 \times \exp(-0.176\ Z)$ and $h = 3.10^{14} \times \exp(0.283\ Z)$. As $h = M/(L \times b)$ is inversely proportional to the dislocation mean free path L [11], the present values of h (15 to 35 μm) can be interpreted as indicating that, at the initiation of flow, L is approximately equal to the grain size (25 μm) specified in reference [4]. Thus the rate of hardening appears to be determined by rate of dislocation accumulation at grain boundaries.

5. DRX kinetic

Three stress-strain curves reconstituted using equation (2) are depicted in Figure 5 together with their respective experimental curves. The Avrami kinetics of DRX were then derived, where the net softening X was defined as the difference between the work hardening and experimental curves. Here the Avrami kinetics are represented by the relation: $\log[\ln(1/(1\text{-}X))] = \log(k) + n \times \log(t)$, where k is the Avrami constant, n the time exponent, and t is the time elapsed since the initiation of DRX. Seven selected Avrami plots are represented in Figure 6, where the kinetics are seen to be approximately linear in the range X = 0.1 to X = 0.9.

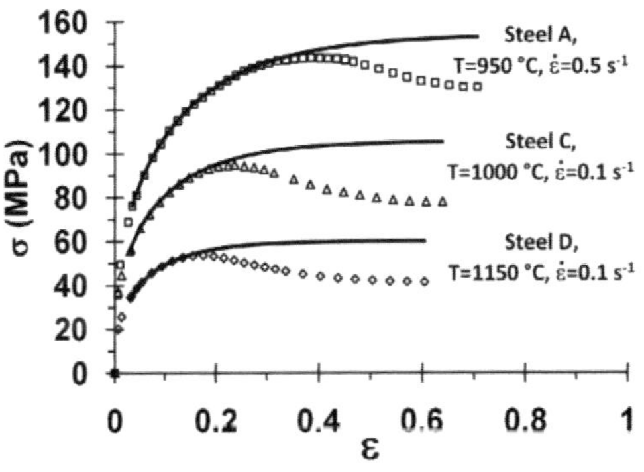

Figure 5. Three experimental flow curves (open symbols) compared to their respective calculated work hardening curves (black lines).

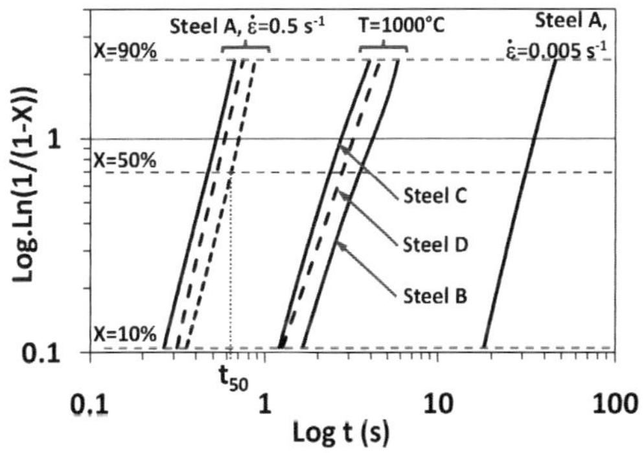

Figure 6. Avrami plots calculated from 7 sets of flow curves showing the effects of temperature, strain rate and chemical composition.

For the present 17 Avrami plots (10 are not displayed in Figure 6), n is about 3 (it ranged from 2.2 to 3.5). It is of interest that an increase of 100°C in the deformation temperature (steel A) only led to an acceleration in the kinetics by a factor of about 1.37. A decrease of 2 orders of magnitude in strain rate, on the other hand, slowed the kinetics down by about the same factor, suggesting that the rate of dynamic softening is approximately proportional to strain rate. From the softening behaviors of steels B, C and D under the same deformation conditions (T = 1000°C, strain rate = $0.1s^{-1}$), it can be seen that Nb (steel B) is the most effective retarder of DRX, followed by the combination of Mo and Cr (steel D), and then by Ti (steel C).

6. Conclusions

The work hardening behavior of the unrecrystallized regions of hot deformed austenite can be determined from $2 \times \theta \times \sigma$ vs. σ^2 plots derived from flow curves depicting the dynamic recovery behavior prior to ε_c, the critical strain for the initiation of DRX. Here $\theta = d\sigma/d\varepsilon$. The work hardening coefficient h increases with Z; values of the former can be interpreted to indicate that the mean free path of the dislocations L at the initiation of plastic flow depends on the grain size. Conversely, the dynamic recovery coefficient r decreases with Z. The net fractional softening attributable to DRX was calculated from the work hardening curves and the Avrami plots were derived in each case. It is evident that strain rate has a strong influence on the DRX kinetics, as the rate of softening is approximately proportional to the strain rate. Among the common alloying elements, Nb has the strongest influence in retarding the rate of DRX.

References

[1] Jafari M and Najafizadeh A 2009 *Mater. Sci. Eng.* A **501** 16
[2] Serajzadeh S and Karimi Taheri A 2002 *Mater. Lett.* **56** 984
[3] Jonas J J , Quelennec X, Jiang L and Martin E 2009 *Acta Mater.* **57** 2748
[4] Dong L, Zhong Y, Ma Q, Yuan C and Ma L 2008 *Tsinghua Sci. Technol.* **13** 609
[5] Ma L, Liu Z, Jiao S, Yuan X and Wu D 2008 *J. Iron Steel Res. Int.* **15** 31
[6] Yue C, Zhang L, Liao S, Pei J, Gao H, Jia Y and Lian X 2009 *Mater. Sci. Eng.* A **499** 177
[7] Estrin Y and Mecking H 1984 *Acta Metall.* **32** 57
[8] Poliak E I and Jonas J J J 1996 *Acta Mater.* **44** 127
[9] Seki I and Nagata K 2006 *ISIJ Int.* **45** 1789
[10] Ghosh G and Olson G B 2002 Acta Mater. **50** 2655
[11] Reichert B and Estrin Y 2007 *Steel Res.* **10-11** 791

A statistical study on tensile characteristics of stainless steel at elevated temperatures

Park SungHo[1], **Park NoSeok**[1], **Kim JaeHoon**[2]

[1] Defense Systems Test Center, Agency for Defense Development, Taean P.O. Box 1, Taean, 357-900, Korea Republic (South)
[2] Department of Mechanical Design Engineering, Choongnam National University, 220 Gung-Dong, Yuseong-Gu, Daejeon, 305-764, Korea Republic (South)

redfox89@dreamwiz.com

Abstract. This study describes the tensile test results of AISI type 304 deformed at room and elevated temperatures. It has been known that the normal distribution fits well with the strength of typical structural materials. The Weibull distribution has many characteristics in the reliability design. In this paper, we showed that the Weibull distribution can be used to describe the scattering of strength data as well as normal distribution by testing goodness of fit. It is found that the parameters of Weibull distribution have decreasing tendency for temperature rise and the coefficient of variation has increasing tendency for temperature rise.

1. Introduction

Stainless steel has become increasingly popular in structural and architectural applications due to its high level of resistance to corrosion, attractive appearance, structural and architectural qualities, as well as its ability to retain its strength at high temperatures [1]. Mechanical properties, such as yield strength, tensile strength, hardness, are not deterministic but statistical. Causes of scattering of the properties may be the incorrectness of test condition (specimen, equipment, etc.), the lack of test technique, chemical variations of test specimen, inhomogeneity of heat treatment conditions, the working method of specimen or the variation of residual stresses. These factors make mechanical properties more statistical. The quantitative distribution characteristics of mechanical properties are used for the reliability design of machine or structure and the basis data for quality control and design validation. Park et al. [2] showed that the normal distribution describes well the tensile and yield strength at various elevated temperatures from normal statistical paper. There are two statistical models of material's strength. One is the weakest link model; the other is the averaged strength model. Generally, the defect dominant strength fits well with the extreme value distribution, and the deformation dominant strength (tensile strength and hardness of ductile material, etc) fits well with the normal distribution. On the normal statistical paper, the tensile strength, yield strength and hardness are shown linear [3]. Recently, many researchers have used the Weibull distribution to acquire a design parameter as well as fatigue life prediction [4-5].

Even though the strength follows normal distribution very well, in this study it is investigated by the Weibull distribution at wide range of temperatures. The aim of this study is to investigate which statistical distribution describes better for strengths of AISI 304 stainless steel and to get various statistical distribution parameters at elevated temperatures and the variation of parameters with temperature.

2. Experimental method

An MTS 810 Universal testing machine of 100kN capacity was used for tensile tests. The MTS Model 653.02 high temperature furnace with a maximum temperature of 1400℃ was used for heating the specimens. The temperature of specimen was measured on the attached thermocouples, type-K, with a maximum measurable temperature of 1550℃ and attached on the specimen at 3 points (at the center of specimen and ±10mm from the center) by strain gage welder.

The material used in this study is AISI 304 (18Cr-8Ni) austenitic stainless steel which is widely used in heat exchanger, transportation vessel, food container, rotor, blade, etc. The test specimens were prepared in accordance with the ASTM Standard E8 and E21 [6-7]. An MTS Model 632.53F-11 of axial extensometer was used to measure the strain of the middle part of the specimen. The gauge length of the extensometer was 25mm.

The temperatures specified in the temperature controller were 21 (R.T.), 100, 200, 350, 500, 650 and 800℃. At each temperature, a steady state tensile test was conducted for 10 specimens. The specimen was heated up to a specified temperature and maintained for 30 minutes. Then the tensile load was applied to specimen. The strain control was used in the tensile testing machine. A tensile loading rate of 2mm/min was used, leading to an initial strain rate of 10^{-3} s^{-1}, and the heating rate of the furnace is 1.9~2.0℃/sec. The material properties, such as yield strength, ultimate tensile strength, modulus of elasticity, elongation, were obtained from the stress-strain curve.

3. Statistical analysis method

By a goodness of fit test, the best fitted model was chosen, which shows the minimum difference between the theoretical model (namely, normal distribution, Weibull distribution, etc.) and the cumulative distribution model. To choose the best fitted model to given strength data, we used three different kinds of goodness of fit methods in the commercial software Weibull++® 7: (1) modified Kolmogorov-Smirnov, (2) Correlation Coefficient γ and (3) Log-likelihood value. In this study, we considered the normal distribution and the (2-parameter) Weibull distribution function only for the competing distribution model for yield strength and tensile strength. Kang [8] used such goodness of fit test models to assess the statistical distribution of flexural strength of woven-fabric laminates with impact-induced damage. In his paper, the author summarized the use of three models briefly.

The coefficient of variation (C.V) is a normalized measure of dispersion of a probability distribution. It is defined as the ratio of the standard deviation (σ) to the mean (μ). For the Weibull distribution, MTTF (mean time to failures) or MTBF (mean time between failures) is used as mean. So, the coefficient of variation of the Weibull distribution is as Eq. 1. Here, Γ is a gamma function and η and β are the scale and shape parameters in the 2-parameter Weibull distribution, respectively.

$$C_v = \frac{\sigma}{\mu} = \eta\sqrt{\Gamma\left(1+\frac{2}{\beta}\right)-\left[\Gamma\left(1+\frac{1}{\beta}\right)\right]^2} \Big/ \eta\,\Gamma\left(1+\frac{1}{\beta}\right) \quad (1)$$

4. Test and statistical analysis results

The details of test results are summarized in Ref. [2]. As the temperature rises, the tensile strength, yield strength and modulus of elasticity are decreased. But the elongation shows a different tendency for temperature. An elongation decreases as the temperature rises at below 650℃, but it increases at above 650℃.

The results of goodness of fit test for each statistical distribution model are shown in Table 1 for tensile strength and in Table 2 for yield strength. The correlation coefficient and Log-likelihood (L-K) value at each temperature for normal distribution are shown in parenthesis for a comparing purpose. Both Modified Kolmogorov-Smirnov (K-S) test and Chi-Squared test return the probability that the respective critical value is less than the value calculated. High values, close to one, indicate that there is a significant difference between the theoretical distribution and this data set. Depending on the sample size used, the K-S test and Chi-Squared test will return different results. For K-S as the sample size increases, $D_{Critical}$ becomes smaller thus increasing the probability that $D_{Critical} < D$. On the other hand the Chi-Squared test is really not valid for smaller sample sizes, i.e. minimum of 35 [9]. In this study, the sample size at each temperature was 10. There might be some errors at modified K-S values and Chi-squared values, but the Weibull distribution seems to be more fitted to the data. The correlation coefficients and L-K values at various temperatures show that both normal and the Weibull distribution fits well with the data, but the Weibull distribution is slightly more fitted to the data than normal distribution.

Table 1 Results of goodness of fit test for tensile strength

Test Method	21°C	100°C	200°C	350°C	500°C	650°C	800°C
Modified K-S value(%)	0.024	0.002	0.216	4.971	4.197	40.456	7.042
Correlation Coefficient(%)	98.743 (97.593)	97.087 (97.555)	96.824 (96.078)	95.464 (90.189)	95.653 (97.891)	91.847 (85.502)	94.427 (87.852)
Chi-Squared value(%)	25.080	27.671	29.708	32.959	25.0119	41.932	33.981
L-K value	-32.128 (-31.951)	-19.649 (-18.689)	-23.949 (-23.438)	-33.115 (-34.153)	-31.110 (-27.749)	-37.228 (-43.419)	-29.701 (35.580)

Table 2 Results of goodness of fit test for yield strength

Test Method	21°C	100°C	200°C	350°C	500°C	650°C	800°C
Modified K-S value(%)	9.3911	90.1507	37.7349	86.7418	85.0706	60.8708	26.9015
Correlation Coefficient(%)	94.7979 (89.163)	85.5912 (79.650)	93.1286 (89.103)	81.6505 (73.998)	86.3306 (86.481)	88.1797 (81.308)	94.7643 (91.327)
Chi-Squared value(%)	36.4749	40.1061	20.9716	46.2782	27.7323	30.5425	34.8022
L-K value	-30.4877 (-31.706)	-21.4977 (-23.526)	-27.2641 (-27.539)	-19.9399 (-22.691)	-25.5514 (-23.484)	-28.1770 (-30.409)	-28.9889 (-28.646)

Fig. 3 is a plot of shape parameter and scale parameter of tensile strength at various temperatures. The shape parameter rises to the maximum at 100℃. Thereafter, it descends rapidly and remains almost constant above 650℃. The scale parameter of tensile strength descends to 200℃, remains almost constant to 300℃ and falls rapidly to 800℃. The shape parameter of yield strength as shown in Fig. 4 rises to the maximum value as in the case of tensile strength. It descends rapidly to 200℃ but rises to near the maximum value but rapidly drops to 650℃. Thereafter, the decreasing rate is low to the 800℃. The scale parameter of yield strength descends gradually as the temperature rises as shown in Fig. 4. The coefficient of variation was plotted against the temperature in Fig. 5 for tensile strength and yield strength. As the temperature rises, the coefficient of variation of tensile strength has tendency of rise. It is peculiar that the coefficient of variation at room temperature (21℃) is higher than that at 100℃. As the temperature rises, the strength decreases but its variation increases.

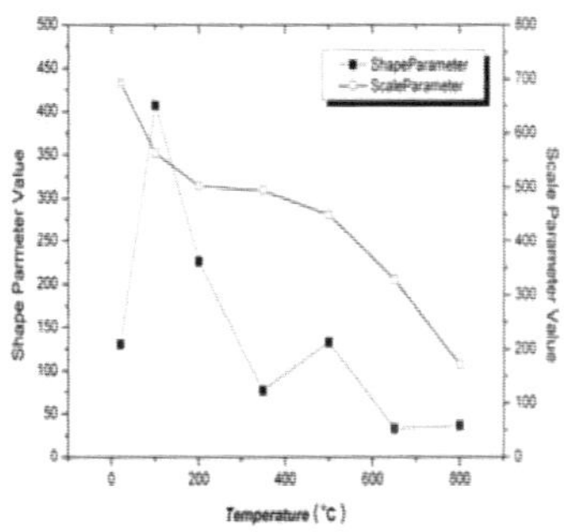

Figure 3. Weibull parameters of tensile strength at various temperatures

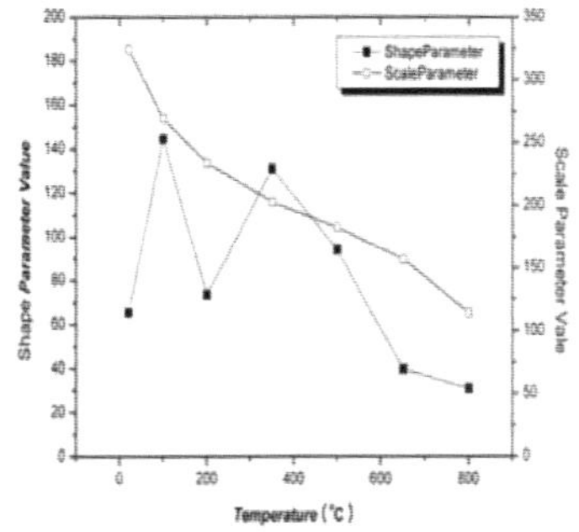

Figure 4. Weibull parameters of yield strength at various temperatures

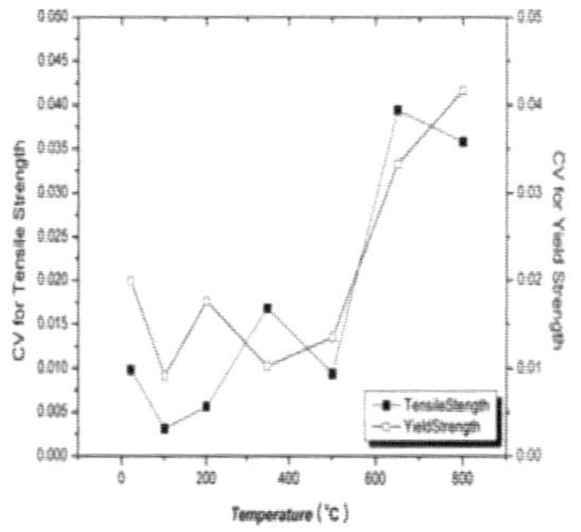

Figure 5. Coefficient of variations of tensile and yield strength at various temperatures

5. Conclusions

This paper aimed to statistically analyze the strength data of AISI 304 stainless steel at elevated temperatures. It is well known that the normal distribution describes well the distribution of tensile strength and yield strength. However, the Weibull distribution has many characteristics to life prediction so it might be used in various reliability design areas. In this study, we investigated the use of the Weibull distribution by a goodness of fit test. The following conclusions were obtained.

(1) On the basis of strength data at various temperatures and use of statistical criteria, the (2-parameter) Weibull distribution can be an appropriate statistical distribution as well as normal distribution.

(2) As the temperature increases, the shape parameter and the scale parameter of tensile strength and yield strength have a tendency of decrease. As the temperature increases, the coefficient of variation which is a normalized measure of dispersion has a tendency of increase.

6. References

[1] K. Abdella, Explicit full-range stress-strain relationships for stainless steel at high temperatures, *J. Construct. Steel Res.* **65**, 2009, pp. 794-800

[2] S. H. Park, N. S. Park, J. H. Kim, A study on tensile characteristics of AISI 304 stainless steel under room and elevated temperatures, *J.Korean Soc. Propulsion Engineers* **12**, 2008, pp. 35-42

[3] J. H. Song, J. H. Park, An introduction to reliability engineering in mechanical design, Intervision (ISBN 978-89-5667-393-6), 2007, pp. 53-55, 217-224

[4] Y. M. Choi, J. D. Lee, J. Ko, Calculation of design parameter of Si_3N_4 for engineering part through the measurement of tensile strength, *J. Korean Ceram. Soc.* **33**, 1996, pp. 709-717

[5] S. J. Kim, M. K. Kwak, S. W. Kwon, U. S. Kong, Probability distribution characteristics for elevated temperature mechanical properties of stainless steels, *J.Korean Soc. Ocean Engineers* **18**, 2004, pp. 64-69

[6] ASTM E8, Standard Test Methods for Tension Testing of Metallic Materials, ASTM International, 2004, pp. 1-24

[7] ASTM E21, Standard Test Methods for Elevated Temperature Tension Tests of Metallic Materials, ASTM International, 2005

[8] K. W. Kang, S. K. Koh, D. K. Kim, K. J. Kim, Assessment of the statistical distribution of flexural strength of woven-fabric laminates with impact-induced damage, *J. Comp. Struct.* **90**, 2009, pp. 60-66

[9] Weibull++® 7 User's manual, Reliasoft Inc., 2008

High temperature creep behaviour of Al-rich Ti-Al alloys

D Sturm[1], M Heilmaier[1], H Saage[1], J Aguilar[2], G J Schmitz[2], A Drevermann[2], M Palm[3], F Stein[3], N Engberding[3], K Kelm[4], S Irsen[4]

[1] Otto von Guericke University Magdeburg, Institute for Materials and Joining Technology, P.O. Box 4120, D-39016 Magdeburg, Germany
[2] ACCESS e.V., Intzestraße 5, D-52072 Aachen, Germany
[3] Max-Planck-Institut für Eisenforschung GmbH, Max-Planck-Str. 1, D-40237 Düsseldorf, Germany
[4] Stiftung caesar, Electron Microscopy, Ludwig-Erhard-Allee 2, D-53175 Bonn, Germany

daniel.sturm@ovgu.de

Abstract. Compared to Ti-rich γ-TiAl-based alloys Al-rich Ti-Al alloys offer an additional reduction of in density and a better oxidation resistance which are both due to the increased Al content. Polycrystalline material was manufactured by centrifugal casting. Microstructural characterization was carried out employing light-optical, scanning and transmission electron microscopy and XRD analyses. The high temperature creep of two binary alloys, namely $Al_{60}Ti_{40}$ and $Al_{62}Ti_{38}$ was comparatively assessed with compression tests at constant true stress in a temperature range between 1173 and 1323 K in air. The alloys were tested in the cast condition (containing various amounts of the metastable phases Al_5Ti_3 and h-Al_2Ti) and after annealing at 1223 K for 200 h which produced (thermodynamically stable) lamellar γ-TiAl + r-Al_2Ti microstructures. In general, already the as-cast alloys exhibit a reasonable creep resistance at 1173 K. Compared with $Al_{60}Ti_{40}$, both, the as-cast and the annealed $Al_{62}Ti_{38}$ alloy exhibit better creep resistance up to 1323 K which can be rationalized by the reduced lamella spacing. The assessment of creep tests conducted at identical stress levels and varying temperatures yielded apparent activation energies for creep of Q = 430 kJ/mol for the annealed $Al_{60}Ti_{40}$ alloy and of Q = 383 kJ/mol for the annealed $Al_{62}Ti_{38}$ material. The latter coincides well with that of Al diffusion in γ-TiAl, whereas the former can be rationalized by the instability of the microstructure containing metastable phases.

1. Introduction

In previous decades Ti-rich Ti-Al-based intermetallic alloys have been developed successfully as materials for high temperature structure applications because of their low density and high specific strength. Favourable creep properties have been achieved for alloys with lamellar microstructures of γ-TiAl and the Ti-rich α_2-phase [1, 2]. Ti-Al alloy systems with increased aluminium content are expected to result in a significant additional weight reduction and a higher oxidation resistance, due to the formation of an adherent Al_2O_3 scale [3]. Furthermore, Al-rich Ti-Al alloys offer adequate strength at high temperatures, which would enable them to be used at significantly higher temperatures than Ti-rich γ-Ti-Al-based alloys. During investigations of the phase equilibria in the Al-rich part of the Ti-Al system Palm et al. [4-6] found that lamellar γ-TiAl + r-Al_2Ti microstructures can be generated by means of heat treatment and that the mechanical properties of these alloys could be improved by the formation of lamellar microstructure. Based on these findings an elaborate investigation has been started which aims at the production and detailed characterization of two-phase microstructures of γ-TiAl + r-Al_2Ti and a subsequent evaluation of the resulting mechanical properties.

2. Experimental

Centrifugal casting with Induction Skull Melting technology has shown high potential for manufacturing γ-TiAl components [7]. Hence, test bars of $Al_{60}Ti_{40}$ and $Al_{62}Ti_{38}$ were produced with this technology by ACCESS in Aachen. In general, the alloys showed good melting and casting behaviour. Nevertheless, some pores have been found by non-destructive x-ray inspection in the centre line of the test bars in the as cast condition, probably caused by uncontrolled solidification initiated by a high thermal gradient. The microstructure of the alloys in the as-cast state and after annealing was studied by light-optical, scanning (SEM) and transmission electron microscopy (TEM) and X-ray diffraction (XRD) analyses. Elevated temperature mechanical properties were characterised with compression creep tests on a screw-driven Zwick Z 100 testing machine at temperatures ranging from 1173 to 1323 K in air. The specimens were 5.5 mm high and 3.1 x 3.8 mm² in cross section and were prepared from the cast bars via electro-erosion and subsequent grinding and polishing of the parallel load-bearing surfaces to minimize friction.

3. Results and Discussion

Optical microscopy and XRD analysis of the $Al_{60}Ti_{40}$ and $Al_{62}Ti_{38}$ alloys in the as-cast condition showed a single-phase γ-TiAl microstructure which would contradict the phase diagram [6]. In-depth characterization of the microstructure of the $Al_{60}Ti_{40}$ alloy by TEM revealed a γ-TiAl matrix with domains of the metastable phase Al_5Ti_3 (5-10 nm domain size). The very small size of these Al_5Ti_3 domains results in a pronounced x-ray diffraction peak broadening. Because of the small size it was, thus, impossible to detect these nano-sized second phase particles by XRD analysis. The $Al_{62}Ti_{38}$ alloy showed a more complex microstructure. There are two different orientations of slabs of the metastable h-Al_2Ti phase in the γ-TiAl matrix. The interfaces between both h-Al_2Ti versions are {310} twin surfaces. The investigation revealed also domains of the metastable Al_5Ti_3 phase in addition to the Al_2Ti phase and the γ-TiAl matrix.

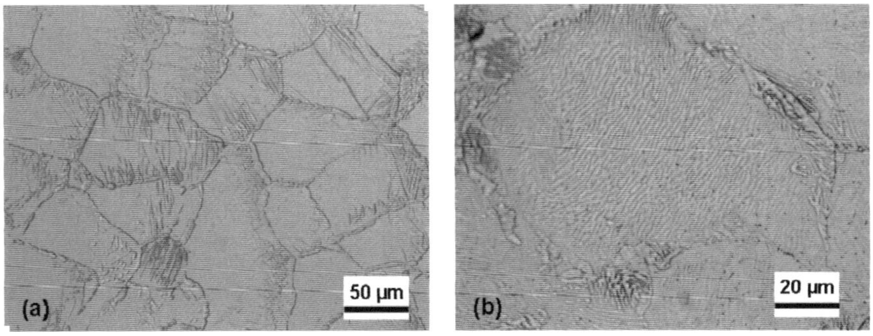

Figure 1. (a) Optical micrograph of the nearly lamellar γ-TiAl + r-Al_2Ti microstructure in $Al_{60}Ti_{40}$ and (b) fully lamellar microstructure of $Al_{62}Ti_{38}$ after annealing

After heat treatment at 1223 K for 200 h and water quenching [4] the microstructure of $Al_{60}Ti_{40}$ was transformed to a relatively coarse, nearly lamellar microstructure of the two phases γ-TiAl and r-Al_2Ti (Figure 1a), which were expected to form in accord with the phase diagram. The $Al_{62}Ti_{38}$ alloy was annealed at 1223 K for 50 h and water quenched [4]. Here the microstructure changed to a fine lamellar microstructure of the same two phases (Figure 1b).

The minimum creep rate obtained from constant true stress tests and the steady state stress obtained from constant strain rate tests were plotted in a double-logarithmic form in Figure 2 for $Al_{60}Ti_{40}$ [8] and $Al_{62}Ti_{38}$, respectively, in the as cast and the annealed condition at 1273 and 1323 K.

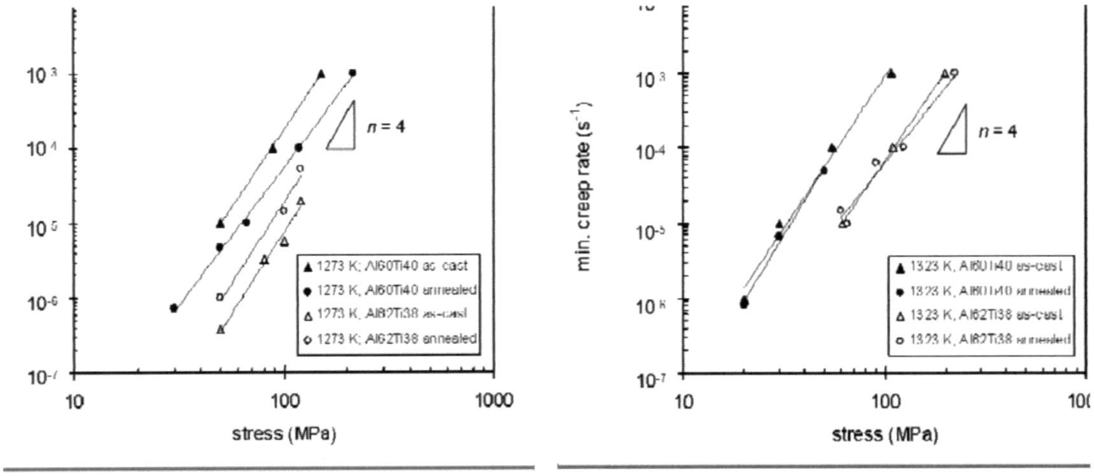

Figure 2. Comparison of secondary creep rates at 1273 K (left) and 1323 K (right) for the $Al_{60}Ti_{40}$ alloy (full symbols) [8] compared with data for the $Al_{62}Ti_{38}$ alloy (open symbols).

Compared with the creep results for the nearly lamellar $Al_{60}Ti_{40}$ alloy [8] it is obvious that both, the as-cast and the annealed $Al_{62}Ti_{38}$ alloy exhibit better creep resistance up to 1323 K. There is no significant difference in the creep behaviour of the as-cast and the annealed condition for both alloys. It should be mentioned that the high Al content of this alloys also provides a very good oxidation resistance which made testing in air possible. Even though the microstructures of all tested conditions are substantially different, the creep strength does not reflect this microstructural variation: with a value of approximately $n = 4$ the stress exponents were found to be relatively constant in the temperature and stress regime investigated. This value is within the range of $n = 3\text{-}6$ which is predicted by various creep models available in the literature [9-12]. This indicates that dislocation climb may be the rate controlling creep mechanism. In order to determine the activation energy of creep utilizing multiple linear regression, additional compressive creep data were plotted together as the logarithm of the minimum creep rate normalized with temperature vs. the stress normalized with shear modulus according to

$$\dot{\varepsilon} = A\left(\frac{\sigma}{E}\right)^n \exp\left(-\frac{Q}{RT}\right) \quad . \qquad (1)$$

The activation energy was varied until an optimal fit (maximum R^2) for the regression was found. The results of these calculations are shown in Table 1. The value for creep of the as-cast $Al_{60}Ti_{40}$ alloy is much higher than the values for interdiffusion of Al or Ti in γ-TiAl found in the literature [12], whereas the value of $Q = 383$ kJ/mol obtained for the annealed $Al_{62}Ti_{38}$ material coincides well with that of Al diffusion in γ-TiAl [12]. The relatively high observed activation energies may be rationalized by the fact that a metastable two-phase microstructure is present in the $Al_{60}Ti_{40}$ alloy.

Table 1: Activation energy for creep

	$Al_{60}Ti_{40}$ as-cast	$Al_{60}Ti_{40}$ annealed	$Al_{62}Ti_{38}$ as-cast	$Al_{62}Ti_{38}$ annealed
Q_{creep}	550 kJ/mol	430 kJ/mol	385 kJ/mol	383 kJ/mol

A TEM bright-field image of the microstructure creep deformed $Al_{60}Ti_{40}$ as-cast alloy is shown in Figure 3a which prooves the formation of a fine subgrain microstructure. This indicates the dissolution of the metastable Al_5Ti_3 phase during creep deformation and is a possible explanation for the observed high activation energy in the as-cast $Al_{60}Ti_{40}$ alloy. By contrast, the activation energy for creep of the $Al_{62}Ti_{38}$ alloys in both conditions are in the range of the activation energies usually observed for fully lamellar Ti- rich Ti-Al alloys (see for example [13]) where interface related creep phenomena as well as the degradation of the microstructure due to recrystallization and phase transformation determine the creep strength and the activation energy. The $Al_{62}Ti_{38}$ alloy shows the same deformed microstructure for the as-cast and the annealed condition with extended low angle grain boundaries formations and dislocations in globular r-Al_2Ti and γ-TiAl (with precipitates of Al_5Ti_3), shown in Figure 3b. Furthermore the investigation revealed areas of lamellar r Al_2Ti + γ-TiAl microstructure. Here the dislocations were found only in the γ-TiAl phase, Figure 3c. The results of the TEM investigation may serve as a plausible explanation for the nearly similar creep behaviour for the as-cast and annealed condition of the alloy.

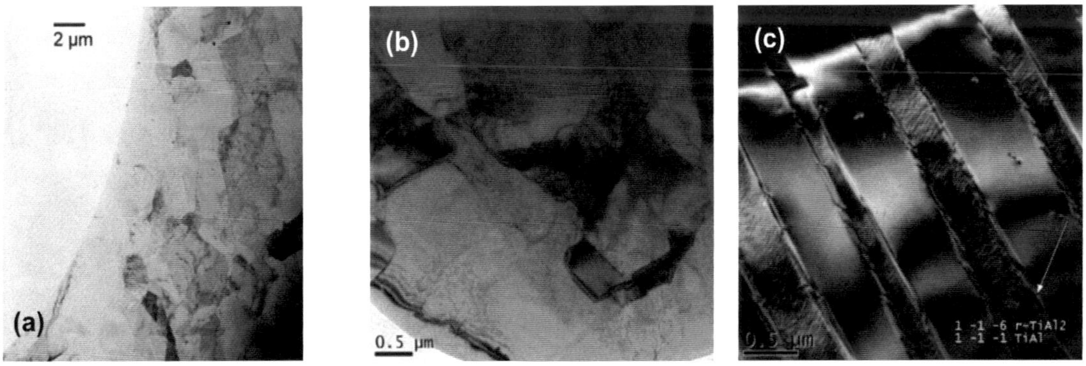

Figure 3. (a) TEM image of the dislocation substructure of $Al_{60}Ti_{40}$, as-cast state, after creep at 1323 K, σ = 60 MPa, ε = 12 % [8]; (b) bright field image of low angle grain boundaries formation of $Al_{62}Ti_{38}$ after creep deformation at 1273 K, σ = 50...100...120 MPa (stress change test), ε = 20 % and (c) dark field image of dislocations in the lamellar γ-TiAl phase in $Al_{62}Ti_{38}$ after creep deformation at 1273 K, σ = 50...100...120 MPa (stress change test), ε = 20 %

Acknowledgements
We gratefully acknowledge financial support from the German Science Foundation (DFG) for funding the group research project Al-rich Ti-Al alloys (PAK 19).

References
[1] Dietrich M 2002 *Titan-Aluminium-Legierungen – eine Werkstoffgruppe mit Zukunft* **1-131**
[2] Clemens H and Kestler H 2000 *Adv. Eng. Mater.* **2** 551-570
[3] Eckert M et al. 1997 *Oxidation of Intermetallics* Wiley-VCH
[4] Zhang L C, Palm M, Stein F and Sauthoff G 2001 *Intermetallics* **9** 229-238
[5] Stein F, Zhang L C, Palm M and Sauthoff G 2001 *Structural Intermetallics* **495-504**
[6] Palm M, Zhang L C, Stein F and Sauthoff G 2002 *Intermetallics* **10** 523-540
[7] Blum M et al. 2002 *Mater. Sci. Eng.* **329-331** 616-620.
[8] Sturm D et al 2009 *Mater. Sci. Eng.* A **510-511** 373-376.
[9] Barrett C R and Nix W D 1965 *Acta Metall.* **13** 1247-1258.
[10] Frost H J and Ashby M F 1982 *The Plasticity and Creep of Metals and Ceramics* (Pergamon Press)
[11] Karthikeyan S, Viswanathan G B and Mills M J 2004 *Acta Mater.* **52** 2577-2589.
[12] Mishin Y and Herzig C 2000 *Acta Mater.* **48** 589-623.
[13] Appel F, Oehring M and Wagner R 2000 *Intermetallics* **8** 1283

Effect of Cr content on the thermal stability of tempered lath structures and precipitates in strength enhanced ferritic steels

H Ghassemi Armaki [1], R P Chen [1], S Kano [1], K Maruyama [1] and M Igarashi [2]

[1] Graduate School of Environmental Studies, Tohoku University, 6-6-02 Aobayama, Sendai 980-8579, Japan

[2] Corporate Research and Development Laboratories, Sumitomo Metal Industry, Ltd., 1-8 Fuso-Cho, Amagasaki 660-0891, Japan

hasan_gm2003@yahoo.com

Abstract. The effect of Cr content on the thermal stability of tempered laths (or elongated subgrains) and precipitates has been studied during long-term aging at 650 °C in three P122 grade steels with increasing Cr content from 9 to 10.5 and 12%. Addition of Cr accelerates the coarsening of subgrains during long-term aging. The number fraction of MX precipitates does change up to 10^4 h aging in 9% and 10.5% Cr steels, whereas it decreases significantly in 12% Cr steel due to the formation of Z phase. The coarsening rate of $M_{23}C_6$ precipitates, mostly located on the subgrain boundaries, increases from 9 to 12% Cr and steel containing 9% Cr has the highest number density of $M_{23}C_6$ after 10^4 h aging. The addition of Cr from 9 to 12% accelerates the coarsening rate of Laves phase particles during aging. As a result, 9% Cr steel shows the most stable tempered lath martesitic structure during long-term aging.

1. Introduction

Advanced 9-12 %Cr heat resistant steels have been used in ultra-supercritical fossil fired power plants. Chromium is a main alloying element of these heat resisting steels to ensure high hardenability and good resistance to oxidation. Addition of Cr from 9 to 12%Cr increases the short-term creep strength, whereas the long-term creep strength decreases from 9 to 12%Cr [1]. High Cr ferritic steels subjected to normalizing and tempering are usually observed to have a tempered lath martensitic microstructure consisting of several types of strengthening mechanisms as following:

-Dislocation hardening: subgrain boundaries and free dislocations within subgrains.
-Precipitation hardening: $M_{23}C_6$ on subgrain boundaries and MX within subgrains.
-Solution hardening: W, Mo and Cr in solution.

All the above mechanisms slow down mobile dislocations [2]. Summation of the contributions from all the mechanisms determines the creep strength of a tempered martensite microstructure in the studied steels. Coarsening of tempered laths and precipitates has been studied during creep [2]. Thermal degradation of strengthening mechanisms proceeds during creep exposure together with its strain induced degradation. For full understanding of the Cr dependence of degradation of microstructure after long-term creep exposure, contributions of the thermal degradation should be separated from those of the strain induced degradation. In the present paper, the Cr dependence of thermal stability of tempered lath structures and MX, $M_{23}C_6$ and Laves phase precipitates have been studied up to 10^4 h aging at 650 °C in three P122 steels with the addition of Cr content from 9 to 10.5 and 12%.

2. Experimental

The chemical composition of the S9, S10 and D12 steels is listed in Table 1. All the chemical compositions given in the present paper are expressed in mass%. The steels marked as S and D are dominantly single phase (α-ferrite) and dual phase (α+δ ferrite), respectively. The Cr concentration increases from 9% (S9) to 12% Cr (D12). They were normalized at 1050 °C for 0.5 h and were cooled in air, followed by tempering at 770 °C for 2 h. These heat treatments make a tempered lath martensitic structure with precipitates of $M_{23}C_6$ and MX. The subgrain width λ of each steel was evaluated in several prior austenite grains covering total area > 80 μm² with TEM (JEOL 2100) operating at 200kV, and $M_{23}C_6$ precipitates were characterized by FESEM. The observation of very fine particles, for example MX particles, demands the replicas on which particles have been extracted. The precipitates were identified from quantification of EDX analysis spectra, and then number fraction of each precipitate was calculated in characterized area. More detailed explanations on the materials and creep rupture data of the same steels are available in Ref. [3].

Table 1. Chemical composition of the steels studied (mass %).

Name	C	Si	Mn	Cu	Ni	Cr	Mo	V	W	Nb	N
S9 (P122-S)	0.097	0.25	0.52	0.87	0.31	9	0.39	0.20	1.87	0.052	0.054
S10 (P122-S)	0.098	0.26	0.49	0.87	0.33	10.5	0.37	0.20	1.87	0.053	0.058
D12 (P122-D)	0.093	0.25	0.49	0.86	0.31	12	0.39	0.20	1.86	0.049	0.066

2. Results and discussion

2.1. Coarsening of Tempered Lath Structures (or Elongated Subgrains)

The subgrain size is plotted as a function of Cr content of the steels in figure 1: after short-term (about 10 h), middle-term (3000 h) and long-term exposure (10^4 h) at 650 °C. The subgrain size decreases with the addition of Cr from 9 to 12% after the short-term exposure. The subgrain size decreases by around 50 nm for the addition of every 1.5% Cr from 9 to 12% Cr. The aging time is very short and this short-term aging can not significantly affect the subgrain size after tempering. So, the decreasing of subgrain size with the addition of Cr is related to the inherent tempered martensitic microstructure of the materials. Such an effect of Cr content on the subgrain size is kept even after middle-term aging (3000 h), though the decrease of subgrain size with the addition of Cr is slightly smaller. The decrease of subgrain size from low to high Cr content in short-term region presents the increase of subgrain boundary strengthening and can cause the increase of creep strength as reported by previous researches [1, 4]. However, the effect of Cr content on the subgrain size changes after long-term exposure as shown by the triangle symbols in figure 1 and subgrain size increases with the addition of Cr content from 9 to 12%. Therefore, steel containing 9% Cr has the most stable tempered laths during long-term aging. The typical TEM images of microstructures after 10^4 h aging at 650 °C for steels S9, S10 and D12 are shown in Figures 2 (a) , (b) and (c), respectively. The subgrain size increases from 9 to 12% Cr and steel S9 shows the lowest subgrain size. The remarked particles in figure 2 are $M_{23}C_6$ carbides. Most of particles are found on the subgrain boundaries.

2.2. Coarsening and Disappearance of Precipitates

Figure 3 shows the number fraction of all the detected precipitates as a function of Cr content of steels after short-term (10 h) and long-term aging (10^4 h) shown as solid and open symbols, respectively. The number fraction of Laves phase particles are less than 2% for all states and not included for more simplify. The number fractions of $M_{23}C_6$ and MX precipitates are highest among the particles. As evident in figure 3, number fraction of MX precipitates does not decrease sensibly after 10^4 h aging for

S9 and S10 steels, though it may decreases in longer aging time for steel S10. Therefore, MX precipitates can not be responsible for the coarsening of tempered lath structures in S9 and S10 steels. But, the number fraction of MX precipitates decreases significantly in steel containing 12% Cr after 10^4 h aging. The formation of Z phase takes place at the expense of MX carbonitrides [5]. As shown in figure 3, the number fraction of Z phase has increased significantly in steel D12. Z phase grow rapidly and can not contribute to strengthening [5]. Therefore, the degradation of tempered lath structures will be more serious in steel D12 after long-term aging.

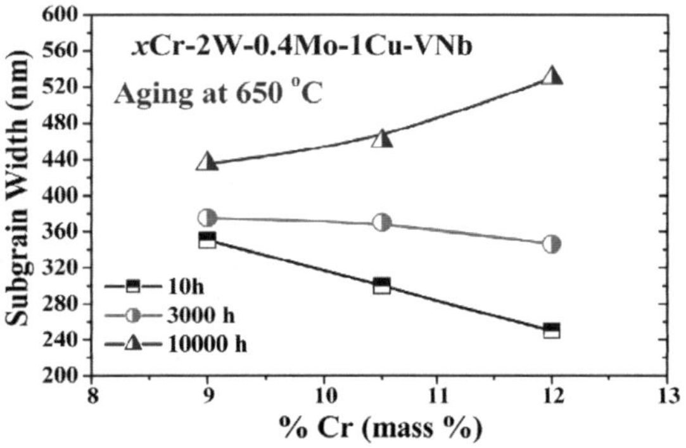

Figure 1. Cr dependence of subgrain coarsening after different aging time at 650 °C.

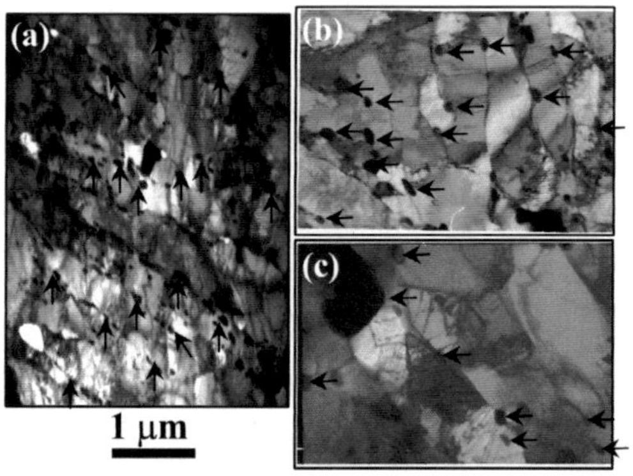

Figure 2. Typical TEM images of (a) S9, (b) S10 and (c) D12, after aging for 10^4 h at 650 °C.

$M_{23}C_6$ carbides are distributed along subgrain boundaries after tempering and act as pining points to migration of subgrain boundaries and consequently recovery of dislocation substructure are controlled by back force due to the particles [2, 6]. Figure 4 shows the mean diameter of $M_{23}C_6$ particles against exposure time at 650 °C. This figure points out that the driving force of coarsening increases by the addition of Cr form 9 to 12% and steel S9 has the lowest coarsening rate. The present authors have reported that the change of Cr concentration in $M_{23}C_6$ precipitates during long-term aging increases with increasing Cr concentration from 9 to 12% Cr. The change in chemical composition causes flow of Cr atoms to $M_{23}C_6$ particles and can assist the particle coarsening in higher Cr steels [7]. Figure 5 shows the frequency of $M_{23}C_6$ particles after 10^4 h aging at 650 °C. The steel S9 has the largest distribution of precipitates and the highest number density of $M_{23}C_6$ particles. Therefore, the more stability of tempered lath structures of steel S9 after long-term aging (as shown in figure 1) is refereed to more pinning points of $M_{23}C_6$ precipitates against migration of subgrain boundaries (as shown typically in TEM image of S9 in figure 2 (a)).

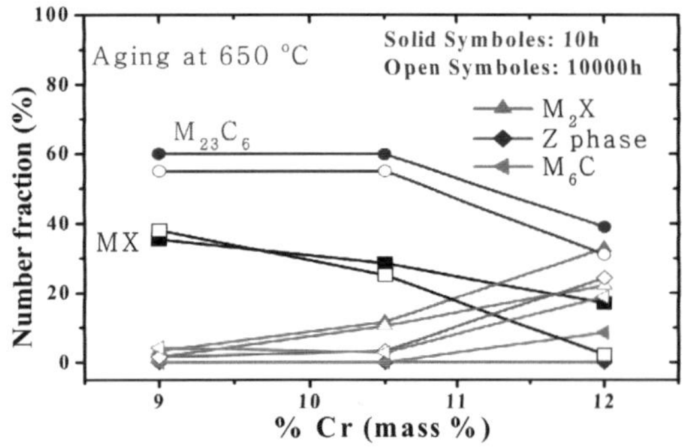

Figure 3. Number fraction of all the precipitates detected by FETEM as a function of Cr content of steels after 10 h and 10^4 h aging at 650 °C.

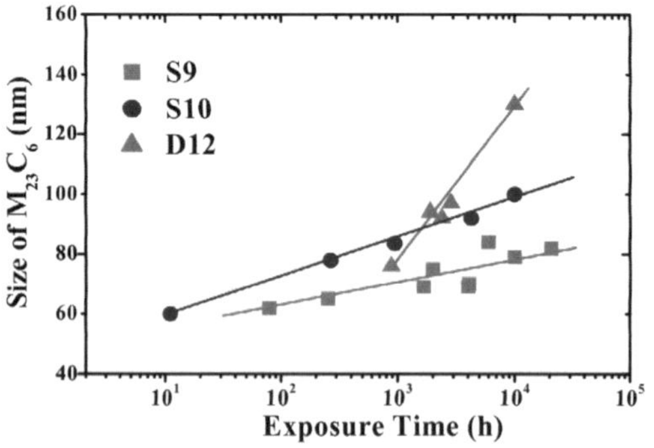

Figure 4. Size of $M_{23}C_6$ precipitates as a function of exposure time at 650 °C,

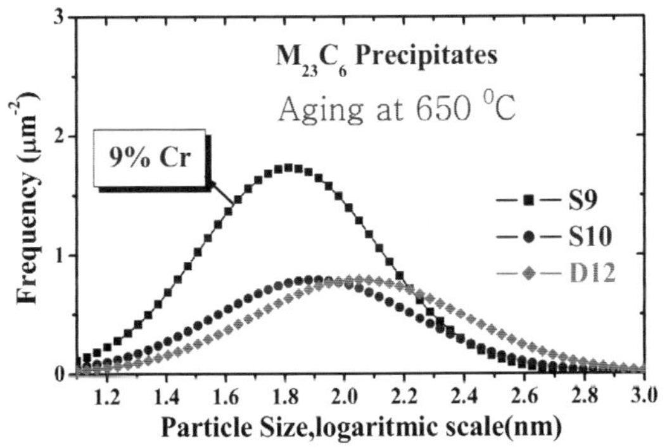

Figure 5. Frequency and size distribution of $M_{23}C_6$ particles in logarithmic scale after 10^4 h aging at 650 °C.

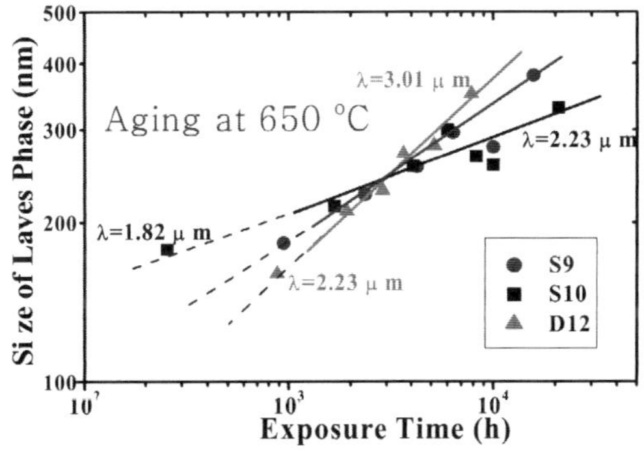

Figure 6. Laves phase particles size as a function of exposure time at 650 °C.

Figure 6 shows the mean diameter of Fe_2 (W, Mo) Laves phase as a function of exposure time at 650 °C. The addition of Cr accelerates the coarsening rate of Fe_2 (W, Mo) Laves phase particles during long-term aging. Laves phase precipitates are found mostly located at tempered laths boundaries [8]. The spacing of Laves phase precipitates has been indicated for shortest and longest aging time of steels S9 and D12 in figure 6. The spacing of Laves phase is much more than those of $M_{23}C_6$ and MX precipitates regardless of Cr content. Therefore, they can not influence sensibly coarsening of subgrains. However, the formation of Laves phase leads to depletion of W and Mo elements and loss of solid solution strengthening. Hence, the Cr dependence of depletion kinetic of W and Mo elements can be more useful in the future study.

4. Summary

1. The addition of Cr from 9 to 12% results in finer subgrain structure after short-term aging. But, the coarsening rate of subgrains accelerates with increasing Cr during aging, and subgrain size increases from 9 to 12% Cr steel after long-term aging.
2. The disappearance of MX precipitates were not found up to 10^4 h aging at 650 °C for steels S9 and S10, and they can not be responsible for coarsening of subgrains in these steels. However, the significant disappearance of MX at steel D12 can accelerates the coarsening of subgrains.
3. Agglomeration of $M_{23}C_6$ is slower at 9% Cr, and their spacing is kept narrow for longer duration aging in the steel. This is the reason for the slow coarsening of tempered lath structures at 9% Cr.
4. The addition of Cr from 9 to 12% accelerates the coarsening rate of Laves phase particles.
5. 9% Cr steel shows the most stable tempered lath martensitic structure during long-term aging.

References

[1] Abe F, Topical Review 2008 *Sci. Technol. Adv. Mater.* **9**
[2] Maruyama K, Sawada K and Koike J 2001 *ISIJ Int.* **41** 641
[3] Ghassemi Armaki H, Maruyama K, Yoshizawa M and Igarashi M 2008 *Mater. Sci. Eng.* A **490** 66
[4] LIU X Y and FUJ ITA T 1989 *ISIJ Int.* **29** 680
[5] Sawada K, Kushima H and Kimura K 2006 *ISIJ Int.* **46** 769
[6] Abe F 2008 *Creep Resistant Steels* (Woodhead publishing, Cambridge) chapter 9 p 279
[7] Ghassemi Armaki H, Chen R P, Maruyama K, Yoshizawa M and Igarashi M *Mater. Lett.* in press
[8] Dimmler G, Weinert P, Kozeschnik E and Cerjak H 2003 *Mater. Charact.* **51** 341

Internal stress model for pre-primary stage of low-stress creep

L Kloc

Institute of Physics of Materials AS CR, Žižkova 22, CZ-61662 Brno, Czech Republic

kloc@ipm.cz

Abstract. Initial transient stage in low-stress creep experiments was observed in all such experiments. Recently, evidences were presented that this stage cannot be considered as a normal creep primary stage, though the shape of the creep curve is similar. The strain reached during this so called pre-primary stage is fully recoverable upon unloading; the internal stresses must play important role in the effect. Model of standard linear anelastic solid was modified by introduction of creeping body instead of viscous dashpot. Both power law and hyperbolic sine creep law were used to fit observed creep curves of model and structural materials. Mainly the model using hyeprbolic sine creep law provides good fit to individual creep curves and sets of creep curves at different stresses.

1. Introduction

Transient stage with decreasing strain rate was observed in almost all low-stress creep experiments. Since the stage resembles the normal primary stage in conventional creep experiments, the interpretation of the transient deformation mechanisms was the same. Nevertheless, this approach met some serious difficulties [1]. Recently, evidences were presented [2] that the observed transient deformation cannot be considered a normal primary stage. One of the main differences is that the strain reached during this stage is almost fully reversible [3, 4], so it should be connected to the internal stresses field building.

The viscoelastic model of the standard anelastic material leads to the McVetty type of the creep equation [5], which is not capable to describe transient creep accurately. In this work, the model was adopted using creeping element instead of viscous dashpot; both power law and hyperbolic sine law creep behavior were considered.

2. Experimental procedure

Some previous experimental results were arbitrarily chosen to test the model capabilities. These were the Ni-15%Cr solid solution [6], the alumina fine grain advanced ceramics [7] and the P-91 type of the creep resistant steel [8, 4].

High strain sensitivity experimental techniques were used, mainly the helicoid spring specimen technique. The experimental procedures were described elsewhere [9] in more details. The important issue is the starting point of the experimental creep curves. Since most of the experiments use the "self-loaded" specimens, no initial strain upon loading is included in the creep curve. The creep curves recording starts in the time instant when the nominal temperature of the test is reached. Since the heating up process is relatively fast, it can be considered as purely elastic initial state.

3. Model

The model consists of two elastic elements A and B, one creeping element C and one viscous element D, as depicted in figure 1. The viscous element D was added to introduce the steady state creep term into the final equations; in fact, no reasonable fit could be achieved without this term.

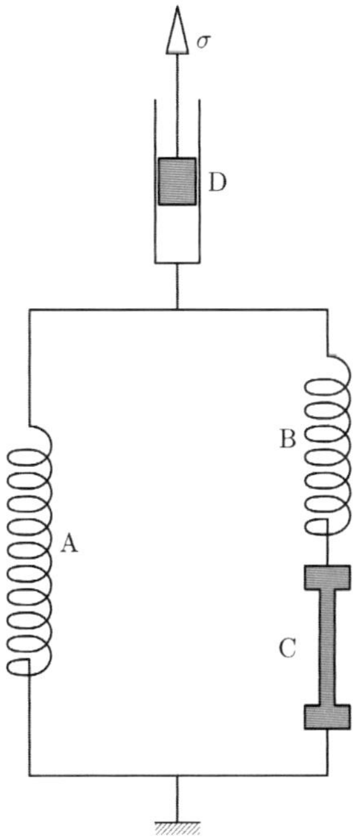

Particular elements are described by equations

$$\dot{\varepsilon}_D = d \cdot \sigma \tag{1}$$

$$\varepsilon_A = a \cdot \sigma_A \tag{2}$$

$$\varepsilon_B = b \cdot \sigma_B \tag{3}$$

$$\dot{\varepsilon}_C = c \cdot (\sigma_B)^n \tag{4a}$$

or

$$\dot{\varepsilon}_C = c \cdot \sinh(g \cdot \sigma_B) \tag{4b}$$

where a, b, c, d, g and $n \neq 1$ are parameters.

Figure 1. Mechanical model of transient creep strain.

Model behavior is then described by differential equations

$$\dot{\varepsilon} = \frac{ac}{a+b}\left(\sigma - \frac{\varepsilon - d\sigma t}{a}\right)^n + d\sigma \tag{5a}$$

or

$$\dot{\varepsilon} = \frac{ac}{a+b}\sinh\left(g\left(\sigma - \frac{\varepsilon - d\sigma t}{a}\right)\right) + d\sigma \tag{5b}$$

for power law or hyperbolic sine law, respectively.

Initial conditions are given by the elastic state, that is at $t = 0$ the elastic strain is $\varepsilon(0) = ab\sigma/(a+b)$ and $\sigma_A = b\sigma/(a+b)$, $\sigma_B = a\sigma/(a+b)$. Since this initial elastic strain is not recorded in the experimental creep curves, it must be subtracted from the final equations:

$$\varepsilon = \frac{a^2\sigma}{a+b} - \frac{a}{\left(\frac{(n-1)c}{a+b}t + \left(\frac{a+b}{a\sigma}\right)^{n-1}\right)^{\frac{1}{n-1}}} + d\sigma t \tag{6a}$$

or

$$\varepsilon = \frac{a^2\sigma}{a+b} + \frac{a}{g}\log\left(\frac{\exp\left(\frac{cg}{a+b}t\right)\left(\exp\left(\frac{bg\sigma}{a+b}\right) + \exp(g\sigma)\right) + \exp\left(\frac{bg\sigma}{a+b}\right) - \exp(g\sigma)}{\exp\left(\frac{cg}{a+b}t\right)\left(\exp\left(\frac{bg\sigma}{a+b}\right) + \exp(g\sigma)\right) - \exp\left(\frac{bg\sigma}{a+b}\right) + \exp(g\sigma)}\right) + d\sigma t \tag{6b}$$

4. Results

For practical reasons, the stress exponent n is limited to values between 3 (Alloy class metals) and 12 (experimental value for the P-91 steel [10]). Since all fitting procedures tend to adhere to one or other of these limits, the model equations were used with values n fixed to both limits. The fits are marked as "PL3" and "PL12, while the model using hyperbolic sine is marked "SH" in what follows.

4.1. Individual creep curves

An example of model creep curves fitted on experimental data for Ni-15%Cr solid solution is depicted in figure 2. It is clear that it is practically impossible to distinguish between results provided by particular model equations, so the fitting residuals are also plotted in figure 3. The statistical errors of the fits are summarized in the table 1

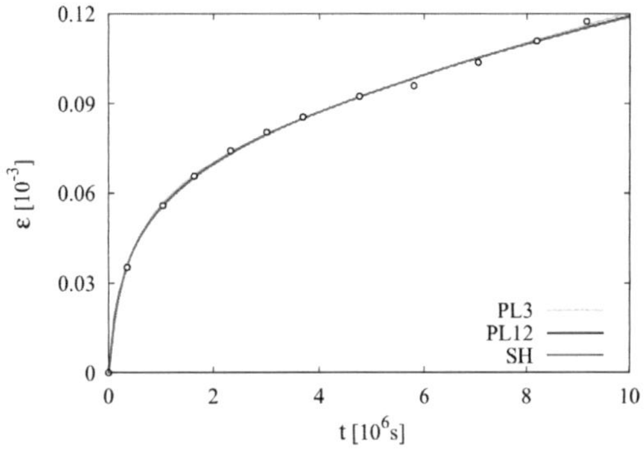

Figure 2. Creep curve of Ni-15%Cr solid solution at 700 °C and 4.9 MPa, fitted by the three model equations

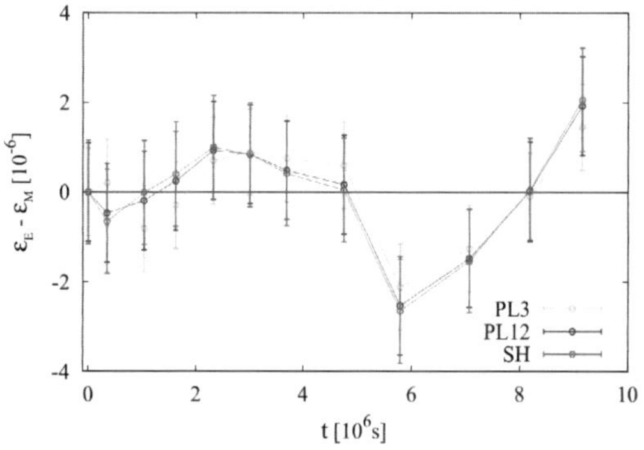

Figure 3. Residuals of the fit from figure 2

Table 1. Average fitting errors for various materials and equations

Material	PL3	PL12	SH
NiCr	9.7×10^{-7}	1.1×10^{-6}	1.2×10^{-6}
Alumina	8.9×10^{-7}	7.7×10^{-7}	7.3×10^{-7}
P-91	6.9×10^{-6}	2.7×10^{-6}	1.2×10^{-6}

It is clear, that while with NiCr and alumina all equations provides comparable results, with P-91 having more pronounced transient stage, the fit provided by PL3 is worse than that provided by PL12 and mainly SH.

4.2. Sets of creep curves

For the physical meaning of the model to be maintained, the parameters a, b, c, d and g should be stress independent. To prove it, the equations were fitted to the sets of creep curves with different stress provided by the helicoid spring specimen technique. Results are shown in figures 4 and 5.

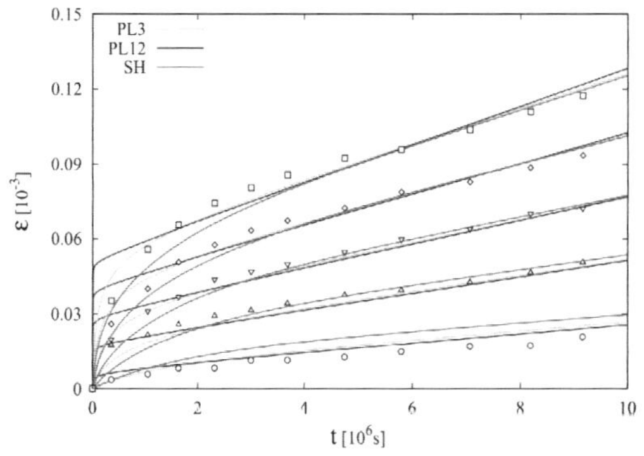

Figure 4. Fit of the creep curves of Ni-15%Cr solid solution at 700 °C and ○ 1.2 MPa, △2.1 MPa, ▽3,0 MPa, ◇4.0 MPa, □4.9 MPa.

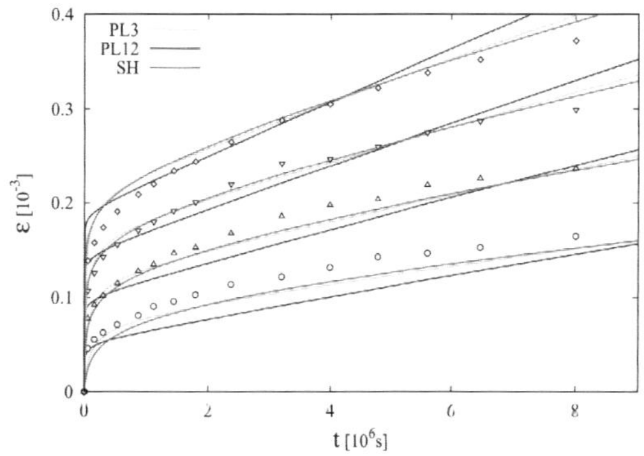

Figure 5. Fit of the creep curves of P-91 steel at 625 °C and ○ 4.1 MPa, △6.3 MPa, ▽8.4 MPa, ◇10.5 MPa.

It is clear that the PL12 equation fail to follow the shape of the creep curves. So the good results of the equation on individual creep curves is paid by strong stress dependence of the parameters. This fact contradicts the physical meaning of the model. Thus, the PL12 model does not work. On the other hand, the PL3 and SH models behave correctly. In most cases, the SH model provides better results than the PL3 one.

5. Conclusion

Simple elastoplastic model using elastic, viscous and creeping elements is able to describe the transient stage of the low stress creep of various material. The best fit is obtained, if the creeping element is governed by the hyperbolic sine creep law (equations 5b, 6b). If the power-law creep rule is used, for the high values of the stress exponent the strong stress dependence of parameters occur, which is not acceptable. On the other hand, model with power-law equation and stress exponent $n = 3$ behaves correctly, but provides generally worse results than the model with hyperbolic sine equation.

Acknowledgments

This work was supported by the Grant Agency of Academy of Sciences of the Czech Republic under contract no. IAA 200410701.

References

[1] Kloc L and Skleničká V 1999 *Creep Behavior of Advanced Materials for the 21st Century* ed R S Mishra et al. (Warrendale: TMS) pp 495–502
[2] Kloc L and Skleničká V 2009 *Mater. Sci. Forum* **604–605** 351–356
[3] Mishra R S and Mukherjee A K 1999 *Creep Behavior of Advanced Materials for the 21st Century* ed R S Mishra et al. (Warrendale: TMS) pp 503–507
[4] Kloc L and Skleničká V 2004 *Mater. Sci. Eng.* A **387–389** 633–638
[5] McVetty P G 1934 *Mech. Eng.* **56** 149
[6] Kloc L, Fiala J and Čadek J 1995 *Mater. Sci. Eng.* A **202** 11–17
[7] Kloc L, Mareček P and Fiala J 2008 *Mater. Sci. Eng.* A **483–484** 541–543
[8] Kloc L and Skleničká V 1997 *Mater. Sci. Eng.* A **234–236** 962–965
[9] Kloc L and Mareček P 2009 *J. Test. Eval.* **37** 53–58
[10] Skleničká V, Kuchařová K, Dlouhý A and Krejčí J 1994 *Proc. Conf Materials for Advanced Power Engineering* ed D Coutsouradis et al. (Dodrecht: Kluwer Academic) p 435

Influence of processing on the microstructure and mechanical behaviour of Mo-Si-B alloys

M Krüger[1], H Saage[2], M Heilmaier[3], M Böning[4], H Kestler[4]

[1] Otto von Guericke University Magdeburg, Institute for Materials and Joining Technology, P.O. Box 4120, D-39016 Magdeburg, Germany
[2] University of Applied Sciences Landshut, Mechanical Engineering, D- 84036 Landshut, Germany
[3] Technical University Darmstadt, Dept. Materials Science, D-64287 Darmstadt, Germany
[4] Plansee SE, Technology Center, A-6600 Reutte in Tyrol, Austria

m.heilmaier@phm.tu-darmstadt.de

Abstract. Mo-Si-B materials consisting of a Mo(Si) solid solution and the intermetallic phases Mo_3Si and Mo_5SiB_2 (T2) were prepared by mechanical alloying (MA) as the crucial step of a powdermetallurgical process. After consolidation via an industrial processing route (cold isostatic pressing, sintering, hot isostatic pressing) the resulting microstructures of Mo-Si-B alloys up to 45% of intermetallic phases reveal a continuous α-Mo matrix with embedded, homogeneously distributed intermetallic particles. Clearly, increasing the amount of Mo solid solution reduces the BDTT (demonstrated by three point bending tests between room temperature and 1200°C), however, values below 900°C could not be obtained due to grain boundary embrittlement caused by Si segregation. Alloying with Zr was proven by Auger analysis in Mo-Si solid solutions to reduce this segregation. Therefore, in a second trial Zr as a (micro-) alloying element was added. The influence of microalloying on ductility and strength is comparatively discussed with reference compositions Mo-6Si-5B and Mo-9Si-8B.

1. Introduction

Since three-phase Mo-Si-B alloys (pioneered by Berczik [1]) combine the excellent high temperature strength of Mo-base alloys and the beneficial oxidation resistance of intermetallic silicide phases as a provider for Si and B necessary for forming a dense coating consisting of SiO_2(B) glass [2], they are considered as candidates for high temperature application in air. It is widely accepted that the ideal microstructure of these alloys is composed of a continuous α-Mo matrix with embedded intermetallic particles [3-7]. A powdermetallurgical (PM) process based on mechanical alloying (MA) as the crucial step may be used to fabricate three-phase Mo-Si-B alloys with this microstructure. In previous work [8] it was demonstrated that while Si is one of the most effective solid solution strengtheners for Mo [9], it also strongly promotes the tendency towards brittle intergranular failure. Amongst other possible elements that potentially ductilize Mo [10], Zr has been shown to improve the fracture toughness of Mo-12Si-8.5B [5]. Miller and Bryhan [11] have demonstrated by atom probe tomography that small amounts of Zr, C and B substitute detrimental oxygen at the grain boundaries of a welded Mo-base alloy. Therefore, in our recent work [12] we studied the mechanical behaviour of mechanically alloyed Mo-1.5Si and Mo-1.5Si-1Zr as compared to Mo-0.3Si and Mo-1.5 Si processed by an industrial PM route (without MA). We have shown that the brittle behaviour of Mo(Si) at room temperature is, at least in part, caused by the reduction of the cohesive grain boundary strength due to segregated Si. As an important outcome Auger analysis showed that additions of Zr dramatically reduce the segregation of Si at the grain boundaries [12]. Furthermore, Zr is responsible not only for the reduction of the grain size of (single phase) Mo-Si solid solutions, additionally it improves the strength level from 500 MPa to about 2 GPa and leads to limited plastic deformation in RT bend tests. Therefore, in this paper we investigate the effect of Zr on the mechanical properties of three phase Mo-Si-B alloys between RT and 1200°C.

2. Experimental

Elemental powder mixtures of Mo, Si and B of 99.95, 99.9 and 98% purity respectively, were used to produce different ternary Mo-Si-B compositions. Mechanical alloying (MA, high energy milling) was carried out under protective argon atmosphere for 20 hours in a planetary ball mill (Retsch® PM 400) with a rotational speed of 200 rpm and a powder to ball weight ratio of 1:13. After milling, the powders were cold isostatically pressed at 200 MPa and sintered under hydrogen atmosphere at 1600°C to reduce the impurity content (mainly oxygen). As a final step the sintered bars were further consolidated by hot isostatic pressing (HIP) at 1500°C and 200 MPa. The microstructures of powder samples were characterized by X-ray diffraction (XRD) with CuK_α radiation. Lattice parameters of α-Mo were calculated on the basis of seven peaks using the Cohen Method [13]. To determine the domain size of the milled powders, the program "breadth" [14] based on the Fourier method was used. The microstructures of consolidated samples were characterized by scanning electron microscopy (FEI ESEM XL30 FEG equipped with EDX). Mechanical properties were assessed by three-point bending tests using rectangular samples with a cross section of about 3.8 x 3 mm^2 which were prepared from HIPed bars by electro-discharge machining, turning and grinding. All tests were carried out using a Zwick electromechanical testing device equipped with a Maytec furnace at a crosshead speed of 0.01 mm/min. Above 500°C the tests were performed under protective argon atmosphere.

3. Results and Discussion

In Table 1 the powder parameters during mechanical alloying are summarized for Mo-9Si-8B. The domain size decreases right from the beginning of MA followed by the decrease of the lattice parameter after 10 h of milling. Since the atomic radius of Si is significantly smaller than that of Mo, it is obvious that substitution of Mo by different proportions of Si atoms leads to the observed decrease of the lattice parameter with increasing Si solution. While MA of Mo-4Si-2B (not shown here, see [6]) leads to a nearly constant lattice parameter after milling times longer than 30 hours, the lattice parameter of powder mixtures with higher Si and B content still decreases up to larger milling times indicating further solution of Si. The formation of a solid solution also becomes apparent in the increasing hardness values during milling.

Table 1. Powder parameters of Mo-9Si-8B in dependence on the milling time.

Alloy	**Mo-9Si-8B**						
Milling time [h]	5	10	20	30	50	70	100
Lattice constant of α-Mo [Å]	3.1471	3.1466	3.1419	3.1372	3.1276	3.1227	3.1144
Domain size of α-Mo [Å]	150	64	-	93	31	21	12
Microhardness (HV 0.01/5)	1089	1176	1177	-	1215	1284	1334

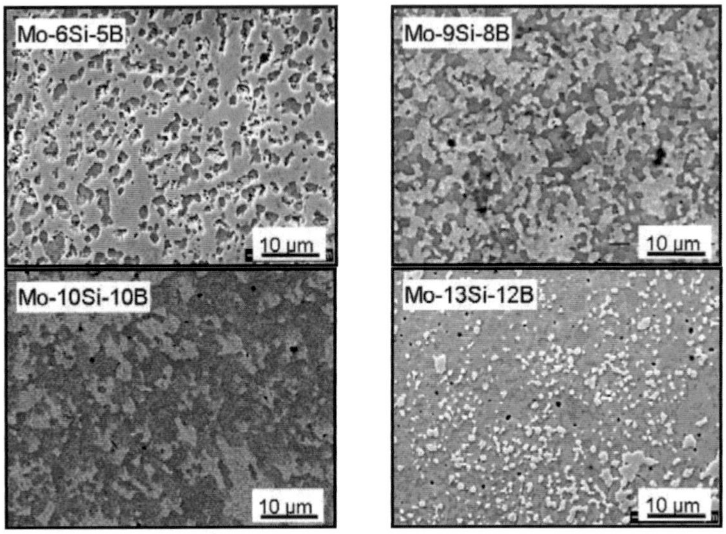

Figure 1. Microstructures of the consolidated Mo-Si-B samples. The α-Mo matrix appears as the brightest phase and the slightly darker grey and the dark grey appearing phases are Mo_3Si and Mo_5SiB_2.

The consolidation as described in the experimental section leads to the formation of an optimal microstructure consisting of a continuous α-Mo matrix (light-grey) and embedded intermetallic particles (dark-grey) for both, the Mo-6Si-5B and Mo-9Si-8B composition, respectively. This is shown in the 2D SEM micrographs in figure 1. In contrast the alloys Mo-10Si-10B and Mo-13Si-12B show a reversed microstructure, namely a continuous intermetallic matrix with islands of α-Mo. These findings have been recently confirmed by 3D SEM FIB tomography [15]. Generally, the materials processed through MA show a very fine microstructure with the average grain sizes of all phases as determined by quantitative stereology to lie between 1 and 1.5 µm [6].

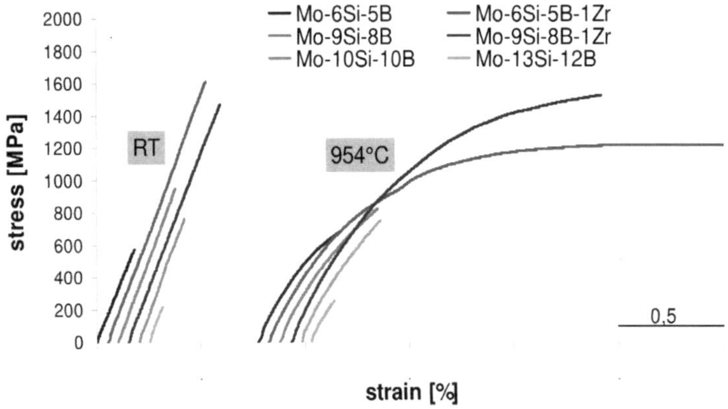

Figure 2. Stress-strain-curves (outer fibre) of Mo-Si-B alloys at RT and 954°C.

Stress-strain (calculated for the outer tensile fibre) curves for the Mo-Si-B base alloys and two alloys with additions of 1% Zr at RT and 954°C are compared in figure 2. All tested alloys behave brittle at RT, though Zr apparently causes an increased strength level of up to 1600 MPa. This is approximately a factor of 2 larger than the values reported in [1] for the same base composition. Furthermore, Zr leads to limited plastic deformation at temperatures as low as 800°C (light blue curve in figure 3), whereas the base alloys remain still brittle up to about 950°C. Consequently, the brittle-to-ductile transition temperature (BDTT) of alloys with a continuous α-Mo matrix could be reduced by about 150°C and by more than 250°C as compared with alloys possessing an intermetallic matrix (orange and green in figure 3). Future work will be directed towards varying the necessary Zr concentration or finding other potentially ductilizing elements to fully exploit the effect on reducing of the BDTT.

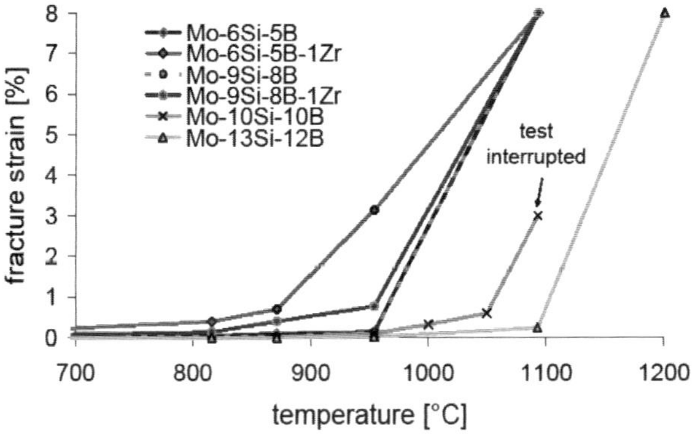

Figure 3. Outer fibre tensile fracture strain of mechanically alloyed Mo-Si-B alloys versus test temperature showing the decrease of BDTT for alloys with continuous α-Mo matrix and the effect of Zr microalloying which decreases the BDTT of these alloys even further. Note that the tests were terminated after reaching 8% plastic strain even if fracture did not occur.

4. Conclusions

Mechanical alloying offers the opportunity to prepare three-phase Mo-Si-B alloys that possess a continuous α-Mo matrix in which uniformly distributed intermetallic particles are embedded. The BDTT decreases by about 150 K when a continuous α-Mo matrix is established instead of an intermetallic matrix. A further improvement of the ductility of Mo-Si-B alloys with continuous α-Mo matrix is reached by the addition of Zr, which in Mo-Si alloys has already been identified as reducing the segregation of Si to the grain boundaries, thus enhancing their cohesive strength dramatically [12].

Acknowledgements

The financial support of the German Science Foundation (DFG) within the framework of the research unit 727 "Beyond Nickelbase Superalloys" is gratefully acknowledged. We thank E.P. George and L. Heatherly (ORNL, Oak Ridge, TN) for the help with Auger analysis and G. Schumacher, J. Banhart and O. Hassomeris from HMI Berlin for fruitful cooperation with 3D SEM/FIB tomography.

References

[1] Berczik D M 1997 *US Patents 5,595,616 and 5,693,156*
[2] Parthasarathy T A, Mendiratta M and Dimiduk D M 2002 *Acta Mater.* **50** 1857
[3] Schneibel J H, Kramer M J, Ünal Ö and Wright R N 2001 *Intermetallics* **9** 25
[4] Nieh T G, Wang J G and Liu CT 2001 *Intermetallics* **9** 73
[5] Schneibel J H, Ritchie R O, Kruzic J J and Tortorelli P F 2005 *Metall. Mater. Trans A* **36** 525
[6] Krüger M, Franz S, Saage H, Heilmaier M, Schneibel J H, Jéhanno P, Böning M and Kestler H 2008 *Intermetallics* **16** 933
[7] Heilmaier M, Saage H, Krüger M, Jéhanno P, Böning M and Kestler H 2009 *Mat. Res. Soc. Symp. Proc.* **1128** U07-07
[8] Sturm D, Heilmaier M, Schneibel J H, Jéhanno P, Skrotzki B and Saage H 2007 *Mater. Sci. Eng. A* **463** 107
[9] Northcott L 1956 *Molybdenum* (New York: Academic Press Inc.)
[10] Geller C B, Smith R W, Hack J E, Saxe P and Wimmer E 2005 *Scripta Mater.* **52** 205
[11] Miller M K and Bryhan A J 2002 *Mater. Sci. Eng. A.* **327** 80
[12] Saage H, Krüger M, Sturm D, Heilmaier M, Schneibel J H, George E, Heatherly L, Somsen C, Eggeler G and Yang Y 2009 *Acta Mater.* **57** 3895
[13] Cohen M U 1935 *Rev. Sci. Instrum..* **6** 68
[14] Balzar D and Ledbetter H 1993 *J. Appl. Crystallogr.* **26** 97
[15] Schumacher G, Hassomeris O and Banhart J 2009 *private communication*

Tensile strength and creep behaviour of austenitic stainless steel type 18Cr - 12Ni with niobium additions at 700°C

V L Sordi and L O Bueno

Federal University of São Carlos, Materials Engineering Department, São Carlos (SP), 13565-905, Brasil

sordi@ufscar.br

Abstract. The effect of niobium additions up to 2.36 wt% on the creep behavior of a series of seven extra low carbon 18Cr-12Ni austenitic stainless steels at 700°C has been investigated. Grain size and hardness measurements, hot tensile tests and constant stress creep tests from 90 to 180 MPa were carried out for each alloy, in the solution treated condition at 1050, 1200 and 1300°C followed by quench in water. The mechanical behavior at high temperature was related to the amount of NbC precipitation occurring during the tests. Solid solution and intermetallic compound effects were also considered. Creep data analysis was done to determine the parameters of the creep power-law equation $\dot{\varepsilon} = A.\sigma^{n}$ and the Monkman-Grant relation $\dot{\varepsilon}.t_R^{m}=K$. Niobium-carbide precipitation in these steels reduces the secondary stage dependence of strain rate with applied stress, resulting in n-values which indicate the possibility of operation of various creep mechanisms. The creep strength during the secondary stage is primarily controlled by the amount of NbC available for precipitation. However, the rupture times increase progressively with niobium content, as the amount of undissolved carbide particles in grain boundaries and the Laves phase precipitation increase.

1. Introduction

Niobium can be added to austenitic stainless steel as to stabilize carbon in austenite, reducing the formation of chromium carbides at the grain boundaries during high temperature service. Thus it is possible to avoid the losses in corrosion resistance, and the tensile strength can be improved by the solid solution strengthening effect of niobium in the austenitic matrix. Creep rupture life can be improved by the precipitation of niobium carbides on dislocations. This effect is generally followed by some ductility loss, since deformation is concentrated at or near the grain boundaries by the high strength of the matrix.. Depending on solution treatment conditions and on chemical composition of the steel, formation of intermetallic compounds, like σ-phase and Fe_2Nb, may occur [1,2,3]. In this work it was considered that niobium carbide solubility in austenite can be described by the relation: Log (Nb.C) = 3.3 – 7200/T, proposed by Minami and Kimura [4,5], and that the composition of niobium carbide is next to Nb(CN), whose stoichiometric relation is Nb/Ceq = 7.74 , where Ceq = (C + 6/7N). Thus, it was possible to estimate the quantities presented in Figure 1. The mechanical properties were evaluated by tensile tests and constant stress creep tests at the temperature of 700°C. Creep data analysis was made by determining the parameters of the creep power-law equation $\dot{\varepsilon} = A\sigma^n$ and the Monkman-Grant [6] relation $\dot{\varepsilon}.t_R^m = K$, where $\dot{\varepsilon}$ is the minimum creep rate, σ is the applied stress and t_R is the rupture time and *A, n, m* and *K* are constants to be determined.

2. Experimental

Seven alloys were considered in this research with chemical composition indicated in Table 1. They were prepared by vacuum melting and submitted to hot rolling to produce plates with about 10 mm thickness. The material was solution-treated at the temperatures of 1050, 1200 and 1300°C, for 1 hour in argon atmosphere followed by quench in water. Hilliard grain size and Vickers hardness measurements, hot tensile and constant stress creep tests were carried out at 700°C. Screw-ended specimens with diameter / gage length ratio of 4 mm / 20 mm and 5mm / 50 mm were used for the tensile and creep tests, respectively. The applied stresses for creep ranged from 90 to 180 MPa, producing rupture times varying from about 0.1 to 5000 h.

Table 1. Chemical compositions of the steels

Alloy	C (%)	Si (%)	Mn (%)	Cr (%)	Ni (%)	Nb (%)	Nb/Ceq
A1	0.020	1.02	1.16	18.24	11.00	zero	0
A2	0.014	0.71	0.91	17.80	11.80	0.22	11.6
A3	0.020	1.00	1.13	18.80	11.20	0.52	16.8
A4	0.026	0.73	0.97	18.13	12.84	0.91	24.6
A5	0.020	1.02	1.21	18.20	11.10	1.14	32.6
A6	0.016	0.92	1.04	17.80	13.20	1.78	55.6
A7	0.020	1.09	1.11	17.99	13.20	2.36	63.8

3. Results and discussion

As shown by Figure 2, hardness increases with niobium content, what could be attributed to the solid solution hardening effect caused by niobium in the austenitic matrix. However, the hardness of steels with high amounts of niobium is more sensible to solution treatment temperature than the hardness of steels with lower additions, in opposition to what happens with grain size. Therefore, the lower

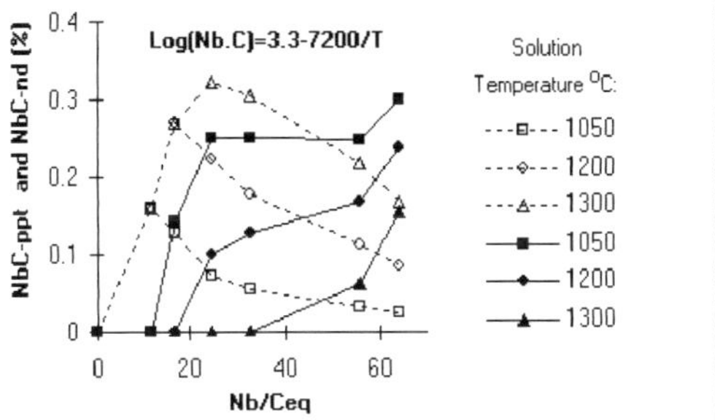

Figure 1 – Amount of niobium carbide available for precipitation (NbC-ppt dashed lines) and undissolved NbC (NbC-nd solid lines).

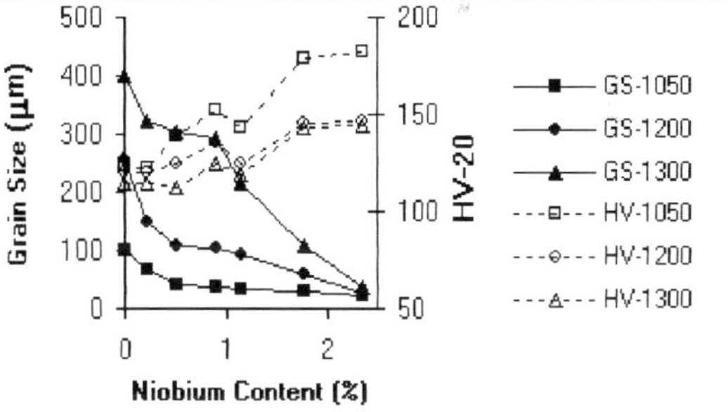

Figure 2. Vickers hardness (dashed lines) and grain size (solid lines) after solution-treatment at 1050, 1200 and 1300°C.

hardness observed for steels solution-treated at 1300°C cannot be justified only by grain size, since grain size presents only a light dependence on the solution treatment temperature for steels with high niobium contents, where hardness varies in a more noticeable way. On the other hand, steels with low niobium contents show great variation in grain size, and practically stable hardness. Increases in hardness for a certain temperature of solution treatment temperature also cannot be explained only by the quantities of niobium in solid solution, because this would imply greater hardness for steels solution-treated at higher temperatures, in opposition to what has been observed. It is possible, therefore, that the presence of undissolved carbides and undissolved intermetallic particles of Fe_2Nb (Laves phase), is the main factor controlling the hardness of these alloys in the solution-treated condition, as already suggested by other authors [7-10].

Figure 3 shows that, for the steels with low niobium additions, the ultimate tensile stress is lower when the solution-treatment is made at 1300 °C, but, as the amount of niobium increases, this situation is reverted. For yield stress this trend does not repeat. Tensile strength seems to result from several effects: the amount of niobium dissolved in the austenitic matrix, the amount of NbC available for precipitation and the amounts of undissolved Fe_2Nb and NbC particles. Comparing Figure 3 to Figure 1, yield stress seems to depend specially on the presence of undissolved particles and on the amount of niobium dissolved in the matrix. Carbide precipitation is strongly accelerated by plastic deformation and, therefore, might have greater influence on the ultimate tensile stress than on the yield stress. It was also observed that the level of deformation at maximum load dropped from 25% for sample A1 (0% Nb) to 15% for sample A7 (2.36% Nb) and that, in general, solution-treated alloys at 1300°C presented the lowest values of final elongation. Denham and Silcock [9] had even observed that the presence of undissolved Fe_2Nb increases the total elongation at 700°C, but they also pointed out that maximum loads were obtained at too low deformations levels, and great part of subsequent deformation was attributed to the formation of cracks at grain boundaries.

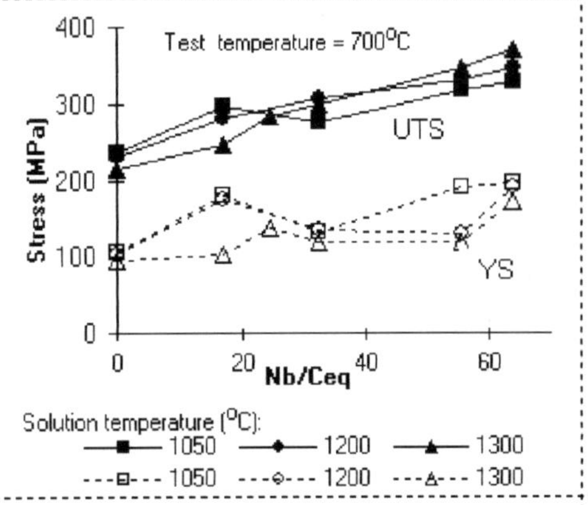

Figure 3 – Effect of Nb/Ceq ratio on tensile strength (solid lines) and yield stress (dashed lines) at 700 °C.

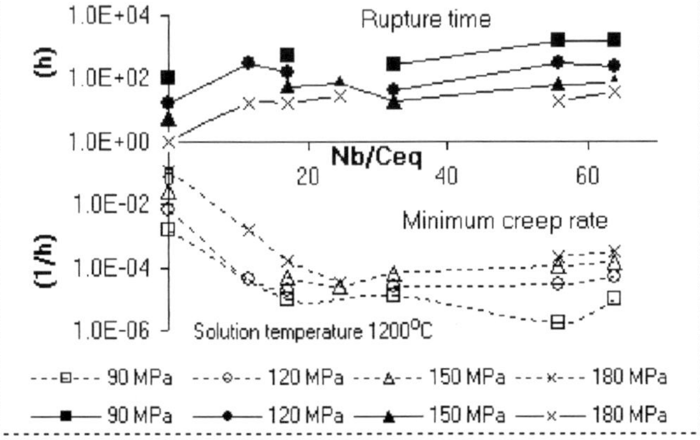

Figure 4. Effect of Nb/Ceq ratio on creep rupture time (solid lines) and minimum creep rate (dashed lines). Solution temperature: 1200 °C. Test temperature:700 °C.

Figure 4 shows the trend of increase in creep rupture time with increase of niobium content. The lowest strain rates at secondary creep stage are obtained for Nb/Ceq relations next to 20, matching with the maximum niobium carbide availability for precipitation indicated in Figure 1. Figure 5 presents the values of the parameters n and m for the equations $\dot{\varepsilon} = A\sigma^n$ and $\dot{\varepsilon}.t_R^{\ m}$, respectively, as function of Nb/Ceq relation. For alloys with low Nb/Ceq ratio, the values of n decreases as the solution-treatment temperature decreases. Increasing the Nb/Ceq ratio means obtaining smaller and smaller grain sizes, less and less dependent on solution-treatment temperature. At the same time, the possibility of niobium carbide precipitation increases, reaching its maximum level around the Nb/Ceq ratio of A4 alloy, and turning down again for greater Nb additions (See Figure 1). From a certain level of addition, creep process is supposed to be controlled by dislocation climb over precipitated particles, reducing the dependence of strain rate on applied stress. Thus, the minimum n-values obtained for the alloys with more niobium carbide available for precipitation can be explained, as well as the increase of n-values with the decrease of solution-treatment temperatures for the alloys with high addition of niobium. The fact that the n-exponent decreases, tending to 1, as the precipitation level increases, may indicate more and more participation of mechanisms which involve diffusion creep in controlling secondary stage strain rates.

A trend of decrease in the values of the m-exponent in Figure 5 can be observed for Nb/Ceq rates that promote niobium carbide precipitation, namely, the A3, A4 and A5 alloys. Low m-values means that very different rupture times can be obtained with small variations of secondary stage creep rates. Decreasing m-values suggest some loss in ductility and major contribution of secondary stage to the total rupture time. The greatest rupture times obtained in this work correspond to the alloys with highest Nb additions, but, according to the adopted solubility relation, these alloys do not present the highest amounts of carbide available for precipitation during the test. In fact, these long rupture times were obtained due to a major participation of tertiary stage in the creep test, because of the presence of undissolved Fe_2Nb and NbC particles at the grain boundaries. These particles are well known to help preventing cracks at the grain boundaries from propagating, resulting in an apparent increase in total creep ductility [1,9].

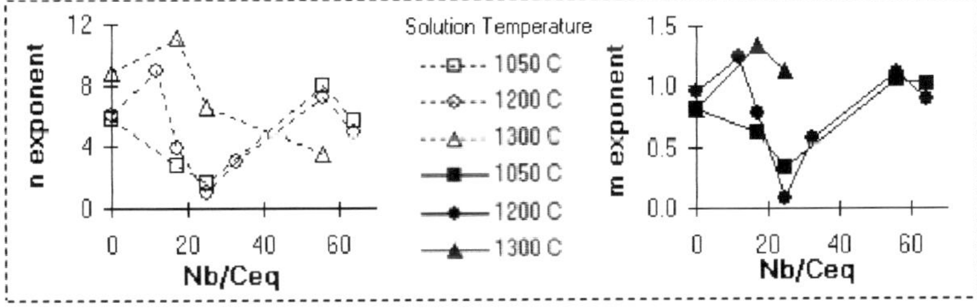

Figure 5. Dashed lines: effect of Nb/Ceq ratio on n exponent of creep power law equation. Solid lines: effect of Nb/Ceq ratio on m-exponent of the Monkman-Grant equation.

4. Conclusions

The reduction in grain size and the increase in the amounts of niobium dissolved in the matrix contribute to the increase in hardness of the studied steels. However, the main hardness-controlling factor is the presence of undissolved NbC and Laves phase particles, which overcome the effects of grain size and niobium into solution.

The lowest secondary stage creep rates occur for the alloys which present higher amounts of niobium carbide available for precipitation during creep test. The greatest rupture times are related to the alloys which presented higher amounts of undissolved phases after the solution treatments.

The presence of undissolved Fe_2Nb and NbC particles makes crack propagation at grain boundaries more difficult, resulting in an increase in rupture time and apparent ductility.

Niobium additions with distinct Nb/Ceq relations and the use of different solution temperatures resulted in a large variations in the n values, in the range from about 1 to 12, which indicate the possibility of operation of various creep mechanism in these steels.

The lower n values were obtained for the conditions where the amounts of niobium carbide available for precipitation are higher, which indicates predominance of the mechanism of diffusion creep, where $n = 1$. A trend of decrease in the values of the m-exponent can be observed for Nb/Ceq rates that promote niobium carbide precipitation, which suggest some loss in ductility and major contribution of secondary stage to the total rupture time.

References

[1] Keown S R and Pickering F B 1981 *Proc. International Symposium Niobium 81* Ed. Harry Stuart San Francisco USA Nov.8-11.
[2] Kallqvist J Andrén H O 1999 *Mater. Sci. Eng. A* **270** 27
[3] Vach M, Kuníková T, Dománková M, Sevc P, Caplovic L, Gogola P, Janovec J 2008 *Mater. Characterization* **59** 1792.
[4] Deardo A J, Gray J M, Meyer L 1981 *Proc. International Symposium Niobium 81* Ed. Harry Stuart San Francisco USA Nov.8-11.
[5] Minami Y and Kimura H 1987 *Trans. Iron Steel Inst. Japan* **27** 299
[6] Monkman F C and Grant N J 1956 *Proc. ASTM* **53** 838.
[7] Sordi V L and Bueno L O 1994 *Proc.VII Congreso Nacional de Metalurgia Conamet'94* 734 Antofagasta Chile.
[8] Pigatin W L and Bueno L O 1990 *Metalurgia-ABM* **46** (388) 242
[9] Denham A W and Silcock J M 1969 *J. Iron Steel Inst.* **May** 585
[10] Padilha A F, Machado I F, Plaut R L P 2005 *J. Mater. Process. Technol.* **170** 89

Phenomenological approach to precise creep life prediction by means of quantitative evaluation of strain rate acceleration in secondary creep

Hiroyuki Sato[1] **and Takaya Miyano**[2]

[1] Graduate School of Science and Technology, Hirosaki University
3 Bunkyo, Hirosaki, Aomori, 036-8561 Japan
[2] Department of Micro System Technology, Ritsumeikan University
1-1-1 Noji-higashi, Kusatsu, Shiga, 525-8577, Japan

g4sato@mech.hirosaki-u.ac.jp

Abstract. A method of creep life prediction by means of Strain-Acceleration-Parameter (SAP), α, is presented. The authors show that the shape of creep curve can be characterized by SAP that reflects magnitude of strain-rate change in secondary creep. The SAP-values, α are evaluated on magnesium-aluminium solution hardened alloys. Reconstruction of creep curves by combinations of SAP and minimum-creep rates are successfully performed, and the curves reasonably agree with experiments. The advantage of the proposed method is that the required parameters evaluated from individual creep curves are directly connected with the minimum creep rate. The predicted times-to-failure agree well with that obtained by experiments, and possibility of precise life time prediction by SAP is pronounced.

1. Introduction

It is widely accepted that creep behaviors are mainly evaluated by minimum or steady state creep rate. The important parameters are its stress dependence, grain size dependence and temperature dependence, and are characterized by the stress exponent, n, the grain size exponent, p, and the activation energy of creep, Q_c. One of the general forms of equations that describe steady-state or minimum strain rates is the equation proposed by Mukherjee, *et al.* [1]:

$$\dot{\varepsilon} = \frac{AD_0Gb}{kT}\left(\frac{b}{d}\right)^p\left(\frac{\sigma}{G}\right)^n \exp\left(-\frac{Q_c}{RT}\right) \tag{1}$$

Here, A, D_0, G, b, k and R are a constant, pre-exponential term of diffusion coefficient, shear modulus, the magnitude of Burgers vector, the Boltzmann's constant and the gas constant, respectively. The values T and σ are the absolute temperature and the applied stress which determine the creep conditions, and d is the grain size.

The minimum creep rates are treated as the most important parameter describing creep behavior. The minimum creep rates, however, are evaluated from just a part of creep curve, and the same minimum creep rates would be observed from creep curves whose shapes are different to each other. For more precise determination and description of creep behavior, not only the minimum creep rates, but also the shape of creep curves should be considered.

Many methods of analysis of whole creep curves have been proposed. For example, the theta projection [2], the Omega projection [3], and the improved methods [4-5] are well known. In these methods, however, the minimum creep rates are expressed indirectly from the parameters estimated from a creep curve. Because the minimum creep rates are treated as the most important parameter that represents creep curve, authors believe the minimum creep rate should be explicitly expressed as an independent parameter, and proposed one parameter that reflects the shape of a creep curve, that is declared independently on the minimum creep rates [6-7].

In this paper, changes of creep rate in secondary creep in magnesium-aluminum solution strengthed alloys are quantitatively evaluated by means of the parameter, and reconstructed creep curves by means of the parameters are shown [7]. The reconstructed creep curves show three stages of creep, and estimated time-to-failure agrees well with that obtained by experiments. The possibility of precise life prediction by evaluating strain rate acceleration in secondary creep is shown.

2. Quantitative evaluation of creep curve

2.1. Evaluation of strain-rate acceleration in secondary creep

Authors propose a characteristic value, Strain-Acceleration-Parameter (SAP), α, that reflects shape of creep curves and reflects acceleration of strain rate in secondary creep [6-7]:

$$\alpha = \frac{d^2}{d\varepsilon^2} \log_{10}\left(\dot{\varepsilon} / \mathrm{s}^{-1}\right)\Big|_{\varepsilon=\dot{\varepsilon}_{\min}} \quad (2)$$

α corresponds to the curvature of the common logarithm of strain rate as a function of strain. The value is defined at a strain of minimum creep rate, ε_{min}, and at a time of minimum creep rate, t_{min}. Because it is the second order differential of strain, the values evaluated by means of simple finite differential vary and spread depending on the precision of calculation. To avoid this difficulty, authors applied least square spline interpolation for calculation of strain rate and related derivatives.

2.2. Reconstruction of creep curve with the Strain-Acceleration-Parameter, α

As the Strain-Acceleration-Parameter, α, is defined as a curvature of the common logarithm of strain-rate as a function of strain, the creep curve can be extrapolated and reconstructed with suitable initial conditions. As α, is defined at minimum creep rate, and conditions at minimum creep rate, *i.e.*, strain and strain-rate are known, the common logarithm of strain rate, $\log\dot{\varepsilon}$, as a function of strain, ε, can be described as equation (3).

$$\log\dot{\varepsilon}(\varepsilon) = \frac{\alpha}{2}\left(\varepsilon - \varepsilon_{\min}\right)^2 + \log\dot{\varepsilon}_{\min} \quad (3)$$

Here, $\varepsilon_{\min}$ and $\dot{\varepsilon}_{\min}$ are evaluated experimentally from individual creep curves. Equation (3) can be solved numerically giving the creep curve, ε(*t*). Required values to reconstruct the whole creep curve are the SAP α, the minimum strain rate $\dot{\varepsilon}_{\min}$, the strain at minimum creep rate $\varepsilon_{\min}$, and the time at minimum creep rate t_{min}. Three of these required four parameters directly correlate with a condition at the minimum creep rate.

3. Application to creep of magnesium-aluminium solid solution alloys

3.1. Creep behaviour of magnesium-aluminium solid solution alloys at 600K.

Magnesium-aluminium solid solution alloys are typical solid solutions whose creep characteristics are divided into two classes termed Alloy-type (Class-I) and Metal-type (Class-II). In case of coarse grained solid solution alloys, the effect of grain size is negligibly small, but solute concentration affects creep rates. As the activation energy of creep is similar to that for diffusion at 600K, one can have modified equation (4) of equation (1).

$$\dot{\varepsilon} = A'\frac{DGb}{kT}N^{-m}\left(\frac{\sigma}{G}\right)^{n} \quad (4)$$

At 600K, Mg-Al solid solutions show transition from Alloy-type to Metal-type at around 20MPa, depending on solute concentration [8].

3.2. Evaluation of the Strain-Acceleration-Parameter(SAP), α

Figure 1 shows examples of evaluation of SAP α. The value defined by equation (2) is evaluated by least square spline fitting of the common logarithm of strain rate as a function of strain at the minimum strain rate. The short vertical lines indicate strain at the minimum creep rate, and the curvatures of the lines are evaluated as shown on Table 1. The α, increases with increasing solute concentration. In the alloys, α, are reasonably evaluated both in the stress range of Alloy-type and Metal-type.

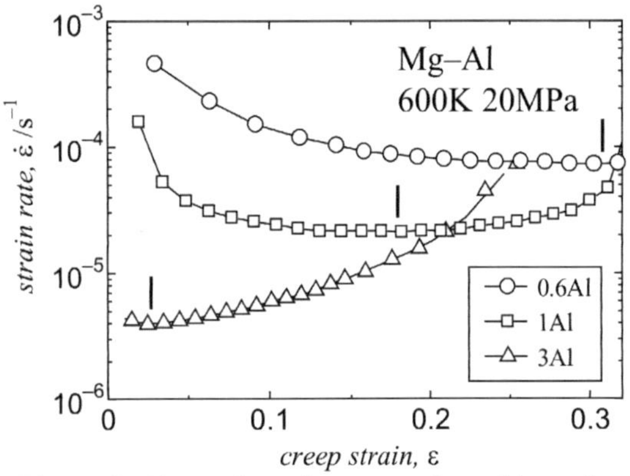

Figure 1. Examples of common logarithm of strain-rate as a function of strain. Curvature at the minimum creep rate is SAP α. [7].

Table 1. Examples of the estimated value of α, for the alloys at 600K, 20MPa.

Alloy	α	Creep behaviour
Mg-0.6Al	15	Metal-type
Mg-1Al	33	Metal-type
Mg-3Al	70	Alloy-type

3.3. Extrapolation and Reconstruction of creep curves

Figure 2 shows an example of reconstruction of creep curves, obtained by numerical integration of equation (3). The curves agree well with the experimental creep curves at the same temperature and the applied stress. At the conditions shown in Figure 2, the times to failure are almost identical in experiment and reconstruction. It is therefore, suggested that the SAP defined at the minimum creep rate reasonably reflects the shape of the whole creep curve. This means that time-to-failure can be predicted by values evaluated quantitatively in secondary creep. The SAP and three values related directly to the minimum creep rate represent the whole creep curve at a given temperature and stress.

3.4. Estimated time-to-failure by SAP α, and experiments

Figure 3 shows a comparison of the time-to-failure estimated by the numerical integration, and experiments. Under the conditions analyzed in this paper, only a limited number of curves show failure in true strain below 0.4. The experimental times-to-failure agree well with that extrapolated from the four values evaluated at a minimum creep rate.

In Mg-1Al alloy, the transition from Alloy-type to Metal-type appears at around 18MPa at 600K. The predicted time-to-failure agrees well with that obtained by experiments in both Alloy-type regime and Metal-type regime. It seems that the quantitative evaluation of creep curve by means of SAP has potential of general application for precise life time evaluation.

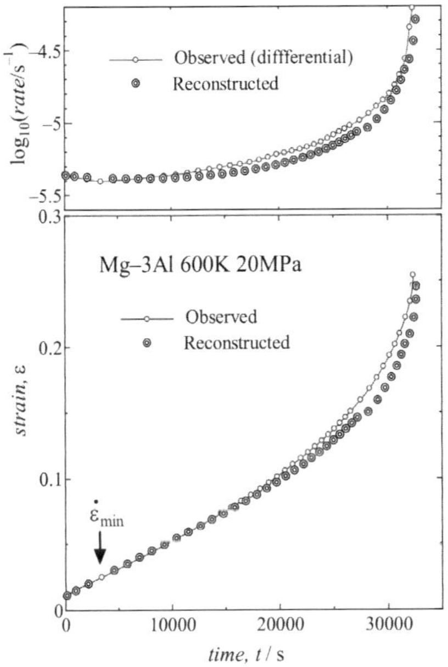

Figure 2. Example of creep curve reconstructed by SAP and experiments. Time-to-failure agree well with experiments [7].

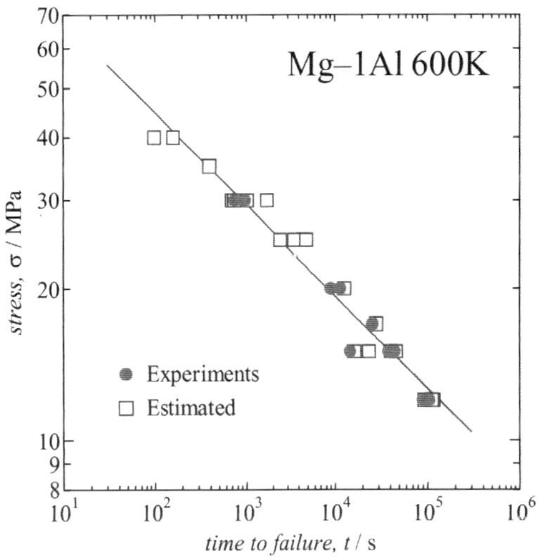

Figure 3. Comparison of time-to-failure extrapolated by SAP at a minimum creep rate (rectangle) and experiments (circle).

4. Conclusions

1. We propose Strain-Acceleration-Parameter (SAP), α, which reflects the shape of a creep curve at a minimum creep rate in secondary creep. The SAP corresponds to the curvature of common logarithm of strain rate as a function of strain.
2. A creep curve can be reconstructed from SAP and three values evaluated at condition of minimum creep rate. The reconstructed creep curve and that obtained by experiments agree well with each other in magnesium-aluminum solution strengthened alloys at 600K.
3. Time-to-failure, extrapolated based on the values evaluated in secondary creep, agree well with experiments. The possibility of precise life time prediction by means of SAP, α, has been demonstrated. The advantage of this method is required values are essentially simple, and the three of four are directly related to a minimum creep rate.
4. In general, micro structural changes during creep deformation cause strain rate change in a complex manner. Applications of this method by means of SAP to a wide variety of alloys are expected to validate the possibility of life prediction from secondary creep.

References

[1] A.K. Mukherjee, *et al.*: *The Interaction Between Dislocations and Point Defects*, Vol. II, Part III, ed. by B.L. Eyre, Atomic Energy Authority, Harwell, UK, (1968), 422-495.
[2] R.W. Evans, *et. al.*:, *Recent Advances in Creep and Fracture of Engineering Materials and Structures*, Ed. by B. Wilshire and D.R.J. Owen, Pineridge Press, U.K., (1982), p135
[3] M. Prager: *Strength of Materials*, Eds. by H. Oikawa, *et. al.*, JIM, (1984), 571-574.
[4] K. Maruyama and H. Oikawa: J. Pressure Vessel Tech., 109, (1987), 142-146
[5] K. Maruyama, *et. al.*: ISIJ International, 37 (1997), 419-423.
[6] H. Sato: Suppl.Proc., Vol.3, 137th Annual Meeting & Exhibition, TMS2008, (2008), 199-204.
[7] H. Sato: Multiple Approach to Fracture of Materials, Composite and Strucutures, 17th European Conference on Fracture, Brno, ESIS, (2008), pp.1103-1110.
[8] H. Sato and H. Oikawa: *Strength of Metals and Alloys*, Eds. by D.G. Brandon, et al., Freund Publishing House, Ltd., U.K., (1991), 463-470.

Compression creep studies of mechanically alloyed nanostructured Fe-12Cr-2W-0.25Y_2O_3 ODS alloy

P Susila[1], D Sturm[2], M Heilmaier[3], BS Murty[1], V Subramanya Sarma[1]

[1] Department of Metallurgical and Materials Engineering, Indian Institute of Technology Madras, Chennai, India-36
[2] Otto von Guericke University Magdeburg, Institute for Materials and Joining Technology, P.O. Box 4120, D-39016 Magdeburg, Germany
[3] Technical University Darmstadt, Dept. Materials Science, D-64287 Darmstadt, Germany

vsarma@iitm.ac.in

Abstract. Nanocrystalline Fe-12Cr-2W (wt.%) and oxide dispersion strengthened (ODS) Fe-12Cr-2W-0.25 Y_2O_3 (wt. %) alloy powders were produced through mechanical alloying in a high energy planetary ball mill. Microhardness studies on individual milled powder particles revealed a clear increase of hardness for the Yttria containing alloy, thus, proving the ODS effect to be operative in the powders. Consolidation was carried out in a uniaxial hot press at 900°C with a pressure of 200 MPa. TEM and XRD analyses consistently revealed the nanocrystalline grain size before (12 to 30 nm) and after consolidation (120 nm), respectively. Compression creep studies in a temperature range between 800 and 1000°C revealed that diffusional creep is obviously suppressed in these materials. However, the creep resistance of the ODS ferritic steel is only marginally higher than its base alloy and both alloys are by far less creep resistant than that of a ferritic steel of comparable baseline composition strengthened by complex and exceptionally thermally stable Ti-Y-O nanoclusters [1]. Both differences can be rationalized by the instability of the microstructure leading to significant particle and grain growth after creep testing.

1. Introduction

Since several decades, oxide dispersion strengthened (ODS) ferritic steels have been discussed for high temperature applications in general and for nuclear applications in particular [2, 3]. This stems from the fact that these nanosized, incoherent and insoluble particles restrict the dislocation motion even at temperatures close to the melting point of the base material. Further, ferritic ODS steels possess a beneficial corrosion resistance [4], oxidation resistance [5] along with the resistance to irradiation damage, especially when alloying with heavy elements such as W [3]. ODS alloy processing is commonly employed by mechanical alloying (MA) as a high energy ball milling process which has the potential to achieve a uniform and controlled dispersion of second phase particles in various metallic matrices [6-8]. The present compression creep study has been motivated by recent reports on the unusually high creep strength of ultrafine grained (UFG) ferritic steels with composition of Fe-(12-14)Cr-(2-3)W-0.4Ti-0.25Y_2O_3 [1,9,10]. It was claimed there that this unparalleled thermally stability is caused by complex and thermally extremely stable Ti-Y-O nanoclusters (with diameters as low as 2 to 4 nm) pinning the grain boundaries as well as lying in the grain interior [3, 10]. Likewise, the observed enrichment of W at grain boundaries was discussed as a possible contributor to the above findings [1]. Therefore, we synthesized ferritic Fe-12Cr-2W-0.25Y_2O_3 ODS steels without titanium as an alloying element in order (i) to shed light on the role of tungsten on creep strength and microstructural stability and (ii) to prove the effect of above Ti-nanoclusters on creep resistance.

2. Experimental

Elemental powders with nominal composition of Fe-12Cr-2W (all in wt.%) were mechanically alloyed (MA) in a Fritsch Pulverisette-5 planetary ball mill with WC vials and balls of 10mm diameter. Nanocrystalline Y_2O_3 was obtained by ball milling separately and about 0.25wt.% of this was added to the elemental powders. Milling was carried out at 300 rpm in all the cases with toluene as process control agent (PCA). The alloys were milled in air atmosphere with a ball to powder ratio (BPR) of 10:1. Powders were collected at different milling times viz. 5, 10, 15, 20, 25 and 30h and were characterized with a Shimadzu X-Ray diffractometer (XRD) using Fe-K_α radiation (λ = 1.9314 Å) to study the milling progress. The hardness of powders was measured using a MHT-10 microhardness tester (load of 5g, dwell time of 10s). Milled powders were consolidated by vacuum hot pressing at 900°C and with a pressure of 200 MPa for 1h under a vacuum of 10^{-6} mbar to obtain about 99% of theoretical density. Compression creep studies were performed on hot pressed samples with a size of 6x3.5x3.5 mm^3 in a screw-driven Zwick Z100 between 800 and 1000°C. Creep testing was either carried out at strain rates from 10^{-3} to 10^{-5} s^{-1} until reaching a constant stress level or at constant creep stress σ to extend the experimental range to lower strain rate levels. Transmission electron microscopy (TEM) analysis of milled powders and consolidated and crept samples was done using a Philips CM20 microscope operating at 200 kV. All bulk samples for TEM analysis were thinned by ion milling.

3. Results and Discussion

As commonly expected during MA, broadening of X-ray diffraction (XRD) peaks with increasing milling time was observed and this is due to the crystallite size reduction and lattice strain increase. The latter were quantitatively evaluated applying Voigt function for XRD peak profile analysis [11] yielding crystallite sizes in Fe-12Cr-2W alloy and the ODS alloy range between 14 nm and 35 nm. These results are in good correspondence with the TEM analysis of the milled ODS alloy powder: quantitative assessment of figures such as 1a yielded an average grain size of 14nm (Figure 1b). Microhardness measurements on milled powders mounted on a resin clearly revealed the direct strengthening of the dispersoids: comparing powders of both the ODS alloy and its particle free counterpart at an equivalent grain size of e.g. 200nm revealed an increase in hardness from 300 $HV_{0.005}$ to about 750 $HV_{0.005}$. After consolidation, (see Figures 1c and d) the ODS ferritic steel is still nanostructured with a mean grain size (intercept lengths measured on TEM micrographs were corrected with a shape factor of 1.5 assuming spherical grains) of 120 nm. Also, the particle free Fe-12Cr-2W alloy remains in the nanostructured state (~ 115 nm nm mean grain size, not shown). From figures like 1c also the mean dispersoid size was estimated to be about 18nm. Note, that this value is already by a factor of 5 larger than those determined for the Ti-Y-O nanoclusters [1]. Calculating the Zener grain size from the above particle parameters according to $d_Z = 4r/(3f)$ where r = 9nm is the mean radius of dispersoids and f = 0.004 their volume fraction (calculated from nominal composition) yields a predicted grain size of ~ 3000 nm. This value is significantly larger than the one measured in Figure 1d (~ 120 nm) and clearly indicates that the consolidated material is still far from thermodynamic equilibrium. The consequences on the creep properties will be discussed next.

Figure 2 summarizes the creep data on the Fe-12Cr-2W alloys with and without Yttria in the temperature range between 800 and 1000°C. As both materials were clearly seen to be nanostructured after consolidation (Figure 1b and d), diffusional creep may be the rate controlling mechanism and this will be checked first. Creep rate as a function of applied stress is plotted in Figure 2 as straight lines for (volume diffusion D_V dominated) Nabarro-Herring (NH) creep according to

$$\dot{\varepsilon} = 40\frac{D_V \Omega \sigma}{kTd^2} \tag{1}$$

where Ω is the atomic volume, k Boltzmann's constant, and d the grain size [12]. Obviously, both materials do not obey this relation, neither in terms of absolute creep rate values nor in possessing a comparable slope of the strain rate – stress relation, i.e. stress exponent n = 5…11 for our alloys vs. unity for NH creep. Rather, their creep resistance is by several orders of magnitude higher. Note, that also the comparison with (grain boundary diffusion controlled) Coble creep would yield similar offsets. Second, in contrast to the hardness measurements discussed before the effect of oxide dispersion strengthening is only moderately visible if at all as the creep curves of both variants seem to overlap at all temperatures tested. Third, compared with the Ti-nanocluster strengthened steel (red line in Figure 2, data taken from [1]) the present ODS alloy is indeed by far less creep resistant. The reason for this peculiar behaviour will be explained next.

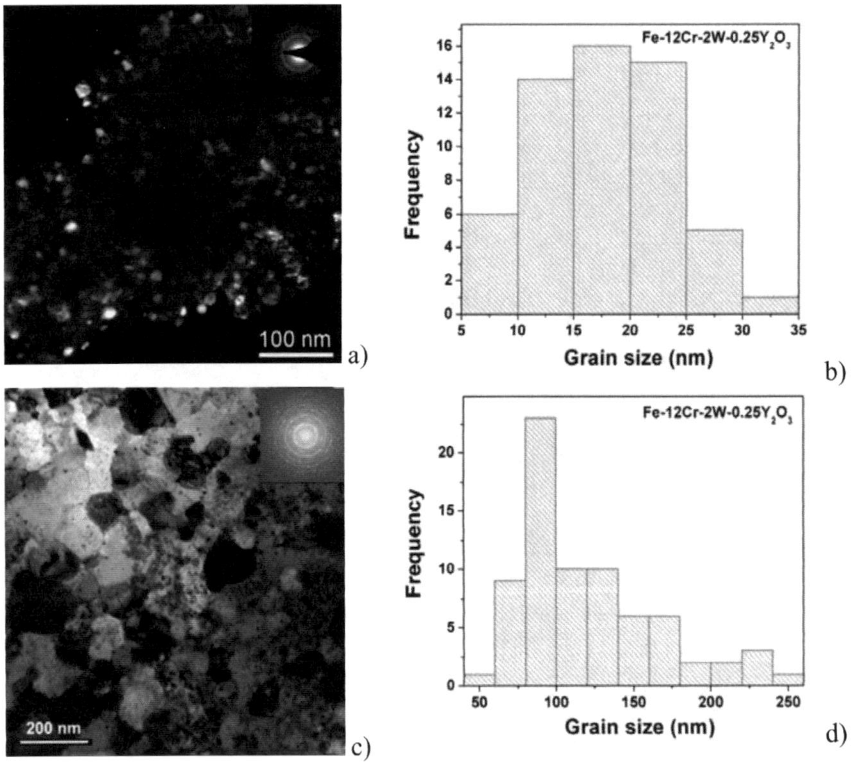

Figure 1. TEM microstructures and grain size distributions of MAed powders (a and b) and consolidated (c and d) Fe-12Cr-2W-0.25Y_2O_3 steel. Average grain sizes after milling and consolidation are 18 and 120 nm, respectively. The mean dispersoid diameter after consolidation is 18 nm.

Post mortem TEM analysis of a sample crept at 800°C is shown in Figure 3. Quantitative evaluation of both, grain size distribution (Figure 3a) and dispersoid sizes reveal that the microstructure of the ODS Fe-12Cr-2W-0.25Y_2O_3 alloy is instable during high temperature creep exposure. This is surprising as ODS alloys are usually known for their thermal stability [6-8]. However, as already critically reviewed in [7, 8] the sole criterion of choosing a compound with the largest (negative) enthalpy of formation for effective dispersion strengthening is obviously not sufficient. Rather, solubility as well as diffusivity of both, cationic and anionic component of the oxide has to be taken into account [7, 8]. Consequently, here the mean dispersoid radius has grown from 9 to r = 28nm after creep. Again, calculating the Zener grain size yields now $d_Z \approx 9\mu m$ which is in far better agreement with the measured mean value of ~ 2.6μm. This indicates that the material is much closer to thermodynamic equilibrium after creep. Hence, from above findings it can be concluded that (i) the role of W is of minor importance and (ii) Ti additions are of paramount importance for obtaining UFG ferritic ODS steels with unparalleled microstructural stability and creep resistance.

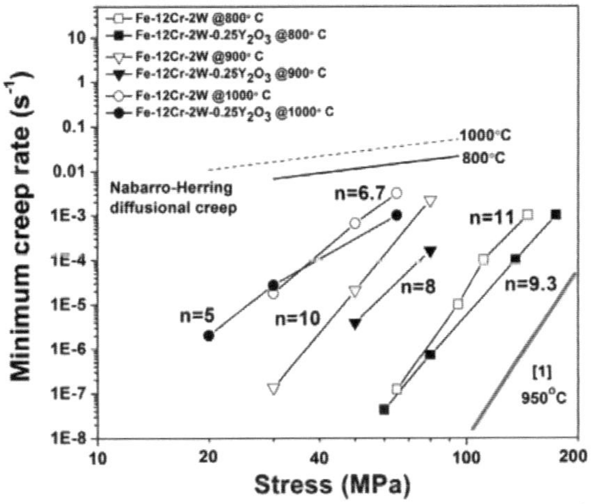

Figure 2. Minimum creep rate vs. applied stress of ODS Fe-2Cr-2W (full), Fe-12Cr-2W (open symbols) and literature data at 950°C for UFG nanocluster strengthened steel [1]

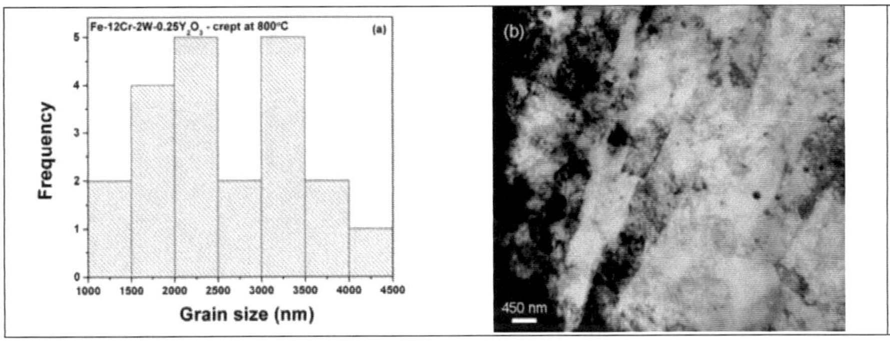

Figure 3. TEM microstructure (b) and grain size distribution (a) of MAed ODS steel crept at 800°C. The mean dispersoid diameter after creep is 56 nm.

Acknowledgements

Financial support of the Indian DST and the German DAAD is gratefully acknowledged. We thank J.H. Schneibel, C.T Liu, M.K. Miller (ORNL, Oak Ridge, TN), and M.J. Mills and P. Sarosi (Ohio State University, Columbus, OH) for fruitful collaboration on the nanocluster steels.

References

[1] Schneibel J H et al. 2009 *Scripta Mater.* **61** 793
[2] Ukai S and Fujiwara M 2002 *J. Nuclear Mater.* **307-311** 749
[3] Klueh R L et al. 2002 *J. Nuclear Mater.* **307-311** 773
[4] Cho H S, Kimura A, Ukai S and Fujiwara M 2004 *J. Nuclear Mater.* **329-333** 387
[5] Pint B A and Wright I G 2002 *J. Nuclear Mater.* **307-311** 763
[6] Benjamin J S 1970 *Metall. Trans.* **1** 2943
[7] Grundmann U, Heilmaier M and Schultz L 1997 *Scripta Mater.* **37** 103
[8] Kudashov D V, Martin U, Heilmaier M and Oettel H 2004 *Mater. Sci. Eng. A* **387-389** 639
[9] Hoelzer D T et al. 2007 *J. Nuclear Mater.* **367–370** 166
[10] Hayashi T, Sarosi P M, Schneibel J H and Mills M J 2008 *Acta Mater.* **56** 1407
[11] Keijser T H, Langford J I, Mittemeijer E J and Vogels A P B 1982 *J. Appl. Cryst.* **15** 308
[12] Poirier J.-P. 1985 *Creep of Crystals* (Cambridge University Press)

Evolution of microstructures during creep in TiAl-base intermetallics with a different Nb content

A Dlouhý, A Orlová and K Kuchařová

Institute of Physics of Materials, AS CR, Žižkova 22, 616 62 Brno, Czech Republic

dlouhy@ipm.cz

Abstract. An evolution of microstructure has been investigated with increasing compression creep strain in intermetallic alloys Ti-48Al-2Cr-2Nb-1B (alloy Nb2) and Ti-46Al-7Nb-0.6Cr-0.2Ni-0.1Si (alloy Nb7) loaded to 350 MPa at 973 K. Scanning electron microscopy and transmission electron microscopy analysis focused on individual deformation modes and on the phase stability of the alloys. Dislocation densities and deformation twinning characteristics were systematically evaluated for strains up to 0.38. Results of these quantitative studies showed that the evolution of total and non-zero <c>-component dislocation densities with strain was similar for all the investigated materials in spite of relevant differences in the initial γ/α_2 microstructures. However, the spacing between deformation twins seemed to scale with the creep strength of the two alloys. These results are discussed in terms of the fundamental deformation modes and their contributions to the strain accumulation kinetics during high temperature creep in TiAl-base intermetallics with a different Nb content.

1. Introduction

TiAl-base intermetallic alloys of the 2nd and 3rd generation are attractive engineering materials for gas turbine and car engine applications [1, 2], but they also represent interesting systems for basic high temperature creep studies [3, 4]. A transition from 2nd to 3rd generation alloys was based on new alloying strategies, which promoted materials with higher Nb content [5]. While the positive influence of Nb on room temperature mechanical properties was soon appreciated [6, 7], the influence of Nb alloying on high temperature strength has not yet been fully understood [4, 8]. There seems to be a general agreement in that the additions of Nb exceeding 5 at% cause an increase of creep strength. Controversies still exist regarding the nature of the strengthening mechanism [4]. Since the solid solution contribution is less important in the creep regime [9], the strengthening effect due to Nb additions was suggested to reflect either a decrease of the stacking fault energy or a limited diffusivity of Nb atoms and/or better thermal stability of γ/γ and γ/α_2 lamellar microstructures [4]. However, the investigations performed so far did not address the evolution of deformation structures with strain in terms of dislocation densities and deformation twinning characteristics and their potential dependence on the Nb content. Therefore, in the present study we focus on two alloys with different Nb content (Ti-48Al-2Cr-2Nb-B and Ti-46Al-7Nb-0.6Cr-0.2Ni-0.1Si) and investigate their high temperature creep and associated microstructural changes in detail. The quantitative assessment of dislocation structures is based on a transmission electron microscopy (TEM) technique in which dislocations with non-zero component of the Burgers vector into the <c>-axis of the $L1_0$ lattice (superdislocations and twinning dislocations) are systematically displayed with a reflection condition $\mathbf{g} = (002)$ [10 – 13].

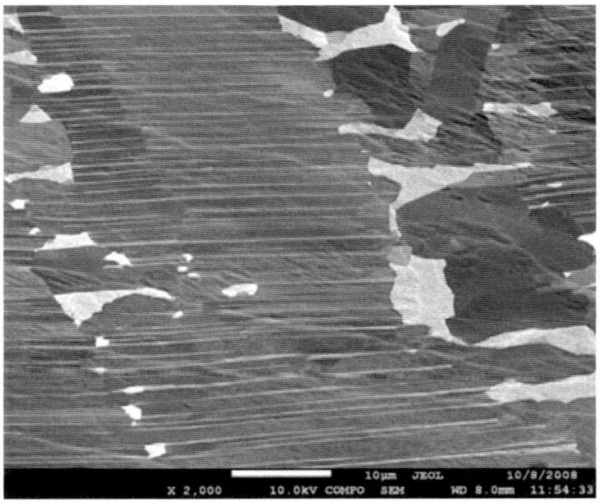

Figure 1. SEM BSE micrograph illustrating prior-to-creep microstructure of the alloy Nb7. The microstructure consists of γ grains and lamellae (dark) and α_2 grains and lamellae (bright).

2. Materials and experiments

The Ti-48Al-2Cr-2Nb-1B (at%) alloy (**Nb2**) was plasma melted at the IRC Birmingham. The alloy was isothermally forged at 1423 K and a strain rate of $5\cdot10^{-3}$ s^{-1} with a height reduction of 70%. Parts of the resulting pancake received three different heat treatments in order to generate different prior-to-creep microstructures: (i) annealing at 1473 K for 4 hours followed by furnace cooling – the resulting microstructure was equiaxed with a low α_2 volume fraction of 1.2% (ELA), (ii) annealing at 1658 K for 1 hour followed by furnace cooling – the resulting microstructure was nearly lamellar (NL) with 4.8 vol.% of α_2 and (iii) after the heat treatment under (i), the alloy was additionally annealed at 1523 K for 1 hour which resulted in Widmanstätten-like distributed α_2 precipitates that occupied high volume fraction of 17.9 % in equiaxed γ grains (alloy version EHA). Further details on the heat treatments and prior-to-creep microstructures of the Nb2 alloy were published elsewhere [10].

The Ti-46Al-7Nb-0.6Cr-0.2Ni-0.1Si (at%) alloy (Nb7) was melted, cast and delivered by Flowserve Dayton in a form of a cylindrical rod the length and diameter of which were 1200 mm and 70 mm, respectively. The rod was HIPped after casting and no additional heat treatment was applied. The volume fraction of the α_2 phase prior-to-creep was close to 5%. Further information on the processing of the Nb7 alloy can be found in [11].

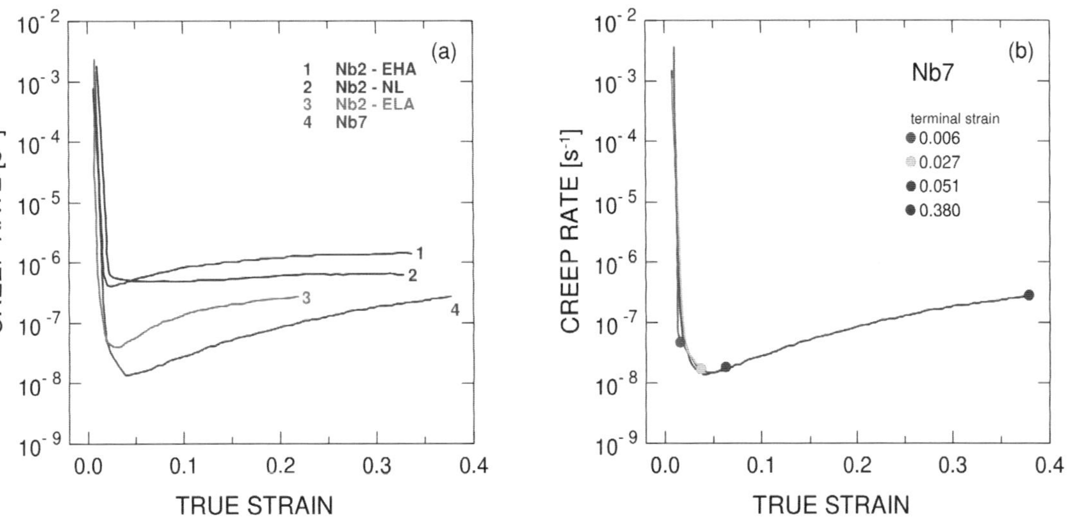

Figure 2. Compression creep curves recorded at 973 K and 350 MPa. (a) A comparison of alloys Nb2 (microstructural variants EHA, NL and ELA) and Nb7 (as cast and HIPped). (b) Interrupted tests of the alloy Nb7.

Compression creep tests were performed in purified argon at 973 K and under the constant applied stress of 350 MPa. The tests were interrupted at pre-selected strains and the corresponding microstructures were investigated. Pre-deformed compression cylinders were cut axially; one surface of the axial cut was electrolytically polished and observed by scanning electron microscopy (SEM). As an example, the SEM image in Figure 1 was taken using back scattered electrons (BSE) and documents the prior-to-creep microstructure of the alloy Nb7, which consists of equiaxed γ ($L1_0$ crystal lattice) and α_2 ($D0_{19}$ crystal lattice) grains surrounded by γ/γ and γ/α_2 lamellar colonies. Thin foils cut from the other half of the pre-deformed cylinders were investigated using (scanning) transmission electron microscopy ((S)TEM). Further details on the creep testing and the quantitative assessment of microstructures were published elsewhere [3, 11, 12].

3. Results

A stress of 350 MPa was applied to the compression cylinders at 973 K. The corresponding creep data are plotted in Figure 2. Creep curves obtained for the alloy Nb2 (three microstructural states ELA, NL and EHA) and the alloy Nb7 (as-cast and HIPped state) are compared in Figure 2a. While the creep

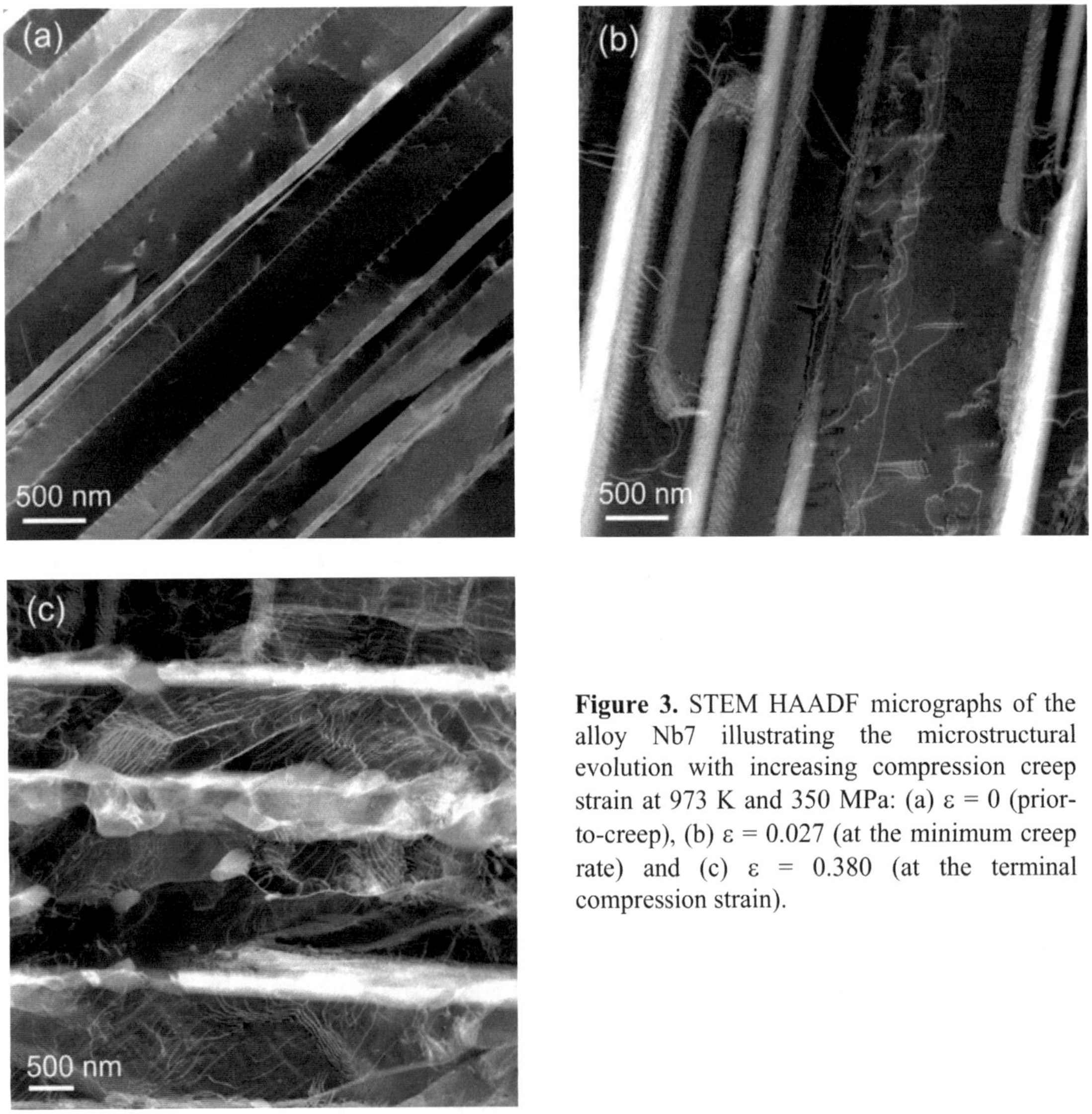

Figure 3. STEM HAADF micrographs of the alloy Nb7 illustrating the microstructural evolution with increasing compression creep strain at 973 K and 350 MPa: (a) $\varepsilon = 0$ (prior-to-creep), (b) $\varepsilon = 0.027$ (at the minimum creep rate) and (c) $\varepsilon = 0.380$ (at the terminal compression strain).

rates of the alloy Nb2 tested in versions NL and EHA are rather similar throughout the investigated strain range, the microstructural version ELA creeps clearly slower. The alloy Nb7 then exhibits even smaller creep rates at all strains. Figure 2b summarizes the interrupted creep experiments performed for the alloy Nb7 and indicates the creep strain levels at which the individual tests were terminated (0.6, 2.7, 5.1 and 38%) in order to analyse corresponding deformation structures.

Examples of high angular annular dark field (HAADF) STEM images are presented in Figures 3a (prior-to-creep microstructure of the alloy Nb7), 3b (after 2.7% of compression strain) and 3c (at the terminal compression strain 38%). The STEM images in Figs. 3a and b clearly show that, while the total dislocation density in the γ–phase increases considerably, the γ/γ and γ/α_2 lamellar interfaces are quite stable and the corresponding interlamellar spacing does not change markedly during primary creep. The total dislocation density (Figure 4a) and the density of dislocations with non-zero <c>-component of the Burgers vector (Figure 4b) seem to evolve similarly with increasing compression creep strain in all investigated material states in spite of relevant differences in their γ/α_2 prior-to-creep microstructures. On the other hand, the distance λ between deformation twins is generally higher for the ELA version of the alloy Nb2 and for the alloy Nb7, as documented in Figure 4c. Since the spacing λ characterizes the intensity of twinning in the individual pre-strained materials, the plots in Figure 4c suggest that the NL and EHA versions of the alloy Nb2 exhibit correspondingly higher twinning activity throughout the investigated range of strain.

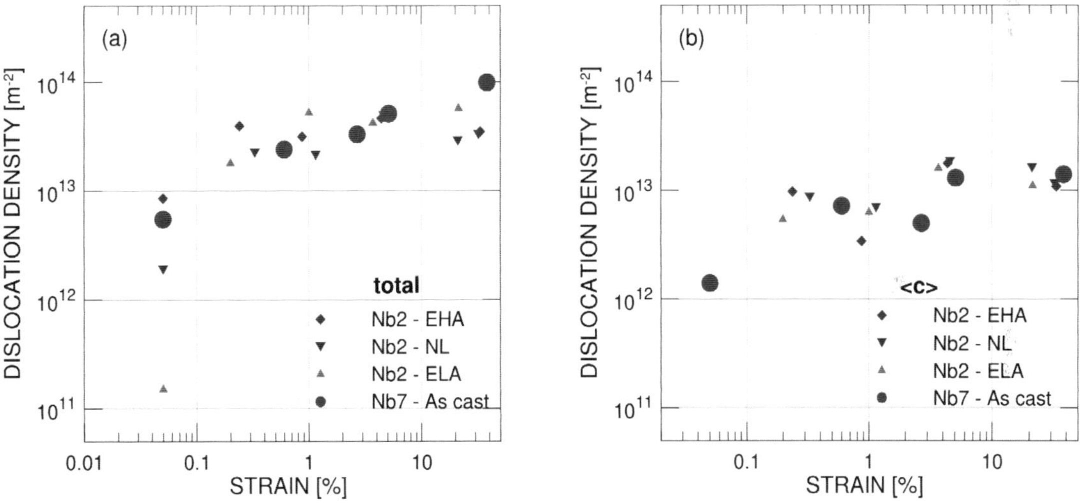

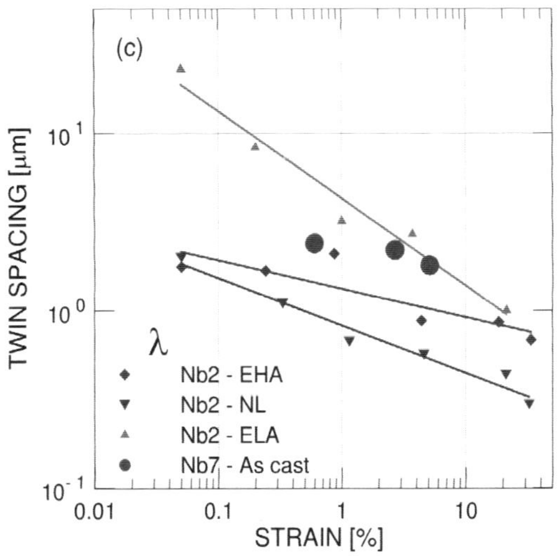

Figure 4. Characteristics of the microstructure after creep in the alloys Nb2 and Nb7 and their evolution with compression creep strain: (a) overall dislocation density, (b) density of dislocations with non-zero <c>-component of Burgers vector and (c) spacing between deformation twins.

4. Discussion

The creep curves presented in Figure 2a show that the creep strain accumulation kinetics is similar for microstructural modifications EHA and ELA of the alloy Nb2 and for the alloy Nb7. In all the three cases, the creep rate decreases by several orders of magnitude in a rather narrow strain range before a minimum creep rate is attained at creep strains of about 3% or less. During this primary stage the total and <c>-component dislocation densities grow by at least one order of magnitude, see Figures 4a and 4b. We note that the <c>-component dislocation density represents approximately one fourth of the total density at comparable strains. On further straining, the hardening-type kinetics changes to an accelerating stage during which creep rate increases and seems to seek for a new steady value, see the curves recorded for material states Nb2-EHA, Nb2-ELA and Nb7. Thus the alloys with considerably different prior-to-creep γ/α_2 microstructures follow the similar type of accelerating stage. Moreover, corresponding γ/α_2 microstructures exhibited no significant changes up to strains well behind the range where the creep rate reaches a minimum. Therefore, we conclude that expected microstructural instabilities like coarsening of γ and α_2 lamellae [4] cannot account for the creep behaviour observed in the present study. Furthermore, similar acceleration stages have also been observed during creep of Al-rich binary alloys that fully consisted of equiaxed γ grains and did not contain any α_2 phase either in the form of lamellae or particles [12]. In the present study, creep experiments were performed at a relatively low temperature of 973 K and, while some recrystallization was detected in the Nb7 specimen subjected to the pre-deformation of 38%, no evidence of recrystallization was found for the other Nb7 microstructures pre-deformed to lower strain levels. These results combined with a fact that strains, at which creep rates exhibit a minimum, are small (between 2 and 3%) strongly support the conclusion that the dynamic recrystallization can be excluded as a reason of the softening during the acceleration stages.

In our earlier studies [11, 12] we have shown that the creep strain accumulation kinetics of the type shown in Figure 2 for the Nb2 alloy modifications EHA and ELA and for the alloy Nb7 can be reasonably well accounted for when combinations of individual deformation modes operating in the $L1_0$ γ-TiAl lattice are taken into account. Mecking and co-workers [13] pointed out that a proper partitioning of the overall strain accumulated in γ-TiAl intermetallics among contributions due to ordinary dislocations, superdislocations and deformation twinning is necessary to fulfil the von Mises criterion. This principle is still important in the creep regime in order to avoid excessive internal stress concentrations. As the critical resolved shear stress for the ordinary slip is small relative to the other two deformation modes, a stress redistribution during primary creep is needed in a majority of cases to promote a sufficient activity of deformation twinning and to activate sources of <c>-component dislocations [11]. Figure 4c shows that a clear difference in the twinning activity exists when the NL modification of the alloy Nb2 is compared to the modification Nb2-ELA or the alloy Nb7. As a consequence, the NL modification of the alloy Nb2 exhibits a different type of creep strain accumulation kinetics (Figure 2a) and the last two material states (Nb2-ELA and Nb7) creep at low creep rates. A contribution to the superior creep strength of the alloy Nb7 may also be due to a drop of diffusion coefficients associated with the Nb alloying [4]. The low twinning activity in the alloy Nb7 is rather unexpected based on the low stacking fault energy argument [4] and needs a further investigation.

5. Conclusions

Creep strength of the TiAl base alloys Nb2 and Nb7 tested at 973 K and 350 MPa increases in a sequence Nb2 - NL, Nb2 - EHA, Nb2 - ELA and Nb7 - as cast. The evolution of the total and <c>-dislocation densities with increasing compression creep strain is similar for all the four investigated material states. Stronger alloy versions exhibit less twinning activity and may thus carry higher internal stresses due to the incompatibility of the plastic strain. Results of this study suggest that neither the overall content of the α_2 phase nor its evolution with the increasing creep strain could account for the observed variations of the creep strength.

Acknowledgement

The authors would like to thank to Dr. R. Ravelle-Chapuis (JEOL Europe), Dr. M. Svoboda (IPM Brno), Ing. T. Bartak (IPM Brno) and Dr. Ch. Somsen (RU Bochum) for the assistance during the SEM and STEM experiments. Financial support has been received from the Czech Science Foundation and from the Grant Agency of the AS CR under contracts nos. 106/07/0762 and 1QS200410502, respectively.

References

[1] Tetsui T, Shindo K, Kobayashi S and Takeyama M 2002 *Scripta Mater.* **47** 399
[2] Tetsui T 2002 *Mat. Sci. Eng. A* **329-331** 582
[3] Dlouhy A and Kucharova K 2004 *Intermetallics* **12** 705
[4] Yamaguchi M, Zhu H, Suzuki M, Maruyama K and Appel F 2008 *Mater. Sci. Eng. A* **483-484** 517
[5] Carneiro T and Kim Y-W 2005 *Intermetallics* **13** 1000
[6] Appel F, Oehring M and Wagner R 2000 *Intermetallics* **8** 1283
[7] Liu Z C, Lin J P, Li S J and Chen G I 2002 *Intermetallics* **11** 653
[8] Zhang W J, Chen G I, Appel F, Nie T G and Deevi S C 2001 *Mater. Sci. Eng. A* **315** 250
[9] Cadek J 1988 Creep *in Metallic Materials* (Amsterdam, Elsevier)
[10] Orlova A, Kucharova K and Dlouhy A 2008 *Mater. Sci Eng. A* **483-484** 109
[11] Dlouhy A, Kucharova K and Brezina J 2001 *Mater. Sci Eng. A* **319-321** 820
[12] Orlova A, Kucharova K and Dlouhy A 2004 *Z. Metallkd.* **95** 1011
[13] Mecking H, Hartig Ch and Kocks U F 1996 *Acta Mater.* **44** 1309

15th International Conference on the Strength of Materials (ICSMA-15) IOP Publishing
Journal of Physics: Conference Series **240** (2010) 012092 doi:10.1088/1742-6596/240/1/012092

A fundamental study of the high temperature deformation mechanisms of magnesium

Adrien Chapuis and Julian Driver

Ecole des Mines de Saint Etienne, Department of Material Science and Structures, UMR CNRS 5146, 158 Cours Fauriel, 42023 Saint Etienne, France

driver@emse.fr

Abstract. Single crystals of Magnesium have been deformed in plane strain compression from room temperature to 450 °C. The deformed crystals have been analysed by EBSD to identify the operative twin variants and parent rotations for comparison with the expected systems. In parallel with these experiments, we have developed a crystal plasticity model based on Taylor principles and the Schmid law, to compute the activated systems together with the twin reorientations and the lattice rotations for large strains. The CRSS values are determined by an iterative procedure that best correlates the experimental data with the numerical simulations for each orientation. From the measured stress/strain data the CRSS for twinning and some slip systems are given over this wide temperature range.

1. Introduction

Since the 1960's there have been very few basic studies of the fundamental mechanisms of plastic deformation in Mg despite their importance for controlling texture, anisotropy and ductility [1-3]. With the advent of modern EBSD it is now possible to follow local reorientations due to both slip and twinning and hence evaluate, with greater accuracy, their respective contributions to the crystal deformation. This paper describes a systematic study of the deformation of Mg crystals as a function of orientation and temperature using EBSD mapping.

Mg single crystals were compressed in a channel die at temperatures from 25 to 450°C. Overall, about 10 initial orientations have been studied, but here we limit the results to 3 significantly different cases. They correspond to the orientations (90 90 30) in which tensile twinning on {10-12} is expected, (0 90 30) where prismatic slip should occur, and the high symmetry (0 0 0) promoting multiple twinning. In parallel with the experiments, a crystal plasticity model has been developed to compute the crystal rotation due to slip and the amount and orientation of twins, as a function of the critical resolved shear stresses (CRSS) for the slip and twinning systems.

2. Experimental procedure

Pure magnesium single crystals were produced by directional controlled solidification in a mould under protective argon atmosphere in our laboratory. Samples were cut by spark cutting.

Plane strain compression tests were performed in a channel die at 25, 150, 250 350 and 450°C and a strain rate of 0.1 s^{-1}. The samples were lubricated with graphite and Teflon for high temperature tests. Just after deformation the samples were immediately quenched in cold water.

EBSD maps were made on a RD-ND plane, both to check the initial orientation and to analyze the lattice rotations and twins. In many cases the inclination of the twin plane on the RD-ND plane gives the twin variant with good accuracy, e.g. {10-12} twins have symmetrical variants of misorientation 8°, but the twin planes can have very different inclinations.

The lattice and twin orientations are given in Euler angles, using the Bunge convention taking the crystal basis as X along [10-10], Y along [-1-120] and Z along [0001].

3. Crystal plasticity analysis

The program describes the plastic deformation by slip and twinning of a rate insensitive material [4] under mixed boundary conditions. It maximizes the imposed external plastic work (1), under condition (2), where M is the generalized Schmidt factor.

$$\dot{W}=\sum_{ij}\sigma_{ij}\dot{\varepsilon}_{ij} \quad (1) \qquad\qquad CRSS\geq\sum_{ij}\sigma_{ij}M_{ij} \quad (2)$$

36 systems are used as follows: 3 basal (0001)<-12-10>, 3 prismatic {10-10}<-12-10>, 6 pyramidal I {10-11}<-12-10> and 6 pyramidal II {-12-12}<-12-1-3> slip systems together with 6 tensile {10-12}<-1011>, 6 compressive {10-11}<10-1-2> and 6 other compressive {10-13}<30-3-2> twinning systems. The orientations of the twinning planes with respect to the RD-ND plane of observation can then be compared with the corresponding experimental observations.

4. Results

4.1. Single crystals 90 90 30 (exact orientation 88 84 32)

Crystals were deformed over the temperature range 25-450°C (Figure 1). The EBSD map of crystal deformed at 350°C (figure 2) is color-coded so that [0001] along ND is red and <10-10> along ND is blue. All {10-12} tensile twin variants (noted A1 to A6) are observed and identified (Figure 3).

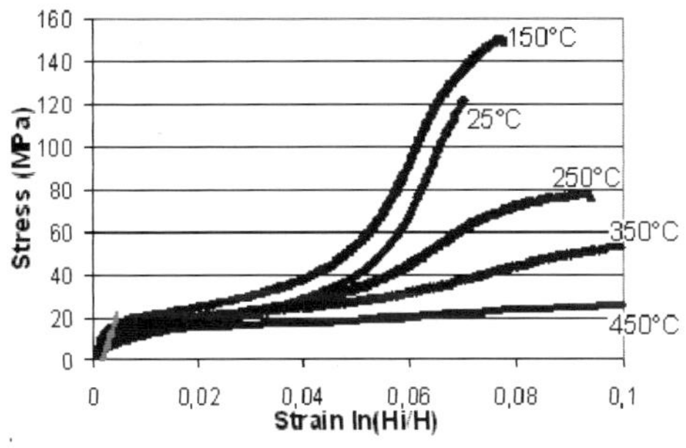

Figure 1. Stress strain curves of (90 90 30) crystals at 5 temperatures

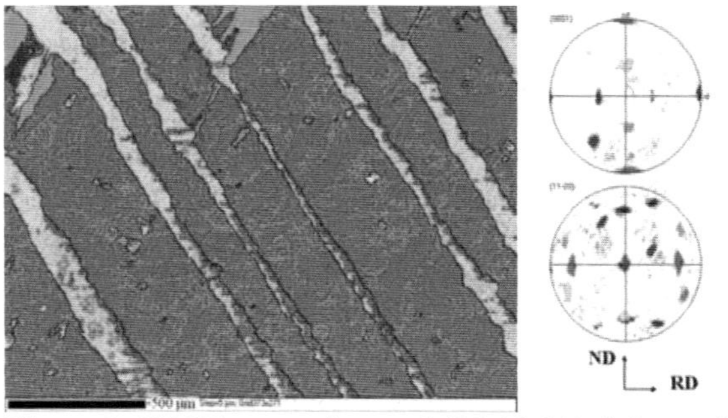

Figure 2. EBSD map of crystal (90 90 30) deformed 0.096 at 350°C, and pole figure

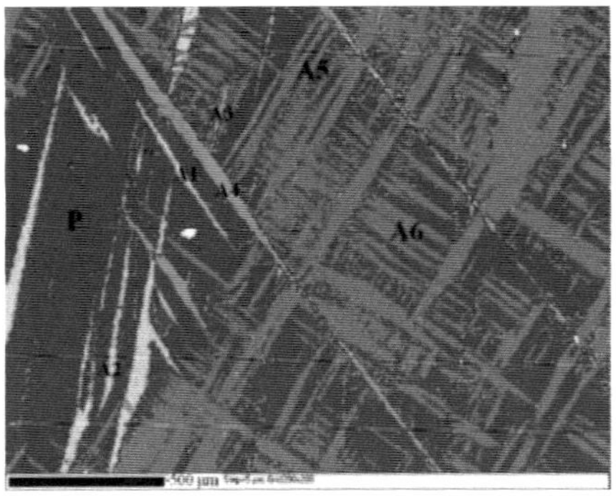

Figure 3a. EBSD map of crystal (90 90 30) deformed 0.04 at RT

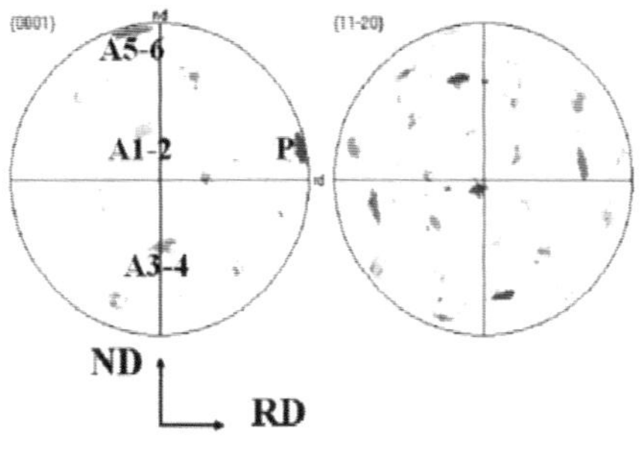

Figure 3b. pole figure of crystal (90 90 30) deformed 0.04 at RT

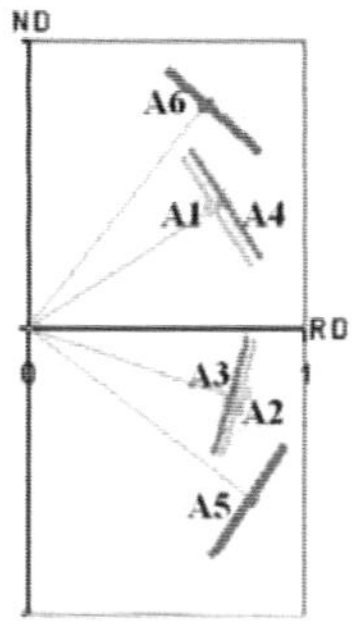

Figure 3c. scheme of twin planes

At all temperatures the dominant deformation mode is primary {10-12} tensile twinning; the two symmetrical twin variants in red dominate, but twins 2 and 3 seem to increase at high temperature.

4.2. Single crystals 0 90 30 (exact orientation 1 89 32)

This crystal is expected to undergo prismatic slip but, as also found by [2], at low temperature there is a superposition of tensile and compressive twinning. We analyzed (Figure 4) compressive {10-11} twins which double twinned into {10-11}{10-12} (noted b-a-) and which cut across {10-12} to form {10-12}{10-13} (noted a-c-). At temperatures higher than 300°C, prismatic slip is observed as the lattice rotates around $\vec{c}$ leading to a softer, stable, orientation close to 0 90 0 (Figure 5).

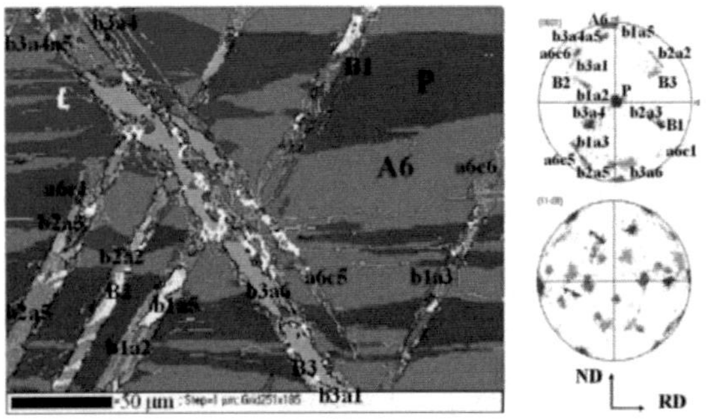

Figure 4. EBSD map of crystal (0 90 30) deformed 0.083 at 150°C and pole figure

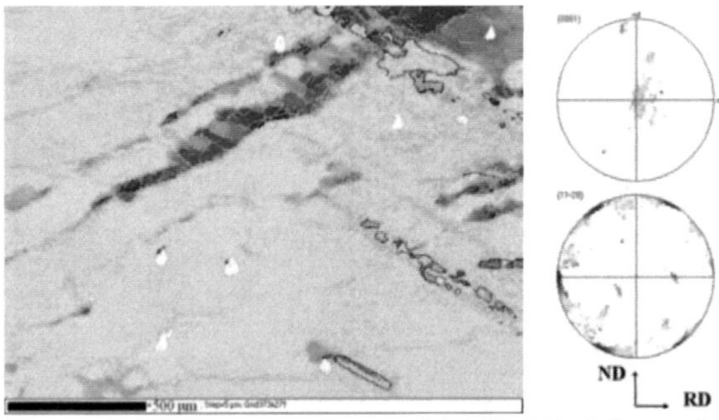

Figure 5. EBSD map of crystal (0 90 30) deformed 0.303 at 350°C and pole figure

Crystals deformed at 250°C shows recrystallized compressive twins, but the parent part does not show any clear rotation, whereas crystals deformed at 350 and 450°C rotated by prismatic or pyramidal I slip, both giving the same rotation around TD.

4.3. Single crystals 0 0 0 (exact orientation 45 1 17)

At 25°C the crystal broke quickly, but in crystals deformed at 150°C and above (Figure 6), we found many twins that could be analyzed (Figures 7, 8) as (-1103)(0-112) double twinning (noted c1a5). On Figure 8c observed twins noted C are {10-13} ; the usual {10-11} twins (noted B) are a little observed at low temperature. At 450°C a few twins were observed but deformation is mainly due to pyramidal II slip.

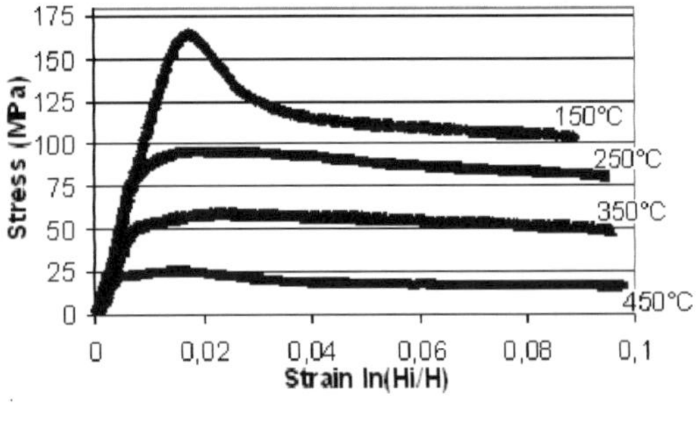

Figure 6. Stress-strain curves of crystals (0 0 0)

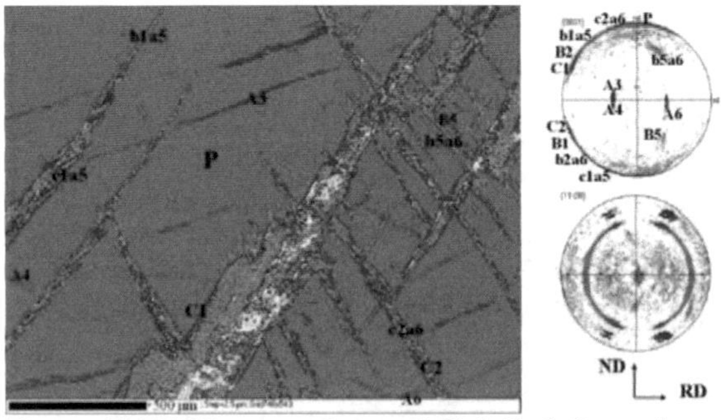

Figure 7. EBSD map of crystal (0 0 0) deformed 0.082 at 150°C and pole figure

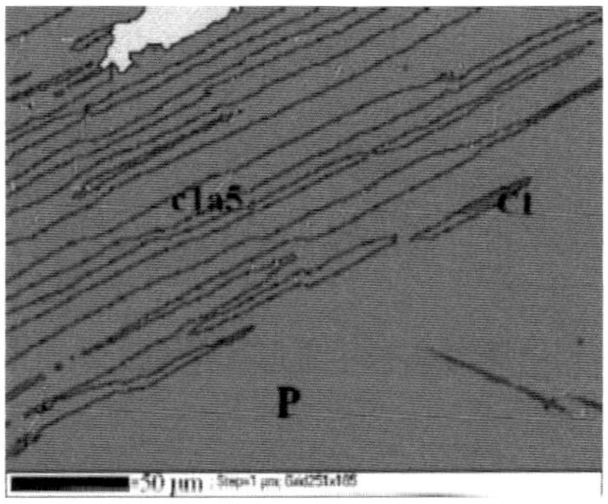

Figure 8a. EBSD map of crystal (0 0 0) deformed 0.088 at 350°C

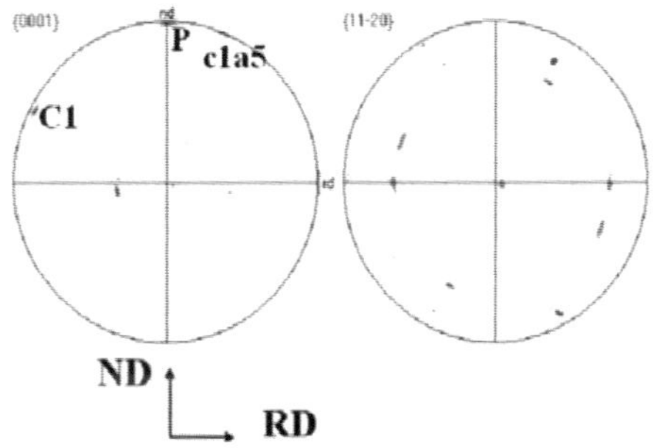

Figure 8b. pole figure of crystal (0 0 0) deformed 0.088 at 350°C

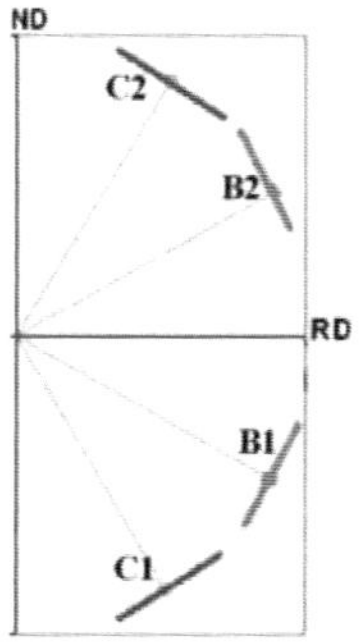

Figure 8c. scheme of twin planes

We conclude that {10-13}{10-12} double twinning occurred at all temperature, whereas usually double twinning is reported on {10-11}{10-12}.

5. Discussion and conclusion

To estimate the CRSS from the flow stress of channel die experiments, we assume proportionality between the CRSS for activated slip or twinning systems and the calculated stress.

Figure 9 summarizes, for each orientation and various temperatures, the experimental yield stress at about 0.2% strain, and the evaluated CRSS.

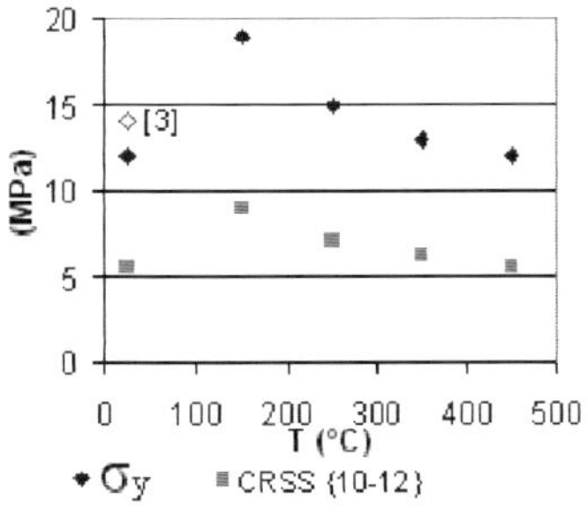

Figure 9a. CRSS of crystal (90 90 30)

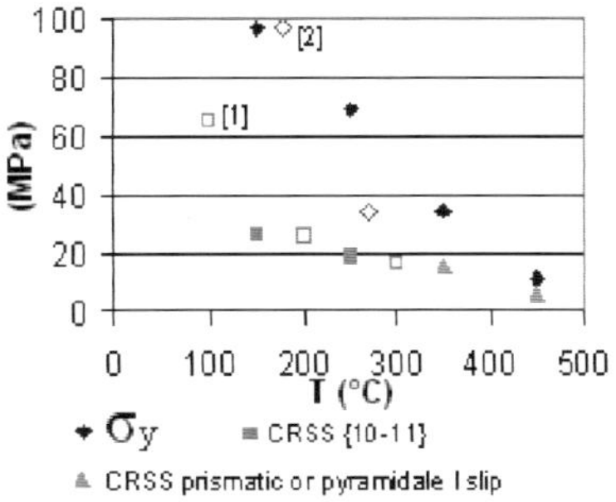

Figure 9b. CRSS of crystal (0 90 30)

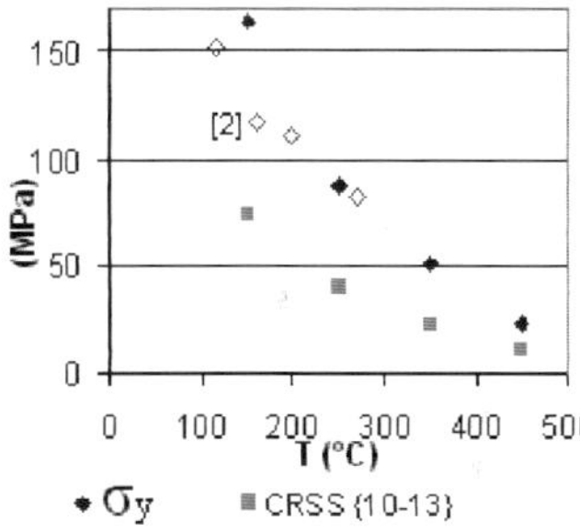

Figure 9c. CRSS of crystal (0 0 0)

For {10-11} twinning our result agree with those of Yoshinaga and Horiuchi [1] (empty points) who used uniaxial compression tests along $\vec{c}$; other channel-die experiments [2-3] also fit quite well. In general the present systematic study of the temperature dependence of slip and twinning in pure Mg gives similar flow stress results to those of previous works, when available, but we have been able to estimate the CRSS values of many systems over a wide temperature range.

References

[1] Yoshinaga H and Horiuchi R 1963 Deformation mechanisms in magnesium single crystals compressed in the direction parallel to hexagonal axis *Trans. JIM* **4**

[2] Wonziewicz B C and Backofen W A 1967 Plasticity of magnesium crystals *Trans. TMS AIME* **239** 1422

[3] Kelley E W and Hosford W F 1968 Plane-strain compression of magnesium and magnesium alloy crystals *Trans. TMS AIME* **242** 5

[4] Tomé C and Kocks U F 1985 The yield surface of h.c.p. crystals *Acta Meatll.* **33-4** 603-21

Mechanical properties of an undercooled aluminium alloy Al-0.6Mg-0.7Si

O Kessler, M Reich

University of Rostock, Faculty of Mechanical Engineering and Marine Technology, Chair of Materials Science, D-18051 Rostock, Germany

olaf.kessler@uni-rostock.de

Abstract. Compression tests of an undercooled aluminium alloy Al-0.6Mg-0.7Si have been performed in a quenching and deformation dilatometer. Samples have been solution annealed and quenched in the dilatometer with varying quenching rates (0.1 K/min to 1000 K/min) and varying quenching temperatures. Immediately after quenching, compression tests on quenching temperature have been performed in the dilatometer. The results have been correlated with the precipitation behaviour of the undercooled aluminium alloy Al-0.6Mg-0.7Si. Stress-strain-curves of quenching rates higher than the critical cooling rate for precipitation differ from those of quenching rates lower than the critical cooling rate. Further, stress-strain-curves of temperatures above and during precipitation differ from those of temperatures below precipitation.

1. Introduction

Age hardening is the most important process to increase the strength of aluminium alloys. It consists of solution annealing, quenching and aging. During quenching a supersaturated solid solution is formed, which is necessary for the following precipitation of strengthening particles. Further, during quenching thermal gradients occur in components, which can generate residual stresses and distortion. Heat treatment simulation is an important tool, to describe these processes. Therefore, thermophysical properties, precipitation behaviour and mechanical properties of the simulated alloys are required depending on temperature and microstructure [1,2]. In the case of quenched aluminium alloys, this means properties of undercooled states. The mechanical properties of these non-equilibrium microstructures are not available. Hot tensile tests of aluminium alloys can be performed at relevant temperatures, but their microstructures are near equilibrium and differ from the undercooled state.

2. Precipitation Behaviour of Al-0.6Mg-0.7Si during Cooling

The precipitation behaviour of aluminium alloys during cooling can be described by continuous cooling precipitation (CCP) diagrams. This method has recently been established by the author [3] using differential scanning calorimetry (DSC). Fig. 1 (top) shows the chemical composition and the CCP diagram of the investigated alloy Al-0.6Mg-0.7Si during cooling from solution annealing 540°C 20min. For cooling rates from some 0.1 K/min to some 100 K/min the temperatures and times of precipitation start and end are given. For this alloy three different precipitation peaks have been detected by DSC. The critical cooling rate to suppress precipitation has been estimated by hardness testing to be about 1000 K/min. The CCP-diagram has been used, to choose relevant cooling rates and temperatures for mechanical testing of undercooled states and to discuss the mechanical properties regarding microstructure and precipitation. Fig. 1 (bottom) shows light microscopy images of Al-0.6Mg-0.7Si microstructures after cooling with 100, 1 and 0.3 K/min. For slow cooling coarse (about 10 μm) rod like and plate like Mg_2Si-precipitates appear on grain boundaries as well as inside the grains. These coarse precipitates lower the particle strengthening potential during aging. With increasing cooling rates, these coarse precipitates become smaller and less before they finally disappear [3].

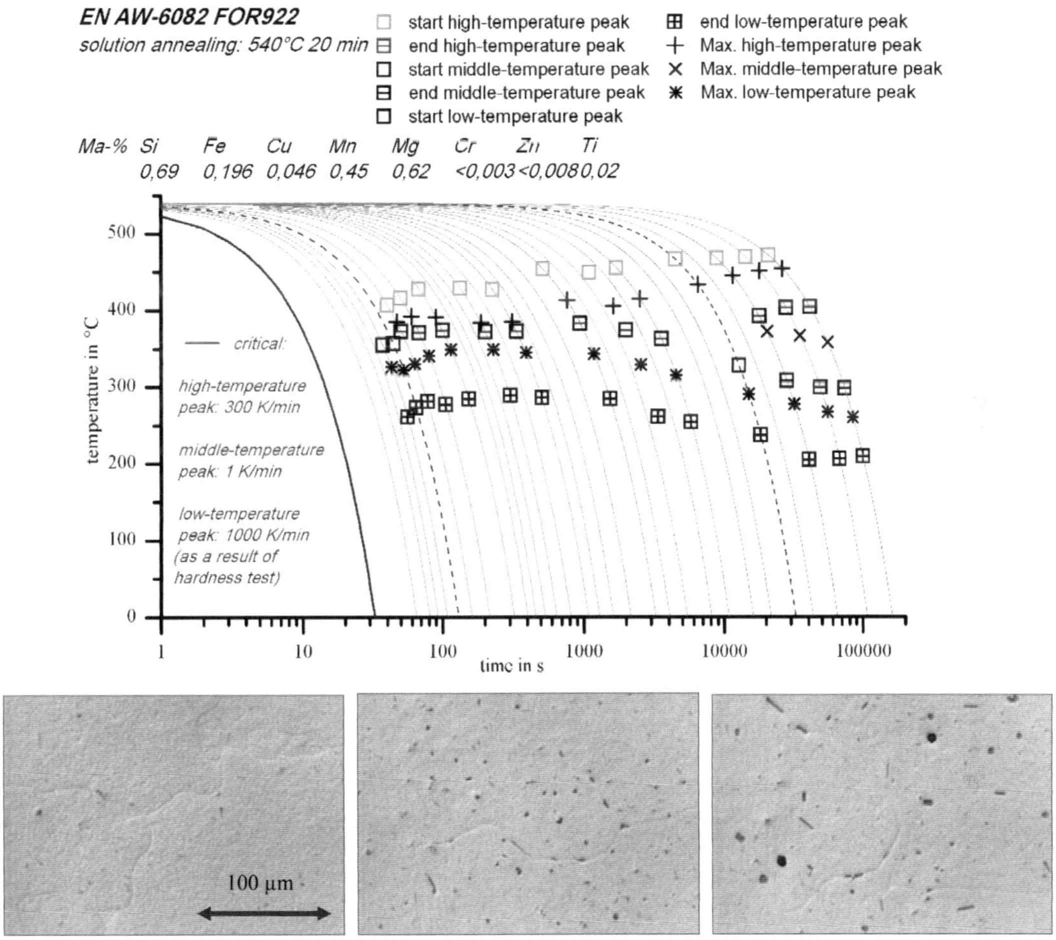

Figure 1. top: Continuous cooling precipitation diagram of Al-0.6Mg-0.7Si, bottom: microstructures after cooling with 100, 1 and 0.3 K/min.

3. Compression Tests in Dilatometer

Solution annealing, quenching with different rates on different temperatures and immediate compression testing on quenching temperature have been performed in a quenching and deformation dilatometer type Bähr 805 A/D. Solution annealing has been done at 540°C for 20min. Quenching rates of 0.1, 1, 10, 100 and 1000 K/min as well as quenching temperatures of 500, 400, 350, 250°C and room temperature (RT) have been investigated. Sample dimensions of ∅5x10mm, a deformation of 3mm and a strain rate of $0.1s^{-1}$ have been chosen. The duration between quenching and deformation amounted 1.4s, the duration of the compression test 3s. Except for the lowest cooling rate of 0.1 K/min all compression tests have been repeated three times and their flow curves have been averaged. Fig. 2 (left) shows the original DSC-curves resulting in the above CCP-diagram. It can be seen, that the quenching temperature of 500°C lays above the precipitation peaks for all cooling rates. Opposite, quenching temperatures of 250°C and RT lay below the precipitation peaks for all cooling rates. Quenching temperatures of 400°C and 350°C are situated in the precipitation interval. Further, with increasing cooling rate the DSC-peak areas (as a measure for the amount of precipitates) decrease [3]. Fig. 2 (right, top) shows exemplarily flow curves (true stress / log. total strain) of the undercooled aluminium alloy Al-0.6Mg-0.7Si at a cooling rate of 100 K/min for different quenching temperatures. The strength increases with decreasing temperature. Fig. 2 (right, bottom) shows exemplarily flow curves (true stress / log. total strain) at a quenching temperature of 250°C for different cooling rates. The strength increases with increasing cooling rate. For further evaluation the yield strength $R_{p0.2}$ has been determined depending on cooling rate and quenching temperature and the particular elastic deformation has been subtracted from all flow curves. The remaining flow curves (σ = true stress above $R_{p0.2}$ / φ = log. plastic strain) have been approximated by different mathematical models. The best fit has been achieved by the hardening model of Hockett-Sherby [4]:

$$\sigma(\varphi) = k_{fs} - (k_{fs} - k_{f0}) \cdot e^{-(N\varphi)^k} \quad (1)$$

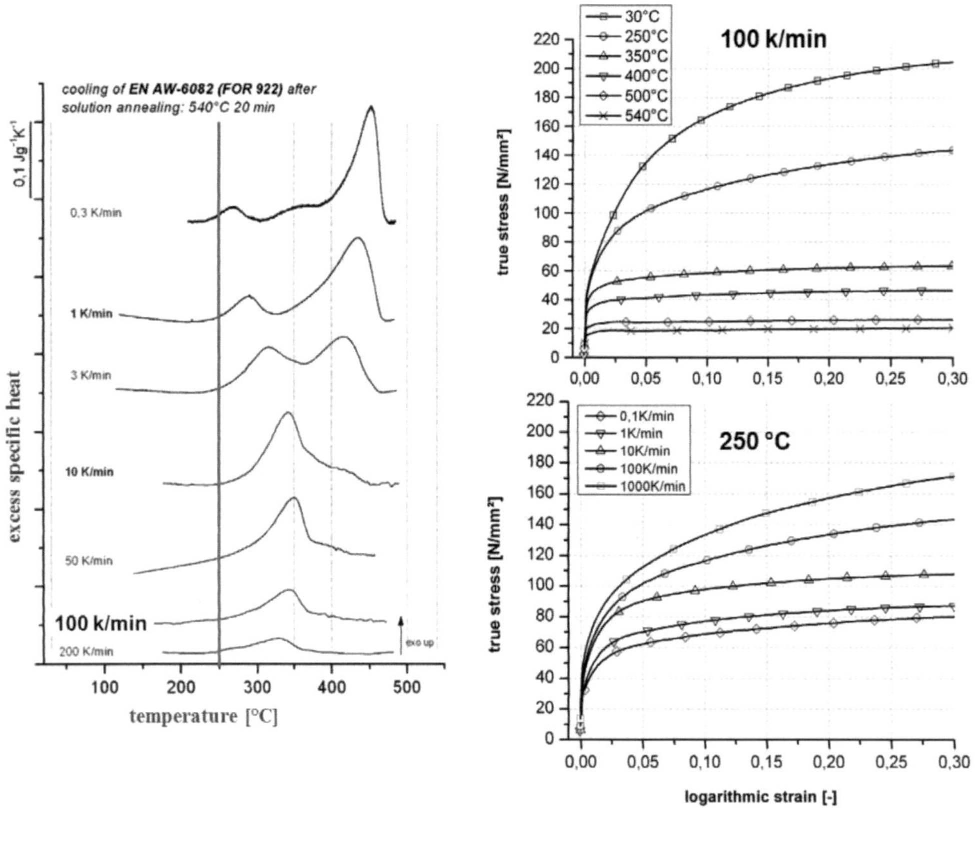

Figure 2. left: DSC curves; right top: flow curves at different temperatures for 100 K/min, right bottom: flow curves for different cooling rates at 250°C.

4. Correlation of Mechanical Properties and Precipitation Behaviour

Fig. 3 shows the yield strength of the undercooled aluminium alloy Al-0.6Mg-0.7Si depending on cooling rate and quenching temperature. For a quenching temperature of 500°C, which lays above the precipitation interval, the yield strength is low (about 20 N/mm²) and does not depend on cooling rate. For quenching temperatures of 250°C and RT, which lay below the precipitation interval, the yield strength is higher (about 40-55 N/mm²) and strongly depends on cooling rate. It increases with increasing cooling rate. Thus it can be concluded, that solid solution strengthening and the possible loss of solute elements during quenching are the mainly responsible mechanisms. For quenching temperatures below the precipitation interval and low cooling rates, the loss of solute elements is high and thereby solid solution strengthening low. Opposite, for quenching temperatures below the precipitation interval and high cooling rates, the loss of solute elements is low and thereby solid solution strengthening high.

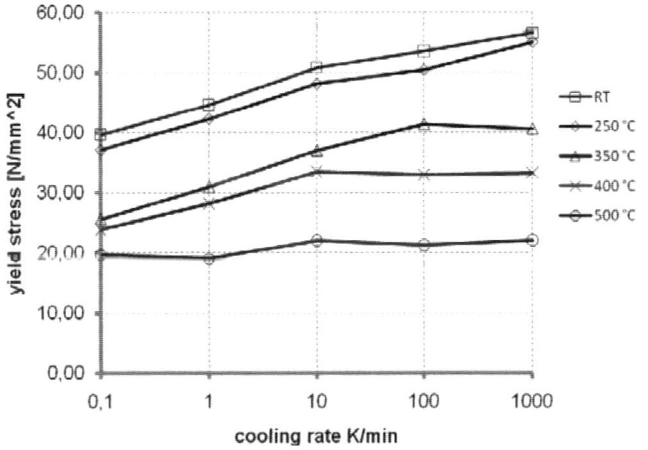

Figure 3. Yield strength depending on cooling rate and temperature.

5. Conclusion and Outlook

Quenching and deformation dilatometry has been proven to be a suitable method for the determination of mechanical properties of undercooled aluminium alloys depending on cooling rate and temperature. For the investigated aluminium alloy Al-0.6Mg-0.7Si a considerable correlation between cooling rate, precipitation behaviour and flow curves has been found. Solid solution strengthening and the possible loss of solute elements during quenching are assumed to be the mainly responsible mechanisms. The best approximation of flow curves has been achieved by the hardening model of Hockett-Sherby. Further, the Hockett-Sherby model has to be implemented into heat treatment simulation and the heat treatment simulation of real aluminium components (e.g. extrusion profiles) has to be performed.

Acknowledgements

The authors gratefully acknowledge funding of this work by the Deutsche Forschungsgemeinschaft, Forschergruppe 922, project KE616/13-1 as well as funding of the quenching and deformation dilatometer by Deutsche Forschungsgemeinschaft and the State Mecklenburg-Vorpommern (DFG INST 264/40-1).

References

[1] Surm H, Kessler O, Hoffmann F, Zoch H, 2008, Modelling of Austenitizing with non-constant heating rate in hypereutectoid steels, *Int. J. Microstructure Mater. Properties* **3** 35-48

[2] Kessler O, Reich M, 2009, Similarities and Differences in Heat Treatment Simulation of Aluminium Alloys and Steels, *Materialwissenschaft u. Werkstofftechnik* **5-6** 473-478

[3] Milkereit B, Kessler O, Schick C, 2009, Recording of continuous cooling precipitation diagrams of aluminium alloys *Thermochimica Acta* **492** 73–78

[4] Hockett J, Sherby O, 1975, Large Strain Deformation of Polycrystalline Metals at Low Homologous Temperatures, *J. Mech. Phys. Solids* **23** 87-98

Estimation and correlation of strengthening components to the evolution of microstructure following cold work and articial aging in AA6111 aluminium

G.K. Quainoo[1], S. Yannacopoulos[2] and C.M. Sargent[3]

[1]Department of Physics, Materials Science & Engineering, Southern Oregon University, 1250 Siskiyou Boulevard, Ashland, OR, 97520, USA
[2]School of Engineering, UBC Okanagan, Art 323 – 3333 University Way, Kelowna BC V1V 1Y7
[3]Department of Mechanical Engineering, 57 Campus Drive, University of Saskatchewan, Saskatoon, SK., S7N 5A9

quainoog@sou.edu

ABSTRACT
In this study, the contributions of the various strengthening components following the application of cold work and precipitation in AA6111 has been evaluated and correlated by means of tensile testing and transmission electron microscopy (TEM). The results show a considerable improvement in yield and tensile strength with increasing level of cold work. The component of strength developed from cold work and precipitation respectively increases with increasing level of cold work. The recovery strength (softening) also increases with increasing level of cold work. TEM showed a strong interaction of strengthening precipitates with dislocations. The density of dislocation tangles is shown to increase with increasing degree of cold work.

1. INTRODUCTION

In the automotive industry, the use of aluminum alloys has increased over the years as manufacturers strive to design lighter vehicles as part of an overall goal to meet the North American Corporate Average Fuel Efficiency (C.A.F.E) Standards. These standards seek to improve fuel efficiencies and reduce vehicular emissions into the atmosphere while meeting governmental and international targets set in regulations such as the Kyoto Protocol. Heat treatable AA6111 has emerged as one of the most prominent materials employed in outer body panels of cars and light trucks.

There has always been the desire to have some way of estimating the yield strength of a part following forming and/or heat treatment operations. For a formed and painted automobile panel, the yield strength involves combining strength components arising from cold work, strain aging, precipitation, and recovery. The challenge over the years has been to find a method of quantifying each of these strengthening components. The conflicting demand of low yield strength alloy and high yield strength components in the 6xxx series alloys containing Cu are partially addressed through the aging response promoted through the automotive paint bake cycle. Unfortunately, only a small fraction of the full aging potential inherent in the alloy is exploited due to the relatively low temperatures as well as short duration of most commercial paint bake cycles. The need for rapid strengthening response in today's finishing lines has led researchers to impose thermo-mechanical process histories along with variation in alloy compositions on 6xxx series aluminum alloys with the view to improving the strengthening characteristics as well as kinetics of these alloys [1-5]. In the present investigation, the contributions of the various strengthening components following the application of cold work and artificial aging in AA6111 has been evaluated with a view to estimating the overall yield strength of a formed part under the paint bake cycle of the alloy.

2. EXPERIMENTAL PROCEDURE

Tensile test specimens of 1mm sheet AA611, (composition in wt % 0.8Mg, 0.6Si, 0.7Cu, 0.25Fe, 0.2Mn, 0.05Cr, 0.06Ti, and balance Al) were machined according to the ASTM E-8 standard and solution heat treated (SHT) at 560°C for 30 minutes in an air furnace and rapidly quenched in 20°C water. The samples were then kept at room temperature for two hours to stabilize them after which some were given 2%, 5% or 10% cold work by stretching and then artificially aged at 180°C for various lengths of time followed by tensile testing measure the yield stress. Samples of AA6111 aluminum at the simulated paint bake cycle of 180°C for 30 minutes as well as overaged samples strained to various levels were examined in a Hitachi 2200 FE STEM at 200kV.

3. RESULTS AND DISCUSSION

Figure 1 (a) shows the variation of yield strength as a function of aging time at 180°C for various levels of cold work. It can be observed that as the aging time increases at each level of cold work, the yield strength values increase up to a peak value after which they decrease with further aging time. The significant increase in yield strength is attributed to the increase in dislocation density, due to cold work, piling up to form tangles and hence increasing the strength of the material [6,7]. The time to attain peak strength for all levels of cold work occurred after about 10 hours of aging.

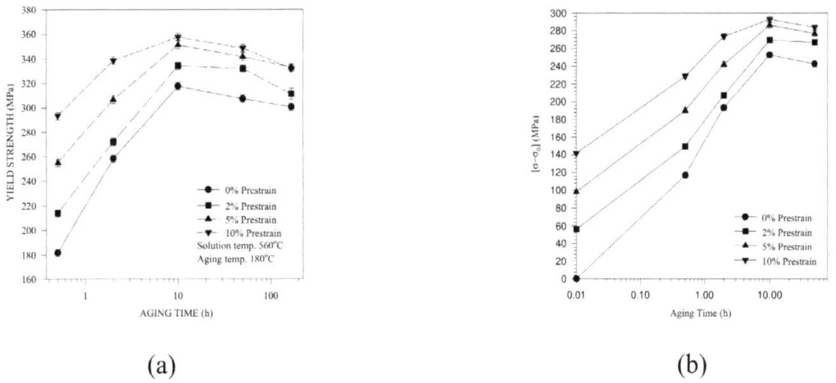

Figure 1: (a) Variation of yield strength with aging time at various levels of cold work on AA6111, (b) change in yield strength following SHT ($\sigma - \sigma_{ss0}$) versus aging time for various levels of cold work on AA6111

3.1 Quantifying the Strengthening Components

Aluminium alloys such as AA6111 strain harden quite rapidly during the forming process, and, for a highly strained panel the yield strength can almost double. However, the increased strength due to forming will recover, and hence soften to some extent due to exposure to the elevated temperatures associated with the paint bake. As such, the strength of the final part is the net effect of cold work, precipitation and recovery. Figure 1 data was used in quantifying the contribution of strength components for cold work, precipitation and recovery under the paint bake temperature. Let us designate σ_y as the net yield stress, σ_{ss0} as the initial strength of the material after solution heat treatment, σ_{cw} as the strength due to cold work, σ_{ppt} as the strength component due to precipitation, and σ_{rec} is the component resulting from recovery. To estimate the strength component following cold work, the yield strength values were normalized with the strength following solution heat treatment σ_{ss0}. Figure 1(b) presents the change in yield strength following SHT, ($\sigma - \sigma_{ss0}$) versus aging time for various levels of cold work at 180°C. The contribution of strength components for 2%, 5% and 10% cold work (σ_{cw}), were respectively 56.2MPa, 98.2MPa and 141.6MPa, showing an increase in strength due to the level of cold work.

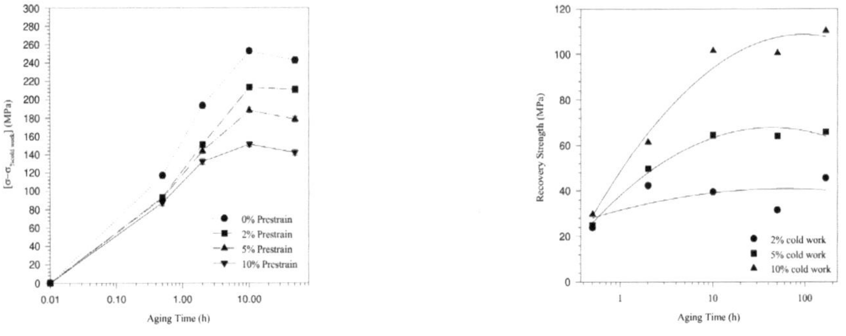

(a) (b)

Figure 2: (a) Variation of ($\sigma - \sigma_{cw}$) with aging time at 180°C (b) Variation of Recovery strength with aging time at 180°C.

For strength components due to precipitation and recovery (σ_{ppt} & σ_{rec}), the change in flow stress with yield strength of the cold worked sample prior to artificial aging, ($\sigma - \sigma_{cw}$) at 180 °C was used. Figures 2(a) presents the variation of ($\sigma - \sigma_{cw}$) with aging time at 180°C. Cold work is known to affect the precipitation process in two ways: (1) heterogeneous nucleation can occur on dislocations [8] and (2) the growth and/or coarsening rate of precipitates on dislocations can be accelerated as a result of dislocation core diffusion [9]. Figure 2(a) shows that there is a decrease in the ($\sigma - \sigma_{cw}$) flow stress values with increasing levels of cold work. This decrease is a result of recovery, a softening process, occurring at elevated temperatures. The difference between the ($\sigma - \sigma_{cw}$) flow stress values of the unstrained sample and those for the various levels of cold work gives an indication of the amount of recovery experienced in the material at the various levels of cold work at 180°C and attributed to the effect of dislocation core diffusion [9], the results of which are presented in Figure 2(b). A close look at the graph shows an increase in recovery with increasing level of cold work at 180°C. Under the paint bake condition of 180°C for 0.5 h, it can be observed that the level of recovery experienced by the material is minimal, indicating that the material retains a good amount of its strength under this condition. Figures 3(a) and (b) show the bright field TEM micrographs of AA6111 samples given 0 and 2% cold work, respectively, following the simulated paint bake cycle. It can be observed in Figure 3(a) that the unstrained sample consists mainly of widely spaced helical dislocations, which are pinned by solute clusters that formed after quenching. Figure 3 (b), on the other hand, shows substantial increase in

(a) (b)

Figure 3: Bright field TEM micrographs of AA6111 samples subjected to the simulated paint bake cycle following (a) 0% cold work and (b) 2% cold work.

dislocation density with increasing level of cold work, thus confirming that the increase in strength of cold worked material is due, in part, to the increased amount of dislocation tangles in addition to precipitates. A closer look at the strained sample again reveals dotted spots in the microstructure which are believed to be needle heads of strengthening precipitate particles. To establish the link between the strength properties of the alloy along with the phases responsible for them, an unstrained sample and a 10% cold worked sample were heavily overaged at 315°C for 2 hrs and TEM analysis done. Figures 4 (a) and (b) show the bright field TEM micrographs of heavily overaged samples of 0% and the 10% cold work respectively. Figure 4(a) shows that the unstrained sample comprises precipitates distributed randomly in the matrix. Figure 4 (b) on the other hand shows precipitates formed in a regular pattern, believed to be previous lines of dislocation generated by cold working. This shows that the increase in strength of the cold worked sample is due, in part, to increased amount of dislocations serving as high energy sites for the nucleation of precipitate phases. As well, the dislocations enhance the diffusion rates due to pipe diffusion by increasing the rate of solute diffusion through short circuit diffusion paths.

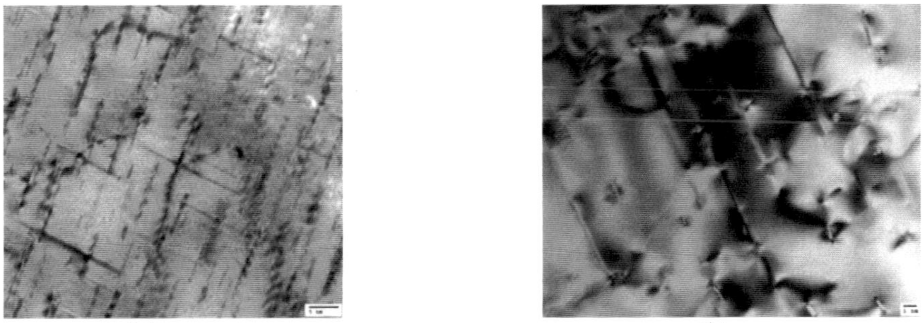

(a) (b)

Figures 4 (a) and (b) show the bright field TEM micrographs of overaged samples at 0% and the 10% cold work respectively.

4. CONCLUSION

The overall strength of pre-strained and aged samples of AA6111 aluminum alloys can be estimated by determining and assembling the various strength components from their precipitation profiles. The component of strength developed from cold work increases with increasing level of cold work. The strength due to precipitation decreases with increasing level of cold work, the loss in strength is attributed softening in the alloy at the paint bake temperature. The loss of strength due to recovery also increases with increasing level of cold work.

ACKNOWLEDGEMENT

The National Sciences and Engineering Research Council of Canada, Engineering Technology Industry Council of Oregon, the Carpenter Foundations and Southern Oregon University are gratefully acknowledged.

REFERENCES

[1] Miao WF, and Laughlin D.E, *Metall. Mater. Trans. A*, 2000, **31A**, 361-371.
[2] Burger G., Gupta AK, Jeffrey PW and Lloyd DJ, *Mater. Char.*, 1995, **33**, p. 23.
[3] Yassar RS, Field DP and Weiland H, *Metall. Mater. Trans. A*, 2005, **36A**, 2059-2065.
[4] Yassar RS, Field DP and Weiland H, *Scripta Mater.*, 2005, **53**, 299-303
[5] Wang X, Esmaeli S and Lloyd DJ, *Metall. Mater. Trans. A*, 2006, **37A**, 2691-2699.
[6] Quainoo GK and Yannacopoulos S, *J. Mater. Sci.*, 2004, **39**, 6495-6502.
[7] Quainoo GK and Yannacopoulos S, *J. Mater. Sci.*, 2004, **39**, 4841-4847.
[8] Russel KC, *Adv. Colloid Interfacial Sci.*, 1980, **13**, 255-261.
[9] Hoyt JJ, *Acta Metall.*, 1991, **39**, 2091.

Mechanical properties of low carbon steel hardened by the Fe_2SiTi phase at high volume fraction

M. Perrier[1, 2], O. Bouaziz[2], Y. Brechet[1], A. Deschamps[1], P. Donnadieu[1]

[1] SIMAP, 1130, rue de la piscine, 38402 Saint Martin d'Hères, France
[2] Arcelormittal Research SA, Voie romaine, 57280 Maizieres les Metz, France

malika.perrier@simap.grenoble-inp.fr

Abstract. The addition of Si and Ti to low carbon steel allows to obtain a nanometric precipitation of an ordered intermetallic phase, Fe_2SiTi, with a volume fraction of the order of 6%; which is much higher than usual for microalloyed steels and more comparable to the situation encountered in aluminum alloys. The resulting precipitation hardening leads to extremely high mechanical properties, depending on the optimization of metallurgical route, which makes these materials very promising for the development of innovative, weight saving automotive solutions. In this contribution the precipitation sequence and kinetics are studied in details by a combination of experimental techniques including small angle neutron scattering and scanning or transmission electron microscopy. Some properties (yield stress, strain hardening, strain to fracture) are discussed in view of the measured precipitate characteristics, notably their size and volume fraction. The fracture mechanism, and particularly the occurrence of brittle fracture, is discussed in view of the state of precipitation.

1. Introduction

The Iron-Silicon-Titanium (Fe-Si-Ti) system is known since the 1930's to show precipitation hardening capability. It has been further assessed by several authors (e.g. [1]), that the Fe_2SiTi phase provided significant precipitation hardening to steels. However, it was also found ([1], [2]) that such precipitation-hardened alloys are rather brittle, limiting the possible applications of this system. Obtaining a useful compromise of properties is only possible if the development of microstructure and related mechanical properties is rationalized. The purpose of this work is therefore to carry out a systematic quantitative investigation of precipitation kinetics and mechanical properties along various heat treatments and to investigate the microstructure / property relationships, on a ternary Fe-Si-Ti alloy.

2. Methodology

The iron-based alloy studied here contains 2,5wt% Silicon and 1wt% Titanium. It was hot rolled between 1250°C and 950°C with long interpass times (30 s), in order to make recrystallization easier. It was then cold rolled up to 75% thickness reduction in the aim to have smaller ferritic grains. The cold-rolled samples, 0,9mm in thickness, were then subjected to heat treatments starting with a quench from 900°C (temperature of solution treatment) followed by ageing treatments between 450°C and 650°C for different times. The matrix remains fully ferritic all along the heat treatment, including the 900°C solution treatment, with a grain size of about 25µm.

The methods used for precipitation characterization were: Scanning Electron Microscopy with a Field Emission Gun (SEM-FEG), Transmission Electron Microscopy (TEM) and Small Angle Neutron Scattering (SANS), carried out at ILL on beamline D11 (Grenoble, France). SANS measurements were carried out under a high magnetic field. Both the nuclear and magnetic contributions could be separated and analyzed to give precipitate size (using a standard Guinier analysis) and volume fraction values (using the Fe_2SiTi composition for nuclear contrast and the magnetic hole assumption for magnetic contrast). The nuclear and magnetic contributions were shown to provide consistent information. Here only the results from the magnetic contribution will be presented (for more details about the method see [3]). For the study of mechanical properties microhardness (Vickers) and tensile tests were performed.

3. Results

3.1. Precipitation characterization

Microscopic observations (Figure 1a, b, c and d) confirm that a nano-scale precipitation occurs during ageing. Depending on the ageing temperature, the precipitates can reach diameters ranging from 3nm to 100nm. The images obtained by SEM-FEG show a large number of precipitates spread homogeneously within the grains (Figure 1 a). Electron diffraction patterns confirmed that the precipitate crystal structure is consistent with the metastable phase Fe_2SiTi (fcc, a = 0.5709 nm) (Figure 1c). More detailed observations made on TEM thin foils confirm the observations made with the SEM-FEG (Figure 1 b and d): radius of about 10nm and density of $1e^{17}$ precipitates per m^3.

Quantification of the precipitate average size was achieved on dark-field TEM images for selected heat treatments (see Figure 2a). More systematic and global information were obtained from the SANS measurements. For heat treatments in the temperature range 500°C-580°C, Figure 2a shows the average precipitate size obtained by SANS and TEM, while the evolution of precipitate volume fraction derived from SANS is given in Figure 2b.

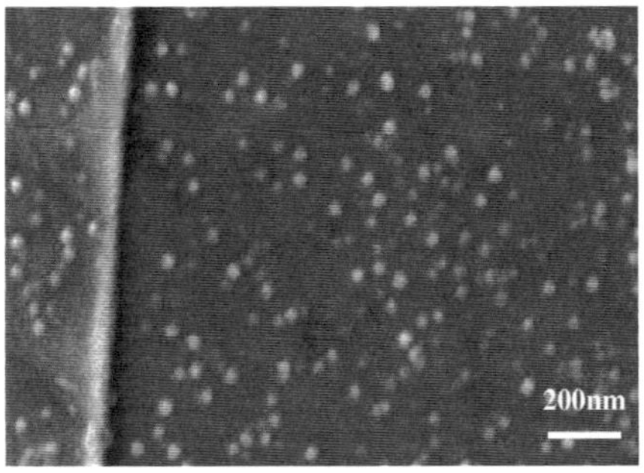

Figure 1a

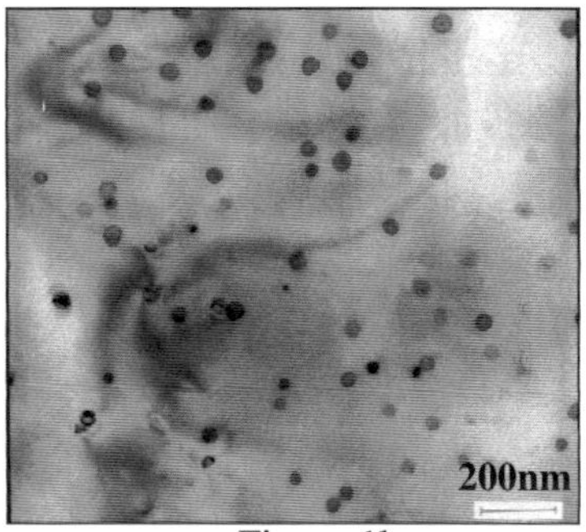

Figure 1b

Figure 1c

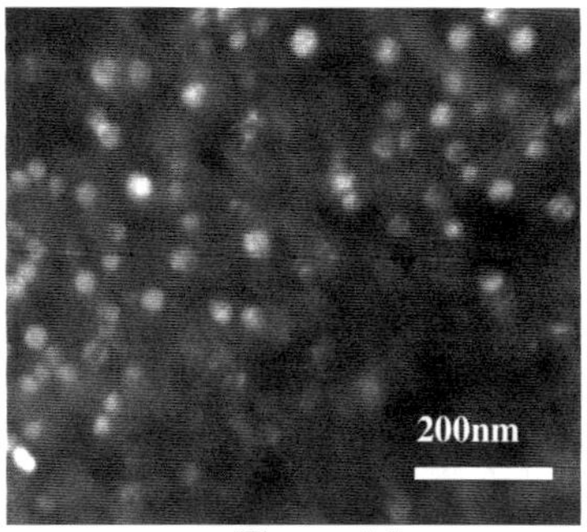

Figure 1d

Figure 1: observations of Fe_2SiTi precipitates with a) SEM-FEG for sample aged at 550°C for 3h; b) TEM bright field for sample aged at 550°C during 3h; c) TEM diffraction pattern in [110] axis (sample aged at 550°C for 6h) and d) TEM dark field for sample aged at 550°C during 15h

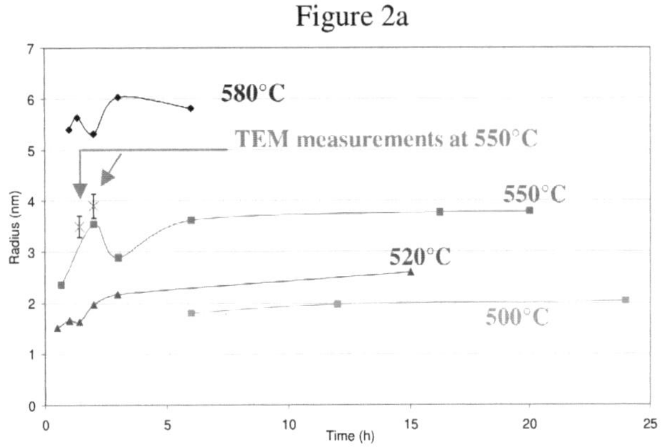

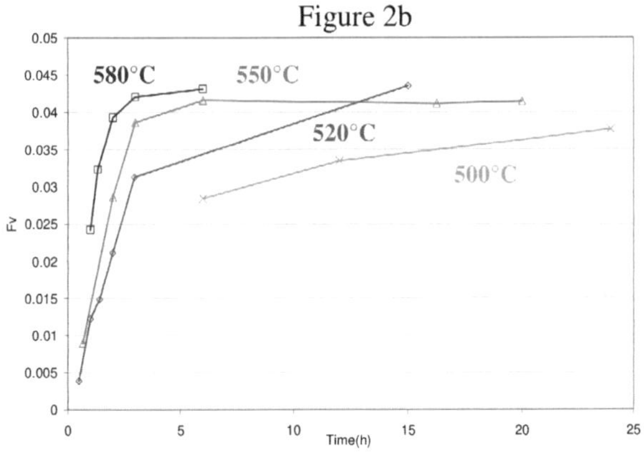

Figure 2: a) Mean radius evolution (nm) and b) volume fraction evolution as a function of time for samples at various temperatures (ranging from 500 to 580°C) obtained by SANS experiments

Figure 2a shows the evolution of the mean radius of the precipitates in Fe-Si-Ti system, obtained by SANS measurements. Two remarkable aspects of the radius evolution can be noticed. First, a reduction of the mean radius is observed after 3 hours ageing. Secondly, the TEM measurements on dark field images, give radius values that are in good agreement with SANS measurements for samples aged under 3hours (1h25 and 2h). Values for samples aged 3 hours and more are not consistent anymore with the ones measured with TEM using picture Figure 1d (3nm vs. 10nm). These two elements could be explained by the formation of a new precipitate population at this stage. In fact, SANS method enables to measure the radius and volume fraction of the phase which is present predominantly. Hence, if we imagine the nucleation of a new phase at about 3 hours ageing with a smaller size, the apparent mean radius would decrease suddenly. Thus, if the new phase nucleates rapidly, we could imagine that in SANS observations these are predominant. This would also explain why the TEM measurements are not in agreement, since with TEM we can only see the bigger particles (3nm is probably too small for TEM resolution here). These observations are to be brought in parallel with the measures of microhardness performed on the same samples (see section 3.2).This possible nucleation of a new phase was already mentioned in former studies ([1]), though it needs to be confirmed by additional observations: atom probe or by high temperature Differential Scanning Calorimetry (DSC) could be used.

3.2. Mechanical properties

Microhardness tests showed a wide range of hardening values depending on the annealing temperature (Figure 3a). The first hardness peak observed at 550°C appears after 2 hours, whereas, at 500°C, hardness still increases after 24 hours, in agreement with the slower precipitation kinetics at lower temperature measured with SANS. Peak hardness is observed to be associated with a precipitate radius of about 3 nm at 550°C.

In order to find a relation between precipitate size, volume fraction and the ductility, tensile tests were performed (Figure 3b). These show the 3.5 nm large precipitates at peak hardness for 550°C give the best compromise between yield stress and ductility. The sample over-aged for 3h has a deformation of 15% but the yield stress decreases suddenly, probably due to a change in the deformation mechanism, namely from shearing to Orowan by passing of the precipitates.

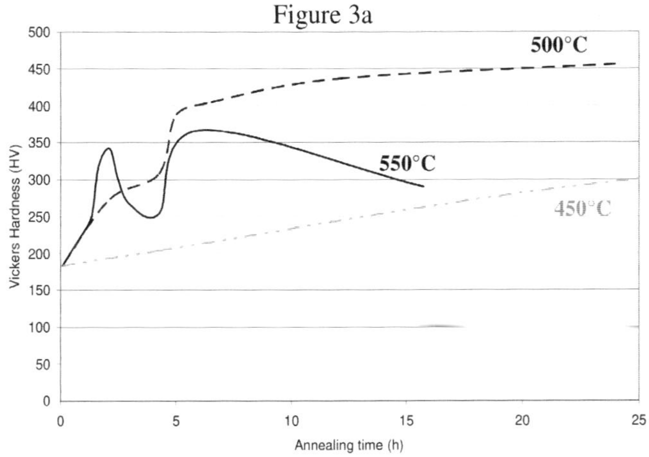

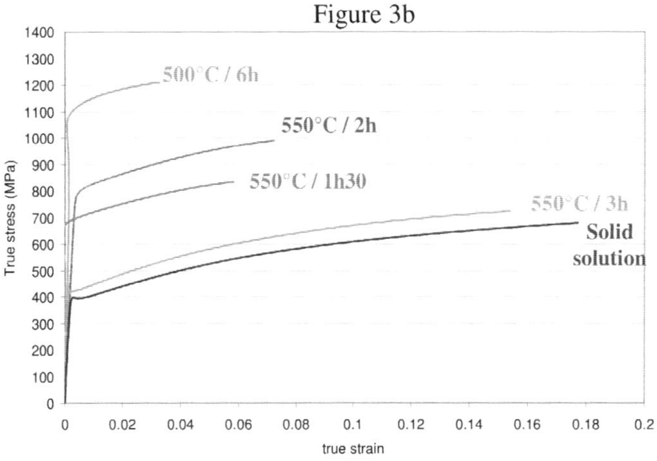

Figure 3: a) Hardness evolution of the Fe-Si-Ti as a function of time at various temperatures and b) tensile test curves showing the evolution of true stress with true strain for different annealing conditions

For samples with very small precipitates (2-3nm) as for example at 500°C on Figure 2a, the hardness reaches very high values (see Figure 3a). However, tensile tests performed on these samples reveal a very brittle behavior (cleavage fracture observed) with a uniform elongation of only 2-3% (Figure 3b). Thus the presence of very small precipitates with high volume fraction does not provide a satisfying strength-ductility compromise. A better solution would be to increase the radius at the cost of a decrease of precipitate density like for example at 550°C. Hardness peak is reached more rapidly, at lower value, but the uniform elongation can reach up to 15%.

4. Discussion

We have seen that, contrary to what has been said in earlier works ([1] and [2]), Fe-Si-Ti alloys can combine remarkable hardening characteristics and interesting deformation capabilities, depending on the heat treatment temperature. This is the result of the choice of an appropriate compromise between mechanical properties and precipitate size and density. Our measurements provide evidence that age hardening at low temperature with a high density of extremely small precipitates leads to embrittlement. At precipitate sizes equal to or above 3nm (corresponding to the ones at peak strength), part of the precipitates become non-shearable, which gives rise to a less pronounced localization of plastic flow and probably explains the ductility improvement.

Acknowledgements
The authors would like to thank C. Dewhurst and all the team of ILL-D11 for help with the SANS experiments.

References

[1] D.H. Jack and R.W.K. Honeycombe, Age hardening of an Fe-Si-Ti alloy, *Acta Metall.* **20**, 1972

[2] R. Papaleo and J.A. Whiteman, The deformation mechanism in ferritic Fe-Si-Ti alloys containing intermetallic precipitates, *Micron* **2**, 1970, 63-70

[3] Perrard, F., Deschamps, A., Bley, F., Donnadieu, P., and Maugis, P., A small-angle neutron scattering study of fine-scale NbC precipitation kinetics in the a-Fe-Nb-C system, *J. Appl. Cryst.* **39**, 2006, 473-482

Understanding the mechanical properties of 2198 Al-Li-Cu alloy in relation with the intra-granular and inter-granular precipitate microstructure

B. Decreus[1,2], A. Deschamps[1] and P. Donnadieu[1]

[1] SIMAP – INPGrenoble-CNRS-UJF, BP75, 38402 Saint Martin D'Hères, France
[2] Alcan CRV, 725 rue Aristide Bergès, BP27, 38341 Voreppe CEDEX, France

benjamin.decreus@simap.grenoble-inp.fr

Abstract. 3rd generation Al-Li-Cu alloys such as AA2198 are known to present extremely high strength levels because of the precipitation state of the material. The final precipitation state is reached via a complex sequence that can involve notably solute clusters, GP zones, θ', T_1 and δ' phases. In this work, we have studied the evolution of mechanical properties during heat treatment. Microhardness and tensile tests have shown a direct link with the evolution of the size of intra-granular T1 precipitates. The evolution of the fracture mode in relation with the precipitation state has been followed by the so-called short bar test that allows to load relatively thin plates in the short transverse direction, which is particularly sensitive to the inter-granular fracture mode. Along the heat treatment, from the naturally aged T351 state and along ageing at 155°C, we show a clear transition from a ductile transgranular fracture mode to a ductile inter-granular fracture mode, controlled by the precipitation of the T_1 phase at the grain boundaries.

1. Introduction

It has been shown in a previous work [1] that the precipitate microstructure developed during heat treatment at 155°C in 2198 alloy (0.8-1.1 Li; 2.9-3.5 Cu; 0.25-0.8 Mg in wt%) was composed mainly of T_1 (Al_2CuLi) plate-shaped precipitates. The morphology characteristic of these intra-granular plates was measured by Small-Angle X-ray Scattering. T_1 precipitates were shown to present a constant thickness around 1nm all along the heat treatment and an increasing mean length, reaching 40nm after 16h of heat treatment.

The aim of the present study is to investigate in parallel the mechanical properties for tensile tests, and relate the compromise between strength and ductility to the precipitate microstructure, both inside the grains and at the grain boundaries. The role of intergranular precipitation on the fracture mode will be investigated in more detail using specific tests where the load is applied along the short transverse direction of the rolled plate, enabling to open the grain boundaries in mode I. This so-called Short Bar test [2] makes also possible to estimate the fracture toughness of the studied material.

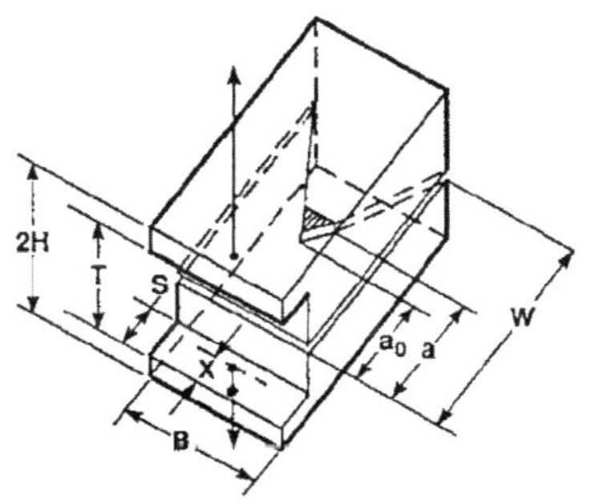

Figure 1. Flow chart of a short bar sample.

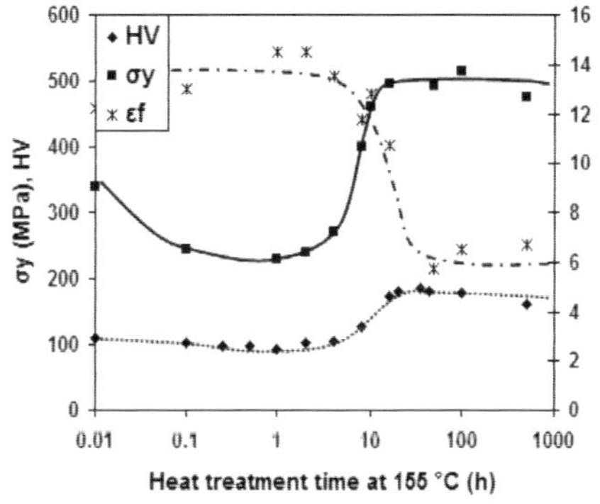

Figure 2. Evolution of yield strength and elongation in 2198 during heat-treatment.

2. Experimental

Short bars samples (see figure 1) were machined in a 12mm thick 2198 plate in T351 state (stretched + naturally aged after solution treatment and quench). They were machined to 19*12.5*11 mm^3 chevron notched samples. Heat treatments were carried out in a Carbolite air furnace with a 20°C per hour heating rate before a plateau at 155°C for various times.

Short bars tests were realized in an Instron 5500R machine with a cross-head speed of 0.5 mm/min. After crack propagation, the fracture of the remaining ligament was achieved using the same Instron tensile machine with manual control.

SEM observations were made in a LEO stereoscan 440 microscope under an accelerating tension of 20 kV. FEG-SEM experiments were made in a Zeiss Ultra-55 with in-lens secondary electrons and back-scattering detectors.

3. Results

Hardness measurements along the same heat treatment show a strength plateau between 20 and 100 hours at 155°C before a drop in hardness from 200 hours onwards (Figure 2). Tensile tests are showing a similar behavior with a yield strength plateau, however, ductility is drops significantly as soon as this plateau begins. The evolution of yield strength is related to the precipitation state and in our case, this plateau can be linked to the slow evolution of T_1 precipitates [2] whose thickness begins to increase significantly only after 100 hours at 155°C.

Short bar tests can be used to estimate the evolution of apparent short-transverse plane-strain fracture toughness during heat treatment. Figure 3 shows that toughness decreases continuously with time during ageing at 155°C whereas yield strength tends to remain rather constant over 10-500h at 155°C. In order to understand this difference, it is necessary to observe the fracture surfaces.

3.1. Fractography analysis

Figure 4 shows a fracture surface of a short bar sample aged 16 hours at 155°C. One can see two different areas; the first at the edge of the fracture surface is flat with well-defined fracture ledges whereas the center of the fracture surface is clearly rougher. In figure 5-a, one can clearly see that the flat area of the fracture surface corresponds to an intergranular mode without the presence of large cavities or intermetallic particles whereas as shown in figure 5-b the central part of the sample consists mainly of micron-size dimples initiated around intermetallics (identified as mainly Al_7Cu_2Fe particles), corresponding to ductile trans-granular fracture.

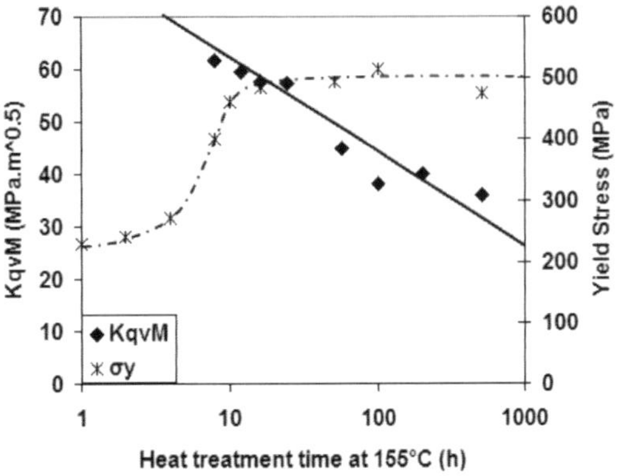

Figure 3. Evolution of plane-strain fracture toughness versus time. Yield strength is also shown as a reminder.

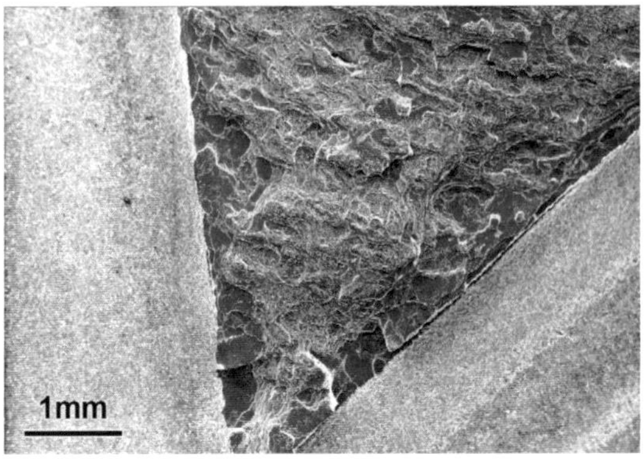

Figure 4. Secondary electrons SEM fractograph of a short-bar sample aged 16h at 155°C. Two different areas can be distinguished.

In short transverse direction, fracture after ageing 16h at 155°C thus occurs according to two very different mechanisms. In the centre of the sample, where plane strain conditions (and associated high stress triaxiality) prevail, cavities nucleate and grow fast around intermetallic particles and the resulting fracture has a ductile, transgranular character. Close to the free surface of the notch, triaxiality is lower, and intergranular fracture is favored, helped by the fact that due to the fibrous structure of the rolled plate the grain boundaries are normal to the loading direction. This general trend is observed all along the heat treatment with an increase in the fraction of planar areas with heat treatment time, see figure 6.

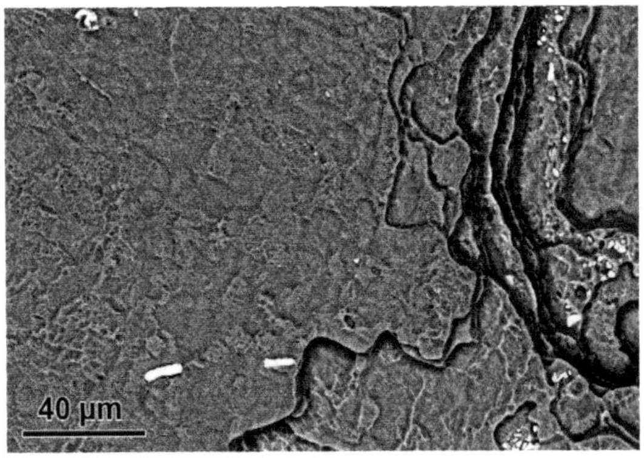

Figure 5-a. Back Scattered electrons image on the plane area of T851 short bar sample

Figure 5-b. Back Scattered electrons image on the ductile area of T851 short bar sample

3.2. Intergranular precipitation

Heat treatment duration has a strong effect on the fraction of intergranular or intersubgranular fracture area, whereas it has little influence on the size of intra-granular precipitates and correspondingly on the alloy yield strength. FEG-SEM was used in order to image more precisely the planar areas of the fracture surfaces, as shown in fig 7. One can clearly see that they are composed of sub-micronic cavities that have initiated around small precipitates. These precipitates seem to be spherical with a diameter around 65nm after 56 hours at 155°C. According to EDS analysis these precipitates contain no zirconium or iron which rules out intermetallics and dispersoïds. Some EDS acquisition carried out at very low accelerating voltage to have a better spatial resolution showed that the precipitates are characterized by a large Cu/Al ratio which is consistent with the T_1 phase.

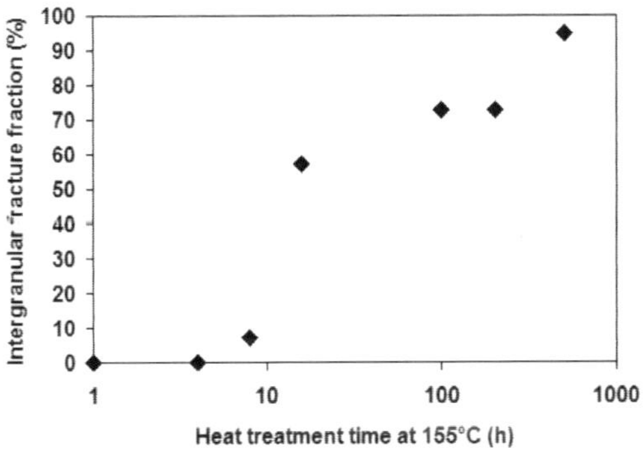

Figure 6. Evolution of the surface fraction of plane areas in 2198 during versus heat treatment time.

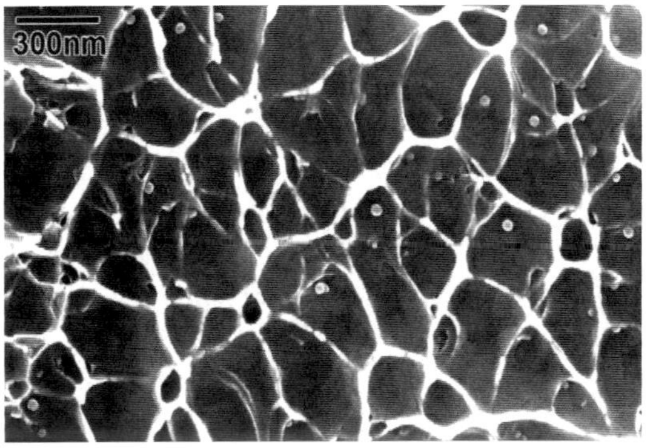

Figure 7. FEG-SEM secondary electrons image on a plane area of 2198 heat treated 56 hours at 155°C. One can see both sub-micronic cavities and spherical precipitates inside those cavities.

4. Discussion and conclusion

The evolution of fracture mode in short-transverse direction in 2198 alloys has been followed using short bars testing and fractography analysis. The main trends of the results are the following:

- Plane-strain fracture toughness quickly decreases during ageing at 155°C, in parallel with an increase of the area fraction of intergranular fracture, and no significant change in the intra-granular microstructure
- Intergranular fracture is ductile, with sub-micron dimples nucleated on precipitates smaller than 60 nm, containing Cu, that are not intermetallics and probably T_1 or T_2 precipitates

Acknowledgements

This work has been partially founded by the ANR project ALICANTDE.

References

[1] Decreus B, Bley F, Deschamps A and Donnadieu P 2008 Quantitative characterization of the precipitation in an AA2198 Al-Li-Cu alloy *Proc. Int. Conf. on Aluminium Alloys 11 (Aachen, DE)*

[2] Cassada W A, Shiflet G J and Starke E A Jr 1991 The effect of plastic deformation on Al_2CuLi (T) precipitation *Met. Trans. A* **22** 299-306

[3] 2002 Standard test method for plain-strain (chevron-notch) fracture toughness of metallic materials. *ASTM-International.* E 1304-97

[4] Kim N J and Lee E W 1993 Effect of T_1 precipitate on the anisotropy of Al-Li alloy 2090 *Acta Metall. Mater.* **41** 941-948

Model for the solid state phase transformation and transformation plasticity during casting in grey cast iron GJL350

V. Runser, J. Hoffmeister and V. Schulze
Institute of Material Science and Engineering 1, University of Karlsruhe (TH), Kaiserstrasse 12, 76131 Karlsruhe, Germany

vincent.runser@iwk1.uni-karlsruhe.de

Abstract. The development of stresses and distortions in solidifying metallic components during casting processes is not only influenced by the temperature field created across the mould but can also be strongly influenced by the transformation kinetics occurring in the solidifying metal. It is challenging to apply numerical methods which are strictly based on a coupled thermo-mechanical model because the models do not account for the inhomogeneous distribution of alloying elements, particularly carbon. Therefore a viable model has to take into account some preselected difference by heated states using short-time austenitizing, which includes a carbon gradient in order to understand the influence of carbon on thermomechanical properties. The aim of the present work is to identify high temperature parameters for a lamellar cast iron GJL-350 in order to include them into a material model. The transformation kinetics for non-isothermal transformations are determined from measured isothermal TTT-diagrams at differently heated states. Doing so, it was also possible to determine the isothermal TTT-diagrams as a function of the carbon content in the matrix.

1. Introduction

The mechanical properties and the fatigue strength of grey cast iron are significantly affected by the thermally induced strains related to the solid state phase transformations occurring during the cooling stages of metallic melts. A clear understanding of these phenomena is helpful for the prediction of crack initiation and microstructure in order to realize improved properties. The development of stresses and distortions in solidifying metallic components during casting processes is not only influenced by the temperature field created across the mould but can also be strongly influenced by the transformation kinetics occurring in the solidifying metal. For cast irons, these influences together with the non steady states at high temperature and the dissolution of graphite particles lead inevitably to a considerable variation in the final properties. These influential states are decisive in the development of microstructure as well as the stress fields in the cast material. For loads below the yield stress of the phases, a linear relation between strain occurring during transformation and the stress applied can be usually observed due to transformation plasticity. The phenomenon of transformation plasticity is often observed for steels during special heat treatments [1]. Therefore the prediction of residual stresses due to the solid phase transformations is incomplete without the consideration of the significant contribution of this additional strain. In the present study, experimental results accounting for the dependence of isothermal solid state transformation behaviour on the dissolved carbon content will be presented and empirically modelled in order to provide a straight forward routine which may be used in thermo-mechanical finite element simulation of casting processes using CASTS. The purpose of this study was the examination of the thermo mechanical and transformation behaviour of a lamellar cast iron GJL-350 using the method of the classification of the carbon gradient into three predefined carbon classes. This method has been successfully applied to a ductile cast iron GJS-600 [3] thus providing the basis to extend it to cast iron of different compositions. The influences of the mold pressure during the solidification of the melt, the produced constraints due to variation of sections combined with the presence of thermal gradient inside the cast component demonstrate the importance of introducing a realistic model which can describe transformation plasticity. In the literature, the consequences of the kinetics of transformation itself have already been demonstrated [2]. Therefore the last part of this work deals with the examination of the thermomechanical behaviour of the cast iron based on measurements on the transformation plasticity.

2. Material and experimental work

The material used was a hypoeutectic lamellar graphite cast iron GJL-350 with the chemical composition of 3.37%C, 1.5%Si, 0.87%Mn and traces of P, S and Cu. The microstructure of the as-cast material is mostly pearlitic (about 87 %). All the experiments have been performed on cylindrical hollow specimens with a wall thickness of 1 mm. Dilatometric experiments for the identification of TTT-diagrams for predefined carbon classes were performed using a quenching device in a servo-hydraulic testing machine under high vacuum (10^{-5} bar), which combines conductive and inductive heating using a welding transformator (Harms & Wende) and an eddy current generator). The temperature was measured with a pyrometer of type IGA. The strain was analysed using a capacitive extensometer. Two independent valves one for water and one for the compressed air allow the efficient quenching from the inside of the specimen and therefore avoid the transformation of the initial captured non steady state. The measurements of carbon content for the classification in three carbon classes of the mixtures M are made using the Electron Probe Microanalysis called EPMA, which was presented in a previous work for a ductile cast iron [3].

3. Results and Discussions

Firstly the specimens were rapidly heated up using short time austenitizing. The parameters used for the short time austenitizing are given in table 1. After a rapid temperature increase followed by quenching down to room temperature, the generated carbon gradient was captured in the martensite, because the cooling rate of 1000 K/s did not allow for further diffusion processes. The measurements of carbon content identified the fractions of the three carbon classes corresponding to the mixtures. The following results present a large variation of carbon classes and are summarized in table 2.

austenitizing condition	heating rate (°C/s)	austenitizing temperature (°C)	austenitizing time (s)
M_1	700	200	2
M_2	250	825	10
M_3	250	930	300

Table 1: Austenitizing conditions applied for the identification of the three mixtures

	C_1	C_2	C_3
M_1	0.28	0.43	0.29
M_2	0.02	0.72	0.260
M_3	0	0	1

Table 2: Fractions of the carbon classes C_1, C_2 and C_3 in the analysed mixtures

Figure 3 shows the thermal strain vs. temperature of a specimen corresponding to the mixture M_3 after heating and quenching. The thermal expansion coefficient and the transformation strain for all phases and all mixtures used for the modelling of thermal stresses, are determined. By quenching to room temperature, martensite forms below martensite start M_s, which is detected in this case from the consequent volume increase at 212°C on the cooling curve. If the quenching is interrupted above M_s at a temperature where diffusive transformations occur such as 420°C, the evolution of the time dependent thermal strain at this temperature is observed and gives precious informations about the kinetic of transformation. Figure 4 gives the variation of the thermal strain vs. the time for different temperatures for M_3. The identification of the start and end of transformation is clearly represented and the interpolation curves t_s and t_f close the domain of transformation. The method was applied for the both other states M_1 and M_2 too. Since the fractions of classes in all mixtures are known, it is simple to calculate the transformation kinetics of the distinct carbon classes and determine the volume of the transformed phase for the three carbon classes using the classical Avrami Model [4] with temperature dependent parameters. The results are summarized in form of the calculated TTT diagrams and reported figure 5. The expansion coefficients and transformation strains are shown in figure 6.

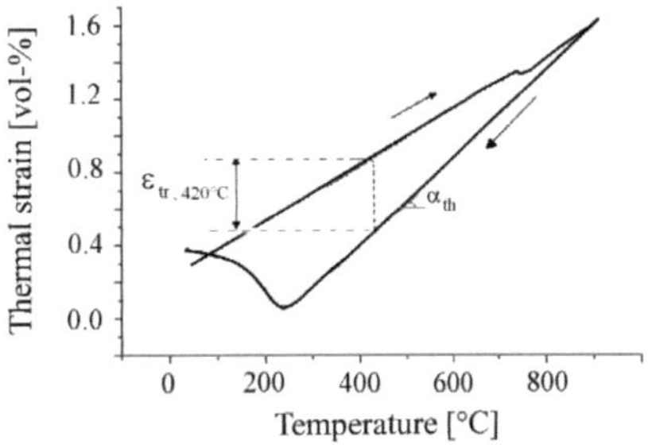

Figure 3: Dilatometric curve for the mixture M3 during heating and cooling

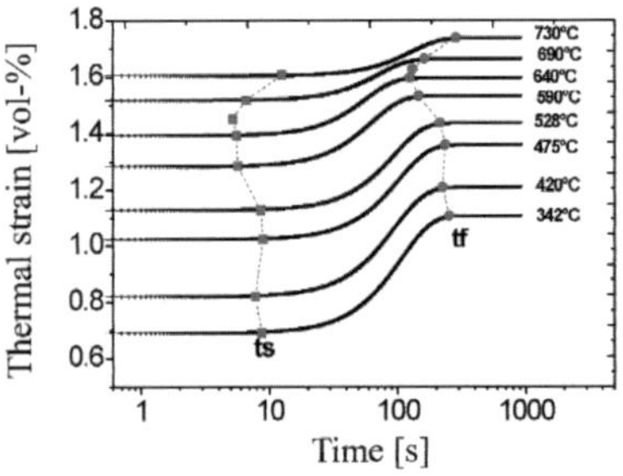

Figure 4: Thermal strain vs. time in dependence to the temperature for the mixture M_3

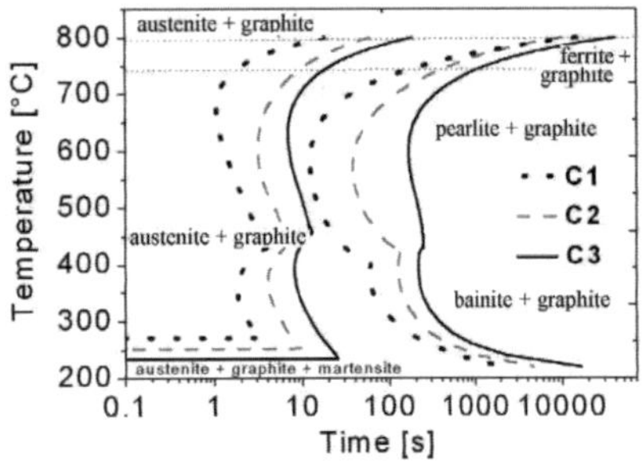

Figure 5: Casting TTT-Diagrams of the three carbon classes C_1, C_2 and C_3

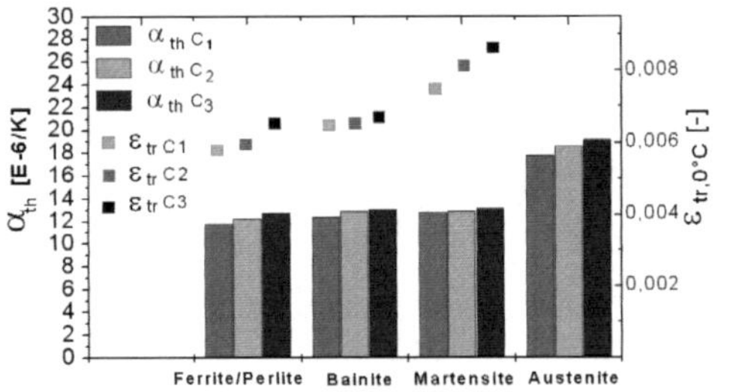

Figure 6: Thermal expansion coefficients and transformation strains of phases associated with a carbon class

For the analysis of transformation plasticity, the samples are loaded with different tensile stresses between 0 and 50 MPa at 600°C. Creep strains appear and must be considered for the calculation of the transformation plasticity strain. The measured total strain could be also divided into four different strains, hence the elastic strain, the transformation strain, the time dependent strain or creep and the transformation plasticity strain, which is expected to be identified. The pure transformation strain is simply determined from the specimen, which has been determined without stress and gives a value of 0.102%. The elastic strain is calculated according to the Hooke's law and leads to an average Youngs modulus of 68300 MPa. The analysis of creep especially at high stresses is more complex. The time dependent strain evolution under stress leads to damage of the specimen. The creep for each stress was experimentally determined and described by four parameters in the Burger's model [6] and extracted from the total strain. Figure 7 shows the strain vs. time during the pearlite transformation at 600°C for the mixture M_1 after removing the elastic and creep strains. A parameter K of transformation plasticity strain according to the law of Greenwood and Johnson [5], which describes the proportionality between the transformation plasticity and the applied tensile stress, is finally determined with a value $4.95*10^{-5}$ mm^2/N. The measurement of the volume change during the transformation under stress allows the determination of the transformation times and the shift of this time as function of the stress. After the extension of the analysis to the both other mixtures, the plot in figure 8 shows the considerable acceleration of the pearlitic transformation kinetic without and with constant stress of 48 MPa in the three non steady states. These results underline the importance of integrating the change of transformation kinetics under constant load in order to simulate realistically the evolution of residual stresses in the solid state.

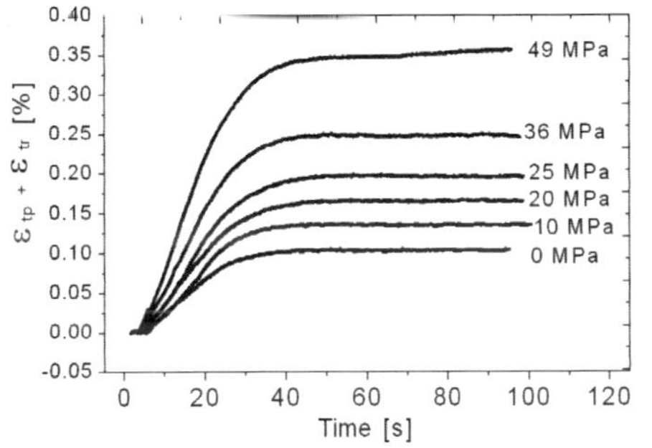

Figure 7: Transformation plasticity and transformation strains vs. time for the mixture M_1 and for different tensile stresses

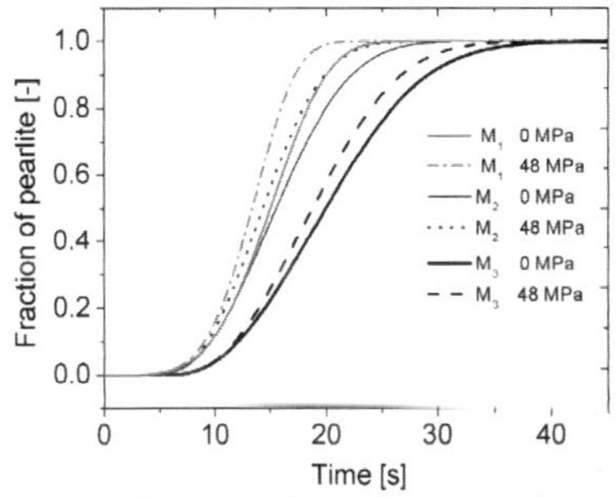

Figure 8: Kinetics of pearlite transformation at 600°C for M_1, M_2 and M_3 with and without stress

4. Conclusions

In this investigation it was shown the importance of considering the repartition of carbon in three carbon classes in the austenite for GJL 350, in order to modelling the prediction of residual stresses and microstructure evolution in the solid state. More carbon in austenite retards the transformation process and plays a major role for the final properties. The transformation plasticity was studied for the mixture M_1 for the pearlitic transformation at 600°C. An applied tensile stress during the transformation accelerates considerably the kinetics of transformation. These results were connected to the coupled simulation model of this cast iron GJL 350 and allow for more realistic predictions for the evolution of the microstructure and residual stresses.

Acknowledgement
The authors express their gratitude for the kind financial support of the German Science Foundation (DFG) within the research project P222.

References

[1] Denis S. et al. 1987, Influence of stresses on the kinetics of pearlitic transformation during continuous cooling, *Acta Metall.* **35** 1621-1632
[2] Maier H.J. and Ahrens U. 2002, Isothermal bainitic transformation in low allow steel, *Z. Metallkd.* **93** 712-718
[3] Runser V and Schulze V 2009, Model of Solid State Transformations of Ductile Cast Iron GJS-600, *ASTM International* **6** Issue 2
[4] Avrami M. 1940, Kinetics of phase change. II. Transformation-time relations for random distribution of nuclei. *J. Chem. Phys.* **8** 212 – 224
[5] Greenwood G.W. and Johnson R.H. 1965, The deformation of metals under small stresses during phase transformations, *Proc. Roy. Soc. London* A **283** 403
[6] Betten J. 2002, *Creep Mechanics*, Springer, 229 - 233

The influence of martensitic transformation on mechanical properties of cast high alloyed CrMnNi-steel under various strain rates and temperatures

L Krüger[1], S Wolf[1], U Martin[2], S Martin[2], P R Scheller[3], A Jahn[3], A Weiß[3]

[1] Institute of Materials Engineering, Technical University Bergakademie Freiberg, Gustav-Zeuner-Str. 5, D-09599 Freiberg, Germany

[2] Institute of Materials Science, Technical University Bergakademie Freiberg, Gustav-Zeuner-Str. 5, D-09599 Freiberg, Germany

[3] Institute of Iron and Steel Technology, Technical University Bergakademie Freiberg, Leipziger Str. 34, D-09599 Freiberg, Germany

swolf@ww.tu-freiberg.de

Abstract. Metastable austenitic steels show excellent mechanical properties, such as high strength combined with excellent ductility and toughness due to martensitic transformation under mechanical loading (transformation induced plasticity effect). A good energy consumption, and, in the case of high-alloyed metastable austenitic steels, a high corrosion resistance, increase the potential of these materials for diverse applications, also in regard of safety requirements. Up to now, numerous wrought alloys were investigated concerning mechanical behaviour, TRIP-effect, martensitic transformation behaviour and modelling of transformation kinetics or stress-strain behaviour. New high alloyed cast CrMnNi-steels, developed at Technical University Bergakademie Freiberg, provide the chance to reduce processing steps, production time and costs. In order to understand the influence of temperature on the martensitic phase transformation behaviour and therefore on mechanical properties and failure, the mechanical response under tensile loading in a temperature range between -70°C and 200°C was investigated. The mechanical behaviour under compressive loading was also examined in a wide range of strain rates between 10^{-4} s^{-1} and 10^{3} s^{-1} to obtain information about the strain rate effect on stress-strain behaviour and microstructural changes.

1. Introduction

Steels using the transformation induced plasticity (TRIP)-effect show high strength and excellent ductility plus high toughness. So far, most investigations deal with wrought alloys, which have to be cold or hot-worked.

In this paper, a recently developed cast high alloyed CrMnNi-steel is investigated. The mechanical behaviour of such steels is closely linked with a martensitic transformation caused by applying a mechanical load. Weiß et al. [1,2] developed stress-temperature-transformation (STT) diagrams for high alloyed wrought steels X5CrNi18-10 (1.4301, AISI304) and X6CrNiTi18-10 (1.4541, AISI 321).

These diagrams characterize typical temperatures and stresses, which define the temperature and stress ranges where transformation occurs. The highest temperature where martensite is formed by mechanical loading is called M_d-temperature [3]. Below this temperature, a critical stress level (triggering stress) is required for activating the transformation process. Several transformation routes are proposed in literature: $\gamma \rightarrow \varepsilon$, $\varepsilon \rightarrow \alpha'$ and $\gamma \rightarrow \alpha'$.

However, it is well known that the transformation behaviour is related to stacking fault energy (SFE). The stacking fault energy is a function of chemical composition and of temperature.

In this paper, the influence of temperature on martensitic transformation and stress-strain behaviour is shown for tensile loading, as well as the influence of strain rate for compressive loading.

2. Materials and methods

After melting in an induction furnace, the material was cast in a copper mould under argon gas atmosphere, using technique of rising casting. Tensile and compressive specimens were heat treated at 1050°C for 30 minutes, to retransform any martensite brought in through mechanical machining. The chemical composition of the cast CrMnNi-alloy is listed in table 1.

Table 1. Chemical composition of cast high alloyed CrMnNi-TRIP-steel.

	Fe and others	C	Cr	Ni	Mn	Si
wt. %	bal.	0.05	15.9	6.1	6.8	0.9

The corresponding initial microstructure is characterized by a dendritic solidification, consisting of coarse austenitic grains (grain size ~ 1mm) and small amounts of δ-ferrite (< 5%), figure 1. Using the empiric equation of Schramm and Reed [4], the SFE is calculated to 17.5mJ/m^2. Many observations on high alloyed metastable austenitic steels have a SFE of 20mJ/m^2 as a critical value governing the kind of deformation mechanism [4-7]. Below this value, the formation of deformation martensite is favoured, instead of mechanical twinning above 20mJ/m^2.

For mechanical testing at moderate values a servohydraulic testing machine was used, whereas the tests at high strain rates under compressive loading were performed with drop impact test and a Split-Hopkinson-pressure-bar (SHPB). After straining, tensile and compressive specimens were cut off to samples for light-optical-microscopy, SEM, EBSD-measurements and magnetic balance testing. With the magnetic balance test the fraction of α'-phase can be measured, e.g. described by Talonen [8].

3. Results and discussion

The TRIP-effect involves an anomaly in elongation values with decreasing temperature, especially in uniform elongation, figure 2. The unusual deformation response is explained by a change in the dominating deformation mechanism of slip, martensite phase transformation (TRIP-effect) or mechanical twinning (TWIP-effect). Weiß correlates the beginning of this anomaly with the formation of ε-martensite and the $M_{d\,\gamma \rightarrow \varepsilon}$ temperature for comparable steels, but with less manganese [1,2].

The maximum elongation value is assumed among 60°C and 100°C, corresponding with beginning formation of α'-martensite, resulting in a strong increase of ultimate tensile strength below this temperature range.

Regarding the stress values at 0.2% plastic strain in figure 2, a second anomaly in mechanical behaviour is observed. At temperatures below 20°C, a decrease in yield strength with a minimum at -40°C occurs. Due to the subsidence of triggering stress for $\gamma \rightarrow \alpha'$-formation below the value of yield stress [3,9] the so called stress-induced martensitic transformation appears. This stress-induced transformation causes a micro plasticity, resulting in lower values for elastic slope and yield strength [9].

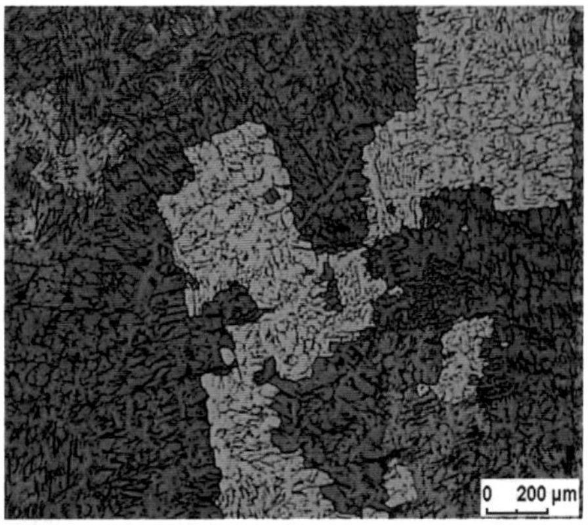

Figure 1. Initial microstructure of cast CrMnNi-alloy.

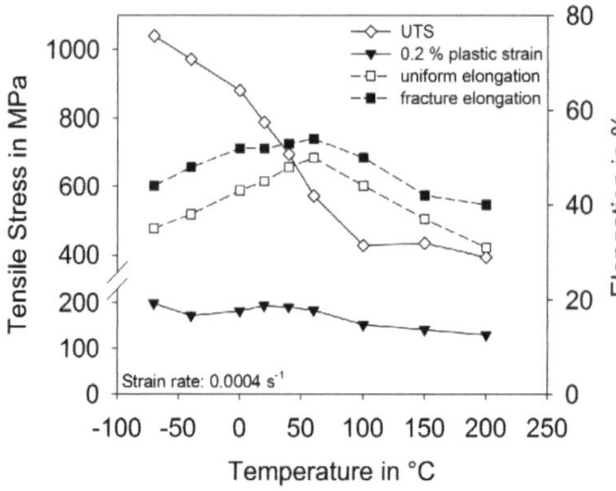

Figure 2. Temperature dependence of strength and ductility values under tensile loading.

The temperature dependence of flow stress and strain hardening behaviour is plotted in figure 3 and figure 4, respectively. The cast CrMnNi-alloy shows a pronounced TRIP-effect, which is indicated by high strain hardening values and microstructural observations. At low temperatures the austenitic phase transforms almost completely to α'-martensite. The formation of α'-phase leads to a sigmoidal shape of the flow curve, especially at low temperatures. The first α'-martensite is measured at 60°C, whereas at higher temperatures mechanical twinning is observed.

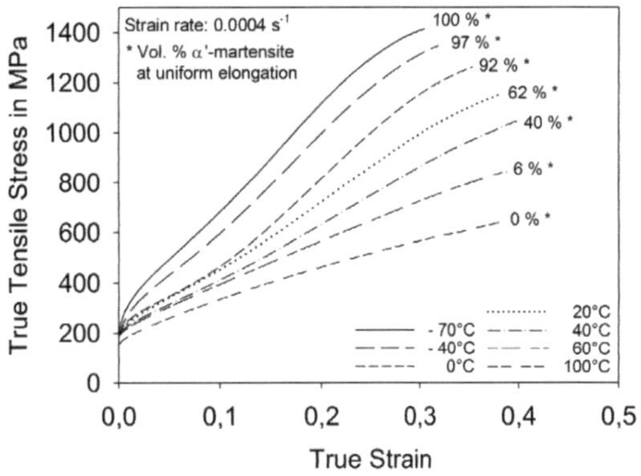

Figure 3. Flow curves and volume fraction α'-martensite after tensile straining at different temperatures and a strain rate of $0.0004s^{-1}$.

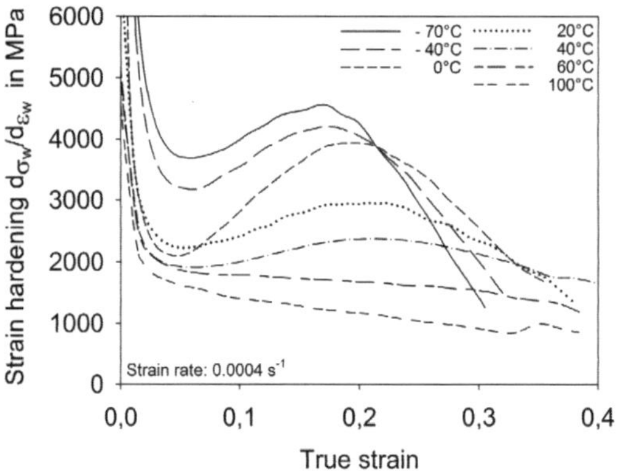

Figure 4. Strain hardening behaviour of tensile straining at different temperatures and a strain rate of $0.0004s^{-1}$.

The martensitic transformation to α'-phase leads to a maximum in strain hardening, corresponding to an inflection point in the flow curve. The lower the temperature, the higher the maximum in strain hardening. With increasing temperature, this maximum is lowered and moved to higher strain values, disappearing at least. As a result of a rise in SFE with increasing temperature, a change in deformation mode to mechanical twinning occurs, resulting in another strain hardening behaviour.

The influence of strain rate on compressive flow stress is shown in figure 5. The yield strength increases slightly with rising strain rate. The strain rate has a strong influence on strain hardening in terms of a transition from isothermal ($0.0004s^{-1}$) to adiabatic behaviour at high strain rates ($2200s^{-1}$). Due to plastic deformation and short deformation time, heat is generated and remains in the specimen. The temperature increase in the sample causes a rise in SFE and therefore inhibits martensitic transformation. For this reason the strain hardening is lowered with increasing strain and strain rate.

The evolution of microstructure during straining at room temperature is explained by the following steps: First of all, thin slip bands emerge, causing a fragmentation of coarse austenitic grains. In these slip bands a high density of stacking faults is generated. If this stacking fault density exceeds a critical value, an indication as ε-martensite is possible. The intersections of slip bands and the shear band-grain boundary intersections build nucleation sites for the formation of α'-phase, figure 6 [6,10-12].

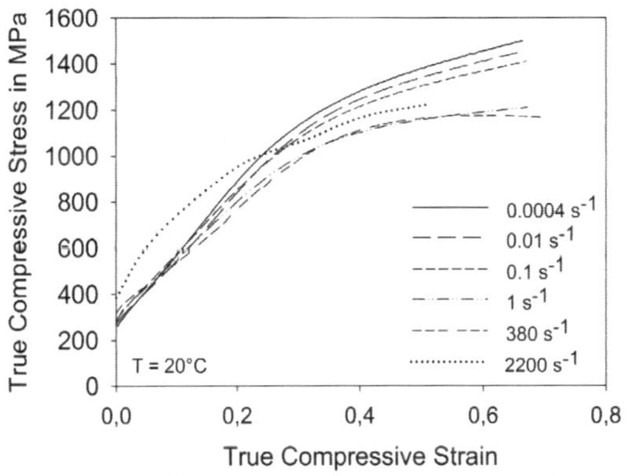

Figure 5. Influence of strain rate on mechanical behaviour under compressive loading at 20°C.

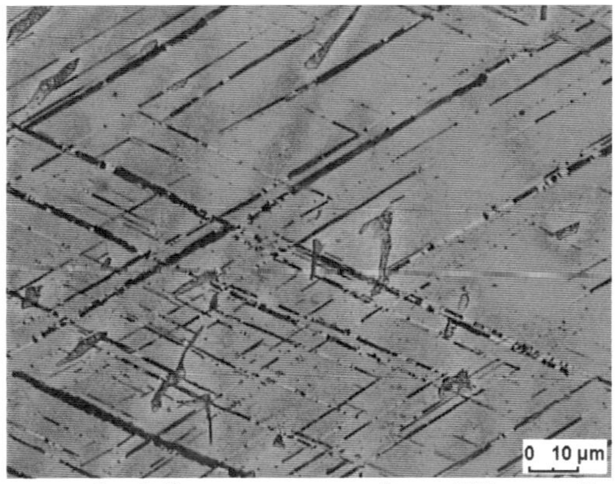

Figure 6. Microstructure after 5% tensile straining at 20°C and a strain rate of $0.0004 s^{-1}$.

4. Conclusions

The examined cast CrMnNi-alloy shows a distinctive TRIP-effect and therefore has excellent mechanical properties. The material undergoes a change in deformation behaviour with varying temperature. While above 100°C twinning occurs, α'-martensite is detectable below this temperature. The formation of α'-martensite during straining is distinguishable in a sigmoidal shape of flow curves and in a pronounced maximum in strain hardening. At temperatures below 20°C an anomaly in yield strength is observed which can be explained by the effect of stress-induced martensitic transformation. At higher strain rates a temperature rise in specimen and a transition from isothermal ($0.0004 s^{-1}$) to adiabatic ($2200 s^{-1}$) material behaviour occurs. Therefore, with increasing strain rate martensitic transformation is inhibited and a lower strain hardening effect is observed.

Acknowledgements
The authors gratefully acknowledge the Deutsche Forschungsgemeinschaft (DFG) for supporting this work carried out within the framework of SFB 799 (Collaborative Research Center B2 and A2).

References

[1] Weiß, A., Scheller, P.R. and Gutte, H., *Steel Grips* **1** (2003) 284-288.
[2] Weiß, A., Gutte, H. and Scheller, P.R., *Steel Res.* **77** (2006) 727-732.
[3] Fahr, D., *Met. Trans.* **2** (1971) 1883-1892.
[4] Schramm, R.E. and Reed, R. P., *Metall. Trans.* A **6** (1975) 1345-1351.
[5] Sato, K., Ichinose, M., Hirotsu, Y. and Inoue, Y., *ISIJ Int.* **29** (1989) 868-877.
[6] Talonen, J and Hänninen, H., *Acta Mater.* **55** (2007) 6108-6118.
[7] Dumay, A., Chateau, J.-P., Allain, S., Migot, S. and Bouaziz, O., *Mater. Sci. Eng.* **483-484** (2008) 184-187.
[8] Talonen, J., Aspegren, P., Hänninen, H., *Mater. Sci. Tech. Ser.* **20** (2004) 1506-1512.
[9] Schumann, H., Wissenschaftliche Berichte des Zentralinstituts für Festkörperphysik und Werkstoffforschung **20** (1979) 137-157.
[10] Olson, G.B., Cohen, M., *Met. Trans.* A **6** (1975) 791-795.
[11] Das, A., Sivaprasad, S., Ghosh, M., Chakraborti, P.C. and Tarafder, S., *Mater. Sci. Eng.* **486** (2008) 283-286.
[12] Schumann, H., *Kristall und Technik* **12** (1977) 363-370.

Journal of Physics: Conference Series **240** (2010) 012099 doi:10.1088/1742-6596/240/1/012099

Microstructure and compression strength of novel TRIP-steel/Mg-PSZ composites

H. Biermann[1], C.G. Aneziris[2], A. Kolbe[1], U. Martin[3], A. Müller[3], W. Schärfl[2] and M. Herrmann[4]

[1]TU Bergakademie Freiberg, Institute for Materials Engineering, Gustav-Zeuner-Strasse 5, D-09599 Freiberg, Germany
[2]TU Bergakademie Freiberg, Institute for Ceramics, Glass and Construction Materials, Germany
[3]TU Bergakademie Freiberg, Institute for Materials Science, Gustav-Zeuner-Strasse 5, D-09599 Freiberg, Germany
[4]Fraunhofer IKTS, D-01277 Dresden, Germany

biermann@ww.tu-freiberg.de

Abstract. A novel steel-based composite material, composed of metastable austenitic stainless steel as matrix and up to 15 % zirconia as reinforcement, is processed by two powder metallurgy routes. The matrix exhibits the so-called TRIP-effect (TRIP: TRansformation-Induced Plasticity) and shows a deformation-induced formation of martensite. Compression tests of rod samples processed by cold isostatic pressing show increased strength compared to the non-reinforced steel matrix up to 20 % strain. Three-point bending tests show, however, reduced ductility for high zirconia contents. Filigree honeycomb structures were produced by a novel extrusion technique with extraordinary high values of specific energy absorption.

1. Introduction

Most Metal-Matrix-Composites (MMCs) have a light-metal matrix such as aluminum or magnesium. MMCs with a steel matrix, however, have not been examined very much up to now, although they may be possible candidates for use as high strength and wear resistant materials [1,2].

The combination of a steel matrix which shows the TRIP effect (TRIP: Transformation-Induced Plasticity) with a metastable ZrO_2-reinforcement also exhibiting a martensitic phase transformation was until now only investigated by the group of Guo et al. with regard to the mechanical properties. The studied composite was produced by the conventional powder metallurgical route and hot pressing [3-5]. Guo et al. thus created TRIP-steel matrix composites with reinforcements made of yttria-partially stabilized zirconia-particles (2Y-PSZ) and obtained an extraordinary high strength in high deformation rate compression experiments.

In the present work results of a new collaborative research center with the title "TRIP-Matrix-Composites – design of tough, transformation reinforced composite materials and structures on Fe-ZrO_2-basis" are presented. Samples of the steel-matrix composite materials were produced by cold-isostatic pressing and a novel extrusion technique which is used conventionally in ceramic technology. First results of a variant of the family of composite materials to be developed were already presented in very recent papers [6-8].

2. Materials and Experimental Details

The composite materials were made of AISI 304 stainless steel powder and zirconia with 3.5 wt.% MgO with particle sizes of 45 µm and 2 µm, respectively. Cylindrical rods without zirconia and with zirconia contents of 5 % and 15 % were prepared by cold isostatic pressing (CIP), square honeycomb structures without zirconia and with zirconia contents of 5 % and 10 % and a dimension of 25.5 mm x 25.5 mm with 196 channels (200 cpsi—channels per square inch) and a wall thickness of 250 µm have been extruded, cf. [6]. The green bodies were dried, debinded and finally sintered at 1350°C for 2 h. The CIPed samples had relative densities of 90 to 93 %. The microstructure of the CIPed material with 15 % zirconia is shown in Figure 1a by an optical micrograph after etching with Beraha II. The fine zirconia grains are clustered due to an insufficient mixing procedure which has to be improved in future. Because of the sintering temperature of 1350°C, the separate zirconia particles are unsintered in these clusters. However, the steel powder particles are sintered to each other and exhibit some binding to the zirconia grains. Figure 1b shows a square honeycomb sample with the filigree macrostructure.

From the rods, cylindrical samples for compression experiments of 6 mm diameter and 9 mm height as well as samples of dimension 4 mm x 6 mm x 40 mm for 3-point bending tests were machined. The CIPed samples as well as the honeycomb samples were tested in a servohydraulic 500 kN testing machine at a displacement rate of 0.016 mm/s ($2x10^{-3}$ s^{-1}).

a)

b)

Figure 1: (a) Microstructure of a CIPed composite sample with 5 % zirconia, optical microscopy, etchant Beraha II; (b) honeycomb sample produced by extrusion.

3. Results and Discussion

The results of compression and bending experiments, respectively, of the CIPed materials are shown in Figure 2. In compression, the non-reinforced sintered steel samples behave comparable to wrought steel of equivalent composition. The composites with 5 % and 15 % zirconia show an increased flow stress. The subsequent stress-strain curves of the two composites, however, differ significantly. The 5 % zirconia composite sample exhibits a comparable strain hardening as the non-reinforced material, however at higher strength. In contrast, the stress-strain curve of the 15 % zirconia composite deviates already at a strain of 8 % due to limited ductility, see the inserted micrographs of the samples after compression to 50 % true compressive strain. Additional investigations showed the deformation-induced formation of α′-martensite in the steel matrix by optical microscopy and EBSD [8].

In bending, the 5 % zirconia sample shows again a higher strength compared to the non-reinforced sample. However, the 15 % zirconia sample fails at small bending stresses in a nearly brittle mode. The fracture surfaces show that the crack path follows the unsintered zirconia clusters [8].

The compressive stress-strain curve of a honeycomb sample is given in Figure 3 together with some micrographs taken at different strains. It shows a large regime of strain hardening, followed by a plateau-like behavior with a flow stress of about 270 MPa at approximately 25 % compressive strain and a subsequent stress decrease. The micrographs show the failure of some walls of the honeycomb structure by bending and first cracks on the surface of the sample.

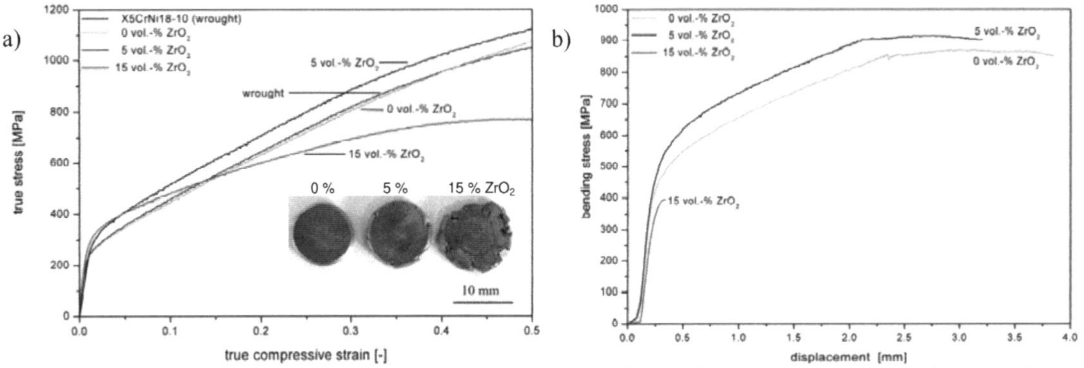

Figure 2: Compressive (a) and bending (b) behavior of CIPed TRIP-matrix composite samples compared to zirconia-free AISI 304 steel.

In addition to the macroscopic failure observed on the surface of the honeycomb samples, SEM investigations have been carried out. Figure 4a shows an overview from the top side of a honeycomb sample containing 10 % zirconia deformed to 50 % compressive strain. The separate walls deform plastically by bending and buckling, respectively. In some cases, rupture of the walls can be found, in accordance to the macroscopic shear band visible in Figure 3, starting at 25 % compressive strain. The deformation and failure modes are in accordance to the model of Gibson and Ashby for cellular materials [10]. The mass and volume specific energy absorption, respectively, observed during compression of the honeycomb samples shows extraordinary high values, cf. [6-8].

A metallographic longitudinal section of the deformed honeycomb sample studied by optical microscopy is shown in Figure 4b. The zirconia particle clusters are deformed because the single zirconia particles remain unsintered. In the steel matrix, a high density of glide bands is visible in the presented micrograph. However, due to the inhomogeneous deformation of the honeycomb samples, other positions of the section showed nearly no traces of deformation (not shown here). A closer view on the deformation bands showed beginning martensitic transformation of the metastable austenitic steel to the α′-martensite, cf. Figure 4b.

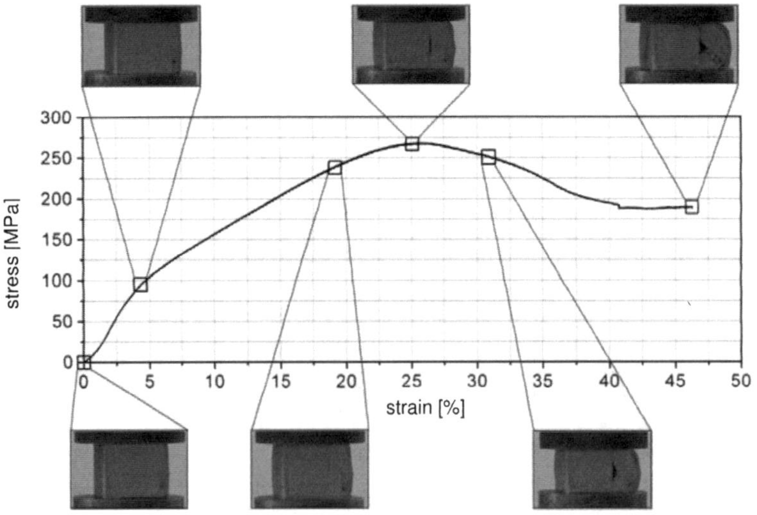

Figure 3: Compressive stress-strain curve of a honeycomb composite sample together with optical photographs of the sample at different deformation states.

a) b)

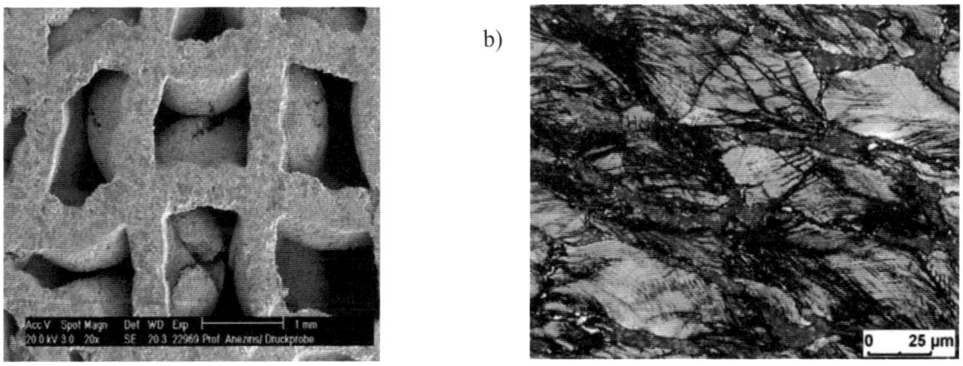

Figure 4: SEM micrograph of a top view [6] (a) and optical micrograph (stress axis vertical, etchant Beraha II) [8] (b) of a honeycomb sample with 10 % zirconia deformed to 60 % compressive strain.

4. Conclusions

In the present paper new composite materials based on austenitic stainless TRIP-steel reinforced with zirconia are presented. In further studies the mixing process of the raw materials has to be improved within the framework of the ongoing research on TRIP-Matrix Composites to avoid the clustering of the zirconia particles. Both composites produced by CIP and by the novel extrusion technique of a plastic feed stock material show a reinforcement effect compared to the reference samples of non-reinforced austenitic steel. In the steel matrix, the deformation-induced formation of α'-martensite was observed (TRIP-effect). It has to be pointed out that in the present investigation the zirconia particles exhibited a low fraction of metastable phases and therefore showed no martensitic transformation.

Acknowledgements

The authors thank German Research Foundation (DFG) for funding of the Collaborative Research Center "TRIP-Matrix Composites" (SFB 799). The authors also acknowledge the support of Dipl.-Ing. T. Mottitschka in experimental work.

References

[1] Karlsohn M, Weber S, Alavi Zaree S R, Müller S, Theisen W, Reimers W and Pyzalla A R, 2008 *Powder Metallurgy* **51** 31
[2] Wittig D, Glauche A, Aneziris C G, Minghetti T, Schelle C, Graule T and Kuebler J 2008 *Mater. Sci. Eng. A* **488** 580
[3] Guo Y, Zhou Y, Li D, Duan X and Lei T 2003 *J. Mater. Sci. Technol.* **19** 137
[4] Guo Y, Zhou Y, Duan X, Li D and Lei T 2004 *Ceram. Int.* **30** 75
[5] Zhou Y, Guo Y, Li D and Duan X 2003 *Trans. Nonferrous Met. Soc. China* **3** 1086
[6] Aneziris C G, Schärfl W, Biermann H and Martin U *Int. J. Appl. Ceram. Technol.* DOI:10.1111/j1744-7402.2008.02321.x, in press
[7] Aneziris C G, Biermann H, Schärfl W, Ballaschk U and Martin U 2009 in *Verbundwerkstoffe*, ed. Krenkel W, Wiley-VCH, Weinheim, 147
[8] Biermann H, Martin U, Aneziris C G, Kolbe A, Müller A, Schärfl W and Herrmann M, *Adv. Eng. Mater.* 10.1002/adem.200900210, accepted for publication
[9] Gibson L J and Ashby M F 1997 *Cellular Solids - Structure and Properties*, 2nd ed. Cambridge University Press

Portevin-Le Chatelier effect in an E911 creep resistant steel with 3%Co additives

A Kipelova[1], R Kaibyshev[1], V Skorobogatykh[2], I Schenkova[2]

[1] Belgorod State University, Laboratory of Mechanical Properties of Nanoscale Materials and Superalloys, Pobeda 85, 308015 Belgorod, Russia
[2] Central Research Institute for Machine-Building Technology, Sharikopodshipnikovskaya 4, 115088 Moscow, Russia

kipelova@bsu.edu.ru

Abstract. The effect of tempering temperature on mechanical properties of an E911+3%Co creep resistant steel was investigated. The mechanical tensile tests were carried out at temperatures from 298 to 1073 K and at strain rates varying from 2.1×10^{-5} s^{-1} to $2.1 \times 10^{-1} s^{-1}$. The Portevin-Le Chatelier (PLC) effect was found in the temperature range of 473 to 623 K. Various attributes of dynamic strain aging (DSA) like serrated flow with an acoustic emission were observed. With increasing temperature the ultimate tensile strength (UTS) and the yield strength (YS) increased while the ductility decreased. The dependences of the critical plastic strain on strain rate and temperature exhibited "inverse" behavior that was associated with concentrated solid solution in the DSA regime.

1. Introduction

The martensitic steels containing 9-12%Cr are of great interest as components for advanced high efficiency power plants [1, 2]. The beneficial combination of mechanical properties of these steels is achieved by a complex alloying to provide both solution and precipitation strengthening. The latter is resulted from homogeneous distribution of stable second-phase nanoscale dispersoids.

Dynamic strain aging (DSA) was shown in martensitic/ferritic steels due to the interaction of interstitial atoms such as C and N with mobile dislocations during plastic deformation under specific conditions of temperature and stain rate. The DSA results in serrated stress-strain curves, promotes a work hardening, which increases an ultimate tensile strength, and reduces a ductility [3, 4]. Concerning applications, the strengthening effect of DSA can be used to improve mechanical properties such as strength, fatigue and creep resistance [5]. On the other hand, the DSA resulted in several detrimental consequences, i.e. decreased ductility and toughness, and increased the fracture appearance transition temperature. The aim of the present work is to examine the influence of tempering temperature on the temperature range of the PLC effect and its influence on mechanical properties of an E911+3%Co creep resistant steel.

2. Experimental procedure

The chemical composition of investigated E911+3%Co creep resistant steel was (C:0.13, Cr:8.6, Co:3.2, W:1.2, Mo:0.9, V:0.2, Cu:0.1, Nb:0.07, Si:0.06, Ni 0.05, N:0.04, Mn:0.02, B:0.005, all in mass%, and the balance Fe). The samples were austenitized at 1323 K for 0.5 h followed by air cooling. Then, the specimens were tempered at temperatures ranging from 473 to 1073 K for 3 hours followed by mechanical tests at the same temperatures. The tensile tests were carried out at strain rates of 2.1×10^{-5} s^{-1} to $2.1 \times 10^{-1} s^{-1}$ by using an Instron model 5882 universal testing machine. Tensile specimens of 3×1.5 mm^2 cross section and 12 mm gauge length were used.

Differential scanning calorimetry (DSC) measurements were conducted on heating and cooling between 298K and 1323 K at the rates of 2 and 10 K/min to analyze a precipitation behavior of the steel. The DSC disk samples with a diameter of 5 mm and a thickness of 0.4 mm were heated and cooled in Al_2O_3 crucibles in an argon atmosphere.

The tempered microstructures was examined by using a JEM-2100 transmission electron microscope (TEM) operating at 200 kV that was equipped with an INCA energy-dispersive X-ray spectrometer (EDS). The TEM foils were prepared by the double jet electro-polishing method using a 10% solution of perchloric acid in glacial acetic acid. Extraction replicas for the microchemical analysis of precipitates were prepared by the evaporation of carbon onto a polished and etched sample surface followed by dissolution of the metallic matrix in a 10%HCl with ethanol at a voltage of 2 V at 293 K. The extraction replicas were caught on small copper grids and cleaned with ethanol.

3. Results and discussion

Serrated stress-strain curves associated with DSA were observed at temperatures of 473 to 623 K and at strain rates varying from 2.1×10^{-5} s^{-1} to $1 \times 10^{-1} s^{-1}$ (figure 1(a), (b)). The type B serrations occurred

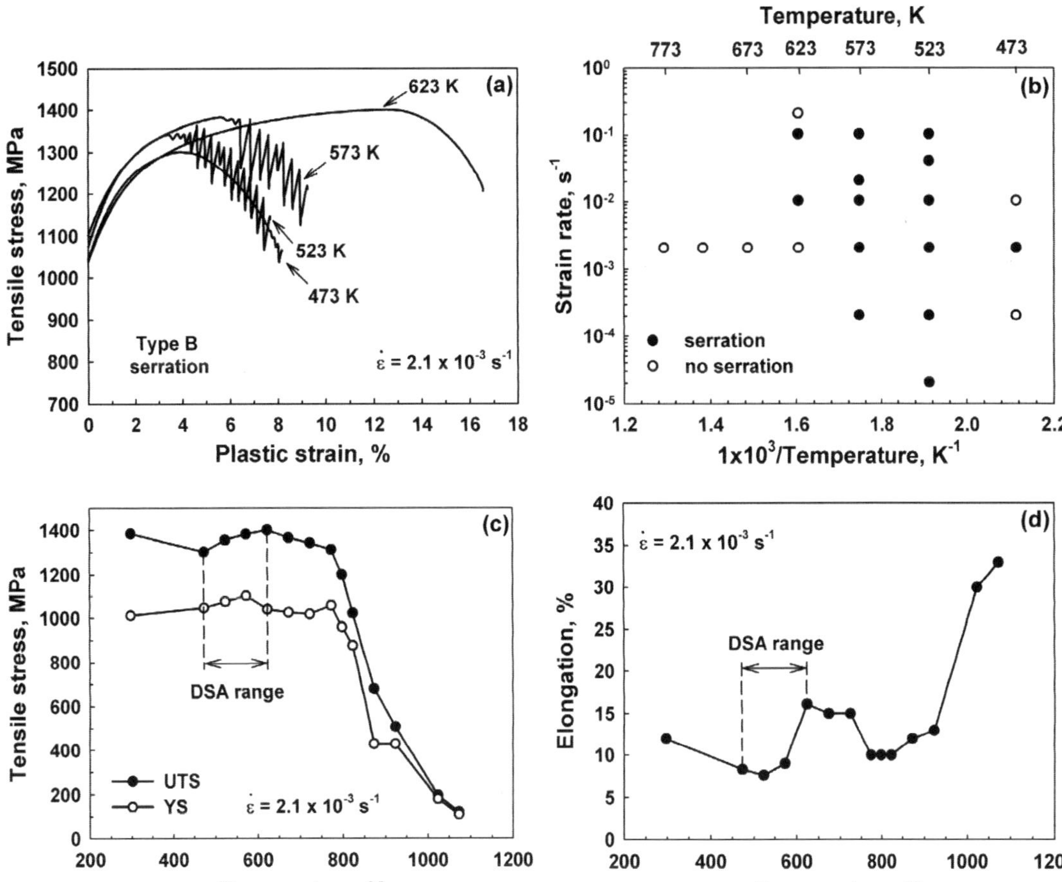

Figure 1. Mechanical properties of an E911+3%Co steel: typical strain-stress curves representing type B serrations (a); temperature-strain rate interval of serrated flow (b); temperature dependence of strength (c) and ductility (d).

under conditions of DSA regime. Type B serrations are oscillations about the general level of the stress-strain curve that occur in quick succession due to discontinuous band propagation arising from the DSA of the moving dislocations within the Luders band [6]. Serrations appear when the carbon or nitrogen concentration on the dislocation exceeds the critical concentration required for the formation of sufficiently large atmosphere. This phenomenon involves the pipe diffusion of carbon or nitrogen from forest dislocations to the waiting mobile dislocations [3].

The temperature dependences of UTS, YS and ductility are shown in figures 1(c), (d). In the DSA range the UTS and the YS gradually increased with increase in the test temperature to 623 K and 573 K, respectively (figure 1(c)). At the same, the ductility exhibited non-monotonous behavior with a minimum at 523 K (figure 1(d)). This suggests that the Luders bands can be potential sites for failure during serrated flow, deteriorating the ductility; and the probability of such failure increases as the material work-hardens. In the DSA range an "inverse" behaviour [7] of the critical plastic strain with strain rate and temperature was found (figure 2(a), (b)). The strain rate and temperature dependencies of critical plastic strain are connected with concentrated solid solution.

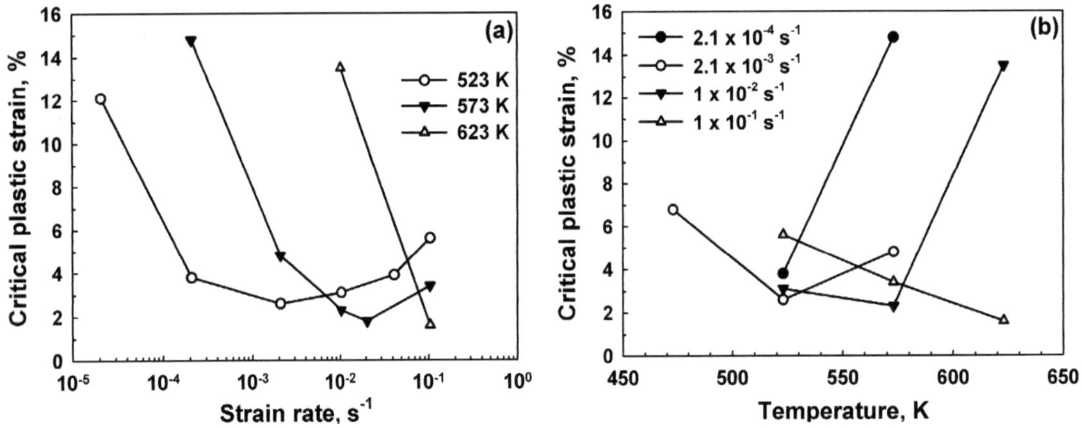

Figure 2. Dependence of the critical plastic strain on strain rate (a) and temperature (b).

The weak exothermic reactions were detected by DSC on heating with the rates of 2 and 10 K/min at 483 and 623 K, respectively. These reactions correspond to the precipitation of second phase particles. The quenched steel samples contained occasional equiaxed particles, approximately 30 nm in diameter; they were identified as NbC by EDS spectra and electron diffraction of the particles extracted from the matrix. Two types of particles, namely, equiaxed NbC particles with an average size of about 18 nm and elongated $(Fe, Cr)_3C$ particles of approximately 80 × 30 nm in size that uniformly distributed within the matrix were observed in the steel tempered at 573 K (figure 3(a)). The ratio of Cr/Fe less than 0.2 was found in the $(Fe, Cr)_3C$ particles. The interaction of these carbides with moving dislocations is clearly seen in the TEM micrograph taken from the deformation substructure after tensile test (figure 3(b)). The volume fraction of homogeneously distributed NbC and $(Fe, Cr)_3C$ carbides was increased with increase in the tempering temperature to 773 K. The amount of Cr in $(Fe, Cr)_3C$ carbides was almost fivefold increased. The increased number of second phase particles led to an increase in the YS and a decrease in the ductility at temperatures of around 773 K (figure 1(c), (d)). Disappearance of DSA at temperatures above 623 K was discussed as a result of the numerous precipitations of fine uniformly distributed carbides acting as extremely effective sinks depleting the interstitial (carbon/nitrogen) atmosphere at dislocations [3]. Further increase in tempering temperatures up to 1073 K led to additional precipitation of $Cr_{23}C_6$ carbides. The intensive carbides precipitation in the temperature range of 773 to 1073 K reduces concentration of alloying elements in solid solution and terminates the DSA. It was also suggested that at elevated temperatures and low strain-rates, the blocked dislocations could drag their atmosphere with them and serrated yielding was unlikely to occur [3].

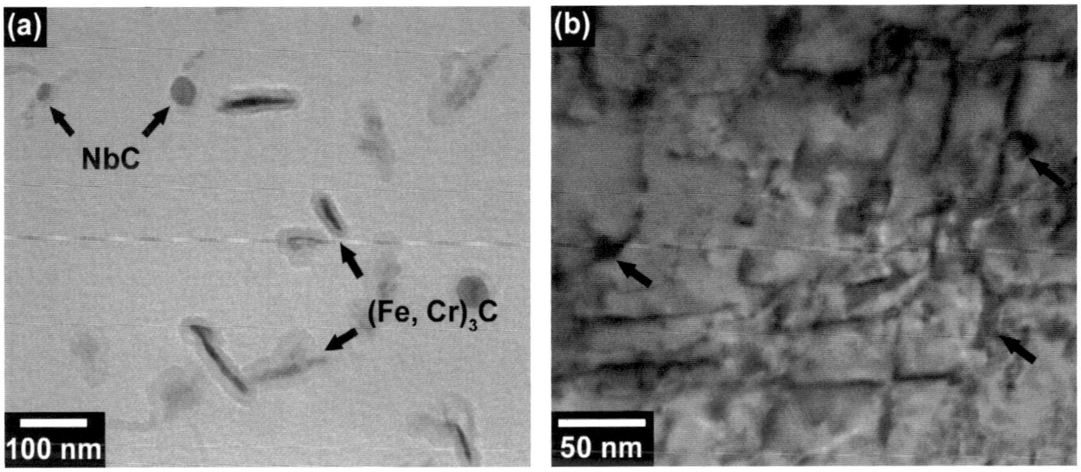

Figure 3. TEM micrographs of the E911+3%Co steel after austenitizing at 1323 K for 0.5 h and tempering for 3 h at 573 K (a) and additionally deformed at 573 K (b).

4. Conclusions

The influence of tempering temperature on the temperature range manifestation of the PLC effect and its effect on the mechanical properties of an E911+3%Co creep resistant steel was examined. The main results are as follows:

1. The E911+3%Co steel demonstrates occurrence of PLC effect in the temperature range of 473 to 623 K due to phenomenon of DSA with the following typical attributes: serrated plastic flow with an acoustic emission, increasing the UTS and the YS along with decreasing the ductility as temperature increases.

2. Intensive carbide precipitation at temperatures above 623 K is responsible for the disappearance of DSA during the tensile tests at elevated temperatures.

Acknowledgements

The study was supported by Federal Agency for Science and Innovations under grant No. 02.523.12.3019. Authors are grateful to Mr. V. Dudko, Center of Common Facilities, Belgorod State University, for his assistance in instrumental analysis of mechanical properties.

References

[1] Ennis P J, Czyrska-Filemonowicz 2002 *Operation Maintenance and Materials Issue* **1** 1
[2] Vaillant J C, Vandenberghe B, Hahn B, Heuser H, Jochum C 2008 *Int. J. Pressure Vessels and Piping* **85** 38
[3] Kishore R, Singh R N, Sinha T K, Kashyap B P 1997 *J. Mater. Sci.* **32** 437
[4] Mirty KL, Hong J, Zou Y, Britt J C Role of Friction and Source Hardening in DSA and Radiation Embrittlement of Ferritic Steels and Iron *Transaction of the 12 th International Conference on Structural Mechanics in Reactor Technology (Stuttgart, Germany, 15-20 August 1993)* 147-1522.
[5] Lee J S, Kim I S, Pack C Y, Kim J W 2000 *J. Korean Nuclear Society* **32** 77
[6] Rodriguez P 1984 *Bull. Mater. Sci.* **6** 653
[7] Brechet Y, Estrin Y. 1995 *Acta Metall. Mater.* **43** 955

Effects of lamellar thickness on misfit dislocation introduction and mechanical properties of γ/α_2 nano-lamellar TiAl alloys

K Maruyama, A Tabata, Y Toriyama, M Suzuki, K Yoshimi

Graduate School of Environmental Studies, Tohoku University, Sendai 980-8579, Japan

maruyama@material.tohoku.ac.jp

Abstract. Stress-strain behavior of lamellar Ti-38Al-3Zr and -3Nb alloys with average lamellar thickness ranging from 10 to 1000 nm were studied at room temperature. Their yield stresses decrease from a high value of coherent lamellar structures to a low value after introducing misfit dislocations onto lamellar boundaries. Their strain hardening rates increase with decreasing lamellar thickness, and then drop to a low level when the misfit dislocations become absent. There is critical thickness of γ lamellae for introduction of misfit dislocations. The thickness decreases with increasing lattice misfit between the constituent phases.

1. Introduction

TiAl Alloys consist of γTiAl and α_2Ti_3Al phases and form a lamellar microstructure. Their yield stresses σ_y increase with decreasing lamellar thickness λ according to the Hall-Petch relation [1-3]:

$$\sigma_y = \sigma_o + k/\sqrt{\lambda} \tag{1}$$

where σ_o is the friction stress, and k is the Hall-Petch slope. Further refinement of their lamellar thickness below 100 nm results in deviation from the Hall-Petch relation [4-6]. Yield stress of polycrystals sometimes exhibits an inverse Hall-Petch relation at very fine grain sizes [7,8].

There is a lattice misfit of 1 to 2 % between the γ and α_2 phases. The misfit is accommodated by elastic deformation (coherent boundary) or by introducing misfit dislocations onto γ/α_2 boundaries (semi-coherent boundary). Hazzledine [9] has predicted a critical lamellar thickness for the introduction of misfit dislocations, and Maruyama et al. [6] have proved the transition from coherent boundaries to semi-coherent boundaries above a critical thickness of γ lamellae. The change in boundary microstructures can affect stress-strain response of lamellar TiAl alloys. The present paper studies how the critical γ thickness for the introduction of misfit dislocations changes with amount of lattice misfit between the constituent phases. Contributions of the misfit dislocation introduction to the lamellar size dependence of yield stress are also studied.

2. Experimental procedure

Ti-38mol%Al-3mol%M (M = Nb or Zr) alloys were used in the present study. The materials used for studying the misfit dislocation introduction were arc-melted and then isothermally forged at 1423 K to nominal strain of 50 %. The other materials for mechanical tests were induction-skull-melted and then hot-extruded at 1473 K to 80 % reduction in area. The alloys were solution treated at 1453 K for 600 s to produce an α_2 single phase state, and then aged in vacuum at 900-1300 K in the $\gamma+\alpha_2$ dual phase field until formation of a fully lamellar microstructure. Since volume fractions of γ phase measured by TEM are 31 and 37 % in the Ti-38Al-3Nb and -3Zr alloys, respectively, most of lamellar boundaries are a γ/α_2 type in the present alloys. The lamellar colony (grain) sizes of the alloys measured by SEM are 75 μm. Compression tests were carried out at room temperature at a strain rate of 3×10^{-3} s^{-1} with specimens of 2×2 mm^2 in cross section and 3 mm height.

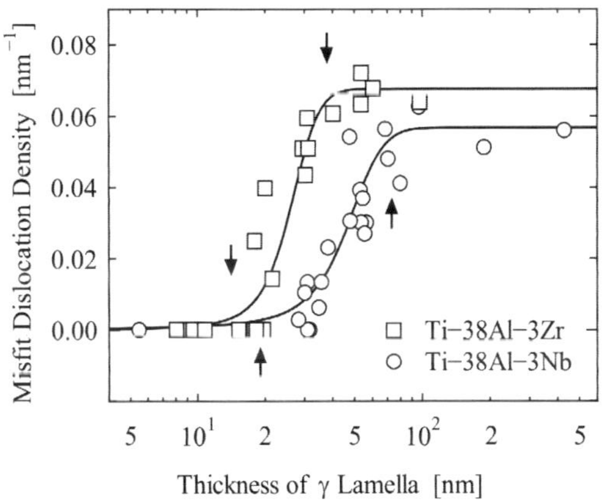

Figure 1. Changes of misfit dislocation density as a function of thickness of γ lamella.

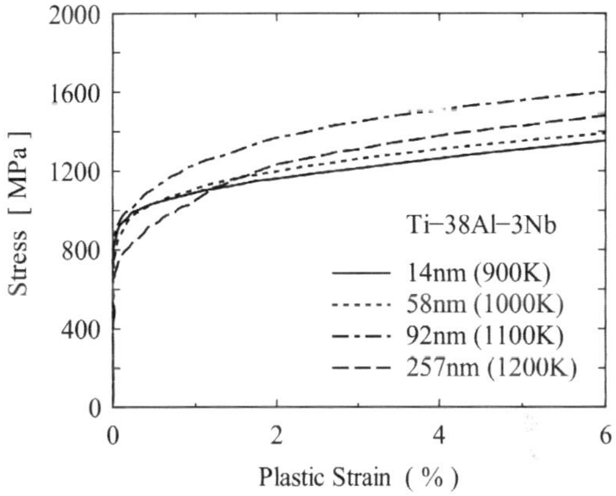

Figure 2. Representative stress-strain curves with different average lamellar thicknesses.

3. Results and discussion

3.1. Introduction of misfit dislocation onto γ/α_2 boundaries

When lamellar boundaries are end on, misfit dislocations make a dot like contrast around them on transmission electron micrographs. Density of misfit dislocations can be measured by counting the dot contrast [6]. Misfit dislocation densities of the two alloys measured by this technique are plotted in Fig. 1 against thickness of the γ lamella in which the density was measured. Misfit dislocation is absent in thin lamellae, and in thick lamellae the dislocation density reaches a stationary level ρ_m typical of the material. The curves in Fig. 1 were determined by regression analyses based on the error function so that the curves give the best fit to the data points. The arrows in the figure indicate the points of $0.05\rho_m$ and $0.95\rho_m$. During aging γ plates precipitate in an α_2 matrix. Lattice misfit $\Delta\varepsilon_m$ is accommodated by elastic deformation, and thereby introducing misfit stress τ_m in γ lamellae. The misfit stress is in proportion to $\Delta\varepsilon_m$. Suppose misfit dislocations are introduced by dislocation motion in γ lamellae, then the critical thickness λ_γ^* of γ lamellae for the introduction of misfit dislocations is given by

$$\lambda_\gamma^* = G\,b / \tau_m \qquad (2)$$

where G is the shear modulus, and b is the length of Burgers vector. The misfit dislocation density ρ_m [m^{-1}] necessary to fully accommodate the misfit strain $\Delta\varepsilon_m$ is given by

$$\rho_m = \Delta\varepsilon_m / b_m \qquad (3)$$

where b_m is the length of Burgers vector of misfit dislocations. The high ρ_m value of Ti-38Al-3Zr alloy points out a larger $\Delta\varepsilon_m$ value in the alloy. The larger $\Delta\varepsilon_m$ results in the lower λ_γ^* value in the Ti-38Al-3Zr alloy.

3.2. Effects of lamellar thickness on mechanical properties

Figure 2 shows representative stress-plastic strain curves of the Ti-38Al-3Nb alloy. The lamellar microstructures were produced by isothermal aging at the temperatures listed in the figure, and their average lamellar thicknesses decrease with decreasing aging temperature. The specimen with the finest thickness (14 nm) has high yield stress but low strain hardening rate, whereas the coarse lamellar material of 257 nm thickness has low yield stress and high strain hardening rate. Their 0.2 % proof stresses increase with decreasing thickness and then reach an upper limit. On the other hand, their 4 % proof stresses take a maximum value at the thickness of 92 nm, and then decrease to a stationary value, indicating the inverse Hall-Petch relation. The lamellar thickness dependence of flow stress varies with location on the stress-strain curves. For more detailed understanding of these variations, we examine the yield stress and the strain hardening rate of the present alloys in Fig. 3.

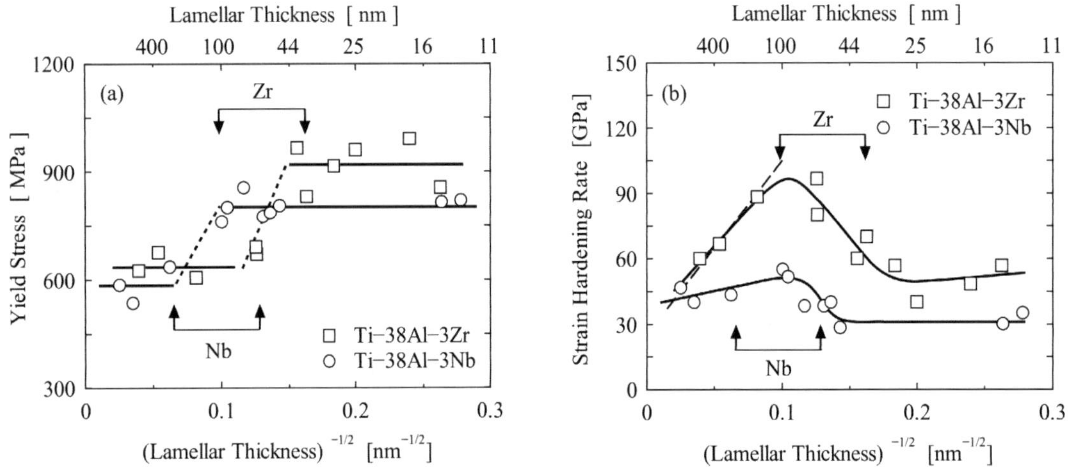

Figure 3. (a) Yield stress and (b) strain hardening rate of Ti-38Al-3Zr and -3Nb alloys as a function of average lamellar thickness.

The yield stress in the present study is defined as follows: A straight line connecting 0.2 % and 0.5 % proof stresses is drawn. The intersection of the straight line and the elastic line gives the yield stress. Strain hardening essentially has no contribution to the yield stress. The strain hardening rate is defined as the average slope of a stress-plastic strain curve between 0.2 % and 0.5 % strain. The yield stress and strain hardening rate are plotted in Fig. 3 against the inverse square root of average lamellar thickness. The arrows in the figure indicate the average lamellar thicknesses that correspond to the γ lamellar thicknesses giving $0.05\rho_m$ and $0.95\rho_m$ in Fig. 1. The yield stress takes a high value at fine lamellar thicknesses, and then decreases stepwise to a low value after introducing the misfit dislocations. Theories on the Hall-Petch relation, for example the dislocation pile-up model [10] and the geometrically necessary dislocation model [11], assume that grain size dependent flow stress is brought about by grain size dependent strain hardening, and that the yield stress defined in the present study is independent of grain size. This is essentially true in the present study. However, it should be pointed out that the lamellar microstructures without misfit dislocation have a higher friction (yield) stress than those having misfit dislocations, since each lamella is elastically deformed in the former for accommodating the lattice misfit. The yield stress levels of the coherent lamellar microstructures increase with increasing misfit strain.

Suppose dislocations glide within a lamella, leaving dislocation segments on both sides of the lamella, and the dislocation segments cause strain hardening. This model gives the following equation for flow stress:

$$\sigma = \sigma_o + 0.4\, M\, G\, b\sqrt{2\, M\, \varepsilon / b\, \lambda} \tag{4}$$

where σ_o is the lattice friction stress, *M* the Taylor factor (=3), *G* the shear modules (61 GPa), *b* the length of Burgers vector (0.28 nm), ε the average plastic strain (0.0035 in Fig. 3(b)), and λ is the average lamellar thickness. This equation provides the following expression on strain hardening rate:

$$d\sigma/d\varepsilon = 0.2\, M\, G\, b\sqrt{2\, M\, /\, b\, \lambda\, \varepsilon} \tag{5}$$

These equations are based on the assumptions that lamellar boundaries are too strong for glide dislocations to pass through, and that annihilation of dislocations cannot occur at the boundaries. In Fig. 3(b), dσ/dε is proportional to $\lambda^{-1/2}$, supporting Eq. (5). The equation gives the slope of the dashed line, and the line agrees well with the experimental data of the Ti-38Al-3Zr alloy in the coarse lamellar microstructures with misfit dislocations. This good agreement points out that the assumptions, namely strong obstacle and no annihilation of dislocations, hold for the lamellar microstructures with a high density of misfit dislocations. The strain hardening rate of the Ti-38Al-3Zr alloy starts to decrease below 100 nm of lamellar thickness, and increases gradually again below 25 nm. The arrows in the figure point out that the disappearance of misfit dislocations from lamellar boundaries brings about the decrease in strain hardening rate. This finding suggests that lamellar boundaries without misfit dislocations are not sufficient obstacles to dislocation glide and/or to annihilation of dislocations left at the boundaries. The strain hardening behavior of the Ti-38Al-3Nb alloy is similar to that of the Ti-38Al-3Zr alloy. However, the slope of the solid line of the Ti-38Al-3Nb alloy for the coarse lamellar microstructures is lower than the prediction from Eq. (5). This fact points out that a sufficient density of misfit dislocations is necessary to keep strain hardening rate at a high level.

4. Conclusions

(a) The critical thickness of γ lamellae for the introduction of misfit dislocations decreases with increasing lattice misfit between the two constitute phases.

(b) Yield stress decreases from a high level of lamellar microstructures with coherent lamellar boundaries to a low level after introducing the misfit dislocations. The yield stress of coherent lamellar microstructures increases with increasing lattice misfit.

(c) Strain hardening rate increases with decreasing lamellar thickness, but drops to lower values after the disappearance of misfit dislocations, namely in coherent lamellar materials. The strain hardening rate increases with increasing lattice misfit.

Acknowledgements

This research was supported by the Ministry of Education, Science, Sports and Culture, Japan (Grant No. 19206066).

References

[1] Maziasz P J and Liu C T 1998 *Metall. Mater. Trans.* **29A** 105
[2] Liu C T and Maziasz P J 1998 *Intermetallics* **6** 653
[3] Dimiduk D M, Hazzledine P M, Parthasarathy T A, Sashagiri S and Mendiratta M G 1998 *Metall. Mater. Trans.* **29A** 37
[4] Parthasarathy T A, Mendiratta M G and Dimiduk D M 1998 *Acta Mater.* **46** 4005
[5] Maruyama K, Suzuki G, Kim H Y, Suzuki M and Sato H 2002 *Mater. Sci. Eng.* **A329-331** 190
[6] Maruyama K, Yamaguchi M, Suzuki G, Zhu H, Kim H Y and Yoo M H 2004 *Acta Mater.* **52** 5185
[7] Misra A, Hirth J P and Hoagland R G 2005 *Acta Mater.* **53** 4817
[8] Shen T D, Schwarz R B, Feng S, Swadener J G, Huang J Y, Tang M, Zhang J, Vogel S C and Zhao Y 2007 *Acta Mater.* **55** 5007
[9] Hazzledine P M 1998 *Intermetallics* **6** 673
[10] Petch N J 1953 *J. Iron Steel Inst.* **174** 25
[11] Ashby M F 1971 *Strengthening Methods in Crystals* ed A Kelly and R B Nicholson (Amsterdam: Elsevier) pp 137-192

Mechanical properties of Cu-Ni-Be system alloys

R Monzen, Y Shimada and C Watanabe

Division of Innovative Technology and Science, Kanazawa University, Kakuma-machi, Kanazawa 920-1192, Ishikawa, Japan

monzen@t.kanazawa-u.ac.jp

Abstract. The effects of addition of 0.17wt%Zr, 0.1wt%Mg and 0.1wt%Co on the mechanical properties of a Cu-1.2wt%Ni-0.2wt%Be alloy have been investigated. Adding Zr, Co or Mg to the Cu-Ni-Be alloy brings about the improvement in strength and stress relaxation property. The Zr, Co or Mg addition decreases the inter-precipitate spacing of γ" precipitates, resulting in the increase in strength. The higher resistance to the stress relaxation of the Zr- or Co-added alloy is attributed to the lower density of mobile dislocations. The improvement of stress relaxation property by the Mg addition is explained by the viscous glide motion of dislocations dragging Mg atoms, in addition to the lower density of mobile dislocations.

1. Introduction

Copper-base alloys are used for electrical parts such as connectors and lead frames because the electrical conductivity of the alloys is very high. The high strength of the copper-base alloys is also required for small devices used in microelectronics. Most of the alloys are usually of the precipitation-strengthened types and are dilutely alloyed with elements of very low solubility to preserve high levels of conductivity. Cu-Ni-Be alloys containing usually about 2wt%Ni and 0.4wt%Be are one of such precipitation-hardenable copper-base alloys. Recently, a Cu-1.27wt%Ni-0.22wt%Be alloy, which contains smaller amounts of Ni and Be than the standard Cu-Ni-Be alloys, has been developed. The strength and electrical conductivity of the Cu-1.27wt%Ni-0.22wt%Be alloy and a standard Cu-2.05wt%Ni-0.35wt%Be alloy aged at 360 to 500 °C after cold rolling have been examined [1]. The former alloy exhibits a lower strength and a higher electrical conductivity than the latter alloy. Both alloys are hardened by disk-shaped coherent precipitates of γ" phase.

In this study, we investigate the strength and stress relaxation property of Cu-1.2wt%Ni-0.2wt%Be alloys with and without 0.17wt%Zr, 0.1wt%Mg or 0.1wt%Co aged at 450°C. It has been reported that the addition of Mg to a Cu-Ni-Si alloy [2] and Zr to a Cu-Cr alloy [3] enhances the strength and resistance to stress relaxation of the Cu-base alloys, and the addition of Co to a Cu-Ni-Be alloy increases the strength of the alloy [4]. This is the reason why Zr, Mg and Co were selected as added elements.

2. Experimental

Cu-1.2%Ni-0.2%Be, Cu-1.2%Ni-0.2%Be-0.17%Zr, Cu-1.2%Ni-0.2%Be-0.1%Co and Cu-1.2%Ni-0.2%Be-0.1%Mg alloys were prepared by melting in an argon atmosphere. The cast alloys were homogenized at 1000°C for 24h in a vacuum, cold-rolled by 50% reduction in thickness and then spark-cut into specimen strips. All the specimen strips were solution-treated at 1000°C for 8h in a vacuum, quenched into water, cold-rolled to 40% reduction in thickness and then aged at 450 °C for various times in an argon atmosphere. The solution-treated specimens had a grain size of 80μm.

Microhardness tests were carried out using the Vickers method. Tensile tests were performed using a static Instron-type testing machine with a constant strain rate of 10^{-3} s^{-1} at room temperature. Electrical resistivity measurements were made by a Hocking AutoSigma 3000 electrical conductivity tester at 20 °C. According to the literature [5], cantilever stress relaxation tests were performed at 180°C for 24h in a nitrogen atmosphere. Transmission electron microscopy (TEM) was carried out using a HITACHI H-9000NAR microscope at an operation voltage of 300kV. Thin foils for TEM observations were prepared by slicing the aged specimens with a spark cutter and by electropolishing using a solution of 67% methanol and 33% nitric acid at -30 °C and 6.5 V in a twin-jet electropolisher.

3. Results and discussion

3.1. Effect of Zr, Mg and Co on strength

The hardness curves on aging at 450 °C for Cu-Ni-Be alloys with and without Zr, Mg and Co showed that the peak hardness effect occurred after aging for 2h. Aging the alloys at 450 °C for 2h after cold rolling to 40% reduction produced disk-shaped precipitates in the Cu matrix of α phase. An analysis of high-resolution TEM (HRTEM) images of the precipitates revealed that the disk-shaped precipitates were a γ"-NiBe phase, which consisted of a two Be-layer structure separated by a Ni layer parallel to the $\{100\}_\alpha$ plane [1]. No other precipitates were observed in Cu-Ni-Be alloys with Zr, Mg and Co.

Table 1 summarizes the tensile properties and electrical conductivity of the present alloys aged at 450 °C for 2h. The addition of Zr, Mg and Co causes an increase in 0.2% proof stress and tensile strength.

Table 1. 0.2% proof stress, tensile stress, elongation and electrical conductivity, and volume fraction, radius, thickness, number density and average inter-tip spacing of disk-shaped γ" precipitates for Cu-Ni-Be system alloys aged at 450 °C for 2h after 40% cold-rolling.

Specimen (wt%)	0.2% proof stress (MPa)	Tensile stress (MPa)	Elongation (%)	Electrical conductivity (%IACS)	Volume fraction	Radius (nm)	Thickness (nm)	Number density ($\times 10^{23}/m^3$)	Inter-tip spacing (nm)
Cu-1.2Ni-0.2Be	490	570	7	65	0.013	5.6	0.5	2.7	15
Cu-1.2Ni-0.2Be-0.17Zr	560	640	7	60	0.016	4.0	0.5	6.3	11
Cu-1.2Ni-0.2Be-0.1Co	510	600	6	60	0.014	5.0	0.5	3.7	13
Cu-1.2Ni-0.2Be-0.1Mg	530	620	6	62	0.014	4.6	0.5	4.6	12

It has been reported that the yield stress of Cu-Ni-Be alloys containing γ" precipitates at room temperature is controlled by the Orowan mechanism at peak-age and over-age conditions [1]. The Orowan stress is inversely proportional to the inter-precipitate spacing λ. The increase in strength due to the addition of Zr, Mg and Co can then be discussed by estimating λ. Nie and Muddle considered the statistical evaluation of the average spacing λ_{ave} between plate tips assuming that the precipitates form as circular disks and arrived at the following expression [6]:

$$\lambda_{ave} = \sqrt{\frac{\sqrt{3}}{2}\frac{1}{\sqrt{2\sin\theta\cdot N\cdot r}}} - \frac{\pi r}{4} - \frac{\sqrt{3}}{2}\frac{h}{\sin\theta}. \quad (1)$$

Here N is the number density of precipitates per unit volume, r is the rdius of precipitates, h is their thickness and θ is the dihedral angle between the plate and the $\{111\}_\alpha$ slip plane. For $\{001\}_\alpha$ precipitate plates, θ=54.74°. The diameter and thickness of γ" precipitates were measured from HRTEM images. To obtain statistically reliable data, more than 100 precipitates were analyzed for each alloy. The volume fraction f for the Cu-Ni-Be alloy was determined by applying the values of electrical resistivity, before and after aging at 450 °C for 2 h, to the experimental data regarding the dependence of electrical resistivity on Ni or Be concentration [7]. For the Cu-Ni-Be alloys with 0.17%Zr, 0.1%Mg and 0.1%Co, all of the trace atoms were assumed to be dissolved in the matrix. Then f was calculated after the resistivity increment caused by 0.17%Zr, 0.1%Mg or 0.1%Co addition was removed. Table 1 lists the values of r and f for the present alloys. The number density N of γ" precipitates was obtained from r, h and f using the equation of $N=f/(\pi r^2 h)$. The estimated values of N and λ_{ave} also are listed in table 1. The values of N for the Zr, Mg and Co-added alloys are larger than that for the Cu-Ni-Be alloy, indicating that the Zr, Mg and Co solutes promote the formation rate of γ" precipitates. Moreover, it is stated that the increase in N or decrease in λ_{ave} by adding Zr, Mg and Co results in an increase in $\sigma_{0.2}$ due to the Orowan looping mechanism.

3.2. Effect of Zr, Mg and Co on stress relaxation property

Table 2 lists the stress relaxation rate for the present alloys tested at 180 °C for 24 h. The stress relaxation rate is defined as the ratio of the deflection upon unloading to the deflection upon loading on a cantilever [5]. A lower stress relaxation rate means a higher resistance to stress relaxation. The addition of Zr, Mg and Co to the Cu-Ni-Be alloy enhances the stress relaxation resistance, and the effect is pronounced for the alloys with Zr and Mg.

Table 2. Stress relaxation rate for Cu-Ni-Be system alloys aged at 450 °C for 2 h. Stress relaxation tests were performed at 180 °C for 24 h.

Specimen (wt%)	Stress relaxation rate
Cu-1.2Ni-0.2Be	0.15±0.02
Cu-1.2Ni-0.2Be-0.17Zr	0.10±0.02
Cu-1.2Ni-0.2Be-0.1Co	0.13±0.02
Cu-1.2Ni-0.2Be-0.1Mg	0.10±0.02

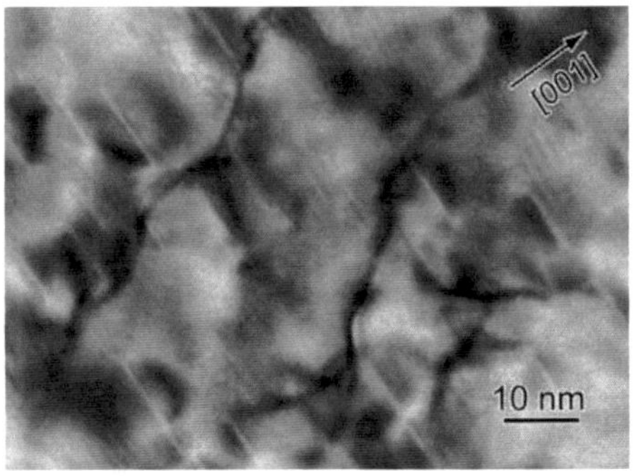

Figure 1. TEM image of disk-shaped γ" precipitates in a Cu-1.2%Ni-0.2%Be alloy aged at 450 °C for 2 h after 40% cold-rolling. The zone axis is $[110]_{\alpha}$.

Since stress relaxation tests were performed at a relatively low temperature of 180 °C in the present study, the stress relaxation is likely to occur by logarithmic creep caused by the relatively short range motion of dislocations [8]. Thus, the stress relaxation depends on the mobility and density of mobile dislocations.

It is well known that the mobility of dislocations decreases when they are dragging their atmospheres of solute atoms behind them. Then the improvement of the stress relaxation property of the Cu-Ni-Be alloy by the addition of Zr, Mg or Co in table 2 might be expected to be attributable to the drag of atmosphere of Zr, Mg or Co atoms. However, we found that the addition of 0.17%Zr and 0.1%Co to pure Cu did not change the stress relaxation rate, although the addition of 0.1%Mg decreased it. There existed no precipitates in the Cu-Zr, Cu-Mg and Cu-Co alloys examined. Thus, it is necessary to discuss another factor responsible for the decrease in stress relaxation rate by the addition of Zr, Mg and Co.

In a previous paper [9], we have found that the present Cu-1.2%Ni-0.2%Be alloy exhibits better stress relaxation property than Cu-1.2%Ni and Cu-0.2%Be alloys. It has been concluded that this result is probably because atoms caused by pairs of Ni and Be chemical bonding are dragged by moving dislocations. Thus, the amounts of Ni and Be atoms dissolved in the Cu matrix affect the stress relaxation property. It may thus be inferred that the development of the stress relaxation property by the addition of Zr, Mg and Co is caused by the increase in amounts of Ni and Be in the Cu matrix. However, it is seen from table 1 that the addition of each element increases the value of *f*: that is, it decreases the amounts of Ni and Be atoms in the Cu matrix.

As mentioned before, the stress relaxation should be affected by the density of mobile dislocations. In order to examine this, stress relaxation tests were carried out at 180°C for 24h, using pure Cu with and without 20% cold-rolling after annealing. The stress relaxation rates of Cu with and without 20% cold-rolling were 0.5 and 0.42, respectively. It is said that a higher density of mobile dislocations brings about a lower resistance to stress relaxation. Therefore, it may be expected that the addition of Zr, Mg or Co lowers the density of mobile dislocations and, as a result, the stress relaxation rate decreases. TEM observations of the present alloys aged at 450 °C for 2 h after 40% cold-rolling revealed that γ" precipitates formed preferentially on dislocations, as exemplified in figure 1. As shown in table 1, the number density of the precipitates increases by the element addition. This increase in number density should cause an increase in the number density of the γ" precipitates preferentially formed on dislocations, resulting in a decrease in the density of mobile dislocations. This explains the development of the stress relaxation property due to the addition of each element.

As mentioned before, the addition of 0.1%Mg to pure Cu improved the stress relaxation property. In the previous paper [2], we have reported that the increase of stress relaxation resistance by the addition of 0.1%Mg to a Cu-2.0%Ni-0.5%Si alloy is caused by the Mg-atom-drag effect on dislocation motion. Therefore, the decrease in stress relaxation rate by adding Mg to the Cu-Ni-Be alloy is attributable to the drag effect of Mg atoms on dislocation motion, in addition to the decrease in the density of mobile dislocations due to the increase in the number density of the γ" precipitates preferentially formed on dislocations.

4. Conclusions

We have found that 0.17wt%Zr, 0.1wt%Co or 0.1wt%Mg added to a Cu-1.2wt%Ni-0.2wt%Be alloy enhances the strength and stress relaxation property of the Cu-Ni-Be alloy. The increase in strength by the addition of each element is ascribed to the decrease in inter-tip spacing of γ" precipitates. The improvement of stress relaxation property by the addition of Zr and Co is explained by the decrease in the density of mobile dislocations due to the increase in the number density of γ" precipitates preferentially formed on dislocations. The increase in stress relaxation resistance by the Mg addition is attributed to the viscous glide motion of dislocations dragging Mg atoms, in addition to the decrease in the density of mobile dislocations.

Acknowledgments

A part of this work was conducted in the Kyoto-Advanced Nanotechnology Network, supported by the "Nanotechnology Network" of the Ministry of Education, Culture, Sports, Science and Technology (MEXT), Japan. This work was also partially supported by a Grant-in-Aid for Science Research from The Ministry of Education, Culture, Sports and Technology under Grant No.19560742.

References

[1] Ota S, Muramatsu N, Watanabe C and Monzen R 2007 *J. Soc. Mater. Sci. Japan* **56** 531
[2] Watanabe C, Hiraide H, Zhang Z and Monzen R 2005 *J. Soc. Mater. Sci. Japan* **54** 717
[3] Watanabe C and Monzen R 2008 *J. Mater. Sci.* **43** 813
[4] Miki M, Ogino Y, Ishikawa S and Natsume Y 1997 *J. JCBRA* **36** 113
[5] 2004 Standard test methods for stress relaxation for materials and structures in ASTM Test Method (ASTM international, West Conshohocken) p 397
[6] J F Nie, B C Muddle and I J Polmear 1996 *Mater. Sci. Forum* **217** 1257
[7] Linde J O 1968 *Helv. Phys. Acta* **41** 1007
[8] Sato E, Yamada T, Tanaka H and Jinbo I 2005 *J. Jpn. Inst. Light Met.* **55** 604
[9] Shimodaira K, Tanabe N, Watanabe C and Monzen R 2009 *J. JRICu* **36** 140

Journal of Physics: Conference Series **240** (2010) 012103 doi:10.1088/1742-6596/240/1/012103

High-strength ultrafine-grained Ti-Fe-Sn alloys with a bimodal structure

L C Zhang[1,2], H B Lu[1], M Calin[3], E V Pereloma[1] and J Eckert[3,4]

[1] Faculty of Engineering, University of Wollongong, Northfields Avenue, Wollongong, NSW 2522, Australia
[2] School of Mechanical Engineering, The University of Western Australia, MDBP MO50, 35 Stirling Highway, CRAWLEY, WA 6009, Australia
[3] IFW Dresden, Institute for Complex Materials, P.O. Box 27 01 16, D-01171 Dresden, Germany
[4] TU Dresden, Institute of Materials Science, D-01062 Dresden, Germany

lczhangimr@gmail.com, laichang@uow.edu.au

Abstract. The microstructure-mechanical properties relationship in ultrafine-grained Ti-Fe-Sn alloys with high strength and large plasticity was investigated. The alloys are mainly composed of a hypereutectic microstructure with micrometer-sized primary dendrites embedded in an ultrafine-grained eutectic matrix. The bimodal composites exhibit a fracture strength higher than 2350 MPa and an enhanced plasticity larger than 7%. The excellent mechanical properties are critically related to the microstructure features of the phase constituents in the alloys.

1. Introduction

Advanced metallic materials with nano-/ultrafine-grained microstructure exhibit much higher strength than their conventional coarse-grained counterparts [1,2]. However, the limited plastic deformability at room temperature of these materials has restricted their technological applications as high-strength structural materials. Therefore, the enhancement of their room temperature plasticity has been an important topic of research in nano-/ultrafine-grained metallic materials [1,2]. Recently, tailoring of the microstructure with a bimodal grain size distribution has been highlighted as a promising approach to significant improvement of the room temperature plasticity in these advanced materials [3-8].

Ti-based alloys have achieved major importance for advanced structural applications due to their high specific strength as well as excellent corrosion resistance. A large number of high-strength Ti-based alloys with a bimodal grain size distribution have been developed in recent years [4,6,8-11], where a high strength nano-/ultrafine-grained eutectic matrix is toughened by micrometer-sized ductile dendrites (such as solid solution and cubic compound). The Ti-based bimodal composites have been reported to have high fracture strength (~2000–2700 MPa) and enhanced plasticity (~2–16%). For example, as-cast $Ti_{60}Cu_{14}Ni_{22}Sn_4Nb_{10}$ rods composed of micrometer-sized β-Ti solid solution dendrites embedded in a nanostructured matrix exhibit a high fracture strength of 2400 MPa and a plasticity of 14.5% [4], and $Ti_{63.375}Fe_{34.125}Sn_{2.5}$ rods containing an ultrafine-grained eutectic matrix and micrometer-sized FeTi dendrites reach a strength of 2650 MPa and a plasticity of 12.5% [6]. Systematic studies on the deformation behavior of the Ti-based bimodal composites have shown that the balance of the mechanical properties in these alloys is critically related to the combined effects of microstructural features (such as size/fraction of each phase and structural compatibility between phases in the microstructure, and their short-range order) [6,8-10]. However, the deformation behavior of the Ti-based bimodal composites is not fully understood due to its native microstructure complexity. For example, Ti-Fe-Sn rods with different alloy compositions display significantly different mechanical behavior due to the alteration of microstructure features in the alloys.

In this work, we report on the enhancement of the room temperature plasticity of Ti-Fe alloys by tailoring of the microstructure features through slight changes of Sn concentration. The improved mechanical properties are attributed to the microstructure modulation of the nano-/ultrafine-grained eutectic matrix and the phase constituents.

2. Experimental

Starting from elemental pieces with purity ≥99.99%, master alloys with a nominal composition of $(Ti_{0.65}Fe_{0.35})_{100-x}Sn_x$ (x = 0, 2.5 and 5%, in atomic percent, the same hereafter) were prepared under a Ti-gettered argon atmosphere. Rods with 3 mm diameter were formed by copper mold casting under argon atmosphere. The structural features of as-cast rods were examined by X-ray diffraction (XRD) using a Philips PW 1050 diffractometer with monochromated Co K_α radiation (λ = 0.178897 nm). The microstructures of transversal cross-sectional surfaces were characterized using a JEOL JSM 6400 scanning electron microscopy (SEM) and a JEOL 2011 analytical transmission electron microscope (TEM) coupled with energy-dispersive x-ray (EDX) analysis. The mechanical properties at room temperature were evaluated by uniaxial compression tests at an initial strain rate of 8×10^{-4} s^{-1}.

3. Results and discussion

Figure 1 shows the room temperature compressive engineering stress-strain curves for the Ti-Fe-Sn alloys. All three alloys display considerable work hardening, high strength and large plasticity. The obtained mechanical properties including yield strength ($\sigma_{0.2}$), fracture strength (σ_f) and plastic strain (ε_p) are summarized in the table shown in the inset of Figure 1. The addition of 2.5% Sn results in a significant increment in fracture strength by ~300 MPa and a noticeable enhancement of plasticity by ~5%. Further addition of Sn to 5% leads to a slight decrease in strength and plasticity compared to the $Ti_{63.375}Fe_{34.125}Sn_{2.5}$ alloy. However, 5% Sn addition still toughens the $Ti_{65}Fe_{35}$ alloy by enhancing the plasticity as well as maintaining a high strength. Therefore, a suitable Sn concentration could both improve the strength and enhance the plasticity of the $Ti_{65}Fe_{35}$ alloy. Among all three alloys, $Ti_{63.375}Fe_{34.125}Sn_{2.5}$ has the best balance of mechanical properties, i.e. a high σ_f of ~2650 MPa and a large ε_p of ~12.5%. This alloy has a much higher strength and a plasticity similar to the data found for the $Ti_{60}Cu_{14}Ni_{22}Sn_4Nb_{10}$ [4] and Ti-Fe-Co alloys [11].

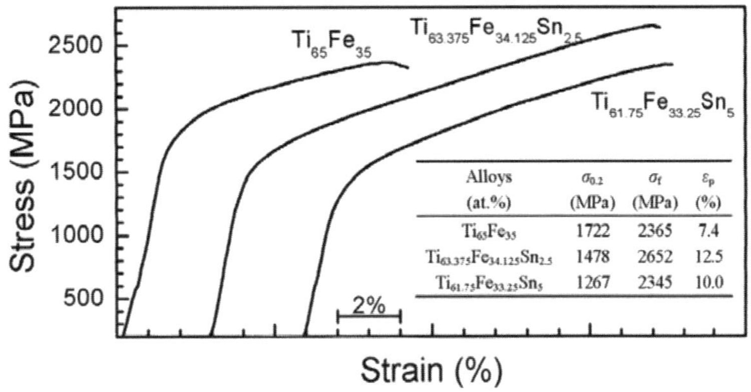

Figure 1. Engineering stress-strain curves of the as-cast Ti-Fe-Sn alloys. The inset summarizes the mechanical properties.

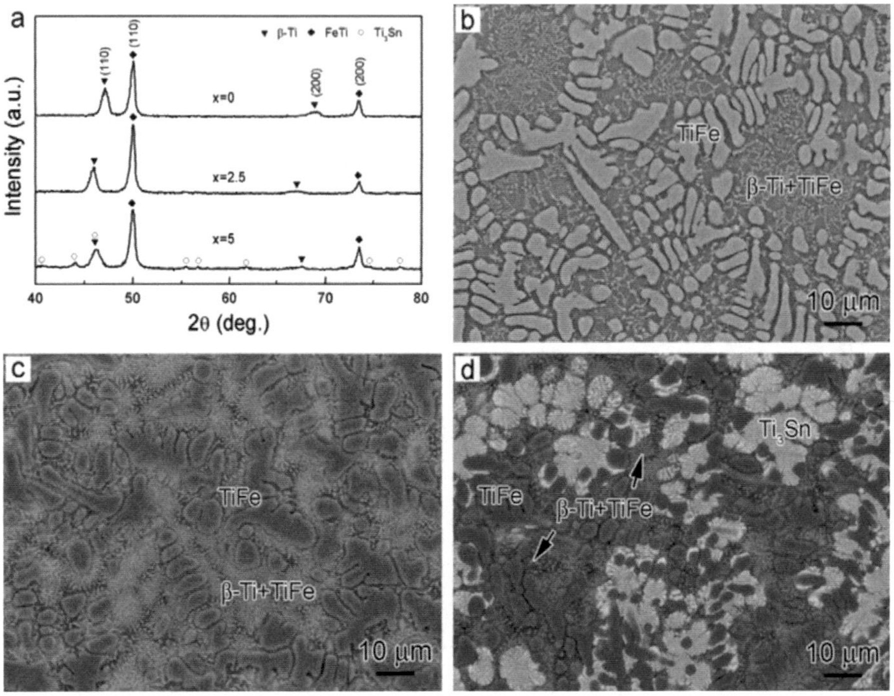

Figure 2. (a) XRD patterns and back-scattered electron SEM micrographs of the as-cast $(Ti_{0.65}Fe_{0.35})_{100-x}Sn_x$ alloys: (b) x = 0, (c) x = 2.5, and (d) x = 5.

XRD patterns (Figure 2(a)) indicate that the $Ti_{65}Fe_{35}$ and $Ti_{63.375}Fe_{34.125}Sn_{2.5}$ alloys mainly consist of body-centered-cubic (b.c.c.) β-Ti and bcc FeTi phases. Further addition of Sn to 5% leads to the appearance of hexagonal Ti_3Sn together with the aforementioned two phases. The microstructures of the Ti-Fe-Sn are illustrated in Figs. 2(b)–(d). $Ti_{65}Fe_{35}$ (Figure 2(b)) and $Ti_{63.375}Fe_{34.125}Sn_{2.5}$ (Figure 2(c)) alloys display a typical hypereutectic microstructure, where FeTi primary dendrites are embedded in an ultrafine (β-Ti + FeTi) eutectic matrix. However, the 2.5% Sn addition slightly decreases the sizes of the primary FeTi dendrites (i.e. length, trunk spacing and arm spacing) from ~50 – 100, 10 – 25 and 2 – 4 μm in the $Ti_{65}Fe_{35}$ alloy to ~20 – 70, 7 – 15 and 2 – 3 μm in $Ti_{63.375}Fe_{34.125}Sn_{2.5}$. Interestingly, the 2.5% Sn addition significantly refines the eutectic matrix, i.e. the width of the FeTi (bright) lamellae decreases from ~0.4 μm ($Ti_{65}Fe_{35}$) to ~200 nm ($Ti_{63.375}Fe_{34.125}Sn_{2.5}$) and that of β-Ti (gray) lamellae decreases from ~2.5 μm ($Ti_{65}Fe_{35}$) to ~160 nm ($Ti_{63.375}Fe_{34.125}Sn_{2.5}$). The volume fraction of the ultrafine eutectic in the two-phase composites is increased from ~65% in $Ti_{65}Fe_{35}$ to ~72% in $Ti_{63.375}Fe_{34.125}Sn_{2.5}$. EDX analysis indicates that the primary FeTi dendrites have a composition of approximately $Ti_{52}Fe_{48}$ in both alloys except for traces of Sn (0.4 ± 0.4 at.%) in the latter alloy. Most Sn dissolves in the eutectic matrix in case of $Ti_{63.375}Fe_{34.125}Sn_{2.5}$, which could further toughen the alloy. As such, the 2.5% Sn addition leads to the refinement of the microstructure constituents (slightly smaller primary FeTi dendrites and considerable finer eutectic matrix) and an increment of the volume fraction of the ductile eutectic matrix by ~7 vol.%), leading to the improvement of the strength and plasticity of $Ti_{63.375}Fe_{34.125}Sn_{2.5}$, as shown in Figure 1. When further increasing the content of Sn to 5% (Figure 2(d)), $Ti_{61.75}Fe_{33.25}Sn_5$ shows a complex microstructure composed of primary Ti_3Sn, FeTi dendrites and (β-Ti + FeTi) eutectic matrix, with volume fractions of ~20, 24 and 56 vol.%, respectively. This alloy has FeTi dendrites and eutectic with a close sizes to those in $Ti_{63.375}Fe_{34.125}Sn_{2.5}$, but contains a lower fraction of eutectic matrix, which is responsible for plasticity. Therefore, $Ti_{61.75}Fe_{33.25}Sn_5$ exhibits a lower plasticity than $Ti_{63.375}Fe_{34.125}Sn_{2.5}$. In addition, the structural incompatibility between Ti_3Sn and b.c.c. FeTi and β-Ti blocks the dislocation transfer across their interfaces, accelerates the failure around Ti_3Sn phase [6].

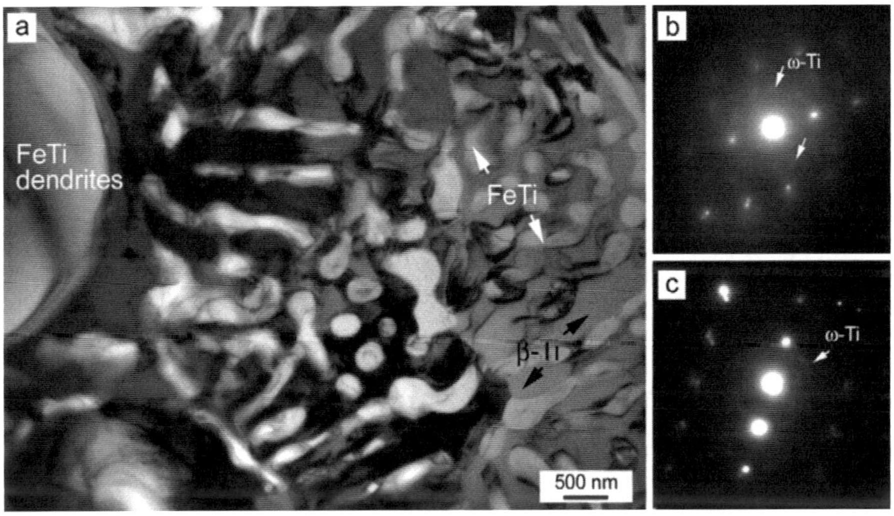

Figure 3. (a) Bright-filed TEM image of the $Ti_{63.375}Fe_{34.125}Sn_{2.5}$ and SAED patterns for the eutectic matrix in the (b) $Ti_{65}Fe_{35}$ and (c) $Ti_{63.375}Fe_{34.125}Sn_{2.5}$ alloys.

TEM observations confirm that $Ti_{63.375}Fe_{34.125}Sn_{2.5}$ (Figure 3(a)) is composed of FeTi dendrites embedded in an ultrafine (β-Ti + FeTi) eutectic matrix and the $Ti_{65}Fe_{35}$ alloy shows a similar microstructure. The selected area electron diffraction (SAED) patterns from the matrix reveal an ω-Ti-like short-range order (SRO) (denoted by arrows) in the matrix of both alloys. Although it has been reported that ω-Ti-like SRO reduces the plasticity of Ti-based alloys [9,10], the SRO does not significantly reduce the plasticity in the current Ti-Fe-Sn alloys. Further work is underway to fully understand the deformation behavior of this alloy system.

4. Conclusions

Ultrafine-grained Ti-Fe-Sn bimodal composites exhibit high strength (over 2350 MPa) and large plasticity (>7%). Their mechanical behavior is associated with the complex microstructure features (i.e., the refinement / fraction of and compatibility between phase constituents).

Acknowledgments

This work was supported by the URC Small Grant and EMI Strategic Grant, University of Wollongong. L.C. Zhang appreciates the support of the Alexander von Humboldt Foundation.

References

[1] Ma E 2006 *JOM* **58** 49
[2] Koch C C, Youssef K M, Scattergood R O and Murty K L 2005 *Adv. Eng. Mater.* **7** 787
[3] Wang Y M, Chen M W, Zhou F H and Ma E 2002 *Nature* **419** 912
[4] He G, Eckert J, Löser W and Schultz L 2002 *Nature Mater.* **2** 33
[5] Fan G J, Choo H, Liaw P K and Lavernia E J 2006 *Acta Mater.* **54** 1759
[6] Zhang L C, Das J, Lu H B, Duhamel C, Calin M and Eckert J 2007 *Scripta Mater.* **57** 101
[7] Park J M, Sohn S W, Kim T E, Kim D H, Kim K B and Kim W T 2007 *Scripta Mater.* **57** 1153
[8] Han J H, Kim K B, Yi S, Park J M, Sohn S W, Kim T E, Das J and Eckert J 2008 *Appl. Phys. Lett.* **93** 141901
[9] Zhang L C, Lu H B, Mickel C and Eckert J 2007 *Appl. Phys. Lett.* **91** 051906
[10] Das J, Eckert J and Theissmann R 2006 *Appl. Phys. Lett.* **89** 261917
[11] Louzguine D V, Louzguina L V, Kato H and Inoue A 2005 *Acta Mater.* **53** 2009

Preparation and internal stress estimation of BN films by ion mixing and vapour deposition technique

S Hanaki, B Leng and H Uchida

Graduate School of Engineering, University of Hyogo
Shosha 2167, Himeji, Hyogo, Japan

hanaki@eng.u-hyoho.ac.jp

Abstract. Boron Nitride (BN) films were synthesized onto silicone wafer by depositing B metal vapour under simultaneous irradiation of N ions. Here, film thickness, ion beam energy and transport ratio (*B*/*N*) were selected as a preparation parameter and they were controlled in the range of 0.2-1μm, 0.2~2keV and 1~5, respectively. The BN films prepared were characterized using several analytical techniques and their internal stresses were estimated using Stoney's equation. From Fourier transform infrared spectroscopy, it was found that use of low energy N ions is effective for the formation of cubic BN (cBN) phase using ion mixing and vapour deposition (IVD) technique. At this condition, high compressive stress is measured and strong correlations were found among crystal structure, internal stress and Knoop hardness of BN films.

1. Introduction

Boron nitride (BN) film has excellent properties such as high hardness, thermal conductivity, thermal stability and so forth [1]. It is very difficult, however, to synthesize the cubic BN (cBN) phase because it is a metastable phase under atmospheric pressure like a diamond. It has been widely recognized that the successful formation of cBN phase requires a bombardment of growing films with energetic ions [2,3,4]. This intensive ion bombardment also produces high compressive stress in the film [4,5,6]. In nucleation and growth of cBN film, however, full understanding of the mechanism has not been established yet. The purpose of this study is to clarify the effects of processing parameters on the internal stress of BN films prepared by ion mixing and vapour deposition (IVD) technique. BN films were synthesized onto silicone wafer by depositing B metal vapour under simultaneous irradiation of N ions. The film thickness, transport ratio (*B*/*N*) and ion beam energy were selected as preparation parameters [7]. The internal stress of obtained films was also estimated by Stoney's equation using the Newton ring method and the correlation between crystal structure of BN film and these processing parameters were discussed.

2. Experimental procedure

Synthesis of the BN films was performed using an IVD machine equipped with two bucket-type ion sources, two evaporators and a pumping system. The BN films were synthesized onto silicon (100) wafer and glass substrate (18^{l}x18^{w}x0.12^{t}mm) by depositing B metal vapour under simultaneous irradiation of N ions. The processing parameters for sputter cleaning and film deposition were summarized in table 1.

Table 1. Conditions of sputter cleaning and film deposition.

Parameter	Sputter cleaning	Film deposition
Acceleration voltage (E)	5keV	0.2~2keV
Acceleration current	20mA	9mA
Arc voltage	80V	80V
Deceleration voltage	1.5keV	1.5keV
Ion incident angle	45°	0°
Transport ratio (B/N)		1~5
Film thickness (d)		0.2~1μm
N_2 gas flow rate	4sccm	4sccm
Substrate temperature	N.T.*	N.T.*
Substrate rotation speed	0.1rev/s	0.1rev/s

* Normal (intentionally 'no-heated') temperature

Internal stress of BN films was evaluated using the Newton ring method [8]. The Newton rings were measured with an optical microscope equipped with a Michelson interferometer that allows determining the radius of curvature of the substrate R. The internal stress σ was estimated using Stoney's equation [8] as shown in eq.(1).

$$\sigma = \frac{E_s h_s^2}{6(1-\nu)Rd} \tag{1}$$

Here, E_s and ν are elastic modulus and Poisson ratio of substrate, h_s and d are thickness of substrate and film, respectively.

The appearance of film surface and cross section was examined by field emission type scanning electron microscope (FE-SEM). Fourier transform infrared spectrometer was also used to determine the crystal structure. The Knoop hardness of BN film on silicon wafer was also measured.

3. Results and discussions

3.1 Morphology and structure

Figure 1 shows examples of FE-SEM micrographs of 1μm-thick BN films prepared at the acceleration voltage E=0.2, 2keV and the transport ratio B/N=4. All BN films prepared under various conditions were of satisfactory quality with very smooth surfaces. From the SEM observations, it became clear that the BN films exhibited the columnar structure with grainy appearance, less independent of acceleration voltage and/or transport ratio.

Figure 2 summarizes FT-IR spectra of 1μm-thick BN films prepared at transport ratio B/N=4 as a function of acceleration voltage E. For the stoichiometric of BN film at 2keV, only two intensive absorption bands around 1380 and 780 cm^{-1} were observed. They are assigned to in-plane and out-of-plane vibrational modes of hBN, respectively. However, as the acceleration voltage is decreased, another absorption band can be observed around 1080 cm^{-1}, which can be assigned to the presence of cBN phase [3]. This suggests that low acceleration energy of N ions is effective for the synthesis of cBN films by IVD technique.

3.2 Internal stress of BN film

Figure 3 shows the relationship between internal stress and film thickness of BN films prepared at the acceleration voltage E=0.2, 2keV and the transport ratio B/N=4. From figure 3, it is confirmed that the compressive stress increases linearly with the increase of film thickness. Besides, the compressive stress tends to became larger at E=0.2keV in comparison with E=2keV. This suggests that the compressive stress increase at low acceleration voltage. It is known that internal stress normally increase at high acceleration voltage due to high impact energy of accelerated ions. However, obtained result does not agree with such tendency and it is considered that the increase of internal stress is related to generation of cBN phase. Figure 4 shows the relationship between internal stress and acceleration voltage of 1μm-thick BN films prepared at the transport ratio *B*/*N*=4. From figure 4, it is also confirmed that the compressive stress increases at the acceleration voltage *E*=0.2keV, which can be assigned to the condition for formation of cBN phase.

Figure 1. FE-SEM micrographs showing surface and cross-sectional views of BN films prepared at acceleration voltage *E*=(a,b)0.2keV and (c,d)2keV (*B*/*N*=4).

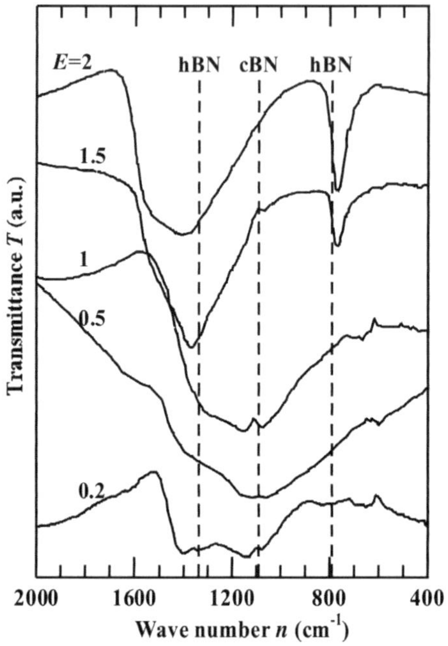

Figure 2. FT-IR spectra of BN films prepared at transport ratio *B*/*N*=4 (*d*=1μm).

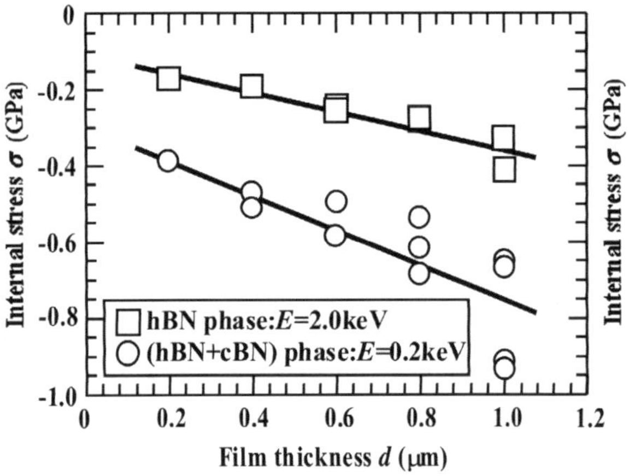

Figure 3. Relationship between internal stress and film thickness of BN films prepared (*B/N*=4).

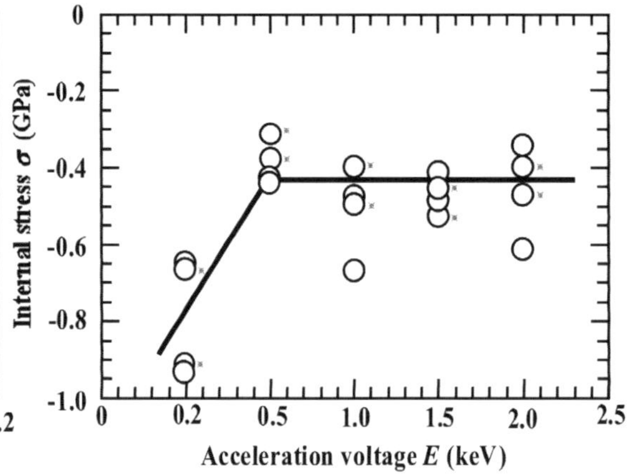

Figure 4. Relationship between internal stress and acceleration voltage of BN films prepared (*B/N*=4, *d*=1.0μm), data points marked (*) represent specimens that were subjected to hardness testing.

Figure 5 shows the relationship between internal stress and transport ratio of 1μm-thick BN films prepared at the acceleration voltage *E*=0.2keV. It is confirmed that the compressive stress increases at transport ratio *B*/*N*=4. It is also confirmed that the mixture phase of hBN and cBN has been formed at this condition [7]. Therefore, it is concluded that the compressive stress of film increases with the increase of cBN content. The high compressive stress is considered to be caused by transition of crystal structure from graphite type hBN to cBN during film growth. However, there is no experimental evidence and further investigations are required for direct evidence.

Figure 6 shows the relationship between internal stress and Knoop hardness of BN films. Knoop hardness was measured for the specimens which have (*) in figure 4 and 5. The solid line in figure 6 is obtained by least-squares method. The coefficient of correlation is -0.843 and it is confirmed that there is a high correlation between internal stress and Knoop hardness. From these results, strong correlations were found among crystal structure, internal stress and Knoop hardness of BN films.

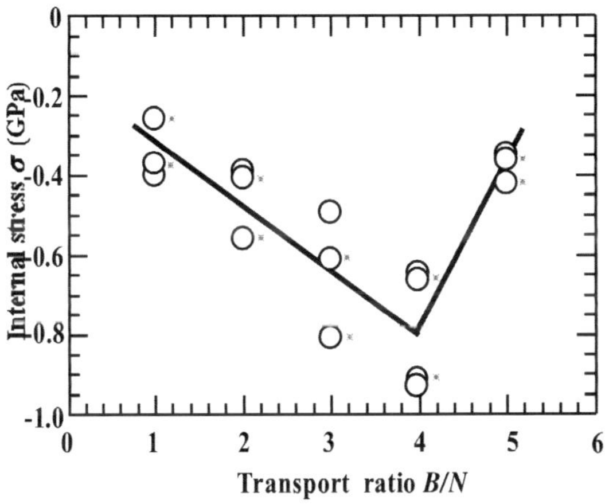

Figure 5. Relatonship between internal stress and transport ratio of BN films prepared (E=0.2keV, d=1.0μm).

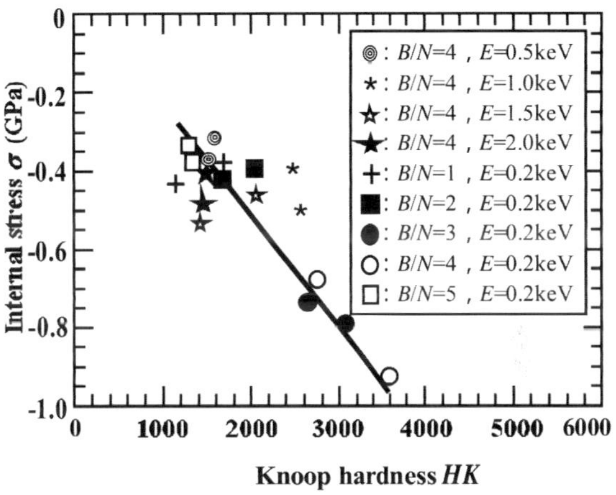

Figure 6. Relatonship between internal stress and Knoop hardness of BN films (d=1.0μm).

4. Conclusions

BN films were prepared by IVD technique and internal stress was measured by the Newton ring method. The effect of processing parameters on the internal stress of BN film can be summarized with the following remarks;

(1) Low acceleration voltage of N ions is effective for the formation of cBN phase.
(2) Compressive stress of BN film increases at the condition which cBN phase is formed.
(3) Strong correlations can be found among crystal structure, internal stress and Knoop hardness of BN films.

Acknowledgements

This work is supported by a Grant-in-Aid for Scientific research (no.20560084) for Promotion of Science.

References

[1] Samantaray C B, Singh R N 2005 *Int. Mater. Rev.* **50** 313-344
[2] Ikebe T, Kawata Y and Hirai Y 1990 *J. Vac. Sci. Technol.* A **8** 3168-3173
[3] Nishiyama S, Kuratani N, Ebe A and Ogata K 1993 *Nucl. Instrum. Meth.* B **80/81** 1485-1488
[4] Djouadi M A, Vasin A, Nouveau C, Angleraud B and Tessier P Y 2004 *Surf. Coat. Technol.* **180-181** 174-177
[5] Mieno M and Yoshida T 1990, *Jpn. J. Appl. Phys.* **29** L1175-L1177
[6] Harms U, Gaertner M, Schutze A, Bewilogua K and Neuhauser H 2001, Thin Solid Films, **385** 281-290
[7] Uchida H, Yamashita M, Hanaki S and Kurihara A 2008, *Mater. Sci. Eng.* A **483-484** 695-697
[8] Hoffman DW 1996 *Physics of Thin Films* **3** 211, New York Academic

A molecular dynamics study of deformation induced phase transformations at fault bands

C W Sinclair

Department of Materials Engineering, The University of British Columbia, Vancouver, BC V6T 1Z4 Canada

chad.sinclair@ubc.ca

Abstract. In many ferrous austenitic alloys a strain induced martensitic transformation is important for determining bulk mechanical response. In this work molecular dynamic simulations have been performed in an attempt to further elucidate the mechanisms by which bcc-α' martensite may form at deformation induced hcp-ϵ martensite bands. The question of critical nucleus size and fault band arrangement are discussed in relation to the Olson-Cohen model for nucleation at fault band intersections as is the resulting orientation relationship.

1. Introduction

The strain induced transformation from austenite (γ) to martensite (ϵ and/or α') is responsible for the strong rate of work hardening in many austenitic steels. Understanding and predicting these bulk mechanical properties requires an ability to predict the rate of phase transformation as a function of plastic strain. This relationship comes from the apparent ability of plasticity to generate potent sites for the nucleation of the phase transformation. Phenomenological models for this transformation consider that the intersection between deformation induced planar faults provides unique conditions for the phase transformation[1, 2]. Olson and Cohen [2] pointed out that if a band of hcp-ϵ martensite (which can be considered as being comprised of stacking faults on every second $\{111\}$ plane) was intersected by a second band of faulted ϵ (comprised of stacking faults on every third atomic plane), then the strain (in a continuum sense) at the intersection of these fault bands would correspond closely to the Bain strain required for the $\gamma \rightarrow \alpha'$ phase transformation (e.g. [4]). In reality, this process of intersection (if it can occur) is atomistic in nature. In order to investigate this transformation from the atomic scale a recent study using molecular dynamics simulations [5] was initiated. The phase transformation was shown to occur at the intersection without the need for externally applied stress and that the product α' was observed to grow into the surrounding fault bands. A Pitsch orientation relationship was found to exist between the two phases. While the MD simulations clearly indicated that the envisioned fault band intersections are potent sites for nucleation of α' several questions remain unresolved. Among these it is important to understand whether the transformation will proceed even in the absence of the perfect intersection geometry described above. For example, will the transformation take place when two perfect ϵ martensite bands intersect? Also not investigated was whether there is a minimum required intersection volume. Finally, it was noted that the Pitsch orientation relationship observed in the MD simulations does not correspond to the the orientation relationship (Kurdjumov-Sachs) found experimentally in these materials. It is the purpose of this note is to expand upon these topics.

The primary aspects of the simulation approach have been outlined elsewhere [5]. Simulations were thermostated to 50 K in an NVT ensemble. The EAM potential developed for Fe by Ackland et al. [6] was used in this study. This potential replicates many of the properties of Fe, but predicts a very low stacking fault energy and, correspondingly small difference in cohesive energy between the fcc and hcp phases. The simulation cell was periodic in the z-direction and fixed in the x and y directions. The ϵ and faulted ϵ were created by introducing stacking faults within the simulation cell. Consistent with our previous work, simulations were performed on a small cell containing 10,835 atoms with dimensions of 10.7 x 9.2 x 1.2 nm in the x, y and z directions. Also presented here are larger scale simulations performed on 234,472 atoms in a box having dimensions of 40.6 x 39.4 x1.83 nm. These latter simulations have been performed using the LAMMPS package(http://lammps.sandia.gov) [7] with visualization performed using Atomeye [8]. The crystallographic geometry of the simulation cell is illustrated in figure 1 where $x = [0\bar{1}1]_{fcc}$, $y = [\bar{1}00]_{fcc}$ and $z = [011]_{fcc}$.

2. Dependence of Transformation on Fault Intersection Characteristics

The Olson-Cohen model for nucleation is specific in the combination of fault bands (ϵ and faulted ϵ) that are best suited for nucleation of α'. Experimental observations, however, routinely show that the transformation can occur at a range of faults. In figure 2(a) the progression of transformation at the fault band intersection is shown for the intersection envisioned by Olson and Cohen. In figure 2(b) a simulation consisting of two intersecting perfect ϵ bands is shown. Red (light) atoms have a coordination number of 12 (fcc or hcp coordination) while blue (dark) atoms have a coordination number of 8 (bcc). Atoms with other coordination numbers are not shown.

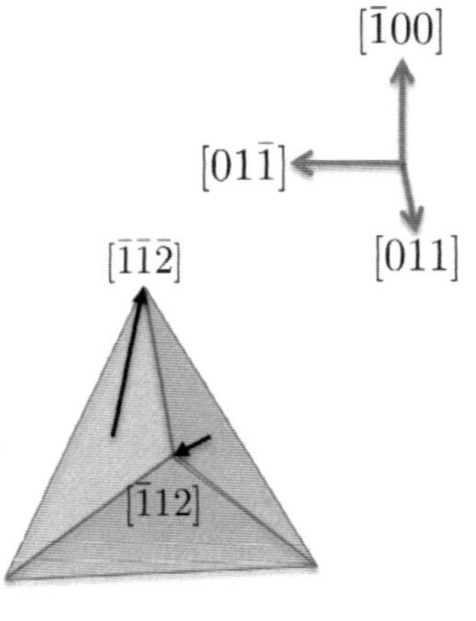

Figure 1. Schematic orientation of simulation cell relative to the Thompson tetrahedron. The Burgers vectors of the two sets of partial dislocations used to generate the ϵ and faulted ϵ bands are also shown.

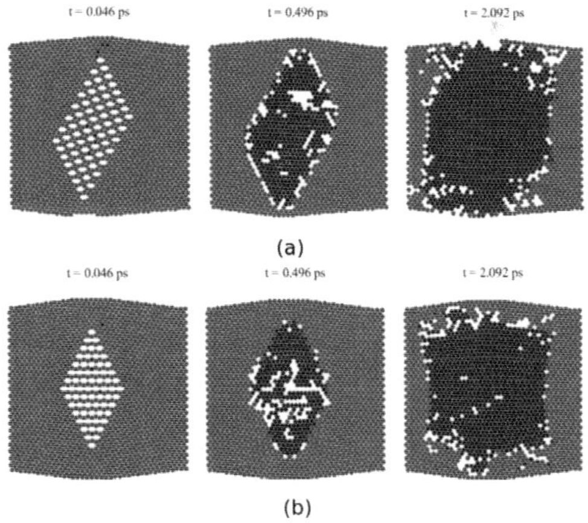

Figure 2. Progression of transformation from $\gamma \rightarrow \alpha'$ in the small simulation cell viewed along the z axis for a) the fault band intersection envisioned by Olson and Cohen and b) intersection between two perfect ϵ bands. Red (light) atoms have a coordination number of 12, blue (dark) atoms a coordination number of 8.

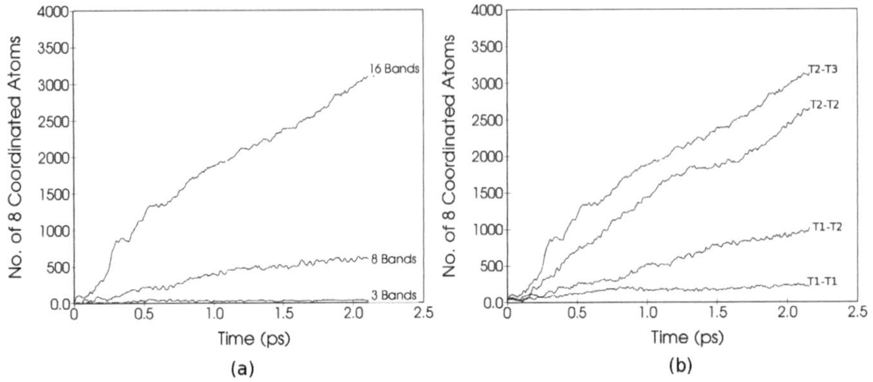

Figure 3. a) The measured rate of transformation in cells containing intersecting ϵ and faulted ϵ with differing intersection volumes defined by the number of stacking faults used to create the ϵ plates. b) The rate of transformation for different types of fault intersection. The nomenclature follows that introduced by Olson and Cohen [2] where $T3$ = faulted ϵ, $T2$ = perfect ϵ and $T1$ = twin.

The rate of transformation for different types of fault intersections is quantified in figure 3(b). Consistent with the images in figure 2, the rates of transformation for intersections formed by a perfect ϵ plate and a faulted ϵ plate (the conditions of Olson-Cohen) and between two perfect ϵ bands are nearly the same. As the intersecting faults are changed the strain in the intersection volume also changes. For the intersection between a perfect ϵ plate and a twin and the intersection between two twins the transformation proceeds much more slowly, consistent with an increasing departure from the Bain strain in the intersection volume.

Figure 3(a) also illustrates how the size of the intersection volume effects the rate of transformation. The results show that there are is a minimum intersection volume required for the transformation to proceed. These results were reproduced with the larger simulation cell where it was observed that when an 8 atomic layer thick ϵ plate was intersected by 1 and 2 stacking faults a small region having bcc coordination was formed but did not grow appreciably. In contrast, the situation of ϵ martensite being intersected by 4 stacking faults results in a rapid transformation with the α' martensite reaching the edges of the simulation cell within 70 ps. For the case of ϵ martensite intersected by 8 stacking faults (figure 4) the transformation proceeds rapidly reaching the edges of the cell after 30 ps.

Another notable feature of the transformation observed in figure 4 is the apparent anisotropic growth of the α' martensite along the perfect ϵ martensite band. This is to be contrasted with the results from the small scale simulation shown in figure 2 (a) where the growth of the α' nucleus was stopped shortly beyond the intersection volume due to the constraints imposed by the simulation cell boundaries.

Careful observation of the local atomic structure at the fcc/bcc interface reveals that the growth of the α' into the prior ϵ band corresponds to a change in orientation relationship between α' and the γ situated outside of the ϵ bands. Consistent with our previous observations, the orientation relationship between the γ and α' formed within the original intersection volume is Pitsch (figure 5(a)). However, the high rate of growth into the prior ϵ band is observed to correspond to the nucleation of a new α' nucleus with a Kurdjumov-Sachs orientation relationship corresponding to a rotation of 5.26° relative to the Pitsch orientation relationship. This observation is significant given that it is the Kurdjumov-Sachs orientation relationship that is most commonly reported between α' and γ in TRIP austenitic steels. Moreover, the morphology of α' martensite in figure 4 corresponds more closely to that observed experimentally

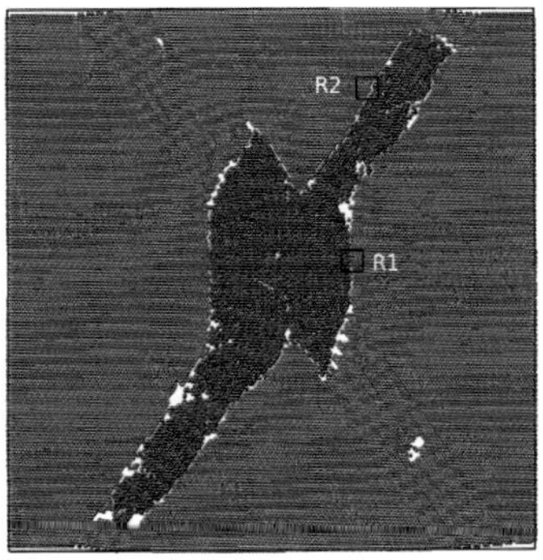

Figure 4. The large simulation cell after 30 ps at 50 K showing the anisotropic growth of the α' martensite (blue/dark) outside of the intersection volume.

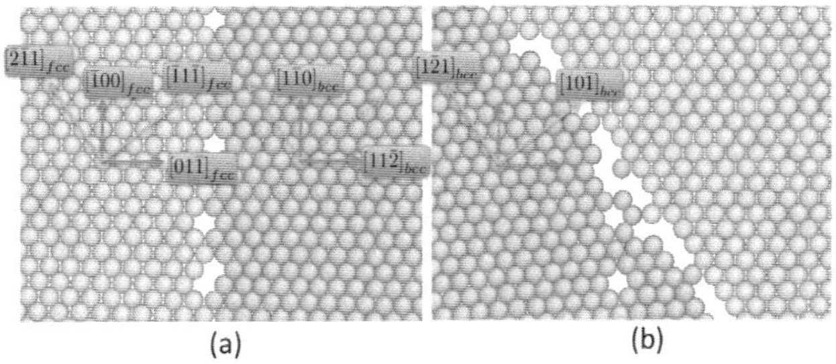

Figure 5. Highlight of the atomic structure at the fcc (yellow/light) and bcc (gray/dark) interface a) near the original intersection volume noted as R1 in figure 4 and b) in the prior ϵ band noted as R2 in figure 4. In the former case the orientation relationship is found to be Pitch (as in [5]) while the α' formed in the prior ϵ band is found to have a Kurdjomov-Sachs orientation relationship.

as α' is often found to grow roughly parallel to and within ϵ bands. As noted before, the strains generated by the ϵ band partially accomplish the Bain strain required for the transformation. The observations reported here likely point to the importance of the interface generated between α' and its surroundings in terms of the mobility of the phase interface, a factor well known to be important for the determination of the crystallography and growth characteristics of thermal martensite [4, 9].

3. Summary

The progression of phase transformation within the context of the Olson-Cohen model has been shown to be sensitive to both the type of fault band intersection as well as fault intersection volume. Moreover, it has been shown that in large simulations, the growth of α' becomes very anisotropic once it grows beyond the original fault band intersection volume. The rapid increase in growth kinetics corresponds to a change in orientation relationship from Pitsch to Kurdjumov-Sachs, the latter being coherent with experiment. Further work continues to elucidate the process of intersection itself.

References

[1] Olson G. B., Cohen M., Metall. Trans. 1975;6A:791.
[2] Olson G B, Cohen M., 1972 *J. Less-Common Metals* **28** 107
[3] Bogers A J, Burgers W G, 1964 *Acta Metall.* **12** 255
[4] Bhadeshia H K D H, 2001 *Worked Examples in the Geometry of Crystals (2nd edition)* (London: Institute of Materials, London)
[5] Sinclair C W, Hoagland R G, 2008 *Acta Mater.* **56** 4160
[6] Ackland G J, Bacon T H D J, Calder A F, 1997 *Phil. Mag. A* **75** 713
[7] Plimpton S J, 1995 *J. Comp. Phys.* **117** 1
[8] J Li, 2003 *Modelling Simul. Mater. Sci. Eng.* **11** 173
[9] Pond, R C, Ma X, Hirth, J P, 2008 *J. Mater. Sci.* **43** 3881

Journal of Physics: Conference Series **240** (2010) 012106 doi:10.1088/1742-6596/240/1/012106

Developing aluminum nanocomposites via severe plastic deformation

S. Goussous, W. Xu, K. Xia

Department of Mechanical Engineering and ARC Centre of Excellence for Design in Light Metals, University of Melbourne, Victoria 3010, Australia

s.al-goussous@pgrad.unimelb.edu.au

Abstract. Severe plastic deformation (SPD) processes were used in mixing and consolidating aluminum based nanocomposites containing various amounts of C nanoparticles. A planetary ball milling machine was used to mix the powders, and back pressure equal channel angular pressing (BP-ECAP) for consolidation. For comparison, Al-Al_2O_3 nanocomposites were processed by the same processing method. TEM observations showed that the combination of the two SPD processes was effective in dispersing the C nanoparticles in the Al matrix, and better dispersion was achieved by increasing the number of ECAP passes. Especially in Al-5 wt% C after ECAP consolidation for 24 passes, the compressive plastic strain was significantly improved. The Al-Al_2O_3 materials appeared to have better particle distribution and ductility compared to the Al-C materials although their strengths were lower.

1. Introduction

Aluminum alloys and composites have attracted considerable attention in structural and functional applications due to their light weight and high specific strength and stiffness. However, precipitation hardened aluminum alloys have not attained the same level of high strength as that achieved in Fe-Fe_3C materials, and they are thermally unstable. Although aluminum matrix nanocomposites have great potential, the challenge remains how to uniformly distribute nanometer sized ceramic particles in the Al matrix, since these nanoparticles have high tendency to agglomerate due to their large surface to volume ratio. Thus, it is critical to solve this problem in order to achieve the desired mechanical properties.

Equal channel angular pressing (ECAP) is a unique SPD process which has found extensive uses in producing ultrafine grained (UFG) metallic materials [1]. The use of back pressure equal channel angular processing (BP-ECAP) is effective in consolidating particles into bulk material due to the high amount of shear involved. In the present investigation, BP-ECAP is used to produce fully dense Al based nanocomposites reinforced with inexpensive ceramic particles (e.g. carbon black). Good dispersion and excellent mechanical properties were obtained in these nanocomposites by combining mechanical milling for dispersion of nanoparticles with ECAP for consolidation.

2. Experimental Materials and Procedures

A detailed description of experimental materials and methods can be found in [2]. The carbon black used in this work consisted of large clusters of an average size of ~500 nm although individual particles had an average size of ~30 nm with a density of 1.7-1.9 g/cm^3. The Al_2O_3 particles had an average size of < 50 nm with a density of 4 g/cm^3.

Different amounts of ceramic particles (2, 5 and 10 wt%) were added to a pure Al powder and the mixtures were processed for 30 hours each using a planetary ball milling machine. BP-ECAP was used to consolidate the mixed powder at 400°C for 8 and 24 passes, respectively, following the C route with the application of a back pressure of 50 MPa. The set up of BP-ECAP is similar to that described in [3].

Density measurements were conducted based on the Archimedes principle. The Vickers microhardness was measured by using a 50 g load for 15 seconds, and an average of 6 readings was taken from each material. Compressive specimens of 4 × 4 × 8 mm were taken along the longitudinal direction of the sample after ECAP and tested at room temperature. An FET Tecnai F20 was used for TEM observations at 200 kV.

3. Results and Discussion

3.1. Density and Hardness

The density and microhardness values are shown in Table 1. It is apparent that the materials produced were fully dense (the density value of 1.87 g/cm^3 was used for C [4] in the calculation of the theoretical densities). There was a significant increase in hardness with an addition of 2 wt% C, although only a moderate further increase was observed in the Al-5 wt% C material. The hardness values of the Al-2 and 5 wt% Al_2O_3 materials were lower than the corresponding Al-C materials, although the effective difference was smaller since the Al-Al_2O_3 composites were lower in volume fractions.

These results showed that the BP-ECAP consolidation is capable of producing bulk metal matrix nanocomposites with full density at temperatures lower than those used in the conventional consolidation processes like HIP and sintering [5-9]. This is desirable since high temperature could cause excessive grain growth and damaging reactions between matrix and reinforcement, resulting in poorer mechanical properties. On the other hand, SPD processes such as ECAP can provide the amount of strain necessary to deform the particles, leading to the filling of pores and strong interfacial bonding without the need of high thermal energy. In addition, fully dense materials with homogeneous particle distribution were produced in this work without resorting to secondary processing such as rolling and extrusion, which are often required following HIP and sintering.

Table 1. Density and microhardness for the Al-C and Al-Al_2O_3 nanocomposites after ECAP consolidation for 8 and 24 passes at 400 °C.

Material	Measured Density (g/cm^3)	Theoretical Density (g/cm^3)	HV (kg/mm^2)
Al (unmilled) (ECAP for 8 passes)	2.7	2.7	33.9
Al (milled) (ECAP for 8 passes)	2.7	2.7	37.1
Al-2 wt% C (ECAP for 8 passes)	2.68	2.68	83.1
Al-5 wt% C (ECAP for 8 passes)	2.65	2.64	96.5
Al-5 wt% C (ECAP for 24 passes)	2.65	2.64	90.5
Al-2 wt% Al_2O_3 (ECAP for 8 passes)	2.72	2.72	61.9
Al-5 wt% Al_2O_3 (ECAP for 8 passes)	2.74	2.74	81
Al-10 wt% Al_2O_3 (ECAP for 24 passes)	2.79	2.79	83.5

Journal of Physics: Conference Series **240** (2010) 012106 doi:10.1088/1742-6596/240/1/012106

3.2. Microstructure and Mechanical Properties

No pores in the nanocomposites produced were found in TEM observations, and this confirms the achievement of fully dense materials. The C clusters in the as-received powder had been dispersed in the Al matrix as individual nanoparticles (< ~100 nm in size), as shown in Figure 1 for the Al-5 wt% C material. The microstructure of pure Al without the addition of reinforcement was totally different, as shown in Figure 2a, having elongated grains with many subgrain boundaries; in contrast, an equiaxed grain structure was observed in Al-2 wt% C as shown in Figure 2b. The nanosized particles in the composites had prevented grain growth and refined the structure during the ECAP process. With increasing number of passes from 8 to 24, as shown in Figure 1 for Al-5 wt% C, more grain refinement was achieved (grain size reduced from ~1 μm after 8 passes to ~300 nm after 24 passes). This phenomenon was also observed by others [10]. On the other hand, the spherical Al_2O_3 nanoparticles (< 50 nm in size) were observed in the Al-Al_2O_3 materials to be homogeneously distributed, as shown in Figure 3.

It is therefore apparent that the combination of mechanical milling and ECAP was effective in dispersing the initially clustered C nanoparticles. A uniform distribution of individual C particles (~50 nm in size) was achieved, and it is believed that most of the dispersion occurred during the mechanical milling process, as also observed by others [11-13]. Following mechanical milling, ECAP ensured that the mixed powder was completely consolidated with full density which was not achieved by conventional compacting and sintering [11, 12]. In addition, continued ECAP led to further structural refinement and more homogeneous dispersion of ceramic nanoparticles in the Al matrix, and this would benefit the mechanical properties of the nanocomposites (see the next paragraph).

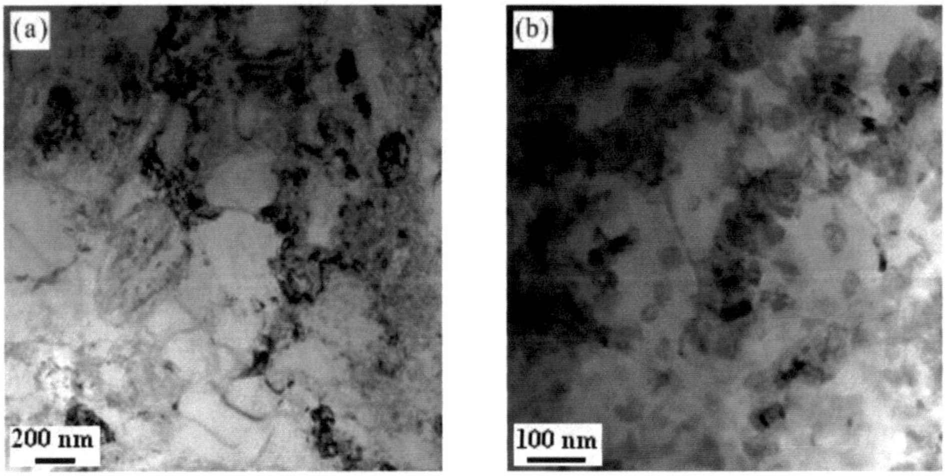

Figure 1. TEM microstructures in Al-5 wt% C after BP-ECAP for a) 8 and b) 24 passes at 400°C.

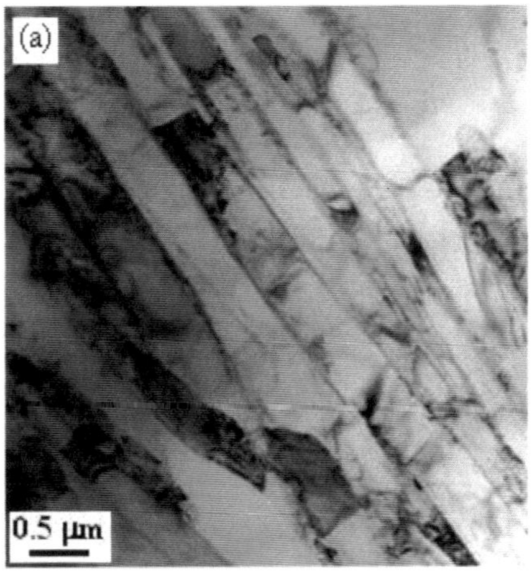

Figure 2. TEM microstructures of (a) pure Al with an elongated grain structure with many subgrain boundaries, and (b) Al-2 wt% C with an equiaxed grain structure, after BP-ECAP for 8 passes at 400°C.

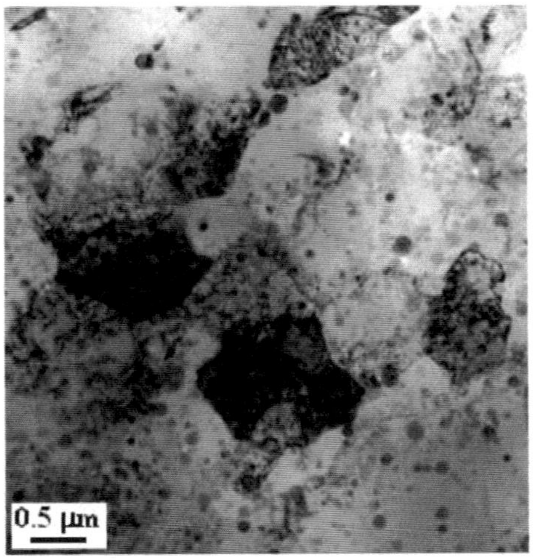

Figure 3. TEM microstructure of the Al-10 wt% Al_2O_3 nanocomposite material, showing the homogeneous dispersion of Al_2O_3 nanoparticles (<50 nm in sizes) in the Al matrix.

The compressive strength and strain of the nanocomposite materials produced showed significant enhancement. The Al-2 wt% C material consolidated for 8 passes showed a 0.2% proof stress ($\sigma_{0.2}$) of 225 MPa and a maximum strength (σ_{max}) of 302 MPa, significantly higher than those of the pure Al after the same processing ($\sigma_{0.2}$ = 58 MPa and σ_{max} = 157 MPa). By the addition of 5 wt% C, these strengths were further increased to 260 and 352 MPa, respectively. Considerable compressive strain (tests stopped at ~35% without failure) was achieved in Al-2 wt% C after consolidation for 8 passes. However, the Al-5 wt% material consolidated for 8 passes fractured at ~10% true strain. The ductility was improved to 20% by increasing the number of ECAP passes to 24, while the compressive strength was almost unchanged. The ductility of the Al-Al_2O_3 materials appeared to be much better, showing true strains of 40-80% without fracture. However, the proof stress in Al-Al_2O_3 materials was lower. For example, $\sigma_{0.2}$ in Al-5 wt% Al_2O_3 processed for 8 passes was ~153 MPa, compared to 225 MPa in Al-2 wt% C at the same conditions (the two having an equivalent volume fraction of ~3 vol%).

The high strength achieved is thought to be resulting from combined contributions from dispersion hardening and grain refinement. The Orowan mechanism becomes more effective with decreasing particle size and increasing homogeneity in particle distribution. On the other hand, the fine and homogeneous dispersion of C nanoparticles had pinned grain boundaries, limiting recrystallization and grain growth and leading to a much refined, equiaxed grain structure, in contrast to the coarse and elongated grain structure in pure Al owing to the dynamic recovery process which occurs at ~400°C [14].

4. Conclusion

Fully dense monolithic Al as well as Al-C and Al-Al_2O_3 nanocomposites were successfully produced by combining two severe plastic deformation processes: mechanical milling to mix the particles and back pressure equal channel angular pressing (BP-ECAP) to consolidate the mixed powder. The BP-ECAP consolidation was carried out at 400°C for up to 24 passes. The clusters (~500 nm in size) in the as-received carbon black powder were broken and dispersed homogenously in the Al matrix following the mechanical milling and ECAP processes, with individual C particles of <100 nm in size.

The nanocomposites displayed significantly higher hardness which increased with increasing reinforcement content. Significantly higher compressive strengths were achieved in the nanocomposites although the Al-C materials appeared to exhibit greater strength and lower ductility than the Al-Al_2O_3 materials. The enhanced mechanical properties were attributed to the fine and homogeneous distribution of the ceramic nanoparticles which provide dispersion strengthening and cause the formation of a finer grain structure. The combination of mechanical milling and ECAP offers a promising method for producing strong metal matrix nanocomposites.

Acknowledgments

This project was partially supported by the Australian Research Council with the ARC Centre of Excellence for Design in Light Metals.

References

[1] Valiev, R.Z. and Langdon, T.G. 2006 *Prog. Mater. Sci.* **51** 881-981.
[2] Goussous, S., Xu, W., Wu, X. and Xia, K. 2009 *Comp. Sci. Technol.* **69** 1997-2001.
[3] Xia, K., Wu, X., Honma, T. and Ringer S.P., 2007 *J. Mater. Sci.* **42** 1551-1560.
[4] Donnet, J.-B. and Voet, A. 1976 *in Carbon Black: Physics, Chemistry, and Elastomer Reinforcement,* New York, Marcel Dekker. 81.
[5] Atkinson, H.V., Zulfia, A., Lima, A., Jones, H. and King, S. 1997 *Mater. Design* **18** 243-245.
[6] Goujon, C. and Goeuriot, P. 2001 *Mater. Sci. Eng.* A **315** 180-188.
[7] Hassan, S.F. and Gupta, M. 2006 *J. Mater. Sci.* **41** 2229-2236.
[8] Jha, A.K., Prasad, S.V. and Upadhyaya, G.S. 1989 *Powder Metall.* **32** 309-312.
[9] Ogel, B. and Gurbuz, R. 2001 *Mater. Sci. Eng.* A **301** 213-220.
[10] Han, B.Q. and Langdon, T.G. 2005 *Mater. Sci. Eng.* A **410-411** 430-434.
[11] Arik, H. 2004 *Mater. Design* **25** 31-40.
[12] Besterci, M. 2006 *Mater. Design* **27** 416-421.
[13] Kamrani, S., Simchi, A., Riedel, R. and Reihani, S.M.S. 2007 *Powder Metall.* **50** 276-282
[14] Humphreys, F.J. and Hatherly, M. 1996 *in Recrystallization and Related Annealing Phenomena*, Oxford, UK, Pergamon. 365.

Research on the drawing process with a large total deformation wires of AZ31 alloy

T Bajor , Z Muskalski, M Suliga

Czestochowa University of Technology, Faculty of Materials Processing Technology and Applied Physics, Al. Armii Krajowej 19, 42-200 Czestochowa, Poland

tbajor@wip.pcz.pl

Abstract. Magnesium and their alloys have been extensively studied in recent years, not only because of their potential applications as light-weight engineering materials, but also owing to their biodegradability. Due to their hexagonal close-packed crystallographic structure, cold plastic processing of magnesium alloys is difficult. The preliminary researches carried out by the authors have indicated that the application of the KOBO method, based on the effect of cyclic strain path change, for the deformation of magnesium alloys, provides the possibility of obtaining a fine-grained structure material to be used for further cold plastic processing with large total deformation.

The main purpose of this work is to present research findings concerning a detailed analysis of mechanical properties and changes occurring in the structure of AZ31 alloy wire during the multistage cold drawing process. The appropriate selection of drawing parameters and the application of multistep heat treatment operations enable the deformation of the AZ31 alloy in the cold drawing process with a total draft of about 90%.

1. Introduction

Magnesium alloys are promising light-weight metal alloys. Owing to their properties, these alloys have a wide range of application, starting from the automotive and the electronics industries, through to household equipment articles [1-5]. Moreover, magnesium alloys are beginning to have increasingly high importance as alloys intended for application in medicine, as they belong to a group of materials with good bio-compatibility with the human body (having the ability to dissolve in the body), thus being able to be utilized in cardiac surgery or orthopaedics. The main applications of medical magnesium alloys include bone implants, stents, as well as surgical threads designed for joining soft tissues [6].

As hexagonal close-packed structure alloys, magnesium alloys have limited plasticity and poor deformability at ambient temperature. They are also characterized by high mechanical anisotropy forming during plastic deformation. The application of the KOBO technology based on the cyclic strain path change effect during the plastic deformation of metal, enables a fine-grained material with good mechanical and plastic properties to be obtained [7]. The advantage of this method is the ability to extrude material with a high extrusion ratio at a low temperature. So obtained material is characterized by a homogeneous structure and can be subjected to a further cold deformation process, e.g. a drawing process, which will result in wire with good plastic properties and strength.

2. Experimental procedure

The AZ31 magnesium alloy ingots in the form of ϕ50 x 150 mm of chemical composition as given in Table 1, extruded by the KOBO method into ϕ2 mm-diameter blank, was used for testing. The extrusion process was carried out under isothermal conditions using two-hole extruding die; the extrusion ratio was $\lambda = 312.5$.

Table 1. Chemical composition of the investigated alloy

Component contents, wt%							
Mg	**Al**	**Mn**	**Nd**	**Sb**	**Zn**	**Fe**	**Si**
96.284	2.58	0.12	0.005	0.017	0.99	0.002	0.002

With the purpose of designation parameter optimally heat treatment was performed annealed in a resistance furnace according to the scheme given in Table 2.

Table 2. Heat treatment parameters

Temperature ^{0}C	**150**				**200**				**250**			
Time [min]	15	30	45	60	15	30	45	60	15	30	45	60

Process of deforming stock carry after selection of optimal interoperative parameters of heat treatment according to scheme placed in table 3. The wire drawing process was carried out in laboratory conditions on a ZWICK Z100 testing machine at a speed of 0.83 [mm/s]. Sintered carbide drawing dies with a drawing angle of $2\alpha = 12°$ were used for the drawing process. A unit draft of G $\approx$10% was applied.

Table 3. Scheme of drawing process

Draw no.	**0-3**	**Heat treatment**	**4-8**	**Heat treatment**	**9-16**	**Heat treatment**	**17-24**	**Heat treatment**
G_T [%]	27.75		42.42		56.58		58.13	

To determine the changes occurring in the microstructure, metallographic examinations were performed on an AXIOVERT 25 optical microscope. The analysis of microstructure development during the cold drawing process and the examination of the effect of draft and heat treatment on structure refinement and mechanical properties of wire were carried out.

3. Experimental results and their analysis

The complex state of stress occurring during the process of drawing hard deformable materials causes an inhomogeneous structure to occur. The application of the interoperative annealing process during the magnesium alloy cold plastic working process is very important for two reasons: Firstly, softening of the material and the restoration of its plastic deformability is observed. Secondly, the recrystallization process is a very effective mechanism of the formation of a structure about small largeness of grain size.

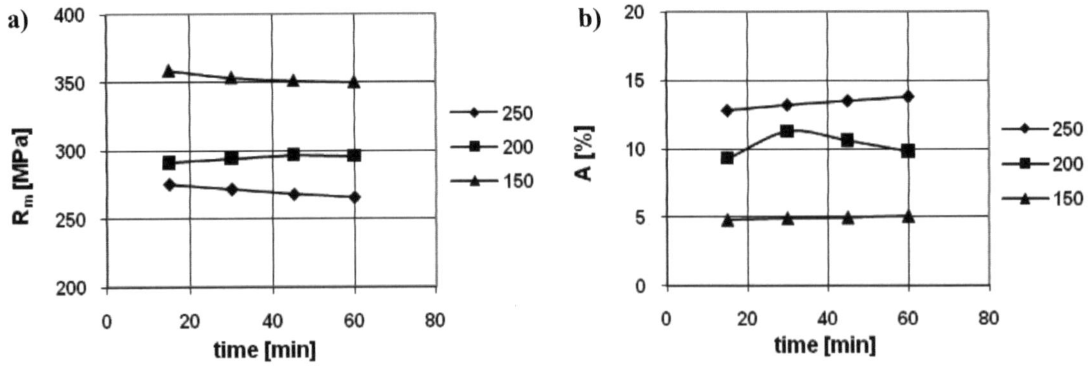

Figure 1. Relationship of mechanical and plastic properties as a function of heat treatment time and temperature **a)** R_m [MPa] = f(t); **b)** A [%] = f (t).

The first stage of the experimental procedure included the determination of the optimal temperature and time of interoperative heat treatment. To this end, the examination of the effect of holding temperature and time on the properties and structure of the material after deformation with a total draft of 27.75% (with a wire diameter of 1.7 mm) was performed. After carrying out experimental tests according to the scheme as shown in Table 2, the following relationships were obtained: R_m = f(t) (Figure 1a) and Ag [%] = f (t) (Figure 1b).

From the obtained testing results it can be observed that with increasing heat treatment temperature an increase in plastic properties (Figure 1b) and a drop in mechanical properties (Figure 1a) follows. Establishment, that: an annealing temperature of 250°C, and a duration of 60 min., are optimal parameter interoperative heat treatment owing to the good mechanical (R_m = 265.9 MPa) and plastic ($\mathbf{A_g}$ = 13.8%) properties of the wire tested. The determined value of the plasticity margin factor, $R_{0.2}/R_m$, is 0.69.

Other points to carry out the heat treatment was chosen on the basis of experimental studies including the analysis of the structure and the plastic and mechanical properties of wires after respective drawing process stages (Figure 2). Decrease of mechanical properties, a significant drop in plasticity and approaching to 1 ratio $R_{0.2}/R_m$ were the considerations that influenced the making of a decision about carrying out subsequent heat treatment.

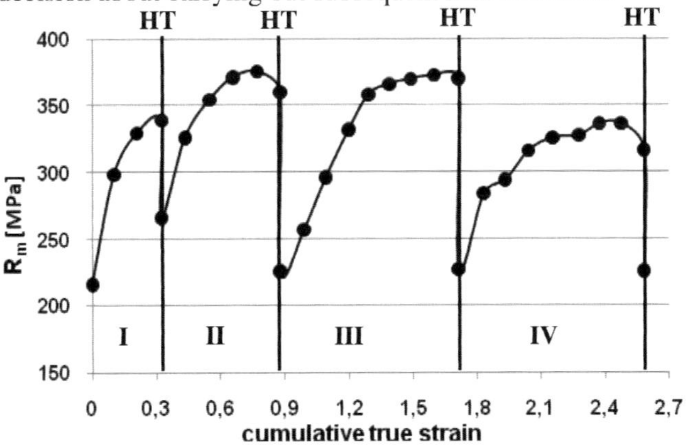

Figure 2. Relationship of the reliability on extension as a function of the cumulated strain at respective deformation process stages (HT – the point of carrying out heat treatment).

The results of detailed tests for the determination of the mechanical properties of the wires under examination after respective drawing process stages and after performed heat treatment are shown in Figure 2. The drawing process was divided into 4 stages: Stage I encompassed 3 draws of a total draft of 27.75%; Stage II – 5 draws of a total draft of 42.42%; Stage III – 8 draws of a total draft of 56.58%; and Stage IV – 8 draws of a total draft of 58.13%. In total, the initial stock was deformed with a total draft of 92.4%.With increasing deformation, a decrease in the mechanical properties of the material under examination was observed at particular deformation process stages. The highest increase was noted at Stages II and III of the deformation process, where the stress level reached a value of 370 MPa, and at Stage IV with a stress level of 340 MPa. An increase in the actual strain value was observed at respective deformation process stages, as follows: Stage I, ε = 0.35; Stage II, ε = 0.55; Stage III, ε = 0.83; and Stage IV, ε = 0.87; which gives a cumulated actual strain value of 2.58.

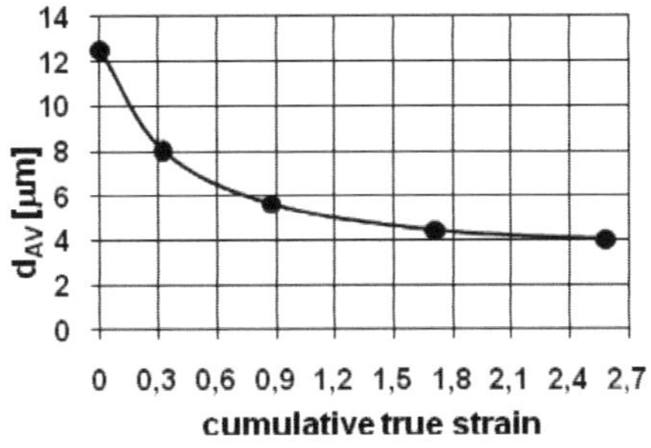

Figure 3. Dependence of average grain size as a function of cumulative true strain after next processes of interoperative heat treatment.

Microstructure observations made after each stage of recrystallization annealing found increment structure refinement, as shown in Figure 3. Example wire structures are illustrated in Figure 4.

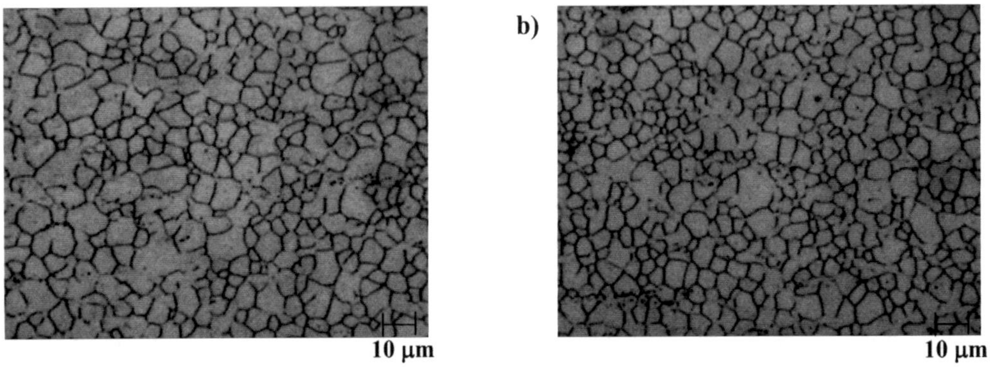

Figure 4. Microstructure of the AZ31 alloy after cold drawing and recrystallization annealing; the magnitude of cumulated true strain is a) 0.87, and b) 1.71.

4. Summary

The investigation presented in the article has found that the factor that has a great influence on the deformability of the AZ31 magnesium alloy in the cold drawing process is the initial coarse-grained material structure. Assuring the correct parameters of the AZ31 alloy drawing process, such as a unit draft of approx. 10%, and optimal heat treatment parameters, enabled a 0.55 mm-diameter wire to be obtained from stock material of an initial diameter of 2mm, as produced by the KOBO method, by carrying out 3 heat treatment operations. The properties of received wire in hard state are R_m = 316 MPa, and the true strain ε = 0.6%; whereas in its soft state after annealing, they are as follows: R_m = 225.5 MPa, and the true strain ε = 10%.

The drawing process carried out following the proposed scheme produced a thin wire of AZ31 alloy with a high level of cumulated actual strain of $\varepsilon = 2.58$.

The performed analysis of structure showed that the main mechanism of material hardening in the drawing process with preset process parameters was the dislocation slip. The application of big deforming accumulated and recrystallization annealing enabled a 3-fold disintegration of structure and fine-grained material about middle-sized 4 μm to be obtained.

References

[1] Yang Z., Li J.P., Zang J.X., Lorimer G.W., Robson J. Review on research and development of magnesium alloys, *Acta Metall. Sinica (Engl. Lett.)* **21** (2008), pp. 313-328,

[2] Xin Wu, Yi Liu: Superplasticity of coarse-grained magnesium alloy, *Scripta M ater.* **46** (2002), pp. 269-274

[3] Watanabe H et al.: Grain size control of commercial wrought Mg-Al-Zn alloys utilizing dynamic recrystallization, *Mater. Trans.* **42** 2001, pp. 1200-1205;

[4] Murai T, Matsuoka S., Miyamoto S, Oki Y.: Effects of extrusion conditions on microstructure and mechanical properties of AZ31B magnesium alloy extrusions, *J. Mater. Process. Technol.* **141** 2003, pp. 207–212,

[5] Su C.W., Lu L., Lai M.O.: Recrystallization and grain growth of deformed magnesium alloy; *Phil. Mag.* **88** 2008, pp. 181-200;

[6] Staiger M.P., Pietak A.M., Huadmai J., Dias G.: Magnesium and its alloys as orthopedic biomaterials: A review, Biomaterials **27**, 2006, pp. 1728-1734,

[7] Pospiech J., Korbel A. Bonarski J, Bochniak W., Tarkowski: Microstructure and texture of Mg-based AZ alloys after heavy deformation under cyclic strain path change conditions, *Mater. Sci. Forum* . **84-86** 2008, pp. 565-570.

Nanomechanical behaviour of Al-Ti layered composites produced by accumulative roll bonding

V Maier[1,2], H W Höppel[2] and M Göken[1,2]

[1] Institute 1, Department of Materials Science and Engineering, University of Erlangen-Nuernberg, Martensstr. 5, 91058 Erlangen, Germany;
[2] Insitute of Advanced Materials and Processes (ZMP), Fürth

verena.maier@ww.uni-erlangen.de

Abstract. In this study Ti-foils were roll bonded together with commercially pure aluminium AA1050. The laminates were produced by using two 1 mm thick AA1050 sheets at the outer side of the stack combined with 100 µm thick Ti-foils as an intermediate layer for each accumulative roll bonding process. The samples were rolled up to 4 ARB cycles. Subsequently the sheets were post-process heat treated at 180°C, 400°C or 600°C, respectively, for 24 hours. The local mechanical behaviour of the Al/Ti intermetallic interfaces have been investigated using nanoindentation experiments. A strong dependence between annealing-temperature, -time and deformation grade is detected. While a heat treatment at 180°C only leads to a weak bonding between Al and Ti with a preservation of the UFG structure, temperatures up to 600°C are causing a complete recrystallisation of the microstructure and formation of diffusion layers with different Al and Ti concentrations.

1. Introduction

Over the last decades, nanocrystalline and ultrafine-grained (UFG) materials with grain sizes of less than 1 µm have gained significant attention. Especially their mechanical properties in terms of strength and ductility show superior behaviour compared to conventionally grained materials [1,2].
Besides the production of UFG sheet materials the accumulative roll bonding (ARB)-process [3-5] provides the very promising possibility to produce composite materials with a layered structure or even laminates. Thus, it is possible to insert different types of particles, fibres and foils to create specific laminates, multistacks and graded materials with tailored properties [6].

Especially the intermetallic compounds of the Al/Ti system have a great potential for industrial applications. In the past, investigations on the diffusion reaction between pure Al and pure Ti were accomplished by Wei et al. [7].

In the current work Ti-Al laminates were produced by ARB and the local behaviour of the Al/Ti-interfaces were systematically studied with nanoindentation.

2. Experimental procedure

2.1. Materials

Aluminium of commercial purity Al 99.5 (equivalent to AA1050A) and Titanium of commercial purity Ti (grade 1) were used. The chemical composition of both materials is listed in Table 1. Prior to the accumulative roll bonding process, the aluminium was solution heat treated at 500°C for 1 h and subsequently quenched in water, while titanium was annealed at 700°C for 15min and air cooled.

Table 1. Chemical composition of commercially pure aluminium AA1050A and pure Ti (grade 1).

wt %	Si	Fe	Cu	Mn	Mg	Zn	Ti	others	Al
AA1050A	0,25	0,40	0,05	0,05	0,05	0,07	0,05	0,03	balance
wt %	O	Fe	N	C	H			others	Ti
Ti (grade1)	0,12	0,15	0,05	0,06	0,013			0,40	balance

2.2. Accumulative roll bonding process (ARB)

The size of the aluminum sheets prior to the roll bonding process were 100 mm x 300 mm x 1 mm, the titanium foils were 20 mm x 300 mm x 100 µm in size. The surfaces of both materials were wire brushed in order to remove oxide layers. The aluminum sheets were stacked on top of each other with the Ti-foil as an interlayer between the two aluminum matrix sheets. Afterwards the stacks were rolled together without lubrication to a 50% thickness reduction using a four high rolling mill (Carl Wezel, Mühlacker). The bonded, laminated metal sheets were halved, folded and once again wire brushed before the next cycle. The sheets were roll bonded up to 4 cycles each time with two interlayer of Ti-foil. The principle is shown in Figure 1. Finally, the produced Al/Ti laminate structures were post-process heat-treated at 180°C, 400°C and 600°C.

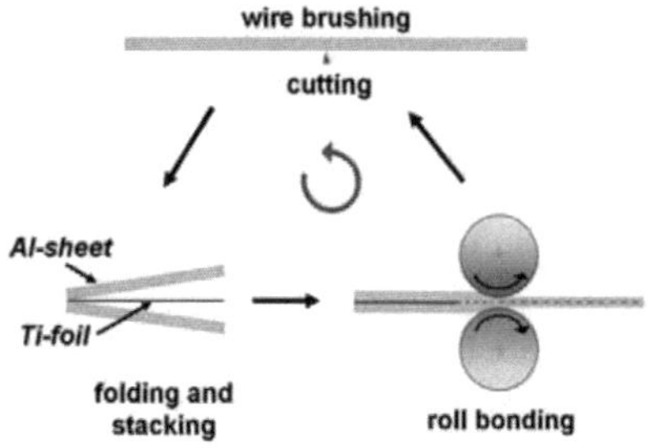

Figure 1. Schematic of adopted accumulative roll bonding process for producing laminate structures

2.3. Investigation of microstructure and mechanical properties

All specimens were prepared for metallographic examination by sectioning transverse and longitudinal to the rolling direction. The specimens were mounted in acrylic resin Technovit 2000 LC, ground, polished with diamond down to 1 µm and electro polished.

Light microscopy, scanning electron microscopy and energy dispersive spectrometry were used to investigate the microstructure and the occurred phases and compounds.

For investigating the local mechanical behavior of the Al/Ti interface as well as the occurred phases, nanoindentation experiments were carried out [8]. The experiments have been performed on a G200 Agilent Nano Indenter with a Berkovich diamond tip and a load controlled method using 20 mN as the maximum load. Arrays of 3 x 12 indents were directly positioned over the Al/Ti interfaces. Neighboring indents were separated by a distance of 20 µm to avoid an influence of the affected plastic zone around an indent.

3. Results and Discussion

3.1. Structure

Figure 2 shows the morphology of Ti fragments in the Al/Ti laminates after 4 cycles of ARB. It is obvious that more fragmentation occurs in the longitudinal section than in the transversal section. For both sections the Ti fragments are homogenously spread in the Al matrix, but depending on the orientation of the cross section the number and size are different. The longitudinal section shows a multiplicity of short, small Ti fragments, whereas in the transversal section very long, thin Ti parts are dominating. Especially in the last bonding layer (middle layer), the transversal Ti fragment is very long and not cracked.

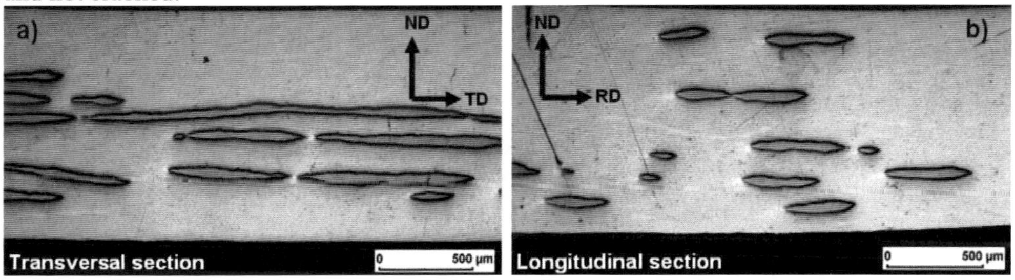

Figure 2. Light microscopic images of the morphology of Ti fragments in Al/Ti laminates after 4 cycles ARB in transversal (a) and longitudinal (b) section

In order to qualify optically the Al/Ti interface behavior, three samples were prepared by using the Focused Ion Beam (FIB) technique. Prior to the preparation, the samples were heat treated at 180°C, 400°C or 600°C, respectively for 24 hours. The prepared cross sections are showing nicely that 180°C and 400°C are not leading to any diffusion reaction between Al and Ti. Only after a 600°C heat treatment for 24 hours an interdiffusion zone occurs consisting of tiny phases, which are embedded in the Al matrix. The newly formed phase was detected to be $TiAl_3$ by EDS analysis (Figure 5).

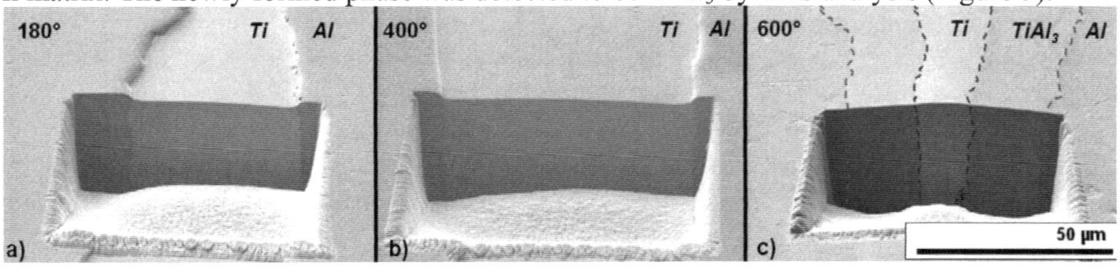

Figure 3. Scanning electron microscopic images of Al/Ti interfaces after various heat treatments at (a) 180°C; (b) 400°C; (c) 600°C prepared by Focused Ion Beam cuts

3.2. Mechanical properties

The different heat treatments are leading to a hardness decrease in the Al matrix, as shown in Figure 4. The local hardness of the ultrafine-grained Al-matrix before annealing averages at 0.6 GPa. First recovery processes and little grain growth occurs during low annealing temperatures of 180°.The local hardness decreases to 0.51 GPa with a relatively high standard deviation. This argues for a less inhomogeneous grain structure, while the UFG structure is mostly preserved. Further increasing of the annealing temperature leads to a continuous decrease in the hardness to 0.42 GPa.

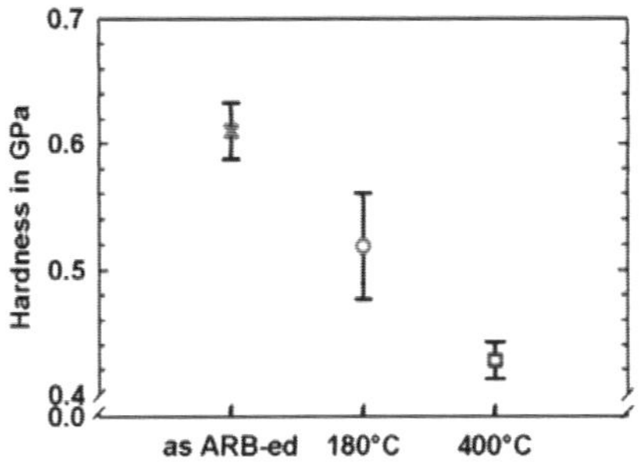

Figure 4. Hardness of Al with increasing annealing temperature (each after 24 hours heat treatment).

The results of the local hardness measurements obtained by nanoindentation experiments through a cross section of the Al/Ti laminate are shown in Figure 5. The compound material was post-process heat treated at 600°C for 24 hours, which led to a formation of an intermetallic $TiAl_3$ phase determined by EDS. Figure 5 shows clearly the dependence between local hardness and chemical composition. The local hardness of the averages at 0.4 GPa for pure Al and respectively 2.1 GPa for pure Ti.

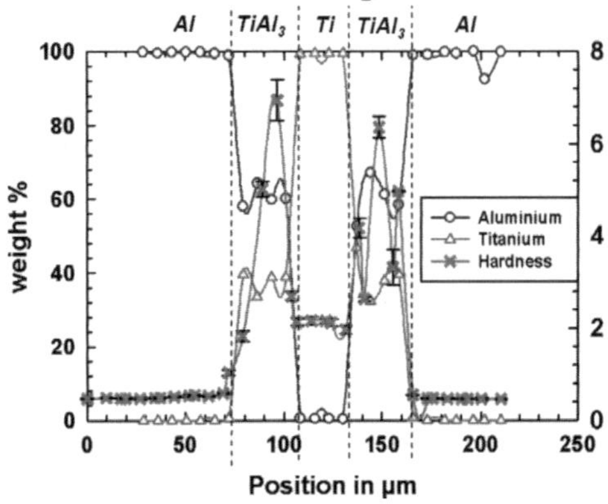

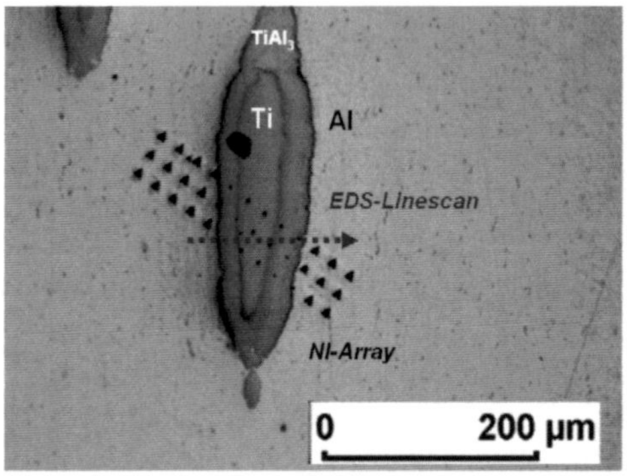

Figure 5. Local hardness evolution and EDS linescan across the Al/Ti multimaterial stack after annealing at 600°C

Comparing the results, it can be shown that the new formed intermetallic phase has a significant higher hardness as pure Al and Ti. The absolute value averages between 4 GPa up to 7 GPa. The value distribution can be explained with the inhomogeneous structures of the formed two phase regions. During further nanoindentation experiments in compact $TiAl_3$, hardness values of almost 9 GPa were observed.

4. Conclusion

In the present study Ti-foils were rolled in pure Al sheets by using the ARB process. The foils broke during the deformation process and were homogenously spread in the Al-matrix. The distribution of those Ti fragments differs between longitudinal and transversal section.

Annealing processes at 180° and 400° led to a weak bonding with plenty of surface defects, while a heat treatment at 600°C led to a formation of an intermetallic $TiAl_3$ phase. Local mechanical properties could be successfully investigated by carrying out nanoindentation experiments.

Acknowledgements

The authors gratefully acknowledge the funding of the German Research Council (DFG), which, within the framework of its 'Excellence Initiative' supports the Cluster of Excellence 'Engineering of Advanced Materials' at the University of Erlangen-Nuernberg.

References

[1] Valiev RZ, Alexandrov IV 2002 *J. Mater. Res.* **17** 5
[2] Höppel HW, May J, Göken M 2004 *Adv. Eng. Mater.* **6** 781
[3] Saito Y, Tsuji N, Utsunomiya H, Sakai T, Hong RG 1998 *Scripta Mater.* **39** 1221
[4] Saito Y, Utsunomiya, Tsuji N, Sakai T 1999 *Acta Mater.* **47** 579
[5] Tsuji N, Saito Y, Utsunomiya H, Tanigawa S 1999 *Scripta Mater.* **140** 795
[6] Min G, Lee JM, Kang SB, Kim HW 2006 *Mater. Lett.* **60** 3255
[7] Wei Y, Aiping W, Guisheng Z, Jiale R 2008 *Mater. Sci. Eng.* **A 480** 456
[8] Oliver WC, Pharr GM 1992 *J. Mater. Res.* **7** 1564

Joint investigation of the local material rotation and lattice spin in a cube {100} <001> oriented single crystal

M Darrieulat, RY Fillit, M Mondon and S Sao-Joao

Ecole Nationale Supérieure des Mines de Saint-Etienne, SMS Centre, UMR CNRS n° 5146, 158 cours Fauriel 42023 Saint-Etienne, cedex 2, France.

darrieulat@emse.fr

Abstract. Cube {100} <001> oriented single crystals of Al 1% Mn were compressed in channel-die. Their lateral faces were covered with transferable carbon grids with a step of 100 μm . At a deformation of about 0.3, the vertical bars of the grids show undulations whose characteristic length is of the order of the millimetre and which become sharper and smaller as the deformation proceeds. Fiducial golden grids with a step of 20 μm remain largely unaffected. This shows that the investigated heterogeneity is typical of the mesoscopic scale and has no directly related patterns at the macroscopic and microscopic level. Microfocussed X-rays were used to measure the crystallographic rotations during the process. The investigated spot was a few 0.1 mm^2. At a deformation of 0.6, the lateral faces of the crystal undergo a split into two Cube orientations each rotated of about 15° around the transverse axis. This is put in relation with the undulations of the bars. At 0.9 an additional rotation around the longitudinal axis appears. The local material rotation and the lattice spin at the mesoscopic scale are interpreted in accordance with previous analyses of the evolution of the Cube texture based on EBSD and the observation of the traces of slip systems.

1. Introduction

Among all the crystallographic orientations, Cube displays a particular behaviour when submitted to plane strains [1]. When compressed in channel-die, four slips systems of a Cube single crystal are equally stressed but locally, only two of them are simultaneously active. So the crystal splits up into deformation bands parallel to the vertical lateral faces, those in contact with the walls of the die. The width of these bands ranges typically from 50 to 200 μm [2] that is, for metals, the mesoscopic scale.

Now, the results above were obtained by observing the traces of the slip systems, electro-litho-depositing fiducial golden grids with a step of 5 μm in a scanning electron microscope and performing EBSD analyses. Other techniques have appeared recently. One of them is the transferable carbon grids made of a polyester film covered by a photographic emulsion on which a laser beam impresses a grid. These grids are applied on to the test piece which has beforehand been covered with a film of glue and deform with it. The step can be 30 μm or wider. Another is the microfocus X-ray texture analysis, in which crystallographic information is gathered from an elliptic spot which ranges from a few tenths to a few hundreds of μm, according to the device, while classical X-rays work on square millimetres. This is documented in various papers, among them [3]. Table 1 sums up, for each scale, the means of investigation of the material rotation of the test piece and the spin of the crystal lattice. The aim of the present paper is to find if, by using carbon grids or microfocussed X-rays, the previous observations can be confirmed.

Table 1. Means of microstructural investigation relevant to each scale.

Scale	Material rotation	Lattice spin
microscopic	Fiducial golden grids	EBSD
mesoscopic	Transferable carbon grids	Microfocussed X-rays
macroscopic	Grids drawn on the test piece	Classical X-rays

2. Experiments

Cube single crystals of Al 1%w Mn were cut into test pieces 10.3 mm high and 8 mm long (unchanging width: 7 mm). Their lateral faces were covered by transferable carbon grids forming initially squares of $100\,\mu m$, the bars being around $35\,\mu m$ thick. They were compressed in channel-die at room temperature by steps of $\overline{\varepsilon} = 0.15$. At each step the test pieces were released from the die, the Teflon ™ wrapping which acts as a lubricant was changed and pictures of the grids were taken with a binocular magnifier, the magnification varying from X = 3 to X = 28.5 . Some of these pictures can be seen in figure 1. The axes have been numbered 1 (elongation), 2 (transverse) and 3 (compression).

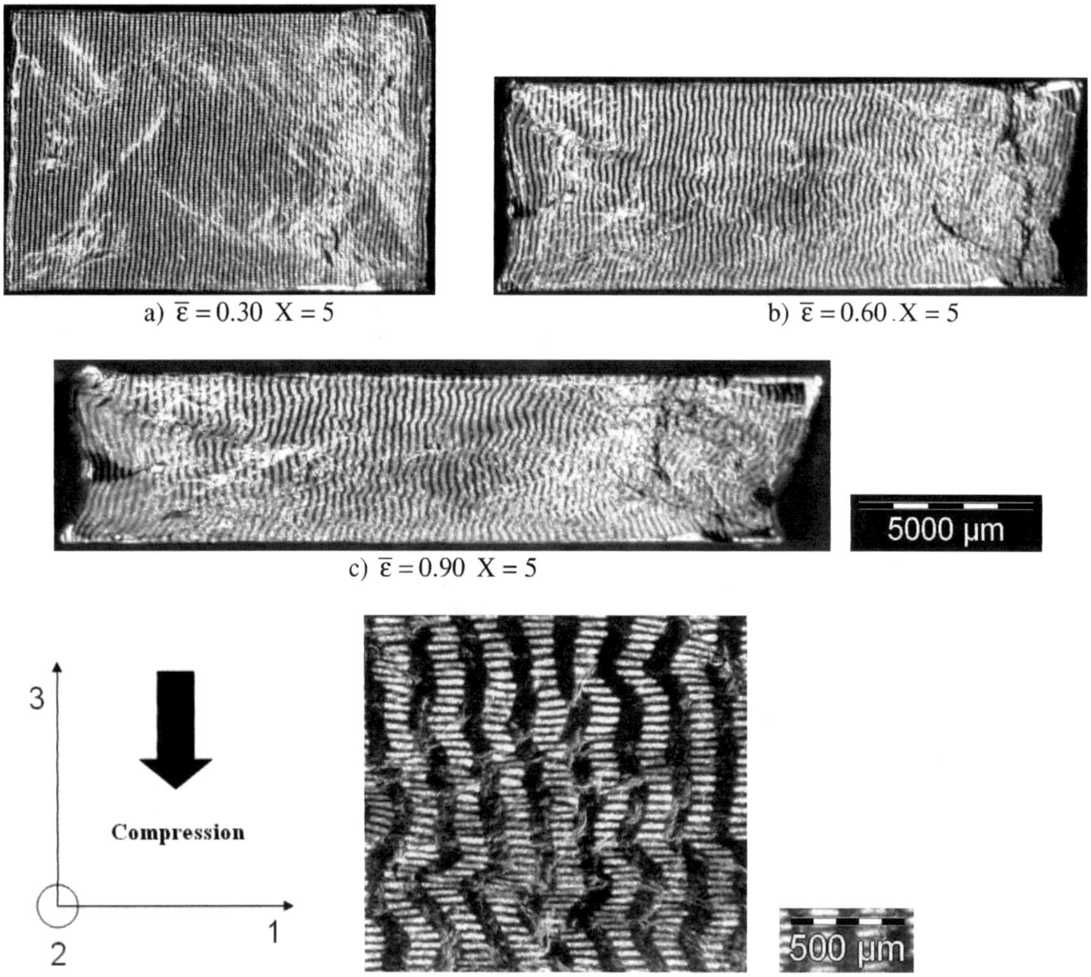

a) $\overline{\varepsilon} = 0.30$ X = 5
b) $\overline{\varepsilon} = 0.60$ X = 5
c) $\overline{\varepsilon} = 0.90$ X = 5
d) $\overline{\varepsilon} = 0.90$ X = 28.5

Figure 1. Carbon grids on the lateral faces of a test piece at various deformations. $\overline{\varepsilon}$: logarithmic deformation. X: magnification.

At $\overline{\varepsilon} = 0.30$, no shear or other heterogeneity can be spotted except on the edges of the test-piece, but the vertical bars show slight undulations whose characteristic length is of the order of the millimetre. At this stage, the strain hardening changes, becomes linear and the abovementioned deformation bands appear. On the lateral faces, at further steps of deformation, the horizontal rungs remain pretty much horizontal and get thinner and thinner. The bars widen and form sharper patterns whose characteristic length lessens as the height of the sample decreases. Such is the case at $\overline{\varepsilon} = 0.60$: see figure 2, b). The phenomenon accentuates at $\overline{\varepsilon} = 0.90$, where it takes in places the aspect of chevrons such as those of insert d). The shears they reveal affect several bars at a time and remain widely horizontal. At larger deformations, the grid deteriorates because of the friction.

Is this an essentially mesoscopic phenomenon? To answer this question, fiducial golden grids with $20\,\mu m$ steps were deposited and similar compressions performed. The results can be seen on figure 2. On stretches of a few hundreds μm, the golden bars remain straight. Figure 1 a), for example, pictures part of one of the abovementioned undulations. Hence, these are typically mesoscopic. Conversely, the fiducial grids reveal heterogeneities of their own, spotted only at the microscopic scale: traces of slip systems, whose inclination with respect to axis 1 lies around 30°; coarse slip bands which form valleys in the initially polished surfaces, as on insert b); a greater activity of the slip systems with the growing deformation on insert c). Occasional shear bands can also be found [4].

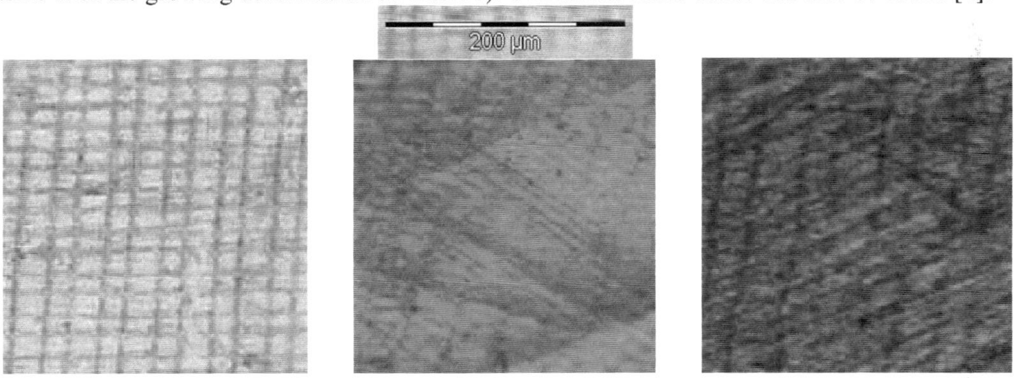

a) $\overline{\varepsilon} = 0.30$ b) $\overline{\varepsilon} = 0.60$ c) $\overline{\varepsilon} = 0.90$

Figure 2. Golden grids on the lateral faces of a test piece at various deformations.

Classical X-ray measurements on the lateral faces show that, after $\overline{\varepsilon} = 0.30$, there is a split in the initial Cube orientation [4]. It rotates clockwise and anticlockwise around axis 2 and the spread gets wider with the deformation. What is the corresponding spatial organisation of the deformation? Microfocuss X-ray texture analysis with $100\,\mu m$ foci was used in an attempt to investigate it.

Such an appliance is easily mounted on a classical rig, but the measurements take time: typically five hours for one point. In the case of the present work, Copper was used as the anode material and the X-rays scanned an ellipsoidal spot whose axes were about 250 μm horizontally and $150\,\mu m$ vertically. Six points were chosen on the face at $\overline{\varepsilon} = 0.60$: four grouped in half a square millimetre at the centre of the sample and two apart. The intensity was collected in a layer less than $30\,\mu m$ deep under the surface of the test piece, hence, in only one vertical deformation band.

The results throughout the face are remarkably uniform, especially at the centre of the test piece. Figure 3, a) shows the split already known through classical measurements. It is $\pm 15°$ around the ideal Cube position. The poles are sharp and the scatter around each of them is of the order of 6°. Measurements were also done for $\overline{\varepsilon} = 0.90$: see figure 3, b).The results are basically the same but an additional rotation around axis 1 is detected. All this shows a repartition between patches rotating clockwise and others counter clockwise, the pattern changing with a characteristic length of about $100\,\mu m$, at least in one direction. At larger deformations, the crystallography becomes confused.

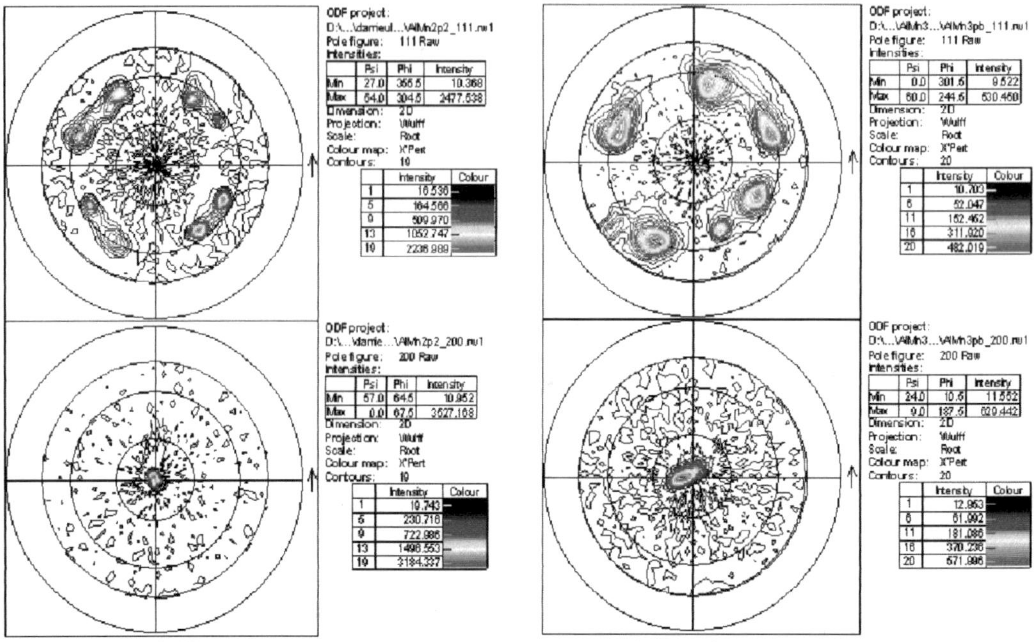

a) $\overline{\varepsilon} = 0.60$ b) $\overline{\varepsilon} = 0.90$

Figure 3. Pole figures obtained by microfocussed X-rays on the lateral {100} faces of the test pieces. Horizontal axis: elongation direction (axis 1, cf. figure 1). Vertical axis: normal direction (axis 3, compression). Upper figures: {111} poles. Lower figures: {200} poles.

3. Discussion

According to the analysis presented in [1], only one pair of slip systems is active locally. The systems work with equal slip rates and provoke a plastic spin $\boldsymbol{\omega}$ in the axes of the crystal such that $\omega_{12} \neq 0,\ \omega_{13} = \omega_{23} = 0$. Now, the material rotation rate $\boldsymbol{\Omega}$ and the lattice spin $\mathbf{r}$ are linked by $\boldsymbol{\Omega} = \mathbf{r} + \boldsymbol{\omega}$ [5]. This implies $\Omega_{13} = r_{13}$. The material rotation $\overline{\Omega}_{13}$ is given by half the slant of the grids. It is determined by the boundary conditions applied to the test piece. The lattice rotation $\overline{r}_{13}$ can be measured on the pole figures. At $\overline{\varepsilon} = 0.60$, it is worthy of note that both are of the order of 15° and that the traces of the slip systems, initially inclined 45° with respect to axis 1, have moved towards 30°, that is, a rotation of 15° due to the tilt of lattice. The observations are much the same at $\overline{\varepsilon} = 0.90$.

This can be interpreted as follows. The test piece is constrained horizontally but has vertical free faces. It evolves as a staking of horizontal layers with $\Omega_{13} = r_{13}$ of opposite signs. Their thickness follows the progress of the compression. The microfocussed X-rays investigate a spot which covers typically two sides of a chevron, hence the splits on figure 3. This scheme is compatible with the activity of either of the pairs of slip systems. No conclusion can be drawn on whether the parting into horizontal layers corresponds to a change in the pairs of active slip systems, but the continuity of the traces of the slip systems suggests that the same pair which is acting on the whole face.

References

[1] Wert JA, Liu Q and Hansen N 1995 *Acta Mater.* **43** 4153-63.
[3] Basson F and Driver J 2000 *Acta Mater.* **48** 2101-15.
[3] Darrieulat M, Zhani K, Fillit RY and Chenaoui A 2009 *J. Eng. Mat. Tech.* **131** 1-1 1-9.
[4] Poussardin JY 2003 PhD dissertation, Ecole des Mines de Saint-Etienne (France).
[5] Havner KS 1992 Finite Plastic Deformation of Crystalline Solids (Cambridge University Press) 34-38.

Mechanical, microstructure and texture properties of interstitial-free steel and copper subjected to equal channel angular extrusion and cold-rolling

A A Gazder[1], S S Hazra[1], C F Gu[2], W Q Cao[2], C H J Davies[2], E V Pereloma[1]

[1] School of Mechanical, Materials and Mechatronics Engineering, University of Wollongong, New South Wales 2522, Australia
[2] Department of Materials Engineering, Monash University, Victoria 3800, Australia

azdiar@uow.edu.au

Abstract. Interstitial-free steel and OFHC copper were subjected to 8 passes, route B_C room temperature ECAE followed by cold-rolling up to 97.5% thickness reduction. Uniaxial tensile tests and Electron Back-Scattering Diffraction were used to characterise the evolution in mechanical properties, microstructure refinement and micro-texture. IF-steel showed continuous increase in strength whereas Cu returned reduced strength and a small gain in ductility at 97.5% reduction. In both metals substructure refinement was accompanied by an increase in high-angle boundary fraction, average misorientation and a slight increase in Σ3 boundaries. An evolution of crystallographic orientations from negative shear to predominantly cold-rolled textures after 95% and 97.5% reduction was observed in both metals.

1. Introduction

In order to make commercial use of ultrafine grained materials their behaviour during subsequent deformation into plate/sheet/foil form via cold-rolling (CR) needs to be understood. To-date, adiabatic shear banding in commercial purity Fe was investigated after 4 passes, route C room temperature (RT) Equal Channel Angular Extrusion (ECAE) followed by cryo-rolling up to 82% [1]. The effect of initial grain size on shear banding in ECAE + 95% CR was also studied in Cu [2]. Superplasticity of Al-3Mg-Sc alloys was investigated either by up to 8 passes, route B_C RT-ECAE and CR up to 70% thickness reduction [3] or 4 passes, route B_C ECAE at 473K followed by up to 85% CR [4]. Hcp Ti has also been subjected to ECAE at 673K and further CR up to 73% [5]. Irrespective of material, transmission electron microscopy (TEM) revealed grains elongated in the rolling direction and the retention of high misorientation angles.

The present study compares the effects of additional CR up to 97.5 % reduction on Interstitial-Free (IF) steel and Oxygen-Free High Conductivity (OFHC) copper previously subjected to 8 passes RT-ECAE. Mechanical properties were determined by RT uniaxial tensile testing while Electron Back-Scattering Diffraction (EBSD) returned subgrain/grain statistics, misorientation, Σ3 boundaries and micro-texture.

2. Experimental Procedure

Billets of 80×20×20 mm^3 Ti IF-steel and 120×20×20 mm^3 OFHC Cu were annealed and subjected to N=8 RT-ECAE via route B_C with 10 kN backpressure at 5 and 2 mms^{-1}, respectively. Following this the billets were further CR for up to 95% (IF-steel) and 97.5% (Cu) thickness reduction. As shown in [6], the rolling direction (RD) was maintained parallel to the prior ECAE extrusion direction (ED).

Flat dog-boned samples of 15(gage)×5×1 mm^3 (IF-steel) and 10×2×1 mm^3 (Cu) were cut from the center of the stable sample length along the ED/RD and subjected to uniaxial tension at strain rates of 8×10^{-4} s^{-1} and 6×10^{-4} s^{-1}, respectively.

ND-ED/RD samples cut from the center of the stable sample cross-section were mechanically ground and electropolished. EBSD was conducted on a JEOL 7001F (IF-steel) and LEO-1350 (Cu) field mission gun-scanning scanning electron microscope fitted with a Nordlys-II detector and the Oxford Channel-5 software package. Operating conditions were 15kV, 1.14nA for IF-steel and 20kV, 5nA for Cu at 20mm working distance. Step sizes of 100, 80 and 25 nm were used for IF-steel and 40 nm for Cu. The maps were initially cleaned using VMAP. Post-processing was undertaken by Channel-5 with a 2° misorientation limit imposed on all maps.

3. Results and Discussion

IF-steel returns greater strength and reducing total elongation (Figure 1, Table 1) with increasing CR reduction. In contrast Cu shows reduced strength and a slight gain in ductility after 97.5% CR. At these imposed strain levels, we expect enhanced dynamic recovery (relative to IF-steel) due to the higher stacking fault energy of Cu [7]. After 95% and 97.5% CR both IF-steel and Cu show a flattening of their tensile curves and reduced rates of decline during Stage-III work hardening. However in contrast to IF-steel, Cu returns a larger period of post-necking elongation.

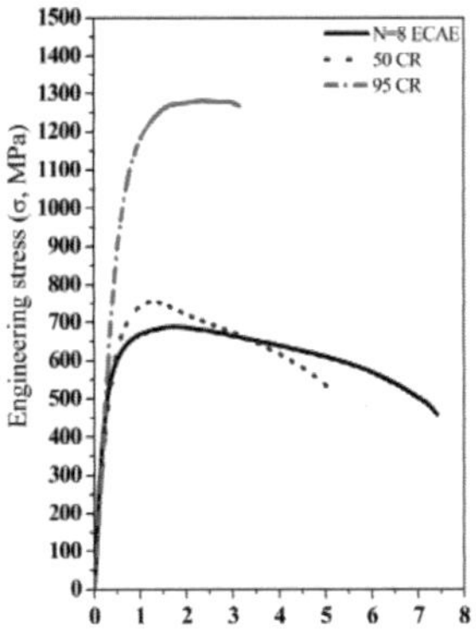

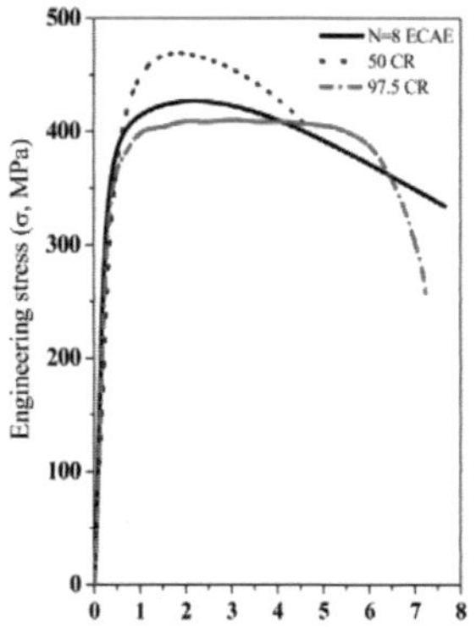

Figure 1. Uniaxial tensile curves of: (a) IF-steel and (b) Cu after N = 8 ECAE and up to 97.5% CR.

Table 1. Mechanical properties of IF steel and Cu after N = 8 ECAE and cold-rolling.

		$\sigma_{0.2}$ (MPa)	σ_{UTS} (MPa)	$\varepsilon_{uniform}$ (%)	ε_{total} (%)
	N=8	607 ± 13	689 ± 19	1.8 ± 0.4	7.5 ± 1.8
IF-steel	50%	639 ± 81	753 ± 58	1.3 ± 0.1	5.1 ± 0.6
	95%	1097 ± 8	1280 ± 3	2.4 ± 0.2	3.2 ± 0.3
	N=8	385 ± 1	427 ± 1	2.2 ± 0.2	7.7 ± 0.2
Cu	50%	419 ± 5	470 ± 4	1.8 ± 0.2	5.0 ± 2
	97.5%	384 ± 2	410 ± 9	3.2 ± 0.2	7.2 ± 2

After N=8 ECAE IF-steel and Cu comprise a mix of equiaxed and elongated substructures (Figures 2(a, d)). Both materials have a large fraction of high-angle boundaries (HABs) (Table 2) with average subgrain (θ≥2°) sizes of 0.43 (IF-steel) and 0.36 (Cu) μm, respectively. Further CR up to 95% and 97.5% produced a convergence of grain (θ≥15°) sizes with subgrain sizes and increase in average misorientation. Importantly, for similar microstructural length-scales 95% CR IF-steel clearly shows shear banding of ribbon-shaped grains whereas 97.5% CR Cu has a mixed morphology of approximately equiaxed grains between layers of elongated substructures and an absence of shear bands (Figure 2(c, f)).

CR produced a rotation of grains such that their long axis coincides with the RD. When the (110) and (111) pole figures of 50% CR IF-steel and Cu are shown in ECAE convention (Figures 3(b, e)) a clockwise rotation around the TD of ~15-20° becomes obvious (compared to Figures 3(a, d)) and corresponds directly to the macroscopic deformation mode.

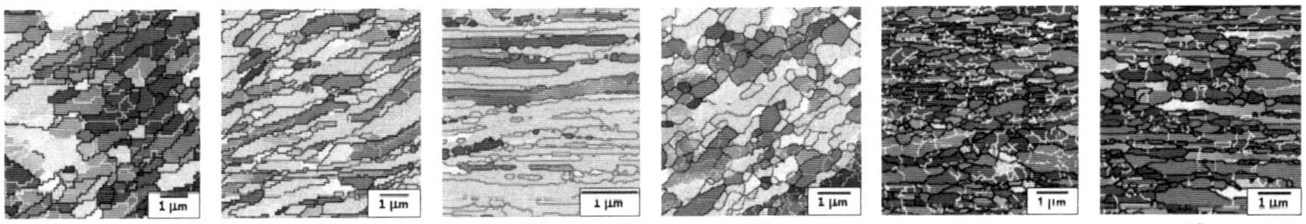

(a) (b) (c) (d) (e) (f)

Figure 2. Representative inverse pole figure maps of (a-c) IF-steel and (d-f) Cu after: (a, d) N=8 ECAE; (b, e) 50% (c) 95% and (f) 97.5% CR. White lines = LAB (2°<θ<15°); black lines = HAB (15°≤θ≤62.8°); green = ⟨101⟩ ; blue = ⟨111⟩ ; red = ⟨001⟩ ; ED/RD = *x-axis*, ND = *y-axis*.

Table 2. Microstructural parameters of IF steel and Cu after N = 8 ECAE and cold-rolling.

		d_{ECD} θ≥2° (μm)	d_{ECD} θ≥15° (μm)	HAB (%)	Aspect ratio θ≥15°	θ_{HAB}	θ_{AVG}	Σ3 (%)
IF-steel	N=8	0.43 ± 0.2	0.64 ± 0.4	63 ± 2	2.6 ± 1.3	37.7 ± 0.1	27.3 ± 0.4	2.1 ± 0.5
	50%	0.39 ± 0.2	0.55 ± 0.4	71 ± 2	3.9 ± 2.4	37.9 ± 0.3	29.8 ± 0.8	2.4 ± 0.2
	95%	0.21 ± 0.2	0.25 ± 0.2	85 ± 1	5.1 ± 4.7	39.3 ± 1.6	35.3 ± 1.6	3.4 ± 1.6
Cu	N=8	0.36 ± 0.2	0.56 ± 0.3	57	2.1 ± 0.9	40.1	25.8	1.7
	50%	0.25 ± 0.2	0.39 ± 0.3	63	3.5 ± 2.3	41.1	28.5	3.0
	97.5%	0.28 ± 0.2	0.36 ± 0.3	79	3.6 ± 3.0	41.8	35.0	3.8

After 95% and 97.5% CR both materials clearly depict typical rolling texture components. In IF-steel the highest intensities are noted for $(115)[1\bar{1}0]$ in the α-fiber (⟨110⟩//RD) and $(111)[1\bar{2}1]$ in the γ-fiber (⟨111⟩//ND). The former along with {114}⟨110⟩ splits the RD-fiber and is a manifestation of alloying additions [8]. In CR Cu the development of typical Copper-type textures (or the β-fiber) from Copper {112}⟨111⟩ through S {123}⟨634⟩ and ending at Brass {110}⟨112⟩ in φ_2 is seen along with Brass-type textures (or the α-fiber) which includes {011}⟨011⟩, Brass {110}⟨112⟩ and Goss {110}⟨001⟩ in φ_1. After 97.5% CR the maximum volume fraction was found for S (44%) followed by the near Brass orientations (24%), Copper (16%) and Goss (3%).

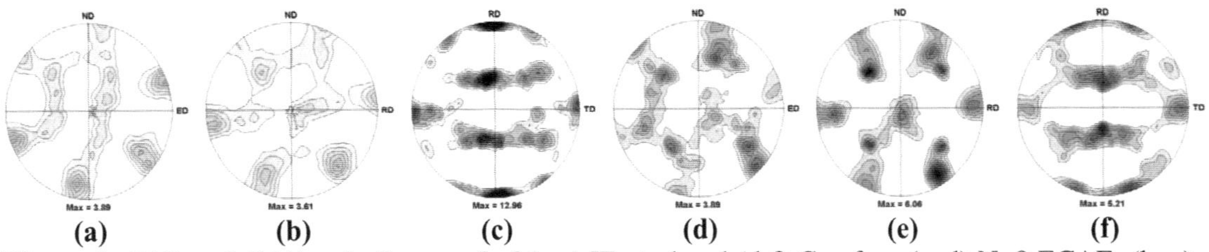

(a) (b) (c) (d) (e) (f)

Figure 3. (110) and (111) pole figures of of (a-c) IF-steel and (d-f) Cu after: (a, d) N=8 ECAE; (b, e) 50% (c) 95% and (f) 97.5% CR. (a,b) and (d,e) in ECAE convention; (c) and (f) in rolling convention.

The textures in 97.5% CR Cu are mainly Copper-type (α-fiber) but contain an unusually prominent Brass component. Pure Brass textures are ascribed to low stacking fault energy via alloying, reduced rolling temperature and high strain rates. It is associated with late stage shear banding occurring within closely spaced deformation twins via latent hardening [9, 10]. While EBSD mapping resolution is a limiting factor in this work, the absence of shear banding in ECAE + 97.5% CR Cu agrees with a previous report that attributed its disappearance to the complex deformation path [2].

Of further interest are the small increases in Σ3 boundary fraction in both metals (Table 2). While the twinning domain reduces in bcc materials [11], the opposite occurs in fcc Cu with deformation twins observed in grain interiors and narrow micro-shear bands via TEM [12]. However Σ3 increase in Cu can also be attributed to boundaries being close to S and Brass variants. Again considering the overall scale of deformation twinning, major changes to bulk texture are not expected [10].

Consequently other aspects of 97.5% CR Cu need discussion. Firstly although the Brass intensity is higher than Copper orientation, the former is shared by both β and α -fibers. Since the volume fraction of β-fiber is greater than α-fiber, it corresponds more to a mixed texture state. Secondly, although the rate of grain refinement is markedly faster in IF-steel, the rate of average misorientation increase is quicker in Cu. A greater level of texture symmetry is more readily observed in 50% CR Cu than IF-steel which suggests the texture subdivision of grains.

While Taylor modelling is unsuitable for explaining Brass textures, the composite Sachs model returns approximate agreement only up to 50% CR [10]. Latent hardening simulations assuming quicker hardening rates for previously inactive slip systems have also shown variations in Brass strength at large CR reductions [13]. Thus an alternative approach involving 'slow latent hardening' wherein active slip systems harden at more rapid rates than inactive systems [14] is suggested. Although such assumptions negate shear banding in fcc Cu (an observation that agrees with EBSD) and promotes cross-slip, it could better simulate Brass orientation development for complex deformation paths, large CR reductions and mixed textures.

4. Conclusions

ECAE followed by cold-rolling leads to increases in strength, grain refinement and average misorientation in IF-steel and Cu. Both metals return typical rolling texture components after highest CR reduction. The nature of shear banding in Cu and strength of the Brass component necessitates further investigation into the contributions from deformation twinning and latent hardening.

Acknowledgements

This work was funded under ARC Discovery Project 0557255.

References

[1] Wei Q, Kecskes L, Jiao T, Hartwig K T, Ramesh K T, Ma E, *Acta Mater.* 2004 **52** 1859
[2] Kusnierz J, Baliga W, Bogucka J 2004 Effect of pre-deformation by ECA pressing on shear banding & texture of cold rolled copper *Applied Crystallography: Proc. of XIX Conf.* (Krakow, Poland, 2003) ed H Moraweic, D Stroz (Singapore: World Sci. Pub. Co. Pte. Ltd.) pp. 181
[3] Akamatsu H, Fujinami T, Horita Z, Langdon T G, *Scripta Mater.* 2001 **44** 759
[4] Park K -T, Lee H -J, Lee C S, Shin D H, *Mater. Sci. Eng.* A 2005 **393** 118
[5] Stolyarov V V, Zhu Y T, Alexandrov I V, Lowe T C, Valiev R Z, *Mater. Sci. Eng.* A 2003 **343** 43
[6] Hazra S S, Gazder A A, Pereloma E V, *Mater. Sci. Eng.* A 2009 *doi: 10.1016/j.msea.2009.06.033*
[7] Humphreys F J, Hatherly M 1995 *Recrystallization and Related Annealing Phenomena* (Oxford: Pergamon)
[8] Toroghinejad M R, Ashrafizadeh F, Najafizadeh A, Humphreys A O, Liu D, Jonas J J, *Metall. Mater. Trans.* A 2003 **34**
[9] Leffers T, Pedersen O B, *Scripta Mater.* 2002 **46** 741
[10] Leffers T, Ray R K, *Prog. Mater. Sci.* 2009 **54** 351
[11] Meyers M A, Vöhringer O, Lubarda V A, *Acta Mater.* 2001 **49** 4025
[12] Huang C X, Wang K, Wu S D, Zhang Z F, Li G Y, Li S X, *Acta Mater.* 2006 **54** 655
[13] Miraglia M, Dawson P, Leffers T, *Acta Mater.* 2007 **55** 799
[14] Beyerlein I J, Tomé C N, *Int. J. Plast.* 2007 **23** 640

Effect of grain size distribution on mechanical properties of ultrafine grained Al severely deformed by ARB process and subsequently annealed

D Terada [1], H Houda [2] and N Tsuji [1]

[1] Department of Material Science and Engineering, Graduate School of Engineering, Kyoto University, Kyoto 606-8501, Japan
[2] Department of Adaptive Machine Systems, Graduate School of Engineering, Osaka University, Suita, Osaka 565-0871, Japan

daisuke.terada@ky7.ecs.kyoto-u.ac.jp

Abstract. A commercial purity aluminum was highly deformed by the accumulative roll-bonding (ARB) process and subsequently annealed. The specimens having various grain size distributions were obtained. In case of the specimen ARB-processed with lubrication, the specimens with mean grain size larger than 3μm showed continuous yielding. On the other hand, in case of the specimen ARB-processed without lubrication, the specimens with mean grain size larger than 3μm showed discontinuous yielding. It suggests that appearance of the yield-drop phenomena can not be decided by the mean grain size. In order to consider effect of grain size distribution, the volume fraction of grains was summed from coarser grains, and the grain size when the summed volume fraction reached 70%, $d_{70\%}$ was estimated from the grain size distribution. it was found that $d_{70\%}$ of specimens which showed continuous yielding were larger than 8 μm while the specimens which showed discontinuous yielding were smaller than 6 μm, regardless of the lubrication condition in the ARB process. The result suggests that the appearance of the yield-drop phenomena depend on $d_{70\%}$.

1. Introduction

Ultrafine grained (UFG) materials having the mean grain size smaller than 1μm can be fabricated by severe plastic deformation (SPD) processes [1-4]. The UFG materials generally exhibit high strength. Additionally, the UFG materials sometimes show particular mechanical properties. For example, yield-drop phenomena similar to that conventionally observed in mild steels can appear in tensile tests of an UFG-Al [5]. The yield-drop phenomena appeared in the specimen having the mean grain size smaller than several micrometers. It is suggested that appearance of the yield-drop phenomena depend on the mean grain size of specimens.

By the way, polycrystalline materials sometimes show wide grain size distribution. The microscopic deformation of the materials having the wide grain size distribution might be inhomogeneous. Thus, it is expected that mechanical properties can not be decided by not only the mean grain size but also grain size distribution. However, effect of the grain size distribution on mechanical properties has not been clarified yet. Therefore, in the present paper, relationship between grain size distribution and yield drop phenomena in an UFG-Al was investigated.

2. Experimental procedures

A commercial pure aluminium (99% purity) was used in this study. The material was severely deformed by ARB process. Surfaces of two sheets of the materials with thickness of 1mm were degreased by acetone and wire-brushed. And then the sheets were stacked to be 2mm thick and roll-bonded by 50% reduction in one pass at ambient temperature. The roll peripheral speed was 2.0m/min using a two-high mill with 310 mm diameter rolls. This procedure defines one cycle of the ARB process. The roll-bonded sheet was cut in half-length, stacked and roll-bonded again.

It is known that microstructure of ARB-processed materials depends on the lubrication condition of the roll-bonding. In order to obtain the specimen with various grain size distributions, the ARB process was carried out with and without lubrication. To obtain same grain size, a number of the ARB cycle under unlubricated condition should be smaller than that under lubricated condition, because grain refinement with lubrication is faster than that without unlubrication. Thus, the ARB process was repeated up to 10 cycles with lubrication and up to 6 cycles without lubrication. Subsequently, the ARB-processed specimens were annealed in air at various temperatures ranging from 150°C to 300°C for 1.8ks.

The mechanical properties of the specimens were characterized by tensile test. Tensile specimens 5mm in gage width and 10 mm in gage length (1/5 of JIS-5 specimen) were tested at an initial strain rate of 8.3 x 10^{-4} s^{-1}. The tensile direction was parallel to the rolling direction (RD) of the sheets.

Microstructures of the ARB processed specimens and heat-treated specimens were characterized by EBSP measurement. The observed section was perpendicular to the transverse direction (TD) of the sheet. The observed plane was electro-polished in a 300ml HNO_3 + 700ml CH_3OH solution at -30°C and voltage of 15V. The EBSP measurements were carried out using the FE-SEM at 15kV. The step size of the EBSP scan was 50nm.

3. Results and discussion

The microstructure of the specimens ARB-processed and subsequently annealed were characterized by the EBSP measurement. The specimen ARB-processed by 10 cycles with lubrication (lubricated specimen) exhibited the lamellar boundary structure having the mean lamellar spacing of 220nm elongated to the RD. After annealing, the lamellar spacing increased with increasing the annealing temperature. The specimen showed comparatively equiaxed grain structure after annealing above 200°C. After annealing above 250°C, the specimen exhibited inhomogeneous microstructure which was composed of fine grains with grain size of several micrometers and coarse grains with several tens micrometers.

The specimen ARB-processed by 6cycles without lubrication (unlubricated specimen) also showed the lamellar boundary structure. The mean lamellar spacing was 290nm. Microstructure change of the specimen during the annealing was similar to that of the lubricated specimens.

Figure 1 shows the stress-strain curves for the specimen ARB-processed and subsequently annealed ((a) the specimen ARB-processed by 10 cycles with lubrication and (b) the specimen ARB-processed

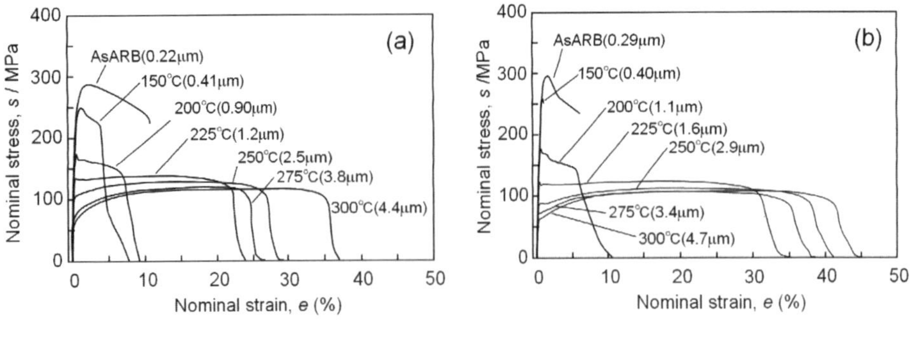

Figure 1. Results of the tensile test of the specimens ARB-processed and subsequently annealed. (a) The specimens ARB processed by 10 cycles with lubrication and (b) the specimens ARB-processed without lubrication

by 6 cycles without lubrication). The annealing temperatures and the mean grain thickness of each specimen are also shown in Fig. 1.

In the lubricated specimen (Fig.1(a)), the flow stress and total elongation decreased after annealing at 150°C. After annealing above 200°C, the total elongation increased and the flow stress decreased with increasing the annealing temperature. The specimens ARB-processed by 6 cycles without lubrication showed the same tendency of the specimen ARB-processed by 10 cycles with lubrication. Both specimens annealed at 200°C showed the yield-drop phenomena.

Figure 2 shows the stress-strain curves of (a) the lubricated specimens and (b) the unlubricated specimens annealed above 225°C in low tensile strain region. In Fig. 2(a), the lubricated specimens annealed at 225°C and 250°C show discontinuous yielding. The specimens annealed above 275°C show continuous yielding. On the other hand, in Fig. 2(b), the unlubricated specimens annealed between 225°C and 300°C show discontinuous yielding. In case of the lubricated specimen, the specimen having the mean grain size larger than 3μm show continuous yielding. However, in case of the unlubricated specimens, the specimens having the mean grain size larger 3μm show discontinuous yielding. The result suggests that the yielding behavior not only depend on the mean grain size.

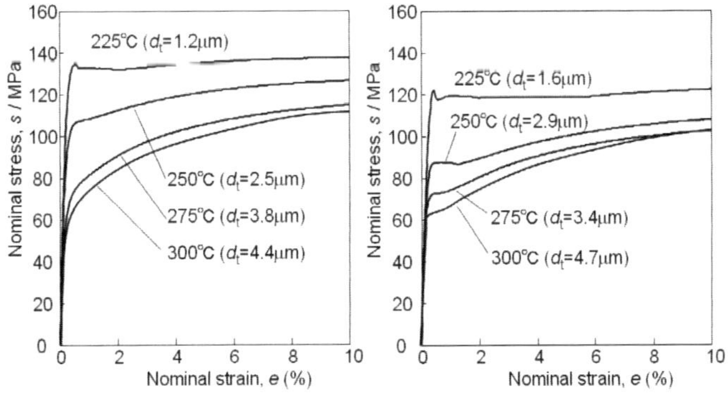

Figure 2. Stress-strain curves of (a) the lubricated specimens and (b) the unlubricated specimens annealed above 225°C in low tensile strain region.

Figure 3 shows the distribution of the grain size (in area) as semi-logarithmic graph for (a) the lubricated specimen annealed at 250°C, (b) lubricated specimen annealed at 275°C, (c) unlubricated specimen annealed at 275°C and (d) unlubricated specimen annealed at 300°C. In the figures, bar chats indicate the measured values obtained from the EBSP data and black lines indicated fitted lognormal distribution curves. The shapes of the grain size distribution for these specimens are similar. However, it seems that the fraction of grain size in area larger than 100 μm^2 of the lubricated specimen annealed at 275°C (Fig.3(b)) which showed continuous yielding is larger than that of other specimens (Fig.3 (a), (c) and (d)) which showed discontinuous yielding. It suggests that the yielding behavior depend on the grain size distribution.

By the way, when a polycrystalline material macroscopically yields, individual grains naturally deform. It is considered that individual grains microscopically yield before the material macroscopically yield. The yielding of the individual grains is called "micro-yielding". It is expected that the materials having the wide grain size distribution such as Fig.3 may deform non uniformly and micro-yielding occur from coarser grains to finer ones. Regarding the relationship between micro-yielding and macroscopic yielding, yielding behavior of an IF steels having various grain size distributions has been reported [6]. In the report, it was suggested that polycrystalline materials including duplex-grained structure materials cause the macroscopic yielding when the grains of 70%-80% vol% are micro-yielded.

Assuming the micro-yielding occur form coarser grains to finer grains, yielding behavior may dependents on the mechanical properties of the grains of 70-80 vol% from coarse-sized grain.

Thus, the volume fraction of grains was summed from coarser grains, and the grain size when the summed volume fraction reached 70%, $d_{70\%}$ was estimated from the grain size distribution as a grain size parameter including an effect of the grain size distribution. $d_{70\%}$ estimated from Fig.3(a), (b), (c) and (d) were 5.1μm, 8.1μm, 4.8μm and 5.5μm, respectively. $d_{70\%}$ of Fig.3(b) which showed continuous yielding is larger than that of other specimens (Fig.3 (a), (c) and (d)) which showed discontinuous yielding. $d_{70\%}$ of the other specimens were also estimated from their grain size distributions. From the results, it was found that $d_{70\%}$ of specimens which showed continuous yielding were larger than 8 μm while the specimens which showed discontinuous yielding were smaller than 6 μm, regardless of the lubrication condition in the ARB process. The result suggests that the appearance of the yield-drop phenomena depend on $d_{70\%}$..

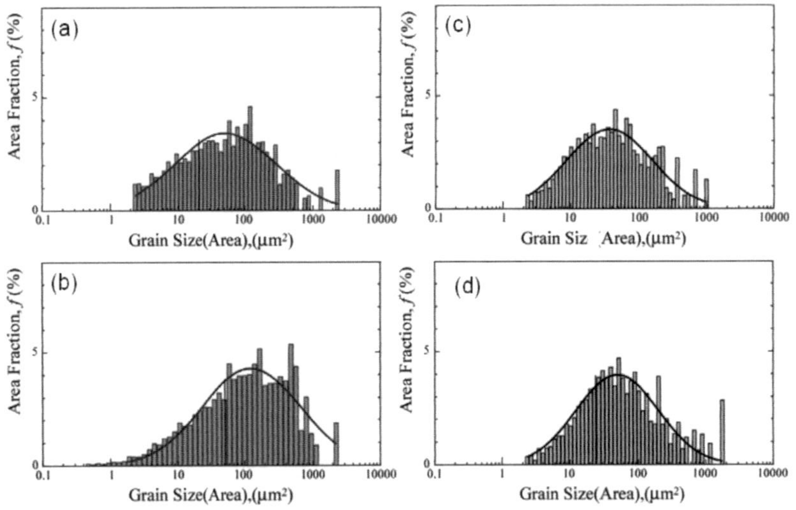

Figure 3. Distribution of the grain size (in area) as semi-logarithmic graph for (a) the lubricated specimen annealed at 250°C, (b) lubricated specimen annealed at 275°C, (c) unlubricated specimen annealed at 275°C and (d) unlubricated specimen annealed at 300°C. The bar chats show the measured value obtained from the EBSP data and black lines indicate fitted lognormal distribution curves.

4. Conclusions

In this paper, the mechanical properties of a commercial purity aluminum highly deformed by the accumulative roll-bonding (ARB) process and subsequently annealed were investigated. The main results are summarized below:

(1)From results of tensile tests, it was found that In case of the specimen ARB-processed by 10 cycles with lubrication, the specimen having the mean grain size larger than 3μm show continuous yielding. However, in case of the unlubricated specimens, the specimens having the mean grain size larger 3μm show discontinuous yielding. The result suggests that the appearance of the yield-drop phenomena can not be decided by the mean grain size.

(2) It was found that the grain size when the summed volume fraction from large grains reached 70%, $d_{70\%}$ of specimens which showed continuous yielding were larger than 8 μm while the specimens which showed discontinuous yielding were smaller than 6 μm, regardless of the lubrication condition in the ARB process. The result suggests that the appearance of the yield-drop phenomena depend on $d_{70\%}$.

References

[1] Langdon T G 2007 *Mater. Sci . Eng.* A **462** 3.
[2] Islamgaliev R K, Yunusova N F, Sabirov I N, Sergueeva A V, Valiev R Z 2001 *Mater. Sci. Eng.* A **319** 877.
[3] Tsuji N, Saito Y, Lee S H, Minamino Y 2003 *Adv. Eng. Mater.* **5** 338.
[4] Song R, Ponge D, Raabe D, Speer J G, Matlock D K 2006 *Mater. Sci. Eng.* A **441** 1.
[5] Tsuji N, Ito Y, Saito Y, Minamino Y, 2002 *Scripta Mater.* **47** 893.
[6] Matoba R, Nakada N, Futamura Y, Tsuchiyama T, Takaki S, 2007 *Tetsu-to-Hanané* **93** 513.

Journal of Physics: Conference Series **240** (2010) 012112 doi:10.1088/1742-6596/240/1/012112

Mechanical behaviour of heavily deformed CuAgZr conductor materials

J Freudenberger, J Lyubimova, A Gaganov, H Klauß, and L Schultz

IFW Dresden, Institute for Metallic Materials, P.O.-Box 270116, D-01171 Dresden, Germany

j.freudenbergerifw-dresden.de

Abstract. The mechanical properties of Cu−7m.%Ag−0.05m.%Zr-alloys have been improved by optimised thermo-mechanical treatments. After processing the conductor material shows an ultimate tensile strength of more than 1.1 GPa, a yield strength of about 1 GPa, a plastic strain of 0.7%, and a conductivity of 60 %IACS (at room temperature). The yield strength of a cold deformed Cu − 7m.%Ag − 0.05m.%Zr-alloy has been assessed on the basis of different hardening mechanisms: solid solution, grain boundary, precipitation and dislocation hardening. The critical shear stress which is determined from the experimentally observed critical shear strength by Schmid's law is between a linear and a quadratic superposition of its individual contributions and hence is in good agreement with the theoretical prediction.

1. Introduction

CuAg-alloys have attracted interest because of their advantageous combination of excellent mechanical strength and relatively high electrical conductivity [1, 2]. The properties of CuAg-alloys originate from their microstructure, which arises from the chemical composition as well as from thermo-mechanical treatments [3, 4]. The microstructure of the alloys is mostly determined by the silver content, which determines the predominantly observed precipitation reaction. The microstructure of a homogenised Cu-7 m.%Ag alloy (mass percent) consists of a Cu-rich solid solution and discontinuous Ag-rich precipitates next to large angle grain boundaries. Adding 0.05 m.%Zr to a Cu-7m.-%Ag alloy suppresses the discontinuous precipitation mode in these alloys and continuous precipitates which are favourable in the sense of maximum strength are observed [5].

2. Experimental Details

Polycrystalline samples of Cu − 7m.%Ag − 0.05m.%Zr (CuAgZr) were molten in an inductance furnace and cast into a graphite mould (D = 32 mm). The CuAgZr alloy has then been homogenised at 850°C for 5 h and subsequently water-quenched. Precipitates were formed during a further heat treatment at 400°C / 4 h. For details please refer to [5].

Vickers hardness HV0.5 (load time 10 s) values were obtained from a minimum of 15 indentations. The ultimate tensile strength σ_{UTS} and yield strength σ_{yp} were measured at room temperature on drawn wire pieces of a total length of 150 mm using an electro-mechanical Instron 8562 testing machine at a displacement rate of 0.2 mm/min.

3. Results and Discussion

The alloys show a reasonable strength after cold deformation up to a logarithmic drawing strain of $\eta = \ln(A_0/A) = 5$ (A_0 and A are the initial and final cross sectional areas). However, there is a demand to enhance the elongation from about 1.4% to 5%. This originates from the design of pulsed high field magnets [6, 7]. Finite Element Analysis simulations show that for achieving the highest possible stress in both, conductor and reinforcement, the conductor needs to be able to tolerate this mentioned strain [7]. A simple way of enhancing the strain in a metallic material is to heat treat the samples [8]. This, however, will at the same time cause a loss in strength. The effect of the temperature of annealing for 2 h on the strength and strain is shown in Fig. 1.

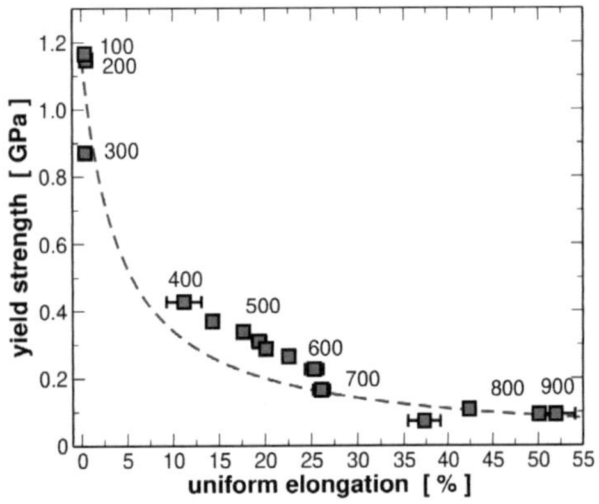

Figure 1. Yield strength of CuAgZr alloys in dependence on the uniform elongation. The values are taken after a heat treatment for 2 h at the depicted temperatures (next to symbols).

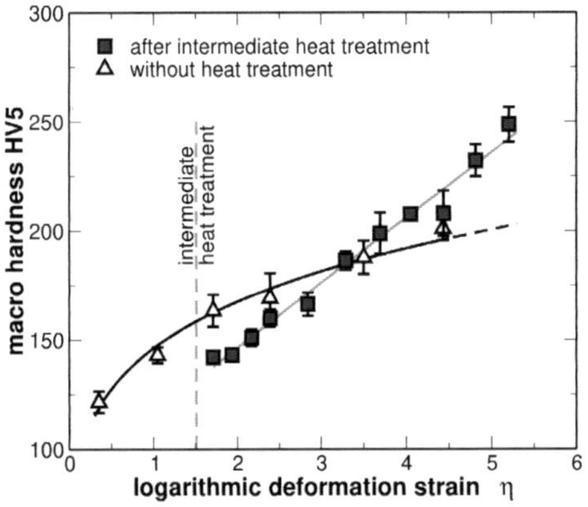

Figure 2. Hardness HV5 in dependence on the logarithmic deformation strain. One set of samples were heat treated at 300°C for 2 h after a cold deformation strain of $\eta = 1.5$ was reached

Figure 1 shows that a temperature of 300°C causes a drop in strength to 800 MPa while the uniform elongation increases by 0.5%. Thus, this treatment must not be applied on the final drawn sample. We have shown earlier [5] that intermediate heat treatments are beneficial for a further enhancement of the strength. As shown in Fig. 2, a heat treatment (300°C / 2 h) of a cold drawn ($\eta = 1.5$) sample causes a drop in hardness. Further cold deformation causes a larger increase of the hardness when compared to the sample without intermediate heat treatment [9]. Thus, at a later stage of deformation, the loss in hardness, that is caused by the thermal treatment, is balanced. For even higher deformation an enhanced hardness is observed when compared to the condition without thermal treatment. The final deformed wire is a compromise of either further enhancing the strength up to 1.4 GPa while keeping the strain to failure at a comparable level or to enhance the strain to failure by 0.5% while keeping σ_{UTS} at 1.2 GPa. Figure 3 summarises the optimisation of the strength in CuAg-alloys. For a Cu-7m.%Ag alloy σ_{UTS} increases with cold deformation as shown. Adding 0.05m.% Zr to this alloy enhances the overall strength level due to the suppression of the discontinuous precipitation reaction [5]. An intermediate heat treatment which has been applied at $\eta = 4$ allows further deformation yielding to $\sigma_{UTS} = 1.4$ GPa.

The final microstructure of CuAgZr following casting, solutionising and precipitation heat treatment bears four different hardening mechanisms: solid solution hardening, precipitation hardening, phase boundary hardening and work hardening. These individual contributions to the total strength are assessed in the following on the basis of general formulas [10]. The increase of the critical yield stress for solid solution strengthening $\Delta\tau_{c,ss}$ is calculated by:

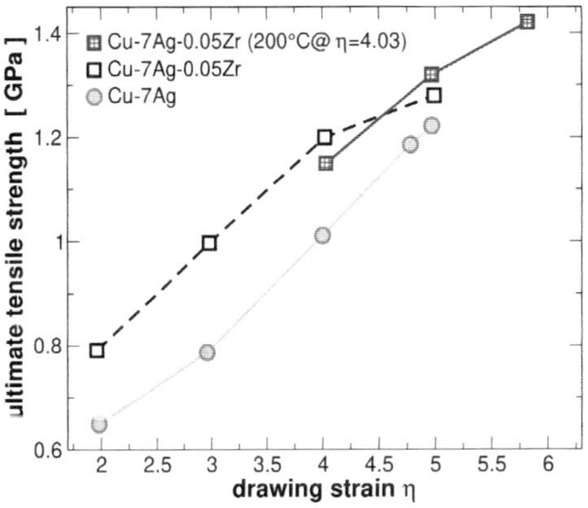

Figure 3. σ_{UTS} in dependence of the drawing strain for Cu-7m.%Ag and CuAgZr. In addition: CuAgZr being intermediately heat treated.

Table 1. Summary of the individual strength contributions (solid solution $\Delta\tau_{\mathrm{c,ss}}$, grain boundaries $\Delta\tau_{\mathrm{c,HP}}$, precipitates $\Delta\tau_{\mathrm{c,precip.}}$, dislocations $\Delta\tau_{\mathrm{c,disloc.}}$) to the total yield strength and their sum according to $\tau_{\mathrm{c,total}} = \Delta\tau_{\mathrm{c}} + \tau_{\mathrm{c,0}}$ and Eq. 5.

$\tau_{\mathrm{c,0}}$	40 MPa
$\Delta\tau_{\mathrm{c,ss}}$	116 MPa
$\Delta\tau_{\mathrm{c,HP}}$	31 MPa
$\Delta\tau_{\mathrm{c,precip.}}$	188 MPa
$\Delta\tau_{\mathrm{c,disloc.}}$	71 MPa
$\tau_{\mathrm{c,total}}$ $(k=1)$	446 MPa
$\tau_{\mathrm{c,total}}$ $(k=2)$	274 MPa

$$\Delta\tau_{\mathrm{c,ss}} = G(|\delta| + \frac{1}{20}|\eta|)^{\frac{3}{2}}\sqrt{\frac{x_{\mathrm{a}}}{3}} = 116\ \mathrm{MPa} \tag{1}$$

In Eq. 1, $\delta = \frac{\partial \ln a}{\partial x} = 0.1302$ is a factor of lattice change and $\eta = \frac{\partial \ln G}{\partial x} = 0.3485$ is a factor describing the change of the shear modulus due to alloying, $x_{\mathrm{a}} = 0.59$ at.% is the Ag fraction within the solid solution. Lattice parameters of $a_{\mathrm{Cu}} = 0.3620$ nm and $a_0 = 0.3618$ nm (solid solution) and shear moduli of $G_{\mathrm{Cu}} = 46$ GPa and $G_0 = 45.91$ GPa have been used for modelling The increase of the yield strength $\Delta\tau_{\mathrm{c,HP}}$ due to the grain refinement is determined by:

$$\Delta\tau_{\mathrm{c,HP}} = k_{\mathrm{HP}}/\sqrt{d} = 31\ \mathrm{MPa} \tag{2}$$

In Eq. 2, $d = 24.6$ μm is the mean grain size and $k_{\mathrm{HP}} = 0.15$ MPa$\sqrt{\mathrm{m}}$ [10, 11], the Hall-Petch constant. The precipitates are aligned along the drawing axis. Shearing of precipitates becomes possible as the slip plane of the dislocations is oriented by an angle of 53° to the long axis of the precipitates. In this case the increase of the strength $\Delta\tau_{\mathrm{c,precip.}}$ is given by:

$$\Delta\tau_{\mathrm{c,precip.}} = \frac{\gamma^{3/2}}{b^2}\sqrt{\frac{rf}{3G}} = 188\ \mathrm{MPa} \tag{3}$$

In Eq. 3, $\gamma = 2.2449$ J/m^2 [12] is the phase boundary energy, $b = 0.2556$ nm the Burgers vector of the solid solution, $r = 15$ nm the radius and $f = 0.07$ the volume fraction of the precipitates. The total number of dislocations can be drastically increased by cold working. The CuAgZr alloys have been cold worked up to a logarithmic deformation strain of $\eta = \ln(A_0/A) = 4.07$. The increase of strength due to work hardening as given by:

$$\Delta\tau_{\mathrm{c,disloc.}} = \alpha G b\sqrt{\rho} = 71\ \mathrm{MPa} \tag{4}$$

In Eq. 4, $\alpha = 0.6$ is a factor according to the geometry and $\rho = 1 \cdot 10^{14}$ m^{-2} is the dislocation density, which has been obtained from TEM analysis and the linear intercept length method [13]. The total critical stress of the CuAgZr alloy is calculated by a superposition of the previously discussed individual contributions. However, there is no generally accepted formula available to assess this superposition. In consequence, an empirical equation has often been used [14, 15]:

$$\Delta\tau_{\mathrm{c}}^{k} = \Delta\tau_{\mathrm{c,ss}}^{k} + \Delta\tau_{\mathrm{c,HP}}^{k} + \Delta\tau_{\mathrm{c,precip.}}^{k} + \Delta\tau_{\mathrm{c,disloc.}}^{k} \quad (5)$$

The exponent k in Eq. 5 is typically between 1 and 2, depending if the individual strengthening mechanisms interact. The critical shear stresses are summarised in Tab. 1 for both k values.

Now we make a Gedankenexperiment: Assuming a linear superposition ($k = 1$) of the individual strengthening mechanisms leads to a critical shear stress of $\tau_{\mathrm{c,total}} = 446$ MPa. For polycrystalline materials the Taylor factor M is used to capture the orientation dependence of the strength from numerous sources of variability. It can be between 2 and 4. The critical shear stress and the experimentally measured shear strength $\sigma_{\mathrm{yp}} = 1064$ MPa match for a Taylor factor of $M = 2.38$. Here we assume the strengthening mechanisms with different strength and interaction distance [15] and thus no interaction. In a second step, we assume a quadratic superposition ($k = 2$) and calculate a critical shear stress of $\tau_{\mathrm{c,total}} = 274$ MPa, which matches to the critical shear strength if a Taylor factor of $M = 3.88$ is assumed. In this case the individual strengthening mechanisms should have a similar strength or have similar obstacle distances, which is not the case. In addition this Taylor factor deviates somewhat from the usual value of 3.06 for polycrystalline materials. However, this can be rationalised by the strong $< 1\,1\,1 >$ fibre texture in the material. This will increase the Taylor factor to about $M = 3.7$. (assuming M being the inverse Schmid factor calculated for a $< 1\,1\,1 >$ oriented fcc single crystal).

The microstructure of the alloys provide reasonable arguments for $k = 1$ as the obstacle distances range from about 2 nm (average spacing between the solute atoms in the case of solid solution hardening) to about 100 nm (distance between dislocations). In contrast to this, a k-value of 2 allows the explanation including texture and an interaction between the strengthening mechanisms. Equation 5 matches for a k-value of 1.76, providing a further argument for the later case. Although there is no physical explanation for a nearly quadratic superposition of the individual contributions to the total shear stress we find an accurate prediction of the actual measured yield strength of the CuAgZr material.

4. Summary

The mechanical properties of CuAgZr have been improved by optimised thermo-mechanical treatments. The optimised material bears a compromise of enhanced strength (plus 200 MPa) or strain (plus 0.5%) while keeping the other properties at a similar level. The yield stress of a cold drawn CuAgZr fits best to a quadratic superposition of the individual shear stresses that arise from the microstructure of the alloy.

This work was funded by the *Deutsche Forschungsgemeinschaft (DFG)*, FR 1714/2 and FR 1714/5. Furthermore, we thank D. Seifert, T. Wolf and M. Frey for experimental support.

References

[1] Heringhaus F, Schneider-Muntau HJ and Gottstein G 2003 *Mater. Sci. Eng.* A **347** 9
[2] Freudenberger J, Grünberger W, Botcharova E, Gaganov A and Schultz L 2002 *Adv. Eng. Mater.* **9** 677
[3] Hong SI and Hill MA 1999 *Mater. Sci. Eng.* A **264** 151
[4] Gaganov A, Freudenberger J, Grünberger W and Schultz L 2004 *Z. Metallk.* **95** 425
[5] Gaganov A, Freudenberger J, Bocharova E and Schultz L 2006 *Mater. Sci. Eng.* A **437** 313
[6] Herrmannsdoerfer T et al. 2003 *J. Low Temp. Phys* **133** 41
[7] Witte H, Magnet design using finite element analysis, Ph.D. thesis, Univ. of Oxford, Dept. of Physics 2007
[8] Lyubimova J, Freudenberger J, Gaganov A, Klauß and Schultz L 2010 *Mater. Sci. Forum* **633** 707
[9] Lyubimova J et al. 2009 *Mater. Sci. Eng.* A DOI: 10.1016/j.msea.2009.08.036
[10] Gottstein G, Physikalische Grundlagen der Materialkunde 2007 Springer
[11] Kozlov E, Zhdanov A, Popova N, Pekarskaya E and Koneva N 2004 *Mater. Sci. Eng.* A **387** 789
[12] Choi KS, Choi Y and Inoue H 2007 *Mater. Sci. Forum* **558** 1295
[13] Heimendahl M, Einführung in die Elektronenmikroskopie 1970 Vieweg
[14] Nembach E and Neite G 1985 *Progress Mater. Sci.* **29** 177
[15] Kocks U, Argon A and Ashby M 1975 *Progress Mater. Sci.* **19** 1

Influence of strain amount on stabilization of ω-phase in pure Ti by severe plastic deformation under high-pressure torsion

Y Todaka [1], H Azuma [1], Y Ohnishi [1], H Suzuki [2] and M Umemoto [1]

[1] Department of Production Systems Engineering, Toyohashi University of Technology, Hibarigaoka 1-1, Tempaku, Toyohashi, Aichi 441-8580, Japan
[2] Quantum Beam Science Directorate, Japan Atomic Energy Agency, Shirakata-Shirane 2-4, Tokai, Naka, Ibaraki 319-1195, Japan

todaka@martens.pse.tut.ac.jp

Abstract. In pure Ti, the influence of shear deformation on the α to ω transformation and the development of texture in the ω-phase under high-pressure torsion (HPT) straining were investigated by means of X-ray and neutron diffractions. The fraction of ω-phase increased with strain in the ω-phase state. Bulk submicrocrystalline ω-Ti was fabricated by HPT-straining under the compressive pressure $P = 5$ GPa with the equivalent strain $\varepsilon_{eq} > 110$ at the rotation speed of 3.3×10^{-3} rev. per sec. (0.2 rev. per min.) at room temperature. The texture of ω-phase evolved by HPT-straining with the prismatic planes parallel to the shear direction of HPT-straining and the basal planes perpendicular to it.

1. Introduction

Severe plastic deformation (SPD) techniques have been applied to the design of bulk ultrafine-grained materials with improved properties, because intense strain can be introduced and high-density lattice defects created in the SPD-processed samples [1]. Among the SPD techniques, high-pressure torsion (HPT) process has a unique characteristic: deformation under high pressure. The principle of the HPT process is based on the procedure developed by Bridgman [2]. A thin disk is placed between two anvils and subjected to torsional straining under high pressure. Thus, the two parameters, *i.e.* the number of turns applied to the disk, N, (an amount of shear strain, γ) and the magnitude of the imposed pressure, P, are important in the HPT process.

The pressure-induced transformations of Ti, Zr, Hf were first observed by Jamieson [3]. It is well-known that pure Ti stabilizes in the hexagonal close-packed α-phase at ambient condition, and under high pressure (≥ 2 GPa) the α-phase transforms into the simple hexagonal ω-phase [4]. Recently, in pure Ti [5, 6] and pure Zr [7] after HPT-straining, the high-pressure ω-phases of these materials were observed at ambient condition. Todaka *et al.* reported that the stabilization of ω-phase at ambient condition occurs *via* the application of shear deformation in the ω-phase state [6].

In this work, to investigate the influence of shear deformation on the α to ω transformation in pure Ti, HPT-straining was applied. The texture evolvement of ω-phase during HPT-straining was also examined by means of X-ray and neutron diffractions.

2. Experimental Procedures

The experiments were conducted using pure Ti (Ti- 0.045O- 0.025Fe- 0.01N- 0.01C- 0.004H in mass %). This material was annealed at 1073 K for 3.6 ks in a pure Ar atmosphere. The annealed Ti was cut into disks of 10 and 20 mm in diameter × 0.85 mm thick for HPT-straining. The HPT process was carried out using anvils with a depression of 0.25 mm in depth and 10 or 20 mm in diameter. The disk was held between the two anvils and was torsion-strained under P = 5 GPa up to N = 10 turns at a rotation speed of 3.3 × 10^{-3} rev. per sec. (0.2 rev. per min.) at room temperature. The shear strain, γ, for HPT-straining is given by the following equations: $\gamma = 2\pi Nr/t$, where r is distance from the disk center and t is the disk thickness; and also the von Mises equivalent strain, ε_{eq}, given by $\varepsilon_{eq} = \gamma/\sqrt{3}$. The disk flowed to the radial direction during HPT-straining due to the lack of side constraint, resulting in the reduction of thickness to 0.6 mm (10 mm diameter disk) and 0.65 mm (20 mm diameter disk). The increment of temperature at the region of anvil around 1 mm (10 mm diameter anvil) away from the surface of disk center was around 5 K during deformation. The HPT process has been described in detail elsewhere [8,9].

The HPT-processed disks were characterized at ambient condition by X-ray diffraction (XRD; Rigaku RINT-2500, Cu K_α, 40 kV, 250 mA, step size 0.02 °, collection time 1 s), neutron diffraction (ND; RESA in the Japan Research Reactor No.3 (JRR-3) at the Japan Atomic Energy Agency, wavelength 0.165 nm, slit size 2 (width) × 15 (height) mm^2) and transmission electron microscopy (TEM; Hitachi H-800, 200 kV). The 10 mm diameter disks for XRD with the center position of r = 5 mm in the HPT-processed 20 mm diameter disks were prepared. The specimens for ND were prepared by stacking the HPT-processed 20 mm diameter disks to 5 mm thick (Figure 1 (a)). The ND was carried out from two different directions: radial and hoop directions (Figure 1 (b), (c)). Diffraction planes in the radial- and hoop- direction measurements were parallel and perpendicular to the shear direction in the HPT-straining, respectively. The TEM observation was performed from the direction parallel to the torsion axis.

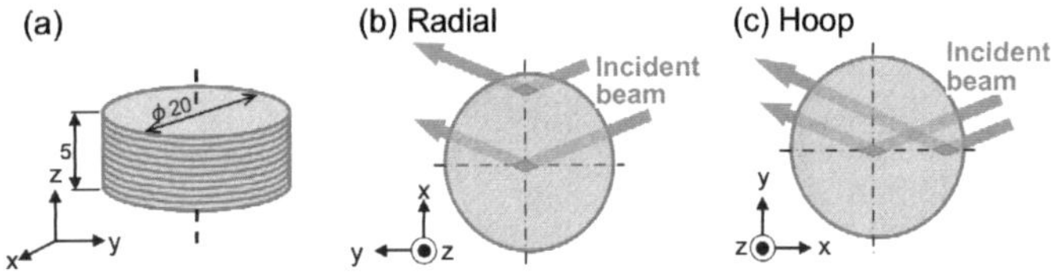

Figure 1. (a) Specimen shape for ND (unit: mm). ND was performed from two different directions: (b) radial and (c) hoop directions.

3. Results and discussion

Figure 2 (a) shows the XRD patterns of the samples after HPT-straining for varying turns. In the sample compressed at P = 5 GPa for 3 ks in the HPT anvils without torsional deformation, no peaks of ω-phase are seen. However, the peak intensity ratio of α-phase changed by compression in the comparison with that of the initial annealed (non-deformed) sample. This change might be caused by reverse-transformation from ω to α-phase after pressure-release and / or by compressive deformation. The increase in the peak intensity of ω-phase and the simultaneous decrease in that of α-phase are observed with increasing numbers of turns, indicating that the α to ω transformation occurred by shear deformation. The volume fractions of ω-phase estimated from the XRD profiles of Figure 2 (a) by using the Rietveld program Quanto [10] were plotted in Figure 2 (b) as a function of ε_{eq} calculated using the following values: t = 0.65 mm and r = 5 mm (the center of XRD-measured disks). The fraction of ω-phase increased rapidly with ε_{eq}, and level off at around 90 vol % with $\varepsilon_{eq} > 100$. This result is consistent with that of ND measurement. The ND profiles at the various radial positions of the sample after HPT-straining for N = 5 are shown in Figure 3. The both peaks of α- and ω- phases were observed near the disk center. However, the peak intensity ratio of α- and ω- phases, I_α/I_ω, decreased with increase in ε_{eq}, and the $(101)_\alpha$ peak disappeared with $\varepsilon_{eq} > 110$ ($r \geq 4$ mm). Figure 4 shows the ND profiles of several diffraction planes at the region of $r = 8$ mm ($\varepsilon_{eq} = 223$) away from the disk center after HPT-straining for $N = 5$. In the radial direction measurements, the $(002)_\omega$ peak of basal plane was not observed, meanwhile the peaks diffracted by the non-basal planes of $(211)_\omega$, $(300)_\omega$ and $(220)_\omega$ could be seen. As well, in the hoop ones, the $(002)_\omega$ peak appeared, and also the peak of the $(112)_\omega$ and/or $(301)_\omega$ pyramidal planes was seen. These ND experiments indicate that the prismatic planes are aligned parallel to shear direction and the basal planes are done perpendicular to it during HPT-straining.

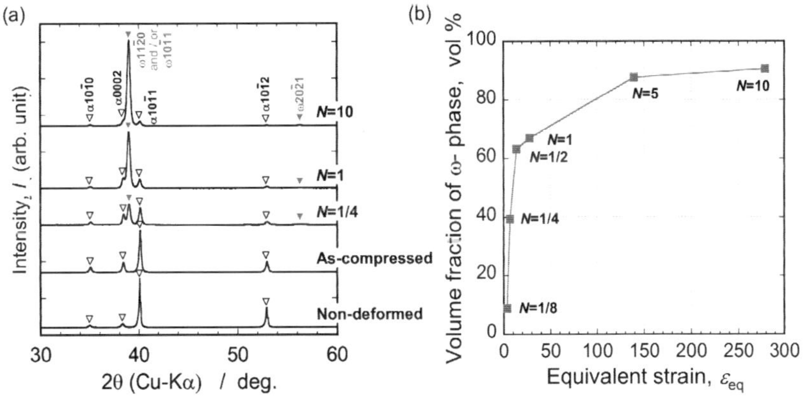

Figure 2. (a) XRD patterns of the samples after HPT-straining for varying turns. (b) Volume fractions of ω-phase estimated from the XRD profiles of (a) as a function of ε_{eq}.

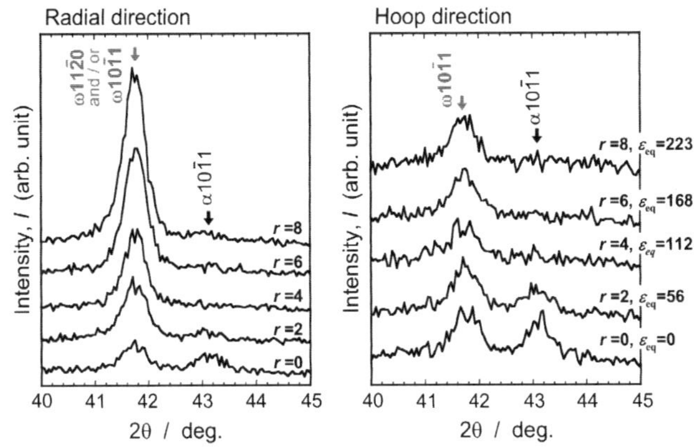

Figure 3. ND profiles at the various radial regions in the HPT-processed sample for $N = 5$. (These profiles are not correct by the length of beam path.)

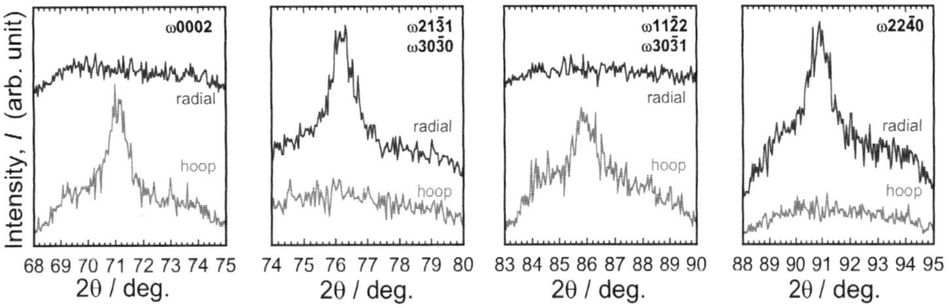

Figure 4. ND profiles at the region of $r = 8$ mm ($\varepsilon_{eq} = 223$) away from the disk center in the HPT-processed sample for $N = 5$.

Figure 5 shows the microstructure at the region of r = 3.2 mm (ε_{eq} = 97) away from the disk center after HPT-straining (N = 5, 10 mm diameter disk). In the dark-field (DF) image, the strongly reflecting regions correspond to the ring of $(110)_\omega$ and/or $(100)_\omega$ marked in the selected-area electron diffraction (SAD) pattern taken from an area 1.2 μm in diameter in the bright-field (BF) image. Equiaxed grains of a few hundred nanometers in diameter with high dislocation density can be seen. The SAD spots almost corresponded to those of ω-phase, which is consistent with the results of X-ray and neutron diffractions. The arrangement of these spots in semicontinuous circles demonstrates that this submicrocrystalline structure contains large misorientations. This is consistent with the DF image where bright regions in grains are non-uniform due to the significant distortions.

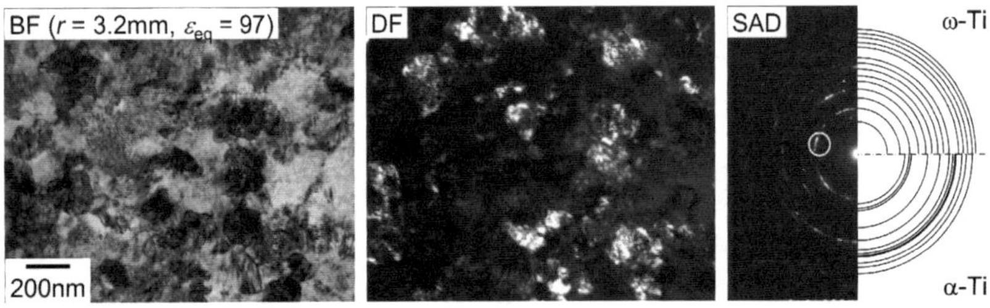

Figure 5. Microstructure at the region of r = 3.2 mm (ε_{eq} = 97) away from the disk center in the HPT-processed 10 mm diameter disk for N = 5.

4. Conclusions

Bulk ω-Ti with grain size of a few hundred nanometers was fabricated by HPT-straining under P = 5 GPa with $\varepsilon_{eq} > 110$ at 3.3×10^{-3} rev. per sec. (0.2 rev. per min.) at room temperature. The fraction of ω-phase increased with strain in the ω-phase state. The prismatic planes of ω-phase were aligned parallel to shear direction during HPT-straining, and also the basal planes were done perpendicular to it.

Acknowledgements

This study was financially supported in part by the Grant-in-Aid for Scientific Research from the MEXT of Japan and the grant for young researchers project of Research Center for Future Technology, Toyohashi University of Technology. The materials for HPT disks (pure Ti) and anvils (DRM2 steel) were provided by Kobe Steel and Daido Steel Co., Ltd. All such support is gratefully appreciated by the authors.

References

[1] Valiev R Z Islamgaliev R K and Alexandrov I V 2000 *Prog. Mater. Sci.* **45** 103
[2] Bridgman P W 1952 *Studies in Large Plastic Flow and Fracture* (New York: McGraw-Hill).
[3] Jamieson J C 1963 *Science* **140** 72
[4] Sikka S K Vohra Y K and Chidambaram R 1982 *Prog. Mater. Sci.* **27** 245
[5] Kilmametov A R Khristoforov A V Wilde G and Valiev R Z 2007 *Z. Kristallogr. Suppl.* **26** 339
[6] Todaka Y Sasaki J Moto T and Umemoto M 2008 *Scripta Mater.* **59** 615
[7] Perez-Prado M T Gimazov A A Ruano O A Kassner M E and Zhilyaev A P 2008 *Scripta Mater.* **58** 219
[8] Todaka Y Umemoto M Yamazaki A Sasaki J and Tsuchiya K 2008 *Mater. Trans.* **49** 7
[9] Todaka Y Umemoto M Yamazaki A Sasaki J and Tsuchiya K 2008 *Mater. Trans.* **49** 47
[10] Altomare A Burla M C Giacovazzo C Guagliardi A Moliterni A G G Polidori G and Rizzi R 2001 *J. Appl. Crystallogr.* **34** 392

Strain hardening and softening in ultrafine grained Al fabricated by ARB process

Genki Horii[1], Daisuke Terada[2] and Nobuhiro Tsuji[2]

[1]Department of Adaptive Machine Systems, Graduate School of Engineering, Osaka University, 2-1 Yamadaoka, Suita, Osaka, 565-0871, JAPAN
[2]Department of Materials Science and Engineering, Graduate School of Engineering, Kyoto University, Yoshida Honmachi, Sakyo-ku, Kyoto, 606-8501, JAPAN

genki.HORII@ams.eng.osaka-u.ac.jp

Abstract. Deformation behavior of a commercial purity aluminium (99% purity) highly deformed by the accumulative roll-bonding (ARB) process was investigated by analyzing the change in the specimen shape during tensile test precisely. True stress obtained by the specimen shape analysis of the conventionally coarse grained material was found to increase continuously even after necking, which indicated that the necking region was work-hardened. On the other hand, true stress of the ARB-processed Al which showed lamellar boundary structure having the mean lamellar spacing of 220 nm decreased after necking, which was quite different from the result of the coarse grained material. This result suggests that the necking region is not hardened but softened by deformation. The softening is considered to be caused by dynamic recovery at grain boundaries.

1. Introduction
Ultrafine grained (UFG) materials fabricated by severe plastic deformation (SPD) processes [1-3] are known to exhibit excellent mechanical properties and unusual deformation behavior [1, 4, 5]. For example, UFG metals fabricated by ARB process have the elongated lamellar UFG structure and show high strength, which is 2-4 times higher than that of coarse grained metals [1]. However, their uniform elongation is often low due to early plastic instability in the tensile test. In contrast, the post-uniform elongation is still large, which is one of the characteristic features of the UFG materials. In the present paper, the post-uniform deformation of an UFG-Al during the tensile test is discussed in detail by analyzing the change in the specimen shapes during tensile test [6].

2. Experimental Procedures
Fully annealed sheets of a commercial purity Al (AA1100) was used as the starting material in this study. The dimensions of the starting sheets were 1 mm in thickness, 55 mm in width and 220 mm in length. Two starting sheets were stacked after degreasing by acetone and wire-brushing the contact surfaces and then roll-bonded by 50% reduction in one pass without lubrication at ambient temperature. The roll-bonded sheet was immediately water-cooled after rolling. The rolled sheet was cut into half-length, trimmed and subjected to the next ARB cycle. This procedure was repeated up to 6 cycles (equivalent strain: ε_{eq} = 4.8) without changing the rolling direction (RD). Electron

backscattering diffraction pattern (EBSD) measurement in a scanning electron microscope equipped with a field emission-type gun (FE-SEM) was conducted for the starting material and the ARB-processed specimen. Observed sections for the EBSP measurement were perpendicular to the transverse direction (TD) of the sheets. The observed plane was electro-polished in a 30 vol% HNO_3 + 70 vol% CH_3OH solution at 243 K and voltage of 20 V. Tensile specimens with 10 mm in gage length and 5 mm in width were cut off from the starting material and the ARB-processed sheet. The tensile direction was parallel to RD. The images of the specimen were recorded continuously during the tensile test from normal direction and transverse direction of the sheet specimens by two digital video cameras. The schematic illustration of the hardware configuration for the specimen shape analysis system is shown in Fig. 1. The shape of the specimen was determined quantitatively by detecting the edge of the specimen in the images using software programmed by ourselves.

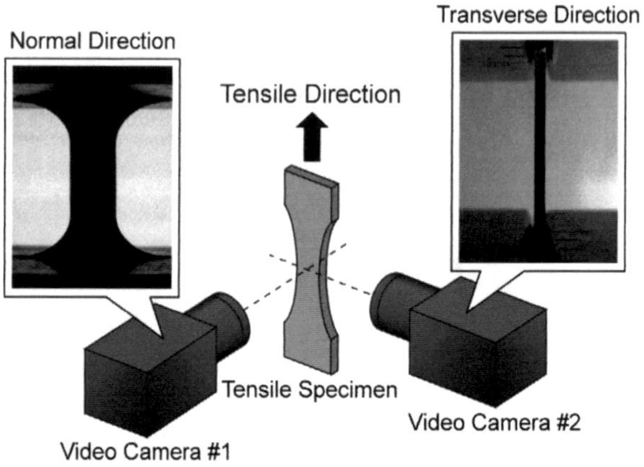

Figure 1 Schematic illustration showing the hardware configuration for the specimen shape analysis.

3. Results and Discussion

3.1. Microstructure

Figure 2 shows the grain boundary maps obtained by EBSD measurement of the starting material and the specimen ARB-processed by 6 cycles. In these maps, high angle grain boundaries (HAGBs) with misorientation angle above 15° are shown as black lines, while low angle grain boundaries (LAGBs) with misorientation angle between 2° and 15° are shown as gray lines.

The starting material shows the equiaxed grain structure having the mean grain size of 24 μm. On the other hand, the ARB-processed specimen showed the lamellar boundary structure elongated along RD. The mean spacing of all lamellar boundaries including low-angle ones were 220 nm.

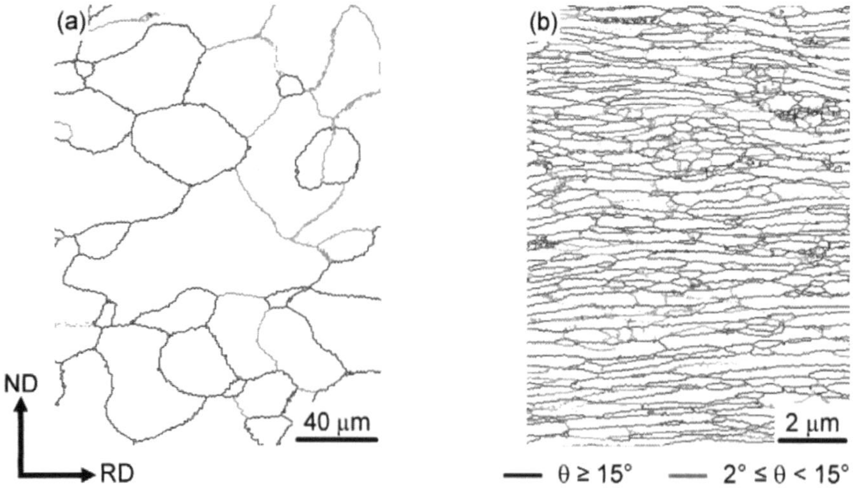

Figure 2 Grain boundary maps obtained by EBSD measurement of (a) the starting material and (b) the ARB-processed specimen. Black lines indicate boundaries with misorientation of $\theta \geq 15°$, while gray lines indicate boundaries with misorientation of $2° \leq \theta < 15°$.

3.2. Mechanical properties

Figure 3 shows the stress-strain curves of the starting material and the ARB-processed specimen. In this figure, gray curves indicate nominal stress while black curves indicate true stress calculated from the minimum cross-sectional area obtained from the specimen shape analysis. The starting material exhibits typical deformation behavior of the coarse grained metals. The ultimate tensile stress (UTS) was 84 MPa at the engineering strain of 33%. The total elongation was 49%. The strain at which necking in the width direction started was estimated as 31% by the specimen shape analysis, which approximately corresponded to the strain of the UTS. Necking in the thickness direction started at 39%. Then, the nominal stress decreased rapidly. True stress calculated from the minimum cross-sectional area was larger than the nominal stress and increased continuously even after necking. This result indicates that the necking region is continuously work-hardened during post-uniform deformation.

The ARB-processed specimen showed peak stress at early stage in the tensile test (Fig. 3 (b)). The UTS of the ARB-processed sample was 290 MPa at the strain of 4.5% and the total elongation was 11%. The strain at which necking in the width direction started was 4.9%, which almost coincided with the strain of the UTS. Necking in the thickness direction started at 6.5%, at which the slope of the nominal stress changed. True stress obtained by the specimen shape analysis decreased after necking in the width direction and increased after necking in the thickness direction, which was quite different from the result of the starting material. This result suggests that the necking region is not hardened but softened by deformation after the peak engineering stress. The softening might be caused by dynamic recovery. It is well known that the pure Al generally undergo dynamic recovery at ambient temperature because of the enhanced cross-slip of screw-dislocations due to its high stacking fault energy. In addition, the UFG materials have a large number of grain boundaries. It is likely that grain boundaries act as a sink for dislocations. This would help annihilation of dislocations and result in enhanced recovery. On the other hand, strain hardening after necking in the thickness direction is thought to be caused by high strain-rate sensitivity of the ARB-processed Al. Significant strain-rate dependence of the flow stress in the UFG materials has been reported [7-10]. In tensile test, strain-rate increases after necking because the deformation proceeds only within the necking region. Thus, the stress might increase when hardening caused by the increase in strain-rate is larger than the softening caused by the recovery in the necking region.

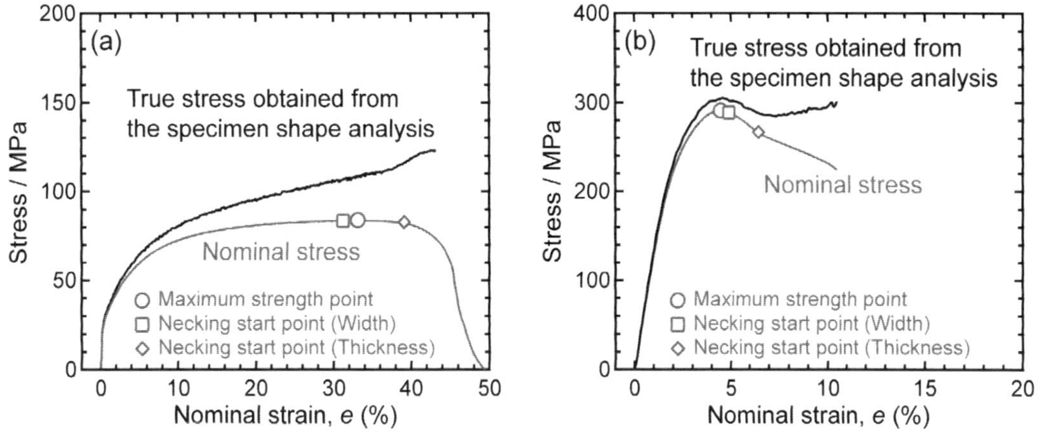

Figure 3 Stress-strain curves of (a) the starting material and (b) the ARB-processes specimen. Gray curves indicate nominal stress while black curves indicate true stress calculated from the minimum cross-sectional area obtained from the specimen shape analysis.

4. Conclusions

The deformation behavior of a commercial purity Al highly deformed by the ARB process was investigated by analyzing the change in the specimen shapes precisely. The major results obtained are summarized below:

(1) Commercial purity Al ARB-processed by 6 cycles showed the ultrafine lamellar boundary structure having mean lamellar spacing of 220 nm.

(2) Specimen shape analysis during the tensile test clarified that the strain at which the necking in the width direction started approximately coincided with the strain at the UTS in both the starting material and the ARB-processed specimen.

(3) True stress of the ARB-processed sample decreased after necking in the tensile test, while that of the starting material with the coarse grains increased. The softening of the ARB-processed Al might be caused by dynamic recovery at grain boundaries.

Acknowledgement

This study was financially supported by a Giant-in-Aid for scientific Research on the Priority Area "Giant Straining Process for Advanced Materials Containing Ultra-High Density Lattice Defects" from the Ministry of Education, Culture, Sports, Science and Technology (MEXT), Japan.

References

[1] Tsuji N, Ito Y, Saito Y and Minamino Y 2002 *Scripta Mater.* **47** 893
[2] Iwahashi Y, Horita Z, Nemoto M and Langdon T G 1997 *Acta Mater.* **45** 4733
[3] Furukawa M, Horita Z, Nemoto M Valiev R Z and Langdon T G 1996 *Acta Mater.* **44** 4619
[4] Huang X, Hansen N and Tsuji N 2006 *Science* **312** 249
[5] Yu C Y, Kao P W and Chang C P 2005 *Acta Mater.* **53** 4019
[6] Martin S 2007 unpublished report, Risø National Laboratory-DTU, Roskilde
[7] Wang Y M and Ma E 2004 *Acta Mater.* **52** 1699
[8] Kim H W, Kang S B, Tsuji N and Minamino Y 2005 *Acta Mater.* **53** 1737
[9] May J, Höppel H W and Göken M 2005 *Scripta Mater.* **53** 189
[10] Li Y J, Zeng X H and Blum W 2004 *Acta Mater.* **52** 5009

Internal stresses in a 15%Cr ferritic stainless steel after large strain unidirectional processing

A Belyakov[1], K Tsuzaki[2], R Kaibyshev[1]

[1] Belgorod State University, Pobeda 85, Belgorod, 308015, Russia
[2] Structural Metals Center, National Institute for Materials Science, Sengen 1-2-1, Tsukuba, Ibaraki 305-0047, Japan

belyakov@bsu.edu.ru

Abstract. An interstitial free ferritic stainless steel was cold worked to a total strain of 4.6. The largely strained steel is characterized by a submicrocrystalline structure consisting of elongated grains/subgrains with the transverse size of about 210 nm; and the fraction of high-angle grain boundaries is about 0.6. Following a rapid rise at an early processing stage, the dislocation density in (sub)grain interiors unusually decreased after total strains of above 2. Nevertheless, the samples are characterized by high residual stresses that result in complex elastic distortions of the crystal lattice within the elongated crystallites. Such internal stresses are shown to be originated from deformation grain boundaries including low-angle subboundaries.

1. Introduction

An interest in nano- and submicrocrystalline structural materials is motivated by a beneficial combination of their mechanical properties. Metallic materials with grain size of tens to hundreds microns were shown offering a lot of superior properties like a high strength with sufficient ductility, enhanced impact toughness, superplasticity at relatively low temperatures and high strain rates, etc. [1-3]. One of the most promising methods for production of submicrocrystalline metals and alloys is based on large strain deformation [4, 5]. Commonly, largely deformed materials are characterized by high residual stresses that affect deformation behavior and mechanical properties. The development of high internal stresses was discussed as a result of non-equilibrium grain boundaries in largely strained materials [5, 6]. However, the sources of high internal stresses in submicrocrystalline metals and alloys have not been studied in sufficient detail. The effect of crystallographic parameters of grain boundaries on the values of internal stresses and their relaxation is still unclear. The aim of the present work is to clarify the origin of high residual stresses and their relationship with dislocation densities and strain-induced grain/subgrain boundaries in a submicrocrystalline ferritic stainless steel processed by large strain plastic working.

2. Experimental

A ferritic stainless steel (Fe – 0.003C – 0.01Mn – 0.001P – 0.001S – 15.0Cr – 0.003N, all in weight%) was used as the starting material. The steel was homogenized and hot forged at 1200°C followed by a hot rolling at 700°C. The large strain deformation was carried out at an ambient temperature by calibre rolling/swaging to a total strain of 4.6. Structural investigations were carried out on longitudinal and transverse sections using a JEM-2100 transmission electron microscope (TEM). Grain/subgrain sizes

Journal of Physics: Conference Series **240** (2010) 012115 doi:10.1088/1742-6596/240/1/012115

were measured on the TEM micrographs by the linear-intercept method, counting the all clear defined (sub)boundaries. The dislocation densities were evaluated by counting the individual dislocations in grain/subgrain interiors. Misorientations across deformation (sub)boundaries and local curvatures within the submicrocrystallites were analyzed by the converged beam Kikuchi diffraction technique. The strain hardening was studied by Vickers hardness tests with a load of 3 N.

3. Results and discussion

Typical cold worked microstructures are shown in figure 1. Unidirectional cold working results in the development of elongated grains/subgrains that aligned along the deformation axis. The deformation (sub)boundaries that arranged parallel to the rolling direction in the sample strained to 2.0 can be considered as dense dislocation walls [7], which separate the cell blocks with high density of interior dislocations. The spacing between the longitudinal (sub)boundaries decreases with increasing the total strain, leading to the formation of lamellar microstructure after a large strain of 4.6.

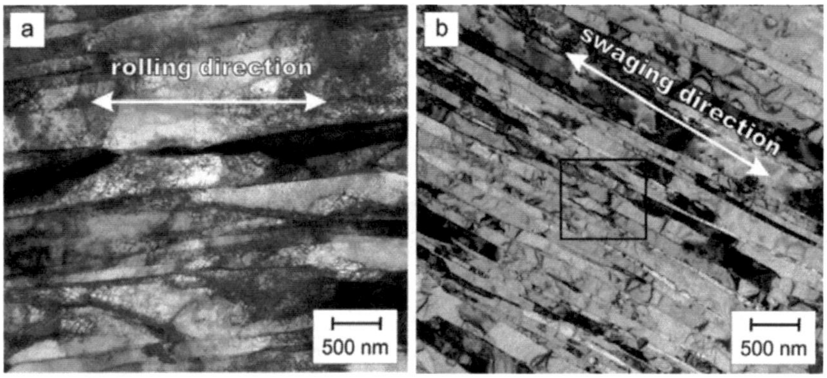

Figure 1. Deformation microstructures in Fe-15%Cr steel after cold working to a total strain of 1.0 (a) and 4.6 (b).

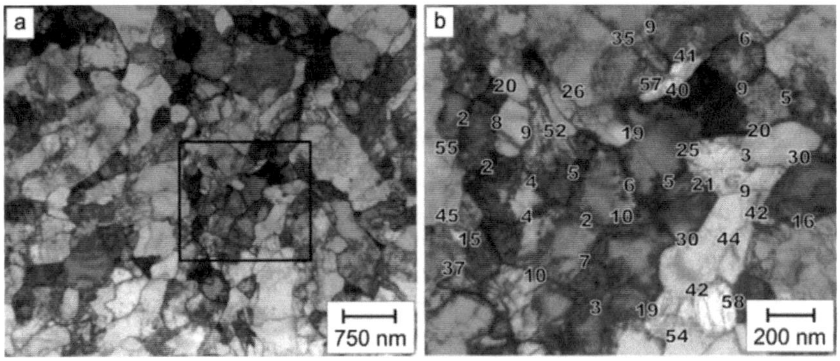

Figure 2. Deformation microstructure in Fe-15%Cr steel after cold working to a total strain of 4.6. Numbers in (b) indicate the (sub)boundary misorientations in degrees.

The decrease in the transverse size of structural elements is accompanied with increase in the (sub)boundary misorientations. Figure 2 presents the cross section microstructure that evolved in the sample processed to a strain of 4.6. The cross section of highly elongated grains/subgrains looks like almost equiaxed microstructure composed of irregular network of (sub)boundaries with different misorientations. The fraction of high-angle grain boundaries comprises about 0.6. Therefore, the deformation microstructure in largely strained sample consists of rod-shaped grains/subgrains separated by various boundary types, i.e. from low-angle dislocation subboundaries to conventional high-angle grain boundaries.

The transverse grain/subgrain size rapidly reduces to about 250 nm in early straining to about 1.0 followed by gradual decrease to 210 nm during further deformation to a strain of 4.6 (figure 3). Correspondingly, the hardness quickly increases at relatively small strains, then, the hardening rate decreases and results in a gradual hardening with increasing the total strain. The structural refinement and the strain hardening at strains of $\varepsilon < 2$ are accompanied with an increase in the dislocation density within grain/subgrain interiors. However, the further straining leads to unusual decrease of the interior dislocation density down to the level comparable to initial state (figure 3).

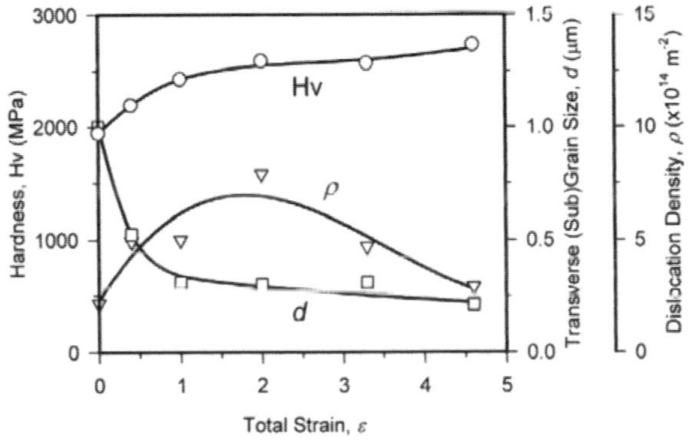

Figure 3. Effect of strain on the hardness, the transverse (sub)grain size and the interior dislocation density.

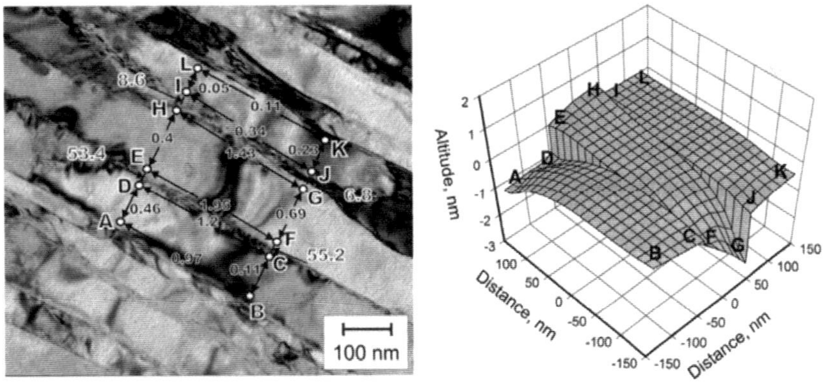

Figure 4. Enlarged portion of fine structure in figure 1b. Numbers indicate the local disorientations in degrees between the pointed micro-regions.

In spite of apparently low density of interior dislocations after large strains, the deformed sample is characterized by high residual stresses. The precise evaluation of local crystallographic orientations revealed large elastic distortions of the fine grains/subgrains. The lattice curvatures attain about 2 over a distance of 300 nm within the (sub)grains, which are free of any interior dislocations (figure 4). The lattice curvatures between the analyzed points are represented more clearly by the bent surfaces in figure 4. The relative altitudes of the surfaces were plotted as deviations of the central electron spots in Kikuchi maps from those of the grain centers, the relative altitudes of which were taken as zero. It is clearly seen that the large elastic distortions are maintained by both the high-angle grain boundaries (misorientation of about 55° between C and F) and low-angle subboundaries (misorientation of about 7° between G and J). This suggests that low-angle subboundaries in largely strained materials serve as effective barriers for dislocation motion, playing a similar role in strengthening as conventional grain boundaries.

The strengthening caused by dislocation substructures should follow a power law function of subgrain size with the exponent of -1.0, while the exponent of -0.5 is attributed to the grain size strengthening mechanism [8, 9]. Figure 5 illustrates the strain hardening as a function of transverse (sub)grain size in various cold rolled/swaged stainless steels. The Hv_0 of 1800 MPa was taken for Fe-18%Cr-7%Ni and Fe-22%Cr-3%Ni [6] and that of 1200 MPa for fully annealed Fe-15%Cr [10]. The size exponent of -0.5 is obtained for all the worked steels. Therefore, the strengthening of largely strained steels obeys the Hall-Petch-type relationship, irrespective of rather large fraction of low-angle subboundaries in strain-induced submicrocrystalline structures.

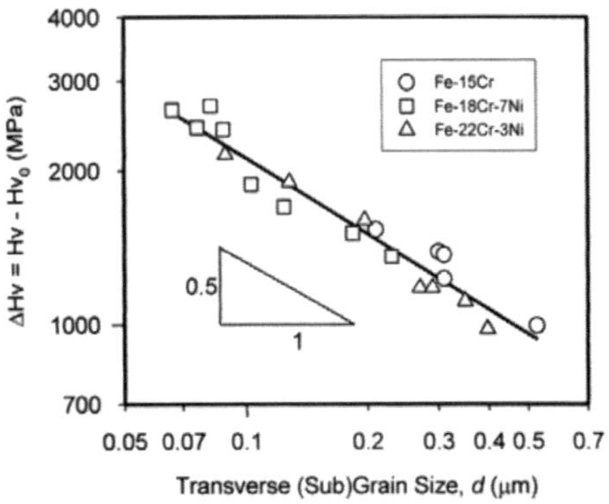

Figure 5. Relationship between the transverse (sub)grain size and the hardening (ΔHv) for bar rolled/swaged steels.

4. Summary

An Fe-15%Cr stainless steel subjected to large strain cold working by bar rolling/swaging is characterized by high residual stresses resulting in complex elastic distortions of strain-induced submicrocrystallites. The internal stresses are maintained by the submicriocrystalline boundaries including low-angle deformation subboundaries. The strain hardening during cold working follows the Hall-Petch-type relationship in spite of wide distribution of grain/subgrain boundary misorientations.

Acknowledgements

The financial support received from the Russian Foundation for Basic Research under grant No. 09-08-97567 is gratefully acknowledged.

References

[1] Wang Y, Chen M, Zhou F and Ma E 2002 *Nature* **419** 912
[2] Musin F, Kaibyshev R, Motohashi Y and Itoh G 2004 *Scripta Mater.* **50** 511
[3] Kimura Y, Inoue T and Tsuzaki K 2008 *Science* **320** 1057
[4] Humphreys F J, Prangnell P B, Bowen R J, Gholinia A and Harris C 1999 *Phil. Trans. R. Soc. Lond.* **357** 1663
[5] Valiev R Z, Islamgaliev R K and Alexandrov I V 2000 *Prog. Mater. Sci.* **45** 103
[6] Belyakov A, Kimura Y, Adachi Y and Tsuzaki K 2004 *Mater. Trans.* **45** 2812
[7] Bay B, Hansen N, Hughes D A and Kuhlmann-Wilsdorf D 1992 *Acta Metall. Mater.* **40** 205
[8] Young C M and Sherby O D 1973 *J. Iron Steel Inst.* **211** 640
[9] Kimura Y, Hidaka H and Takaki S 1999 *Mater. Trans JIM* **40** 1149
[10] Belyakov A, Tsuzaki K, Kimura Y and Mishima Y 2007 *J. Mater. Res.* **22** 3042

Grain refinement of coarse grained gold by combined thermo-mechanical process of severe plastic deformation and low temperature annealing

H. Miura[1], I. Itabashi[2], G. Yu[1] and T. Sakai[1]

[1] Department of Mechanical Engineering and Intelligent Systems,
UEC Tokyo (The University of Electro-Communications), Chofu, Tokyo 182-8585, Japan
[2] Technical Department, Tanaka Kinzoku Kogyo. K.K.
2-73 Shin-machi, Hiratsuka, Kanagawa, 254-0076, Japan

miura@mce.uec.ac.jp

Abstract. A high purity gold composed of coarse grains (d = 1.7 mm) was repeatedly thermo-mechanically processed by multi directional forging (MDF) at room temperature and annealing at warm temperature range. The effects of small addition of Ca (180 at. ppm) to Au on the static recrystallization (SRX) behavior and ultrafine grain evolution were investigated. Extensive SRX was observed in the MDFed Au sample to cumulative strain of $\Sigma\Delta\varepsilon = 1.2$ even at 453 K. After 2 cycles of the repeated TMP, fine grains of about d = 10 μm was obtained in the Au sample. The Au-Ca alloy MDFed to $\Sigma\Delta\varepsilon = 1.2$, however, was hardly SRXed even by annealing at 543 K. Subsequently, the fine-grained Au and Au-Ca alloy were further MDFed to $\Sigma\Delta\varepsilon = 8.0$ at maximum at room temperature. By the prolonged MDF, average (sub)grain size of about 200 nm was achieved in the Au sample. The hardness increased with increasing cumulative strain. Although the tensile strength was also raised with cumulative strain, large loss of ductility did not appear.

1. Introduction

Recently, special attention is paid to fabrication of ultrafine grained (UFGed) structure by severe plastic deformation (SPD). It is because the grain refinement induces both of strengthening and formability of metallic materials. The former effect is brought by reduction of mean-free path of dislocations by high density of grain boundaries, which is well-known as Hall-Petch relation.

By SPD, multidirectional forging (MDF) [1-3], equal channel angular pressing (ECAP) [4] and accumulative roll-bonding (ARB) [5], UFGs from 20 to 200 nm in diameter were easily and homogeneously evolved. Actually, such UFGed structures evolved by MDF were revealed to possess quite high tensile strength, i.e., 920 MPa and 2.2 GPa for Cu-Zn alloy [2] and SUS316 stainless steel [3], respectively. However, such UFGed metallic materials obtained by SPD tend to possess poor ductility. Even though the above MDFed alloys exhibited superior ductility of about 20 % in nominal strain, most of the SPD alloys show quite poor ductility, less than 10 %, in nominal strain [5].

It is known traditionally that Au and its alloys possess superior nature of extensive deformability. For example, deep drawing to fabricate Au wire of 20 μm in diameter is industrially common technique. The Au wire is composed of ultrafine fiber structure, where the direction is parallel to the drawing axis [6]. However, studies about the mechanical behavior of the UFGed Au and Au alloys processed by SPD are quite few as far as the authors know. The purpose of the present study is to investigate the evolution UFGs in Au and Au alloys during MDF and to reveal their mechanical properties.

2. Experimental

Hot extruded high purity Au (99.999%) and Au-180 at. ppm Ca alloy were cut into rectangular shaped samples of dimensions of $9.8 \times 7.8 \times 6.5$ mm^3. Hereafter, these samples are referred as 5N-Au and Au-Ca, respectively. The initial grain size of the as-hot extruded samples was so coarse (1.7 mm and 1.2 mm, respectively) that MDF to obtain UFGs is not efficient, they were preliminary fine grained by repeated cycles of combined thermo mechanical process (CTMP) of MDF to $\Sigma\Delta\varepsilon = 1.2$ and low temperature annealing. This is because that grain refinement during MDF is more accelerated when the initial grain size is smaller [1]. In the MDF method, the samples were pressed with changing loading direction 90 degrees pass by pass (i.e. *z* to *x* to *y* to z ...). A pass strain of $\Delta\varepsilon = 0.4$ and strain rate of 3.0×10^{-3} s^{-1} were employed for MDF.

After the sufficient grain refinement by CMTP, the samples were further MDFed up to cumulative strain of $\Sigma\Delta\varepsilon = 8$ at maximum (i.e., 20 passes) at room temperature. The evolved microstructure was examined by optical microscopy (OM) and transmission electron microscopy (TEM). The grain size was estimated by the line-intercept method. Mechanical properties were investigated by micro Vickers hardness test and tensile test. Tensile test was carried out on an Instron-type mechanical testing machine at an initial strain rate of 2.5 x10^{-3} s^{-1} at room temperature. The dimensions of the gage size of the tensile samples were 3 x 3 x 0.8 mm^3.

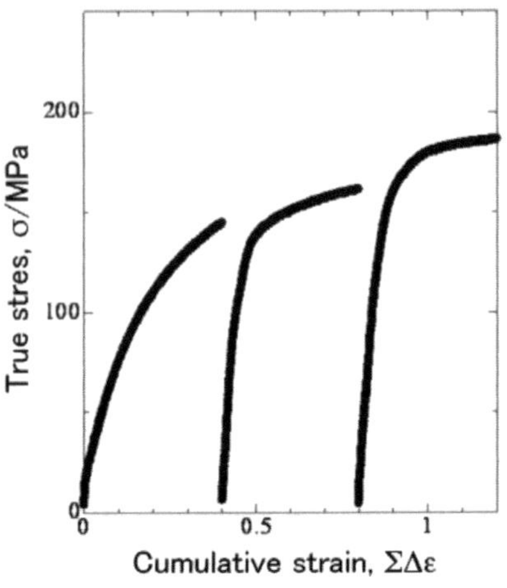

Fig. 1 True stress vs. cumulative strain curves of the as-extruded 5N-Au during multi directional forging at room temperature.

3 Results and discussion

3.1. Grain refinement of coarse grained initial microstructure

The true stress-cumulative strain curves obtained during MDF of the coarse grained as-hot extruded 5N-Au are shown in Fig. 1. The rapid work hardening occurred at the first pass of forging. The work hardening appears to be gradually saturated with increasing cumulative strain. This would be closely related to the saturation of dislocation density, but not to evolution of UFGs. For the UFG evolution by MDF, much higher cumulative strain more than $\sum\Delta\varepsilon = 3$ is required [1-3].

The above MDFed 5N-Au to $\sum\Delta\varepsilon = 1.2$ was subsequently annealed at warm temperature range. Figure 2 summarized the results of hardness change during annealing. The hardness decreases with increasing annealing time. The softening appears more rapid as the temperature becomes higher. The softening was induced mainly by occurrence of static recrystallization (SRX), as will be shown later. In contrast, softening did not appeared in the MDFed Au-Ca alloy even at 483 K, while the result is not shown here. The low temperature annealing of the MDFed Au-Ca was, therefore, carried out at 543 K. This result indicates that addition of quite small amount of Ca to Au causes drastic improvement of thermal stability.

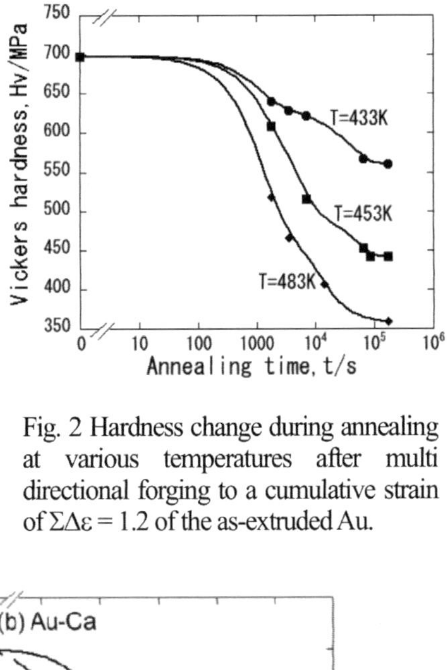

Fig. 2 Hardness change during annealing at various temperatures after multi directional forging to a cumulative strain of $\Sigma\Delta\varepsilon = 1.2$ of the as-extruded Au.

After the first cycle of CTMP, the samples were further processed by the secondary cycle of CTMP. Figure 3 summarized the grain size change during low temperature annealing after secondary MDF. The gain size of 5N-Au reduced down to about 900 μm after the first cycle of CTMP, and then, it became to be about d = 10.5 μm at minimum after second CTMP (Fig. 3 (a)). When compared Figures 2 and 3 (a), it is evident that the time to onset SRX was much shortened. This is because that density of grain boundaries, where

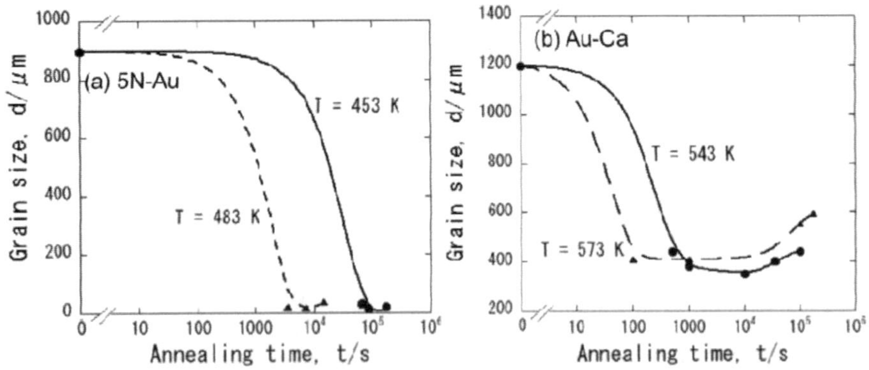

Fig. 3 Change in the average grain size during 2ndary annealing after 2ndary multi directional forging to $\Sigma\Delta\varepsilon = 1.2$ ($\Sigma\Delta\varepsilon_{Total} = 2.4$): (a) for 5N-Au and (b) for Au-Ca.

Fig. 4 Microstructural change in 5N-Au; (a) before MDF (i.e., just after 2cycles of CMTP), MDFed to (b) $\Sigma\Delta\varepsilon = 2.4$ (or $\Sigma\Delta\varepsilon_{Total}$ (= 1.2 + 1.2 + 2.4) = 4.8) and (c) $\Sigma\Delta\varepsilon = 6.0$ ($\Sigma\Delta\varepsilon_{Total} = 8.4$).

heterogeneous deformation microstructure can evolve along them to raise stored energy, affects much on the SRX behavior. That is, SRX was much stimulated by the reduction of grain size from 1.7 mm to 900 μm. Actually, it is known that ultra grain refinement by SPD is accompanied with loss of thermal stability [2]. It should be also noted that SRX took place at relatively low temperature around 433 K; 0.32 Tm, where Tm the melting temperature. Such low SRX temperature in 5N-Au should be caused by the effects of high purity and high energy stored by MDF. On the other hand, the minimum grain size of Au-Ca obtained after 2 cycles of CTMP was 260 μm (Fig. 3 (b)). The coarser grain size of Au-Ca obtained was induced by relatively higher annealing temperature compared with that of 5N-Au.

Fig. 5 Evolved microstructure in the 5N-Au sample MDFed to $\Sigma\Delta\varepsilon = 6.0$ ($\Sigma\Delta\varepsilon_{Total} = 8.4$).

3.2. Multi directional forging and the evolved microstructure

The fine-grained 5N-Au and Au-Ca by 2 cycles of CTMP were further MDFed at room temperature to cumulative strain of $\Sigma\Delta\varepsilon$ = 8.4 at maximum, while the grain sizes were largely different (10.5 μm and 260 μm). The microstructural change was observed by OM and the typical photographs are exhibited in Fig. 4. It can be seen in Fig. 4 that, with increasing cumulative strain, the initial straight grain boundaries having sharp contrast gradually curved and became unclear accompanied with evolution of substructures in grain interiors. After MDF to $\Sigma\Delta\varepsilon = 6.0$ ($\Sigma\Delta\varepsilon_{Total} = 8.4$), it is difficult to identify the initial grain boundaries among the evolved granular-look structure.

Figure 5 shows the TEM microstructure of 5N-Au MDFed to $\Sigma\Delta\varepsilon$ = 6.0. It is evident that the 5N-Au was already composed of fine (sub)grains raging from 100 to 400 nm. Suzuki et al. estimated the grain sizes in Au and Au-Ag alloy ECAPed from the surface morphology to be 200 to 400 nm [7]. The above grain sizes are almost comparable with that observed at the cross section of deeply drawn Au wire [6].

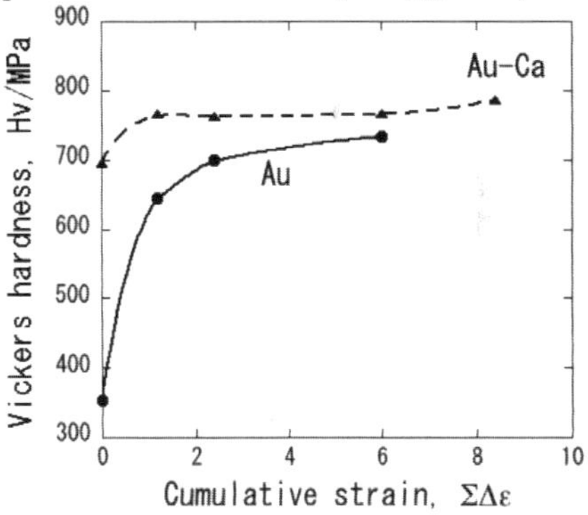

Fig. 6 Hardness change during MDF.

Belyakov et al. have revealed that UFG evolution during MDF is more accelerated with decreasing initial grain size [1]. It can be estimate, therefore, the grain size in Au-Ca must be much coarser than that in 5N-Au. Sakai et al. proposed a mechanism of UFG evolution during MDF as low temperature dynamic recrystallization [8]. In this mechanism, introduced dislocations form dislocation walls at first and, then, their misorientation gradually increase to evolve high angle boundaries with increasing cumulative strain. Therefore, the microstructure obtained by MDF inevitably contains a certain fraction of low angle boundaries.

3.3. Mechanical properties of the multi directional forged Au and Au-Ca alloy

Figure 6 represents the hardness changes during MDF. The higher hardness of Au-Ca before MDF is resulted mainly by higher dislocation density because of incomplete occurrence of SRX. It is apparent that the hardness of both samples gradually increased with cumulative strain and looked saturated at high cumulative strain region.

The lower hardness even though the expected finer grain size in 5N-Au may be affected by recovery at ambient temperature [9]. Suzuki et al. measured the hardness of 4N purity Au ECAPed to equivalent strain of 4 by route Bc and reported it to be about 940 MPa [7]. The results obtained by MDF in the present study seem to be comparable or slightly lower than that by ECAP, although the purity is much different.

The MDFed 5N-Au and Au-Ca samples were tensile tested and the true stress vs. nominal strain curves are shown in Fig. 7. It can be seen that the ultimate tensile strength (UTS) was gradually raised with increasing cumulative strain. The UTS of 330 MPa was achieved in the 5N-Au sample. The slightly higher value of the UTS in the MDFed 5N-Au than that of the ECAPed 4N-Au (245 MPa) [7] would be resulted from more uniform evolution of UFGs. The increment of UTS of the Au-Ca is rather smaller than that of the 5N-Au and it looked soon saturated over $\Sigma\Delta\varepsilon = 1.2$. This difference would be resulted from much delayed evolution of UFGs owing to the coarse initial grains in Au-Ca. It is also notable in Fig. 7 that ductility was not so reduced, more than 40 %, even at high cumulative strain region around $\Sigma\Delta\varepsilon = 6.0$. Similar tendency of such large ductility of the ECAPed Au and Au alloys was reported [7]. UFGed metallic materials fabricated by SPD process tends to lose its ductility [5]. The observed large ductility of Au and Au-Ca even after MDF should be due to the inherent nature of extended ductility in Au.

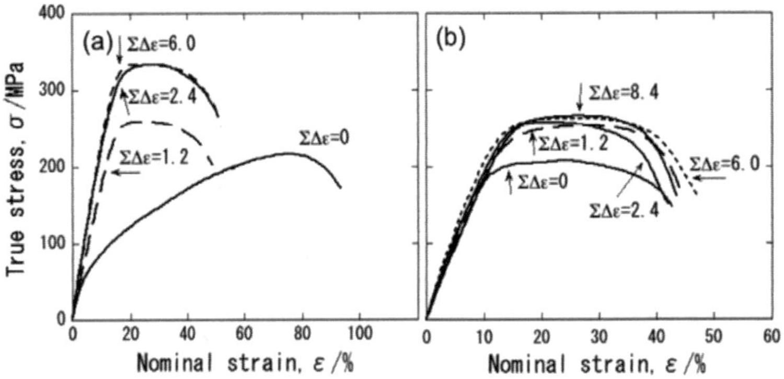

Fig. 7 True stress vs. nominal strain curves for (a) Au and (b) Au-Ca alloy MDFed

4. Summary

Combined thermo mechanical process (CTMP) of multi directional forging (MDF) and low temperature annealing was applied to high purity Au (5N-Au) and Au-Ca alloy composed of coarse grains. Uniform evolution of fine grains of about 10.5 μm was obtained in 5N-Au after 2 cycles of CMTP. Addition of small amount of Ca to Au improved thermal stability, and therefore, recrystallization temperature was much raised. The higher thermal stability resulted coarser grain size of the Au-Ca alloy even after 2 cycles of CMTP.

Ultrafine (sub)grains (UFGs) were almost uniformly evolved by the prolonged MDF of the above fine grained 5N-Au. The UFGed 5N-Au exhibited quite high ultimate tensile strength (UTS) and ductility. The excellent balance of mechanical properties of the UFGed 5N-Au must be related with the inherent nature of Au. In contrast, the coarse grained Au-Ca alloy showed comparably lower UTS even though MDFed to higher cumulative strain. This was assumed that initially coarser grain affected to retard UFG evolution.

References

[1] A. Belyakov, T. Sakai, H. Miura, R. Kaibyshev, K. Tsuzaki, 2002 *Acta Mater.* **50** 1547
[2] H. Nakao, H. Miura and T. Sakai, 2008 *J. Japan Inst. Metals* **72** 397
[3] H. Miura and Y. Nakao, 2009 *Mater. Sci. Forum* **633-634** 577
[4] Y. Iwahashi, Z. Horita, M. Nemoto and L. G Langdon, 1997 *Acta Mater.* **45** 4733
[5] N. Tsuji, Y. Ito, Y. Saito and Y. Minamino, 2002 *Scripta Mater.* **47** 893
[6] S. S.Kang, H. S.Jung, W. H.Bang, J. H.Cho, K. H.Oh, D. S.Kim, J. S.Cho, Y. J.Park and S. H.Park., 2005 *Mater. Sci. Forum* **495-497** 907
[7] T. Suzuki, A. Vinogradov and S. Hashimoto, 2004 *Mater. Trans.* **45** 2200
[8] T.Sakai, H.Miura, K.Usui and T.Koba, 2004 *Proc. 25th Risoe Int. Symp. Mater. Sci.* Denmark, 509
[9] L. E.Murr, 1975 Interfacial Phenomena in Metals and Alloys, Addison-Weseley, Massachusetts, USA, 377

Mechanical properties of Fe -10Ni -7Mn martensitic steel subjected to severe plastic deformation via cold rolling and wire drawing

H. Ghasemi-Nanesa[1], M. Nili-Ahmadabadi[1,2*] H. Shirazi[1]

[1] School of Metallurgy and Materials Engineering, Faculty of Engineering,
[2] Center of Excellence for High Performance Materials, University of Tehran, P.O. Box14395-731, Tehran, Iran

sut.caster.81710018@gmail.com

Abstract. Fe-Ni-Mn martensitic steels are one of the major groups of ultra-high strength steels that have good mechanical properties and ductility in as annealed condition but they suffer from severe inter-granular embitterment after aging. In this paper, the effect of heavy shaped cold rolling and wire drawing on the mechanical properties of Fe-Ni-Mn steel was investigated. This process could provide a large strain deformation in this alloy. The total strain was $\varepsilon \sim 7$. Aging behavior and tensile properties of Fe-10Ni-7Mn were studied after aging at 753 K. The results showed that the ultimate tensile strength and ductility after cold rolling, wire drawing and aging increased up to 2540 MPa and 7.1 %, respectively, while the conventional steels show a premature fracture stress of 830 MPa with about zero ductility after aging.

1. Introduction

Fe-Ni-Mn Martensitic steels are one of the major groups of ultra high strength steels. They have excellent age hardenability by the formation of nanometer sized inter-metallic precipitates such as f.c.t θ-NiMn [1-3]. They suffer from intergranular embrittlement after aging. Discontinuous coarsening of grain boundary precipitates was found to be the main source of embrittlement [4-9]. It is known that grain refinement is an effective way to improve strength and toughness of metallic materials. At present, great attention is paid to the processes of severe plastic deformation (SPD) due to the formation of nano (sub micron grain size) structures upon deformation [9-13]. SPD consists of various methods such as equal channel angular pressing (ECAP) [10-12], accumulative roll banding (ARB)[11-13], high pressure torsion (HPT)[11] and etc. Combination of wire drawing and cold rolling leads to a new method of severe plastic deformation suitable for industrial process, which promotes the work hardening by increasing dislocation density. Most ductile metals can be drawn into ultrafine wire by pulling them through thinner and thinner dies. In most cases, drawing conditions have to be chosen with care: low drawing speeds, low reduction for each pass and diamond dies with low approach angles are recommended [14,15]. Structures obtained during SPD have specific features such as small grains down to submicron and nano level, low density of free dislocations, high angle misorientaion of these grains and high energy and non-equilibrium state of grain boundaries [11]. These structures lead to changes in the mechanical properties: a significant increase in the strength at good ductility. Effect of ECAP and heavy cold rolling of aged martensitic steels have been studied and

* corresponding author: nili@ut.ac.ir

the resulted microstructure and mechanical properties were discussed [16-17]. It was shown that the aforementioned SPD processes have a major impact on the microstructure and mechanical properties of steels. However, the aim of this paper is to consider the features of nanostructure formation during SPD via shaped cold rolling and wire drawing with focusing on mechanical properties of Fe-10wt % Ni- 7wt % Mn.

2. Experimental procedures

Fe-10wt%Ni- 7wt%Mn was prepared by vacuum induction melting (VIM) and vacuum arc remelting (VAR) routes. The chemical composition of the studied alloy is given in Table.1 The remelted ingot was forged by 50% reduction at 1423 k then homogenized in vacuum furnace at 1473 K for 43200 sec. Heavy shaped cold rolling and wire drawing were carried out to investigate the effect of the large strain plastic deformation on the mechanical properties. Shaped rolling process was carried out in 24 passes. Then the drawing process was carried out with 19 passes with 0.05 mm reduction in diameter at each pass. A final diameter of wire was 0.45 mm. The total applied strain was $\varepsilon \sim 7$. Aging behavior was studied at 753 K for different holding time. Microhardness measurements were carried out by Vickers method of 200 g load. Tensile properties were measured using a conventional tensile testing machine with a crosshead speed of 1mm/min. The fracture surface was studied by a scanning electron microscope equipped with EDAX[2].

Table .1 Chemical composition of the alloy (Wt %)

Fe	Ni	Mn	C	S	P	N	Al
base	10.50	7.00	0.006	0.007	0.005	0.005	0.003

3. Results and discussion

Figure 1 shows a scanning electron microscopic (SEM) micrograph of as annealed sample with fully lath martensite microstructure.

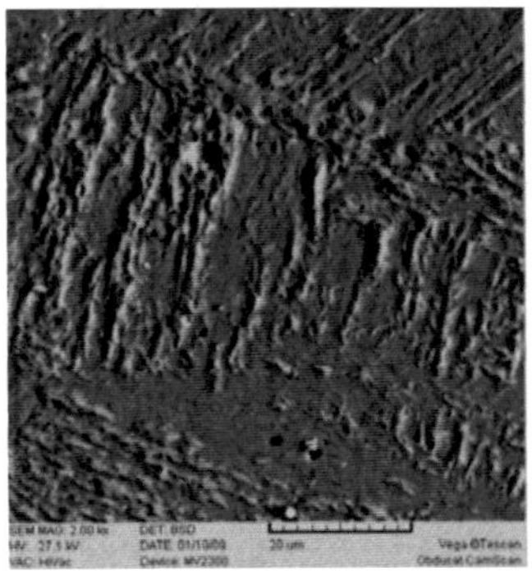

Figure1. SEM Micrograph of as annealed ample

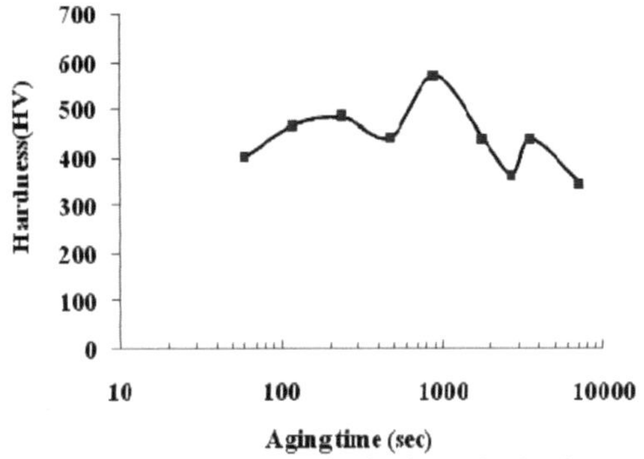

Figure 2. Hardness changes during isothermal aging in sample A

Figure 2 shows the hardness changes in sample A during isothermal aging. The maximum hardness of about 573 HV was obtained after 900 sec. Based on a hardness-aging time curve, the precipitation of f.c.t-θ (NiMn) inter-metallic particles is believed to be the main source of hardening [1-3, 19-20]. In two aging time intervals, softening occurred. This phenomenon reported in [21] is attributed to with three main reasons. First, the changes in the dispersion of clusters cause softening. Second, it comes

2 Energy Dispersive Spectroscopy

from the reverse transformation of martensite to austenite. Thirdly, softening comes from the rearrangement of dislocation substructure leading to a reduction in the total dislocation density. Therefore, for this alloy the hardness decreased to a value of 442 HV after 480 sec and of 362 HV after 2700 sec supposedly due to some of the above phenomena. Figure 3(a) shows mainly brittle intergranular fracture along prior austenite grain boundaries (PAGBs) in the aged-free strain sample where the fracture path is denoted by a dashed white line. The ductile part in fracture surface of the aged wire was cut and shown in Figure 3(b). Small dimples in the fracture surface are perceivable in the magnified picture so the fracture mode is rather defined as the mixed mode of ductile and brittle fracture. The ductile part is responsible for enhanced ductility during the tensile test.

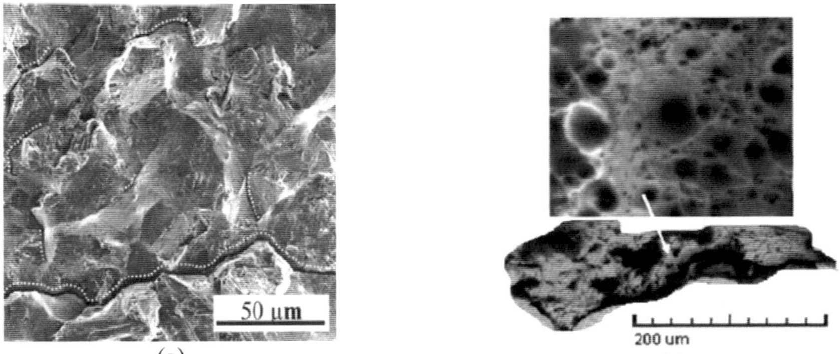

(a) (b)

Figure 3. SEM micrographs (a) brittle fracture in free strain and aged sample (b) ductile part - fracture surface of aged wire

Table 2 shows tensile properties of both samples A and B in as drawn and aged conditions. A total elongation of about 6.9 % and a tensile strength of about 1900(MPa) were obtained for sample A. After aging, there was no important change in mechanical properties. As shown in 1 for sample B, there was a reduction in the UTS with further deformation that could be related to reverse transformation of martensite to austenite [22]. After aging because of more work hardening and more grain refining and formation of finer precipitates, there was a recognizable increase in the UTS and it was the maximum UTS that had been reported for Fe-Ni-Mn alloys yet. A ductility of about 7.1 % was obtained. Heavy cold deformed wire with a total measured strain about $\varepsilon \sim 7$ showed a tensile strength of about 1740 MPa in sample B while as annealed steel shows a tensile strength of about 830 MPa. Sample B revealed a tensile strength of about 2540 MPa after aging with a tensile ductility of ~ 7.1 %.

Table.2 Tensile properties of Fe-Ni-Mn wires

	Wire diameter (mm)	UTS (MPa) - as drawn	UTS (MPa) -as aged	Ductility (%)
Sample A	0.60	1900	2000	6.9
Sample B	0.45	1740	2540	7.1

According to investigations of grain boundary formation during large strain deformation, two different kinds of grain boundaries can be produced: incidental dislocation boundaries (IDBs) and geometrically necessary boundaries (GNBs). These definitions are based on grain boundaries scaling behavior and rate of increase in the average misorientation angle across the boundaries with increasing strain especially in larger strains. Also, the average misorientation angle increases more rapidly with strain for GNBs than that for IDBs. Dislocations in IDBs are redundant whereas in GNBs are nonredunant. Generally presence of GNBs is more effective than IDBs in hardening of microstructure. During wire drawing of bcc materials, grains not only elongate in the drawing direction but also get a kind of folding. In order to fold the grains over each other, some extra dislocations have to be generated. These dislocations are not annihilated and are defined as geometrically necessary boundaries (GNBs). Formation of GNBs during wire drawing, precipitation density and size can be explained by a significant difference between tensile strengths of samples A and B after aging. It can be estimated that higher GNBs density along with more uniform and fine precipitates in 0.45 mm than 0.60 mm diameter aged wire leads to higher tensile strength in the former than the latter wire.[13-15, 23-24]

4. Conclusions

Combination of cold rolling and drawing is a kind of severe plastic deformation to fabricate nanostructured materials.

Maximum tensile strength was obtained after aging for 900 sec and this tensile strength is the highest value of strength that reported until now in all articles dealt with the examination of Fe-Ni-Mn martensitic steels. The wire exhibits ductile fracture. After aging the fracture surface showed the combination of brittle and ductile mechanisms. It can be concluded that the mechanical properties and ductility were simultaneously improved by severe plastic deformation via cold rolling and wire drawing.

Aging treatment at 753 K for 480 sec and 2700 sec leads to softening.

References

[1] M. Yodogawa, *Trans J.I.M* **17**, 1976, 799–808
[2] V.M. Kardonski, M.D. Perkas, *Metall. Sci. Heat Treat.* **3**, 1966, 254–256.
[3] J. Singh, C.M. Wayman, *Mater. Sci. Eng.* **94**, 1987, 233–242.
[4] M.D. Perkas, *Metall. Sci. Heat Treat.* **4**, 1968, 415–425.
[5] M. Tanaka, T. Suzuki, M. Yodogawa, *Trans. I.S.I.J.* **8**, 1968, 236–244.
[6] D.R. Squires, E.A. Wilson, *Metall. Trans.* **3**, 1972, 575–581.
[7] L.T. Shiang, C.M. Wayman, *Metallography* **21**, 1988, 399–423.
[8] L.T. Shiang, C.M. Wayman, *Metallography* **22**, 1989, 15–36.
[9] B. Q. Han, E. J. Lavernia, F. A. Mohamed, *Metall. Mater. Trans.* **34A**, 2003, 71.
[10] R. Z. Valiev, R. K. Islamgaliev, I. V. Alexandrov, *Prog. Mater. Sci.* **45**, 2000, 103.
[11] S. V. Dobatkin, J. Zrnik, I. Mamuzic, *Metalurgika* **45**, 2006, 313-321.
[12] V.M. Segal et al, Russ. *Metall. (Metally)*, 1981, 99.
[13] Y. Saito, *Acta Mater.* **47**, 1999, 579.
[14] M. Zidani, M. Messoudi, T. Baudin, D. Soals, C. Derfouf, N. Bounar, *Metal* **13**, 2008.
[15] Bert Verlinden, Julian Driver, Indradev Samajdar, Roger D. Doherty, *Thermo-mechanical Processing of Metallic Materials, Elsevier Pergamon Material Series*, 2007.
[16] M. Nili-Ahmadabadi, H. Shirazi, A. Fatehi, S. Hossein Nedjad, *Int. J. Modern Phys.* B **22** 2008, 2814-2822.
[17] M. Iranpour Mobarakea, M. Nili-Ahmadabadia, B. Poorganji , A. Fatehi, H. Shirazi, T.Furuhara, Habibi Parsa, S. Hossein Nedjad, *Mater. Sci. Eng.* A **49**, 2008, 172–176.
[18] N.H. Heo, *Sci. Mater.* **34**, 1996, 1517–1522.
[19] S. Hossein Nedjad, M. Nili Ahmadabadi, R. Mahmudi, T. Furuhara, T. Maki, *Nanostructured Materials by High-Pressure Severe Plastic Deformation*, 2006, 265-269.
[20] S. Hossein Nedjad, M. Nili Ahmadabadi , R. Mahmudi, T. Furuhara, T. Maki, *Mater. Sci. Eng.* A **438– 440**, 2006, 288–291.
[21] E.V. Pereloma, A. Shekhter 1, M.K. Miller, S.P. Ringer, *Acta Mater.* **52**, 2004, 5589-5602.
[22] H. Ghasemi-Nanesa, M. Nili-Ahmadabad, H. Shirazi, S.Hossein Nedjad, submitted for Scripta Mater., 2009.
[23] D. A. Hughes, N. Hansen, D. J. Bammann, *Scripta Mater* **48**, 2003, 147–153.
[24] H. Mughrabi, *Acta Mater.* **54**, *2006*, 3417–3427.

Mechanical properties of nanostructured nickel based superalloy Inconel 718

Sh Mukhtarov, A Ermachenko

Institute for Metals Superplasticity Problems RAS, 39, Khalturina, Ufa, 450001, Russia

shamil@anrb.ru

Abstract. This paper will describe the investigations of a nanostructured (NS) state of nickel based INCONEL® alloy 718. This structure was generated in bulk semiproducts by severe plastic deformation (SPD) via multiple isothermal forging (MIF) of a coarse-grained alloy. The initial structure consisted of γ-phase grains with disperse precipitations of γ″-phase in the forms of discs, 50-75 nm in diameter and 20 nm in thickness. The MIF generated structures possess a large quantity of non-coherent plates and rounded precipitations of δ-phase, primarily along grain boundaries. In the duplex (γ+δ) structure the grains have high dislocation density and a large number of nonequilibrium boundaries. Investigations to determine mechanical properties of the alloy in a nanostructured state were carried out. Nanocrystalline Inconel 718 (80 nm) possesses a very high room-temperature strength after SPD. Microcrystalline (MC) and NS states of the alloy were subjected to strengthening thermal treatment, and the obtained results were compared in order to determine their mechanical properties at room and elevated temperatures.

1. Introduction

It is well known [1] that by SPD an average grain size of metals and alloys can be refined to a NS state (<1 μm). In nickel-iron superalloy Inconel 718, that is widely used in aircraft engine manufacture, ultra-fine grain and NS states allow the fabrication of components by superplastic deformation at higher strain. Components operating at elevated temperatures are subjected to heat treatment to have strengthening coherent disperse particles of second phase precipitated. Mechanical properties of NS Inconel 718 are reported elsewhere [2]. But the data on mechanical properties of NS Inconel 718 after heat treatment are missing, and that is why such results are important.

The goal of the present paper is to investigate structure and mechanical properties of NS Inconel 718 subjected to strengthening heat treatment.

2. Material and methods

The material selected for this study was hot-deformed and precipitation hardened with strengthened phase γ″ (bct, Ni_3Nb) nickel based alloy Inconel 718 (Ni-18Cr-0.6Al-1.1Ti-18Fe-5Nb-2.9Mo-0.1Co, %Wt.) produced by Pratt & Whitney. The testing samples of alloy for SPD via MIF were machined from 250 mm diameter billets. MIF is a process in which a billet is compressed between flat dies on all three orthogonal directions during a sequence of forging passes with gradual decrease of temperature.

Bulk samples were produced on 630 ton-force press with decreasing deformation temperature from 950 to 800°C and to 575°C for processing, MC and NS states, respectively.

Deformation at high temperatures results in the transformation of γ''-phase into orthorhombic δ-phase. MC and NS forged billets were subjected to annealing at 980°C/1 hour, aging at 720°C/8 hours, furnace cooling to 620°C, hold at 620°C for total aging time of 18 hours. [3]. Tensile properties were determined at room temperature and 650°C, using sheet samples, with gauge dimensions of 5 mm x 2 mm x 15 mm. Stress rupture life and fatigue strength were determined using samples with gauge dimensions ∅3 mm x 18 mm and ∅4 mm x 25 mm, respectively.

Optical microscopy and transmission electron microscopy (JEM-2000EX) were used to investigate the microstructure.

3. Experimental results and discussion

3.1. Initial microstructure

MIF of bulk samples from investigated alloy carried out to various strains resulted in the formation of MC and NS states with mean grain sizes: 1 μm and 0.08 μm (figure 1). The TEM and SEM [4] studies have shown that the duplex structure consists of γ-phase and high temperature modification of Ni_3Nb - δ-phase. Plates of δ-phase with non-coherent boundaries are distributed uniformly. The NS state is characterized by higher dislocation density and non-equilibrium γ/γ boundaries [5]. Carbides with a mean size of 5 μm are present.

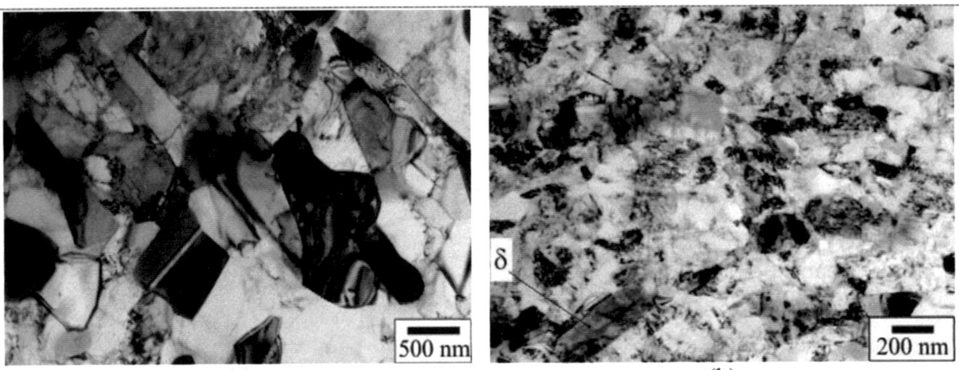

(a) (b)

Figure 1. TEM image of Inconel 718 produced by MIF with mean grain size: (a)-1 μm; (b)-0.08 μm.

3.2. Mechanical properties of alloy after MIF at room temperature

According to the Hall-Petch relationship the strength of metals and alloys increases with decreasing grain size [1]. Tensile properties of Inconel 718 after MIF with various grain sizes are shown in table 1. The NS (γ+δ) alloy exhibits very high room-temperature strength, which is much higher than that of the (γ+γ'') alloy subjected to the strengthening thermal treatment. At the same time, increase in strength of this alloy in a NS state is accompanied by some reduction of ductility.

3.3. Mechanical properties after heat treatment

The studies on microstructure after carrying out the standard heat treatment [3] of MC and NS alloys showed that it resulted in the generation of uniform structure, the average grain sizes of γ-phase being 3.9 and 4.6 μm, accordingly. The difference of initial structural states is inherited after carrying out thermal treatment. From figure 2 it is evident that after heat treatment (HT) some quantity of globular δ-phase is present in grain boundaries, which retards grain growth during annealing. Volume fraction of δ-phase is 2.9% for MC state after heat treatment and 3.3% for NC state.

Table 1. Mechanical properties of alloys after MIF at room temperature.

Grain size (μm)	Phases	Ultimate strength (MPa)	Yield strength (MPa)	El (%)	RA (%)
-	$\gamma+\gamma''$	≥1276[a]	≥1034	≥12	≥15
1[b]	$\gamma+\delta$	1184	920	21.9	22.2
0.3[b]	$\gamma+\delta$	1560	1300	5.1	11.0
0.08[b]	$\gamma+\delta$	1920	1845	4.8	6.1

[a] Mechanical properties after heat treatment [6].
[b] [4].

The data on mechanical properties of the heat treated alloy are presented in table 2. These data evidently correspond to the material specification requirements. It is obvious that the initial nanostructure leads to maximum strength for the alloy at room temperature.

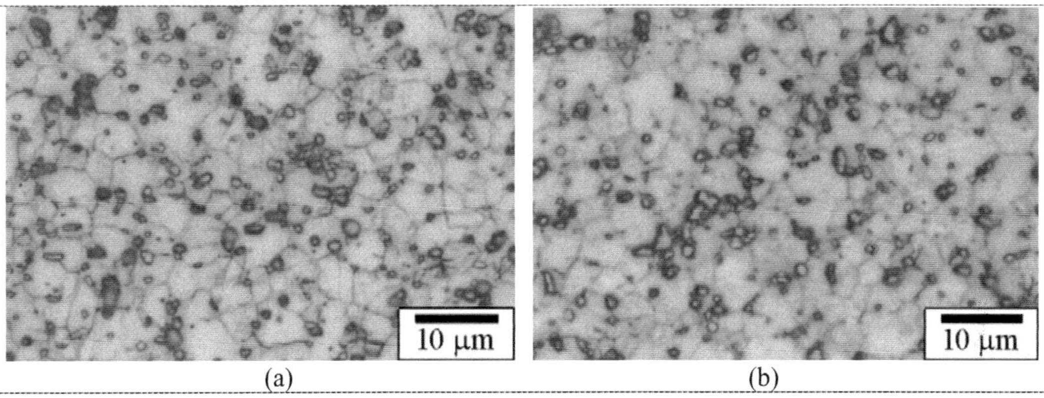

Figure 2. Microstructure of Inconel 718 after MIF and heat treatment: (a) – MC+HT; (b) – NS+HT.

Table 2. Mechanical properties of Inconel 718 after MIF and heat treatment.

State	Ultimate strength (MPa)	Yield strength (MPa)	El (%)	RA (%)	Grain sizes (μm)	Fatigue strength (MPa)	Fatigue strength (Cycles)
AMS 5662[a]	≥1276/1000[b]	≥1034/862	≥12/12	≥15/15	-	-	-
CG+HT [a]	1428/1176	1180/976	19/18	20/32	11-22	910	10^5 [c]
MC+HT	1488/1169	1234/995	17/23	23/45	4.6	910	$>1.6\cdot10^5$
NS+HT	1520/1164	1252/993	19/21	35/42	3.9	910	$>1.6\cdot10^5$

[a] [6], coarse-grained (CG).
[b] At room temperature/at 650°C.
[c] At grain size 19.6 μm [7].

Comparative fatigue tests of the samples at room temperature have shown that the properties of NS alloy on the scale of 10^5 cycles are higher by the factor of 1.6 than those stipulated elsewhere [7].

Stress rupture data shown in table 3 indicate that with decreasing a mean grain size of γ-phase one observes the tendency towards decrease of stress rupture and increase of ductility. All presented conditions meet the material specification requirements.

Table 3. Stress rupture data at 650°C of Inconel 718 after MIF and heat treatment.

State	Stress (MPa)	Time (hrs)	El (%)	RA (%)
AMS 5662	689	≥23	≥4	-
MC+HT	710	29.8	20.2	61.0
NS+HT		27.0	25.3	73.5

The investigation results show that for increasing alloy's strength properties at operating temperatures it is appropriate to have δ-phase totally dissolved in order to increase the quantity of strengthening γ″-phase precipitates during aging.

4. Summary

With decreasing grain size to NS state of alloy 718 occurs a significant increase of strength properties at room temperature, in particular ultimate strength increased up to 1920 MPa.

It is highly appropriate to employ NS Inconel 718 after heat treatment when it possesses high strength and sufficient ductility.

For increasing service life of NS Inconel 718 components it is reasonable to develop a special heat treatment.

Acknowledgements
This study was carried out with the financial support of the Russian Foundation for Basic Research, Projects # 07-08-00287a, 09-08-08126z.

References

[1] Valiyev R Z and Alexandrov I V 2000 *Nanostructured Materials Processed by Severe Plastic Deformation* (Moscow: Logos)
[2] Mukhtarov Sh, Valitov V, Gigliotti M F X, Subramanian PR, Marte J S and Dudova N 2008 *Mater. Sci. Forum* **584-586** **1** 458-63
[3] Chandler W T, Ghosh A K and Mahoney W M 1984 Superplastic forming and diffusion bonding of Inconel 718 *J. Spacecraft* **21** **1** 61-4
[4] Mukhtarov Sh Kh 2010 *Mater. Sci. Forum* **633-634** 569-75 in press
[5] Valitov V A, Mukhtarov Sh Kh and Raskulova Yu A 2006 *The Physics of Metals and Metallography* **102** **1** 97-104
[6] Bernshtein L M and Matevosian A P 1982 *Proc. Int. Conf. Deformation and Properties of Aerospace Material Technique* ed V S Sandler (Moscow: Metallurgia) p 376
[7] http://www.specialmetals.com/documents/Inconel%20alloy%20718.pdf

High performance of mechanical and electrical properties of Cu-Cr-Zr alloy sheets produced by ARB process and additional thermo-mechanical treatment

T Akita[1], K Kitagawa[1], K Kita[1], M Gotoh[1], Y Hirose[1] and N Tsuji[2]

[1] Graduate School of Natural Science and Technology, Kanazawa University, Kakuma, Kanazawa, Ishikawa 920-1192, Japan
[2] Department of Materials Science and Engineering, Graduate School of Engineering, Kyoto University, Yoshida Honmachi, Sakyo-ku, Kyoto 606-8501, Japan

masahide@cs.s.kanazawa-u.ac.jp

Abstract. The additional cold rolling and the aging process were applied to Cu-0.85Cr-0.07Zr alloy sheets processed by ARB, and mechanical properties and structural information were investigated for the purpose of further improvement of the mechanical properties and the electrical conductivity. From the results of the tensile test and the measurement of electrical conductivity, ARB/aged/CR was most appropriate processing in order to achieve technical advantages. The high tensile strength of 745 MPa and the high electrical conductivity of 68 %IACS were obtained simultaneously. In addition, the improvement of incomplete boundaries generated during ARB processing was possible by thermo-mechanical treatment.

1. Introduction

Fine-grained metals achieve very high yield strength in accordance with Hall-Petch relation. Accumulative roll bonding (ARB) processing [1] was taken up in this study to get fine-grained thin sheets, which are expected to be used as electrical parts. On the other hand, conventional thermo-treatment like age hardening has been used in order to improve some mechanical properties of metallic materials until today. It is expected that such materials would obtain more excellent properties with thermo-mechanical treatments, i.e., the severe plastic deformation (SPD) and the thermo-treatment. However, materials given huge strain by SPD have a problem of unstabilization of thermal stability in the viewpoint of recrystallization and recovery.

In our previous study, the improvement of mechanical properties of ARBed Cu-Cr-Zr alloy after extra annealing processes was reported [2]. As a result, the high tensile strength of 625 MPa, the large total elongation of 15 % and the high electrical conductivity of 79 %IACS were obtained by ARB and subsequent aging treatment. However the average crystal grain size of 240 nm in ARB/aged material was larger than that of 210 nm in ARB/5cycles materials due to recrystallization by heat treatment. Therefore subsequent mechanical plastic deformation processing would affect a refinement of grains, and higher strengthening is expected simultaneously.

In this study, the additional cold rolling and the aging process were applied to Cu-Cr-Zr alloy sheets processed by ARB, and mechanical properties and structural information were investigated for the purpose of further improvement of the mechanical properties and the electrical conductivity.

2. Experimental

The Cu-0.85Cr-0.07Zr (in mass %) alloy was used in the present investigation. It was quenched in the 5 % water solution of NaCl after heating at 1000 °C for 2 hours for the purpose of solution heat treatment. Samples having dimensions of 50 x 300 x 1 mm^3 for the ARB processing were cut from the 80 % cold-rolled plate (As-rolled material). The ARB processing was carried out up to 5 cycles (As-ARBed material). In the ARB processing, a plate of 1 mm thick was cut into two pieces and they were stacked with brushed surface. Stacked plates were subjected into the roll bonding process with reduction of 50 % per one cycle. In addition, the aging treatment to ARBed materials was conducted at 450 °C for 2ks in the salt bath (ARB/aged material). The most excellent strength-electrical conductivity combination could be accomplished by this aging time and temperature [3]. As mentioned in the introduction section, further improvement about strength is expected by additional mechanical treatment to the ARB/aged material. Therefore, the additional cold roll (CR) processing was carried out to the As-ARBed (ARB/CR material) and the ARB/aged (ARB/aged/CR material) samples respectively. The thickness of two kinds of CRed materials was reduced from 1.0 mm to 0.2 mm by the additional cold rolling. An explanation drawing of each processing is shown in Fig.1.

Mechanical tensile test with the specimens of 3 mm in width and 10 mm in gage length was performed by tensile test machine at the strain rate of 10^{-3} sec^{-1}. Vickers hardness on the ND-RD section was also measured at the load of 2.94 N for 15 sec. Electrical conductivity of all specimens was measured by using eddy-current method.

Electron backscatter diffraction pattern (EBSD) images on the ND-RD section were obtained by the orientation imaging microscopy (OIM) system attached to FE-SEM. The average grain size, the fraction of high angle grain boundaries (HAGBs) and the grain boundary image were analyzed from EBSP data sets. Then, SEM micrographs on fracture surface of each specimen after tensile test were compared.

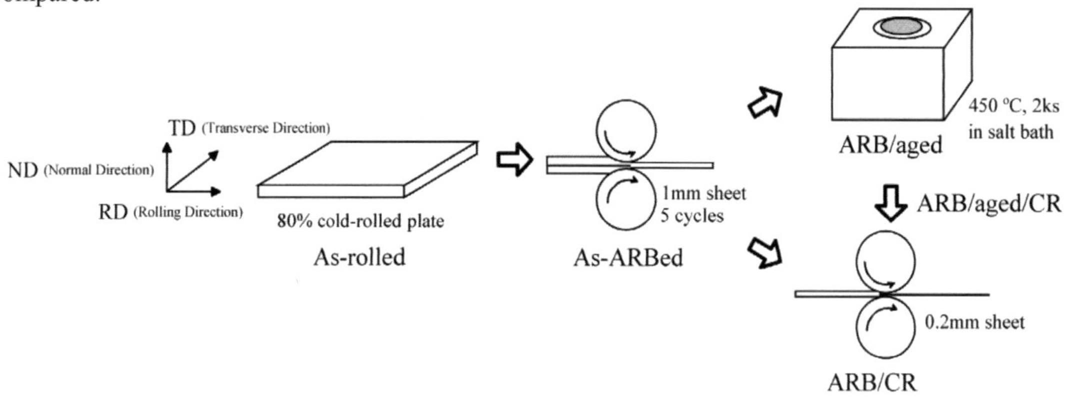

Figure 1. Materials preparation in this study.

3. Results and Discussion

3.1. Mechanical and Electrical Properties

Some mechanical properties obtained from present specimens are summarized in Table 1. The ARB processing improved the strength of the Cu alloy without deterioration of the electrical conductivity. Moreover, the elongation was improved simultaneously. In general, it is known that metallic materials have a trade-off relation between strength and ductility. Nevertheless, the result in this experiment was different from our knowledge. It is very interesting phenomenon. Subsequently, both the strength and the elongation were improved in the ARB/aged processing. It is noted in particular that high electrical conductivity could be achieved at the same time. Ordinarily, the dissolution of precipitation from solid solution and the elimination of dislocation with recovery occur by heat treatment, and they would affect the strength, the elongation and the electrical properties of this material. On the other hand, the ARB/CR processing could not improve the strength, and besides, it decreased the elongation and the electrical conductivity. Finally, although the electrical conductivity of the ARB/aged/CR processing diminished slightly in comparison with that of the ARB/aged processing, the most superior result about the tensile strength in all specimens was obtained by the ARB/aged/CR processing.

Table 1. Mechanical properties and electrical conductivity of all specimens.

Specimen (Thickness)		Vickers hardness (HV)	Yield stress (MPa)	Tensile strength (MPa)	Total elongation (%)	Electrical cond. (%IACS)
As-rolled	(1.0 mm)	141	460	464	7.5	35
As-ARBed	(1.0 mm)	170	540	585	10.0	35
ARB/aged	(1.0 mm)	209	605	625	15	79
ARB/CR	(0.2 mm)	170	600	626	6.9	39
ARB/aged/CR	(0.2 mm)	217	700	745	7.0	68

3.2. Microstructure and Related Data by EBSD

Boundary maps by EBSD are shown in Fig.2. The average grain sizes and the fraction of HAGBs analyzed by EBSD are summarized in Table 2. The lamellar structure consisting of grains like pancakes was created by the ARB processing, and this structure hardly changed by subsequent aging treatment. Refined grains about 200nm by the ARB contribute to the strength of specimens. Figure 3 shows a TEM micrograph of the ARB/aged specimen. Some precipitations that seem compounds Cu_5Zr in a grain can be recognized. The additional aging can induce the dissolution of precipitate without serious coarsening of grain size of matrix materials. Since grain refinement was promoted by additional cold rolling after additional aging treatment, the ARB/aged/CR specimen obtained the highest strength of all specimens. It is supposed that the ARB/aging/CR processing would induce extreme refinement of crystal grains due to the pile up of dislocations associated with small precipitations generated by the aging treatment ahead of the cold rolling. In addition, the strain-induced solid solution formation could occur because the electrical conductivity after additional CR was lower.

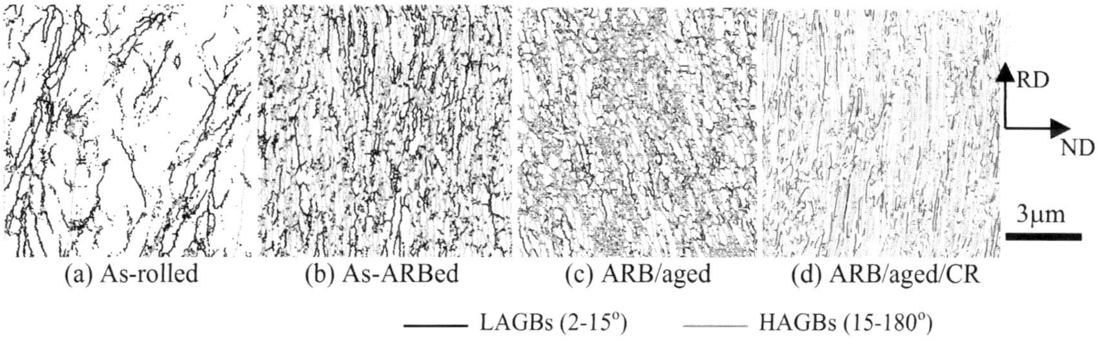

Figure 2. Boundary maps of specimens obtained by EBSD. Gray lines and black lines indicate High angle grain boundaries (HAGBs: 2-14°) and Low angle grain boundaries (LAGBs: 15-180°) respectively.

Table 2. Average grain size (lamella intervals) and fraction of HAGBs.

Specimen	Lamella intervals (nm)	Fraction of HAGBs (%)
As-rolled	~ $6x10^3$	31
As-ARBed	230	75
ARB/aged	260	81
ARB/aged/CR	130	84

3.3. Influence of Bonding Boundary in ARB on Strength

There is a possibility that the sheets processed by ARB include some incomplete bonding boundaries. Indeed, some such faults were found in all specimens from the observation of the fracture surface in tensile specimens. Two SEM pictures of fracture surface around bonding boundaries are shown in Fig.4. There were equiaxial dimples on the fracture surface, and it means indicating the typical ductile fracture. Some faults caused by incomplete bonding were found out among dimples. It could be observed that faults in the ARB/aged/CR specimen were pretty smaller than that in the ARB/aged specimen. The additional thermal and mechanical treatments were able to improve the condition of incomplete bonding boundaries. It is supposed that atomic migration takes place by additional heat treatment and subsequent plastic deformation segmentalizes remained faults. Although it is unclear whether the crack initiation corresponds with the incomplete bonding boundary, reduction of such faults would improve the mechanical properties.

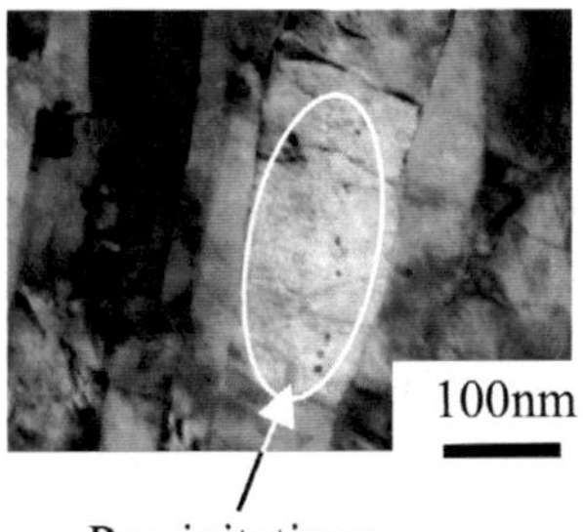

Figure 3. TEM micrograph of ARB/aged specimen.

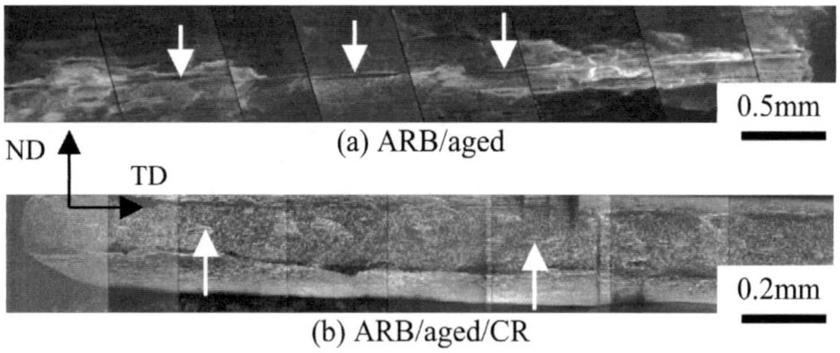

Figure 4. Incomplete bonding boundaries on fracture surface by tensile test. White arrows point to some of faults in specimens.

4. Conclusions

The ultrafine-grained Cu-Cr-Zr alloy processed by ARB was investigated with respect to the strength, the electrical conductivity and the grain size after additional aging and subsequent cold rolling. Major improvements from As-ARBed material are summarized as follows;

- The ARB/aged processing extremely improved not only the mechanical properties but also the electrical conductivity.
- The ARB/aged/CR processing achieved the highest strength of all specimens without sacrificing much of the electrical conductivity.
- The additional aging and CR processing was able to reduce incomplete bonding boundaries generated during ARB processing.

Acknowledgments

The authors are very grateful for the assistance provided by Mr. S Adachi of TSL solutions K.K., Japan on EBSD analysis. The assistance of Mr. K Saitoh and Mr. M Katoh at Kanazawa University is greatly appreciated. This work was supported in part by a Grant-in-Aid from Japan Science and Technology Agency and Japan Copper and Brass Association.

References

[1] Saito Y, Tsuji N, Utsunomiya H, Sakai T and Hong R G 1998 *Scripta Mater.* **39** 1221
[2] Kitagawa K, Akita T, Kita K, Gotoh M, Takata N and Tsuji N 2008 *Mater. Sci. Forum* **584-586** 791
[3] Otake Y, Kitagawa K, Kita K, Takata N, Tsuji N and Aoki S 2007 *J. JRICu* **46** 142

Effect of intense plastic straining on microstructure and mechanical properties of an Al-Mg-Sc alloy

R. Kaibyshev[1], E. Avtokratova[2], O. Sitdikov[2]

[1] Belgorod State University, Pobeda 85, Belgorod, 308015, Russia
[2] Institute for Metals Superplasticity Problems, Khalturina 39, Ufa, 450001, Russia

rustam_kaibyshev@bsu.edu.ru

Abstract. An Al-5%Mg–0.18%Mn–0.2%Sc-0.08%Zr-0.002%Be was subjected to equal-channel angular extrusion up to true strains of ~3 and ~8, that resulted in the formation of partially recrystallized and fully recrystallized structure, respectively. It was shown that the alloy with partially recrystallized structure exhibits highest strength and ductility. The material with fully recrystallized structure showed lowest fatigue crack growth rate and highest value of fracture toughness. Reasons of this unusual effect of microstructure on crack propagation resistance under fatigue are discussed.

1. Introduction

The formation of micron scale grained structure in Al-Mg-Sc alloys by equal-channel angular extrusion (ECAE) technique leads to significant increase in their strength [1] and extraordinary superplastic ductilities [2]. These properties, however, are not sufficient for structural aluminum alloys. For commercial use as a material for critical components of an aircraft the knowledge of fatigue properties and fracture toughness of these alloys processed by ECAE is very important. Unfortunately, the limited number of studies were dealt with examination of these properties in Al-Mg-Sc alloys [3-6]. The aim of the present work is to report effect of ECAE on strength, ductility, fatigue crack growth rate and fracture toughness. It is known [1] that it is possible to produce different microstructure by ECAE varying strain. In this study the mechanical properties were examined for an Al-Mg-Sc alloy with partially recrystallized structure and fully recrystallized structure produced by ECAE.

2. Experimental

The 1570C Al with a chemical composition of Al-5%Mg–0.18%Mn–0.2%Sc-0.08%Zr-0.002%Be (in weight %) was used as the starting material [7]. The ingot was cut into plates with dimensions of 152x152x25 mm^3. These plates were subjected to ECAE with rectangular shape of channels up to true strains of ~3 and ~8 by route B_{cz} at a temperature of 325°C. Details of this ECAE procedure were described in previous works [3,4]. Tensile specimens of 6 mm gauge length and 1.4 x 3 mm^2 cross-section were machined from these plates with tension axis lying parallel to direction of last extrusion. These samples were tensioned to failure at ambient temperature and a strain rate of 5.6 x 10^{-3} s^{-1} using an Instron 1185 testing machine. The tests for cyclic crack resistance were conducted using a Schenk hydropulse PSA hydraulic machine. Compact tension (CT) specimens 20 mm thick were cut in accordance with ASTM Standard E399-06. The specimens were loaded using a sinusoidal cycle with a frequency of 5 Hz and an asymmetry coefficient R = 0.1. The fatigue crack growth rate as a function of the stress-intensity factor range, ΔK, was obtained in accordance with ASTM standard E647-08. Examinations of microstructure and fractured surface were carried out using optical microscope Nikon L-150 and scanning electron microscope JSM-840 equipped with EBSD attachment. Details of structural characterizations were reported previously [3,4,7].

3. Results and discussion

No coarse second-phase particles with a size higher than 1μm were found in initial structure [7]. Coherent dispersoids of $Al_3(Sc,Zr)$-phase having equiaxed shape and a size ranging from 5 to 10 nm are uniformly distributed within the grains [7]. ECAE up to true strains of ~3 and ~8 resulted in the formation of partially (fig.1a) and fully (fig.1b) recrystallized structure. Fraction of recrystallized grains is 0.25 and 0.95; their average size is ~1.6 μm and ~1.2 μm, respectively. Highly elongated unrecrystallized grains aligned along the last extrusion direction can be clearly distinguished in the samples strained up to ε~3 (fig.1a). Careful inspection of Fig.1a shows that subgrains locate within deformation bands.

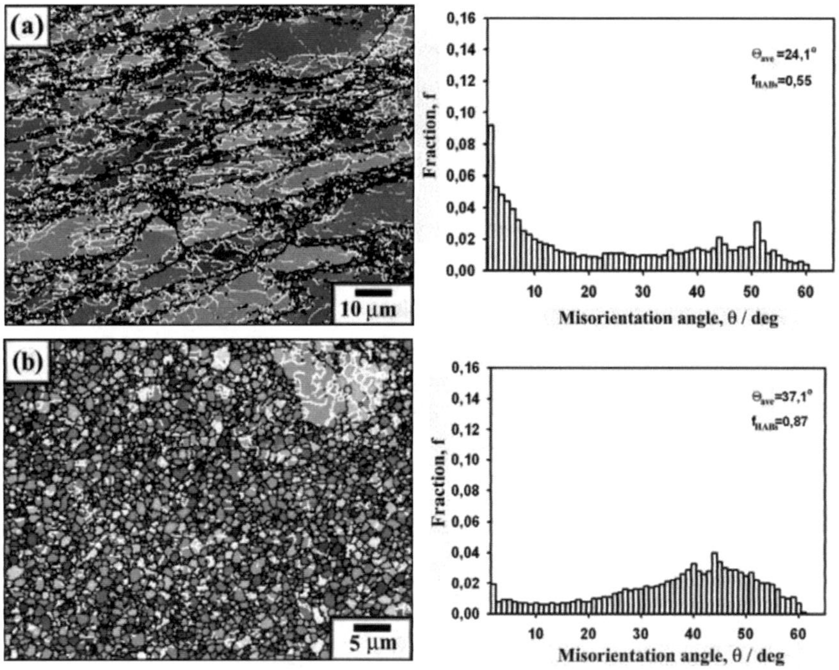

Figure 1. Microstructures in the 1570C after ECAE to a total strain of ~3 (a) and ~8 (b).

Mechanical properties of the 1570C alloy in initial state and after ECAE with different strains are summarized in Table 1. It is seen that intense plastic straining leads to a significant increase in yield stress; total elongation remains virtually unchanged. Ultimate stress increment is minor. Samples strained to ε~3 exhibit highest strength and ductility. Therefore, the formation of partially recrystallized structure in the 1570C provides highest strength that can be attributed to highest density of lattice dislocations.

In contrast the recrystallized state of the 1570C alloy showed lowest values of fatigue crack growth rate (Fig.2). At stage I, the fatigue crack growth rate for the 1570C alloy with partially recrystallized structure is higher by a factor of 3.75 than that for the alloy with fully recrystallized structure. At

ΔK=6 MPa·√m, fatigue crack propagation growth rates are 3×10^{-8} and 8×10^{-9} m/cycle, respectively. At stage II and ΔK=30 MPa·√m, the values of fatigue crack growth rates are 5.8×10^{-6} and 3.4×10^{-6} m/cycle for the 1570C subjected to ECAE up to true strains of ~3 and ~8, respectively.

Table 1. Mechanical properties of the 1570C alloy in initial state and after ECAE

State	YS (0.2%) (MPa)	UTS (MPa)	Elongation (%)
Initial state	238±3	354±6	28±1
ECAE with a total strain of ε~3	309±4	390±2	32±3
ECAE with a total strain of ε~8	300±4	379±6	31±2

As a result, the fracture toughness for the 1570C alloy with fully recrystallized structure attained a value of 43 ± 2 MPa·√m, while this value for the alloy with partially recrystallized structure was found to be 33 ± 2 MPa·√m. Therefore, the formation of fully recrystallized structure provides achieving higher crack propagation resistance in spite of slightly lower strength.

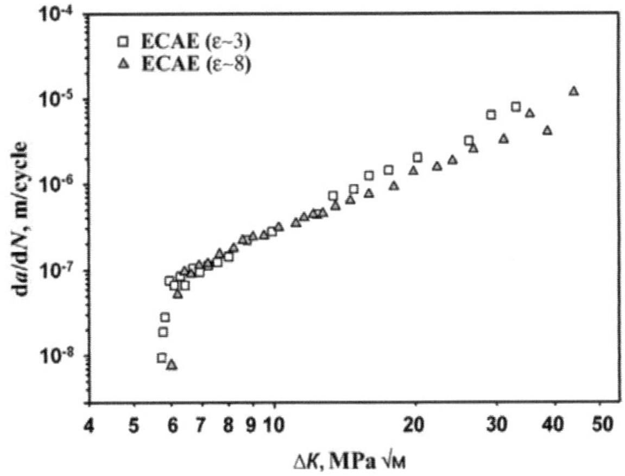

Figure 2. Relationships between crack growth rate and stress intensity factor range.

Crack paths and fractography are represented in Fig.3 and Fig.4, respectively. Notably, no significant difference in fatigue crack growth paths for two states of the 1570C alloy was found at low and moderate values of the stress intensity factor range. However, at larger values ΔK close to ~30 MPa·√m (Fig.3), the crack propagation surface for the sample strained up to ε~3 is more smooth (Fig.3a) than that deformed to ε~8; that induces a higher fatigue crack growth rate. Examination of fatigue crack surfaces (Fig.4) supports this conclusion. At high values of stress intensity factor range, ΔK, both samples exhibit coarse fracture surface, where the fracture crack growth seems to occur by classical crack-tip blunting mechanism that gives rise to striations formation accompanied by opening secondary cracks [3]. At the same time, in the samples with partially recrystallized structure many cracks that grow along deformation bands are also observed. These can give an additional contribution in acceleration of fatigue crack growth rate.

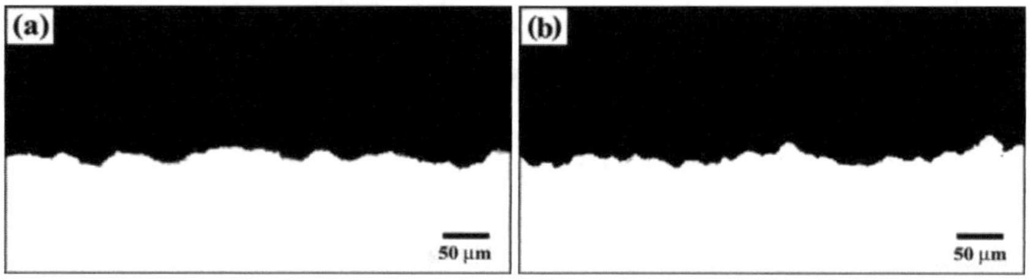

Figure 3. Optical micrographs for fatigue crack growth paths at 13 mm from the straight through notch for the 1570C subjected to ECAE up to a total strain of ~3 (a) and ~8 (b).

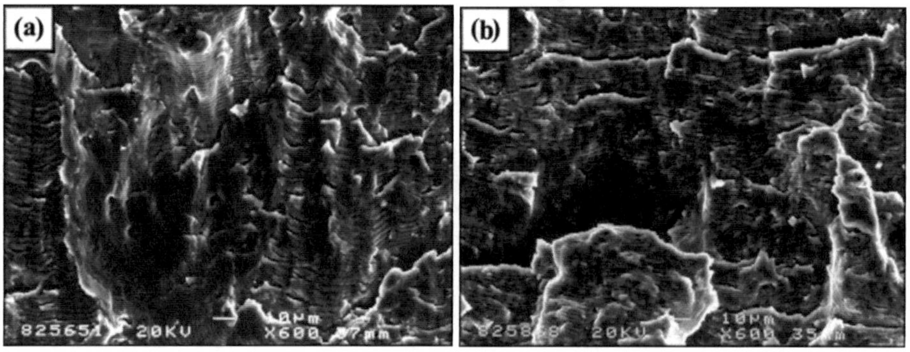

Figure 4. Fatigue crack surfaces at ΔK~30 MPa·√m for the 1570C subjected to ECAE up to a total strain of ~3 (a) and ~8 (b).

4. Summary

The 1570C alloy belonging to Al-Mg-Sc system was subjected to large strain by ECAE. This processing resulted in fully recrystallized structure with an average grain size of ~1.2 μm and enhancement of strength and crack propagation resistance. Values of fatigue crack growth (da/dN= $3,4\times10^{-6}$ m/cycle at ΔK =30 MPa·√m) and fracture toughness (ΔK_{max}=43±2 MPa·√m) are close to these properties of AA2324 alloy that allow considering the 1570C alloy for aircraft applications.

Acknowledgements
This work was supported by the International Science and Technology Center under grant No.2011 and Federal Agency of Education under grant no.P1065 .

References

[1] Valiev R Z, Islamgaliev R K and Alexandrov I V 2000 *Prog. Mater. Sci.* **45** 103
[2] Musin F, Kaibyshev R, Motohashi Y and Itoh G 2004 *Scripta Mater.* **50** 511
[3] Avtokratova E.V., Sitdikov O.Sh., Kaibyshev R.O., Watanabe Y. 2009 *The Physics of Metals and Metallography* 2009 **107** 291
[4] Avtokratova E.V., Kaibyshev R.O., Sitdikov O.Sh. 2008 *The Physics of Metals and Metallography* **105** 500
[5] Vinogradov A., Washikita A., Kitagawa k., Kopylov V.I., 2003 *Mater. Sci. Eng.* **349** 318
[6] Vinogradov A. 2007 *J. Mater. Sci..* **42 1797**
[7] Kaibyshev R.,Avtokratova E, Apollonov A., Davies R. 2006 *Scripta Mater.* **54** 2119

Physical foundations of crystal structure refinement under severe plastic deformation

V. E. Panin and V. E. Egorushkin

Institute of Strength Physics and Materials Science SB RAS, Tomsk, 634021, Russia

paninve@ispms.tsc.ru

Abstract. The problem of crystal structure refinement under severe plastic deformation has been examined in the framework of the multiscale approach used in strength physics and nonequilibrium thermodynamics. This approach assumes that fragmentation of a solid in the entire hierarchy of the structural-scale levels (nano-, micro-, meso- and macroscale levels) plays a crucial role in the plastic deformation and fracture of materials. Formation of submicrocrystalline or nanocrystalline structure in solids subjected to severe plastic deformation is associated with fragmentation of the strain-induced band structures in the bending internal stress field. Flowing defect boundaries at the lower scale accomodate fragmentation of any curvature in a highly nonequilibrium material under SPD.

1. Introduction

Translation invariance is known to assure the thermodynamic stability of crystals. Violation of the translation invariance in a defect crystal causes its internal energy to increase. Because of this, the crystal undergoes continuous fragmentation, and defects are displaced toward fragment boundaries. Processes of this kind obey laws of nonequilibrium thermodynamics. The nonequilibrium thermodynamics of a deformed solid as a multiscale system and the corpuscular-wave dualism of plastic shear were discussed elsewhere [1]. In this work, the thermodynamical approach is applied to a description of fragmentation of solids under severe plastic deformation (SPD).

2. The nonequilibrium thermodynamics of fragmentation of solids in external force fields

Metastable mesoscopic substructures formed in a crystal fragmented at different scale levels are responsible for the local minima seen in the plot of the Gibbs thermodynamic potential against the molar volume, $F(v)$, shown in figure 1. To describe the local minima in terms of the thermodynamics of nonequilibrium states, we will consider the effective potentials $U(v,\alpha)$ in local defect-structure fields α [2], because this parameter accounts for the local fall in the magnitude of $F(v,\alpha)$ due to the entropy production during defect phase formation.

The relation for the Gibbs thermodynamic potential is of the form:

$$F(v) = U - TS + pv - \sum_{i=1}^{n} \mu_i c_i , \tag{1}$$

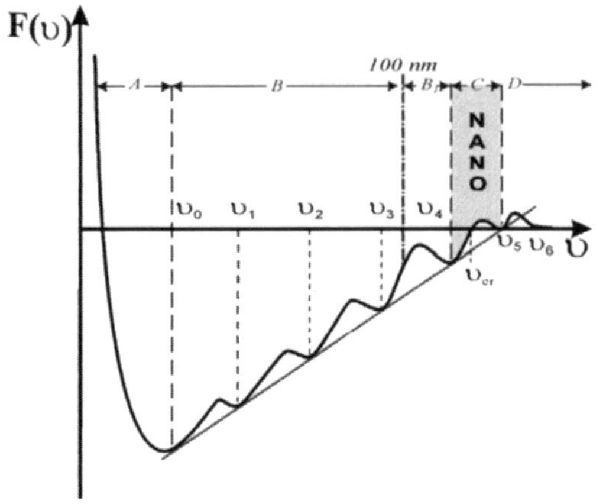

Figure 1. Dependence of the Gibbs thermodynamic potential F on the molar volume v with allowance made for local hydrostatic tension zones of different scale levels.

where U is the internal energy, T is the temperature, S is the entropy, p is the pressure, v is the molar volume and μ_i is the chemical potential of the ith element with concentration c_i.

Since defect phases occur in local hydrostatic tension zones [1], v is used as an independent variable for the Gibbs thermodynamic potential. In the equilibrium state of a crystal for $v = v_0$, $F(v_0)$ has a minimum value. For any change in v, the value of $F(v)$ for the examined crystal is increased. By virtue of (1) the local minima in the ascending branch of the $F(v, \alpha)$ curve are attributable to the entropy production and redistribution of alloying elements between the basic crystal and the defect phases because of the difference in the chemical potentials of each of the ith components.

Following [3], the entropy production σ_s is written as

$$\sigma_S = \frac{\chi(\nabla T)^2}{T^2} + \rho\frac{\overline{v}\overline{\sigma}}{T} - \frac{\rho}{T^2}\left(\left[\overline{\alpha v}\right], \nabla T\right), \tag{2}$$

where χ is the thermal conductivity, ρ is the density of the material, $\overline{v}$ is proportional to the density of the defect flux and $\overline{\sigma}$ is the hydrostatic tension. The augend in (2) is the entropy production due to heat release. The addend defines the work done by the stress field on the defect flux to provide the motion of defects. The third term is related to the energy flux of the mechanical field of the crystal through the surface bounding the local structural transformation zone during the formation of the defect phase. The first two terms always have positive values and specify the entropy production. The third term has a negative value and reduces the entropy production. This term accounts for the relaxation of the disturbance of the mechanical field of the strained crystal, as the local structural transformation takes place in the local hydrostatic tension zone.

The fact that easy local structural transformation may take place during generation of dislocation nuclei near the ground state of a solid (the average molar volume $v = v_0$) determines a high value of the negative term in (2) for σ_s. In a thermodynamically stable crystal, the splitting of dislocations is limited by the energy flux of the mechanical field. As a consequence the entropy production is small during generation of dislocations and shallow minima appear in the $F(v)$ curve for $v = v_1$. For this reason, dislocations propagate along the slip plane as solitons without leaving any traces. The generation of dislocation nuclei is classified as local structural transformations in the hydrostatic tension zones.

As the degree of nonequilibrium of a solid is increased, it moves away from the ground state when $F(v_0)$ is at a minimum. For a certain range of values of v_i, local structural-phase transformations take place in the local hydrostatic tension zones, as seen in the ascending branch of the $F(v) = f(v)$ curve to the right of figure 1. The transformations are not suppressed by weak energy fluxes of the mechanical field of a nonequilibrium crystal. In this case, the negative term plays a secondary role in the entropy production, whereas the positive terms are growing in importance. The depth of the local minima in the $F(v) = f(v)$ curve increases. Irreversible defect structures (twins, bands of different scales and strain-induced martensite) are formed in the strained material.

As evident from figure 1, the condition $F(v) < 0$ is met in the range v_0–v_{cr}. This means that all crystal fragmentation stages under SPD in this range can be described in the framework of existing theories of deformation defects and their self-organization.

Notably, a nonequilibrium solid under SPD demonstrates a new peculiar quality where the Gibbs thermodynamic potential for the strained material approaches zero, and a two-phase pre-transition state is formed in the material (region *C* in figure 1). In this state, some of the nanocrystals with molar volume v_4 are surrounded by quasi-amorphous interlayers with molar volume v_5. We emphasize that the Gibbs thermodynamic potential for the nanocrystals is negative, even though their molar volume is dramatically increased (which is why the material is in a highly nonequilibrium state), whereas the quasi-amorphous interlayers have $F(v) > 0$. The shear instability of the quasi-amorphous interlayers accounts for the anomalous dynamic behavior of the Hall-Petch relation for nonequilibrium two-phase nanocrystals-amorphous phase systems. The pre-transition two-phase state is referred to as the nanostructural state of a solid for a grain-size range $d < 30$ nm. In addition to that, a solid may undergo fragmentation of nanosized structures with the proviso that $F(v) < 0$ (region B_1 in figure 1). Mechanical behavior of solids of this kind can be described using Hall-Petch relation. Finally, local zones formed for $v > v_5$, wherein $F(v) > 0$, are evidence of microporous structure developing in these regions of the material [1]. To prevent the porosity formation, SPD must be accomplished under hydrostatic compression

3. The influence of solid nonequilibrium degree on mechanisms of its refinement at SPD

Two kinds of SPD were used to refine structure of specimens under study and, by doing so, to change their nonequilibrium degree: cold rolling and high cycle fatigue testing

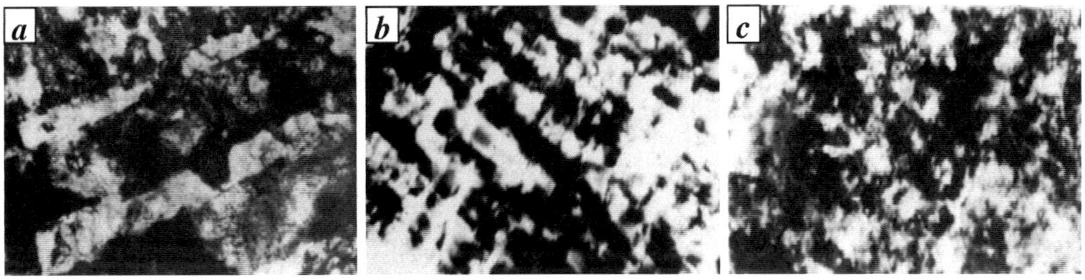

Figure 2. Formation of macroscopic localized-deformation bands at different stages of cold rolling of high-nitrogen austenitic steel Cr20NMn20V: $T = 293$ K and $\varepsilon > 95$ %: (a) macrobands as a system of contacting subbands of crystal and defect phases, × 18160; (b) macroband fragmentation under superplastic flow, ×16000; (c) cellular substructure, ε=99%,x30000.

As an example, figure 2 shows the evolution of the mesoband structure in high-nitrogen austenitic steel Cr20NMn20V subjected to very high degrees of cold rolling. In the initial stage of cold rolling, a typical mesoscopic structure is formed in the material However, very high degrees of cold rolling give rise to macroscopic bands evident as a defect phase (figure 2(a)). Notably, the macrobands experience severe bending. The macroband length increases dramatically, and continuous and discrete misorientations as high as several tens of degrees are accumulated at the macroband boundaries. The electron diffraction patterns from the regions within the bands exhibit a system of reflections arranged in pairs, which is evidence of two-phase macroband structure. One type of plastic-shear subbands has the same crystal structure as the initial material with a certain density of dislocations characterized by chaotic distribution. Another type of subbands is a defect phase in its own right. On further increase in the degree of cold rolling the bent crystal subbands are fragmented. The space between the fragments is filled with the defect phase (figure 2(b).). Increasing of ε up to 99% is accompanied by formation of

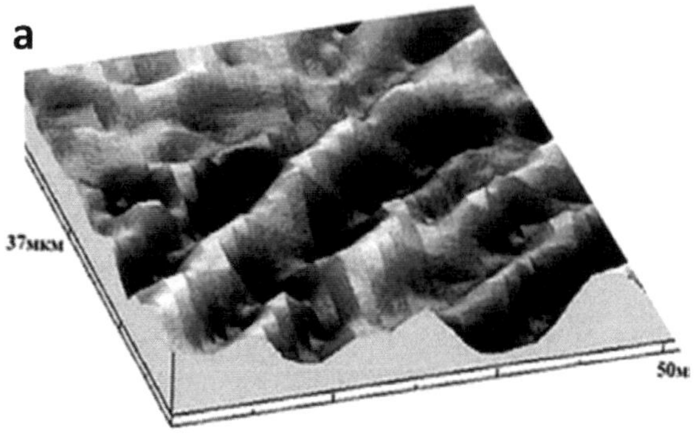

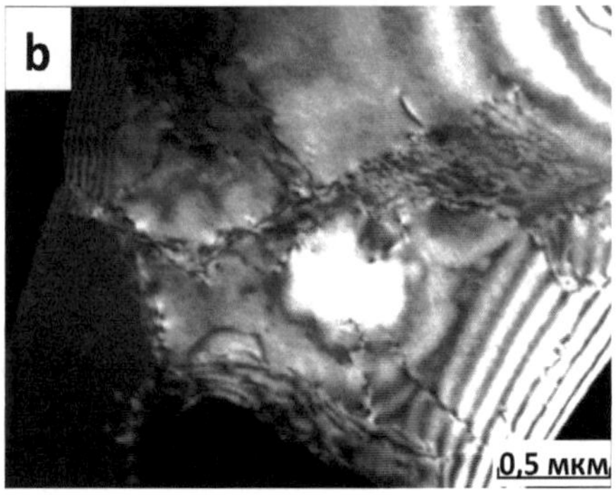

Figure 3 Laser profilometry (a) and TEM (b) of a high-purity 35 μm thick Al film glued to a commercial Al substrate; alternating bending at T=293R with the number of cycles $N=2,9\cdot10^6$

cellular substructure with the characteristic cell size of 100 nm (figure 2(c)). Cold-working dynamic recrystallization develops in some zones. In cold-rolled materials, structures of this kind are formed in the superplasticity regime.

An ingenious method of SPD was offered in [4] making it possible to obtain high severe plastic deformation. Thin film of high-purity Al glued to a commercial Al substrate is subjected to fatique testing. It allows to realize very high degree of SPD within Al-film under alternating bending with the number of cycles N up to 18×10^6. This method was also used in this investigation.

Figure 3(a) shows the 3D micrograph to characterize extrusion of material from surface layer of Al-film at $N=2,9\cdot10^6$. It is easily seen that quasiperiodical extruded mesobands of localized plastic deformation appear. Each mesoband consists of a set of lamellae being displaced along each other. Very high curvature of lamellae indicates about high nonequilibrium state of the extruded material. As it is seen in figure 3(b), plastic flow of extruded lamella is realized by propagation of flowing spiral – wise defect boundaries, whose trajectory and curvature depend upon internal stress field. The results shown in figure 3 testify that the refinement of a very nonequilibrium solid develops as multiscale process where at the lower scale the flowing defect boundaries accommodate fragmentation of any curvature. Sophisticated description and discussion of plastic flow mechanisms of highly nonequilibrium solid will be published soon in a separate paper.

Acknowledgement
The work was supported by the SB RAS (Project № 3.6.1.1 and 1).

References
[1] Panin V E and Egorushkin V E 2008 *Phys. Mesomech.* **11** 105
[2] Leontovich M A 1938 *Zh. Exp. Teor. Fiz.* No. 7 844
[3] Egorushkin V E 1990 *Russ. Phys. J.* **33** 135
[4] Panin V E, Elsukova T.F, Egorushkin V E, Vaulina O Y, Pochivalov Y I 2009 *Deformation and Fracture of Materials* No. 6 2 (*in Russian*)

Influence of laser-induced microstructure notches on the configuration of residual stresses in a steel plate

S Pawlak[1], **K Dutka**[2], **J Bonarski**[1]

[1] Institute of Metallurgy and Materials Science Polish Academy of Sciences, 25 25 Reymonta Str., 30-059 Kraków, Poland
[2] University of Warmia and Mazury, 2 Michała Oczapowskiego Str., 10-719 Olsztyn, Poland

nmpawlak@imim-pan.krakow.pl

Abstract. Among technological problems related to manufacturing materials with controlled and localized properties, the effect of a microstructure notch is still unsatisfactorily recognized. Laser modification of selected areas of material structures is one of the most promising technologies in the field. The work presents selected variants of generating the intended microstructure changes in the near-surface areas of steel plates by means of a laser beam. A field of residual stress was generated in the examined steel as a result of phase transformations accompanying the laser treatment. Both the microstructure changes and configuration of residual stresses were identified by means of the optical microscopy, and X-ray diffraction technique. The results, additionally related to crystallographic texture, allowed to characterize the microstructure modifications and to interpret a potential change of mechanical properties. It was found, that the level of the stresses and its range can vary essentially with parameters of the laser treatment, which gives one a chance to control the behaviour of a material in the conditions of mechanical loading, by exploitation of a structure element.

1. Introduction

A generally understood decay of constructional elements can be defined as a process of multi-aspectual degradation of material structure. Progress of the degradation process depends on the nature of destructive factors and on the resistance of a material. From the point of view of the efficiency of a construction and the safety its exploitation, the possibility of controlling the profile of decay zone spreading in a material constitutes a interesting question. One way of achieving a control of that sort is to introduce a definite disorder of material microstructure which makes it possible to control both the size of the destruction area, and the propagation route of a destructive process. Among many methods of generating microstructural notches, the laser melting/hardening of a surface is one of the most effective methods, which is due to a comparatively large range of possible variants and close control over the location of generated notches [1,2]. This work presents a texture and stress profile of the microstructure of steel with a microstructural notch made by controlled laser treatment.

2. Experimental procedure

Structural steel 45H (0.45%C, 0.65%Mn, 0.27%Si, 0.95%Cr) with maximum content of P and the content of S within the range of 0.025-0.035% was used in the experiments. The sample in the form of plates(ca. $55 \times 10 \times 4$ mm^3) was cut off a steel-rod perpendicular and parallel regarding the rolling direction(RD). Before cuting the samples, laser remelting in the form of straight paths was performed on a selected surface of the rod. As a result of thermal treatment (hardening at 1050°C and then annealing at 510°C), the investigated steel reached the hardness of 350 HV and the yield stress of 1200 MPa. In laser processing were used: high-power fibrous laser HPFLSPI 200C and high-power diode laser HPDL ROFIN SINAR DL 200. The direction of remelted paths was parallel or perpendicular to the rolling direction (RD). Both a diode and fiber-type laser with various combinations of power and velocity travel were applied. The treatment was carried out in nitrogen or argon protective atmosphere. The research area of each sample was prepared in the form of metallographic specimens, where cross-sections of laser-remelted paths were revealed. Electron and optical microscopy were applied to characterize the microstructure and morphology of laser-modified areas. Residual stresses and crystallographic texture in selected sub-areas surrounding laser-treated volumes of material were analyzed by X-ray diffraction techniques. Topography of changes in the above characteristics of microstructure of the samples in question was obtained (figure 1).

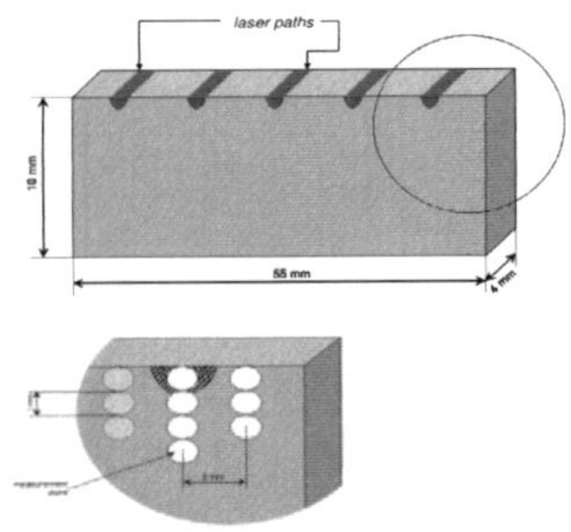

Figure 1. Scheme of investigated sample with cross-section of the laser paths (top). Details of area marked by the circles present the location of X-ray measurement points (solid-line circles filled in white color) on sample surface (left-bottom). Data for points marked by the dashed-line circles filled in gray color were assumed the same as the counterpart "white" points located symmetrically with respect to the laser path position. Orthogonal reference system assumed in stress analysis is presented too (right-bottom)

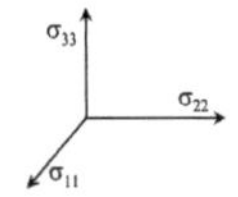

The presented investigations concern four chosen samples with paths remelted by means of lasers configured for various parameters of treatment. The samples are denoted as DA22_1, DN22_1, WA22_3,and WN22_3. The notation reads: type of laser (D– diode-type, W– fibrous-type), protective atmosphere (N - nitrogen, A - argon), temperature of hardening (2 corresponds to 1050°C) as well as the direction of laser paths (2 - perpendicular to RD). The last digit in a sample symbol indicates the number of the path (definite power and relative velocity of the laser head). Experimental data for texture and stress analyses were registered in the areas (measurement points) marked in figure 1.

3. Results and discussion

3.1. Microstructure

The pictures obtained by means of optical microscopy were used to identify the size of the heat-affected zone (HAZ), the distribution of martensite and ferrite in modified areas of steel. The results show that the size of the HAZ, remelting area and hardened volume increased almost threefold when the diode laser was used, as compared to the treatment with the fibrous laser.

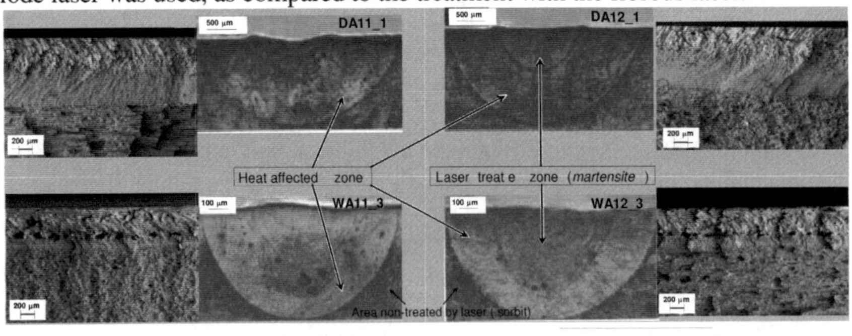

Figure 2. Cross-sections of samples with selected laser paths performed by the diode and fibrous laser. Nearby each of the cross-section, a related longitudinal-break counterpart registered by means of electron microscopy is presented. The optical and electron microscopy pictures have been made in the IMIM PAN in Krakow and in the WITPiS in Sulejowek, respectively.

Optical and electron microscope pictures of the cross-sections of laser paths reveal a strong inhomogeneous microstructure of the modified areas of the steel samples (figure 2). Apart from the intentionally introduced disturbances of the microstructure, distinct zones of gas pores are observed. The reaction of the examined steel to localized laser beam was also reflected in the shape of the microstructure irregularities (elongated sub-areas) in the path region. Depending on the parameters of the laser treatment, a specific morphologic pattern of the path can be observed. Visible differences between the tracks result from differences in the power and velocity of laser travel. Individual laser paths are characterized by considerable heterogeneity produced during the melting process followed by subsequent crystallization (difference in size and shape of dendrites). A distinct diversity in the size of the HAZ accompanying each of the paths is visible in its surroundings. The morphology and configuration of individual phases with a defined preferred orientation of crystallites and the post-processing morphological objects (dendrites, pores) create a kind of microstructure notch in the sample space which generates a field of residual stresses in laser-modified steel.

3.2. Crystallographic texture

Texture analysis of investigated samples reveals the preferred orientation of crystallities related to the cross-section of laser paths. The coordinate system (ND, TD, RD) assumed in texture analysis is related to the one assumed for stress analysis and presented in figure 2: ND ≡ σ_{11}, TD ≡ σ_{22}, RD ≡ σ_{33}. Experimental (110), (100) and (211) pole figures of ferritic-martensitic phase were registered for the round-shape surface areas with the diameter of 1.0 mm, covering the remelted region and partially HAZ. The X-ray diffraction technique was applied by the Schulz's method of back-reflection[4]. Based on the experimental pole figures, an orientation distribution function (ODF) was calculated by discrete the arbitrarily defined cells method [5], which allows one to reproduce complete pole figures and inverse pole figures of the normal direction (ND) for the ferrite-martensite phase. LaboTex software was used in texture analysis [6]. As it can be observed in the inverse pole figures of ND for selected paths presented in figure 6, the texture of the material bulk treated by the diode laser differs from the material treated by the fiber laser. Apart from orientations of ND dominating in all of the considered paths and situated in the range of <216> ÷<116> directions, an additional, relatively strong component close to <321> orientation was identified in the case of the diode laser. Another important result concerning the RD direction informs about its dominant space orientation near <100> in the case of diode-laser paths and close to <410> orientation in the case of fiber-laser ones. The third direction (TD) connected to the investigated sample surface (cross-section of the laser paths) does not show any stronger diversification depending on the kind of laser and parameters of treatment.

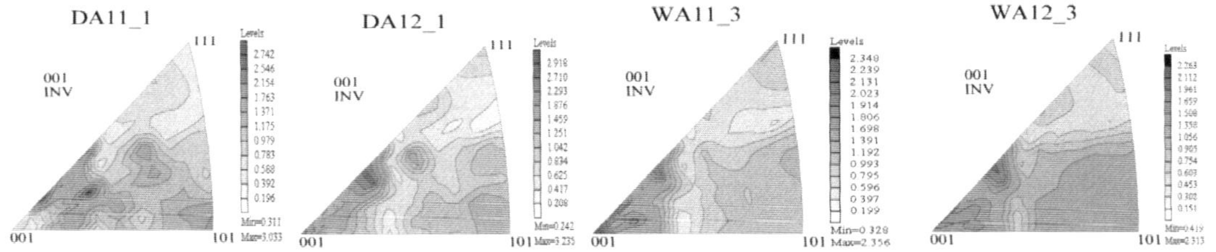

Figure 3. Inverse pole figures of normal direction (ND) for selected laser paths performed by the diode laser: DA11 and DA12, and by fibrous one: WA11 and WA12

3.3. Residual stresses

The distribution of residual stresses (σ_{22} component in figure 2) in the central zone of the laser paths and in its surroundings was analyzed by the X-ray $\sin^2\psi$ method [7]. The diffraction profiles were measured by means of Rigaku MSF X-ray Stress Analyzer. Young's modulus E = 223,3 GPa and Poisson coefficient ν= 0.28 have been assumed in the calculations. The results are shown in figure 4. The revealed distributions of the residual stresses depend essentially on the parameters of the performed laser treatment. The highest compressive stresses were generated by the fiber laser in nitrogen atmosphere and the lowest ones by the diode laser in the same protective atmosphere. Along with compressive stresses, tensile stresses also were changed. The highest compressive stresses (ca. –250 MPa) were identified in sample DA12 for path No. 1, at the distance of about 1.0 mm from the central point of the cross-section (figure 4). In sample DA11 (path No. 1) for the same distance, tensile stresses (up to 60 MPa) were identified. The larger the distance from the central point (~3.0 mm), the greater the tensile stresses become in both samples treated by the diode laser. The distribution of the tensile stresses in sample DA12 passes much more softly into the distribution of compressive ones, as compared to sample DA11. In the case of the fiber laser (sample WA11 and WA12), an analogous tendency in the distribution and level of the stresses is observed, regarding mutual orientation of the paths direction and rolling direction of matrix material before laser treatment. Essential difference in the stress distribution are observed for samples DA11_1 and WA11_3. Among the analyzed variants of the laser treatment, the most diverse stresses (from compressive $\sigma_{22} =$ 200 ±140 MPa to tensile $\sigma_{22} = 310 \pm 240$ MPa) were identified in the case of paths generated by fibre laser in argon atmosphere and for laser paths perpendicular to RD.

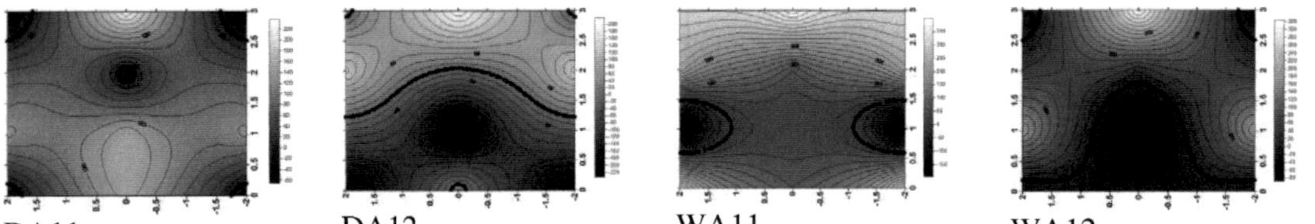

DA11 DA12 WA11 WA12

Figure 4. Distribution of values of the σ_{22} component of stress field in the area of the cross-section of the diode laser path No.1 and of the fibrous laser path No.3.

4. Conclusions

Based on the performed analyses of texture, stress analyses and microscopic observation, the laser modified areas of investigated steel can be qualified as a poly-phase microstructural notch. The morphology of remelted areas surrounded by the heat affected zones create specific conditions for maintaining a specific field of residual stresses in the modified material. In the case of the fiber laser treatment, the modified microstructure is characterized by greater dimensions and exhibits more homogeneous and weakly shaped texture, as compared to the case of diode laser treatment. The configuration of the stress field (its σ_{22} component) is essentially influenced by the parameters of the laser treatment. The investigated variants of laser modification make it possible to identify the areas of potential weakness in the material in response to external mechanical loading. It is also possible to foresee the road of crack propagation in examined steel plates after applied laser modification.

Acknowledgements

The authors would like to acknowledge the National Centre for Research and Development (NCBiR) for the financial support of project N R15004804, the Polish Ministry of Education and Science for supporting the work within the framework of project PBZ-MNiSW-3/3/2006 as well as by the allocation of founds designed for science in the years of 2007-2010.

References

[1] Yang S, Wang Z, Kokawa H, Sato Y 2008 *Mater. Sci. Eng.* A **474** 112-119
[2] Grum J 2007 *J. Achievements Mater. Manufact. Eng.* **24** 17-25
[3] Durand-Charre M 2004 Microstructure of Steels and Cast Irons (Berlin: Springer)
[4] Schulz L G 1949 *J. Apply. Phys.* **20** 1030
[5] Pawlik K 1986 *Phys. Stat. Sol.* **134** 477.
[6] LaboTex 2006 The Texture Analysis Software by LaboSoft s.c.
[7] Noyan I C and Cohen J B 1987 Residual Stresses –Measurement by Diffraction and Interpretation

Increasing strength, ductility and impact toughness of ultrafine-grained 6063 aluminium alloy by combining ECAP and a high-temperature short-time aging

L. W. Meyer , R. Schönherr , M. Hockauf

Chemnitz University of Technology, Materials and Impact Engineering, D-09107 Chemnitz, Germany

robert.schoenherr@mb.tu-chemnitz.de

Abstract. Since fully-dense ultrafine or nanocrystalline bulk materials can be processed, there has been an increasing scientific interest in several plastic deformation (SPD) procedures, particularly in the last decade. Especially the equal-channel angular pressing (ECAP) has widely been investigated due to its ability of producing billets sufficiently large for industrial applications in functional or structural components. The significant strength increase based on grain refinement is typically accompanied by a significant decrease in ductility and toughness. Within this work, a new methodology was applied for combining ECAP with a subsequent high-temperature short-time aging for the 6063 aluminium alloy. An increase in strength, ductility as well as impact toughness regarding its coarse grained counterparts was reached. More precisely, ultimate tensile strength, elongation to failure and impact toughness were increased by 46%, 21% and 40% respectively. This was observed after only one run of ECAP at room temperature in a solid-solution treated condition and an aging at 170° C for 18 minutes. The regular aging time for maximum strength at 170° C is around 6 hours. Longer exposure times lead to recrystallisation and, as for regular aging, it leads to overaging, both causing a decrease of properties. The work demonstrates a strategy for an efficient processing of commercial Al-Mg-Si alloys with outstanding mechanical properties.

1. Introduction

For property optimization of metallic materials by creating an ultra-fine grained microstructure, the ECAP deformation has gained acceptance during the last years [1-3]. During this procedure, a rod-shaped semi-finished product is pressed through an angled channel. When passing the deformation zone (cutting angle of channels), a large shear strain is implemented, which is, compared to extrusion pressing, completely homogenous above the cross-section with a very small fringe.

2. Experimental procedure

The deformation of materials took place in an ECAP prototype tool which is equipped with movable parts on the channel walls to reduce friction. The basic structure is shown in Figure 1 . The shear of the used channel angle of $\Phi = 90°$ is $\gamma \sim 2$ which is equivalent to a comparable deformation of $\varphi \sim 1.1$ [4]. Since the tool's entry and exit channels have the same cross-section, repetitive compressions for cumulating the deformation amount can be attached. In the following text, the number of passes is indicated by the continuous index *N*.

Journal of Physics: Conference Series **240** (2010) 012123 doi:10.1088/1742-6596/240/1/012123

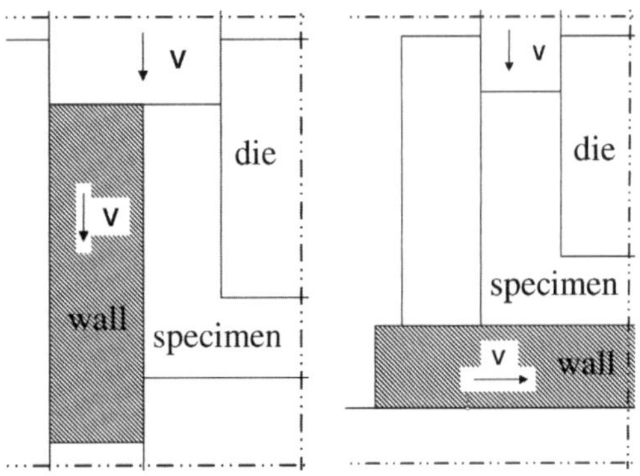

The majority of current researches deals with unalloyed materials or model alloys which are examined in a way that is irrelevant for practical usage (e.g. soft-annealed or highly-aged states of precipitation-hardened alloys, see [5] as an example). In the present paper, a precipitation-hardened aluminium alloy of technical purity in a completely aged state is accounted for. Table 1 shows the chemical composition of the aluminium alloy used for the ECAP deformation.

Figure 1. Movable tool walls in the first channel (left) and in the second channel (right) [3]

Table 1: Chemical composition (wt. %) of the investigated aluminium alloy

Denomination	Si	Fe	Cu	Mn	Mg	Cr	Zn	Ti	Al
EN AW-6060*	0.43	0.16	<0.01	<0.01	0.52	<0.01	<0.01	<0.01	balance
EN AW-6063	0.68	0.16	0.01	0.01	0.70	0.002	0.04	0.01	balance

*material for metallographic researches

The characteristic mechanical behaviour in the form of quasistatic technical stress-strain curves, was determined by A3-cylindrical-tension test specimen. Furthermore, the notched-bar impact work was determined by an instrumented notched-bar-impact bending test in compliance with DIN EN ISO 14556 using miniature Charpy V-notch specimen (4 x 4 x 40 mm^3). The microstructure was valuated using images from an optical and electron microscope of the low-alloyed aluminium EN AW-6060. The technical basis formed a field-emission scanning electron microscope (*Neon40*) from Zeiss.

3. Results and discussion

3.1. Development of Strength and Ductility

For the researches, material EN AW-6063 was solution heat-treated at 530°C for one hour (index -W), and then deformed up to six times by ECAP route E (see [3]).

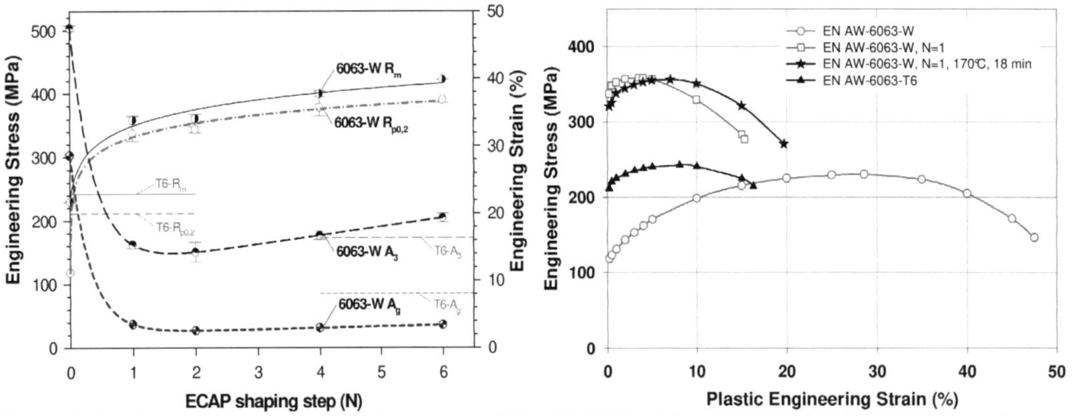

Figure 2. Mechanical behaviour of the investigated EN AW-6063 subject to the ECAP shaping steps and stress-strain curves for solid-solution annealed EN AW-6063-W, simple ECAP shaped EN AW-6063-W, N = 1 with post ECAP heat-treated condition 170°C, 18 minutes compared to the reference EN AW-6063-T6

The development of the mechanical values yield point $R_{p0,2}$, tensile strength R_m, uniform elongation A_g, and elongation to failure A_3 in Figure 2 show a strong increase of strength values with a high loss of ductility. The properties of the maximum-aged heat-treatment state (index –T6) of the undeformed alloy are used as reference values. After six deformations, yield point and tensile strength of the ECAP aluminium are 84% and 74% above the T6 reference state. However, uniform elongation, which is important for constructive purposes, drops to 42% of the undeformed state.

This lost ductility shall be regained by a downstream heat treatment at 170° C. The consequences of the downstream heat treatment on the simple-shaped material are shown at the right side of Figure 2. Annealing for 18 minutes at a temperature of 170°C leads to a retrieval of ductility whereas the uniform elongation increases from 3.5% to respectable 7%, which almost matches the value of the reference state T6 (8%). Strength does not change. In accordance with the study on AA6060 (presented in ref. [6]) the maximum ductility achieved during aging decreases also for the AA6063 with increasing number of passes and/or decreasing aging temperature, while the maximum strength behaves vice versa. Since the billets are designated for safety relevant applications the focus was to optimize the damage tolerance and thus only one pass is sufficient.

Compared to the reference T6, this optimized state shows an increase in tensile strength of 46% as well as in elongation to failure of 21%.

3.2. Development of Charpy Toughness

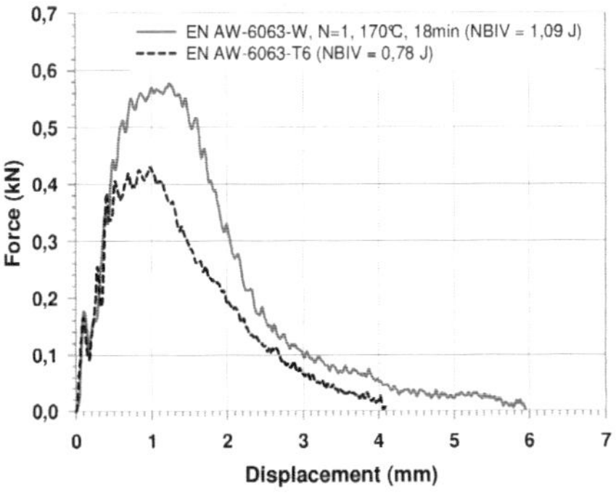

The notched-bar impact value (NBIV), which results from integrating the load-deflection curve of the instrumental notched-bar-impact bending tests as shown in Figure 3, features an increase of absorbed energy of 40% (compared to the reference state T6) which is due to ECAP deformation and the downstream heat treatment.

Figure 3. Force-displacement curves received from Charpy impact tests of simple ECAP shaped EN AW-6063-W, N = 1 with post ECAP heat-treated condition 170°C, 18 minutes compared to the reference EN AW-6063-T6

3.3. Microstructure

The intense decrease of grain size due to the implemented shear strain explains the significant increase in strength due to ECAP (see left side of Figure 2).

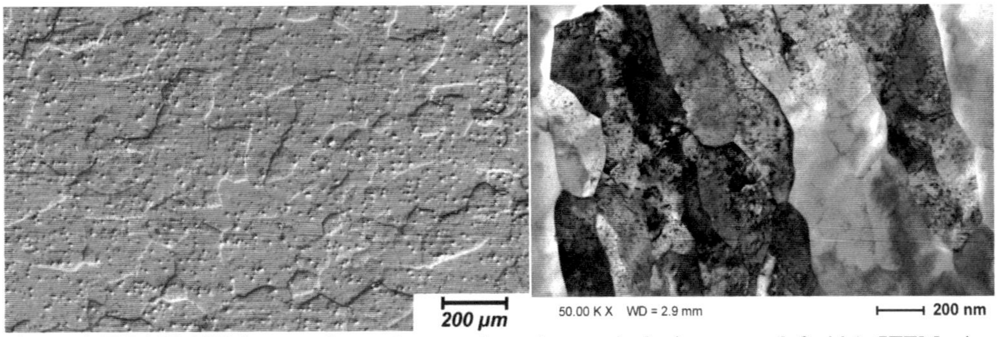

Figure 4. EN AW-6060 in normal granular state through an optical microscope (left side), STEM micrograph of ultrafine grained state after N = 8 deformations (right side)

The initial microstructure in Figure 4 (left side) shows an average grain size of 120 µm, which decreased to 310 nm (Figure 4; right side) by an eightfold ECAP deformation. Hereafter, the optimizing heat treatment after the first ECAP deformation shall be discussed. The blurry dark areas in the once-only ECAP shaped state (a) shown in the EBSD image in Figure 5 (left side) indicate strong inner tensions. These dark areas represent the newly formed dislocations which accumulate at the grain boundaries and inside the grains, and increase the strength of the material. Light areas indicate no tensions inside the grain. The more the aluminium strengthens due to dislocation accumulations the blurrier is the black-white contrast. With the optimizing heat treatment (b), tensions are partially reduced, and the picture contrast is increased. The dislocations can shift, and as a result, the developing small-angle grain boundaries contribute the grain refinement. This microstructure recovery allows a retrieval of ductility.

The high-resolution TEM image of the once ecaped state (a) in Figure 5 (right side) clearly shows the dislocation accumulations (white arrows) which partially form small-angle grain boundaries. These boundaries match the blurry lines running through the grain in the EBSD image in Figure 5 (left side). After the heat treatment, the TEM image (b) shows small-angle grain boundaries which developed due to the shifting of the dislocations (white arrows). Furthermore, there are finest precipitations close to the free dislocations, acting as nuclei, which preserve the strength of the material during heat treatment.

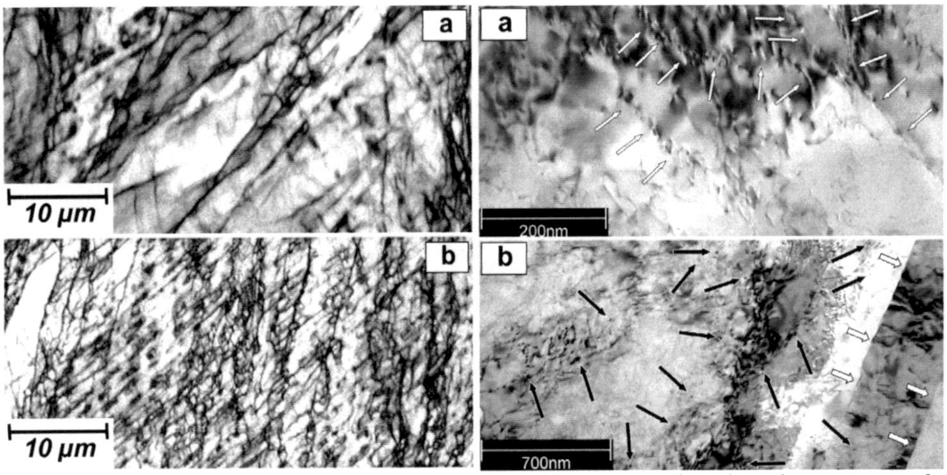

Figure 5. EBSD quality (band contrast) maps (left side) representing the microstructure after N = 1; TEM bright-field images (right side) showing the microstructure after N = 1 for (a) As-processed condition without subsequent aging and (b) high-temperature aged at 170 °C for 18 minutes [6]

4. Summary and conclusion

The presented results demonstrate the high potential of the combined process that induces high plastic strain like ECAP for EN AW-6063 with an appropriate aging treatment for achieving high strength, high charpy impact toughness and high ductility, compared to the commercial coarse grained counterparts. The effects occur after a single pressing and a short high-temperature aging, making the process attractive for practical applications.

Acknowledgements

The authors gratefully acknowledge the German Research Foundation (DFG) for supporting this work carried out within the framework of Sonderforschungsbereich 692.

References

[1] Hall E O 1951 *Proc. Royal Soc. London* **64** 474f
[2] Petch N J 1953 *J. Iron Steel Inst.* **174** 25
[3] Segal V M 2004 *Mater. Sci. Eng.* A **386** 269
[4] Iwahashi Y 1997 *Acta Mater.* **45** 4733
[5] Horita Z 2001 *J. Mater. Process. Technol.* **117** 288
[6] Hockauf M, Meyer L W, Zillmann B, Hietschold M, Schulze S, Krüger L 2009 *Mater. Sci. Eng.* A **503** 167

Journal of Physics: Conference Series **240** (2010) 012124 doi:10.1088/1742-6596/240/1/012124

Strain hardening and microstrains during cyclic incremental forming of carbon steel and pure iron

L W Meyer, C Kuprin and P Frint

Chemnitz University of Technology, Materials and Impact Engineering, Erfenschlager Str. 73, D-09125 Chemnitz, Germany

lothar.meyer@mb.tu-chemnitz.de

Abstract. The evolution of subsequent yield loci after large plastic strains has been studied for pure iron and steel with ferrite-pearlite structure. Large strains have been applied by torsion of thin walled tubes with different strain paths: monotonic and cyclic incremental torsion. Points on the yield loci have been probed by biaxial torsion-tension or torsion-compression load using a single-probe method. The results show strong influence of the microstructure towards anisotropic hardening. Microstrain values from XRD profile analysis are added and confirm the findings for hardening behaviour.

1. Introduction

In continuum mechanics, strain hardening is described by the model of the yield surface and its strain-dependent evolution. Translation, expansion and rotation or shape changes are denoted as kinematic, isotropic and distortive hardening, respectively. In complex strain paths with reversed strain, plastic yield can be strongly anisotropic, dependent on material and its microstructure [1]. A strong Bauschinger effect was reported for multiphased materials, and attributed to back stresses caused by microstrains. These microstrains can exist between grains and different phases (2nd order) and inside grains and phases due to dislocation arrangement (3rd order).

This study intends to compare strain hardening evolution of steel with a ferrite-pearlite structure and pure iron during different strain paths under large plastic shear strain. The influence of a second hard phase – the Fe_3C-lamellae in pearlite grains – towards path-dependent anisotropic hardening is to be demonstrated.

2. Experimental work

2.1. Material and sample shape

For the investigation of subsequent yield loci two materials have been choosen: steel 42CrMo4 with normalised ferrite-pearlite structure and commercial pure iron (0.009%C; 0.090%Mn; 0.005%P; 0.001%S) with grain sizes of 15μm and 30μm, respectively. The samples were thin walled tubes with inner and outer diameter of 10 and 12.5 mm, respectively and a deforming length of 10 mm.

2.2. Plastic pre-strain in torsion

Large plastic strains have been applied by free end torsion test at room temperature. Two different strain paths have been applied: Monotonic torsion and a cyclic incremental path with forward strain amplitude $\gamma_+ = 0.26$ and backward strain amplitude $\gamma_- = 0.15$ at a shear strain rate of $0.08s^{-1}$. The maximum shear strain γ was calculated from the rotation angle ϕ, the outer sample diameter d_o and the length l to $\gamma = \phi d_o/2l$. The maximum shear stress was calculated from the measured torque M, assuming linear stress distribution versus the radius: $\tau = M/(\pi(d_o^4-d_i^4)/16d_o)$. The torsional strain paths were conducted up to certain strain stages, as indicated by points in figure 1a for steel.

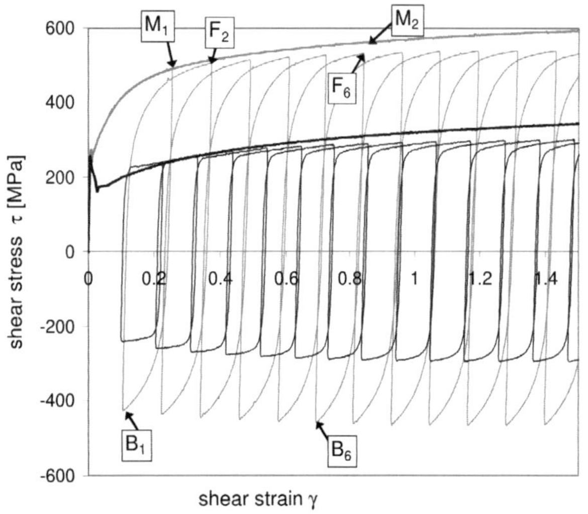

Figure 1a. Shear stress versus shear strain for steel (light curves) and pure iron (dark curves).

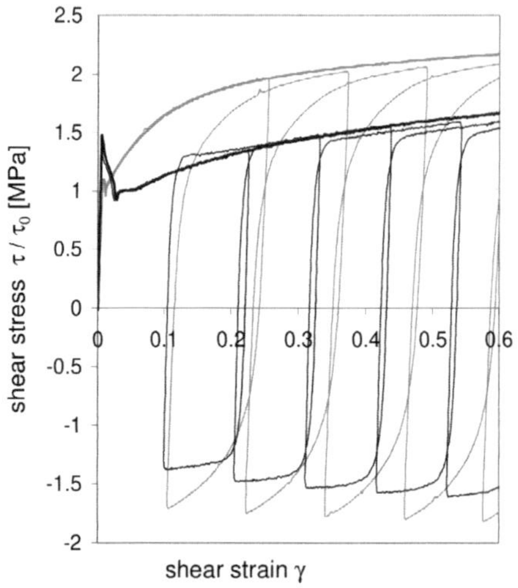

Figure 1b. Shear stress normalised with the first yield shear stress τ_0 for steel (light curves) and pure iron (dark curves).

2.3. Biaxial testing

At certain strain stages, the samples were partially unloaded before biaxial testing. Only partial unloading was essential steel to prevent plastic flow during complete unloading, as shown in [2]. Biaxial testing was done by combined torsion-tension or torsion-compression test, with torsion in prestrain and in opposite direction. With this procedure, subsequent yield points in the axial stress-shear stress space could be detected. A single-probe method was applied to exclude influences of yield probing procedure, because pre-strained material is very sensitive towards any further deformation. The method for yield point determination has been described in detail elsewhere [2]. Here, for initial yield points the instability in shear stress – axial stress plot was used as yield criterion and for subsequent yield points a 0.0005 (0.05%) v. Mises offset strain. Therefore, the strains had to be determined by optical strain field analysis using an Aramis system (by GOM company). Strains and stresses have been calculated using the current sample geometry. The equivalent strain rate for biaxial testing was $0.01s^{-1}$.

For prestrain and biaxial testing a hydropuls tension-torsion testing machine SchenckPTT250 was used.

2.4. Microstrain determination

Microstrains have been calculated from XRD profile Rietveld-analysis using the Diffrac plus-TOPAS (4.2) software from Bruker axs. Diffraction profiles were measured on flat grinded and polished mantle surfaces of the thin walled tube samples using a Siemens D5000 diffractometer and CoKα radiation. Microstrains have been derived using only Gaussian contribution to the line profile adjustment.

3. Results

3.1. Strain hardening

Torsional flow curves show the dependency of flow stress on the strain path, figure 1a. The cyclic forming paths lead to reduced flow stresses compared to the monotonic deformation. For steel, strain reversal results in early re-yielding and considerable strain hardening (as slope of the flow curve, $\Delta\tau/\Delta\gamma$) during each deformation step. Pure iron, however, shows a much smaller slope $\Delta\tau/\Delta\gamma$ and yield stresses, which are almost not influenced by strain reversal. The shape of the hystereses can be compared more clearly in figure 1b, where the stresses are normalised with the first yield shear stress.

The yield loci of the two materials in shear stress-axial stress space are depicted in figures 2-5.

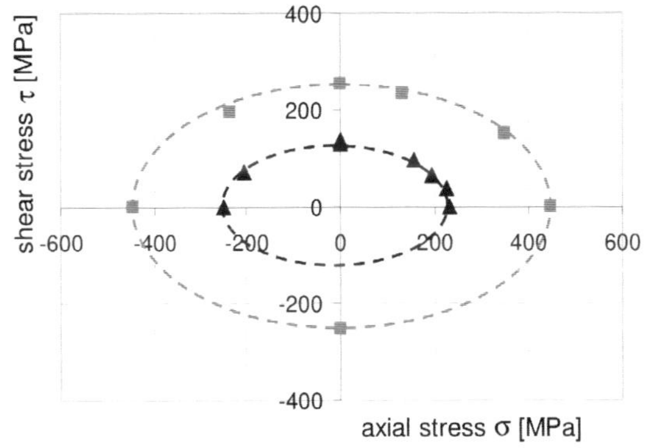

Figure 2. Initial yield points of steel (squares) and pure iron (triangles).

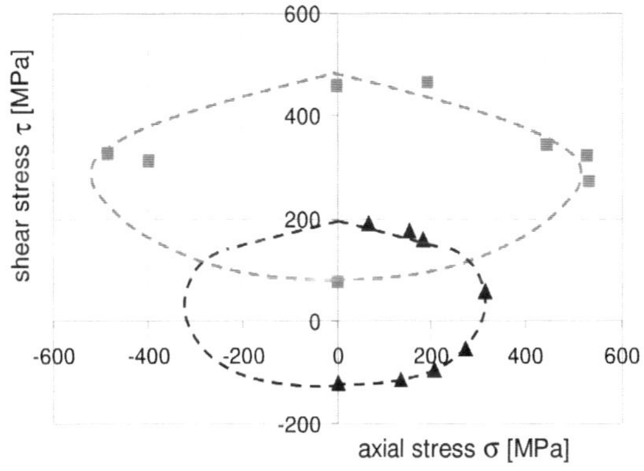

Figure 3. Subsequent yield points of steel (squares) and pure iron (triangles) after forward shear strain up to point M_1.

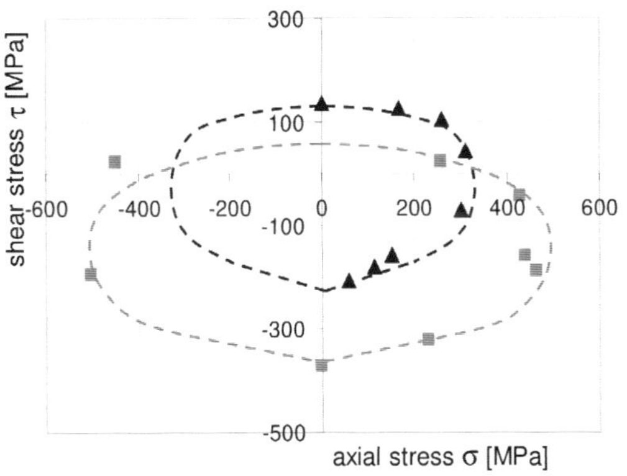

Figure 4. Subsequent yield points of steel (squares) and pure iron (triangles) after cyclic incremental deformation up to point B_1.

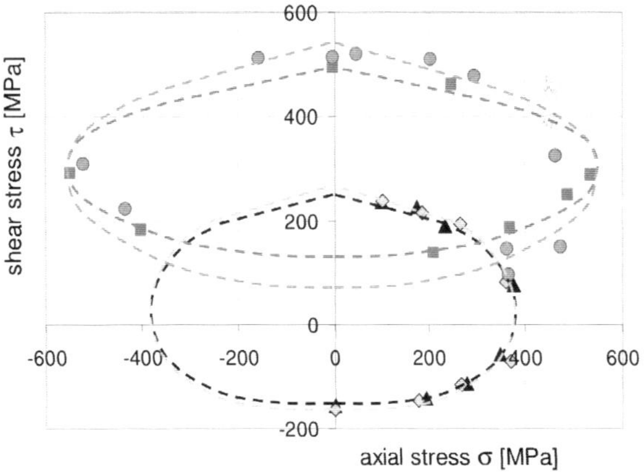

Figure 5. Subsequent yield points of steel (above) and pure iron (below) after monotonic torsion up to point M_2 (dots; diamonds) and cyclic incremental torsion up to point F_6 (squares; triangles).

Figure 2 shows the initial yield loci. The experimental points allow the construction of ellipses with nearly ideal v. Mises axial ratio. When a torsional prestrain was applied, expansion, distortion and translation of the yield loci into last prestrain direction appear, the latter especially distinct for steel. The shapes of the yield loci have been constructed by means of the experimental points, using simple geometric forms. During the cyclic incremental strain path, the subsequent yield locus expands, translates according to the strain path, and generates a higher curvature towards the last prestrain direction, figures 2-5. All hardening components are visible in both materials, whereas isotropic hardening dominates in pure iron and anisotropic hardening in steel. The influence of strain path at a certain resulting deformation is shown in figure 5. The cyclic strain paths lead to a shrinking dimension of the yield loci regarding the pre-strain axis by early re-yielding, especially for steel.

3.2. Microstrains

Microstrains are increasing during deformation, whereas backward strain lowers them in the beginning of the cyclic incremental strain path, figure 6. For both materials, the strongest increase of microstrains occurs in the first strain step towards M_1, where the flow stress increases remarkably. Afterwards, alterations are smaller. For steel the dependency of microstrain value on the forming path is more pronounced than for iron, forward compared to backward strain stages as well as cyclic compared to monotonic deformation are clearly different. For iron, the strain states B_6, F_6 and M_6 lead to similar microstrains.

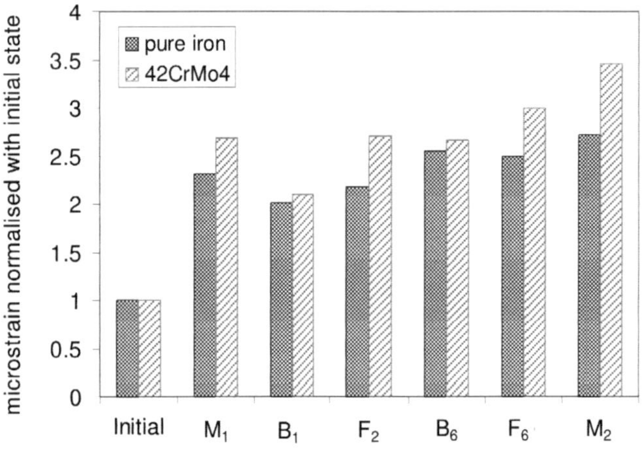

Figure 6. Microstrains of steel and pure iron at different strain stages (see figure 1a). Microstrains have been normalised to the value for the initial state (heat treated). The initial values are 0.024% and 0.030% for iron and steel, respectively.

4. Conclusions

During large plastic deformation of steel and pure iron with reversed strain all hardening components have been found: Isotropic and anisotropic hardening, which comprises translation and distortion of yield loci. Cyclic incremental deformation of steels leads to strong anisotropic hardening. This is due to the second hard phase in steel, which cause strong path-dependent microstrains. Subsequent yield loci of the single-phased iron show mainly isotropic hardening, anisotropic hardening portions are small, but visible. The less path-dependent microstrains, which are present in pure iron, therefore may arise from dislocation pattern or inter-grain strains. Transmission electron microscopy has to substantiate these findings.

Acknowledgement

The authors acknowledge the DFG for supporting this work carried out within the framework of project SPP 1146.

References

[1] Hecker S S 1976 Experimental studies of yield phenomena in biaxially loaded metals *Constitutive equations in viscoplasticy - Computational and engineering aspects: Proc. ASME Winter Ann. Meet.* (New York City, NY, USA, 5-10 Dec 1976) ASME AMD **20** 1-33

[2] Meyer L W, Kuprin C and Shutov A V 2009 Experimental yield point determination in metals after large plastic strains *Constitutive Modeling of Kinematic and Anisotropic Hardening Effects for Ductile Materials: Proc. 3rd Forming Technology Forum Zurich* (Zurich, Switzerland, 5-6 May 2009) Institute of Virtual Manufacturing ETH Zurich, 71-76

Texture development of ARB processed Mg/Al multilayers

H.-G. Brokmeier[1], H. Chang[2], W.M. Gan[2], M.Y. Zheng[2], K. Wu[2]

[1] Clausthal University of Clausthal, Institute of Materials Science and Engineering, 38678 Clausthal- Zellerfeld, Germany
[2] Harbin Institute of Technology, School of Materials Science and Engineering, Harbin 150001, P.R. China

heinz-guenter.brokmeier@tu-clausthal.de

Abstract. Our work deals with accumulative roll bonding (ARB) of pure Mg sheet (0.9mm thickness) and of Al5052 sheet (0.5mm). A stacking of Al-Mg-Al was firstly rolled to 50% reduction at 400°C and secondly ARB has been processed up to 3 cycles. In such multilayers as well as highly mixed composites of two-phased system texture development, phase reactions and strain accumulation are of basic interest, which needs a combination of different experimental methods for characterization. The present paper deals with the global texture evolution measured by thermal neutrons to average always over the whole sample thickness, SEM and optical microscopy indicates the macroscopic development of Mg and Al layers. The initial materials show typical and strong basal plane texture of hexagonal Mg (17.9mrd) and a recrystallization texture of cubic Al (8.5mrd). Co-deformation of Al/Mg/Al leads to strong decrease of both textures, whereas Mg has always a much stronger texture than Al5052. ARB processing produces only weak Al-textures. After sandwich-rolling and 1 cycle ARB rotated cube is observed in Al5052, which does not exist after 2 and 3 cycles of ARB.

1. Introduction

Accumulative roll bonding (ARB) is one kind of severe plastic deformation (SPD) process which can produce bulk ultra-fine grained (UFG) metallic materials [1]. The ARB process consists of multiple cycles of rolling, cutting, stacking and solid-state deformation bonding, so that a large strain can be accumulated in the metallic sheet during the ARB process without any sheet geometrical changing [2]. ARB with excellent quality has been processed at many single phase materials, but also some two-phase materials were tested. Al/Mg composites up to 3 ARB cycles were produced by Chen, Hsieh and Wu [3] at 300°C. Experimental problems in the texture analysis of lamellar two-phase systems restrict the investigation of texture evolution. ARB processed Al/Al(Sc) was reported by Quadir [4]. The motivation to investigate ARB processed multilayers is the combination of mechanical properties and damping behaviour (Mg - low density and high damping capacity; Al - good formability and high strength). Due to the high penetration power, neutron diffraction is excellent for the investigation of multi-phased systems and global textures [5]. This contribution is focused on texture evolution during ARB. In addition X-ray diffraction (constitution of intermetallic phases), SEM including EBSD, optical microstructure, micro-tomography (crack distribution, boundary conditions), microhardness and mechanical tests have been carried out.

2. Experiment

The initial materials in this study were commercially pure magnesium and Al 5052 alloy. Pure Mg slabs with the dimension of 120 × 50 × 4 mm^3 were cut from the cast ingot and then rolled into sheets with the thickness of about 0.90 mm at 400°C. The Al alloy was received in the form of a rolled sheet with the thickness of 0.5 mm and annealed at 400°C for 30 min. After degreasing and wire-brushing, the pure Mg sheet and the two annealed Al 5052 alloy sheets were stacked by the sequence of Al-Mg-Al. It has to be noticed that the former rolling direction of Mg and the former transverse direction of Al 5052 were oriented parallel. Thereafter, warm roll-bonded with 50% reduction at 400°C was carried out to obtain a sandwich structure. The primary sandwich sheet was processed by ARB at 400°C up to 3 cycles to produce Mg/Al multilayers. During ARB processing, the samples were kept at 400°C for 5min to reach the rolling temperature before each cycle and air-cooled after rolling. All the experiments were conducted on a 110mm diameter roll-mill at a rolling speed of 7m/min without lubricate. The microstructure of the Mg/Al composite was observed by HITACHI S3000 scanning electron microscope (SEM). The global texture of the two constitute metals were measured by the texture diffractometer TEX-2 at GKSS Research Centre, Germany [6]. The texture measurements were done on a stack of several sheets with a size of 10 × 10 × 11 mm^3 for the penetration depth of neutron beam. Due to the high penetration power of thermal neutrons anisotropic absorption problems described by Bunge et al. [19] did not occur. Four complete Mg pole figures ((10.0), (00.2),(10.1) and (11.0)) and two complete Al pole figures ((111) and (200)) were measured . The orientation distribution function (ODF) of the cubic Al and the hexagonal Mg was calculated using the iterative series expansion method with a maximum series expansion coefficient of l_{max}=23.

3. Microstructure

Figure 1 shows the SEM micrographs of the Al/Mg/Al composite for the different states of ARB. The layer thickness of Mg as well as of Al decreases gradually with increasing number of ARB cycles. The deformation of the two constituting metals was almost homogenous till the second ARB cycle. After 2 cycles intermetallic phases were determined by X-ray diffraction. After 3 cycles necking was observed, see figure 1d. Layer thickness distribution along the rolling direction (RD) was inhomogeneous. Plastic instabilities caused by different flow properties of constituent phases makes the harder phase to neck and finally rupture during the co-deformation of dissimilar metals [7].

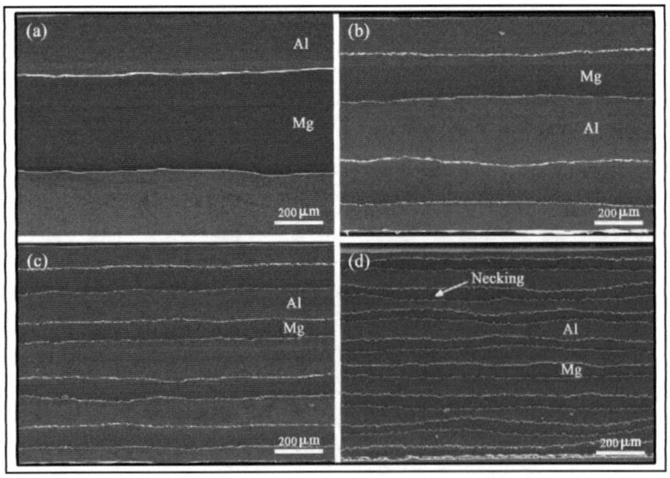

Figure 1. SEM microstructure of Mg/Al ARB processing
a) sandwich - co deformation of Mg and Al; b) 1 ARB cycle;
c) 2 ARB cycles; d) 3 ARB cycles

4. Texture

The quantitative texture shows that both starting materials have their typical textures well known as Mg deformation texture and Al recrystallization texture. In the case of hexagonal Mg a relatively strong basal plane orientation exists (17.9mrd – multiples of a random distribution), see figure 2. Due to annealing, Al shows as main texture component the cube orientation {100} <100> with 8.5mrd, see figure 3.

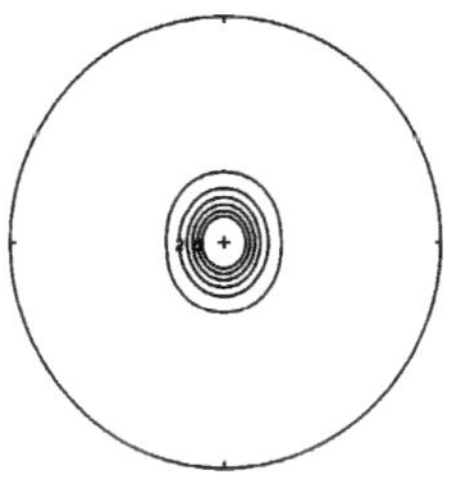

Figure 2. Starting texture of Mg
Levels 1.0, 3.0,5.0,7,0, 9.0, 11.0 mrd

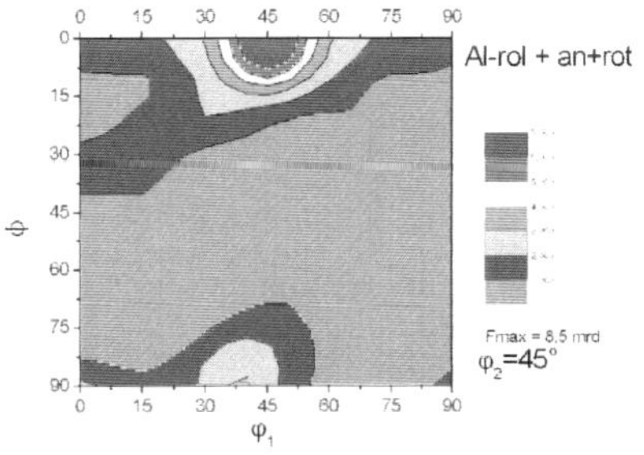

Figure 3. Al5052 texture, starting texture

After sandwich and ARB rolling the texture type of Mg is not changed. One obtains only changing texture sharpness and changing elongation of the central maximum in the basal pole figure (00.2). First step of sandwich rolling leads by a co-deformation of the Mg sheet covered by two Al sheets to decrease the texture sharpness from 17.9mrd to 12.8mrd (figure 4). This evolution has been observed for many composites, that texture type is identical to the deformation of pure materials but orientation densities are much lower [8]. 1 ARB cycle gives a stronger texture as also bee seen for Al. Both phases carry part of the deformation and not only the cubic Al. After 2 and 3 cycles Mg-textures decreases continuously, which can be explained by the wavy microstructure of the Mg- layers, see figures 1d and 4. Hexagonal Mg prefers an orientation of the basal plane (00.2) parallel to macroscopic glide plane. The elongation of the central maximum in RD of the (00.2) pole figure increases and the maximum orientation in ND decreases.

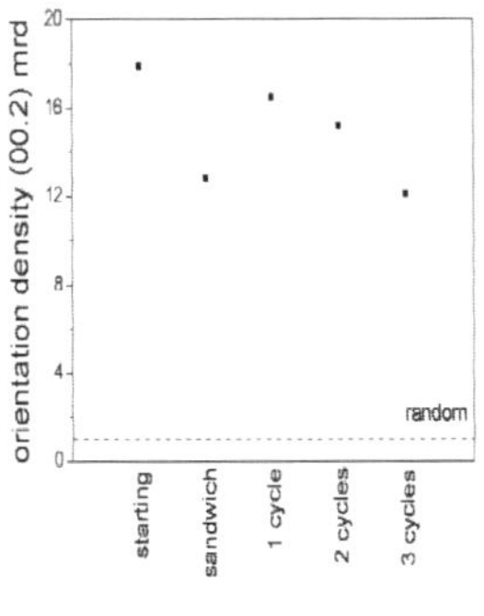

Figure 4. Texture evolution of Mg demonstrated by (00.2) orientation density

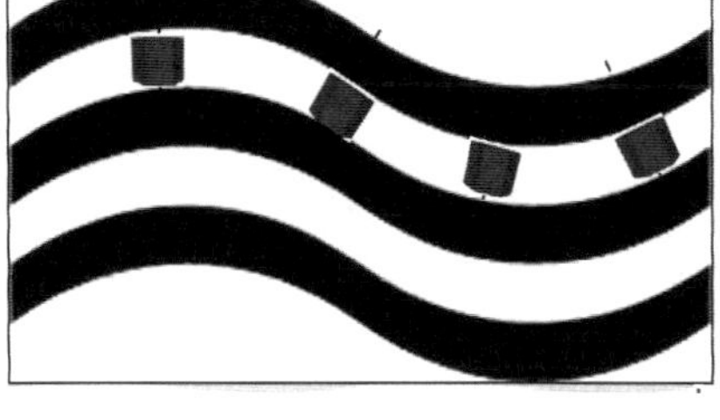

Figure 5. Schematic orientation of hexagonal Mg-grains related to the microstructure after 3 ARB cycles

The initial texture after rolling and annealing is dominated by the cube component {100} <100> with 8.5mrd, see figures 3 and 6. Rotating the sheet by 90° leads in addition to weak {110} <111> and {110} <110> orientations. Co-deformation by sandwich rolling of 0.5mmAl/1mmMg/0.5mmAl results in a strong decrease of the Al-texture mainly of the cube component. Brass {110} <112> and rotated cube {100} <110> are growing on a low level, see figures 6 and 7. The evolution of the brass component describes also the evolution of the copper- and the S-component. After 1 cycle ARB rotated cube is stable, brass is growing and cube becomes unimportant. The existence of rotated cube indicates shear deformation, which we expect on the boundary between Mg and Al. The microstructure after 2 ARB cycles shows beginning wavy structure in RD. Brass, copper and S get a little sharper but no rotated cube is observed. As already seen for the development of Mg texture the sharpness of all existing Al texture components decrease after 2 ARB cycles.

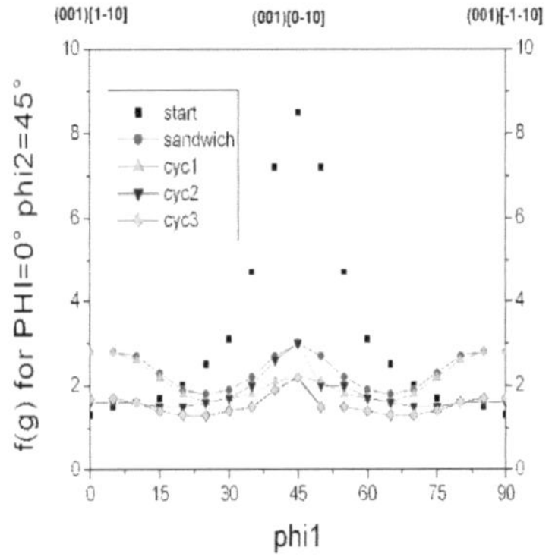

Figure 6. Al-texture evolution

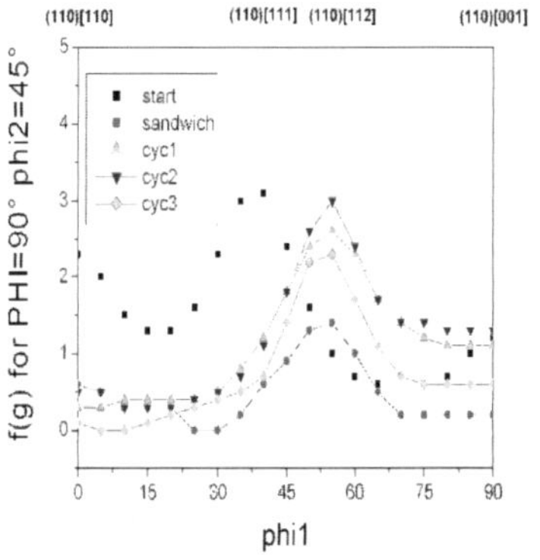

Figure 7. Al-texture evolution (α- fiber)

5. Conclusion

Mg/Al multilayers were produced with sufficient bonding conditions. After 2 ARB cycles at 400°C intermetallic phases (Al_3Mg_2, $Mg_{17}Al_{12}$) form and the layers become wavy. Necking is observed after 3 ARB cycles. Neutron diffraction is excellent to describe the global texture of co-deformed multilayers without any problems in anisotropic absorption. Mg forms the typical basal plane texture with (00.2) planes parallel to the rolling plane. The texture sharpness varies due to the deformation condition and decreases from 1 cycle to 3 cycles of ARB processing. The strong initial texture of the annealed Al sheet is destroyed by sandwich rolling. Compared to Mg the Al texture is much weaker and develops on a low level. Co-deformation by sandwich rolling and 1 ARB cycle ARB show rotated cube together with brass, copper and S-component, which indicates the presence of shear. For 2 and 3 ARB cycles ARB no rotated cube is observed, but formation of intermetallic phases. The deformation texture of Al increases from sandwich rolling to 2 ARB cycles and decreases during 3rd cycle.

Acknowledgments

This work was supported by the National Nature Science Foundation of China (No. 50571031) and the German DAAD-PPP project D/08/01772.

References

[1] Tsuji N, Saito Y, Lee S H and Minamino Y 2005 *Adv. Eng. Mater.* **5** 338
[2] Saito Y, Utsunomiya H, Tsuji N and Sakai T 1999 *Acta Mater.* **47** 579
[3] Chen M C, Hsieh H C and Wu W 2006 *J. Alloys Comp.* **416** 169
[4] Quadir M Z, Ferry M, Al-Buhamad O and Munroe P R 2009 *Acta* Mater. **57** 29
[5] Brokmeier H-G 1999 *Textures Microstr.* **33** 13
[6] Brokmeier H-G, Zink U, Schnieber R and Witassek B 1998 *Mater. Sci. Forum* **273-275** 277
[7] Min G H, Lee J M, Kang S B and Kim H W 2006 *Mater. Lett.* **60** 3255
[8] Brokmeier H-G 1994 *Mater. Sci. Eng.* **A175** 131

15th International Conference on the Strength of Materials (ICSMA-15) IOP Publishing
Journal of Physics: Conference Series **240** (2010) 012126 doi:10.1088/1742-6596/240/1/012126

Characterization of metal-forming processes with respect to non-monotonity

K. Bobor, Gy. Krallics

Department of Materials Science and Engineering, Budapest University of Technology and Economics, 1521 Budapest, Hungary

kristof.bobor@gmail.com

Abstract. The method of severe plastic deformation (SPD) is often used to produce bulk ultra-fine grained materials. SPD is a procedure, whereby the structure of the material changes from the initial coarse grained state to ultra-fine grained structure. For these techniques shear deformation as well as the so-called non-monotonic deformation are characteristic. The aim of this paper is to compare different metal-forming techniques with respect to non-montonity of deformation. A quantity and its calculation for the characterization of non-monotonity are introduced. Study of different metal-forming processes (simple shear, ECAP, upsetting) concerning non-monotonity and the comparison of these results are presented.

1. Introduction

The method of severe plastic deformation is often used to produce bulk ultra-fine grained (UFG) materials, whereby the structure of the material changes from the initial coarse grained state to UFG structure. By this microstructure special material properties can be obtained. Essentially, for these SPD techniques the shear deformation as well as the so-called non-monotonic deformation is characteristic. The paper introduces the calculation of a degree of non-monotonity (DNM), which can be a measure of technology planning of SPD processes. On the other hand, the mechanical study of different metal-forming techniques will be presented, with calculation and comparison of DNM.

2. Monotonic deformation

The mechanical concept of monotonity was introduced by Smirnov-Aljajev [1]. A forming process develops monotonically if no component of the rate of deformation tensor **d** changes its sign, namely the eigenvectors of the rate of deformation tensor are parallel to the same material lines during the whole deformation process and the Lode parameter remains constant. The two necessary conditions for a monotonic deformation that the relative state of the material lines and the eigenvectors of the rate of deformation tensor are not changing during the process, and the Lode parameter remains constant. Investigating the monotonity of the forming processes, can be distinguished which are monotonic processes and which are further or closer to the monotonic state. Through the inspection of the firs condition (the type of the deformation) can be ascertained more about the evolving microstructure, than through the analysis of the stress state (second condition). Through the investigation of the type of the deformation, a measure can be created, which is appropriate to describe the degree of non-monotonity of a forming process.

In this chapter the definition of a degree of non-monotonity (DNM) is shown. To define the monotonity, or rather the non-monotonity, the separate analysis of the rigid body rotation of a material particle and the rotation of the eigenvectors of **d** is necessary. Both values can deduced from the velocity gradient tensor, hence an appropriate description of the velocity field of the forming process is required (equations (1)(2)). The velocity gradient tensor **L** can be decomposed into a symmetric and an anti-symmetric part using equation (3). The first describe **d**, the second shows the rigid body rotation of a material point. The vorticity tensor **ω** can transcribe to vector-form using equation (4).

$$\boldsymbol{v}_{(x_1,x_2,x_3)} = \left[\boldsymbol{v}_1(\boldsymbol{x},t), \boldsymbol{v}_2(\boldsymbol{x},t), \boldsymbol{v}_3(\boldsymbol{x},t)\right] \quad (1) \qquad \mathbf{L}_{kl} = \boldsymbol{v}_{k;l} = \boldsymbol{v}_{k,l} + \boldsymbol{v}_s \Gamma^s_{kl} \quad (2)$$

$$\begin{aligned} \mathbf{d}_{kl} &= (\mathbf{L}_{kl} + \mathbf{L}_{lk})/2 \\ \boldsymbol{\omega}_{kl} &= (\mathbf{L}_{kl} - \mathbf{L}_{lk})/2 \end{aligned} \quad (3) \qquad \boldsymbol{\omega} = \begin{bmatrix} 0 & -\omega_3 & \omega_2 \\ \omega_3 & 0 & \omega_1 \\ -\omega_2 & -\omega_1 & 0 \end{bmatrix} \quad \boldsymbol{\omega} = \begin{bmatrix} \omega_1 \\ \omega_2 \\ \omega_3 \end{bmatrix} \quad (4)$$

The $\boldsymbol{\omega}$ vector specified defines the axis and the angular velocity of the rotation of a material point at the given point of space and at the given time. Furthermore the rotation of the eigenvectors of the rate of deformation tensor is required. The eigenvectors $\boldsymbol{n}_i$ can calculated from **d**. The relation between the eigenvector $\boldsymbol{n}_i$ and its rate of change $\dot{\boldsymbol{n}}_i$ with respect to time can be written as follows

$$\dot{\boldsymbol{n}}_i = \boldsymbol{\Omega} \times \boldsymbol{n}_i \quad (5)$$

Where $\boldsymbol{\Omega}$ is the vector of rotation and **Ω** is the anti-symmetric tensor form of $\boldsymbol{\Omega}$, similar to equation (4). **Ω** can be calculated from the system of equations (5). The $\dot{\boldsymbol{n}}_i$ vector is a material derivative of the eigenvector $\boldsymbol{n}_i$, which can be expressed with the formula (6). One of the necessary conditions of monotonity is, that the eigenvectors of the rate of deformation coincide with the same material line through the whole deformation. This condition is fulfilled if $\boldsymbol{\Omega}$ and $\boldsymbol{\omega}$ vectors are equal. If the difference between the vectors $\boldsymbol{\beta}=\boldsymbol{\omega} - \boldsymbol{\Omega}$ differ from 0 then the process is non-monotonic in the given time for the given material point. In order to characterize the deformation of a chosen material particle through the forming process, the integration of $\boldsymbol{\beta}$ over the time and along a particle trajectory is needed. The DNM μ is the integral of the absolute value of $\boldsymbol{\beta}$ according to equation (7).

$$\dot{\boldsymbol{n}}_i = \frac{d\boldsymbol{n}_i}{dt} + \boldsymbol{v}_k \boldsymbol{n}_{i;k} \quad (6) \qquad \mu = \int_0^t |\boldsymbol{\beta}| dt \quad (7)$$

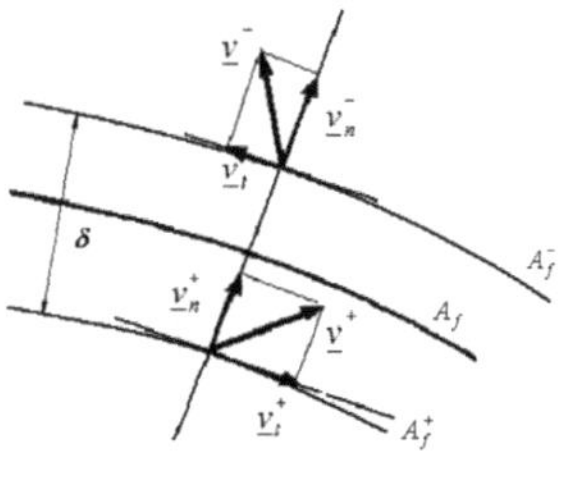

Figure 1. Discontinuity surface

Figure 2. Simple shear

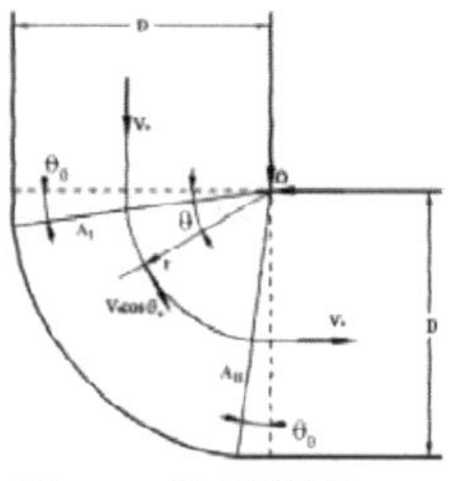

Figure 3. ECAP.
Cylindrical coordinate system [2]

The starting point of the calculation of the DNM is the known velocity field. For that purpose an appropriate model of the forming process is necessary, which describes the material flow. Analytical models are often used to modeling a forming process. In this paper analytical models were also used for the calculations. These contain mostly simplification in order of faster and easier calculus. They are not so accurate as a finite element computation, but provide a continuous velocity field.

Discontinuity surfaces are often used to simplify the flow model. The velocity of a particle which is passing through this surface has a stepwise change (figure 1). The normal component to the surface A_f

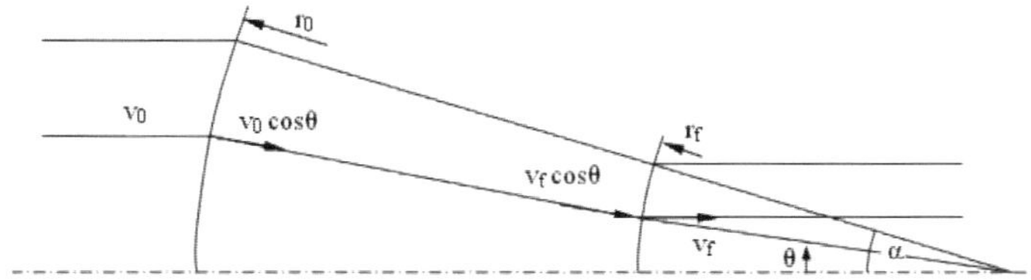

Figure 4. Forward rod extrusion. spherical coordinate system [3]

remains continuous ($\boldsymbol{v}_n^+=\boldsymbol{v}_n^-$) but the tangential changes are abrupt. The material particle is deformed infinitely rapid; hence the calculation of the DNM on these surfaces is not trivial. Assume that the tangential velocity near to the surface A_f in the δ wide region between A_f^+ and A_f^- changes continuously. The non-zero components of **L**, **ω** and **d** are

$$\mathbf{L}_{21}=\frac{\Delta v_t}{\delta} \qquad \mathbf{d}_{12}=\mathbf{d}_{21}=\frac{\Delta v_t}{2\delta} \qquad \boldsymbol{\omega}_{21}=-\boldsymbol{\omega}_{12}=\frac{\Delta v_t}{2\delta} \tag{8}$$

This shows that a rigid body rotation occurs in the δ wide region, without any rotation of the eigenvectors of the rate of deformation. The value of the expression $\Delta v_t/\delta$ becomes infinitesimally high if $\delta \to 0$. If a material particle goes through this region in t_0 time, then $\delta = v_n \cdot t_0$. From this calculation results that the DNM and equivalent strain are the following

$$\mu=\int_0^{t_0}\frac{\Delta v_t}{2\delta}dt=\int_0^{t_0}\frac{\Delta v_t}{2v_n t_0}dt=\frac{\Delta v_t}{2v_n t_0}\int_0^{t_0}dt=\frac{\Delta v_t}{2v_n}\frac{t_0}{t_0}=\frac{\Delta v_t}{2v_n} \qquad \varphi - \frac{\Delta v_t}{\sqrt{3}v_n} \tag{9}$$

5. Characteristic deformation of metal forming processes

5.1. Simple shear

The velocity field for simple shear is given in a Cartesian coordinate system using equation (10) with a parameter k. The DNM is increasing with the deformation, according to the equation (11). Simple shear (figure 2) is the way of deformation - strongly non-monotonic - in the idealized process of ECAP and HPT.

$$\boldsymbol{v}_{(x_1,x_2,x_3)}=\begin{bmatrix}k\cdot x_2 & 0 & 0\end{bmatrix} \tag{10}$$

$$\mu=\sqrt{3}/2\cdot\varphi \tag{11}$$

5.2. Forward rod extrusion

The material flow in a constricted channel (figure 4) can be described in a spherical coordinate system using equation (14). Five points of the cross section at θ/α=0;0,3;0,5;0,8;1 were investigated during the extrusion from the initial to the final diameter in four steps (ø10 to ø8, ø8 to ø6, ø6 to ø5, ø5 to ø4).

$$\boldsymbol{v}_{(r,\vartheta,\theta)}=\begin{bmatrix}v_0 r_0^2\cos\theta\cdot x_1^{-2} & 0 & 0\end{bmatrix} \tag{12}$$

5.3. ECAP

The flow model of [2] was used (figure 3). According to this model, material particles in the die's deformation zone are moving in a circular way between two discontinuity surfaces. The velocity field is given in a cylindrical coordinate system according to equation (13). The DNM on the entry and exit surface is $\mu=\sqrt{3}/2\varphi_{AI,II}=1/2\tan\theta_0$ Between the surfaces the equivalent strain and DNM are $\mu=\pi/4-\theta_0$ and $\varphi_{chanell}=(\pi/2-2\theta_0)/\sqrt{3}$ The relation between the equivalent strain and DNM is $\mu=\sqrt{3}/2\varphi$, which shows that the deformation is simple shear-like non-montonic.

$$\boldsymbol{v}_{(r,\theta,z)}=\begin{bmatrix}0 & v_0\cdot\cos\theta & 0\end{bmatrix} \tag{13}$$

$$\boldsymbol{v}_{(r,\theta,z)}=\begin{bmatrix}v_0 r_0^2\cos\theta\cdot x_1^{-2} & 0 & 0\end{bmatrix} \tag{14}$$

5.4. Simultaneous torsion, tension and frictionless upsetting of a cylindrical specimen

The velocity of the deformation can be described in a cylindrical coordinate system as follows

$$\boldsymbol{v}_{(r,\theta,z)} = \begin{bmatrix} \frac{r v_0}{2h} & \frac{\omega_0 r z}{h} & -\frac{z v_0}{h} \end{bmatrix} \tag{15}$$

where $h=h_0-v_0t$ is the actual height of the specimen and ω_0 the angular velocity of torsion. By uniaxial tension or frictionless upsetting (v_0=const, $\omega_0=0$) μ remains zero and the process is pure monotonic (the lode parameter is also constant). If no tension (or upsetting) occurs ($v_0=0$, ω_0 =const) then the type of deformation is simple shear, the relationship described by equation (11) is valid. If both types of forming are present, then the DNM can change between simple shear type and pure monotonic type deformation (figure 5).

6. Results

In figure 5 the results of the DNM calculation are shown for the above presented processes. As it is shown in the diagram, the simple shear, ECAP, and the torsion differ most from a monotonic deformation. By forward extrusion, the DNM is not constant in the cross section: in the symmetry axis the deformation is monotonic, and with the distance from the centre the DNM is increasing, but remains far from ECAP or simple shear. By uniaxial tension and frictionless upsetting the value of the DNM remains zero according to the requirements.

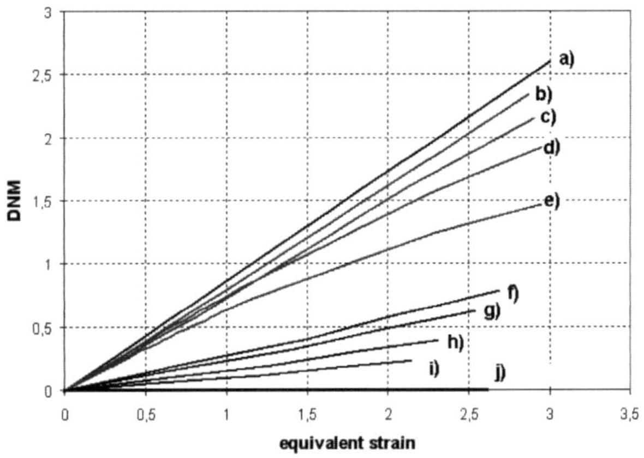

Figure 5. Results
a) ECAP, torsion, simple shear
b) torsion+upsetting* (v=-20 mm/s)
c) torsion+upsetting* (v=-30 mm/s)
d) torsion+tension* (v=20 mm/s)
e) torsion+tension* (v=30 mm)
f) forward extrusion ($\alpha=\theta=10^o$)
g) forward extrusion ($\alpha=10^o$ $\theta=8^o$)
h) forward extrusion ($\alpha=10^o$ $\theta=5^o$)
i) forward extrusion ($\alpha=10^o$ $\theta=3^o$)
j) Tension, upsetting, forward extrusion ($\alpha=10^o$ $\theta=0^o$)

*ω=1 rad/s; h_0=10mm

7. Conclusion

Several authors made experiment and investigation about the grain refinement by the mentioned forming processes (ECAP[4][5], HPT[6], extrusion[7][8]). The experiments show that through the procedures like ECAP and HPT stronger grain refinement occurs than through the other forming processes. The results show that the DNM for the former processes is higher than for the latter. For processes which differ more from pure monotonic deformation, an UFG microstructure evolves sooner. The above mentioned method can be suitable for planning of production of UFG metals, enhancement of existing methods, or planning new technologies.

Reference

[1] G. A. Smirnov-Aljajev 1978 *Сопротивление Материалов Пластическому Деформированию, Машиностроение* (Leningrad)
[2] A.R. Eivani, A.K. Taheri, *J. Mater. Process. Technol.* **182** (2007) 555–563
[3] B. Avitzur 1968 *Metal Forming* (McGraw-Hill Book Comp)
[4] J.-Y. Chang, A. Shan *Mater. Sci. Eng.* A **347** (2003) 165-170
[5] U. Chakkingal, A. B. Suriadi, P.F. Thomson *Scripta Mater.* **39** (1998) 677-684
[6] C. Xu, Z. Horita, T. G. Langdon *Acta Mater.* **56** (2008) 5168–5176
[7] S. Kanekoa, K. Murakamib, T. Sakaic *Mater. Sci. Eng.* A **500** (2009) 8-15
[8] J. Swiostek, J. Goken, D. Letzig, K.U. Kainer *Mater. Sci. Eng.* A **424** (2006) 223–229

Journal of Physics: Conference Series **240** (2010) 012127 doi:10.1088/1742-6596/240/1/012127

Strength and ductility of medium carbon steel after equal channel angular pressing

S. Dobatkin[1,2] J. Zrnik[3], S. Nikulin[2],T. Kovarik[3]

[1] A.A.Baikov Institute of Metallurgy and Materials Science, Russian Academy of Sciences, Leninsky prospect 49, 119991 Moscow, Russia
[2] Moscow Steel and Alloys Institute (Technological University), Leninsky prospect 49, 119049 Moscow, Russia
[3] COMTES FHT, Prumyslova 995, 33441 Dobrany, Czech Republic

dobatkin@imet.ac.ru

Abstract. Upon the ECA pressing at 400°C of the AISI 1045 steel with 0.45% C, the processes of dynamic recovery are developed with the formation of subgrains of ~320 nm in size and isolated grains of submicron size with high-angle boundaries. The fragmentation and partial spheroidization of the cementite lamellae is observed within the pearlite colonies. Deformation in pearlite grains is not sufficient, either to break down the cementite lamellae structure or to produce the refinement of ferrite grains. Annealing of the deformed steel with 0.45%C after ECA pressing causes the perfection of the recovered structure in ferrite and further processes of fragmentation, spheroidization, and coagulation of carbide phase. Annealing at T = 550°C for 5 h leads to almost complete spheroidization of carbides with an average size of ~280 nm and increases to some extent the average ferrite grain (subgrain) size to ~410 nm. The submicron grain-subgrain structure of the steel with 0.45%C causes a significant strengthening (YS = 960 MPa) at a retention of satisfactory elongation (EL = 8%). Subsequent annealing at T = 550°C for 5 h increases ductility to EL = 14% and decreases the yield strength to YS = 745 MPa.

1. Introduction

Equal-channel angular (ECA) pressing allows one to obtain submicro- and even nano-crystalline state in bulk billets of different metals and alloys including commercial ones. Now the application of this method for the achievement of ultrafine-grained structure in iron-based alloys, in particular, in steels attracts increasing interest. A series of articles on low-carbon steels are published up until now [1-6], but there is virtually no data on the ECA pressing of medium-carbon steels.

The aim of the present work was to study the structure formation in ferrite and carbide phases of the medium carbon steel (0.45% C) upon ECA pressing and subsequent annealing purposely for the improvement of mechanical properties.

2. Experimental Procedure

Commercial 0.45%C steel – AISI 1045 (0.45% C; 0.63% Mn; 0.23% Si; 0.18% Cr; 0.043% Al; 0.02% Cu) was taken for the study; in the initial state, the steel was air-quenched from 960°C after holding for 2 h. The ECA pressing of the steel samples of dia.10 x 60 mm in size was performed at T = 400°C by N = 4-6 according to route Bc at an angle of 120° between the channels. The initial and deformed structure was examined with an "Olympus PME 3"optical microscope, JEM 1000 transmission electron microscope, and an LEO 430i scanning electron microscope. The tensile properties were measured with an INSTRON 3382 testing machine using the specimens of 20 mm in length and 3 mm in diameter.

3. Results and Discussion

The size of the initial ferrite grains and pearlite colonies before deformation was about 15 μm. After ECA pressing by 4 and 6 passes, the examination by optical microscopy and scanning electron microscopy (SEM) revealed the fragmentation of the pearlite colonies. Their average size decreased to 8 μm. The SEM analysis showed that both discontinuous and wavy morphologies of cementite lamellae are presented in deformed structure. The beginning of lamellae fragmentation is noticeable after ECA pressing for 4 passes, and the beginning of spheroidization is revealed after 6 passes. The transmission electron microscopy (TEM) examination of the thin foils after four passes (N = 4) showed the imperfect subgrained structure with rather thick subgrain boundaries, high density of lattice and grain-boundary dislocations, and an average subgrain size of about 0.4 μm. The beginning of the fragmentation processes is clearly seen inside the pearlite

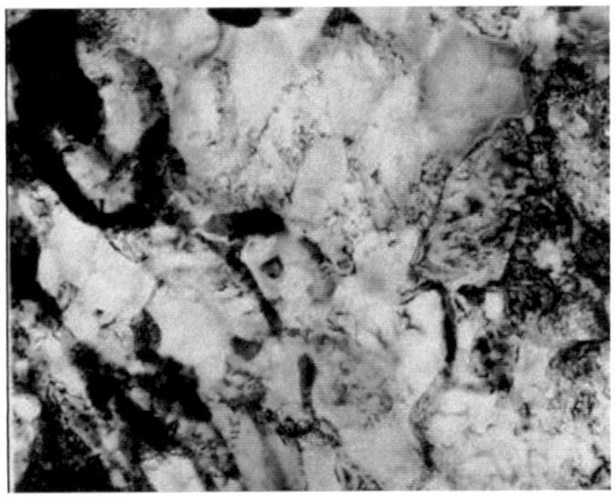

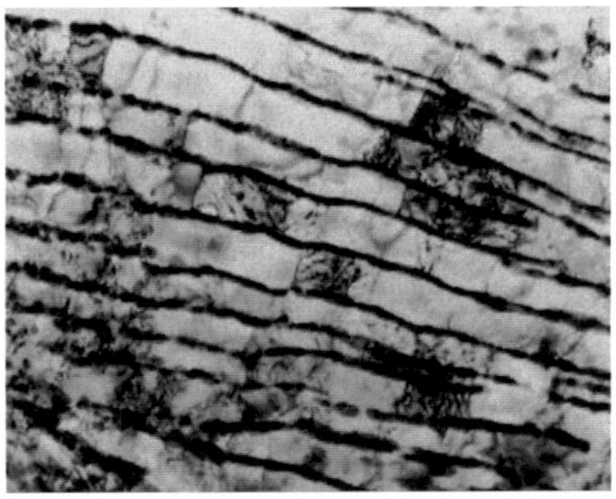

Figure 1. TEM microstructure of 0.45%C steel after ECAP at 400°C and N=6 passes.

colonies. As the number of passes increases to 6, the average subgrain size decreases to ~0.3 μm, and the substructure becomes more perfect due to the dynamic recovery, which leads to subgrain-boundary thinning and to density decrease of free dislocations (Fig. 1a). A banded contrast at some boundaries indicates the local appearance of high-angle boundaries. This evidences the beginning of the transformation of the subgrained structure into submicrocrystalline one, i.e., the dynamic recrystallization proceeds. The beginning of the fragmentation and spheroidization processes is clearly observed in the pearlite colonies (Fig. 1b).

ECA pressing by N = 6 increased the yield strength and ultimate tensile strength by a factor of about 2.5 (up to 960 MPa), and 1.5 (up to >1 GPa), respectively (Table 1). The elongation in this case decreased more than by a factor of 2 (to ~8%), and the yield ratio became 0.95. For the improvement in ductility, an post-deformation heat treatment was performed.

Annealing at 500°C for 3 h after ECA pressing did not virtually change the structure: the processes of the carbide spheroidization were expressed very weakly. Even after annealing for 50 h, the spheroidization was not completed, and the carbide distribution remained nonuniform.

The SEM analysis showed that, upon annealing at 500°C after the ECA pressing by N = 6 compared to that by N = 4, the cementite lamellae fragmentation was more pronounced, the spheroidization both inside and outside the pearlite colonies occurred more rapidly; and the carbide size outside the pearlite colonies after annealing for 50 h was larger, 0.4 μm vs. 0.3 μm (the average carbide size inside the pearlite colonies after annealing for 50 h was 0.15 μm for both states) (Fig. 2).

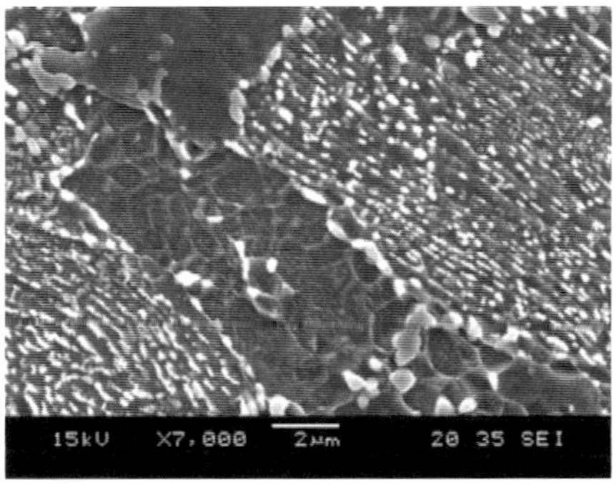

a.

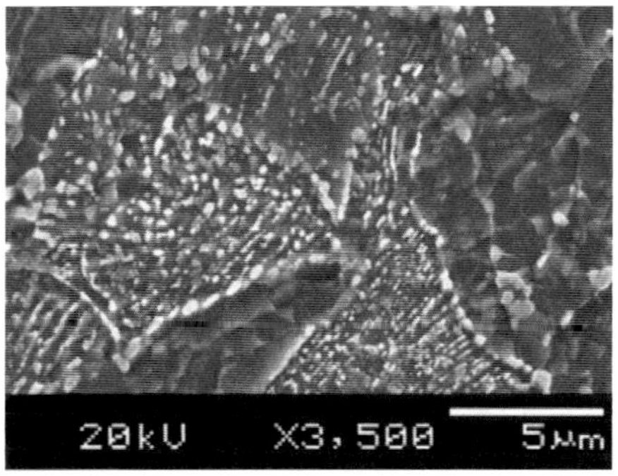

b.

Figure 2. Structure of 0.45%C steel after ECAP at 400°C and heating (500°C, 50 h.): a-N=4; b-N=6. SEM

Table 1. Mechanical properties of 0.45%C steel after ECAP and heating

State	YS, MPa	UTS, MPa	EL, %	RA, %
Normalization	402	725	18.1	22
ECAP, N=4	892	981	6.3	43
ECAP, N=6	960	1013	7.7	31
ECAP,N=6+heating at 500°C(10 h.)	798	829	12.6	51
ECAP, N=6+heating at 550°C(5 h.)	746	830	13.8	45

At steel annealing at temperature of 550°C, the carbide spheroidization processes occurred more rapidly. After annealing for 3 h, the lamellar morphology of the pearlite colonies was still noticeable, but, already after annealing for 5 h, a virtually completely spheroidized structure was observed (Fig.3). In this case, the spheroidization inside the pearlite colonies was completed. As a whole, the distribution of spherical carbides with an average diameter of ~0.3 μm in the ferrite matrix was substantially more uniform than that was after annealing for 50 h at 500°C. The ferrite had predominantly grained structure with an average grain size of ~0.4 μm. Annealing at 550°C (5 h) decreased YS to ~750 MPa, but the elongation in this case increased to 14% and the yield ratio decreased from 0.95 to 0.90 (Table 1, Fig. 4).

4. Conclusions

1. Upon the ECA pressing at 400°C of the steel with 0.45% C, the processes of dynamic recovery at the unsteady stage are developed in ferrite with the formation of subgrains of ~320nm in size and isolated grains of submicron size with high-angle boundaries. The fragmentation and partial spheroidization of the cementite lamellae are observed in the pearlite colonies.
2. Annealing of the steel with 0.45%C after ECA pressing causes the perfection of the recovered structure in ferrite (decrease in the density of free dislocations, thinning of the subgrain boundaries, and insignificant subgrain growth) and further processes of fragmentation, spheroidization, and

coagulation of carbide phase. Annealing at T = 550°C for 5 h leads to almost complete spheroidization of carbides with an average size of ~280 nm and somewhat increases the average ferrite grain (subgrain) size to ~410 nm.

3. The submicron grain-subgrain structure of the steel with 0.45%C causes a significant strengthening (YS = 960 MPa) at a retention of satisfactory elongation (EL = 8%). Subsequent annealing at T = 550°C for 5 h increases ductility to EL = 14% and decreases the yield strength to YS = 745 MPa.

Figure 3. Structure of 0.45%C steel after ECAP at 400°C (N=6) and heating (550°C, 5h.). TEM

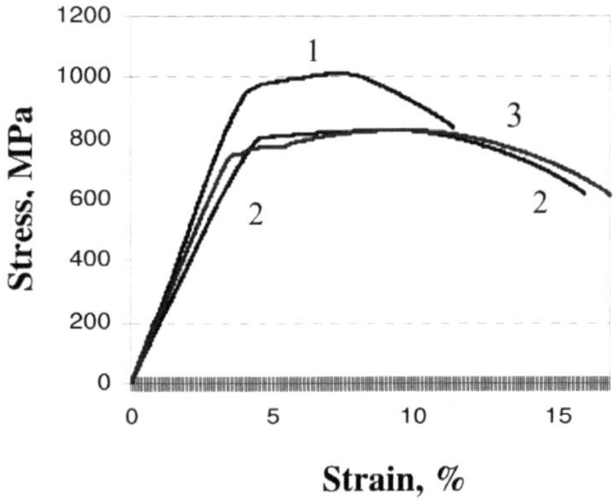

Figure 4. Mechanical properties of 0.45%C steel: 1-ECAP; 2-ECAP + heating 500°C, 10 h.; 3-ECAP + heating 550°C, 5 h.

References

[1] Shin D H, Kim I, Kim J and Park K T 2001 *Acta Mater.* **49** 1285
[2] Fukuda Y, Oh-ishi K, Horita Z and Langdon T G 2002 *Acta Mater.* **50** 1359
[3] Wang J T, Xub C, Duc Z Z et al 2005 *Mater. Sci. Eng.* A **410–411** 312
[4] Dobatkin S V, Odessky P D, Pippan R et al 2004 *Metals* **1** 110 (in Russian)
[5] Dobatkin S V, Zrnik J and Mamuzic I 2006 *Metallurgy* Zagreb Croatia **45** 313
[6] Zrnik J, Mamuzic I, Dobatkin S V et al 2007 *Metallurgy* Zagreb Croatia **46** 21

Microstructure and mechanical properties of accumulative roll bonded aluminium alloy AA5754

T Hausöl, H W Höppel and M Göken

Department of Materials Science and Engineering, Institute I
University Erlangen-Nürnberg, Martensstr.5, 91058 Erlangen, Germany

tina.hausoel@ww.uni-erlangen.de

Abstract. The aluminium alloy AA5754 is used for many technical applications. In this work, the accumulative roll bonding process is applied to this alloy in order to investigate the potential of an ultrafine-grained structure on the mechanical properties of this Al-Mg alloy. Sheets from AA5754 (AlMg3) were successfully processed by accumulative roll bonding in order to obtain an ultrafine-grained microstructure. The ARB process was performed at 230 °C or 250 °C up to 7 or 8 cycles respectively. Thus the grain size decreased from 10 µm (initial state) to approximately 80 nm (ultrafine-grained state, normal direction). The microstructural evolution and the mechanical properties have been investigated by means of scanning electron microscopy, hardness measurements and tensile testing. After one ARB cycle the samples showed an increase in hardness by a factor of almost 2 in comparison to the as-received material. Further processing causes a linear increase of hardness with each additional cycle. Yield strength and tensile strength of the roll bonded specimens are highly increased in comparison to the as-received samples whereas the ductility declined. A considerable increase in ductility is obtained by heat treatment of the ARB specimens at 250 °C, but on the expense of a moderate decreased strength. The deformation behaviour is also influenced by the ultrafine-grained structure. The occurrence of the Portevin-Le Chatelier effect is manifested by serrated stress-strain curves. The amplitude of serrations increases with increasing number of ARB cycles but can be reduced by the appliance of a higher strain rate. Lüders strain only occurs at the as-received, i.e. not strained, samples.

1. Introduction

For reason of weight reduction aluminium and its alloys become more and more important in the automotive industry. While aluminium alloys of the 6xxx series are mostly used for outer panels for which besides strength, a good formability and a high surface quality are required, 5xxx alloys are mostly used for structural panels where strength, good deep drawing and stretch behaviour are important issues [1]. The enhancement of strength and stiffness provides the possibility of scaling down the dimensions and consequently an improvement in light-weight design. For this purpose an ultrafine-grained microstructure which is coincided with extraordinary mechanical properties affords high potential. In this context, accumulative roll bonding (ARB) is a well known process for the production of ultrafine-grained sheet materials [2]. Aluminium of commercial purity and numerous aluminium alloys have already been subjected to the ARB process and show promising properties for prospective engineering application, see for example [3-6]. In this work, the accumulative roll bonding process is applied to the aluminium alloy AA5754 in order to investigate the potential of an ultrafine-grained structure on the mechanical properties of this Al-Mg alloy.

2. Experimental

The material investigated was the aluminium alloy AA5754 often used in the automotive industry for structural panels [1]. The chemical compostition is listed in table 1.

Table 1. Chemical compostion of aluminium alloy AA5754.

Wt.%	Si	Fe	Cu	Mn	Mg	Cr	Zn	Ti	other	Al
AA5754	0.4	0.4	0.1	0.5	2.6-3.6	0.3	0.2	0.15	0.15	balance

The as-received sheets were in the recrystallized (soft) state. The ARB-process was applied to sheets with 1 mm x 100 mm x 300 mm in size. Prior to roll bonding the surfaces were wire brushed in order to remove the oxide layer, stacked on top of each other and pre-warmed to 230 and 250 °C respectively for 210 s in a furnace. The sheets were immediately after pre-warming roll bonded in a four high rolling mill (Carl Wezel, Mühlacker) with a nominal thickness reduction of 50 % per cycle. The bonded aluminium sheets were air cooled, halved, wire brushed and stacked again before rolling the next cycle. For more details concerning ARB see [2, 3]. The sheets were roll bonded up to 7 or 8 cycles respectively.

The microstructure was observed using the electron channeling contrast (ECC) in a scanning electron microscope (SEM, Crossbeam 1540 EsB, Zeiss) in order to charactarise microstructural changes during ARB. The specimens were mechanically grinded and polished and electropolished using the standard Struers electrolyte A3 at 25 V for 20 s at room temperature.

The hardness depending on the number of ARB cycles was determined using a Vickers hardness tester (V-100 A, Leco) at a load of 49.05 N. Flat tensile specimens with a gauge length of 33.5 mm were machined in rolling direction from the as-received and the roll bonded sheets after 2, 4, 6 and 7 or 8 cycles respectively. For monotonic testing a universal testing machine Instron 4505 with a clip-on extensometer was used. All tensile tests were performed at room temperature using three strain rates of 10^{-2} s^{-1}, 10^{-3} s^{-1}, 10^{-4} s^{-1}.

Furthermore ARB specimens after 4 cycles were investigated after an annealing heat treatment at 250 °C for 10 minutes and 60 minutes and a heat treatment at 450 °C for 10 minutes to achieve a recrystallized state.

3. Results and discussion

3.1. Microstructure

The development of the microstructure during the ARB process is shown in figure 1. The mean grain-size in the as-received soft condition of the aluminium alloy AA5754 is 10 μm.

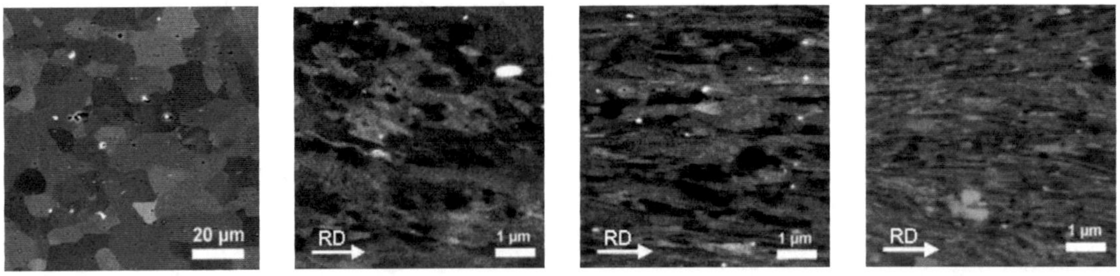

a) b) c) d)

Figure 1. SEM electron channelling contrast micrographs of AA5754 a) in the as-received state (AR) and after b) two, c) four and d) eight ARB cycles processed at 250 °C (RD: rolling direction).

After 2 cycles the grains show an irregular morphology due to deformation (figure 1b). Further rolling leads to an ultrafine-grained microstructure after 3-4 ARB cycles with elongated grains in rolling direction (figure 1c-d). After 8 ARB cycles the median grain size perpendicular to the rolling direction is 80 nm whereas there is no distinction between small and high angle grain boundaries.

3.2. Mechanical Properties

The evolution of the hardness with regard to the number of ARB cycles is shown in figure 2. There is no influence of the process temperature on the hardness of the sheets. However processing of ARB sheets by 8 ARB cycles has only been possible at 250 °C. After the first ARB cycle the hardness is nearly twice as high as in the initial state. Further processing leads to a linear increase in hardness up to 7 or 8 cycles respectively for both processing temperatures. Compared to the results on AA6016 [4] it becomes obvious that for the non-precipitation hardened alloy AA5754 the introduction of an ultrafine-grained microstructure by ARB is more effective in terms of strengthening the material by fine grains. In combination with the findings for the AA1050 alloy [3] the high degree of solute atoms in AA5754 appears to be the key-issue for the continuous increase in strength with increasing ARB cycles.

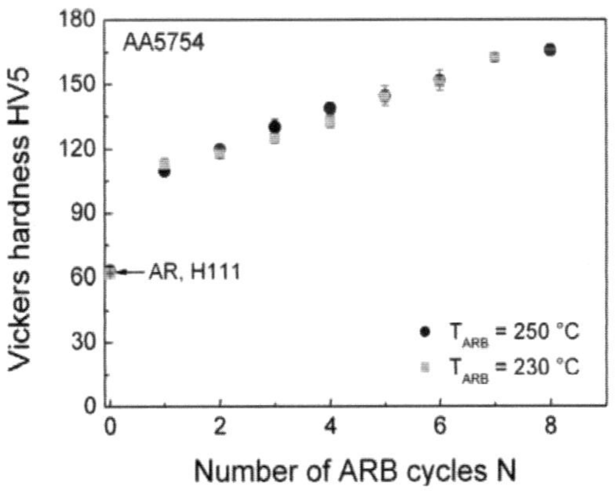

Figure 2. Hardness evolution at different process temperatures of the aluminium alloy AA5754 with an increasing number of ARB cycles.

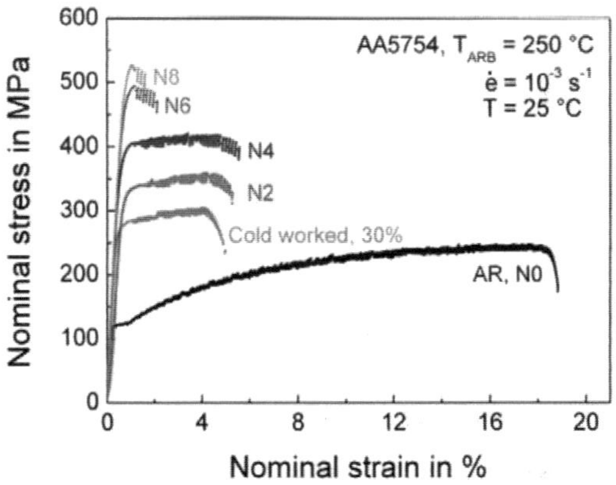

Figure 3. Stress-strain curves of as-received, cold rolled and ARB processed AA5754 performed at room temperature and a strain rate of 10^{-3} s^{-1}.

Representative tensile stress-strain curves of the as-received, the cold rolled (30 %) and the ARB samples can be seen in figure 3. With increasing number of ARB cycles the yield and the ultimate tensile strength (UTS) progressively increase. After 4 ARB cycles, a remarkable increase in yield strength of 330 % and in UTS of 170 % in comparison to the soft condition can be achieved. The elongation to failure is compromised by 75 % in comparison to the as-received condition but is increased slightly in comparison to the cold rolled CG condition. The gain in strength by further processing is on the expense of the ductility.

In order to improve the ductility, ARB specimens rolled-up to 4 cycles were heat treated at 250 °C for 10 minutes and 60 minutes and at 450 °C for 10 minutes to get a fully recrystallised condition. The mechanical properties received by tensile testing are shown in figure 4. Even though the heat treatment of 10 minutes at 250 °C causes a drop of 83 MPa in yield strength and UTS the values are still above the ones of the hard conventional grain-sized (CG) condition. Furthermore the elongation to failure is doubled in comparison to the not annealed ARB sample. In the case of the specimens heat treated for one hour at 250 °C the strength of the hard CG state and additionally almost the ductility of the recrystallized state are reached.

A Lüders strain behaviour is only observed in the case of the soft as-received condition (figure 3, black curve) which fulfils the precondition of the absence of mobile dislocations. The serrated yielding due to the Portevin-Le Châtelier (PLC) effect is observed for all conditions. The magnitude of serrations increases with increasing strain and as well with increasing number of ARB cycles. The decreasing grain size of the ARB processed samples is simultaneously attended by an increasing number of grain boundaries which act as obstacles for mobile dislocations [7].

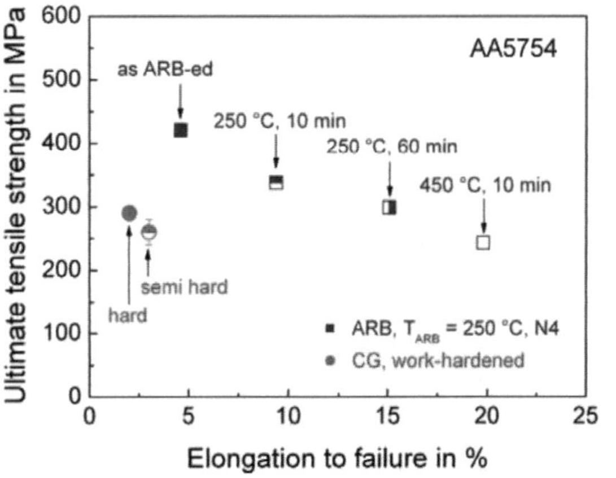

Figure 4. Mechanical properties of AA5754 with 4 ARB cycles after different annealing treatments in comparison to CG AA5754 sheets. Data for CG materials from [8].

The results of tensile testing at different strain rates of the as-received samples and ARB specimens roll bonded up to 4 cycles show that the serrations form with different intensity and morphology. More detailed, serrations of type B are found for the strain rates 10^{-3} s^{-1} and $10^{-4}s^{-1}$ and serrations of type A for 10^{-2} s^{-1}, compare [9]. For the ARB specimens, an inverse strain rate sensitivity is qualitatively observed, as for higher strain rates lower strength levels are revealed. Due to the occurrence of the serrations a quantitative analysis of the strain rate sensitivity exponent m does not make sense.

4. Conclusions

Aluminium alloy AA5754 has been processed up to 7 or 8 cycles respectively by using accumulative roll bonding producing an ultrafine-grained microstructure down to 80 nm. This favourably influences the mechanical properties like highly increased hardness, yield and ultimate tensile strength. Ductility is lost in comparison to the as-received state but is, up to 4 ARB cycles, increased in comparison to the cold rolled CG condition. A considerable increase in ductility can be reached by a short heat treatment at 250 °C, but on the expense of a moderately decreased strength. The deformation behaviour is also influenced by the ultrafine-grained structure. The occurrence of the Portevin-Le Châtelier effect is manifested by serrated stress-strain curves. The amplitude of serrations increases with increasing number of ARB cycles but can be reduced by the appliance of a higher strain rate. Lüders strain only occurs in the as-received, i.e. not strained, samples. The ultrafine-grained condition of AA5754 shows a high potential for light-weight constructions in automotive applications.

References

[1] Miller W S, Zhuang L, Bottema J, Wittebrood A J, De Smet P, Haszler A. and Vieregge A 2000 *Mater. Sci. Eng.* **A280** 37
[2] Saito Y, Utsunomiya H, Tsuji N and Sakai T 1999 *Acta Mater.* **47** 579
[3] Höppel H W, May J and Göken M 2004 *Adv. Eng. Mater.* **6** 781
[4] Topic I, Höppel H W and Göken M 2007 *Int. J. Mater. Res.* **98** 320
[5] Topic I, Höppel H W, Staud D, Merklein M, Geiger M and Göken M 2008 *Adv. Eng. Mater.* **10** 1101
[6] Lee S H, Saito Y, Sakai T, Utsunomiya H 2002 *Mater. Sci. Eng.* **A325** 228
[7] Wen W and Morris J G 2003 *Mater. Sci. Eng.* **A354** 279
[8] DIN EN 485-2: Aluminium and aluminium alloys –Sheet, strip and plate – Part 2: Mechanical properties *DIN Deutsches Institut für Normung e.V.* 2008.
[9] Balík J and Lukáč P 1993 *Acta Metall. Mater.* **41** 1447.

Quantitative determination of a change of dominant slip system in tensile FCC single crystals

M.S. Szczerba and P. Pałka

Department of Metallic Materials and NanoEngineering
University of Science and Technology, al. Mickiewicza 30, 30-059 Kraków, Poland

E-mail: szczerba@agh.edu.pl

Abstract. A change of dominant slip system (CDSS) in tensile FCC single crystals, as illustrated by properties of Cu-11at.%Al alloy single crystals, is quantitatively analyzed. The analysis shows that before and after the changeover cumulative shear of the dominant slip system is of the order of a magnitude larger than that of any other active secondary system. It was also found that at the critical condition for the CDSS to occur the rate of work hardening of the previously dominant slip system becomes equal to that one which is taking over.

1. Introduction

The changeover of dominant slip of FCC single crystals from one octahedral plane to another was probably first observed by Elam, as illustrated by room temperature tensile properties of Cu-Zn single crystals [1]. By means of the shape changes analysis of a distorted crystal and X-ray measurements of crystal lattice rotation Elam proved undoubtedly that the CDSS occurs at the end of overshoot, where the primary dominant slip system is replaced by another dominant system operating on the conjugate slip plane. Since then, the overshoot phenomenon has been the subject of extensive studies that confirmed its universality and importance in plasticity of FCC crystals. However, the experimental data were insufficient to be quantitatively discussed in terms of the CDSS effect (see review paper [2]). Very systematic qualitative studies of the CDSS effect in tensile FCC single crystals are more recently reported by Basinski, Szczerba and Embury [3]. Examination of a large body of stress-strain (σ–ε) and rate of strain hardening ($d\sigma/d\varepsilon$) curves of Cu and single phase copper alloys (Al, Ni, Si) has additionally shown that classical behaviour predicted by the Considère construction is never exhibited by soft oriented tensile single crystals. Instead the CDSS causes a sudden effective decrease in strain hardening which is accompanied by strain localization and the specimen fails prematurely, which makes it impossible to perform a complete quantitative analysis of the CDSS effect. On the other hand, very recently systematic quantitative studies of the amount of shear of primary and secondary slip systems in copper single crystals deformed in tension at room temperature are reported in [4]. The data show that prior to the entry of the CDSS the amount of cumulative shear of primary slip is usually of the order of a magnitude larger than that associated with secondary slip systems. However, the complete quantitative analysis of the CDSS effect has not yet been performed. This is probably because of the fact that in most cases the amount of slip in the new dominant slip system was either impossible to determine due to the neck formation - soft oriented crystals, or the measurements could be affected by inhomogeneous distribution of crystal strain due to Lüders deformation - orientations near the $[001]$-$[\bar{1}11]$ symmetry line of the basic stereographic triangle. In this paper the properties of Cu-Al alloy single crystals, which allow studying efficiently the amount of dominant and secondary slip before and after the CDSS, are analyzed. To perform the analysis, the full gradient deformation matrix was determined from current orientation and shape measurements at successive strain intervals and the magnitude of shears in eight independent slip systems was calculated by means of the Chin, Thurston and Nesbitt method [5]. Moreover, a mechanical condition for the CDSS effect to occur and its influence on deformation mechanisms at large strain plastic flow of FCC crystals will also be discussed.

2. Experimental procedure

Cu-11at.%Al alloy single crystal samples of prismatic shape and dimensions 4·4·100 mm^3 were used in the studies. The initial tensile crystal orientation was less than 2 degrees away from the $[001]$-$[\bar{1}11]$ symmetry line of the standard stereographic triangle and close to the $[\bar{1}12]$ crystallographic direction. The solute concentration and the initial crystal orientation were chosen to ensure that the tensile single crystals will survive the overshoot instability/strain localization and they sustain further homogeneous deformation large enough to perform the complete and reliable analysis of the CDSS effect. The single crystals were grown in a natural temperature gradient furnace in a vacuum better than 10^{-5} hPa. The samples were mechanically and chemically polished with Mitchell reagent in order to obtain lateral surfaces of great smoothness. Then, the grids of measurement points were placed on the crystal surfaces by the indenter of an Instron Wolpert 2100 microhardness tester with a value of the indentation load within the range 0.1 – 0.2 N. Taking advantage of a micrometric X-Y table coupled with an optical microscope and the Bruker D8 Advance X-ray diffractometer, the measurements of the mutual placement of the grid points were conducted and the crystallographic orientation of both of lateral crystal surfaces was determined. The samples were then step by step deformed in tension at room temperature with an initial strain rate of 10^{-3} s^{-1} and strain intervals of about 0.05 of the tensile logarithmic strain. The distorted samples were placed again in a specially-prepared holder that was adapted to fasten it both at the X-ray diffractometer and the micrometric X-Y table. Sets of experimental data describing the shape changes of the samples in terms of geometric crystallography after each deformation step were obtained in the form of 3D base vectors and a magnitude of the shear strains in eight independent slip systems could be calculated with a resolution of the measurement method better than $5 \cdot 10^{-3}$. Additionally the stress–strain and rate of work hardening curves of the Cu-11at%Al single crystals were determined. More details about the experimental procedure and the method of calculation of shears in the independent slip systems have been given elsewhere [4-6].

3. Experimental results and discussion

The stress-strain (б-ε) and rate of work hardening curves of the Cu-11at.%Al single crystals are shown in figure 1. These curves are limited to the range of tensile strain where the total crystal deformation is produced by slip only - it is well known that these single crystals deform at larger strains by mechanical twinning. The б-ε curve can be divided into two distinct stages with a sharp transition somewhere in the middle, where catastrophic drop from 1000 MPa down to about 200 MPa of the rate of strain hardening (dб/dε) is observed. Watching lateral surfaces of a deforming crystal it was easy to see that the abrupt drop in work hardening was followed by a discrete plastic front travelling along the specimen. However, during this period of plastic flow ($\varepsilon \approx 0.3$-0.4) a specimen was hardened enough to sustain further homogeneous deformation and to restore the high rate of strain hardening. The X-ray measurements show that the transient local decrease in dб/dε corresponds to the end of overshoot of the crystal tensile axis as illustrated in figure 2. The figure shows the angular distance from the $[001]$-$[\bar{1}11]$ symmetry line as a function of strain. Since the primary slip system, $(111)[\bar{1}01]$ dominates at the very beginning the deformation process, the tensile axis moves from the initial position towards the operating slip direction $[\bar{1}01]$ crossing the $[001]$-$[\bar{1}11]$ symmetry line during the first strain intervals. Further crystal deformation proceeds in the state of the overshoot which at the critical point (ε=0.26) reaches the maximal value of about 7 degrees. Starting from the critical point the reverse process occurs. The tensile axis moves backward to the initial stereographic triangle crossing again at ε=0.47

the symmetry line, so the state of second overshoot is achieved. It is clear that at the end of overshoot the CDSS must take place which consists of the changeover of the dominance from primary (**P**) into conjugate (**C**) slip system, $(\bar{1}\bar{1}1)[01\bar{1}]$. Corresponding changes of Schmid factors of the dominant **P** and **C** slip systems are shown in figure 3.

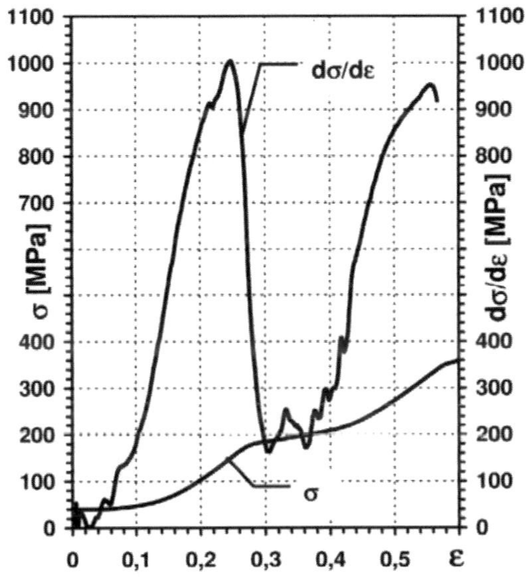

Fig.1. Stress-strain and rate of work hardening curves of the Cu-11at.%Al single crystals

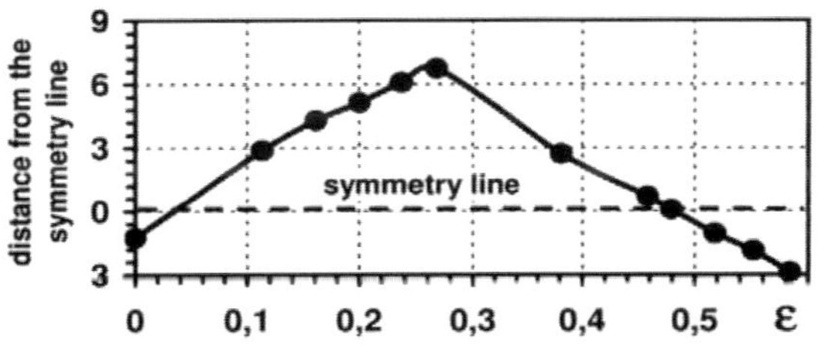

Fig.2. Changes of the tensile axis position during deformation process

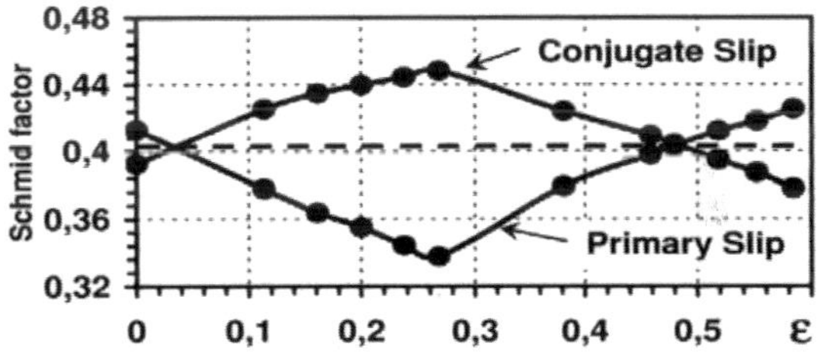

Fig.3. Schmid factor changes of the dominant primary **P** and conjugate **C** slip systems.

Now, taking the product of the tensile stress and the current Schmid factor value one can calculate the resolved shear stress τ operating in the dominant **C** and **P** systems. The evolution of τ_C and τ_P values during the deformation process is shown in figure 4. It is to note that at the end of overshoot the difference between τ_P and τ_C is the largest and the ratio τ_C/τ_P at this point reaches the value of 1.3. It means that the entry of the new dominant **C** system is due to the latent hardening well over-stressed what may result in the sharp transient local decrease in dσ/dε and the temporary strain localization. The key experimental data concerning the amount of shears operating in all four octahedral slip planes of the crystal lattice, or alternatively, in eight independent slip systems, is shown in figure 5. The two important points are: (i) prior to the CDSS the crystal deformation is totally dominated by the **P** slip system and the amount of shear in this system is of the order of a magnitude larger that any other secondary slip system including the **C** slip system; (ii) after the CDSS crystal deformation is totally dominated by the **C** slip system and the amount of shear in this system is of the order of a magnitude larger that any other secondary slip system including the **P** slip system. So, at the end of the overshoot the CDSS effect consists of the dominant slip changeover of the **P→C** type. Moreover, figure 6 shows that the changeover does occur when the gradually decreasing strain hardening rate (dτ/dγ) of **C** system approaches the same numerical value as the gradually increasing strain hardening rate of **P** system. It is important to note that operation of the new dominant **C** system leads quickly to the state of the end of the second overshoot and another CDSS effect might occur where the opposite **C→P** type of the changeover can take place. Such multiple CDSS effect can occur in copper [4] and less concentrated Cu-Al alloy single crystals [6] where the sequence of **P→C→P→C** have been observed before the tensile specimen necked to failure. However, at the end of the second overshoot of the investigated Cu-12at.%Al crystals mechanical twinning is observed instead. Such the dominant slip – mechanical twinning transition in the case of Cu-8at.%Al single crystals oriented for single and double glide was discussed in detail by Szczerba, Bajor and Tokarski [7]. Summarising the obtained results and the performed discussion it seems reasonable to support strongly an idea that the CDSS phenomenon might be of great importance in large strain plasticity of FCC materials. The phenomenon should be also seriously considered as a process which is transformational in nature.

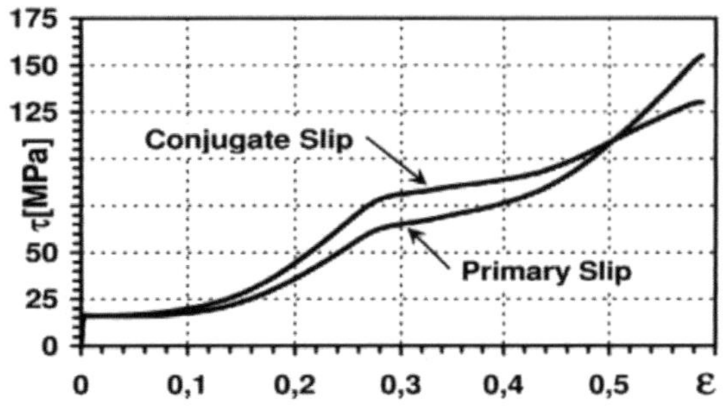

Fig.4. The evolution of resolved shear stress of the dominant **P** and **C** slip systems

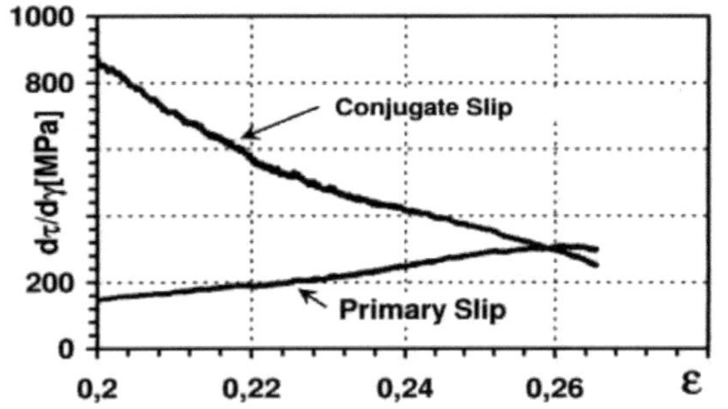

Fig.6. The rates of work hardening of the dominant **P** and **C** slip systems

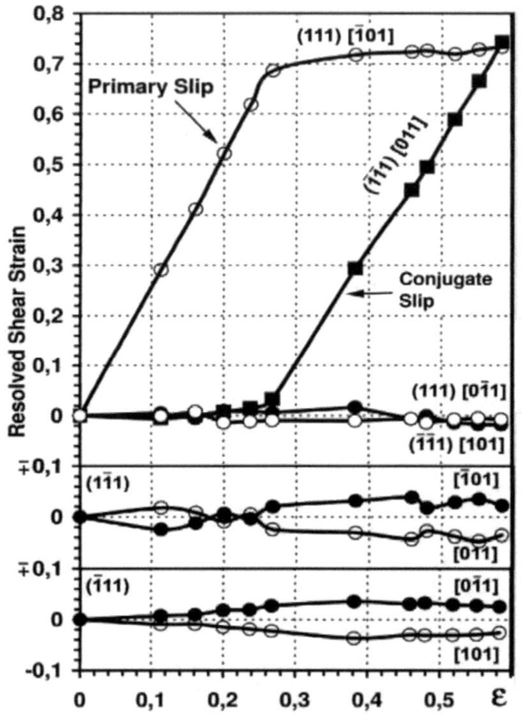

Fig.5. The amount of shear in eight independent slip systems versus the tensile deformation

4. Conclusions

On the basis of the obtained results and the performed discussion the following conclusions can be drawn: (i) – the complete quantitative analysis of the CDSS effect occurring in the investigated Cu-11at.%Al single crystals was successfully done; (ii) - the amount of shear of the dominant slip system prior to and after the CDSS is of the order of a magnitude larger than for any other secondary slip system; (iii) - The CDSS does occur at the end of overshoot when the strain hardening rates dτ/dγ of the dominating slip system and the system which is taking over the crystal deformation become equal.

Acknowledgments

The financial support of the Ministry of Higher Education and Science / AGH University of Science and Technology, Kraków, Poland, grant number 11.11.180.255 is greatly acknowledged.

References

[1] Elam C F 1927 *Proc. Roy. Soc.* A **113** 148
[2] Basinski S J and Basinski Z S 1979 *Plastic Deformation and Work Hardening, Dislocation in Solids*, ed. F R N Nabarro (North – Holland Publishing Company) p 333
[3] Basinski Z S, Szczerba M S and Embury J D 1997 *Phil. Mag.* A **76** 743
[4] Basinski Z S and Basinski S J 2004 *Phil. Mag.* **84** 213
[5] Chin G Y, Thurston R N and Nesbitt E A 1966 *Trans. Metall. Soc. AIME* **236** 69
[6] Szczerba M S and Pałka P 2009 *Arch. Metall. Mater.* **54** 51
[7] Szczerba M S, Bajor T and Tokarski T 2004 *Phil. Mag.* **84** 481

The effect of equal channel angular pressing on the tensile properties and microstructure of two medical implant materials: ASTM F-138 austenitic steel and Grade 2 titanium

A de A Mendes Filho, VL Sordi, AM Kliauga and M Ferrante

Federal University of São Carlos, Materials Engineering Department, São Carlos, 13565-905, Brasil

ferrante@ufscar.br

Abstract. Titanium and F-138 stainless steel are employed in bone replacement and repair. The former material was ECAP-deformed at room temperature and at 300°C, followed in some cases by cold rolling. The steel was ECAP-deformed at room temperature only. Work-hardening behavior was studied by making use of the Kocks-mecking plots and microstructural evolution was followed by TEM. Conclusions show that for Ti, ECAP combined with cold rolling gives the best strength-ductility combination, whilst room temperature ECAP increases the tensile strength of the steel but caused substantial ductility loss.

1. Introduction

One of the most promising application fields of ECAP technology is that of orthopedic items, such as plates, screws, and implants. Presently, only three metallic materials are employed: ASTM F-138 austenitic stainless steel and titanium, both the metal and the Ti6Al4V alloy. The steel is widely used due to its good performance and accessible cost, whilst the Ti alloy is the best solution. However, there is an intense search for new alloy formulations in which the potentially cytotoxic Al and V are absent. Thus, complex Ti-based alloys are under development [1] but cost and complexity do not favour this strategy. Instead, the upgrading of the mechanical properties of commercial purity materials so as to approach those of the alloy is a better solution [2, 3], well within the scope of ECAP.

This paper presents results on tensile behaviour of ECAP - deformed (with and without cold rolling finish) Grade 2 Ti and F-138 stainless steel, and examines the feasibility of room temperature deformation for both. As for the latter material, the very preliminary study here reported was motivated by the scarcity of data regarding its behaviour under severe plastic deformation (SPD).

2. Experimental

2.1. Materials: commercial purity Grade 2 Ti and AISI F-138 austenitic stainless steel were supplied as extruded rods, which were subsequently annealed (Ti at 710°C, steel at 1100°C) and air cooled.

2.2. ECAP- deformation: performed in a H-13 steel die ($\Phi = 120^\circ$ and $\Psi = 20^\circ$) with 10 mm diameter cross section channels. Heating was supplied by five resistance heaters embedded in the die and thermocouples were located close to the channel angle. The experiments were performed at 25 and 300°C ± 20°C on 70 mm long specimens, pressed using 5 mm s^{-1} ram speed; The Ti specimens were subjected to one (1X) and four (4X) passes, following Route B_C. The hot-deformed samples are identified by the letter H. Two Ti samples (0X and 4XH were subjected to cold rolling (CR) up to 70% reduction without intermediate anneals. Cold deformation and two passes were applied to the steel.

2.3. Characterization: the microstructure was observed by transmission electron microscopy (TEM) on polished sections parallel to the specimen axis, employing a 5 perchloric-95 acetic acid electrolyte. Vickers hardness measurements preceded tensile tests which were performed on subsize specimens (12 mm^2 and 12 mm gauge length) at room temperature and strain rate equal to 10^{-3} s^{-1}.

3. Results and discussion

Tensile data are summarized in Table I, and the changes in σ_u and ε_t resulting from SPD are illustrated in Figure 1 for Ti. Data clearly show that for Ti those properties have been optimized, a remarkable result for conflicting properties, whilst the conventional behaviour is followed by the steel. Dealing with the tensile behaviour of Ti first, comparison with literature data shows that, although the present samples were submitted to a relatively low shear strain (γ = 0,66 and 2,64 for one and four ECAP passes respectively) they exhibit similar when not higher strength than specimens much more heavily deformed [3, 4, 5]. More interesting, in all comparisons the ductility of the samples ECAped at 300°C was higher. On this respect, an instance of concurrent increase of strength and ductility observed on nanostructured Ti after post-deformation anneal at 300°C was interpreted by Valiev et al. as due to grain boundary structure change, thus leading to grain sliding and rotation [6]. Clearly, more data would be necessary to better analyze the present results

Table I. Hardness and tensile properties of hyper deformed Ti and F-138 steel.

Material	Condition	σ_y (MPa)	σ_u (MPa)	ε_t (%)	ε_u (%)	HV
Titanium	0X	320	408	21	9.0	140
	0X + CR	585	702	17	2.5	236
	1X	564	589	12	3.7	196
	1X H	530	573	25	3.2	205
	4X H	636	678	23	4.0	229
	4X H + CR	796	876	21	3.3	271
F-138	0X	374	640	32	23	130
	1X	889	907	12	2.8	321
	2X	1055	1095	9	2.8	339

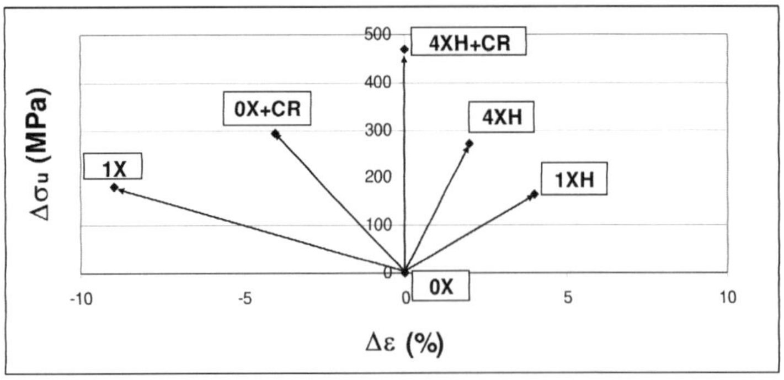

Figure 1. Tensile properties map for Ti samples plotted in the space $\Delta\sigma_u$ *versus* $\Delta\varepsilon_t$. Differences are taken in relation to the 0X sample in terms of σ_u and ε_t.

Figure 1 also shows that ECAP+CR gives the best strength-ductility combination and the highest strength. As for the stainless steel, yield strength reached a large increment (182%), but ductility losses were close to 70%.

Work hardening (W-H) is a very important phenomenon that controls the first stage of plastic deformation, that is, uniform elongation (ε_u), possibly a parameter more important that total elongation (ε_t) since no internal defects are yet formed. Table I shows values of ε_u while Figure 2 describes the W-H behaviour of the samples in terms of Kocks-Mecking plots (K-M). This type of presentation quantifies two important parameters: the initial W-H rate (θ_{max} or $[d\sigma/d\varepsilon]_{max}$), obtained by extrapolation to $\sigma - \sigma_y = 0$ of the linear part of the curve, and the evolution of θ with flow stress. The former parameter is associated to the dislocation accumulation whilst the latter is related to restoration processes.

The analysis of the present K-M plots runs as follows: for microcrystalline Ti sample 1X exhibits the smaller θ_{max} followed by a relatively fast initial decrease of W-H rate; also, this specimen has the smallest saturation stress (i.e. $\sigma - \sigma_y$; indicated by the dotted line), indicating a very high initial dislocation density. As for the samples deformed at 300°C, they all show similar behavior, irrespective if CR follows ECAP - deformation or not. Furthermore, saturation stresses are higher, and in contrast with the non-deformed sample, no stage IV is observed. The K-S plots also show that $\theta_{max} \approx \mu_{Ti}/300$, that is, the initial rate of hardening is small, a characteristic shown by nanocrystalline metals, and also by HC metals due to the difficulty of nucleating dislocations in secondary slip systems. As for the steel, SPD caused a very high θ_{max} followed by a sharp decrease of W-H rate; again, only stage III is observed whilst the non-deformed sample shows the presence of stage IV. The initial W-H rate (i.e. θ_{max}) of sample 1X is larger than that of sample 2X, ($\mu/100$ and $\mu/400$, respectively) again suggesting that the latter has a higher defect density. Also, its restoration rate is slower, suggesting lower defects mobility. Finally, for both steel and Ti the hardening capacity H_C, defined as σ_u/σ_y is close to 1.1, thus explaining the relatively low uniform elongation observed.

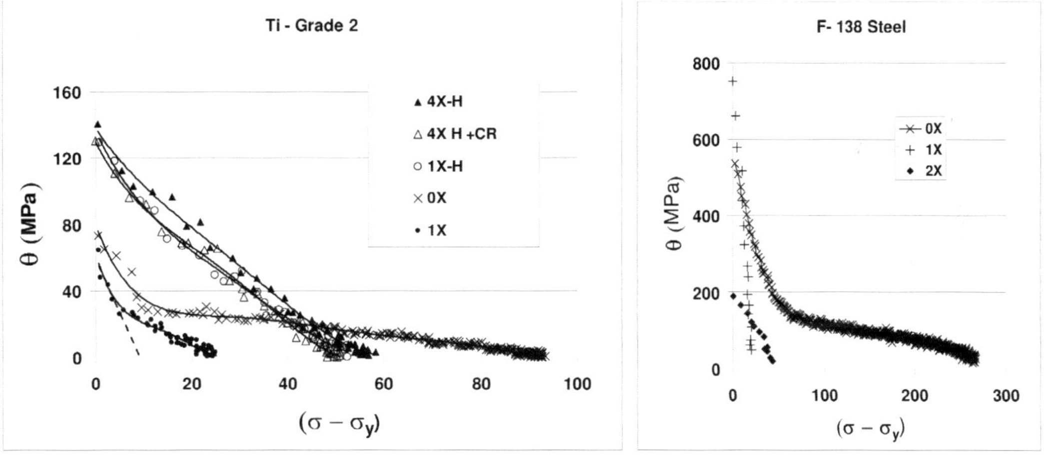

Figure 2. Kocks-Mecking plots for Ti and for the F-138 steel. Dotted line exemplifies how saturation stress is determined

The microstructural evolution is shown in Figure 3, and the TEM micrographs a and b refer to the F-138 steel; twinning is the major deformation mechanism, and inside twins stacking fault contrast is clearly seen (figure 3a). After the second pass shear bands develop and grain fragmentation begins. As for Ti, most of the plastic deformation seems to have been assisted by dislocations and twinning is not the major deformation mechanism, an observation supported by a theoretical calculation on slip/twinning transition as a function of grain size [7]. Final grain size is close to 200 nm and 100 nm for samples 1X and 1X+CR, respectively.

4. Final remarks

It was shown that ECAP-deformation of Grade 2 Ti produces a substantial increase of both strength and ductility, and that the best results are obtained by that SPD technique followed by cold rolling. However, when performed at room temperature, ECAP leads to substantial ductility loss. Work-hardening rate was not very pronounced: initial value was close to μ/300 for the specimens deformed at 300°C and μ/600 for sample 1X. On the other hand, the F-138 steel tensile behaviour is typical of material submitted to SPD, that is, high strength but very low ductility. The K-S plots showed very efficient restoration processes acting in sample 1X.

As for the steel microstructural response to SDP, deformation bands with internal stacking faults were seen after the first pass; further deformation reduced the bands thickness and grain fragmentation is initiated. For Ti, deformation seems to be mainly dislocation-assisted and the main difference between samples 1X and 1XH seems to be the dislocation density, which is larger in the 25°C deformed sample. Finally, CR appears to be very efficient in decreasing the size of the microstructural features.

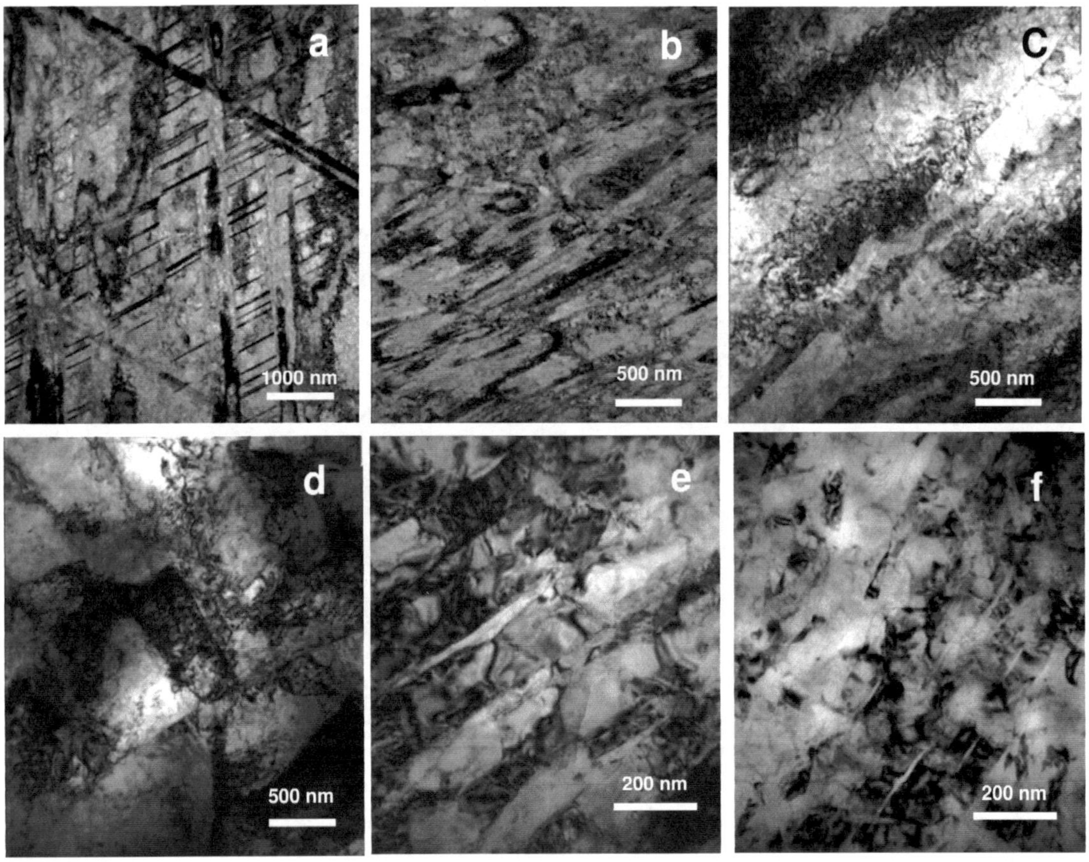

Figure 3. TEM micrographs from longitudinal sections of deformed samples: (a) F-138 sample 1X; (b) F-138 sample 2X; (c) Ti-1X; (d) Ti-1XH; (e) Ti-4XH (f) Ti-4XH+CR

References

[1] Geetha M, Singh AK, Asokamani R and Gogia AK 2009 *Progr. Mater. Sci.* **54** 397
[2] Ko YG, Shin DH, Park KT and Lee CS 2006 *Scripta Mater.* **54** 1785
[3] Zhao X, Fu W, Yang X, Langdon TG 2008 *Scripta Mater.* **59** 542
[4] Stolyarov VV, Zeipper L, Mingler B and Zehetbauer M 2008 *Mater. Sci. Eng.* A **476** 98
[5] Stolyarov VV, Zhu Y, Alexandrov IV, Lowe TC and Valiev R 2003 *Mater. Sci. Eng.* A **343** 43
[6] Valiev, RZ, Sergueeva, AV, Mukherjee, AK 2003 Scripta Mater. **49** 669
[7] Meyers MA, Benson DJ, Vöhringer O, Kad BK, Xue Q and Fu H-H 2002 *Mater. Sci. Eng.* A **322** 194

Microstructure evolution and strength response of ultrafine grain medium carbon steel processed by high pressure torsion

Jozef Zrnik[1,2], Reinhard Pippan[3,4], Stephan Scheriau[3,4]

[1] COMTES FHT Ltd, Prumyslova 995, 33441 Dobrany, Czech Republic
[2] West Bohemian University, Universitni 22, 301 01 Plzen, Czech republic
[3] Erich Schmid Institute of Materials Science, Austrian Academy of Science, Jahnstrasse 12, A-8700 Leoben, Austria
[4] Christian Doppler Laboratory for Local Analysis Deformation and Fracture, Jahnstrasse 12, A-8700, Leoben, Austria

jzrnik@comtesfht.cz

Abstract. High pressure torsion (HPT) deformation method at increased temperature of 400°C with varying strain was applied to refine microstructure of the medium carbon steel (AISI 1045). To investigate the deformation behaviour of the ferrite-pearlite two phase structure the different shear deformation and constant high hydrostatic pressure of 7 GPa were applied. The shear stress evolution during deformation and measurement of the torque were recorded and related to structure development and mechanical strength. In order to characterize microstructure development the transmission electron microscopy (TEM) was accomplished. By execution of the first turn the grain refinement was observed at deformed disc periphery. In the centre of the disc, no matter what strain was introduced, the ferrite grain structure showed moderately deformed features and pearlite colonies were preserved. With further equivalent strain increase, equilibrium between the refinement of coarse phases and new grains restoration processes led to saturation of the grain refinement process. Upon tensile properties testing, the yield strength and ultimate strength increased with increasing equivalent strain (ε_{eq}) and only short region of strain hardening period prior failure appeared, regardless the strain applied. A small drop in the hardness across the disc was measured after execution of N= 4 and 6 turns, which may be related to formation of fine grain structure and structure recovery in disc centre.

1. Introduction

During the recent decade, bulk nanostructered materials produced by severe plastic deformation (SPD) have been investigated intensively. The production of fine grained materials by SPD, leads to a large number of investigations focusing on the substructure development and related mechanical properties. It has been already well known that SPD of metallic materials, involving processes such as equal angular pressing (ECAP), accumulative roll bonding (ARB) and high pressure torsion (HPT) is capable of producing ultrafine grained (UFG) materials with submicrometer or nanometer grain size [1,2]. The wide range of results available for various materials deformed by HPT confirmed a saturation in microstructural refinement and the strength and hardness as strain increases [3, 4]. The results point out that there are differences in the development of the microstructure between the pure and multiphase steels [5,6].

In this work two phase medium carbon steel has been deformed at increased temperature of 400°C by high pressure torsion to different strain (executing different numbers of turns) at hydrostatic pressure of 7 GPa. The underlying relationship between microstructure and mechanical properties was analyzed with respect to the local axial position of analyses on the deformed disc.

2. Material and experimental procedure

A commercial medium carbon steel (Fe-0.45%C, 0.42%Mn, 0.23%Si 0.18%Cr, 0.043%Al in wt. %) was supplied in form of rod with a diameter of 20 mm. Prior HPT deformation the steel was the annealed for 1.5 h at temperature of 850°C followed by air cooling in order to obtain uniform normalized structure. The mixture of quite coarse ferrite-pearlite structure was obtained as showed in SEM micrograph in Fig. 1. Following the annealing treatment, the size of pearlite colonies was of ~ 30 μm and of ~ 15 μm was size of ferrite grains.

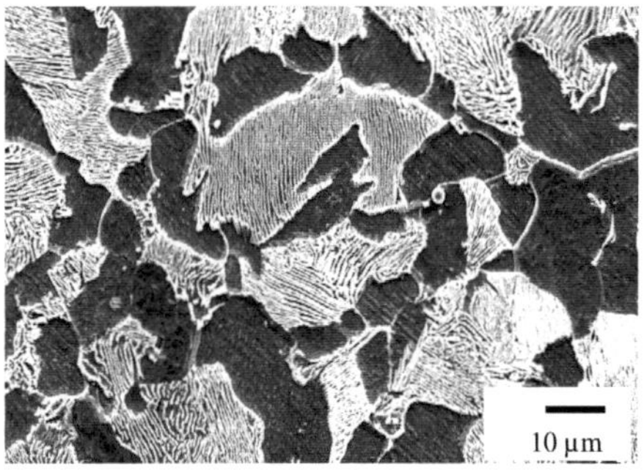

Figure 1. SEM micrograph of initial ferrite – pearlite structure of steel.

The discs of 8 mm in diameter and ~ 0.95 mm in thickness have been cut off from the rod were deformed by torsion up to 6 turns at temperature of 400°C and at a pressure of 7 GPa. The equivalent von Mises strain ε_{eq} as a function of the number of turns ***n*** was calculated according to the relation: $\varepsilon_{eq} = 2\pi nr / t\sqrt{3}$. The size of effective strain conducting N- 1, 2, 4, and 6 turns was ε_{eq} ~ 15, 30, 60 and 90. The changes in mechanical properties in dependence on straining (number of turns) were determined by microhardness measurement across the deformed disc with a load of 300 g using Vickers microhardness tester, by tensile test using subsize tensile pieces cut off at discs periphery (in radial direction) and by in-situ measurement of the torque. The reliable and accurate possibility to measure the torque during deformation allowed to quickly to determine changes in mechanical strength (shear stress τ) without applying other method afterwards.

The transmission electron microscopy (TEM) was used to evaluate the microstructural evolution with respect to disc site and the number of turns corresponding to shear stress. The TEM micrographs were obtained by using JEOL JEM 2000FX operating at 200 kV. The purpose of different site selection was to evaluate an effect of shear strain magnitude across the disc (at peripheral and axial position of disc) on ultrafine grain microstructure development.

3. Results and discussion

Investigating material comprised of the coarse ferrite-pearlitic steel AISI 1045, as shown in Fig.1. Refining coarse steel structure to nanostructure size the application of large strain, usually with an equivalent strain ε_{eq} more than 10, is required. Electron microscopy analysis of thin prepared from

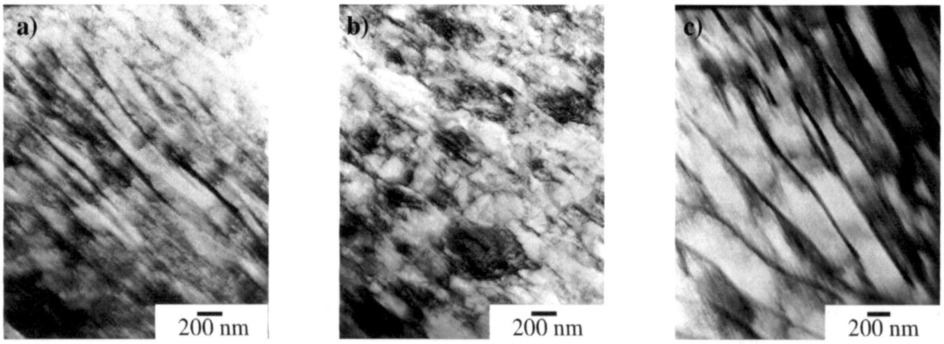

Fig. 2. TEM micrographs of microstructures observed at disc periphery after HPT conducting 1 turn, (a & b) and corresponding to the disc centre (c) experienced the ε_{eq} ~ 15.

deformed discs exposed to different strain revealed formation of various structure characteristics in dependence of shear strain applied and localisation.

3.1. Microstructure

In order to compare the size of torsion straining at different position of deformed disc, Fig. 2 shows the microstructure observed in steel after exposure to the first turn at temperature of 400°C. Finishing the first turn ($\varepsilon_{eq} \sim 15$) the heterogeneity in structure development was still evident across the disc. At disc periphery alignment of smeared (partially dissolved) lamellae in pearlite next to the area of banded subgrain-like structure is documented in Fig. 2a,b. In the centre of the disc the lamellae remained, but some dissolution of the cementite lamellae, can be observed with smeared-like features, as presented in Fig. 2c. Conducting 4 and 6 turns ($\varepsilon_{eq} \sim 60$ and ~90), the dual phase structure was significantly modified with respect to analysis site as presented in Fig. 3. Regardless the strain size, successfully refined structure with small grains (grain size about of ~ 200 nm) having random high angle orientation was found in disc edge area as presented in Fig. 3a. Selected area diffraction method (SAED) confirmed high angle disorientation of small grains. In the centre of the disc the structure has moderately deformed features. In ferrite grains tangled dislocations are present, and pearlite colonies were only slightly deformed and/or locally fragmented, Fig. 3b,c. From the TEM micrographs cementite lamellae thinning in central disc area is evident in structure. This fact is attributed to partial dissolution of cementite during shearing and results in decreasing of the lamellae thickness.

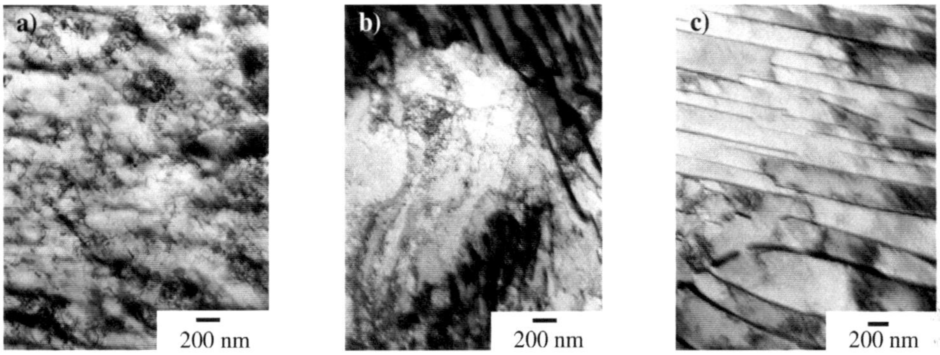

Fig. 3. TEM micrographs of deformed microstructures conducting 4 turns found at disc periphery and at the centre of disc experienced the $\varepsilon_{eq} \sim 60$: a) edge, b, c) centre.

3.2. Mechanical properties

In order to characterize the changes in mechanical properties due to the large shear deformation the various methods were applied. Vickers microhardness and tensile properties were evaluated at room temperature. The possibility to measure the torque in-situ during deformation provided a method to determine mechanical strength. The torque value includes contribution from disc deformation inside the cavity and contribution from outside the cavity where material is impressed btw anvils (burr) [5].

Microhardness evaluation after different straining (N turns) was convenient choice to evaluate the mechanical strength and is often applied [6,7]. In present work the method was mainly used to determine the strength change across the disc (at disc periphery and the centre) after application of different strain ε_{eq} and compared with hardness of the initial steel. The measured Vickers hardness data (HV30) for different number of turns are stated in Tab.1. The results show that some decrease in hardness appeared across disc, performing the 4th turn. In deformed microstructure analyzed by TEM of thin foils there were not observed any specific structure diversities, that could match the results, and

Tab. 1 Microhardness as function of N

N turns	edge	centre	edge
1	530	364	509
2	524	342	528
4	467	308	426
6	536	517	491

then returning again to previous higher hardness values after 6 turns.

To obtain mechanical strength values, the specimens for tensile tests were machined out of HPT samples [8]. Two specimens (see in fig. 4) were cut off from the deformed disc. Tensile behaviour was evaluated at room temperature and records in form of load and sample elongation are documented in Fig. 4. The strength was increasing with increasing strain (turn numbers) and maximum load value obtained was higher than 1700 N. The development of the ultimate tensile strength is similar to the in-situ measured torque. The records of in-situ measured torque and shear stress (τ) for turn's number of N1 and N6 are presented in Fig. 5. The possible explanation could be that the onset of steady state for both values coincidences with saturation in the decrease of the grain structure size and microhardness values. The strain necessary to reach this state deformation depends on the materials structural state.

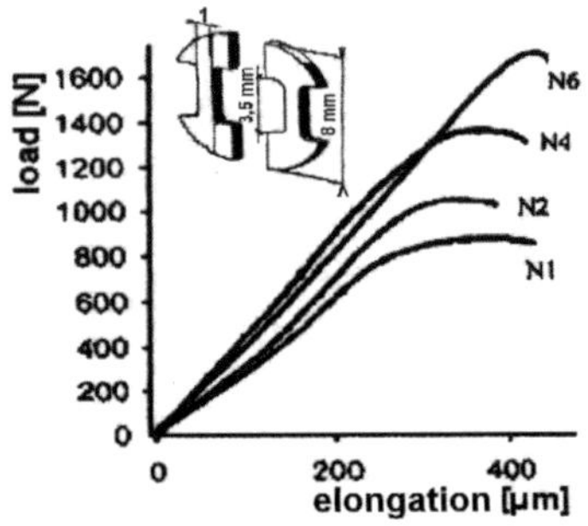

Fig. 4. Tensile records of after different HPT straining.

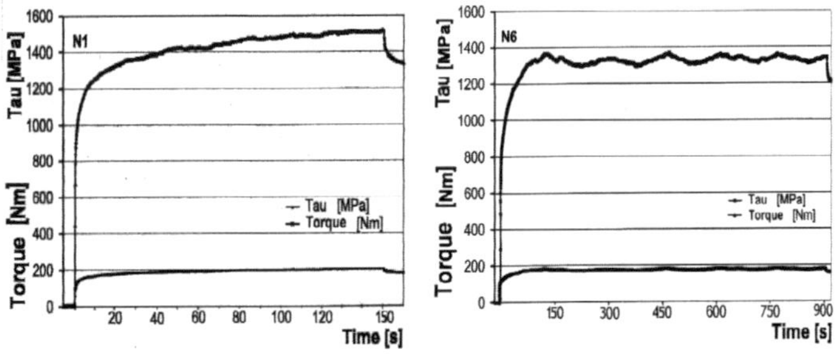

Fig. 5. Comparison of the in-situ torque and shear stress curves for N1 and N6

4. Summary and conclusions

High pressure torsion method at increased temperature of 400°C was applied to refine microstructure in AISI 1045 steel. In dependence of strain applied there are markedly differences in the development of the microstructure as regards the site of analysis. Grain refinement and cementite lamellae dissolution was observed after the first turn ($\varepsilon_{eq} \sim 15$) at disc periphery. In disc centre, the deformed structure of ferrite with dislocation tangles had moderately deformed features of coarse cementite lamellae in pearlite grains. At higher strain applied cementite lamellae are aligned, thinner and fragmented in dependence of the strain introduced. The effective strain increase, set an equilibrium between the fragmentation cementite phase and new fine grains restoration processes and led to saturation of the grain refinement process. Upon tensile properties, the yield strength and ultimate strength increased with increasing ε_{eq}. A small decrease in the hardness across the disc was measured across the disc after execution of 4 turns, which may be related to formation of fine grain structure in the disc and dislocation structure recovery at the disc centre.

Acknowledgement: The support from the Erich-Schmid-Institute of Materials Science of the Academy of Sciences in Leoben, Austria and form the Ministry of Education of the Czech Republic (MSM 2621691901) is greatly appreciated.

References

[1] V.M. Segal, V.I. Reznikov, A.E. Drobyshevski, Russian Metallography, 1, 1981, 99.
[2] R.Z. Valiev, A.V. Korznikov, R.R. Mulyukov, Mater. Sci. Eng. A **168** 1993, 99.
[3] F. Wetscher, A. Worhauer, R. Stock, R. Pippan, Mater. Sci. Eng. A **387-389** 2004, 809.
[4] N.Q. Chinh, G. Horvath, Y. Horita, T.G.Langdon, Acta Mater. **52** 2004, 3555.
[5] F. Wetscher , Doctoral thesis, Leoben 2006, Austria.
[6] M. Richert, Q. Liu, N. Hasen, Mater. Sci. Eng. A **260** 1999, 275.
[7] R.K. Islamgaliev, W. Buchgraber, Y.R. Kolobov, N.M. Amirkhanov, A.V. Sergueva, K.V. Ivanov, G.P. Grabovetskaya, Mater. Sci. Eng A **319-321** 2000, 872.
[8] A. Vorhauer, R. Pipan, Scripta Mater. **51** 2004, 921

Sequence effects of twist extrusion and rolling on microstructure and mechanical properties of aluminum alloy 8112

Sh Ranjbar Bahadori, S A A Akbari Mousavi and A R Shahab

School of Metallurgy and Materials Engineering, University College of Engineering, University of Tehran, Tehran, Iran, P.O. Box: 11155-4563.

shrb1984@yahoo.com

Abstract: Recently severe plastic deformation processes have been the essence of metal forming researches to produce ultrafine-grained materials. Twist Extrusion (TE) is one of the most unprecedented methods developed in recent years, but remained in laboratory scales. The main reason for this matter refers to some deficiencies like microstructure heterogeneity occurring after TE. However, employing conventional forming techniques could make TE industrial and, moreover, reduce the bulk structure grain size. Using a conventional forming process such as rolling after twist extrusion has been suggested by this paper. T.E process of Al 8112 samples was carried out using a twisted die with 60° die angle and the samples were processed through rolling subsequently. The results demonstrated that implementation of rolling not only reduced heterogeneity but also decreased the grain size and, consequently, enhanced the bulk strength.

1. Introduction

Severe plastic deformation (SPD) is a family of metal forming processes which uses hydrostatic pressure to impose high strain values on the material in order to generate ultrafine-grained microstructure without any changes in material geometry [1]. It includes several different techniques such as equal channel angular pressing (ECAP) [2], high pressure torsion (HPT) [3], accumulative roll bonding (ARB) [4], multi-directional forging (MDF) [5] and repetitive corrugation and strengthening (RCS) [6]. Another developed technique which is called twist extrusion (TE) has been examined by some investigations. TE process ensures accumulation of large strain deformation in the material, refinement of microstructure and enhancement of mechanical properties and hardness, but these exceptional properties concur with heterogeneity. The novel solution for this problem is to use a conventional metal forming technique such as rolling process subsequent to TE. Park et al. [7] examined the effect of post-rolling after ECAP on the deformation behavior of Al-Mg alloy. They found that post-rolling after ECAP is capable of enhancing the superplastic behavior of the materials significantly. Moreover, Stolyarov et al. [8] investigated on the twist extrusion of CP-titanium processed by subsequent rolling. It was discovered that the combined process of TE and rolling, not only decreases the heterogeneity but also reduces the grain size and increases the strength properties.

In this investigation, the effects of post rolling after the twist extrusion process on commercially pure aluminum were considered.

2. Experimental procedure

Aluminum alloy 8112 with the composition tabulated in table 1 was selected for the present study. Its initial grain size was about 600 μm and an annealing treatment was performed to the as-received material (500°C for 2h), giving a grain size of about 859 μm.

Table 1. Chemical composition of commercially Al 8112.

Al	Fe	Si	Mg	Cu	Zn	Cr	Ti	Mn	Other
96.03	0.90	0.86	0.73	0.44	0.43	0.26	0.23	0.11	Bal.

The circular Al 8112 billets were machined to achieve twist samples with the dimensions of 70×32×18 mm^3. The samples were extruded in a twisted channel with the rotation angle (α) of 90° and the slope angle (β) of 60° as illustrated in figure 1. TE process was performed for one pass at room temperature and then a partial annealing treatment was conducted on the twist extruded billet at 300°C for 0.5 h to decrease the hardness of sample and to prevent sample cracking during post rolling. Finally, cross cold rolling was carried out in longitudinal and transversal directions up to a total reduction of 37% subsequently to decrease the microstructure heterogeneity.

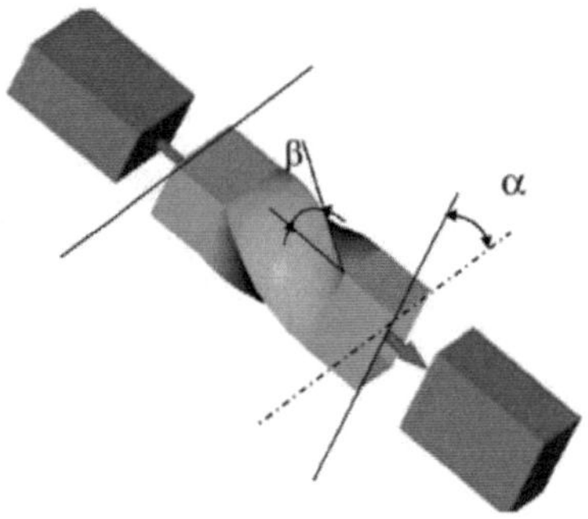

Figure 1. Schematic of the twist extrusion process [9].

The microstructures of the TEed samples with and without subsequent rolling were examined by scanning electron microscopy (SEM) of CamScan MV2300 making a comparison between center and edge regions in the outer longitudinal section to examine the heterogeneity evolutions. This cross-section was selected because during TE process the maximum strains occur in peripheral regions of workpiece [10]. Moreover, grain size measurements were undertaken by the linear intercept method on light optical micrographs obtained with an Olympus light microscopy, where the grain size is the chord length from the line intersections.

Hardness tests were performed by an ESEWAY hardness tester on the samples prepared for microstructure observations. The Brinell hardness tests (HB) were employed by using the load of 30 kg/mm^2. At least five separate measurements were performed on each sample.

3. Results and discussions

3.1. Microstructure

Figures 2(a) and 2(b) illustrate the microstructures of twist extruded sample without partial annealing, selected from center and edge regions of outer longitudinal plane (red dots show the position of the selected region for the SEM examinations). Both the center and edge regions of TEed sample consist of equiaxed grain sizes of 151 μm and 128 μm, respectively and both with tremendous microstructure heterogeneity. After employing one pass TE process, the range of grain size in outer longitudinal cross-section specified at different positions was between 90 μm and 190 μm. Because of the die geometry, the edge regions underwent more deformation than the center region and as a consequence, the edges consisted of finer grains. These results are in accordance with the previous experiments [11, 12]. Figure 2(c) and 2(d) show the microstructure of center and edge regions after the rolling process; hereafter, the TE and the TE + cold rolling processes are denoted as the TE and TE + CR, respectively. The TE + CR samples comprised finer grains of 80 μm and 74 μm at the centre and edges, respectively, demonstrating the effect of CR on reducing the heterogeneity. After this stage the range of grain size throughout the cross-section was between 60 μm and 90 μm.

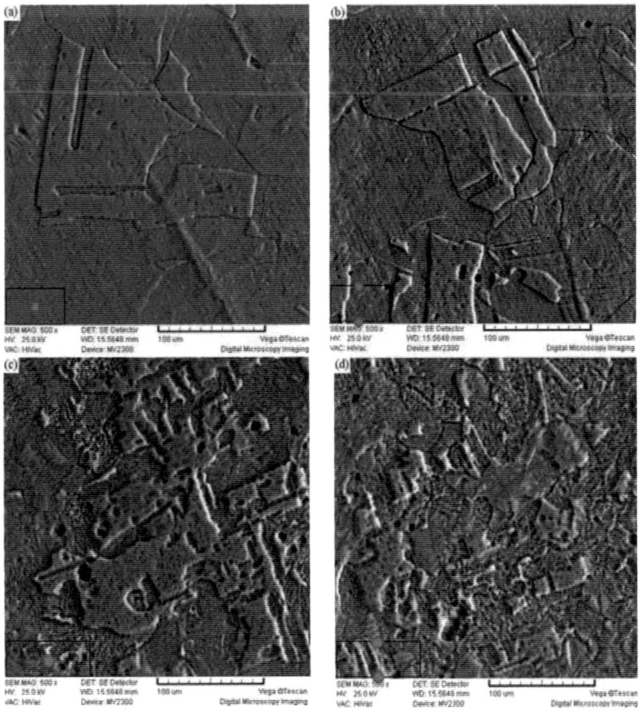

Figure 2. SEM microstructures of Al 1100 after TE in (a) center and (b) edge regions and after TE + CR in (c) center and (d) edge regions. (Red dots show the position of the selected regions).

3.2. Hardness

Table 2 summarizes the hardness values and the mean grain sizes of Al 8112 samples for center and edge regions in different stages. The as-annealed hardness of alloy is 39 +/- 2 HB. After employing the one pass TE, the hardness increases to 57 +/-3 HB and it enhances to 68 +/- 2 HB after CR in the center region. The edge region hardness increases in a similar sequence from 65 +/-3 HB after twist to 72 +/-2 HB with subsequent rolling. These differences refer to microstructure heterogeneity discussed previously.

Table 2. The hardness values and grain sizes of Al 8112 in different stages.

Sample	As-received	Anneal	TE (Center)	TE (Edge)	TE + CR (Center)	TE + CR (Edge)
Hardness (HB)	46 +/-2	39 +/-2	57 +/-3	65 +/-2	68 +/-3	72 +/-2
Mean Grain size (μm)	600	859	151	128	80	74

Figure 3 shows that after TE, the hardness values decrease from the edges towards the center encompassing a huge heterogeneity across the longitudinal section. In contrast employing CR resulted in an enhancement of the hardness and a reduction in the heterogeneity. After CR the hardness values almost approached an average value of 70 +/-2 HB throughout the longitudinal section.

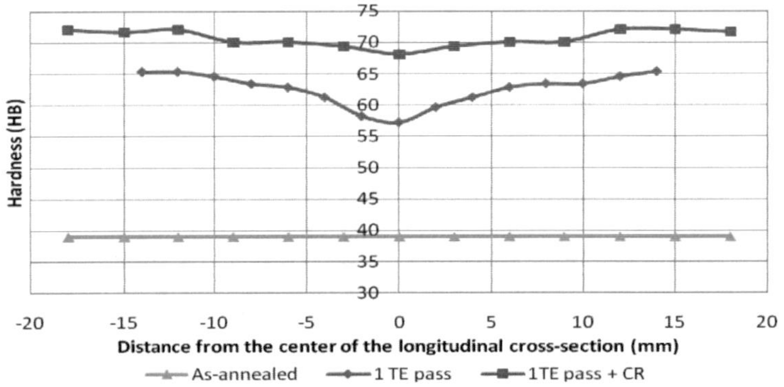

Figure 3. Hardness variations across the middle of the outer longitudinal cross-section of Al 8112 after TE and TE + CR.

4. Conclusions

1. An Al 8112 sample was subjected to twist extrusion for one pass with and without subsequent cold rolling. The microstructure evolutions showed that CR not only reduces the grain size but also reduces the microstructure heterogeneity across the longitudinal section.
2. The hardness variations reflected the enhancement of strength properties by employing the subsequent cold rolling after TE. In addition, hardness profiles showed the reduction of mechanical heterogeneity according to the microstructure evolutions.

References

[1] Valiev R Z, Estrin Y, Horita Z, Longdon T G, Zehetbauer M J and Zhu Y T 2006 *JOM* **58 (4)** 33-39.
[2] Segal V M, Reznikov V I, Drobyshevskiy A E and Kopylov V I 1981 *Metally* **1** 99-105.
[3] Smirnova N A, Levit V I, Pilyugin V I, Kusnetsov R I, Davydova L S and Sazonova V A 1986 *Fiz. Met. Metalloved* **61** 1170-1177.
[4] Saito Y, Utsunomiya H, Sakai T and Hong R G 1999 *Mater. Sci. Forum* **304-306** 73-78.
[5] Salishchev G A, Valiakhmetov O R and Galeyev R M 1993 *J. Mater. Sci.* **28** 2898-2903.
[6] Huang J Y, Zhu Y T, Jiang H G and Lowe T C 2001 *Acta Mater.* **49** 1479-1505.
[7] Park K T, Lee H J, Lee C S and Shin D H 2005 *Mater. Sci. Eng.* A **393** 118-124.
[8] Stolyarov V V, Beygelzimer Y E, Orlov D V and Valiev R Z 2005 *Phys. Met. Metall.* **99** 204-211.
[9] Varyukhin V, Beygelzimer Y E, Synkov S and Orlov D V *2005 The 3rd International Conference on Nanomaterials by Severe Plastic Deformation (NanoSPD3)*, Japan.
[10] Akbari Mousavi S A A, Shahab A R, Mastoori M 2008 *Mater. Des.* **494** 1316-1329.
[11] Shahab A R 2007 M. Sc. Thesis, School of Metallurgy and Materials Engineering, University College of Engineering, University of Tehran, Iran.
[12] Beygelzimer Y E, Varyukhin V, Synkov S, Orlov D 2009 *Mater. Sci. Eng.* A **503** 13-17.

Effects of recrystallization on texture, microstructure and mechanical properties in HPT-deformed pure Mg

Bartlomiej J. Bonarski[1,2], Erhard Schafler[1], Borys Mikulowski[2] and Michael J. Zehetbauer[1]

[1] Research Group Physics of Nanostructured Materials, Faculty of Physics, University of Vienna, Boltzmanngasse 5, A-1090 Wien, Austria
[2] Department of Metallic Materials and Nanotechnology, Faculty of Non-Ferrous Metals, AGH – University of Science and Technology, Al. Mickiewicza 30/A-2, PL-30059 Krakow, Poland

bartlomiej.bonarski@univie.ac.at

Abstract. Mg of purity 99.8 wt% was deformed by High Pressure-Torsion at hydrostatic pressures 1 to 4 GPa and RT, up to plastic shear strains of 120. X-ray texture analysis showed up deviations from expected shear texture, which increased with increasing shear strain and hydrostatic pressure. According to TEM and SEM investigations these deviations can be understood in terms of recrystallization. The current paper aimed at the differences of the recrystallization processes which occur *during* HPT deformation and unloading (*dynamic* recrystallization, DRX), and those *after* deformation (*static* recrystallization, SRX). For this purpose, two sorts of samples were investigated: (i) such being stored at RT immediately after HPT deformation, and (ii) such being stored at 77 K immediately after HPT deformation, and stored at RT for a minimum and constant time needed for preparation. The results show that SRX brings the texture closer to the ideal shear texture and to higher strength values, but to smaller ductilities than DRX does. The mechanical properties can be attributed to changes of texture rather than to those of grain size.

1. Introduction

Magnesium is considered to be a promising material for engineering in times when energy consumption reduction achieved by weight reduction gains much importance. By wide-range application of Mg and its alloys mostly in automotive and aircraft industry [1, 2] a lot of attention has been devoted to investigations of the physical deformation behavior of magnesium. An efficient way to improve the mechanical properties of Mg is a significant reduction of grain size, which enhances its strength and can be achieved by methods of Severe Plastic Deformation (SPD) [3].

2. Experimental methods

Polycrystalline magnesium of technical purity (99.8 wt.%) samples were hot extruded (350°C) into rods of 8 mm diameter and then deformed by High-Pressure Torsion (HPT, [4]) at hydrostatic pressures 1-4 GPa up to shear strains $\gamma=120$ [5]. During HPT processing a Cu-ring has been attached in order to achieve homogeneity (for further details see [6]).

X-ray texture measurements of HPT-deformed samples were performed on radial cross sections of the samples (RD, Fig. 1a) using a Cu$K\alpha$ beam collimated to a spot size of 300 μm, being part of a BRUKER-AXS DISCOVERY D-8 diffractometer equipped with a GADDS area detector. Pole figures (0002), $(10\bar{1}0)$, $(10\bar{1}1)$ and $(10\bar{1}2)$ were measured with Bruker-AXS D-8 Discovery diffraction system supplied with GADDS area detector. Complete pole figures were calculated from ODF by the Arbitrary Defined Cells (ADC) method [7] using monoclinic symmetrization using LaboTex 3.0 software.

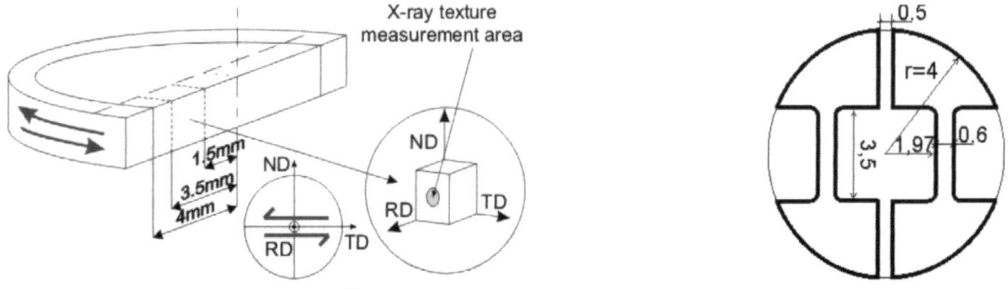

(a) (b)

Fig. 1. (a) Setup of X-ray texture measurement with respect to HPT sample geometry and shear directions (b) tension samples as prepared from HPT-deformed specimen. ND =Normal Direction = torsion axis, RD =Radial Direction, TD = Tangential Direction.

Tension tests were performed on samples cut from HPT material by spark erosion (as shown in Fig. 1b) at RT, at a constant strain rate $\dot{\varepsilon} = 1.4 \cdot 10^{-3}[s^{-1}]$ and RT. TEM imaging was carried out with thin foils being prepared from the radial cross sections used for the texture investigations, within a TEM Philips CM200 operated at a voltage of 200kV (for further details see [5, 8]). EBSD investigations were carried out by a system HKL-OXFORD INSTRUMENTS being part of a SEM Hitachi S-3400N operated at 20keV.

3. Results and discussion

As reported in previous works [5, 8], the texture of pure magnesium after HPT reveals components expected for deformation by shear ('B-fiber', see [9]), but the positions of these components seems surprising – they are deviating from the position of ideal shear (by a certain angle α (Fig. 2cd).This deviation is apparently caused by recrystallization during and after the HPT deformation , as it was confirmed by TEM investigations reported in [5, 8]. The deviation has been found to increase with the hydrostatic pressure applied during HPT which suggested but not proved that the deviation is arising from the *dynamic* part of recrystallization: While this dynamic recrystallization (DRX) is assumed to operate during HPT and the unloading process, the part of static recrystallization (SRX) is to occur afterwards, i.e. in entirely load-free condition. Special experiments have been performed in order to clarify the specific effects of DRX and SRX to microstructure and mechanical properties. For this purpose two sets of samples have been prepared investigated: (i) those which have been stored at RT after HPT processing for a few months as long as no change of microstructure has been observed (called 'Recrystallized – RX' samples in what follows), and (ii) those which have been stored at 77 K (in liquid nitrogen) immediately after HPT processing (called 'Frozen – F' samples in what follows). While the first set was expected to show mainly effects of static recrystallization (SRX), the second one should reflect mostly influences from dynamic recrystallization (DRX). Both groups of samples were subjected to the same investigations, such as X-ray diffraction, SEM and TEM imaging as well as tension tests, with a minimum and approximately constant time of RT storage required for sample preparation and investigations. The texture investigations of the F-type samples showed that deviations α of the B-fiber from ideal shear texture get even larger than that of the RX-type samples, confirming the suspicion that it is mainly the dynamic part of recrystallization which is responsible for the deviation α of B-fiber from its ideal shear position (see Fig. 3a, and [9]). From the same results, however, it can be concluded also that the static part of recrystallization (SRX) exhibits a clear tendency to move the position of B-fiber towards the ideal shear position.

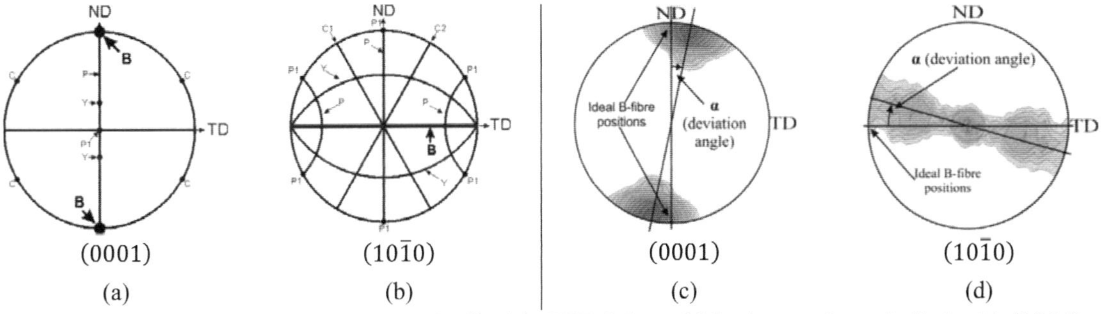

Fig. 2. 'Ideal' shear deformation components (B-fiber) in HPT deformed Mg shown schematically by (a) (0001)and (b) $(10\bar{1}0)$ pole figures [9], and experimental pole figures of (c) (0001)and (d) $(10\bar{1}0)$ planes, with angle α indicated which describes the deviation of the position of B-fiber from the ideal position.

These findings of effects specific to DRX and SRX seem to be correlated with the results of the tension tests. Fig. 3b compares the stress-strain characteristics obtained for both the RX-type and the F-type samples. The RX-type samples show a much higher strength, while at the same time their ductility is lower than that of the F-type samples. It seems that the static part of recrystallization (SRX) increases the strength whereas the dynamic one (DRX) is beneficial to the ductility. This can be explained if one takes a closer view on the microstructure of differently treated samples (please see [8] for complete results).

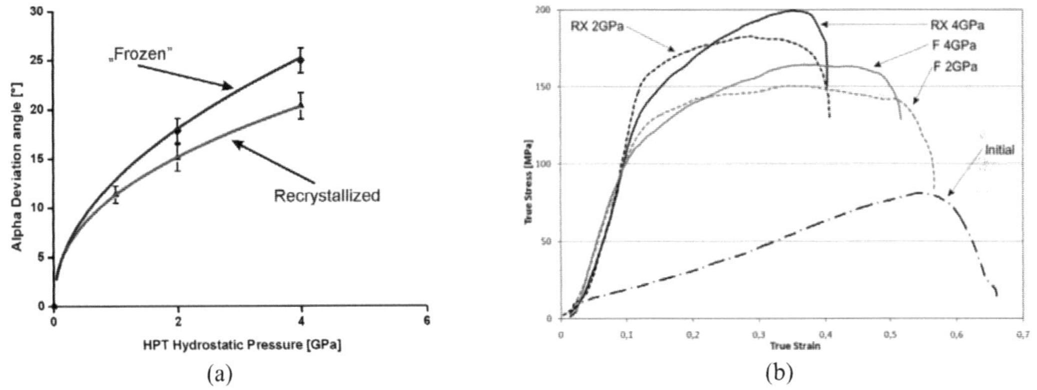

Fig. 3. (a) Deviation angle α indicating the deviation of B-fiber from its ideal (shear) position and (b) stress-strain characteristics of HPT-deformed Mg samples. Recrystallized (RX) samples show higher strength but lower ductility than 'Frozen' (F) ones. Data of initial Mg before HPT deformation are shown for comparison.

All HPT deformed samples of Mg reveal a strong bimodal microstructure with large, recrystallized grains beyond micron-size coexisting with submicron-size grains produced during severe plastic deformation (see [5]) However, irrespective of the pressure applied, RX-type samples show a much larger volume fraction of the recrystallized grains than the F-type samples, and their resulting average grain size is markedly larger than that of the F-type samples (see [8]). From both of those tendencies, one would expect a lower strength as well as a lower ductility of the RX-type samples compared to the F-samples, but the tensile tests tell a contrary story. From that it is concluded that the grain size and/or distribution does play only a minor role for the observed stress-strain characteristics and that it is the grain orientation rather than the grain size which govern the mechanical properties observed. Also if one considers the microstructures resulting from HPT deformation performed at different pressures (Fig. 4), these conclusions are confirmed. The average grain size d and its distribution is markedly larger in case of pressure relaxation after HPT at 4 GPa (d=1.8 μm) than after HPT at 2 GPa (d=2.4 μm), although the strength values demonstrated in Fig. 3b is larger and the ductility is lower.

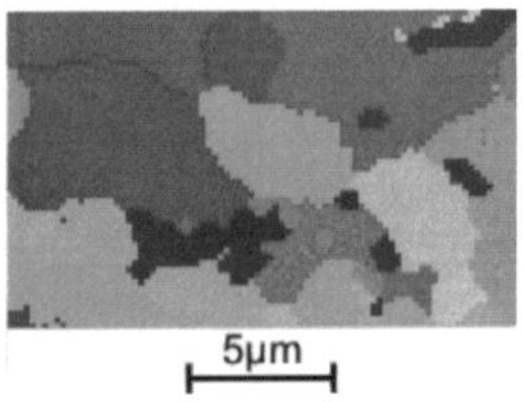

Fig. 4. EBSD images of RX-type samples after HPT deformation at different hydrostatic pressures: Left hand side: 2GPa; right hand side: 4GPa.

4. Summary and Conclusions

Commercial purity magnesium subjected to HPT deformation at room temperature develops textures close to those expected for pure shear deformation, but distinct deviations occur due to recrystallization. Particular experiments have been performed which showed that it is mainly the dynamic part of recrystallization which is responsible for the deviation from ideal shear texture, and that the static part taking place after dynamic recrystallization turns the texture closer to the ideal shear position. The behaviour of mechanical properties are in line with that picture: While the dynamic recrystallization lowers the strength but increases ductility, the static recrystallization after HPT processing enhances the strength and decreases the ductility which can be explained by the large differences in texture during dynamic and static recrystallization rather than the variations in grain size and/or distribution.

Acknowledgements. One of the Authors, Bartlomiej Bonarski acknowledges his special thanks to PhD Małgorzata Perek from Department of Metallic Materials and Nanotechnology at the AGH – University of Science and Technology in Cracow, Poland for great help with SEM EBSD investigations. The authors are grateful for financial support from the Focus Project "Bulk Nanostructured Materials" of University of Vienna, from the PhD Program "I022-N Experimental Materials Science - Nanostructured Materials" of University of Vienna, and from the Austrian Science Fund, Project NFN- S10403. They kindly acknowledge further supports by Grant No. 11.11.180.255 and Researcher's Promoter Grant No. N N507 402435 provided by the Polish Ministry of Science and Higher Education.

References

[1] K. U. Kainer, Magnesium Alloys and Their Applications,Wiley-VCH Weinheim, Germany (2000).
[2] M. Hakamada, T. Furuta, Y. Chino, Y. Chen, H. Kusuda and M. Mabuchi, Energy **32** (2007) 1352.
[3] R. Z. Valiev, Nature Mater. **3** (2004) 511
[4] R. Pippan, in 'Bulk Nanostructured Materials', eds M. J. Zehetbauer and Y. T. Zhu, Wiley-VCH Weinheim (2009)
[5] B. J. Bonarski, E. Schafler, B. Mingler, W. Skrotzki, B. Mikułowski and M. J. Zehetbauer, J. Mater. Sci. **43** (2008) 7513
[6] B. J. Bonarski, B. Mikułowski, E. Schafler, C. Holzleithner and M. J. Zehetbauer, Arch. Metall. Mater. **53** (2008) 117
[7] K. Pawlik and J. Pospiech, Proc. Workshop Theoretical methods of texture analysis, Clausthal-Zellerfeld, Germany (1986)
[8] B. J. Bonarski, N. A. Enikeev, E. Schafler, B. Mikulowski, R. Z. Valiev and M. J. Zehetbauer, Metall. Mater. Trans. A, submitted for publication (2009)
[9] B. Beausir, L. S. Toth and K. W. Neale, Acta Mater. **55** (2007) 2695.

Strength of nanoporous Ni-based superalloy membranes

J. Rösler, O. Näth

Technische Universität Braunschweig, Institut für Werkstoffe, Langer Kamp 8, D-38106 Braunschweig, Germany

j.roesler@tu-bs.de

Abstract. Nanoporous Ni based superalloy membranes are a new material class. They are fabricated from the two phase γ/γ' base material by thermomechanical processing, followed by selective phase extraction. Compared to other metallic membrane materials, they stand out due to an extremely regular and fine channel-like porosity on the nanoscale. This allows for particularly interesting applications in areas such as particle filtration, catalysis of chemical reactions or heat exchange. However, fundamental understanding of the mechanical behaviour is a prerequisite in all these cases. Thus, the microstructure property correlation of these novel materials is analyzed here, examining a γ'-membrane (where the γ-phase is leached out) by tensile testing. It will be demonstrated that nanoporous superalloy membranes are remarkably strong materials, provided the processing parameters are properly selected.

1. Introduction

Porous metals having pore dimensions on the nanoscale are receiving considerable attention as they are of particular interest for functional applications (e.g. phase separation or catalysis of chemical reactions) where small pore dimensions and large specific surface areas are required in combination with metallic properties [1-6]. To date, two processing routes have emerged, allowing to produce cellular metals with porosity on the nanoscale. Dealloying, where a less noble alloying element such as silver is leached from a solid solution strengthened nobel metal such as gold, is one potential route [1-3]. The second route, developed by the authors and discussed here, uses two-phase alloys, in particular Ni-based superalloys, containing cube shaped γ'-precipitates embedded in the Ni-rich γ-matrix [4-6]. The "trick" is to alter the microstructure by thermomechanical processing such that an interpenetrating network of both phases is formed, whereby each phase is in itself interconnected. Then, one phase can be (electro-)chemically dissolved while the other remains rigid. As a result, a porous and permeable membrane material is obtained, containing channel-like interconnected pores, some hundred nanometers wide, at the location of the dissolved phase.

The described process has been demonstrated successfully in [5,6] on the example of single crystalline Ni-based superalloys, where directional coarsening of the γ'-particles by uniaxial creep deformation in tension as described, for example, in [9,10] was used to form the required interpenetrating and interconnected microstructure. However, satisfactory mechanical properties are also essential for practical applications. For this reason, we report here on the strength and elastic stiffness of a nanoporous superalloy membrane material, using tensile testing.

2. Material and processing

The single crystalline Ni-based superalloy CMSX-4 was used here as starting material. Its chemical composition (in wt-%) is given by: Ni-6.5Cr-9.6Co-0.6Mo-6.4W-3.0Re-5.6Al-1.0Ti-6.5Ta-0.1Hf. The microstructure of the material after solution and precipitation heat treatment is depicted in fig. 2a. It consists of cube shaped γ'-particles embedded in the γ-matrix. The material was subsequently creep deformed in tension, whereby the loading direction was essentially parallel to the crystallographic [001]-orientation. While stress and temperature were set to 170MPa and 1273K, respectively, the creep duration was varied. Typical creep curves are shown in fig. 1, displaying a minimum of the strain rate at a critical strain ε_c of about 0.2%. Sheets, about 16mm wide and 140mm long, were then machined from the deformed creep specimen and, after polishing, the γ-phase was electrochemically dissolved, using the parameters given in [5]. Typically, the final thickness of the sheets was 0.3mm. Their ends were masked so that solid end pieces for easy clamping in the tensile test machine resulted.

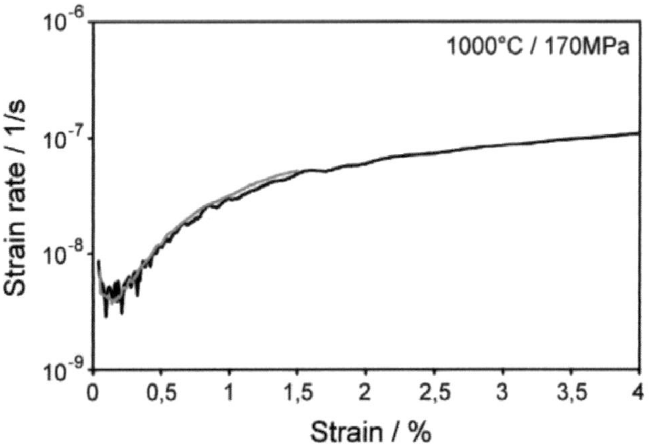

Fig. 1: Creep curves (strain rate versus strain) obtained for CMSX-4 at 1000°C / 170MPa. Note the existence of a strain rate minimum at a strain of about 0.2%.

3. Microstructure and mechanical properties

The microstructure of the sheets after creep deformation and selective phase extraction is depicted in fig. 2b-d as a function of the prior creep strain. Channel-like porosity, a few hundred nanometers wide, is clearly visible at the location of the dissolved γ-phase, while the ligaments are composed of the γ′-phase. The tensile stress during creep deformation was applied in vertical direction and directional growth of the γ′-particles in the plain perpendicular to the applied stress is apparent. At the smallest creep strain of about 0.1%, individual γ′-particles are connected with their neighbors along side faces (see fig. 2b). However, an interconnected network has not yet been established. Consequently, the sample is disintegrating upon handling and mechanical testing is not possible. In fact, this behavior is generally observed when the creep strain ε is less than ε_c. In contrast, when ε is larger than ε_c, the samples maintain their integrity upon handling and display significant mechanical strength (see below). This finding suggests that interconnectivity of the prior γ′-precipitate particles to a rigid, three-dimensional network is established at the point in time when the creep rate minimum is attained, i.e. when the material is at its largest creep resistance. In fact, it is qualitatively understandable why the material reaches it larges flow strength when this particular microstructural state is reached: Beforehand, the γ′-phase is still existing in the form of isolated precipitates (even though their morphology is no longer cube shaped but elongated) and the imposed deformation of the sample can be accommodated by creep deformation of the softer γ-phase alone. Once interconnectivity of the γ′-phase is established, this stronger phase must be deformed as well so that the flow strength increases until interconnectivity is established. Upon further deformation, the microstructure coarsens (compare fig. 2c with 2d), leading to softening of the material. As a result, a creep rate minimum at $\varepsilon = \varepsilon_c$ is obtained.

The ultimate tensile strength R_m and elastic stiffness of the membrane materials, obtained by tensile testing at ambient temperature, is displayed in fig. 3 as a function of the prior creep strain. Hereby, the loading direction was again perpendicular to the γ′-ligaments. Note that R_m increases initially with the creep strain. This is so because more cross-links between γ′-ligaments are formed in the course of creep deformation so that the stability of the γ′-network increases beyond the critical creep strain ε_c. However, the microstructure of the nanoporous material becomes also coarser and less regular at large prior creep strains (compare fig. 2c with 2d), leading to softening. Consequently, a strength maximum of about 100MPa at about 4% creep strain results. This is a remarkably large value for a porous material and a consequence of (i) the high strength of the base material and (ii) the regularity of the microstructure.

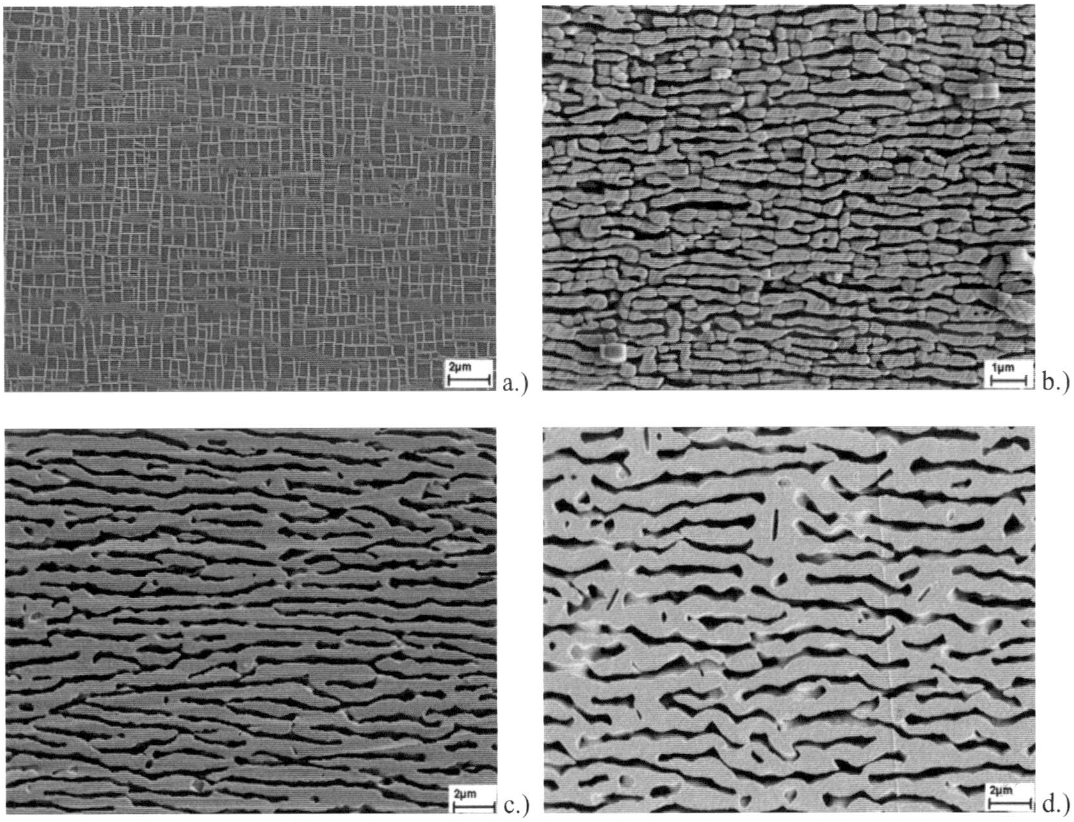

Fig. 2: Microstructures of the CMSX-4 base material (a) and the membrane materials after creep deformation to a strain of approx. 0.1% (b), 1,5% (c) and 11,5% (d).

The elastic stiffness follows the same trend with the creep strain as the ultimate tensile strength (see fig. 3), which can be explained analogously. Note, however, that the elastic modulus of single crystalline $Ni_3(Al, Ti)$ in [001]-direction is 133GPa [7]. Thus, the values obtained here are about one order of magnitude less even though the pore volume fraction is only about 26%. This comes about because perpendicular loading causes substantial shear and bending of the ligaments. In fact, it is shown in [8] that the obtained results can be quantitatively explained when axial, shear and bending displacements are taken into account and it turns out that bending is the most significant contribution, even though the other two contributions cannot be neglected.

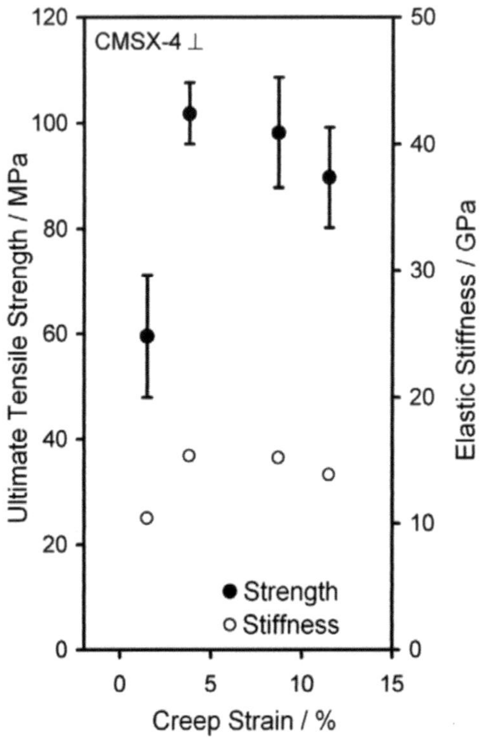

Fig. 3: Ultimate tensile strength and elastic stiffness of the nanoporous membrane materials as a function of prior creep strain. The loading direction is perpendicular to the γ'-ligaments.

Conclusion

Nanoporous superalloy membranes stand out as they combine remarkably fine and regularly arranged porosity on the nanoscale with excellent mechanical strength. The results demonstrate that creep deformation beyond the creep rate minimum is a prerequisite for the fabrication of these membrane materials and that best strength is achieved after creep strains of about 4% as a well interconnected but still very fine and regular microstructure is attained at the same time. The elastic stiffness, which is about one order of magnitude less than that of the bulk material when loading is perpendicular to the γ'-ligaments, is a result of ligament bending and shear.

Acknowledgements
We would like to thank the Deutsche Forschungsgemeinschaft for financial support of this work.

References

[1] Y. Ding and J.J. Erlebacher, *Am. Cem. Soc.* **125** (2003) 7772.
[2] R. Li and K. Sieradzki, *Phys. Rev. Lett.* **68** (1992) 1168.
[3] J. Erlebacher, M.J. Aziz, A. Karma, N. Dimitrov and K. Sieradzki, *Nature* **410** (2001) 450.
[4] J. Rösler and D. Mukherji, *German Patent* (2008) 10231577.
[5] J. Rösler, O. Näth, S. Jäger, F. Schmitz and D. Mukherji, *Acta Mater.* **53** (2005) 1397.
[6] J. Rösler and D. Mukherji, *Adv. Eng. Mater.* **5** (2003) 916.
[7] H. Yasuda, T. Takasugi and M. Koiwa, *Acta Metall. Mater.* **40** (1992) 381.
[8] J. Rösler and O. Näth, *Acta Mater.* **58** (2010) 1815.
[9] R.A. MacKay and L.J. Ebert, *Metall. Trans.* **16A** (1985) 1969.
[10] H.-A. Kuhn, H. Biermann, T. Ungar, H. Mughrabi, *Acta Metall. Mater.* **39** (1991) 2783.

Journal of Physics: Conference Series **240** (2010) 012135 doi:10.1088/1742-6596/240/1/012135

Bulk nanoscale materials in steel products

B Chehab[1], X Wang[1], J-P Masse[1], O Bouaziz[2], H Zurob[1] and D Embury[1]

[1] Department of Materials Science and Engineering, McMaster University, Hamilton, Ontario, L8S 1X3 Canada

[2] Arcelor Research, Voie Romaine-BP30320 57283 Mainz'res-le'-Metz Cedex, France

emburyd@mcmaster.ca

Abstract. Although a number of nanoscale metallic materials exhibit interesting mechanical properties the fabrication paths are often complex and difficult to apply to bulk structural materials. However a number of steels which exhibit combinations of plasticity and phase transitions can be deformed to produce ultra high strength levels in the range 1 to 3 GPa. The resultant high stored energy and complex microstructures allow new nanoscale structures to be produced by combinations of recovery and recrystallisation.

The resultant structures exhibit totally new combinations of strength and ductility to be achieved. In specific cases this also enables both the nature of the grain boundary structure and the spatial variation in structure to be controlled. In this presentation both the detailed microstructural features and their relation to the strength, work-hardening capacity and ductility will be discussed for a number of martensitic and austenitic steels.

1. Introduction

There is a very extensive literature on the production of nanoscale materials using a wide variety of techniques including vapour deposition, electro deposition, severe plastic deformation (SPD), ECAP and other methods. In a number of cases materials with very high strength and attractive combinations of strength and ductility can be produced. In addition the dominant length scales such as the grain size suggest are such that we need to modify the conventional descriptions of yielding and work hardening to account for the events which occur at these reduced length scales.

However few of these investigations yield materials which can be considered as bulk materials which can be readily utilized for current engineering applications. An alternative approach is to consider a more conventional material such as a heat treatable structural steel and exploit the rich variety of structures which can be produced by using combinations of deformation and phase transitions and a number of authors have produced ultrafine scale structures by this approach using cold rolling and subsequently tempering low carbon lath martensites [1-5]. It should be noted that cold rolling reduction of 50% i.e. a true strain of 0.8 was sufficient to produce ultra fine grained materials by this route whereas in SPD processing the strain levels are often of the order of 4. Other researchers have used warm rolling and Poorganji et al [6] have used dynamic recrystallisation of an Fe-2%Mn-0.1%C steel to produce ultrafine grain (UFG) ferrite.

In the current work we have used the concept of combining cold rolling with a subsequent heat treatment but starting with a compositionally graded steel with a C content in the central region of 0.4%c which was deformed by cold rolling to a true strain of 1.6. The objective was to use a range of heat treatments at temperatures between 300°C and 900°C to explore the range of fine scale structures which can arise from subsequent heat treatment. The essential objective is that this approach produces not only a refinement in one important length scale e.g. ferrite grain size but a concomitant refinement in related length scales such as carbide size or martensite island size.

Thus the approach yields not only nanoscale structures in bulk materials but provides the opportunity to control both yield strength and competitive processes such as damage initiation and ductility in both ferritic steels containing carbides and in dual phase structure composed of ferrite plus martensite.

2. Experimental methods

The current work was performed on a steel of composition Fe-2.07%Mn-0.4%C-0.4%Cr-0.36%Si-.0.12%Mo. The material was partially decarburised at 1075°C and then water quenched to form a completely martensitic structure at the centre as shown in Fig. 1. It was then rolled using a modified sandwich rolling technique applied to the compositionally graded to produce a highly deformed martensitic which is illustrated in Fig 2.It should be noted that under these conditions the martensite is completely plastic resulting in a reduction of the lath size as can be seen from a comparison of Fig. 1 and Fig. 2.

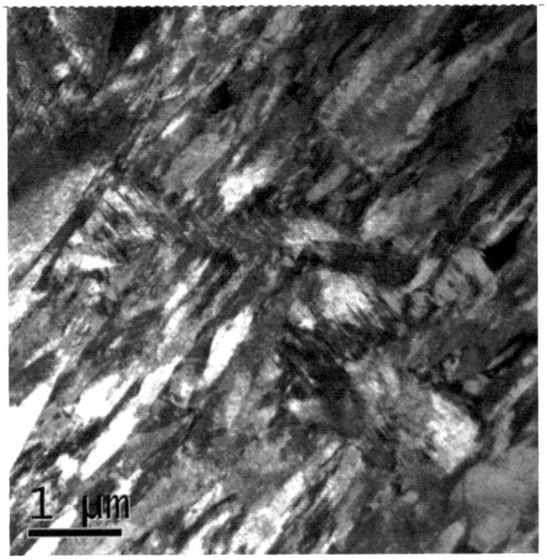

Figure 1. Martensitic structure produced on quenching the compositionally graded steel. Some martensite laths are twinned.

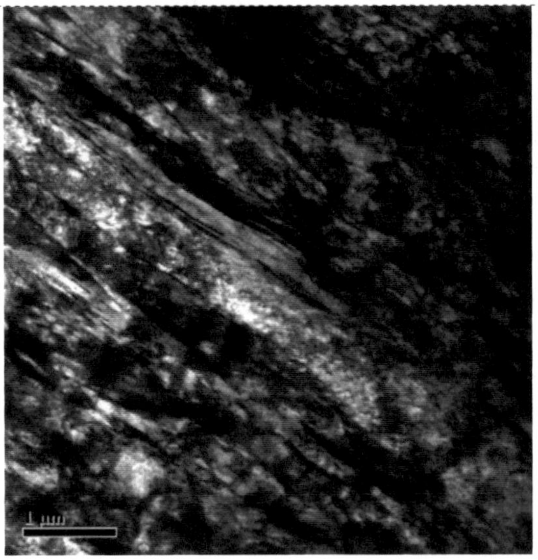

Figure 2. Martensitic structure after rolling to a true strain of 1.0. Note the narrow lath size produced by deformation of the martensite.

This structure was then heat treated over a range of temperatures from 300°C to 900°C. The resulting microstructures were characterized using optical SEM and TEM methods to establish the nature, length scale, and distribution of the various components. Microstructures of as-quenched, cold rolled and heat treated at different temperatures and times were characterized using PHILIPS CM12 TEM, operated at 120 KV. The plane-view TEM foils were prepared as follows. Thin slices were first cut off from samples and ground down (both sides) to a thickness of approximately 80 µm; 3 mm diameter discs were punched followed by twin-jet electro-polishing in a solution of 10 pct. perchloric acid and 90 pct. methanol, at -35°C and 15V.

The mechanical properties of the structures produced by heat treatment on the rolled martensitic structure were assessed both by hardness measurements and by tensile tests performed at a strain rate of 0.005 sec^{-1} and the data plotted in the form of true stress–true strain data in the range of uniform strains and the ductility was assessed from measurements of the final reduction in area.

3. Results

The heat treatments of the rolled material in the temperature range 300 to 600°C produced a range of structures with very fine scale ferritic structures an example is shown in Fig. 3 It should be noted that the initial scale of the recrystallised ferritic structure is determined by the scale of the martensite lath size. Thus the recrystallisation process appears to have a very high nucleation frequency and by varying the time and temperature of recrystallisation. It is possible to produce bulk nanoscale structure composed of fine scale spherodised carbides and fine scale ferrite in which the grain size is in the range 10 to 200 nm.

In samples recrystallised at 775°C on quenching from 775°C these showed dual phase structures of the type illustrated in Fig. 4 The scale of the martensitic phase reflects the fine scale of the austenitic phase formed during the intercritical anneal and it is interesting to note that the small scale martensitic phase appears to exhibit a limited number of variants.

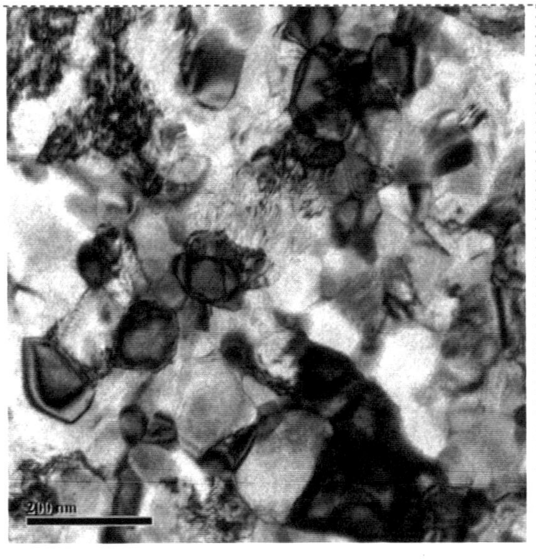

Figure 3. Nanoscale structure of ferrite grains and carbides after annealing cold rolled martensite 4 hours at 500°C.

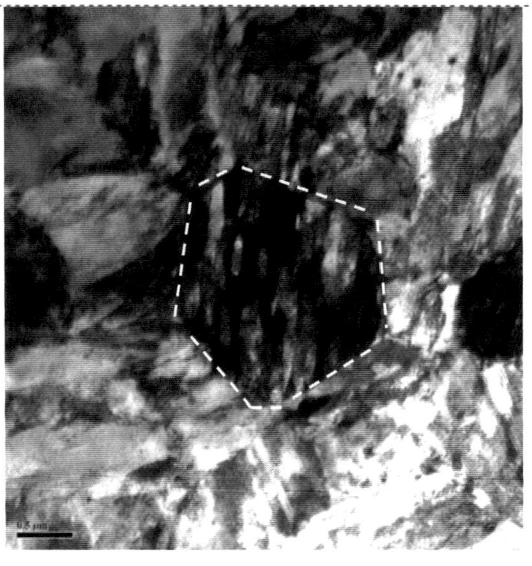

Figure 4. Dual phase structure produced by annealing cold rolled martensite 2 min at 775°C. The dotted line denotes the size of the prior austenite grain size.

The results of the tensile tests performed in the present study are summarised in Fig 5. This shows the true stress true strain curves for the initial martensitic structure formed in the compositional graded structure, the stress strain response after cold rolling to 100% and the stress strain response for the fine scale ferritic /carbide structure and the fine scale dual phase structure.

All the materials are compositionally graded but both the initial as quenched structure and the fine scale dual phase structure exhibit very pronounced elastic-plastic transitions.

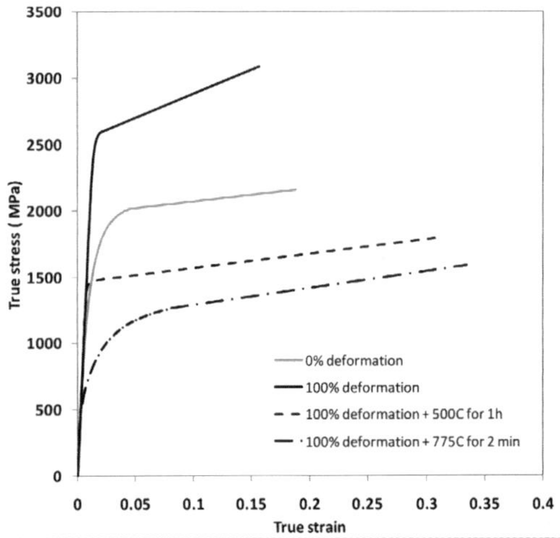

Figure 5. True stress-true strain for various treatments applied to a compositionally graded martensitic structure. The fracture strains are based on the reduction in area at fracture. It is of value to note the combinations of strength and ductility and the form of the elastic plastic transition.

4. Discussion

It is clear from the present study that bulk nanostructured steels can be produced even in steels with C levels as high as 0.4% In the present work we have utilised compositionally graded or architectured materials to achieve high ductility in the original as quenched martensitic structures. The combinations of strength and ductility produced both in the nanoscale ferrite/carbide steels and the fine scale dual phase steels are very attractive and represent an alternative processing route to produce new combinations of properties.

There are many aspects of the nanoscale materials which warrant further detailed study. The recrystallisation on the heavily deformed martensitic structure appears to take place by having a very high nucleation frequency coupled with a restriction of the ferrite grain size to the scale of the initial width of the elongated martensite laths .This is in part due to the simultaneous formation of a fine scale spherodised carbide structure which pins the recrystallised ferrite structure.

In the case of the fine scale dual phase structure this is much finer in scale than dual phase steels produced by conventional intercritical annealing. The simplified martensitic islands which contain only one or two variants are of interest and these also appear to undergo large plastic strain prior to damage initiation resulting in attractive fracture properties. A detailed account of the fracture processes in these nanoscale materials will be reported in a subsequent publication.

Acknowledgements
The authors are grateful to Arcelormittal and to NSERC Canada for research support during the period of this study.

References
[1] Valiev R Z, Islamgaliev R K and Alexandrov I V, 2000 Bulk Nanostructures from Severe Plastic Deformation *Progress Mater. Sci.* **45** 103
[2] Tsuji et al 2002 *Scripta Mater.* **46** 305
[3] Ueji et al 2002 *Acta Mater.* **50** 4177
[4] Zhao X et al 2005 *Mater. Sci. Eng.* **A 397** 117
[5] Tianfu et al 2006 *Mater. Sci. Eng.* **A 432** 216
[6] Lan H F et al 2007 *ISIJ Int.* **47** 1652
[7] Poorganji B et al 2008 *Scripta Mater.* **59** 279

A simple dislocation model of the influence of high-angle boundaries on the deformation behavior of ultrafine-grained materials

W Blum[1] and P Eisenlohr[2]

[1] Department of Materials Science and Engineering, Institute I: General Materials Properties, Martensstraße 5, 91058 Erlangen, Germany
[2] Max-Planck-Institut für Eisenforschung, Max-Planck-Straße 1, 40237 Düsseldorf, Germany

Wolfgang.Blum@ww.uni-erlangen.de, p.eisenlohr@mpie.de

Abstract. The deformation resistance of ultrafine-grained (UFG) materials is modelled on the basis of the evolution of the average dislocation density with strain in the course of glide and recovery of dislocations. In contrast to materials with conventional grain size (CG), dislocations are stored and annihilated solely at the high-angle boundaries, where screw dislocations glide and edge dislocations climb towards annihilation sites. The high-angle boundaries enhance both the rates at which dislocations are stored and recovered. Depending on the spacing of high-angle boundaries, temperature and strain rate, UFG materials are softer or harder than their CG counterparts.

1. Introduction

In [1] a simple dislocation model was presented which describes the maximum deformation resistance of UFG and nanostructured materials. The model adapts classical dislocation theories of steady-state creep of CG materials at elevated temperatures to the case of UFG materials. UFG materials are defined here not on an absolute grain size scale, but on a microstructural basis as materials where the spacing d of high-angle boundaries is so small that new low-angle boundaries do not form, implying that dislocations are no longer stored in the grain interior in the form of a subgrain structure, but at the high-angle boundaries limiting the mean free dislocation path. This means that the stored edge dislocations find a relatively easy exit from the material by grain boundary diffusion. Screw dislocations are relatively easily absorbed by gliding parallel to the boundaries to sinks (like triple junctions of grains); consequently steady-state deformation of UFG materials is directly linked with grain boundary sliding.

The steady-state deformation resistance was determined analytically in [1] by solving the condition of dynamic equilibrium, $\dot{\rho}^+ = \dot{\rho}^-$, for the average dislocation density ρ and neglecting the thermal activation of glide by relating the flow stress σ at given temperature T and strain rate $\dot{\varepsilon}$ to the inverse of the average dislocation spacing $\delta \approx \rho^{-0.5}$ in the form $\sigma = \alpha' G b \sqrt{\rho}$ where G is the shear modulus, b is the Burgers vector length and α' is the dislocation interaction constant [1] for UFG materials.

In the present work, the model presented in [1] is extended by including the evolution of the dislocation structure during straining and the thermal activation of glide. The results are compared to experimental data for UFG Cu produced by equal channel angular pressing (ECAP) [2] and nanocrystalline Ni produced by pulsed electro deposition (PED) [3].

2. The model

2.1. UFG materials

Most of the model equations for the UFG case have appeared in [1]. Here we describe their physical meaning.

The microstructural definition of UFG materials given above can be expressed as $\Lambda = 0.6d$ where Λ is the mean free path of dislocations, in other words, $1/\Lambda$ is the dislocation length which is stored at the high-angle boundaries per slipped area. The rate at which dislocation length is stored per volume of UFG material with resolved shear strain γ follows as $\mathrm{d}\rho^+/\mathrm{d}\gamma = 1/(b\,\Lambda)$.

Edge dislocations generate single-sided steps at the high-angle boundaries which are addressed in the literature as extrinsic grain boundary dislocations. They are annihilated by extrinsic grain boundary dislocations (steps) of opposite sign. The annihilation occurs either spontaneously when the spacing d_{dip} of the dipole partners is less than a certain value $d_{\mathrm{spon,b}}$ or, else, by climb along the high-angle boundaries with the aid of grain boundary diffusion, which is assumed to be driven by the self stresses of edge dipoles. Spontaneous annihilation simply reduces the rate of dislocation storage by the factor $1 - d_{\mathrm{spon,b}}/d_{\mathrm{dip}}$ and thereby sets a low-temperature limit to the dislocation density which can be stored in the material. Climb yields the rate of loss of edge dislocations with time, $\dot{\rho}^-_{\mathrm{climb}}$.

The net rate of change of ρ in UFG materials results as the difference of storage and recovery rates, $\dot{\rho} = (\mathrm{d}\rho^+/\mathrm{d}\gamma)\,(1 - d_{\mathrm{spon,b}}/d_{\mathrm{dip}})\,\dot{\gamma} - \dot{\rho}^-_{\mathrm{climb}}$, where γ is the resolved shear.

Several numerical parameters were introduced [1] which quantify the relative contribution of a given slip system to the total strain, the fractions of dislocations which are stored directly at the boundaries and in front of them inside the grain, the relative factor of relaxation of the stress field when an edge dislocation reaches a high-angle boundary and the fraction of dislocations at the boundaries which find a dipole partner there. These factors were used to fit the model to the experimental results [1].

The generation and glide of dislocations are supported by thermal activation. The rate $\dot{\gamma}$ of thermally activated glide is expressed as the density of gliding dislocations, set to be a fraction $f_{\mathrm{mob}} \approx 10^{-3}$ of the dislocations inside the grains, and the velocity of glide formulated according to [4]. The parameters of the thermal obstacles were chosen such that the flow stress is diminished by less than 20% compared to the maximum shear stress $\hat{\tau}$ which the dislocation obstacles can withstand. This means that the dependence of the flow stress on temperature T and strain rate $\dot{\gamma}$ is mainly controlled by recovery, not by glide of dislocations as proposed for instance in [1].

With these expressions the evolution of the deformation resistance from yielding to the steady state can be modelled by integrating the set of ordinary differential equations for ρ (structure evolution) and γ (glide).

2.2. CG materials

For the sake of comparison with the UFG case a simple model of deformation of CG materials was also set up. It uses expressions for thermally activated glide and for dipole capture and spontaneous as well as climb-controlled annihilation of edge dislocations similar to the ones in [5]. The model parameters were adjusted with the only aim to bring the model into reasonable agreement with the behavior of CG Cu which, in the temperature and stress range considered, is characterized by an exponential stress dependence of the steady-state deformation rate (so-called power law breakdown behavior), but is not well documented by experimental results.

Journal of Physics: Conference Series **240** (2010) 012136 doi:10.1088/1742-6596/240/1/012136

3. Results and discussion

Fig. 1 shows examples of the modelled evolution of the deformation resistance of pure UFG Cu produced by severe plastic deformation (i.e., with initial high-angle boundary spacing $d =$ 0.35 μm). It is postulated that d increases in the course of coarsening of the subgrain structure towards the stress-dependent steady-state value (see [2]). The deformation resistance quickly attains a maximum and declines with further straining. The agreement with experimental curves (grey) is good.

The steady-state deformation resistance of UFG Cu is given in Fig. 2 for constant d; this corresponds to the maximum of the deformation resistance where the effect of subgrain coarsening is still negligible. When the stress becomes higher than indicated by the vertical dashed line, the steady-state subgrain size $14\,b\,G\,/\sigma$ becomes smaller than d. This means that subgrains can form in the grains so that the UFG behavior is expected to break down. For the nanostructured Cu with d of 0.1 μm no such breakdown is predicted implying that the UFG behavior should prevail. It is seen that for the lower two temperatures the UFG Cu is stronger than the CG material. In the UFG Cu with $d = 1$ μm a transition to CG behavior is predicted

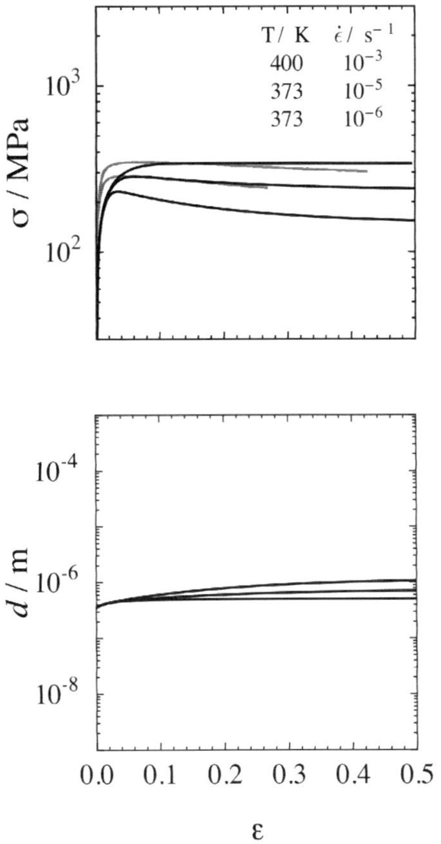

Figure 1. Evolution of flow stress σ and high-angle boundary spacing d with strain ϵ for UFG Cu (black lines: model, grey lines: experiment).

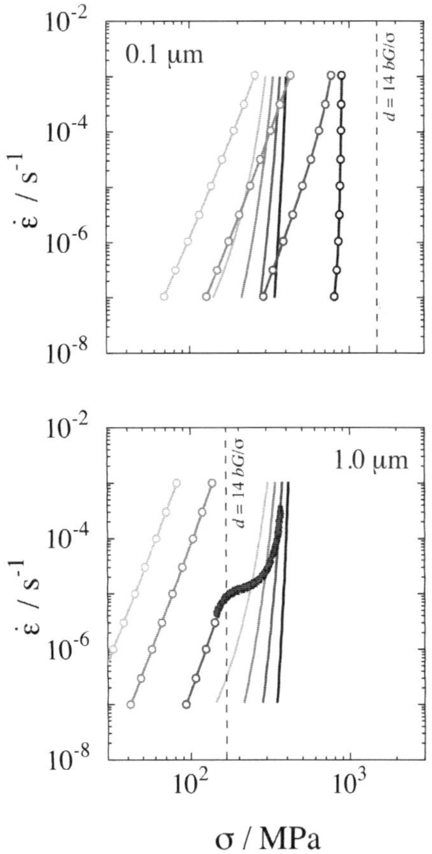

Figure 2. Steady-state deformation resistance at 600, 500, 400, and 300 K (in sequence from light grey to black, i.e., left to right) for $d = 0.1$ and 1.0 μm for UFG (lines with symbols) and CG Cu.

as indicated by the heavy S-shaped connecting line for 400 K. The UFG material is softer than the CG material due to the faster recovery at high-angle boundaries compared to recovery in the grain interior.

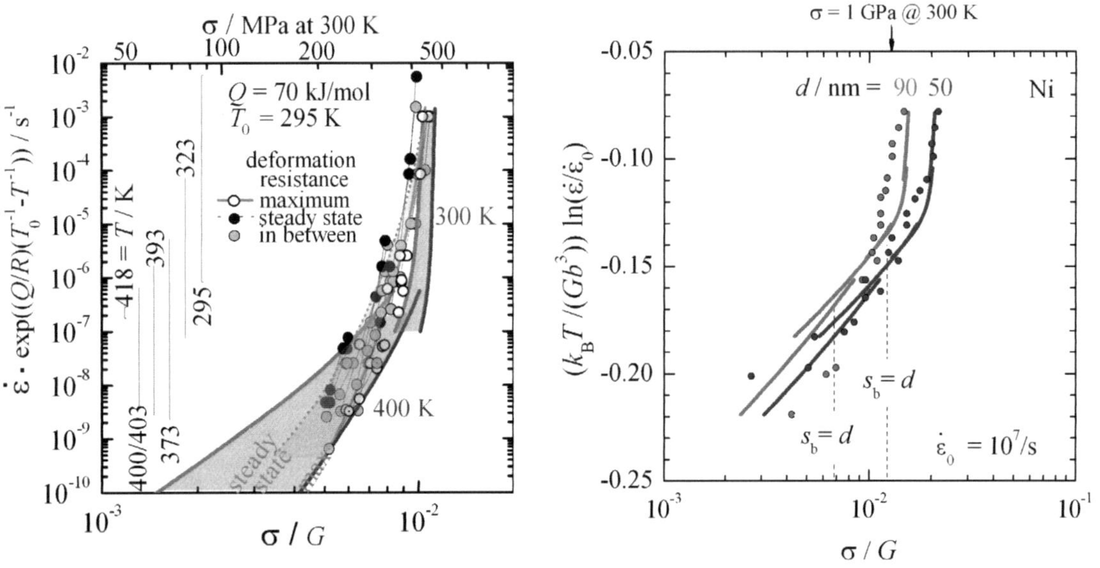

Figure 3. Comparison of modelled steady-state deformation resistance with experimental data from [2, 3]; left: UFG Cu with initial d = 0.35 µm without (blue) and with (red) d-evolution during subgrain coarsening, right: nanocrystalline PED Ni with d = 90 (red) and 50 nm (blue). Note difference in T-normalization; k_B and R are Boltzmann constant and gas constant, respectively.

Fig. 3 shows the steady-state deformation resistance of UFG Cu prepared by ECAP and nanostructured Ni prepared by pulsed electro deposition. At constant d the modelled curves for UFG Cu (blue lines, left) are in reasonable agreement with the data for the measured maximum deformation resistance (open circles). When subgrain coarsening is included, the model results (red lines, left) are consistent with the experimentally observed softening during strain-induced subgrain coarsening up to the steady state (black symbols). For nanostructured Ni the maxima of the modelled deformation resistance are consistent with the experimental data, at least up to the point where the spacing s_b of extrinsic edge dislocations at high-angle boundaries is less than the boundary spacing d. The influence of d is captured by the model.

4. Concluding remark

The present work confirms that the simple dislocation model proposed in [1] is capable of describing the behavior of UFG metals. This means that the deformation resistance and its strain rate sensitivity can be understood in terms of dislocation recovery at high-angle boundaries occurring after thermally activated glide through the grain interiors.

References

[1] Blum W and Zeng X H 2009 *Acta Mater.* **57** 1966–1974
[2] Blum W, Li Y J and Durst K 2009 *Acta Mater.* in print
[3] Li Y J, Mueller J, Höppel H, Göken M and Blum W 2007 *Acta Mater.* **55** 5708–5717
[4] Kocks U F, Argon A S and Ashby M F 1975 *Progr. Mater. Sci.* **19** 1–291
[5] Blum W, Eisenlohr P and Breutinger F 2002 *Metall. Trans.* **33A** 291–303

In situ TEM observations of reverse dislocation motion upon unloading of tensile-deformed UFG aluminium

F. Mompiou[1], M. Legros[1], D. Caillard[1], H. Mughrabi[2]

[1] CEMES-CNRS, 29 rue Jeanne Marvig, BP4347, F-31055 TOULOUSE, France.
[2] Department of Materials Science and Engineering, University Erlangen-Nürnberg, Martensstrasse 5, D-91058 Erlangen, Germany.

caillard@cemes.fr

Abstract. An *in situ* transmission electron microscopy study has been performed on ultrafine-grained (UFG) aluminium during tensile loading and unloading in the microyield regime. The goal was to assess the reasons for the unusually large inelastic backflow that had been observed earlier during unloading on UFG material, as compared to that of conventional grain size material. It was noted that in particular edge dislocations emitted by sources within the grains during loading run back into the dislocation sources and disappear during unloading, explaining at least semi-quantitatively the rather large inelastic backflow in UFG material.

1. Introduction

Bulk ultrafine-grained (UFG) metals produced by equal channel angular pressing (ECAP) have attracted much attention during the last decade because of their extraordinary strength properties [1]. In the present study, the tensile loading and unloading behaviour during microyielding of UFG aluminium specimens was investigated by *in situ* transmission electron microscopy (TEM). The motivation for this work stems from the observation that UFG copper [2] and aluminium [3] specimens exhibited unusually large inelastic reverse deformations during unloading, whereas conventional grain (CG) size material showed almost completely elastic behaviour during unloading. Figures 1a and 1b show as an example the microyielding loading/unloading behaviours observed on UFG and CG aluminium specimens, respectively [3].

2. Experimental procedure

UFG aluminium of commercial purity with a 300 nm average grain size was obtained by ECAP-processing with 8 route B_c passes, with a rotation of 90° in the same sense around the specimen axis after every pass (courtesy of Johannes May, Erlangen). Rectangular microsamples, of size 3mm×1mm, were prepared by spark cutting, followed by mechanical and electro-polishing. They were fixed on a Gatan room-temperature straining device, and strained in a 2010HC transmission microscope [4]. The fixation was such that the stress could be released without inducing any compression. The images were recorded by a Megaview III video camera, and analyzed frame by frame.

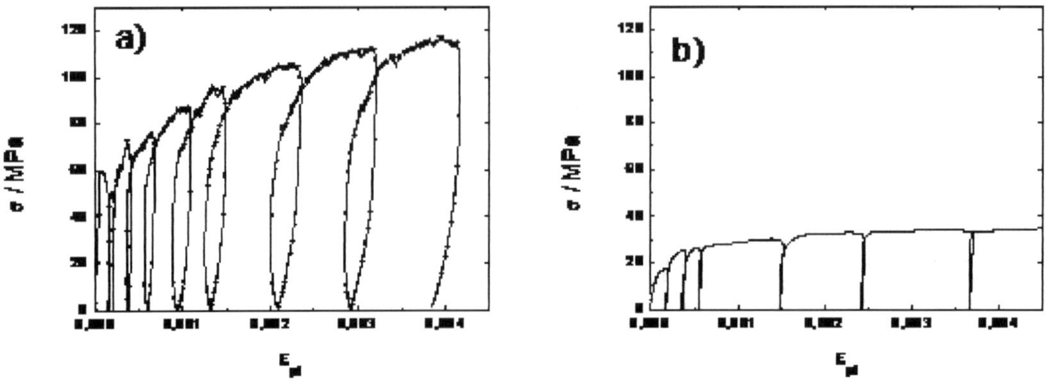

Fig. 1: Tensile microyielding stress-strain curves (plot of stress σ versus plastic strain ε_{pl}) with repeated unloadings and reloadings. a) UFG aluminium, as-ECAP. b) CG aluminium. From [3].

3. Results

As soon as the stress is applied, sources start to operate in some grains. They all consist of one dislocation segment rotating around a single anchoring point, generally a dislocation junction or a super-jog. This type of source is sometimes referenced as "single-armed source" or "spiral source". Only a few grains deform plastically at yield, presumably those with the largest sizes, the most favourable orientations, and the most appropriate dislocations and anchoring points. All emitted dislocations glide to the neighbouring grain boundaries (GBs), and almost never annihilate inside the grains. Then, more and more grains are deformed till the plastic deformation becomes homogeneous.

In thin foils, the character of the dislocations interacting with GBs is strongly determined by the sample orientation, which can induce different behaviours, as shown in figs. 1 and 2.

Fig. 2 shows a source rotating anticlockwise and emitting 4 dislocation loops in a 600nm grain. The anchoring point S is a super-jog. The slip traces at the foil surface are stair-shaped, which indicates that the dislocations have cross slipped in front of the left GB, as a result of their strong screw component. One consequence of this behaviour is that the 4 dislocations have entered the left GB at different places. Upon unloading, no large reverse dislocation motion could be observed, except for the unbowing of the source dislocation (not shown in the figure).

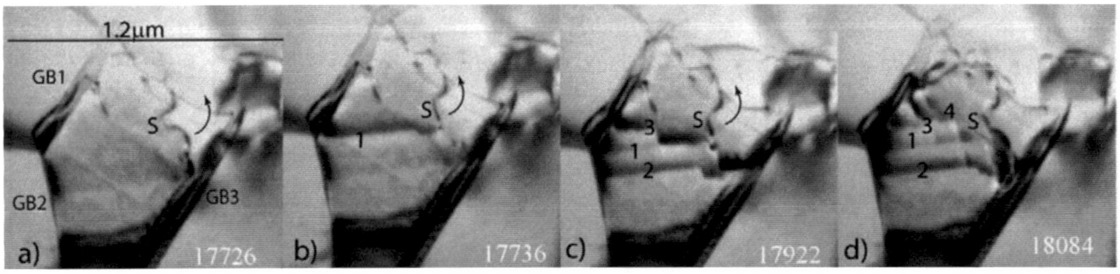

Fig. 2: Single-armed dislocation source (anchoring point S) emitting 4 near-screw dislocation loops. As a result of their cross-slip (see stair-shaped traces, noted 1 to 4), the dislocations enter the left GBs (GB1 and GB2) at different places. No reverse motion of dislocations (that characterizes inelastic behaviour) has been observed upon subsequent unloading.

Fig. 3 shows another grain containing a single-armed source, anchored in S', rotating anticlockwise, and emitting 5 dislocation loops. At variance from the preceding case, no cross slip takes place in front of the two neighboring GBs, because of the strong edge character of the dislocation segments. Then, all dislocations enter the GBs at almost the same place. Small "pile" ups are formed. In aluminium, this may appear surprising but can be explained by two observations: i) pile ups are not perfect because dislocations do not lie in the same exact plane, and ii) the pile-up length is short on account of the small grain size. As shown in what follows, the inelastic reverse dislocation flow effect is large.

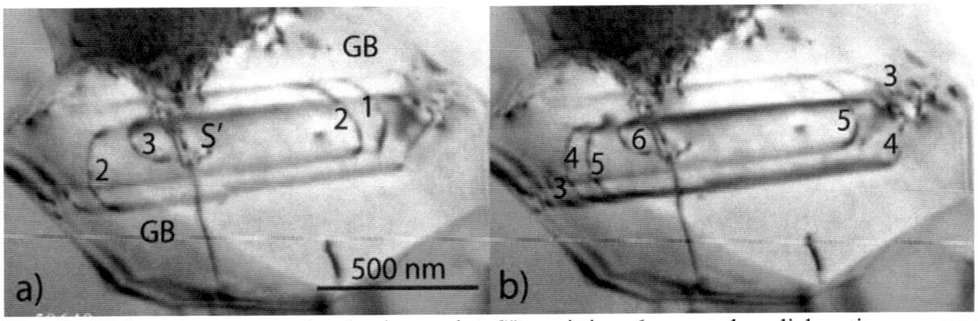

Fig. 3: Dislocation source (anchoring point S') emitting 6 near-edge dislocation segments. The emitted dislocations form short pile-ups and enter the neighbouring GBs at almost the same place.

Fig. 4 shows the same grain as in fig. 3. On its left side, the source S' emits 2 dislocations which immediately enter the left GB, in (b) and (d). The contrast of the inserted dislocations rapidly vanishes, between (b) and (c), and between (d) and (e). No reverse motion of these dislocations is observed upon unloading, in (e) and (f), although this has occurred at different times of the experiment (see fig. 5). On the right side of the source, a strong broad contrast, noted C, can be seen at the intersection of the active slip plane and the right GB. This contrast results from the cumulated strain contrasts of all previously-inserted dislocations. Dislocation 1 is not completely inserted in the right-side GB, in (b-d), and dislocation 2 is piled up against dislocation 1, in (d), which shows that the broad contrast corresponds to a repulsive elastic stress. Upon unloading, as shown in (e) and (f), this stress is sufficient to extract dislocation 1 from the GB, and to induce the back motion of dislocations 1 and 2 over a fairly large distance.

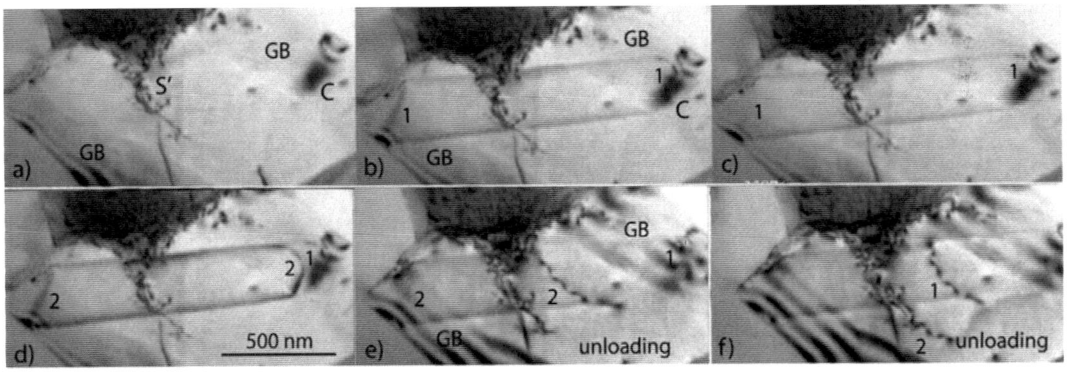

Fig. 4: Inelastic effect in the grain of fig. 2 during loading-unloading. a) The broad contrast C is due to previously inserted dislocations. b) During loading, he first emitted dislocation (1) is completely inserted in the left GB, and partly inserted in the right one. c) The contrast of 1 vanishes in the left GB. d) A second dislocation has been emitted, which is repelled by the GB on the right side. e) After unloading, dislocation 2 moves back on the right side. f) Dislocation 1 is extracted and also moves back (nothing occurs on the left side).

Fig. 5 shows the same behaviour at the left GB. The same source emits a dislocation loop, noted 1, which enters the left GB, as in fig. 4. The inserted part on the left side is noted 1_a. This insertion is however incomplete, because one portion noted 1_b remains out of the GB (fig. 5b). Upon unloading, the portion 1_b moves back, (fig. 5c), and extracts the inserted segment 1_a (fig. 5d) till the whole dislocation disappears at the source (not shown). This sequence shows that the left GB contains the same repulsive stress field as the right one, although it does not exhibit the same strong broad contrast (note however a distortion in the fringe spacing of the left GB). In other words, repeated insertions at the same GB place generate a strong repulsive stress field, which can sometimes (but not always) appear as a broad TEM contrast.

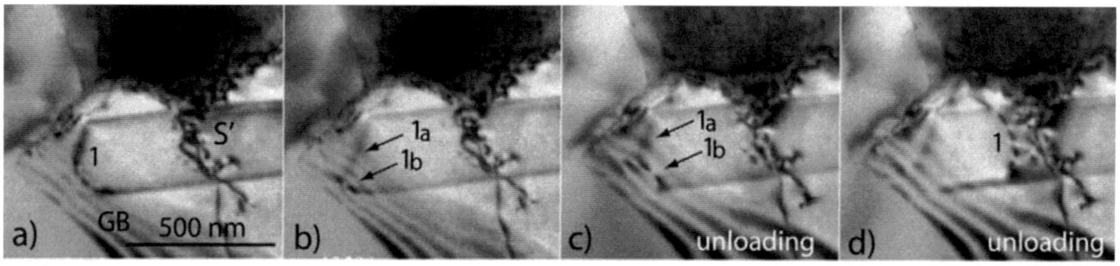

Fig. 5: Same behaviour as in fig. 4, at the left GB. a) After emission of loop 1. b) On the left-side, dislocation 1 is partly inserted in the GB (1a) and partly out of the GB (1b). c) Upon unloading, 1b bows in the back direction. d) Dislocation 1 is completely extracted and moves back to the source.

4. Discussion and conclusions

When dislocations that are emitted by sources during loading have a strong edge character, their interaction with GBs shows a strong inelastic reverse motion when the stress is reversed. Using the Orowan law, and considering that 2 dislocations on average move back to the source upon unloading, the inelastic reverse strain in the grain studied is about 5×10^{-4}. This is equivalent to what has been measured in macroscopic tests, indicating that in this case the inelastic effect is well reproduced in the thin foil.

The internal stress at the origin of the back motion of dislocations originates from the cumulated elastic stress fields of the dislocations inserted previously in the GBs. This explains why the inelastic reverse flow effect is more pronounced for edge dislocations (which all enter the GBs at the same place) than for screw ones (which are more dispersed by cross slip). In the bulk material, the inelastic reverse flow effect should however be present in each grain containing a source, because the emitted dislocation loops always exhibit a near-edge character somewhere at one of the surrounding GBs.

It is noted that the observations made in the earlier mechanical microyielding studies [2,3] and in the present *in situ* TEM investigation indicate that the dislocations generated during loading disappear almost completely during unloading, implying that there is no net increase of dislocation density after a loading/unloading cycle. This matches nicely the results obtained by Van Swygenhoven's group in an *in situ* X-ray broadening study on tensile-deformed nanocrystalline nickel [5]. In this work, the X-ray peak broadening observed during loading was found to disappear completely during unloading.

References

[1] Valiev RZ, Islamgaliev RK and Alexandrov IV 2000 *Progr. Mater. Sci.* **45** 103-189.
[2] Höppel HW, Xu C, Kautz M, Barta-Schreiber N, Langdon TG and Mughrabi H 2004 *Proc. 2nd Internat. Conference on Nanomaterials by Severe Plastic Deformation Vienna* ed M Zehetbauer and RZ Valiev (Weinheim: WileyVCH) 677-683.
[3] Mughrabi H, Höppel HW and Kautz M 2004 *Scripta Mater.* **51** 807-812.
[4] Pettinari F, Couret A, Caillard D, Molénat G, Clément N and Coujou A 2001 *J. Microscopy* **203** 47-56.
[5] Budrovic Z, Van Swygenhoven H, Derlet PM, Van Petegem S and Schmitt B 2004 Science **304** 273-276.

Can we use kinetic analysis for investigation of deformation processes in nanocrystalline materials?

Yu Ivanisenko[1] , L Kurmanaeva[1], K Yang[2] and H-J Fecht[1,2]

[1] Institute of Nanotechnology, Karlsruhe Institute of Technology, Hermann-von-Helmholtz-Platz 1, 76344 Eggenstein-Leopoldshafen, Germany
[2] Institute of Micro and Nanomaterials, University of Ulm, Alber-Einstain-Allee 47, 89081 Ulm, Germany

julia.ivanisenko@kit.edu

Abstract. Present paper represents an examination of strain rate response of nanocrystalline (NC) Pd with the aim to shed a light at deformation mechanisms operating in this material with extremely small grain size of 14 nm. Specimens were prepared by inert gas condensation method, and studied using conventional and in situ in SEM compression tests, and instrumented high pressure torsion (HPT). NC Pd demonstrated very high compression strength and good ductility, furthermore, strain rate jump tests revealed high strain rate sensitivity of 0.05, and corresponding activation volume was only $5b^3$. Microstructure investigation in the gauge sections revealed signatures of various deformation mechanisms: dislocation slip, twinning, cooperative grain boundary sliding and shear banding. These results raise a question about the applicability of kinetic analysis for nanocrystalline materials.

1. Introduction

The mechanical behaviour (strength, ductility, mechanisms of deformation, strain hardening) of nanocrystalline materials is not yet understood and is one of the key topics of modern material science. A major impediment to studies of the deformation of nanocrystalline metals is their typically very limited ductility in tension, which restricts the experimentally accessible parameter space. Compression tests and torsion testing under high pressure [1,2] offer an important opportunity, since they allow practically unlimited shear strain without fracture.

In this paper we study the deformation behaviour of NC Pd at low strains (using ex situ and in situ in SEM compression tests) and large strains (using instrumented high pressure torsion (HPT)). We also perform the analysis of plastic deformation kinetics in NC materials by means of strain rate change (jump) tests.

2. Experimental

Nanocrystalline Pd with a mean grain size of 14 nm (figure 1a) was obtained by inert gas condensation. Uniaxial compression tests were performed using a dedicated testing machine for miniature specimens with dimensions 0.9 mm×0.9 mm×1.3 mm. The strain was precisely measured by means of a laser extensometer P-50 from Fiedler Optoelectronics. The instrumented HPT tests were performed in a compression-torsion mode under a pressure of 4.5 GPa in a Schenck™ testing machine using three strain rates ($\dot{}$): 1 s^{-1}, 10^{-1} s^{-1} and 10^{-2} s^{-1}. All experiments were performed at room temperature. For conventional bright and dark field imaging, a Phillips CM 30 operated at 300 kV was used.

3. Results

3.1. Deformation behaviour of NC Pd in compression tests

As-prepared igc Pd with a homogenous nanocrystalline microstructure demonstrated pronounced strain hardening, very high compression strength and good ductility (figure 2). The compression stress (σ) is significantly rate-dependent: strain rate sensitivity coefficient m, obtained from the jump tests, is estimated as:

$$m = \partial \log(\sigma) / \partial \log(\dot{\varepsilon}) \tag{1}$$

yielding 0.05, and activation volume defined as:

$$\Delta V = \sqrt{3}kT/m\sigma \tag{2}$$

where k is the Boltzmann constant and T is the temperature, is equal to $5b^3$ (b=Burgers vector).

3.2. In situ observation of the development of deformation relief during compression test

Using the special technique of grey-scale correlation analysis, it was revealed that NC Pd specimens were deformed homogenously only until the yield point, and after that the sample was sheared along planes inclined at around 45° to the compression axis. Further compression led to greater shearing and buckling in the middle of the gauge length. Careful inspection of the sample surface revealed great number of shear bands emerging and multiplying during loading. These shear bands are seen as very narrow but distinct lines distributed with a spacing of 10μm (figure 3).

3.3. Instrumented HPT test

The mechanical response and the microstructure development during HPT in a very wide range of strain and strain rate in NC Pd was studied in [2]. Typical Torque vs. shear strain curves for the strain range $\gamma<15$ are shown in figure 4. Very similar to the results of compression test, rapid strain hardening, high strength and strain rate sensitivity is observed. The microstructure investigation of the HPT processed samples revealed a larger grain size at the strain value $\gamma = 15$, where the mean grain size was $D = 40$ nm, more than twice the initial value. At an aspect ratio of 1.1, the grains appear still equiaxed at this strain. Dislocations and their pile-ups - spreading throughout entire grains – can be found in larger grains with diameter >50 nm. Importantly, however the fraction of grains with dislocations is rather low, most grains are visibly free of dislocations. Some amount of twins in the form of thin lamellae is also observed inside of many grains. Remarkably, a formation of planes of cooperative grain boundary sliding was observed very frequently. A typical example is shown in Fig. 1b.

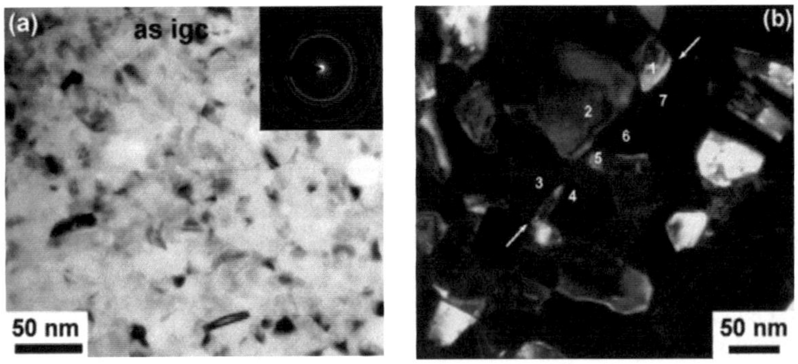

Figure 1.

Microstructure of NC Pd in the as-prepared igc state (a) and after HPT for γ =15 showing formation of planes of cooperative grain boundary sliding. Reprinted from [2] with permission of Elsevier.

4. Discussion

We start our discussion with the analysis of strain rate sensitivity and activation volume revealed in compression test. The values of m and ΔV are consistent with those obtained for electrodeposited Ni with a grain size of 30 nm [3,4,5,6,7] and point to a thermally activated control of plastic deformation in NC Pd. On the other hand they are inconsistent with values typical for all known deformation processes: dislocation glide, creep, grain boundary sliding. This fact led many scientists to conclude that some special kinds of dislocation mediated mechanisms are operating in NC materials [6-8,9]. For example it was shown that dislocation nucleation on grain boundaries should be thermally activated [7,8]. Furthermore a propagation of a dislocation loop through the grain would require thermal assistance to overcome various obstacles such as dislocation tangles in a narrow area next to a grain boundary [10], pinning at grain boundary ledges [8], or at grain boundary impurities [7]. All these "obstacles" are localised to a very small spacing at/in the grain boundary with typical dimensions of 5-10 b^3, and calculation of corresponding m using Eq.2 would result in high m values close to those obtained experimentally.

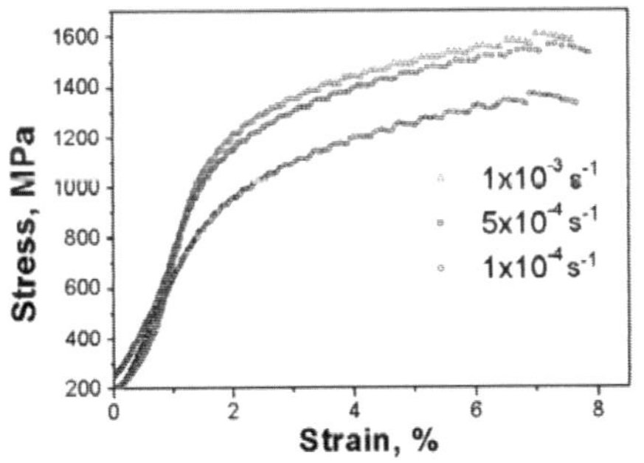

Figure 2. Stress vs. strain compression curves of NC Pd with a mean grain size of 14 nm.

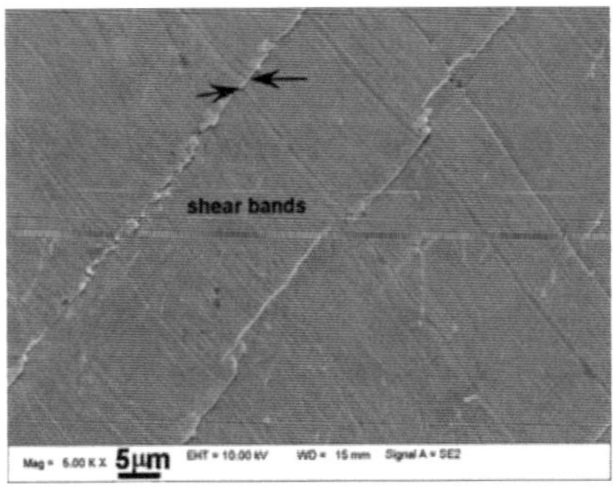

Figure 3. Shear bands revealed in NC Pd during in situ compression test. Arrows indicate the step produced by the shear band on the polishing scratch line.

In this respect we would like to emphasise that our investigations revealed signatures of various deformation processes occurring in NC Pd with a grain size of 14 nm: from dislocation slip and twinning to cooperative grain boundary sliding (Fig. 1b) and shear banding (Fig. 3).

This raises the question about the applicability of kinetic analysis for NC materials. This analysis is valid for the case when the deformation mechanism **uniquely** determines the strain rate $\dot{\gamma}$. This strain rate is determined by the Arrhenian equation [11,12]:

$$\dot{\gamma} = \dot{\gamma}_0 \exp(-\frac{\Delta G(\tau_e)}{kT}). \tag{3}$$

Here ΔG the Gibbs free energy of activation for the stress-assisted thermally activated process and τ_e the thermal component of the applied shear stress, i.e. the stress necessary to overcome the short range barrier responsible for the temperature and strain rate dependence. The pre-exponential factor $\dot{\gamma}_0$ depends on the volume fraction of fertile material that can take part in the configurational transformation, the characteristic frequency of the cluster or other configuration along the activation path (a cluster can be a dislocation line segment, or a twin embryo, etc.) and finally on a freestanding transformation shear strain [12], and represents therefore the rate of operating deformation process. Activation energy ΔG and activation volume ΔV can be found from following equations based on Eq.3:

$$\ln[\dot{\gamma}(T_1)/\dot{\gamma}(T_2)] = \frac{\Delta G}{k}(\frac{1}{T_2} - \frac{1}{T_1}) \tag{4}$$

when stress is constant, and

$$\ln[\dot{\gamma}(\tau_1)/\dot{\gamma}(\tau_2)] = \frac{\Delta V}{kT}(\tau_1 - \tau_2) \tag{5}$$

when temperature is constant.

Obviously, these equations cannot be used in cases where **more than one** deformation mechanism controls the stress-strain relationship, because $\dot{\gamma} = \dot{\gamma}_{01}\exp(-\frac{\Delta G_1(\tau_e)}{kT}) + \dot{\gamma}_{02}\exp(-\frac{\Delta G_2(\tau_e)}{kT})$ is not an exponent, and taking its logarithm would yield an apparent activation energy $\Delta G_1<\Delta G<\Delta G_2$ gradually changing in the measurement ranges of temperature and stress [13]. The same is valid for the activation volume.

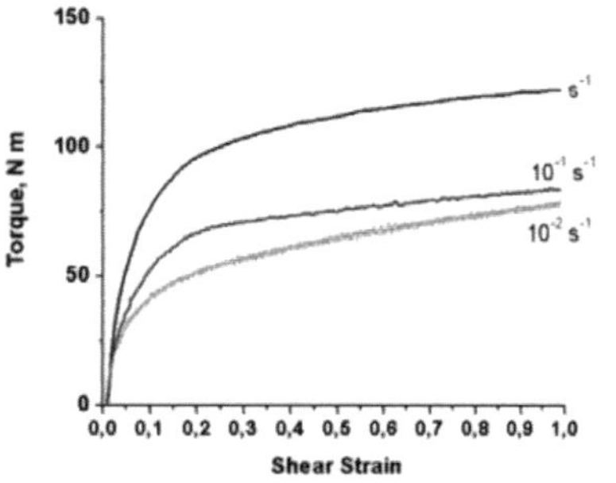

Figure 3. Torque vs. Shear Strain curve of NC Pd deformed at different shear strain rates. Only portion of the diagram up to shear strain $\gamma = 1$ is shown.

5. Conclusions

A variety of deformation mechanisms are operating in NC Pd with a grain size of 14 nm during plastic strain. In case of several dominating mechanisms the application of the kinetic analysis may be hindered and may lead to confusing results. Careful analysis of microstructure in the gage section after test or/and in situ observation should be performed to justify the conclusions yielding from the kinetic analysis.

Acknowledgment

Funding support of Deutsche Forschungsgemeinschaft under contract FOR714 is gratefully acknowledged.

References

[1] Pippan R, Wetcher F, Hafok M, Vorhauer A, Sabirov I. 2006 *Adv. Eng. Mat.* **8** 1046.
[2] Ivanisenko Yu, Kurmanaeva L, Weissmueller J, Yang K, Markmann J, Rösner H, Scherer T, Fecht H-J 2009 *Acta Mater.* **57** 3391.
[3] Dalla Torre F, Van Swygenhoven H, Victoria M 2002 *Acta Mater.* **50** 3957.
[4] Kumar K S, Van Swygenhoven H, Suresh S 2003 *Acta Mater.* **51** 5743.
[5] Dalla Torre F, Spatig P, Schaublin R, Victoria M 2005 *Acta Mater.* **53** 2337.
[6] Li Y J, Mueller J, Höppel H W, Göken M, Blum W 2007 *Acta Mater.* **55** 5708.
[7] Wang Y M, Hamza A V, Ma E 2006 *Acta Mater.* **54** 2715.
[8] Van Swygenhoven H, Derlet P M, Frøseth A G 2006 *Acta Mater.* **54** 1975.
[9] Cheng S, Spenser J A, Milligan W W, 2003 *Acta Mater.* **51** 4505.
[10] Conrad H 2003 *Mater. Sci. Eng.* A **341** 216.
[11] Calliard D, Martin J L 2003 *Thermally Activated Mechanisms in Crystal Plasticity* (Amsterdam: Pergamon) 13.
[12] Argon A S 2008 *Strengthening Mechanisms in Crystal Plasticity* (New York: Oxford University Press Inc.) 45-52.
[13] Shtremel M A 1997 *Strength of Alloys. Part II, Deformation.* (Moscow: MISIS) 114-117.

The influence of severe plastic deformation by high pressure torsion on structure and mechanical properties of Hadfield steel single crystals

G G Zakharova , E G Astafurova

Institute of Strength Physics and Materials Science, Siberian Branch of Russian Academy of Sciences, Akademichesky prospect 2/4, 634021, Tomsk, Russia

galinazg@yandex.ru

Abstract. Hadfield steel single crystals have been deformed by high pressure torsion at room temperature (P=5GPa) for 1, 2, 3 revolutions. The resulting microstructure has been studied by means of transmission electron microscopy (TEM) and X-ray analysis. The size of fragments decreases with increasing number of revolutions due to interaction of slip dislocations, microbands and thin twins. As a result of severe plastic deformation, the microhardness of the Hadfield steel has been increased, and a portion of ε, α' martensite has been found.

1. Introduction

In recent years, much attention is paid to nanocrystalline (NC) materials (10nm<d<100nm) [1, 2], because of their unique properties cased by small grain size. The influence of severe plastic deformation (SPD) on structure and properties of plastic metals, such as Cu, Ni, Al and low-strength alloys based on it, was widely investigated. The works devoted to investigation of nanocrystalline steels are not enough.

Austenitic carbon steels are an important class of industrial materials but they are hard-to-deform by the methods of SPD. Therefore, fabricating of NC states in cheap low-alloyed steels allows to extend the class of construction materials, because SPD extremely increases the mechanical properties of steels [3,4]. The nanostructuring of steels creates a new class of high-strength materials with prospect of further application.

Hadfield steel is austenitic manganese steel which possesses high strain hardening attributed to the formation of twin boundaries and dynamic strain ageing [5-8]. Twin boundaries provide strong barriers to dislocation motion, and Mn-C couples in the cores of dislocations lock them. Hence, Hadfield steel is a perspective material for hardening under SPD, as quick fragmentation by twinning and fast increase in dislocation density are supposed at room temperature. The main object of this paper is the characterization of microstructure occurring under high pressure torsion of single crystalline Hadfield steel.

2. Experimental procedures

The single crystalline Hadfield steel (Fe-13Mn-1.3C, wt. %) specimens were chosen for investigation as a model material without any inner boundaries which can effect the fragmentation process. Single crystals were grown with the Bridgman technique in a He atmosphere. They were then homogenized for 24 h at 1373K. Electro-discharge machining was utilized to cut discs with diameter of 10 mm and height of 0,6 mm. All specimens were solution-treated and water-quenched from 1373K after 1h. They then were mechanically ground, chemically etched to remove damaged surface layer. Specimens were subjected to high pressure torsion (HPT) at room temperature (P=5GPa) for number of revolutions from one to three. The structure analysis was performed using Philips CM30 transmission electron microscope and Olympus GX-71 optical microscope. Shimadzu XRD-6000 diffractometer with Cu-K_α radiation was utilized for X-ray research. The microhardness was defined in the middle of specimen radius with Duramin 5 device with load of 200g. True logarithmic strain was calculated in accordance with e=ln(υr/h) [1] in the middle of the radius, where υ-rotation angle; r, h- disk radius and thickness.

3. Results and discussion

The mechanical properties (microhardness) and TEM images of single-crystalline Hadfield steel after high pressure torsion (HPT) are presented in Table 1 and Figure 1. Fragmentation of single-crystalline Hadfield steel causes the essential increase in steel microhardness after HPT. Initially, steel was in low-strength state with microhardness of 2.2 GPa. In comparison with initial state, the microhardness of Hadfield steel after one revolution increases more than three times up to the value of 7.1 GPa (Table 1). The following revolutions do not rise the microhardness considerably, and after three revolutions a value of 7.8 GPa is reached.

Table 1. The influence of HPT on microhardness of Hadfield steel and corresponding true logarithmic strain

state	initial	upset	1 revolution	2 revolutions	3 revolutions
H_μ, GPa	2.2	4.1	7.1	7.6	7.8
e	0	0.2	4.0	4.7	5.0

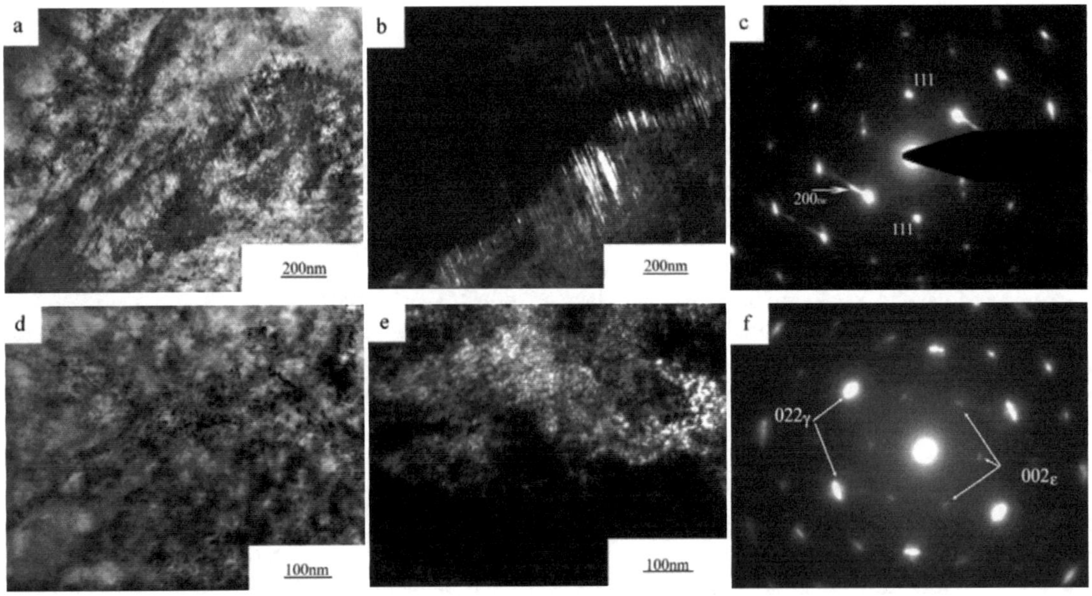

Figure 1. TEM images of Hadfield steel after HPT for one (a-c) and three revolutions (d-f): (a, d) bright field images; (b) dark field image (twins, corresponding reflex is depicted on (c)), (e) dark field image (ε-martensite, corresponding reflexes are depicted on (f)); (c,f) selected area diffraction patterns from areas of 0.1 μm^2 of (a) and (d), respectively

With increasing number of HPT revolutions the initial structure is refined. TEM bright field images of the Hadfield steel after HPT and corresponding selected area electron diffraction (SAED) patterns reveal colonies of thin twins with mean width of ~5-10 nm (Figs. 1 a-c) and fine-grained ε-martensite (Figs. 1 d-f). Fragment boundaries can be hardly recognized. So, HPT of Hadfield steel at room temperature does not give a structure with well-developed high-angle boundaries. SPD results in fragmentation of an initial single-crystalline structure with micro-, nano- twins and shear bands. The new boundaries have mainly twin or band nature. The refinement of the initial structure by SPD was also confirmed by X-ray analysis. Size of the coherently scattering regions decreases, fraction of martensite grows with increasing number of HPT revolutions (Fig. 2, Table 2). The amount of ε-martensite after 1 revolution is very small (less than 2%). Three HPT revolutions lead to the appearance of α'- martensite and an increase of ε-martensite fraction as revealed by X-ray diffraction (Fig. 2, Table 2). The small fractions of ε, α' -martensite are the result of severe plastic deformation by HPT.

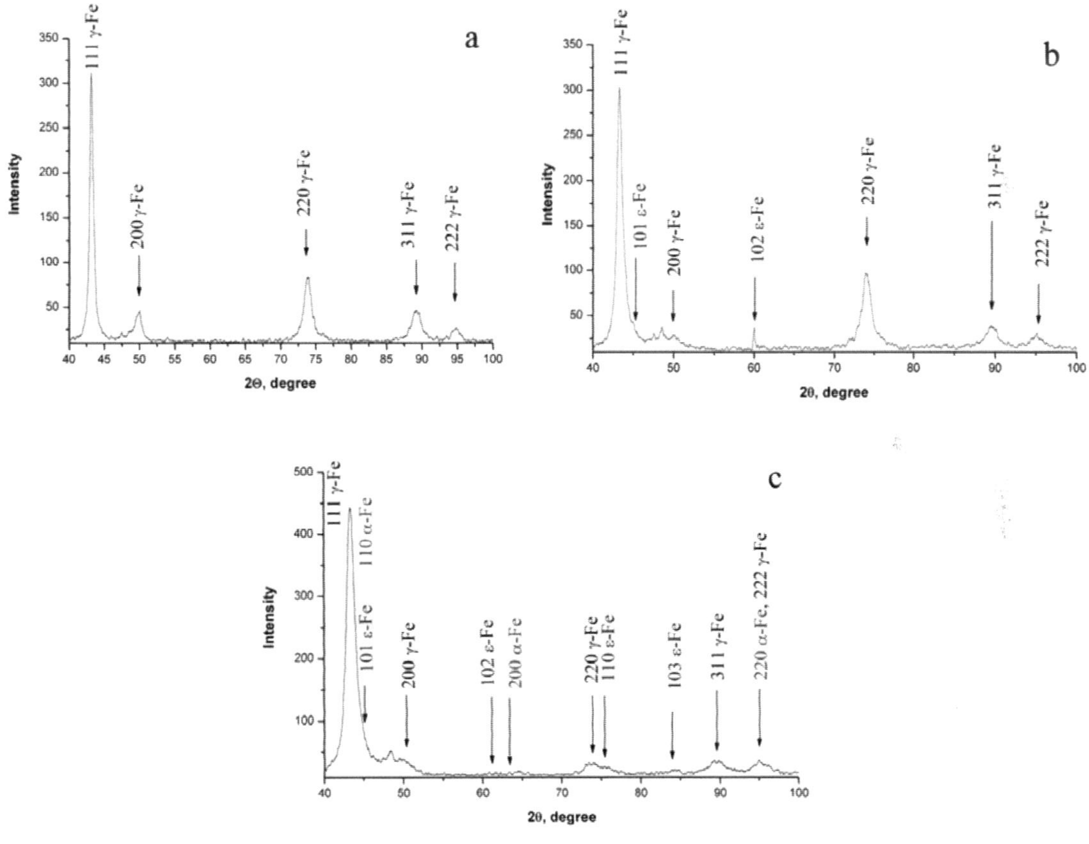

Figure 2. X-rays line profiles of Hadfield steel after HPT for one (a), two (b), and three revolutions (c)

Table 2. The influence of HPT on values of Δd/d, coherently scattering regions (CSR) and phase composition in Hadfield steel

	initial	1 revolution	2 revolutions	3 revolutions
CSR, nm	-	16.9	11.16	10.36
Δd/d	-	$3.49*10^{-3}$	$4.23*10^{-3}$	$4.53*10^{-3}$
Phases	γ	γ, ε	γ, ε, α'	γ, ε ,α'

More investigations are needed to unravel the nature of the martensitic transformation under SPD of stable Hadfield steel. The high carbon concentration lowers the martensite start temperature and suppresses the martensitic transformation in Hadfield steel at room temperature. Early works attributed high strain hardening of Hadfield polycrystals to the formation of ε and α' martensite due to local decarburization of bulk crystals. However, later studies show that the volume fraction of martensite is too low to give a high contribution to strain hardening. This is in good agreement with the present study shown that the increase in ε and α' martensite fraction with the number of revolutions does not provide a substantial increase in microhardness (Tables 1 and 2).

Mechanical twinning is supposed to play a dominant role in strength of Hadfield steel during HPT. As twinning starts in a lattice supersaturated with local barriers (due to high carbon and Mn content), it has non-homogeneous character. Twins are thin, and even a small twin volume fraction gives high strain hardening. A network of twin boundaries forms from the very beginning of deformation and corresponds to the fastest hardening effect (Table 1). As revealed by TEM and X-ray diffraction, further deformation corresponds to the formation of a polycrystalline structure that is textured and consists of twins, bands and martensite.

4. Conclusions

As a result of high pressure torsion at room temperature, the microhardness of Hadfield single-crystalline steel was increased. HPT results in refinement of the initial structure and formation of ε and α' deformation martensite. The size of structural fragments decreases and the fraction of α'-, ε-martensite rises with increasing number of revolutions. The formation of substructure in Hadfield steel predominantly determined by twinning and consequently by the formation of special type boundaries minimizes the contribution to microstructure refinement by slip.

Acknowledgments

This research was supported by the RFBR (projects 09-08-99062) and by Russian Ministry of education and science (contract No. P2366).

References

[1] Valiev R.Z., Alexandrov I.V., 2007. *Bulk Nanostructured Metallic Materials,* IKC Akademkniga, Moscow: 1-398, in Russian.

[2] International Symposium on Bulk Nanostructured Materials: from Fundamentals to Innovation, BNM 2007, edited by Valiev R.Z., Langdon T.G., Alexandrov I.V., Zhu Y.T., Estrin Y. and Kostorz G., 2009. *Mater. Sci. Eng.* **A 503** 1–208.

[3] Astafurova E.G., Dobatkin S.V., Naydenkin E.V., Shagalina S.V., Zakharova G.G., 2008. *Mater. Sci. Forum* **584-586** 649-654.

[4] Qu S., Huang C.X., Gao Y.L., Yang G., Wu S.D., Zang Q.S., Zhang Z.F., 2008. *Mater. Sci. Eng.* **A 475** 207–216.

[5] Karaman I., Sehitoglu H., Beaudoin A.J., Chumlyakov Y.I., Maier H.J. and Tomé C.N., 2000. *Acta Mater.* **48** 2031-2047.

[6] Astafurova E.G., Kireeva I.V., Chumlyakov Yu.I., et al., 2007. *Int. J. Mater. Res.* **98** 144-149.

[7] Raghavan K.S., Sastri A.S., Marcinkowski M.J., 1969. *Trans. Met. Soc. AIME* **245** 1569-1575.

[8] Dastur Y.N., Leslie W.C., 1981. *Met. Trans.* **12A** 749-759.

Journal of Physics: Conference Series **240** (2010) 012140 doi:10.1088/1742-6596/240/1/012140

Crystal plasticity modelling and experiments for deriving microstructure-property relationships in γ-TiAl based alloys

Claudio Zambaldi , Dierk Raabe

Max-Planck-Institut für Eisenforschung GmbH, Max-Planck-Str. 1, 40237 Düsseldorf Germany

c.zambaldi@mpie.de

Abstract. Single-crystals of γ-TiAl cannot be grown for the compositions present inside the two-phase γ/α_2-microstructures that show good mechanical properties. Therefore the single crystal constitutive behaviour of γ-TiAl was studied by nanoindentation experiments in single phase regions of these microstructures. The experiments were extensively characterized by a combined experimental approach to clarify the orientation dependent mechanical response during nanoindentation. They further were analyzed by a three-dimensional crystal plasticity finite element model that incorporated the deformation behaviour of γ-TiAl. The spatially resolved activation of competing deformation mechanisms during indentation was used to assess their relative strengths. On the length-scale of multi-grain aggregates two kinds of microstructures were investigated. The lamellar microstructure was analyzed in terms of kinematic constraints perpendicular to densely spaced lamellar boundaries which lead to pronounced plastic anisotropy. Secondly, the mechanical behaviour of massively transformed microstructures was modelled by assuming a lower degree of kinematic constraints. This resulted in less plastic anisotropy on a single grain scale and lower compatibility stresses in a 64-grain aggregate. On the macroscopic length scale, the results could possibly explain the pre-yielding of lamellar microstructures.

1. Introduction
Gamma-TiAl based alloys are currently investigated for application in turbine aero-engines. Due to their low density of about 4g/cm³ they exhibit excellent specific mechanical properties [1].

In this paper two approaches are presented to analyze the micromechanics of γ-TiAl based alloys on different length scales. First the single phase behavior of tetragonal, $L1_0$-ordered γ-TiAl was studied by a combined analysis employing nanoindentation experiments and crystal plasticity finite element modeling (CPFEM). Second, on a larger length scale, the two-phase micromechanics of a lamellar and a refined microstructure were simulated by a homogenized CPFEM-model that takes into account the combined effect from single crystal plasticity and the interface-dominated flow behavior.

2. Analysis of the single phase properties by nanoindentation, EBSD, AFM and CPFEM
The inelastic deformation behavior of γ-TiAl is highly anisotropic [2]. Bulk single crystals cannot be produced for the near-stoichiometric compositions that are present in two-phase microstructures with balanced properties, e.g. [3]. Therefore an alternative approach was developed to analyze the

intrinsic properties of the constitutive phases. Nanoindentation experiments were performed in single phase regions of a two-phase microstructure. For the discrimination of the three order variants in γ-TiAl a newly developed indexing scheme for electron backscatter diffraction (EBSD) patterns was applied [4]. It was used to measure the crystallographic direction of the indentation axis. The pile-up formation around the indent was measured by atomic force microscopy (AFM). Finally, three-dimensional CPFEM simulation of the indentation process was employed to assess the deformation contributions from each deformation system.

Nanoindentation of TiAl alloys has previously been studied [5,6]. For the analysis of plastic anisotropy, axis-symmetric indenter shapes are well suited because they do not interfere with the plastic anisotropy of the crystal under consideration [6]. Figure 1 gives a view of the combined results from a topographic measurement and orientation mapping for several indentations. The Berkovich indenter geometry inhibited formation of pile-up in the directions of the edges.

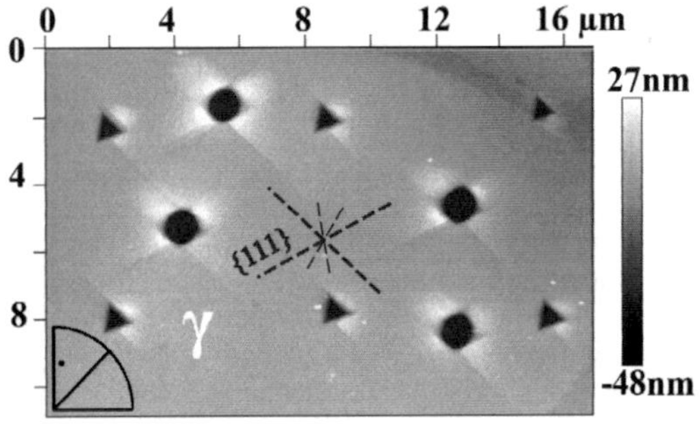

Figure 1) Nanoindents in γ-grains of nearly lamellar Ti-46Al-8Nb (at.-%); AFM topographic map (greyscale) combined with an EBSD orientation map (001 inverse polefigure colouring). The {111}-traces coincide with the observed surface steps. Cono-spherical (tip radius < 1 μm) and Berkovich indentations were performed with maximum loads of 6 mN and 3 mN, respectively. The indentation axes were oriented along [1 10 6]. Remaining indent depths were about 250 nm for the 6 mN load.

2.1. Crystal plasticity modelling of the indentation deformation

An existing elasto-viscoplastic crystal plasticity formulation [7] based on the works of Kalidindi et al. [8] was changed to use specific strength and hardening characteristics for deformation systems of different types. Additionally, a coarse implementation of the four twin deformation modes in TiAl was carried out by the installation of uni-directional slip systems. Reorientation through twinning was neglected. The constitutive law was coupled with a finite element solver as a material subroutine. Four glide systems by ordinary dislocations, b=1/2 ⟨110⟩, eight by superdislocations, b=⟨101⟩, and four true twinning modes, b=1/6 $\langle 11\bar{2} \rangle$ were taken into account on the {111} planes of the γ-phase.

A three-dimensional finite element model of indentation was generated to simulate the nonlinear deformation during indentation of an axis-symmetric rigid body. The formation of hillocks was analyzed after unloading. The formation of characteristic pile-up patterns during the indentation could be related to the existence of distinct lobes of slip activity below the surface. The topography of the pile-up shape was a characteristic feature of the indented crystallographic direction as has been previously reported for indentation into high-symmetry orientations of copper [9].

Simulations were performed for indentation axes throughout the crystallographic unit triangle of γ-TiAl. To this end, the unit triangle was discretized at an approximate resolution of 9° (Δ=9°, cf. [10]) by a method after Helming et al. [10]. The simulated pile-up characteristics were found to be in good agreement with experimentally measured pile-up topographies, figure 2.

The measured pile-up patterns were analyzed and compared with the simulation results for different sets of constitutive parameters for several orientations. It was found that mostly ordinary and, to a minor extent, twinning deformation is operating during the indentation of single phase γ-TiAl. These two modes could explain the observed pile-up patterns throughout orientation space. The adjusted values for the critical resolved shear stress (CRSS) for ordinary dislocation glide and twinning were 150 MPa and for superdislocation glide a CRSS of 300 MPa was assumed which lead to almost complete suppression of this mode. Load-displacement curves showed good agreement with the experimental data for loading and unloading.

Figure 2) Predicted pile-up patterns evolving in γ-TiAl. The topographic data was obtained from 51 CPFEM simulations. The individual patterns are placed in an inverse polefigure projection of the respective indentation axes. The impressions are shown in bluish colour, the upheaval in greyscale.
Topographies in the 001–110–010 triangle can be obtained by a mirroring operation of the given topographies through the $(\bar{1}10)$ plane. The number of dominant hillocks ranges from one to four. Two experimental pile-up topographies are also shown.

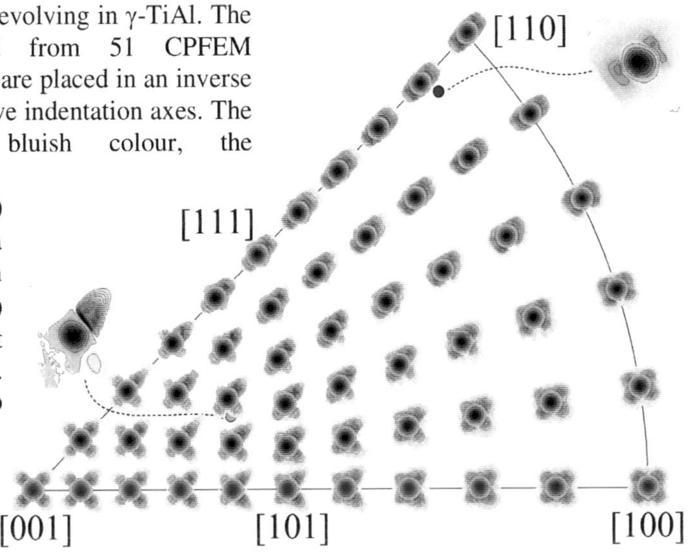

3. A homogenized model for two-phase γ/α_2 microstructures

3.1. Anisotropic yielding of lamellar bundles

The second length scale on which the micromechanics were studied was on the order of grain sizes. The yielding behaviour of lamellar poly-synthetically twinned (PST) samples that contain only one lamellar orientation, shows a pronounced dependency of the angle between the loading axis and the lamellar interfaces [11-14]. The micro-mechanics are dominated by the high density of parallel α_2/γ and γ/γ interfaces. A homogenized CPFEM model was developed that reflects the kinematic constraints resulting from the interfaces, figure 3.

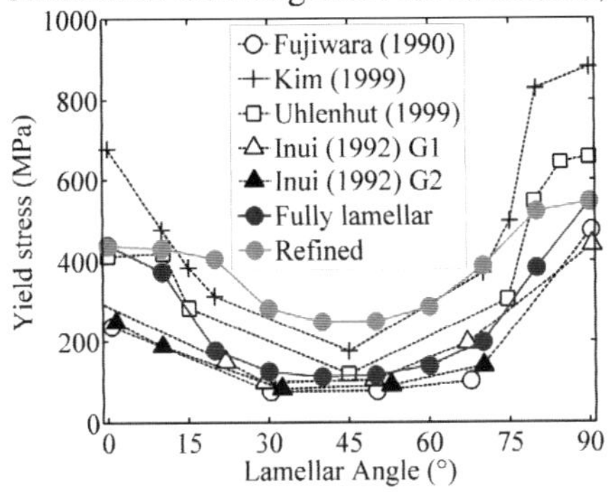

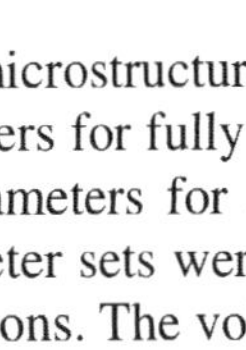

Figure 3) Anisotropic yielding of a simulated PST crystal in compression compared to literature data [11-14]. The assumed yielding for a refined micro-structure is shown in comparison to the yield points from the fully lamellar model. The definition of the lamellar angle, Φ, is shown for a vertical deformation axis.

3.2. Polycrystal mechanics

To simulate the tensile response of real microstructures, a 64-grain aggregate was constructed with periodic boundary conditions. The parameters for fully lamellar (FL) microstructures were taken to be identical to the PST parameters. The parameters for refined (RF) microstructures were changed to reflect less plastic anisotropy. Both parameter sets were used for simulated tensile deformation of the 64-grain model with random grain orientations. The volume averaged response of the grain aggregates is shown in figure 4-a. The simulated curves were generally lying below the data from tensile tests on polycrystalline microstructures. That could be explained by different compositions of the PST crystals (binary) and the FL and refined microstructures that were additionally alloyed with 8 at.-% Nb or Ta and show a smaller lamellar spacing which is known to increase the overall strength. More important, however, is the observation that the simulated fully lamellar aggregate shows a flow curve in excellent agreement with the pre-yielding behaviour of the experimental data. The extended pre-yielding of the

FL microstructure is absent in the refined microstructure. Therefore the strong plastic anisotropy might explain the pre-yielding observed in fully lamellar microstructures [15-17].

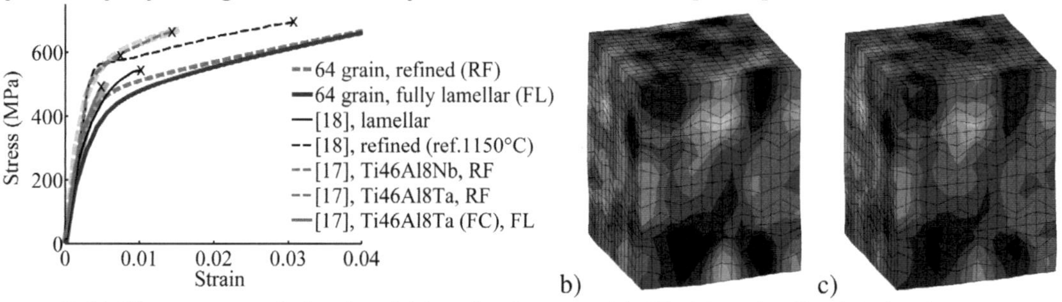

a) b) c)

Figure 4) (a) Flow curves of simulated 64-grain clusters with FL (–) and refined (- -) microstructures compared to literature data [17,18]; the bold, semi-transparent curves are the scaled results from the respective simulations and match the experimental curves from [17] almost perfectly; (b) Distribution of axial stress for the FL and (c) refined microstructure simulations, greyscale from 0 MPa (white) to 1600 MPa (black).

4. Conclusions

The presented nanoindentation analysis combines experimental and computational methods to form a coherent picture of the single phase mechanics. Easy activation of ordinary dislocation glide in the γ-TiAl phase was found to be an intrinsic property, i.e. not mainly related to the interfaces.

The analysis of a 64-grain cluster with two different sets of constitutive parameters revealed the strong plastic anisotropy in lamellar microstructures as the probable source of the experimentally observed pre-yielding behavior.

Acknowledgements

Dr. F. Roters is acknowledged for valuable discussions. Dr. G. Behr and Dr. W. Löser (IFW Dresden) are thanked for providing the sample material. Support by the EU FP6 integrated project IMPRESS (NMP3-CT-2004-500635) is gratefully acknowledged.

References

[1] Appel F and Wagner R 1998 *Mater. Sci. Eng.* ***R22*** 187–268
[2] Mecking H, Hartig C and Kocks U F 1996 *Acta Mater.* **44**, 1309-1321
[3] Bird N, Jiao S and Taylor G 2000 *Intermetallics* **8** 133–41
[4] Zambaldi C, Zaefferer S and Wright S I *J. Appl. Cryst.*, accepted
[5] Göken M, Kempf M and Nix W 2001 *Acta Mater.* **49** 903–11
[6] Kempf J, Göken M and Vehoff H 2002 *Mater. Sci. Eng.* **A329** 184–9
[7] Roters F 2005 *Comput. Mater Sci.* **32** 509–17
[8] Kalidindi S R, Bronkhorst C A and Anand L 1992 *J. Mech. Phys. Solids* **40** 537-569
[9] Wang Y, Raabe D, Klüber C and Roters F 2004 *Acta Mater.* **52** 2229–38
[10] Helming K, Tamm R, Fels B 1998 *Mater. Sci. Forum* **273-275** 119–24
[11] Uhlenhut H 1999 *PhD Thesis* (Aachen: Shaker)
[12] Fujiwara T, Nakamura A, Hosomi M, Nishitani S R, Shirai Y and Yamaguchi M 1990 *Phil. Mag.* **A61** 591–606
[13] Inui H, Oh M and Nakamura A 1992 *Acta Metall.* 1992 **40** 3095–104
[14] Kim M-C, Nomura M, Vitek V and Pope D P 1999 *Proc. Mat. Res. Soc. Symp.* **552** ed George E, Yamaguchi M *et al.*, KK3.1.1
[15] Hu D, Huang A, Jiang H, Mota-Solis N and Wu X H 2006 *Intermetallics* **14** 82–90
[16] Wu X H, Jiang H, Huang A, Hu D, Mota-Solis N and Loretto M H 2006 *Intermetallics* **14** 91
[17] Saage H, Huang A, Hu D, Loretto M and Wu X 2009 *Intermetallics* **17** 32–8
[18] Couret A, Molénat G, Galy J and Thomas M 2008 *Intermetallics* **16** 1134–41

The effect of dopants on the brittle-to-ductile transition in silicon single crystals

Youn Jeong Hong, Masaki Tanaka, Keiki Maeno, Kenji Higashida

Department of Materials Science and Engineering, Kyushu University,
744 Motooka, Fukuoka, 819-0395, Japan

hong@kyudai-mse.org

Abstract. The brittle-to-ductile transition (BDT) in boron, antimony and arsenic doped Cz silicon crystals has been experimentally studied, respectively. The BDT temperatures in antimony and arsenic doped silicon wafers are lower than that in a non-doped wafer while the BDT temperature in a boron doped wafer is almost the same as that in the non-doped wafer. The activation energy was obtained from the strain rate dependence of the BDT temperature. It was found that the values of the activation energy in the antimony and arsenic doped wafers are lower than that in the non-doped and boron doped wafers, indicating that the dislocation velocity in the antimony and arsenic doped silicon is faster than that in the non-doped while the dislocation velocity in the boron doped is the same as that in the non-doped. The effect of increasing in dislocation velocity on the BDT temperature was calculated by two-dimensional discrete dislocation dynamics simulations, indicating that the increasing in dislocation velocity decreases the BDT temperature in silicon single crystals.

1. Introduction

Silicon crystals have been used as model crystals to understand the mechanism behind the brittle-to-ductile transition (BDT) behaviour since St. John [1] demonstrated that they show a very sharp BDT. Many studies after his work revealed that the controlling process of the BDT is dislocation glide, deducing from the fact that the values of the activation energy obtained were close to those for dislocation glide [2-4]. Roberts *et al* [5] showed that the BDT temperature in phosphorous doped silicon was lower than that in the non-doped one and the activation energy obtained from the strain rate dependence of the BDT temperature was close to that for dislocation glide. It implied that the BDT behaviour was affected by doped elements via change of dislocation velocity. Since several kinds of dopants are added in silicon crystals for commercial use, it is necessary to investigate the effect of dopants on the BDT behaviour. In this study, the effect of p-type dopant (boron) and n-type dopants (arsenic or antimony) on the BDT behaviour will be discussed based on the dislocation shielding theory [6]. The effect of the change in dislocation velocity on the BDT was also investigated by two-dimensional discrete dislocation dynamics simulations.

2. Experimental Procedure

Specimens were prepared from one-side mirror polished (001) wafers grown by Czochralski technique without dopants (1×10^{15}atoms/cm^3 of boron remained) and those with dopants of boron (3.6×10^{19} atoms/cm^3), antimony (2.2×10^{18}atoms/cm^3) and arsenic (2.8×10^{19}atoms/cm^3). Rectangular specimens of $8\times35\times0.62$mm^3 were cut out from the wafers for subsequent three-point bending tests. Fracture tests were carried out at temperatures between 770K and 1120K. The cross-head speed (CHS) was set to be 0.002mm/min, 0.02mm/min and 0.2mm/min, respectively. Apparent fracture toughness was obtained by following equations.

$$K_Q = \sigma_f\sqrt{2\pi a}f(\xi),\ \text{with } \xi = \frac{a}{w} \text{ and } f(\xi) = 1.090 - 1.735\xi + 8.20\xi^2 - 14.18\xi^3 - 14.57\xi^4, \quad (1)$$

where σ_f is the fracture stress, a is the notch depth and w is the specimen height. The BDT temperature was determined as one at which the highest value of apparent fracture toughness in each CHS was achieved. The activation energy for the controlling process of the BDT was obtained from Arrhenius plots of the logarithm of the CHS against the inverse temperature.

3. Results and Discussions

Figures 1 (a)-(d) show temperature dependence of the apparent fracture toughness in boron doped, non-doped, antimony doped and arsenic doped silicon single crystals respectively. The apparent fracture toughness increases with increasing temperature, from the lowest valued of 1.3MPa.m$^{1/2}$. The BDT temperature, T_{BDT}, is defined as the highest apparent fracture toughness, in this study. The relationship between the BDT temperatures in each specimen was found to be: $T_{BDT,\,As} < T_{BDT,\,Sb} < T_{BDT,\,B} = T_{BDT,\,non}$ at any strain rate. It was thus found that p-type dopant (boron) does not affect the BDT temperature while n-type dopants (arsenic and antimony) decrease the BDT temperature. The same kind of result that n-type dopants show lower BDT temperature was obtained by Roberts *et al* although the dopant is not the same as those employed in the present study but phosphorous [5]. In order to investigate the influence of doping on the rate determining process, *i.e.*, dislocation mobility, the activation energy was measured from the deformation rate (CHS) dependence of the BDT temperature obtained in this study.

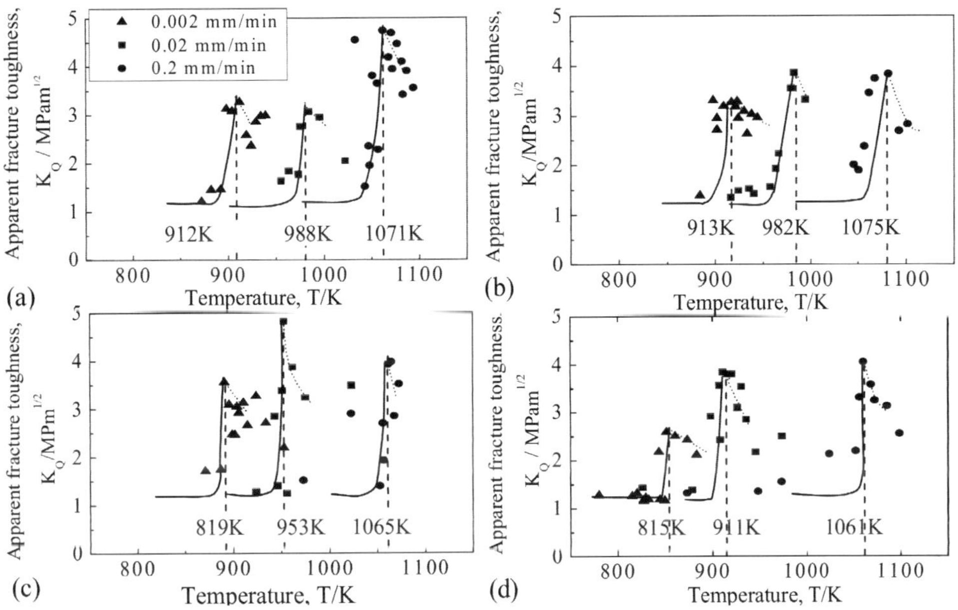

Figure 1. Temperature dependence of apparent fracture toughness of (a) boron doped, (b) non-doped , (c) antimony doped and (d) arsenic doped silicon single crystals.

Figure 2 shows Arrhenius plots of the logarithms of the CHS against the inverse BDT temperature. The slopes of each Arrhenius plot indicate the activation energy of 1.6eV, 2.0eV, and 2.3eV for arsenic doped, antimony doped and boron/non-doped silicon crystals, respectively. The relationship between the activation energy, E, in each specimen is: $E_{As} < E_{Sb} < E_B = E_{non}$. The tendency of the activation energy is the same as that of the BDT temperature. The results indicate that the activation energy for dislocation glide is decreased by the antimony and arsenic doping but not changed by the boron doping. In order to investigate the effect of the decrease in the activation energy for dislocation glide on the BDT temperature, the temperature dependence of fracture toughness was calculated by two-dimensional discrete dislocation dynamic simulations. The detail of the simulation is shown elsewhere [7,8].

Figure 3 shows the temperature dependence of the fracture toughness calculated by using the two-dimensional model. The symbols of triangle, square and circle demonstrate results from the activation energy of 1.6eV, 2.0eV, and 2.3eV for dislocation glide, respectively, corresponding to that for arsenic, antimony and boron/non-doped silicon crystals. It is to be noted here that the BDT temperature decreased with decreasing in the activation energy for dislocation glide.

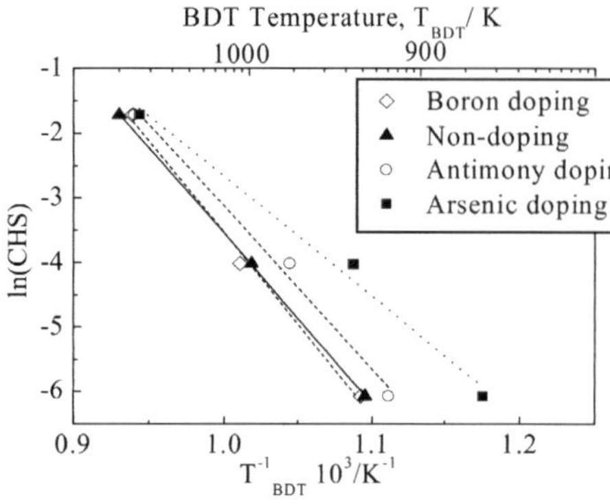

Figure 2. Arrhenius plots of logarithms of the cross-head speed (CHS) against the inverse BDT temperature.

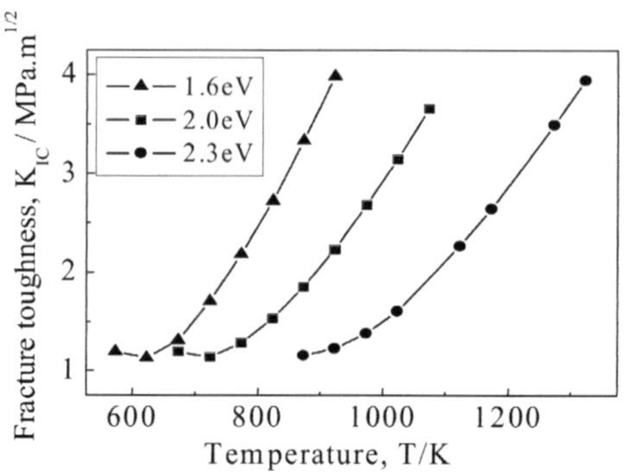

Figure 3. Calculated fracture toughness against temperature where the values of activation energy for dislocation glide were set to be 1.6eV, 2.0eV, and 2.3eV respectively.

Therefore, it is concluded that the decrease in the BDT temperature in n-type doped silicon is due to the increase in the dislocation velocity in silicon single crystals.

4. Conclusions

The effect of dopants on the BDT behaviour in silicon crystals was investigated. The BDT temperature decreased with n-type dopants. The activation energy obtained from the rate dependence of the BDT temperature indicates that the dislocation velocity in n-type (antimony or arsenic doped) silicon are faster than that in non-doped silicon. Two-dimensional discrete dislocation dynamics simulations indicate that increasing in dislocation velocity decreases the BDT temperature. It is concluded that the decrease in the BDT temperature in n-type silicon is due to the increase in the dislocation velocity.

References

[1] John St C1972 *Phil. Mag.* **32** 1193-1212.

[2] Samuels J, Roberts S G 1989 *Proc. R. Soc. Lond.* A, **421** 1-23.

[3] Serbena F C, Roberts S G 1994 *Acta Metall. Mater.* **42** 2505-2510.

[4] Imai M, Sumino K 1983 *Phil. Mag.* A **47** 599-621.

[5] Samuels J, Roberts S G 1989 *Proc. R. Soc. Lond.* A **421** 1-23.

[6] Thomson R 1986 Solid State Physics Ehrenreich H and Turnbull D (London: Academic press INC) **39** 2-462.

[7] Hong Y J, Tanaka M and Higashida K 2009 *Mater. Tans. JIM* doi: 10.2320/matertrans. MAW200919.

[8] Hartmaier A, Gumbsch P 2002 *Phil. Mag.* A **82** 3187-3200.

Journal of Physics: Conference Series **240** (2010) 012142 doi:10.1088/1742-6596/240/1/012142

Crack tip dislocations observed by TEM-tomography in silicon single crystals

Sunao Sadamatsu, Masaki Tanaka, Masaki Honda, Kenji Higashida

Department of Materials Science and Engineering, Kyushu University

744 Motooka, Fukuoka 819-0395 JAPAN

sxdxmx@kyudai-mse.org

Abstract. 3D observations of dislocations at a crack tip were attempted by transmission electron microscopy and computed tomography in order to reveal the 3D structure of dislocations emitted around a crack tip. {011} cracks were introduced into a (001) silicon single crystal wafer by using an indentation method at room temperature. The specimens indented were heated and kept at high temperatures to introduce dislocations from the crack tip. The specimen holder was tilted ±31° by 2° step and dislocation images were taken at every step. The diffraction vector was kept nearly 220 during the tilting operation. The Burgers vectors of the dislocation segments were determined, which included the signs of Burgers vectors. The dislocations observed here were those which accommodate mode II stress intensity around the crack tip. 3D observations using electron tomography reveal these complex crucial processes around the crack tip, which should contribute to understanding the dislocation process improving fracture toughness of crystalline materials.

1. Introduction

Bcc metals such as ferrite steel or materials with a diamond structure such as silicon crystals have remarkable temperature dependence of plastic deformability. They fractured without plastic deformation at low temperatures, but they are easily deformed plastically when temperatures rise high enough for dislocation glide. The change of the fracture mode from a ductile manner to brittle one is called "brittle-to-ductile transition (BDT)". Because fracture of engineering materials has often caused historic disasters or fatal accidents due to structure failures, it is essential to understand the fundamental mechanism behind the BDT in order to keep safety and reliability of human life in this century. Several significant investigations using silicon crystals have been made to understand the fundamental mechanism of the BDT [1-5], because silicon is a suitable model crystal for studying dislocation processes controlling the BDT. According to their results, the BDT is controlled not only by the mobility of each dislocation but also by the complicated process of dislocations multiplied in the 3-D space around a crack tip. Therefore, it is essential to reveal the three-dimensional structure of crack tip dislocations in order for further understanding of the BDT mechanism.

In our previous works, crack tip dislocations in silicon crystals were observed by high voltage electron microscopy (HVEM) [6, 7]. However, they were limited to the observation of simple arrays of dislocations with low-density. Generally, dislocation multiplication around a crack tip suddenly occurs, and their structure during or after the multiplication becomes too complex to be analysed. Recently, in order to overcome this problem, a new technique, called TEM-tomography, has been developed in the field of transmission electron microscopy, which enables to reveal the complex manner of 3-D structure of lattice defects in crystalline materials [8-12]. In the present study, crack tip dislocations were observed by TEM-tomography in order to reveal their 3-D complex structures.

2. Experiment

P type (001) silicon wafers were employed. Cracks were introduced at room temperature by a Vickers indenter with a load of 200g and duel time of 5 seconds. In order to introduce dislocations from the crack tip, the sample was kept at 873K for 2 hours. The crack tip area was thinned by an ion milling machine (Fischione, model-1010). In TEM observations, a series of images of crack tip dislocations were acquired by tilting the specimen with the angle of 2° in each tilting step, and the maximum tilt angle was ±31°. The diffraction vector $\boldsymbol{g}(hkl)$ was maintained close to be 220 during the acquisition of series of images by adjusting the [110] direction of the sample to be parallel to the tilt axis of the holder. The observation was made by using a high-voltage electron microscope (HVEM), JEOL JEM-1300NEF, operating at 1250 kV at the Research Laboratory of High Voltage Electron Microscopy, Kyushu University. The Burgers vector of observed dislocations was determined by the invisibility criterion of $\mathbf{g}\cdot\mathbf{b} = 0$. The sign of Burgers vector was determined by the direction of oscillating contrasts of dislocation lines. The definition of the Burgers vector in this study is following the finish to start / right handed (FS/RH) convention.

3. Results and discussion

Figure 1 shows a bright field image of dislocations emitted from a crack tip. The incident beam direction and foil normal were both [001]. The diffraction vector was set to be [220]. The crack is extending from the right to the left in the figure. The crack tip is surrounded by dislocations with a complex structure, the configuration of which is too complicated to analyse only from this figure.

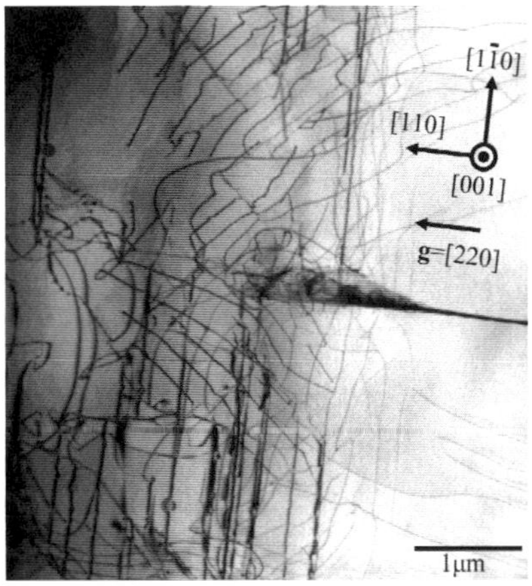

Figure 1 Bright-field image around a crack tip. The incident beam direction and foil normal are both [001].

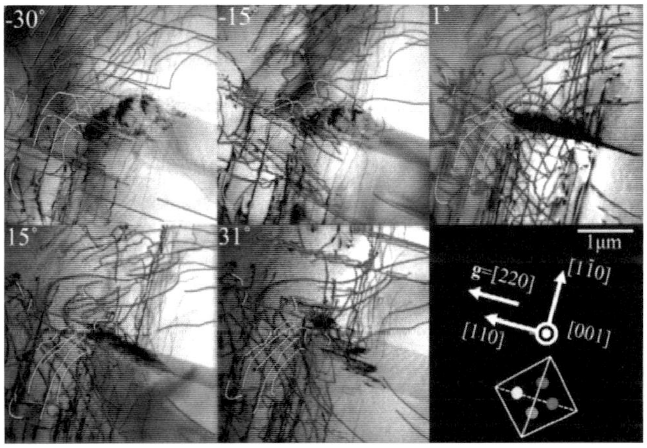

Figure 2 Snapshots from a series of tilting images around a crack tip in Fig. 1. Each dislocation is coloured according to the {111} planes where the dislocations are. Thompson's tetrahedron is shown in the right bottom of the figure.

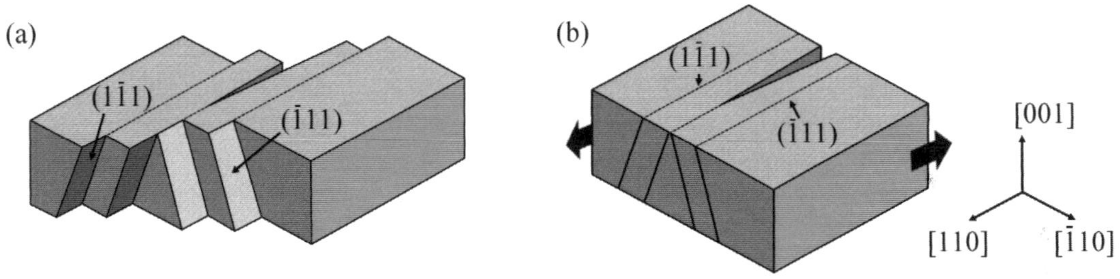

Figure 3 Schematic drawing of the deformation due to the glide of red and blue dislocations in Fig.2. (a) $(1\bar{1}1)$ and $(\bar{1}11)$ planes are glided, (b) mode II stress is applied on the crack plane .

Figure 2 shows snapshots from a series of titling images taken in the same area as shown in Fig.1. The specimen thickness was determined to be approximately 1.4μm by using the contrast width of the crack. This indicates that HVEM is an effective method to observe such a thick specimen including a large number of dislocations in the view field, which is essential to clarify the multiplication process of dislocations. As was described, the titling axis and diffraction vector were set to be [220], and the specimen holder was tilted up to ±31°. The slip plane of each dislocation was determined by the tilting images. Dislocations coloured in red, blue, yellow and green are lying on $(\bar{1}11)$, $(1\bar{1}1)$, (111) and $(11\bar{1})$ plane, respectively. The geometry of the slip planes and slip directions is shown with a Thompson's tetrahedron in the bottom of the figure. The Burgers vector of each dislocation was determined by the invisibility criterion of **g**·**b**=0. The sign of the Burgers vector was also determined by the direction of oscillating contrasts of dislocation images[13]. It is to be noted here that red and blue dislocations are dominant, the Burgers vector of which was determined to be **b**=a/2[110]. The

Burgers vector of the dislocations indicates that those dislocations were emitted in order to accommodate mode II stress intensity around the crack tip. Figure 3 shows schematic drawings of the morphology resulting from the glide of dislocations with the Burgers vector of **b**=a/2[110]. When dislocations coloured in red on $(\bar{1}11)$ and dislocations coloured in blue on $(1\bar{1}1)$ are emitted and gliding away from the crack tip, they should induce displacements around the crack tip as shown in Fig. 3 (a), which corresponds to the displacement due to mode II stress intensity as shown in Fig. 3(b). Thus, the crack mode observed in the present study is determined to be mainly mode II. Further analyses are necessary to clarify the multiplication process of these crack tip dislocations.

4. Conclusion

The dislocation structures emitted from the crack tip were observed by using TEM-tomography in silicon crystals. The method of tomography enables to reveal a complex structure of crack tip dislocations, which cannot be attained only by conventional 2D TEM observations. The slip planes and the Burgers vectors of the dislocations were determined. The dislocations were found to be accommodating mode II stress intensity around the crack tip. TEM-tomography is a powerful tool to clarify such complex 3D structures of dislocations around the crack tip, which should contribute to the further understanding of the interaction between dislocations and the crack tip.

References

[1] St. John C 1975 *Phil. Mag.* **32** 1193-1212
[2] Samuels J and Roberts S G 1989 *Proc. R. Soc. Lond. A* **421** 1-23
[3] Hirsch P B, Roberts S G and Samuels J 1989 *Proc. R. Soc. Lond. A* **421** 25-53
[4] Michot G, de Oliveira M A L and Champier G 1999 *Mater. Sci. Eng. A* **272** 83-89
[5] Brede M and Haasen P 1988 *Acta Metall.* **36** 2003-2018
[6] Higashida K, Narita N, Onodera R, et al. 1997 *Mater. Sci. Eng. A* **237** 72-78
[7] Higashida K, Narita N, Tanaka M, et al. 2002 *Phil. Mag. A* **82** 3263-3273
[8] Barnard J S, Sharp J, Tong J R, et al. 2006 *Science* **313** 319
[9] Sharp J H, Barnard J S, Kaneko K, et al. 2008 *J. Phys. Conf. Ser.* **126** 012013
[10] Tanaka M, Higashida K, Kaneko K, et al. 2008 *Scripta Mater.* **59** 901-904
[11] Kaneko K, Inoke K, Sato K, et al. 2008 *Ultramicroscopy* **108** 210-220
[12] Kimura K, Hata S, Matsumura S, et al. 2005 *J. Electron Microscopy* **54** 373-377
[13] Ruvimov S S and Scheerschmidt K 1994 *Phys. Stat. Sol.* **(a)141** 269-284

Pre-strain direction effect on microstructure evolution and energy storage process during uniaxial tension of austenitic steel

M Maj, W Oliferuk

Institute of Fundamental Technological Research Polish Academy of Sciences, Pawińskiego 5b, 02-106 Warsaw, Poland

mimaj@ippt.gov.pl

Abstract. In the present paper the influence of pre-strain direction on energy balance during deformation of austenitic steel was investigated and the analysis of microscopic phenomena responsible for this influence was performed. The specimens with different pre-strain directions were prepared and the ratio of the stored energy increment to plastic work increment, called energy storage rate, as a function of plastic strain was experimentally determined. At the initial stage of plastic deformation of annealed materials this quantity vs. plastic strain has a maximum. It has been shown that for specimens strained in the same direction as pre-strain the energy storage rate decreases monotonically with deformation while for specimens where strain path was changed, the maximum of the energy storage rate is observed (as in case of annealed material). The study of slip and microstructure evolution at meso- and micro-scales have shown that the change in pre-strain direction leads to the redistribution of internal stresses generated by incompatible slip in neighbouring grains of different orientation. Just after change in strain direction the accommodation of these stresses takes place not only by generation of geometrically necessary dislocations but also by micro-shear banding.

1. Introduction

The deformation process always leads to change in the temperature of deformed material what is a macroscopic manifestation of phenomena proceeding on the level of its microstructure. Thermal effects accompanying the deformation process were investigated by many researchers in last years [1]. The plastic work w_p equal to total energy expended on irreversible deformation can be presented as a sum of energy stored in the material e_s and energy dissipated as a heat q_d. Thus

$$e_s = w_p - q_d, \tag{1}$$

where e_s, w_p, q_d are the specific quantities.

The energy stored during deformation process is defined as a change in internal energy measured after unloading and describes the state of the cold worked material. As a measure of the energy conversion at each instant of straining the energy storage rate $\Delta e_s / \Delta w_p$ was used. This macroscopic quantity is influenced by many micro-scale mechanisms. Both the microstructure evolution and energy storage process during monotonic deformation were studied extensively in many papers [2]. In accordance with LEDSs hypothesis developed by Hansen and Kuhlmann-Wilsdorf, the evolution of dislocation structures during monotonic deformation tends towards the configurations of lower energy [3]. In recent years the rapid development of methods of material properties improvement using strain path change is observed. Therefore, it is important to study the pre-strain direction effect on the energy storage process. Thus, the aim of this paper is to study the effect of pre-strain direction on energy storage rate and to find the interpretation of obtained results on the basis of microstructure analysis.

2. Experimental procedure and results

2.1. Determination of energy storage rate

The experiments were performed on austenitic stainless steel 316L. The steel sheet was initially annealed at 1100°C and water quenched. Firstly, large specimens (500mm x 105mm x 1.5mm) were cut out and pre-strained to $\varepsilon^p = 0.068$. From such prepared material, three groups of specimens (*L*, *S* and *T*) were cut out using electro-erosion machine (figure 1).

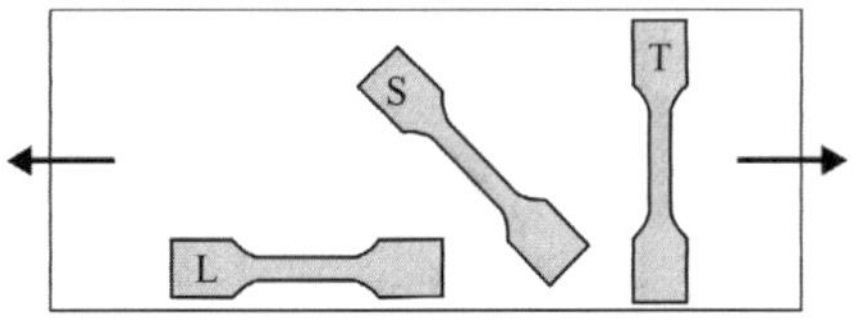

Figure 1. Orientation of the specimens (pre-strain direction marked).

These specimens and the *0* ones, made of non-deformed sheet, were strained with constant strain rate $\dot{\varepsilon} = 7\times10^{-3}s^{-1}$. During tension the temperature distribution on the surface of the specimen was measured using IR thermography. Simultaneously the stress and strain were determined. Like in the previous reports, a method of the stored energy determination without interrupting the deformation and without using a calorimeter was employed [4]. The stored energy e_s was determined as a difference between plastic work w_p and energy dissipated as a heat q_d. The plastic work was derived directly from the load-displacement curve under assumption that elastic properties do not change for this range of deformation. The energy dissipated as a heat q_d was determined by simulating the process of specimen heating during deformation using a controlled electrical power supply in such a way, that the temperature increase over time during simulation was identical to that measured by IR camera during tensile testing. When straining and simulation are conducted under identical conditions then the heat, which would have been transferred to the surroundings if the temperature of the unloaded sample had returned to the initial value, is the same in both cases and can be calculated from electrical parameters.

The energy storage rate as a function of plastic strain is presented in Figure 2. It is seen that for L specimens, strained in direction of pre-strain, this quantity decreases monotonically with deformation while for specimens where strain path was changed (S and T) the maximum of $\Delta e_s / \Delta w_p$ is observed (as in case of annealed material).

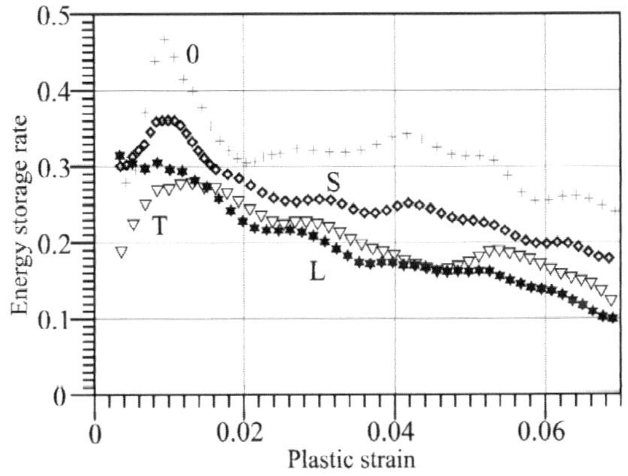

Figure 2. Energy storage rate vs. plastic strain.

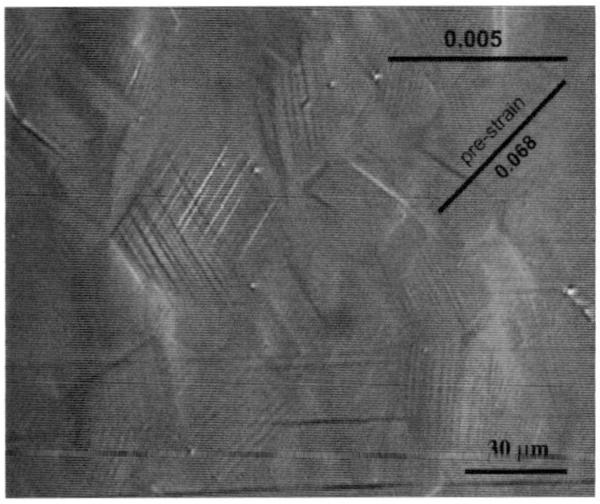

Figure 3. The surface of the *S* specimen after polishing and subsequent straining.

2.2. *Analysis of material structure*

The study of material structure was performed in two scales: the slip evolution was studied using optical microscopy with Differential Interference Contrast (DIC) whereas an evolution of dislocation structures was investigated using Transmission Electron Microscopy (TEM).

2.2.1. Slip evolution. The specimens with different pre-strain direction (*L*, *S* and *T*) and that at annealed state 0 were electropolished in order to obtain mirror-like surface. Such prepared specimens were strained to small amount of plastic strain $\varepsilon^p = 0.005$. Then, their surfaces were analysed using optical microscopy with DIC at magnification up to 600x. The advantage of this method is that it reveals effects generated during last stage of deformation. In figure 3 the surface of the S specimen after polishing and subsequent straining is shown. In particular grains the slip bands are seen. It is characteristic that for the L specimen strained in direction of pre-strain, only one set of slip bands dominate in particular grains, whereas when strain path was changed, more than one family of slip band appears. The quantitative analysis of percentage of grains $\eta = \frac{L^*}{L_t} \cdot 100\%$ with more than one set of slip bands was performed. L^* and L_t are the number of grains with more than one set of slip bands and the total number of grains, respectively. The results of the analysis are presented in table 1.

Table 1. Percentage of grains with more than one set of slip bands.

Specimen	L^*	L_t	$\eta = \frac{L^*}{L_t} \cdot 100\%$
$L_{+0,005}$	39	2000	1,9
$S_{+0,005}$	189	2000	9,4
$T_{+0,005}$	174	2000	8,7
$0_{+0,005}$	106	2000	5,3

It is seen that the η value is significantly lower in case of *L* specimens than for *S* and *T* ones where an additional set of slip bands were observed just after change in strain direction. This extra set of slip bands is necessary to accommodate intergranular stresses generated between neighboring grains after change in direction of external load with respect to pre-strain direction.

2.2.2. Microstructure evolution. The thin foils for TEM studies were cut out from the same specimens for which the slip analysis was performed. The observations were carried out at magnifications up to

20000x. In Figure 4 the microband crossing the dislocation structure is presented. This element of microstructure is typical for *S* and *T* specimens and is not observed in case of *L* one. Micro-shear bands cut across the grain boundary are seen in Figure 5. This deformation mechanism is characteristic for advanced stages of deformation during monotonic straining. The micro-shear bands were observed in *T* specimens only (strained perpendicularly to pre-strain direction).

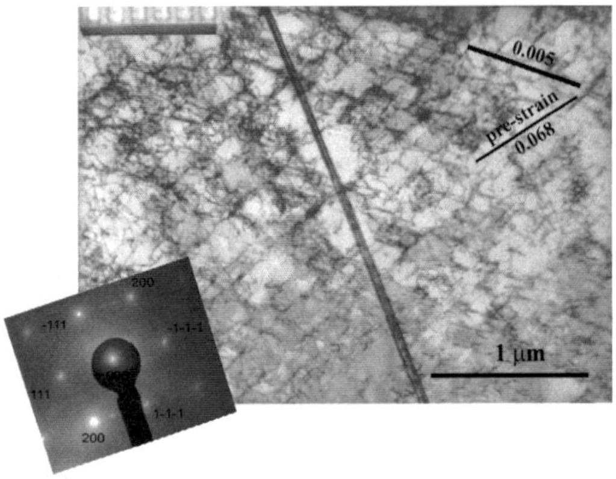

Figure 4. Microband crossing the dislocation structure in the *S* specimen.

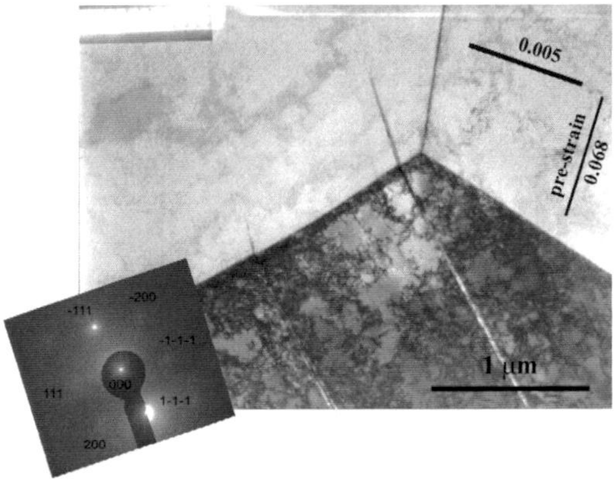

Figure 5. Micro-shear bands crossing the grain boundary in the *T* specimen.

Identification of energy storage rate components. Theoretical and experimental approach

W Oliferuk[1,2], **M Maj**[1]

[1] Fundamental Technological Research, Polish Academy of Sciences, Pawinskiego 5b, 02-106 Warszawa, Poland
[2] Białystok Technical University, Wiejska 45A, 15-351 Białystok,

wolif@ippt.gov.pl

Abstract. The subject of the present paper is decomposition of energy storage rate into terms related to different mode of deformation. The stored energy is the change in internal energy due to plastic deformation after specimen unloading. Hence, this energy describes the state of the cold-worked material. Whereas, the ratio of the stored energy increment to the appropriate increment of plastic work is the measure of energy conversion process. This ratio is called the energy storage rate. Experimental results show that the energy storage rate is dependent on plastic strain. This dependence is influenced by different microscopic deformation mechanisms.
It has been shown that the energy storage rate can be presented as a sum of particular components. Each of them is related to the separate internal microscopic mechanism. Two of the components are identified. One of them is the storage rate of statistically stored dislocation energy related to uniform deformation. Another one is connected with non-uniform deformation at the grain level. It is the storage rate of the long range stresses energy and geometrically necessary dislocation energy. The maximum of energy storage rate, that appeared at initial stage of plastic deformation is discussed in terms of internal micro-stresses.

1. Introduction

When metals deform plastically an energy conversion occurs; a part of the mechanical work w_p done during the single cycle of plastic straining is converted into a heat q_d. The rest of it remains in the strained material and it is known as the stored energy e_s.

$$e_s = w_p - q_d, \tag{1.1}$$

where e_s, w_p and q_d are specific quantities.

The stored energy characterises the cold-worked state of the deformed material and it represents the change in internal energy measured at external stress free state. Thus, the stored energy depends on deformation history. The measure of energy conversion at each instant of the deformation process is the rate of energy storage de_s/dw_p. It has been shown experimentally, that at the initial stage of plastic deformation of annealed polycrystalline materials the dependence of the de_s/dw_p on the work w_p expended during plastic deformation has a maximum [1, 2]. Such results can be related to different influence of particular micro-scale deformation mechanisms on the energy stored rate at different deformation stage. Thus de_s/dw_p is macroscopic quantity that depends on number of internal parameters H_1, H_2, ..., H_n. Each of them is related to individual mechanism or deformation mode. Since, there is a coupling between different mechanisms, the function can not be, in general, written in the form: $\sum_{i=1}^{n} e_s^{(i)}(H_i)$. The similar remark concerns also the energy storage rate. The question appears. Is it possible to distinguish the influence of the given internal parameter on the rate of energy storage? In order to find the answer, the theoretical analysis of energy storage rate, on the basis of phenomenological thermodynamics of plastic deformation has been performed. The results of the analysis have been used to decompose the total energy storage rate, obtained for the initial stage of uniaxial tension of austenitic steel, into two components. The results of our previous work [4] and work by Szczepiński [5] were used for such decomposition.

2. Decomposition of the energy storage rate

In the course of loading plastic strain is a monotonic function of time t, $\varepsilon^P(t)$. Hence, the time t can be replaced by the plastic strain ε^P. In effect, all internal parameters H (independently of their physical meaning) are some functions of ε^P,

$$H = H\left(\varepsilon^p\right). \tag{2.1}$$

The difference between incremental plastic work and stored energy can be presented in the form:

$$dw_p - de_s\left(H\right) = dw_p\left(1 - Z\right),\ Z = \frac{de_s\left(H\right)}{dw_p}, \tag{2.2}$$

where Z is the total energy storage rate. It is quantity describing energy conversion at a given instant of deformation. For infinitesimal increment of stored energy:

$$de_s\left(H\right) = \sum_{i=n}^{n}\left(\frac{\partial e_s}{\partial H_i}\right)dH_i. \tag{2.3}$$

Taking into consideration that H is function of ε^p and $dw_p = \frac{1}{\rho_0}\tau d\varepsilon^p$, the energy storage rate Z, during uniaxial tensile deformation, can be written in the form:

$$Z = \frac{de_s}{dw_p} = \frac{\rho_0}{\tau}\cdot\frac{de_s}{d\varepsilon^p} = \frac{\rho_0}{\tau}\cdot\sum_{i=1}^{i=n}\frac{\partial e_s\left(H\right)}{\partial H_i}\cdot\frac{dH_i}{d\varepsilon^p} = Z_1 + Z_2 + \ldots Z_n, \tag{2.4}$$

τ is Kirchhoff stress.

From (2.4) it follows that the rate of energy storage can be presented as a sum of particular components. Each of them describes the contribution of some microscopic mechanism (that is represented by H_i) to the energy conversion.

3. Determination of the energy storage rate

The energy storage rate can be obtained differentiating the stored energy e_s as a function of the plastic work w_p.The experimental method of stored energy determination, as in the previous works by Oliferuk et al. [2 - 4], was employed. The work w_p was derived from the load-elongation curve. The dissipated energy (equal to q_d, see Eq. 1.1) was determined by simulating the process of specimen heating during deformation using a controlled electrical power supply (related to the unit mass) in such a way that the temperature increase over time during the simulation was identical to that measured during tensile testing. When the straining and the simulation are conducted under identical conditions then the heat, which would have been transferred to the surroundings if the temperature of the unloaded sample has returned to the initial value, is the same in both cases and can be calculated from the measurable electrical quantities. The method gives the same order of magnitude of the stored energy as that obtained by X – ray and calorimetric measurements [6 - 8].

3.1. Components of the energy storage rate
The stored energy is the energy of stress field that remains in the deformed material after unloading. It can be divided into at least two parts: the energy of stress field connected with uniform and non-uniform deformation at the grain level. Thus, the energy storage rate can be also divided into two appropriate terms. The stored energy E_{s1} connected with non uniform deformation is a sum of a lattice stretch energy, energy of geometrically necessary dislocations and the energy of long-range internal stresses due to heterogeneous distributions of dislocations in the gauge part of the specimen. In our previous work [4] it has been shown that during uniaxial tension this part of the stored energy can be estimated from experimentally obtained load–displacement curve and it is equal to the AEB area (Figure 1). This energy can be calculated also from stress-strain curve. The specific energy e_{s1} is equal to the product of the $A'E'B'$ area and mass density ρ_0 of a tested material (Figure 2).

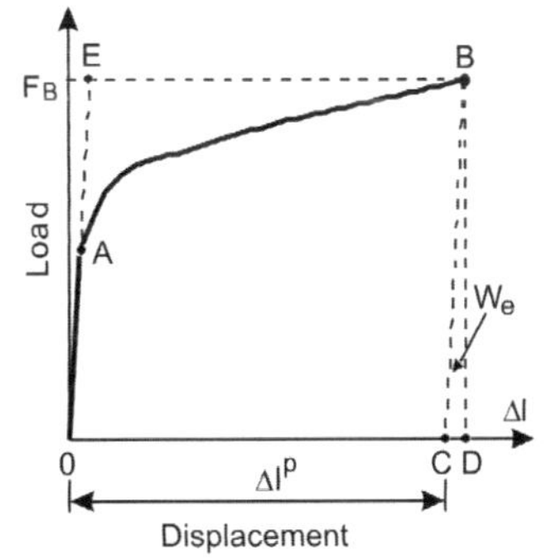

Figure 1. Generalized load versus generalized displacement curve. W_e is elastic energy.

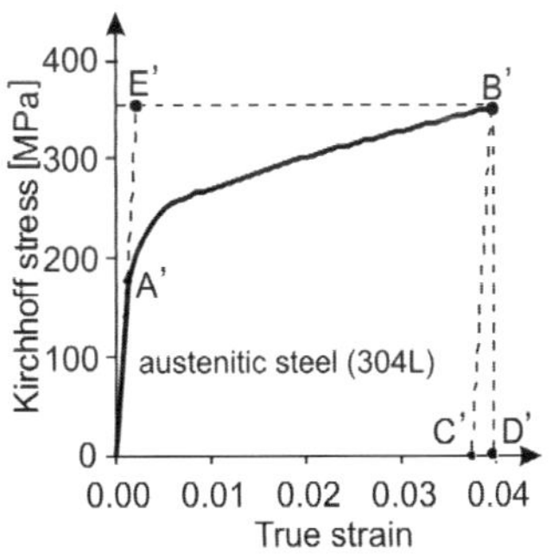

Figure 2. Stress-strain curve at the initial stage of tensile deformation of austenitic steel.

From the load-displacement dependence the E_{s1} as a function of plastic work can be determined and then $Z_1 = de_{s1} / dw_p$ can be calculated. Therefore, the remaining term $Z_2 = de_{s2} / dw_p$ of the total energy storage rate is related to uniform deformation. e_{s2} is the specific energy of statistically stored dislocations accumulated by their mutual trapping. It can be determined as a difference between experimentally measured total energy storage rate de_s / dw_p and the energy storage rate de_{s1} / dw_p related to non-uniform deformation.

4. Experimental procedure and results
The experiments were performed on the 304L austenitic stainless steel with mean grain size 7 µm. All specimens were strained using the MTS testing machine at the constant strain rate $\dot{\varepsilon} = 4{,}3 \cdot 10^{-3}\ s^{-1}$. During the tensile test the temperature distribution on the surface of the specimen was measured by IR camera. Simultaneously, the stress and strain were determined. Figure 2 shows the typical stress-strain curve for the tested steel. On the basis of the thermo-mechanical characteristics, the stored energy as a function of the true strain was determined. It was assumed that before plastic deformation, at annealed state, the energy stored in the specimen is zero. The experimentally determined total stored energy as a function of the plastic work for the tested material is presented in Figure 3. In the same figure the part e_{s1} of the stored energy calculated from the stress-strain curve is shown. Differentiating the total stored energy e_s, and the stored energy e_{s1} as a function of plastic work, the total energy storage rate

$Z = de_s / dw_p$ and the $Z_1 = de_{s1} / dw_p$ are determined. Results of such operation is shown in Figure 4. According to result of the theoretical analysis (see paragraph 3.1) the rate Z_2 of energy accumulated in statistically stored dislocations was calculated as the difference between Z and Z_1.

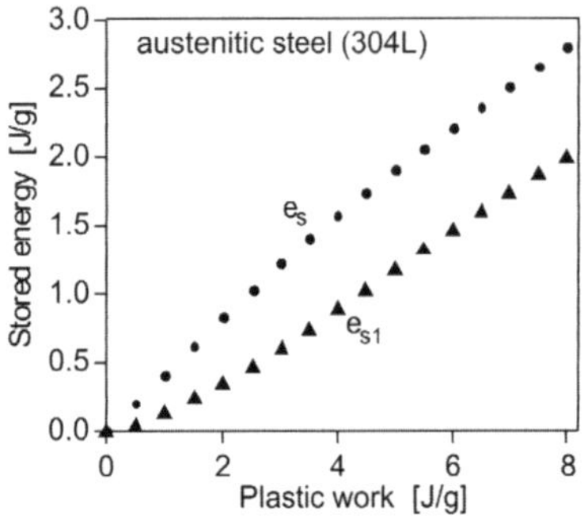

Figure 3. The total stored energy e_s and the energy e_{s1} connected with non-uniform deformation as a function of plastic work at the initial stage of tensile deformation.

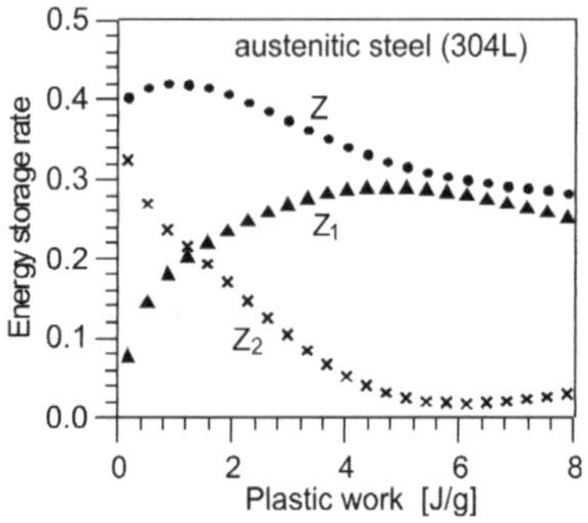

Figure 4. Decomposition of the total energy storage rate Z into two components Z_1 and Z_2. $Z = Z_1 + Z_2$.

5. Discussion

It is seen that the maximum of the total energy storage rate (Z) is mainly determined by the energy storage rate (Z_1) related to non-uniform deformation at the grain level (Figure 4). Just after loading, different orientation of grains leads to the incompatible slip, what generates long-range internal stresses and Z_1 increases. To ensure compatible deformation of different parts of plastically non-uniform material geometrically necessary dislocations (GND-s) are generated, what reduces the existing long-range internal stresses and creates a new ones, because distribution of GND-s is non-uniform [9, 10]. This means that plastic deformation is a continuous process ensuring compatible strain of different material part until deformed specimen will be broken. Microstructure evolution during plastic deformation tends to formation of low energy dislocation structures (LEDSs) what causes a decline of the energy storage rate [2, 11]. At the initial stage of plastic deformation of the tested steel, Z_2 manifests such tendency. But it should be noticed that formation of LEDSs concerns also GND-s.

Acknowledgements

This work has been supported by Council of Science (Poland) under Grant No. N N501 0094 33 (2007-2010).

References

[1] Chrysochoos A, Maisonneue O, Martin G, Caumon H 1985 *Nucl. Eng. Design* **114** 323
[2] Oliferuk W, Świątnicki W A, Grabski M W 1991 *Mater. Sci. Eng.* A **161** 55
[3] Oliferuk W, Świątnicki W A, Grabski M W 1995 *Mater. Sci. Eng.* A **197** 49
[4] Oliferuk W, Maj M 2009 *Eur. J. Mech. A/Sol.* **28** 266
[5] Szczepiński W 2001 *Arch. Mech.* **53** 615
[6] Mandal D, Baker I 1995 *Scr. Metall. Mater.* 4 **40** 645
[7] Borbély A, Driver J H, Ungár T 2000 *Acta Mater.* **48** 2005
[8] Cao W Q, Gu C F, Pereloma E V, Davies C H J 2008 *Mater. Sci. Eng.* A **492** 74
[9] Ashby M F 1970 *Phil. Mag.* **21** 399
[10] Bay B, Hansen N, Hughes D A, Kuhlmann-Wilsdorf D 1992 *Acta Metall. Mater.* 2 **40** 20
[11] Kuhlmann-Wilsdorf D 1989 *Mater. Sci. Eng.* A **197** 1

Cavitation failure during superplastic tensile loading in alumina-base ceramics

K Hiraga[1], B-N Kim[1], K Morita[1], H Yoshida[1] and M Tabuchi[1]

National Institute for Materials Science, Ibaraki 305-0047, Japan

hiraga.keijiro@nims.go.jp

Abstract. Tensile failure in superplastic alumina-base materials shows a close similarity to ductile failure in metallic materials containing voids or hard inclusions. The occurrence of microcracks, which is followed by macroscopic cracking towards final failure, cannot be explained from the brittle propagation of preexistent defects, but from the coalescence of intergranular cavities formed and grown by superplastic deformation. The cavity coalescence is shown to occur as the cavity separation distance decreases to a certain level owing to an increase in the number and size of the cavities.

1. Introduction

Although earlier studies on superplastic ceramics have often reported that tensile failure occurs through cracking in a direction perpendicular to the stress axis [1-7], the mechanisms of such cracking have still been unclear. For example, the causes of cracking assumed in the studies are essentially different from each other. Kim *et al.* [1] proposed that cracking is caused by the brittle propagation of preexistent defects and the propagation occurs from the beginning of tensile deformation. From microstructural features after deformation, Schissler *et al.* [2] and Chokshi *et al.* [3] associated the cracking with extensive interlinkage among adjacent cavities. There have, however, been few quantitative studies on these assumptions and on the details of tensile failure in superplastic ceramics. The present study was undertaken to examine this issue by using alumina-base materials under different combinations of the initial grain size, tensile rate and deformation temperature. Emphasis was placed on the onset conditions of cracking.

2. Experimental methods

The following materials were synthesized from high purity oxide powders (> 99.97%) by methods described elsewhere [5]: a monolithic alumina doped with 0.2 wt% MgO and a composite alumina dispersed with 10 vol% ZrO_2 particles. For relative densities larger than 99%, the grain size of the alumina phase (defined as 1.56 times the average intercept length) was adjusted to d_0 = 0.7-2.8 μm and 0.45-2.0 μm for the MgO-doped and the ZrO_2-dispersed materials, respectively. From these materials, flat tensile specimens with gauge dimensions of ^{l0}10-^{t0}3-^{w0}3.5 mm were machined and pulled to failure or to prescribed strains with an Instron-type machine. The ranges of the loading temperature and the strain rate were 1400-1550 °C and 2×10^{-4}-3×10^{-3} s^{-1}, respectively. After loading, a surface layer of 300-500 μm in thickness was removed from the gauge portion for polishing. By using quantitative optical microscopy [5] of which resolution limit was set to 0.5μm, the area-equivalent circle diameter, the major and minor axis lengths and the aspect ratio (the major to the minor axis ratio) of each cavity and the volume fraction of cavity damage were measured at the center of the as-polished gauge portion.

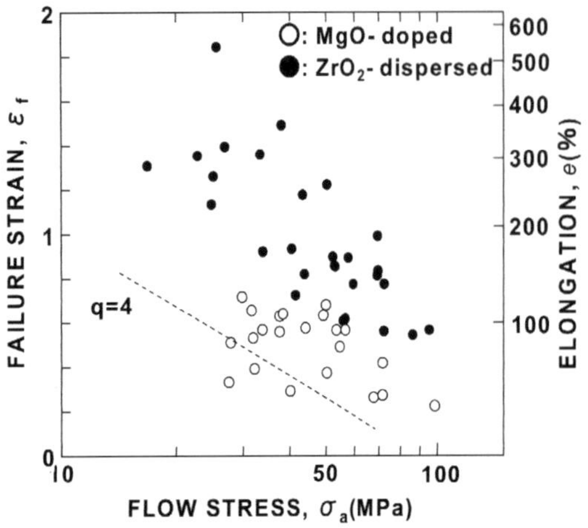

Figure 1 Tensile elongation as a function of the average flow stress σ_a, showing large scattering against a given level of flow stress.

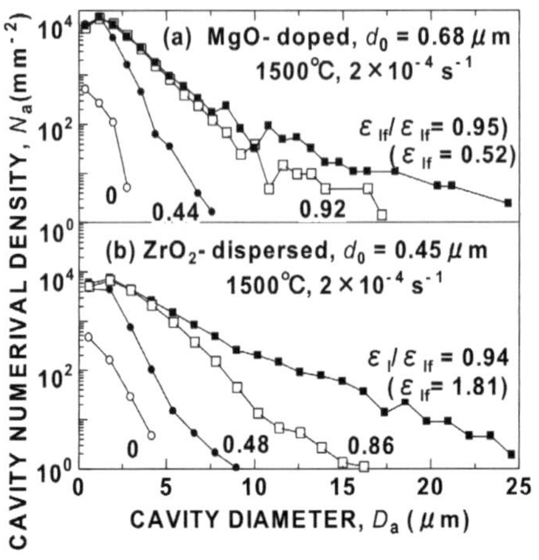

Figure 2 Typical evolution of cavity-size distribution with strain. The interval of cavity size is 0.8 μ m.

3. Examination of the brittle crack-propagation model

Figure 1 summarizes the tensile ductility of the present material as a function of the average flow stress, $\int\sigma(\varepsilon)d\varepsilon/\varepsilon_f$, where $\sigma(\varepsilon)$ is the stress at a strain of ε and ε_f is the failure strain. There is a trend of decreasing tensile ductility with increasing flow stress. Such a trend that has often been noted for superplastic ceramics [1-4] is reasonable since higher stresses may generally accelerate primary processes of failure. However, tensile ductility shows a large scatter for a given flow-stress level that can be attained by different combinations of the grain size and the loading conditions. Thus, the data show that tensile ductility cannot be related simply to the flow-stress level.

Figure 1 also indicates that the tensile failure of the present materials cannot be explained from the brittle crack-propagation model [1] given by equation (1).

$$\varepsilon_f = \ln\left\{\left[\frac{(2\gamma_s - \gamma_{gb})E}{\pi C_0}\right]^{1/q}\sigma^{-2/q}\right\}, \qquad (1)$$

where ε_f is the fracture strain, γ_s is the surface energy, γ_{gb} is the grain-boundary energy, E is the elastic modulus, C_0 is the initial crack size related to the initial grain size and q is the rate constant of a crack-growth law given by $dC/d\varepsilon = qC$ and it takes a value of about 4 [1]. For a given chemical composition, namely a given combination of γ_s, γ_{gb} and E, Eq. (1) predicts a linear relationship between ε_f and $\ln\sigma$ with a slope of $-2/q$. However, the ε_f–σ_a data given in figure 1 show a large scatter on the logarithmic scale. A q-value of about 4 that appears for a certain set of the data does not support the brittle cracking model. A study on cavity growth in the present materials [7] showed that q is smaller than 2.0 except for a last stage of deformation where the tensile strain exceeds 90% of the failure strain. This fact is contrary to such a premise of the model [1] that steady crack growth with $q \sim 4$ occurs from the beginning of loading.

4. Evolution of cavitation damage relating to tensile failure

Figure 2 shows the evolution of cavity size distribution as a function of the local strain, ε_l, at which the measurement was made. The size distribution shifts towards higher densities and larger sizes with increasing strain, showing continuous cavity formation and growth with loading. The evolution of cavity damage in the present materials was similar for a given value of $\varepsilon_l/\varepsilon_{lf}$, where ε_{lf} is the local strain at the failed tip. For $\varepsilon/\varepsilon_f < \sim 0.9$, cavity coalescence was limited or localized and most cavities were rather equiaxed with diameters of < 10 μm and aspect ratios of < 2.0. Microcracks (figure 3(a)), which

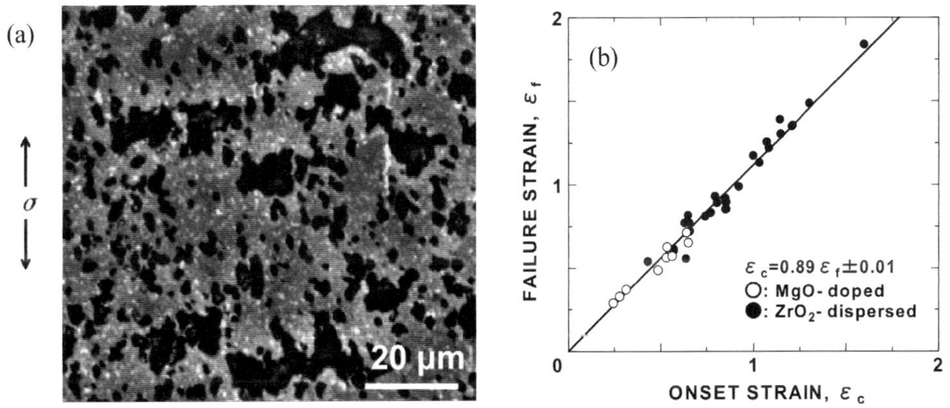

Figure 3 (a) Microcracking observed in the ZrO_2-dispersed material loaded at 1500 °C and at $2\times10^{-4}s^{-1}$ to $0.93\varepsilon_{lf}$ and (b) the relationship between the onset strain of microcracking and the failure strain.

were defined as cavities having major axis lengths of ≧ 20 μm and aspect ratios of ≧ 2.0 in this study, did not appear until the value of $\varepsilon/\varepsilon_f$ exceeded ~ 0.9. For $\varepsilon/\varepsilon_f$ > ~ 0.95, the size and the number of microcracks increased steeply, and some of them interlinked or grow in the direction normal to the tensile axis, resulting in macroscopic cracks for final failure. Such a close relationship between microcracking and tensile failure is explicitly confirmed in figure 3 (b), where the failure strain is proportional to the onset strain of microcracking with a slope of about 0.9.

5. Microcracking through cavity coalescence

We examined the onset of microcracking using a model for cavity coalescence through necking of the matrix among adjacent cavities [8], since the microstructural observation strongly suggests that this mechanism may work in the present material (figure 3(a)). The model predicts that cavity coalescence, namely microcracking occurs when the following condition is fulfilled:

$$\frac{P(1-f)}{1-f_0}=\frac{\sigma_m}{Y}+\frac{3+\nu}{3\sqrt{\nu^2+3}}, \qquad (2)$$

where P is the plastic constraint factor, σ_m is the mean normal stress, Y is the uniaxial yield stress of the matrix and ν is the Lode variable and f and f_0 are the current and the initial volume fractions of cavities, respectively. In the present materials, f_0 is in a range of 6×10^{-4} - 2×10^{-3}. The plastic constraint factor can be calculated from cavity data as $P = 0.1/(D_{av}/\lambda)^2+1.2/(D_{av}/L_{av})^{1/2}$, where D_{av} and L_{av} are the average diameter and the center to center spacing of cavities, respectively, and $\lambda = L_{av} - D_{av}$ is the cavity-separation distance. The equation shows that the P-value decreases with increasing cavity number and size.

As typically shown in figure. 4, there is a close relationship between the evolution of the plastic constraint factor and the onset of microcracking. That is, microcracking starts to occur as the value of P decreases from the initial range of 209-216 to a narrow range of 2.1-2.9 owing to an increase in the number and size of cavities (figure 2). This relationship was found both in the MgO-dope and in the ZrO_2-dispersed materials independently of the combination of the initial grain size, temperature and strain rate. From this result, we examined equation (2) as follows. Assuming the isotropic growth of spherical cavities as a first-order approximation, and considering that little necking occurs in the gauge portion of superplastic ceramics and ν is unity for uniaxial tension, equation (2) is simplified as [9]:

$$P(1-f)\approx 1. \qquad (3)$$

The values of $P(1-f)$ at the onset of microcracking.were calculated by using the cavity-size distributions and f = 0.11-0.33. The calculated values are plotted in figure 5 as a function of the failure strain, ε_f.

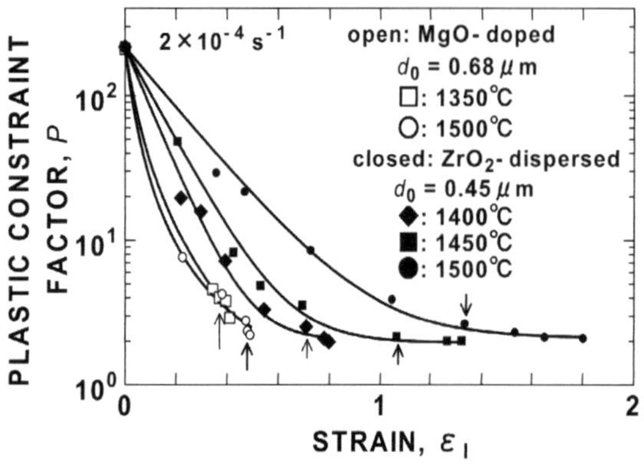

Figure 4 Typical evolution of the plastic constraint factor as a function of strain. The arrows indicate the onset points of microcracking.

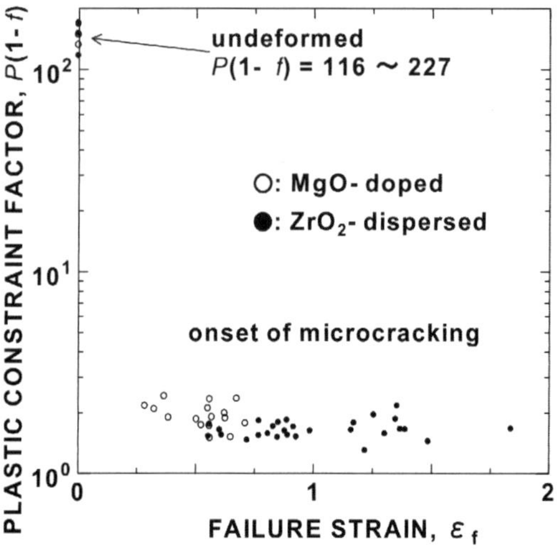

Figure 5 Plastic constraint factor for the onset of microcracking as a function of the failure strain.

Figure 5 indicates that the cavity-coalescence model works well for the onset of microcracking in the present materials. The value of $P(1\text{-}f)$ is decreased from the initial range of 116-227 to a narrow range of 1.9 ± 0.3 (MgO-doped) and 1.7 ± 0.2 (ZrO_2-dispersed) at the onset of microcracking, independently of the combination of the initial grain size, strain rate, temperature and failure strain. The $P(1\text{-}f)$ values slightly larger than unity at the onset of microcracking may arise mainly from the simplification of the cavity shapes. The result shown in figure 5 is in contrast to the large scatter appearing in the ε_f- σ_a relationship (figure 1) and leads to a conclusion that the primary process of tensile failure in the superplastic alumina is cavity coalescence due to internal necking of the matrix among adjacent cavities.

6. Conclusions

1. Tensile ductility in superplastic alumina cannot be explained only from the level of flow stress: failure strain ε_f shows a large scatter for a given flow stress level attained by different combinations of the strain rate, loading temperature and grain size.
2. Microcracking, which appears as the strain exceeds $\sim 0.9\varepsilon_f$ and leads to macroscopic cracking for final failure, does not occur through a brittle process, but through a plastic process of cavity-coalescence. Analysis of cavitation data shows that the microcracking starts to occur as the cavity-separation distance decreases to a certain level owing to an increase in the number and size of cavities.

Acknowledgement

This work was supported by a Grant-in-Aid for Scientific Research (B) 21360328.

References

[1] Kim W-J, Wolfenstine J and Sherby O D 1991 *Acta Metall. Mater.* **39** 199.
[2] Schissler J D, Chokshi A H Nieh T G and Wadsworth J 1991 *Acta Metall. Mater.* **39** 3227.
[3] Chokshi A H, Nieh T G and Wadsworth J 1991 *J. Am. Ceram. Soc.* **74** 869.
[4] Yoshizawa Y and Sakuma T 1992 *Acta Metall. Mater.* **40** 2943.
[5] Hiraga K, Nakano K, Suzuki T S and Sakka Y 2002 *J. Am. Ceram. Soc.* **85** 2763.
[6] Hiraga K, Kim B-N, Morita K, Yoshida H, Suzuki T S and Sakka Y 2007 *Sci. Technol. Adv. Mater.* **8** 578.
[7] Hiraga K, Nakano K, Suzuki T S and Sakka Y 1999 *Mater. Sci. Forum* **304-306** 431.
[8] Thomason 1985 *Acta Metall.* **33** 1087.
[9] Hiraga K and Nakano K 2004 Z. Metallkde. 95 559.

Plasticity and X-ray Line Profile Analysis of the Semicrystalline Polymer Poly(3-hydroxybutyrate)

F. Spieckermann[1], H. Wilhelm[1,2], E.Schafler[1], M. Kerber[1], S. Bernstorff[3] and M.J. Zehetbauer[1]

[1] Research Group Physics of Nanostructured Materials, Faculty of Physics, University of Vienna, Boltzmanngasse 5, Wien, Austria
[2] Laboratory of Polymer Engineering LKT-TGM, Wexstrasse 19-23, 1200 Wien, Austria
[3] Sincrotrone ELETTRA Trieste, Strada Statale 14 - km 163,5 in AREA Science Park, 34012 Basovizza, Trieste, Italy

florian.spieckermannunivie.ac.at

Abstract. The evolution of the microstructure during compressive deformation of the biodegradable polymer poly(3-hydroxybutyrate) (P3HB) was investigated in-situ via X-ray diffraction using synchrotron radiation. Flow curves were measured in-situ together with X-ray profiles for several degrees of deformation. The profiles were analysed using Multi-Reflection X-ray Line Profile Analysis (MXPA) adapted by the authors for semicrystalline polymers providing lamella thickness, crystallinity, and the presence and density of dislocations as a function of the deformation. In contrast to previous investigations in α crystallised isotactic polypropylene (α-iPP), P3HB does not exhibit a deformation induced increase of the dislocation density which suggests mechanisms other than dislocations to be involved in plastic deformation of P3HB.

1. Introduction

In a previous paper on α crystallised isotactic polypropylene (α-iPP) [1] it has been demonstrated that the use of Multi-Reflection X-ray Profile Analysis (MXPA) as developed by Ungár and co-workers [2, 3] allows for the proof of the existence of dislocations in semicrystalline polymers. It was also possible for the first time to measure the density of dislocations quantitatively, which in the case of α-iPP was found to increase with increasing plastic deformation.

In general, the mechanical properties and particularly the yield stress of semicrystalline polymers seem to be governed by the size of the crystalline lamellae rather than by the crystallinity [4]. For the mechanism related, Flory and Yoon [5] and Balta-Calleja and Cruz [6] suggested that plastic deformation in semicrystalline polymers is driven by localised melting and recrystallisation under adiabatic conditions. However, this concept is not able to account for the specific change in yielding behaviour for larger lamella thicknesses as it was observed experimentally in semicrystalline polyethylene [4]. Contrary to this concept, Argon et al. [7] suggested that the mechanisms in question involve dislocations i.e. by the thermally activated generation of dislocations on the surfaces of nano-sized lamellae, and by a dislocation half-loop mechanism for larger lamellae. One strategy to shed more light on this problem is to investigate other types of semicrystalline polymers especially by means of MXPA which is capable of analysing the presence and density of dislocations on the one hand, and the lamella thickness on the other.

Journal of Physics: Conference Series **240** (2010) 012146 doi:10.1088/1742-6596/240/1/012146

Recently the polyester poly(3-hydroxybutyrate) (P3HB) produced by microorganisms experienced some attention of the industrial community due to its high biodegradability. As the mechanical properties and the processibility are comparable to those of polypropylene (PP) this material qualifies for a number of future applications especially for medical and agricultural purposes.

From a scientific point of view, P3HB is particularly interesting for MXPA measurements because unlike many other semicrystalline polymers (a) its crystal system is simple orthorhombic and (b) its wide angle X-ray scattering (WAXS) pattern shows a large number of high intensity peaks [8], which is important for the efficiency of the MXPA. Moreover, the lamella thickness of P3HB is as small as 7 nm which should strongly affect the mechanical properties, and/or the generation dynamics of dislocations in case of their existence. Furthermore, there is only little data in the literature on the evolution of the microstructure of P3HB during deformation. Therefore the experiments presented in this paper aim at the presence of dislocations and potentially related deformation mechanisms in P3HB.

2. Experiments

2.1. Sample Preparation

P3HB rods were purchased from Goodfellow Cambridge Limited. For compression tests cylindrical samples with a height of 10 mm and a diameter of 6 mm were machined.

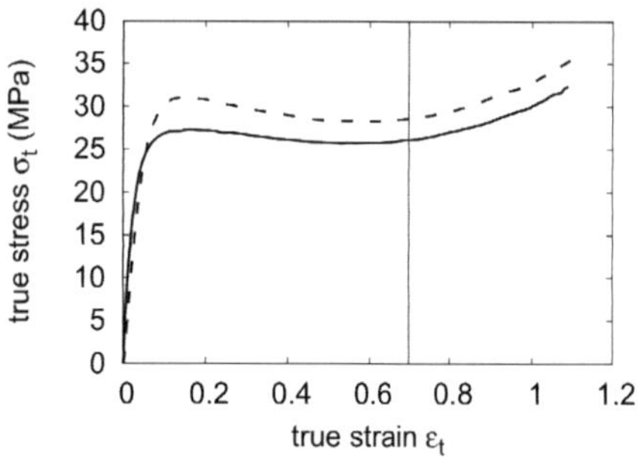

Figure 1. Compression true stress-true strain curve of P3HB for strain rates of $\dot{\epsilon}_t = 4 \cdot 10^{-3}$ s^{-1}(- - -) and $\dot{\epsilon}_t = 4 \cdot 10^{-4}$ s^{-1}(——). After yielding a small softening is followed by a strong hardening beyond $\epsilon_t \approx 0.7$ (vertical line).

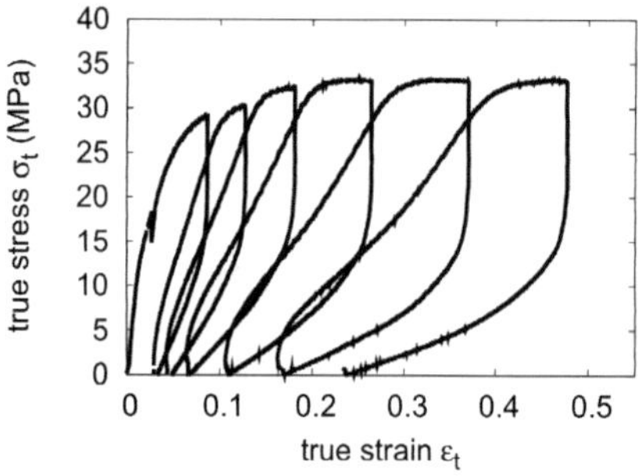

Figure 2. Compression true stress-true strain curve of P3HB for cyclic loading-unloading of an in-situ MXPA-measured sample. The large opening of the cycles is an indication for strong activity of the amorphous phase.

2.2. Mechanical Experiments and In-situ Diffraction Experiments

Uniaxial compression tests were carried out at room temperature on a Shimadzu AG50 deformation machine. Geometric stress-strain correction of the compressed samples was ensured by a video extensometer capturing the maximum bulging diameter with a digital camera. The curves resulting from deformation at two different strain rates are shown in figure 1.

In-situ WAXS measurements during plastic deformation were performed in transmission setup at the SAXS-Beamline 5.2L of Sincrotrone ELETTRA Trieste. For this purpose a specially designed miniature compression machine was used. The stress-strain curve of a cyclic loading unloading experiment is depicted in figure 2. The photon energy used was 8 keV which corresponds to a wave length of CuK_α radiation of 0.154 nm. The incident beam had a spot size of 200 μm × 500 μm on the sample. The photon flux amounted to $5 \cdot 10^{11}$ photons $mm^{-2}s^{-1}$. The WAXS spectra were recorded with a linear position-sensitive detector (1024 channels, type PSD 50 of Braun, Munich, Germany) at a distance of 370 mm between the detector and the specimen. To ensure sufficient statistics for a reliable evaluation of the WAXS profiles, at least 10^4 counts were collected in the maxima of the diffraction peaks. The recorded profiles

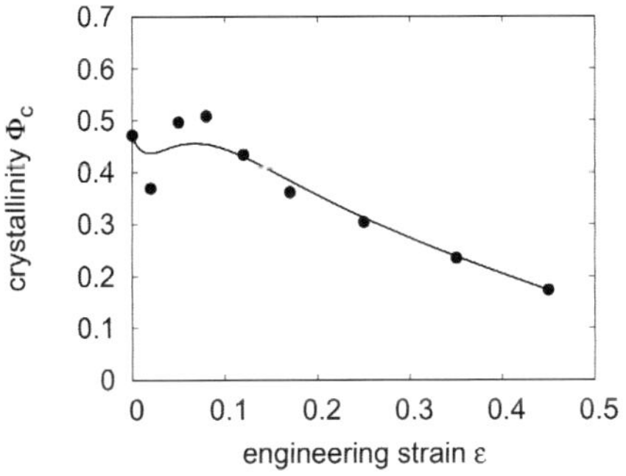

Figure 3. The evolution of crystallinity (●) as a function of the engineering strain as derived from the deconvolution of the amorphous and the crystalline WAXS signals under compressive load. The line (——) is added to guide the eye.

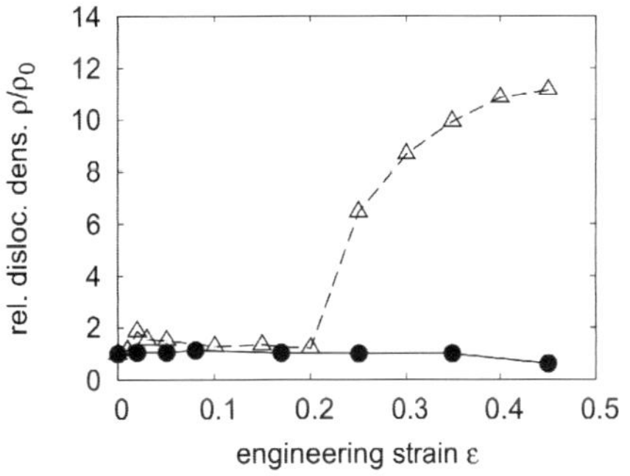

Figure 4. Strain characteristics of the relative dislocation density in uniaxial compression for P3HB (●) and α-iPP (△) as a function of the engineering strain. In-situ diffraction patterns were recorded under load.

were evaluated by means of the Multi-Reflection X-ray Line Profile Analysis (MXPA) [2, 3]. The evaluation procedure followed three steps, starting with the background determination and peak separation. After this a pre-fit of the components of the average dislocation contrast factor was done by evaluation of the modified Williamson-Hall plots. These values were then used as starting values for the evaluation with the program CMWP-fit [9, 10]. Details of this procedure are given in the papers by Wilhelm et al. [1] and Kerber et al. [11].

3. Results and Discussion

Surprisingly the change of the profile broadening induced by plastic deformation found in P3HB seems to be very small. Earlier investigations of X-ray line broadening in α-iPP [1] revealed a strong deformation induced increase of the broadening. This could be related to the formation of dislocations while the coherently scattering domain size (CSD) (which is also responsible for an important part of the broadening) remained nearly constant. Nevertheless the CSD was found to correspond very well to the lamella size of α-iPP as determined by DSC or SAXS [12, 13]. In the present investigation of P3HB, a stronger scattering for the CSD in comparison with α-iPP is found which results in a CSD between 6-11 nm. The broadening related to dislocations, however, does not seem to be affected by deformation (figure 4). Although in absolute terms the initial dislocation density turns out to be even higher in P3HB ($6.5 \cdot 10^{16}$ m^{-2}) than in α-iPP ($3.8 \cdot 10^{16}$ m^{-2}[1]) it does hardly change as a function of the compressive strain applied (figure 4). This contrast to the observations in α-iPP indicates that in P3HB a different deformation mechanism is active. Different explanations may account for the unexpected behaviour of P3HB:

(i) The dislocation density is high enough to accommodate the crystallographic slip, and generation and annihilation of dislocations may be in balance.

(ii) Intralamellar slip occurs with a mechanism other than dislocation motion, e.g. localised adiabatic melting [5] without mobilisation of dislocations.

(iii) Interlamellar processes are responsible for the main part of the deformation due to a high stiffness of the crystallites. In this case the lamellar crystallites would fragment before crystallographic slip occurs. Like mechanism (ii), also this mechanism would not involve dislocation motion.

As the interaction between dislocations in polymers is expected to be low [4] the hardening curve in figure 1 cannot be used to decide between the three processes. The opening of the cycles shown in figure 2 increases with deformation. According to Hiss et al. [14] this can be attributed to the viscoelastic properties of the amorphous phase. The strong reduction of the crystalline volume fraction plotted in figure 3 is in accordance with this observation, since it suggests a stronger influence of the amorphous phase on the relaxation and deformation behaviour of the material. Both observations support the occurrence of mechanisms (ii) and (iii). In particular, the considerable amount of kinematic hardening which can be seen in figure 2, and the marked brittleness of the bulk material hint at the formation of cracks as suggested by mechanism (iii).

Nevertheless, the real reasons for the occurrence of a deformation mechanism in P3HB, which is different to that operating in α-iPP remain unclear. Future measurements of the deformation induced texture might prove useful in order to identify possibly activated slip systems. Transmission and/or scanning electron microscopy should give insight in how far fragmentation of the crystallites occurs with deformation.

4. Summary and Conclusions

With the present experiments, using the method of Multiple X-ray Profile Analysis, for the first time the existence of dislocations in melt-crystallised P3HB was shown. A considerable initial dislocation density of $6.5 \cdot 10^{16}\text{m}^{-2}$ was found. However, a deformation-induced change in the dislocation density has not been observed in P3HB which suggests one of the following microstructural deformation mechanisms to occur: (i) a balance of dislocation annihilation and generation, (ii) deformation-induced adiabatic melting, and (iii) brittle fragmentation of the lamellae.

Aknowledgements

Support by the EU contract RII3-CT-2004-506008(IA-SFS) is gratefully acknowledged. Thanks are given to the University of Vienna for providing support within the PhD Program 'Experimental Materials Science – Nanostructured Materials' and the Focus Project 'Bulk Nanostructured Materials'.

References

[1] Wilhelm H, Paris A, Schafler E, Bernstorff S, Bonarski J, Ungar T and Zehetbauer M 2004 *Mater. Sci. Eng. A* **387-389** 1018
[2] Ungár T and Borbély A 1996 *Appl. Phys. Lett.* **69** 3173
[3] Ungár T, Gubicza J, Ribárik G and Borbély A 2001 *J. Appl. Cryst.* **34** 298
[4] Galeski A 2003 *Prog. Polym. Sci.* **28** 1643
[5] Flory P and Yoon D 1978 *Nature* **272** 226
[6] Balta-Calleja F and Cruz C 1996 *Acta Polym.* **47** 303
[7] Argon A, Galeski A and Kazmierczak T 2005 *Polymer* **46** 11798
[8] Bruckner S, Meille S, Malpezzi L, Cesaro A, Navarini L and Tombolini R 1988 *Macromolecules* **21** 967
[9] Ribárik G, Ungár T and Gubicza J 2001 *J. Appl. Cryst.* **34** 669
[10] Ribárik G, Gubicza J and Ungár T 2004 *Mater. Sci. Eng. A* **387-389** 343
[11] Kerber M, Schafler E, Wieczorek A, Ribárik G, Bernstorff S, Ungár T and Zehetbauer M 2009 *Int. J. Mater. Res.* **100** 770
[12] Spieckermann F, Wilhelm H, Schafler E, Aifantis E and Zehetbauer M 2009 *J. Eng. Mater. Technol.* **131** 011109–5
[13] Wilhelm H, Spieckermann F, Schafler E, Kerber M and Zehetbauer M *in preparation*
[14] Hiss R, Hobeika S, Lynn C and Strobl G 1999 *Macromolecules* **32** 4390

Bimodal grain size distributions in UFG materials produced by SPD: Their evolution and effect on mechanical properties

H.W. Höppel[1], M. Korn[1], R. Lapovok[2] and H. Mughrabi[1]

[1]University of Erlangen-Nürnberg, Department of Materials Science and Engineering, Institute I, Martensstr. 5, 91058 Erlangen, Germany
[2]Monash University, Materials Engineering, Building 19, Clayton 3800, Australia

mughrabi@ww.uni-erlangen.de

Abstract. The mechanical properties of bulk ultrafine-grained materials produced by severe plastic deformation can be modified (sometimes enhanced) by a mild annealing treatment which leads, in some cases, to a bimodal grain size distribution, characterized by a good combination of strength and ductility. Bimodal grain size distributions can also evolve during cyclic deformation at rather low homologous temperature. Here, the conditions under which bimodal grain size distributions evolve and how they affect the mechanical properties, as studied by the authors and as reported so far in the literature, will be reviewed and discussed .

1. Introduction

Bulk ultrafine-grained (UFG) metals produced by equal channel angular pressing (ECAP), the most widespread severe plastic deformation (SPD) technique, have attracted much attention during recent years because of their extraordinary strength properties [1]. As will be discussed in detail below, *bimodal grain size distributions* in UFG metals, consisting of regions with coarsened grains amidst the original UFG grain structure, have been found repeatedly in bulk ECAP-processed UFG material of copper, nickel and α-brass under different, frequently not well-defined circumstances, for example

- after cyclic deformation at room temperature,
- after long-time anneals at room temperature or short-time anneals at higher temperatures, and
- shortly after ECAP-processing.

The effect of bimodal grain size distributions on the monotonic and the fatigue strength properties has been beneficial in some cases but not in others. The aim of the present work is to present some new results, to clarify some seemingly contradictory findings, to identify in more detail the conditions under which bimodal grain structures evolve and to discuss their effects on the mechanical properties.

2. Occurrence of bimodal grain size distributions, effects on monotonic and fatigue strength

2.1. Evolution of bimodal grain size distributions during strain- and stress-controlled fatigue

In an early fatigue study on UFG copper (99.98%), an interesting observation was made, namely that, during fatigue in strain control at room temperature (RT), a strong cyclic softening occurred. At the same time, it was noted that the originally strongly ECAP-hardened and more or less homogeneous UFG grain structure had undergone a marked coarsening in locally confined regions, whereas the original UFG microstructure was retained in the remaining regions [2]. Similar results were obtained

Figure 1. Examples of bimodally coarsened UFG copper. a) TEM micrograph of "optimal" bimodal grain size structure in annealed high-purity (99.99%) UFG copper [3,4]. b) SEM micrograph of bimodal grain size structure in commercial purity (>99.95%) UFG copper, < 3 weeks after ECAP.

on fatigued high-purity (99.99%) UFG copper [3,4] in which clearly *bimodal grain size distributions* were found and, less clearly, on fatigued high purity (99.99%) nickel [5]. It follows that, in the as-ECAP-processed UFG copper, the grain structure is microstructurally unstable at RT, i.e. at a rather low homologous temperature of ca. 0.2. The process of *local* grain coarsening during deformation has been identified as thermally activated temperature- and strain-rate-dependent *dynamic recrystallization* [3,4]. It is remarkable that, in stress-controlled fatigue of high-purity UFG copper, almost no cyclic softening occurred and local grain coarsening was almost negligible [3,4].

Normally, in material of lower *purity*, grain coarsening and cyclic softening would be expected to be impeded. This seemed to be confirmed in work on commercial purity UFG copper [6], however after fatigue in stress control. Only very recently, the same authors found softening and bimodal grain coarsening in strain control. Thus, final conclusions concerning the effect of purity remain vague.

2.2 Introduction of bimodal grain size distributions by annealing to enhance the fatigue properties

Fatigued UFG copper has always been found to have larger fatigue lives, compared to those of conventional grain size (CG) copper in a Wöhler (S-N) plot but not in the Coffin-Manson plot that is typically used for LCF (Low Cycle Fatigue) representations [2,3,4]. Since it is known that good LCF performance requires a sufficiently high ductility [7] and in view of the lowered ductility of the strongly ECAP-hardened UFG material, the rather unsatisfactory LCF fatigue performance is understandable. In an attempt to remedy this deficiency, Höppel et al. [3,4], compare also [8], subjected their UFG copper of high purity (99.99%) to a mild annealing treatment in order to enhance the ductility at the expense of a moderate loss of strength. In a systematic search for the most suitable annealing treatment with respect to enhancing fatigue life, a 2 h anneal at 170°C was found to be *optimal*. In the annealed material, a *bimodal grain size distribution* was found, with coarse grains of some μm grain size embedded in the original UFG grain structure. With increasing annealing time, the volume fraction of coarse grains increased. Figure 1a shows, as an example, a TEM micrograph of the annealed UFG copper with the optimal bimodal grain size distribution. The annealed UFG copper with the bimodal grain size distribution exhibited a remarkable enhancement of fatigue lives by a factor of 7 in the Coffin-Manson plot and an overall excellent fatigue performance, implying that the optimal bimodal grain size distribution provides a good compromise between fatigue strength and ductility.

2.3 Evolution of bimodal grain size distributions during ageing at room or higher temperatures

It is not always possible to obtain bimodal grain structures by annealing. So far, in the case of bulk SPD-processed UFG metals, bimodal grain size distributions have only been encountered or produced in the case of copper, nickel and α-brass, while considerable efforts to introduce a bimodal grain structure by annealing in commercial purity UFG aluminium failed [8,10], since grain coarsening always occurred in a more or less homogeneous manner. The crucial factor seems to be the competition between recovery and recrystallization [9]. After strong and rapid initial recovery, favoured by high stacking fault energy, not enough driving force is left for subsequent recrystallization. This is probably the case encountered in aluminium.

For both UFG high-purity copper and UFG commercial purity aluminium, grain coarsening was noted after resting at room temperature for some months (up to a year) after ECAP-processing. However, whereas the typical bimodal grain size structure developed in the case of UFG copper, grain coarsening in UFG aluminium occurred more or less homogeneously as noted previously above.

2.4 Evolution of bimodal grain size distributions (shortly) after ECAP

Commercial purity copper, subjected to 8 ECAP passes, route B_c, at rather high ECAP deformation rate (5 mm/s and 10 mm/s) with back pressure [11], and inspected after a rest of some weeks at RT after ECAP-processing, revealed regions with bimodal grain structures. This was most marked in the centre of the billet (cross section 20 mm × 20 mm) and progressed with further RT-resting. An example of the locally coarsened grain structure is shown in Fig. 1b. The coarsening observed probably resulted from an *in situ* anneal due to the deformation heating during (rapid) ECAP-processing, being stronger in the centre than at the periphery because of less heat loss. A simple estimate predicts a temperature rise that is proportional to the ECAP deformation rate (unpublished).

3. Exploitation of bimodal grain size distributions to enhance monotonic and fatigue strength

As already discussed, beneficial effects of bimodal grain size distributions have been noted in some cases. Shortly after the work of Höppel et al. [3,4], Ma and his group succeeded in producing copper with an excellent combination of good monotonic strength and ductility by cold rolling at liquid nitrogen temperature, followed by an annealing treatment which again led to a bimodal grain size distribution [9]. Based on this success, Ma and others have since advocated the advantages of using material with a bimodal grain size distribution for obtaining good strength and ductility.

However, this expectation is not always borne out in practice. For example, in the case of UFG α-brass, a bimodal grain size distribution was introduced successfully by annealing, and although the parameters of the annealing treatment were varied systematically, only a marginal improvement of the fatigue performance was achieved [8,10]. It appears that the success of such annealing treatments depends not only on finding the optimal annealing conditions but also on the material.

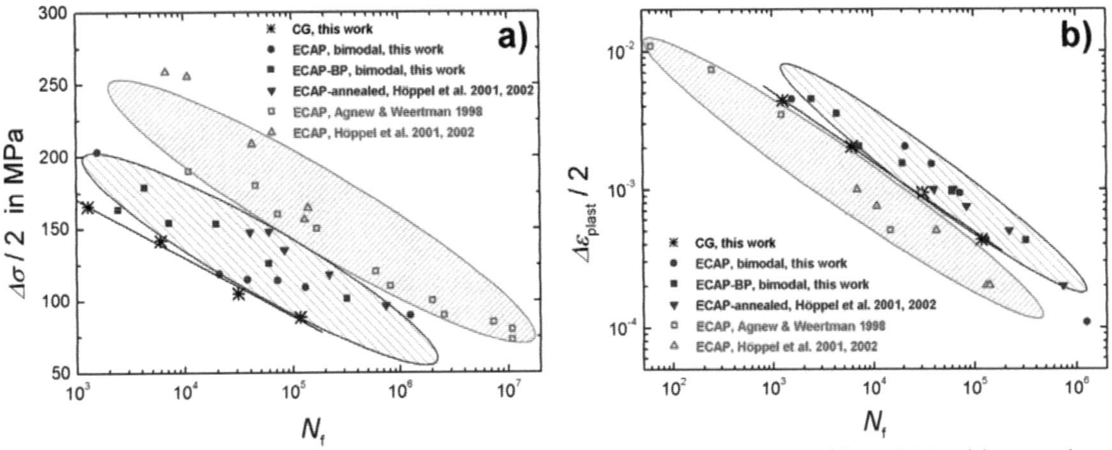

Figure 2. Fatigue life data of commercial purity UFG copper, processed by ECAP without and with back pressure (ECAP-BP). a) Wöhler (S-N) plot, stress amplitude $\Delta\sigma/2$ vs. number of cycles to failure, N_f, showing roughly: $N_{f,\,CG} \leq N_{f,\,bimodal} \leq N_{f,\,UFG}$. b) Coffin-Manson plot, plastic strain amplitude $\Delta\varepsilon_{pl}/2$ vs. N_f, showing roughly: $N_{f,\,UFG} \leq N_{f,\,CG} \leq N_{f,\,bimodal}$.

In order to assess the effects of ECAP-processing with back pressure [11], commercial purity copper which had been ECAP-processed (12 passes, route B_c) with and without back pressure (BP) was fatigued in strain control. Aroused by an unusually large scatter of the fatigue lives, the authors discovered more or less large regions of a bimodally coarsened grain distribution in the initial UFG grain structure, inherited either from ECAP (see Section 2.4) or from resting at RT after ECAP (Section 2.3). The fatigue life data obtained in this study are displayed in Figures 2a and 2b in Wöhler (S-N) and Coffin-Manson plots, respectively, with some other published data. It should be noted that, while there is a broad scatter band, most ECAP-processed specimens exhibited larger fatigue lives than the CG material, not only in the Wöhler (S-N) but also in the Coffin-Manson plot. This quite remarkable enhancement of fatigue performance is similar to the improvement achieved in the case of high-purity UFG copper by annealing, discussed in Section 2.2. In the present case, the initial bimodally coarsened grain structure of admittedly ill-defined origin obviously had a beneficial effect.

4. Summary and Conclusions

Summarizing and confining ourselves to the well investigated cases, mainly on UFG copper, the following conclusions are drawn with respect to the factors governing the occurrence of coarsening, the evolution of bimodal grain size distributions and their effects on the mechanical properties:

- Some UFG metals (e.g. copper or nickel), exhibit a strong tendency to undergo *local* grain coarsening by *dynamic recrystallization* and to develop *bimodal grain size structures* during mild annealing or mild cyclic deformation at rather low temperature (RT). Other UFG materials, e.g. aluminium, also exhibit grain growth at RT but in a rather uniform manner.
- Bimodal grain size structures can very probably evolve by *in situ* annealing during ECAP, due to the heat production and temperature rise, especially at rather high ECAP deformation rates.
- In strain-controlled fatigue and, to a much smaller extent, in stress-controlled fatigue, bimodal grain size distributions can evolve, accompanied by cyclic softening.
- Ageing for short times at elevated temperatures or at RT for longer periods (months, years) can lead to *local* grain coarsening and formation of bimodal grain size distributions.
- Local grain coarsening is expected to be impeded in less pure UFG material. However, bimodal grain size distributions have been found in commercial and high purity materials. Hence, convincing evidence of such an effect of (im)purity is currently lacking.
- Introduction of bimodal grain size distributions by an annealing treatment can sometimes (not always) enhance the monotonic and fatigue strength properties of UFG materials to different degrees, depending on the annealing conditions, degree of bimodality and the material.
- Tailored enhancement of mechanical properties by annealing in order to introduce bimodal grain size distributions is not always possible and requires "optimal" annealing conditions.

References

[1] Valiev RZ, Islamgaliev RK and Alexandrov IV 2000 *Prog. Mater. Sci.* **45** 103-189.
[2] Agnew SR, and Weertman JR. 1998 *Mater. Sci. Eng.* A **244** 145-152.
[3] Höppel HW, Brunnbauer M, Mughrabi H, Valiev RZ and Zhilyaev AP 2001 In: *Proc. Werkstoffwoche 2000*, http://www.materialsweek.org/proceedings.
[4] Höppel HW, Zhou ZM, Mughrabi H and Valiev RZ 2002 *Phil. Mag.* A **82** 1781-1794.
[5] Thiele E, Holste C and Klemm, R 2002 *Z. Metallkd.* **93** 730-736.
[6] Lukáš P, Kunz L and Svoboda M 2005 *Mater. Sci. Eng.* A **391** 337-341, and personal communication, Sept. 2009.
[7] Mughrabi H 2000 In: Lowe TC, Valiev RZ, editors. *Investigations and Applications of Severe Plastic Deformation*, Kluwer Academic Publishers, 241-253.
[8] Mughrabi H, Höppel HW, Kautz M and Valiev RZ 2003 *Z. Metallkd.* **94** 1079-1083.
[9] Wang Y, Chen M, Zhou F and Ma E 2002 *Nature* **419** 912-914.
[10] Mughrabi H, Höppel HW and Kautz M 2004 *Scripta Mater.* **51** 807-812.
[11] Lapovok RY 2005 *J. Mater. Sci.* **40** 341-346.

High strength and high ductility in as-deposited nanocrystalline Ni

Pinqiang Dai[1,2], Weichang Xu[1], Dian Tang[1]

[1] School of Materials Science and Engineering, Fuzhou University, Fuzhou 350108, PR China

[2] Department of Materials Science and Engineering, Fujian University of Technology, Fuzhou 350002, PR China

pqdai@126.com

Abstract. In the present study, an electrodeposited nanocrystalline (nc) Ni sample with high strength and superior ductility relative to many other electrodeposited nc-Ni was prepared. The superior ductility in the present nc-Ni sample free of defects was ascribed to mixed grains, the size of which spanned nano- and sub-micro scales at its as-deposited state with a grain size distribution ranged from 5 to 120nm. Obvious dislocation motion happening in coarse-grained polycrystalline was observed in large grains of nc-Ni matrix resulting in a remarkable enhanced ductility without a decrease in the strength. The present nc-Ni with an average grain size of 27.2nm prepared by direct current electrodeposition shows the average ultimate tensile strength of 1200MPa and the average elongation to failure of 10.4%.

1. Introduction

Nanocrystalline (nc) materials with high strength and high hardness, are expected to be new generation engineering structural materials, but to obtain the higher ductility must be the premise of practice applications. In the past decade, deformation mechanisms in nc materials were extensively investigated through experimental studies and molecular dynamics (MD) simulations. Some new deformation mechanisms including partial dislocations emission from grain boundaries (GB) [1], deformation twin [1-2], grain rotation [3], GB migration [4] have been well-proposed. However, the direct relationship between deformation mechanisms and associated mechanical properties, especially for the ductility, is not fully understood yet. Nonetheless, some effective routes have been suggested to improve the ductility of nc materials [5-8]. One of these routes was focused on the microstructure design, e.g. a composition of nanoscale grains and ultrafine grains (UFGs) (100-500nm) [5,7]. This method relies on the nc matrix to hold the high strength but increases the ductility by introduction of the conventional dislocation sliding to large grains. According to the characteristic of nc materials having abnormal grain growth, the bimodal microstructure can be obtained by thermal annealing. For electrodeposited nc metals, however, low temperature annealing always causes embrittlement [9]. In the present study, a nc-Ni sample prepared by electrodeposition with a broad grain size distribution (BGSD nc-Ni) was reported. The nc-Ni sample has the upper limit of grain size over 120nm but with an average grain size of 27.2nm in the as-deposited statement. The combination of high strength and good ductility was obtained for the BGSD nc-Ni sample. The intracrystalline dislocation network was observed in the post-deformed microstructures confirming the conventional intracrystalline dislocation sliding mechanism.

2. Experimental procedure

In the present study, an aqueous sulfamate-based electrolyte was used to produce BGSD nc-Ni by direct current electrodeposition. The electrolyte was made of 500g L^{-1} $Ni(NH_2SO_3)_2 \cdot 4H_2O$, 20g L^{-1} $NiCl_2 \cdot 6H_2O$, 30g L^{-1} boric acid, 15g L^{-1} $MnCl_2 \cdot 6H_2O$, 2.5g L^{-1} soluble saccharin and 0.1g L^{-1} lauryl sodium sulfate . The main electrodeposition conditions are listed in table 1. Under these conditions, BGSD nc-Ni, the thickness of which was 180-200μm, was deposited on the stainless steel substrate, which had been polished to a mirror-like finish surface prior to electrodeposition. After electrodeposition, the deposits can be mechanically stripped from the stainless steel.

The composition of the BGSD nc-Ni was detected by X-ray Fluorescence (XRF). The microstructures of BGSD nc-Ni were observed by transmission electronic microscope (TEM). TEM observations were performed using a Tecnai G2 F20 S-TWIN operated at 200KV. TEM samples were prepared by double-jet electropolishing using an electrolyte consisting of 5 vol.-% perchloric acid and 95 vol.-% ethanol at a temperature below -20℃. The average grain size was determined from dark field TEM images using the image analysis software (Photoshop 7.0) to count at least 500 grains for each sample.

The dog bone shaped tensile specimens of BGSD nc-Ni with a gauge length of 10mm and gauge width of 6mm were cut using the electro-discharging machine. Room temperature (RT) tensile tests were carried out with the strain rate varying from $5\times10^{-5}s^{-1}$ to $10^{-2}s^{-1}$.

Table 1. The main condition for preparing BGSD nc-Ni

Electrolyte volume (L)	1
Electrolyte temperature (℃)	60
pH	3.5
Current density (A/dm^2)	8
Anode material	Electrolytic Ni (purity > 99.9 mass.%)
Ratio of cathode area to anode area in the plating bath	1:4
Cathode substrate	Stainless steel sheet (1Cr15Mn8Ni5Cu2)

3. Results and discussion.

The TEM bright field image of as-deposited nc-Ni sample is shown in figure 1(a). It is observed that isolated large grains with the size of 120nm are surrounded by small grains with the size below 30nm forming an inhomogeneous microstructure, which is similar to grains undergone abnormal growth in the low temperature annealing. Some defects e.g. dislocation, twin or sub-grain boundary, marked by black arrows, are clearly seen within large grains. The inset is a typical ring pattern further confirming that the global grain size is in the nanoscale. The statistical diagram (figure 1(b)) shows that the BGSD nc-Ni sample has an average grain size of 27.2nm with a grain size distribution ranging from 5 to 120nm. No Mn was detected in the nc-Ni though there was $MnCl_2$ in the electrolyte. The $MnCl_2$ affects the deposition process and results in a broad grain size distribution in the deposits.

The engineering stress-strain curves for BGSD nc-Ni are shown in figure 2(a). The average 0.2% offset yield strength, $\sigma_{0.2}$, is 638MPa and the average ultimate tensile strength, σ_{UTS}, is 1176MPa. The average elongation to fracture, δ_{ETF}, is 10.6%. In comparison with nc-Ni with a relatively narrow grain size distribution (NGSD) of 10-50nm and an average grain size of 15-25nm, which has a σ_{UTS} of 1200-1300MPa and a common ductility with δ_{ETF} of 4-5% [9-10], apparently, the present BGSD nc-Ni samples possess the markedly enhanced ductility and the similar strength.

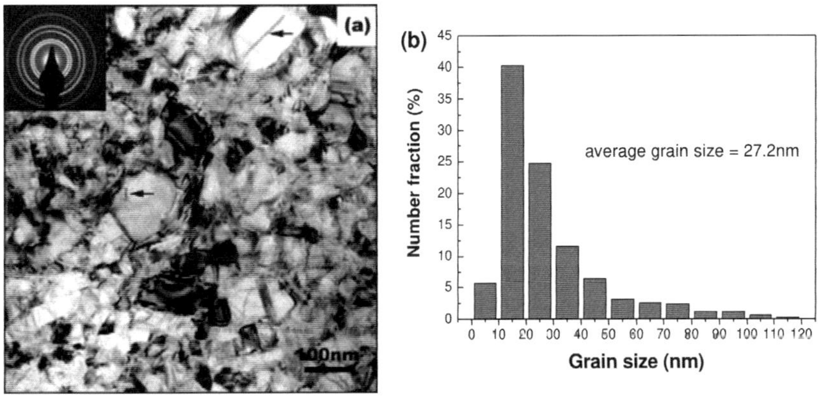

Figure 1. TEM bright field image (a) and grain size statistical distribution (b) for electrodeposited nc Ni with a broad grain size distribution.

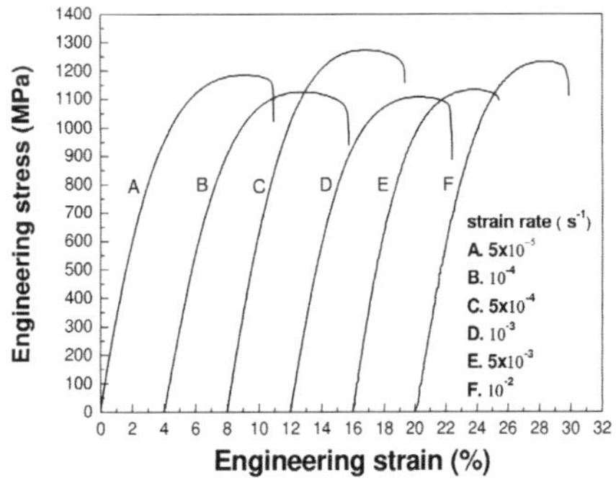

Figure 2. Engineering stress-strain curves

Figure 3(a) shows a TEM bright field image from the tensile fracture of BGSD nc-Ni. Apparent grain growth is observed in contrary to the as-deposited grain size in figure 1(a). In comparison with figure 1(a), the statistical analysis in figure 3(b) displays that the average grain size after deformation is increased by 19.3nm and that there is a sharp decrease in the number fraction of sizes below 30nm. In addition, few large grain coalescences with sizes over 200nm are also observed in the post-deformed microstructures, as shown in figure 3(c). These observations indicate the propensity of grain growth driven by applied stresses in BGSD nc-Ni is stronger than that of NGBD nc-Ni [3,4,11]. It seems that the large grains can act as cores to merge neighboring small grains by grain rotation [3,11] forming the grain coalescence during plastic deformation. In addition, dislocations trapped in the grain coalescence and dislocation pile-ups at GBs are observed. One of these phenomena was captured in figure 3(c). These findings offer compelling evidence to support the conventional intracrystalline dislocation-sliding mechanism dominated in BGSD nc-Ni.

It was pointed that dislocations are hardly to move in small nanograins due to the need of very high stress to activate dislocation sources, e.g. the Frank-Read source [12]. The critical stress, τ, to activate the Frank-Read source can be expressed as $\tau = Gb/L$, where G is shear module (80GPa for Ni), b the smallest Burgers vector in face-centered cubic metals (0.249nm for Ni) and L the distance between two pinning points (typical L=d/3 or d/4 and d is the grain diameter). According to the equation, for grains of sizes of 80-120nm, τ is estimated about 498-746MPa, which is close to the present experimental yield stress. Accordingly, the Frank-Read source can be activated in the large grains of sizes over 80nm in the practical tensile deformation. Moreover, grain growth in the plastic deformation process may further favour the activation of the Frank-Read source with increasing deformation strains.

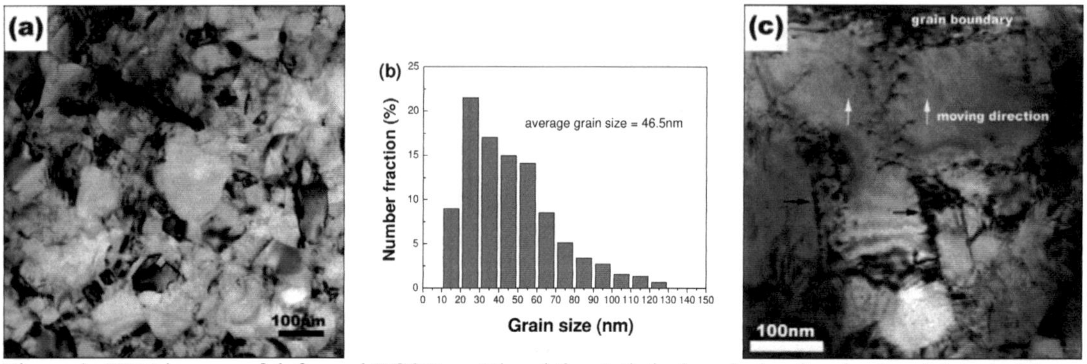

Figure 3. TEM images of deformed BGSD nc-Ni and the statistical analysis of grain size. (a) typical image showing pronounced global grain growth, (b) statistical grain size distribution after deformation, (c) image showing significant grain growth and pile-ups of dislocation ahead of grain boundaries after deformation.

Conclusion

(1) In comparison with typical NGSD (10-50nm) nc-Ni, with an average grain size of 15-25nm, has a common ductility with δ_{ETF} of ~4% and a high strength with σ_{UTS} of 1200-1300MPa at room temperature and strain rates of 3×10^{-3} or $10^{-4}s^{-1}$. However, the present BGSD (10-120nm) nc-Ni sample prepared by direct current electrodeposition with an average grain size of 27.2nm exhibits a considerable ductility (over 10% of δ_{ETF}) along with the similar strength.

(2) The intracrystalline dislocation network was observed in the post-deformed microstructure confirming the conventional intracrystalline dislocation sliding mechanism in BGSD nc-Ni.

Acknowledgements

The financial support provided by the Innovation Project of Fuzhou University, the Foundations of Fujian provincial natural science research programme (No. 2008J0145-E0810006) and the National 863 projects of China (No. 2007AA03Z325) is gratefully acknowledged.

References

[1] Froseth A G, Derlet P M and Van Swygenhoven H 2005 *Adv. Eng. Mater.* **7** 16-20.
[2] Chen M, Ma E, Hemker K J, Sheng H, Wang Y and Cheng X 2003 *Science* **300** 1275-7.
[3] Shan Z, Stach E A, Wiezorek J M K, Knapp J A, Follstaedt D M and Mao S X 2004 *Science* **305** 654-7.
[4] Farkas D, Frosteh A and Van Swygenhoven H 2006 *Scripta Mater.* **55** 695-8.
[5] Wang Y, Chen M, Zhou F and Ma E 2002 *Nature* **419** 912-5.
[6] Lu L, Shen Y, Chen X, Qian L and Lu K 2004 *Science* **304** 422-5.
[7] Zhang X, Wang H, Scattergood R O, Narayan J, Koch C C, Sergueeva A V and Mukherjee A K 2002 *Acta Mater.* **50** 4823-30.
[8] Guduru R K, Murty L L, Youssef K M, Scattergood R O and Koch C C 2007 *Mater. Sci. Eng.* **A 463** 14-21.
[9] Wang Y M, Cheng S, Wei Q M, Ma E, Nieh T G and Hamza A 2004 *Scripta Mater.* **51** 1023-8.
[10] Wu X L, Zhu Y T and Ma E 2006 *Appl. Phys. Lett.* **88**(121905) 1-3.
[11] Wang Y B, Li B Q, Sui M L and Mao S X 2008 *Appl. Phys. Lett.* **92**(011903) 1-3.
[12] Legros M, Elliott B R, Rittner M N, Weertman J R and Hemker K J 2000 *Phil. Mag.* **A 80** 1017-26.

Crystallographic characterization of catastrophic shear in submicron nickel at low temperatures

SR Dey[1,2], L Hollang[2], K Reuther[2], R Hübner[3], R Chulist[2], W Skrotzki[2]

[1] Research Department: Integrity of Small Scale Systems/High Temperature Materials, Institut für Werkstoffe, Ruhr-Universität Bochum, Universitätsstraße 150, D-44801 Bochum, Germany
[2] Institut für Strukturphysik, Technische Universität Dresden, D-01062 Dresden, Germany
[3] Center for Complex Analysis, Global Foundries, Wilschdorfer Landstraße 101, D-01109 Dresden, Germany

suhashranjandey@gmail.com

Abstract. Nickel produced by additive-free pulsed electro-deposition with grain size of about 150 nm was deformed in tension at 320 K and 4 K. At both temperatures the specimens exhibited strong parabolic hardening and good ductility. However, at 4 K a sudden transition from homogeneous deformation to "repeated catastrophic shear" took place. Microstructural characterization of the sheared region showed deformed grains elongated in between tensile axis and growth direction. The texture information from the sheared region was obtained using conical dark field measurements in the transmission electron microscope. The pole figures of the deformed 4 K specimen show a typical shear texture which clearly differs from the starting texture.

1. Introduction

Pure metals with grain sizes between the nanocrystalline (< 100 nm) and microcrystalline (> 1 μm) state are very promising objects for investigations of the transition from "conventional" deformation behavior to "nano"-behavior. To study this temperature dependent transition, additive-free pulsed electrodeposited (PED) nickel was chosen as model material for face-centred cubic metals [1]. Additive-free PED nickel was pure, fully dense and contained a large fraction of high angle grain boundaries with a narrow grain size distribution.

It was the aim of the present work to extend investigations of grain size induced changes of deformation mechanism already performed on pure polycrystalline nickel with large grain size [2].

2. Experimental

A rectangular plate of PED nickel (70 × 40 × 2 mm^3) was produced using a sulfamate bath without additives (e.g. saccharin). Two cylindrical tensile specimens ("dog bone" shaped, gauge length 10 mm and diameter 0.8 mm) were cut from the PED plate by spark-erosion with the tensile axis (TA) aligned perpendicular to the growth direction (GD). Subsequently, the specimens were electropolished to remove the disturbed surface layer. Tensile deformation was carried out in an open continuous flow cryostat (He coolant) permitting to keep the temperature constant within less than 0.05 K [3]. The cryostat was mounted in an Instron 4502 load frame. The experiments were performed at constant crosshead displacement rate providing an initial strain rate of 4×10^{-4} s^{-1}. Electron backscatter diffraction (EBSD) and transmission electron microscopy (TEM) were performed in a scanning electron microscope Zeiss Ultra 55 and a Technai, respectively. TEM foils were cut by focused ion beam (FIB). Conical dark field measurements in the TEM after processing through ACT software yield orientation imaging maps and related pole figures. The pole figures were obtained using LABOTEX software.

3. Results and Discussion

The initial microstructure and texture of the additive-free PED nickel plate is given in figure 1. Few grains are elongated along GD. The average grain size obtained from the EBSD scan is 150 nm. There is a weak <110> fiber texture along GD.

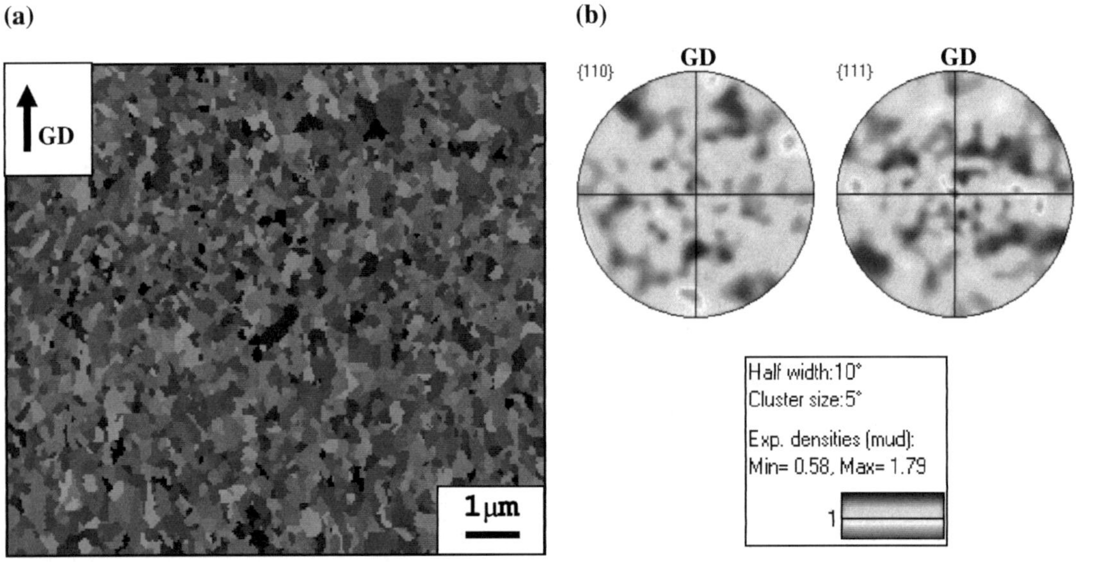

Figure 1. **(a)** EBSD map of the initial microstructure, acquisition step size 40 nm, **(b)** {110} and {111} pole figures from (a)

Figure 2 shows the engineering stress (force with respect to the initial cross section) vs. crosshead displacement for the specimens deformed at 320 K and 4 K. The two stress drops at 320 K are the results of stress relaxation tests performed in order to study the deformation mechanisms [2]. Apart from that, the 320 K curve shows homogeneous continuous deformation, but at a rather high stress level of more than 1500 MPa at the end of deformation. Reports in the literature (e.g. [4, 5]) claim (X-ray) grain sizes between 20 nm and 50 nm for such stress levels in PED nickel. Initially, the curve at 4 K behaved similar to that at 320 K, but at 2400 MPa the deformation mode suddenly changed towards repeated shear events. After a first stress drop of 16.9% the deformation continued elastically before a second stress drop of the same order appeared. Also the second event was not fatal for the specimen. Again the deformation continued elastically, but then the specimen failed. The shear events were accompanied by loud acoustic emission.

Similar shear events are reported by Niewczas et al. [6, 7] on copper single crystals deformed in tension at 4.2 K, and are attributed to "adiabatic deformation" (see e.g. [8]). In this case, a dramatic load drop of 16.5 % marks the sudden transition from homogeneous slip on the primary and conjugate slip systems at the beginning of deformation towards twinning alone. Niewczas et al. [6, 7] argue that the onset of twinning acts as trigger for the adiabatic deformation in the way that after a substantial overshoot of the crystal axis the resolved stress in the twinning system reaches the stress necessary to nucleate twins. It is interesting to note that the relative depth of the load drops in copper single crystals is almost the same as in PED nickel polycrystals.

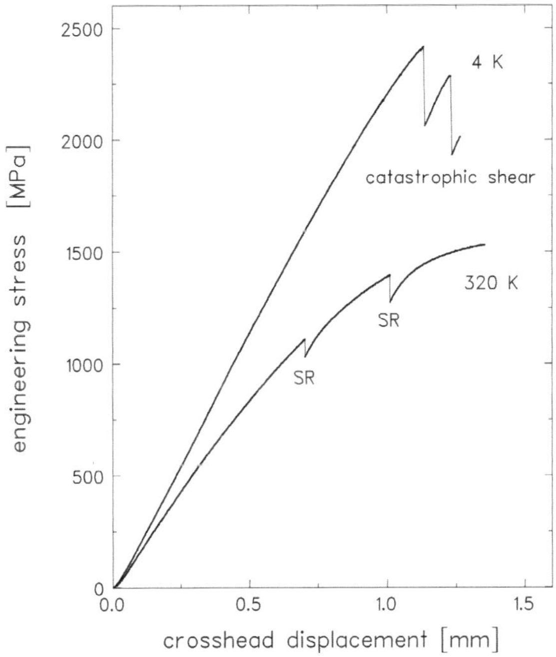

Figure 2. Engineering stress vs. crosshead displacement at 320 K and 4 K (stress relaxation tests are marked by SR)

After dismounting the 4 K tensile tested specimen no distinct shear band, but a necked region of about 200 μm extension in tensile direction around the crack was recognized which probably represents the sheared region. For microstructural observations of the sheared region, a thin foil cut through FIB parallel to GD was studied by TEM (figure 3). Elongated grains are directed in between TA and GD. While initially the grains were almost equiaxed (figure 1a) this indicates deformation-induced grain shape changes. Obviously, during the shear event localized deformation took place also clearly changing the texture. As figure 4 shows, the texture evolved towards a typical shear texture produced by shear on {111}<110> slip systems [9].

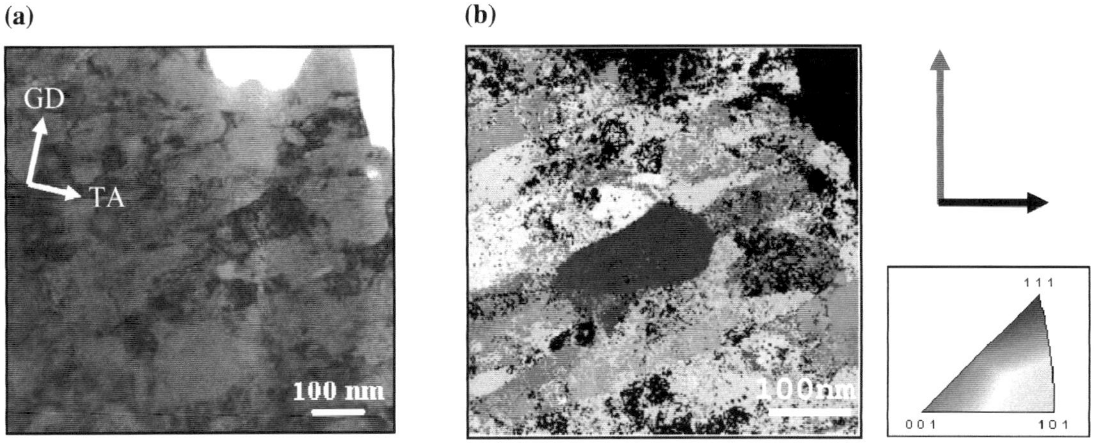

Figure 3. (a) Bright field TEM micrograph of the sheared region of the 4 K specimen after fracture, **(b)** orientation map of part of the region shown in (a), color code with respect to vertical upward direction (red colored), black dots not indexed

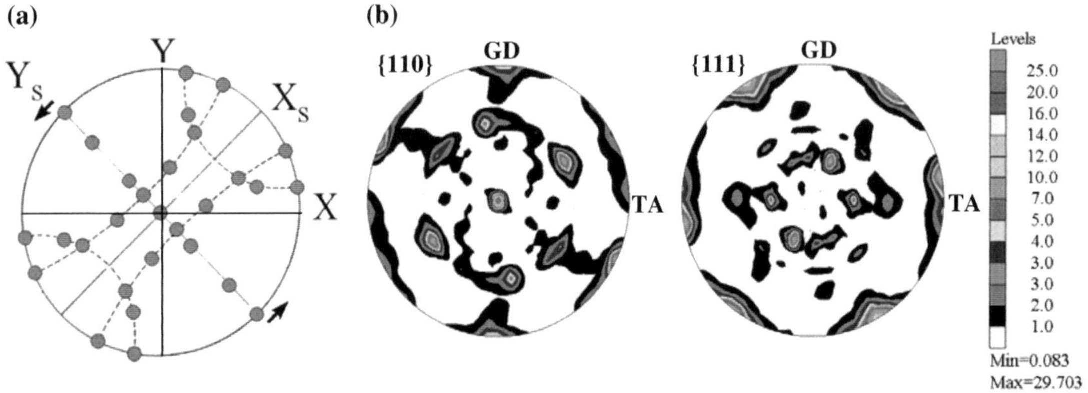

Figure 4. **(a)** Schematic representation of an ideal {111} pole figure for simple shear along X_S on the plane normal to Y_S, **(b)** experimental {110} and {111} pole figures related to figure 3(b)

Interestingly, the pole figures of catastrophic shear at cryogenic temperature are comparable with those obtained for cyclic deformation at room temperature [10]. This indicates that the localized shear observed most probably is carried by dislocations. So far, no sign is found about the nature of the triggering mechanism for catastrophic shear.

4. Conclusions

During uniaxial tension at 4 K, additive-free pulsed electrodeposited nickel containing nearly equiaxed grains of 150 nm in size at an engineering stress of 2.4 GPa changed from homogeneous deformation to localized catastrophic shear. TEM of the sheared region reveals substantial elongation of the grains in between TA and GD. Pole figures show a typical shear texture produced by slip on {111}<110> systems. Thus, localized shear deformation of sub-microcrystalline nickel at low temperatures is carried by dislocation slip. The trigger mechanism for the onset of strain localization is probably related to twinning.

Acknowledgements

S.R. Dey would like to thank the Alexander von Humboldt Research Foundation for financial support during his stay at the Technische Universität Dresden. This work was also supported by the Deutsche Forschungsgemeinschaft through contract Sk21/23-1. Thanks are due to H.-T. Reiter for sample cutting.

References

[1] Thompson AW 1975 *Acta Metall.* **23** 1337
[2] Hollang L, Hieckmann E, Brunner D, Holste C and Skrotzki W 2006 *Mater. Sci. Eng.* A **424** 138
[3] Brunner D and Diehl J 1992 *Z. Metallkde.* **83** 828
[4] Yin WM, Wang SH, Mirshams CH and Xiao CH 2001 *Mater. Sci. Eng.* A **301** 18
[5] Schwaiger R, Moser B, Dao M, Chollacoop N and Suresh S 2003 *Acta Mater.* **51** 5159
[6] Niewczas M, Basinski ZS, Basinski SJ and Embury JD 2001 *Phil. Mag.* A **81** 1121
[7] Niewczas M, Basinski ZS and Embury JD 2001 *Phil. Mag.* A **81** 1143
[8] Poirier JP 1980 *J. Struct. Geol.* **2** 135
[9] Bacroix B 1986 PhD Thesis, McGill University, Montreal
[10] Dey SR, Beausir B, Hollang L and Skrotzki W 2009 *Scripta Mater.*, DOI: 10.1016/j.scriptamat.2010.02.010.

Creep in TiO_2 doped Zirconia: Implications for high strain rate superplasticity

Maya Kini and Atul H. Chokshi,

Department of Materials Engg., Indian Institute of Science, Bangalore 560 012, India

achokshi@materials.iisc.ernet.in

Abstract In ceramics, dopants offer the possibility of higher creep rates by enhancing diffusion. The present study examines the potential for high strain rate superplasticity in a TiO_2 doped zirconia, by conducting creep experiments together with microstructural characterization. It is shown that both pure and doped zirconia exhibit transitions in creep behaviour from Coble diffusion creep with n~1 to an interface controlled process with n~2. Doping with TiO_2 enhances the creep rate by over an order of magnitude. There is evidence of substantial grain boundary sliding, consistent with diffusion creep.

1. Introduction

Superplasticity refers to the ability of some fine grained polycrystalline materials to exhibit large elongations to failure [1]. The process is well established in metallic alloys, and it is being utilized to form components with complex shapes. It is now recognized that even nominally brittle intermetallic compounds and ceramics can exhibit large ductility under some conditions. A 3 mol% yttria stabilized tetragonal zirconia (3YTZ) has become a standard material for examining different aspects of superplasticity in ceramics [1-4]. Experimental data reveal that superplasticity occurs in fine grained materials tested at high temperatures. The superplastic strain rate is given as

$$\dot{\varepsilon} \propto D d^{-p} \sigma^{n} \quad (1)$$

where D is the diffusion coefficient, d is the grain size, σ is the stress and p and n are constants termed the inverse grain size and stress exponent, respectively. For typical superplastic conditions, n~p~2. The diffusion coefficient can be expressed as $D=D_0\exp(-Q/RT)$, where D_0 is the pre-exponential term, Q is the activation energy for the rate controlling process, R is the gas constant and T is the absolute temperature.

Conventional superplastic Ti and Al alloys with grain sizes of ~10 μm exhibit superplasticity at relatively slow strain rates of $\sim10^{-4}$ s^{-1} [1]. For economic viability, it is desirable to displace optimum superplasticity to higher strain rates. Equation (1) suggests that it is possible to obtain high strain rate superplasticity by refining the grain size, and this is a popular approach in metallic alloys where a refinement in grain size from ~10 to ~1 μm leads to an enhancement in strain rate from $\sim10^{-4}$ to $>10^{-2}$ s^{-1}[1].

In contrast to metallic alloys where superplasticity requires a grain size of <10 μm, superplasticity in ceramics requires a finer grain size of < 1 μm. Further refinement in grain size to <0.1 μm is possible, but difficult. Equation (1) also suggests that an increase in temperature can lead to an increase in strain rate, by enhancing D. However, experimental data on superplastic zirconia reveals that an increase in testing temperature to 1823 K leads to drastic grain growth [5], so that initial grain sizes of ~0.3 μm increase rapidly to values > 1 μm. These results suggest that it is not possible to sustain superplasticity at higher temperatures in nominally single phase ceramics. Kim et al. [6] have used the strategy of mutiphase alloying to limit grain growth at higher temperatures to obtain high strain rate superplasticity in zirconia-spinel composites.

In contrast to metallic alloys, where diffusion is controlled by the thermal concentration and migration of defects, diffusion in ceramics can also be modified by doping. Experimental data in 3YTZ indicate that there are many transitions in stress exponents from ~7 at high stresses (related to intragranular dislocation creep) to ~1 and ~2 and ~3 with a decrease in stress corresponding to Coble diffusion creep, and interface controlled diffusion creep [7]. These transitions are sensitive to impurity content as well as concentrations of dopants [1-3,7-9]. Experiments have shown that elements such as Al, Si, Ti and Ge influence both the flow stress and the transitions in creep behavior [8-13]. A recent detailed study has shown that silica additions do not influence the diffusion coefficients, but they retard the transition from Coble creep to interface controlled creep and possibly enhance ductility by limiting concurrent cavitation [14]. Sakuma and co-workers have shown that Ti and Ge either individually or together also enhance ductility, and these have been attributed to changes in bonding [10,11,13]. However, the influence of Ti additions on deformation mechanisms has not yet been examined.

We report below some experimental results on pure 3YTZ and TiO_2 doped 3YTZ, and demonstrate that TiO_2 addition enhances diffusion giving rise to the possibility of high strain rate superplasticity in such materials.

2. Experimental Materials and Procedure

Pure 3YTZ powders were obtained from Tosoh corporation in Japan. A wet-chemistry approach was utilized to dope 3YTZ powders with TiO_2; precursor of TiO_2 was formed by the hydrolysis of Titanium n-Butoxide [15]. Based on earlier promising results by Sakuma, we produced a composition of 3YTZ doped with 4.8 wt% TiO_2. Dense compacts of pure 3YTZ and TiO_2 doped 3YTZ were prepared by cold compaction and sintering at 1723 K for 2 hours. A range of grain sizes between ~0.6 and 1.4 μm were obtained by annealing sintered samples at various temperatures for different times. Parallelepiped specimens with nominal dimensions of 3x3x5 mm were machined for compression testing; only samples with densities >99% were used for the creep study. X-ray diffraction was used to determine the crystalline phases present in the materials, and to also calculate the theoretical densities.

Constant load and load change experiments were conducted for evaluating the creep characteristics. Selected deformed specimens were examined after creep testing for changes in the grain size and shape. Surface offsets at grain boundaries, indicative of grain boundary sliding, were measured using AFM data.

3. Experimental Results and Discussion

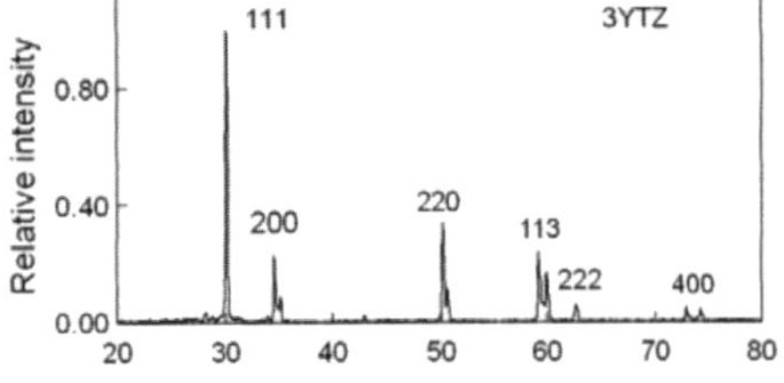

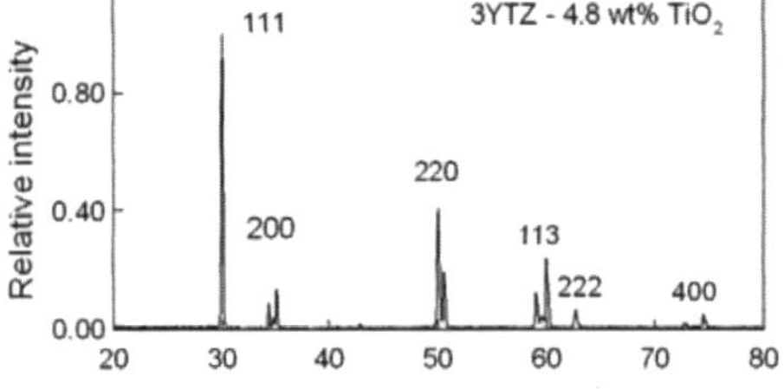

Fig 1: XRD patterns for annealed (a) 3YTZ (b) 3YTZ – 4.8 wt% TiO_2

Figures 1a and 1b are x-ray data obtained from pure 3YTZ and TiO_2-doped 3YTZ: both materials are largely tetragonal, with small {200} and {311} cubic zirconia peaks in annealed specimens with coarse grain sizes. There was no evidence of any separate TiO_2 peaks in the doped material, indicating that all the TiO_2 was dissolved in the 3YTZ. All specimens exhibited an equiaxed grain shape.

The variation in creep rate with strain is depicted in Fig. 2 for the TiO_2-doped specimens tested at either a constant load or by the load change method. There is almost no primary region, and most of the data correspond to steady state creep. There are very small transients accompanying load changes. These results suggest that there no significant internal microstructural changes, consistent with experimental observations on selected deformed specimens of the retention in an equiaxed grain shape and no grain growth.

Figure 3 is a logarithmic plot of the variation in strain rate with stress for the pure and the doped specimens with a grain size of ~0.7 to 1.0μm. It is clear that the data from load change tests are consistent with those from individual constant load tests. Furthermore, Fig. 3 depicts transitions in stress exponents from n~1 to n~2 with a decrease in stress, for both materials. Comparison of the data indicates also that TiO_2 doping enhances the creep rates by over an order of magnitude and reduces the stress for a transition from n~1 to n~2.

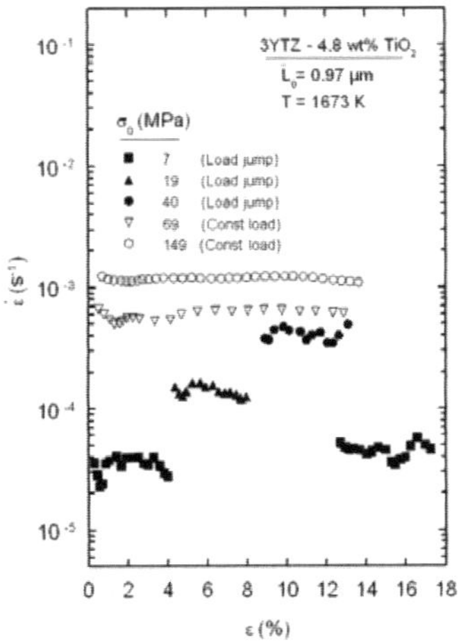

Fig 2: Strain rate vs strain curves for 3YTZ – 4.8 wt% TiO_2

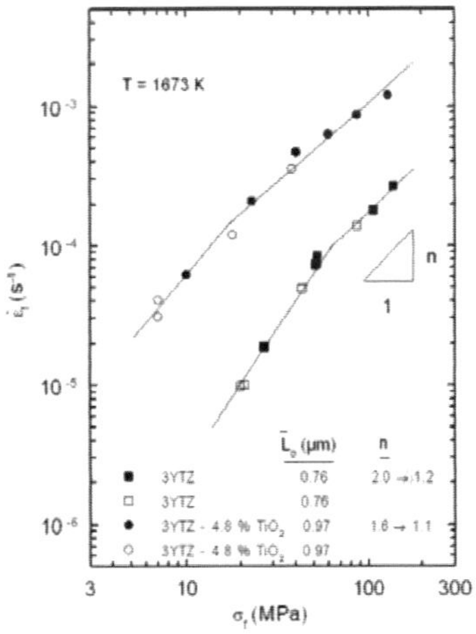

Fig 3: Variation of strain with stress for 3YTZ and 3YTZ – 4.8 wt% TiO_2 showing n~ 2→ 1

Earlier detailed studies on pure and SiO_2-doped 3YTZ [7,14] indicate that the n~1 region is related to Coble diffusion creep which leads to the following creep rate $\dot{\varepsilon}_{CO}$:

$$\dot{\varepsilon}_{CO} = \frac{33\delta D_{gb} G}{kT}\left(\frac{\sigma}{G}\right)\left(\frac{b}{d}\right)^3 \tag{2}$$

where δ is the grain boundary width, D_{gb} is the grain boundary diffusion coefficient, G is the shear modulus, b is the magnitude of the Burgers vector and k is Boltzmann's constant. Experimental data show that p=3 when n=1, consistent with Coble diffusion creep. For a constant stress and temperature, eqn. 2 suggests that an increase in creep rate is related to an increase in D_{gb}. Thus, the present experimental data indicate that titania doping enhances grain boundary diffusion in 3YTZ, offering the potential for high strain rate superplasticity.

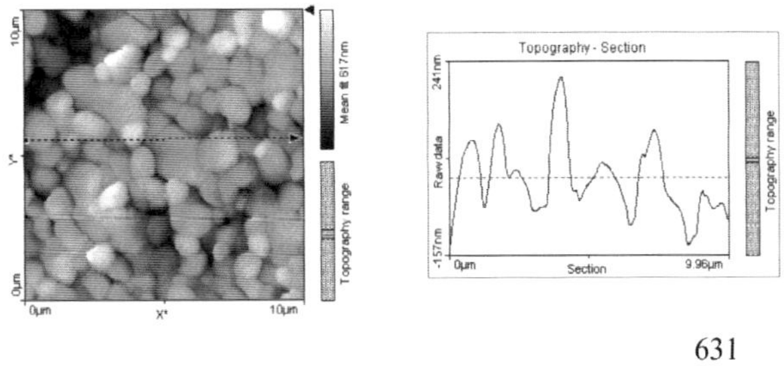

Fig 4: AFM image of deformed 3YTZ- 4.8 wt% TiO_2 and surface profile showing vertical offsets.

It is recognized that diffusion creep must involve GBS to maintain specimen coherency, and this is referred to as Lifshitz sliding [18, 19], and measurements of sliding should yield high values for the GBS contribution. Figure 4 shows the surface topology of a deformed TiO_2 doped specimen; the offsets along grain boundaries are recorded, as shown on the right had side of Fig. 4. Such measurements of offsets along many boundaries can be used to calculate the GBS contribution to creep, following procedures described by Langdon. The experimental results yield a value of the GBS contribution to creep of ≥50%, which is consistent with Coble diffusion creep.

Coble creep and Lifshitz sliding are expected to lead to an elongation of grains along the tensile axis [16, 17]. However, the grains retained their equiaxed shapes in the present study. It is possible to account for this observation by referring to a model for diffusion creep developed by Lee [18], which involves grain switching and the retention of an equiaxed grain shape. It has also been pointed out that grain growth will also lead to grain switching which enables the retention of an equiaxed grain shape [19].

The transition from n~1 to n~2 at lower stresses is consistent with the operation of an interface controlled diffusion creep process, involving grain boundary sources and sinks that do not operate perfectly. The segregation of TiO_2 to grain boundaries, reported elsewhere [20], is likely to modify the efficiency of grain boundaries, and this can lead to a decrease in the stress for a transition from Coble diffusion creep to interface controlled Coble diffusion creep.

4. Summary and Conclusions

Compression creep experiments on pure and TiO_2-doped zirconia reveal a transition in stress exponents from n~1 at high stresses to n~2 at low stresses. The data are consistent with a transition from Coble diffusion creep to an interface controlled process at lower stresses. Titania doping enhances the creep rate by one order of magnitude, indicating that the dopant enhances grain boundary diffusion. There is considerable grain boundary sliding, consistent with Coble diffusion creep.

Acknowledgment

This work was supported by a grant from CSIR.

References

[1] A. H. Chokshi, A.K. Mukherjee and T.G.Langdon, *Mater. Sci. Eng.* **A10** (1993), p. 237
[2] F. Wakai, S. Sakaguchi and Y. Matsuno, *Adv. Ceram. Mater.* **1** (1986), p. 259
[3] A. H. Chokshi, *Mater. Sci. Eng.* **A166** (1993), p.119
[4] M. Jimenez-Melendo, A. Dominguez Rodriguez and A. Bravo- Leon, *J. Am. Ceram. Soc.* **81** (1998), p. 2761
[5] D.J. Schissler , A.H. Chokshi, T.G.Nieh and J. Wadsworth, *Acta Metall. Mater.* **39**, p. 3227
[6] B. N. Kim, K. Hiraga, K. Morita and Y. Sakka., *Nature* **413** (2001), p. 288
[7] I. Charit and A. H. Chokshi, *Acta Mater.* **49** (2001), p. 2239
[8] D. M. Owen and A. H. Chokshi, *Acta Mater.* **46** (1998), p. 667
[9] J. A. Hines, Y. Ikuhara, A. H. Chokshi and T. Sakuma, *Acta Mater.* **46** (1998), p. 5557
[10] K. Tsurui and T. Sakuma, *Scripta Mater.* **34** (1996), p. 443
[11] A. Kuwabara, M. Nakano, H. Yoshida, Y. Ikuhara and T. Sakuma, *Acta Mater.* **52** (2004), p. 5563
[12] E. Sato, H. Morioka, K. Kuribayashi & D. Sundararaman, *J. Mater. Sci.* **34** (1999), p. 4511
[13] H. Ýoshida, K.Morita, B. N. Kim, K. Hiraga and T. Tamamoto, *Acta Mater.* **57** (2009), p. 3029
[14] S. Ghosh and A. H. Chokshi, *J. Am. Ceram. Soc.* (2010), in press
[15] J. Chen, L. Gao, J. Huang and D. Yan, *J. Mater. Sci.* **31** (1996), p. 3497
[16] W. R. Cannon, *Phil. Mag.* **25** (1972), p. 1489
[17] T. G. Langdon, *J. Mater. Sci.* **41** (2006), p. 597
[18] D. Lee, *Metall. Trans.* **1** (1970), p. 309
[19] A. H. Chokshi, *Scripta Mater.* **42** (2000), p. 241
[20] H. Yoshida, *J. Ceram. Soc. Japan* **114** (2006), p. 155

Preparation and mechanical property of polymer-based biomaterials

P. Zhang, G. Chen, X. F. Zheng

School of Materials Science and Engineering, Harbin Institute of Technology at Weihai, Weihai 264209, China

pzhang@yeah.net

Abstract. The porous polymer-based biomaterial has been synthesized from PLGA, dioxane and tricalcium phosphate (TCP) by low-temperature deposition process. The deformation behaviours and fracture mechanism of polymer-based biomaterials were investigated using the compression test and the finite element (FE) simulation. The results show that the stress-strain curve of compression process includes linear elastic stage I, platform stage II and densification stage III, and the fracture mechanism can be considered as brittle fracture.

1. Introduction

Biomaterial is essentially a material that is used and adapted for a medical application. For instance, biomaterials can be used in joint replacements, bone plates, bone cement, artificial ligaments and tendons, skin repair devices, etc. Up to now, many types of biomaterials have been developed and used widely [1-6]. In general, materials options for synthesis biomaterial include metals, ceramics and polymers. Unfortunately, metallic and ceramic biomaterials are not suitable to replace soft tissues because of markedly different mechanical properties. Polymer-based biomaterials may be a good candidate for replace soft tissue, because synthetic polymers that have various properties can be produced by changing the amount or kind of reactant and used according to its purpose. Poly-lactic acid is a kind of polymer which make from lactic acid as the main raw material and has good thermal stability, solvent resistance, mechanical properties, physical performance, biodegradablity. So it is an ideal green polymer material [7-11].

In this paper, the polylactic-co-glycolic acid based biomaterials were fabricated by the low-temperature deposition forming process. For investigating the mechanical properties of the porous material, the deformation behavior in load process of the prepared biomaterial and the mechanism of fracture were analyzed by using compression experiment and FE simulation.

2. Experiments

2.1. Experimental materials

Experimental materials were polylactic-co-glycolic acid (PLGA) and dioxane. In the polylactic-co-glycolic acid, the ratio of LA and GA is 5 to 5, molecular weight is 25 myriad. The ratio of weight of PLGA and volume of dioxane is 0.10. Solid powder is tricalcium phosphate (TCP). The ratio of TCP and PLGA is 3 to 7.

2.2. Experimental equipment and forming process

The low-temperature deposition equipment developed by Tsinghua University was used [12]. The flow chart of forming process can be shown as Fig.1.

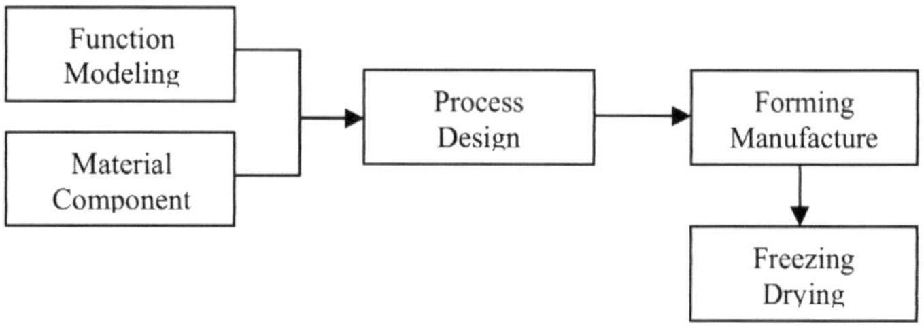

Figure 1. Flow chart of forming process

Firstly, according to the structure design of parts, CAD model can be built. Then, relevant process file and NC code file used to numerical control machining can be generated through data processing software. Secondly, according to the proportion of materials components, the selected polymer materials were dissolved in liquefier. Thirdly, NC code file can control the sprayer ejecting the slurry and scanning movement. The extrusion materials can solidify and bond rapidly in low temperature room. The designed part can be formed through the pile-up process. At the same time, porous structure can be obtained due to the phase separation. Fourthly, the formed parts were dried in dehydrator. The structure of formed part can transform to microporous structure due to the sublimation of dissolvent.

2.3. Mechanical property test

The compression mechanical properties of test samples were measured using Instron 5569 electronic universal testing machine. The size of the round sample was ¢16 mm×10 mm. The load velocity was 1 mm/min.

3. Results and Analysis

3.1. Porous patterns

Fig.2 shows that in the prepared porous materials, there are lots of pores with uniform thickness of wall and most of the pores are connected and distributed homogeneously. It can also be found that TCP powders have been distributed homogeneously on the surface of wall.

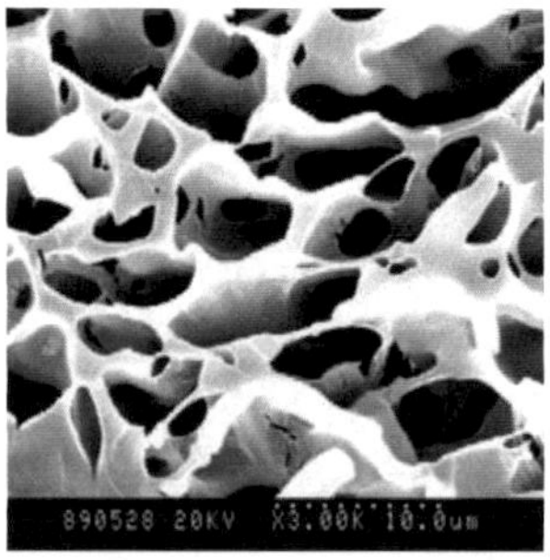

Figure 2. SEM image of porous material

3.2. Stress-strain curve

The stress-strain curve of compression process shown in Fig.3 includes linear elastic stage I, platform stage II and densification stage III, which is the typical compression curve of porous structure.

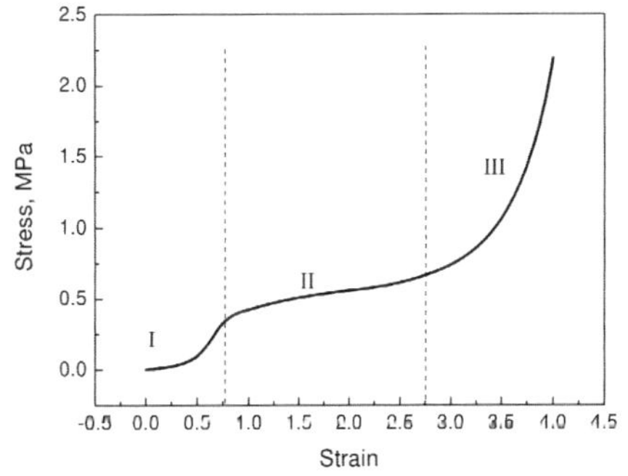

Figure 3. Stress-strain curve of compression process

Flow stress is the most important material parameter. Usually, flow stress is the function of equivalent effective strain and equivalent effective strain rate, and which can be described by the expression (1).

$$\bar{\sigma}=\bar{\sigma}(\bar{\varepsilon},\dot{\bar{\varepsilon}}) \qquad (1)$$

Where $\bar{\sigma}$, $\bar{\varepsilon}$, $\dot{\varepsilon}$ are flow stress, equivalent effective strain and equivalent effective strain rate respectively.

According to difference of the yield rule, the flow stress datum of porous materials can be considered as the yield stress of porous body or compact body. Then, the flow stress regression model of porous materials can be fitted as

$$\bar{\sigma}=0.314+0.007\varepsilon^{3.841} \qquad (2)$$

3.3. Analysis of compression process

FE simulations of porous materials can disclosure the flow and densification rule, and predict the defects generated during the plastic deformation process. Fig.4 shows the equivalent effective strain nephograms of compression process. It can be seen that the pore walls of porous materials can be crumbled under the pressure with increasing of the compression strains, then the block porous material become more and more compact. Porous materials have the characteristics of variability of volume, lateral flow, inhomogeneous deformation and compaction due to the porous pattern distinction, which leading to the complexity of deformation regularity.

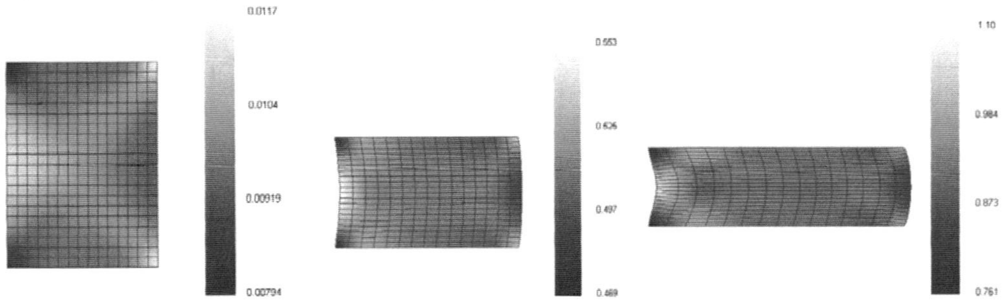

(a) 1step (b) 40step (c) 60step

Figure 4. Equivalent effective strain nephogram of compression process

3.4. Analysis of fracture mechanism

Fig.5 shows the damage nephogram of compression process. It can be seen that with increasing of compression the surfaces of porous materials contact with the dies were collapsed firstly, then the collapse range enlarged gradually. During the compression process, porous materials generate certain lateral deformation. SEM image of fracture pattern shown in Fig.6 indicate that external section has been compacted and inner section of fracture is still with porous structure. The fracture pattern is tearing shape, so the fracture mechanism can be considered as brittle fracture.

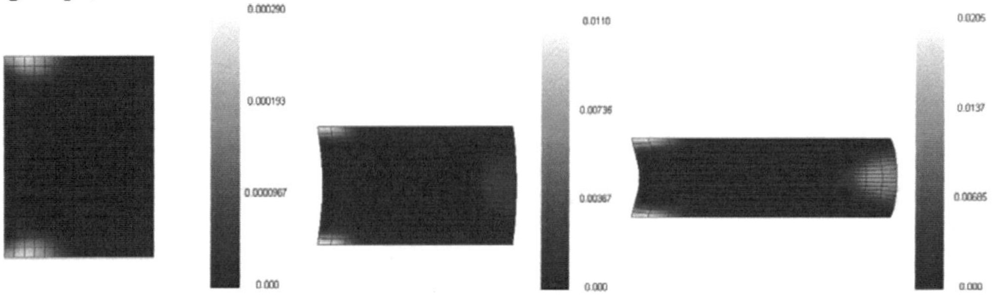

(a) 1step (b) 40step (c) 60step

Figure 5. Damage nephogram of compression process

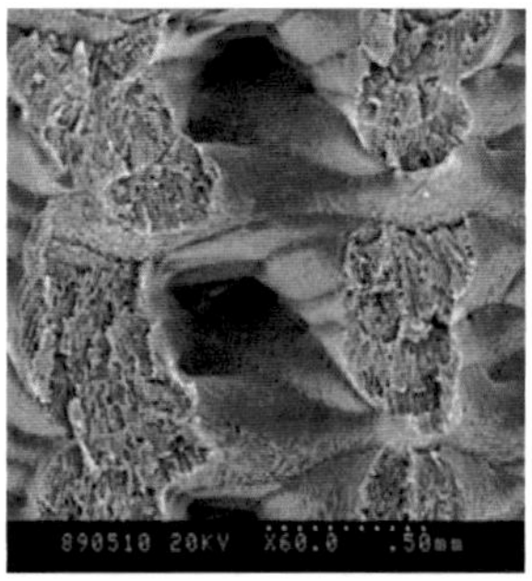

Figure 6. SEM image of fracture pattern

4. Conclusions

(1) The compression curve of the prepared porous biomaterial includes linear elastic stage, platform stage and densification stage. The porous materials have the characteristics of variability of volume, lateral flow, inhomogeneous deformation and compaction due to the porous pattern distinction, which leading to the complexity of deformation regularity.

(2) After compression test, the external section of fracture is compacted and the inner section of fracture is still porous.

(3) The fracture pattern is tearing shape and the fracture mechanism can be considered as brittle fracture.

Reference

[1] D.V. Andreeva and D.G. Shchukin, *Materials Today* **11** (2008) 24-30.
[2] R. Stoop, *Injury* **39** (2008) 77-87.
[3] K.X. Wei, G. Meng and S. Zhou, *J. Sound Vibration* **298** (2006) 1150-1153.
[4] G. Song, V. Sethi and H.N. Li, *Eng. Struct.* **28** (2006) 1513-1524.
[5] S. Sharma, P. Kaur and A. Jain, *Biomacromolecules* **4** (2003) 330-336.
[6] M. Sardar, I. Roy and M.N. Gupta, *Biotechnology Progress* **19** (2003) 1654-1658.
[7] M. Yokoyama, *Drug Discovery Today* **7** (2002) 426-432.
[8] D. Oupicky, T. Reschel and C. Konak, *Macromolecules* **36** (2003) 6863-6872.
[9] K. Haraguchi, T. Takehisa and M. Ebato, *Biomacromolecules* **7** (2006) 3267-3275.
[10] S. Kim and K.E. Healy, *Biomacromolecules* **4** (2003) 1214-1223.
[11] M.P. Lutolf and J.A. Hubbell, *Nature Biotechnology* **23** (2005) 47-55.
[12] Z. Xiong, Y.N. Yan and S.G. Wang, *Scripta Mater.* **46** (2002) 771-776.

Journal of Physics: Conference Series **240** (2010) 012153 doi:10.1088/1742-6596/240/1/012153

High pressure torsion of Cu-based metallic glasses

S Hóbor[1], Zs. Kovács[1] A. P. Zhilyaev[2], L. K. Varga[3], P. J. Szabó[4] and Á. Révész[1]

[1] Department of Materials Physics, Eötvös University, Budapest, H-1518, P.O.B. 32, Budapest, Hungary
[2] Centro Nacional de Investigaciones Metalúrgicas, 28040 Madrid, Spain and Institute for Metals Superplasticity Problems, RAS, 450001 Ufa, Russia
[3] Research Institute for Solid state Physics and Optics, Hungarian Academy of Sciences. H-1525 Budapest, P.O.B. 49, Hungary
[4] Deparment of Materials Science and Engineering, University of Technology and Economy, Budapest, H-1111 Hungary

hobors@metal.elte.hu

Abstract. Cu-Zr-Ti metallic glass was subjected to high pressure torsion applying different revolution times (180s, 120s, 60s). Both deformation and deformation rate dependent microstructural and thermal properties were characterized by scanning electron microscopy, X-ray diffraction and calorimetry, respectively. In order to estimate the temperature rise in the metallic glass during high pressure torsion, quasi three-dimensional heat conduction equation with a source term was considered. Solutions indicate that the saturation temperature strongly depends on the revolution time, i. e. on the deformation rate.

1. Introduction

Recently high pressure torsion (HPT) based on severe plastic deformation [1] was successfully applied to induce nanocrystals in amorphous alloys [2-4]. As known, shear deformation in metallic glasses is localized in very thin shear bands (~10 nm) below the glass transition temperature (T_g), characterized by enhanced atomic mobility and local temperature rise [5-7]. Approaching T_g, the plastic deformation shows a transition from localized to homogenous viscous flow [8].

In a recent study, the observed morphological and microstructural features introduced by HPT of an amorphous $Cu_{60}Zr_{30}Ti_{10}$ alloy was interpreted by temperature profiles obtained from a simple model based on one-dimensional heat-conduction [9]. Later a quasi three-dimensional model was developed and the effect of the instrumental parameters on the temperature rise in the disk-shaped samples was investigated in detail [10]. In this paper the role of the most important parameter, i. e. the revolution time on the microstructure and thermal stability of amorphous $Cu_{60}Zr_{30}Ti_{10}$ alloys is presented.

2. Experimental

An ingot of nominal composition of $Cu_{60}Zr_{30}Ti_{10}$ was prepared by induction melting of high purity (99,9 %) Cu, Zr and Ti metals. Ribbon was obtained by using single roller melt spinning technique in inert atmosphere with a rotating Cu-wheel. Pieces of the ribbon were placed between HPT-anvils and compressed into several porosity free, round shaped disks with an average diameter of 8 mm and an average thickness of about 150-250 μm under an applied pressure of 6 GPa, with 5 whole turns and three different revolution times (t_{rev}= 180 s, 120 s and 60 s). Hereafter the corresponding samples are denoted as #180, #120 and #60, respectively. The accumulated shear strain for HPT deformation at a radius r can be represented by $\varepsilon(r,t) = 2\pi \cdot t \cdot r / L \cdot t_{rev}$, where L and t are the thickness of the disk and the time, respectively. Microstructure was examined by a special high resolution double crystal diffractometer with negligible instrumental broadening equipped with a fine focus rotating copper anode (Nonius, FR 591). The cross-section of the beam impinging the specimen was 0.1x2 mm^2. Diffraction patterns were registered by FUJI Imaging Plate (BAS MS2025). Morphology studies were performed on a Philips XL 30 scanning electron microscopy (SEM) in backscattered electron (BSE) mode on the mechanically polished cross-section of the disks, while the compositional changes were quantitatively determined by energy dispersive X-ray (EDX) analysis with a relative accuracy of 3 %. A Perkin Elmer power compensated differential scanning calorimeter (DSC) was applied to investigate the thermal behavior and crystallization applying continuous heating experiments performed at scan rate of 40 K/min. Transformation enthalpies were obtained as the area of the exothermic peaks.

3. Results

As seen, the XRD patterns taken on both the centre and perimeter of the #180 sample are dominated by a broad, featureless amorphous halo at around 40.5 deg, however, severe deformation induce a couple of faint Bragg-peaks for the perimeter of #120 disk (see Fig. 1). The largest shear rate (sample #60) results in the formation of intense peaks corresponding to hcp $Cu_{51}Zr_{14}$ (p6/m a= 1.13 nm and c = 0.82 nm) and to hcp Cu_2ZrTi (p6$_3$/mmc, a = 0,505 nm, c=0,82 nm) phases even for the centre, while the amorphous halo becomes almost negligible for the perimeter of the disks.

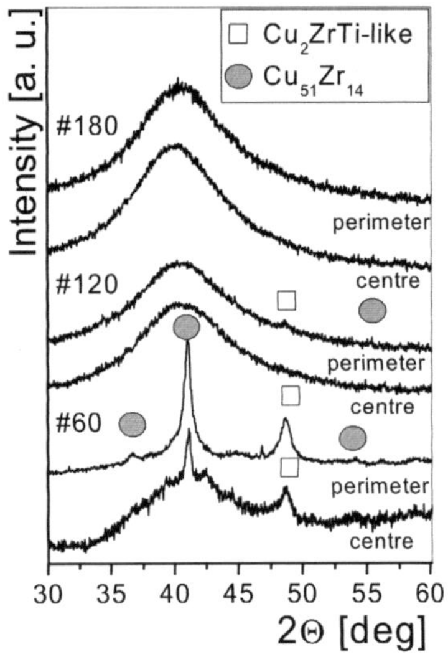

Figure 1. *X-ray diffractograms taken at the centre and perimeter of the HPT-disks.*

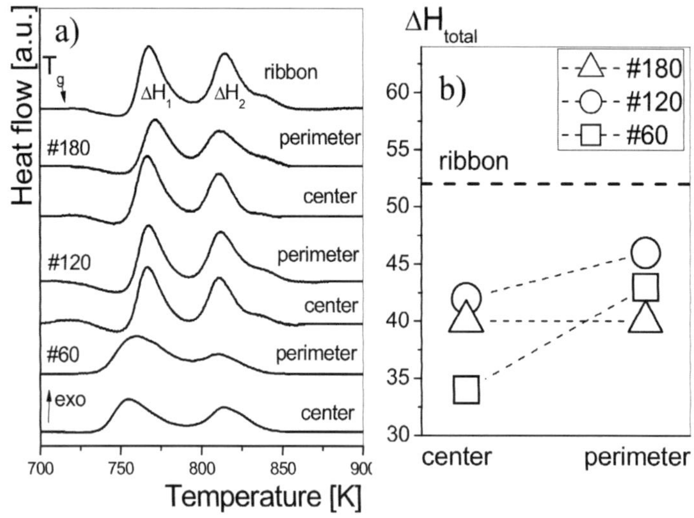

Figure 2. (a) *DSC curves of the fully amorphous alloy and the centre and perimeter of the disks.* **(b)** *Total crystallization enthalpies obtained from the DSC curves.*

The DSC thermograms of the #180 and #120 samples exhibit similar features as of the as-quenched ribbon, i. e. T_g is followed by two exothermic peaks (Fig 2a). For each disk the glass transition become less pronounced with increasing deformation, furthermore it totally disappears for the largest deformation rate (#60). At the same time the exothermic peak shift to lower temperatures and considerably broaden. The variation of the amorphous content with different shear rates can be characterized by the total enthalpy release ($\Delta H_{TOT} = \Delta H_1 + \Delta H_2$). For the slowest t_{rev} (#180) a 23% drop in ΔH_{TOT} is observed both for the centre and perimeter, independently from the magnitude of deformation, in line with the XRD data (see Fig 2b). In the case of samples #120 and #60 the decrease in ΔH_{TOT} for the centre is 19% and 35%, respectively, nevertheless a remarkable increase occurs for the perimeter.

The microstructural differences associated with various shear rates are visualized in the SEM BSE images (Fig. 3a-c). As seen in Fig. 3a, the entire sample #180 is characterized by several bunches of elongated dark grey bands of 20-40 µm slightly enriched in Ti ($Cu_{59}Zr_{27}Ti_{14}$) dispersed in the amorphous matrix ($Cu_{61}Zr_{29}Ti_{10}$). On contrary, the morphology of sample #120 (Fig. 3b) obeys a stronger deformation dependence, dominated by elongated particles of about 10-30 µm in the centre and much smaller ones (~5 µm) at the perimeter. Furthermore the central region of sample #60 contains several robust, dark grey blocks of 15-25 µm, homogenously dispersed in the matrix (Fig 3c). However, the BSE image of the perimeter exhibits only contrastless homogenous microstructure without any detectable crystalline blocks.

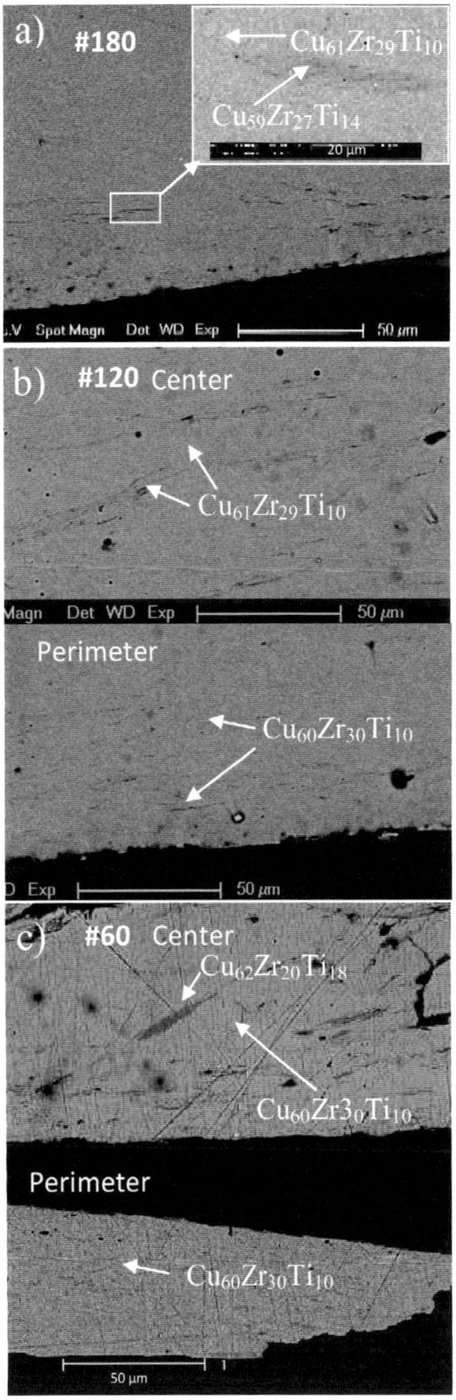

Figure 3. *SEM BSE images of the cross-section of HPT-disks with different revolution time* ***(a)*** *with t_{rev}=180 s where the inset is the enlargement of an elongated block,* ***(b)*** *with t_{rev}=120 s* ***(c)*** *with t_{rev}=60 s*

4. Discussion

The contrast between the microstructure and thermal behaviour of the HPT-disks processed by different deformation rates can be explained in terms of a three-dimensional thermoplastic model based on heat conduction [10]. In brief, the temperature evolution in the disk can be derived from the heat conduction equation

$$c(\mathbf{r})\cdot\rho(\mathbf{r})\cdot\frac{\partial T(\mathbf{r},t)}{\partial t} - \Delta(k(\mathbf{r})\cdot T(\mathbf{r},t)) = S(\mathbf{r},t),$$

where $c(\mathbf{r})$, $\rho(\mathbf{r})$ and $k(\mathbf{r})$ are the heat capacity, the density and the heat conductivity, respectively. $S(\mathbf{r},t)$, the plastic deformation induced heat release, can be calculated as

$$\begin{array}{ll} \beta\cdot\sigma\cdot d\varepsilon/dt, & \text{if } T<T_g \\ \beta\cdot\eta\,(d\varepsilon/dt)^2, & \text{if } T>T_g, \end{array}$$

where β, ε and η are the fraction of plastic work converted to thermoplastic heating, the shear strain and the viscosity that can be obtained from the Vogel-Fulcher equation [11].

The calculated temperature curves clearly show that the rate of the temperature rise as well as the saturation temperature drastically depends on the t_{rev} (Fig. 4). According to the model the temperature increases by ~180K for both samples #180 and #120, during the entire process of five whole revolutions, indicating that it remains below T_g, corresponding to extreme deformation in the narrow shear bands [7]. As a consequence plenty of nanocrystalline nuclei can form in the amorphous matrix near these locally heated deformation zones [9], that partly coalesce in the plane of shear deformation, forming bunches of elongated crystalline bands, as seen in Fig. 3a-b. For sample #180 the shear rate is the lowest, explaining the lack of deformation dependent microstructure confirmed by XRD and SEM. However, higher shear rate for sample #120 moderately limits the coalescence of the nuclei in the outer regions of the disk (Fig. 3b), in line with the observed slight increase in ΔH_{TOT} (Fig. 2b) corresponding to the excess grain boundary enthalpy of these nanocrystals. As seen in Fig. 4, for the sample deformed by the highest shear rate (#60) the temperature reaches T_g in a relatively short time (~55s), thereafter the rate of the temperature rise decreases abruptly, yielding a temperature saturation at around 715 K. Accordingly the HPT-process may be divided into two distinct stages according to the deformation mode. During the first stage of deformation plenty of nuclei form in the in the entire sample similarly to samples #180 and #120. However, after the temperature exceeds T_g the growth and coalescence of these nanocrystals is strongly limited by the persistent shear in the perimeter of the disk. Nevertheless this effect is less pronounced in the centre, moreover the devitrification is enhanced by the high temperature, yielding the formation of robust crystalline blocks seen in Fig. 3c, and the reasonable drop in ΔH_{TOT} in Fig. 2b.

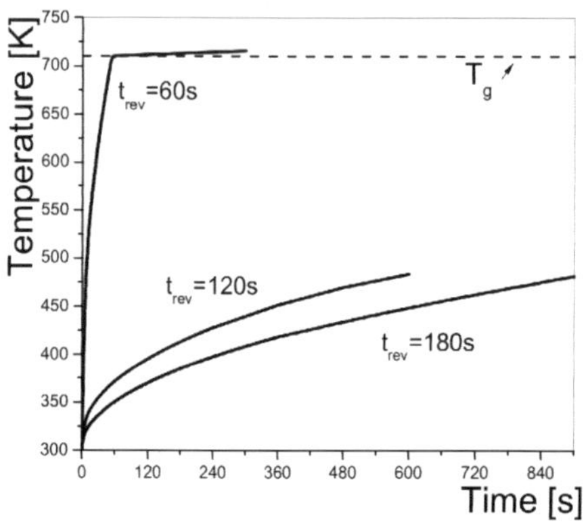

Figure 4. *Temperature evolution as a function of the processing time at the centre of the HPT-disk with different revolution time.*

5. Conclusions

Numerical calculations based on quasi three-dimensional heat conduction equation have satisfactorily described the microstructural features and thermal behaviour of amorphous $Cu_{60}Zr_{30}Ti_{10}$ alloy processed by HPT with different shear rate.

Acknowledgement

We appreciate the support of the Hungarian Scientific Research Fund (OTKA) under grant No. 67893. Á. R. is indebted for the Bolyai Scholarship of the Hungarian Academy of Sciences.

References

[1] Valiev R Z, Islamgaliev R K and Alexandrov I V, 2000 *Prog. Mater. Sci.* **45**, 103.
[2] Boucharat N, Rösner H and Wilde G, 2007 *J. Non-Crsyt. Solids* **354,** 592.
[3] Kovács Zs, Henits P, Zhylaev A P and Révész Á, 2006 *Scripta Mater.* **54,** 1733
[4] Révész Á, Hóbor S, Szabó P J, Zhilyaev A P, Kovács Zs, 2007 *Mater. Sci. Eng. A* **460-461**, 459.
[5] Chen H, He Y, Shiflet G J and Poon S J, 1994 *Nature* **367,** 541.
[6] Zhang Y and Greer A L, 2006 *Appl. Phys. Lett.* **89,** 071907.1
[7] Lewandowski J J, Greer A L, 2006 *Nature Mater.* **5** 15.
[8] Schuh C A, Hufnagel T C, and Ramamurty U, 2007 *Acta Mater.* **55,** 4067.
[9] Hóbor S, Révész Á, Szabó P J, Zhilyaev A P, Kovács Kis V, Lábár J L and Kovács Zs, 2008 *J. Appl. Phys.* **104** 033525
[10] Hóbor S, Kovács Zs and Révész Á, 2009 *J. Appl. Phys.* **106**, 023531
[11] Fulcher G S, 1925 *J. Am. Ceram. Soc.* **8,** 339.

Al-based metal matrix composites reinforced with nanocrystalline Al-Ti-Ni particles

S Scudino[1], F Ali[1], K B Surreddi[1], K G Prashanth[1], M Sakaliyska[1], J Eckert[1,2]

[1] IFW Dresden, Institut für Komplexe Materialien, Postfach 27 01 16, D-01171 Dresden, Germany
[2] TU Dresden, Institut für Werkstoffwissenschaft, D-01062 Dresden, Germany

s.scudino@ifw-dresden.de

Abstract. Al-based metal matrix composites containing different volume fractions of nanocrystalline $Al_{70}Ti_{20}Ni_{10}$ reinforcing particles have been produced by powder metallurgy and the effect of the volume fraction of reinforcement on the mechanical properties of the composites has been studied. Room temperature compression tests reveal a considerable improvement of the mechanical properties as compared to pure Aluminum. The compressive strength increases from 155 MPa for pure Al to about 200 and 240 MPa for the samples with 20 and 40 vol.% of reinforcement, respectively, while retaining appreciable plastic deformation with a fracture strain ranging between 43 and 28 %.

1. Introduction

Among the advanced engineering materials, Al-based metal matrix composites (MMCs) have been attracting considerable attention due to their remarkable mechanical properties, including low density, high strength and good fatigue and wear resistance [1,2]. In addition, MMCs offer the possibility to tailor their properties to meet specific requirements, which renders this type of materials quite unique in comparison to conventional unreinforced materials [1,2].

Different types of materials, ranging from the typical ceramic reinforcements, such as Al_2O_3 and SiC [1-3], to more unconventional reinforcements, such as quasicrystals [4,5] and complex metallic alloys (CMAs) [6], have been successfully used as reinforcements in MMCs. Other possible candidates as reinforcing agents in MMCs are amorphous, partially amorphous and nanocrystalline Al-based alloys, which have attracted widespread attention as potential candidates for structural as well as functional applications due to their high strength combined with low density [7-10].

To test the effect of the nanocrystalline particulate reinforcements on the mechanical properties of Al-based composites, in this work, Al-based MMCs reinforced with nanostructured $Al_{70}Ti_{20}Ni_{10}$ particles have been produced by powder metallurgy methods. The results reveal encouraging room temperature mechanical properties consisting of high strength combined with considerable plastic deformation.

2. Experimental

Milling experiments starting from pure elemental powder mixtures (purity >99.9 wt.%) with nominal compositions $Al_{70}Ti_{20}Ni_{10}$ were performed using a Retsch PM400 planetary ball mill and hardened steel balls and vials. No process control agent was used. The powders were milled for 100 h with a ball-to-powder mass ratio (BPR) of 10:1 and a milling intensity of 150 rpm. To avoid or minimize possible atmosphere contamination during milling, vial charging and any subsequent sample handling was carried out in a glove box under purified argon atmosphere (less than 1 ppm O_2 and H_2O). Al-based metal matrix composites consisting of elemental Al powder blended with different amounts (V = 20 and 40 vol.%) of $Al_{70}Ti_{20}Ni_{10}$ powders were synthesized through powder metallurgy methods. Consolidation was done by uni-axial hot pressing followed by hot extrusion under argon atmosphere at 773 K and 530 MPa. The extrusion ratio was 4:1. The density of the consolidated samples was evaluated by the Archimedes principle revealing a relative density of about 98 %. The phases and the microstructure were characterized by X-ray diffraction (XRD) using a Philips PW 1050 diffractometer (Co K_α radiation) and by scanning electron microscopy (SEM) using a Hitachi TM-1000 tabletop microscope. The Young's Modulus was evaluated by ultrasonic measurements using an Olympus 5900 PR ultrasonic pulser-receiver. According to the ASTM standard for compression testing [11], cylinders with a length/diameter ratio of 2.0 (10 mm length and 5 mm diameter) were prepared from the extruded samples. The specimens were tested with an INSTRON 8562 testing facility under quasistatic loading (strain rate of $8x10^{-4}$ s^{-1}) at room temperature. Both ends of the specimens were polished to make them parallel to each other prior to the compression test.

3. Results and discussion

The XRD patterns of the composites reinforced with different amounts of $Al_{70}Ti_{20}Ni_{10}$ particles are shown in Figure 1(a) along with the pattern of the as-milled $Al_{70}Ti_{20}Ni_{10}$ reinforcing powder. The pattern of the as-milled reinforcement is characterized by the presence of few broad diffraction peaks which can be identified as the Al_3Ti phase (space group Pm-3m). The diffraction peaks in are extremely broad, indicating that the phase formed is of nanometer dimensions.

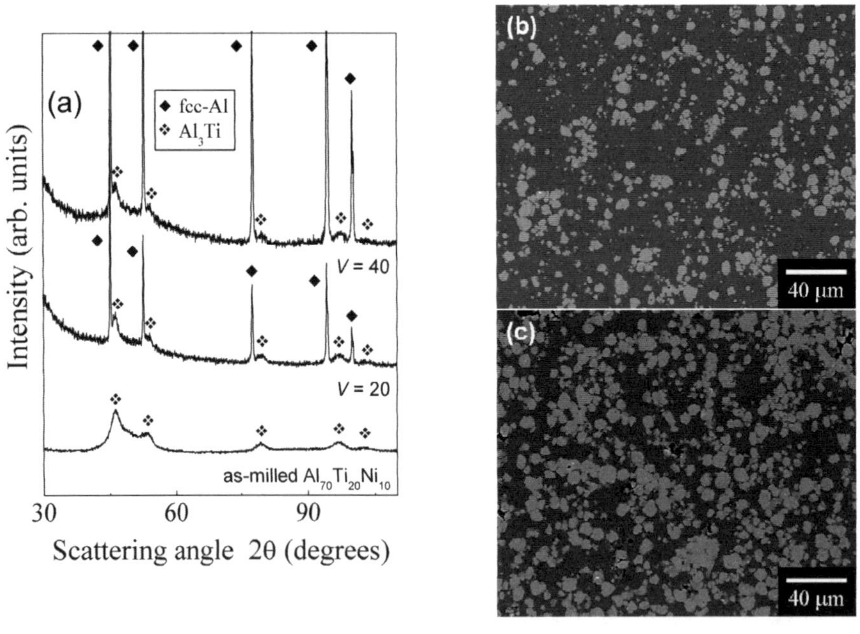

Figure 1. (a) XRD patterns (Co K_α radiation) of the as-milled $Al_{70}Ti_{20}Ni_{10}$ powder and composites with 20 (V = 20) and 40 vol.% (V = 40) of $Al_{70}Ti_{20}Ni_{10}$ reinforcement. SEM micrographs for the consolidated composites with (b) 20 vol.% and (c) 40 vol.% of $Al_{70}Ti_{20}Ni_{10}$ reinforcement.

The pattern of the composite with V = 20 shows the presence of fcc Al along with the Al_3Ti phase. No sharpening of the diffraction peaks belonging to the Al_3Ti phase can be detected, indicating that none or negligible grain growth of the reinforcing phase has occurred in the current consolidation conditions (extrusion at 773 K and 530 MPa). Similar results can be observed for the sample with 40 vol.% reinforcement, where the intensity of the $Al_{70}Ti_{20}Ni_{10}$ diffraction peaks are increased with respect to the composite with V = 20 due to the higher volume fraction of reinforcement. The detection of only two phases in the patterns of the composites with V = 20 and 40 indicates that no reaction between matrix and reinforcement to form additional phases took place during consolidation.

Figures 1(b) and 1(c) show the SEM micrographs taken from the cross-section of the consolidated composites. The images display a microstructure consisting of bright particles (the $Al_{70}Ti_{20}Ni_{10}$ reinforcement) homogeneously dispersed in the Al matrix (the dark regions). The $Al_{70}Ti_{20}Ni_{10}$ particles become more interconnected with increasing amount of reinforcement. Only few pores are visible, further corroborating the high density of the consolidated specimens.

Typical room temperature uni-axial compression true stress-true strain curves of the tests under quasistatic loading for the composite materials are shown in Figure 2(a) together with the curve for the extruded pure Al. The addition of the nanostructured reinforcement is very effective for improving the mechanical properties of the Al matrix. The compressive strength (the maximum compressive stress which the material is capable of sustaining [12]) increases from 155 MPa for pure Al to about 200 for the samples with 20 vol.% of reinforcement, while retaining appreciable plastic deformation reaching an ultimate strain of 43 % before fracture occurs. With increasing the volume fraction of reinforcement to 40 vol.%, the compressive strength further raises to 240 MPa, and the strain at break is about 28 %. These results indicate that the addition of the nanostructured reinforcement leads to composite materials with compressive strengths exceeding that of pure Al by 30 – 50 %, while retaining appreciable plastic deformation. Besides the enhancement of the mechanical properties, the addition of the nanocrystalline reinforcement has a positive effect on the Young's modulus of the composites, which increases from about 70 GPa for pure Al to 75 and 85 GPa for the composites with 20 and 40 vol.% of reinforcement, respectively [Figure 2(b)].

The strengthening effect of particles in metal matrix composites is generally attributed to two factors [13]: (*i*) the load bearing effect of the reinforcement, in which the reinforcement can share the applied stress directly by stress transfer from the matrix [14-16] and (*ii*) the dislocation strengthening in the matrix, which is related to the nucleation of additional dislocations in the matrix due to the introduction of the reinforcement [17,18]. Recently, it has been found for composites with large

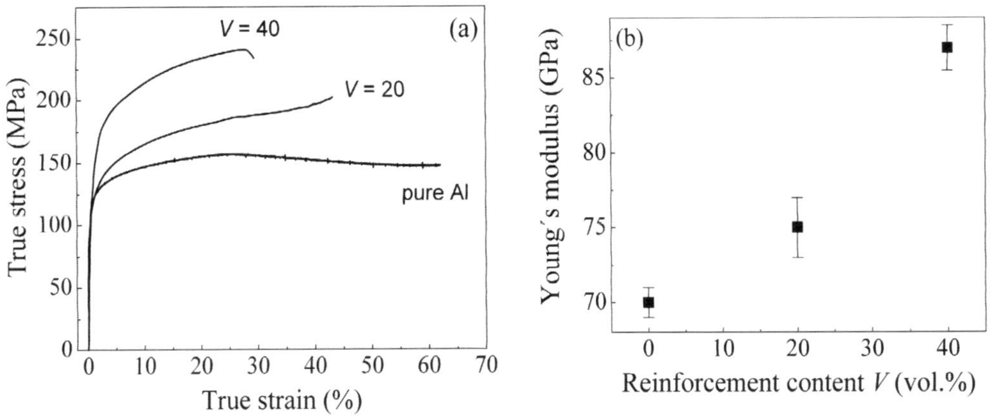

Figure 2. (a) Room temperature compression true stress-true strain curves for the hot pressed and hot extruded pure Al, composites with 20 vol.% (V = 20) and 40 vol.% (V = 40) of nanostructured $Al_{70}Ti_{20}Ni_{10}$ particles; (b) Young's modulus as a function of the reinforcement content.

amount of reinforcement (≥ 20 vol.%) that the mean free path of the matrix (i.e. the characteristic matrix ligament size λ) plays a dominant role in the strengthening of Al-based composites [6]. This size effect, similar to the strengthening by grain refinement, can significantly contribute to the strength of the material because the matrix/particle interface can effectively inhibit dislocation movement. The matrix ligament size can be remarkably reduced by using small reinforcing particles [6]. Due to their small size (< 20 μm), the present nanocrystalline particles are very effective in reducing the matrix ligament size λ, which decreases from 15 μm for the sample with V = 20 to about 8 μm for the composite with 40 vol.% of reinforcement, therefore, explaining the observed increase of strength with increasing reinforcement content.

4. Summary

To test the validity of the nanocrystalline particulate reinforcements on the mechanical properties of composites, Al-based MMCs consisting of pure Al reinforced with different volume fractions of nanostructured $Al_{70}Ti_{20}Ni_{10}$ particles have been produced by powder metallurgy methods. The results reveal encouraging room temperature mechanical properties. The compressive strength increases from 155 MPa for pure Al to about 200 and 240 MPa for the samples with 20 and 40 vol.% of reinforcement, respectively, while retaining appreciable plastic deformation with a fracture strain ranging between 43 and 28 %.

Acknowledgments

The authors would like to thank B. Bartusch and H. Schulze for technical assistance, and J.S. Kim for stimulating discussions. This work was supported by the German Science Foundation under grant Ec 111/16-2. F. Ali acknowledges the PIEAS for financial support. K. B. Surreddi is grateful for the financial support provided by the DAAD.

References

[1] Clyne T W and Withers P J. 1993 *An Introduction to Metal Matrix Composites* (Cambridge University Press Cambridge)
[2] Kainer K U 2006 *Metal Matrix Composites. Custom-made Materials for Automotive and Aerospace Engineering* (WILEY-VCH Weinheim)
[3] Slipenyuk A, Kuprin V, Milman Y, Goncharuk V and Eckert J 2006 *Acta Mater.* **54** 157
[4] Schurack F, Eckert J and Schultz L 2003 *Phil. Mag.* **83** 1287
[5] El Kabir T, Joulain A, Gauthier V, Dubois S, Bonneville J and Bertheau D 2008 *J. Mater. Res.* **23** 904
[6] Scudino S, Liu G, Sakaliyska M, Surreddi K B and Eckert J 2009 *Acta Mater.* **57** 4529
[7] Inoue A 1998 *Progess Mater. Sci.* **43** 365
[8] Inoue A and Kimura H 2001 *J. Light Metals* **1** 31
[9] Scudino S, Surreddi K B, Sager S, Sakaliyska M, Kim J S, Löser W and Eckert J 2008 *J. Mater. Sci.* **43** 4518
[10] Prashanth K G, Scudino S, Murty B S and Eckert J 2009 *J. Alloys Comp.* **477** 171
[11] ASTM E9-89aR00 2000 *Standard test methods for compression testing of metallic materials at room temperature* (West Conshohocken, PA: ASTM)
[12] ASTM E6-03 2003 *Standard terminology relating to methods of mechanical testing, Annual Book of ASTM Standards* vol 03.01 (West Conshohocken, PA: ASTM)
[13] Lloyd D J 1994 *Inter. Mater. Rev.* **39** 1
[14] Piggot M R 1980 *Load-Bearing Fiber Composites* (Oxford: Pergamon Press)
[15] Nardone V C and Prewo K M 1986 *Scripta Mater.* **20** 43
[16] Nardone V C 1987 *Scripta Mater.* **21** 1313
[17] Vogelsang M, Arsenault R H and Fisher R M 1986 *Met. Trans. A* **17** 379
[18] Shi N, Wilner B and Arsenault R J 1992 *Acta Metall. Mater.* **40** 2841

Production of high-strength $Al_{85}Y_8Ni_5Co_2$ bulk alloy by spark plasma sintering

K B Surreddi[1], V C Srivastava[2,3], S Scudino[1], M Sakaliyska[1], V Uhlenwinkel[2], J S Kim[4] and J Eckert[1,5]

[1] IFW Dresden, Institut für Komplexe Materialien, Postfach 27 01 16, 01171 Dresden, Germany
[2] Institute für Werkstofftechnik, Universität Bremen, Badgasteiner Str.3, D-28359 Bremen, Germany
[3] Metal Extraction & Forming Division, National Metallurgical Laboratory, Jamshedpur-831007, India
[4] Research Center for Machine Parts and Materials Processing, University of Ulsan, San 29, 680749 Ulsan Korea
[5] TU Dresden, Institut für Werkstoffwissenschaft, D-01062 Dresden, Germany

k.b.surreddi@ifw-dresden.de

Abstract. Highly dense bulk samples were produced by spark plasma sintering (SPS) through combined devitrification and consolidation of partially amorphous $Al_{85}Y_8Ni_5Co_2$ gas atomized powders. The microstructure of the consolidated samples shows a mixed structure containing crystalline, ultrafine-grained and amorphous/nanocrystalline particles. The sintered sample exhibits a remarkable high strength of about 1050 MPa combined with 3.7 % fracture strain.

1. Introduction

In recent years, significant efforts have been dedicated to the production of amorphous, partially amorphous, nanocrystalline and ultra-fine grained (UFG) Al-based alloys [1-5]. Due to their high strength combined with low density, these alloys are very attractive candidates for structural and functional applications [1-5]. For example, Al-based amorphous alloys with more than 85 at.% Al, such as $Al_{85}Y_8Ni_5Co_2$, show a tensile fracture strength exceeding 1200 MPa combined with good bending ductility [6]. Partially crystallized Al-based amorphous ribbons with a microstructure consisting of an uniform dispersion of fcc Al nanoparticles within the residual amorphous matrix exhibit even higher fracture strengths and hardness with respect to fully amorphous alloys of the same composition [7-11]. However, in spite of their promising mechanical properties, the maximum scale of these materials is generally limited to a thickness of less than 100 micrometers, which strongly restricts their engineering application.

To overcome this limitation, powder metallurgical methods, including powder consolidation, have been used to produce bulk high strength Al-based materials [4]. Among the consolidation techniques used, spark plasma sintering (SPS) is particularly attractive due to many advantages, including good densification, a rapid consolidation process and reduced grain growth [12-13]. Accordingly, in this work, partially amorphous $Al_{85}Y_8Ni_5Co_2$ gas atomized powders were consolidated by SPS. Consolidation by SPS results in a heterogeneous microstructure consisting of crystalline, nanocrystalline and amorphous phases, which leads to promising room temperature mechanical properties, i.e. high strength combined with good plastic deformation.

2. Experimental Details

Gas atomized partially amorphous powder with nominal composition of $Al_{85}Y_8Ni_5Co_2$ (at.%) were used as starting material. The Phase formation and the microstructure were studied by X-ray diffraction (XRD) using a Philips PW 1050 diffractometer (Co Kα radiation) and by scanning electron microscopy (SEM) using a JEOL JSM 6400 microscope. The thermal stability of the as-atomized powders was studied by differential scanning calorimetry (DSC) at a heating rate of 40 K/min with a Perkin-Elmer DSC7 under a continuous flow of purified argon. The powders were consolidated into cylindrical specimens of 10 mm diameter under high vacuum by spark plasma sintering at a temperature of 673 K (heating rate of 10 K/min) and an applied pressure of 500 MPa using a Sumitomo SPS-515S sintering machine. The density of the consolidated samples was evaluated by the Archimedes principle and was found to be 99 % of the theoretical density. Cylindrical specimens with a length/diameter ratio of 2 (6 mm length and 3 mm diameter) were prepared from the SPS samples by wire erosion method. The specimens were tested with an INSTRON 8562 testing facility under quasistatic compressive loading (strain rate of $1x10^{-4}$ s^{-1}) at room temperature. Both ends of the specimens were polished to make them parallel to each other prior to the compression test.

3. Results and discussion

Figure 1 displays the XRD pattern of the as-atomized $Al_{85}Y_8Ni_5Co_2$ powder. The pattern shows a weak broad maximum characteristic of an amorphous structure centered at about 45° along with the diffraction peaks from fcc Al, and Al_3Y and Al_9Co_2 intermetallic phases. This indicates that the cooling rate during gas atomization is not high enough to suppress the formation of crystalline phases [14]. Figure 2 shows the isochronal DSC scan (40 K/min) of the as-atomized powder and the shrinkage rate behavior during the SPS process. The DSC scan reveals a clear glass transition (T_g) at 550 K followed by three exothermic peaks with onset temperatures T_{x1} = 570 K, T_{x2} = 610 K and T_{x3} = 660 K due to crystallization of the glassy phase. A similar multi-step crystallization behavior has been observed by several authors [6,15-18] and corresponds to the formation of fcc Al in the first crystallization event followed by the formation of the Al_3Y and Al_9Co_2 phases at higher temperatures [6,16-17].

During consolidation by SPS a constant pressure of 500 MPa was applied from room temperature through the crystallization events to the final sintering temperature (673 K to avoid complete

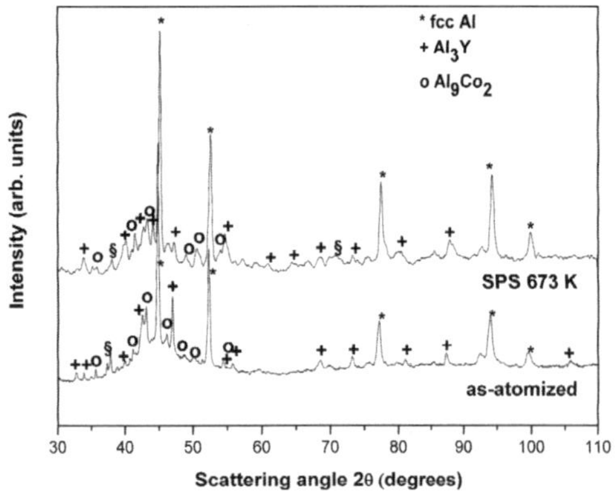

Figure 1. XRD patterns of the as-atomized powder and the bulk sample consolidated by SPS at 673 K.

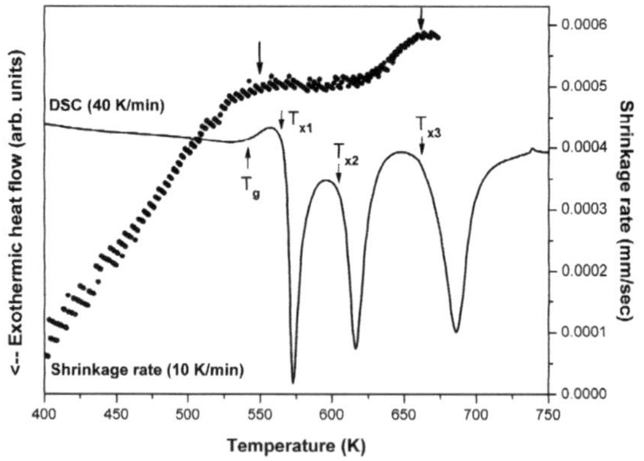

Figure 2. Isochronal DSC scan (40 K/min) of as-atomized powder and shrinkage rate of the powder during SPS (data points).

crystallization), which was held for about 3 minutes. The consolidation of the $Al_{85}Y_8Ni_5Co_2$ powder at high temperatures can thus be considered as a combined *in-situ* (partial) devitrification and densification of the powder. The shrinkage rate during SPS (data points in Figure 2) reveals two maxima at about 550 and 670 K, respectively, which are associated with the highest densification rates. These two main shrinkage events take place at the glass transition region (where the glassy phase transforms into the supercooled liquid) and in the temperature range between the second and the third crystallization events, where the residual amorphous phase with a different composition may undergo the glass transition, as observed for the $Al_{87}Ni_8La_5$ glassy powder [19].

Figure 3. SEM micrographs of the sample sintered at 673 K.

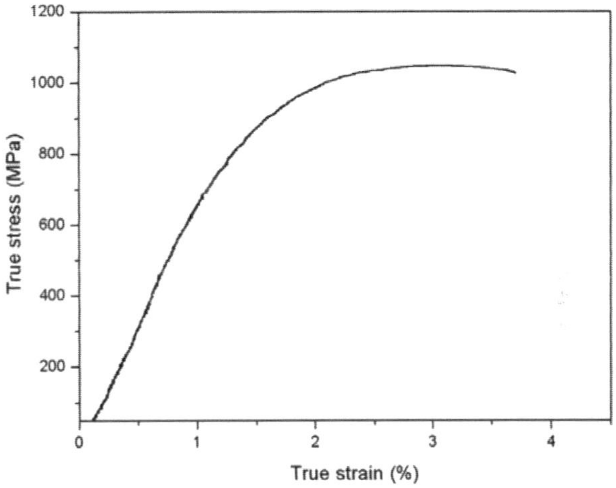

Figure 4. Room temperature true stress-true strain curve of the consolidated sample.

The XRD pattern of the $Al_{85}Y_8Ni_5Co_2$ powder sintered at 673 K is also shown in Figure 1. The structure is similar to the as-atomized powder and consists of fcc-Al together with the two intermetallic compounds Al_3Y and Al_9Co_2. The pattern still shows a weak broad maximum at about 45°, indicating that an amorphous phase is retained after sintering. The diffraction peaks in Figure 1 are rather broad, indicating that the phases formed are of nano sized or ultra-fine dimensions.

Figure 3 shows SEM images of the sample sintered by SPS at 673 K. The microstructure is rather heterogeneous, consisting of particles with different characteristics: small featureless particles (most likely with nano-crystalline or amorphous structure), particles with large crystalline features, ultrafine-grain structured particles and particles with mixed crystalline/nano-crystalline/amorphous features. EDX investigations reveal that the dark areas between the particles in Figure 3 correspond to fcc Al.

A typical room temperature uni-axial compression stress-strain curve for the sintered sample is shown in Figure 4. The material exhibits a yield strength (0.2 % offset) of about 800 MPa followed by a region with strain hardening up to the maximum stress of 1050 MPa and fracture strain of 3.7 %.

Spark plasma sintering of the $Al_{85}Y_8Ni_5Co_2$ powder leads to highly dense specimens displaying high strength combined with good plastic deformation. Such a behavior is presumably due to the multi-phase microstructure consisting of soft fcc-Al and high-strength intermetallic compounds (see Figures 1 and 3). However, the observed room temperature plastic deformation is in contrast to what was reported for other Al-based alloys produced by consolidation of gas-atomized powders [4], which display a similar microstructure. For example, although fully crystallized Al-Ni-Y-Co samples exhibit an extremely high compressive strength of 1420 MPa [4], the plastic strain is only about 1 % [20]. This is similar to recent results of Sasaki *et al.* [21] on nanocrystalline $Al_{85}Ni_{10}La_5$ samples produced by SPS of gas-atomized amorphous powders, which show a compressive strength exceeding 1200 MPa, but no plastic deformation. A possible explanation for the larger plastic deformation of the present material may be linked to the heterogeneous microstructure characterizing the consolidated sample, consisting of crystalline, nanocrystaline and amorphous phases (Figure 3), which might provide the right balance between soft and hard phases explaining the observed high strength combined with good plastic deformation.

4. Summary

In-situ devitrification and consolidation of gas-atomized $Al_{85}Y_8Ni_5Co_2$ powders into highly dense bulk specimens was carried out by spark plasma sintering. Room temperature compression tests of the consolidated bulk material reveal remarkable mechanical properties, namely, a high compression strength of 1050 MPa combined with a fracture strain of about 3.7 %. These findings demonstrate that the combined devitrification and consolidation of glassy precursors by spark plasma sintering is a suitable method for the production of Al-based materials characterized by high strength and good plastic deformability.

Acknowledgements

The authors would like to thank B. Bartusch, M. Gründlich and H. Schulze for technical assistance and F. Ali and M. Stoica for stimulating discussions. This work was supported by the German Science Foundation under grant Ec 111/16-2. K.B. Surreddi is grateful for the financial support provided by DAAD.

References

[1] Inoue A, Ohtera K, Tsai A P and Masumoto T 1988 *Jpn. J. Appl. Phys.* **27** 479
[2] Inoue A 1998 *Progress Mater. Sci.* **43** 365
[3] Schurack F, Eckert J and Schultz L 2001 *Acta Mater.* **49** 1351
[4] KawamuraY, Mano H and Inoue A 2001 *Scripta Mater.* **44** 1599
[5] Watanabe M and Inoue A 1993 *Mater. Trans. JIM.* **34** 162
[6] Inoue A, Matsumoto N, Matsumoto T 1990 *Mater. Trans. JIM* **31** 493
[7] Yoshizawa Y, Oguma S and Yamaauchi K 1988 *J. Appl. Phys.* **64** 6044
[8] Suzuki K, Kataoka N, Inoue A and Masumoto T 1990 *Mater. Trans. JIM* **31** 743
[9] Inoue A *Mater. Sci. Forum* 1995 **179-181** 691
[10] Kim Y H, Inoue A and Masumoto T 1990 *Mater. Trans. JIM* **31** 747
[11] Eckert J 2002 *Nanostructural Materials: Processing, Properties, and Applications*, ed. Koch C C (Norwich, NY: William Andrew Publishing), 423-525
[12] Tokita M 1998 *Mater Sci. Forum* **308–311** 83
[13] Omari M 2000 *Mater. Sci. Eng. A* **287** 183
[14] Suryanarayana C 1991 *Processing of Metals and Alloys, Materials Science and Technology: A comprehensive treatment*, ed. R W Cahn, P Haasen and E J Kramer (VCH, Weinheim) Vol. 15
[15] Yewondwossen M, Dunlap R A and Lloyd D J 1992 *J. Phys.: Condens. Matter.* **4** 461
[16] Dong P, Hou W L, Chang X C, Quan M X and Wang J Q 2007 *J. Alloys Comp.* **436** 118
[17] Bassim N, Kiminami C S, Kaufman M J, Oliveira M F, Perdigao M N R V and Botta Filho M J 2001 *Mater. Sci. Eng. A* **304-306** 332
[18] Schumacher P and Greer A L 1993 *Key Eng. Mater.* **81-83** 631
[19] Scudino S, Surreddi K B, Nguyen H V, Liu G, Gemming T, Sakaliyska M, Kim J S, Vierke J, Wollgarten M and Eckert J *J. Mater. Res.* (in press)
[20] Inoue A, Kawamura Y, Kimura H M and Mano H 2001 *Mater. Sci. Forum* **360-362** 129
[21] Sasaki T T, Hono K, Vierke J, Wollgarten M and Banhart J 2008 *Mater. Sci. Eng. A* **490** 343

A statistical physics consideration about the strength of small size metallic glass pillars

Changqiang Chen, Yutao Pei, Jeff.Th.M. De Hosson

Department of Applied Physics, Materials Innovation Institute, University of Groningen, Nijenborgh 4, 9747 AG Groningen, Netherlands

cqchen01@gmail.com

Abstract. We have fabricated micro-/nano- pillars of a Zr-based metallic glass, $Zr_{50}Ti_{16.5}Cu_{15}Ni_{18.5}$, with pillar tip diameters ranging from ~750 nm to ~110 nm. These pillars were mechanically tested quantitatively in-situ in a Transmission Electron Microscope (TEM). Due to a slight tapering of ~3°, the deformation accommodated with shear bands is driven downwards from the top. However, the real time monitoring of shear bands evolution in-situ in TEM makes it possible to measure the effective load-bearing diameter, i.e., the diameter ahead of the most forefront shear band, which enables the precise measurement of yield strength of these pillars. Consequently, it turns out that the yield strength is essentially size independent and close to the bulk value. Statistical physics description of strength of small size metallic glasses is discussed, and it is concluded that Weibull Statistics is inappropriate in describing strength of metallic glasses at micro and nano length scales.

1. Introduction

Size effect on strength of metallic glasses receives recently increasing attention, due to their unique mechanical properties and potential application in microelectromechanical systems. Compared to single crystals which exhibit strengths times, or even orders, higher than their bulk counter part at micro- or nano- scale due to a dislocation source limited deformation [1-3], MGs obviously lack a dislocation involved mechanism due to the intrinsically disordered structure. Nevertheless, a couple of recent publications reported more or less size effects on strength of micrometer scaled specimens [4-8]. Lee et al [4] and Lai et al [5] reported up to 100 and 86 per cent increment of yield strength, respectively, of micrometer-diameter pillars of a Mg-based and a Zr-based MG over the values of bulk glasses, and interpreted the size effect according to Weibull statistics. Cheng et al [6] measured a yield stress of 2.25 GPa of micropillars of another Zr-base glass whose bulk value is 1.9 GPa. Schuster et al [7] reported modest increase (~9%) of the yield strength of $Pd_{40}Ni_{40}P_{20}$ micropillars over the bulk one, and later [8] attributed this effect to measurement artifacts. In contrast, experiments by Volkert et al [9] on micropillars of PdSi showed a slight decrease of strength with decreasing size, but it is still hard to conclude about a possible size effect because of experimental errors [9,10].

Apparently, there is at present no general consensus on how size effects influence the strength of metallic glasses. Accurate measurement of individual small metallic specimens is still a challenge mainly because of the difficulties in achieving uniform gauge lengths of pillars with diameters smaller than 1 μm. This is in fact one of the origins of the discrepancy in size effects reported in literature. Also the positioning of the specimen and alignment of the test system is a critical issue for extremely small specimens. In this letter we present quantitative in situ TEM compression of micro-/nano- MG pillars, which enables measuring pillars of diameters down to tens nm scale. Quantitative load response is correlated with dynamic observation of yielding evolution, which allows real time measurement of effective diameter bearing the externally applied load.

2. Experimental

Amorphous micropillars were fabricated by FIB (focused ion beam) from metallic glass ribbon which is Zr-based $Zr_{50}Ti_{16.5}Cu_{15}Ni_{18.5}$, by using An FEI Strata DB235 dual beam FIB-SEM system operating at 30 KV and 20 PA for final shaping. More than 40 pillars with tip diameter ranging from 110 nm to 750 nm and having well defined gauge lengths ranging from 1.2-2.0 μm, aspect ratios from 3 to 8, and small taper angles between 2.0-3.5°, are fabricated and studied.

In situ TEM compression experiments were performed using a recently developed Hysitron picoindenter TEM holder (Hysitron Inc., Minneapolis, MN) equipped on JEOL 2010F TEM, with a diamond flat punch of 2 μm diameter. It is integrated with a miniature capacitive load–displacement transducer that permits high resolution load and displacement measurements (resolution of ~0.3μN in load, ~1 nm in displacement) [1113]. A nominal strain rate of $\sim 10^{-2}$/s is used.

3. Results

The deformation of all specimens is dominated by intermittent shear banding. With the small tapering, the deformation is driven downwards from the top, leaving the lower part purely elastically deformed. The effective diameter was measured just ahead of the most forefront shear band, which is the minimum diameter at the purely elastically deformed region. Although dynamic measurement was continuously performed, to ensure accurate measuring of effective diameter which is the main source of error, a compression test was normally interrupted at a certain stage to acquire images still after unloading. The recorded peak load before unloading divided by measured diameter was taken as yield strength. A typical example is illustrated in Figures. 1. It can be seen that the engineering stress taking into account the diameter change (taper angle) during compression still overestimates the results. However, the yield stresses calculated by measuring the effective diameter in two successive test cycles (indicated in the curve) are essentially consistent (1.73GPa), indicating an accurate measurement.

The values of the yield strength measured from a large number of Zr-based micropillars are plotted in Figure. 2 as a function of effective pillar diameter. With the accurate measurement of load-bearing diameter, the error bars indicate very small uncertainties in the measurement of the yield strength. It turns out that there is no dependence of the yield stress on pillar diameter over the current size scale. The result is understandable. As well known the high yield strength of metallic glass is due to the lack of an ‘easy’ flow mechanism like ‘dislocation sliding’ in crystalline materials. As a consequence, in metallic glass one would not expect a “smaller is stronger” phenomenon observed in nanocrystalline materials due to dislocation starvation mechanism. Strength of BMGs is essentially controlled by interatomic bonding and has a roughly linear relationship with elastic modulus as $\sigma_f = 0.02E$ or $\tau_y = 0.027G$ [14], where σ_f and τ_y are normal and shear fracture strength, E and G, Young’s modulus and shear modulus, respectively.

4. Discussion

This finding is different from the refs [4-7], but is consistent with refs [8,9]. When inspecting literature about the yield strength of pillars it should be kept in mind that, by assuming a tip diameter or an average diameter as the pillar size without taking into account the dynamical evolution of the effective diameter during compression, an artificial size effect will always be attained, since a constant tapering angle has larger influence on thinner pillars. This kind of error however is circumvented here by in-situ dynamically measuring the effective diameter as mentioned above.

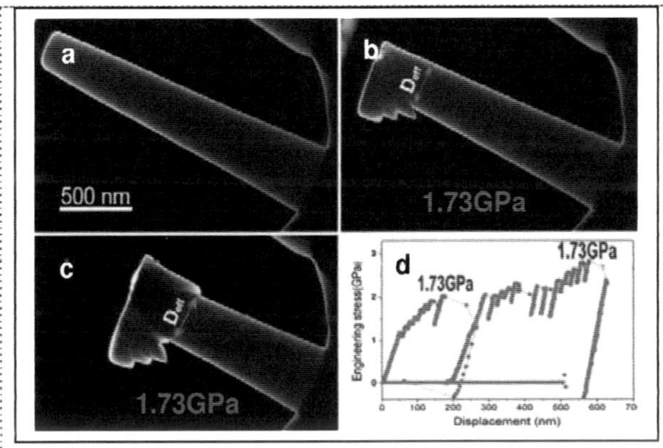

Figure 1. TEM micrographs showing the deformation morphology of a pillar of 270 nm tip diameter subjected to two loading-unloading cycels: (a) before compression and (b) and (c) after unloading of each compression cycle with the yield stress indicated; the effective load bearing diameter is measured just ahead of the most forefront shear band as illustrated in (b) and (c); (d) engineering stress vs displacement curve, with the yield stress measured in-situ in TEM is also indicated.

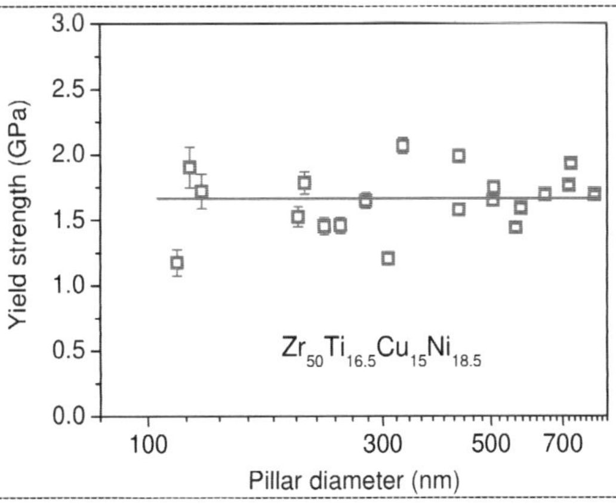

Figure 2. Yield stress versus pillar diameter for both metallic glasses. The error bars reflect the uncertainties in the stress due to actual pillar diameter, which are minimized by measuring the effective diameter in the in situ TEM images.

It would be also interesting to make a connection to a statistical physics description of strength by investigating the dependence of the failure stress on the effective size. The reported small size induced increase in yield stress has been interpreted according to Weibull statistics (WS). In WS [15] the probability of failure at a stress σ is expressed by $P_y(\sigma)=1-\exp[-V(\sigma/\sigma_0)^m]$, where σ_0 is a scale parameter, m Weibull modulus. This equation is widely used in fitting of $\ln[-\ln(1-P_f)]$ versus $\ln\sigma$, with m being the slope. A size dependence of the mean value of σ can be derived, which is described by $\sigma_1/\sigma_2=(V_2/V_1)^{1/m}$ or generally $\sigma\propto V^{-1/m}$. If WS would be continuously extrapolated from macro- to micro-/nano- scale specimens, the strength of submicrometer MGs would be exceptionally high. The reported size independent strength in this work indicates that the Weibull statistics cannot describe the nature of strength data of submicrometer scale MGs.

WS does not apply due to the fact that it does not make a distinction between nucleation, propagation and overlap of stress fields. WS is derived based on the weakest link hypothesis that fracture occurs once the fracture of the weakest link occurs and there is no interaction among flaws. The size effect described is due to the fact that a WS material has high probability of having a larger destructive defect in a larger volume [15]. Together with a Griffith failure criterion, sizes of the destructive flaws $a_{c,1}$ and $a_{c,2}$ in V_1 and V_2 respectively are correlated by $a_{c,1}/a_{c,2}=(V_1/V_2)^{2/m}$. Clearly, very small flaws become fracture origins in small specimens. In fact, an inverse-power-law distribution of flaw size (with the exponent being $-m/2-1$) is a necessary condition of WS [16]. This tendency of a large amount of smaller flaws acting as fracture (or yield) origin in smaller specimens may put the applicability of WS in small specimens questionable. It has been recognized that materials become insensitive to flaws when the flaw size is smaller than 30 nm [17], since the widely used engineering concept of stress concentration is no longer valid for such small flaws.

Although WS may be useful in describing the strength of BMGs [18,19] when the strength is limited by large flaws (>micrometer), like as-casting pores, inclusions, surface irregularities, which act effectively as stress-concentrators, it is not applicable to sub-micrometer MG pillars as shown above.

5. Conclusions

We have fabricated micro-/nano- pillars of a Zr-based metallic glass, $Zr_{50}Ti_{16.5}Cu_{15}Ni_{18}$, with pillar tip diameters ranging from ~650 nm to ~90 nm. These pillars were mechanically tested in-situ in TEM as a function of pillar diameters. For other details see [20]. It turns out that the yield strength of the metallic glass is close to those of the bulk glass and is essentially size independent. Statistical physics description of strength for small size metallic glass is discussed, and Weibull statistics is found inappropriate in describing the strength of metallic glasses at micro and nano length scales.

Acknowledgements
This research was carried out under the Project No. MC7.06246 in the framework of the research programme of the Materials Innovation Institute M2i (www.M2i.nl), the former Netherlands Institute for Metals Research (NIMR), Delft, the Netherlands.

References

[1] M. D. Uchic, D. M. Dimiduk, J. N. Florando and W. D. Nix, *Science* **36** 4 (2003).
[2] C. A. Volkert and E.T. Lilleodden, *Phil. Mag.* **86** 13 (2006).
[3] B. Wu, A. Heidelberg and J.J. Boland, *Nature Mater.* **4** 525 (2005).
[4] C. J. Lee, J. C. Huang and T. G. Nieh, *Appl. Phys. Lett.* **91** 161913 (2007).
[5] Y. H. Lai, C. J. Lee, Y. T. Cheng, H. S. Chou, H. M. Chen, X. H. Du, C. I. Chang, J. C. Huang, S. R. Jain, J. S. C. Jang and T. G. Nieh, *Scripta Mater.* **58** 890 (2008).
[6] S. Cheng, X L Wang and P. K. Liaw, *Appl. Phys. Lett.* **91** 201917 (2007).
[7] B. E . Schuster, Q. Wei, M.H. Ervin, S. Q. Hruszkewycz, M. K. Miller, T.C. Hufnagel and K. T. Ramesh, *Scripta Mater.* **57** 517 (2007).
[8] B. E. Schuster, Q.Wei, T.C. Hufnagel and K. T. Ramesh, *Acta Mater.* **56** 5091 (2008).
[9] C. A. Volkert, A. Donohue and F. Spaepen, *J. Appl. Phys.* **103** 083539 (2008).
[10] C. A. Volkert, Private communication.
[11] W. A. Soer, J.Th.M. De Hosson, A.M. Minor, J.W. Morris Jr. and E.A. Stach, *Acta Mater.* **52** 5783 (2004).
[12] W. A. Soer, J. Th. M. De Hosson, A. M. Minor, Z. W. Shan, S. A. Syed Asif and O. L. Warren, *Appl. Phys. Lett.* **90** 181924 (2007)
[13] J. Th. M. De Hosson, W. A. Soer, A. M. Minor, Z. W. Shan, E. A. Stach, S. A. Syed Asif and O.L.Warren, *J. Mater. Sci.* **41**, 7704 (2006).
[14] W. L. Johnson and K. Samwer, *Phys. Rev. Lett.* **95** 195501 (2005).
[15] W. Weibull, *J. Appl. Mech.* **18** 293 (1951).
[16] R. Danzer, P. Supancic, J. Pascual and T. Lube. *Eng. Fract. Mech.* **74** 2919 (2007)
[17] H. Gao, B. Ji, I. L. Jager, E. Arzt and P. Fratzl. *Proc. Natl. Acad. Sci. U S A* **100** 5597 (2003).
[18] W. F. Wu, Y. Li and C.A. Schuh. *Phil. Mag.* **88** 71 (2008).
[19] Y. Y. Zhao, E. Ma and J Xu. *Scripta Mater.* **58** 496 (2008).
[20] C.Q.Chen, Y.T.Pei and J.Th.M. De Hosson, *Phil.Mag. Lett.* (2009), in press.

Experimental validation of microstructure evolution in crystalline materials

C Rehrl[1], S Kleber[2], T Antretter[3] and R Pippan[1]

[1] Erich Schmid Institute of Materials Science, Austrian Academy of Sciences, Leoben, Austria
[2] Böhler Edelstahl GmbH, Kapfenberg, Austria
[3] Institute of Mechanics, University of Leoben, Leoben, Austria

christian.rehrl@mu-leoben.at

Abstract. The influence of initial grain orientation and geometrical restrictions (grain boundaries, sample geometry) on the microstructural evolution was investigated at ambient temperature. Single- and polycrystalline austenitic steel samples, strained under uniaxial tension, show pronounced orientation changes which are strongly dependent on initial orientation. The numerical simulation results from a crystal plasticity FEM model match the experimental ones.

1. Introduction

Under load, the crystal anisotropy of polycrystalline metals and geometrical restrictions (grain boundaries or forming dies) cause strong inhomogeneities of stress and strain and the crystal orientation changes on the local scale. Especially texture evolution and grain fragmentation during the plastic deformation are of technical interest for the metal forming industry.

A recent work [1] supported the plan to examine this structural evolution by identifying the physical processes in coarse grained crystals. With this motivation uniaxial tension tests on single (sc) and poly crystals (pc), actually a bi-crystal, were performed. The analysis was performed on a free sample surface by using EBSD-technique and stereophotogrammetric deformation analysis. Particular attention has been paid to clarify the influence of initial orientation on the structural evolution. For this study a really important advantage of coarse grained samples is the nearly columnar grain structure, which provides a bridge to validate numerical simulation results. For the simulation, a crystal plasticity finite element model was used [2, 3].

2. Sample preparation and experimental setup

The analysed material was a coarse pc austenitic steel, which consists of millimeter-sized grains. The chemical analysis is given in Table 1.

Table 1. Chemical contents (wt%) of the analysed austenitic steel Böhler A220.

Fe	Cr	Ni	Mo	Mn	Si	N	C
63.2	17.5	14.5	2.7	1.7	0.3	0.07	0.03

Figure 1. Tensile test sample fabrication out of a coarse grained sheet (a,b) and Kammrath&Weiss test unit (c) with highlighted sample position.

Tension test samples were machined out of sheets as shown in Figure 1(b). The coarse grained structure makes it possible to cut out samples with a cross section of 1x1mm^2 being either sc or pc over the whole inner gauge length of 5mm (Figure 1(a)). Subsequently, these samples were mechanically ground, polished and electrolytically polished (electrolyte A2) to remove the deformation layer from the cutting and mechanical grinding process.
The samples were deformed in an open displacement controlled mode to engineering strains of about 30% using a tension test unit (Kammrath&Weiss) as shown in Figure 1(c). This equipment utilized two moveable crossheads, which held the center of the sample at a fixed position during the elongation. The experiments were conducted at a constant crosshead speed of 10μm/s and at ambient temperature.

The analysis procedure consists of a combination of the measurements of the in-plane strains and the local crystal orientation during and after deformation test. Measurements on the crystal orientation were carried out with an EBSD-SEM system (a TSL EBSD system interfaced to a LEO 1525). The initial crystallographic orientations and the grain boundary profile of the sc and the pc sample are plotted as Inverse Pole Figures (IPFs) related to the tension direction in Figure 2. The $\langle 11\ 7\ 20\rangle$ axis of the sc sample was parallel to the loading direction. For the pc sample the two largest grains show a soft $\langle 7\ 3\ 17\rangle$ and a hard $\langle 5\ 5\ 7\rangle$ orientation parallel to the loading direction. In-plane strains were calculated from local displacement fields by means of an in-situ stereophotogrammetric deformation analysis [4, 5]. To support the measurement of the local strain, a randomly distributed pattern was applied to the sample surface by a fine mist of paint. To demonstrate the potential of crystal plasticity FEM simulations, the incremental strain in tension direction is compared to the experimental one.

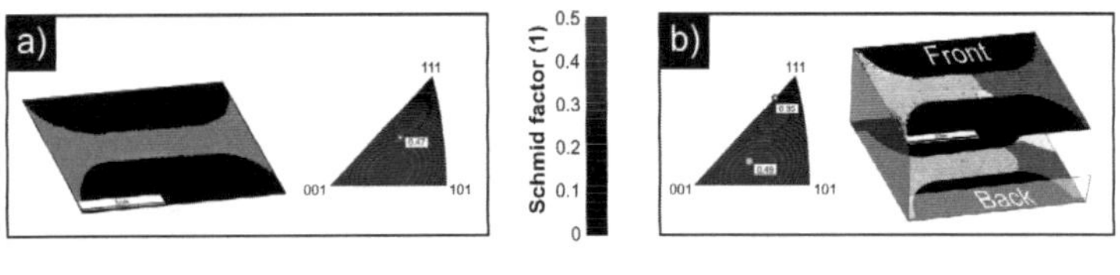

Figure 2. Initial Orientation of sc and pc sample in IPF Tensile test sample fabrication out of a coarse grained sheet (a,b).

3. Results and discussion

3.1. Macroscopic stress strain response

In Figure 3 the engineering stress vs. displacement data of the sc and pc samples are plotted. The sc sample was strained without any interruption up to an ϵ of 34%. In the case of the pc sample, the experiment was conducted incrementally in two steps to correlate the crystal orientation changes with the local strain evolution.

Under normal circumstances a stage I zone is not typical for a pc sample however in our case we are analyzing a bicrystal. For both investigated samples the three hardening stages appear during the deformation. A plausible explanation for this response might be the initial crystallographic orientation and the grain boundary profile. As shown in Figure 2(b), the left coarse grain - having a soft single slip orientation with a Schmid factor of 0.48 - stretches over the whole inner gauge length without any grain boundary in between. In absence of constraining effects it follows that the pc sample deforms like a soft single crystal. The smaller stage I zone, the higher hardening moduli in the stage II and the more pronounced softening at stage III in comparison to the sc sample (in spite of the higher Schmid factor) is due to the constraint from the grain boundary.

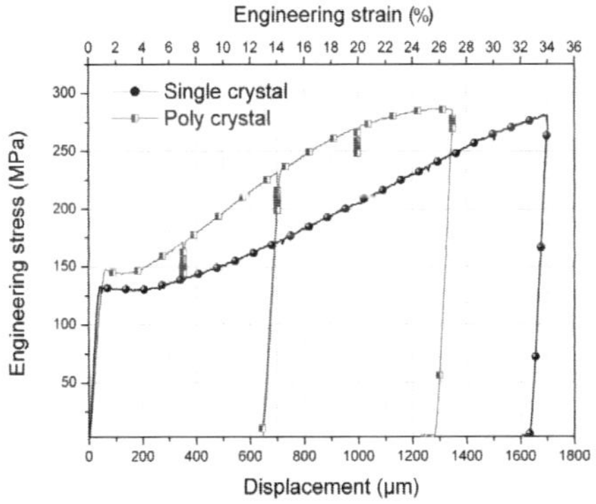

Figure 3. Engineering stress strain curves for sc and pc sample.

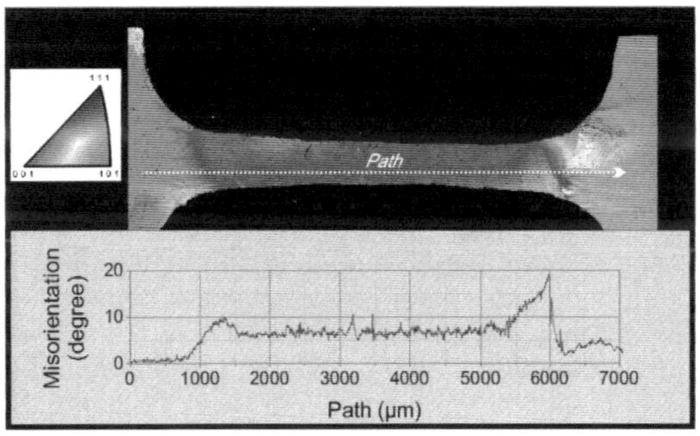

Figure 4. Inverse pole figure in tension direction with a point to origin misorientation plot of a 34% strained single crystalline sample.

3.2. Local microstructure evolution

Single crystalline sample

Figure 4 shows the crystal orientation related to the tension direction and a point to origin misorientation plot of the sc sample after ϵ=34%. Between the sample flanks the dislocation can move freely to the sample surface and no orientation gradient is found. Conversely, pronounced orientation changes occur at the sample flanks due to geometrical restrictions which cause a rigid body rotation of the inner area.

Polycrystalline sample

As already mentioned, the strongest deformation in the pc sample is localized in the soft crystal (Figure 5). From comparison of incremental strain data to crystal orientation data, visualized as path plots in Figure 6, the following results can be presented: The first strain increment of 14% leads to a strong localization and a very heterogenous distribution of strain in the soft grain without any effect on the misorientation gradient. In spite of the small strains, which appear in the hard crystal, the fragmentation process is more developed than in the soft one. With increasing ϵ the crystal fragmentation occurs in both crystals. This phenomenon motivates to investigate patterns of different slip system activation on the sample surface which are illustrated as close-up scans in Figure 5. A fading procedure is used to combine IPFs and digital images. In case of the soft crystal strain heterogeneities and, for the hard crystal, different activation of slip systems cause the fragmentation process. For the soft crystal there is one major slip system with a maximum Schmid factor and, therefore, it is hard to activate a second slip system. In the hard crystal two slip systems have the same Schmid factor. That means, the resolved shear

stress on each slip system is equal. As a result small orientation fluctuations or geometrical restrictions, leading to a more complex strain tensor, can favour one slip system very easily. For the crystal fragmentation process the initial orientation has a strong influence [6].
As shown for the pc sample in Figure 6 the predicted local strains from the simulation model match the experimental ones.

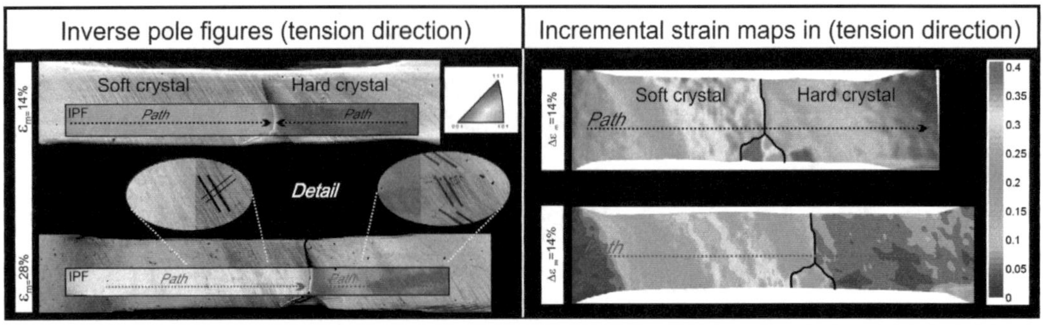

Figure 5. Inverse Pole Figures and incremental in-plane strain maps in tension direction illustrate the local plastic behaviour of the pc sample. The results along the paths indicated by dotted lines are compared in Figure 6.

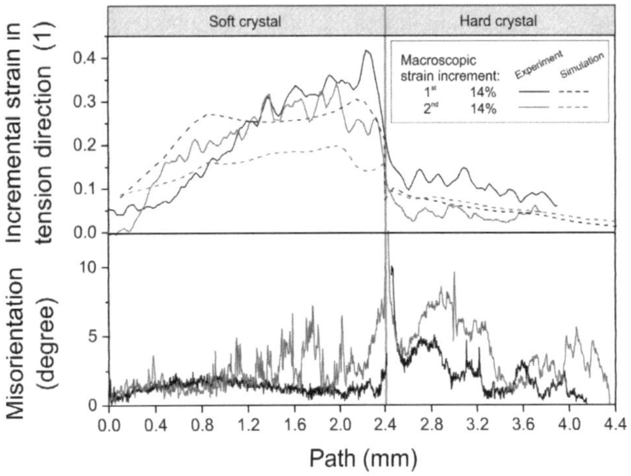

Figure 6. A comparison of experimentally determined and simulated incremental strain paths (dashed lines are numerical results) with misorientation paths of the pc sample. The position of the paths is plotted in Figure 5.

Conclusion

Dislocation slip and shear constraints are the physical mechanisms which control the texture evolution. In double slip orientated grains, the fragmentation is more pronounced as in single slip orientated ones.

References

[1] Zhao Z, Ramesh M, Raabe D, Cuitio A.M. and Radovitzky R 2008 *Int. J. Plast.* **24** 2278-2297
[2] Wu T Y, Bassani J L and Laird C 1991 *Proc. R. Soc. London* A **435** 1-19
[3] Bassani J L and Wu T Y 1991 *Proc. R. Soc. London* A **435** 21-41
[4] Tatschl A and Kolednik O 2003 *Mater. Sci. Eng.* A **339** 265-280
[5] Tatschl A and Kolednik O 2004 *Mater. Sci. Eng.* A **364** 284-399
[6] Raabe D, Zhao Z and Mao W 2008 *Acta Mater.* **50** 4379-4394

Numerical verification of three point bending experiment of magnetorheological elastomer (MRE) in magnetic field

Danuta Miedzinska[1], Anna Boczkowska[2], Konrad Zubko[3]

[1] Military University of Technology In Warsaw, Faculty of Mechanical Engineering
[2] Warsaw University of Technology, Faculty of Materials Science and Engineering
[3] Military University of Technology In Warsaw, Faculty of Advanced Technologies and Chemistry

dmiedzinska@wat.edu.pl

Abstract. In the article a method of numerical verification of experimental results for magnetorheological elastomer samples (MRE) is presented. The samples were shaped into cylinders with diameter of 8 mm and height of 20 mm with various carbonyl iron volume shares (1,5%, 11,5% and 33%). The diameter of soft ferromagnetic substance particles ranged from 6 to 9 μm. During the experiment, initially bended samples were exposed to the magnetic field with intensity levels at 0,1T, 0,3T, 0,5T, 0,7 and 1T. The reaction of the sample to the field action was measured as a displacement of a specimen. Numerical calculation was carried out with the MSC Patran/Marc computer code. For the purpose of numerical analysis the orthotropic material model with the material properties of magnetorheological elastomer along the iron chains, and of the pure elastomer along other directions, was applied. The material properties were obtained from the experimental tests. During the numerical analysis, the initial mechanical load resulting from cylinder deflection was set. Then, the equivalent external force, that was set on the basis of analytical calculations of intermolecular reaction within iron chains in the specific magnetic field, was put on the bended sample. Correspondence of such numerical model with results of the experiment was verified. Similar results of the experiments and both theoretical and FEM analysis indicates that macroscopic modeling of magnetorheological elastomer mechanical properties as orthotropic material delivers accurate enough description of the material's behavior.

1. Introduction

Magnetorheological elastomers (MREs) are materials with rheological properties which can be changed in a continuous way, rapidly and reversibly by the applied magnetic field. They are the solid analogues of magnetorheological fluids (MRFs), consisting of magnetically permeable particles (such as iron) added to a viscoelastic polymeric material prior to crosslinking. Before the curing process of the polymer a strong external magnetic field is applied. The field induces dipole moments within the particles, which seek minimum energy states. The chains of particles with collinear dipole moments are formed and curing of the polymeric matrix material locks the chains in place. In this orientation, the particles can form separate chains in the three-dimensional simple lattice structures or even more complicated structures, where particles have multiple interaction points [4].

The microstructure of MRE is shown in figure 1.

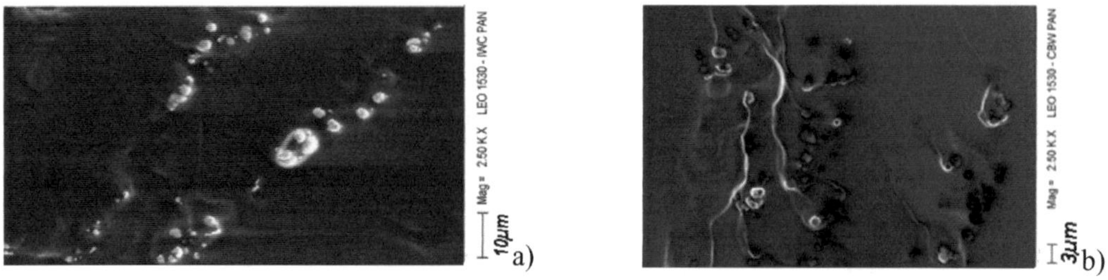

Figure 1. SEM images of MRE obtained from PU 80/20 [1] filled with 11,5 vol. % of carbonyl-iron particles, cured under magnetic field of: a) 100mT, b) 300mT [1].

2. Experimental results

The experiments have been conducted to evaluate the magnetorheological material response to the applied magnetic field. An electromagnet has been used to produce the magnetic field. The field strength has been constantly monitored using the Hall-probe method. The cylindrical samples (ø=8mm, h=18mm) have been used for the experiments. The experimental setup is shown in figure 2.

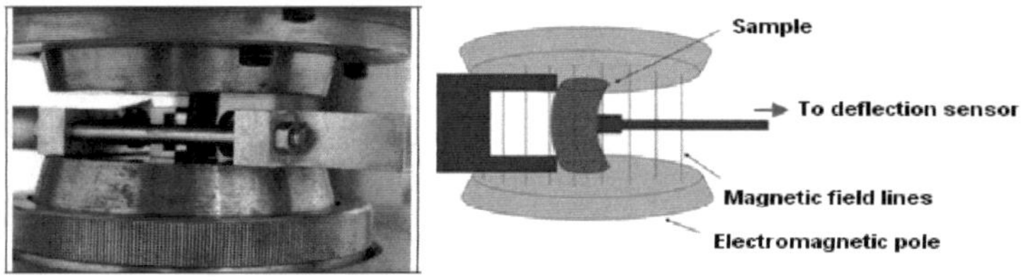

Figure 2. Experiment setup for MRE examination in the magnetic field [3].

The experiment objective was to measure the response of samples to a deflection under the magnetic field. The samples were placed parallel to the magnetic field lines and deflected before the application of a magnetic field. A deflection, which is analogous to a three point bending, was applied to change the orientation of the particles chains in the material. After the application of the magnetic field the samples tend to straighten which was measured by a deflection sensor. The magnetic field within the range of 0-0,9 T has been applied. The samples response became stronger with rising intensity of the field in all the cases. A deflection value was changing depending on the content of iron in a sample. The highest deflection was observed for samples with 11.5% of iron particles. The lower deflection of samples with 33 vol. % Fe may be caused by a higher stiffness of samples with a greater amount of particles [2].

3. Numerical model and analysis

A numerical model was developed to verify the results the experiment described above. The model was based on the assumption that MRE behaves like an orthotropic material with the material properties of MRE on the direction along the iron chains - and of a pure elastomer - on the other directions. Such an assumption can be made for the small deformations of a sample, what took place in the considered experiment.

The material properties, especially the Young modulus were taken from previous experiments [1] and were presented in figure 3. The assumed value of Poisson ratio was 0,46.

The FE analysis consisted of the two stages: the first one was to initially deflect a cylindrical sample (ø=8 mm, h=18 mm) by a nodal force to the deflection value taken from the experiment described before. Then the deflected sample was stretched by the external force simulating the

[1] PU 80/20 – a polyurethane elastomer made of 80% of an initial polymer and 20% of a hardener

influence of the mass forces appearing in the iron chains under the magnetic field. A scheme of the numerical experiment stages is presented in figure 4.

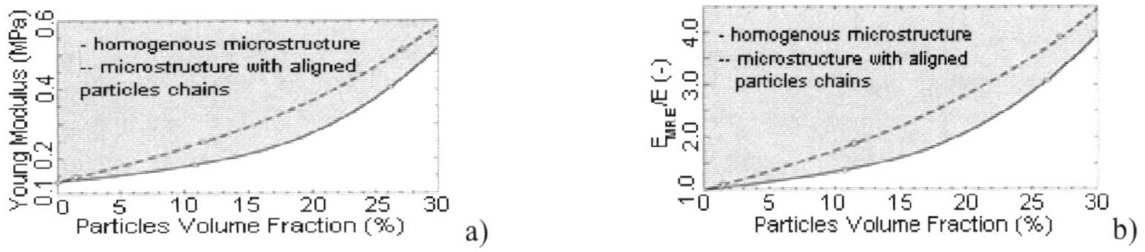

Figure 3. Influence of the iron volume fraction on the Young's module of the magneorheological elastomer (a). An increase in the Young's modulus value in comparison with a pure elastomer (b).

A numerical analysis was carried out for the MRE sample with 11,5 vol. % Fe and the iron chains parallel to the sample axis. The considered magnetic fields were 0,1, 0,3, 0,5. 0,7 and 1T. The number of Fe dipoles in 1 mm^3 was about 273058.

A static FE analysis was accomplished with MSC Patran/Marc computer code.

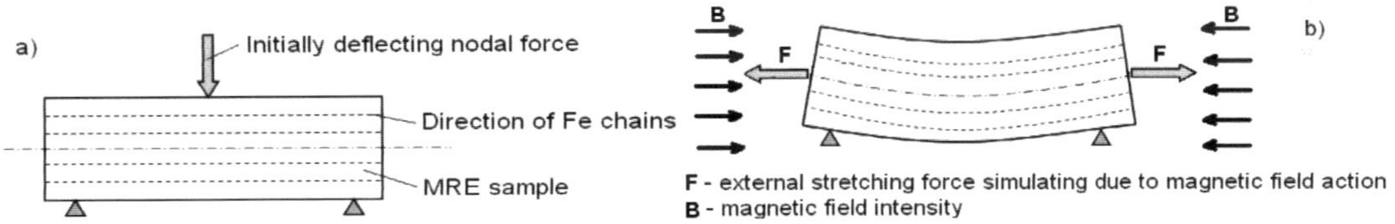

Figure 4. Stages of FE analysis: initial bending (a), magnetic field action consideration (b).

4. Analytical calculations

The external stretching force, which simulated the action of magnetic field on Fe chains, was calculated according to the equation describing a dipole interreaction force in iron chains along the sample axes. The equation is as follows [5]:

$$F_0 = \left(\frac{3\mu_0}{4\pi}\right) \cdot (m^2) \cdot (5\cos^2\alpha - 1) \cdot \left(\frac{1}{r^4}\right) \qquad (1)$$

where: α - angle between the external magnetic field and the dipoles chain axes, $\mu_0 = 4\pi \cdot 10^{-7}$ H/m – a magnetic permeability of a void, r=12 μm – the distance between two interreacting dipoles, m – magnetic momentum of MRE, considered in accordance to the volume iron share and magnetic field intensity (taken from previous researches). The resulted value describes the force per 1 gram of the elastomer – iron composite. For FE analysis the force F_0 value was multiply by the value of the sample mass, what is given with the equation (2):

$$M = V \cdot (U \cdot \rho_{Fe} + (1 - U)\rho_{elastomer}) \qquad (2)$$

where: V – MRE volume, U – volume share of iron particles in elastomer, ρ_{Fe} = 7874 kg/m^2 – iron density, $\rho_{elastomer}$ = 1030 kg/m^3– pure elastomer density.

5. FE analysis results

The response of the initially bended MRE sample to the external stretching force simulating the action of a magnetic field was analyzed. In agreement with the experiment during the FE analysis the sample should exhibit a complete return to the initial state. It means that the final deflection of the numerical sample first bended and then subjected to the external force simulating magnetic field action should approach zero. A comparison between the experimental and FE analyses is presented in Table 1. The values of the magnetic field intensity and of the initial deflection were taken from the experiment, the external force simulating the action of a magnetic field on the iron chains was calculated with the equation (1). The initially bending force and the sample response values are the results of the numerical analysis.

Table 1. Comparison of the results of the experimental and the FE analysis.

Magnetic field (experiment)	Initially bending force (FE analysis)	Initial deflection (experiment)	External force simulating magnetic field action (analytical calculations)	Total final deflection (FE analysis)
[T]	[N]	[mm]	[N/kg]	[mm]
0,1	0,059	0,028	1,6	$1{,}55\cdot10^{-3}$
0,3	0,44	0,2085	4,4	$1{,}87\cdot10^{-3}$
0,5	0,59	0,279	6,3	$1{,}61\cdot10^{-3}$
0,7	0,63	0,295	10,1	$1{,}44\cdot10^{-3}$
1	0,72	0,34	12,7	$1{,}39\cdot10^{-3}$

It can be concluded that a response of the bended sample to a force simulating magnetic field action corresponds to the experimental findings that the sample should vanish. The values of the final deflection vary from zero by only about 0,15%, what means that the initial bending deflection almost disappeared.

6. Conclusions

The three point bending experiment of a MRE sample in a magnetic field was numerically simulated.

The results of the applied modeling method of magnetorheological elastomers for small deformations agrees with experiments. The considered material model was orthotropic with the material properties of MRE in the direction along the iron chains and – of a pure elastomer – for the other directions. The FE modeling resulted in about 0,15% of correspondence between the numerical and experimental analyses.

It also allows to consider the influence of the stiffer iron chains in elastomer. The model shows a distribution of the material properties in MRE. It takes into consideration a direction of Fe chains. It can be also modified if the tested MRE sample has iron chains directed not parallel to the sample axis (for example if an angle between axes and chains reached 45 degrees the orthotropic material properties local coordinate system are rotated in the same way – see fig. 5).

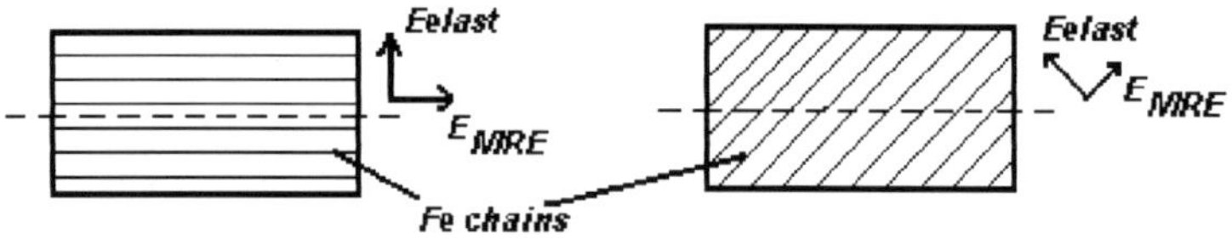

Figure 5. Local axes of the orthotropic material properties

The model can be used to predict a behavior of MRE working in a domain of small deformations. It is a simple way how to assess the influence of a magnetic field, a volume share of iron particles in an elastomer and a direction of Fe chains in a tested sample.

Acknowledgement

The paper is supported by a grant No 515R/1090/6391/008, financed in the years 2008-2011 by Ministry of Science and Higher Education, Poland.

References

[1] Boczkowska A, Awietjan S, Babski K, Wróblewski R, Leonowicz M 2006 *Proc. SPIE* **6170**, 31700R1-11
[2] Boczkowska A, Awietjan S 2008 *Mater. Sci. Forum* **587-588** 630-634
[3] Boczkowska A, Awietjan S. Wróblewski R. 2007 *Smart Mater. Struct.* **16** 1924-1930
[4] Boczkowska A, Awietjan S 2009 *Solid State Phenomena* **154** 107-112
[5] ShenY, Golnaraghi M F and Heppler G R 2004 *J. Intelligent Mater. Syst. Struct.* **15** 27-35

Journal of Physics: Conference Series **240** (2010) 012159 doi:10.1088/1742-6596/240/1/012159

Consideration of pinning and dissolution with dividing sites in Monte Carlo simulation of grain growth

C. Sekkak[1], N. Rouag[1] and R. Penelle[2]

[1] Laboratoire Microstructures et Défauts, Université Mentouri Constantine, 25000, Algérie.
[2] ICMMO, LPCES, Université de Paris Sud, UMR CNRS 8182, Bt410, 91405 Orsay Cedex, France.

nadrou@wissal.dz

Abstract
Various distributions of particle size are considered in the present work to study the effect of pinning by a bimodal distribution of particles on grain growth simulation in comparison with that of the real distribution of MnS (30-50 nm) and AlN (5-8 nm) in a Fe3%Si alloy grade HiB. A dividing site technique is used to make easy consideration of different size of particles together without increasing matrix size for calculation. The dissolution of precipitates introduced after the stagnation stage allows grain growth to start again.The dividing sites technique allows the consideration of particle dissolution by gradual vanishing of peripheral layers of particles, in agreement with the experimental evolution.

1. Introduction

The second phase particles which act on the grain boundary migration rate permit to control the grain size and orientation during normal growth. So, it is possible to seek the most appropriate matrix for the abnormal growth of a given orientation such as the Goss grains in Fe-3%Si steels. The grain boundary dragging by a second phase depends on its shape, size, distribution and volume fraction. Zener [1] was the first to propose a model permitting to calculate this dragging force, then, several models were proposed to explain the effect of a homogeneous distribution of particles on recrystallization and grain growth [2-5]. Simulations of materials containing particles generally take into account only one particle size [6-14]; the introduction of several particle sizes was performed in an analytical way [15,16]. The case of bimodal distribution was considered by simulation of interactions between grain boundary migration and precipitates [17]. The aim of the present work concerns the influence of the bimodal distribution of MnS (30-50 nm) and AlN (5-8 nm) inhibitors on normal grain growth of Fe-3%Si alloy, grade Hi-B that presents an inhomogeneous microstructure, where the size factor is not required condition for grain growth.

2. Dragging force in simulations procedures

As mentioned, Zener [1] was the first to propose a relation between the dragging force and grain size, volume fraction of the particles and the grain boundary energy, i.e.:

$$Pz = \frac{3}{2}\left(\frac{f\gamma}{r}\right) \qquad (1)$$

Pz is the pinning force, γ is the grain boundary energy, f is the volume fraction of the particles and r is their average radius. The introduction of particles is performed by randomly allocating them in the matrix [6-8]. The pinning study is performed by introducing the Zener dragging force Pz [18]. In the present work, the initial matrix used in the simulation is the microstructure of a Fe-3%Si sheet grade Hi-B after primary recrystallization. This microstructure was characterized by EBSD on a grid 200x230 sites with a step scanning equal to 2 µm, (figure 1-a).

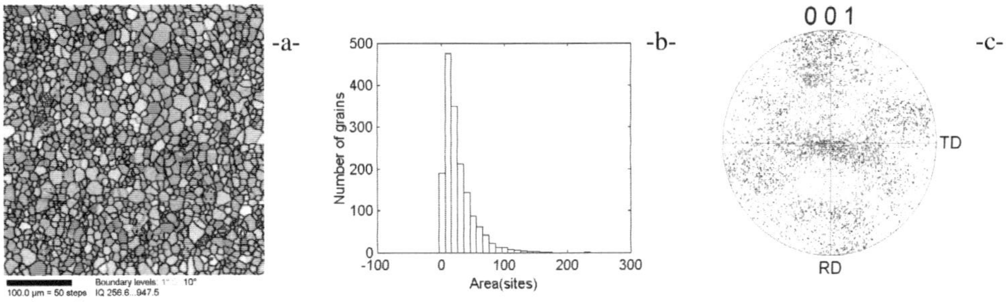

Figure 1: Initial matrix used in the simulation (a), grain size distribution (b) and {100} pole figure (c).

A deterministic approach of a Monte Carlo simulation technique was adopted from analysis of the experimental parameter evolution. Inside a grain each site of the studied zone is characterized by an orientation given by the three Euler angles (φ_1, φ, φ_2). An orientation number different from all the orientations present in the matrix is assigned to precipitates: (-1) for the precipitates inside the grains and (-2) for those at grain boundaries. Four particle sizes (1, 7, 13 and 19 sites) have been considered. This choice corresponds to a node (1site) with its first (7sites), second (13 sites) and third (19 sites) neighbours, in a hexagonal grid of simulation. The use of grid with low resolution (200x200) [6] by allocating one site by particle does not allow to simultaneously consider particles with different sizes. Figure 2 shows the random distribution of a 2% volume fraction of particles with 1, 7, and 19 sites by particle. Fig.2 reveals the presence of grains whose size is similar to that of precipitates having 7 sites; the grain size distribution of the initial matrix is presented figure 1-b. The first size class shows an abnormality: the particles are bigger than many grains of the matrix. The effect is obviously more pronounced with 13 or 19 site particles. Then, consideration of these precipitates in the calculation of the pinning effect does not have any more sense; it is rather two-phase microstructure, in which one can not consider pinning effect.

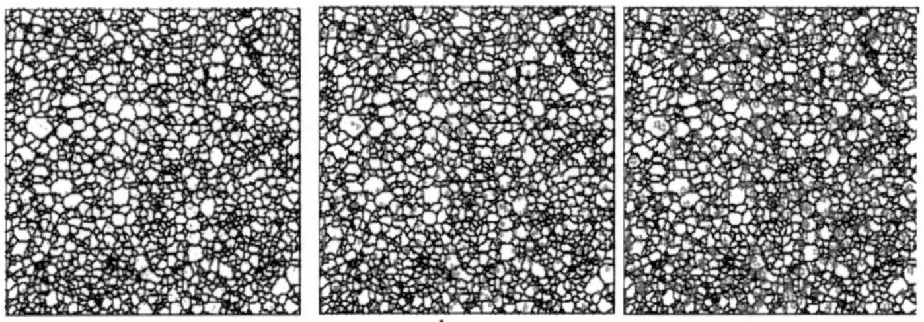

-a- -b- -c-

Figure 2
Microstructure 200x200 with random distribution of 2% particles affecting: 1 site (a), 7 sites (b) and 19 sites (c). (by particle)

3. Simulation procedure with dividing sites

Generally, the real microstructures used in simulation are characterized by EBSD [19]. The precipitates whose size is less than 1 μm are difficult to characterize because of the low resolution of W filament-SEM, a technique of dividing sites was used in the simulation to bypass the difficulty [17]. At the particle scale, each site is divided in a given number of sites permitting the introduction of different size of particles while respecting the ratios between particle and grain sizes. Double-grid method was also used for modeling stressed grain growth related to grain boundary migration [20]. The choice of initial matrix size and that of dividing coefficient G must be in agreement with the experimental values of the ratio between biggest precipitate sizes and smallest grain sizes. In the present matrix (HiB Fe3%Si with MnS and AlN inhibitors), that ratio is about 5%. With the initial matrix (200x230 sites) 4 sites are attributed to the smallest grains (5μm), therefore it is impossible to reasonably consider particle effect in simulation by assigning them an entire site. With a dividing site coefficient G=64, one site is treated as 64 sub-sites, a matrix with (1600x1840) sites is used instead of the initial matrix (figure 3). The introduction of different sizes of particles becomes easy, one sub-site is attributed to AlN particle and 13 sub-sites are attributed to MnS particle.

With that dividing technique, a finer grid is locally used without increasing the calculation time, since the memory is not occupied by big matrixes. The dividing procedure is only performed during computation; the calculation time is practically the same as that with 1 site/particle and a matrix 240X240. To achieve the same effect of growth with adequate size reports for (largest particle/smallest grain) and (smallest particle/largest particle) it would be necessary to start the simulation with initial matrixes with a minimal size 10^5x10^5. After dividing sites, the first class is equal to about 200 sites instead of 10 sites in the initial matrix (Fig.1 b) without dividing; this smaller size among the grains of the matrix is clearly higher than that of the biggest precipitates (13 sub-sites) considered in the present work.

The grain boundary energy was firstly considered as isotropic i.e. $\gamma = J = 1$. With anisotropy of the grain boundary energy, J involves different values in the simulation [21]. The initial microstructure was drawn with a step size equal to 2 μm. After dividing sites the grid parameter becomes equal to

0.25 μm. The particle diameter for a hexagonal grid is given by $d = \left(\frac{4A}{\pi}\right)^{1/2}$, where $A = \frac{na^2\sqrt{3}}{2}$, n is the site number of the particle and 'a' is the grid parameter [10].

Table 1 presents the values used in the simulation. The figure 3 shows the dragging force for different implicit values of Pz = 0, 1, 3, 5.

Particle size (sites)	Particle diameter d (μm)
1 site	0.2625
7 sites	0.6946
13 sites	0.9465
19 sites	1.1443

Table 1:
Values of pinning force used in simulation.

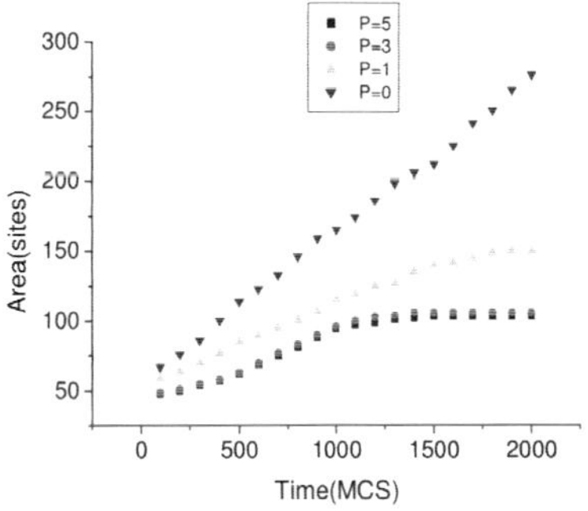

Figure 3: Normal grain growth kinetics for different values of the dragging force Pz.

The figure 4 shows that when the dividing coefficient G increases the pinning effect is more marked and the maximal size of particles to stabilize pinning effect increases; one changes from 5 sites for G=0 to ~ 30 sites for G = 64. The technique permits to consider different size of particles with a bimodal distribution and different layers of dissolution without increasing calculation matrixes.

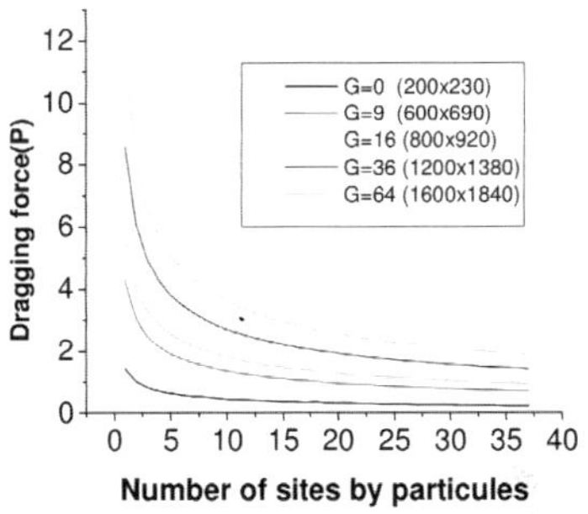

Figure 4: Dividing site effect on dragging force Pz.

4. Bimodal distribution

The choice of the site number allocated to particles is based on the neighbor number for the grid; for a hexagonal grid 1, 7 and 19 sites [10, 12, 13, 16] are generally considered. The second neighbors being constituted of 6 sites, it was chosen to work with 1, 7, 13 sites because this distribution is more in agreement with the experimental matrix. The bimodal distribution of MnS (30-50 nm) and AlN (8-9 nm) precipitates in the Fe-3%Si Hi-B gives a size ratio of about 3.5 between the two types of precipitates [22]. To study the effect of the bimodal distribution of these precipitates on the normal grain growth, the average size of each population was taken into consideration, that is: 1 sub-site (d=0.26) for AlN and 13 sub-sites (d=0.94) for MnS, respecting together the experimental diameter ratio and the true neighborhood of the hexagonal grid. The two populations of particles have been identically allocated in the microstructure, the volume fraction of the two populations of precipitates was equal 2 and 1% respectively, the dragging force is calculated with values of table 1.

6. Simulation of the precipitate dissolution

The dissolution of the precipitates begins by the diffusion of the peripheral atoms. In first approximation, that behavior can be described in simulation procedure by gradual vanishing of precipitate volume, the peripheral sites become sites of the matrix. The dissolution of precipitate occurs by successive layers; the sites initially allocated to particles (that is not submitted to reorientation) are gradually submitted to reorientation, layer after layer. Figure 5 shows the isotropic case with 2% particles in the initial matrix. In a first approach the precipitates dissolution was considered at the end of stagnation, about after 1300 MCS. The dividing sites technique allows the consideration of particle dissolution by gradual vanishing of peripheral layers of particles, the results are most in agreement the experimental evolution.

Figure 5
Dissolution with and without dividing site in isotropic case of grain growth with 2% particles.

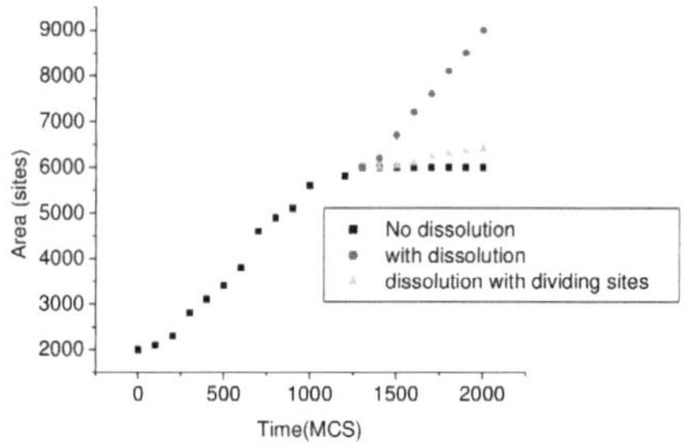

Conclusion

This work was focused on the introduction of a bimodal distribution of precipitates similar to the experimental one that control normal growth in a Fe-3%Si Hi-B. The study was performed in a determinist way from the evolution parameter analysis obtained experimentally. A high resolution matrix (1600x1840) coupled to a technique of dividing sites of the experimental matrix was used to respect the ratio between the size of biggest particles and that of smallest grains. The dissolution of the precipitates at the end of growth is an important step because the purification of the matrix by precipitate dissolution enables to get a secondary matrix without residual primary matrix. In the procedure simulation, the dividing sites technique allows the consideration of particle dissolution by gradual vanishing of peripheral layers of particles, in agreement with the experimental evolution.

References

[1] C. Zener, C.S. Smith, *Trans. Met. Soc. AIME* **175** 47 (1948)
[2] T. Gladman, *Proc. R. Soc. Lond.* **A294** 298 (1966)
[3] M. F. Ashby, J. Harper, J. Lewis, *Trans. AIME* **245** 413 (1969)
[4] P. Hellman, M. Hillert, *Scand. J. Metals* **4** 211 (1975)
[5] T. Gladman, *Scripta Metall. Mater.* **27** 1569 (1992)
[6] D. J. Srolovitz, M. P. Anderson, G. S. Grest, P. S. Sahni, *Acta Metall.* **32** 1429 (1984)
[7] R. D. Doherty, D. J. Srolovitz, A. D. Rollet, M. P. Anderson, *Scripta Metall.* **21** 675 (1987)
[8] M. P. Anderson, G. S. Grest, R. D. Doherty, D. J. Srolovitz, *Scripta Metall.* **23** 753 (1989)
[9] J. Gao, R. G. Thompson, B. R. Patterson, *Acta Mater.* **43** 3653 (1997)
[10] B. K. Kad, P. M. Hazzeldine, *Mater. Sci. Eng.* **A238** 70 (1997)
[11] M. Soucail , R. Messina , A. Cosnuau , L.P. Kubin, *Mater. Sci. Eng.* **A27** 1 (1999)
[12] A. Miyake, *Scripta Mater.* **45** 1009 (2001)
[13] G. S. Thompson, J. M. Rickman, M. P. Harmer, *J. Mater. Res.* **11** 1520 (1996)
[14] Xiaoyan Song, Guoquan Liu, Nanju Gu, **Mater. Sci. Eng. A270** 178 (1999)
[15] R. Elest, J. Vanhumbeeck, L. Delaey, *Acta Metall.* **36** 1723 (1988)
[16] P. Rios, *Acta Metall. Mater.* **40** 649 (1992)
[17] Y. Tsuzuki, Y. Suzuki, *Mater. Sci. Forum* **204-206** 233 (1996)
[18] I. Ohnuma,H. Ohtani, K. Ishida, T. Nishizawa, *Mater. Sci. Forum* **204-206** 313 (1996)
[19] T. Baudin, P. Paillard, R. Penelle, *Scripta Mater.* **36** 789 (1997)
[20] J.S. Chen , V. Kotta, H. Lu , D. Wang , D. Moldovan , D. Wolf, *Comput. Methods Appl. Mech. Eng.* **193** 1277 (2004)
[21] G. Abbruzzese, A. Campopiano, S. Fortunati, *Textures Microstruct.* **775** 14 (1991)
[22] N. Rouag, G. Vigna, R. Penelle, *Acta Metall. Mater.* **38** 1101 (1990)

In situ measurements of X-ray peak profile asymmetry from individual grains

C Wejdemann [1], U Lienert [2] and W Pantleon [1]

[1]Center for Fundamental Research: Metal Structures in Four Dimensions, Materials Research Division, Risoe National Laboratory for Sustainable Energy, Technical University of Denmark Frederiksborgvej 399, 4000 Roskilde, Denmark

[2]Advanced Photon Source, Argonne National Laboratory, 9700 South Cass Avenue, Argonne, IL 60439, USA

chwe@risoe.dtu.dk

Abstract. Two copper samples, pre-deformed in tension to 5% plastic strain, are subjected to an in situ tensile deformation of 1% plastic strain while X-ray peak profiles from individual bulk grains are obtained. One sample is oriented with the in situ tensile axis parallel to the pre-deformation axis, and peak profiles are obtained with the scattering vector parallel to this direction. The profiles show the expected asymmetry explained by the composite model as caused by intra-grain stresses. The other sample is oriented with the in situ tensile axis perpendicular to the pre-deformation axis, and peak profiles are obtained with the scattering vector parallel to the in situ tensile axis. In this case the profiles initially show an opposite asymmetry, but during the in situ deformation the asymmetry reverses sign as the deformation under new loading conditions leads to changes in the intra-grain stresses.

1. Introduction

A heterogeneous dislocation structure with almost dislocation-free regions (subgrains) enclosed by dislocation-rich regions (dislocation walls) is a characteristic result of plastic deformation in many metals. According to the composite model [1] the heterogeneous dislocation density leads to a heterogeneous stress distribution. Unidirectional tensile deformation results in backward stresses in the subgrains and forward stresses in the walls (compared to the applied external stress). The intra-grain stresses can be studied by X-ray peak profile analysis. Peak profiles from a sample deformed in tension may have an asymmetric shape depending on the orientation of the scattering vector with respect to the tensile axis. The two extreme cases are when the scattering vector is parallel to the tensile axis (the axial case) and when it is perpendicular to the tensile axis (the side case). In these two cases the asymmetry is observed to be of different signs in accordance with the composite model [2, 3].

Traditionally peak profiles have been obtained from single crystals or by averaging over many grains in a polycrystalline sample. In the present study a setup designed to study the dynamics of dislocation structures in individual bulk grains [4, 5] is used to study the evolution of peak profile asymmetry during in situ deformation of two different copper samples. One sample (the 90° sample) is subjected to a 90° rotation of the tensile axis which makes it possible to observe changes in peak profile asymmetry as the initial side case develops into an axial case. The other sample (the 0° sample) serves as a reference sample with no change in tensile axis for which the axial case is realized throughout the experiment.

2. Experimental procedure

Recrystallized, weakly textured sheets of 99.99% pure copper with a mean grain size of ~20 μm and dimensions 300 mm x 50 mm x 0.31 mm are pre-deformed in tension to 5% plastic strain with a strain rate of $1.4 \cdot 10^{-4}$ s^{-1}. Two smaller dog-bone shaped samples with a gauge length of 8 mm and a gauge width of 3 mm are spark cut from the sheets for further tensile deformation. For the 0° sample the new and the old (pre-deformation) tensile axes are parallel and for the 90° sample they are perpendicular to each other.

At beamline 1-ID-XOR at the Advanced Photon Source (APS, Argonne National Laboratory) a special setup for obtaining three-dimensional reciprocal space maps of individual bulk grains with high resolution is established [4, 5]. The setup combines a narrow (~30 μm), highly monochromatic ($\Delta E/E = 7 \cdot 10^{-5}$), 52 keV X-ray beam with a two-dimensional detector (pixel size 79 μm) placed ~3.6 meters behind the sample. While rocking the sample in steps of 0.015° around the horizontal axis perpendicular to the scattering plane (the ω-axis in Figure 1) two-dimensional images are acquired, which are then combined to produce a three-dimensional reciprocal space map. The resolution in the 2θ-direction is ~0.0012°. For in situ deformation the samples are mounted with the in situ tensile axis (almost) parallel to the scattering vector in a custom made load frame with a strain gauge glued onto the gauge area.

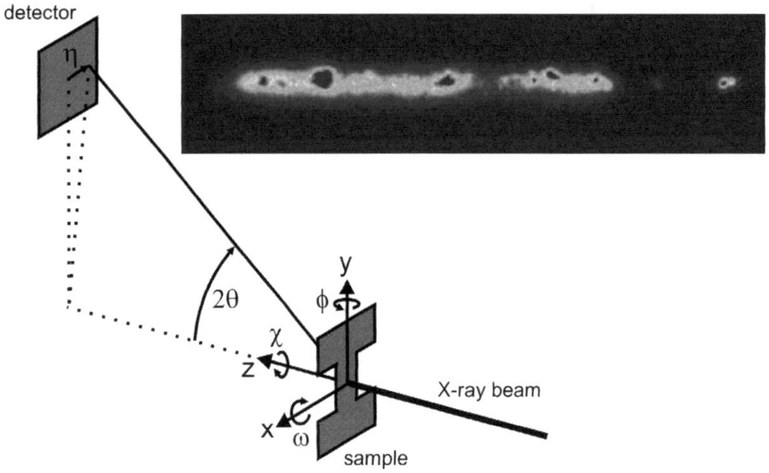

Figure 1. Sketch of the setup and part of a two-dimensional detector image of a 400 reflection.

The samples are mounted in the load frame one at a time and a small load corresponding to a plastic strain of ~0.03% is applied for stabilization purposes. The 0° sample is oriented with the scattering vector parallel to the pre-deformation tensile axis and the 90° sample is oriented with the scattering vector perpendicular to the pre-deformation tensile axis. For both samples the scattering vector is parallel to the in situ tensile axis. For each sample four grains are selected based on criteria used in a recent similar experiment [6], and a reciprocal space map of one particular 400 reflection is obtained for each grain. The sample is then deformed in unidirectional tension with a strain rate of ~10^{-4} s^{-1}. At 7 different strain levels the deformation is paused (allowing the load to relax slightly) while additional reciprocal space maps of the same 400 reflections are obtained.

The raw data show characteristic features related to the dislocation structure in the grains. As shown in Figure 1 the images contain localized areas with high intensity superimposed on a slowly varying cloud of lower intensity. The high-intensity areas are due to scattering from almost dislocation free regions of the crystal (the subgrains) while the spread-out cloud is due to scattering from regions with a high dislocation density (the dislocation walls) [4].

3. Data analysis and results

Before the individual two-dimensional images are combined, the background level around the 400 reflection is estimated for each image and subtracted from the intensity, the intensity in the image is normalized with the beam intensity and cosmic rays are removed.

The three-dimensional reciprocal space maps contain information about different aspects of the dislocation structure of the grains [3,6,7]. This paper focusses on the asymmetry of the radial peak profile. The reciprocal space maps are intensity distributions on a three-dimensional space which can be parameterized by the angles 2θ, η and ω (see Figure 1). The peak profile is the one-dimensional intensity distribution that results when the diffracted intensity is integrated over the azimuthal angles η and ω and the intensity is considered as a function of the scattering angle 2θ only. To calculate the peak profile, the values of 2θ spanned by each voxel are determined and the intensity in the voxel is distributed on a 2θ-axis accordingly. Two examples of peak profiles from the same grain from the 90° sample are shown in Figure 2. The figure shows that as the sample is deformed, the profiles shift to lower values of 2θ (due to the average elastic strain caused by the imposed external load), widen (due to dislocation accumulation) and change asymmetry (from having the slower-decaying tail on the high-2θ side to having it on the low-2θ side of the intensity maximum).

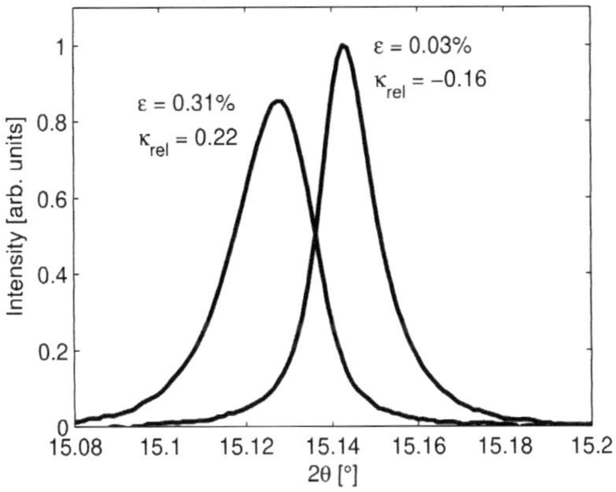

Figure 2. Radial peak profiles from the same grain from the 90° sample. The strain (ε) along the in situ tensile axis and the asymmetry (κ_{rel}) is indicated.

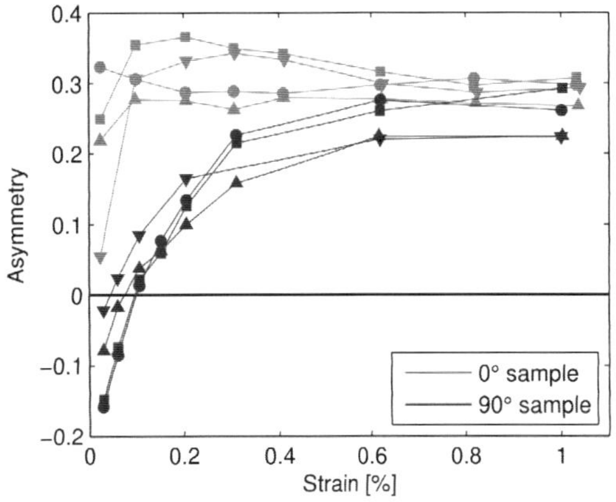

Figure 3. Peak profile asymmetry characterized by $\kappa_{rel} = (\Delta_l - \Delta_h)/(\Delta_l + \Delta_h)$.

To quantify the asymmetry of the profiles a split pseudo-Voigt function (e.g. [8]) is fitted to each profile and the half-widths at half maximum on the low-angle (Δ_l) and the high-angle (Δ_h) sides (of the profile maximum) are determined. The parameter $\kappa_{rel} = (\Delta_l - \Delta_h)/(\Delta_l + \Delta_h)$, then gives a measure of the relative asymmetry of the profiles which assigns a positive asymmetry to profiles having a slower-decaying tail on the low-2θ side of the intensity maximum and a negative asymmetry to profiles having a slower-decaying tail on the high-2θ side of the intensity maximum. Using this measure the peak profile asymmetry is determined for the 8 different grains as a function of strain along the in situ tensile axis and shown in Figure 3. The curves generally fall in two groups. The grains from the 0° sample show a more or less constant peak profile asymmetry, while the grains from the 90° sample (subjected to change in deformation axis) start with a negative peak profile asymmetry which becomes positive and reaches the level from the 90° sample grains with increasing deformation.

4. Discussion

In deformed metals with a dislocation structure consisting of dislocation-rich walls and almost dislocation-free subgrains, the composite model predicts peak profiles to be possibly asymmetric. The peak profiles are a superposition of two sub-profiles, one arising from the walls and the other from the subgrains [3]. Due to their different deformation resistances, a compatible deformation of hard walls and soft subgrains requires the building-up of tensile stresses in the walls and compressive stresses in the subgrains. Such internal stresses in the direction of the tensile axis develop when the applied load causes plastic deformation in the softer subgrains but is insufficient to cause yielding of the harder walls still deforming elastically.

Applying these arguments to the present experiment, the results in Figure 3 can be understood. For the 0° sample, which is observed in the axial case at all times, the intra-grain stresses that have developed during the pre-deformation do not change significantly during the in situ deformation, and a positive peak profile asymmetry is observed for all strains. The somewhat smaller peak profile asymmetry seen for some of the grains at the first point on the curves ($\varepsilon \sim 0.03\%$) might be a result of the unloading after the pre-deformation.

For the 90° sample, which is mounted for observation in the side case (with respect to the pre-deformation axis), the intra-grain stresses developed during the pre-deformation give rise to a negative asymmetry before the in situ deformation. As the sample is deformed along the in situ tensile axis, the situation shifts to the axial case, causing the intra-grain stresses to change and leading to a positive peak profile asymmetry of the same level as for the 0° sample.

The internal stresses induced by the pre-deformation in the 90° specimen act perpendicular to the in situ tensile axis and cause a triaxial stress state when loading along the new tensile axis. Transversal tensile stresses lower the equivalent stress in the walls, whereas the compressive stresses along the transversal direction in the subgrains increase the equivalent stress for them. When loading along the new tensile direction, dislocation-free subgrains yield first while the walls still comply elastically. The plastic deformation of the subgrains leads to an accumulation of dislocations at the interface to the walls producing again forward stresses in the walls and backward stresses in the subgrains with respect to the new tensile direction.

Acknowledgments

This work was supported by the Danish National Research Foundation and the Danish Natural Science Research Council. Use of the Advanced Photon Source was supported by the U.S. Department of Energy, Office of Science, Office of Basic Energy Sciences, under Contract No. DE-AC02-06CH11357.

References

[1] Mughrabi H 1979 *Proc. of the Fifth Conf. on the Strength of Metals and Alloys* vol 3 ed Haasen P *et al.* (Oxford: Pergamon Press) 1615
[2] Ungár T, Mughrabi H, Rönnpagel D and Wilkens M 1984 *Acta Metall.* **32** 333
[3] Mughrabi H, Ungár T, Kienle W and Wilkens M 1986 *Phil. Mag. A* **6** 793
[4] Jakobsen B, Poulsen H F, Lienert U, Almer J, Shastri S D, Sørensen H O, Gundlach C and Pantleon W 2006 *Science* **312** 889
[5] Jakobsen B, Poulsen H F, Lienert U and Pantleon W 2007 *Acta Mater.* **55** 3421
[6] Wejdemann C, Poulsen H F, Lienert U and Pantleon W 2009 *IOP Conf. Series: Mater. Sci. Eng.* **3** 012003
[7] Jakobsen B, Lienert U, Almer J, Poulsen H F and Pantleon W 2008 *Mater. Sci. Eng. A* **483-484** 641
[8] Enzo S and Schiffini L 1999 *Profile fitting and analytical functions* in *Defect and Microstructure Analysis by Diffraction* ed Snyder R L *et al.* (Oxford: Oxford University Press) 29

Texture evolution during tensile necking of copper processed by equal channel angular extrusion

W Pantleon[1], S Richter[2], S Martin[2] and JR Bowen

Risoe National Laboratory for Sustainable Energy, Technical University of Denmark, Frederiksborgvej 399, 4000 Roskilde, Denmark

pawo@risoe.dtu.dk

Abstract. Polycrystalline copper is severely plastically deformed by equal channel angular extrusion to an equivalent strain of 10. Cylindrical ASTM standard tensile specimens machined from the extruded material are deformed in tension until failure. The forming neck develops an anisotropic cross section with its long axis along the original transversal direction. Microstructure and texture are investigated by electron backscatter diffraction in the as-deformed condition and after additional tensile deformation. The texture is dominated by two components close to <110>{112} directly related to the shear deformation in the intersection plane of the channels, weakens with tensile deformation and is finally replaced by a restricted <111>+<100> fibre texture characteristic for tensile deformation in the neck region.

1. Introduction

Severe plastic deformation causes refinement of the grain structure and subdivision of original grains by deformation-induced boundaries. Caused by their microstructure, severely plastically deformed metals have higher yield and ultimate tensile strength, but lower ductility (cf. [1,2]). During tensile tests their work-hardening rate is rather low and plastic deformation localizes at an early stage causing necking and finally failure. For an understanding of the mechanical response of severely plastically deformed metals, characterization of the microstructural evolution during homogeneous tensile deformation is insufficient due to the small uniform elongation and the microstructure in the neck must be investigated as well.

2. Experimental details

Polycrystalline OFHC copper is deformed by equal channel angular extrusion (ECAE) [3] to an equivalent strain of 10. Billets are extruded 15 times through a die with an angle of 120° between the equivalent cylindrical channels without any rotation of the specimen (route A) (cf. [4]). From the core of the billets cylindrical ASTM standard tensile specimens are spark-eroded with a diameter of 4 mm, a gauge length of 20 mm, and the new tensile direction corresponding to the extrusion direction (ED).

[2] Now at: Institute of Materials Science, Freiberg University of Mining and Technology, Gustav-Zeuner-Str. 5, D-09596 Freiberg, Germany

Using an INSTRON testing machine, the tensile samples are pulled to failure with a cross head speed of 0.5 mm/min (which equals an initial strain rate of $4 \cdot 10^{-4}$ s^{-1}). The resulting engineering stress engineering strain curves are presented in figure 1 (with the engineering strain obtained from fiducial markers on the specimen surface). The maximum force required is reached already at 1% strain (uniform elongation), and the ultimate tensile strength is about 436±5 MPa. The ECAE deformed specimens fail after pronounced necking by a 45° shearing of a large section. Cross sections in the homogeneously deformed region remain circular, but the neck develops anisotropically with an elliptical cross section as sketched in figure 2. The long axis of the elliptical neck shape is aligned with the transversal direction (TD) of the extrusion process. An aspect ratio of about 0.7 between shorter and longer axis is determined in a cross section close to the fracture surface. From the area of this cross section, an equivalent strain $\varepsilon = \ln(A_0/A) = 1.2$ is obtained locally in the neck.

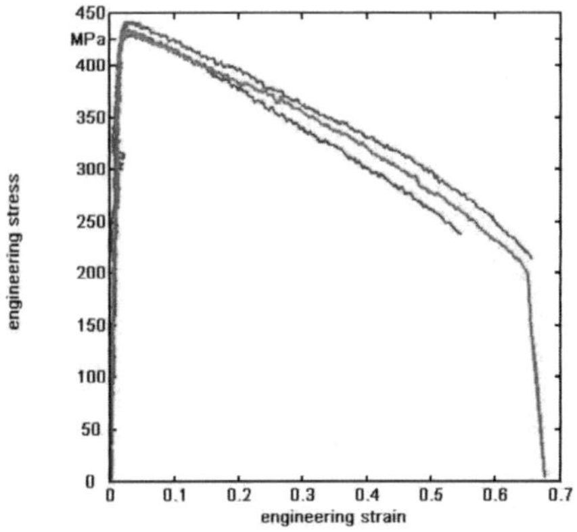

Figure 1. Engineering stress versus engineering strain during tensile loading of ECAE deformed copper for three specimens from core of different billets.

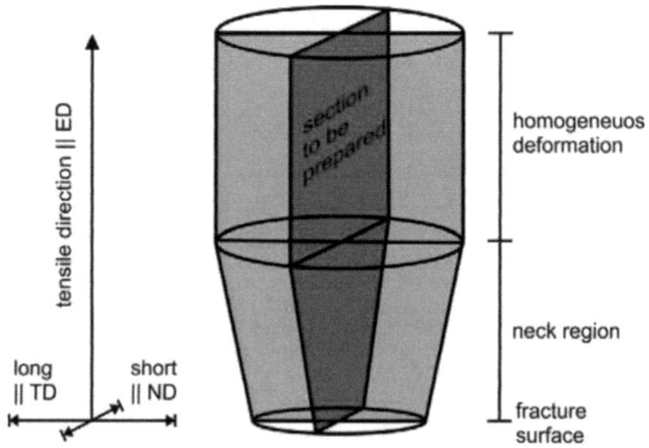

Figure 2. Sketch of upper part of a failed tensile specimen with homogeneously deformed region of circular cross section and neck region with elliptical cross section.

The microstructure is investigated by electron backscatter diffraction (ESBD) with a Zeiss Supra 35 FEGSEM using a Nordlys2 detector and data acquisition software HKL Channel5 by Oxford Instruments. EBSD requires a clean planar surface for investigating the microstructure and several preparation techniques for examining the neck region are tested [5]. The microstructural investigations are performed on a midsection through the centre of the tensile deformed specimen comprising the tensile axis and the short axis of the neck (corresponding to normal direction, ND) as sketched in figure 1. Successful preparation is achieved by coating the sample with dissolvable carbon cement prior to embedding it in epoxy. The relevant midsection is then exposed by grinding, extracting the sample manually from the epoxy, mechanical diamond polishing by hand and final electro-polishing.

On the prepared surface large orientation maps are collected on square grids with step size of 0.1 μm between individual measuring points. Two maps are gathered at different positions on the prepared midsection, one in the region of homogeneous deformation, the other in the neck close to the fracture surface. An additional orientation map is collected on a transversal section in the centre of an ECAE deformed billet not subjected to any tensile deformation.

3. Results and discussions

Figure 3 presents the spatial arrangement of orientations for the homogeneously deformed region and the neck. The homogeneously deformed region resembles closely the quite heterogeneous microstructure after ECAE [6] with regions of apparently small grains caused by grain refinement and regions with rather large grains which did not fragment severely during ECAE. The elongated grains in the deformation structure of the neck reflect the large tensile deformation to an equivalent strain of more than 1.

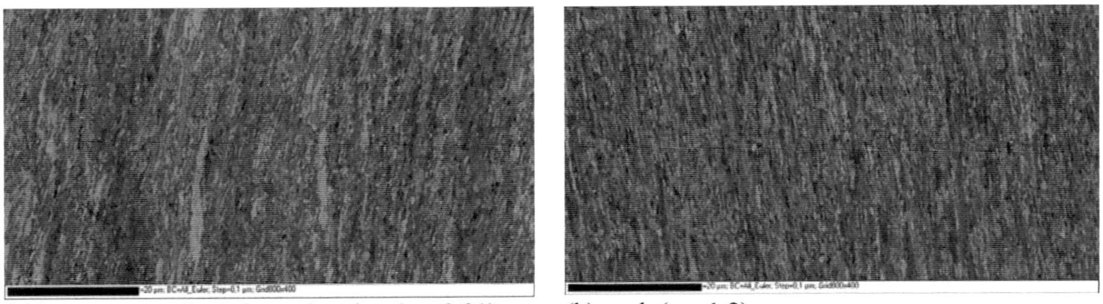

(a) homogeneously deformed region ($\varepsilon = 0.01$) (b) neck ($\varepsilon \approx 1.2$)

Figure 3. Orientation maps (80 µm × 40 µm with different colours representing different orientations) from tensile deformed ECAE specimen. The tensile direction (and extrusion direction) is vertical. The horizontal direction corresponds to the normal direction and the short axis of the elliptical neck.

From the measured orientations the texture of the deformation structures is inferred. In figure 4 the determined 111, 110 and 100 pole figures are summarized. A 110 pole close to the extrusion and tensile direction is obvious for the ECAE deformed specimen (figure 4a) and for the additionally homogeneously deformed region (figures 4b). Such a preferred [110] direction along the extrusion direction has been observed e.g. by [7] and can be related to the shear deformation along the channel intersection plane during ECAE [4]. The 111 pole figure of the ECAE deformed specimen in figure 4a reveals a 111 pole pointing along the transversal direction. Taking into account the multiplicity of four for 111 poles, the existence and position of six other 111 poles indicate the presence of two distinct preferred orientations. Both preferred orientations have a common 111 pole and are rotated from each other by 180° around the crystallographic [111] axis or TD. Combining the information of ED being parallel to [110] and TD to [111], ND must be along a crystallographic [112] direction and the texture of the ECAE deformed specimen an orthorhombic <110>{112} texture. Taking into account the geometry of the 120° die and the inclination by 60° of the intersection plane of the two channels with respect to ED, the rotated texture in the intersection plane is also a <110>{112} texture with two components. This confirms occurrence of simple shear on the intersection plane as these two texture components are commonly observed after shear deformation and designated B and $\overline{\mathrm{B}}$ orientation [8].

Orientations with a [110] direction along the tensile axis are instable under the following tensile deformation and replaced by a <111>+<100> fibre texture characteristic for unidirectional tension as obvious from the 111 and 100 pole figures of the neck regions: The orthorhombic texture developed during ECAE has a maximum pole density of 6.1 with respect to a random distribution of orientations. This texture weakens due to the slight tensile deformation to a tensile strain of 0.01 in the homogeneously deformed region to 5.1 and strengthens again to 7.4 in the neck due to formation of a dominant <111> fibre texture – with an additional weak <100> fibre. The fibre textures are not completely isotropic perpendicular to the tensile direction and variations of the pole density along the diffraction rings observed for both specimens are inherited from the orthorhombic ECAE texture.

Further investigations are required for resolving, if the small inclinations of the poles from the ideal directions are an intrinsic effect of the processing conditions as in the case of ECAE (cf. [6,9]) or an artefact of the chosen preparation technique. For a second ECAE processed specimen, an analogous texture development is observed. In both cases, the elliptical neck shape forms with its long axis along the original transversal direction and hence along an initial crystallographic [111] direction.

4. Summary

The microstructure of deformed copper after equal channel angular extrusion is analyzed in the as-deformed state and after additional tensile deformation to failure. After ECAE an orthorhombic texture consisting of two components close to <110>{112} is observed. The small plastic deformation in the homogeneously deforming regions causes a slight weakening of the texture. In the neck a (restricted) <111>+<100> double fibre texture characteristic for tensile deformation is formed. The neck develops an anisotropic cross section with its long axis along the transversal direction of the ECAE processing.

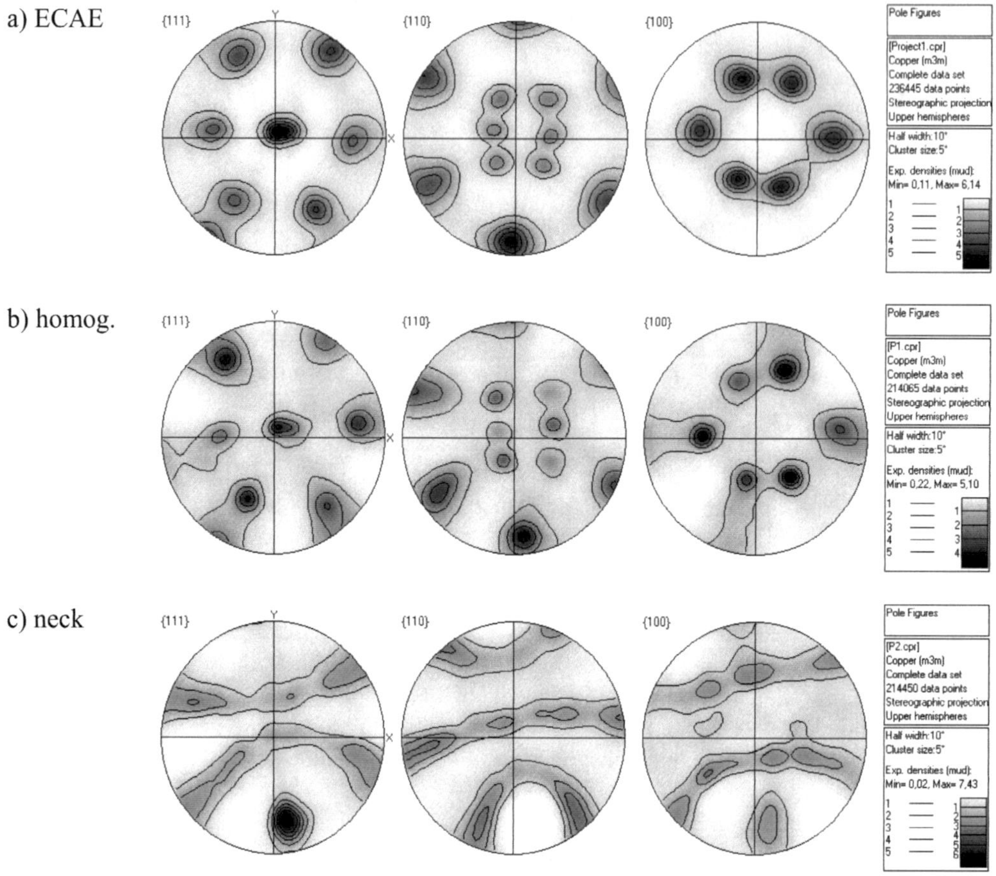

Figure 4. Pole figures: (a) as ECAE deformed and after additional tensile deformation: from (b) the homogeneously deformed region and (c) the neck. Extrusion and tensile direction are vertical; the horizontal direction is the normal direction.

Acknowledgements
The authors wish to thank the Danish National Research Foundation for financial support through the Center for Fundamental Research: Metal Structures in Four Dimensions and Prof. P.B. Prangnell, Manchester Materials Science Centre, for access to the ECAE processing facilities.

References
[1] Bowen J R, Prangnell PB, Juul Jensen D and Hansen N 2004 *Mater. Sci. Eng. A* **387-389** 235-9.
[2] Valiev R Z and Langdon T G 2006 *Progr. Mater. Sci.* **51** 881-981.
[3] Segal V M 1995 *Mater. Sci. Eng. A* **197** 157-64.
[4] Bowen J R, Gholinia A, Roberts S M and Prangnell P B 2000 *Mater. Sci. Eng. A* **287** 87-99.
[5] Richter S, Bowen J R and Pantleon W 2009 Preparation of tensile test fracture surface cross sections for high resolution electron backscatter diffraction, *in preparation for Mater. Char.*
[6] Mishin O and Bowen J R 2009 *Metall. Mater. Trans. A* **40** 1684-92.
[7] Huang W H, Chang L, Kao P W and Chang C P 2001 *Mater. Sci. Eng. A* **307** 113-8.
[8] Toth L S, Arruffat Massion R, Germain L, Baik S C and Suwas S 2004 *Acta Mater.* **52** 1885-98.
[9] Agnew S and Weertman J R 1998 *Mater. Sci. Eng. A* **242** 174-80.

15th International Conference on the Strength of Materials (ICSMA-15) IOP Publishing
Journal of Physics: Conference Series **240** (2010) 012162 doi:10.1088/1742-6596/240/1/012162

Thermal Shock Behaviour of Ceramics under Different Testing Conditions

R. Pulz , B. Rehmer

Federal Institute for Materials Research and Testing
Unter den Eichen 87, 12205 Berlin, Germany

Robert.Pulz@bam.de

Abstract. Beside the presentation of interesting parts of the experimental equipment the thermal shock behaviour of two ceramics will be discussed. Specimens of silicon carbide and zirconia were investigated in air and vacuum. The influence of the testing conditions on the strength will be addressed.

1. Introduction

The use of ceramic materials is of particular interest due to the excellent strength at high temperatures, the resistance to wear and corrosion as well as the low density. In most potential fields of applications ceramic materials are also exposed to thermal shock and thermal fatigue. Due to their poor thermal conductivity thermal shock leads to high thermal stresses and to creation of micro cracks. The material or unit fails when reaching a critical level of damage. This damage may depend on the material itself or on the environmental media.

Current publications often report about quench tests to investigate the thermal shock behaviour of ceramics e.g. [1-3]. The great disadvantage of this test method is the influence of test conditions like sample shape and size, quenching media and test method used to measure residual strength on results. So the quench test is mostly of interest for qualitative comparison of materials in the field of materials research but never for dimensioning of constructions.

To overcome this limitation, different methods were developed to carry out thermal shock tests under repeatable conditions heating up by infrared lamps or by laser e.g. [4-6].

The presented thermal shock tests were carried out using laser beam heating by helical beam guidance across thin sample. The resulting temperature distribution was recorded by high-frequency infrared (HF-IR) camera. Using a vacuum-sealed test chamber, tests under different environmental media are possible. This is of great importance due to lack of publications addressing this influence.

2. Experimental

2.1 Experimental setup

Tests were carried out using thermal shock test equipment with laser beam heating. The main feature of the test equipment is an extremely fast heating-up thermal shock resulting in high temperature gradients in thin disks using a Nd:YAG-Laser. Focusing optical system enabled helical beam guidance across the sample. Resulting temperature fields were recorded by HF-IR camera. Fig. 1 presents a sketch of the used test chamber with all important components and Fig. 2 shows the construction of the sample holder.

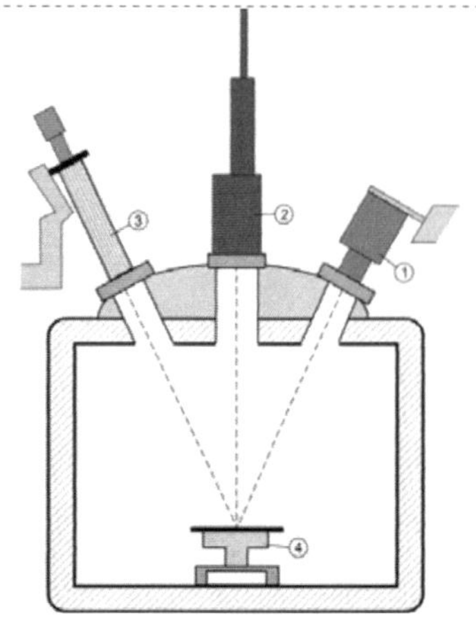

Figure 1. Sketch of the test chamber:
1- HF-IR-camera,
2- focusing optical system with laser,
3- optical window with adapted CCD-camera,
4- sample holder

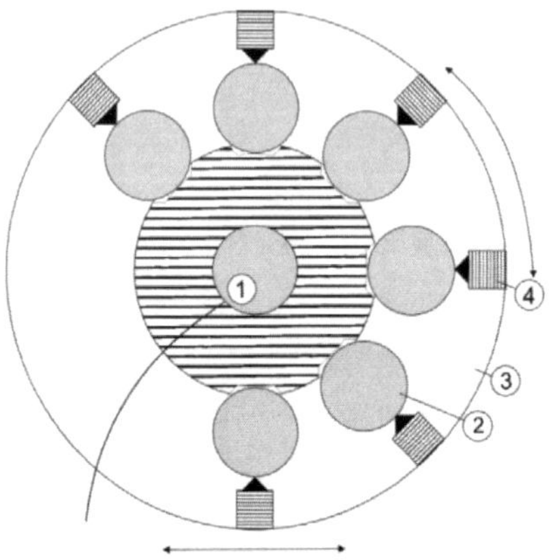

Figure 2. Sketch of the sample holder:
1- accoustic emission sensor,
2- sample,
3- sample holder,
4- contact pin

Some details of the test equipment are listed below:

- Nd:YAG laser with focusing optical system
  - 1 kW maximum laser power output
  - wave length $\lambda = 1064$ nm
  - continuous radiation
  - user-defined laser beam control
- HF-IR camera
  - InSb focal array $\lambda = 3$ to 5 µm
  - exposure time 2 µs up to 14 ms
  - 256 x 256 pixel, resolution 12 Bit
  - 1500 Hz
- Acoustic emission sensors
  - 150 kHz sensor with sampling rate 5 MHz
  - 32 MB memory depth
- Test chamber
  - hermetically sealed up to 4 x 10^{-3} mbar
  - CCD-camera up to $\lambda = 1.1$ µm
  - Contact pin for sample fixing without obstruction of strain

2.2 Investigated materials
Investigations were carried out with the commercial ceramics zirconia (MgO-PSZ) and sintered silicon carbide. Ground samples were prepared with a diameter of 20 mm and a thickness of 0.3 mm.

2.3 Calculation of critical thermal shock strength
Laser beam heating up by helical beam guidance across the thin sample starting in the middle induces temperature profiles presented in Figure 3. Accurate space- und time resolved determination of temperature demands accurate calibration of the HF-IR camera for each material and each

environmental media so that the correlation between the readout digits of the camera and the surface temperature of the sample is known. A resulting calibration curve is presented in Figure 4.

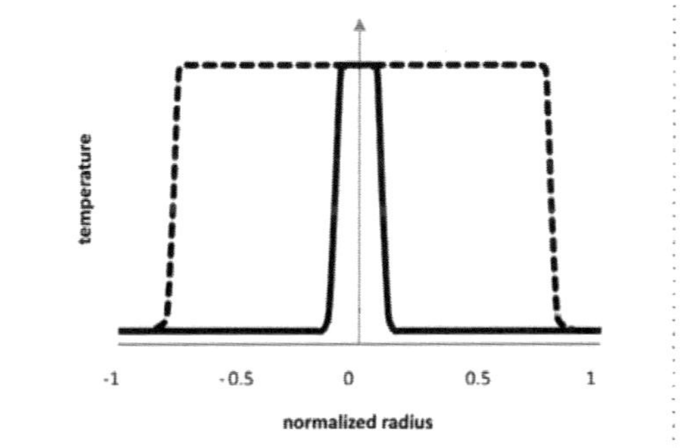

Figure 3. Schematic temperature distribution at the beginning (continuous line) and the end (dashed line) of thermal shock test.

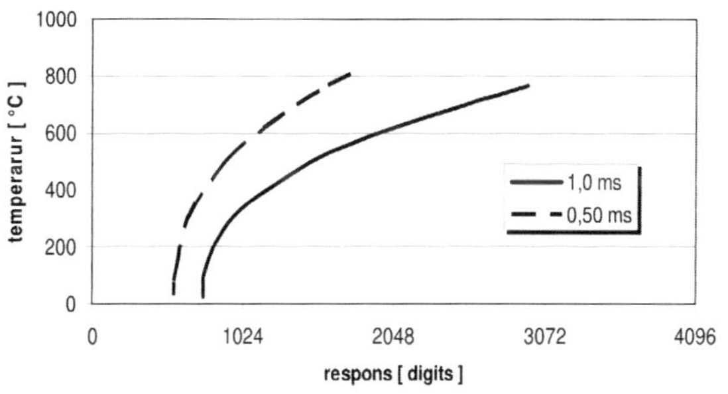

Figure 4. Calibration curve of SSiC for different integration times of the HF-IR camera.

Thermal shock strength calculations are based on the knowledge of space- und time resolved temperature fields. Provided that there is no temperature gradient through thickness direction and only elastic material behaviour occurs, the distribution of tangential stresses is calculated according to equation (1) and distribution of radial stresses according equation (2).

$$\sigma_{\Theta}(r,t) = E(T(r))[-\alpha(T(r))T(r) + \int_0^{r_0} \alpha(T(r))T(r)r \cdot dr + \frac{1}{r^2}\int_0^{r} \alpha(T(r))T(r)r \cdot dr] \quad (1)$$

$$\sigma_{rad}(r,t) = E(T(r))[\frac{1}{r_0^2}\int_0^{r_0} \alpha(T(r))T(r)r \cdot dr - \frac{1}{r^2}\int_0^{r} \alpha(T(r))T(r)r \cdot dr] \quad (2)$$

Where σ is the stress, T the temperature, r the radius, t the time, E(T(r)) Young's Modulus as a function of T and $\alpha(T(r))$ the coefficient of thermal expansion as a function of T.

The resulting schematic trend of the distribution of calculated tangential and radial stresses at the beginning and the end of thermal shock test is shown in Fig. 5 and 6. It is supposed that failure is caused by the maximum of tangential stresses at the moment of failure so the maximum of tangential stress equates with the thermal shock strength. Radial stresses were neglected.

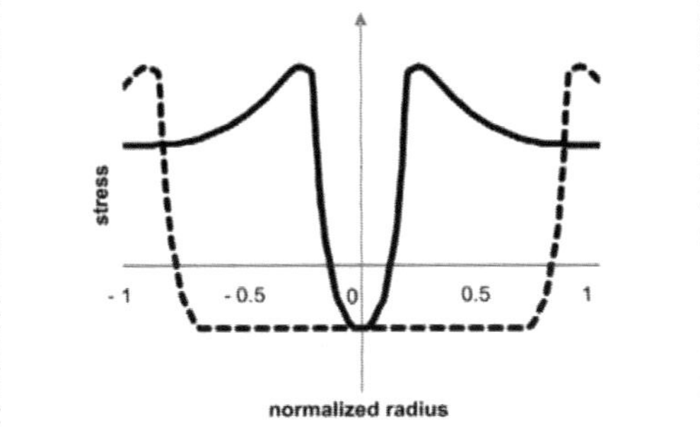

Figure 5. Schematic trend of distribution of tangential stresses at the beginning (continuous line) and the end (dashed line) of thermal shock test.

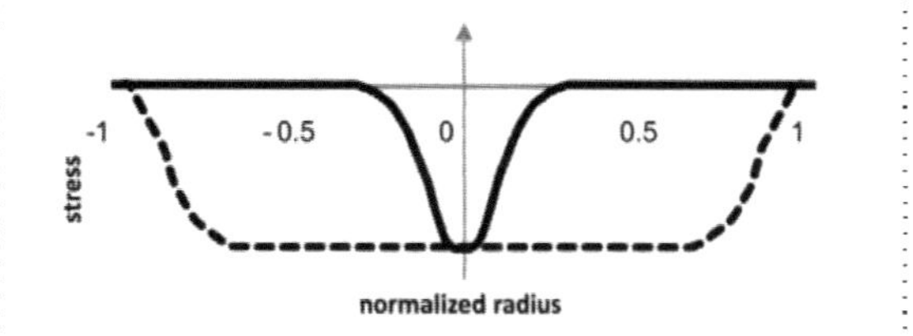

Figure 6. Schematic trend of distribution of radial stresses at the beginning (continuous line) and the end (dashed line) of thermal shock test.

3. Results and Discussion

For silicon carbide and zirconia and each environmental media about 30 samples were tested. The maximum tangential stress in the moment of failure was determined as thermal shock strength. The strength distribution of ceramic samples is described by the Weibull material parameters m and σ_0 and the presentation of measured strength results is given in form of a Weibull plot for silicon carbide (Fig. 8) and zirconia (Fig. 9).

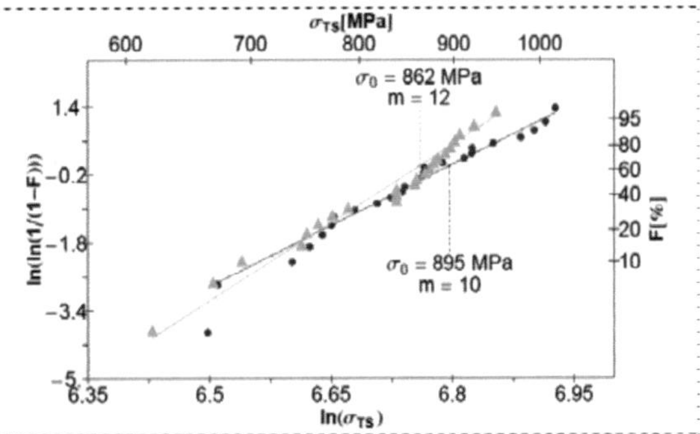

Figure 8. Weibull plots for silicon carbide, • air, ▲ vacuum.

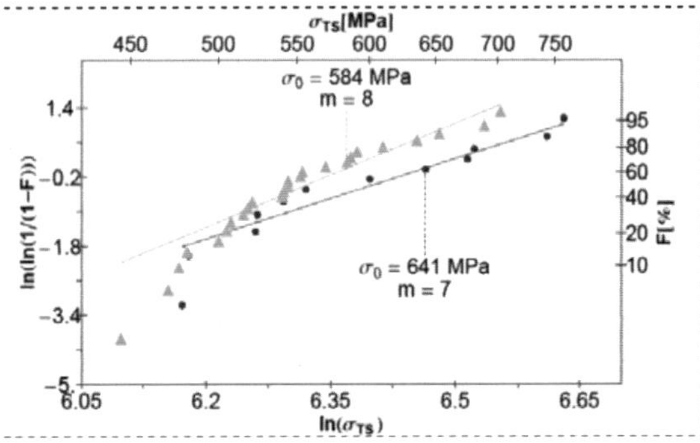

Figure 9. Weibull plots for zirconia, • air, ▲ vacuum.

Compared to test results for experiments in air there is no significant effect for testing in vacuum. One reason can be the short duration of tests, typically less than one second. It will be interesting to investigate the influence of environmental media again in fatigue tests or for materials which are known for their sensitivity to subcritical crack growth.

4. Conclusions

The presented test equipment offers the possibility to carry out repeatable thermal shock tests in different environmental media. Based on the time and space resolved measurement of temperature, failure stresses can be determined.

Thermal shock tests on silicon carbide and zirconia did not show an influence of environmental media on thermal shock strength.

References

[1] Colombo,P., Hellmann,J.R., Shelleman,D.L.: Thermal shock behavior of silicon oxycarbide foams. *Journal of the American Ceramic Society* **85** (2002), 2306-2312.

[2] Liu,N., Shi,M., Yuan,Y, Chao,S.,Feng, J.,Majewski, P.,Aldinger,F.: Thermal shock and thermal fatigue study of Sr- and Mg-doped lanthanum gallate. *International Journal of Fatigue* **28** (2006), 237-242.

[3] Meiheng,L., Xiaofeng,S., Wangyu,H., Hengrong,G.: Thermal shock behaviour of EB-PVD thermal barrier coatings, *Surface & Coating Technology* **201** (2007)7387-7391.

[4] Knoblauch,V.: Thermoschockprüfung und bruchmechanische Charakterisierung hochtemperaturbelasteter Keramiken und Composite. Thesis, Hamburg (2000)

[5] Rettig,U.: Laserschock-Messmethode-Ein quantitatives Verfahren zur Charakterisierung der Temperaturwechselbeständigkeit hochfester keramischer Werkstoffe. Thesis, Dresden (1997)

[6] Munz,D., Fett,T.: Ceramics- Mechanical Properties, Failure Behaviour, Materials Selection, Springer Series in Materials Science, Vol. 36, Berlin, Spinger, (1999).

Investigation of mechanical properties of gold thin films using a bulge test technique

A.Hémel[1], A.Jacques[1], T.Schenk[1], O.Ferry[1], T.Kruml[2]

[1]Institut Jean Lamour (UMR 7198 CNRS-INPL-UHP), Nancy université, Ecole des mines, Parc de Saurupt, 54042 Nancy, France
[2]Institute of Physics of Materials, Academy of Sciences, Zizkova 22, Brno, Czech Republic

audrey.hemel@mines.inpl-nancy.fr

Abstract. A bulge test device has been built, with the aim to perform mechanical tests on membranes with a thickness in the 100 *nm* to 10 *µm* ranges, between room temperature and 900°*C*. Our set up a differential atmospheric pressure is applied across a membrane, while the deflection is measured using a laser interferometer. We show first results obtained from gold membranes with different thicknesses.

1. Introduction

Thins films are used in many application such as microelectronics and since a few years in MEMS devices. As these films are subjected to high stresses, either during industrial process or during device operation, their mechanical properties are particularly relevant.

However, these materials generally have microstructures (grain size, shape, texture...) and chemical compositions which differ from these of bulk materials. Moreover, their properties cannot be directly extrapolated from these of bulk materials: well-known scaling laws such as the Hall-Petch [1] equation ($d^{-1/2}$ dependence of the yield stress on the grain size of polycrystalline materials) can be expected - and are indeed observed - to fail somewhere in the sub-micron range.

Among the different possible experimental techniques, from nanoindentation to direct testing of micron size specimens, the bulge test, first built by Beams in 1959 [2], offers a good compromise between two difficulties: specimen preparation and analysis of experimental data. Free-standing membranes are prepared from thin films deposited on a silicon substrate by opening a window within the substrate by classical wet or dry etching methods. The specimens are then loaded by use of a differential atmospheric pressure on both sides of the film.

The temperature dependence of the mechanical behaviour is of major importance. However, the range of currently existing bulge test devices does not exceed 450°*C* [3, 4, 5]. Other high temperature mechanical tests may be done by substrate curvature measurement, but require temperature variations during the test.

The aim of the present paper is to describe a bulge test apparatus currently being built, in order to test ductile and brittle films in a temperature range from room temperature to ~ 900°*C*, and to report first results obtained at room temperature on ductile gold thin films with a submicron thickness.

2. Description of the bulge test device.

A diagram of the device in room temperature configuration is shown in figure 1. A sample is clamped on a sample holder using a graphite screw and two graphite rings for an air-tight sealing. A differential pressure is applied using a syringe connected to a computer-controlled motor. A differential pressure gauge measures the applied pressure with a resolution of 0.1 *Pa* and a pressure range from 0 to 0.5 *MPa*. A graphite rod glued to a silicon mirror (total mass: 4 *mg*) is placed on the sample. Deflection of films induced by the applied pressure is measured using a Michelson interferometer with a λ=770 *nm* laser beam reflecting on the mirror. Pressure and deflection are measured continuously (figure 2), in order to command the pressure syringe for constant stress rate loading or tests under constant stress (i.e. creep tests).

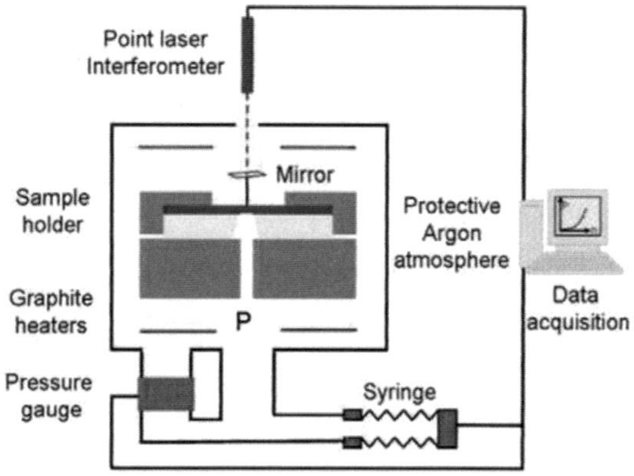

Figure 1. Diagram of the bulge test device.

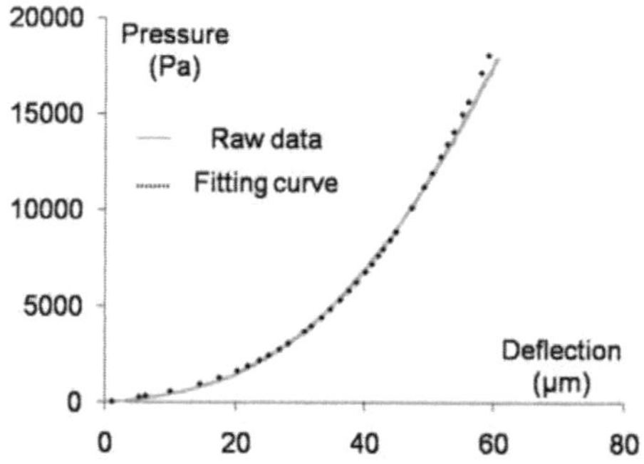

Figure 2. Bulge test raw data (grey line) obtained for a rectangular gold membrane (7*1.5 mm^2) with a 200 *nm* thickness, and fit using (1) (dots)

The specimen can be heated using a pair of circular graphite resistors placed on top and bottom of the sample holder. The temperature is measured by two type K thermocouples and controlled by a PID system. The furnace is water-cooled and thermally insulated from the rest of the device. The whole is shut in a vacuum vessel which can be put under a secondary vacuum, in order to be under high-purity conditions before floating the vessel with pure argon. During pumping the device and heating, the specimen must be carefully kept at zero differential pressure, in order to avoid loading of the film.

2.1. Sample preparation

One of the major difficulties of this technique is the sample preparation. The films are deposited on 380 μm thick silicon substrates. A gold mask with a rectangular (7*1.5 mm^2) window is deposited by PVD on the substrate opposite to the film. The substrate is etched in a KOH solution at 80°*C* (wet etch) or by Deep Reactive Ion Etching. Both etch techniques leave perfect rectangular freestanding thin films.

2.2. Data analysis

The elastic deflection h of a square or rectangular membrane subject to a uniform pressure p is given by the following simplified equation [6, 7, 8]:

$$p = C_1 \frac{Mt}{a^4(1-\nu)} h^3 + C_2 \frac{\sigma_0 t}{a^2} h \tag{1}$$

Here C_1 and C_2 are two coefficient depending on the shape of the sample, M is the plane strain modulus, a the half side length, t the thickness of the film, ν the Poisson's ratio and σ_0 the residual stress of the film. The experimental data are fitted according to (1) and allow the extraction of basic parameters such as E and σ_0. Figure 2 provides an example of a pressure-deflection curve from a rectangular gold thin film (t = 200 *nm*). The fluctuations of the pressure and deflection remain within the thickness of the curve. The dots are the fit of the initial data points (up to a 45 μm deflection) using (1).

For a rectangular membrane, we can assume a plane strain state in the centre of the membrane, the p, h data can be translated into σ, ε curves using the following relation [6, 7]:

$$\sigma_{xx} = \frac{pa^2}{2ht} \qquad \varepsilon_{xx} = \frac{2h^2}{3a^2} + \varepsilon_0 \tag{2}$$

3. Experimental results

3.1. Microstructure

The microstructure of the film was analysed by X-ray diffraction, electron backscattering diffraction and transmission electron microscope.

The morphology and the grain size of both as deposited and annealed thin film were observed and determined from TEM observation. All films show a bimodal grain distribution with large grains (~ 1 μm) surrounded by smaller grains (~ 30 *nm*). The cross sectional view figure 3 e) reveals a columnar grain structure and shows that there is more than one grain stacked in the film thickness.

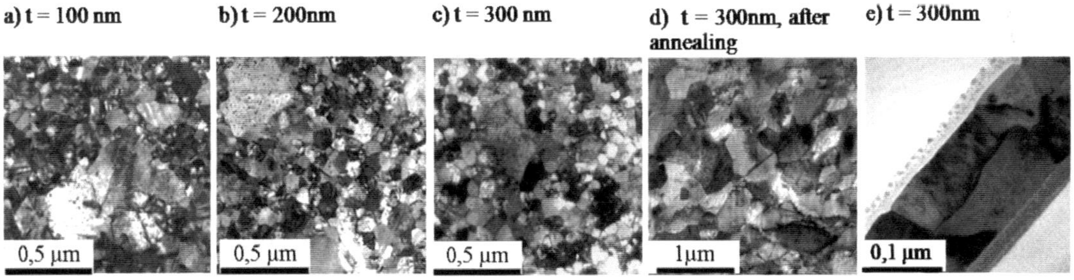

Figure 3. a)-d) Microstructure observed using TEM of gold thin films with different thickness deposited on silicon wafer, e) Cross section view of a 300 *nm* thick gold thin film.

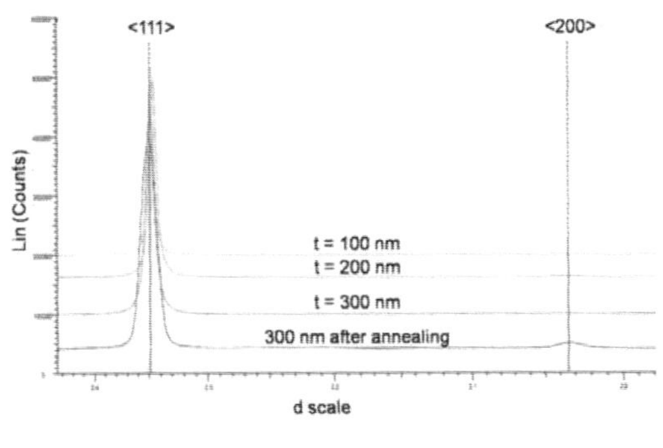

Figure 4. X-ray diffraction from a θ/2θ set up. The scans of Au films deposited on Si showing a strong <1 1 1> texture for the raw film, and a dual <1 1 1>/<2 0 0> texture after annealing.

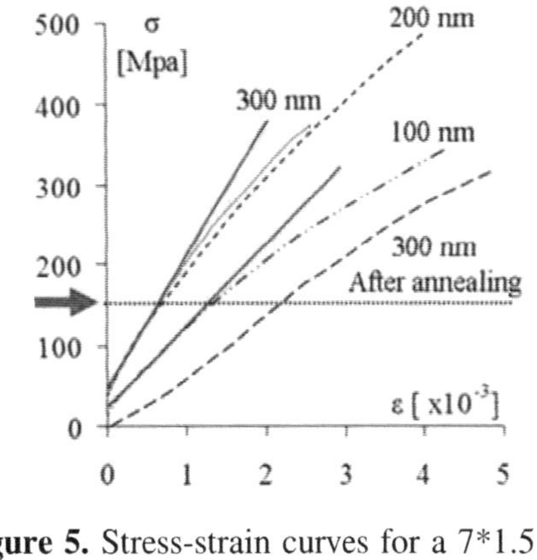

Figure 5. Stress-strain curves for a 7*1.5 mm^2 rectangular gold thin films with different thickness. The red arrow shows the beginning of microplasticity and the plane strain modulus was evaluated from the red lines.

Figure 3 a)-d) shows typical plane views from gold thin film with different thicknesses before and after annealing. The micrographs reveal the presence of defects such as growth twins and dislocations which can appear during the growth process. The average grain size was measured using the intercept method and varies from 50 to 70 *nm* for the as-deposited films. Annealing at 260°*C* of the free standing film for 20 *h* (after etching of the window) increases the average grain size to 200 - 300 *nm* [9].

The texture was determined by means of X ray diffraction. The results are shown in figure 4 and reveal a strong <111> texture [10]. A low (200) peak appears after annealing.

3.2. Stress strain curves

The free-standing gold thin films were initially buckled. The clamping of the specimens made them flat, except at their edges which were still wrinkled. Figure 5 shows typical stress vs. strain curves before and after annealing. The apparent initial stress (50 *MPa* for raw specimens and -2 *MPa*) for the annealed one thus do not represent the initial state of the film. The plane strain modulus M calculated from the slope of the elastic part (red line figure 5), lies between the value of M <111> (145 *GPa*) and M <100> (53 *GPa*) [10] (138, 122 and 90 *GPa* for thin films with a thickness of 300, 200, and 100 *nm* respectively). For the thinner film the reduced stiffness can be attributed to the film deposition technique. It can be related to a change in microstructure (porosity) of the film instead of a thin film effect. After annealing, the plane strain modulus is about 60 *GPa*. The decrease of M after a heat treatment is due to a change of texture (see the appearance of a <200> peak Fig 4).

A major source of systematic experimental error is the initial value of the deflection: For a film under a low initial stress, or a buckled film, the initial position of the mirror on rod device is very sensitive to its weight or to friction. This might explain the low modulus of the 100 *nm* film and the apparently erroneous behaviour of the annealed film.

Microplasticity begins approximately at $\sigma = 150$ *MPa (arrow Fig 5))*. This is probably the result of the heterogeneous microstructure, as the largest grains are expected to yield first, while the smaller ones still deform elastically. For each grain, we expect a different local stress depending on its size and orientation. The plastic domain of the curves is very short (less than 10^{-3}), and does not allow defining a classical 0.2% yield stress, as transition from purely elastic to plastic behaviour is smooth. The fracture strain happens approximately at $\varepsilon = 0.3\%$ for as-deposited films while fracture was not reached for the annealed thin film. This difference may be the result of the increase in the average grain size.

4. Conclusion

A new bulge testing device has been built, aiming at strain membranes of brittle materials from room temperature up to 900°*C*. The first room temperature tests on gold films give results consistent with literature data. The variation of Young's modulus can be attributed to changes in the volume fraction of the <111> fibre components. The yield stress remains insensitive to the film thickness. Further improvements will focus on non contact, full 2D measurements of the film shape, in order to get a realistic view of the stress and strain distribution within plastically strained membranes.

References

[1] Hirth J P and Lothe J 1968 *Theory of dislocations* (McGraw-Hill ed., New York)
[2] Beams J W 1960 *The Structure and Properties of Thin Films*, 183 (C.A. Neugebauer et al. eds., Wiley, New York)
[3] Cieslar M, Oliva V, Karimi A and Martin J L 2004 *Mater. Sci. Eng.* **387** 734
[4] Cieslar M, Oliva V, Karimi A and Martin J L 2004 *J. Alloys Comp.* **378** 312
[5] Kalkman A J, Verbruggen A H and Janssen G C A M 2003 *Rev. Sci. Instru.* **74** 1383
[6] Vlassak J J and Nix W D 1992 *J. Mater. Res.* **7** 3242
[7] Small M K, Vlassak J J, Powell J B, Daniels B J and Nix W D 1993 *Mat. Res. Soc. Symp. Proc.* **308** 159
[8] Timoshenko S P and Woinowsky-Krieger S 1959 *Theory of plates and shells* (McGraw-Hill, ed., New York)
[9] Greiser J, Müller D, Müllner P, Thompson C V and Arzt E 1999: *Scripta Mater.* **41** 709
[10] Espinosa H D, Porok B C and Peng B 2004 *J. Mech. Phys. Solids* **52** 667

Pulsed low-energy positron beams: A useful tool to investigate defect structures in deformed metals and alloys

W Egger[1], P Sperr[1], G Kögel[1], H-J Gudladt[2]

[1]Institut für Angewandte Physik und Messtechnik, [2]Institut für Werkstoffkunde, Universität der Bundeswehr München, D-85577 Neubiberg, Germany

werner.egger@unibw.de

Abstract. To understand in more detail the behaviour of deformed metallic materials, the knowledge of defects and defect distributions at an atomistic level is important. To investigate the plastic zone in front of a crack tip, the positron annihilation lifetime spectroscopy with a pulsed monoenergetic positron beam of variable energy allows the detection of vacancies, dislocations, vacancy clusters and micro voids in the crack surface near region. Moreover, for a given defect type it is possible to determine its concentration. The positron lifetime measurements in samples of different materials (aluminium, copper) showed different defect profiles for crack surfaces produced by monotonic and cyclic deformation. In addition, this technique was able to characterize the kind of damage (monotonic or cyclic) by analysing the different positron lifetimes.

1. Introduction

The lifetime prediction of cyclically deformed constructions used for airplanes and cars becomes increasingly important in order to minimize the amount of very expensive and time consuming mechanical tests. Typically, linear damage accumulation models were used for the fatigue lifetime prediction, but in most cases the damage evolution is non-linear, and, consequently, microstructural aspects of damage have to be taken into account.

In general, two stages can be distinguished in the fatigue lifetime: The early stage of deformation, accompanied by the formation of the crack, and the stage characterized by the cyclic growth of the fatigue crack. Considering the crack as a kind of damage phenomenon it is necessary to understand the formation as well as the propagation of the crack. The latter can be measured by conventional techniques like compliance measurements or by the DC potential drop method.

In case of monotonic or cyclic crack propagation of ductile materials it has been found that the process zone (monotonic or cyclic plastic zone) in front of the crack tip determines the behaviour of the crack expansion. Consequently, the knowledge of defects and defect distributions at an atomistic level within this zone is highly desirable.

Positron annihilation lifetime spectroscopy with pulsed monoenergetic positron beams of variable energy allows for depth profiling of vacancies, dislocations, vacancy clusters and micro voids. Moreover, for a given defect it is possible to determine its concentration. Therefore, this technique offers a unique tool for investigations in deformed materials.

Journal of Physics: Conference Series **240** (2010) 012164 doi:10.1088/1742-6596/240/1/012164

In this paper we will give a short overview of defect characterizations with pulsed low energy positron beams in cracked specimens of pure copper and a technical aluminum alloy. Most of the results have been published and discussed in more detail elsewhere [1,3-5]. Our aim is mainly to draw the attention on a method of potentially great interest to a wider scientific community of materials scientists.

2. Positron lifetime spectroscopy with pulsed positron beams

The positron is the anti particle of the electron. In condensed matter it will annihilate with an electron into two or more gamma quanta with a lifetime characteristic for the local electron density at the annihilation site. A positron with energies in the keV-range, implanted in a metal, thermalizes rapidly within a few picoseconds. This time is short compared to the time of hundreds of picoseconds typically spent before annihilation. In a defect free crystalline material, the positron survives for a time which is characteristic for the material. In the presence of open volume defects, due to the missing repulsive ionic potential, transition to a localized state, trapping, may occur. The electron density at the site of the defect is locally reduced and, consequently, the positron lifetime is prolonged with respect to the lifetime in the defect free bulk. Trapping occurs also at regions with different positron affinity than the crystalline matrix, e.g. at precipitates. Again, the lifetime is characteristic of the material and the type of defect. The high trapping rate into open volume defects and regions of high positron affinity is the reason for the high sensitivity of positrons as defect probes: Defect concentrations in the ppm-range may still be detected, from the trapping rate of the positron in a given defect the defect concentration may be inferred.

In positron lifetime spectroscopy the time spent by the positron between its implantation in the solid and its annihilation is measured. In a pure, defect free material, this yields an exponential decay spectrum which may be characterized by a single lifetime. In the presence of defects, additional exponential components with characteristic defect lifetimes may be detected. The relative intensities convey information on the defect concentration. Another useful but less specific parameter to characterize the lifetime spectrum is the intensity-weighted sum of all lifetimes, the mean lifetime. A mean lifetime longer than the lifetime in the defect free material points to the presence of defects.

To depth resolve defect structures one has to employ monoenergetic positron beams of variable energy. The implantation profile of such a beam may be approximated by the derivative of a Gaussian centred about a mean implantation depth. The mean implantation depth $\overline{z}$ scales with the density ρ of the material and the beam energy E as $\overline{z} \propto E^{1.6} / \rho$. Positrons implanted with a given energy sample a region extending from the surface down to two times the mean implantation depth.

To perform positron lifetime measurements with a monoenergetic beam, the beam has to be pulsed. A pulsed low energy beam system of variable energy for depth profiling of laterally homogeneous defect distributions, PLEPS [6], was developed at our institute and operates nowadays at the high intensity positron source NEPOMUC at the Munich research reactor FRM-II [7]. Also, a Scanning Positron Microscope (SPM) [8] has been developed, which offers additional lateral resolution down to 1 micrometer. After demonstrating its usefulness with a weak conventional positron source in the laboratory, which entailed however excessively long measurement times, it is currently installed at NEPOMUC and will be operative soon. Further technical details of both beam systems can be found in the references [6, 8].

3. Experimental results

3.1. Sample preparation

To gain more information about the crack evolution, fatigue crack propagation tests have been undertaken with single edged notched (SEN) specimens. For crack initiation, a starter notch of 1 mm depth was cut into the small side of the specimens. The specimens were cyclically deformed in a fatigue testing machine under symmetric push pull conditions. The actual crack length was measured by a potential drop method with a detection limit of 1 µm /cycle. Both signals, the stress amplitude as well as the crack length, were used to control the stress intensity factor, ΔK, and, as a consequence, it was possible to keep it constant. This procedure allows keeping the crack growth rate constant, too. The latter was important to obtain a constant defect configuration in front of the crack tip. Now the defect configuration within the crack tip near region was measured with PLEPS and SPM as it can be seen in the Figure 1. At first, the crack flanks that are visible on the specimen surface and the surrounding plastic zone were scanned by the SPM in X- and Y- direction with a lateral resolution of 5 µm. Thereafter, the specimen was broken and the crack surface was investigated by PLEPS. To obtain a depth profile of positron lifetimes the monoenergetic beam was varied between 0.5 keV and 18 keV.

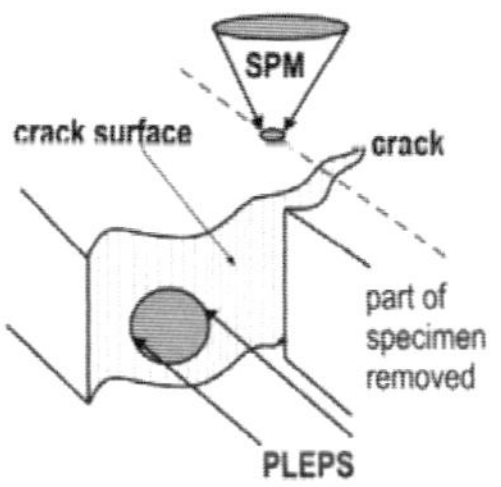

Figure 1. Employing PLEPS and the SPM on cracked specimen [2]

3.2. Results in pure copper

At first, the surface of the SEN specimen of a high purity copper alloy as seen by the SPM at 16 keV implantation energy is shown in Figure 2 [1]. The fatigue crack propagates form the lower right hand side to the upper left hand side of the picture. Different greyscales in the map correspond to different mean positron lifetimes. Close to the crack flanks the longest mean positron lifetimes have been measured. The band of enhanced lifetime is 100 µm wide, nearly symmetrical, and can be correlated with the cyclic plastic zone. The typical positron lifetime in plastically deformed copper is 172 ps [2]. Therefore, the maximum value of 250 ps in the mean positron lifetime indicates that defects with a larger free volume than mono vacancies and dislocations must exist close to the crack flanks.

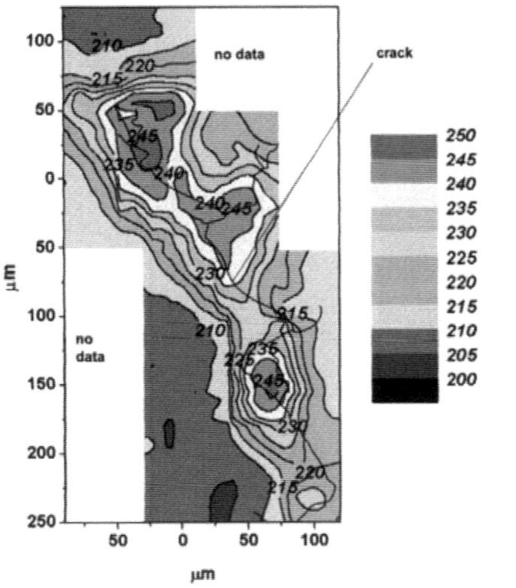

Figure 2. Mean lifetime map of a fatigue crack tip in pure copper [1].

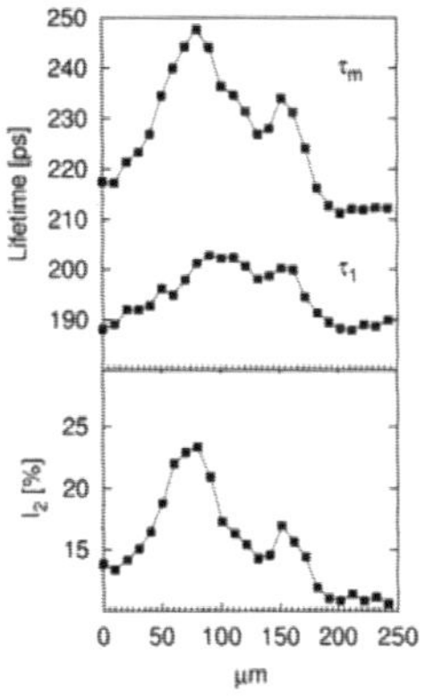

Figure 3. Line scan over the crack tip in pure copper. The mean lifetime, and the shorter lifetime are shown, the longer lifetime was fluctuating around 400 ps. Also shown is the intensity of the second lifetime [1].

To get additional information, a line scan was performed perpendicular to the crack as can be seen in Figure 3 [1]. The spectra could be decomposed into two positron lifetimes. The shorter one of about 190 ps was slightly larger than the 172 ps. This indicates the presence of small vacancy clusters close to the crack flanks. The longer lifetime of about 400 ps can be attributed to larger vacancy clusters. Taking the intensity of the second lifetime into account on can see that positron trapping in vacancy clusters becomes most pronounced close to the crack flanks. Since the crack developed into two branches, two pronounced maxima are visible in the plots.

Now, the specimen was cracked and the fatigue crack surface was measured by the depth sensitive PLEPS. Figure 4 shows the mean lifetime vs. the mean implantation depth [3]. The solid line follows the prediction of a diffusion trapping model for positrons in pure copper without any deformation. The squares represent the lifetime measurement of a specimen of pure copper, carefully polished after the annealing at 800 °C in high vacuum. A good correlation between the model and the measurement had been found and the specimen was used as a reference. Lifetime measurements on the cracked surface of a monotonously fractured specimen showed that positron trapping at higher implantation energies occurs mainly in dislocations. In the case of fatigue fracture, a mean lifetime of 270 ps was found at higher implantation energies. Decomposition of the lifetime spectra into two components pointed to the presence of dislocations and of open volumes of 10 to 20 vacancies. This indicates that within the cyclic plastic zone the defect structure must differ from the defect structure induced by monotonic deformation [3].

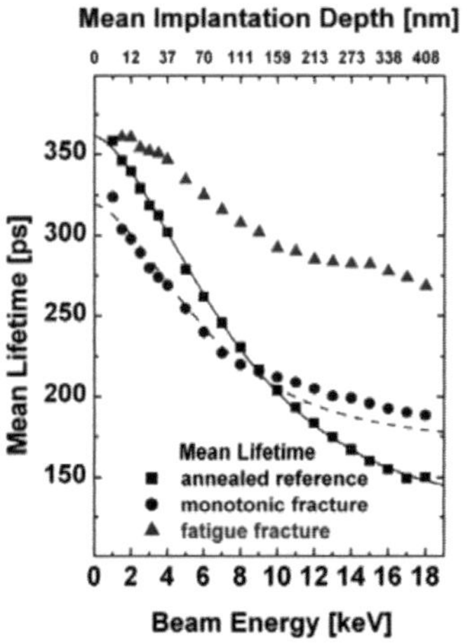

Figure 4. Mean lifetime in an annealed reference, on a monotonously and on a fatigue frac tured surface in pure copper [1].

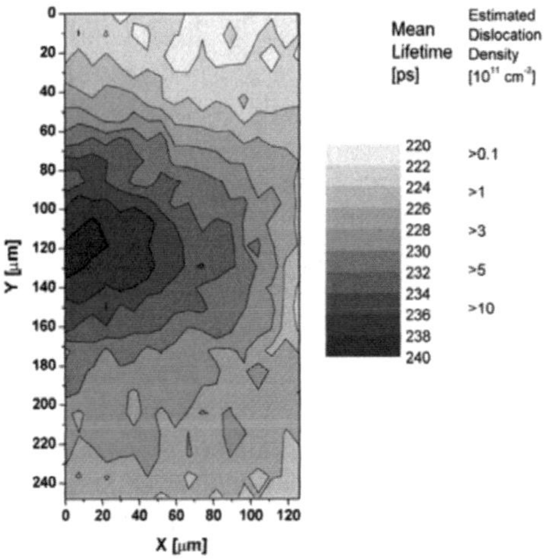

Figure 5. Mean lifetime map $\tau(x,y)$ of a crack tip in an AlMgSi-alloy [4].

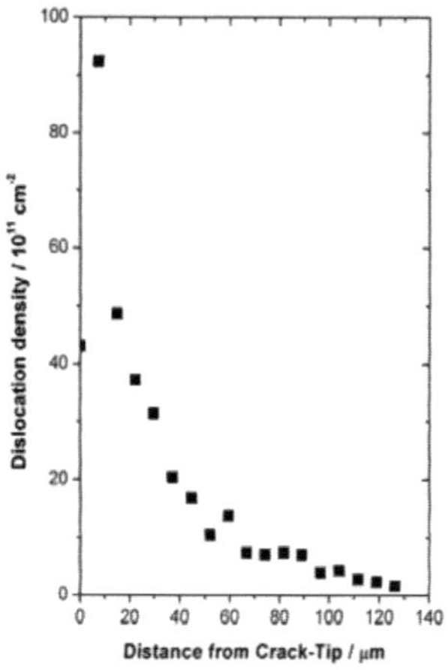

Figure 6. Dislocation density as a function of distance from the crack tip.

3.3. Results in the AlMgSi-alloy AA6013

Starting from these results, a technical AlMgSi-alloy (AA6013) has been investigated by using SPM and PLEPS. At first the SEN-specimen was cyclically deformed at K = 12 MPa$\sqrt{m}$ in ambient air. After localizing the crack tip region of the fatigue crack the latter was investgated by the SPM as can be seen in Figure 5 [3]. The fatigue crack tip was surrounded by a plastic zone with a high dislocation density. The mean lifetime value of 240 ps points to a high density of dislocations as it is expected for plastically deformed material. At a radial distance of more than 60 µm from the crack tip, where the deformation becomes less pronounced, the positron lifetime was measured to be about 220 ps, corresponding to total trapping in precipitates, typical for the untreated material. It becomes thus possible to estimate the dislocation density c_{disl} from the SPM-diagram to [6]:

$$c_{\text{disl}}(x,y) = 4 \cdot 10^{11} \frac{\tau(x,y) - 220\text{ ps}}{240\text{ ps} - \tau(x,y)} \ [\text{cm}^{-2}].$$

In addition, the dislocation density close to the crack tip was found to be as high as in heavily deformed metallic materials where a dislocation density of several 10^{12} cm^{-2} can be expected. With increasing distance from the crack tip c_{disl} decreases, as it can be seen in Figure 6. At a distance of 120 μm from the crack tip the density was found to be more than two orders of magnitude smaller than in the crack tip near region.

In addition, measurements with PLEPS on the cracked surface of fatigued samples were performed. Besides a lifetime corresponding to dislocations, a second lifetime, associated to small vacancy clusters of up to 20 vacancies was observed. Those vacancy clusters vanished within a distance of 2 μm from the crack surface [3]. As in the case of copper, on the cracked surface of monotonously fractured specimens, total trapping of positrons in dislocations, but no vacancy clusters were observed.

4. Summary and conclusions

The experiments have shown that positron lifetime spectroscopy with pulsed low energy beams of variable energy is able to elucidate complex defect structures at an atomistic level and to distinguish between different kinds of defects close to the crack tip for pure copper and aluminium alloys. The measurements clearly show that the defect configuration is quite different in the crack tip near area for specimens fractured under cyclic or monotonic loading conditions. The mean lifetime in the specimens fractured under cyclic loading conditions is enhanced with respect to the mean lifetime in those fractured under monotonic loading conditions. The presence of a second lifetime was only found in fatigued specimens and indicate a complex defect structure consisting of dislocation networks and small vacancy clusters. Those defects are located within the cyclic plastic zone in front of the crack tip.

The presence of vacancy clusters of less than 1 nm in diameter confirms earlier results from Page et. al. [9]. They found an increasing volume fraction of voids by SANS, in OFHC copper cyclically deformed at 405°C. Obviously, vacancy clusters are characteristic for fatigued copper, even at room temperature. It is well known that ordered dislocation structures develop during cyclic deformation under stress as well as under strain control. The latter had been studied with single crystals and polycrystals of Cu and Al-based alloys, cyclically deformed under strain as well as under stress control [10-12]. In case of low cycle fatigue dislocation networks as well as persistent slip bands (PSB) have been found. The kind of dislocation structure depends on the degree of the strain amplitude. For single edge-notched specimens fatigued under stress control, the formation of a persistent slip band has been found in front of the notch, where the highest stress fields will occur. In the early stage of crack propagation the fatigue crack follows the slip band and, in single crystals, oriented for single slip a crystallographic stage I crack has been observed by TEM [12]. It should be mentioned that these kind of dislocation structure has been only observed under cyclic deformation. In addition, the existence of the PSB and corresponding micro cracks has been detected by Katagiri et al. in the late 70s by using TEM [13].

Now, it can be summarized that in the light of fracture mechanics the dislocation arrangement has to be taken into account in order to understand the crack extension under cyclic deformation. It is assumed that the kind of the plastic zone in front of the crack tip determines the crack growth behaviour. The existence of the dislocation network was confirmed for a copper bulk material by ECCI measurements undertaken with SEM in front of the fatigue crack tip [14]. In case of monotonic deformation, however, this kind of dislocation arrangement cannot be found.

The extension of the cyclic plastic zone in front of a fatigue crack can be determined by positron lifetime measurements with the SPM, taking the dislocation density into account as it is seen in Figure 6. The difference of the monotonic and cyclic plastic zones must be reflected in the monotonic and cyclic crack growth rate, as special fatigue tests in aluminium alloys, in ferritic and in austenitic steel indicate [15]. Currently, cracked surfaces of ferritic and austenitic steels are investigated with PLEPS at NEPOMUC.

References

[1] W Egger , G Kögel, P Sperr, W Triftshäuser, S Rödling, J Bär, H-J Gudladt, *Appl. Surf. Sci.* **194** (2002) 214
[2] T Lepistö, J Yli-Kauppila, P Kettunen, P Hauttojärvi, *Phys. Stat. Sol.* (a), **67** (1981) K93
[3] W Egger , G Kögel, P Sperr, W Triftshäuser, S Rödling, J Bär, H-J Gudladt, *Z. Metallkd.* **94** (2003) 687
[4] W Egger , G Kögel, P Sperr, W Triftshäuser, J Bär, S Rödling, H-J Gudladt, *Mater. Sci. Eng.* A **387-389** (2004) 317
[5] W Egger, G Kögel, H J Gudladt, *Int. J. Mater. Res.* **97** (2006) 12
[6] W Egger, P Sperr, G Kögel, and G Dollinger, *Phys. Stat. Sol.* (c) **4** (2007) 3969
[7] C Hugenschmidt, G Dollinger, W Egger, G Kögel, B Löwe, J Mayer, P Pikart, C Piochacz, R Repper, K Schreckenbach, P Sperr, M Stadlbauer, *Appl. Surf. Sci.* **255** (2008) 29
[8] A David, G Kögel, P Sperr, W Triftshäuser, *Phys. Rev. Lett.* **87** (2001) 067402
[9] R Page, J R Weertman, M Roth, *Acta Metall.* **30** (1982) 1357
[10] J M Finney, C Laird, *Phil. Mag.* **31** (1975) 339
[11] H Mughrabi, *Mater. Sci. Eng.* **33** (1978) 207
[12] W Vogel, M Wilhelm, V Gerold, *Acta Metall.* **30** (1982) 31
[13] K Katagiri, A. Omura, K Koyanagi, J Awatani, T Shiraishi, H Kaneshiro, *Met. Trans.* **8A** (1977) 1769
[14] J Bär, C Schumacher, C Kröber to appear in *Prakt. Metallographie* (2009)
[15] J Steinbock, H-J Gudladt to appear in *Int. Conf. of Fracture 12* (2009)

Research on dynamic wear behavior at elevated temperature of HVOF sprayed nanostructured WC-17Co coating

Yan Liu[1], Hui Chen[1], Guoqing Gou[1], Mingjing Tu[2]

[1] Institute of Materials Science and Engineering, Southwest Jiaotong University, Chengdu, 610031,China,
[2] Institute of Materials Science and Engineering, Sichuan University, Chengdu, 610041, China

xnrpt@swjtu.edu.cn (Hui Chen)

Abstract. Nanostructured WC-17Co coatings have been prepared by means of High Velocity Oxy-fuel (HVOF) technique. The wear resistance at the elevated temperature (500℃) of nanostructured coatings was compared using GCr15 steel as counterpart in sliding wear tests. The results show that when at the temperature of 500℃, the wear failure mechanism turns from plastic deformation to fracture resulted from crack propagation and adhesive wear. With the wear going, abrasive wear dominate in the coating, then turns into adhesive wear with changes of microscope.

1. Introduction

Many engineering components are subjected to elevated temperature in tribological conditions during service. High contact surface temperatures also arise due to the generation of friction heating, particularly for large sliding speeds and contact loads. Under these extreme environments, the wear resistance strongly diminishes, which may lead to the rapid failure of the coating [1, 2]. Thus, it starts to be a priority to know the tribological behavior of hard coatings with temperature.

WC-Co materials have been widely applied in cutting tools and wear resistant components for its high hardness, strength and wear resistance.

Yang et al. [4] researched the sliding wear behavior of HVOF sprayed WC-12%Co coatings at elevated temperatures. Previous work [5] researched the wear behavior of nanostructured and ultra-fine HVOF Spraying WC-17Co coatings at 600℃. The results show that the wear resistance and the fracture toughness of the nanostructured coatings at elevated temperature are better than that of the ultra-fine coatings. Ref [6] investigated the friction and wear behavior at 600℃ of plasma sprayed nanostructured WC-Co coatings.

A lot of researches have carried out on HVOF spraying nanostructured WC-Co coatings. However, the dynamic wear failure mechanism at high temperature of HVOF spraying nanostructured coating is little understood. In the present study, nanostructured WC-17Co coatings have been prepared by means of High Velocity Oxy-fuel (HVOF) technique. The dynamic wear failure mechanism at elevated temperature of HVOF spraying nanostructured WC-17Co coating was studied using the high temperature sliding wear tester.

2. Experimental detail

One commercial nano powder was used as feedstock in the present study. The powder was agglomerated into sprayable sizes. The powder was obtained from Inframat Ltd., Famington, USA. The powder characteristics are shown in Table 1.

Table 1. Powder characteristics

Feedstock powder	Feedstock size/μm	WC grain size
Nanostructured WC-17Co	5~45	50-500nm

The Q235 was used as substrate. The HVOF spraying system was D-3000 (Zhengzhou Dayang Thermal Spraying Co. Ltd.). Airplane kerosene was used as a fuel. The optimized HVOF parameters are summarized in Table 2.

Table 2. HVOF spraying parameters

Spraying material	Oxygen（m^3/h）	Fuel（m^3/h）	Ar（m^3/h）	Powder-feed rate (g/min)
Nanostructured WC-17Co	35	25	0.7	1.0

Sliding wear test was carried out using MMU-5G sliding tester (Yihua Tester Co. Ltd., Jinan, China), which described in Ref [6]. Sliding wear tests were performed using GCr15 steel as the counterpart, at sliding speed of 200r/min. And wear circumstance temperature is 500℃, under constant load of 200N. Weight losses of the specimens were measured using an electronic balance with an accuracy of ±0.1mg. The worn traces were observed with SEM.

3. Results and discussion

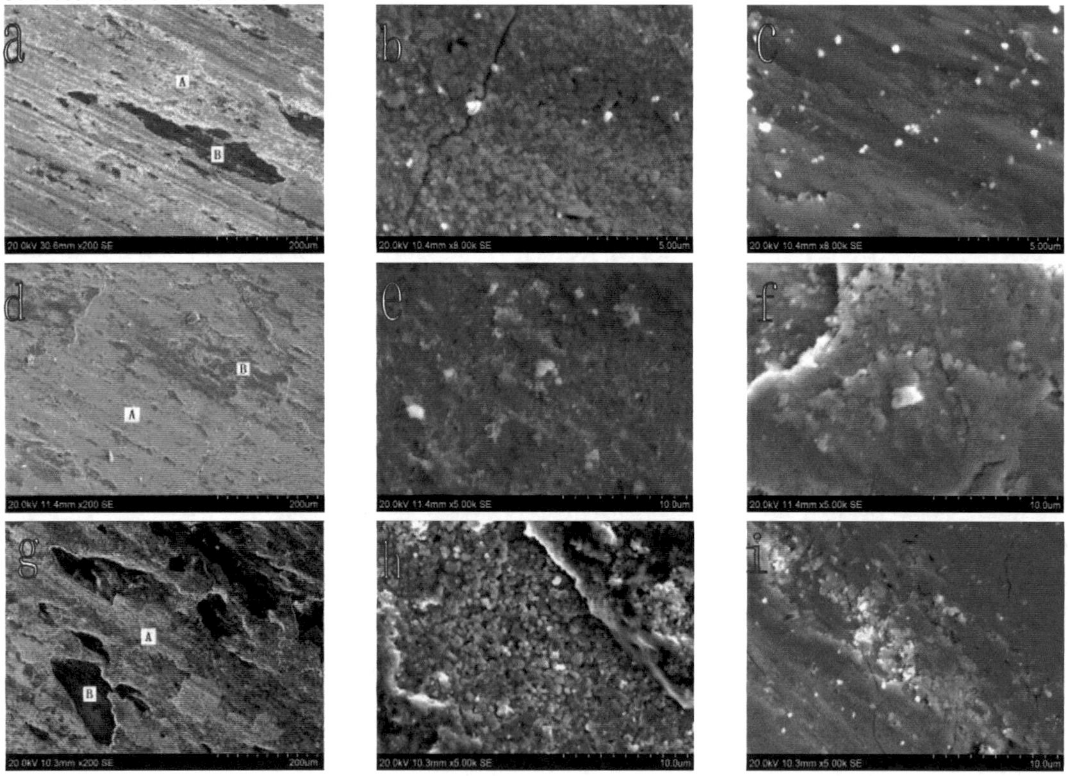

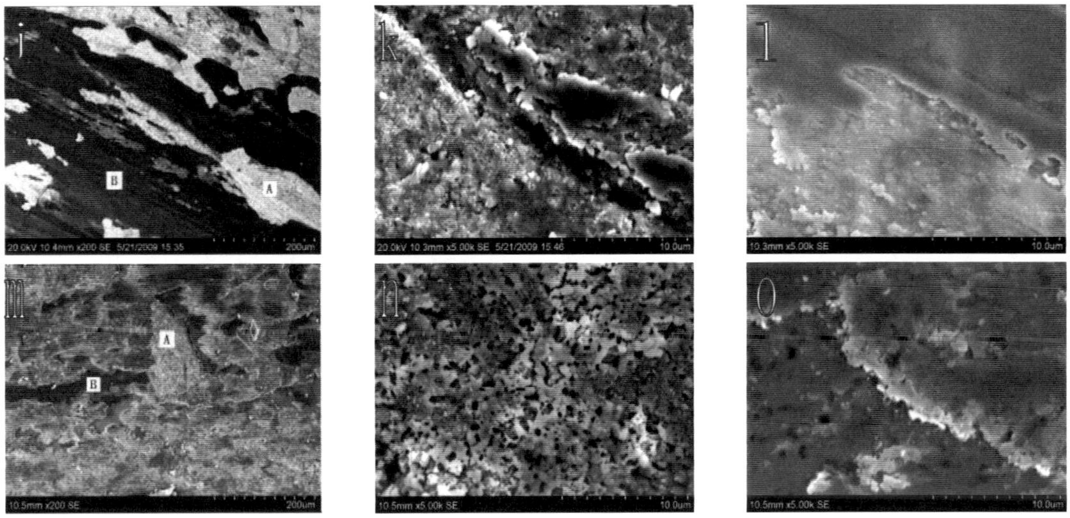

Figure 1. Worn trace after different wearing time: (a) (b) (c) worn trace after 20min wearing; (d)(e)(f)worn trace after 40min wearing; (g)(h)(i)worn trace after 60min wearing; (j)(k)(l) worn trace after 80min wearing; (m)(n)(o) worn trace after 100min wearing;

The worn trace after different wearing time all consist of bright region (A) and dark region (B). EDS analysis indicates that there is more element Fe in region B than in region A except the worn trace after 40min wearing, which illustrates that adhesion wear is more severe in region B.

It can be seen from Figure 1 (a) that after wearing 20min, the wear of the coating is slight. From Figure 1 (b), magnification of region A, it can be found that the morphology is similar to the coating before wearing. WC particles distribute uniformly and dispersive in the Co matrix. Figure 1 (c) is magnification of distinct B in Figure 1 (a), from where it can clearly be seen traces of plastic deformation, which could be rational since the local flash temperature at the rubbing interface might be sufficiently high to soften the asperities in contact and cause deformation of the asperities under initial high contact stress [7].

Figure 1 (d) is the worn trace of coating undergoing 40min wearing. From the EDS result, it can be concluded that severe adhesion wear occurs. Once adhesive between counterpart materials takes place, the adhesive force actually acts as a strain stress perpendicular to the lamellar direction. The adhesive force probably caused micro-cracking and fracturing along the inter-lamellar contact and resulted in flake off of flat plate-like lamella [8]. Figure 1 (e) is the magnified image of distinct A of Figure 1 (d). Oxide of Fe forms a tribofilm covering the coating. WC particles began to become broken. Because Co-based begins to melt and lost, the WC particles fracture in the process of wearing at high temperature. And the WC particles act as abrasive involved in the process of wear and being grounded. Figure 1 (f) is a magnified image of region B of Figure 1 (d). Cracks are clearly seen in the worn trace and fracture has occurred in some parts of the worn trace.

It can be seen from Figure 1 (g) that furrows form after 60min wearing. Figure 1 (h) indicates that the oxides formed in the surface detached from the coating, exposing a fresh surface. From Figure 1 (i) it can be found the cracks propagate, forming net-like shape. The contact points bear two kinds of load in the process of wearing: mechanical load and thermal load. Rapid increase of heat load leads to over-temperature in local surface of counterpart, resulting in larger temperature gradient within the objects, causing thermal stress. Under the dual roles of thermal stress and mechanical stress, surface and internal fracture by duplicating fever. Plastic deformation occurs with wearing. When deformation accumulates to a certain amount, cracks occur under the cyclic stress in the contact area. The expansion of cracks causes spalling in the region of cracking [9].

Furrows are visible in Figure 1 (j). And so are traces of adhesive wear. Figure 1 (k) is the enlarged image of region A in Figure 1 (j). It can be seen that WC particles has been broken and bonded together, losing the original morphology of WC particles, which illustrates that the coating morphology has already begun to change with wearing. From Figure 1 (k) it can be found the region of plastic deformation becomes smooth. The particles in Figure 1 (h) tear off down and become abrasive. In the grinding tablets wear process, the abrasive edges and corners will penetrate the material surface in the role of force, while pierce the material surface in the role of tangential force, resulting in furrows. The existence of particles will reduce the penetrated depth into the matrix and prevent further abrasive plowing, thus improving the wear resistance of materials. But after hard particles falling, they not only lose the role of anti-wear, but also become abrasive themselves, further exacerbate wearing and reduce the wear resistance of composite materials significantly. Usually rigid particles are uprooted under the impact instead of being refined gradually.

From Figure 1 (m) it can see that the coating experienced severe adhesive wear. Figure 1 (n) is the enlarged image of region A in Figure 1 (m). It can be found that the original morphology of WC can not be seen. The particles are connected together, which may due to extremely high flash point temperature. In the process of grinding, only a part of protruding particles bear pressure. The stress is so large and the flash point temperature is so high that may result in the melting of Co-phase, WC particles flow in the melting Co phase, and bond together under the role of friction force, forming the topography of Figure 1 (n). Figure 1 (o) is the enlarged image of region B in Figure 1 (m). It can be seen clear traces of plastic deformation and fracture of the coating.

From the analysis above, it can be concluded that plastic deformation occurs at the beginning. When plastic deformation accumulates to a certain degree, cracks occur under the dual roles of thermal stress and mechanical stress. The expansion of cracks causes delaminating in the region of cracking.

A part of the abrasives produced by spalling of coating are taken out of the worn surface, and some remain in the coating acting as abrasives participating wearing. The abrasives cut the coating, forming furrow morphology. The occurrence of abrasive wear deteriorates the wear process. With the processing of wearing, the transfer of materials has taken place between the coating and the counterpart and morphology of the coating changed dramatically.

4. Conclusions

(1) The worn trace after different wearing time all consist of two distinct regions. The dark region experiences severe adhesion wear.

(2) The wear mechanism is different after different wearing time.

(3)The wear failure mechanism of the coating turns from plastic deformation to fracture resulted from crack propagation and adhesive wear. With the wear going, abrasive wear dominate in the coating, then turns into adhesive wear again with changes of microscope.

References

[1] N.M.G. Parreira, N.J.M. Carvalho, F. Vaz, A. Cavaleiro. Surf. Coat. Technol., 2006, 200: 6511.
[2] T. Polcar, N.M.G. Parreira, A. Cavaleiro. Wear, 2007, 262: 655.
[3] I. Hussainova, I. Jasiuk, M. Sardela, M. Antonov. Wear, 2009, 267:152 - 159
[4] Q. Yang, T. Senda, A. Hiroe, Surface & Coatings Technology, 2006, (200):4208 -4212.
[5] Hui Chen, Guoqing Gou, Mingjing Tu, Yan Liu. Surface Engineering, 2009, 25(7):502-506.
[6] Hui Chen, Guoqing Gou, Mingjing Tu, Yan Liu. Journal of Materials Engineering Performance (JMEP), Published on line: 28 March 2009.
[7] K. Jia, T.E. Fisher, Wear, 1997, 203–204: 310–318.
[8] J.F. Li, C.X. Ding. Wear, 2000, 240: 180–185.
[9] Zhang Yongzhen et al. Science Press, Beijing, 2007

Journal of Physics: Conference Series **240** (2010) 012166 doi:10.1088/1742-6596/240/1/012166

Residual strength of GFR/POM as a function of damage

G Zachariev[1,3] , H-V Rudolph[2], H Ivers[2]

[1]Central Laboratory of Physico-Chemical Mechanics, Bulgarian Academy of Sciences, Acad. G Bonchev Str., Block 1, 1113 Sofia, Bulgaria.
[2] Federal Institute of Materials Research and Testing (BAM), Unter den Eichen 87, D-12205 Berlin, Germany

meziodo@gmail.com

Abstract. A relation between the residual strength and the dispersed damage accumulated in a short fiber reinforced polyoximethylene (GFR/POM) samples under tension is found. For that purpose dependencies of damage and residual strength on loading percentage are used. Damage as a function of loading percentage is known for the material under study. To find the dependency of residual strength on loading percentage a subsidiary function is introduced and a method is proposed for determination of the parameters in the dependency on the basis of the experimental data. Both damage and residual strength are measured after unloading samples that have been loaded applying different loading percentages. Damage is the accumulation of new internal surfaces that arise under mechanical loading in the whole volume of the material. They are registered by a new original method of X-ray refraction. The analytical relation between the residual strength and damage accumulated is compared to the experimental results found for the residual strength under different damage degrees.

1. Introduction

Mechanical loading triggers processes of material deformation and fracture. While the process of deformation can be followed by means of various methods, recording the macrodeformation of a specific basis, the registration of fracture (its kinetics) is a far more complex problem. This is the reason why, performing mechanical investigations, one can not directly measure fracture but such integral quantities as strength and time to fracture. Meanwhile, serious progress in fracture record was marked applying direct physical methods (X-ray treatment, acoustic emission etc.). Yet, the problem of relating the measured quantities to material strength capacity still persists.
The process of fracture in our case is a process of accumulation of dispersed damage in the whole material volume. Accumulated damage determines the degree of fracture.
The aim of the present study is to find a relation between the residual strength and the degree of fracture (damage).

2. Experimental

2.1. Material

A material in which mechanical loading induces accumulation of dispersed (delocalized) damage till macrofracture was studied – injection moulded short glass fiber reinforced polyoximethylene. Samples

[3] Corresponding author.

Journal of Physics: Conference Series **240** (2010) 012166 doi:10.1088/1742-6596/240/1/012166

were fabricated by Ticona GmbH, Frankfurt a.M. according to DIN EN ISO 527. The glass fiber part (E – glass) is 26 wt.%, the mean fiber length is 320 μm (max. 800μm, min. 60μm) and the fiber diameter – 10 ± 1μm. Silan compounds were used for fiber adhesion dressing.

2.2. Mechanical loading

The samples were loaded in tension by an Instron testing machinery with a three constant loading rates equal to 0.22, 1.1 and 5.5 kN/min. After reaching a certain loading percentage (part of the force at fracture) the samples were unloaded instantly. The unloaded samples were loaded till fracture with loading rate 1.1 kN/min to find the residual strength. Nine loading percentages were used – 50, 70, 80, 85, 90, 93, 95, 97 and 98% of the force at fracture. Three samples were loaded at each step and loading rate. Experiments were carried out at room temperature and 50% humidity.

2.3. Measure of damage process

The damage process has a clear physical meaning – it is the accumulation of new internal surfaces in the whole volume of the material: micro and mezzo cracks (called cracks in what follows). They are oriented perpendicularly to the loading direction and to the scattering plane (the plane determined by the collimated and refracted X-ray beams) (figure 1). Damage accumulation proceeds relatively homogeneously in isolated volumes until there occurs interaction between these zones as well as damage transfer to the macrovolume leading to collapse.

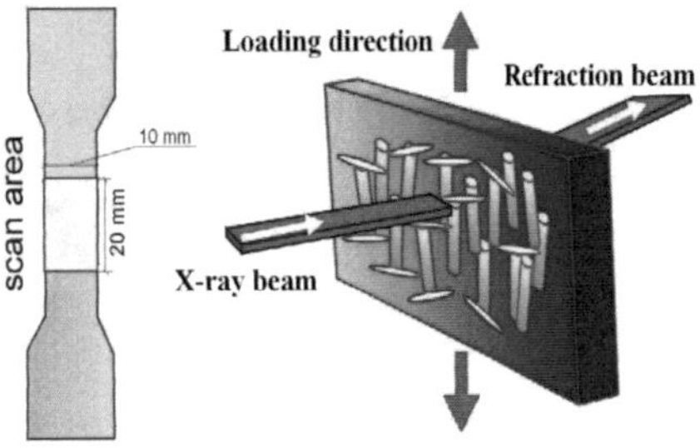

Figure 1. Damage as a result of cracking. Fiber parallel to the scattering plane.

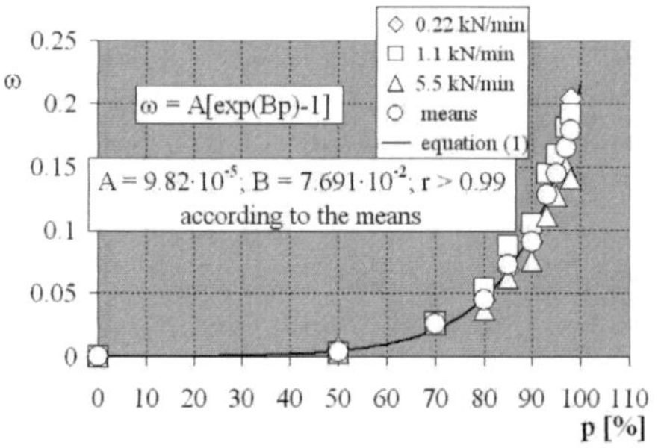

Figure 2. Damage accumulation ω (cracks) as a function of loading percentage p. Description according to (1).

The new internal surfaces in the volume which occur as a result of mechanical loading correspond to the accumulated damage. They are registered by a new original method of X-ray refraction developed in the Federal Institute of Materials Research and Testing in Berlin (Prof. M. Hentschel – [1]). A measure of damage ω is the quantity $\Delta c_m d$ – the mean difference between the refractions cd measured on loaded-unloaded and virgin samples in the scanned area (figure 1). Quantity c = kS/V denotes the refraction factor proportional to the inner surface density (ratio "surface S / volume V"), k is a constant of the instrument, d is sample thickness and m denotes a mean value.

3. Relation residual strength - damage

We are going to find the relation between the residual strength R_{rs} and damage ω on the basis of their dependencies on the loading percentage p. The dependence $\omega - p$ is shown in figure 2. Damage values ω are in relative units. Their analytical description based on the idea for the distribution of weak spots in the material [2] is shown there, too. Parameters in the equation are determined according to the mean values of ω for the three load rates. It was ascertained in [2], that for the material investigated and for the applied loading rates, damage depends on the loading percentage, only, but not on the loading rate. The mean values of ω are shown in figure 2 as circles. The dependence $\omega - p$ reads:

$$\omega(p) = \mathrm{A}[\exp(\mathrm{B}p) - 1] \qquad (1)$$

with A = 9,82.10^{-5} and B = 7,69.10^{-2} (a correlation coefficient r ≈ 1).

The specification of the relation between the residual strength R_{rs} and p faces difficulties due to the character of function $R_{rs}(p)$: the residual strength varies slightly up to p = 97% and sharply (almost jump-wise) in the range 99 – 100% (figure 3). The problem is to find an analytical description of R_{rs} for the whole range of loading percentage p from 0 till 100%, on the basis of experimental data before the "jump". We get a "smooth run" of R_{rs} choosing a function f(p) and specifying R_{rs} as depending on f(p), i.e. R_{rs}[f(p)]:

$$R_{rs}[f(p)] = R_0 - \Delta R_{rs}[f(p)] \qquad (2)$$

where R_0 is the initial strength of the undamaged material, ΔR_{rs} is its decrease caused by damage.

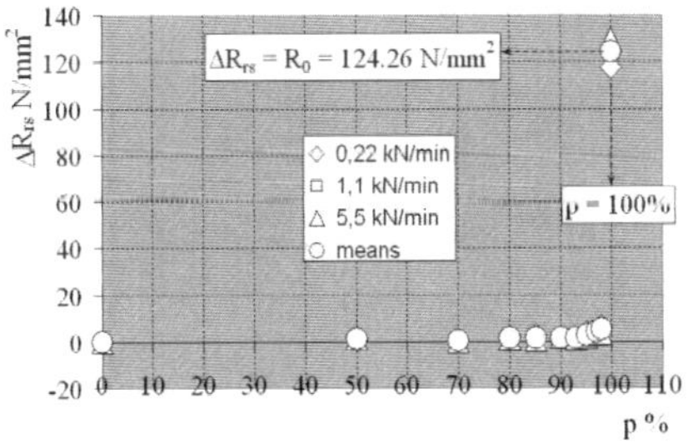

Figure 3. The decrease of residual strength ΔR_{rs} caused by damage as a function of loading percentage p.

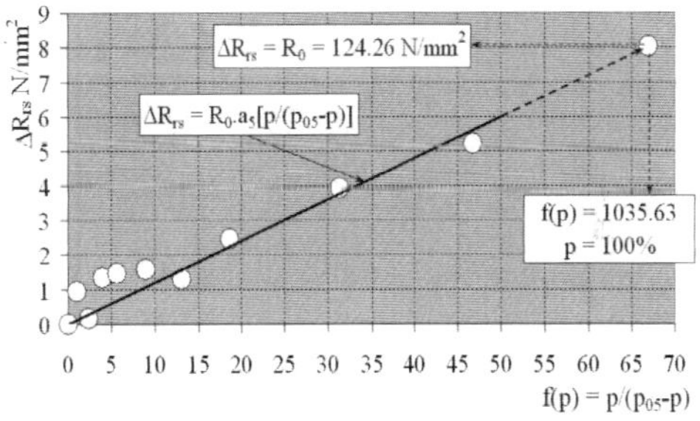

Figure 4. ΔR_{rs} as a function of f(p).

Function f(p) is chosen in the form:

$$f(p) = p/(p_0 - p) \qquad (3)$$

where p_0 is a parameter.
It is assumed, that function ΔR_{rs} is proportional to f(p):

$$\Delta R_{rs} = k.f(p) = k[p/(p_0 - p)] \qquad (4)$$

with a coefficient of proportionality k = R_0.a, and a is a parameter.
Parameters a and p_0 in equation (4) are found applying a linear regression with iteration. As a first step it is assumed that $p_0 = p_{01}$ = 100. The respective parameter a = a_1 is found on the basis of the method of least squares:

$$a_1 = (1/R_0)[p_{1i}\Delta R_{rsi}]/[p_{1i}^2] \qquad (5)$$

where $p_{1i} = p_i/(p_{01} - p_i)$, p_i and ΔR_{rsi} (mean values) correspond to the experimental data at point i (nine points altogether – for the loading percentages 50, 70, 80, 85, 90, 93, 95, 97 and 98%). Square brackets denote sums.
As a second step it is assumed that $p_0 = p_{02} = (1 + a_1)$.100. Repeating the procedure in (5) with p_{02}, we calculate parameter a_2. It is assumed at the third step that $p_0 = p_{03} = (1 + a_2)$.100 and according to (5) we calculate parameter a_3. The procedure is repeated until $a_j = a_{j-1}$, where j is the j-th step. This condition is fulfilled at the 5-th step, and for the parameters a and p_0 follows that: a = a_5 = 0,0009656 and $p_0 = p_{05}$ = 100,09656.
So, equation (4) reads:

$$\Delta R_{rs} = aR_0[p/(p_0 - p)] \tag{6}$$

with a and p_0 given above.
The line according to eq.(6) satisfies three conditions: it crosses point $\Delta R_{rs} = 0$, p = 0%, point $\Delta R_{rs} = R_0$, p = 100% and passes maximally close to the experimental points of the loading percentages.
The "smooth run" of ΔR_{rs} as a function of f(p) is shown in figure 4. It is seen, that the sharp jump in figure 3 near p = 100% disappears in figure 4. This "smooth run" of the experimental data allows the determination of parameters a and p_0 in equations (6) and (2).
The dependence ΔR_{rs} (R_{rs} respectively) on p is shown in figure 5.

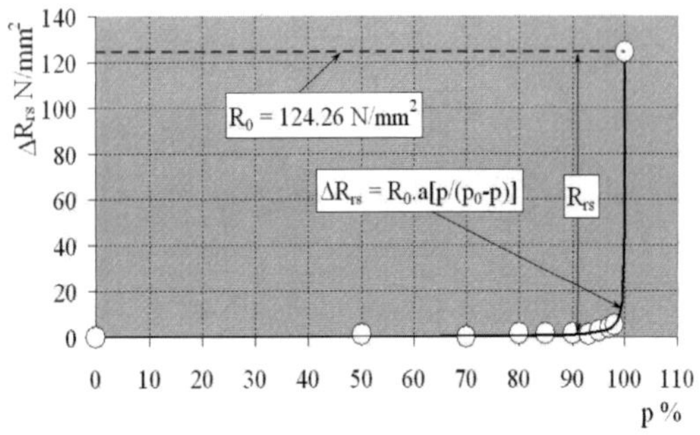

Figure 5. ΔR_{rs} (and R_{rs} respectively) as a function of p.

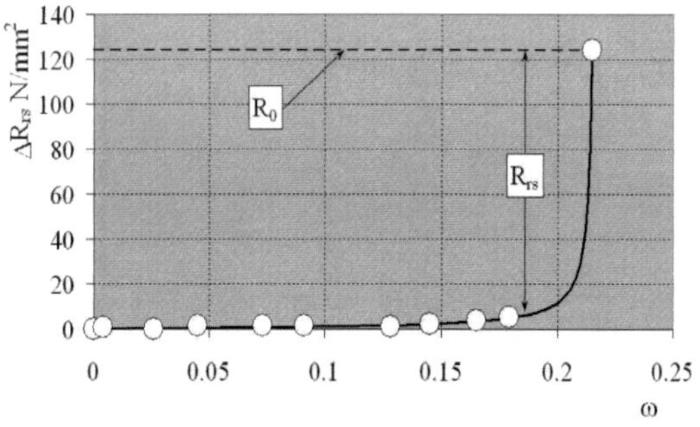

Figure 6. Residual strength R_{rs} and its decrease ΔR_{rs} as a function of damage ω.

Residual strength R_{rs} as a function of the degree of fracture (damage) ω is found for known functions $\omega(p)$ and $R_{rs}[f(p)]$ according to equations (1), (2) and (6):

$$\Delta R_{rs}(\omega) = aR_0\ln(1 + \omega/A)/[Bp_0 - \ln(1 + \omega/A)] \tag{7}$$

The decrease of residual strength ΔR_{rs} and the residual strength R_{rs} , respectively, as a function of damage ω are shown in figure 6. Initial strength R_0 = 124,26 N/mm^2, a, p_0, A and B are given above. The comparison of the experimental data with the theoretical dependence according to the equation (7) in figure 6 shows a good coincidence.

4. Conclusions

- A relation between the residual strength and the dispersed damage at short-term loading is found – equation (7).
- The relation found is verified for short glass fiber reinforced polyoximethylene.
- A method is proposed, which allows the determination of the parameters in the equation that relates the residual strength to the loading degree.
- A proof of the above dependencies is needed for other materials that undergo dispersed damage until fracture.

Acknowledgements
The investigation was carried out with the financial support of Alexander von Humboldt Foundation (Bonn, Germany).

References

[1] Hentschel M P, Ekenhorst D, Harbich K-W, Andersen S B 1994 *Proc. Sixth European Conf. on Nondestructive Testing (Nice)* J.Ph.Berge Paris La Cofrend 167-172
[2] Zachariev G, Rudolph H-V, Ivers H 2004 *Composites Part A* **35** 1119-1123

Journal of Physics: Conference Series **240** (2010) 012167 doi:10.1088/1742-6596/240/1/012167

Effect of an applied stress on discontinuous precipitation in a Cu-Ag alloy

R Monzen[1], T Terazawa[2] and C Watanabe[1]

[1] Division of Innovative Technology and Science, Kanazawa University, Kakuma-machi, Kanazawa 920-1192, Ishikawa, Japan
[2] Division of Mechanical Science and Engineering, Kanazawa University, Kakuma-machi, Kanazawa 920-1192, Ishikawa, Japan

monzen@t.kanazawa-u.ac.jp

Abstract. The effects of an applied tensile stress on the growth rate and morphology of discontinuous precipitation (DP) product have been studied for a Cu-5wt%Ag alloy aged at 300 °C. The DP cell consists of lamellae of the rod-shaped Ag precipitates and solute-depleted Cu matrix. The tensile stress accelerates the growth of DP cells along both the loading direction (LD) and transverse direction (TD) but the cell growth rate along the TD is faster than that along the LD. Transmission electron microscopy has revealed that the tensile stress is apt to produce the Ag precipitates elongated in a <110> direction nearly perpendicular to the LD in a cell, irrespective of the cell growth direction. The observed morphology of the Ag precipitates and the promotion of the cell growth, namely precipitate growth under tension can be understood through the interaction energy between the external stress and the misfit strains of precipitate. The growth of Ag precipitates toward the direction perpendicular to the LD explains the faster cell growth along the TD.

1. Introduction

The effect of an applied tensile stress on the growth rate of discontinuous precipitation (DP) cells has been first examined for six binary alloys of Cu-Ag, Cu-Mg, Cu-Cd, Zn-Cu, Ag-Cu and Pb-Sn by Sulonen [1, 2]. For the six alloys, the growth rates of DP cells at grain boundaries aligned parallel and transverse to the loading direction were different from each other. An explanation for this effect, which was based on an elastic strain energy interaction between the applied stress and the coherency stress field, was proposed. Later, Hillert [3] has provided a quantitative treatment of the experimental results by Sulonen, based on the elastic interaction between the misfit stress field around the solute atoms ahead of the reaction front and applied tensile stress. However, the effect of external stress on the DP morphology has not yet been metallographically studied.

In this study, the effects of an applied tensile stress on the growth rate and morphology of DP cells are investigated for a Cu-5wt%Ag alloy aged at 300 °C. The DP cell in Cu-Ag alloys consists of lamellae of the Ag-rich β phase and solute-depleted α phase [4, 5]. We have found that the growth of DP under the tensile stress is accelerated along the loading direction (LD) and transverse direction (TD) compared with the growth of DP under no applied stress, but the growth rate of DP cells along the TD is faster than along the LD. In addition, from transmission electron microscopy (TEM) observations, it has been shown that aging under tension tends to produce the rod-shaped Ag precipitates elongated in a <110>$_\alpha$ direction nearly perpendicular to the LD in a cell, independent of the growth direction of DP cells. This observation accounts for the anisotropy in the cell growth rate.

2. Experimental

Ingots of a Cu-5wt%Ag alloy were prepared by melting 99.99%Ag and 99.99%Cu. The alloy ingots were homogenized at 800 °C for 24 h in a vacuum, cold-rolled to 50% reduction in thickness and then spark-cut into specimen strips. The specimens had a cross-section of 3 mm × 6 mm and a gage length of 20 mm. All the specimens were solution-treated at 780 °C for 1 h, quenched into water and subsequently aged at 300 °C for various times either under an applied stress of 20 MPa (stress aging) or under no stress (free aging). The applied stress of 20MPa is about one third of the yield strength of the solution-treated specimen at 300 °C.

Thin foils for TEM observations were prepared by slicing the aged specimens with a spark cutter and by electropolishing using a solution of 67% methanol and 33% nitric acid at -30 °C and 6.5V in a twin-jet electropolisher. Microscopy was carried out using a HITACHI H-9000NAR or a JEOL 2000EX microscope at an operation voltage of 300 or 200 kV.

3. Results

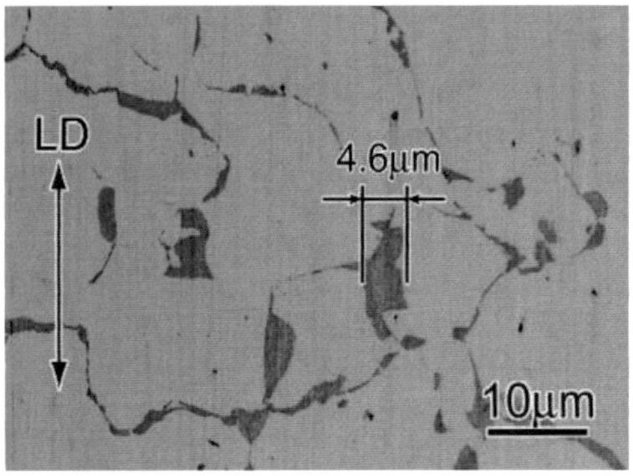

Figure 1. Optical micrograph of a tensile-stress-aged specimen. Aging was carried out at 300 °C for 1 h. An arrow indicates the loading direction (LD).

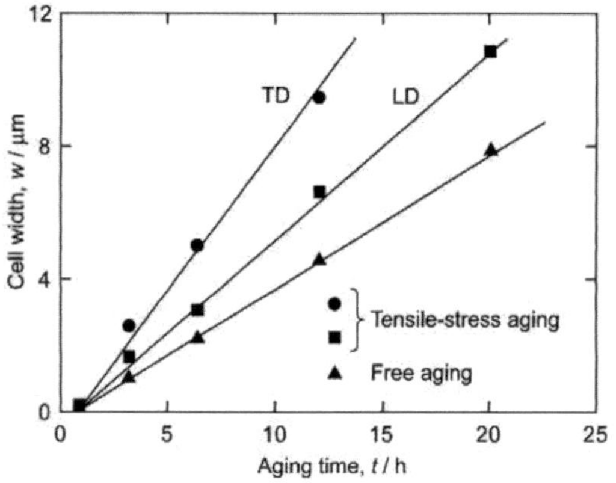

Figure 2. Variation in the width of DP cells during free aging and tensile-stress aging at 300 °C. LD=Loading Direction, TD=Transverse Direction.

DP cells under no stress grew randomly to all directions. In the specimen, tensile-stress-aged, the width of cells in the TD is larger than that in the LD, as exemplified in figure 1. Figure 2 shows the cell width w against aging time t for the specimens, free-aged (FA) and tensile-stress-aged (TSA) at 300 °C. A linear relationship exists between w and t for these specimens. The growth of DP in the TSA specimen is promoted along the LD and TD in comparison with that in the FA specimen, but the growth rate of DP cells along the TD is faster than along the LD. The latter is in agreement with the observation by Sulonen [2] for a Cu-5wt%Ag alloy aged at 500 °C for 30 min under tension.

Interpretations of the results of Sulonen [1, 2] on the effect of applied stress on DP have been put forward by Sulonen [1, 2] and Hillert [3]. When DP occurs in an elastically isotropic solid under an external stress σ and with a coherency strain δ in the solute diffusion zone abutting a grain boundary, the elastic strain energy ΔG in the coherent zone at the grain boundary perpendicular to the LD is written as [6]

$$\Delta G = \frac{E}{1-\nu}\delta^2 - \frac{2\nu}{1-\nu}\delta\sigma \,. \tag{1}$$

Here E is the Young's modulus and ν is the Poisson's ratio. If the grain boundary is parallel to the LD, the driving force for moving it (elastic strain energy) is expressed as

$$\Delta G = \frac{E}{1-\nu}\delta^2 + \delta\sigma\,. \qquad (2)$$

Since the sign of δ is plus in the present work [7], these equations predict that application of tensile stress during aging enhances the growth rate of DP cells in the TD but lowers it in the LD. However, this prediction is in conflict with the result of figure 2.

DP reactions in the FA and TSA specimens were always observed to occur in two directions from one boundary. Individual DP cells gradually grew laterally together and formed a double seam with reaction fronts. TEM observations of various DP cells in the FA specimen revealed that Ag precipitates in a cell had an elongated shape along a $<110>_\alpha$ direction of the Ag-depleted Cu matrix. This direction is in agreement with the elongated $<110>_\alpha$ direction of rod-shaped Ag precipitates in disk-shaped aggregates parallel to $\{001\}_\alpha$ planes in Cu-5.7wt%Ag single crystals [7]. The discontinuous Ag precipitates exhibited a cube-on-cube orientation relationship to the Cu matrix, in accordance with the previous studies on Cu-Ag alloys [8, 9]. In the FA specimen, rod-shaped Ag precipitates behind an advancing grain-boundary grew in a $<110>_\alpha$ direction nearly perpendicular to the boundary.

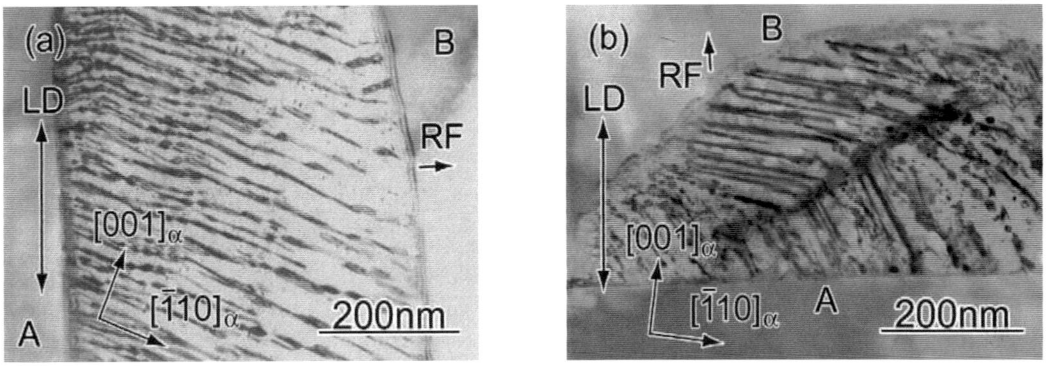

Figure 3. TEM images of rod-shaped Ag precipitates in DP cells after movement of grain boundaries, aligned nearly (a) parallel and (b) perpendicular to the loading direction (LD), in the specimen, tensile-stress-aged at 300°C for 30min. RF=Reaction Front.

Figures 3(a) and 3(b) depict TEM images of rod-shaped Ag precipitates in DP cells after migration of grain boundaries aligned nearly parallel and perpendicular to the LD in the specimen, tensile-stress-aged at 300 °C for 30 min. The zone axes in these figures are parallel to the $[110]_\alpha$ direction. In figure 3(a), the Ag precipitates are elongated in the $[\bar{1}10]_\alpha$ direction of grain A nearly perpendicular to the grain boundary. It should be noted in figure 3(b) that the Ag precipitates initially grow along the $[101]_\alpha$ direction of grain A relatively perpendicular to the advancing boundary, but their growth direction is changed into the $[1\bar{1}0]_\alpha$ direction nearly perpendicular to the LD. This change in growth direction was often observed in the TSA specimen, irrespective of the cell growth direction. It was confirmed that the initial or final growth direction of Ag precipitates in some DP cells was a $<110>_\alpha$ direction nearly perpendicular to the grain boundary or the LD. The initial growth direction is unaffected by the applied stress.

4. Discussion

As mentioned in section 3, the result of figure 2 that the cell growth velocities along the LD and TD under tension are faster than that under no stress cannot be explained from equations (1) and (2). The origin of the promotion of the cell growth velocity (figure 2) and the observed morphology of Ag precipitates (figure 3) under the tensile stress can be understood to arise through the interaction energy due to the presence of positive misfit strain ε_{ij} (stress-free transformation strain) between a Ag precipitate and an external stress σ_{ij}. The interaction energy means the work done by the external stress during the growth of discontinuous Ag precipitates. The interaction energy ΔE per unit volume of the Ag precipitate is expressed as [10, 11]

$$\Delta E = -\sigma_{ij}\varepsilon_{ij} \quad (3)$$

This equation predicts that the growth of Ag precipitates, namely DP cells is promoted by applied tensile stress. This prediction is in agreement with the result of figure 2. Also, it may be expected from equation (3) that the rod-shaped Ag precipitates elongated in a $<110>_\alpha$ direction nearly perpendicular to the LD in a cell are produced under tension, since the misfit strains along the elongated $<110>_\alpha$ direction of a rod-shaped Ag precipitate and along the direction normal to the elongated direction are, respectively, ε_{33}=0.015 and ε_{11}=ε_{22}=0.12 [7]. This also is coincident with the observed morphology of Ag precipitates in figure 3. Therefore, it can be concluded that, as a consequence of the change in growth direction of the Ag precipitates toward the direction perpendicular to the LD as shown in figure 3(b), the cell growth rate in the TD is faster than that in the LD.

5. Conclusions

Application of a tensile stress during aging a Cu-5wt%Ag alloy at 300°C promotes the growth of discontinuous precipitation cells in both the loading direction (LD) and transverse direction (TD) but the cell growth along the TD occurs more rapidly than along the LD. The tensile stress has no influence on the initial growth direction of discontinuous rod-shaped Ag precipitates in a cell, which is a $<110>_\alpha$ direction nearly perpendicular to the advancing grain-boundary, but is prone to change into a $<110>_\alpha$ direction nearly perpendicular to the LD. The change in growth direction of Ag precipitates toward the direction normal to the LD and the acceleration of the cell growth, namely precipitate growth under tension can be understood through the interaction energy due to the presence of positive misfit strains between the applied stress and Ag precipitate. As a result of the change in precipitate growth direction toward the direction perpendicular to the LD, the cell growth along the TD occurs more rapidly.

Acknowledgments

A part of this work was conducted in the Kyoto-Advanced Nanotechnology Network, supported by the "Nanotechnology Network" of the Ministry of Education, Culture, Sports, Science and Technology (MEXT), Japan. This work was also partially supported by a Grant-in-Aid for Science Research from The Ministry of Education, Culture, Sports and Technology under Grant No.19560742.

References

[1] Sulonen M S 1964 *Acta Metall.* **12** 749
[2] Sulonen M S 1964 *Acta Polytec. Scand.* (a) **38** 3
[3] Hillert M 1972 *Metal. Trans.* **3** 2729
[4] Gust W, Beuers J, Steffen J, Stiltz S and Predel B 1986 *Acta Metall.* **34** 1671
[5] Gupta S P 1998 *Can. Metall. Q.* **37** 141
[6] Chung Y H, Shin M C and Yoon D Y 1992 *Acta Metall. Mater.* **40** 2177
[7] Watanabe C, Monzen R, Nagayoshi H and Onaka S 2006 *Phil. Mag. Lett.* **86** 65
[8] Liu J B and Meng L 2008 *J. Mater. Sci.* **43** 2006
[9] Han K, Vasquez A A, Xin Y and Kalu P N, 2003 *Acta Mater.* **51** 767
[10] Eshelby J D 1957 *Proc. R. Soc.* (a) **241** 376
[11] Eshelby J D 1959 *Proc. R. Soc.* (a) **252** 561

15th International Conference on the Strength of Materials (ICSMA-15) IOP Publishing
Journal of Physics: Conference Series **240** (2010) 012168 doi:10.1088/1742-6596/240/1/012168

Compressive strength and hardness of metal matrix syntactic foams

I N Orbulov[1], Á Németh[1] and J Dobránszky[2]

[1] Department of Materials Science and Engineering, Budapest University of Technology and Economics, Goldmann Gy. tér 3., Budapest, H-1111, Hungary
[2] Research Group for Metals Technology of the Hungarian Academy of Sciences, Goldmann Gy. tér 3., Budapest, H-1111, Hungary

orbulov@gmail.com

Abstract. Six types of metal matrix syntactic foams (MMSFs) were produced by pressure infiltration technique. The foams were investigated by upsetting tests at increased (220°C) and at room (25°C) temperature. The parameters were the constituents of the composites and the aspect ratio (height-diameter ratio, H/D) of the specimens. The characteristic properties were: the compressive strength, the fracture strain, the structural stiffness of the foams and the absorbed energy. The strength, the strain and the energy were decreased while the stiffness was increased by increasing the H/D. Increased temperature caused ~25 % drop in the strength and in the stiffness. Macrohardness, depth sensitive and dynamic hardness tests were also performed on MMSF blocks: macrohardness is a structural property and independent from the matrix material. The depth sensitive hardness is sensitive to the deformation capability of the matrix and to a possible change reaction. The dynamic hardnesses of the MMSFs were higher than the hardness of the matrices and this is a microballoon related property.

1. Introduction

Metal matrix syntactic foams (MMSFs) are particle reinforced composites containing ceramic microballoons. The most common production methods are pressure infiltration [1] and blending method [2]. MMSFs have been investigated by many groups: for example Rohatgi et al. and Wu et al. investigated the effect of the microballoon volume fraction on the compressive strength, which is increased by increasing the volume fraction of the particles [3, 4]. Some groups investigated the coefficient of thermal expansion (CTE) and the creep resistance of aluminium-fly ash composites. They proved that, the CTE decreased while the creep resistance increased by increasing the microballoon content in the composites [5, 6]. Balch and Dunand produced MMSFs by pressure infiltration. The volume fraction of the microballoons was maximized and they investigated the compressive failure modes of the foams [7, 8]. Palmer et al. made tensile, upsetting and bending tests on pressure infiltrated MMSFs, too [9]. Dou et al. proved that, the foams show strain rate sensitivity: both the compressive strength and the energy absorption capacity increased by increasing the strain rate [10]. Kiser et al. performed upsetting tests on MMSFs under both uniaxial and multiaxial loadings [11]. It can be concluded that, MMSFs have been widely investigated, but the lack of increased temperature is a motivating reason to perform upsetting tests at increased temperature.

The common matrix materials (Al alloys) have usually low hardness and wear resistance; this can be improved by the ceramic microballoons: the wear resistance and the hardness increased by increasing the microballoon content [2, 12]. MMSFs wear by microcutting and delamination caused by crack propagation below the rubbing surface [13] and in general the conventional Al foams have lower wear resistance compared to MMSFs having the same Al volume fraction [14]. MMSFs also take part in industrial applications. They are good energy absorbers, mechanical dampers. They have low weight, outstanding specific properties, localized failure, etc. They can be used as low weight moving components and energy absorbers. Because of the wide application range their compressive properties and hardness should be characterized.

2. Materials and specimens

Al99.5 or AlSi12 was used as matrix matrial. Three types of ceramic microballoons were used as fillers (Envirospheres Pty. Ltd.). Their constituents were the same (36-40 wt% Al_2O_3, 55-60 wt% SiO_2, 1-5 wt% other oxides); their name and geometrical properties are listed in table 1. Overall six MMSF blocks were produced by pressure infiltration. The microballoon content was ~64 vol%. The infiltration pressure (0.4 MPa) was ensured by Ar gas for 30 s. The infiltration temperature was 600°C and 710°C in the case of AlSi12 and Al99.5 matrix respectively. A 10×35×55 mm specimen for hardness tests and Ø14 cylindrical specimens with H/D=1, 1.5 and 2 were machined from each block.

Table 1. Average geometrical parameters of the microballoons

Type	Outer diameter (μm)	Wall thickness (μm)	Specific surface area (μm^{-1})
SL150	100	3.7	0.060
SLG	130	5.4	0.046
SL300	150	6.7	0.040

3. Experimental

The upsetting tests were done by a Zwick 50 testing machine. The die had four guide bars, polished and lubricated surfaces. The strain rate was 0.01 s^{-1} and the tests were performed at increased (220°C) and at room (25°C) temperature until 50% of engineering deformation was reached. Overall 216 specimens were compressed (2 matrices×3 microballoon types×3 H/D×2 temperatures×6 specimens).

Brinell macrohardness tests were done by a VPM HPO 13/6062 testing machine with an Ø10 mm hardened steel ball. The loading force was 1840 N or 2450 N in the case of Al99.5 and AlSi12 matrix respectively and they were applied for 10 s. In the depth sensitive hardness tests [15, 16] an Ø2.5 mm hard metal mandrel with hemisphere ending was pushed into the specimens by 0.02 mm/s. The load was increased by a MTS 810 type hydraulic testing machine until a previously set force level (1600 N) was reached, and then the specimen was unloaded. Three measurements were done on each specimen. Dynamic hardness tests were done by Equotip tester. The device shots an indenter at the surface and measures its velocity at 1 mm distance from the surface before and after impact. The dynamic hardness can be calculated from the velocities. Ten measurements were done on each specimen.

4. Results and discussion

As main strength properties the compressive strength and structural stiffness should be analyzed. Figure 1 and 2 shows the strength as the function of H/D in the case of the two matrices. In the case of Al99.5 matrix strength strongly decreased between H/D 1 and 1.5 at room temperature. The strength decreased further by increasing the H/D in the case of SL150 filler, but remained the same in the case of SLG and SL300 fillers. This indicates that at higher H/D the load was carried by the microballoons and SL150 microballoons were simply too weak. As the H/D increased the shearing like stress also increased in the specimens because the specimens became higher and the small deviations of the specimen and the stress field resulted in an increasing shearing effect. Due to their small wall thickness SL150 microballoons were the most sensitive. At increased temperature this effect appeared only in the case of higher H/D (between 1.5 and 2).

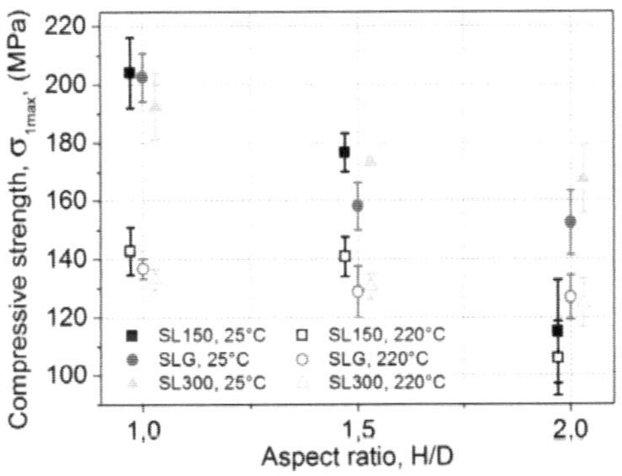

Figure 1. The strength of Al99.5 matrix MMSFs

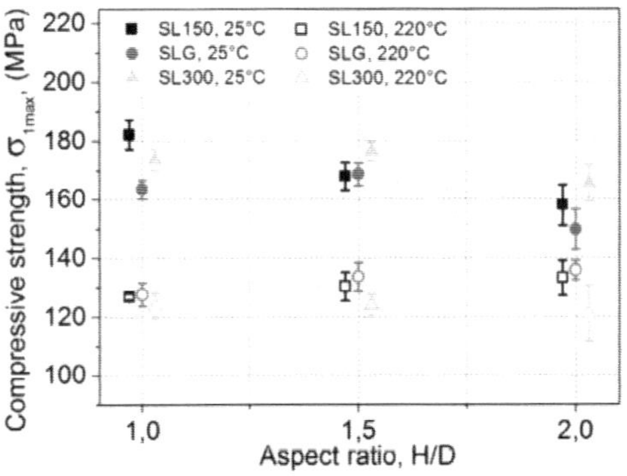

Figure 2. The strength of AlSi12 matrix MMSFs

In the case of AlSi12 matrix the situation was reversed. In the case of SLG and SL300 fillers the strength was almost constant at lower H/D; the microballoons and the matrix carried the load together. After that – at higher H/D – the strength decreased. This means AlSi12 matrix tolerated higher shear load than Al99.5 matrix, because the effect of shearing (the decrease of strength) appeared in the case of higher H/D only. In general, increased temperature caused ~25% drop in the compressive strength in every case. Figure 3 and 4 show the structural stiffness of the MMSFs as the function of H/D in the case of the two matrices; their gradient was almost the same in every cases.

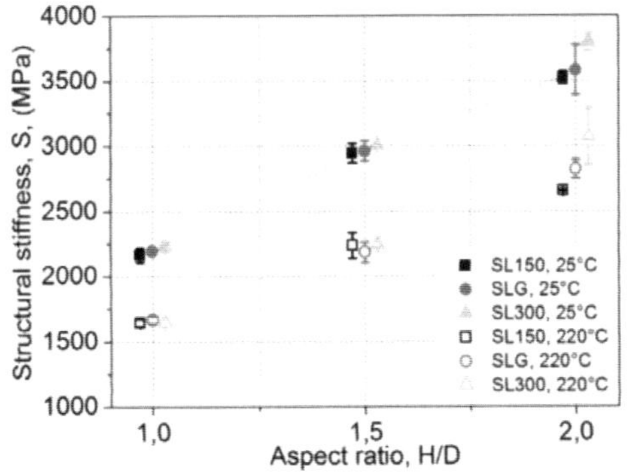

Figure 3. The stiffness of Al99.5 matrix MMSFs

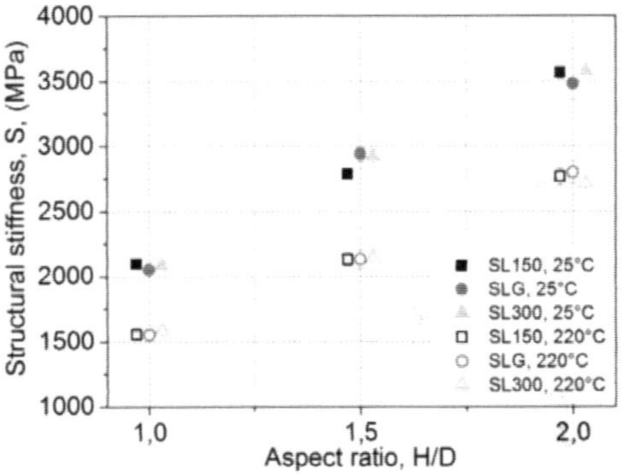

Figure 4. The stiffness of AlSi12 matrix MMSFs

The structural stiffness increased linearly in all cases and was independent from the type of microballoons. The gradient of the stiffness was the same at room and at increased temperatures also. The microballoons and matrices had a small effect on the stiffness and it was determined by the H/D. Increased temperature caused ~25% drop in the structural stiffness in the case of both matrices.

The measured global hardness is shown in figure 5. In the case of Al99.5 and AlSi12 the incorporation of ceramic microballoons hardened and softened the base material respectively. The largest MMSF hardness can be ensured by SL150 microballoons, having the largest specific surface area (surface area/volume ratio). In general the macrohardness is a structural property and proved to be independent from the matrix material. The depth sensitive hardness measurements gave force-depth diagrams as results. The area under the registered curve is proportional to the energy absorbed during the test (figure 6). In the case of Al99.5 matrix the energy was decreased by incorporating SL150 and SLG type microballoons because a change reaction ($4Al+3SiO_2=2Al_2O_3+3Si$) took place between the microballoons and the matrix during the manufacturing, which resulted in lower energy absorbing capability. With SL300 fillers the MMSFs showed the same energy absorbing as pure Al99.5 matrix material because the change reaction mentioned above did not take place [17]. In the case of AlSi12 matrix the energy absorption increased, so the material "softened", the levels of dissipated energy were almost the same in every case. The dynamic hardness of AlSi12 matrix was approximately two times higher than Al99.5. The foams were harder than matrices, but the original hardness difference between the matrices (~70%) decreased to ~20% in the case of MMSFs, so the matrix materials have small effect on the dynamic hardness of MMSFs and therefore this is a microballoon related property. Similarly to Brinell hardness testing, syntactic foams with SL150 type fillers showed the highest hardness.

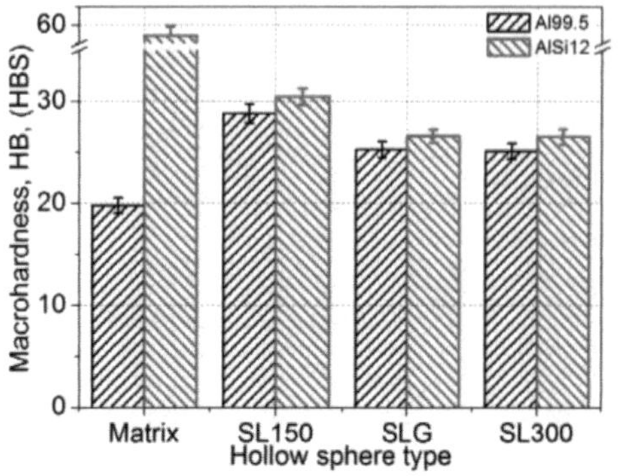

Figure 5. Brinell like macrohardness of MMSFs

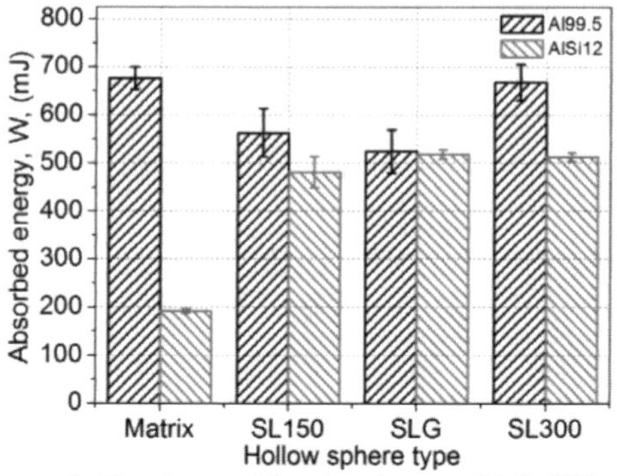

Figure 6. Depth sensitive hardness of MMSFs

5. Conclusions

The compressive strength decreased by increasing the H/D. SL150 type microballoons were the most sensitive because of their smallest wall thickness. The structural stiffness increased almost linearly by increasing the H/D and the gradient was always the same. The effect of microballoon type was negligible. Increased temperature caused ~25% drop in compressive strength and structural stiffness.

The macrohardness is a structural property and independent from the matrix; it cannot demonstrate the change reaction. The depth sensitive hardness tests showed sensitivity to the deformation capability of the matrix and to the presence of the change reaction. The dynamic hardness of the MMSFs was higher than the hardness of the matrices and this is a microballoon related property.

Acknowledgements

Our Metal Matrix Composites Laboratory is supported by grant # GVOP 3.2.1-2004-04-0145/3.0 and by Hungarian Research Fund, NKTH-OTKA K69122. Special thanks to Róbert Tóth and C. H. Erbslöh Hungária Speciality and Industrial Minerals Ltd. for providing microballoons.

References

[1] Rohatgi P K, Guo R Q, Iksan H, Borchelt E J and Asthana R 1998 *Mater. Sci. Tech.* A **244** 22
[2] Ramachandra M and Radhakrishna K 2005 *J. Mater. Sci.* **40** 5989
[3] Rohatgi P K, Kim J K, Gupta N, Alaraj S and Daoud A 2006 *Composites* A **37** 430
[4] Wu G H, Dou Z Y, Sun D L, Jiang L T, Ding B S and He BF 2007 *Scripta Mater.* **56** 221
[5] Rohatgi P K, Gupta N and Alaraj S 2006 *J. Compos. Mater.* **40** 1163
[6] Coteau O and Dunand D C 2008 *Mater. Sci. Eng.* A **488** 573
[7] Balch D K, O'Qwyer J G, Davis G R, Cady C M, Gray III. G T and Dunand D C *Mater. Sci. Eng.* A **391** 408
[8] Balch D K and Dunand D C 2006 *Acta Mater.* **54** 1501
[9] Palmer R A, Gao K, Doan T M, Green L and Cavallaro G 2007 *Mater. Sci. Eng.* A **464** 85
[10] Dou Z Y, Jiang L T, Wu G H, Zhang Q, Xiu Z Y and Chen G Q 2007 *Scripta Mater.* **57** 945
[11] Kiser M, He M Y and Zok F W 1999 *Acta Mater.* **47** 2685
[12] Ramachandra M and Radhakrishna K 2007 *Wear* **262** 1450
[13] Rohatgi P K and Guo R Q 1997 *Tribol. Lett.* **3** 339
[14] Zhang L P and Zhao Y Y 2007 *J. Compos. Mater.* **41** 2105
[15] Gubicza J, Juhász A, Arató P, Szommer P, Tasnádi P and Vörös G 1996 *J. Mater. Sci. Lett.* **15** 2141
[16] Gubicza J, Juhász A, Tasnádi P, Arató P and Vörös G 1996 *J. Mater. Sci.* **31** 3109
[17] Orbulov I N, Dobránszky J, Németh Á 2009 *J. Mater. Sci.* **44** 4013

The effects of some laser parameters on the surface and near surface region of laser treated cast iron cylinder bore

K Májlinger and P J Szabó

Budapest University of Technology and Economics Department of Materials and Engineering, Goldmann György tér 3., Budapest 1111, Hungary

pmgpwo@hotmail.com

Abstract. The environmental and pollution materials emission standards in Europe are going to be more and more strict. In order to keep the standards, a large European automotive manufacturer makes a laser treatment on the cast iron cylinder bores of the V-engine blocks. Samples of laser treated cast iron cylinder bore with lamellar graphite were investigated. Four samples were treated with Nd-YAG laser and Yb-fiber laser sources in three different configurations. Microhardness measurements were made to evaluate the hardness profile of the treated layer. In order to evaluate the microstructure and grain size of the laser treated layer, scanning electron microscopic (SEM) images were taken in cross section with a SEM/focused ion beam (FIB) dual beam electron microscope. The opened graphite area percent were also determined by image analysis method on the surface after laser treatment with a SEM in backscattered electron (BSE) mode, because the outburned graphite holes are the oil reservoirs for lubrication during operating conditions of the engine.

1. Introduction

The automotive engine development is defined by three main factors: (I) to make the engines more cost effective, (II) to keep the environmental regulations and (III) the increasing fuel prices, which means that by even higher engine power the fuel and oil consumption should not increase, and the engines should have a long lifetime [1]. Both the consumption reduction and the power increase of Otto- or Diesel-engines cause higher pressures in the combustion space resulting higher tribological load and wear rate for the piston rings and to the cylinder bore [1].

To improve mechanical and tribological properties of cast iron cylinder bores it is possible to use special mechanical [2] and laser structuring treatments [3], or – which is used by a large European automotive manufacturer – a special laser treatment on the V-block engines to alter the properties of the cylinder bores. Due to the laser treatment, the near surface area of the cylinder bore becomes harder and more wear resistant, furthermore, due to the inhomogenity of the perlitic matrix and carbon lamellae, oil reserving holes are formed. This treatment results in an increased power output and reduced oil consumption [1, 4].

2. The Laser Treatment and Samples

The goal of the laser treatment is to melt a thin layer on the surface and at the same time the laser induced plasma over the surface evaporates the graphite lamellae, deeper than the metallic surface itself. So these "holes" are practically non-communicating oil reservoirs [1]. This treatment is patented, the patent number is: EP 1 738 859 A1.

Recently the cylinder bores in the mass-production are treated with a Xe-Cl excimer laser which works in the UV wavelength region (308 nm) at a repetition rate of 300 Hz. During treatment the workpiece is lasered with a rectangular laserspot with 4 times overlap in a definite raster. The adjustable parameters during the laser treatment are: (I) scale of the overlap (scanning grid), (II) mean laserpower on the surface and (III) the speed of the scanning.

There are further experiments to find alternative laser sources, because the excimer lasers have high maintenance requirements and the maintenance and servicing costs are higher than those of the solid states lasers. We investigated four samples, treated with Nd-YAG laser and Yb-fiber laser sources in three different configurations. The material of the samples is grey cast iron with lamellar graphite (GJL-250). The main laser parameters for the laser treatment are shown in table 1.

Table 1. Main laser parameters for laser treatment

Sample	Lasering strategy	Laser type	Wavelength	Pulse duration	Repetition frequency	Energy density on the surface
A	1x4	Nd-YAG	1064 nm	40 ns	6000 Hz	52 mJ/mm^2
B	2x2	Nd-YAG	1064 nm	40 ns	6000 Hz	52 mJ/mm^2
C	2x2	Nd-YAG	1064 nm	160 ns	6000 Hz	50 mJ/mm^2
D	2x2	Yb-Fiber	1065 nm	140 ns	12500 Hz	49 mJ/mm^2

The only difference between sample A and B is just in the lasering strategy; sample A was treated with four times overlap of the laser spot (four laser pulses) during the insertion of the laser output tube into the cylinder bore. The samples B, C and D were treated with two times overlap during the insertion as well as during removal (also four laser pulses).

3. Experimental

The hardness profile of the surface and the near surface layer of the laser treated surface were measured with a Fisherscope® HM2000 XY microhardness measurement system which utilized the load-indentation depth method according to ISO 14577-1. The indenter was equipped with a Vickers pyramid. The hardness measurements were performed on the laser treated surface.

The FIB investigations were made by a FEI Nova Nanolab 600 SEM/FIB dual beam scanning electron microscope. Half of a regular cross section for TEM lamellae cutout were made and then images were taken – with lowered ion current – of the wall of the obtained cavity (figure 3).

Because the outburned graphite holes are the oil reservoirs for lubrication during operating conditions of the engine [1, 4, 5]; it is important to know how many free graphite holes are going to influence the sliding behaviour. The graphite area of the base material and the laser treated surface were determined by Image Pro Plus software on SEM images in BSE mode. Twenty images were pasted together to cover a 1900×1860 μm^2 area approximately.

4. Results

Figure 1 and 2 shows the hardness vs. indentation depth curves of the four samples. Every curve is the mean value of seven separate measured curves on the same sample. The curves start at 0.1 μm indentation depth, because at the first 100 nm of the indentation – most likely because of the surface roughness of the treated surface – curves have a standard deviation of many order of magnitude. The standard deviation of the mean curves (for every sample) is approx. ± 500 N/mm^2.

The hardness profiles obtained by 150 mN load (figure 1) show that sample A was much harder then the other samples until 400 nm of depth. The hardness values decrease continuously until that point, than tend to the value of approx. 3700 HM. Samples B and D are actually very similar, the "softest" hardness profile was obtained for sample C.

The hardness profiles obtained by 1000 mN load (figure 2) shows that the hardness values also decrease continuously until 400 nm of depth and samples B and D are actually very similar, but the "softest" hardness profile – unlike with 150 mN load – was obtained for sample A.

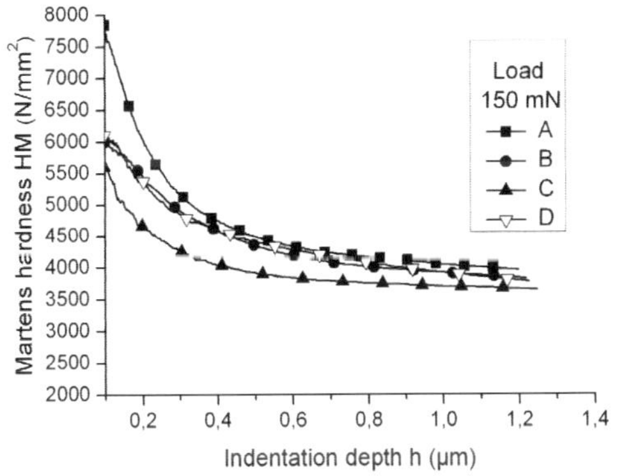

Figure 1. Hardness profiles by 150 mN load

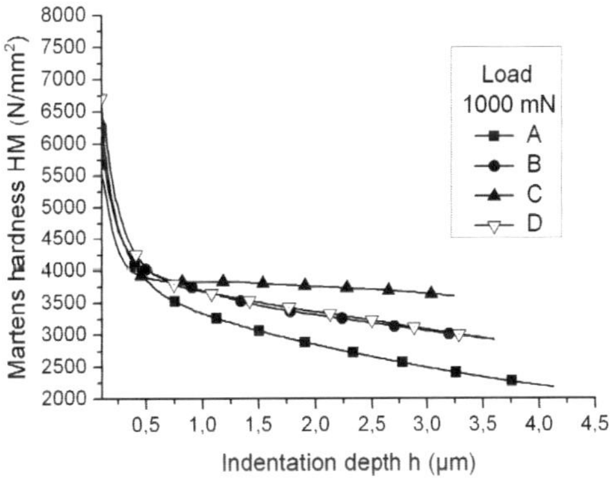

Figure 2. Hardness profiles by 1000 mN load

On the FIB cross section image (figure 3) and on the transmission electron microscope (TEM) image (figure 4) it is well seen that the laser treated layer is ultra fine grained. Below that the base material of perlitic cast iron has many larger grains. The ultra fine grained structure was most likely formed due to the large heat gradient during solidification of the molten surface.

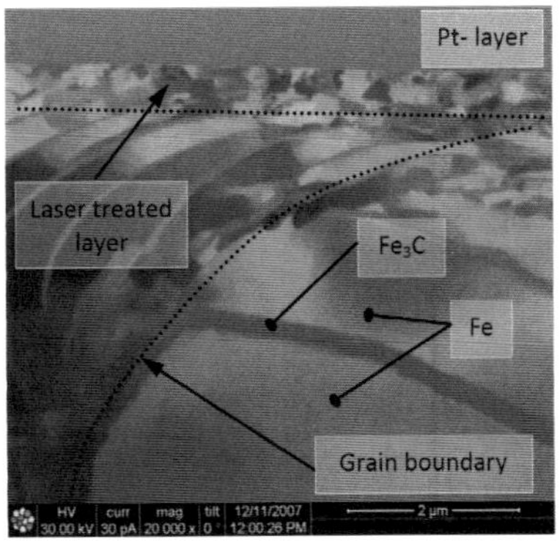

Figure 3. FIB image of the cross section of sample „B" and the mechanically heavily deformed grains near the surface

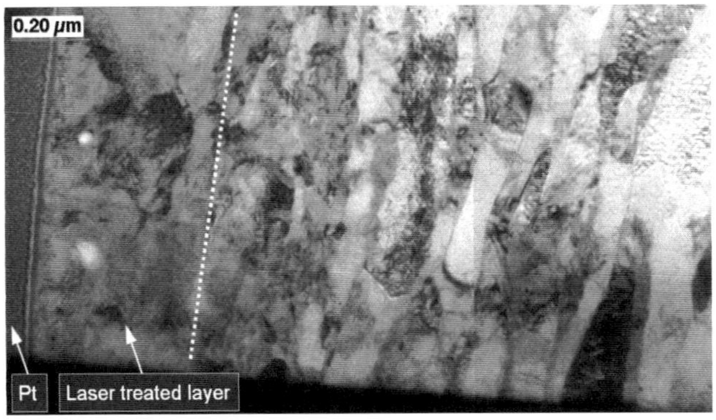

Figure 4. TEM image of the laser treated surface of sample „C" and the mechanically heavily deformed near surface layer

Table 2. Thickness (R) and grain size (d) of the laser treated layer of the cylinder bores measured on FIB, TEM and SEM images

Sample	Thickness of the laser treated layer			d_{FIB} (nm)	d_{TEM} (nm)
	R_{FIB} (µm)	R_{SEM} (µm)	R_{TEM} (µm)		
A	0.76 - 0.93	0.52 - 1.26	---	386	150-560
B	0.65 - 0.71	0.57 - 1.01	0.5	179	160-360
C	0.85 - 1.18	0.65 - 1.19	1.1	222	100-400
D	0.53 - 1.00	---	0.5 - 0.8	250	---

The thickness of the laser treated layers on the FIB images (R_{FIB}) was determined with Image Pro Plus image analysis software. The thickness values were measured also by SEM (R_{SEM}) on cross section samples prepared by conventional mechanical way, and on TEM (R_{TEM}) samples prepared by

FIB and classical TEM sample preparation. The layer thicknesses obtained by different methods are listed in table 2. To determine the average grain size several FIB images were pasted together, so the pasted images covered an area of approx. 20×7 μm^2. The average grain sizes (d_{FIB}) and the maximal and minimal values of grain sizes (d_{TEM}) measured from TEM images with Image Pro Plus program are also listed in table 2.

After mechanical finishing (honing), a part of the graphite lamellae on the bore surface are buried by a thin ferrite smear of the base material. After laser treatment a part of these lamellae will be opened. The ratio between graphite area percent of the base material ($G_{BaseMat.}$) and the graphite area percent of the opened graphite lamellae on the treated surface ($G_{LaseredSurf.}$) defines an effectivity ($G_{Opening}$) of the laser process. The graphite area ratios are listed in table 3 which indicates that the lasering strategy and the repetition frequency have a great effect on the graphite opening.

Table 3. Graphite area ratios

Sample	$G_{BaseMat.}$ (%)	$G_{LaseredSurf.}$ (%)	$G_{Opening}$ (%)
A	10.53	6.75	64.1
B	8.92	4.98	55.8
C	9.33	4.97	53.2
D	8.60	6.70	77.9

Additional x-ray diffraction measurements were made on samples treated with Yb-fiber laser and Xe-Cl excimer laser. The measurements showed that after laser treatment retained austenite was formed in every sample, as it has been reported in many other cases and for different laser types [6-9].

Additional x-ray photoelectron spectroscopy measurements indicated nitrogen incorporation of about 1.5 at% in the laser treated surface and after Ar-ion bombardement at 25nm depth of the samples. Nitrogen solution into metallic surfaces is not unusual even in case of other laser sources induced plasma [9-12].

5. Conclusions

Due to the laser treatment on the cylinder bore the top layer of the surface was molten and at the place of outburned and opened graphite lamellae oil reserving holes were formed. The top layer showed considerably increased hardness which is most likely caused by the ultra fine grained surface layer and the nitrogen incorporation into the surface. Retained austenite was also found in the treated layer.

References

[1] Lindner H, Bergmann H W, Brandenstein C, Lang R, Queitsch A, Reichstein S and Stengel E 2003 *VDI-Berichte* **1764** 73
[2] Knoll G, Lagemann V, Lechtape-Grüter R, Robota A Schlerege F 2003 *VDI-Berichte* **1764** 63
[3] Andersson P, Koskinen J, Varjus S, Gerbig Y, Haefke H, Georgiou S, Zhmudd B and Buss W 2007 *Wear* **262** 369
[4] Herbst L, Lindner H, Heglin M and Hoult T 2004 Targeting diesel engine efficiency, *Industrial Laser Solutions for Manufacturing* CDT (2004)
[5] Duffet G, Sallamand P and Vannes A B 2003 *Applied Surface Science* **205** 289
[6] Benyounis K Y, Fakron O M A, Abboud J H, Olabi A G and Hashmi M J S 2005 Surface *Journal of Materials Processing Technology* **170** 127
[7] Alabeedi K F, Abbouda J H and Benyounisb K Y 2009 *Wear* **266** 925
[8] Lian S and Chenglao L 1991 *Wear* **147** 195
[9] Schaaf P 2002 *Progress in Materials Science* **47** 1
[10] Carpene E, Landry F, Han M, Lieb K P and Schaaf P 2002 *Hyperfine Interactions* **139/140** 355
[11] Carpene E and Schaaf P 2002 *Applied Surface Science* **186** 100
[12] Höche D, Rapin G, Kaspar J, Shinn M and Schaaf P 2007 *Applied Surface Science* **253** 8041

Effect of yttrium addition on grain growth of α, β and α+β titanium alloys

B. Poorganji, A. Kazahari, T. Narushima, C. Ouchi and T. Furuhara
Institute for Materials Research, Tohoku University, Japan

behrang@uwaterloo.ca

Abstract. In this study effects of small amount of yttrium addition on grain growth of C.P titanium (single α), Ti-14Mo-3Nb-1.5Zr (single β) and a newly developed Ti-4.5Al-6Nb-2Mo 2Fe (α+β) titanium alloy are investigated. By adding yttrium nano-sized Y_2O_3 particles are formed in 'in-situ' mode. These particles lead to significant suppression of grain growth in single α and β alloys rather than in the α+β alloy. The results are discussed with regard to conventional grain growth and grain boundary pinning models. It is concluded that yttrium is a potential micro alloying element in titanium alloys and could be used as a grain refining agent.

1. Introduction

The titanium and titanium alloys are used in a wide variety of industrial applications such as aerospace, marine, automotive, and so on [1]. Generally, hot working for breakdown of the cast structure of titanium ingots is conducted at high temperatures which is known as β-processing. On the other hand, the final stage of hot working is mostly conducted in the two-phase region at lower temperatures [2]. The main reason for selection of such a kind of processing route is that heating to temperatures above the β transus temperature of titanium alloys causes a very coarse microstructure due to rapid grain growth of the β phase, which cannot be refined by the final heat treatment [2].

Recently, Hotta et al. [3] have shown that small amounts of yttrium addition to SP-700 alloy resulted in a marked refinement of β grain size as well as retardation of β grain growth at elevated temperatures. Yttrium has a strong affinity to oxygen in titanium, and in-situ Y_2O_3 oxide particles are formed by a very small amount of yttrium addition around 0.02 to 0.05 mass%. These particles are able to control and suppress β grain growth by exerting a pinning effect on the grain boundaries (Zener Effect) [4]. The Zener equation, which was modified by Nishizawa et al. [5], described by Eq. (1), shows the relationship between the average grain radius R and the particle radius r, and its volume fraction f_p :

$$R = \beta r / f_p^{\,m} \tag{1}$$

where m and β are 2/3 [5] and 4/3 [4], respectively.

In the two-phase region of α+β titanium alloys with the volume fraction of the phases being comparable, the rate of coarsening is controlled by inter-diffusion between the phases over distances which are in the order of the phases' size [6]. Generally, it is known that the grain growth rate in deformed metallic materials is much faster compared with un-deformed ones. This behaviour is known as dynamic or deformation grain growth [7-11].

Recently, some studies were carried out to control grain growth in titanium alloys using particles like TiB [12-15]. These studies are related to titanium matrix composites reinforced with ceramics. Such materials show superior high temperature strength. However, some disadvantages of these materials restrict their application. One is high production cost through conventional composite production methods like adding TiB particles into the molten material. The other is deterioration of elongation during superplastic deformation because of using a high fraction of hard particles [16, 17]. Production of in-situ formed Y_2O_3 particles by adding Y during the melting process showed to be very effective in controlling β grain size during high temperature annealing [3]. Furthermore, Y is much cheaper than B and the production process is simpler.

In this study, the influence of yttrium on grain growth was investigated in single α, C.P titanium, new β, Ti-14Mo-3Nb-1.5Zr, and new α+β type titanium alloy, Ti-4.5Al-6Nb-2Mo-2Fe, in particular the pinning of grain boundaries by Y_2O_3 particles during isothermal heating at 1073K of all alloys and also during hot deformation of the α+β two-phase structure.

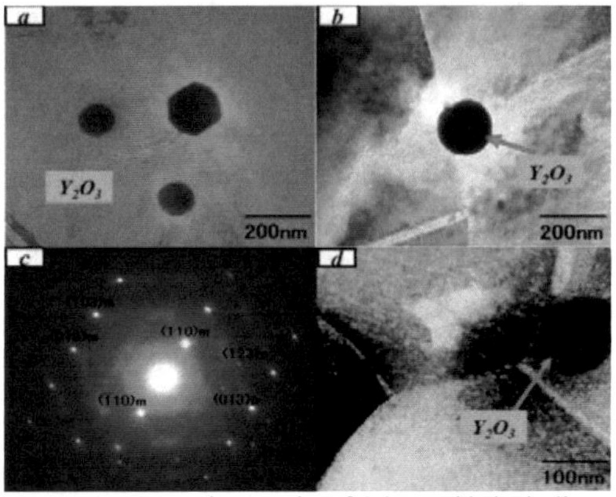

Figure 1. TEM micrographs of 0.05Y added Ti alloys subjected to heat treatment at 1073K for 3.6ks. (a, b and d) BF images and (c) SAD pattern of single β alloy.

2. Experimental procedure

Button ingots of alloys were prepared by arc melting with a non-consumable electrode and a water-cooled copper mould. Raw materials used in this study were titanium sheets containing 0.02% oxygen, high purity metal sheets and yttrium powder. Static grain growth in these alloys was examined by isothermal heating of cylindrical specimens with a diameter of 10 mm and a length of 10 mm at 1073 K for time periods between 3.6 to 40 KS, followed by water quenching. Grain growth behaviour during hot deformation of α+β alloy was investigated by hot compression using a hot working simulator of THERMEC-Master Z. Specimens were compressed at a temperature of 1073K to a true strain of 0.7 followed by He-gas quenching. The deformation was conducted at strain rates of $2 \times 10^{-4}s^{-1}$. The microstructure was observed by optical microscopy (OM) and scanning electron microscopy (SEM, PHILIPS XL30FEG). Average α grain size was measured from micrographs using the line intercept method. The presence and size of Y_2O_3 particles were identified by thin foil observation using transmission electron microscopy (TEM, JEOL-2000EX2) at 200kV.

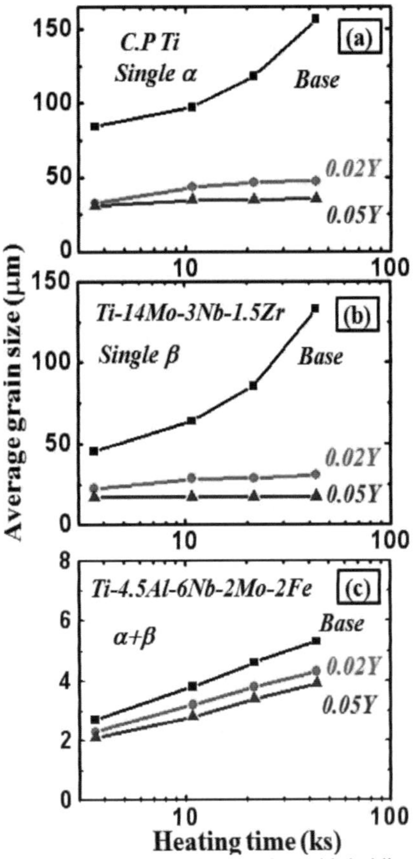

Figure 2. Changes of grain size with holding time of (a) single α, (b) single β and (c) α+β Ti alloys.

3. Results and discussion

Figure 1 shows TEM micrographs of 0.05Y added alloys subjected to heat treatment at 1073K for 3.6ks (a, b and d) and a selected area diffraction pattern (SAD) of single β alloy (c). The average size of Y_2O_3 particles was around 100 nm with a spread from 20 to about 150 nm. The Y_2O_3 particles seem to have spherical shape. Figure 2 shows changes of grain size in single α, single β and α+β alloys with holding time at 1073K. It is obvious that the addition of yttrium to single phase alloys leads to extensive refinement of the microstructure. Higher amount of yttrium addition causes lower average α and β grain size. By adding 0.05% Y grain growth is fully suppressed in these alloys.

On the other hand, in the two-phase alloy, which has a fine initial microstructure, grain growth is not suppressed. However, as shown in Fig. 2 (c), yttrium results in a slightly finer grain size. The reason could be attributed to the size and fraction of the Y_2O_3 particles. Figure 3 shows the grain size as a function of particle size for three different levels of particle volume fractions according to the Zener equation. It should be noted, that this figure shows the effect of particle fraction and size on grain size of the single phase microstructure. Here, it is assumed that if the particles could completely suppress grain growth in the single phase, this will result in complete retardation of grain growth in the two-phase microstructure. From Fig. 3 it is clear that for suppressing grain growth in the two-phase alloy with a volume fraction of 0.05% the size of the Y_2O_3 particles should be decreased to about 10nm. Meanwhile, it is possible to control the grain size by increasing the volume fraction of the Y_2O_3 particles with size of 0.1μm from 0.05 to 1%. This shows that for both controlling the grain size and complete suppression of grain growth optimization of the particle size and fraction is necessary.

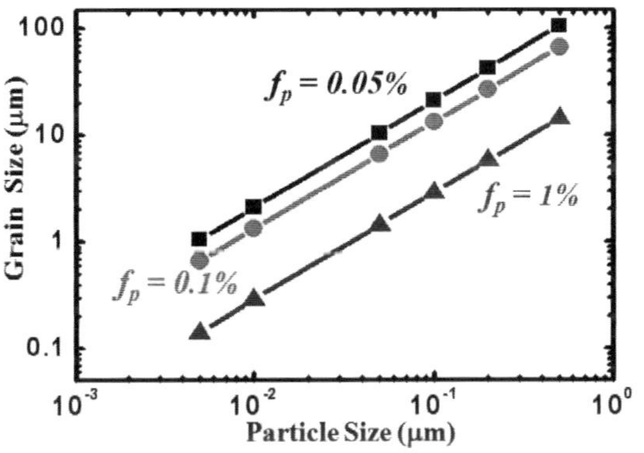

Figure 3. Grain size against particle size for three different levels of particle volume fractions based on the Zener equation.

The effect of deformation on the microstructure of the two-phase alloy used could be cleared more when comparing the microstructure after static and dynamic grain growth at the same heating time. Microstructures of the base and 0.05Y alloy after heating for 3.6ks at 1073K are shown in Fig. 4 (a) and (b), respectively. After 50% compression of the alloy at a strain rate of $2 \times 10^{-4}s^{-1}$ which corresponds to the same heating time (3.6ks), the microstructure obviously is coarser as shown in Figs. 4 (c) and (d). This difference in grain size is due to the difference between dynamic and static grain growth. Dynamic grain growth results in about two times larger grain size compared to static grain growth. Superplastic elongation of both alloys was obtained by hot tensile test. The base alloy yields very high superplastic elongation of about 2500%. However, addition of yttrium to the alloy reduces this value to 1500%. The achievement of large superplastic elongations in new alloys is dedicated to extremely fine two-phase microstructures as well as an optimum design of chemical composition. It is important to note that here very high elongation was obtained at much lower temperature than observed for conventional superplastic titanium alloys like Ti-6Al-4V [16]. Addition of yttrium caused a reduction in superplastic elongation which may be attributed to the presence of Y_2O_3 particles. These particles may act as strain accumulation sites during hot tensile deformation enhancing void nucleation and growth during superplastic deformation and resulting in a reduction of ductility [14, 15]. Therefore, it seems that increasing the volume fraction of Y_2O_3 particles is not an appropriate way to control grain growth of the

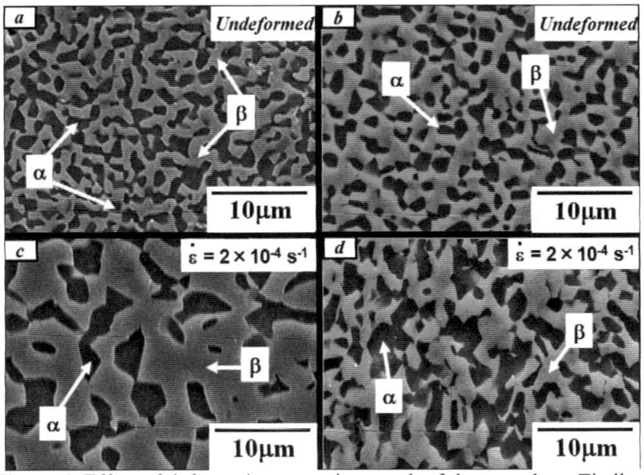

Figure 4. Effect of deformation on grain growth of the two-phase Ti alloy: undeformed base (a) and Y added (b), deformed at a strain rate of $2 \times 10^{-4}s^{-1}$ base (c) and Y added (d).

two-phase alloy. The only procedure remaining to control grain growth seems to be the decrease of particle size. In this regard, further studies are needed to achieve finer particle sizes.

4. Conclusions

In this study, the effect of yttrium addition to single α, C.P titanium, single β, Ti-14Mo-3Nb-1.5Zr, and new α+β type titanium alloy, Ti-4.5Al-6Nb-2Mo-2Fe, was investigated and the following results were obtained. Y_2O_3 particles with an average size around 100 nm are formed in an in-situ mode. Grain growth is suppressed by pinning of the grain boundaries by Y_2O_3 particles during isothermal heating of single α and single β alloys at 1073K while adding Y has a small effect on grain growth of the two-phase structure. Grain growth is accelerated by deformation, but 0.05% yttrium addition showed to be effective in suppression of dynamic grain growth. Y_2O_3 particles have detrimental effect on superplastic elongation.

References

[1] M. Ninomi and J.C. Williams, *Proc. 10th World Conf. on Titanium* (Ti-2003 Science and Technology), G. Lütjering and J. Albrecht, Eds., Wiley-VCH, Weinheim, (2004) Vol. 1 p. 95.
[2] R. Tricot, *Proc. 6th World Conf. on Titanium*, P. Lacombe *et al.*, Societé Francaise de Metallurgie, (1988), p. 23.
[3] S. Hotta, K. Yamada, T. Murakami, T. Narushima, Y. Iguchi and C. Ouchi, *ISIJ Int.* **46** (2006) 129.
[4] C. Zener, quoted by C. S. Smith, *Trans. AIME* **175** (1984) 15.
[5] T. Nishizawa, I. Ohnuma and K. Ishida: *Mater. Trans. JIM* **38** (1997) 950.
[6] F.J. Humphrys and M. Hatherly, *Recrystallization and Related Annealing Phenomena*, 2nd edition, Pergamon Press, Oxford, (2003), p. 365.
[7] D.S. Wilkinson and C.H Caceres, *Acta Metall.* **32** (1984) 1335.
[8] P.S. Bate, K.B. Hyde, S.A. Court and F.J. Humphreys, *Mater. Sci. Forum* **447-448** (2004) 61.
[9] E. Sato and K. Kuribayashi, *ISIJ. Int.* **33** (1993) 825.
[10] M.K. Rabinovich and V.G. Trifonov, *Acta Mater.* **44** (1996) 2073.
[11] P. Bate, *Acta Mater.* **49** (2001) 1435.
[12] T. Nomura, N. Yamamoto, T. Narushima, Y. Iguchi and C. Ouchi, *Proc. 10th World Conf. on Titanium* (Ti-2003 Science and Technology), G. Lütjering and J. Albrecht, Eds., Wiley-VCH, Weinheim, (2004), Vol. 2, p. 1235.
[13] K. Geng, W. Lu and D. Zhang, *Mater. Sci. Eng. A.* **360** (2003) 184.
[14] K. Gen, W. Lu, Y. Qin and D. Zhang, *Mater. Res. Bull.* **39** (2004) 873.
[15] M. Kobayashi, K. Funami, S. Suzuki and C. Ouchi, *Mater. Sci. Eng. A.* **243** (1998) 284.
[16] T. Seshachayulu, S.C. Medeiros, W.G. Frazier and Y.V.R.K. Prasad, *Mater. Sci. Eng. A.* **284** (2000) 184.

AUTHOR INDEX

AUTHOR INDEX

AUTHOR INDEX

AUTHOR INDEX

AUTHOR INDEX